INTERNATIONAL SYMPOSIUM ON HYDROMETEOROLOGY

June 13-17, 1982

Denver, Colorado

AWRA

WMO

Edited by

AMS

A. Ivan Johnson

Robert A. Clark

Weather Modification Association

WMA

AGU

IAHS/AISH

USNC/SH

American Water Resources Association

5410 Grosvenor Lane, Suite 220
Bethesda, Maryland 20814

AMERICAN WATER RESOURCES ASSOCIATION TECHNICAL PUBLICATION SERIES

TPS-82-1

LIBRARY OF CONGRESS CATALOG CARD NUMBER: 83-72705

PREFACE

The fields of hydrology and meteorology have become increasingly interdisciplinary and close working relations between these two major divisions of science have become more and more necessary, resulting in the field of hydrometeorology. The International Symposium on Hydrometeorology was organized by AWRA to provide information on the current state-of-the-art of hydrometeorology and to discuss future directions in hydrometeorology. The papers are diverse in nature ranging from acid rain to probable maximum precipitation and weather modification.

The papers in this *Proceedings* volume provide background material for the many applications of hydrometeorology by both meteorologists and engineers. They illustrate the diverse interests of the several participating societies and of the numerous scientists from the Federal, state, university, and private sectors. The volume is organized into sections that deal with various program areas related to hydrometeorology:

- instrumentation and data acquisition,
- hydrologic modeling and forecasting,
- water supply and water resources,
- long-term planning for water quality,
- drought and climate,
- weather modification,
- acid rain, and
- probable maximum precipitation and frequency for design.

This *Proceedings* volume on hydrometeorology is a unique reference volume for all professional hydrologists, meteorologists, and engineers as well as students who are interested in the current state-of-the-art and future directions in hydrometeorology. It is a valuable reference with more than 90 papers by scientists and engineers from 12 countries.

Co-Editors:

A. Ivan Johnson

Robert A. Clark

A. IVAN JOHNSON is currently Water Resources Consultant to Woodward-Clyde Consultants. He completed undergraduate degrees in Civil Engineering and in Mathematics, and his graduate study was in soil mechanics and soil physics, all at the University of Nebraska, Lincoln. Following three years at a steel works, two years in the Navy Sea Bees, and two years with the Nebraska Department of Roads and Irrigation, he spent 31 years as a hydrologist with the U.S. Geological Survey's Water Resources Division, retiring in 1979 to enter into private practice. He has authored over 80 published works and since 1963 has been well-known for his direction of applied hydrologic training courses for international participants and also has served as a consultant in many foreign countries.

ROBERT A. CLARK is currently the Director of the Office of Hydrology of the National Weather Service. The Office of Hydrology is responsible for the river and flood forecasting program of the NWS. Prior to becoming Director in 1974 he was a Professor in the Department of Meteorology at Texas A&M University. Earlier he served as a hydrologist and hydrometeorologist with the Bureau of Reclamation and as an aviation forecaster with the Air Force. He has also been a hydrometeorological consultant for several engineering firms operating in the United States and in foreign countries. He served as Chairman of the Technical Program Committee for this symposium.

ACKNOWLEDGMENTS

A Symposium and *Proceedings* like this draws upon the skills of many active people. The organizing committees are listed on the following page. The Symposium had six active co-sponsoring organizations – World Meteorological Organization, American Meteorological Society, Weather Modification Association, American Geophysical Union, International Association of Hydrological Sciences, and the U.S. National Committee on Scientific Hydrology. Their assistance in developing the program and in publicizing the Symposium is gratefully acknowledged.

We are indebted to the authors who presented and prepared their papers for publication. The session moderators not only served ably in charge of their sessions but also served as reviewers, along with the members of the Technical Program Committee, of all published papers. Each paper was reviewed and evaluated by two to three reviewers.

Our special thanks goes to Dr. Wilnot N. Hess, Director of the National Center for Atmospheric Research, Boulder, Colorado, for his excellent opening talk, "Rain – A Blessing and a Menace."

Kenneth D. Reid, AWRA Executive Director, was especially helpful with all arrangements, and Charlene Young, AWRA Publications Manager, produced these *Proceedings* and was extremely patient with late authors, reviewers, and editors. All of the AWRA officers were very supportive of the organizing committees.

A. Ivan Johnson
Symposium General Chairman

Robert A. Clark
Technical Program Chairman

INTERNATIONAL SYMPOSIUM ON HYDROMETEOROLOGY

Sponsored by

AMERICAN WATER RESOURCES ASSOCIATION

Co-Sponsors

WORLD METEOROLOGICAL ORGANIZATION

AMERICAN METEOROLOGICAL SOCIETY

WEATHER MODIFICATION ASSOCIATION

AMERICAN GEOPHYSICAL UNION

INTERNATIONAL ASSOCIATION OF HYDROLOGICAL SCIENCES

U.S. NATIONAL COMMITTEE ON SCIENTIFIC HYDROLOGY

TECHNICAL PROGRAM COMMITTEE

CHAIRMAN

ROBERT A. CLARK
National Weather Service, NOAA
Silver Spring, Maryland

JOHN J. CASSIDY
Bechtel Corporation
San Francisco, California

JOSEPH S. CRAGWALL, JR.
U.S. Geological Survey
Reston, Virginia

RAYMOND HERRMANN
U.S. National Park Service
Fort Collins, Colorado

JOHN W. JAMES
Weather Modification Association
Reno, Nevada

JOHN F. MILLER
National Weather Service, NOAA
Silver Spring, Maryland

J. NEMEC
World Meteorological Organization
Geneva, Switzerland

BERNARD A. SILVERMAN
U.S. Bureau of Reclamation
Denver, Colorado

C. RUSSELL WAGNER
U.S. Geological Survey
NSTL Station, Mississippi

LOCAL ARRANGEMENTS COMMITTEE

CHAIRMAN

RICHARD O. HAWKINSON
U.S. Geological Survey
Lakewood, Colorado

Exhibits
RICHARD H. PEARL
Colorado Geological Survey
Denver, Colorado

Tours
SHERMAN R. ELLIS
U.S. Geological Survey
Lakewood, Colorado

Publicity
DONALD T. JENSEN
U.S. Bureau of Reclamation
Denver, Colorado

Visual Aids
RICHARD F. HADLEY
U.S. Geological Survey
Lakewood, Colorado

AWRA BOARD OF DIRECTORS

OFFICERS

President .... PHILLIP E. GREESON
Past President .... KENNETH D. SCHMIDT
President-Elect .... MELVIN W. ANDERSON
Vice President .... DAVID A. RICKERT
Secretary .... WILLIAM R. BOGGESS
Treasurer .... MARY H. MARSH
Director-at-Large .... JERRY R. ROGERS

DIRECTORS

Pacific Northwest .... CHARLES R. COLLIER
Pacific Southwest .... MICHAEL M. MOORADIAN
Mountain .... J. PAUL RILEY
West North Central .... DAVID J. ANDERSON
West South Central .... TROY L. LOVELL
East North Central .... N. EARL SPANGENBERG
East South Central .... VARLEY E. WIEDEMAN
Middle Atlantic .... PETER E. BLACK
Chesapeake .... DONALD R. WIESNET
South Atlantic .... THOMAS N. DEBO
New England .... LARRY R. JOHNSTON

TABLE OF CONTENTS

Section 1
INSTRUMENTATION AND DATA ACQUISITION

Section 2
HYDROLOGIC MODELING AND FORECASTING

Section 3
WATER SUPPLY AND WATER RESOURCES

Section 4
LONG-TERM PLANNING FOR WATER QUALITY

Section 5
DROUGHT AND CLIMATE

Section 6
WEATHER MODIFICATION

Section 7
ACID RAIN

Section 8
PROBABLE MAXIMUM PRECIPITATION AND FREQUENCY FOR DESIGN

Section 1

INSTRUMENTATION AND DATA ACQUISITION

U.S. GEOLOGICAL SURVEY'S WATER DATA PROGRAM NEW APPROACHES IN THE 1980'S

Thomas J. Buchanan and Bruce K. Gilbert[2]

ABSTRACT: The water resources issues of the 1980's will require hydrologic data at many new locations, more data in real-time, and additional parameters to be measured. A refocus of program and improved technology in data collection is necessary to meet these expanding needs. The U.S. Geological Survey provides much of the hydrologic information collected in our Nation. The Survey is evaluating its water data network and developing plans to meet future needs. Network evaluation will provide the opportunity to shift efforts in many areas to effective data collecting activities of greatest need to the Nation. Satellite data transmission is expected to play a most important role in providing the real-time information in water management and flood hazard protection. Modernization of hydrologic instrumentation will make it possible to collect and analyze data in a highly automated fashion, and provide the opportunity to measure more parameters on a real-time basis with a high degree of accuracy.
(KEY TERMS: hydrologic data; data acquisition; data transmission; data storage and retrieval; network design.)

INTRODUCTION

That this Nation has a wide variety of water problems is not a particularly profound observation. It is important, however, to note that in addition to increased stresses of supply and demand, the many complexities of water quality characteristics, and the impacts of natural and man-made hazards, there is a growing impatience for solutions to the problems. In part, this results from development of an enhanced public awareness and understanding of hydrologic processes, and in part from expectations for technological and management breakthroughs to enhance supplies and to handle wastes.

Briefly, the principal problems to be addressed include:

- *Growing demands for water for municipal, industrial, agricultural, mining, energy development, and other uses.*
- *Deterioration of the quality of ground and surface waters.*
- *Hydrologic hazards, both natural and those resulting from man's activities.*

Some of the solutions to these problems likely will include interbasin transfers, weather modification, water conservation, genetic improvement of crop species, implementation of water quality protection strategies, and large scale shifts of the economic base from region to region. It seems clear that new and different types of water data will be required for design purposes and to evaluate the environmental, economic, social, and political impacts of alternatives.

CURRENT U.S. GEOLOGICAL SURVEY DATA PROGRAMS

The U.S. Geological Survey is the principal Federal agency responsible for the collection of hydrologic data (Gilbert and Buchanan, 1982). To meet local, State, and national needs, the Survey cooperates with State and local governments (Buchanan and Gilbert, 1982) and other Federal agencies in collecting information on the occurrence, quality, quantity, distribution, utilization, movement, and availability of surface and ground water resources.

The Geological Survey maintains a nationwide system of stream gaging stations, ground water observation wells, and water quality sampling stations. The resulting data provide the foundation for hydrologic appraisals, environmental impact assessments, engineering design, and energy related studies, as well as establishing a baseline description of the Nation's water resources.

Surface Water

The Geological Survey collects continuous stage (water level) records at about 8,000 stream sites and 1,000 lake and reservoir sites throughout the Nation, and computes water discharge (flow) of most of the streams from the stage records. Discharges at almost 8,000 additional stream sites are measured periodically or at peak stages, low flows, or in response to other selected hydrologic conditions. Discharge has been measured at thousands of stream sites on a one-time basis. These stream gaging sites include naturally flowing streams, streams or canals that are at times affected

[1] Respectively, Assistant Chief Hydrologist and Deputy Assistant Chief Hydrologist; U.S. Geological Survey, 441 National Center, Reston, Virginia 22092.

by man's activities, and waterways where the discharge is completely controlled.

Ground Water

The ground water data collection network maintained by the Geological Survey consists of almost 27,000 public and privately owned observation wells. Measurements are made of the depth to water below land surface, or the height above land surface in flowing wells. Continuous data recorders are used on more than 2,000 of these wells; the remainder are measured on a periodic basis. Additional wells are measured manually as part of specific investigations, and typically some of these are added to the observation well program.

Quality of Water

The Geological Survey also monitors the quality of surface and ground water. Water samples collected at selected surface water stations are analyzed for a large variety of organic and inorganic constituents depending on requirements. Temperature is measured on site at nearly every station. Commonly pH, specific conductance, and dissolved oxygen are measured on site also. Major inorganic constituents, the major nutrients, nitrogen and phosphorus, and often trace metals and organic compounds, are determined in most samples. Samples for determination of the quality of suspended sediment are taken at some of the same stations. In recent years, analyses for pesticide and radiochemical constituents have been increased in response to the growing awareness and concern about sediment particles acting as carriers of contamination. Fecal coliform and fecal streptococcus counts are obtained at several hundred stations, and samples for phytoplankton identification and counting are commonly collected from lakes and reservoirs.

Chemical constituents in ground water samples also are measured as part of the observation well network, although far fewer wells are sampled for water quality than are measured for water level. Wells typically are sampled only once or twice a year.

Sediment

Suspended sediment data, such as concentration, load, or particle size, are collected at approximately 2,800 stream gaging sites. Quantities of sediment transported are determined by measuring sediment concentration in streams and relating these measurements to concurrent stream flow. Such data are needed, for example, to plan for storage of sediment in reservoirs, to identify areas of erosion for possible remedial treatment, and to serve as a basis for evaluation of land-use change.

Water Use

A water-use information program was started in 1978 to provide for the comprehensive and systematic collection, storage, analysis, and dissemination of water-use data and information throughout the United States. Statistics on domestic, industrial, and agricultural water use are required for the planning, management, and development of the Nation's water resources and to provide information necessary to identify and resolve critical water problems.

CHALLENGES AHEAD

Although the Survey collects a great deal of hydrologic data, there are some deficiencies in the geographical distribution and in the kinds and frequency of data obtained. In part, this reflects some lag in responding to growth and shifts in population and depletion of some local sources of water. The development and management of water resources on a regional and national basis, which require more data, obtained on a more timely basis, also have contributed. In the past, expanded needs for information have been accommodated by modifying data collection operations, adding funding resources and staff, or by substitution of one station for another.

Realistically, it must be assumed that additional funds and staff will be forthcoming in the 1980's only to the extent they can be made available from other Survey activities. Considering the potential for some continuing level of inflation and little, if any, prospect for budget increases, possibilities for any real funding growth are few. During the past five years, funding and staff constraints have driven the data collection activity to the point where frequency of visits to hydrologic stations has decreased to the uppermost limits of available instrumentation, and data losses are becoming of concern. Essentially the same limits have been reached with respect to the capacity of the existing number of staff to perform work. And because of increased capital costs, it is no longer possible to install a new data collection station by discontinuing one in current operation.

Thus, the challenges to provide new and additional water data are indeed great. In response, some fresh and innovative approaches are under development for the utilization of recent technological improvements, definition of enhanced sampling strategies, and application of unique statistical techniques for hydrologic network analysis. Simply stated, the goal is to furnish timely information, where needed by the user, with no compromise on scientific integrity. Given the anticipated needs for water data within the potentially available resources, this is no small job.

TECHNOLOGICAL IMPROVEMENTS

The Geological Survey has embarked on a program of technological improvement to keep pace with the needs for water data in the 1980's. The efforts currently in progress are to develop an improved data acquisition system, to develop new sensors for data collection, and to make operational a network of real-time data stations.

At present, a wide variety of outdated electronic and electromechanical systems are used to sense and record various hydrologic parameters. Punched paper tape recorders collect water stage and quality data automatically at fixed time intervals. The tapes are manually retrieved from the stations and are then processed for data computation, analysis, retrieval, and use. Although many water samples continue to be collected manually, some stations are equipped with automatic sampling or on-site monitoring devices. The monitors can automatically measure and record information such as temperature, pH, specific conductance, and dissolved oxygen. On an individual basis, each piece of equipment has served its intended purpose satisfactorily. However, it has become evident that added efficiency would result from using a more coherent data acquisition system with a single power source, a timer to synchronize sensing and recording functions, and a central recorder from which data could be retrieved electronically. Such a system would be easier to support logistically than the current diverse configurations now in operation. The most pressing areas for technology improvement in the 1980's clearly will be the measurement of water quality parameters, together with automation of field station operation.

The increased need for hydrologic data, the numerous chemical constituents continually being discovered in our waters, and the increased costs of salaries and travel, make it mandatory that new sensors be developed. For example, an improved method of determining water levels is needed, preferably by development of a noncontact type of sensor. In addition, new devices are needed to measure velocity or discharge at a stream cross-section. There also are emerging needs for new sensors for a variety of water quality characteristics, especially for toxics, synthetic organics, heavy metals, nutrients, and biological parameters.

An Adaptable Hydrologic Data Acquisition System (AHDAS), custom designed to fit the Survey's particular requirements, is being developed based on the results of planning studies recently completed. The system will consist of a core unit that will supply the power and programing to operate all the sensors at a station and to record the information obtained. The technology exists today to combine and automate the collection of various data under a central control system, which is the key to development of this equipment. Ideally, all the systems should operate from the same power source and timer. The required design will permit newly developed sensors to be added and operated through the central control system.

The Geological Survey has also conducted real-time hydrologic data collection tests using Government and commercial satellites since 1972. Evaluation of these tests shows that expanded use of satellite data relay technology can be invaluable for management and monitoring of critical hydrologic parameters. Telemetry of data from remote data collection sites can greatly reduce time for delivery of the data to the user and may provide benefits to the data collecting agency including reduction of field travel costs and remote monitoring of instrumentation functions, as well as automated data handling. Figure 1 shows the principal elements of a satellite system for the relay of hydrologic data.

The Geological Survey currently operates over 225 data collection platforms using the Geostationary Operational Environmental Satellite (GOES) operated by the National Oceanic and Atmospheric Administration (Shope and Paulson, 1981). Satellite data telemetry has already proven to be a reliable and, for many application, a cost-effective tool for the acquisition of remote hydrological data. A significant expansion of the existing satellite telemetry network now depends on a more complete evaluation of the needs for, and benefits of, having real-time hydrologic data for water resources management. In concert with development of AHDAS, as well as improved equipment interfaces, and the addition of geographically dispersed satellite receive sites, it is anticipated that increasing reliance will be placed on telemetry in the next few years.

INTELLIGENT SAMPLING STRATEGIES

Current systems collect data at fixed and intentionally high frequencies in order to define adequately extreme hydrologic events, which often are of short duration. Consequently, large amounts of essentially superfluous data are normally collected.

In conjunction with development of AHDAS, various intelligent sampling strategies are being evaluated for recording data on the basis of specific information requirements rather than solely on the current method, elapsed time. Such strategies can be easily implemented by AHDAS because of the microprocessor technology it will employ.

For example, stream stage must be recorded only once or twice in a 24-hour period, unless commands for more frequent sampling are triggered by the microprocessor as it detects water level changes or rates of change that exceed a prescribed amount. This may enable the Geological Survey to compute hydrologic records with fully acceptable reliability and equivalent or enhanced information content, but with fewer data points than are now being recorded. This also would markedly decrease the data processing burden, particularly on telemetry systems.

In addition, tests are being conducted of procedures that can be used by AHDAS to detect and flag critical hydrologic data, to minimize the amount of data needed for reconstruction of the original time series record, to identify data for subsequent review, and to recognize errors. Research is being conducted to measure the relative performance between the several candidate sampling strategies and the current system. The measures of performance include comparisons of various differences in peak discharge, mean daily discharge, and stage that may be introduced by using

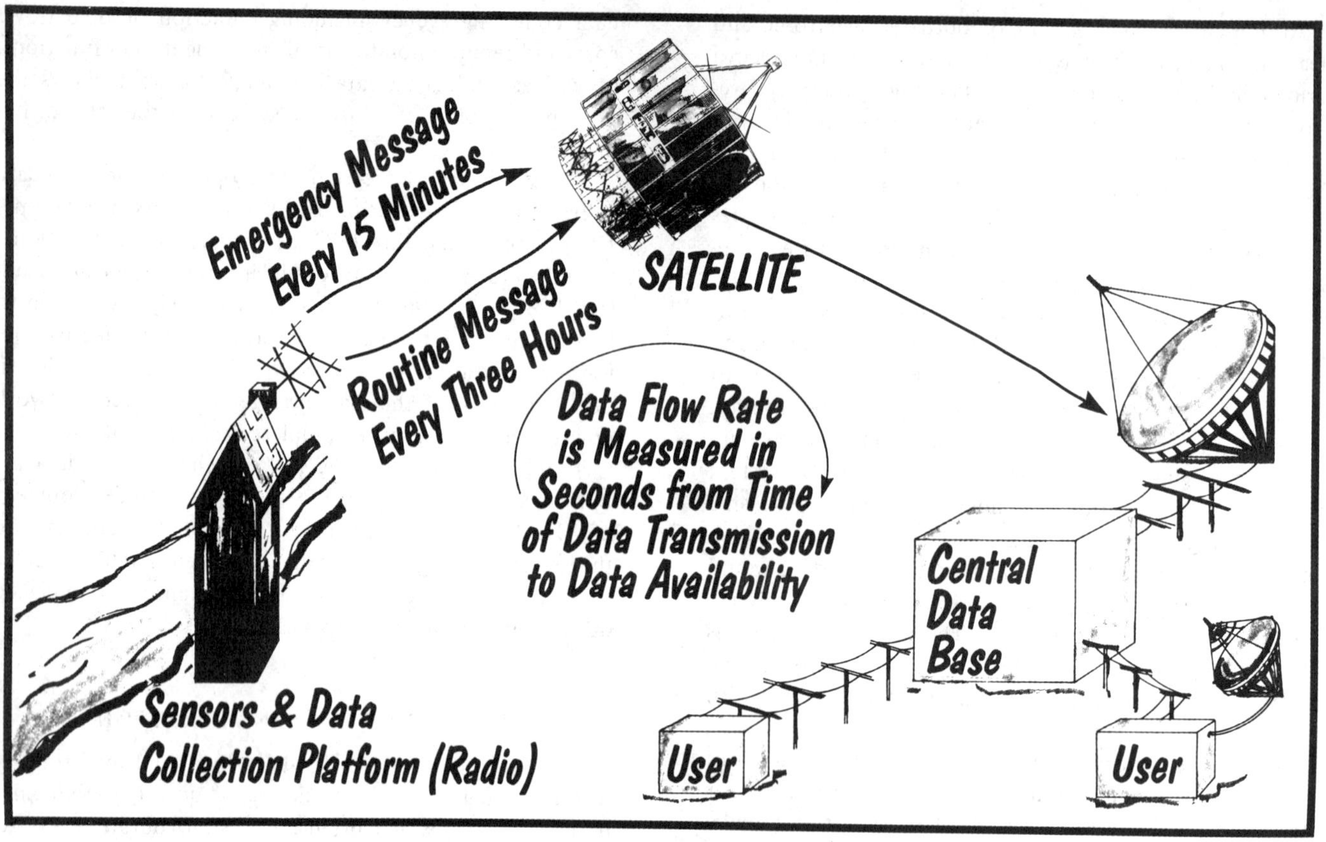

Figure 1. Principal Elements of a Satellite Data Relay System.

the intelligent sampling rather than the traditional time dependent records. The effectiveness of intelligent sampling performance will vary with the characteristics of individual streams. Thus, one technique may be found to perform best for a steep gradient stream and another for a low gradient stream. For this reason, intelligent sampling research is being conducted with data from numerous types of streams, in various physiographic, geologic, and climatological settings. It is essential to determine which strategies are most promising for future implementation and which are most suitable for various streams. It will, of course, be possible to implement the intelligent sampling strategies at well sites more easily than on streams.

NETWORK ANALYSIS

The collection of hydrologic data is the most vital and intrinsic component of the U.S. Geological Survey's mission to provide the information needed for the optimum utilization and management of the Nation's water resources. Along with this is the responsibility to ensure that data collection activities are carried out as economically and effectively as possible – in other words, to provide the most essential information with the funds available.

The U.S. Geological Survey's data network is continually examined to make certain that this goal is being met. For example, during 1981, the Geological Survey conducted an analysis of the stream gaging network in each of the four States, using statistical techniques to assess the cost effectiveness of the operation. This type of analysis is designed to minimize the sum of variances of errors of estimates of stream flow within a network of gages, subject to given operational and budgetary constraints (Moss and Gilroy, 1980).

The technique described above provides for cost-effective resource allocations geared to specified standards for data reliability. However, in addition to generating information on baseline conditions, long term trends, and hydrologic processes, the stream gaging program responds to many factors, including legislative, judicial, and administrative directives. Also of fundamental importance are data requirements for flood and water supply forecasting, current purpose water management, and for input to various other hydrologic investigations and research.

The analysis tested in 1981 is being modified to accommodate these criteria and along with other statistical techniques will then be applied to the Survey's stream gaging programs on a State-by-State basis, nationwide, in the early 1980's. It is anticipated that some data collection sites may be discontinued, others will need to be established, and at selected existing stations, the parameters measured and the frequency of measurement may be altered on the basis of the network analysis.

SUMMARY

The Nation will continue to face a variety of water problems in the 1980's and will be seeking new alternatives as well as currently available solutions. The U.S. Geological Survey, as the principal Federal agency responsible for hydrologic data collection, will be called on to provide additional data for water resources planners and managers. To accomplish this will be a major challenge because constraints are expected to continue on funding resources and staff.

In anticipation of these increasing demands, the Survey is working on a number of technological improvements so as to furnish as much information as possible with the resources available. An Adpatable Hydrologic Data Acquisition System, currently under development, is expected to serve as the central core of data collecting instrumentation for the future. The system would supply the power and programing to operate all the sensors at a station, as well as to record or telemeter the information collected. A vigorous program of sensor development is underway to provide the capability to measure current parameters more accurately and dependably, and to measure other parameters, critical to various aspects of water resources planning and management.

Data relay stations enable water managers to make responsive decisions regarding the effective use and protection of our water resources in real-time. Current investigations are designed to determine when and how much additional real-time data are needed. Rather than calling for data collection based on standard time intervals, it may be possible to develop instrumentation that will collect data in response to preselected key changes in river stage, water quality, or other hydrologic characteristics.

The Geological Survey is also reviewing its networks of data stations to determine where it may be possible to discontinue activities, to change the frequency of measurement or the number of parameters measured, and to determine sites where additional data may be needed. In addition, the analysis will indicate the most cost-effective means of collecting the necessary data within prescribed limits of accuracy.

LITERATURE CITED

Buchanan, T. J. and B. K. Gilbert, 1982. U.S. Geological Survey Federal-State Program. Water Resources Planning and Management Journal, American Society of Civil Engineers 108(WR1): 37-45.

Gilbert, B. K. and T. J. Buchanan, 1982. Water-Data Program of the U.S. Geological Survey. U.S. Geological Survey, Circular 863, 16 pp.

Moss, M. E. and E. J. Gilroy, 1980. Cost Effective Stream-Gaging Strategies for the Lower Colorado River Basin: The Blythe Field Office Operations. U.S. Geological Survey Open-File Report 80-1048, 112 pp.

Shope, W. G. and R. W. Paulson, 1981. Data Collection Via Satellite for Water Management. Transportation Engineering Journal, American Society of Civil Engineers 107(TE4):445-455.

BUREAU OF RECLAMATION'S AUTOMATED HYDROMETEOROLOGIC DATA COLLECTION ACTIVITIES

Richard H. Ives[1]

ABSTRACT: In a variety of programs, the Bureau of Reclamation employs automated data collection platforms (DCP's) to collect hydrometeorologic data and then transmit these data via the Geostationary Operational Environmental Satellite (GOES) system to ground receiving sites. These real-time hydrometeorologic data contribute significantly to improved Bureau program operations in weather modification research, wind and solar energy potential site evaluations, and in water resources project operations.

The establishment of a Meteorological Surface Observation Network (MESONET), consisting of 128 solar powered DCP's, enables Bureau scientists in weather modification research to more accurately monitor weather developments. The deployment of 22 solar powered DCP's at sites with high wind energy potential will permit Bureau engineers to assess the feasibility of installing wind turbine generators at selected sites throughout the West. In addition to being fitted with an anemometer, two of the wind monitoring DCP's are also measuring solar insolation for solar energy potential site evaluations in the Southwest. The Bureau has also installed 66 DCP's and has plans for an additional 110 DCP's in a hydromet system in the Pacific Northwest to provide real-time hydrometeorologic data for making operational decisions to maximize benefits of multipurpose reservoir systems.

Real-time hydrometeorologic data are inexpensive and the data are readily available to users in the Bureau and in other agencies, the data can be immediately integrated with a variety of predictive models, and results are obtained far sooner than would be possible with manual data collection techniques. In addition, for Bureau water resources projects, real-time data can provide the basis for making prompt adjustments in the operational requirements and enhance the ability of Bureau officials to monitor and respond to either drought or flood situations.

(KEY TERMS: data collection; hydrometeorological data; monitoring; networks.)

INTRODUCTION

The Bureau of Reclamation is responsible for multipurpose projects to develop, manage, and conserve our limited water and related land resources to meet present and future requirements in the 17 Western States. Bureau projects consisting of such features as dams, reservoirs, canals, and power plants currently satisfy a significant portion of the water and energy needs of the West.

Accompanying the evolution of the Reclamation program has been a dramatic increase in the need for hydrometeorologic (hydromet) data in diverse geographical areas. The increasing complexity of multipurpose project planning and operations and specialized research needs, in conjunction with significant recent advances in telecommunications, has resulted in the Bureau's use of a satellite linked data collection system (DCS) in several applications to obtain real-time hydromet data. This technology has proved to be cost-effective, reliable, and extremely adaptable for meeting the differing data collection needs of various Bureau programs.

BUREAU OF RECLAMATION PROGRAM OF TODAY

The Reclamation Act of 1902 authorized the Secretary of the Interior to administer the Federal Reclamation program to reclaim the arid and semiarid lands in the West. At the turn of the century, private irrigation development had reached a plateau, and an active role for the Federal Government was seen as the key to continued development of irrigation in the West. In the 1920's an era of multipurpose water resources development began.

Reclamation projects now supply water for irrigation, municipal and industrial use, hydroelectric power generation, environmental enhancement, and recreation, as well as provide storage for flood control. Reclamation projects are indeed more complex today than what could have ever been envisioned when irrigation was the sole driving force. In addition, the great web of Reclamation projects supplying water and power to the West is part of an integrated system of interdependent activities which requires considerable coordination with other agencies and organizations.

Accompanying the evolution of the multipurpose project have been some significant changes in and additions to those activities which comprise the Reclamation program. An essential thread which weaves its way through many Bureau activities is the need for reliable hydromet data. While the

[1] Staff Assistant for Research, Bureau of Reclamation, Washington, D.C. 20240.

U.S. Geological Survey (USGS), Soil Conservation Service (SCS), and the National Weather Service (NWS) are the primary providers of such data, in some cases, the Bureau is required to collect hydromet data where data either are not available in an area of interest or are inadequate for the specialized needs encountered. The focus of this paper is upon those Bureau activities including weather modification research, special studies in evaporation reduction, water resources project operations, dam safety, and solar and wind energy site evaluations, which have incorporated state-of-the-art satellite linked hydromet data collection techniques to meet specific project requirements.

SATELLITE DATA COLLECTION SYSTEMS

In only the past decade we have seen the development of a cost-effective satellite DCS made possible by the availability of earth orbiting satellites. The DCS technology involves the collection in real time of hydromet or other environmental data from remotely located sensors and the transmission of those data via satellite to a ground receiving station. The real-time data are then conveniently available as direct input into an automatic data processing system.

A satellite DCS consists of three essential components including: (1) one or more sensors connected to a field radio or data collection platform (DCP) which transmits the data to the satellite; (2) a radio transponder on the satellite which can both receive and retransmit data from many widely distributed DCPs; and (3) an earth receiving station or downlink capable of receiving and disseminating data to a processing facility for operational use or post-analysis.

Early satellite DCSs in the 1960's were developed for navigational experiments to determine the positions of mobile DCPs as the stations were interrogated. The Nimbus satellites were instrumented with a navigational and data relay system to collect data from remote DCPs to track mobile platforms located on balloons, buoys, and ships by interrogating selected DCPs during successive orbital passes.

The first widespread access to DCS technology was provided with the July 1972 launching into near polar orbit of the first LANDSAT satellite, formerly known as Earth Resources Technology Satellite (ERTS). LANDSAT-1 was a low altitude satellite primarily designed to demonstrate the feasibility of remote sensing, however, it was also capable of relaying environmental data collected from remotely sited platforms. The orbit of the satellite and range of the transponder generally allowed relay of data from a specific DCP only twice a day. The LANDSAT linked DCP typically emitted a brief data transmission or data burst every 90 or 180 seconds and the satellite would receive the data when it passed within approximately 2000 kilometers of the DCP and then relay the data to a downlink. The LANDSAT DCS is a random access system, with one radio frequency channel, meaning DCP transmissions can overlap or interfere with one another. Thus, at a predictable rate there could randomly occur mutual interference of two or more transmissions resulting in no data relay.

The launching in 1974 of the first prototype geostationary satellite resulted in a significant advance in satellite linked data acquisition systems. A geostationary satellite's position remains relatively fixed with respect to some point on the earth's surface permitting a DCP within range to continually transmit to the transponder. Currently, there are several Geostationary Operational Environmental Satellites (GOES), including primary ones positioned at 75° (GOES-East) and 135° (GOES-West) west longitude, and others including a backup satellite at 107° (GOES-Central), which can relay data transmitted from a DCP located nearly anywhere in the western hemisphere.

For the automatic collection of hydromet data measured by *in-situ* sensors, there are a variety of data collection systems available, including telephone, radio, or satellite linked systems as well as meteor burst communications. The Bureau has recognized that the satellite DCS has unique characteristics which make it the preferred choice for meeting many of its in-house data collection needs.

The satellite DCS can provide data from remote, often times inaccessible, sites where no telephone lines are available or several expensive repeaters would be required for a radio linked system, and provides continuous knowledge of the status of the system. In addition, it is extremely flexible due to the availability of various sensor arrangements and transmission modes. An important feature of the satellite DCS is the ability to acquire real-time data and automatically provide input directly to the data processing facility to enhance operational capability and eliminate the need for manual keypunching or data handling. The portability of the DCP also permits installations for short-term projects (e.g., research), thus eliminating the need for costlier, more permanent systems. While the satellite linked system may initially cost more to install than other systems, the operation and maintenance costs have generally been considerably less. The ability to transmit via satellite also permits the establishment of a hydromet DCS which can provide coverage of a broad geographical area and be easily accessed by others.

The current GOES system, with 80 active channels, can receive data from DCPs operating in self-timed, interrogate, or random reporting modes. The more common self-timed DCP transmits a brief data burst at a programmed time interval through the satellite to a ground receiving station. The GOES system has the capability to transmit data on one channel from many DCPs reporting at different times, such that 180 DCPs, each reporting every three hours, can transmit in one minute intervals on the same channel.

In the interrogate reporting mode, a specific DCP's transmission is elicited by the National Oceanic and Atmospheric Administration/National Environmental Satellite Service's ground receiving station at Wallops Island, Virginia, through the satellite.

WATER RESOURCES PROJECT OPERATIONS

The essence of the Bureau of Reclamation program continues to be the multipurpose water projects which supply municipal and industrial water to more than 19 million people or 30 percent of the West's population. In addition, nearly 148,000 farms with about 10 million acres of irrigated land receive water from these projects. The Bureau's 50 hydroelectric power plants annua-ly generate more than 40 billion kilowatt-hours of energy. Although the major portion of the hydromet data collection for the operation of these projects is fulfilled by other agencies, the Bureau has responsibility for an extensive data acquisition system in the Pacific Northwest.

The Columbia River basin contains about 80 multiple and single purpose water resources projects of the Bureau, U.S. Army Corps of Engineers (COE), and other non-Federal entities. The coordinated operation of these interdependent projects has become exceedingly complex since flood control, hydroelectric power generation, navigation, recreation, environmental considerations and water supply for irrigation, municipalities, and industry must all be taken into account. Thus, the Columbia River Operational Hydromet Management System (CROHMS) was formed, in part to provide for a coordinated and ultimately, automated, data collection and management program.

To prevent unnecessary duplication and ensure uniformity and reliability in hydromet data collected, participating CROHMS agencies, including the Bureau, Bonneville Power Administration, COE, USGS, and NWS agreed to assume compatible data collection functions in assigned subwatersheds in the Columbia River basin. All agencies involved, including the SCS which collects snow course data in the watershed for runoff forecasting purposes, agreed to upgrade and automate data collection networks. All data are to be provided to the CROHMS data bank in Portland, Oregon, which disseminates real-time information for decisionmaking relating to flood control power and irrigation scheduling, recreation, and water supply forecasting.

The Bureau has responsibility for data collection in the Upper Deschutes River basin in Oregon, and Yakima River basin in Washington, and the Middle and Upper Snake River basin, primarily in Idaho and Oregon, with a small portion in Wyoming, Utah, and Nevada.

The Bureau installed in 1976 and 1977 a radio linked system of 18 remote data stations in the Yakima River basin and completed in 1981, phase one of a program to install an extensive satellite DCS in the Snake River basin. By the end of 1982, the Snake River basin DCS, covering the Bureau's Boise and Minidoka Projects, will consist of 118 DCPs measuring stream flow, precipitation, ambient air temperature, snow water content, and reservoir elevation. The Snake River network will relay data to a Bureau downlink in Boise which inputs the data to a central processing facility which also transmits to the CROHMS data bank hourly.

Each DCP will have the capability to transmit through two channels on the GOES system – one for self-timed data transmissions and the second channel for adaptive random transmissions. In the self-timed mode, the DCP will obtain values from each sensor at 15-minute intervals and then transmit all stored data (12 values per sensor) via satellite every three hours. In the adaptive random reporting mode, data transmissions are unscheduled and elicited if either a sensor output or the rate of change between successive sensor readings exceed programmed threshold values. This will alert the Bureau to significant changes in the watershed that may require immediate attention.

WEATHER MODIFICATION RESEARCH

While the United States has an abundant water supply, its geographical distribution and availability often times do not coincide with demand. Thus, the Bureau has been investigating weather modification as an augmentation technique since 1962. Successful demonstration of a viable precipitation enhancement technology would result in valuable additional water supplies and, ultimately, more effective utilization of Bureau reservoir systems.

Scientists in the Bureau's weather modification research program (Project Skywater), aware that few surface weather stations were located in field project areas, recognized around 1970 the potential for using satellite technology in transmitting essential hydromet data. It was thought that satellite linked data acquisition systems could assist in monitoring weather conditions to control cloud seeding operations as well as evaluating snowfall increases resulting from project seeding.

A 22-month field test begin in March 1973 to evaluate the feasibility and reliability of using LANDSAT linked DCPs to collect data from remote unattended sites in the San Juan Mountains of southwestern Colorado. Seven battery operated DCPs were installed which relayed temperature, precipitation, humidity, wind speed and direction, solar radiation, stream flow, and snowpack water content observations via satellite to the Goddard Space Flight Center in Greenbelt, Maryland, from where they were transmitted to the headquarters of Project Skywater in Denver. The study revealed that satellite relayed data could reliably and cost-effectively be obtained from nearly inaccessible sites in mountainous terrain where severe wintertime conditions were encountered.

A follow-on study in 1975, investigating the operational utility of the DCS in cloud seeding operations, successfully showed that the availability of near-real-time data could aid in making daily decisions and provide a quality check of system performance. However, due to the polar orbit of the LANDSAT satellite, real-time data, available at approximately 12-hour intervals, weren't always sufficiently current to use in operational decisionmaking. Subsequently, the opportunity to continuously obtain reliable real-time data from a geostationary satellite has greatly enhanced the ability of Project Skywater scientists to monitor weather conditions in cloud seeding project areas.

To support field project operations, Project Skywater currently has 150 self-timed DCPs, although some are on loan for other Bureau activities in evaporation research and wind and solar energy site evaluations. Each DCP measures sensor output every 5 minutes and then transmits the stored data (12 values per sensor) hourly via satellite. The DCPs, known as Portable Remote Observation of the Environment (PROBE) stations, are powered by batteries charged by photovoltaic panels and programmed to measure standard meteorologic parameters which include: ambient air temperature, precipitation, humidity, barometric pressure, and wind speed and direction. A Bureau downlink at Denver receives the data and they are input into the Project Skywater computer for utilization in models for quasi-real-time predictions for cloud seeding operations.

WIND AND SOLAR ENERGY SITE EVALUATIONS

The Bureau has been investigating ways to provide optimum utilization of the hydroelectric resources and power transmission network in the West. It has been concluded that other renewable energy supplies, including both wind and solar power, could be integrated with the Bureau's existing hydroelectric system. The availability of sufficient quantities of either wind or solar energy, which are both an unpredictable and intermittent resource, could permit the storage of water and, therefore, hydroelectric energy, in reservoirs. While the Bureau is currently evaluating the potential wind resource at numerous sites throughout the West, the integration of solar power is only being considered in the Bureau's Lower Colorado Region where the hydropower network consists of Hoover, Davis, and Parker Dams and Powerplants, and an ample supply of solar energy is available.

For the evaluation of either wind or solar energy potential at remote and nearly inaccessible sites, the Bureau has opted to install satellite linked DCPs, each powered by batteries charged by a solar photovoltaic array, to obtain meteorologic data. A total of 22 DCPs have been deployed; at 20 sites in eight states, standard meteorologic parameters are collected to evaluate the potential wind resources; and at two sites (Yuma, Arizona, and Hurricane Mesa, Utah), additional solar data are collected.

SPECIALIZED RESEARCH PROJECTS OR STUDIES

The Bureau is interested in investigating opportunities to conserve precious water supplies by optimizing project operations. A study has been initiated to examine the potential for reducing evaporation at Lake Mead in the arid Southwest. It may be possible to reduce evaporation by withdrawing warmer surface water from the reservoir, however, extensive modeling of the thermal and chemical processes involved is required. An evaporation study must look at the entire Lower Colorado River system to ensure that savings in one area are not offset by losses elsewhere inasmuch as Lake Mead evaporation reductions could result in increased evaporation downstream due to warmer water entering Lake Mohave and lower reaches of the river. In addition, the cost, impacts, and physical possibility of modifying Hoover Dam intake structures to withdraw warmer surface water must be assessed.

A critical part of the Bureau's study is the installation six PROBE stations to collect the required hydromet data for the modeling activities. Three DCPs will be situated on floating platforms at selected locations on Lake Mead; two similarly positioned DCPs will be placed on Lakes Havasu and Mohave; and one DCP will be land based near Lake Mead. The lake positioned DCPs will be instrumented to obtain a variety of data including: wind speed and relative humidity at both the 2-meter and 30-foot elevations, ambient air and water surface temperatures, barometric pressure, and precipitation. In addition, in the Lake Mead and Lake Havasu areas, solar energy influx will also be measured by a solar pyranometer. The data will ultimately be utilized as input to mass transfer and energy budget equations for Lake Mead to determine if it will be feasible to reduce evaporation.

DAM SAFETY ACTIVITIES

A dam safety hazard relates to a structural, operational, or geographically induced deficiency which could cause or contribute to a dam failure, either with the passage of time or by a specific natural event. Deficiencies may be identified in a Bureau structure because older dams were not designed to meet current standards, cannot safety pass newly calculated maximum hydrologic events or cannot sustain estimated maximum credible seismic events, or due to design inadequacies.

In some instances, real-time hydromet data may be extremely valuable in monitoring those conditions or circumstances which could potentially jeopardize the structural integrity of a Bureau dam. For example, where a dam safety hazard exists, the Bureau may impose a temporary reservoir operating restriction until corrective action can be successfully completed. An automated DCS enables the Bureau to improve reservoir operations and thereby reduce the loss of the reservoir resource and potential revenues resulting from limiting reservoir levels.

The Bureau will begin installing satellite linked DCPs in the summer of 1982 for the purpose of improving operations at Twin Buttes Dam, San Angelo Project, in west-central Texas, where a dam safety hazard exists. Due to concern about foundation seepage at the earthfill structure, the Bureau has instituted a restriction upon the reservoir filling level until an effective relief well program is constructed and operating. An operating criteria has been imposed that balances concern for the structural safety of the dam with concern for abrupt releases downstream and limited utilization of the reservoir.

A satellite DCS will provide real-time hydromet data to aid in forecasting major inflows to permit improved operations using more reliable forecasts in lieu of operating under emergency guidelines which may result in a considerable loss of the reservoir resource. A total of 14 DCPs will be deployed – 12 DCPs to measure reservoir stage, stream flow, and precipitation at the dam or in the 2000 square miles of the watershed upstream, and two DCPs to obtain piezometer readings from observation wells located below the downstream face of or in the dam. The DCPs will be similar to those installed in the Pacific Northwest and have the capability to transmit through two channels on the GOES system – one for self-timed data transmissions and the second channel for adaptive random transmissions. This flexibility will permit the Bureau to recognize significant changes at the dam site or in the watershed and take immediate action.

SUMMARY AND CONCLUSIONS

The Bureau of Reclamation currently employs satellite linked data collection platforms to obtain real-time hydrometeorologic data for a variety of program uses. The opportunity to automatically acquire and quickly process essentially real-time data has been helpful in the Bureau's water project operations – including dam safety activities – as well as in research projects or special studies in weather modification, evaporation reduction, and renewable energy. The pressure to optimize the use of our valuable resources has, with limited manpower and budgets, encouraged the use of new technologies which enable us to improve our decisionmaking process.

ACKNOWLEDGMENTS

The author would like to express his appreciation to Messrs. Bill Harrison, Stan Hightower, Joe Kitchen, Dan Lute, Dean Newkirk, and Marv Willis for their assistance in providing information on their programs within the Bureau.

LITERATURE CITED

Kahan, A. M., 1974. Use of the ERTS-1 Satellite Data Collection System in Monitoring Weather Conditions for Control of Cloud Seeding Operations. Bureau of Reclamation.

Kahan, A. M., 1976. Use of the LANDSAT-2 Data Collection System in the Colorado River Basin Weather Modification Program. Bureau of Reclamation.

Kallio, N. A., 1978. Operational Hydrometeorological Data Collection System for the Columbia River. U.S. Geological Survey Circular 756.

Paulson, R. W., 1978. Use of Earth Satellites for Automation of Hydrologic Data Collection. U.S. Geological Survey Circular 756.

ADVANCES IN GAUGING OPEN CHANNELS AND RIVERS USING ULTRASONIC AND ELECTROMAGNETIC METHODS

John D. Newman[1]

ABSTRACT: This paper describes two electronic methods of flow measurement which have been developed to overcome some of the limitations of traditional methods of continuous flow measurement in open channels. The first method relies on accurately timing ultrasonic pulses traversing the river at an angle to the flow. The time varies according to the velocity of the water and the discharge can be inferred. The second method is based on electromagnetic induction. Here electrodes set into the river banks pick up a voltage generated by the water flowing through a vertical magnetic field set up by an electromagnet coil. Both methods basically measure the velocity of the water and include a water level gauge to enable the measurement of the cross-sectional area of the river which is combined with the velocity to indicate discharge. The velocity and level measurements are also available as additional river management parameters. The new methods are suitable for gauging flow in rivers and artificial channels. The choice of technique will usually be decided on a cost premise but generally the electromagnetic technique is more suited to channels up to 25 meters wide and the ultrasonic technique is more suited to wider channels.
(KEY TERMS: data collection; surface water; monitoring; networks.)

INTRODUCTION

One of the most important variables to be considered for water management is volumetric flow. This paper describes two electronic methods of flow measurement which have been developed to overcome some of the limitations of traditional methods of continuous flow measurement in open channels.

Flow data are used for various purposes including irrigation, flood warning, resource planning, river regulation and the design of major capital works including reservoirs, pumping stations, and flood alleviation schemes. Traditional methods of gauging discharge include the establishment of a relationship between water level and flow by sample measurements with current meters or by dilution gauging. A natural stable relationship occurs on some rivers to give a reasonable accuracy but where bed conditions vary due, for example, to accretion and weed growth, the use of level to infer discharge can be grossly inaccurate and backup from tidal effects or a confluence with another river makes the method unacceptable.

An artificial relationship can be established by the installation of a measurement weir or flume but the cost of this technique is becoming prohibitively high in all but the smallest of streams and artificial channels. Structures may be susceptible to the buildup of accretion, silt, and backup is again a problem. The afflux caused by a structure can create additional flooding problems or pumping costs. The two methods of measuring flow in open channels have been developed to overcome some of these problems.

The first method relies on accurately timing ultrasonic pulses traversing the river at an angle to the flow. The time varies according to the velocity of the water. The basic method has been used for some years and is also known as the accoustic flowmeter.

The second method is based on electromagnetic induction. Here electrodes set into the river banks pick up a voltage generated by the water flowing through a vertical magnetic field set up by an electromagnet coil. Both methods basically measure the velocity of the water and include a water level gauge to enable the measurement of the cross sectional area of the river which is combined with the velocity to indicate discharge. The velocity and level measurements are also available as additional river management parameters.

The choice of method will usually be decided on a cost premise but generally the electromagnetic method is more suited to channels up to 25 meters wide and the ultrasonic method is more suited to wider channels. Where the water is inhomogeneous due to poor mixing from two sources or the channel is obstructed by silt, weeds, accretion, and debris, then the electromagnetic method is more suitable. Application of the electromagnetic method is being found for gauging underground sewer pipes.

Both of the gauges are the result of sponsorship by the Water Data Unit, a branch of the Department of the Environment and the Water Research Centre in the United Kingdom. The early work on the ultrasonic gauge was carried out by staff of the United Kingdom Atomic Energy Authority Laboratories at Harwell. The electromagnetic

[1] Director, Environmental Systems Division, Sarasota Automation Ltd., King's Worthy, Winchester, Hampshire, S023 7QA England.

gauge has been developed by 'SARASOTA' – formerly 'Redland Automation.'

THE MULTIPATH ULTRASONIC RIVER GAUGE

This method has been used for several years with varying degrees of success. The principle is briefly as follows: multiple pairs of sonar transducers are mounted on opposite banks of the river at a number of defined depths with each transducer acting alternatively as a transmitter and receiver of pulses of ultrasound (see Figure 1). Pulses of sound are transmitted from an opposite pair of transducers which form a path (this path is angled to the direction of water flow). The pulses are received by the alternate transducers of the pair after a period of time depending on the path length, the velocity of sound along the path, and the resolved component of the velocity of the water.

Figure 1. Ultrasonic Transducers on the River Severn at Bewdley.

If the water is still, the transit times will be equal, but if the water is moving, one pulse will arrive earlier and one pulse later. The transit times are t_1 and t_2 for the particular path (refer to Figure 2). For one ultrasonic flight path the resolved component of the mean velocity of the water along the path can be computed as follows:

$$t_1 = \frac{L}{(c+Vp)} \qquad t_2 = \frac{L}{(c-Vp)}$$

where:

L = the length of the path,

c = the velocity of sound in water, and

V_p = the resolved mean velocity of the water along the path.

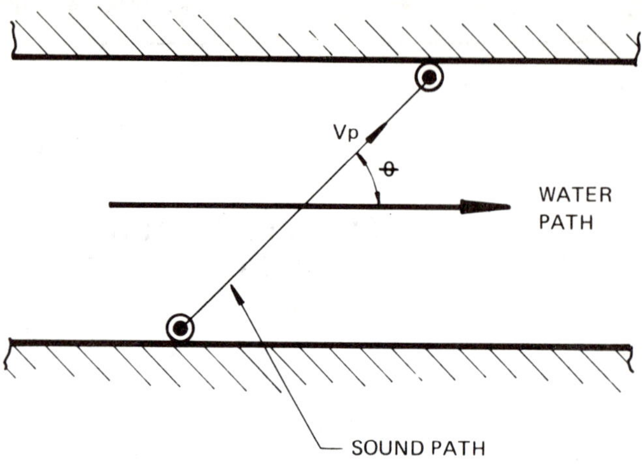

Figure 2. Schematic Plan of River Showing Angled Flight Path.

The equations combine to give:

$$Vp = \frac{L}{2}\left(\frac{1}{t_1} - \frac{1}{t_2}\right) \qquad (1)$$

(which is independent of c)

$$(\text{also } c = \frac{L}{2}\left(\frac{1}{t_1} + \frac{1}{t_2}\right)) \qquad (2)$$

The average velocity of the water V_1 is given by:

$$V_1 = \frac{V_p}{\cos\theta}$$

where θ is the angle between the direction of flow and the flight path.

Multipath Sequencing

On completion of one measurement of t_1 and t_2 at a particular depth, corresponding measurements of time are made at other depths in a similar manner, in sequence, starting at the transducer pair nearest the bed and continuing up to the valid pair, nearest to the surface (see Figure 3).

From the various measurements of transit times, transducer depths, angles to the flow, the parameters can be combined with a depth measurement to calculate flow using a microprocessor. The processor decides which transducers are at valid depths based on a measurement of the water surface level and determines whether to excite the upper transducers.

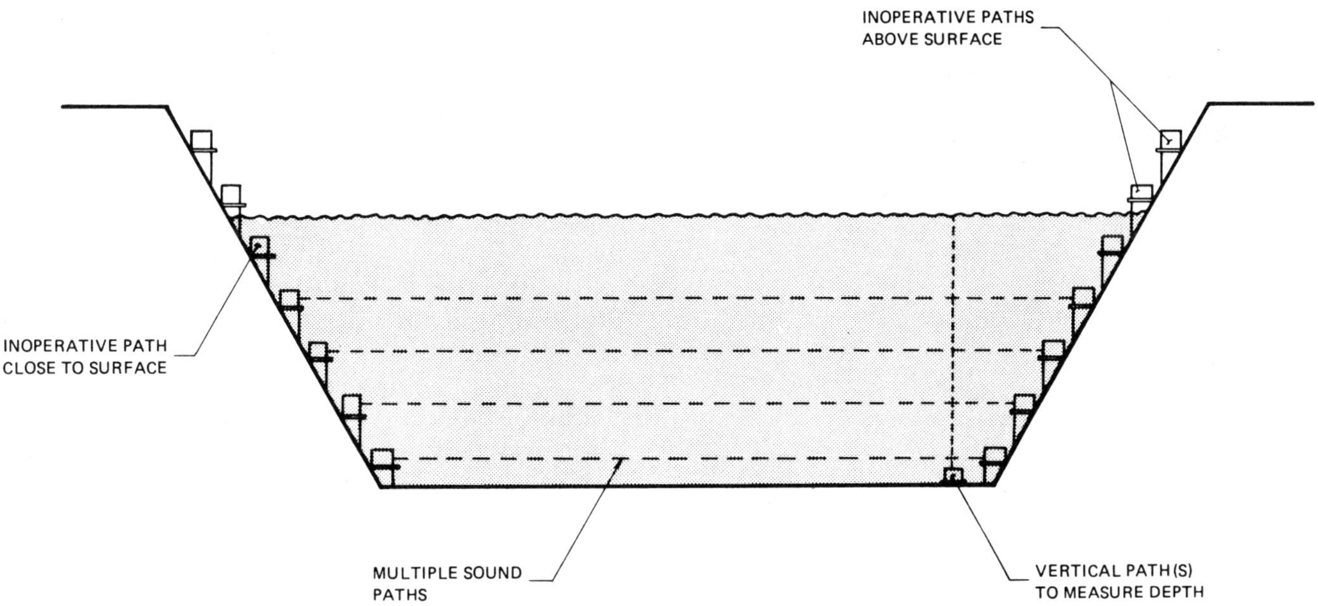

Figure 3. Schematic Section of River Showing Multiple Paths.

Depth Measurement

Depth is measured with similar transducers mounted under the water with beams pointing vertically towards the surface (see Figure 3).

The pulse transmit time to the surface and back to the same transducer gives a measure of river depth. We use the speed of sound already calculated from the horizontal path transit times t_1 and t_2 (see Equation 2). Multiple transducers are used to give several independent measurements of depth and a 'majority vote' is taken to improve on the accuracy and reliability of the measurement. A new feature is to set the 'level' transducers at different datum levels to give a separate computation of the speed of sound and hence improve the accuracy and reliability.

Depth Requirements

Special problems arise for the paths nearest to the surface and the river bed. If the transducers are too near to these boundaries, reflected signals from the boundary can interfere with the direct transmission and cause gross inaccuracies in the time measurement. The path length of the reflected beam is

$$Lr = \sqrt{L^2 + (2d)^2} \qquad (3)$$

where d is the path distance from the boundary layer and L is the length of the direct path.

The minimum difference in path lengths Lr – L must be greater than half a wavelength of the ultrasonic beam, i.e.,

$$Lr - L > \frac{c}{2F}$$

where c is the velocity of sound and F is the carrier frequency of the ultrasonic pulse

$$\sqrt{L^2 + (2d)^2} - L > \frac{c}{2F} \qquad (4)$$

which reduces to:

$$d > \sqrt{\frac{Lc}{4F}} = 0.0274\sqrt{L}\,\text{meters} \qquad (5)$$

for F = 500,000 Hz and L measured in meters. This is a minimum requirement and in practice it may need to be higher than the above figure due to obstructions on the bed or temperature stratification.

Temperature Stratification

Temperature stratification can be a particular problem with low velocity rivers in either severe heat or cold conditions. Here the temperature of the river can be stratified at different levels and, due to the dependence of the speed of sound on water temperature, the beam will bend and in extreme conditions it will not reach the corresponding transducer on the far bank. This problem limits the use of the technique to rivers where stratification does not occur, or the user may have to accept a loss of data at certain times when severe conditions occur. There is a danger that inaccurate results can be obtained under marginal conditions. Similar problems occur in estuarine waters, confluences, or at inflows where unequal mixing causes refraction.

Cross Flow

Further difficulties can arise if the angle θ between the flight path and the direction of the flow is not known accurately. However, the problem can be overcome by using a "crossed path array" (see Figure 4; Va is the actual velocity of the water, V_1 is the assumed velocity of the water, and Vp is the measured velocity component along the beam).

Now

$$Va = \frac{Vp}{\cos(\theta \pm \delta)}$$

The discharge in one "slice" of water depth H is

$$\frac{Vp}{\cos(\theta \pm \delta)} \times L \sin(\theta \pm \delta) \;\; H = Vp.L.H. \tan(\theta \pm \delta)$$

If δ is assumed zero then the assumed discharge is:

$$Vp.L.H. \tan \theta$$

which gives a percentage error:

$$\left(\frac{\tan \theta}{\tan(\theta \pm \delta)} - 1\right) \times 100$$

The error is +19 percent or −16 percent, depending on the skew direction for $\theta = 45^o$ and $\delta = 5^o$. A second crossed path gives corresponding errors in the opposite direction and hence the errors can be combined to give a small residual error.

Signal Processing

The degree of complexity of measuring the times of the multiple paths and combining the measurements has only been made economically viable by the use of a microprocessor to carry out the necessary computations.

Further use of the processor can be made to ensure that data fits a sensible pattern by ensuring that the velocity profile with depth fits a near logarithmic relationship and that the velocity at any level does not change significantly compared with other levels. Pulse arrivals must fit into defined "windows" in time to ensure the raw data is valid. The expected arrival time is known from sound velocity measurements and path lengths since t_1 and t_2 do not differ by more than 0.5 percent.

If a path fails it is ignored and, by interpolation or extrapolation, the next best calculation of discharge is made. Many time measurements are made over the measurement section for each period of several minutes. Flow pulsations are averaged out and any temporary loss of transmission (caused by, say, flotsam in the beam) allows for a partial measurement during the period.

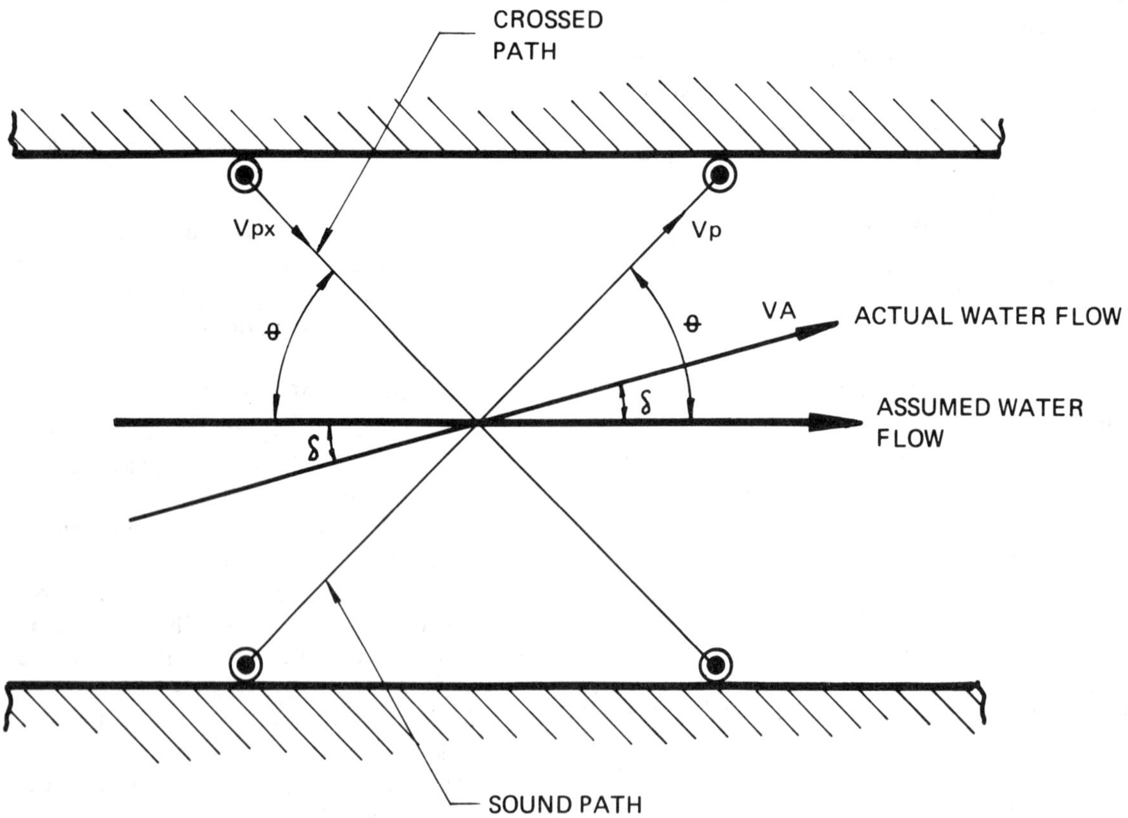

Figure 4. Schematic Plan of River Showing Crossed Paths.

Transducer Electronics

The transducers are driven from head amplifiers and multiplexers mounted near the transducer arrays. The multiplexers route the control signals to the appropriate amplifiers under the control of the microprocessor. Any number of paths from 1 to 32 can be used (including the ultrasonic depth gauges). Electronic counters measure from the time the transducers are driven to the time the two receive pulses are detected. The counters then transfer onto the following paths using the processor to control the multiplexer. The use of electronic switches to route the signals, enables rapid switching between channels and a time and spacial integration across the river section. A feature of the gauge is that additional paths can be added at a later date, without great difficulty, by adding further head amplifiers and transducers. With narrower rivers, particularly if reflectors are used, the head amplifiers may be mounted with the main control unit.

Reflectors

When using an ultrasonic flow meter by transmitting pulses of ultrasound at an angle across a channel there are advantages in using a reflector(s) on the far bank.

(a) No cable has to cross the river.

(b) There is no active element on the far bank to require power.

(c) The main and reflected beams make different angles to the direction of flow and give a first order cancellation of errors caused by skew or crossed flow.

The preferred type of reflector is a corner reflector in the vertical direction in the form of a length of angled steel. This reflector is the only type found to be noncritical to align. These are shown in Figure 5, a photograph of an installation using 14 transducer pairs and reflectors on the River Severn.

Figure 5. Multiple Corner Reflectors on the River Severn.

The reflectors may need regular cleaning to maintain their reflective properties. The cleaning period depends on the quality of the water.

For an Ultrasonic River Gauge Site

(a) The channel should be straight for at least ten times the river width.

(b) The channel width can be in the range 2 to 200 meters. Greater widths are possible but the factory should be consulted. Narrower channels may favor the E. M. method.

(c) The channel cross-section should preferably be symmetrical and the bed should be level.

(d) The channel should be free of weed growth which can otherwise disperse the beam.

(e) Entrained air caused by 'rapids,' waterfalls, etc., must be avoided as this will disperse the beam.

(f) Upstream inflows to the river at different temperatures or laden with high salt content should be avoided as the beam can be refracted. Similarly, gauging should not be attempted just downstream from a confluence since uneven mixing can cause beam refraction.

(g) River sites with an unstable bed should be avoided.

(h) Wide, shallow rivers should be avoided otherwise the beam may reflect from the surface or the bed and cause gross inaccuracies.

(i) Rivers with low velocity should be avoided where possible since temperature stratification can cause refraction of the ultrasonic beam.

THE ELECTROMAGNETIC RIVER GAUGE

History

Faraday, in 1832, was the first to realize that water, flowing in earth's magnetic field, generates a potential, which is proportional to the average velocity of the water and an attempt was made using the method to gauge the River Thames. In 1953-54 the tidal flow through the Straits of Dover was gauged and other large estuaries have since been gauged. The application of the technique using the earth's field for gauging the flow in small fresh water rivers was considered, but the very small undirectional potentials induced in small channels cannot be measured in the presence of interferring signals. Workers have used an electromagnetic flow meter with an electromagnet excited by an alternating current to generate the magnetic field instead of using the earth's magnetic field. This technique has been successfully applied to pipe flow meters and point current meters, and a similar technique has been used for the small channels. Figure 6 gives a schematic illustration of the technique. A coil is buried below the river bed or bridged across the channel to generate a vertical magnetic field across the full channel width. Electrodes, set into opposite banks of the channel, pick up an induced potential

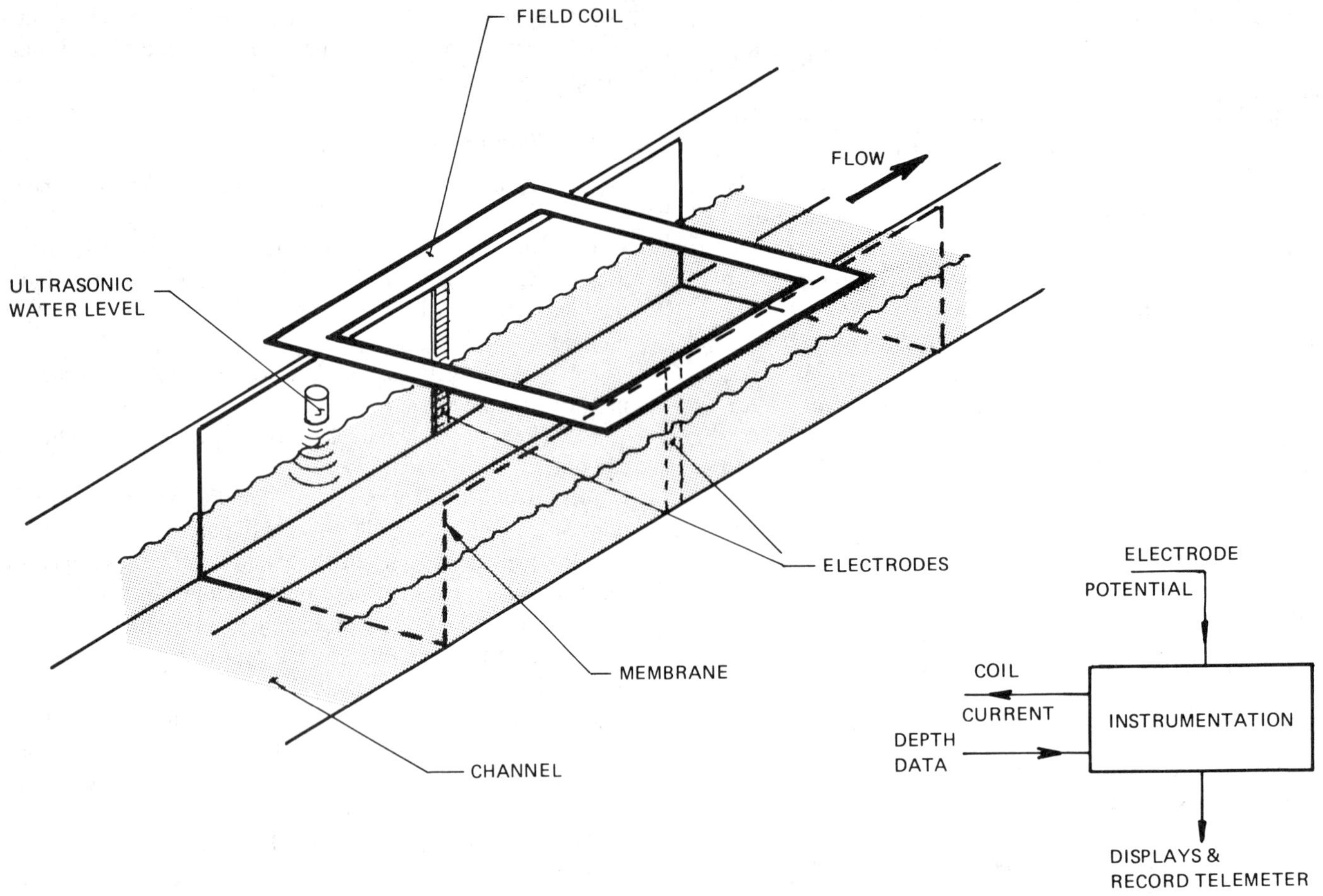

Figure 6. Schematic Diagram of Open Channel Electromagnetic Flow Meter.

generated by the water flowing through the magnetic field. The site photograph (Figure 7) shows an example of the coil, electrodes, and membrane installed on a small river.

Figure 7. Electromagnetic Flow Meter on Swillbrook.

Control equipment drives current through the coil and measures the induced potential which is proportional to the average velocity of the water. This combines with a measurement of river cross-section to give an output of discharge. An insulated membrane isolates the induced potential from the bed of the river which otherwise would attenuate the signal due to the bed acting as an electrical conductor.

In certain situations the bed attenuation can be measured and a compensation made of this effect. However, the compensation is difficult to achieve and at low velocities the attenuated signal becomes very difficult to measure. This problem limits the use of the gauge, with no membrane, to rivers with high velocities and small variations in depth and conductivity. This is typical of a chalk stream with a steady base flow.

Theory

Faraday's law of electromagnetic induction relates the potential generated to the length of a conductor moving in a magnetic field by the equation:

$$E = Tv.b \text{ volts}$$

where:

E = e.m.f. generated transverse to the flow (volts),

T = magnetic field intensity (Tesla),

v = average velocity of the river water (meters per second), and

b = river width (meters).

The minimum measurable direct potential would be of the order of 10 millivolts since there are interfering polarization voltages. Assuming some bed losses, this results in a minimum river width of the order of 1000 meters and a velocity of 4 meters per second to generate a sufficiently large potential for accurate measurement. This would limit the technique to the very largest rivers such as the Mississippi, the Colorado, the Nile, the Danube, etc. and estuaries, and the method could not be used near the Equator where the Earth's field is horizontal. For smaller rivers an alternating field is generated with the coil which is bridged across the river and this enables the measurement of much smaller voltages.

The effect of the bed attenuation can be avoided by lining the channel with an insulating membrane. The advantages of the membrane are to increase the probe voltage and avoid inconsistency caused by inaccurate compensation of the bed attenuation. The disadvantage of the membrane is the cost factor, both for the basic material and installation difficulties, and this limits the maximum river width.

Electromagnet

Depending on the river dimensions and the range of flow to be measured the electromagnet coil will be wound with insulated copper cable using between 100 and 300 turns. The coil will normally be installed in ducts laid in troughs in the river bed, excavated with a drag line dredger. Manholes in the banks of the river at each corner of the coil will give access for installation. The coil is excited with a current of between 3 and 5 amps from the coil switching unit, a part of the instrumentation. The magnitude of the magnetic field will be between 300 and 1000 ampere turns to produce an oscillating magnetic field (this in turn produces an alternating potential of the same frequency at the voltage electrodes). The coil is rectangular in plan and extends just beyond the river width. The coil may be bridged across the channel for smaller channels up to about 5 or 6 meters wide. This may be cheaper than burying the coil in the river bed.

Coil Power Supply

The normal mains supply is transformed, rectified, and filtered to produce a direct potential with a low ripple content. The current is fed to the electromagnet coil and reversed at the required switching rate using semiconductor switches in the processing unit. The voltage applied to the coil will vary according to the width of the channel from about 24 volts to 240 volts, and the power fed into the coil will vary from about 50 watts to 1 kilowatt, again depending on the width of the river.

Electrodes

Figure 8 shows a photograph including a voltage electrode fixed to the side of a 6-meter wide channel where the section has been made rectangular using brick walls. (The electromagnetic flow meter at this site has replaced a previously installed weir which had insufficient range.) The probe material is 20 s.w.g. stainless steel sheet. The length of the probe is approximately 1 meter and the width of the probe is 125 mm. The surface of the probe is covered with a porous polyethelene membrane to minimize movement of water against the probe which would otherwise cause variable polarization effects. At other sites stainless steel rod electrodes are used, again covered with a porous material to avoid movement of water at the probe.

Figure 8. Electromagnetic Flow Meter on the River Little Test.

Instrumentation

The signals from the voltage electrodes are measured in the instrumentation. Particular care has been taken with the electronics to achieve a large measurement range approaching 140 db so that the signal can be filtered from noise picked up by electrodes. The signal voltage, which is proportional to the velocity of the water, is digitized, combined with depth and site data, and finally processed in the instrumentation using a Z80 microprocessor with associated memory and computer program to provide an output of discharge on one or more output recording devices.

The instrumentation is based on a microprocessor bus bar system which gives a high degree of flexibility in providing output options. The same system is also used on the ultrasonic flow gauge and the features of the processor are described later in this article.

Performance

The electromagnetic river gauge has been evaluated against several check methods in artificial channels and natural rivers. Accuracies of the order of ±2 percent are being achieved.

Applications

The electromagnetic river gauge is particularly suited to rivers where there is considerable weed growth or buildup of debris, such as accretion in mountain streams. There is no problem with entrained air or temperature stratification or high sediment content. It can also be used to gauge effluent flow in open channels. The method tolerates a large variation in stage, measures reverse flow, is unaffected by backup, and causes no change in water level or constriction to navigation or fish migration. There is no visual indication of the presence of the station (apart from an instrumentation hut) if the membrane is covered and the coil is buried.

Ideal Sites

The following paragraphs list the major parameters of sites that are to be considered when making the decision on whether a site is suitable for an Electromagnetic Gauge:

(a) The channel should be straight for at least three times the river width.

(b) The channel width should be less than 25 meters on a price criteria.

(c) The section of the channel should be symmetrical about the river center line.

(d) The banks should preferably be near vertical or have a one-in-two slope, otherwise field or scale model calibrations will be necessary.

(e) There should be no major source of electrical interference near the site. Electrified railways, fish screens, electrical power grid lines, and radio transmitters are the most likely sources of interference. A site survey to measure the noise is necessary where doubt arises.

(f) Entrained air, high sediment load and temperature stratification effects, weed growth and variable accretion are unimportant and a feature of the electromagnetic technique.

(g) Heavily reinforced concrete or steel piling can distort the magnetic field and should preferably be avoided, otherwise on-site calibration will be necessary.

(h) Data is necessary on the range of flows, depths, and velocities to be measured to determine the necessary field strength of the electromagnetic. Sections with higher velocities are preferable and should be more than 0.03 meters/second to achieve high accuracies.

(i) The channel needs to be lined with an insulating membrane. There may be difficulties in carrying out this exercise. The membrane is heavy and an area is necessary on the side of the river to maneuver the membrane and possibly temporarily divert the channel. A degraded version of the electromagnetic gauge is possible with no membrane.

(j) The depth to width ratio of channel area and the variation in depth are not important which is a feature of the electromagnetic technique.

ELECTROMAGNETIC PIPE FLOW METER

A special development of the Electromagnetic Flow Meter has been used to gauge flow in part filled sewer pipes and culverts. The gauging of flow in part filled pipes presents a number of problems not normally associated with the established 'full pipe' flow meters. The main differences are as follows:

(a) The water has a free surface.

(b) Gravity flow is usally much slower than in full pipes where the flow is usually pumped.

The main impact of these differences has been the necessity to use instrumentation which is capable of measuring much smaller voltages than existing pipe flow meters. The instrumentation is very similar to that used for river gauging.

The other important aspect of the free surface flow meter is the need to measure surface level. The instrumentation has been designed to interface to several water level sensors, including float gauges, air reaction manometers, pressure gauges, and ultrasonic level gauges. The preferred method depends on the range of level and the application. We mainly prefer the ultrasonic level gauge which is a non-contract method.

The Operational Flow Meter

A Glass Reinforced Plastic (G.R.P.) pipe section is installed in an existing chamber or a specially constructed manhole (see Figure 9). This pipe section is sealed to an existing pipe to form a continuous line. The G.R.P. section contains prefitted electrodes and an electromagnet coil is installed above the pipe.

Where access to the chamber is restricted, the G.R.P. pipe is cut into segments which are taken through the existing access and the segments reassembled inside the chamber. A cover in the pipe allows for access to the electrodes for maintenance.

A vertical pipe is "Teed" into the G.R.P. section and used as a sonic wave guide for a level sensor.

The principle of operation is identical to the open channel flow meter described above, and similar instrumentation is used.

The coil is again excited with a low frequency square wave current to set up a vertical magnetic field across the

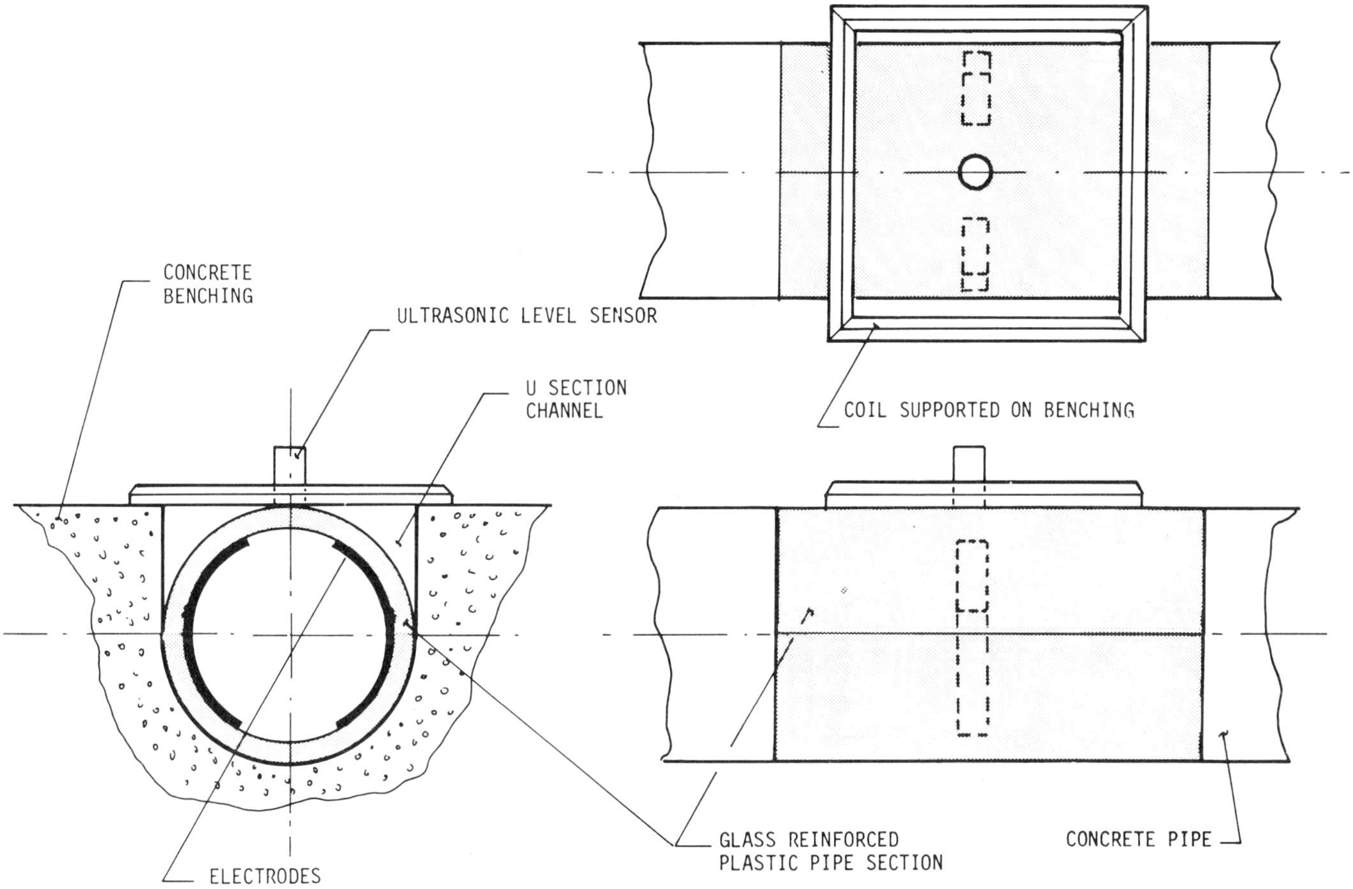

Figure 9. Schematic Diagram of a Glass Reinforced Plastic Sewer Pipe Flow Meter.

pipe. The coil cable consists of 100 turns of 2.5 mm^2 and a current of about 3 amps is used to excite the coil. The flowing water again generates a small voltage between opposite sides of the pipe, which is approximately proportional to the average velocity of water in the pipe. This voltage is picked up by the electrodes at the sides of the pipe. A porous polythene membrane is again used to cover the electrodes to limit wave movement against the electrodes and this membrane also reduces greasing problems at the electrodes. Strip electrodes are used made from stainless steel. If the section were rectangular with a uniform magnetic field, we would obtain a true average of the velocity of the water in the section by measuring the induced voltage. The flow meter would then be independent of varying velocity profiles. However, with circular sections and spatially limited magnetic fields, the averaging is less than ideal. Practical experiments, supported by calibration tests at the National Engineering Laboratories, show that accuracies of the order of ±4 percent can still be achieved with typical variations in flow profiles and backup effects in the pipe.

The G.R.P. section is a naturally insulating material and overcomes the problems of the loss of signal due to leakage currents in the bed of the channel or pipe.

Applications

The main application of this new part filled electromagnetic pipe flow meter is to gauge domestic and industrial effluent and storm water in underground sewer pipes. The flow meter will operate from the dry weather flow condition to the surcharge condition.

Where there is a requirement to measure flows at various points in either a storm sewer or a foul sewer network a number of G.R.P. pipe inserts can be installed fitted with electrodes, coils, and mountings for sonic level sensor. The instrumentation may then be temporarily (or permanently) connected to the pipe section and continuous flow measurements taken. When flow records at this point have been obtained over the required period of time, the instrumentation may then be transferred to a second site.

It should be noted that the improvements to the electronics for both the part filled pipe flow meter and the river flow meter have made the technique suitable for pipe diameters beyond those currently available for full pipes. The maximum diameter would be set by the pipe technology, and hence application can be made to large pipes under pressure, or to tunnels used for hydroelectric power systems, etc. The maximum pipe diameter would be similar to that of rivers, i.e., about 25 meters.

PROCESSING INSTRUMENTATION

The instrumentation is based on a microprocessor bus bar system which gives a high degree of flexibility in providing output options. This same system is used on the SARASOTA Electromagnetic Flowmeter and the Ultrasonic Flowmeter to give a number of common modules, with consequent advantages of simple manufacture and common spares holdings. Figure 10 shows a common block diagram of the two flow gauges. Special modules for each gauge are of course necessary and these are highlighted on the block diagram. Photographs show the control units for each gauge.

The common modules of the two systems are:

(a) The central processor module, which holds the Z80 microprocessor and the operating system program.

(b) The processor timer to provide timing signals to the processor.

(c) An optimal real-time clock for tagging data.

(d) An addressing module which feeds signals to either the coil switching unit for an electromagnetic gauge or to the transducer multiplexer in an ultrasonic gauge.

(e) A station data module which stores the necessary site constants and random access memory for storing data.

(f) An operator switch module which interfaces to controls on the instrument panel and a keyboard.

(g) A programmable read only memory module where the main system program is stored.

(h & i) Two modules to drive numerical displays on the instrument panel.

(j) A parallel output driver which can either drive a special telemetry system or alarm indicators on the instrument panel.

(k) An analogue module to drive a chart recorder.

(l) A cassette module to drive a magnetic cassette recorder.

(m) An RS232 module to communicate with a variety of serial data outputs.

(n) A depth gauge interface module.

(o) A paper tape punch driver module.

The main control unit is also common to the two systems and consists of a printed circuit card frame, power supplies, and a motherboard printed circuit board with

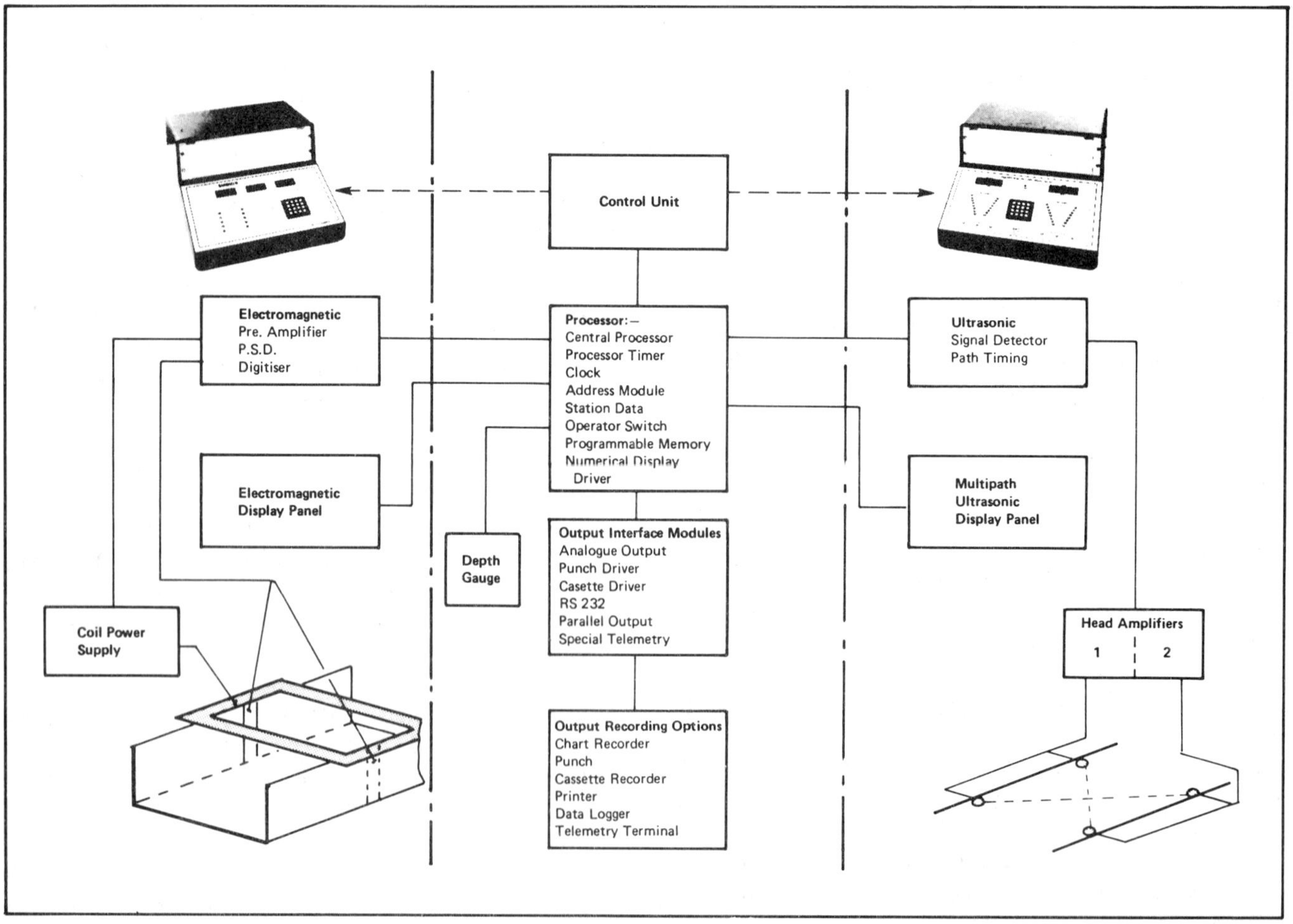

Figure 10. Block Diagram of Ultrasonic and Electromagnetic Instrumentation.

defined interconnections in the form of a bus bar structure to interconnect the various modules. A special bus bar structure has been used to allow for the digital signaling and low level analogue signals with special earthing arrangements.

Special modules for the electromagnetic gauges are:

(a) a preamplifier,
(b) a phase sensitive detector,
(c) a digitizer, and
(d) a separate coil power supply.

Special modules for the ultrasonic gauge are:

(a) a signal detector,
(b) a path timing unit, and
(c) a separate head amplifier unit for the transducers.

The output options for both flow gauges can be arranged to display or record flow, depth, and various other parameters associated with the instrumentation and in particular the L.E.D. displays, give data on engineering parameters, system constants, stored data, etc.

The control unit can also accept inputs from other sensors and act as a data logger or a telemetry outstation.

Software

A software operating system is used for the microprocessor for both gauges and a feature is the use of distributed software and hardware subroutines so that where the system extensions are required this is achieved by plugging in additional hardware modules which include any necessary software subroutines.

This feature has enabled considerable flexibility so that the software can be extended on a modular basis without interfering with the main program.

CONCLUSIONS

The two new electronic methods of river flow gauging are showing advantages over traditional methods. Rivers can now be gauged without creating afflux, difficult sites with weed growth can be gauged, variable backwater is unimportant, and most significantly the cost is lower, particularly on the larger rivers.

The electromagnetic method of measuring flow has been applied to part filled pipes for gauging effluent and storm water in underground pipes.

Full advantage has been taken of the latest microelectronic computing techniques to give significant commonality between the two new gauges and the modular technique employed allows for considerable output recording options.

An Annex, written by Dr. S. T. Walker of the Severn-Trent Water Authority, gives a user's reaction to the new methods.

Annex

EXPERIENCE WITH ADVANCED TECHNOLOGY OPEN CHANNEL FLOW GAUGES IN A TYPICAL BRITISH WATER AUTHORITY

S. T. Walker[2]

Severn-Trent was among the last British Water Authorities to install first generation single-path ultrasonic gauges but has been in the forefront of development and use of both multipath ultrasonic and second generation electromagnetic gauges. Both have become available at opportune times for resource hydrometry and have provided solutions to some hitherto intractible hydrometric problems. From a quickly established base in open channel measurement, electromagnetic technology is now being tested in another difficult area – sewer flow measurement – and promises high performance at reasonable cost.

At the time of writing, Severn-Trent is in the process of commissioning four multipath ultrasonic systems that have been designed to give an extremely high level of performance. Considerable information redundancy has been provided through path and water level sensor replication, to achieve both high accuracy of measurement and low vulnerability to damage or sensor malfunction. Crossed path and signal reflection techniques have both been utilized to productive effect and, in different applications, the systems provide both significant economy and much improved performance over conventional equivalents – where these would have been viable at all. Of the four systems referred to, the first (the reflected signal variant at Bewdley, the control gauge for the regulated River Severn) has been operational for two months and is performing well.

The Authority also has five operational electromagnetic systems at open channel sites that vary from 3 meters to 20 meters in width. The longest running of these, on a twin effluent channel at Derby Water Reclamation Works has functioned with virtually no down time for 2½ years, and repeated checks continue to confirm the stability of its rating – this despite a deliberate "no maintenance" policy. This gauge provides highly accurate measurement in an extremely complex hydraulic environment of rapidly varying flows under varying degrees of drowning by the receiving river. At other natural river sites, lack of head, weed infestation, and differential drowning are all in the process of being overcome, although calibration work is still in

[2]Principal Officer, Measurement and Instrumentation, Severn-Trent Water Authority, Abelson House, 2297 Coventry Road, Sheldon, Birmingham, B26 3PU England.

progress on the sites that have trapezoidal cross-sections. Measurement sensitivity and accuracy has been good and electronic reliability excellent.

ORGANIZATION OF INSTRUMENTATION CONTROL AND DEVELOPMENT WITHIN THE U.S. GEOLOGICAL SURVEY WATER RESOURCES DIVISION

C. Russell Wagner[1]

ABSTRACT: The U.S. Geological Survey, Water Resources Division (WRD) has made a major commitment to upgrading hydrologic instrumentation in the 1980's. The Hydrologic Instrumentation Facility (HIF) was established in 1980, and given nationwide responsibility for research, development, testing, evaluation, procurement, warehousing, distribution, repair, and calibration services for hydrologic instrumentation. The reorganization of these activities into a single facility has resulted in improved hydrologic instrumentation services to the Survey.
(KEY TERMS: instrumentation; hydrology; hydrologic data; data collection equipment.)

HYDROLOGIC INSTRUMENTATION FACILITY

The U.S. Geological Survey's Water Resources Division (WRD) is responsible for assessing the quality and quantity of the nation's water resources. It develops scientific knowledge for understanding these resources. The Survey, as a strictly scientific organization, has no management or regulatory authority. The WRD is field oriented with approximately 90 percent of its staff operating from District offices located in nearly every state of the Nation. The vast water data collection activities require a large amount of specialized hydrologic instrumentation. The recently created U.S. Geological Survey, NSTL Station, MS Hydrologic Instrumentation Facility is a service facility to support the Survey's use of these specialized instruments.

Early in the history of the Survey, hydrologic instrumentation was either invented by innovative field personnel or they used modified commercial items. Many of the items currently used by Survey personnel bear the names of early engineers within the Survey. The first organized hydrologic instrument development efforts were initiated in Columbus, Ohio, in the early 1930's as a project of the Survey's Columbus District office.

During recent years, limited procurement, warehousing, distribution, and repair functions were developed under the jurisdiction of the Survey's Administrative Division, and were located at various sites around the Washington, D.C., area. These activities were integrated with other Geological Survey logistical support programs. The need for more advanced hydrologic instrumentation became apparent as the Survey became involved in complex hydrologic studies, money, and personnel resources became scarce, and centralized guidance was needed to provide such instrumentation.

In 1977 Survey senior staff decided to reorganize several groups supporting instrumentation into a single organizational unit. This unit, named the Hydrologic Instrumentation Facility (HIF), was located at the U.S. Geological Survey's Gulf Coast Hydroscience Center at NSTL Station, Mississippi. The facility became operational in March 1980.

National Space Technology Laboratories (NSTL) is a National Aeronautical and Space Administration (NASA) site that is being converted from a space vehicle engine test center to a host facility for numerous Federal agencies conducting high technology research. It is located approximately 50 miles northeast of New Orleans, Louisiana. One of the advantages of the HIF being situated at NSTL is that NASA, the host agency, provides on-site contractual support services, including broad consulting engineering expertise, drafting services, photographic services, publication services, and a variety of other technical support specialties.

The HIF has been given nationwide responsibility for all aspects of hydrologic instrumentation, in support of Survey programs. Program authority is under the guidance of the Instrumentation Coordinator, a staff position in the Office of Assistant Chief Hydrologist for Operations. An Instrumentation Committee composed of field managers and technical specialists was created to provide long term planning and oversight to the program.

The HIF has a staff of approximately 50 professional, technical, and clerical personnel organized into six sections. To ensure that the HIF maintains responsiveness to field needs, most management positions have been staffed with personnel having previous Survey field experience.

[1]Chief, Hydrologic Instrumentation Facility, U.S. Geological Survey, Water Resources Division, Gulf Coast Hydroscience Center, Bldg. 2101, NSTL Station, Mississippi 39529.

HIF has 36,000 square feet of space, laboratory, and warehouse space.

Instrument Development Laboratory

The Instrument Development Laboratory (IDL) is the principal research and development resource of the Water Resources Division (WRD) in the area of hydrologic instrumentation. It has three broad assignments:

1) Updating and modifying existing instrumentation to take advantage of improved materials or technology;

2) Continued development of instrumentation compatible with equipment owned by the WRD at the existing level of technology, and

3) Research and design of totally new instrumentation that will lead the state of the art for instrument packages of the future.

Research and development is conducted both in-house and by contract with public and quasi-public research groups and private enterprise. Personnel have responsibility for developing criteria, statements of work, and monitoring contracts for outside research. This section has staff members with field experience in surface water, water quality, ground water, and hydrometeorological data needs, as well as a technical staff of electronic, electrical, and mechanical engineers and technicians.

The demand for hydrologic instrumentation normally is not large enough to stimulate private sector development of complete systems that meet WRD requirements. For many years WRD has purchased individual components from a variety of manufacturers and either modified the units or built custom interface units to connect them together. The result is a variety of power supplies, performance expectations, and lack of interchangeability. Also, the Survey is subject to the individual manufacturer's model changes and is forced into sole source procurements. The HIF is developing a systems approach to families of instruments that will be interchangeable and compatible in terms of power sources, inputs, and outputs and will retain patent rights to this equipment so that competitive procurement will be possible.

Some of the technologies that the HIF is exploring includes measurement of flow by electromagnetics and acoustics. A variety of noncontact means of recording water levels, as well as pressure transducers, are also being explored.

Test and Evaluation Section

The Test and Evaluation Section establishes minimum performance standards for instrumentation to ensure that WRD data collection requirements are met and that legal and scientific credibility is maintained. The section establishes acceptance standards, develops quality control procedures and test methods. The Section works closely with other Federal agencies so that its evaluation criteria are in compliance with all applicable legal and operational requirements.

The Section designs and conducts engineering tests and analysis of both USGS developed and commercial hydrologic instrumentation. It operates a variety of testing equipment, including a large environmental test chamber, that simulates worst case conditions of temperature, humidity, and vibration. Results of temperature and humidity failures has aided engineers in improving the design of many pieces of equipment. The HIF is now able to drive a prototype unit to failure to determine which components are most vulnerable.

A "qualified products list" (QPL) is being developed as a guide to the procurement of commercial products meeting USGS requirements. Federal Regulations allows the government to require prospective bidders to have their product tested and qualified for the QPL before bids are submitted in response to a procurement solicitation. After a QPL is established via an announcement in the Commerce Business Daily, only vendors whose products have met the requirements of the published specification will be invited to bid. The only published QPL at this time relates to stage measuring devices, but others will be prepared in the future.

The combination of tight accuracy standards and exhaustive testing should result in the delivery of more reliable instrumentation to field users.

Field Service and Supply Section

The Field Service and Supply Section operates a warehouse, initiates procurement actions to purchase equipment, monitors contracts, and fills field equipment orders from Districts. In most cases field equipment is sold to field offices. In the case of widely used and complicated monitoring equipment, the HIF maintains a rental program to the field and retains the responsibility of procurement and maintenance of the equipment. The processing of field orders is supported by an interactive Computerized Support System (HIF/CSS). This system allows field users to scan the computerized catalog and place orders directly from their terminal. The HIF/CSS also generates accounting documents that credit the HIF account and debit the ordering District accounts within the Survey. The computerized system tracks inventory and procurement actions and generates management reports on sales patterns that will improve future stocking and procurement decisions.

The Field Service and Supply Section maintains close liaison with field offices on equipment needs and adds or deletes items to the inventory in response to field needs. Presently there are over 700 items in the inventory. With the exception of some convenience items such as boots and waders, all of the items either are built to USGS plans and specifications, or require tight quality control procedures to ensure that Survey data collection standards are met.

1 The HIF warehouse sells a significant volume of hydrologic instrumentation to other Federal agencies. This section responds to frequent calls from other Federal agencies regarding the whole range of instrumentation problems and possible solutions.

Repair and Calibration Section

The Repair and Calibration Section provides maintenance, repair, and calibration services for both District owned and rental program equipment. The section operates a complete machine shop and a modern electronics repair shop. Capabilities exist, either in-house or by contract, to repair, adjust, or calibrate virtually any type of USGS owned equipment. Charges to field offices are assessed on a straight time and materials basis.

On request, work is performed for other Federal agencies and state and municipal agencies that are cooperators of WRD Districts.

This section fabricates and modifies many unique and specialized pieces of equipment which are sold through the Field Service and Supply Section, and is responsible for building prototype devices developed by the Instrument Development Laboratory.

Routine stock items requiring quality control testing or precalibration are processed by the Repair and Calibration Section before being stocked in the warehouse.

Technical Services Section

The Technical Services Section provides in-house support by preparing, maintaining, and updating technical documentation for all Water Resources Division equipment. A file of selected industry standards for common components is maintained to assist design engineers in standardizing WRD equipment to the greatest extent possible. This section prepares specifications and engineering drawings as requested by other HIF sections; maintains and controls changes to these specifications and engineering drawings, provides copies of specifications, engineering drawings, and electronic artwork to contractors; and assists in monitoring contracts for procurement of equipment.

This section is responsible for operating a management computer data base. The present support consists of the Computerized Support System (HIF/CSS), which directly supports a variety of inventory, accounting, and administrative functions. The system improves cost accounting and controlled property transactions to both the HIF, ordering field offices, and Administrative Division personnel.

Administrative Services Section

The Administrative Services Section is responsible for the formulation and execution of the budget, accounting, personnel, procurement, and office management necessary to suport the operating needs of the HIF. It provides procurement services, maintains supply source and procedure information, including an automated microfilm system of GSA supplies, open market supplies, industrial specifications, Federal specifications, and military specifications.

This section also manages several rental programs, wherein commonly used instrumentation is procured and owned by the HIF, and rented to the field offices on a monthly basis. This concept is highly successful in maximizing the use of field instruments. Currently, the rental program includes 14,000 Analog Digital Water Level Recorders, 500 Water Quality Monitors, and 40 microprocessor controlled programmable water quality monitors. A new program is being established to rent Satellite Data Platforms.

SUMMARY

The consolidation of numerous hydrologic instrumentation support activities into a single organization has provided the mechanism for an efficient and effective approach to total hydrologic instrumentation services in support of Survey programs. Specialized work of the Sections of HIF range from development of new instrumentation to maintenance and calibration of instruments time honored in Survey use.

INTERNATIONAL SYMPOSIUM ON HYDROMETEOROLOGY
JUNE AMERICAN WATER RESOURCES ASSOCIATION 1982

A CRITICAL ANALYSIS OF FIVE YEARS OF SNOTEL PERFORMANCE IN THE ROCKY MOUNTAINS

Garry L. Schaefer and Bernard A. Shafer[1]

ABSTRACT: In 1978, the USDA Soil Conservation Service began receiving hydrometeorological information through the first operational meteorburst communication system called SNOTEL (snow telemetry). This paper provides a comparative analysis of snow pillow sensors and precipitation gage catch in the variable snowpack environments in the Colorado and New Mexico Rockies for a five-year period (1978-82). Telemetered data are compared to "ground truth" measurements and an explanation given for some anomalous events. Telemetry performance for the 1981-82 period is evaluated. During this five-year period, there were several extreme years of record providing a wide range of hydrometeorological conditions under which to examine system performance. Telemetered data are found to be of sufficient accuracy for water supply forecasting applications.

A discussion of second generation units incorporating a microprocessor programmed to perform numerous functions, including computing minimum, maximum, and average air temperature, as well as event recording is presented.

A subtelemetry network has been developed to add more remote sites to the SNOTEL system without adding more costly meteorburst transceivers.

(KEY TERMS: telemetry; SNOTEL; meteorburst communication; snow pillow.)

INTRODUCTION

For the past 50 years, the USDA Soil Conservation Service (SCS) has coordinated the cooperative snow survey program in the Western U.S. excluding California for the principal purpose of forecasting water supplies. The surveys are conducted by sending teams of snow surveyors into the mountains monthly during the winter to manually sample snow depth and snow water equivalent with a coring device called a Federal sampler. As the management of limited surface water resources in the West became more sophisticated, demand mounted for more timely and comprehensive information from remote mountain areas. To help satisfy this demand, SCS pioneered the development and implementation of a snowpack telemetry (SNOTEL) system covering ten western states and Alaska. The system is unique in that it uses meteor burst radio technology to enable communication with remote sites. A comprehensive description of the SNOTEL system configuration has been given by Shafer and Burke (1978), and systemwide operational experience has been presented by Barton and Crook (1980).

The purpose of this study is to analyze the performance of the SNOTEL network in Colorado and New Mexico for the period 1978-1982 and to report the outlook for further enhancements to the system. Performance was analyzed on two levels: 1) determining the efficiency and accuracy of the radio telemetry component, and 2) verifying telemetered sensor values compared with ground truth. Observed annual fluctuations in system responses are related to the variation in number of available transmission paths. Sensors studied were the precipitation gage and snow pillow. Standard statistical tests were used to analyze the influence of systematic and random errors as well as to establish levels of confidence in the transmitted data.

Inconsistencies in telemetered pillow and precipitation gage data are explained as a function of observed physical phenomena. The evolution in the system is described including addition of dedicated microprocessors to specific channels, upgrading of the transceiver component with second generation designs, inclusions of event reporting capability, and the possible use of a subtelemetry system to expand the number of remote sites.

RADIO TELEMETRY (RF) PERFORMANCE STATISTICS

At 5:00 a.m. Pacific standard time, the systemwide poll (SWP) is initiated for 470 remote sites throughout the West. The response to the SWP exhibits a diurnal and annual periodicity as predicted by meteor burst communication theory. The ratio of available meteor communication paths at 6:00 a.m. compared with 6:00 p.m. is approximately 4:1; the ratio between August compared with February is also approximately 4:1. The effect of

[1] Respectively, Assistant Snow Survey Supervisor and Snow Survey Supervisor; USDA, Soil Conservation Service, P.O. Box 17107, Denver, Colorado 80217.

this annual variation on system performance is demonstrated in Figure 1, a plot of the Colorado-New Mexico SNOTEL system response to the SWP versus time. As anticipated, system performance is highest in summer and lowest in winter. Performance was poorest in February 1981 when only 76 percent of the sites responded to the SWP. This low performance is heavily influenced by hardware problems that were subsequently corrected, so it is not necessarily indicative of a normal condition. The best performance was observed in July 1981: a 96 percent response to the SWP. Average system performance from February 1981 through May 1982 was 89 percent.

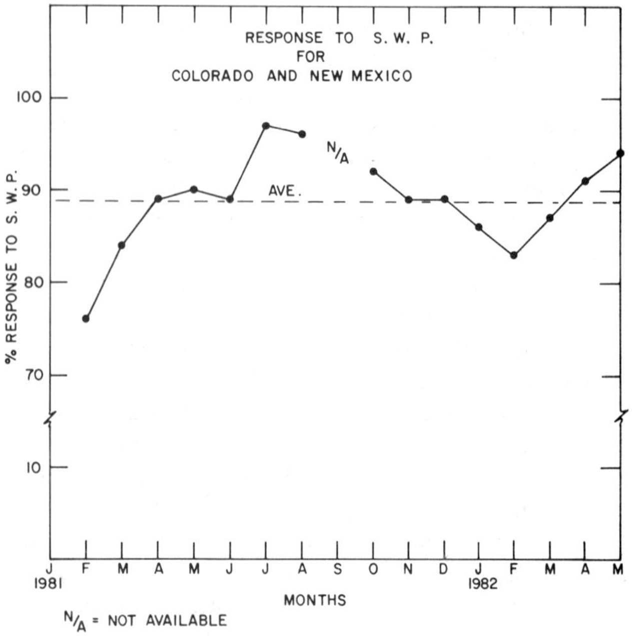

Figure 1. Percent Response to Systemwide Pole (S.W.P.) by Month.

Sites that fail to respond to the SWP can be interrogated on an *Ad Hoc* (demand) basis. Most sites respond to this poll within one hour. Using *Ad Hoc* polls in combination with the SWP improves system performance as measured by a site report in a 24-hour period. The improvement is several percentage points higher than that with only SWP data.

DATA ANALYSIS

Hydrologic Variability

The 1978-82 period included several extreme years and several near average years. The year of 1979 was very wet with winter snowpacks having a return interval of 25 years. On the other hand, 1977 was an extremely dry year with precipitation amounts having return intervals of from 50 to 100 years. The climatological diversity evident in the period was optimum for evaluating sensor performance in a wide variety of conditions.

Precipitation Gage

The standard SNOTEL precipitation gage is a storage device 12 or 16 feet high with a 12-inch diameter orifice. The gage is initially charged with an antifreeze solution and oil near October 1; the oil is used to inhibit evaporation. Precipitation is continually accumulated in the gage for the entire year. Fluid levels in the gage are read visually by means of a manometer in an adjacent shelter house. The static hydraulic head represented by the fluid in the gage is also sensed in the shelter house by a pressure transducer that converts the pressure to an analog signal. The analog signal is subsequently converted to a digital value for transmission.

An analysis was performed to establish the relationship between telemetered precipitation readings and observed values. Observed readings are made manually at periodic intervals by survey teams who record the fluid level in a manometer. Telemetered values are generally those reported in the 5:00 a.m. SWP.

A typical relationship between telemetered and observed precipitation values at the Upper San Juan, Colorado remote site is shown in Figure 2. A highly linear relationship is evident, as expected. A least squares fit expresses the observed value (Y) as a function of the telemetered value (X) plus a constant. For Upper San Juan the equation of best fit is

$$Y = 1.0344X - 0.93 \text{ (inches)}.$$

This equation was derived from 34 observations taken over four years of operation (1979-82). The equation explains 99.8 percent of the variation in the data. The standard error of the estimate is ±0.80 inches.

This type of analysis is performed annually for each remote site to evaluate sensor performance and to adjust for obvious systematic bias in the data acquisition system. However, the equations developed for each site reflect both systematic and random errors. One source of systematic error is in setting the zero offset at the beginning of each water year; this procedure subtracts the initial hydraulic head from the transducers. Other possible systematic errors are malfunctions of transducers and microprocessors at the remote sites.

Random error cannot be accounted for so easily. Sources of random error include observer error, temperature effects on the transducers or other temperature sensitive electronic parts, and occurrence of precipitation between the time data are transmitted and the time an observer reads the manometer. Regression analysis of the type shown helps identify the magnitude of both random and systematic error and indicates when corrective action is warranted.

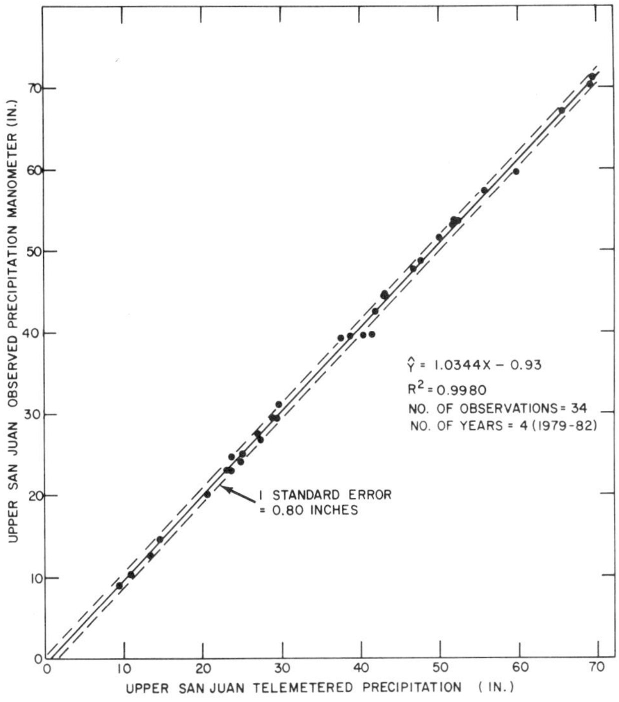

Figure 2. Upper San Juan, Colorado, Observed Precipitation Manometer (value) vs. Telemetered Precipitation (value).

A simple linear regression was performed for the aggregated Colorado-New Mexico system consisting of 58 sites over a five-year period to obtain a "best fit" equation as shown in Figure 3. The equation, obtained from evaluation of 1,077 precipitation observations over a range of 0 to 70 inches, is

$$Y = 1.004X - 0.04 \text{ (inches)}.$$

This equation explains 99.6 percent of the variation in the data and has a standard error of ±0.71 inches.

The hypothesis that observed and telemetered values are identical implies an equation with slope of 1.000 and a constant of 0.00 inches. Comparison of the above derived equation with the hypothesized values indicates a close correspondence, and statistical analysis indicated that the hypothesis of equality could not be rejected at the 2 percent level of significance. A 95 percent confidence interval for the slope was found to be 1.000-1.008. Since the hypothesized slope of 1.000 lies at the lower end of this interval, it is possible that the population slope is really 1.000.

In addition to fitting a linear equation to the data by means of least squares, another approach was taken: the actual difference between observed and telemetered precipitation values was analyzed. The mean difference was calculated by taking the absolute value of the precipitation gage manometer minus the telemetered reading and dividing the number of observations. For 1,077 observations the mean difference was 0.48 inches. The standard deviation of the differences between the observed and telemetered values was 0.71 inches. The average difference expressed as a proportion of the observed value was 4.3 percent.

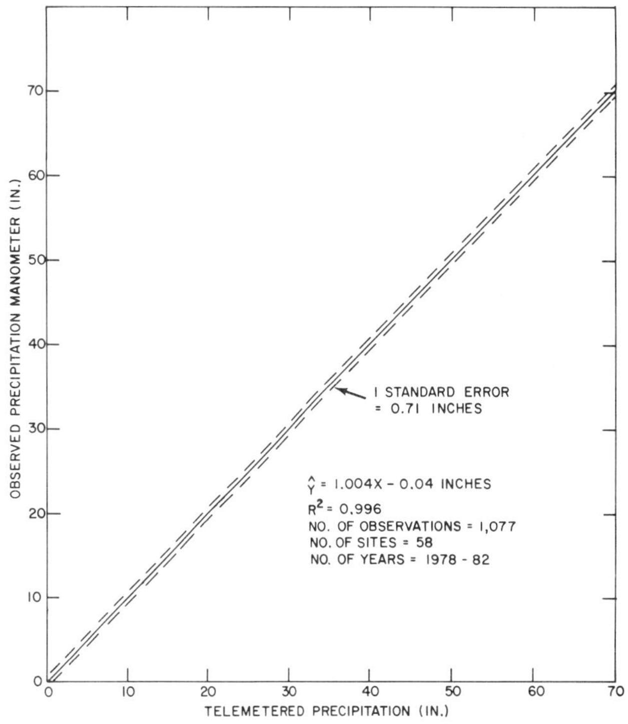

Figure 3. Relationship Between Observed Precipitation Manometer Values and Telemetered Precipitation Values Aggregated for Colorado-New Mexico SNOTEL Sites (1978-82).

The following conclusions can be drawn from these analyses: 1) there appears to be minimal systematic error inherent in telemetering of precipitation data; 2) telemetered values can be expected to be within ±0.71 inches of a simultaneous observation 68 percent of the time; and 3) telemetered precipitation data are accurate enough for use in stream flow forecasting but may be of limited value for applications such as weather modification studies, where resolution of 0.01 inches frequently is required.

Although telemetered precipitation data may correspond to observed manometer readings, they may not in fact reflect reality, as demonstrated in the following description of an anomalous event recorded at the Upper San Juan SNOTEL site in the winter of 1979-80. Figure 4 shows the plot of the precipitation gage daily values and the snow pillow daily values with their respective, periodic observed manual readings. During February, a major storm in the San Juan mountain range deposited 11 inches of snow water equivalent in 10 days as recorded by the snow pillow. This caused a snow plug to form in the precipitation gage. Because of this plug the precipitation gage showed no increase while the snow pillow continued to increase. On March 28 an 11-inch increase in the precipitation gage reading was recorded. This increase was a result of the snow plug dislodging from the side of the gage and partially melting and mixing with the antifreeze solution. Throughout the period in question, telemetered precipitation readings corresponded closely with observed manometer levels.

Occurrences such as this demonstrate the desirability of having dual sensors for measuring winter precipitation.

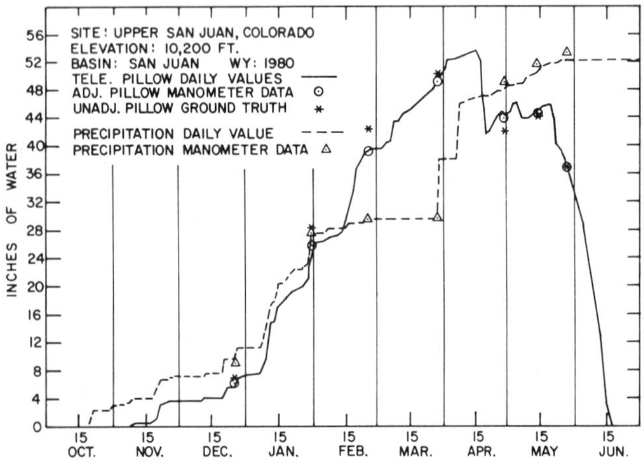

Figure 4. Upper San Juan 1980 Water Year Daily Plot of Precipitation and Snow Pillow Readings Showing an Anomaly in the Precipitation Channel During Mid-February to Mid-April Period.

Snow Pillow

A snow pillow is a hydraulic weighing platform used to measure the water equivalent of the snow resting upon it. Snow pillows in the SNOTEL system are normally 4 ft. x 5 ft. x 1 inch stainless steel tanks plumbed together in configurations of three or four and filled with a 50 percent by volume solution of water and methanol. The pillows are laid flush with the ground surface on a level pad. The hydraulic pressure exerted by the weight of snow on the pillow is transmitted into the shelter house by means of copper tubing. In the house the pressure is sensed by a pressure transducer as well as manifest as static head observable in a manometer. The pressure transducer converts the static hydraulic head to an analog signal, which in turn is converted to a digital value for transmission.

Observed pillow manometer readings and the corresponding telemetered values were analyzed as in the precipitation gage study. A simple regression analysis was employed to check the accuracy of the telemetered readings vs. the snow pillow manometer readings adjusted for the specific gravity of the fluid. Figure 5 shows a typical analysis for a four pillow configuration at the Lake Irene site. The fitted equation is

$$Y = 0.9717X + 0.01 \text{ (inches)},$$

where Y is the pillow manometer and X is the telemetered value. The equation shown in Figure 5 is derived from 31 observations taken over a four-year period. This equation accounts for 99.9 percent of the variation in the data set and has a standard error of ±0.25 inches.

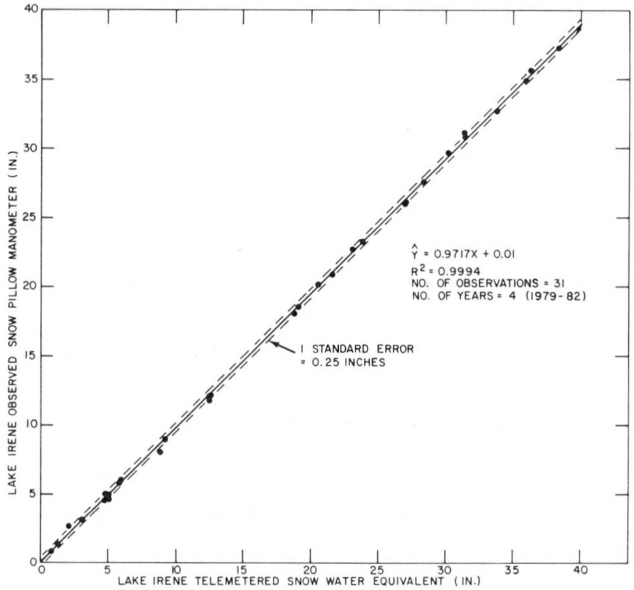

Figure 5. Relationship Between Lake Irene, Colorado, Observed Manometer Values and Telemetered Snow Water Equivalent Values.

As in the case of precipitation data, variation in the relationship between observed and telemetered values can be attributed to both systematic and random errors. Some sources of systematic errors include: 1) incorrect designation of the zero offset used to eliminate the initial static head on the transducer in a no snow condition, 2) malfunctioning of the transducer, 3) application of improper excitation voltage to the transducer, 4) effects of temperature on electronic components, and 5) malfunctioning of transceiver. No attempt was made to adjust any values for such errors prior to analysis.

Sources of random errors include: 1) expansion and contraction of air trapped inside the snow pillow and plumbing lines, 2) transducer malfunctions, 3) misreading of the pillow manometer, and 4) lack of synchronization between transmission and observation times.

A procedure analogous to the precipitation gage analysis was followed to evaluate the performance of the snow pillow telemetered data. Observations from 54 sites in Colorado and New Mexico ranging from 0 to 65 inches of snow water equivalent (SWE) were pooled to form a single sample. A simple linear regression equation was developed from 693 observations (Figure 6). The resulting equation is

$$Y = 0.997X - 0.07 \text{ (inches)},$$

where X = telemetered SWE and Y = observed pillow manometer value. This equation accounts for 98.9 percent of the variation in the data and has a standard error of 1.07 inches.

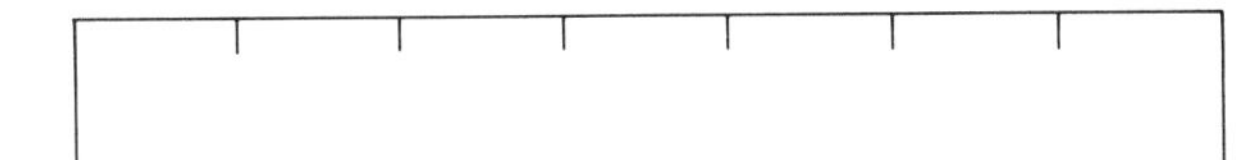

Figure 6. Relationship Between Observed Snow Pillow Manometer Values and Telemetered Snow Water Equivalent Values Aggregated for Colorado-New Mexico SNOTEL Sites in the Period 1978-82.

Again, the hypothesis that the observed and telemetered values are equal implies an equation with a slope of 1.000 and a constant of 0.00. The hypothesis that the calculated slope of 0.997 is really equal to unity was tested and was not rejected at the 20 percent level of significance. The 95 percent confidence interval for the slope was calculated to be 0.989-1.005. The hypothesized slope lies well within this interval which lends credence to the argument that the population slope is equal to unity.

The mean difference between observed manometer readings and the telemetered values was found to be ±0.50 inches. The standard deviation of the absolute value of the differences was 1.08 inches. This figure is substantially higher than the 0.71 inches in the precipitation gage analysis. The average difference expressed as a proportion of the observed value was 9.1 percent.

The foregoing analysis leads to the following conclusions: 1) no appreciable systematic error is apparent in the telemetry of observed pillow manometer readings; and 2) the mean difference between pillow manometer observations and telemetered values is about the same as for the case of precipitation gages, but the influence of random errors, and therefore the dispersion, is greater in the snow pillow data.

Although the telemetry of pillow manometer values appears to be reliable, one must still exercise caution in accepting the reported readings unconditionally as a true representation of physical reality. The possible consequences of such a course of action can be demonstrated by the data recorded at the Middle Creek, Colorado, SNOTEL site in 1982. Figure 7 shows the telemetered pillow and precipitation data for this period. A serious disparity is evident between control samples taken with a Federal sampler and telemetered data. It was first believed that a pillow failure was involved, but further field investigation revealed several very strong ice layers in the snowpack. These ice layers resulted in a condition known as "bridging," which prevented the full weight of the snowpack from registering on the snow pillows and caused the observed manometer readings and telemetered readings to decrease markedly. Throughout this period the precipitation gage continued to show increases during storm events. The discrepancy again reinforces the desirability of having dual sensors at remote sites.

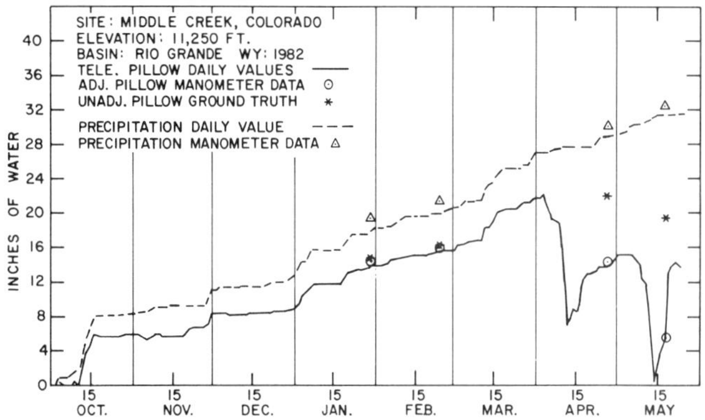

Figure 7. Middle Creek 1982 Water Year Daily Plot of Precipitation and Pillow Channels Showing an Anamolous Event Occurring in the Pillow Channel.

The previous systemwide analysis substantiated the equivalence of telemetered and observed pillow manometer values. Next, the correlation was sought between pillow control samples taken manually (grouth truth) and the telemetered values. At the four corners of the pillows a Federal sampler was used to measure the SWE. The four individual samples were averaged to obtain a single estimate of the SWE on the pillow.

Possible sources of systematic error in this analysis are similar to those in the previous analysis involving the adjusted pillow manometer and telemetry, with the additions of: 1) the tendency of the Federal sampler to overweigh by 10 percent (Farnes, *et al.,* 1982), 2) the nonuniform distribution of snow on the pillows, and 3) the tendency of the pillow to weigh more than what is physically on the pillows (California Cooperative Snow Survey, 1976; Cox and Zuzel, 1978; Smith and Boyne, 1981).

Random error not only can result from the electronic equipment but also can be introduced in the control sampling by: 1) not sampling in the correct locations, 2) obtaining poor samples because of ice layers in the snowpack, 3) misreading the weighing scales, and 4) using different Federal samplers.

A simple regression analysis was performed between pillow control samples (Y) and telemetered readings (X). In this analysis three and four pillow configurations were lumped together. The linear equation derived from a sample set of 510 observations from 48 sites was:

$$Y = 0.993X + 0.22 \text{ (inches).}$$

This equation explains 96.9 percent of the observed variation and has a standard error of ±1.99 inches. The mean difference between control samples and telemetered pillow values was ±1.32 inches. In this analysis no attempt was made to correct the data set to account for the systematic error induced by Federal sampler overweighing.

A further analysis was conducted to determine whether the accuracy of four pillow and three pillow systems differed significantly. As a general rule, three pillow systems have been installed in shallower snowpack areas with average April 1 SWE's of less than 15 inches; four pillow systems were installed in all other cases. Ground truth vs. pillow manometer data were stratified into three pillow and four pillow systems. A regression analysis was performed on the pillow control samples (Y) and the pillow manometer readings (X).

The regression equation for the three pillow systems derived from a sample set of 277 observations from 32 sites was

$$Y = 0.953X + 0.64 \text{ (inches).}$$

The equation for the four pillow systems was

$$Y = 1.007X + 0.44 \text{ (inches).}$$

This equation was derived from 248 observations from 16 sites. The standard error of the estimate was 1.33 inches and 2.16 inches, respectively, for the two equations. The equation for the three pillow system accounts for 96.2 percent of the variation in the sample set, and that for the four pillow system accounts for 97.2 percent of the observed variation. No compensation for the inherent 10 percent overweighing characteristic of the Federal sampler was made in the foregoing analysis.

The results of the regression analysis described above lead to the following conclusions: 1) the three pillow system tends to overweigh by 5 percent compared to the control samples, 2) the four pillow system measures close to what the control sample records, and 3) the equation for the four pillow system accounts for slightly more of the variation in the data set than does the equation for the three pillow system.

SECOND GENERATION EQUIPMENT

As SNOTEL evolves advancing technology has produced a new generation of equipment that embodies substantial improvements over the original models. Some of the improvements are: 1) reduced size; 2) inclusion of a random number generator concept for polling, which improves response when sites are in close proximity to one another; 3) added microprocessor capability; and 4) event recording capability.

With microprocessors, it is possible to program different functions into the transceiver. For example, the transceiver can compute minimum, maximum, and average temperatures as well as store sensor values at specific times for later transmission; this was not possible with the earlier models. Event reporting enables equipment at a remote site to automatically transmit a message to the master stations when the sensors have exceeded certain limits, e.g., precipitation in excess of 1.0 inch in 1 hour. This capability is functional around the Mount St. Helens area in Washington.

Another enhancement is equipment to link several nearby data collection sites to a single station with meteorburst capabilities or to a Geostationary Operational Environmental Satellite (GOES) platform. The outlying stations communicate with the meteorburst or GOES station using conventional, low power VHF telemetry. This subtelemetry system can transmit eight channels of data. It relies on the meteorburst radio or a GOES satellite to transmit the data to a central facility. The cost per site to add a subtelemetry network onto the existing SNOTEL system is less than that of installing individual meteorburst transceivers. A limitation to the subtelemetry system is the requirement for line-of-sight to the receiver. This disadvantage can be overcome with the addition of a repeater.

SUMMARY AND CONCLUSIONS

Telemetry and sensor performance in the Colorado-New Mexico portion of the Westwide SNOTEL network was analyzed for the period 1978-1982. From February 1981 to May 1982, the efficiency of the telemetry system in acquiring data during scheduled systemwide polls exhibited a characteristic annual variation with minimums in midwinter and maximums in summer. Efficiency ranged from 76 to 96 percent, with an average of 89 percent.

Evaluation of the relationship between observed snow pillow and precipitation gage readings and their corresponding telemetered values shows an absence of any appreciable systematic error, i.e., observed and telemetered values are equivalent. However, random sources of error result in an average expected difference of about ±0.5 inches between observed and telemetered data. This accuracy is acceptable for applications in water supply forecasting, the purpose for which the system was designed.

Both three and four pillow configurations are currently deployed as standard snow sensors. Comparison of ground truth data taken with a Federal sampler and telemetered snow water equivalents reveals a nearly 1:1 correspondence in four pillow systems, but a 5 percent systematic overweigh in three pillow systems. Ground truth data were not adjusted for the 10 percent systematic bias of the Federal sampler in this analysis.

Inconsistencies in telemetered data for two cases involving both snow pillow and precipitation gage readings were found to be related to physical causes. Such occurrences reinforce the desirability of having dual precipitation sensors at remote sites.

Innovations in miniaturization and microprocessor applications have led to improvements in remote data acquisition in the SNOTEL system. These innovations will be phased in over a period of years as original equipment is replaced, and should improve performance and data quality. Future enhancements include event reporting capability, elementary manipulation of data at remote sites, and inclusion of subtelemetry systems.

LITERATURE CITED

Barton, Manes and Arthur G. Crook, 1980. Operational Experience with the SNOTEL System. *In:* American Society of Civil Engineers Annual Convention.

California Cooperative Snow Survey, 1976. Snow Sensor Evaluation in the Sierra Nevadas, California. State of California, Department of Water Resources, Division of Planning.

Cox, L. M. and J. F. Zuzel, 1978. SNOTEL Sensor Evaluation and Forecast Development Study. Northwest Watershed Research Center, Agricultural Research, Boise, Idaho.

Farnes, Phillip E., Barry E. Goodison, Neal R. Peterson, and Robert P. Richards, 1982. Metrication of Manual Snow Survey Equipment. 50th Annual Meeting, Western Snow Conference, April 20-23, 1982.

Shafer, B. A. and M. Burke, 1978. SNOTEL. Proceedings of the First Membership Conference of the National Water Data Exchange, U.S. Geological Survey Open File Report 79-206, pp. 110-121.

Smith, F. W. and H. S. Boyne, 1981. Snow Pillow System Behavior for SNOTEL Application. Colorado State University Final Report, pp. 22-26.

INTERNATIONAL SYMPOSIUM ON HYDROMETEOROLOGY
JUNE AMERICAN WATER RESOURCES ASSOCIATION 1982

AN ADAPTABLE HYDROLOGIC DATA ACQUISITION SYSTEM

Richard H. Billings and Vito J. Latkovich[1]

ABSTRACT: Some programs of the U.S. Geological Survey require collecting hydrologic and hydrometeorologic data in a continuous and unattended manner. Presently, a system of automatic instruments is used to collect these data throughout the United States at an estimated 15,000 sites that were established in piecemeal fashion over more than two decades. The system is no longer able to keep pace with the demand for increased performance.

An Adaptable Hydrologic Data Acquisition System (AHDAS), custom designed to the particular requirements of the Survey, is under development. AHDAS will be a system of complementary instruments and modules which may be arranged to fit any practical data collection application. It will incorporate intelligent, programmable microprocessor features with solid state data memory and be capable of operation with a full spectrum of existing and even future sensors, communications systems, and peripherals. AHDAS will be a flexible system that may be altered in size and shape to provide total support for an estimated 20 years.
(KEY TERMS: data collection; data transmission; instrumentation; equipment; hydrologic.)

AN ADAPTABLE HYDROLOGIC DATA ACQUISITION SYSTEM

The U.S. Geological Survey (USGS), through its Water Resources Division, is the principal Federal organization responsible for providing water resources information. Through a system of interrelated data collection programs, measurements of surface water, ground water, and hydrometeorological parameters are made at over 15,000 individual remote field locations throughout the nation. These data are entered into a computerized data base system for analysis, storage, and dissemination.

This massive field data collection program requires a substantial investment of manpower. Consequently, field automation is highly desirable as a means of increasing overall organizational efficiency. However, technologies suitable for this applicable have developed slowly; previous attempts at automation have met with limited success. A growing disparity is evident between the limited capabilities of present field instrumentation and the need for expanded program support.

The *Adaptable Hydrologic Data Acquisition System (AHDAS)* will not only close the aforementioned gap, but also anticipate the needs of future programs. The ADHAS Project, initiated in 1981, will provide a system of complimentary instruments and modules which may be arranged to provide support to any given practical set of data collection requirements. Development of such versatile instrumentation is made possible through the emergence of microprocessor technology.

Automatic field data acquisition requires the performance of four basic tasks: (1) data measurement, (2) data processing/recording, (3) data transmittal, and (4) process control. The degree of complexity of each task is related to the specific application.

- *Data Measurement.* Parameters of interest, or some aspect of the physical medium which can be conveniently related to them, must be sensed and translated to an alternate useable form. No standard technique or technology exists for performing this task; specific approaches are dictated by the nature of the parameters and the conditions under which they must be measured. Microprocessors are employed in measurement, however typically in highly dedicated, nonsystematic roles. Given this limitation, AHDAS is not expected to play a significant role in data measurement.

- *Data Processing/Recording.* The outputs of measurement systems must be appropriately scaled, conditioned, or otherwise processed into desired formats and then entered into the data record. If measured values are converted to digital formats, then a microprocessor system is more than adequate for this task. In fact, the data record can be maintained in the solid state memory normally associated with such a system.

- *Data Transmittal.* Field records must be transmitted and entered into the network data base. USGS programs emphasize two basic methods: (1) data are retrieved by courier in a batch mode and manually transported to local offices for entry, and (2) data are transmitted in real-time (or near real-time) directly from the field site by satellite or other telemetry links. A significant portion of this function can be handled by a microprocessor system.

[1] Respectively, Electronic Engineer and Chief, Instrument Development Laboratory, U.S. Geological Survey, WRD, Hydrologic Instrumentation Facility, Bldg. 2101, MS 49-H (Billings), 6-H (Latkovich), NSTL Station, Mississippi 39529.

• *Process Control.* The basic elements which together provide the functions of measurement, recording, and data transmittal must be coordinated into an overall synergistic process. Similar control may be required by external devices which, although not directly involved in the data collection activity, must be coordinated with it. Process control functions are entirely within the scope of microprocessor system capabilities.

The optimum role for the AHDAS thus encompasses data processing/recording, data transmittal, and process control. Data measurement, being highly individualized, is excluded as are the actual telemetry communication links. The AHDAS will, however, be capable of operation with the full range of these measurement systems and with a variety of telemetry systems (satellite, radio, land-line, telephone, etc.). This proposed role is illustrated in Figure 1.

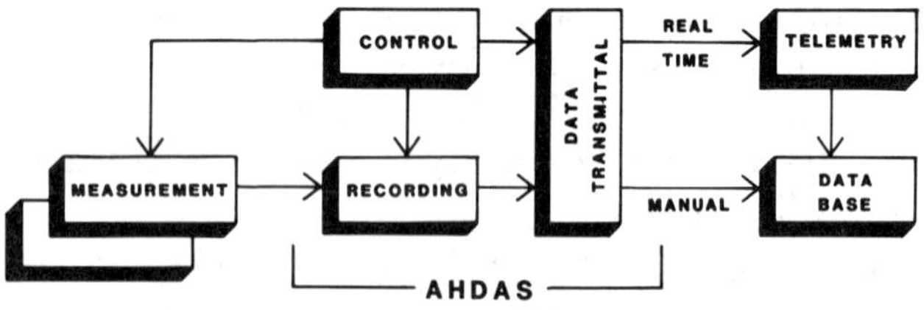

Figure 1. AHDAS Conceptual Role.

The USGS has contracted with the Mitre Corporation to perform the initial system design of the AHDAS. Mitre is conducting in-depth studies of USGS programs and requirements as well as an assessment of candidate technologies. Their final detailed AHDAS configurational concept will be provided through a set of definitive system performance specifications suitable for use in follow-on equipment design, development, and production activities.

Two basic elements are planned: a stationary field recorder and a portable field interrogator. The recorder will operate continuously and unattended in the field and the interrogator will be carried by the field user to provide him with access to the recorders.

The field recorder will be a versatile, general purpose data acquisition system. It will be housed in an environmental enclosure whose dimensions will vary with the specific configuration of instruments. Power will be supplied by an external battery which may be kept under constant charge from commercial line power, solar power, etc. The recorder will provide the user with the freedom to configure an overall system which fits his particular data collection needs.

• *Hardware Interface.* External devices (measurement systems, communications systems, etc.) will be connected through dedicated instrumentation modules which will provide hardware and circuitry to establish physical and functional compatibility with the AHDAS microcomputer. The system may be adjusted to fit a wide variety of applications by adding or removing these modules as required.

• *Data Memory.* Solid-state data memory, designed for modular expandability, will be provided. This type memory is easily interfaced and accessible to real-time data systems.

• *Operating Software.* A set of concise, user oriented programming instructions will be provided, each specifying unique functions basic to the total data acquisition task. By selectively invoking these instructions, the user may program the recorder, in the field, for a wide variety of operating schemes.

The portable field interrogator will provide the user with the ability to service the recorders in his charge. It will operate intermittently from internal, rechargeable batteries. Its physical characteristics (size, shape, weight, durability, etc.) will be optimized for this highly portable use. The interrogator will furnish the single means of access to the recorders in the field, providing three essential service functions: operational programming, data handling, and diagnostic testing.

• *Operational Programming.* Using the specified instruction set, recorder operating routines may be installed, removed, modified, or examined. Programs may be inserted manually by keyboard or automatically in a bulk mode from internal memory.

• *Data Handling.* The interrogator will provide operational support for the manual data collection and transport process. Recorder data records may be extracted electrically and examined to determine their validity, detect trends, etc. These records may then be appropriately annotated and stored temporarily in memory within the interrogator for transport from the field. Data base entry may be accomplished directly at the local office terminal or remotely through a telephone modem.

• *Diagnostic Testing.* The interrogator will perform operational status tests in the field to detect faulty recorders.

This proposed AHDAS concept will provide the cornerstone for a systematic overall automatic data collection program. The basic form of the recorder will address the simplest and most prevalent program application: water level measurement (approximately 10,000 present installations). In this form, the recorder will operate with an external water level sensor (a shaft encoder driven by a float wire assembly or a servo-manometer) and essentially duplicate the operation of the present USGS water level recorder. With little programming effort, this recorder may be upgraded to provide advanced features such as data record compaction, event detection, error flagging, data averaging, etc. If data are desired in real-time, an appropriate interface module may be added and the recorder may be programmed for operation with a specified external telemetry system.

From this basic form, many more complex configurations may be constructed. For example, with the appropriate programming and an added interface module, the recorder may be made to control an external sediment sampler and key its operation to predefined water level events. Or, a multiparameter water quality monitoring system may be configured, again through added programming and interface modules, as well as expanded data memory capacity.

This proposed concept is being developed within a complex framework of design goals. Several organizational issues are being studied in-depth to assess their impacts upon the ability of the AHDAS to provide cost-effective support to the mission and organizational objectives of the USGS. Results of this analysis will be reflected in the system performance specifications as design requirements, and will include the following:

- *Integrity of Data.* The basic product of the USGS is data which are made freely available for public use. These data must be of a known, high degree of accuracy. The design of the AHDAS will incorporate measures which will ensure the basic integrity of data throughout the collection process, from the point of measurement to the data base.

- *System Reliability.* AHDAS design, development, and fabrication will conform to standards of quality and reliability which are an acceptable compromise between the need for uninterrupted service and controlling economic factors.

- *Environmental Compatibility.* AHDAS will be designed for operation within a range of environmental conditions which encompass an acceptable majority of those found at USGS field locations.

- *Operating Characteristics.* Operating characteristics of the AHDAS will be in harmony with the capabilities of the USGS, including: manpower, skill levels, logistics, and economics. A user interface which will permit straightforward operation and remain essentially constant over the life of the system will be provided.

Microprocessor technology has become very well established around accepted industry standards. There is reason to expect that this technology will remain well supported over time and that an AHDAS life cycle of 20 years or more is possible. The ground work necessary to achieve this lifespan is being laid in the system design. Emphasis is being placed on the development of system performance specifications which are forward looking and capable of supporting equipment design, development, and production through periods of evolving technologies and changing program needs.

The system performance specifications, scheduled for completion during the first quarter of calendar year 1983, will be the first major product of this project. The specifications will be invaluable in all further phases of the program over the life cycle of the AHDAS. They will provide the substantive system description from which equipment design and development may proceed and will include mechanisms with which performance of finished products may be verified. They will assure continuity of products over successive production cycles, allowing for the incorporation of new or alternate technologies which may emerge, while maintaining essential performance characteristics. Properly used, these specifications will provide the basis of an effective logistics support program.

LITERATURE CITED

Buchanan, T. J. and B. K. Gilbert, 1983. Water Data Program of the U.S. Geological Survey – New Approaches in the 1980's. Proceedings of the American Water Resources Association, International Symposium on Hydrometeorology, pp. 3-7.

A REAL-TIME HYDROLOGIC DATA ACQUISITION NETWORK FOR THE ARKANSAS RIVER BASIN, COLORADO

Robert Tatnall and Paul Kirshen[1]

ABSTRACT: A commercial real-time hydrologic data collection network of 81 sites has been set up in the Arkansas River Basin in Colorado by Environmental Research & Technology, Inc. (ERT), a subsidiary of COMSAT. Stage is monitored at all sites; salinity and water temperature are monitored at several sites. The data from the sensors are transmitted via the GOES satellite to ERT's downlink in Concord, Massachusetts, and then relayed to users in the field by dedicated telephone lines.

The stage sensors are either maintained by a cooperating agency or ERT. At all sites, however, ERT owns, operates, and maintains the transmitting and associated equipment. ERT also provides quality control for the data collection. Users have access to the data and associated software products as needed for a monthly fee.

Examples of standard software products include archived records, emergency transmissions, hydrographs, basin schematics showing stages and flows at particular times, and excedences of threshold values. Site specific products such as forecast and operational models are also available.

The advantages of the commercial system to users are that large capital expenditures are avoided, only necessary data are purchased, manpower requirements are kept low, many of the research and development efforts in data collection and system operation are undertaken at no cost by ERT, and finally, the economic nature of the system encourages cooperation between various kinds of data users.

(KEY TERMS: data collection; data transmission; hydrologic; meteorologic.)

INTRODUCTION

Environmental Research & Technology Inc. (ERT), a subsidiary of COMSAT, has recently set up a hydrological monitoring system in the Arkansas River Basin, Colorado, in cooperation with the United States Geologic Survey (USGS), the Colorado Division of Water Resources, and local water users. Stage or reservoir elevation is monitored at 81 sites in the basin. The data from the sensors are transmitted via the GOES satellite to ERT's receive station in Concord, Massachusetts, and then relayed to users in the field by dedicated telephone lines. This project is demonstrating the value of real-time data in data collection activities and analysis, and in administration and management of the river for irrigation, flood control, water supply, and other purposes. ERT has installed and is operating and maintaining the system. The data and related software products are sold to users at a monthly or annual cost.

Presented in this paper are a brief description of the Arkansas River Basin, the monitoring equipment and system, some of the data products that have been developed, and future plans for the project.

Arkansas River Basin, Colorado

The river originates in the eastern slopes of the Rocky Mountains in central Colorado. It then flows south and east through a mountainous area of steep gradients until it enters the Colorado Piedmont near Canon City. It then flows east through the broad rolling plains until it enters Kansas. The drainage area in Colorado is 28,970 square miles. The major tributaries are Fountain Creek, St. Charles River, Huerfano River, Aphishapa River, Purgatoire River, and Big Sandy Creek.

Precipitation in the upper elevations of the basin occurs mainly as snowfall. The average annual precipitation at Leadville is 18.5 inches. Most of the precipitation in the Piedmont is rainfall and averages 11.7 inches annually at Pueblo, 16.1 inches at Lamar. Eighty percent of the precipitation occurs between April and October at irregular intervals and sometimes with a very high intensity.

The average annual discharge at Pueblo is approximately 800 cfs, the maximum discharge of record (up to 1969), 103,000 cfs. Most of the average annual runoff occurs in the April to July period. The average river discharge actually decreases downstream because of irrigation withdrawals.

Major aquifers are found in the valley fills of the basin, the Ogallala formation, and the Dakota Sandstone.

The population of the basin is approximately 700,000. Major water uses include irrigation (both from surface water and ground water), municipal and industrial water supply, and recreation. Of these, irrigation is the greatest consumer of the water. The water in the basin is presently over appropriated and late in the water year usually only

[1] Respectively, Manager, Systems Engineering, and Water Resources Consultant; ERT, 696 Virginia Road, Concord, Massachusetts 01742.

senior water rights can be supplied. In 1981 over $5 million was spent in litigation over water rights.

Floods are a major problem in the basin particularly between Canon City and John Martion Reservoir. Floods have damaged both irrigation projects and urban areas.

To meet these water needs, the river is carefully administered by a state engineer who administers the available water in the river basin to users based upon the priorities of their decrees. There are also extensive water management facilities. These include eight transbasin diversions, three major reservoirs, many irrigation diversions and canals (several as long as 40 miles and one over 100 miles long), and a major industrial water supply canal (46 miles).

Monitoring Equipment and System

Figure 1 shows a diagram of the overall operation of the system. Each field station monitors stream stage or reservoir elevation, stores data temporarily, and then transmits the data via the GOES satellite every three hours to a central satellite receiver and data collection facility located at ERT in Concord, Massachusetts. At ERT the data are screened for transmission errors and anomalous values,

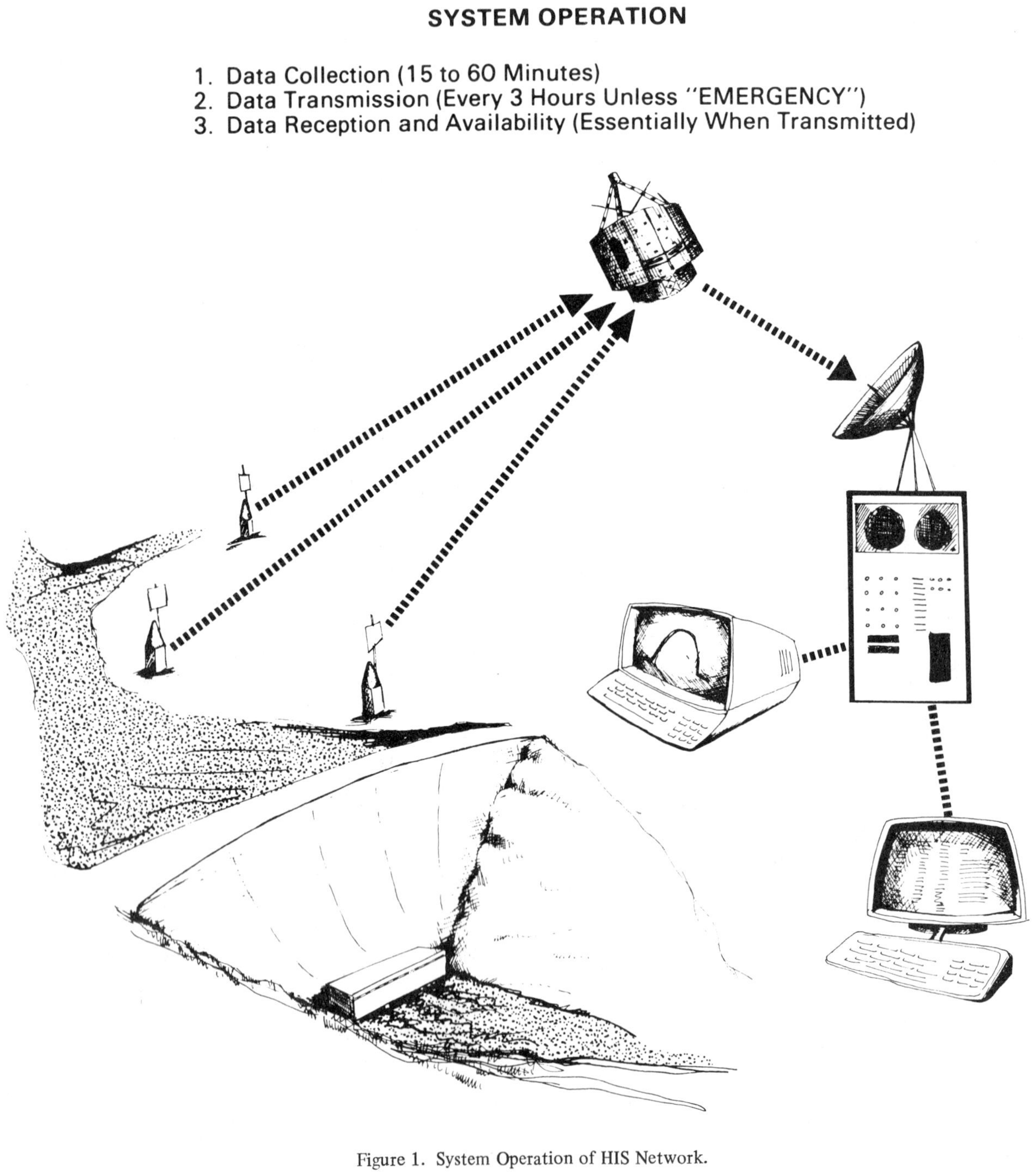

Figure 1. System Operation of HIS Network.

flagged, and then archived in a data base for interactive access. A number of programs (some of which are described subsequently in this paper) have been developed which enable interactive users to access and display data in a convenient fashion.

Stage is measured by either the movement of a float in a stilling well or with a manometer. The linear motion or pressure are converted into shaft rotation, which, in turn, encodes the data on a four-decade BCD shaft encoder. An automatic digital recorder (ADR) samples the encoder position on command from the data collection platform (DCP), a microprocessor controlled device, at each station. Sampling is done at 15-minute intervals. Readings are stored temporarily by the DCP and are transmitted to the satellite (and thus to ERT) every three hours. During emergency situations (for example, high stage), the data are sent at the sampling interval. A DCP can be programmed to sample up to four ADRs and up to eight other analog inputs. Station power is supplied through a combination of battery and solar panel; the panel supplies enough power to meet the long term average requirement of the system and keep the battery charged.

The sites presently being monitored include 57 stream stations, 13 irrigation and municipal ditches, 3 reservoirs, and 8 transmountain diversions. The sites are either USGS, state, or ditch stations. They were selected to best meet data collection, monitoring, and administrative needs.

Data Products

There are three categories of data products; quality assurance (Q/A), tabular and graphic reports, and system and software maintenance programs. A Data General S250 minicomputer is used for all the data products. The system includes a dual computer configuration with a primary and backup system. Failover can be accomplished within minutes for a near 100 percent availability of the central facility.

Q/A is a major component of ERT's project. As mentioned earlier, data are screened and flagged as they are received at Concord. They are screened for print errors, and then for stage values that are high, high-high, low, low-low, or show rapid rise, rapid fall, no change over a certain time interval – all according to standard USGS procedures. Battery voltage is also checked. Shown in Figure 2 is the output from part of a system reliability check. The command for execution of the program is shown in the upper left-hand corner, followed by the desired time period. For each station shown the USGS station ID is listed, followed by the number of sample points expected, number actually received, percent recived, and the number of exceedances for each threshold.

A program has also been developed that will recognize when there is a sustained increase or decrease in stage height or there is excessive oscillation in the stage. Experience indicates that when such situations are being recorded either

Count_DCP_Data Report   Based on Preliminary Data

For Period From 05/06/1982 09:01:54
Through 05/07/1982 09:01:53 (Time Zone 6):

Sta. ID	Exp	Rec	% Rec	Zero	Non	LL	L	H	HH	R+	R-	NC	Volts
APISHAPA RIVER NEAR FOWLER,CO													
07119500	87	87	100.0	0	0	14	87	0	0	0	0	0	218.
ARKANSAS R. CATLIN DAM,CO													
07119700	174	174	100.0	0	0	0	0	22	0	0	0	0	217.
COLORADO CANAL AT MILE 3.8 NR BOONE													
99990010	89	89	100.0	0	0	0	0	0	0	0	0	0	220.
ROCKY FORD-HIGLINE AT MILE 4.9 NR BOONE													
99990011	89	89	100.0	0	0	0	0	89	0	0	0	0	226.
OXFORD FARMERS DITCH AT MILE 1.0 NR BOONE													
99990008	89	89	100.0	0	0	0	0	64	0	0	0	0	223.
ARKANSAS RIVER NEPESTA,CO													
07117000	88	86	97.7	0	0	0	0	0	0	0	0	0	225.
HOLBROOK CANAL AT MILE 3.4 NR. ROCKY FORD													
99990003	88	88	100.0	88	0	88	88	0	0	0	0	88	217.
ROCKY FORD CANAL AT MILE 1.2 NR. MANZANOLA													
99990001	87	87	100.0	0	0	0	0	61	0	0	0	0	222.

Figure 2. System Reliability Report.

a hydrologic or man-made event is occurring (which may, for example, require the rating curve to be updated) or some of the equipment has failed or is clogged, etc. In any case, the user is alerted to potential data quality problems at that site.

One of the most frequently used data reports is a tabulation of stage and discharge or volume at a station over a specified period of time. A station report is shown in Figure 3. The screening flag columns can indicate the result of the data screening process (for example, 0040 indicates "significant data," 0020 indicates "less than lower limit," etc.). It is possible to plot the discharge data as a hydrograph (see Figure 4). Stage is converted to discharge using a computerized rating curve. As discussed subsequently, software has been developed that automatically updates the rating curve according to USGS procedures.

Another product is a schematic of all or part of the network showing the time, stage, discharge or volume, and the changes over a time interval for each station in the schematic. Part of a schematic is shown in Figure 5. The schematic can also be produced in color graphics with different parts and conditions in different colors and patterns.

Software has also been developed that will produce a daily flow report for each station in accordance with USGS requirements and format. Such a report is shown in Figure 6. This allows users to continuously examine and validate data and should decrease publication time for water data. All of these programs run essentially in real-time.

The parameters describing each station (for example, longitude, latitude, equipment numbers, limits for screening thresholds, etc.) are set using the system maintenance programs. A particularly complex maintenance program is that used to update the variation diagram for each station. A variation diagram shows how much stage values of the base rating curve for each station need to be adjusted to account for shifts in the rating curve due to channel bed changes, etc. The standard USGS procedures only allows the diagram to be defined by three points; ERT's procedures allows nine points. Datum correction maintenance programs also exist.

All of the programs are user-friendly. As an example, shown in Figure 7 is the sequence of commands necessary to display the historical and present flows at a station.

Future Plans for Demonstration Project

In a previous pilot program with the USGS, ERT achieved average monthly data capture rates that exceeded 98 percent. Stage data accuracy was ±0.03 to ±0.4 feet of USGS instruments. In this program, capture rates of 99 percent with an accuracy of ±0.02 feet appear achievable.

```
                                          05/07/1982 10:17:41 Eastern Standard Time

HIS List Discharge Report                Based on Preliminary Data

USGS Station I.D. 07094500                 DCP I.D. 1622B5DC
ARKANSAS RIVER AT PARKDALE, CO.
     Last Transmission Received 05/07/1982 07:27:21 (Time Zone 6)
     (This was 1 hour, 50 minutes ago.)


For Period From 05/06/1982 09:17:39
        Through 05/07/1982 09:17:38 (Time Zone 6):

Device Number  1  Parameter Code 00065
Discharge data based on a Rating Curve dated 10/19/79
                               Stage   Discharge
           Date      Time       (ft)     (cfs)          Flags

     05/06/1982 09:32:21        3.56      697.
     05/06/1982 10:02:21        3.56      697.
     05/06/1982 10:47:20        3.55      691.
     05/06/1982 11:17:20        3.55      691.
     05/06/1982 11:47:20        3.55      691.
     05/06/1982 12:17:20        3.55      691.
     05/06/1982 12:47:20        3.54      685.
     05/06/1982 13:17:20        3.55      691.
     05/06/1982 13:47:21        3.54      685.
     05/06/1982 14:17:21        3.54      685.
     05/06/1982 14:47:21        3.54      685.
     05/06/1982 15:17:21        3.55      691.
     05/06/1982 15:47:21        3.70      803.
     05/06/1982 16:17:21        3.81      893.
     05/06/1982 16:47:21        3.82      905.
```

Figure 3. Station Stage/Discharge Report.

```
                                  05/07/1982 10:20:02 Eastern Standard Time

HIS Plot Discharge Report          Based on Preliminary Data

USGS Station I.D. 07094500            DCP I.D. 1622B5DC
ARKANSAS RIVER AT PARKDALE, CO.
     Last Transmission Received 05/07/1982 07:27:21 (Time Zone 6)
     (This was 1 hour, 53 minutes ago.)

For Period From 05/06/1982 09:19:24
        Through 05/07/1982 09:19:23 (Time Zone 6):

 Device Number   1 Parameter Code 00065

Discharge data base

                                  05/07/1982 10:30:41 Eastern Standard Time

HIS Plot Discharge Report          Based on Preliminary Data

USGS Station I.D. 07094500            DCP I.D. 1622B5DC
ARKANSAS RIVER AT PARKDALE, CO.
     Last Transmission Received 05/07/1982 07:27:21 (Time Zone 6)
     (This was 2 hours, 3 minutes ago.)

For Period From 05/06/1982 09:30:14
        Through 05/07/1982 09:30:13 (Time Zone 6):

 Device Number   1 Parameter Code 00065

Discharge data based on a Rating curve dated 10/19/79

                  Plot of Discharge(cfs)  Versus  Time(hours)

                                                (Data is compressed)
```

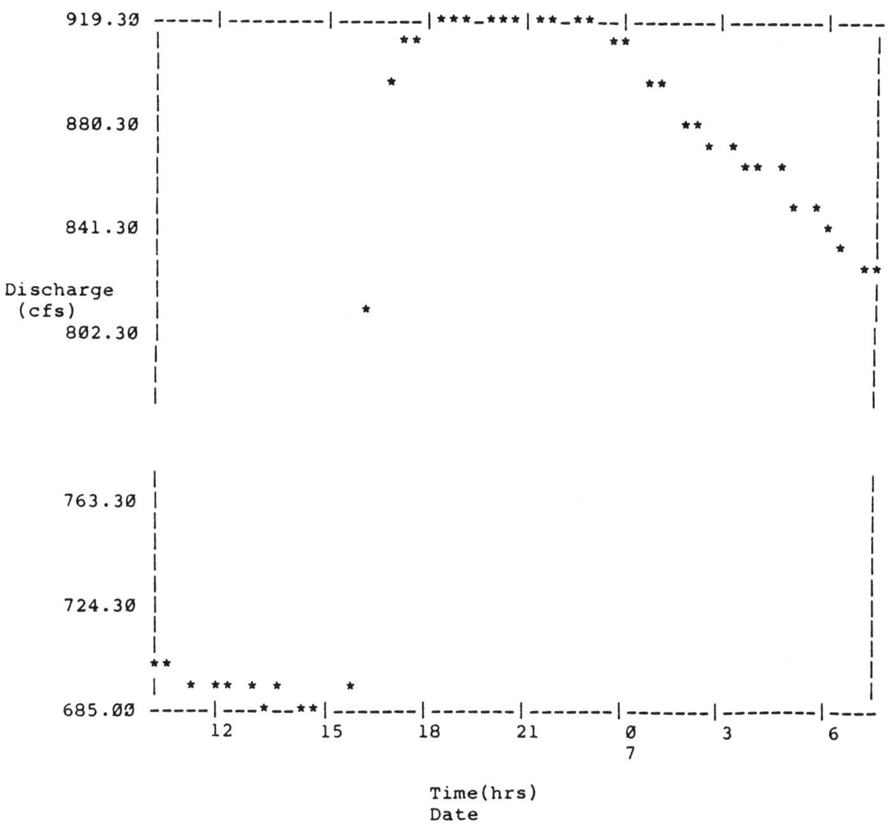

Figure 4. Station Hydrograph.

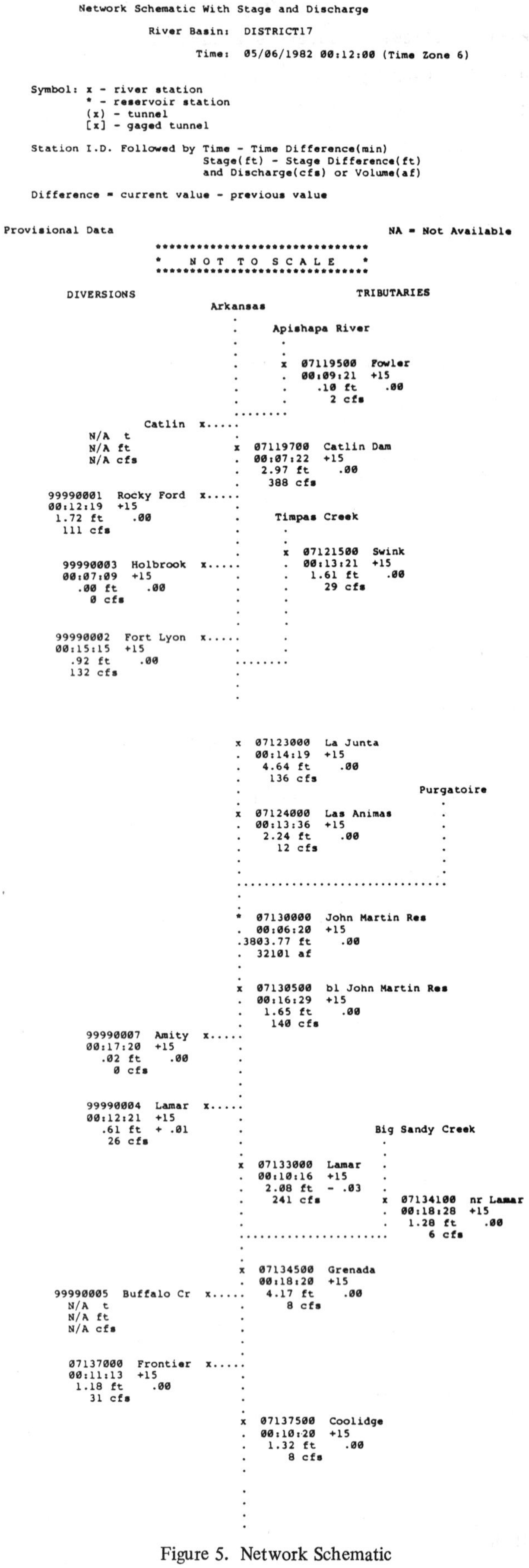

Figure 5. Network Schematic

The program is also being expanded to the other river basins in Colorado so that the total number of stations will be 150. At several stations, additional monitoring equipment will be added for solar radiation, precipitation, and water quality.

Finally, ERT will continue to work closely with water users in the system to further increase the commercial value of the real-time network. Presently, the data are used primarily for collection and administration activities. It is expected that these activities will continue to grow but that the real-time data will also be used to better estimate and model transit losses and other hydrologic occurrences so that flows will be more predictable and used more efficiently.

ACKNOWLEDGMENTS

The authors would like to acknowledge the support and cooperation ERT has received on this program from Jerry Hughes and Ron Steeger of the USGS, Pueblo; and Robert Jesse and Jim Kasic, Colorado Division of Water Resources, Pueblo. Our colleagues in the Information Services Group at ERT are also acknowledged for their programming skills.

```
Daily flow Report                          Based on Preliminary Data

USGS Station I.D. 07094500                 DCP I.D. 1622B5DC
ARKANSAS RIVER AT PARKDALE, CO.


For Period From 05/06/1982 00:00:00
        Through 05/06/1982 23:59:59 (Time Zone 6):

Test Difference:    .02
    DATE   MAX GH   MIN GH    MEAN  EQUIV      MEAN       SHIFT       DATUM
           (TIME)   (TIME)     GH     GH      DISCH        ADJ         CORR

 A  5- 6     3.83     3.54    3.66  3.66        779       -.08W         .00
           (2232)   (1502)

            Stage, in hundreds of feet, at indicated hours (LST)

       1      2      3      4      5      6      7      8      9     10     11     12
       -      -      -      -      -      -      -      -      -     --     --     --

AM   3.60   3.61   3.61   3.60   3.60   3.60   3.60   3.58   3.57   3.56   3.55   3.55
PM   3.54   3.54   3.54   3.78   3.82   3.83   3.83   3.83   3.83   3.83   3.82   3.82

 NOTE.     Symbols used above have the following meanings:
 A - Successive adjusted punch readings differ by more than the specified
     allowable test difference
 P - Daily summary is for an incomplete day
 R - One or more input value is outside the range of the rating table in use
 W - Shift is a value weighted by discharge where shift varies with stage
     during the day
```

Figure 6. USGS Daily Flow Report.

```
Enter time zone you wish to use: 6

Do you want data for a network of stations
or for only one station or DCP i.d.?
(Enter "net", "stn", or "dcp"): stn

Enter Station I.D.: 07094500

Do you want (1) the one latest reading, or
            (2) data over a given time span, or
            (3) data for the last number of hours?

Enter Code ("1", "2", or "3"): 3

Enter amount of time (hhmmss): 240000

Enter minimum reporting interval (hhmmss): 003000

Enter screening flag mask (4 hex chars):

You have selected data for station 07094500
  for time from 05/06/1982 09:17:39 to 05/07/1982 09:17:38 (Time Zone 6)
  for parameter code 65
  with a minimum reporting interval of 30 minutes, 0 seconds
  with no selection by screening flags.

Is this correct (y or n)? y
```

Figure 7. Example of User Commands.

DATA TELEMETRY BY METEOR BURST

Don Sytsma and Ray E. Leader[1]

ABSTRACT: This paper provides an introduction to meteor burst technology and its general application for the telemetry of hydrometeorology data. Two specific applications will be featured: (1) Coastal measurement system; and (2) Event reporting for flood forecasting. In the fall of 1981, a Coastal Measurement System, using meteor burst as the principal telemetry link, was installed off Kodiak Island, Alaska. The purpose of the system is to gather wave data from two buoys, wind data from a nearby island, telemetering a synopsis of this data directly to the engineering offices of the Corps of Engineers in Anchorage. The data is being used in the final design of a breakwater to be constructed in Chiniak Bay. Event reporting for flood forecasting at Mt. St. Helens is described as the second application. Seven meteor burst sites were installed on and around Mt. St. Helens, Washington, to provide event reporting for rapid changes in precipitation and snowfall. Upon the occurrence of an event, the appropriate SCS offices in Portland, Oregon and Spokane, Washington, are notified immediately.
(KEY TERMS:

INTRODUCTION TO METEOR BURST COMMUNICATIONS

Data transmission by bouncing signals off ionized meteor trails provides a reliable means of short and long range communications. Data collection systems presently exist that consist of as many as several hundred remote stations reporting to one master station with data rates up to 100 words per minute. The system is called meteor burst communications and, while it has existed for years, the modern computer technology has made this technique a cost effective alternative to phone lines, ordinary radio, and satellite. The basic propagation mode of meteor burst is reflection off the ionized trails left by micrometeors as they enter the earth's atmosphere, are heated and vaporized. Billions of meteors large enough to give usable trails enter the atmosphere each day.

The operational principle is as follows: the Master Station transmits a continuous, coded signal, usually in the 40 to 50 MHz region. When a meteor appears in the proper location, it reflects that signal to a receiving Remote Station. The remote station decodes the signal, recognizes the signal path, and turns on its transmitter to reflect a signal back along the same path to the master station.

Information can be sent in either direction until diffusion reduces the electron density in the trail to a value too low to sustain reflection. The maximum length of a single hop link is about 2000 kilometers, a distance determined by the height of the meteor burn-up region and the curvature of the earth.

The typical meteor trail has a useful duration of a few hundred milliseconds, while wait times between suitably located trails can range from seconds to minutes depending on time of day, time of year, and system design factors. Hence the transmission consists of bursts of high-data rate transmissions of tens to hundreds of characters, separated by a period of silence. One important by-product of this characteristic is the ability of many links to share a common transmission frequency, a feature important in data collection systems.

The exchange of information can be in either direction; it can consist of short messages such as sensor data readout, coded messages of up to several hundred characters, text messages of a few words, or long messages achieved by splicing together the transmissions of successive bursts. In such an application, average throughputs of up to several hundred words per minute can be achieved with relatively simple equipment. A meteor burst system is ideally suited for operation in harsh environments for the collection of hydrological, meteorological, and oceanographic data. Figure 1 illustrates the basic principles of a meteor burst communications system.

ADVANTAGES OF THE METEOR BURST SYSTEM

Meteor burst systems have many unique features that make it a cost effective alternative to phone lines, ordinary radio, microwave and satellite:

1. 0-1200 mile range without the need for repeaters;
2. Automatic time sharing of single frequency;
3. Multiple remote stations served by one master station;
4. Independent ownership of communication network;

[1]Respectively, General Manager and President, Meteor Communications Corporation, 22419 72nd Avenue South, Kent, Washington 98032.

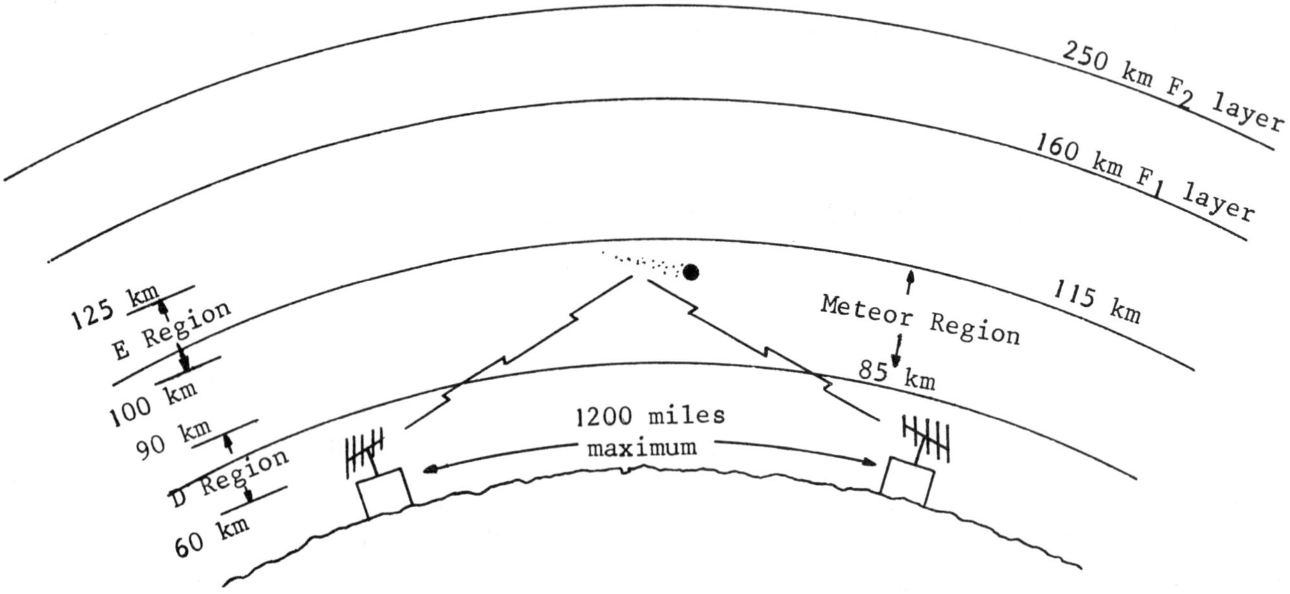

Figure 1. Meteor Burst Communication System.

5. Provides two-way message communications;

6. Remote stations may be added, relocated or deleted with no changes required at the master station; and

7. The system may be expanded to several thousand remote stations.

EXAMPLES OF MBCS USE

Meteor burst communications systems (MBCS) can be employed effectively for both point-to-point services and multiple station networks for ranges up to 2000 km. For extended range, relay stations may be employed using data store-and-forward techniques. Meteor burst systems today are fully automated and simple to operate. Training is required only for familiarization with the terminal operation. It has proven to be a cost-effective, dependable communications service – without recurring leased circuit costs.

Communication services can be readily established for a variety of purposes, such as remote oil and mineral exploration activities, airborne platforms, shipboard use, and communications with work crews in remote locations. Its rapid deployment capability also makes it ideal for disaster and emergency communications. The equipment is available in physically compact terminals, making it ideal for mobile and buoy applications.

Two major data acquisition networks in operation today are the U.S. Department of Agriculture's SNOTEL system and the Alaskan Meteor Burst Communication System (AMBCS).

The SNOTEL system contains over 500 remote data sites spread over 11 western states. The sites, which are unattended and powered by a solar panel and battery, are polled daily for accumulated snow, precipitation, and temperature data. These data are then used for streamflow forecasting and for making assessments of the water resources in the Western United States. The remote sites are microprocessor controlled and provide data averaging and event reporting. Within the SNOTEL system, a network of event reporting sites are situated around Mt. St. Helens to provide immediate reporting of sudden moisture and temperature changes to alert people of impending floods. The SNOTEL network is planned to be expanded to over 1000 sites over the next few years. The Alaska Meteor Burst System, with a master station located in Anchorage, is owned and operated by six federal agencies. Figure 2 is a block diagram of the Alaskan Meteor Burst Communications System. It performs multiple applications in both data acquisition and message communications. These applications include:

1. Air temperature, windspeed and direction, precipitation, snowpack, humidity, and river stage;

2. Message communications from remote BLM survey camps;

3. Flight weather data from remote air fields; and

4. Gulf of Alaska wave data from buoys located near Kodiak.

The Federal Emergency Management Agency owns several portable master stations and a number of remote communications terminals for rapid deployment of emergency communication to disaster areas.

Another application that has found repeated use is the combination of a Loran C receiver with meteor burst for transmitting position data. Ships have been tracked from

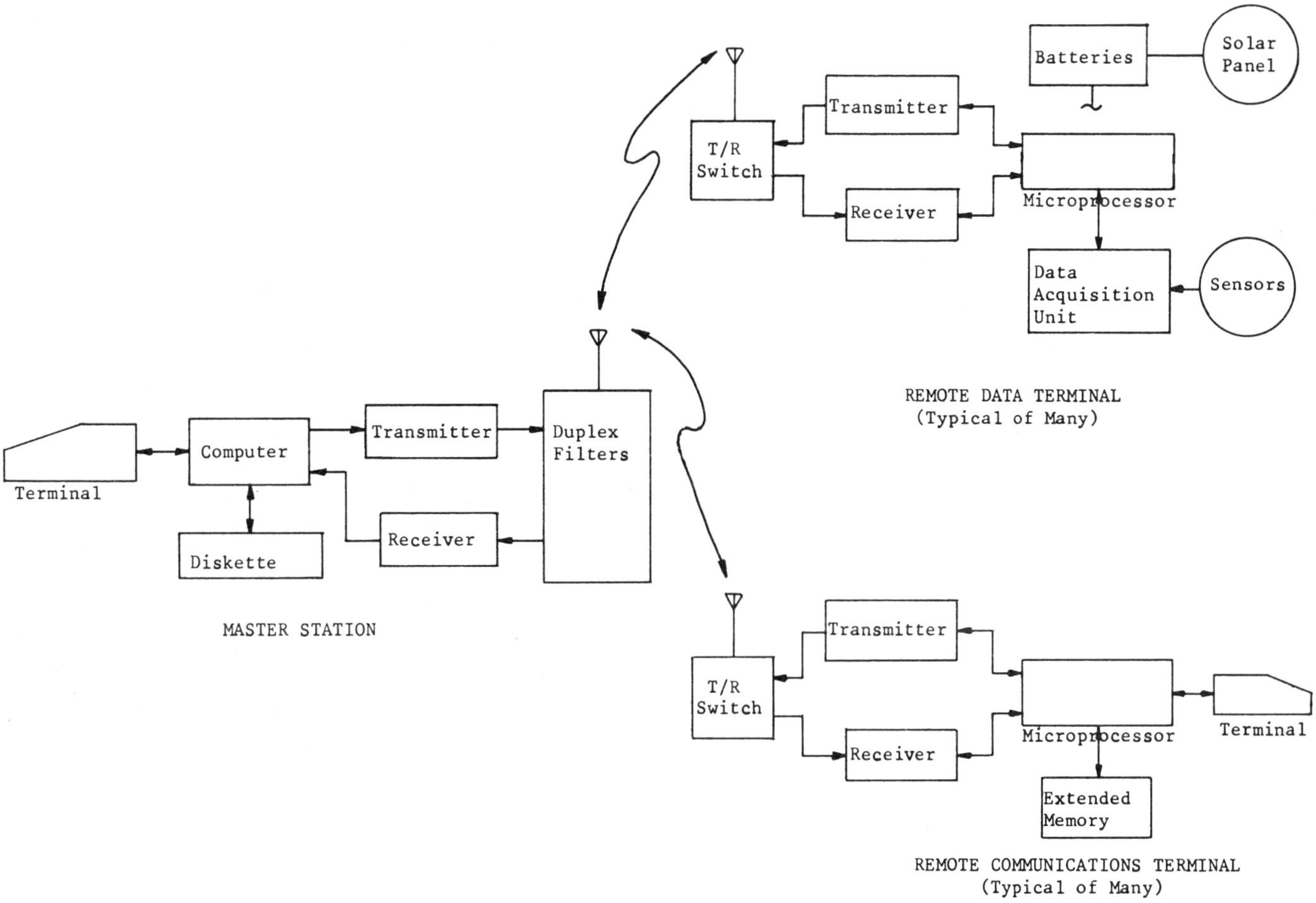

Figure 2. Alaskan Meteor Burst Communications System.

Valdez to Seattle and from Seattle to San Diego. An extension of this application will be the position monitoring of icebergs and drifting buoys.

PERFORMANCE CHARACTERISTICS

The performance of a meteor burst link is defined in terms of data throughput, or as the "waiting time", required to transfer a message between two stations at a specified reliability. For event reporting, such as at Mt. St. Helens, the "waiting time" to get a message through is the critical parameter. For telex or facsimile transmission, the amount of data throughput is usually more important. The primary parameters that affect system performance are as follows:

Operating frequency:	30-50 MHz
Transmitter power:	100 to 1000 watts
Antenna gain:	0 to 2 db
Data rates:	2 Kbps to 8 Kbps
Mode of operation:	half duplex or full duplex

As a rule of thumb, the "waiting time" to transmit a 30 character message is about 1.5 minutes at a probability of .95. The operation of a 100 WPM circuit is practical using full duplex master stations.

EVENT REPORTING AT MT. ST. HELENS

Within the SNOTEL system, a network of event reporting sites is located around Mt. St. Helens to provide immediate reporting of sudden changes in snowpack conditions and precipitation to alert people to impending floods. Figure 3 depicts such an event reporting site. These sites will automatically begin transmitting special "alert messages" to the master station when an out-of-range condition occurs. When the master station receives an "alert message" from any one of the event sites, it will immediately poll all the sites assigned to that particular "event group."

The out-of-range conditions are programmable for each application. For Mt. St. Helens, they are as follows:

1. An increase in snow pillow pressure, equivalent to one (1) inch of water, during a one hour time period, or

2. An increase in precipitation pressure, equivalent to one (1) inch of water, during a one hour time period, or

Figure 3. Event Reporting Site on Mt. St. Helens.

Figure 4. Base Station, Kodiak, Alaska.

3. A decrease in snow pillow pressure, equivalent to 1.5 inches of water, during an eight hour time period.

COASTAL MEASUREMENT SYSTEM IN KODIAK, ALASKA

In the fall of 1981, a Coastal Measurement System, using meteor burst as the principal telemetry link, was installed off Kodiak Island in Alaska. Figure 4 depicts the base station at the Corps of Engineers Office in Kodiak. The purpose of the system is to gather wave data from buoys located in Chiniak Bay as shown in Figure 5. A wind station, showin in Figure 6 was also installed on nearby Puffin Island to gather data on both wind speed and wind direction. This wind and wave data was telemetered directly to the offices of the Corps of Engineers located in Anchorage. The data is being used in the design of a breakwater to be constructed in Chiniak Bay.

Figure 5. Buoys in Chiniak Bay.

SUMMARY

Data telemetry using meteor burst technology has matured to where "off-the-shelf" equipment is available. MBCS has proven to be a cost effective alternative to phone lines, ordinary radio, and satellite communications. Performance is predictable and a system can be designed and custom tailored to user requirements. Typical cost for a complete remote station is from $5,000 to $10,000 and the cost of a master station will range from $40,000 to $100,000, depending on area of coverage, performance and data processing requirements.

Figure 6. Wind Station on Puffin Island.

Section 2

HYDROLOGIC MODELING AND FORECASTING

SUITABILITY OF REMOTE SENSING CAPABILITIES FOR USE IN HYDROLOGIC MODELS

Eugene L. Peck, Thomas N. Keefer, and Edward R. Johnson[1]

ABSTRACT: The results of a study on the suitability of remote sensing capabilities for use in hydrologic models is reported. Seven hydrologic models (API, CREAMS, NWSRFS, STORM, STANFORD, SSARR, and NWSRFS Snowmelt) are reviewed. Information on the primary and secondary roles of parameter and state variables with respect to their function in the models is used to determine the usefulness of six remote sensing capabilities with the seven hydrologic models.

The results indicate remote sensing information has only limited value for use with the hydrologic models in their present form. The usefulness of the remote sensing information would be greatly enhanced with minor modification to the models.

General characteristics of a hydrologic model needed to make maximum use of remotely sensed data are discussed as are recommendations for incorporating remotely sensed data in specific models.

(KEY TERMS: remote sensing; hydrologic models; water resources management.)

INTRODUCTION

The results of a study on the suitability of remote sensing capabilities for hydrologic modeling are presented in this paper. The study was accomplished in two phases. Phase one consisted of a review to determine the suitability of hydrologic models commonly used by federal agencies to use remotely sensed data. The objectives of the second phase were to evaluate current strategies for using remotely sensed information in the models and to develop recommendations for improved use of such information.

Remote sensing techniques provide direct measurement of land characteristics, vegetative cover, and the states of water in the hydrologic cycle. To date, however, use of remote sensing techniques has been of limited value.

REVIEW OF HYDROLOGIC MODELS

The following five hydrologic models commonly used by federal agencies were selected for review:

- Antecedent Precipitation Index (API),
- National Weather Service River Forecast System (NWSRFS),
- Storage, Treatment, Overflow, Runoff Model (STORM),
- Stanford Watershed Model IV (SWM), and
- Streamflow Synthesis and Reservoir Regulation (SSARR).

Two other hydrologic models were reviewed. The Chemical, Runoff and Erosion from Agricultural Management System (CREAMS) model was included because of its extensive use in the field of agriculture. The NWSRFS Snow Accumulation and Ablation model was also selected since it is commonly used with several of the basic hydrologic models.

The following terms and definitions were used in the review of the hydrologic and snowmelt models.

Inputs – The set of driving forces required periodically by the model.

Parameters – The set of values that are changed to make a general hydrologic model apply to a particular location. Parameters are constant with time or at most, vary only slightly with time as compared to inputs.

States – The set of internal model values sufficient to start the model. The states of the model completely define the past history of inputs.

Outputs – Variables of interest that can be computed from knowledge of the states and inputs. Usual examples are streamflow and actual evapotranspiration.

The selected hydrologic models were generally developed for use with ground measurements for input. The descriptions of the hydrologic models in the literature are not adequate for the purpose of evaluating the usefulness of remote sensing capabilities in the operational use of the model.

The review conducted on the hydrologic models gave special emphasis to those parts of the models for which

[1]Respectively, President and Associate, Hydex Corporation, 11150 Main St., Suite 502, Fairfax, Virginia 22030; and Assistant Professor of Civil Engineering, School of Civil Engineering, Georgia Institute of Technology, Atlanta, Georgia 30332.

remote sensing capabilities could be of greatest value. These were the soil moisture accounting, snowmelt and channel inflow computation. The complete findings of this phase were presented in an interim NASA report (Peck, *et al.*, 1981a). Due to limitations on the length of this paper only the most pertinent information and data are presented herein.

The interim report contains a schematic diagram for each model. For illustrative purposes only the schematic diagram for the SSARR model is shown in this paper (Figure 1). The diagram shows the inputs, processes, states, decision points, and outputs of the model as indicated by the following:

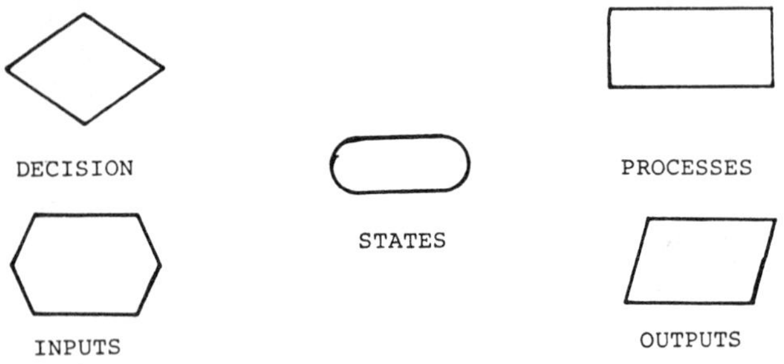

All parameters of each model are indicated by bold type and states by light type. The location of various components indicates the various levels of moisture zones (when a model has more than one) and the relative positioning with depth of the location of states and operating processes.

To assist in understanding the role of each parameter and state variable, various tables are present in the interim report. For each of the basic soil moisture accounting models a set of tables present the definitions of all parameters and state variables. A second set identifies the primary and secondary role of each parameter (states variable) relative to runoff components, soil moisture horizons, and hydrologic processes. As an example, Table 1 is the tabulation showing the primary and secondary role of parameters in the SSARR model. A list of definitions of parameters and states of the SSARR model is given in Table 2.

The tables in the interim NASA report can be used to identify which parameters (state variables) are related to specific runoff components, soil moisture horizons, or hydrologic processes.

The information also gives an immediate indication of the overall complexity of the model and which runoff components, soil moisture horizons, and processes are modeled most precisely.

REMOTE SENSING CAPABILITIES

For this study it has been assumed that the user is a knowledgeable modeler of the behavior of the hydrologic system and seeks to collect information on current inputs to his model and the current states of the model. Sensing is used in a narrow definition. It is taken to mean estimating the average value of a variable over some areal extent by examining the characteristics of the radiation from

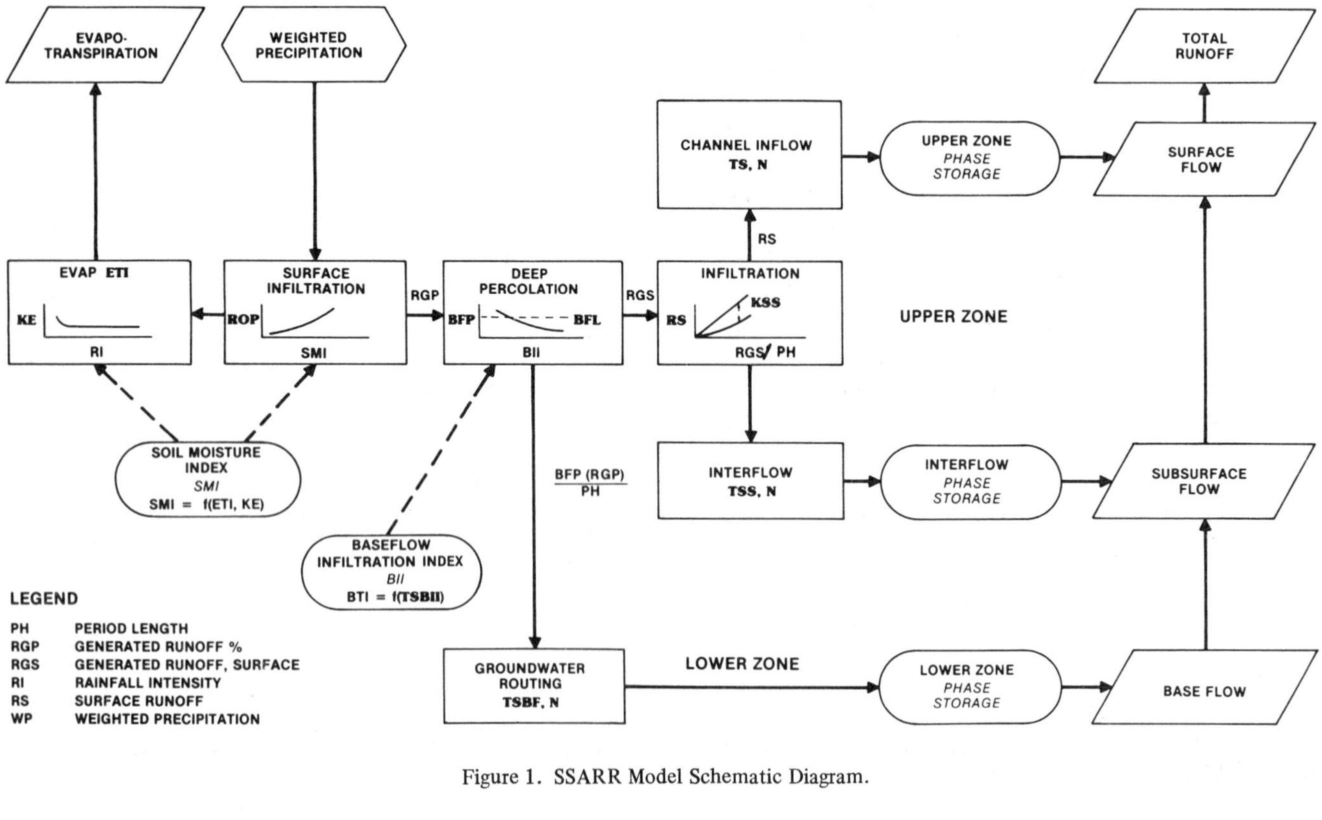

Figure 1. SSARR Model Schematic Diagram.

TABLE 1. Role of Parameters – SSARR Model.

Categories	Parameter Assignment	
	Primary	Secondary
RUNOFF COMPONENTS		
Immediate Runoff		
Surface Runoff	RS/RGS table, TS, N*	ROP/SMI table, KE
Interflow	RS/RGS table, TSS, N* KSS	
Baseflow	BFP/BII table, TSBF, N*	BFL
SOIL MOISTURE HORIZONS		
Upper Zone	ROP/SMI table	KE
Lower Zone	TSBII, BFL	
PROCESSES		
Infiltration	KSS	RS/RGS table, ROP/SMI table
Percolation	BFP/BII table, BFL	TSBII
Evaporation	KE	ETI, optional
Interception		
Losses		

*N is listed as a single variable name; there are, however, distinct N's for surface flow, interflow, and base flow.

TABLE 2. Definition of Parameters and States of SSARR Model.

	PARAMETERS
BFL	Base flow infiltration limit.
BFP	Base flow, percent.
ETI	Evapotranspiration index.
KE	Percent effectiveness of ETI (function of rainfall intensity, RI).
KSS	Limiting subsurface infiltration rate.
N	Number of routing phases (surface flow),
N	Number of routing phases (subsurface flow),
N	Number of routing phases (baseflow).
ROP	Runoff percent.
RS	Surface runoff percent, function of RS/RGS table.
TS	Time of storage; surface flow.
TSS	Time of storage; subsurface flow (interflow).
TSBF	Time of storage; baseflow.

	STATES
SMI	Soil Moisture Index.
BII	Base Flow Infiltration Index.
PHASE STORAGE	Phase storage (discharge or stage) for surface flow.
PHASE STORAGE	Phase storage (discharge) for subsurface flow.
PHASE STORAGE	Phase storage (discharge) for baseflow.

that area. The major emphasis is placed on the use of satellite sensors for measuring the radiation.

Emphasis has been placed on those variables that have at least an intuitive connection to portions of existing hydrologic model and that experience (or research) has shown can be remotely sensed with some degree of success. Since precipitation is an input to all models and no modification to a model would be required for its use it was excluded from consideration.

Remote sensing capabilities of six variables were reviewed for use with the operational hydrologic models. These were:

- Areal extent of snow cover,
- Areal extent of frozen ground,
- Impervious area,
- Land cover,
- Soil moisture, and
- Water equivalent of snow cover.

Of the six variables only three (areal extent of snow cover, land cover, and impervious area) can be considered to have operational measurement techniques at the present time. All three may be obtained through analysis of LANDSAT images.

None of the other three variables (areal extent of frozen ground, water equivalent of snow cover, and soil moisture) with the possible exception of frozen ground, can currently be measured effectively from satellites. All three are awaiting further research on combined active/passive microwave measuring techniques.

USEFULNESS OF REMOTE SENSING IN HYDROLOGIC MODELS

In spite of the recognized potential value of remotely sensed data for water resource management, the federal agencies responsible for river forecasting and water supply prediction are not using such data as a primary operational data base. To examine the reasons for this and to suggest improvements in remote sensing application, it was necessary to couple together the information from the review of the hydrologic models and the review of remote sensing capabilities.

There are several strategies for using remotely sensed data, or, indeed, any type of data, in hydrologic models. The first is to estimate the inputs to the models.

A second strategy for using remotely sensed data in hydrologic models is to update a state of the model to be consistent with the observed data.

A third strategy for using remotely sensed data is for calibrating the parameters of the model. In traditional applications, parameters are estimated only once based on current topographic and land cover data and hydrometeorological data for some calibration interval. It is certainly possible to recalibrate a model based on new data. For example, remote sensed observations of changes in land cover can be used for recalibration of models, thus, blurring the distinction between updating and calibrating the model.

Each of the seven hydrologic models was analyzed to determine the usefulness of the six selected remotely sensed variables for the present configurations of the model and with minor modification or adaptation. The analysis of the SSARR model is given in Table 3 as an example.

The most obvious conclusion from the analyses for the basic soil moisture accounting models is that in their present configurations there is not a significant potential for using remotely sensed observations for input, updating, or calibration. With minor modifications, some of the models can take advantage of the significant potential for applying remotely sensed data to hydrologic modeling.

POTENTIAL USEFULNESS

The greatest beneficial use of remote sensing can be realized for operational hydrology with major modification of the hydrologic models. Such improvements must consider the combined use of remotely sensed and ground measurements. Some of the factors that should be considered are discussed below.

A major feature of any hydrologic model is the scale of the model. Involved are the horizontal scale (basin size), the vertical scale (soil and snow horizons), and the time scale (time step). It is desirable to match the scale of the model

TABLE 3. Use of Remotely Sensed Data in SSARR Hydrologic Model.

Remotely Sensed Variable	Present Configuration*	Minor Modification or Adaptation
Soil Moisture	1. No 2. No 3. No	1. No 2. Update soil moisture index (SMI) 3. Define ROP/SMI table
Impervious Area	1. N/A 2. No 3. No	1. N/A 2. Could be used to modify or select alternate tables 3. Could infer shape of ROP/SMI, BFP/BII, RS/(RGS/PH) tables
Land Cover	1. N/A 2. No 3. No	1. N/A 2. To modify or substitute the infiltration table, RS/(RGS/PH) and evapotranspiration table, KE/RI 3. No
Areal Extent Snow Cover	1. N/A 2. Can be used to update snow covered area 3. Might be used to check model depletion curve calibration if available frequently enough	1. Could be used as input if available frequently enough 2. Update snow covered area 3. Model calibration of depletion curve
Areal Extent Frozen Ground	1. No 2. No 3. No	1. No 2. Update soil moisture index, SMI. Adjust phase storage for subsurface flow, to modify or substitute the ROP/SMI, BFP/BII, and RS/(RGS/PH) tables 3. No
Water Equivalent Snow Cover	1. No 2. No 3. No	1. No 2. No 3. No

*1. Input; 2. Update; 3. Calibrate.

to the scale of the observations since this will make the observations more directly useable.

The natural horizontal scale for lumped parameter models is the basin. Each model has some range of appropriate basin size. The usual approach is to derive areal average values of precipitation and other input items. This will continue to be the major approach to dealing with mismatched horizontal scales of model and observations. The data processing techniques to estimate the appropriate areal average values need to be much more sophisticated in order to combined observations that have various spatial sampling scales. An alternative approach is to match the horizontal scale of the model to the observations.

The vertical scale of the model should match the observation. If a remotely sensed soil moisture measurement represents the top 10 cm of soil, the model should ideally have a single state variable that represents the moisture content of the top 10 cm of soil. There is no guarantee that the most appropriate vertical scale for hydrologic purposes will match the vertical scale of observations.

The time step of the model must be short enough to identify significant variations in observed quantities. If observed quantities undergo significant diurnal variation, it must be possible to identify the modeled hydrologic state at the time of day of the observation.

An inherent tradeoff exists between the accuracy and timelessness of observations and the complexity of a model. It is important to remember both the accuracy and measurement frequency requirements.

Certain structural features of a model can make it more difficult to update the model states. It is helpful to avoid nonfunctional (table driven) components and highly nonlinear components. These features make it difficult to identify the relationship between observations and model states.

There is a need for those engaged in the design and operation of remote sensing systems to obtain feedback from hydrologic modelers. For example, the accuracy of measuring the water equivalent of the snow cover may be much less than the requirement set by hydrologists. However, measurements of less accuracy indicating incremental changes (e.g., by 1 or 2 cm intervals) may be of value for those responsible for forecasting snowmelt floods.

Continued research is needed in discrimination problems. Very few variables can be reliably identified from space without extensive ground truth. Multisensor systems that use several spectral bands combined with point observations on the ground may be needed for accurate determination of variables.

The greatest potential for increasing the usefulness of present models is the development of procedures for using remotely sensed measurements of the snow cover (areal extent and water equivalent) for input or updating the NWSRFS snowmelt model.

The second most promising use would be changes in state variables of soil moisture accounting models (SSARR, NWSRFS, and STORM) to provide for objective use of soil moisture as input or updating of the models.

The impervious area and land cover measurements can be used to some extent with existing models. Modification of some models (or development of procedures to incorporate remotely sensed data) could enhance the usefulness of the impervious area and land cover information and provide means for using the areal extent of frozen ground.

The final report of the NASA study (Peck, *et al.*, 1981b) presents detailed information on recommended strategies for using remotely sensed data in hydrologic models.

CONCLUSIONS

The basic conclusions from the study are:

- Hydrologic models in their present configuration do not have a significant potential for using remotely sensed information.
- The use of remotely sensed data could be significant with minor modification of existing models.
- The greatest potential for remote sensing could be realized by high resolution, passive, microwave sensors of satellites.
- Hydrologic modeling can be improved through the development of a new generation of models or subroutines for existing models which recognize the characteristics of the new remote sensing capabilities.

ACKNOWLEDGMENTS

Special thanks are due to Dr. Albert Rango, NASA, for his support and contribution to the study. It was accomplished under Goddard Space Flight Center, NASA Contract No. NAS5-26446.

LITERATURE CITED

Peck, E. L., R. S. McQuivey, T. N. Keefer, E. R. Johnson, and J. L. Erekson, 1981a. Review of Hydrologic Models for Evaluating Use of Remote Sensing Capabilities. NASA CR 166674, Goddard Space Flight Center, Greenbelt, Maryland, 99 pp.

Peck, E. L., T. N. Keefer, and E. R. Johnson, 1981b. Strategies for Using Remotely Sensed Data in Hydrologic Models. NASA CR 66729, Goddard Space Flight Center, Greenbelt, Maryland, 84 pp.

RUNOFF CURVE NUMBER ESTIMATION USING REMOTE SENSING TECHNIQUES

Kim Johnson, S. Dhamotharan, and Steve Allred[1]

ABSTRACT: Soil Conservation Service Runoff curve numbers are being extensively used by hydrologists to predict runoff values, especially in ungaged watersheds. The reliability of this method depends heavily on the determination of the individual runoff curve numbers. This becomes more critical in extremely nonuniform watersheds such as those in western river basins where energy development projects are being planned. Accurate information on curve number parameters on the subbasin level are a prerequisite to reliable runoff values. Detailed manual surveys to obtain necessary information is always expensive. It is demonstrated herein that remote sensing techniques can be used for runoff curve number estimation to the required level of precision with a significant savings of time and money.

Ground cover distribution, within the basin being studied, is determined by spectral classification to a map base to produce a ground cover map. Using digital image processing techniques, digital maps of soil hydrologic groups, and individual drainage subbasins are prepared to produce a basinwide geographic data base. An estimated runoff curve number map is produced by combining field observations of cover density, the ground cover map, and the soils map. By digitally overlaying the curve number map and drainage basin maps, an average curve number estimate for each drainage basin is determined.

This methodology can provide a cost-effective means of obtaining comprehensive curve number parameters with a minimum of field observations and with a statistically significant result.

(KEY TERMS: runoff forecasting; remote sensing; runoff curve numbers.)

INTRODUCTION

Development of surface flow values for the watersheds in energy development and mining project areas can be a time consuming and expensive task. Frequently these projects are located in remote areas where hydrologic information such as recorded stream flows are lacking. In those cases where recorded data are lacking extensive use is made of computer watershed models for estimating surface flow values. These models, such as the Soil Conservation Service (SCS) TR-20 Program on the HEC-1 Program (Corps of Engineers, 1981), define the subject basin as a series of interconnected subbasins. Physical features such as subbasin size and geometry, channel characteristics, and surface runoff potential are used to define the expected flows given specific precipitation events. The precision with which the basin parameters are defined will control the modeling results. Some basin features, such as size and channel length, may be determined using topographic maps which are generally available. A significant parameter that cannot be well defined using topographic maps is the surface runoff potential for a basin which is frequently expressed as the Soil Conservation Service Curve Numbers (CN's). Normally extensive field work and mapping are necessary to determine the distribution of land cover and soils for developing CN's. Landsat digital data has been demonstrated as a suitable source of information for mapping basin ground cover for determining CN values (Jackson and Rawls, 1982; Ragan and Jackson, 1980; Slack and Welch, 1980; Bondelid, *et al.,* 1980).

In addition to supporting the analysis of Landsat data, digital image processing is also well suited for geographic information processing. Slack and Welch (1980) noted the potential of using digital processing for directly developing model information based upon Landsat imagery. Smith and Blackwell (1979) have demonstrated the application of image processing for basin water quality monitoring and Cermak, *et al.* (1979), have used Landsat classified data in conjunction with a digital geographic information system to support basin modeling.

Morrison-Knudsen's Hydrology Department has implemented procedures for developing CN information using remote sensing and geographic data processing. The procedures are based on the use of digital image processing for the analysis of Landsat digital data in conjunction with other basin information.

DATA CHARACTERISTICS

The image processing techniques used to determine CN distribution are based upon a raster image processing system. A raster data set, the basic data component, consists

[1] Respectively, Senior Engineer, Remote Sensing Applications; Manager, Hydrologic Analysis; and Director, Hydrologic and Environmental Services; Morrison-Knudsen Company, Inc., P.O. Box 7808, Boise, Idaho 83729.

of a matrix of values which have a ground or map equivalence of small cells or grids. An individual cell is referred to as a "pixel," a contraction of the term picture-element. Within an image data set, each pixel has a single value or data number (DN). Since the DN values are maintained in a matrix form, the geometric location, x-y coordinate, of each pixel is a function of the pixel's location in the matrix. Specific x-y coordinate values for each DN value are not maintained. For geometric registration purposes the entire matrix may be referenced to a map projection.

In the case of Landsat image data, the DN values represent the reflectance of the earth's surface recorded in one of four spectral bands, by the satellite's multispectral scanner. Each of the four individual digital images consists of a matrix of DN values, low for dark areas and high for light areas. These digital images may be transformed to a photograph using a film recorder or used to drive a video display device. Digital image data need not be limited to data recorded by remote sensing systems. Any data that has a spatial distribution may be represented by a matrix of DN values.

PROJECT AREA

The project area consists of the upper tributary reaches of the Parachute Creek drainage in Garfield County, Colorado. The basin is characterized by a drainage system that is deeply incised into the shales and sandstones of the Roan Plateau. The drainage area is 139 square miles and the elevations range from a low of 5790 feet msl to a high of 9281 feet msl. Within this region of the basin no recorded stream flow information is available. The project objective was to determine flood flows at several points in the basin for 1, 6, and 24-hour precipitation events and for various recurrence intervals. The SCS TR-20 hydrologic model which requires subbasin average CN values was used for estimating the flood flows.

CURVE NUMBER ESTIMATION

As defined by the SCS (1972) a combination of a hydrologic soil group and a land use or cover class is a hydrologic soil cover complex for which a CN may be assigned. The conventional approach is to measure the homogeneous areas created by each combination of a mapped soil complex and land cover unit. A hydrologic unit or basin weighted CN is then determined based upon the percentage distribution for each hydrologic soil complex. These same procedures are followed except Landsat imagery and digitized soil maps are used to determine the distribution of hydrologic soil cover complexes. Figure 1 depicts the analysis sequence. In addition to the Landsat data, advanced soil survey maps, limited aerial photography coverage, and vegetation field sampling data were available.

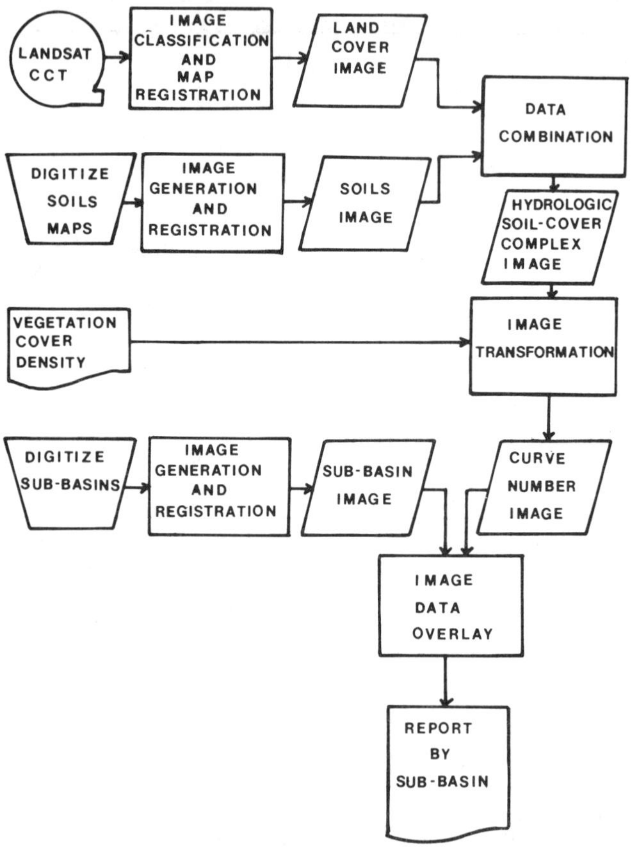

Figure 1. Image Processing Flow.

The distribution of land cover was determined by classification of Landsat imagery recorded on September 30, 1980 (scene ID 22078-17131). The land cover classification was developed using modified clustering (Flemming, *et al.*, 1975) procedures and maximum likelihood classification. The Landsat data for a series of limited areas were clustered using a guided clustering algorithm. The clustering areas are selected to obtain a representative selection of ground cover conditions throughout the basin. The clustering/classification results were evaluated and class assignments were made using the available aerial photography and vegetation survey information. The Landsat classification identified and mapping the land use and cover for the entire basin. The land use and cover classes are listed in Table 1.

The classified Landsat imagery was then geometrically registered to the UTM map base for the region so that the classified data could be directly combined with other mapped information.

The distribution of the hydrologic soil groups was determined using advanced soils survey maps obtained from the SCS. Two hydrologic soil groups B and D, soils were identified within the basin. A third hydrologic group representing the talus slopes was also defined. The soils information was drafted on USGS 1:24,000 scale topographic maps and digitized. The digitized data were converted to a UTM registered image-map in which the DN values indicated the specific hydrologic soil groups. The two, co-registered, data sets, land cover and soils, were combined to produce a composite image-map in which the DN values indicate each unique occurrence of a ground cover class and hydrologic soil group, the hydrologic soil-cover complexes.

TABLE 1. The Curve Numbers Assigned to Each Hydrologic Soil-Cover Complex Defined Using Landsat Data.

Cover Class	Hydrologic Soil Groups B	Hydrologic Soil Groups D
Coniferous Forest	50	72
Deciduous Forest	42	57
Mountain Brush	65	86
Sage Brush	60	79
Grasslands	79	89
Shadscale – Salt Brush	76	92
Juniper	77	90
Farmland	58	78
Bare Soil	83	97

The vegetation field survey data was used to determine an average cover density for each cover class. Using the tables and information provided in "Procedures for Determining Peak Flows in Colorado" (SCS, 1977) a specific CN was assigned for each hydrologic soil-cover complex, based upon an average cover density. Table 1 is a summary of the CN's values by hydrologic soil-cover complex. By transforming hydrologic soil-cover complex DN values to the appropriate CN value a digitan CN image map for the basin was generated. The average CN value for the basin is 63.

In order to determine the average CN for each subbasin, the subbasins defined for the TR-20 model were digitized from USGS 1:24,000 scale maps. The digitized boundaries were converted to another map registered image in which each DN value represented a specific subbasin. A total of 174 subbasins were digitized. Using an overlay and summary reporting program, the CN image map and subbasin map were overlayed so that the distribution of CN's within each subbasin were reported in a tabular file. The tabular file was then processed so that the mean CN value for each basin was determined and printed out for inclusion in the TR-20 hydrologic model. The subbasin CN's vary between 53 and 79. These CN's compare favorably with CN values developed for other western Colorado basins, using standard mapping techniques.

CONCLUSIONS

Landsat data, other available data, and digital image processing techniques have been used to develop CN information at the subbasin level for direct input to hydrologic models. The Landsat imagery allows one to identify the composition and distribution of ground cover throughout the basin. Digital image processing facilitates the production of basin resource maps for cover and soils and supports the development of composite resource maps such as hydrologic soil cover complex information. The applied procedure has resulted in substantial ccost and time saving for developing CN information in areas where only limited information is available. Another advantage of the procedure is that each level of information development is documented by the individual image maps which facilitates modeling documentation and quality control.

LITERATURE CITED

Army Corps of Engineers, 1981. HEC-1 Flood Hydrograph Package. Hydrologic Engineering Center, Davis, California.

Bondelid, T. R., T. J. Jackson, and R. H. McCuen, 1980. Comparison of Conventional and Remotely Sensed Estimates of Runoff Curve Numbers in Southeastern Pennsylvania. Proceedings of the American Society of Photogrammetry Annual Meeting, St. Louis, Missouri.

Cermak, R. J., A. D. Feldman, and R. P. Webb, 1979. Hydrologic Land Use Classification from Landsat. *In:* Satellite Hydrology, M. Deutsch, D. R. Wiesnet, and A. Rango (Editors). American Water Resources Association, Minneapolis, Minnesota, pp. 262-269.

Flemming, M. D., J. S. Berkebile, and R. M. Hoffer, 1975. Computer Aided Analysis of Landsat-1 MSS Data; A Comparison of Three Approaches Including "Modified Clustering Approach." LARS Information Note 072475, Laboratory for Applications of Remote Sensing, Purdue University, West Lafayette, Indiana.

Jackson, T. J. and W. J. Rawls, 1981. SCS Urban Curve Numbers from a Landsat Data Base. Water Resources Bulletin 17(5):857-862.

Ragan, R. M. and T. J. Jackson, 1980. Runoff Synthesis Using Landsat and the SCS Model. Journal of the Hydraulics Division, ASCE, 106(HY5):657-658.

Slack, R. B. and R. Welch, 1980. Soil Conservation Service Runoff Curve Number Estimates from Landsat Data. Water Resources Bulletin 16(5):587-593.

Smith, A. Y. and R. J. Blackwell, 1979. Applications of Digital Image Processing Techniques and Information Systems to Water Quality Monitoring of Tahoe Lake. *In:* Satellite Hydrology, M. Deutsch, D. R. Wiesnet, and A. Rango (Editors). American Water Resources Association, Minneapolis, Minnesota, pp. 662-672.

Soil Conservation Service, 1972. National Engineering Handbook Section 4:Hydrology.

Soil Conservation Service, 1977. Procedures for Determining Peak Flows in Colorado.

INTERNATIONAL SYMPOSIUM ON HYDROMETEOROLOGY
JUNE AMERICAN WATER RESOURCES ASSOCIATION 1982

A SIMULATION STUDY OF THE RECESSION COEFFICIENT FOR ANTECEDENT PRECIPITATION INDEX

Bhaskar J. Choudhury, Bruce J. Blanchard, and Clarence W. Richardson[1]

ABSTRACT: The antecedent precipitation index (API) has been a useful indicator of soil moisture conditions for watershed runoff calculations and recent attempts to correlate this index with spaceborne microwave observations have been fairly successful. We show that the prognostic equation for soil moisture used in some of the atmospheric general circulation models (GCM) together with Thornthwaite-Mather parameterization of actual evapotranspiration leads to API equations. The recession coefficient for API is found to depend on climatic factors through potential evapotranspiration and on soil texture through the field capacity and the permanent wilting point. Climatological and daily weather data for Wisconsin are used to simulate the annual trend and day-to-day fluctuations of the recession coefficient. Good quantitative agreement is shown with the observed trend at Fennimore and Colby watersheds in Wisconsin. This study suggests that API could be a unifying vocabulary for watershed and atmospheric general circulation modelers.
(KEY TERMS: API; GCM; soil moisture.)

INTRODUCTION

Soil moisture condition just before a rainfall has long been recognized as a major factor in watershed runoff predictions, although actual runoff depends upon many other factors, e.g., physical land features and vegetation. It is readily understood that when the soil is wet, runoff is greater than when it is dry. Soil moisture condition is also a major consideration in general circulation models (GCM) used in global models of the atmosphere because over the land the atmosphere exchanges heat with the surface and derives atmospheric moisture at a rate depending upon soil wetness. Local weather such as sea breeze and heat islands are also known to depend upon the state of the soil moisture conditions.

Saxton and Lenz (1967) expressed the API in the following recurrence form:

$$API_j = K(API_{j-1} + R_{j-1}) \tag{1}$$

where K is the recession coefficient and R_{j-1} is the amount of rainfall, both for (j−1)th day. Once the recession coefficient and an initial value of API are given, the API equation forms a powerful model for simulating the moisture conditions. The model to date remains conceptual, since no formal basis for this equation has been provided. Starting with the prognostic equation for soil moisture used in some of the atmospheric GCMs we derive the API equation. Using climatological and soil texture data we simulate the recession coefficient and show agreement with observations.

API MODEL

In formulating the atmospheric GCM, Manabe (1969) considered the following prognostic equation for moisture conditions:

$$Z \frac{d\theta}{dt} = P(t) - E_a(t) \tag{2}$$

where Z is the thickness of a fairly deep soil surface zone, θ is volumetric moisture, P(t) and $E_a(t)$ are, respectively, the rates of precipitation and actual evapotranspiration, and t is time.

There is no rigid prescription for choosing the thickness of the surface layer, Z. For bare soils a thickness of 0.2m may be appropriate, but for vegetated soils, moisture extraction may occur from a deeper soil layer. (Note that actual root depth varies with plant species, climatic factors, and soil factors.) Manabe (1969) considers the maximum water holding capacity of the surface layer, $Z\,\theta_{fc} = 0.15$m (θ_{fc} being the field capacity), which for Colby (θ_{fc} = 0.482) and Fennimore (θ_{fc} = 0.364) watersheds to be studied here correspond to Z values of 0.32 and 0.42m. The root depth in these watersheds is not known. Noting that the analysis of recession coefficients for these watersheds is based on the soil moisture in the surface 0.31m layer, a compromised value of Z = 0.35m is thought to be reasonable.

Actual evapotranspiration depends upon evaporative demand of the atmosphere, transpiration through the plants

[1] Respectively, Physical Scientist and Hydrologist, Hydrological Sciences Branch, Code 924, Goddard Space Flight Center, Greenbelt, Maryland 20771; and Hydrologist, Grassland, Soil and Water Research Lab., U.S. Department of Agriculture, Temple, Texas 76501.

and soil water availability. The Thornthwaite-Mather relation is chosen for this study

$$E_a = \left(\frac{\theta - \theta_w}{\theta_{fc} - \theta_w}\right) E_p \tag{3}$$

where θ_w is moisture at the permanent wilting point and E_p is potential evapotranspiration.

With Equation (3), the solution of Equation (2) can be written as

$$W(t+\Delta t) = \phi(t+\Delta t, t)\, W(t) + \int_t^{t+\Delta t} \phi(t+\Delta t, t')\, P(t')dt' \tag{4}$$

where

$$W(t) = Z[\theta(t) - \theta_w]$$

$$\phi(t+\Delta t, t) = \exp\left\{-\int_t^{t+\Delta t} \frac{E_p(t')dt'}{Z(\theta_{fc} - \theta_w)}\right\}$$

Δt = time interval.

If time interval Δt is chosen as one day and j is used as the day index then Equation (4) can be expressed in several alternate forms.

If rain intermittently occurred during the previous sunshine period then Equation (4) can be approximated as (compare with Equation 1).

$$W_j = K(W_{j-1} + R_{j-1}) \tag{5}$$

where the average value of $\phi(t + \Delta t, t)$ appears as the recession coefficient, K, given by

$$K = \exp\left(-\frac{E}{Z(\theta_{fc} - \theta_w)}\right) \tag{6}$$

where E = daily total potential evapotranspiration for the previous day (j−1).

SIMULATION OF RECESSION COEFFICIENT

The recession coefficient as given by Equation (6) can be simulated knowing soil texture and climatic data. From Saxton and Lenz (1967) the maximum available water, $Z(\theta_{fc} - \theta_w)$, is 0.0735m and 0.105m, respectively, in the top 0.35m soil layer for Fennimore and Colby watersheds. Climatological data, actually observed for these watersheds are unavailable, and were therefore synthesized from published sources. Global insolation is a major factor determining potential evapotranspiration, and the quality of Weather Service data for insolation is questionable. We used a model developed by Choudhury (1982) to calculate the daily total insolation. The daily total potential evapotranspiration is calculated using Penman equation (Thom and Oliver, 1977).

As a check of the climatological data and model equations, the *simulated* potential evapotranspiration and net radiation are shown in Figure 1 together with the regression equation based on *observed* evapotranspiration and net radiation during July through September at Hancock, Wisconsin (Tanner and Pelton, 1960). Although simulated values are more-or-less within the standard error of estimate for the regression equation, we see that the calculated evapotranspiration are generally lower than what would be expected from the regression equation. Tanner and Pelton (1960) had noted that the Penman equation somewhat underestimates the evapotranspiration.

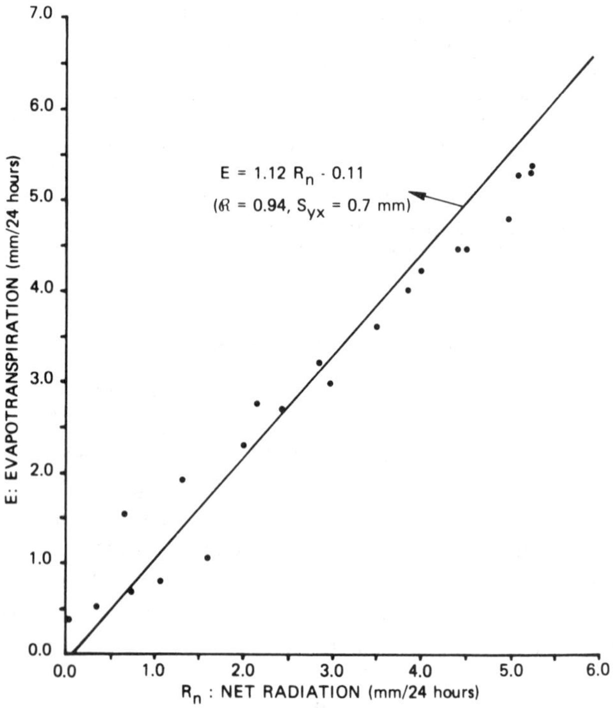

Figure 1. Comparison of Simulated (●) Penman Evapotranspiration and Daily Net Radiation with the Regression Equation from Tanner and Pelton (1960).

The simulated annual trend of the recession coefficient for Fennimore watershed together with observations (Saxton and Lenz, 1967) is shown in Figure 2 using both Penman equation and the regression equation from Tanner and Pelton (1960). Except for April-May period, the simulated trend is within the observed values. In addition to calculations using the climatological mean weather data, we calculated the daily recession coefficients using simulated daily weather for Wisconsin (Richardson, 1981). Figure 3 illustrates that day-to-day variations in weather lead to fluctuations in the simulated recession coefficients, which to a large extent mimic the observed scatter. The higher recession coefficient for April-May period indicates that the rate of soil drying was slower than the calculated rate.

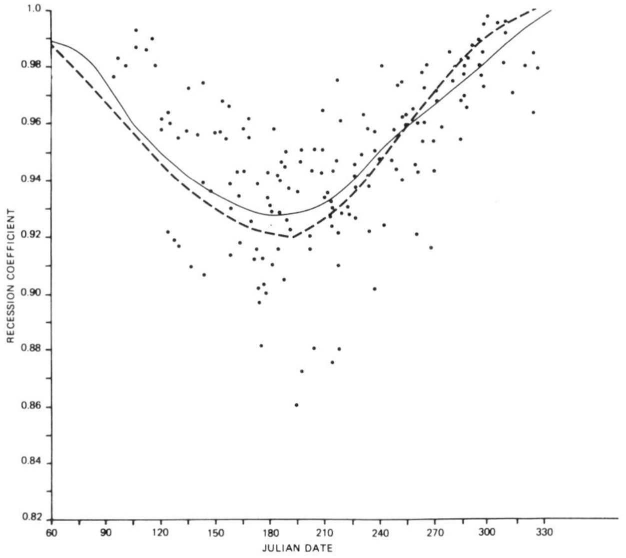

Figure 2. Comparison of Observed (●) and Simulated Annual Trend of the Recession Coefficient (solid line based on Penman equation, dashed line based on Tanner-Pelton regression equation). The observed data points are from Saxton and Lenz (1967).

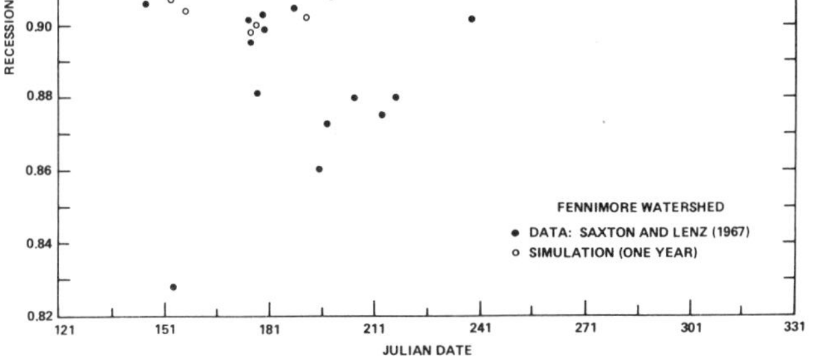

Figure 3. Simulated Daily Recession Coefficients (x) Based on Tanner-Pelton Regression Equation. Day-to-day variation in the recession coefficient is due to net radiation, which affects potential evapotranspiration.

Saxton and Lenz (1967) observed that recession coefficients for the Colby watershed were generally higher than those for Fennimore. Since these two watersheds are only 240km apart it is reasonable to assume that the climatological parameters would be about the same. Assuming this climatological equivalence, the difference in the recession coefficients should then arise from the difference in the soil texture which determines the available water, $Z(\theta_{fc} - \theta_w)$. From Equation (6) one would obtain

$$K_{Colby} = K_{Fennimore}^{0.7} \qquad (7)$$

The observed mean recession coefficient is 0.92 during July at Fennimore, which, according to Equation (7), would imply the corresponding value for Colby should be 0.943. This calculated value for Colby agrees well with the observed value of 0.95.

CONCLUSION

We discussed a formal basis for the API model and a rational method for calculating the recession coefficient. To the extent that one can use the climatological (or pan evaporation) data to obtain potential evapotranspiration and soil texture information, it will be possible to calculate a good first approximation for the recession coefficient, and thus simulate moisture conditions in large areas. Microwave remote sensing of API would be useful in updating watershed models and in monitoring global climate as an indicator of ground wetness.

LITERATURE CITED

Choudhury, B. J., 1982. A Parameterized Model for Global Insolation Under Partially Cloudy Skies. Solar Energy 29:479-486.

Goldberg, B. and W. H. Klein, 1980. A Model for Determining the Spectral Quality of Daylight on a Horizontal Surface at any Geographic Location. Solar Energy 24:351-357.

Manabe, S., 1969. Climate and Ocean Circulation, 1, The Atmospheric Circulation and the Hydrology of the Earth's Surface. Mon. Weather Rev. 97:739-774.

Richardson, C. W., 1981. Stochastic Simulation of Daily Precipitation, Temperature, and Solar Radiation. Water Resour. Res. 17: 182-190.

Saxton, K. E. and A. T. Lenz, 1967. Antecent Retention Index Predict Soil Moisture. Proc. Am. Soc. Cil Eng. J. Hydraulics Div. 93:223-241.

Tanner, C. B. and W. L. Pelton, 1960. Potential Evapotranspiration Estimates by the Approximate Energy Balance Method of Penman. J. Geophys. Res. 65:3391-3413.

Thom, A. S. and H. R. Oliver, 1977. On Penman's Equation for Estimating Regional Evaporation. Quart. J. Roy. Met. Soc. 103: 345-357.

Thornthwaite, C. W. and R. R. Mather, 1957. Instructions and Tables for Computing Potential Evapotranspiration and the Water Balance. Publ. in Climatology 10:185-311.

RELATIONSHIP OF ANTECEDENT FLOW RATE TO STORM HYDROGRAPH COMPONENTS

James A. Lynch and Edward S. Corbett[1]

ABSTRACT: The importance of antecedent flow rate (AFR) as an index of soil moisture conditions on a forested watershed was quantitatively evaluated and relationships between AFR and individual storm hydrograph components were developed. The relationship between AFR and antecedent soil moisture (ASM) was also obtained. This study was carried out using a simulated rainfall facility installed on a 19.59-acre watershed. The irrigation system consisted of 5 miles of aluminum and plastic pipe and utilized 520 rotating sprinklers mounted on 5-foot risers. Water pumped from a nearby lake was used as the source for the simulated rainstorms. Eight, 1.50-inch storms were replicated at various AFR and ASM conditions. Rainfall intensity was 0.24"/hr. Quickflow, total storm runoff, peakflow, time of recession, and duration of quickflow resulting from these rainstorms were significantly correlated with AFR. Time to peakflow was not significantly correlated with AFR. AFR was an excellent index of ASM.

(KEY TERMS: antecedent flow; antecedent soil moisture; storm hydrographs; peak flows; quick flow; storm duration.)

INTRODUCTION

One of the most important and fundamental needs in watershed research is to fully understand watershed behavior – that is to understand the entire hydrologic system which affects the disposition of precipitation enroute through the drainage basin to its ultimate discharge as streamflow. This conversion of precipitation into streamflow is the result of complex interactions between hydrologic processes and numerous climatic and physiographic factors. One of these factors is antecedent soil moisture (ASM).

A common characteristic of most studies aimed at quantifying rainfall-runoff relationships is the absence of actual soil moisture data; ASM is generally estimated. Methods used for predicting ASM are many and varied. Some are extremely complex involving many parameters. Others are rather simple relationships between climatic parameters.

The relationship of baseflow at the beginning of a storm to the observed hydrograph has not received the attention it deserves. This parameter represents an "average" watershed condition since it is an integration of all the hydrologic processes and physiographic factors involved in the conversion of precipitation to streamflow. In addition, it varies from month to month and reflects the effects of evapotranspiration and other seasonal and climatic factors. Consequently, the use of antecedent flow rate (AFR) as an index of average ASM content of a watershed seems to be a logical approach.

The purpose of this study was to quantitatively assess the use of AFR as an index to ASM and to develop relationships between AFR and storm hydrograph components and characteristics.

PROCEDURE

The Rainfall Simulation System

The objectives of this study were accomplished using an irrigation system installed on a 19.59-acre forested watershed in central Pennsylvania. The rainfall simulation facility allows for control of watershed conditions prior to the application of simulated rainfall as well as rainfall intensity, duration, and amount.

The irrigation system was installed on a watershed that is essentially elliptical, with symmetrical sides facing north and south. Average sideslope steepness is 30 percent. The system consisted of approximately 5 miles of aluminum and plastic pipe. Two heavy-duty centrifugal pumps supplied water from a nearby lake. Five hundred and twenty rotating sprinklers with 5/32-inch nozzles mounted on 5-foot risers distributed the water over the basin. The sprinklers were located on the watershed in a 40-foot grid spacing. Flow control valves on each sprinkler allowed a known quantity of simulated rainfall to be applied once the system was pressurized. The flow control valves had a maximum application rate of 0.24 inch per hour.

Separation of quickflow from delayed flow was based on the technique described by Hewlett and Hibbert (1967).

[1]Respectively, Associate Professor of Forest Hydrology, School of Forest Resources, The Pennsylvania State University, 311 Forest Resources Laboratory, University Park, Pennsylvania 16802; and Principal Hydrologist, Northeastern Forest Experiment Station, U.S. Forest Service, 104 Academic Projects Bldg., University Park, Pennsylvania 16802.

A line was projected from the beginning of each storm hydrograph at a slope of 0.05 csm per hour until it intersected the recession side of the hydrograph. The duration from the initiation of runoff to the point of intersection of the delayed flow separation line with the recession limb was termed duration of quickflow. This point also marked the end of the storm recession which was initiated at the point of maximum peakflow. In this analysis, base flow volumes resulting from antecedent storms were excluded from delayed flow. This was accomplished by extending a horizontal line of constant discharge equal to antecedent flow rate from the beginning of storm runoff until it intersected the recession limb of the storm hydrograph. Total storm runoff was the sum of the delayed flow and quickflow volumes.

Soil Moisture Determination

Soils on the watershed are mostly shaly silt loams ranging in depth from 1 to 3 feet on the upper side slopes and ridge top to 3 to 9 feet on the lower side slopes and channel area. Soil infiltration and percolation rates are greater than the maximum expected rainfall intensity for the area. Field capacity for the entire soil mantle (average depth 5 feet) has been estimated at 12.59 inches of water. Soil moisture was measured using the neutron scattering technique. Both surface and depth probes were used.

Affects of Antecedent Flow Rate on Storm Hydrographs

In order to quantitatively evaluate the effects of AFR on stormflow volumes and timing, eight 1.5-inch simulated rainstorms were applied to the entire watershed under various AFR and ASM conditions. Rainfall intensity for all storms was 0.24 inch per hour; rainfall duration was 6.25 hours. The storms were applied between August 1 and September 19 under as similar climatic conditions as possible.

Quickflow, total storm runoff, maximum peakflow, time to maximum peakflow, time of recession and duration of quickflow from each simulated rainstorm were compared with AFR using correlation analysis. Empirical relationships between the individual storm hydrograph parameters and AFR were derived using regression analysis. These relationships were compared with similar models developed with the storm hydrograph parameters and ASM. In addition an empirical relationship between AFR and ASM was derived.

RESULTS AND DISCUSSION

Affects of Antecedent Flow Rate on Storm Hydrographs

The lowest antecedent flow rate used in this study was 0.11 csm; the highest AFR at which a storm was simulated was 2.00 csm (Table 1). A comparison of the storm hydrograph parameters following each simulated rainfall application is given in Table 1. It is evident from these data that antecedent flow conditions were closely related to the individual storm responses. Total storm runoff ranged from 0.179 inch to 1.286 inches. Peak flows ranged from 9.1 to 57.5 csm and the duration of quickflow ranged from 25.4 to 49.3 hours. The percentage of rainfall occurring as storm runoff ranged from 12 percent at the lowest AFR to over 86 percent at the highest AFR.

Quickflow (QV) and Total Storm Runoff (TSR)

The relationship between AFR and QV and TSR for the eight simulated rainstorms are shown in Figures 1 and 2. The mathematical function that best described these curvilinear relationships was $Y = b_o + b_1 \ln X$, where Y = QV or TSR and X = AFR. Using this curvilinear relationship,approximately 97 percent of the variation in QV and TSR could be accounted for on the basis of the logarithm of the independent variable. Although a strong linear relationship existed between these variables, these curvilinear relationships were more realistic. Storm discharge tends to level off as AFR approaches levels indicative of watershed saturation. Intuitively, storm runoff reaches a maximum equal to rainfall amount. Although the curvilinear relationships in Figures 1 and 2 do not completely level off, they do estimate storm discharge more closely than do straight-line relationships. The curvilinear estimates are also more accurate at lower AFR than are linear estimates.

In comparison to linear and curvilinear relationships based on ASM, AFR curvilinear relationships were slightly better for QV and TSR. The coefficients of determination for these storm runoff parameters using ASM as the independent variable were 0.95; the coefficients of determination using AFR as the independent variable were approximately 0.97.

Maximum Peakflows (MP)

Approximately 96 percent of the variation in MP could be explained by the linear model $Y = b_o + b_1 (1/X) + B_2X$, where Y = MP and X = AFR. This curvilinear relationship is illustrated in Figure 3. Although this relationship is only slightly better than the simple linear relationship (r^2 = 0.95), it gives a better estimate of MP at low AFR. Furthermore, the coefficient of determination for the AFR-MP curvilinear relationship was better than the ASM-MP coefficient ($r^2 = 0.89$).

Time to Maximum Peakflow (TMP)

The relationship between AFR and TMP was very poor. AFR explained only 49 percent of the variation in TMP assuming a linear relationship existed. No curvilinear function could explain the variations in TMP over the range of AFR studied. In comparison, ASM appeared to be a better estimator of TMP. However, straight-line linear relationships for TMP are unrealistic in that they assume a

TABLE 1. Storm Hydrograph Characteristics and Components for Eight 1.50-Inch Simulated Rainfall Events at Different Antecedent Flow Rates.

	Hydrograph Characteristic and Component							
Date	ASM (inches)	AFR (CSM)	QV (inches)	TSR (inches)	MP (CSM)	TMP (hours)	TR (hours)	DQ (hours)
8/1	10.95	0.109	0.104	0.179	9.10	6.38	19.05	25.43
8/7	11.32	0.187	0.329	0.480	18.36	6.45	32.63	39.08
8/14	11.42	0.279	0.470	0.611	23.02	6.35	36.13	42.48
8/19	11.62	0.441	0.483	0.626	21.86	6.37	37.42	43.79
8/23	12.00	0.697	0.670	0.817	27.77	9.87	37.58	47.45
8/27	12.22	0.879	0.771	0.919	29.99	9.12	41.60	50.72
9/16	12.33	1.303	1.015	1.171	48.13	8.82	40.38	49.20
9/19	12.45	2.000	1.137	1.286	57.46	8.93	40.35	49.28

NOTES:
ASM – Antecedent Soil Moisture
AFR – Antecedent Flow Rate
QV – Quickflow Volume
TSR – Total Storm Runoff
MP – Maximum Peakflow
TMP – Time to Maximum Peakflow
TR – Time of Recession
DQ – Duration of Quickflow

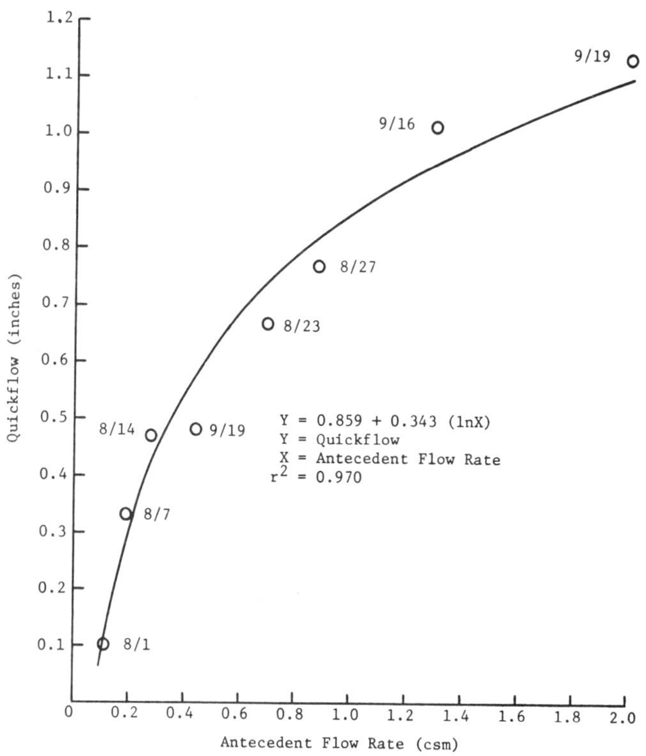

Figure 1. Curvilinear Relationship Between Antecedent Flow Rate and Quickflow Volume.

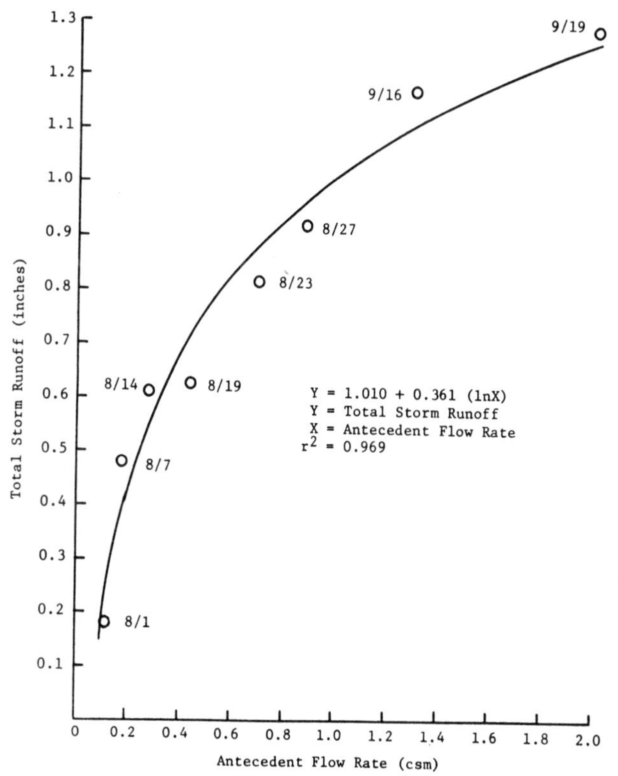

Figure 2. Curvilinear Relationship Between Antecedent Flow Rate and Total Storm Runoff.

gradual transition over the range of AFR studied. This gradual transition was not supported by the data. The times to maximum peakflow were very similar for the first four storms (6.38 hours) and then abruptly increased to 9.2 hours at higher AFR.

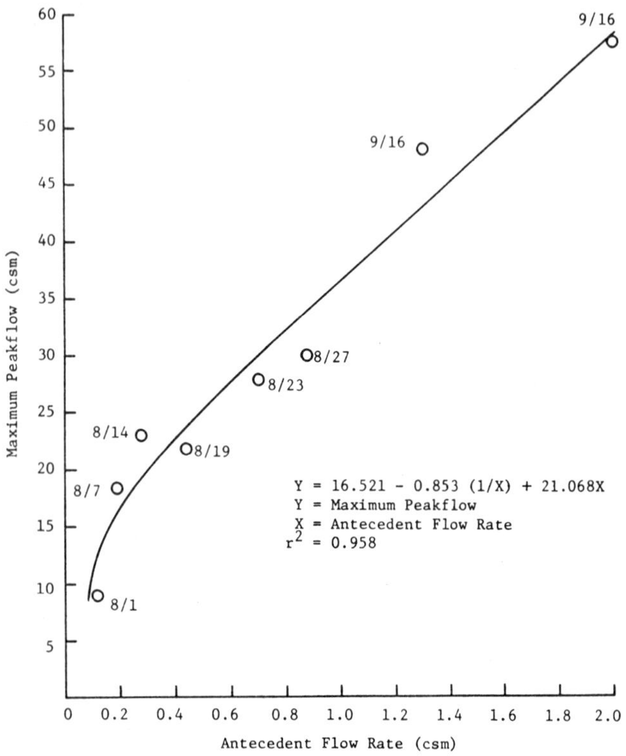

Figure 3. Curvilinear Relationship Between Antecedent Flow Rate and Maximum Peakflow.

Time of Recession (TR) and Duration of Quickflow (DQ)

AFR, TR, and DQ were curvilinearly related. These curvilinear relationships (Figure 4) take the form $Y = b_o + b_1 (1/X)$, where Y = TR or DQ and X = AFR. Using this model, 94.3 percent and 97.0 percent of the variation in TR and DQ, respectively, could be explained by AFR.

In comparison, the relationships between AFR, TR, and DQ were better, although in the case of TR only slightly better than the ASM relationships. Coefficients of determination between AFR, TR, and DQ were 0.3 and 4.3 percent higher, respectively, than for the ASM relationships.

Relationship Between Antecedent Soil Moisture and Antecedent Flow Rate

A good correlation between AFR and ASM further supports the use of AFR as an index to moisture condition of a watershed. Thirty-six ASM measurements and corresponding AFR were used to determine this relationship. ASM measurements ranged from 9.21 to 12.98 inches; AFR ranged from 0.04 to 4.31 csm. Although the linear relationship between ASM and AFR was significant at five percent, it was not very good (r^2 0.36). This lack of strong linear association between these variables was primarily due to the variability of streamflow at moisture contents above 12.00 inches or 95 percent field capacity. This variability is evident in Figure 5. It is also obvious from Figure 5 that a curvilinear function best described the relationship between ASM and AFR. This curvilinear function, a hyperbola, is represented by the model $Y = b_o + b_1 (1/X)$, where Y = ASM, and X = AFR. Using this model, 89.1 percent of the variation in ASM could be explained on the basis of AFR.

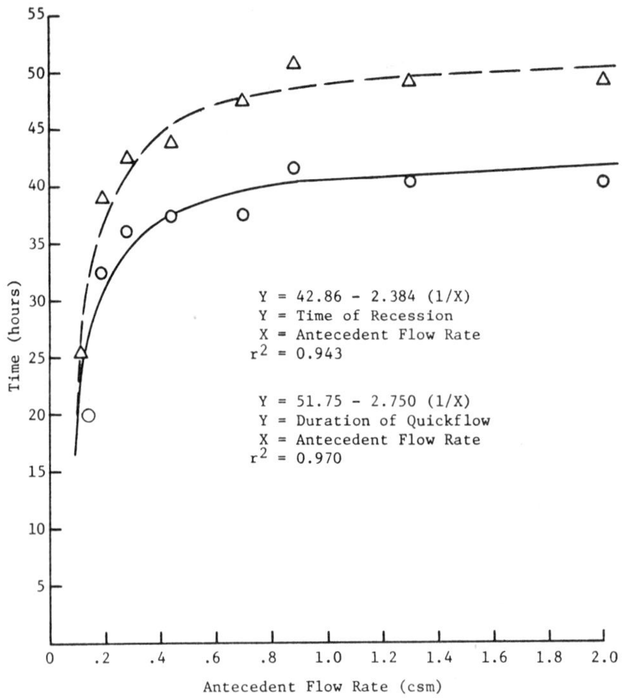

Figure 4. Curvilinear Relationships Between Antecedent Flow Rate and Time of Recession and Duration of Quickflow.

The level of the horizontal asymptote (b_o) occurred at a moisture content of 12.36 inches. Although a leveling off of the curve is intuitive as soil moisture approaches field capacity, the asymptote occurred 0.23 inch below the estimated field capacity. This may be associated with the fact that during the summer months, soil moisture contents above 12.36 inches occur infrequently and then only following large storms. Since only those moisture measurements and corresponding flow rates in which no rainfall had occurred for at least 24 hours were used in this study, only seven soil moisture measurements above 12.36 inches were involved in the analysis. Of these seven, four were above 12.50 inches.

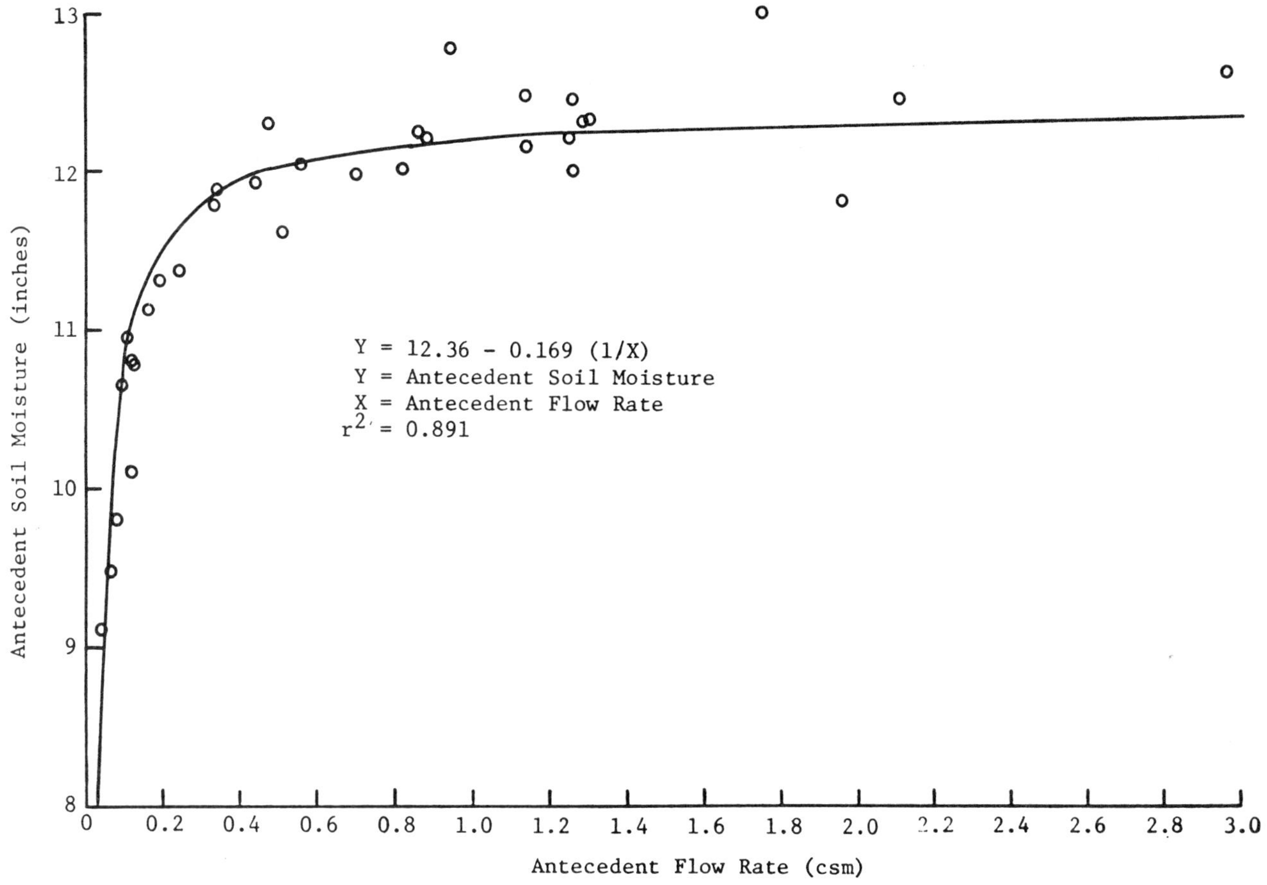

Figure 5. Curvilinear Relationship Between Antecedent Soil Moisture and Antecedent Flow Rate.

CONCLUSION

Based on the results of this study, AFR was found to be an excellent indicator of average soil moisture condition on the site, except at low flow rates when small changes in AFR are often associated with large changes in ASM. A curvilinear relationship existed between ASM and AFR. All storm hydrograph parameters except time to maximum peakflow were significantly related to AFR. These AFR curvilinear models were better than ASM relationships.

LITERATURE CITED

Hewlett, J. D. and A. R. Hibbert, 1967. Factors Affecting the Response of Small Watersheds to Precipitation in Humid Areas. International Symposium on Forest Hydrology, Pergamon Press, New York, New York, pp. 275-290.

INTERNATIONAL SYMPOSIUM ON HYDROMETEOROLOGY
JUNE AMERICAN WATER RESOURCES ASSOCIATION 1982

EVAPORATION TIME SERIES ANALYSIS FOR MILFORD LAKE IN KANSAS

H. Vernon Knapp, Yun-Sheng Yu, and Ernest C. Pogge[1]

ABSTRACT: A weekly evaporation time series generated by the mass transfer method for the Milford Lake displays a predominant annual cycle with low evaporation in spring and winter and high evaporation in summer and fall. A mathematical model for the time series id developed. The model consists of a mean weekly component represented by a Fourier approximation and a stationary random residue by an autoregressive model. From the synthetic evaporation series, the frequency distribution is obtained. A sensitivity analysis shows that the estimated evaporation is most sensitive to changes in water temperature.
(KEY TERMS: lake evaporation; mass transfer; time series analysis; stochastic process.)

INTRODUCTION

Evaporation is a stochastic process whose characteristics must be determined through an analysis of the evaporation time series. Whereas the mean annual evaporation estimated from the measured pan evaporations is generally accepted in engineering practice, the time series for weekly evaporation cannot be obtained accurately from pan evaporation because of uncertainties involved in selecting the pan coefficients. For the Milford Lake with a surface area less than 50 square miles, weekly evaporations can be calculated with sufficient accuracy by using the mass transfer method and available climatologic and water temperature data.

The objective of this paper is to determine the statistical characteristics of the weekly evaporation and to develop a mathematical model for the evaporation time series. The sensitivities of the estimated evaporation to errors in measurements are also determined.

ANALYSIS OF CLIMATOLOGIC AND WATER TEMPERATURE DATA

Mass Transfer Equation

The mass transfer equation developed from the Lake Hefner studies (USGS, 1954) and later modified by Harbeck (1962) is

$$E = 0.00338\, A^{-0.05}\, u_2(e_w - e_a) \quad (1)$$

in which E is daily evaporation in inches, A the lake area in acres, u_2 the wind speed in mph at two meters above the water surface, e_w the saturation vapor pressure in mb, e_a the air vapor pressure unaffected by the water surface in mb. The values of e_w and e_a can be calculated by substituting the water temperature and dewpoint temperature in ^oC, respectively, for T in the following equation (Linsley, *et al.*, 1975)

$$e_w \text{ and } e_a = 33.86[(0.00738T + 0.8072)^8 - 0.000019\,|\,1.5T + 48\,| + 0.001316] \quad (2)$$

If the climatologic data is not observed in the immediate vicinity of the lake, Equation (2) may be modified through the introduction of lake-land ratios for the wind speed and air vapor pressure (Derecki, 1975). For the Milford Lake, the estimated lake-land ratios are close to unity and Equation (1) was used directly to compute evaporation.

Water Temperature and Saturation Vapor Pressure

Water temperature in the Milford Lake has been measured once a week at a location just upstream of the Milford Dam since the reservoir began operation in June 1967. The records however show occasional gaps of greater than four weeks without measurements. In analyzing the data, the following regressive equation was developed and used to fill the missing data.

$$T_t = 0.670\, T_{t-1} + 0.0532\, a_t \quad (3)$$

in which T_t and T_{t-1} are water temperatures of the current and the preceding week; a_t is a normally distributed random variable with zero mean and unity standard deviation. The measured water temperature is assumed to represent the weekly average temperature.

[1] Hydrologist, Illinois State Water Survey, Box 5050, Station A, Champaign, Illinois 61820; and Professors, Department of Civil Engineering University of Kansas, 1039 Learned Hall, Lawrence, Kansas 66045.

The saturation vapor pressure can be computed from Equation (2) using the weekly average temperature. A sample of the computed saturation vapor pressure from 1968 to 1971 is shown in Figure 1. A dominant seasonal component with a period of one year is clearly seen in Figure 1. This trend generally reflects the variation of water temperature.

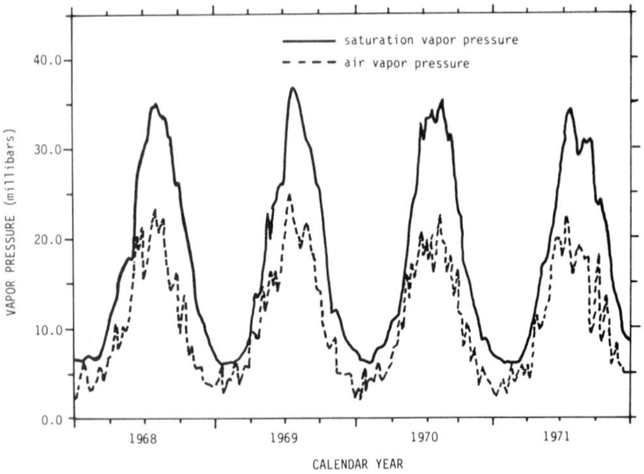

Figure 1. Milford Lake Saturation Vapor Pressure and Air Pressure, 1968-1971.

Dewpoint Temperature and Air Vapor Pressure

Daily average dewpoint temperatures measured by Concordia, Kansas, were used. Concordia is located about 40 miles northwest of the Milford Lake. The weekly dewpoint temperature series was obtained from the daily values. The weekly air vapor pressure computed from Equation (2) for the period from 1968 through 1971 is also shown in Figure 1 as a dashed curve. The air vapor pressure likewise has a predominant annual cycle. High frequency components of significant amplitudes are also present. However, a spectral analysis of both saturation and air vapor pressure series does not show any significant difference in their spectral contents (Yu, *et al.*, 1981).

Wind Speed

Wind speeds measured at Concordia, Kansas, were converted according to the 1/7-power law to give the wind speeds at a height of 2 meters above the water surface. An analysis of the wind speed series leads one to conclude that the wind speed appears to be random short-term fluctuations with a weak seasonal component. Therefore, for each week in a year, the mean of the observed values can be used adequately in the mass transfer equation.

WEEKLY AND MEAN MONTHLY EVAPORATIONS

A sample of the weekly evaporation series obtained from Equation (1) for 1968-1971 is shown in Figure 2. The general pattern of the evaporation series is very similar to those of vapor pressures shown in Figure 1 with irregular components superimposed on a dominant seasonal component. The modeling of the time series will be discussed in the next section. The mean monthly evaporations calculated from the weekly values are presented in Table 1.

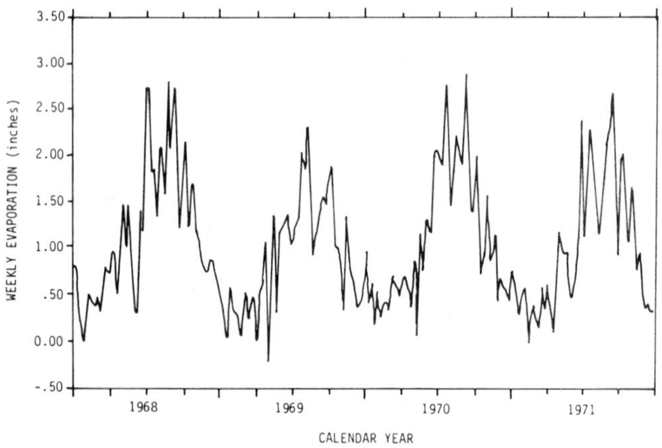

Figure 2. Milford Lake Weekly Evaporation, 1968-1971.

The mean monthly evaporations for November through May are all below four inches. But the evaporation increases substantially in June and maintains a high level through September. The mean annual evaporation is 53.70 inches about 3 percent higher than the value given by Kohler, *et al.* (1959). The total evaporation from June through September exceeds 50 percent of the mean annual value.

The pan coefficients computed from the values listed in Table 1 and the measured pan evaporations for April through October are shown in Table 2. The pan coefficient increases steadily from a low of 0.39 for April to 1.16 for October with an average value of 0.73. The total evaporation for these months amounts to 78.4 percent of the mean annual evaporation.

MATHEMATICAL MODEL FOR WEEKLY EVAPORATION SERIES

The weekly evaporation time series may be considered as composed of a deterministic periodic component and a stationary random component. The former is the mean weekly evaporation and the latter the residue. The respective models are developed as follows.

TABLE 1. Mean Monthly Evaporation for Milford Lake.

	Month											
	January	February	March	April	May	June	July	August	September	October	November	December
Evaporation (inches)	1.69	1.50	2.21	2.78	3.91	6.12	8.07	8.11	7.08	5.84	3.96	2.44

Mean Annual Evaporation = 53.70 inches.

TABLE 2. Pan Coefficients for April Through October.

	Month						
	April	May	June	July	August	September	October
Pan Coefficient	0.39	0.52	0.64	0.75	0.83	1.01	1.16

Mean Weekly Evaporation and Standard Deviation

The mean weekly evaporation is deterministic and may be represented by a Fourier approximation. The number of Fourier components required to adequately simulate the series can be determined from a spectral analysis of the given series. The computed spectral density function (Yu, *et al.*, 1981) shows significant spectral peaks at one and two cycles per year. The removal of these two components essentially eliminates all significant peaks for frequencies equal to or greater than one cycle per year. The spectral values for frequencies less than one cycle per year however were not substantially reduced. Such low frequency components are believed to be the effect of sampling fluctuation for the short time series and not an inherent property of the series.

The equation deveioped for the mean weekly evaporation is

$$\bar{E}_j = 1.02 - 0.79 \sin\left(\frac{2\pi j}{52} + 0.98\right) - 0.12 \sin\left(\frac{4\pi j}{52}\right) \quad (4)$$

in which $\bar{E}_j$ is the mean evaporation in inches for the jth week of a year.

A residual series is obtained from the differences between the original series and the mean weekly evaporation given by Equation (4). The standard deviation of the residual series is also cyclic and may be expressed as

$$s_j = 0.38 + 0.18 \sin\left(\frac{2\pi j}{52} - 0.15\right) \quad (5)$$

in which s_j is the standard deviation of the jth week of a year in inches.

For comparison, the mean weekly evaporations computed from Equation (4) and from the original series are shown in Figure 3a and those for the standard deviation in Figure 3b. Significant discrepancies exist between the two mean evaporation curves, especially for the months of February, March, April, and September. For the standard deviations, the maximum discrepancy between the two curves is less than one standard deviation for a given week. In view of the brevity of the evaporation series, Equation (4) is considered a proper estimate of the mean weekly evaporation. The annual evaporation calculated from the mean weekly values given by Equation (4) is 53.56 inches, very close to 53.70 inches of the original series.

Residual Series

The residual series is given by

$$E'_{i,j} = (E_{i,j} - \bar{E}_j)/s_j \quad (6)$$

in which $E'_{i,j}$ is the dimensionless residue of weekly evaporation for the jth week of ith year, $E_{i,j}$ the original weekly evaporation.

The residual series is assumed to be a stationary stochastic process. Its model can be identified based on the properties of its autocorrelation function (ACF) and partial autocorrelation function (PACF) as discussed by Box and Jenkins (1976). An analysis of the ACP and PACF of the residual series, Yu, *et al.* (1981), suggests an autoregressive model with the following general expression

$$\begin{aligned} E'_{i,j} = {} & \beta_1 s_j (E_{i,j-1} - \bar{E}_{j-1})/s_{j-1} \\ & + \beta_2 s_j (E_{i,j-2} - \bar{E}_{j-2})/s_{j-2} \\ & + \beta_3 s_j (E_{i,j-7} - \bar{E}_{j-7})/s_{j-7} + \epsilon_{i,j} \end{aligned} \quad (7)$$

The first, second, and third terms on the right side of Equation (7) are residues for time lags equal to 1, 2, and 7 weeks and $\epsilon_{i,j}$ is a random variable. The parameters β_1, β_2, and β_7 can be estimated by using either the least squares method or the stepwise regression technique as presented elsewhere (Yu, *et al.*, 1981).

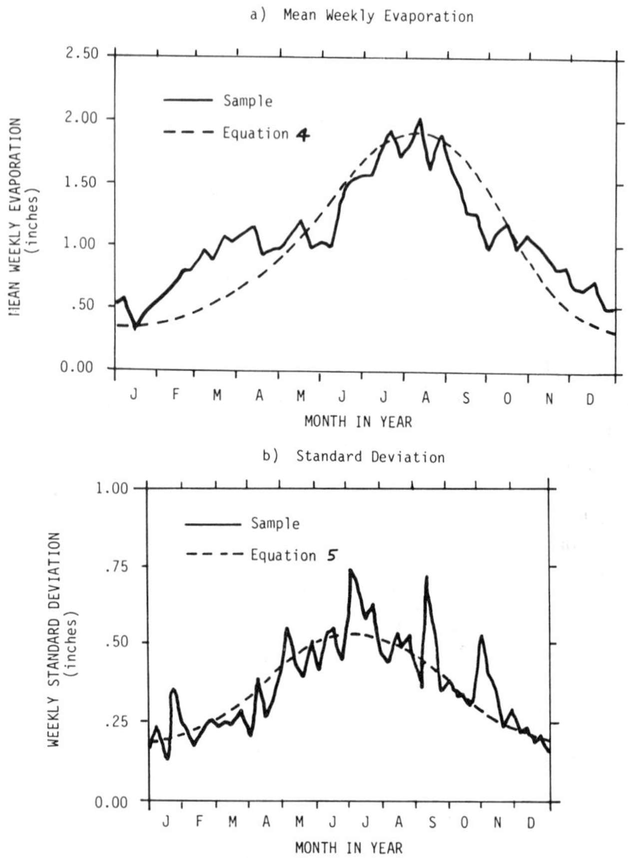

Figure 3. Mean Weekly Evaporation and Standard Deviation.

In order to select the most favorable and parsimonious model, the effects of neglecting lag-2 and/or lag-7 terms on the results were determined based on the values of the following parameters for each model: the explained variation, the standard error of estimate, the lag-one correlation coefficient of the model output, the variance and extreme values of the annual evaporation series. The final model selected is

$$E_{i,j} = \bar{E}_j + 0.139\, s_j\, (E_{i,j-1} - \bar{E}_{j-1})/s_{j-1} + 0.948\, s_j\, a_{i,j} \qquad (8)$$

in which $a_{i,j}$ is a random variable with standard normal distribution. A plot of the computed weekly evaporation series from Equation (8) for four years is shown in Figure 4 with a general pattern similar to that of the original series, Figure 2.

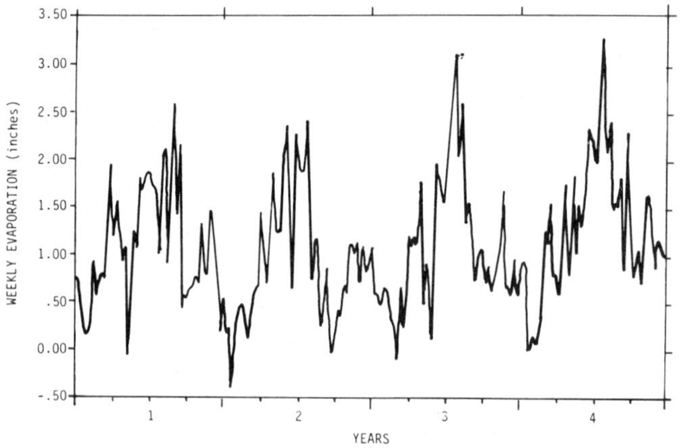

Figure 4. Milford Lake Simulated Weekly Evaporation.

The synthetic series has an explained variance of 78 percent and a standard error of estimate of 0.33 inch. The standard error of estimate is equal to the standard deviation between the original evaporation series and the synthetic series. The lag-one correlations of the synthetic series for 11 years and 1000 years are 0.361 and 0.315, respectively. The lag-one correlation of the original series is 0.319. The 1000-year series has a minimum annual evaporation of 39.51 inches and a maximum 67.42 inches. The extremes of the original 11-year series are 45.82 inches and 60.96 inches.

Frequency of Lake Evaporation

From the 1000-year weekly evaporation series generated by Equation (8), the probabilities of exceedence of the lake evaporation over a consecutive 3, 6, and 12 months are obtained and shown in Table 3.

TABLE 3. Lake Evaporation Frequency.

Probability of Exceedence (percent)	Number of Consecutive Months		
	3	6	12
50	21.90	40.44	53.56
10	24.12	43.60	58.65
5	24.69	44.55	60.17
2	25.45	45.64	61.78
1	25.89	46.25	62.95
0.5	26.30	46.80	64.15

For the 12 consecutive months, the expected value is 53.56 inches and the value at a 2 percent chance of exceedence is 61.78 inches.

SENSITIVITY ANALYSIS

The sensitivity of the estimated evaporation from the mass transfer equation to errors in wind speed, water temperature, and dewpoint temperature and the sensitivity of the model to changes in estimated model parameters are determined. For the mass transfer method, the percent changes in annual and seasonal evaporations per unit change in water temperature, dewpoint temperature, and wind speed, respectively, are computed and shown in Table 4.

TABLE 4. Sensitivity of Evaporation.

Independent Variable	Spring	Summer	Fall	Winter	Annual
Water Temperature (^oC)	20.8	13.1	13.9	20.3	15.6
Dewpoint Temperature (^oC)	14.9	7.7	7.6	12.9	9.8
Wind Speed (mph)	7.8	8.8	8.5	7.8	8.2

The evaporations in spring and winter are usually low and the relative sensitivities as shown in Table 4 are high. On both annual and seasonal basis, evaporation appears to be most sensitive to changes in water temperature. This result is contrary to the finds of Phillips (1978) who found evaporation to be most sensitive to dewpoint temperature. Evaporation is more sensitive to water temperature than to dewpoint temperature because the former is higher than the latter, i.e., at higher temperatures the vapor pressure given by Equation (2) increases more rapidly. On annual basis, an increase in average annual water temperature by 1^oC would increase 15 percent of the annual evaporation or 8.3 percent per ^oF which is higher than 3.8 percent per ^oF given by Coleman and DeCoursey (1976) for Chickasha Lake, Oklahoma.

The sensitivity of the model to the error in the estimated autoregressive coefficient β_1 is determined by computing the percent change in the standard deviation of annual evaporation per relative change in β_1. A 10 percent error in β_1 would cause a 5 percent change in the standard deviation of the annual evaporation. A 100 percent error in β_1 would result in 76 percent change in the annual standard deviation.

CONCLUSIONS

The weekly evaporation time series estimated by the mass transfer method for the Milford Lake displays an annual cycle with low evaporation in winter and spring and high evaporation in fall and summer. On the average, the total evaporation for the four months from June through September exceeds 50 percent of the mean annual evaporation. An analysis of the weekly evaporation series leads to a mathematical model which includes a mean weekly component represented by a Fourier approximation and a stationary random residue by an autogressive model. The estimated evaporation is shown to be most sensitive to changes in water temperature.

ACKNOWLEDGMENTS

The work upon which this publication is based was supported in part by funds provided by the Office of Water Research and Technology, U.S. Department of the Interior, Project No. B-053-KAN.

LITERATURE CITED

Box, G. E. P. and G. M. Jenkins, 1976. Time Series Analysis: Forecasting and Control. Holden-Day, San Francisco, California.

Coleman, G. and D. G. DeCoursey, 1976. Sensitivity and Model Variance Analysis Applied to Some Evaporation and Evapotranspiration Models. Water Resources Research 12(5):873-879.

Dereck, J. A., 1975. Evaporation from Lake Erie. NOAA Technical Report ERL-GLERL3.

Harbeck, G. E., Jr., 1962. A Practical Field Technique for Measuring Reservoir Evaporation Utilizing Mass-Transfer Theory. USGS Professional Paper 272-E.

Kohler, M. A., T. J. Nordenson, and R. D. Baker, 1959. Evaporation Maps for the United States. U.S. Weather Bureau Technical Paper No. 37.

Linsley, R. K., Jr., M. A. Kohler, and J. L. H. Paulhus, 1975. Hydrology for Engineers. McGraw-Hill, St. Louis, Missouri.

Phillips, D. W., 1978. Evaluation of Evaporation from Lake Ontario During IFYGL by a Modified Mass Transfer Equation. Water Resources Research 14(2):197-206.

U.S. Geological Survey, 1954. Water-Loss Investigation: Lake Hefner Studies. USGS Professional Paper 269.

Yu, Y. S., E. C. Pogge, C. T. Field, H. V. Knapp, and S. Esmaeil-Beik, 1980. A Study of Low-Flow Operation of Milford and Elk City Lakes in Kansas. Project Completion Report, the Kansas Water Resources Research Institute at the University of Kansas, Contribution No. 217.

MEASURED EVAPORATION IN HIGH RAINFALL AREAS, LEEWARD KOOLAU RANGE, OAHU, HAWAII

Paul C. Ekern[1]

ABSTRACT: Small evaporimeters for high rainfall areas required a factor of 0.55 for conversion to Californiagrass water use. Elevated Class A pan evaporation was 0.63 and surface pan evaporation 0.54 evaporimeter water loss. Evaporimeter and meteorological parameters were measured along a 14 km transect in central Oahu to the 7,620 mm rainfall zone at 610 m, and along a 6.5 km transect in southern Oahu to the 4,191 mm rainfall zone at 153 m on the leeward slopes of the Koolau Range. Monthly evaporation was highly correlated with r^2 values of 0.663 to 0.936 between sites beneath the cloud base, but with only r^2 of 0.02 with the highest site frequently within the cloud. Daily evaporation correlation with sunlight over a 12-month period for the sites had r^2 values between 0.55 to 0.84. Priestley-Taylor α was nearly 2 for the dry Holmes site. Values of α were 10 percent more than 1.26 for the wet valley sites and 10 percent less than 1.26 on the ridges when net radiation was calculated from long wave radiation with an unimpeded sky. Clear sky net long wave radiation was 20 percent greater in the valley than on the ridge. Runoff was 22 percent, evaporation 35 percent, and percolation 43 percent of the 2,835 mm of rainfall over a 12-month period. Direct interception of cloud water was less than 305 mm/yr. The 984 mm of evaporation was 30 percent greater than that often extrapolated for the critically water short Pearl Harbor ground water basin.
(KEY TERMS: evaporation; rainfall; Hawaii.)

Pumpage from the Pearl Harbor ground water basin, equivalent to 704 mm/yr of rainfall on the 322 km^2 watershed, is a major source of water for the irrigation and urban needs of central Oahu (State Water Commission, 1979). Direct seepage from the basin into the ocean of 333 mm/yr is partially offset by recharge from return irrigation percolate of 164 mm/yr so the net withdrawal becomes equivalent to 872 mm/yr of rainfall over the entire watershed. The artesian heads in the partially confined Ghyben-Herzberg lens have steadily declined under the last 100 years of pumpage from about 9 m to a current 4.6 m and a projected 3.3 m by the year 2000 (Mink, 1980). Alarmed over the storage withdrawal and potential rise of the salt water transition zone at the base of the lens, the Department of Land and Natural Resources in 1979 designated this the first regulated Hawaiian ground water basin (Reg. 9 DLNR Groundwater Use Act 1961).

Surface runoff from the entire watershed is only 18 percent of the rainfall of 1,870 mm/yr (Hirashima, 1971). Runoff hydrographs are steeply triangular with the time to peak 2.58 hr for Waikele Stream, the largest in the shed (Wu, 1969). Runoff from the high rainfall aquifer sources on the leeward Koolau range are about 22 percent of their average 2,972 mm annual rainfall with time of concentration 1.7 and time to peak 0.95 hr. These flood flows carry a heavy sediment load, are difficult to store, and not easily developed for potable water (Ekern, 1976).

Florisitic and ecologic studies in Kipapa Gulch included rainfall and evaporation measurements with Livingstone bulb atmometers for a nine-month period in 1932 (Hosaka, 1935). A rudimentary assessment of evapotranspiration was made to accompany the rainfall and stream flow measurements in the Kipapa sector for 1957, 1958, and 1959 (Mink, 1962). Evapotranspiration in the high rainfall aquifer source region has been subsequently extrapolated from Class A pan evaporation measurements on plains planted to irrigated sugarcane where rainfall is low and positive advection of sensible heat large (Takasaki, 1969; Ekern, 1972). Only two pan evaporation sites were available in the high rainfall areas and lysimetric measures of water use by grass and fern were nearly double the value for the pan (Stearns and Vaksvik, 1935). Pan evaporation records are notoriously poor when rainfall occurs and would not at all be suited for the high rainfall areas (Bloemen, 1978).

The premise of equilibrium evaporation into a saturated layer of air propounded by Priestley-Taylor (1972) must surely be met in these high rainfall sectors. The rain forest (Carlquist, 1980) with continuously wetted leaf canopy should further modify the combination energy budget aerodynamic equation (Shuttleworth and Calder, 1979).

The rate of decrease in evaporation from areas of positive advection drops abruptly at the point where rainfall exceeds evaporation and the radiation energy budget alone dominates further evapotranspiration (Brutsaert and Stricker, 1979). A more precise appraisal of the withdrawal by

[1]Hydrologist, Water Resources Research Center, University of Hawaii at Manoa, Holmes Hall, Room 283, 2540 Dole St., Honolulu, Hawaii 96822.

evapotranspiration from the potential recharge into the vital Pearl Harbor aquifer is urgently needed.

METHODOLOGY

Evaporation was measured along two transects across the leeward Koolau Range. The southern Oahu transect, which extended normal to the ridge from 6,440 m to within 4,800 m of the crest, had four sites on the floor of Manoa Valley and one site on the adjacent ridge (Table 1). The second, in central Oahu, which extended normal to the ridge from 14,000 m to within 800 m of the crest, had seven sites along Kipapa ridge and one on the valley floor at the gaging station on Kipapa Stream. Water loss was measured from a 152 mm diameter carborundum stone evaporating surface which wicked water from a 100 mm deep reservoir of schedule 40 white, 152 mm ID PVC pipe mounted 1.1 m above the soil surface. Clear Plexiglass 3 mm thick and 0.1 m^2 formed a rain shield 100 mm above the stone surface. The dark, continuously wet, porous carborundun with an evaporational plane at or near the surface more nearly simulated evaporation from a wetted leaf than any open pan evaporimeter might (Davenport, 1967). The medium to fine textured stones (Wilcox, 1963) had low total porosity so that the distinctive end point was readily detected as the reservoir was refilled and the pores became saturated. The volume required to refill the reservoir after periods of 1 to 4 days was recorded. Reservoirs of duplicate evaporimeters at the Manoa Valley sites were insulated with 50 mm of white styrofoam and water temperatures recorded for a time to check on heat gain or loss through the reservoir walls.

TABLE 1. Transect Sites, Elevation, Distance from Crest, and Annual Rainfall, Oahu, Hawaii.

Transect Site	Elevation (m)	Distance From Crest (m)	Annual Rainfall (mm)
	KIPAPA		
Mililani	122	14,057	1,016
Bench Mark	183	11,734	– – –
Koa Ridge	366	7,315	1,544
Traveler's Palm	457	5,182	2,263
Site No. 4	488	3,658	2,845
Site No. 5	518	2,286	3,868
Site No. 6	625	732	4,730
Stream Gage	210	7,468	– – –
	MANOA		
Holmes	23	5,578	912
Mauka Campus	31	4,724	1,471
Huelani	107	3,505	2,642
Lyon	104	1,753	3,556
St. Louis Heights	311	3,505	– – –

The evaporimeters were calibrated against Class A pan evaporation at Holmes and Mauka sites in Manoa Valley and at Mililani, Bench Mark, and Koa Ridge on the Kipapa transect. Consumptive use of water by Californiagrass was measured at the Mililani site in eight large percolate lysimeters adjacent to the pan and evaporimeter stations (Handley and Ekern, 1981).

Auxiliary meteorological parameters of sunlight and rainfall were measured at each site. An Eppley PSP at the Holmes site was used to standardize the Eppley 848 and Licor sensors at the field sites (Ekern, 1978). Soil temperature at 150 and 300 mm depths was measured by thermistors at the ridge sites. Wind velocity was measured at the Holmes, Mauka, Bench Mark, and No. 6 site. Since the No. 6 site was often in cloud, a fog drip interceptor measured direct interception of cloud water here (Juvik and Ekern, 1978). The USGS stream flow records from station (16212800) on the Kipapa Stream were used to determine the surface runoff from the watershed. Stream gage records were taken on Manoa Stream adjacent to the Holmes site.

The fraction of global radiation converted into net radiation was calculated based on a reflectance of 0.10 (Ekern, 1965a). Daily net long wave exchange of 150 cal/cm^2 for clear sky was prorated for cloudy sky on the actual fractional receipt of the maximum (approximate clear day) global sunlight at Holmes (Schooley, 1969; Reed, 1976). Net radiation was recorded for brief intervals with a Funk miniature net radiometer at valley and ridge sites on the Manoa transect. Extensive study of net radiation adjacent to the Kipapa Ridge Mililani site has been reported (Ekern, 1965).

RESULTS

At Mililani, comparison between water use by grass and evaporimeter measurements was more reliable in summer when rainfall was light and uncertainties in rainfall catch on the lysimeters was less. During the summer of 1981, water use by full canopy (1 m tall) grass from the four large lysimeters averaged 0.55 of the 7.6 mm/day loss from the evaporimeters. The evaporation from individual lysimeters varied with biomass from 0.4 to 0.7 of the evaporimeter rate. In winter, ratios of grass use to evaporimeter were generally greater but more variable and more uncertain, with an average of 0.65. Overall water use by grass from the spring of 1979 through the summer of 1980 had latent energy equivalent to 0.56 the global sunlight, a quite reasonable relationship indicative of a minimum of positive advection of heat immediately downwind from the irrigated grass plots some 12 x 24 m in area. Water use by fully transpiring vegetation in these water budget studies has been assessed as 0.55 the amount measured by the evaporimeters. This value is in accord with prior measurements of the calibration factor of 2.65 for 76 mm Bellani plates and 0.85 for the 1.2 m diameter Class A pan and 0.99 or better for 3.05 m diameter lysimeters (Ekern, P. C., 1964, The Evapotranspiration of Pineapple in Hawaii, Res. Report No. 109, Pineapple Res. Institute of Hawaii, 233 pp., Private Publication).

Linear correlations of daily evaporimeter and Class A pan evaporation at the Mauka campus for the summer months of 1981 had coefficient of determination, r^2, values of 0.716 with the highest individual monthly value of 0.805 for July. Summer 1981 ratios of pan to evaporimeter were 0.63 for Mililani and Californiagrass use 10 percent greater than pan evaporation. In the summer, Bench Mark pan evaporation (surface level rather than 1 m elevation) was 0.54 the evaporimeter. These values are in general accord with relations reported for sugarcane (Jones, 1980) and sod (Ekern, 1965b; Lau, *et al.*, 1975).

Insulation of the Holmes evaporimeter from July 1980 through June 1981 made only small differences in water loss though water temperatures in the reservoir were 2.6°C and in the uninsulated reservoir 2.1°C below air temperature in October 1981. Insulation increased winter evaporation slightly and reduced summer evaporation. The insulated reservoir at Mauka in November 1981 was 1.8°C below and the uninsulated reservoir the same as air temperature.

Cumulative evaporation among the sites was highly correlated (Table 2). Correlation between Huelani and the other sites for individual monthly evaporation in 1981 had r^2 values from 0.663 to 0.936 for all but site No. 6. Site No. 6 frequently within the cloud had the lowest evaporation poorly correlated with the other sites. These high correlations among the stations beneath the cloud means the uniformity of the evaporation pattern allows a single station to effectively monitor evaporation changes over much of the high rainfall portion of the leeward watershed.

Correlations of evaporation with solar radiation gave higher r^2 values during the summer trade winds when the aerodynamic factor was least variable. At Huelani the average r^2 was 0.641 for May through August 1981; 0.505 for winter 1980-1981, and 0.628 for winter 1981-1982. Correlations of daily values of evaporation with sunlight had r^2 values 0.61 for Huelani, 0.547 for Holmes, and 0.838 for Mauka for the 12 months April 1980 through March 1981. Separation into periods with daily wind run above and below 120 km/day for November 1980 through April 1981 gave at Huelani r^2 0.584 for low wind, 0.461 for high wind, compared to 0.529 for the combined period. The high values of 0.78 and 0.80 for r^2 in January and February 1982 reflect the occurrence of the lowest winter sunlight values in 50 years of record for Honolulu.

Soil temperatures suggest an annual value of 23°C for the lower and 20°C for the upper valley sites and 21°C for the 305 m and 18°C for the 610 m ridge sites. The Priestley-Taylor multiplier to convert net radiation into equilibrium evaporation $(S/S + \gamma)$ with S the slope of the saturation water vapor pressure with temperature and γ the psychrometric constant, a function of barometric pressure, was 0.726 at sea level, 0.69 at the valley head, 0.70 to 305 m, and 0.673 at 610 m. With α of 1.26, the fractions of net radiation converted into actual evaporation, $\alpha\ (S/[S+\gamma])$, became 0.91 for the lower, 0.87 for the upper valley, 0.88 for the lower, and 0.85 for the upper ridge sites.

The high rate of evaporation with an actual α of 1.77 was evidence for positive advection and an aridity effect for the low rainfall Holmes site (Table 3) (Brutsaert and Strucker, 1979). Valley sites had evaporation 10 percent greater than that calculated for net radiation with net long wave radiation from an unobstructed horizon and α of 1.26 while ridge sites had evaporation 10 percent less than that

TABLE 2. Cumulative Evaporation Measured for Manoa and Kipapa Transect Sites, Oahu, Hawaii, 1981.

	Manoa Sites				Kipapa Sites						
1981	Holmes (mm)	Mauka (mm)	Huelani (mm)	Lyon (mm)	Mililani (mm)	Koa Ridge (mm)	Puu (mm)	Traveler's Palm (mm)	Site No. 4 (mm)	Site No. 5 (mm)	Site No. 6 (mm)
January	142	104	113	75	80	74	82	71	67	69	61
February	258	200	220	155	135	141	140	137	129	130	102
March	436	350	351	255	211	226	231	211	202	206	146
April	655	502	482	343	344	319	335	307	276	282	192
May	883	677	623	438	502	414	439	404	364	370	241
June	1,131	863	763	539	647	520	556	518	468	464	287
July	1,382	1,068	920	656	– –	629	676	641	567	565	347
August	1,616	1,260	1,077	775	– –	738	798	764	669	645	384
September	1,840	1,435	1,219	886	– –	838	902	861	756	740	441
October	2,046	1,596	1,360	995	– –	940	1,004	957	839	819	493
November	2,226	1,746	1,485	1,096	– –	1,019	1,068	1,012	895	872	523
December	2,377	1,856	1,599	1,184	– –	1,083	1,140	1,075	958	931	572
Ratio*	1.49	1.16	1.0	0.74	– –	0.68	0.71	0.67	0.60	0.58	0.36

*Compared to Huelani annual value.

TABLE 3. Measured andPriestley-Taylor Calculated Evaporation for Manoa and Kipapa Transects, Oahu, Hawaii, 1981.

Transect Site	Net Radiation (mm)	Calculated Evaporation (mm)	Measured Evaporation (mm)	Actual α	Rainfall (mm)
		Manoa			
Holmes	1,863	1,696	2,428	1.77	800
Mauka Campus	1,862	1,657	1,856	1.40	1,279
Huelani	1,607	1,414	1,353	1.44	2,035
Lyon	1,278	1,112	1,184	1.34	3,580
		Kipapa			
Koa Ridge	1,460	1,285	1,083	1.06	1,931
Puu	– – –	– – –	1,140	– –	– – –
Traveler's Palm	1,332	– – –	1,073	1.16	2,712
Site No. 4	1,320	1,172	958	1.05	3,574
Site No. 5	1,132	973	931	1.21	4,197
Site No. 6	894	760	572	0.95	5,147

calculated on the same basis. Measured net long wave radiation was 20 percent greater on the ridge than in the valley. This entrapment of long wave radiation would increase the energy available for evaporation in the valleys and the enhancement of long wave radiation loss on the ridge would reduce the energy available for evaporation. Evidence for a high α value from the postulated enhancement of advective exchange to the continuously wetted leaf surfaces was not found for the highest rainfall site No. 6 (Shuttleworth and Calder, 1979).

A unique pattern of evaporation with distance from the ridge or rainfall amount must exist for the valley transect of southern Oahu and the ridge transect of Central Oahu. For comparable distances from the crest or rainfall, the Manoa sites have evaporation as much as 30 percent more than the Kipapa sites. However, for the winter months the differences in evaporation between adjacent valley and ridge sites on each transect were quite small, St. Louis Heights was 1.01 times Huelani and Koa Ridge was 1.04 the Stream Gage evaporation.

The water balance for the high rainfall (> 1,524 mm/yr) portion of the Kipapa watershed for the 12 months October 1980 through September 1981 (Table 4) showed that 22 percent of the 2,835 mm of rainfall left as stream flow, 35 percent evaporated, and 43 percent percolated for recharge. These values correspond closely to the averages of 22, 38, and 40 percent developed in 1980 based on an advective net radiation approach to evaporation (Broadbent, E. W., 1980, An Estimate of the Geohydrologic Boundaries and the Hydrologic Budgets of the Pearl Harbor and Waialua Ground Water Basins, Amer. Factors, Inc., 119 pp.) but differ from the 20, 30, and 50 percent suggested in 1962 (Mink, 1962). Recharge of 1,219 mm from this 32.6 percent high rainfall portion would be equivalent to 0.326 x 1,219 = 397 mm of water over the entire 322 km^2 watershed. This 397 mm would supply only 46 percent of the average annual 872 mm of net pumpage and seepage from the basin. Recharge can occur from other areas in winter when rainfall exceeds evaporation or where nontranspiring pineapple fields occur (Ekern, 1965c). Still this 1980-1981 12-month period recharge was considerably less than the average annual recharge from the high rainfall area equivalent to 63 percent of the net removal in the 1980 study (Broadbent, 1980, *ibid*).

Direct interception of cloud water contributed no more than about 305 mm of additional water to a very limited portion of the watershed. Though such interception on the windward slopes can be important as a water source, these lee slopes of the Koolaus received insignificant amounts of fog dip.

SUMMARY

An evaporimeter, shielded from rainfall, with a continually wetted porous evaporation surface required a 0.55 calibration factor to represent unrestricted water use by vegetation. Correlation r^2 values of 0.7 to 0.9 for monthly evaporation from sites beneath the cloud would allow use of a single monitoring site for this high rainfall portion of the watershed. As expected in high rainfall areas the dominant component affecting evaporation was the net radiation balance. Positive advection of heat sharply increased actual evaporation in dry areas where evaporation exceeded rainfall. Particularly during the trade wind period, sunlight was an effective index of evaporation and the Priestley-Taylor adaptation for nonadvective evaporation was appropriate for the high rainfall sectors when adjusted for net long wave balances on the ridge and in the valley. In this major recharge part of the Kipapa watershed for a 12-month period (1980-1981) runoff was 22 percent, evaporation 35 percent, and ground water recharge 43 percent of the 2,835 mm of rainfall. Recharge from this one-third

TABLE 4. Water Balance of Upper Kipapa Watershed, Oahu, Hawaii.

Year/Month	Rainfall (mm)	Evaporation (mm)	Runoff (mm)	Percolate (mm)	Evaporation:Rain (percent)	Percolate:Rain (percent)
1980						
June	308	84	105	119	27	39
July	320	96	92	121	30	40
August	227	101	58	62	48	27
September	139	113	43	–17	95	–12
October	315	82	47	186	26	59
November	105	91	13	1	87	1
December	205	60	44	102	29	50
1981						
January	173	68	29	76	39	44
February	179	61	26	92	34	51
March	294	76	68	150	26	51
April	359	80	88	191	22	53
May	305	89	101	115	28	38
June	126	94	9	22	73	17
July	307	96	77	134	31	44
August	316	93	79	144	29	46
September	151	94	38	19	63	12
Annual						
October 1980 through September 1981	2835	984	619	1232	35	43

portion of the total watershed was only one-half that needed to meet the average net withdrawal by pumpage and seepage from the Pearl Harbor basin.

ACKNOWLEDGMENTS

Instruments and support came in part from DOE Project 1897 and Honolulu Board of Water Supply and Dept. of Land and Natural Resources Grants 337 and 347.

LITERATURE CITED

Bloemen, G. W., 1978. A High-Accuracy Recording Pan-Evaporimeter and Some of its Possibilities. Tech. Bul. 107. Inst. for Land and Water Management, Wageningen, The Netherlands. J. Hydrol. 39:159-173.

Brutsaert, W. and H. Stricker, 1979. An Advective-Aridity Approach to Estimate Actual Regional Evapotranspiration. Water Resources Research 15:443-450.

Carlquist, S., 1980. Hawaii; A Natural History, 2nd Edition. Pac. Trop. Bot. Garden, S. B. Printers, Inc., Honolulu, Hawaii, 468 pp.

Davenport, D. C., 1967. Variations of Evaporation in Time and Space II. Study of Spatial Changes Using Evaporimeters. J. Hydrol. 5:329-350.

Ekern, P. C., 1965a. The Fraction of Sunlight Retained as Net Radiation in Hawaii. J. Geophys. Res. 70:785-793.

Ekern, P. C., 1965b. Evapotranspiration of Bermudagrass Sod in Hawaii. Agron. J. 58:387-390.

Ekern, P. C., 1965c. Evapotranspiration of Pineapple in Hawaii. Plant Physiol. 40:736-739.

Ekern, P. C., 1972. Use of Water by Sugarcane in Hawaii Measured by Hydraulic Lysimeters. Proc. 14th Congress Int. Soc. Sugar Cane Technologists, pp. 805-812.

Ekern, P. C., 1976. Turbidity and Sediment-Rating Curves for Streams on Oahu, Hawaii. *In:* Soil Erosion: Prediction and Control. Soil Conservation Society of America, Ankeney, Iowa, pp. 242-254.

Ekern, P. C., 1977. Drip Irrigation of Sugarcane Measured by Hydraulic Lysimeters, Kunia, Oahu. Technical Report No. 109, Water Resources Research Center, University of Hawaii, Honolulu, Hawaii, 99 pp.

Ekern, P. C., 1978. Variation in Sunlight Induced by Topography Under the Tradewind Regime on Oahu, Hawaii. Preprint, Conf. on Climate and Energy, American Meteorological Society, 8-12 May 1978, Asheville, North Carolina.

Handley, L. L. and P. C. Ekern, 1981. Irrigation of Californiagrass with Domestic Sewage Effluent: Water and Nitrogen Budgets and Crop Productivity. Technical Report No. 141, Water Resources Research Center, University of Hawaii, 29 pp.

Hirashima, G. T., 1971. Availability of Streamflow for Recharge of the Basal Aquifer in the Pearl Harbor Area, Hawaii. Water-Supply Paper 199-B, Geological Survey, Washington, D.C., 26 pp.

Hosaka, E. Y., 1935. Floristic and Ecological Studies in Kipapa Gulch, Oahu. M.S. Thesis, University of Hawaii, Honolulu, Hawaii, 586 pp.

Jones, C. A., 1980. A Review of Evapotranspiration Studies in Irrigated Sugarcane in Hawaii. Hawaiian Planters' Record 59:195-214.

Juvik, J. O. and P. C. Ekern, 1978. A Climatology of Mountain Fog on Mauna Loa, Hawaii Island. Technical Report No. 118, Water Resources Research Center, University of Hawaii, Honolulu, Hawaii, 63 pp.

Lau, L. S., P. C. Ekern, P. C. S. Loh, R. H. F. Young, N. C. Burbank, Jr., and G. L. Dugan, 1975. Water Recycling of Sewage Effluent by Irrigation: A Field Study on Oahu. Technical Report No. 94, Water Resources Research Center, University of Hawaii, Honolulu, Hawaii, 151 pp.

Mink, J. F., 1962. Rainfall and Runoff in the Leeward Koolau Mountains, Oahu, Hawaii. Pac. Sci. 16:147-159.

Mink, J. F., 1980. State of the Groundwater Resources of Southern Oahu. Report to Board of Water Supply, City and County of Honolulu, 83 pp.

Priestley, C. H. B. and R. J. Taylor, 1972. On the Assessment of Surface Heat Parameter Flux and Evaporation Using Large-Scale Parameters. Mon. Wea. Rev. 100:81-92.

Reed, R. K., 1976. On Estimation of Net Long-Wave Radiation from the Oceans. J. Geophys. Res. 81:5793-5794.

Schooley, A. H., 1969. Radiation Measurements at Sea. J. Geophys. Res. 74:958-961.

Shuttleworth, W. J. and I. R. Calder, 1979. Has the Priestley-Taylor Equation Any Relevance to Forest-Evaporation? J. Appl. Meteor. 18:639-646.

State Water Commission, 1979. Hawaii's Water Resources. Directions for the Future. A Report to the Governor of the State of Hawaii by the State Water Commission, 40 pp.

Stearns, H. T. and K. N. Vaksvik, 1935. Geology and Ground-Water Resources of the Island of Oahu, Hawaii. Haw. Div. Hydrography Bull. 1, 479 pp.

Takasaki, K. J., 1978. Summary Appraisals of the Nation's Ground-Water Resources. Hawaii Region. Geol. Survey Prof. Paper 813-M, USGS, Washington, D.C., 29 pp.

Takasaki, K. J., G. T. Hirashima, and E. R. Lubke, 1969. Water Resources of Windward Oahu. Hawaii Geol. Survey Water-Supply Paper 1894, USGS, Washington, D.C., 119 pp.

Wilcox, J. C., 1967. A Simple Evaporimeter for Use in Cold Areas. Water Resources Research 3:433-436.

Wu, I., 1969. Flood Hydrology of Small Watersheds: Evaluation of Time Parameters and Determination of Peak Discharge. Trans. Amer. Soc. Ag. Engr., ASCE 12:655-663.

CALIBRATING A WATERSHED ON DEEP GLACIAL TILL

N. Earl Spangenberg[1]

ABSTRACT: The Little Plover River Watershed, located on deep glacial till in central Wisconsin, shares calibration problems with watersheds in similar environments. Efforts to develop reasonable calibration equations included the use of climatic data and discharge from other basins. The best calibration for annual flows seems to involve the use of precipitation data lagged four months to account for flow through the glacial mantle. Low flows and annual peak flows were also calibrated with the same values from nearby streams. No outstanding calibration equations were developed. It was concluded that calibration of watersheds developed on glacial deposits is still an annoying problems for the watershed scientist attempting to develop baseline data in glacial landscapes.
(KEY TERMS: annual peak discharge; low flows; glacial soils; watershed; calibration.)

INTRODUCTION

The Little Plover River is located in an area of changing land use in central Wisconsin. Although urban expansion has made some inroads, the growth of irrigated agriculture has put more and more land under the wheel of the center-pivot irrigation rig. The source of water for irrigation is plentiful ground water which is available at relatively shallow depths. Several reports (Weeks and Stanglund, 1971; Weeks, *et al.*, 1965) have expressed concern that growing groundwater use in this basin, and throughout the region will have significant impact on streamflow, stream depth, and stream temperature. This study was carried out in order to establish a calibration standard for the Little Plover River to document changes in annual streamflow characteristics.

THE AREA

The Little Plover River is a tributary to the Wisconsin River in central Wisconsin. It is an effluent stream draining part of an extensive glacial outwash called the Central Sands.

The major aquifer in the basin consists of glacial outwash deposits of Quantenary age overlying Precambrian crystalline rocks and limited outcrops of Cambrian sandstone. Saturated thickness of the deposits in the Little Plover basin varies from 90 feet to approximately 10 feet. Groundwater flow rates in the aquifer vary from 0.2 feet per day to about 3 feet per day (Weeks, *et al.*, 1965).

Average annual precipitation is 31.5 inches. Monthly temperatures vary from 16.4^{O}F in January to 71.8^{O}F in February. Vegetation consists of native hardwood scrub pine forests, cropland and some noncultivated grasslands. Suburban development associated with the city of Stevens Point occupies a small part of the basin. Soils in the basin consist of loamy sands and sandy loams developed in the outwash areas, and coarser soils developed on the morainal material deposited on the east side of the basin. Permeability ranges from 6 to 20 inches per hour (Soil Conservation Service, 1978).

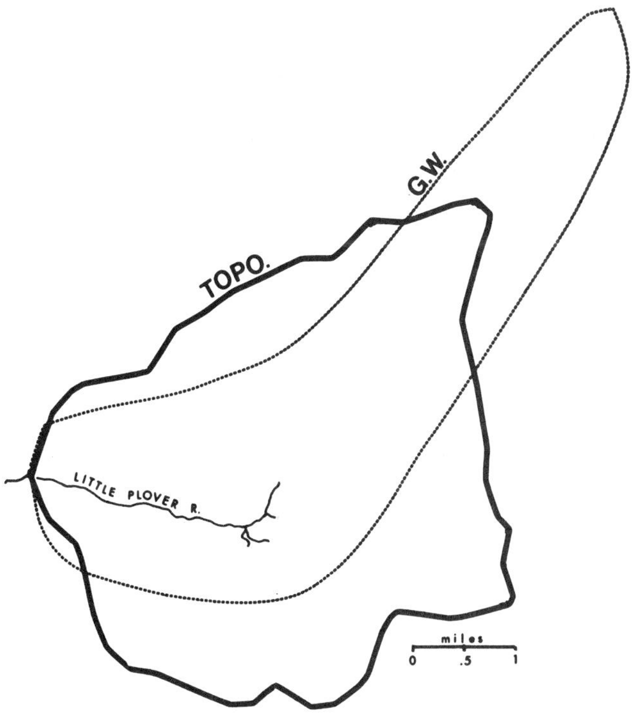

Figure 1. Groundwater and Topographic Divides of the Little Plover River Basin.

[1] Associate Professor, Water and Forestry, College of Natural Resources, University of Wisconsin, Stevens Point, Wisconsin 54481.

The topographic basin has an area of about 21 square miles, a large part of which the U.S. Geological Survey considers "noncontributing." The groundwater basin has an area of about 13 square miles and forms a comma shape to the north and east (Weeks, et al., 1965).

Annual stream flow averages 3872.7 cfs. Daily flow varies from a high of 99.0 cfs to a low of 1.4 cfs. The flow duration curve for the basin is gentle, and reflects a high degree of groundwater control over the flow. The U.S. Geological Survey report for the basin indicates that 91 percent of the flow in the Little Plover River is derived from groundwater (Weeks, *et al.,* 1965).

CALIBRATION

It is difficult to calibrate basins which are not "tight," because underflow which cannot be measured will allow for significant unaccountable variation in discharge.

The Little Plover River represents a calibration challenge since it is not a classical tight basin. In spite of possibilities of frustration, there is a need to establish reference for this river so that we can eventually assess impacts of use in the basin. There is some hope in the fact that a buried sandstone ridge near the gauge forces much of the flow to the surface. The authors of the U.S. Geological Survey report on the basin indicate that ". . . underflow from the basin is small" (Weeks, *et al.,* 1965).

Calibration was attempted using regression equations with total water year discharge as the dependent variable. A variety of independent variables were used both singly and in combination. Precipitation at Stevens Point and total water year discharge at nearby and/or similar rivers were the primary independent variables used.

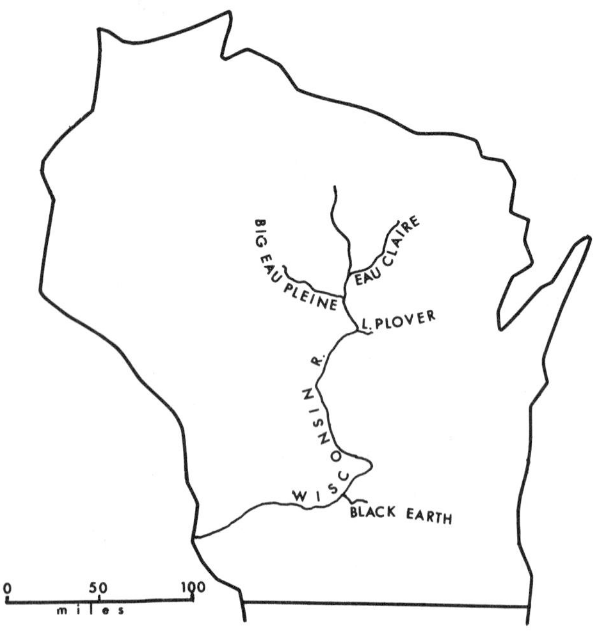

Figure 2. Locations of Rivers and Streams Used for Calibration in Wisconsin.

Other variables included average annual depth to groundwater, and precipitation at Stevens Point averaged on a four month lagged water year. This variable was used becuase there seemed to be a four month lag between peak rainfall accumulation and streamflow peak.

In addition to total discharge the Little Plover River peak flow was used as independent variable in regression equations employing yearly peak flows on the associated streams as independent variables.

RESULTS

Results of the regression analyses are shown in Tables 1 and 2. While October-September precipitation at nearby Stevens Point did not yield any exciting results, the lagged (June-May) precipitation gave a far better equation.

Although the rivers used as calibration streams were not adjacent to the Little Plover River, they are as close as possible, considering the density of gauging in the state. It is interesting that the two (Eau Claire and Black Earth Creek) which are on deeper glacial till show better agreement with the Little Plover River than the Big Eau Pleine which drains thin till.

When multiple regression analysis was used, the best precipitation equation was improved, but only slightly, by the addition of Black Earth Creek as an independent variable.

The equations involving mean annual depth to groundwater failed to produce useful equations. There were some promising indications, but duration of record was not sufficient to generate enough data for reliable equations.

No regression analysis developed any satisfactory equations for accounting for variation in peak flows.

DISCUSSION

Although the general problem of calibrating a "leaky" watershed using available surface water and climatic data remains unsolved, two possible lines of effort which may help in calibrating glacial till basins have been indicated in this study. The first is suggested by the higher coefficients of determination associated with the two equations involving the Eau Claire River and Black Earth Creek, as opposed to the smaller value associated with the shallow basin of the Big Eau Pleine. The flow duration curves for the two deeper glacial till basins resemble the flow duration curve of the Little Plover River far more than does the curve for the Big Eau Pleine. Future work in general calibration would save time by developing calibration equations using streams whose flow characteristics resemble one another in this way. These kinds of comparisons certainly will not provide the best results, but they provide a starting point to which other data may be added.

TABLE 1. Selected Calibration Equations for Little Plover River Annual Flow.
(Dependent Variable: Y = Little Plover River October-September Discharge)

Independent Variable	Equation*	r^2 (percent)	F
1. Eau Claire Discharge (October-September)	$Y = 1791 + 0.021x$	32.4	9.65**
2. Black Earth Discharge (October-September)	$Y = 1905 + 0.163x$	23.1	6.41***
3. Big Eau Pleine Discharge (October-September)	$Y = 2819 + 0.014x$	11.2	3.28
4. Stevens Point Precipitation (October-September)	$Y = 2116 + 51.3x$	–0.5	0.89
5. Stevens Point Lagged Precipitation (June-May)	$Y = -360 + 131x$	51.1	20.8**
6. Eau Claire Discharge and Stevens Point Lagged Precipitation	$Y = -183 + 0.0044x_1 + 113x_2$	46.2	8.7**
7. Black Earth Creek Discharge and Stevens Point Lagged Precipitation	$Y = -963 + 0.106x_1 + 112x_2$	57.3	13.16**
8. Big Eau Plaine Discharge and Stevens Point Lagged Precipitation	$Y = -300 + 0.001x_1 + 128x_2$	45.5	8.5**
9. Well 15 Mean Monthly Depth to g.w.	$Y = 20070 - 503x$	16.5	2.29
10. Well 35 Mean Monthly Depth to g.w.	$Y = 13013 - 2068x$	31.4	3.75

*Equations 1-8, n = 18; Equations 9 and 10, n = 7.
**F significant at 99 percent level.
***F significant at 95 percent level.

TABLE 2. Selected Calibration Equations for Little Plover River Peak Flows.
(Dependent Variable: Y = Little Plover River Peak Flow)

Independent Variable	Equation*	r^2 (percent)	F**
1. Eau Claire River Peak Flow	$Y = 33.2 + 0.0048x$	8.4	2.66
2. Black Earth Creek Peak Flow	$Y = 36.6 + 0.0301x$	3.0	1.56
3. Big Eau Pleine Peak Flow	$Y = 47.3 + 0.002x$	–5.2	0.12
4. Well 15 highest ground water level	$Y = 280 - 7.57x$	15.4	2.28
5. Well 35 highest ground water level	$Y = 85.4 - 14.2x$	11.6	1.92

*Equations 1-3, n = 18; Equations 4 and 5, n = 7.
**No F-values significant at 95 percent or 99 percent level.

The second possible approach to developing better calibration equations for the deep glacial till basin is to develop a method of allowing for the effects of delayed transmission of water through the aquifer. The four month precipitation lag used in this study was suggested by visual examination of precipitation and streamflow records. Where these records might not be available, it may be possible to develop appropriate lag factors using known characteristics of the aquifer.

Calibration of a watershed on deep glacial till remains a frustrating exercise. Further work must be done to identify other physical characteristics which will allow for better accounting for variability in flow.

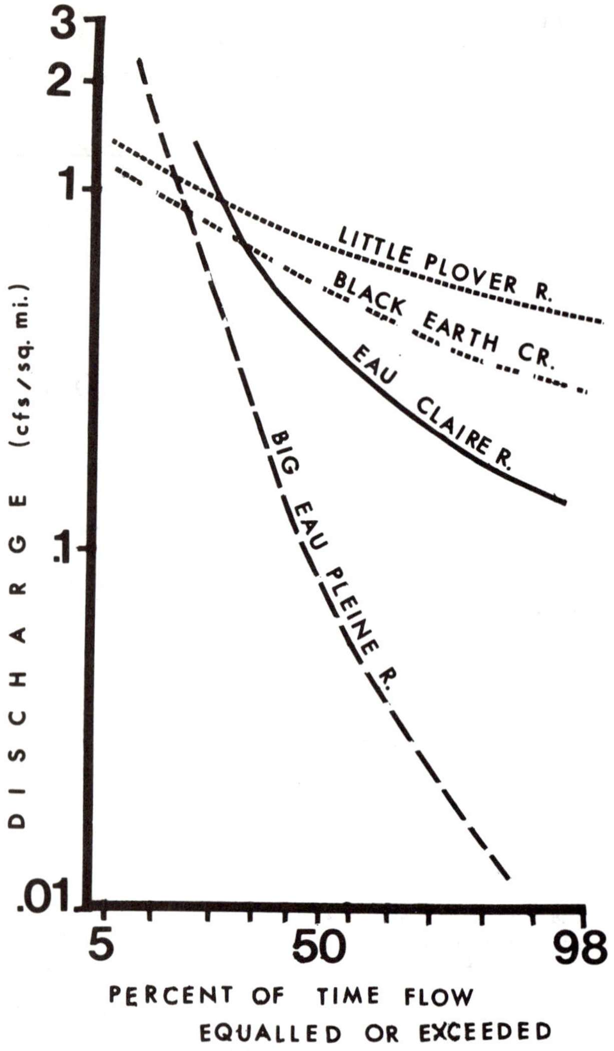

Figure 3. Comparison of Flow Duration Curves for Streams Used in Calibration with the Little Plover River.

LITERATURE CITED

Devaul, R. W. and J. H. Green, 1971. Water Resources of Wisconsin, Central Wisconsin River Basin. Hydrologic Investigations Atlas HS-367, U.S. Geological Survey, Washington, D.C.

Hindall, S. M. and R. G. Borman, 1974. Water Resources of Wisconsin, Lower Disconsin River Basin. Hydrologic Investigation Atlas HA-479, U.S. Geological Survey, Reston, Virginia.

Soil Conservation Service, 1978. Soil Survey of Portage County, Wisconsin. Soil Conservation Service.

U.S. Geological Survey (yearly). Water Resources Data for Wisconsin. U.S. Geological Survey, Madison, Wisconsin.

Weeks, E. P. and H. G. Stangland, 1971. Effects of Irrigation on Streamflow in the Central Sand Plain of Wisconsin. U.S. Geological Survey, Water Resources Division, Open File Report, Madison, Wisconsin.

Weeks, E. P., D. W. Ericson, and C. L. R. Holt, Jr., 1965. Hydrology of the Little Plover River Basin, Portage County, Wisconsin, and the Effects of Water Resource Development. U.S. Geological Survey Water Supply Paper 1811, 78 pp.

HYDROMETEOROLOGICAL SIGNIFICANCE OF RIME ICE DEPOSITS ON TREES IN THE COLORADO ROCKIES

Edward E. Hindman, Randolph D. Borys, and Paul J. DeMott[1]

ABSTRACT: Observations and measurements were made in liquid water clouds during snowfall periods at mountain top sites in the Colorado Rockies during the winter of 1981-1982. Liquid clouds were observed to occur frequently during snowfall periods at most sites. On the average, the rime rates were measured to be ten times the precipitation rates. The rime deposits were estimated to contribute 4-11 percent to the water content of the snowpack and to contribute up to 55 percent to the trace element composition of the snowpack.
(KEY TERMS: rime ice deposits; icing deposits; water content of ice deposits; rime rates; precipitation rates; trace elements; snowpack chemistry.)

INTRODUCTION

Hindman and Grant (1981) have shown that, for the period February-April 1981 at Vail and Steamboat Springs, Colorado, rime ice deposits on high elevation trees occurred during most snowfall periods. Further, they report that, on the average, at Steamboat Springs, the rime rates were 10 times greater than the precipitation rates, a result consistent with values from Finland (Makkonen, 1981).

As a result of these findings, a more extensive study was initiated of the rime ice phenomenon in the Colorado mountains during the winter of 1981-82. Rime measurements were made at Vail (Eagle's Nest Ridge, 3048 m MSL), Steamboat Springs (Storm Peak, 3156 m MSL), Berthoud Pass (chairlift top, 3662 m MSL), and Wolf Creek Pass (poma lift top, 3581 m MSL). Further, a mountain top laboratory was established at the summit of Storm Peak at Steamboat Springs to conduct detailed rime ice investigations. The facility is referred to as the Storm Peak Laboratory (SPL).

In this paper, it is shown from the 1981-82 data that 1) liquid clouds occurred frequently during snowfall periods; 2) the rime rates, on the average, were ten times the precipitation rates; and 3) the rime deposits were estimated to contribute 4-11 percent to the water content of the snowpack and contribute up to 55 percent to the trace constituent composition of the snowpack.

INSTRUMENTATION

The rime measurements at Wolf Creek Pass (WCP) in the Southern Colorado Rockies, at Vail and Berthoud Pass (BP) in the Central Colorado Rockies and at Steamboat Springs (SBS) in the Northern Rockies were made using a procedure detailed by Hindman and Grant (1981). Briefly, a 6.2 mm diameter wood dowel was exposed at eye level to the cloud at a location where the airflow was unobstructed. The width and depth of rime ice deposited over a period were recorded (see Figure 1). The wind speed and temperature values at the beginning and end of the period were also recorded. The observer recorded whether or not it was snowing and if liquid cloud was present. These measurements and observations were taken daily at each site by cooperative observers.

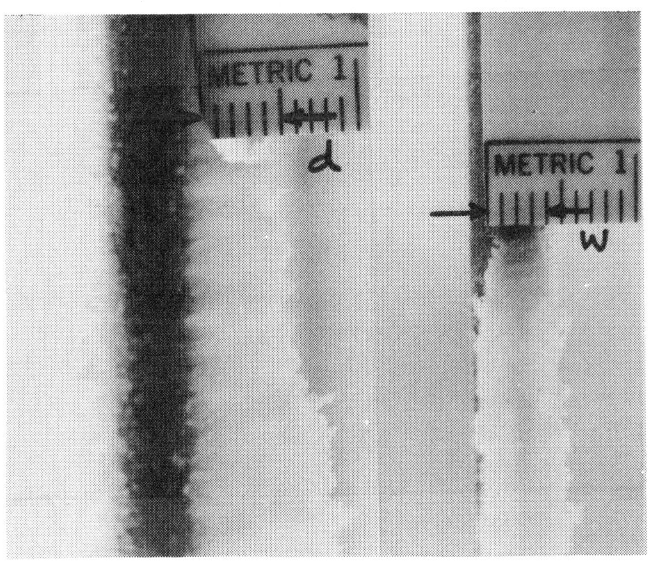

Figure 1. Rime Ice Depth (d=5 mm) and Width (w=4 mm) Measurements Made on 2 December 1981 for a Four-Hour Period, 9 ms^{-1} Average Wind Speed, $-11C$.

[1] Respectively, Research Associate, Ph.D. Candidate, and Ph.D. Student; Department of Atmospheric Science, Colorado State University, Fort Collins, Colorado 80523.

More detailed and extensive measurements and observations were recorded at the SPL including the dowel measurements. The dowels were placed at the SPL site and at three other sites at lower elevations to determine the vertical extent of rime deposits on the windward slopes of the Park Range. Rime densities were estimated at SPL by dividing the mass of rime accumulated on a preweighed dowel by the measured volume of the deposit.

Cloud water measurements were made at the SPL using the roto-rod device described by Rogers, *et al.* (1983). A spinning wire accumulated rime for a short period. The rime deposit was carefully removed from the wire and weighed. The mass of rime is related to cloud liquid water content in gm^{-3} (LWC). According to Rogers, *et al.* (1983). LWC values from the roto-rod vary by ±50 percent from LWC determined from a more nearly accurate device independent of droplet characteristics. The relationship for roto-rod LWC was derived in winter mountain clouds in the Wyoming Rockies and, therefore, should be valid for use at SPL, where clouds with similar microphysical properties occur.

A small branch was cut off a nearby Engelman spruce (*Picea Engelmannii*) to estimate the amount of rime a natural obstacle would collect. The branch was weighed prior to exposure to the airstream and reweighed, approximately every hour, to measure the integrated rime accumulation. Rime ice and snowfall were collected separately using plastic collection devices for later chemical analyses. These analyses assessed the deposition of trace elements to the snowpack via direct cloud water interception by the mountain and via indirect deposition through ice crystal riming.

The cloud physical and chemical studies reported here were part of a large winter mountain cloud investigation conducted by Colorado State University. More detailed physical and chemical properties of the rime deposits at SPL are reported by Hindman, *et al.* (1982).

PROCEDURES

The answers to the questions posed to the cooperative observers (Is a water cloud present; and Is it snowing?) were tabulated. Three categories were developed: coexistence of cloud and snowfall, existence of only cloud, and lack of cloud but snowfall occurring.

The rime and precipitation rates were developed from the cooperative observer data, the data from SPL, and 24-hour snowfall water equivalent measurements from the U.S. Forest Service (WCP, Vail, BP), from the Steamboat Springs Ski Company, and from Feng and Grant (1982). First, the rime data were examined for periods that had both cloud and snowfall occurring at the beginning and end, and an average wind speed $\geqslant 2.7\ ms^{-1}$ (below this value the collection efficiency of the dowel is poorly defined). Second, the 24-hour snowfall records were checked to verify that new snow was reported at the beginning and end of the 24-hour period. Third, the rime rates (R_1, $mm\ hr^{-1}$) were calculated using the following expression:

$$R_1\ (mmhr^{-1}) = \frac{d_1\ P}{\Delta t\ e} \qquad (1)$$

where d_1 is the depth of the rime deposit (mm), P is the average porosity of the deposit (fraction of the deposit estimated to be ice, typically 0.8), Δt is the collection period (hr), and e is the collection efficiency of the dowel as reported by Hindman and Grant (1981). A value of 10 μm was used as the droplet median-volume diameter in e determinations. Equation (1) is accurate to ±50 percent due to uncertainties in P and e values. Finally, the precipitation rates (R_2, $mm\ hr^{-1}$) were calculated using the following expression:

$$R_2\ (mm\ hr^{-1}) = \frac{d_2}{24\ hr} \qquad (2)$$

where d_2 is the water equivalent (mm) of the 24-hour snowfall accumulation. Values of R_2 are good to ±20 percent.

The tree branch was exposed to the cloudy air and also to conditions when the cloud vanished. First, the exposed branch was photographed, the dowels were scraped clean of rime and the wind speed, direction, and air temperature were recorded. Second, the branch was carefully weighed. Third, the branch was reexposed and the roto-rod device was deployed. Fourth, the rime deposit on the roto-rod was weighed. Finally, the wind speed, wind direction, temperature, and rime deposits on the dowels were measured. The rime deposit on the preweighed dowel was measured and weighed.

RESULTS

The results of the cloud and snowfall observations from the cooperative observers are given in Table 1. The large number of observations at Steamboat Springs were primarily made by CSU personnel at SPL. It can be seen from the table that large differences occurred in the incidence of liquid cloud at the various sites. At WCP and SBS, 95 percent and 90 percent of the observations during storm periods reported liquid cloud present, while at Vail and BP, 53 percent and 60 percent of the observations reported liquid cloud present. In contrast, Hindman and Grant (1981) report that 95 percent and 71 percent of the observations at Vail and SBS, respectively, during the winter of 1980-81 reported liquid cloud present. The data are too limited to draw firm conclusions. It is apparent from these results that rime ice is a frequent occurrence during snowfall periods at exposed, high elevation locations at WCP, Fail, BP, and SBS. Further, Judson (1971) reports frequent

TABLE 1. Percent of Observations Which Reported the Indicated Conditions.

Site	Period	Number of Observations	Cloud and Snowing (percent)	Cloud and Not Snowing (percent)	No Cloud and Snowing (percent)
Wolf Creek Pass	11/24/81 to 4/7/82	42	76	19	5
Vail	11/26/81 to 4/10/82	53	28	25	47
Berthoud Pass	11/26/81 to 4/1/82	52	58	2	40
Steamboat Springs	11/30/81 to 4/7/82	241	78	12	10

episodes of rime on Mines Park (3808 m MSL) just 2 km downwind of the BP site.

The results of the rime rate and precipitation rate measurements are shown in Figure 2. Three significant features are apparent in the figure. First, the average rime rates for WCP, Vail, and SBS are 10 times the precipitation rates: 4.7 ± 1.5 vs 0.46 ± 0.13 mm hr^{-1}. This result is consistent with the result from Hindman and Grant (1981) for data from 1980-81 at SBS. Further, the precipitation rate values are consistent with measurements made by Feng and Grant (1982) at SBS. Second, although the regression equation in Figure 2 does not have a large correlation coefficient ($r = 0.39$) due to the large scatter in the data, it appears that the rime and precipitation rates are positively correlated. Finally, observations were made typically once per day at WCP, Vail, and BP. It was necessary to assume that the rime and snowfall episodes occurred for the entire 24-hour period. Thus, the R_1 and R_2 values from these sites may be conservative. At SBS, the measurements by CSU personnel were nearly every hour, so the durations of the rime and snowfall episodes were accurately determined. The SBS R_1 and R_2 values are considered accurate. It can be seen, in Figure 2, that the WCP, Vail, and SBS R_1 and R_2 values are not significantly different because of the overlapping uncertainties in the mean values. Consequently, the R_1 and R_2 values from WCP and Vail are considered representative values. The BP values are significantly different and hence are considered conservative values.

The results from one of the two test branch experiments are given in Figure 3. (Results from the 27-28 February 1982 experiment were similar to those in Figure 3.) Inspection of the figure reveals the following patterns. First, the branch accumulated rime and the maximum value was reached when the cloud dissipated (the cloud period did not last long enough to determine if the branch became saturated with rime and shed rime). Second, the branch lost rime after the maximum value was reached due to wind buffeting and sublimation. Third, the rime rates were quite variable with initial high values. Fourth, the rime densities were initially high then decreased through the cloud period. The test branch was approximately 100 cm^2 in cross-section area and weighed 33 g.

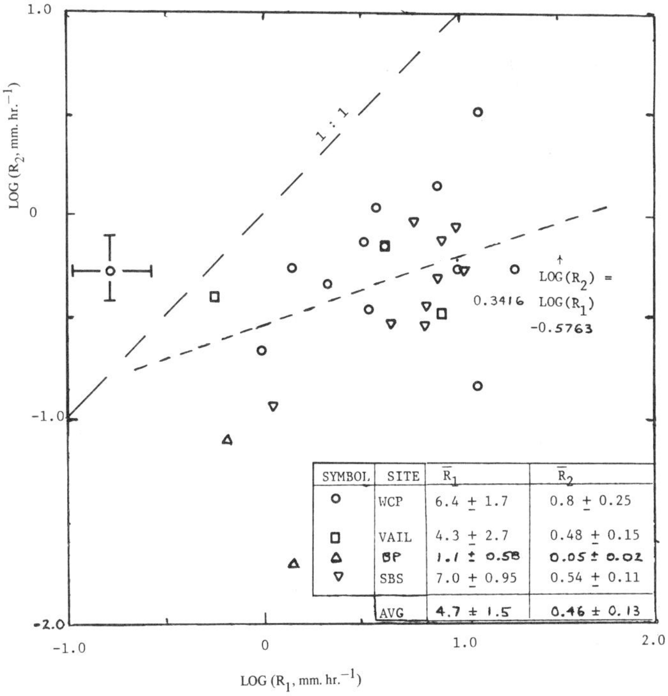

Figure 2. Rime Rates (R_1) and Precipitation Rates (R_2) from Wolf Creek Pass (WCP), Vail, Berthoud Pass (BP), and Steamboat Springs (SBS) for the Winter of 1981-82. The uncertainty in the measurements and the linear regression equation are given.

DISCUSSION

The basic information has been assembled to estimate the hydrometeorological significance of the rime ice deposits (i.e., do they contribute to the water content of the snowpack). The amount of water deposited to the snowpack by rime and precipitation can be estimated in the following manner.

The rime measurements as a function of elevation at SBS reveal that, on the average, the upper 400 m of the Park Range is enveloped in cloud. Further, according to climatological records, the majority of the snowfall occurs within 25 km upwind of the Range. Consequently, the flux of water to the Park Range from rime and from precipitation

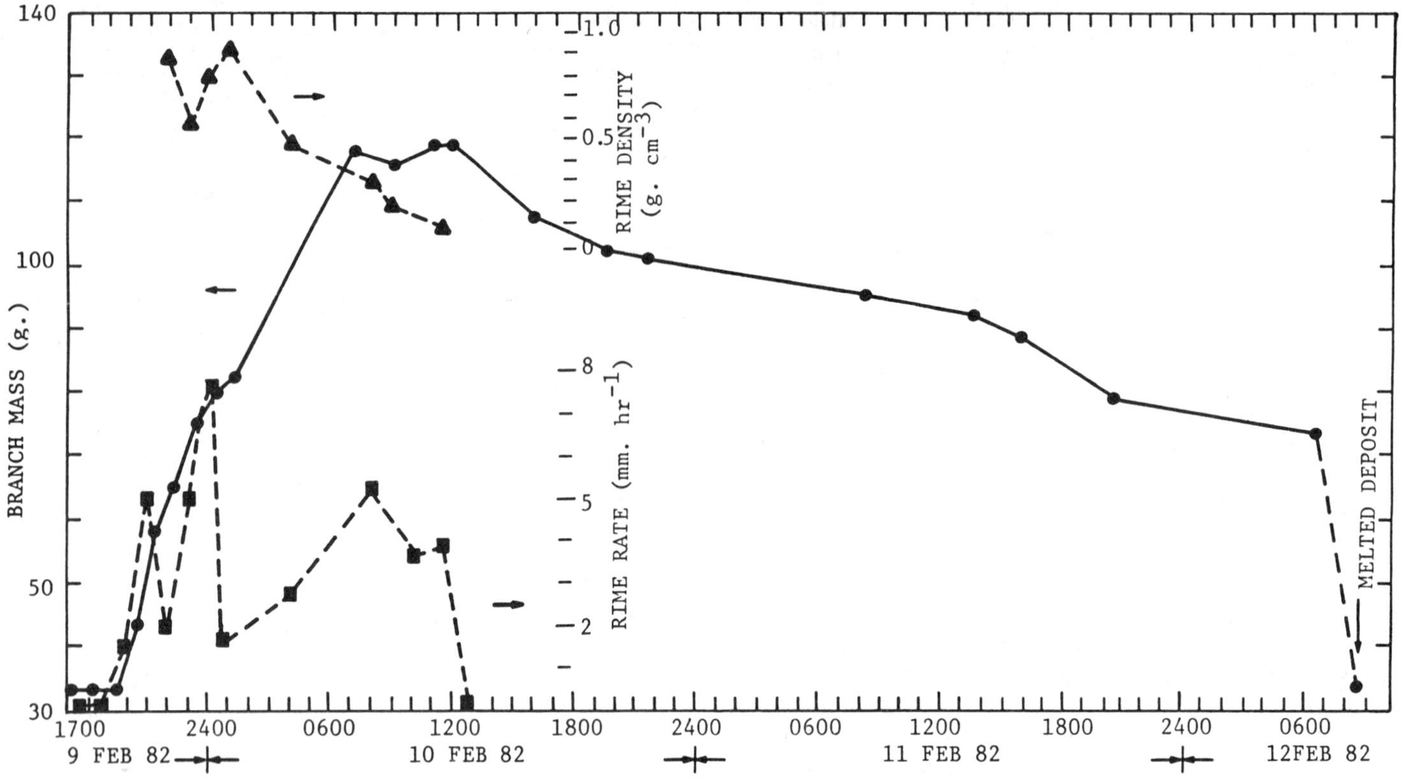

Figure 3. Rime Accumulation by a Test Branch on 9 and 10 February 1982 at SPL.
The corresponding rime rates and rime densities are given.

can be estimated using the R_1 and R_2 values from Figure 2. Assuming a unit width, the fluxes are as follows:

Rime Flux = 7.0 mm hr^{-1} x 0.4 km x unit width x 0.9 g cm^{-3} = 2.5 x 10^4 g hr^{-1}

Precipitation Flux = 0.54 mm hr^{-1} x 25 km x unit width x 1 g cm^{-3} = 1.4 x 10^5 g hr^{-1}.

Thus, according to this estimate, the flux of water due to precipitation is 5.6 times greater than the flux due to direct interception of cloud liquid water by the mountain barrier. Consequently, 15 percent of the water content of the snow could be due to rime deposition and 85 percent due to snowfall (assumed to be unrimed).

The rime accumulation on the test branch shown in Figure 3 was used to estimate the rime flux: 85 g/ 12 hrs/ 100 cm^2 = 0.071 g hr^{-1} cm^{-2}. The rime flux to the mountain was 0.071 g hr^{-1} cm^{-2} x 0.4 km x cm = 2.8 x 10^3 g hr^{-1}. The precipitation rate for the same period was 11.9 mm/24 hr = 0.50 mm hr^{-1} corresponding to a precipitation flux of 0.50 mm hr^{-1} x 25 km x 1 g cm^{-3} = 1.3 x 10^5 g hr^{-1}. Thus, according to this estimate, the flux of water due to precipitation is 46 times greater than the flux due to rime deposits. Consequently, 2 percent of the water content of the snowpack could be due to rime deposition and 98 percent due to snowfall (assumed to be unrimed).

The two estimates of the contribution of rime ice deposits on trees (assuming the deposits shed to the snow) to the snowpack ranged from 2 to 15 percent. The large variation is due to uncertainties with the collection efficiency of the rimed dowel, the cross-section areas of the test branches, and the limited data set. The collection efficiencies could be low by a factor of 1.5 causing the rime contribution to decrease from 15 percent to 11 percent. The cross-section areas could be high by 50 percent causing the rime contribution to increase from 2 percent to 4 percent. It appears that a range from 4 to 11 percent is justified based on the limited data available.

Recent rime ice and snowfall trace element measurements by Hindman, *et al.* (1982), at SPL have revealed the following. Preliminary results from the winter of 1981-82 show that, overall, the rime ice collections contained 2-3 times higher trace element concentrations than did the snow collections. During a single snowfall event with no snow crystal riming, the rime ice collections contained 8 to 10 times higher concentrations than did the snow collections. Our analysis of rime and precipitation fluxes reveals that between 4 and 11 percent of the water content of the snowpack is due to rime deposits. Assuming then, that for 100 g of water in the snowpack, 11 g is due to rime deposits. Using the maximum trace element concentration difference between rime deposits and unrimed snowfall (factor of 10), the rime water contributes up to 55 percent of the trace element content of the snowpack while the snowfall contributes 45 percent. If the amount of rime on snow crystals is considered, the contribution of riming processes to the trace element content of the snowpack could be significantly greater. In addition, the degree of crystal riming may play a role in determining the relative amounts of various trace elements in a given snowfall (Borys, 1982).

CONCLUSIONS

Observations and measurements were made in liquid water clouds during snowfall periods at mountain top sites in the Colorado Rockies during the winter of 1980-81. The observations indicated that liquid clouds occurred frequently during snowfall periods at most of the sites. The measured rime rates averaged 10 times the precipitation rates. When the watershed areas affected by the rime deposits and snowfalls were considered, the calculated contribution of rime to the water content of the snowpack ranged between 4 and 11 percent. When the trace element content of rime ice and snowfall collections were included, up to 55 percent of the trace element content of snowpack was estimated to come from rime deposits and 44 percent from snowfall deposits (assumed to be unrimed snow). It appears from these limited data, that rime deposits on trees, when they shed to the snow, may contribute about as much water to the snowpack as appears possible from cloud seeding efforts ($\approx$ 15 percent, Dennis, 1980). The deposits are estimated to contribute significantly to the trace element content of the snowpack.

ACKNOWLEDGMENTS

The cooperative rime ice measurements were made by professional ski patrol personnel at Wolf Creek Pass, Vail, Berthoud Pass, and at Steamboat Springs. The Steamboat Ski Corporation is acknowledged for logistics support for the Storm Peak Laboratory and the other CSU instrument sites on the ski area. Snowfall data were provided by the Avalanche Warning Center of the U.S. Forest and Range Experiment Station, Ft. Collins, Colorado. This study was a portion of the research sponsored by NSF Grant ATM 8109590.

LITERATURE CITED

Borys, R. D., 1982. The Interaction of Aerosol Particles, Cloud Water and Ice Crystals and Their Relationship to Precipitation Scavenging in the Arctic. Paper presented at the 4th Conf. on Precipitation Scavenging, Dry Deposition and Resuspension, November 29-December 3, 1982, Santa Monica, California.

Dennis, A., 1980. Weather Modification by Cloud Seeding. Academic Press, New York, New York.

Feng, D. and L. O. Grant, 1982. Correlation of Snow Crystal Habits, Number Flux and Snowfall Intensity from Ground Observations. Paper presented at the Conf. on Cloud Physics, 15-17 November 1982, Chicago, Illinois.

Hindman, E. E. and L. O. Grant, 1981. Utility of Mountaintop Rime-Ice Measurements. Ppts. 2nd Mt. Meteorology Conf., Am. Meteor. Soc., Boston, Massachusetts, pp. 404-408.

Hindman, E. E., R. D. Borys, and P. J. DeMott, 1982. Physical and Chemical Properties of Rime Ice Deposits at Storm Peak Laboratory, Steamboat Springs, Colorado. Presented at VI Intl. Symposium on Physics and Chemistry of Ice, 2-6 August 1982, Rolla, Missouri.

Judson, A., 1971. An Infrared De-icing Unit for Cup Anemometers. USDA Forest Service Research Note, RM-187, 4 pp.

Makkonen, L., 1981. Estimating Intensity of Atmospheric Ice Accretion on Stationary Structures. J. Appl. Meteor., 20:595-600.

Rogers, D. C., D. Baumgardner, and G. Vali, 1983. Determination of Supercooled Liquid Water Content by Measuring Rime Rate. J. Climate and Appl. Meteor. 22:153-162.

INTERNATIONAL SYMPOSIUM ON HYDROMETEOROLOGY
JUNE AMERICAN WATER RESOURCES ASSOCIATION 1982

A COMPARATIVE ANALYSIS OF VARIOUS RAINFALL-RUNOFF MODELS

Ricardo A. Smith and Roger A. Amisial[1]

ABSTRACT: The rainfall-runoff models are classified as Parametric and Black Box (linear and nonlinear) models. At the same time the models in each of these categories are classified as generalization models, in which rainfall-runoff data are necessary for parameter estimation, and regionalization models, where this kind of data is not necessary. Several models, covering all the spectrum for the Black Box Generalization models, were implemented. All these models are used to simulate the rainfall-runoff process for the Potomac River near Cumberland, Maryland. A comparative analysis is then made between all the models using a wide variety of fitting criteria. The results are presented in tabular and graphical form and conclusions are drawn from the study.
(KEY TERMS: rainfall-runoff; models; Potomac River.)

INTRODUCTION

The rainfall-runoff models are classified in a general form as generalization and regionalization models. The generalization models are those which allow the estimation of an hydrologic variable for a fictitious condition using historical data of the same variable. For example, the estimation of the maximum annual discharges for a return period greater than the historical records is done by a generalization process when data are available at the point of interest. When the estimation procedure is concerned with a variable for which there is no available records, a regionalization model needs to be used.

This paper is concerned with a comparative analysis of the conventional generalization models of the linear type for the rainfall-runoff process, thereafter referred to as linear generalization models. The conventional models use a black box rainfall-runoff relationship to estimate runoff at a point of interest. That is, the input to the system, excitation, is converted to output, response, by the action of the system, the model, of which only the transform function is known. These models are estimation techniques of the watershed system that transform rainfall in runoff, assuming in this case that the system is a linear, stationary initially relaxed system. This representation of the watershed system is known in Hydrology as Unit Hydrograph, and the comparative analysis presented here is concerned with the different estimation techniques of such hydrograph.

The Unit Hydrograph concept is based on the unit element concept: if the effect due to a unit action is known, the effect due to any other action is given by the combination of that action and the effect due to the unit action. This combination can be of a multiplicative, combinatorial, convolutive, or any other form. The Unit Hydrograph was defined by Sherman (1932) as the direct hydrograph resulting from a unit depth effective rainfall, uniformly distributed over the watershed and of a specific duration. This definition combined with the initial assumption of linearity, stationarity and initially relaxed system constitutes what is known as the Unit Hydrograph Theory (Chow, 1964).

Since the introduction of the Unit Hydrograph concept (Sherman, 1932), a great deal of estimation techniques have been presented in the literature. These techniques can be classified in two groups (Eagelson, *et al.*, 1966): truly linear techniques and optimal linear techniques. The truly linear techniques estimate the Unit Hydrograph assuming that the convolution equation that relates direct runoff with rainfall excess and the unit hydrograph is the only equation that characterizes the system. The following techniques fall in this category: the direct solution of the convolution integral (Blank, *et al.*, 1971); the use of orthogonal functions (O'Donnell, 1960; Dooge, 1973; Papazafiriou, 1975); the use of transformations (Levi and Valdes, 1964; Delleur and Rao, 1971; Blank, *et al.*, 1971) and least squares procedures in matrix form (Newton and Vinyard, 1967; Singh, K. P., 1976). The optimal linear techniques estimate the unit hydrograph by an approximate solution to the convolution equation mentioned above using a computational procedure that defines the "optimal" Unit Hydrograph in some sense (least squares, minimum absolute deviations, etc.). Techniques such as the use of time series (Eagelson, *et al.*, 1966); the use of conceptual models (Nash, 1958; Nash, 1960); the use of linear programming (Deininger, 1969; Mays and Coles, 1980); the use of search procedures

[1] Respectively, Researcher and Professor, Interamerican Center for Integral Development of Land and Water Resources, Apartado Postal 219, Merida, Venezuela.

(Diskin and Boneh, 1975; Singh, V. P., 1976; Amisial, *et al.*, 1981), fall in this group.

A comparative analysis of all this Unit Hydrograph estimation techniques is presented in the next section. This analysis uses as fitting criteria some indicators suggested in the literature. A description and discussion of these indicators is not under the scope of this paper.

COMPARATIVE ANALYSIS

Several comparative analysis have been made in the literature (Singh, V. P., 1976; Singh, K. P., 1976; Aitken, 1973; Dooge, 1977). Several authors, when introducing their estimation procedure for the first time, include a brief comparative analysis with other forms of estimation (Singh, K. P., 1976; Singh, V. P., 1976; Diskin and Boneh, 1975; Dooge, 1977). These comparisons go from just visual adjustment between observed and computed flows to comparisons using complex indicators or coefficients between observed and computed flows such as the residual mass curve coefficient. The hydrograph characteristics normally used for comparisons are: peak discharges, time to peak, hydrograph volume, prediction errors for the peak discharge and the hydrograph volume, and the mean, the standard deviation, and the coefficient of variation of the hydrograph. In general, the coefficients r used for the comparative analysis measure the degree of association between predicted and observed hydrographs. The correlation coefficient ($0 \leq r \leq 1$), the coefficient of efficiency ($r \leq 1$), the coefficient of determination ($r \leq 1$), the special correlation coefficient ($0 \leq r \leq 1$) and the residual mass curve coefficient ($r \leq 1$) are used in this analysis. Other indicators used in this analysis are the mean square error and the integral square error.

The comparative analysis is made using four storms over the Potomac River watershed near Cumberland, Maryland (Singh, K. P., 1976). This watershed has an area of 2,266 square kilometers, and the storms are given as effective rainfall-direct runoff events. For each storm the four-hour unit hydrograph was estimated using all the techniques described above. When possible, the unique four-hour unit hydrograph was also derived. When this was not possible, the average unit hydrograph was estimated. Each one of the estimated unit hydrograph was used to predict all the direct runoff hydrographs. The comparative analysis between the different estimation techniques is then performed using the indicators mentioned above.

In the cases where estimation techniques do not allow the estimation of a unique unit hydrograph, a visual average unit hydrograph or the unit hydrograph from one of the storms should be used in the prediction. The results obtained using a mathematical average unit hydrograph were poor. The comparative analysis presented here used the unique unit hydrograph for prediction. When this is not possible the prediction is made using the unit hydrograph the produces the best results according to the indicators.

Due to space limits not all the results can be presented here. Only some tables and figures are presented as examples of the results obtained in this comparative analysis. Table 1 presents the prediction results for some truly linear techniques for storm 1. Table 2 presents those results for the optimal linear techniques. Figure 1 presents the observed and predicted direct runoff hydrograph for storm 1 using the truly linear and the optimal linear techniques. From the results obtained the following conclusions can be drawn:

1. The least squares in matrix form, the Laguerre functions, the Fourier transform, and the unit weight functions are superior unit hydrograph estimation techniques than the others truly linear techniques.
2. In the optimal linear techniques group, search procedure, linear programming, and the use of time series produced the best results.
3. The optimal linear techniques showed more consistency and produced better results than the truly linear techniques.
4. Although all the optimal linear techniques mentioned above produced almost identical results, the use of search procedures without constraints is recommended here due to its simplicity and flexibility. This technique has been criticized because it does not guarantee realizable results (nonnegative unit hydrograph ordinates and unit volume unit hydrograph). In all the storms studied by the authors this condition occurred naturally with this technique.

LITERATURE CITED

Aitken, A. P., 1973. Assessing Systematic Errors in Rainfall-Runoff Models. Journal of Hydrology 20:121-136.

Amisial, R. A., R. A. Smith, T. A. Bandes, and M. Castro, 1981. Incidencia de los Mētodos de Generalizaciōn y Regionalizaciōn sobre la Selecciōn de las Medidas de Control de Inundaciones. Informe de Avance al CONICIT, CIDIAT, Mērida, 250 pp.

Blank, D., J. W. Delleur, and A. Giorgini, 1971. Oscillatory Kernel Functions in Linear Hydrologic Models. Water Resources Research 7(5):1102-1117.

Chow, V. T., 1964. Runoff. *In:* Handbook of Applied Hydrology, V. T. Chow (Editor). McGraw Hill, New York, New York, pp. 14-1–14-54.

Deininger, R. A., 1969. Linear Programming for Hydrologic Analysis. Water Resources Research 5(5):1105-1109.

Delleur, J. W. and R. A. Rao, 1971. Linear Systems Analysis in Hydrology – The Transform Approach, the Kernel Oscillations and the Effect of Noise. *In:* Systems Approach to Hydrology, V. Yevjevich (Editor). Water Resources Publications, Fort Collins, Colorado, pp. 116-139.

Diskin, M. H. and A. Boneh, 1975. Determination of an Optimal Instantaneous Unit Hydrograph for Linear Time Invariant Systems for Multi-storm Records. Journal of Hydrology 24:57-76.

Dooge, J. C. I., 1973. Linear Theory of Hydrologic Systems. U.S. Dept. of Agriculture Technical Bulletin 1468, Washington, D.C., 327 pp.

TABLE 1. Performance Indices for Storm 1 Using Truly Linear Techniques.*

Statistic	Volumetric Model	Direct Solution	Least Square	Fourier Transformation	Unit Weight	Laguerre	Harmonics
Peak	2.09	2.91	2.27	2.33	2.25	2.24	2.16
Time to Peak	32.00	24.00	24.00	20.00	28.00	28.00	24.00
Volume	0.64	0.70	0.64	0.72	0.64	0.64	0.63
Mean	0.70	0.76	0.70	0.78	0.70	0.69	0.69
Standard Deviation	0.66	1.00	0.76	0.83	0.77	0.76	0.75
Coefficient Variation	0.95	1.30	1.09	1.06	1.10	1.10	1.10
Predicted Error for Peak	6.41	–30.00	–1.40	–4.45	–0.90	–0.40	3.50
Predicted Error for Volume	0.00	–9.84	–0.20	–12.50	–0.40	0.40	1.40
Correlation Coefficient	0.91	0.94	1.00	0.98	1.00	1.00	0.98
Mean Square Error	0.11	0.14	0.00	0.04	0.00	0.00	0.02
Coefficient Efficiency	0.82	0.75	1.00	0.94	1.00	1.00	0.97
Integral Square Error	9.80	11.40	0.47	5.67	0.49	0.50	4.00
Residual Mass Curve Coefficient	0.75	0.50	1.00	0.90	1.00	1.00	0.97

*Observed values on Table 2.

TABLE 2. Performance Indices for Storm 1 Using Optimal Linear Techniques.

Statistic	Observed	Time Series	Linear Program	Search No Restriction	Search Restriction	Nash's Model
Peak	2.24	2.28	2.24	2.24	2.23	1.83
Time to Peak	24.00	24.00	24.00	24.00	24.00	20.00
Volume	0.64	0.64	0.64	0.64	0.64	0.43
Mean	0.70	0.69	0.70	0.70	0.69	0.46
Standard Deviation	0.77	0.76	0.76	0.76	0.77	0.61
Coefficient of Variation	1.10	1.10	1.10	1.09	1.11	1.32
Predicted Error for Peak		–2.00	0.00	–1.20	0.15	18.20
Predicted Error for Volume		0.20	0.00	–0.24	0.00	33.50
Correlation Coefficient		1.00	1.00	1.00	1.00	0.75
Mean Square Error		0.00	0.00	0.00	0.00	0.31
Coefficient Efficiency		1.00	1.00	1.00	1.00	0.47
Coefficient Determination		1.00	1.00	1.00	1.00	0.47
Integral Square Error		0.56	0.20	0.43	0.28	16.70
Special Correlation Coefficient		1.00	1.00	1.00	1.00	0.84
Residual Mass Curve Coefficient		1.00	1.00	1.00	1.00	0.66

Dooge, J. C. I., 1977. Problems and Methods of Rainfall-Runoff Modelling. *In:* Mathematical Models of Surface Water Hydrology, T. A. Ciriani, V. Maione, and J. R. Wallis (Editors). John Wiley and Sons, New York, New York, pp. 71-108.

Eagelson, P. S., R. Mejia-R, and F. March, 1966. Computation of Optimum Realizable Unit Hydrograph. Water Resources Research 2(4):755-764.

Levi, E. and Valdes, 1964. A Method for Direct Analysis of Hydrographs. Journal of Hydrology 2(2):182-190.

Mays, L. W. and L. Coles, 1980. Optimization of Unit Hydrograph Determination. J. of Hydraul. Div. ASCE 106(HY2):85-97.

Nash, J. E., 1958. The Form of the Instantaneous Unit Hydrograph. Inter. Assoc. Hydro. Sciences Publication 42:114-118.

Nash, J. E., 1959. Systematic Determination of Unit Hydrograph Parameters. Journal Geophys. Res. 64(1):111-115.

Newton, D. W. and J. W. Vinyard, 1967. Computer Determined Unit Hydrograph from Floods. J. of Hydraul. Div. ASCE 93(HY5):219-235.

O'Donnell, T., 1960. Instantaneous Unit Hydrograph Derivation by Harmonic Analysis. Int. Ass. Hydr. Sci. 51:546-557.

Papazafiriou, Z. G., 1975. Polynomial Approximation of the Kernel of Closed Linear Hydrologic System. Journal of Hydrology 27: 319-329.

Sherman, L. K., 1932. Streamflow from Rainfall by the Unit-Graph Method. Eng. News-Record 108:501-505.

Singh, K. P., 1976. Unit Hydrograph, A Comparative Study. Water Resources Bulletin 12(2):381-392.

Singh, V. P., 1976. Comparison of Two Mathematical Models of Surface Runoff. Hydrol. Sciences Bull. 21(2):285-299.

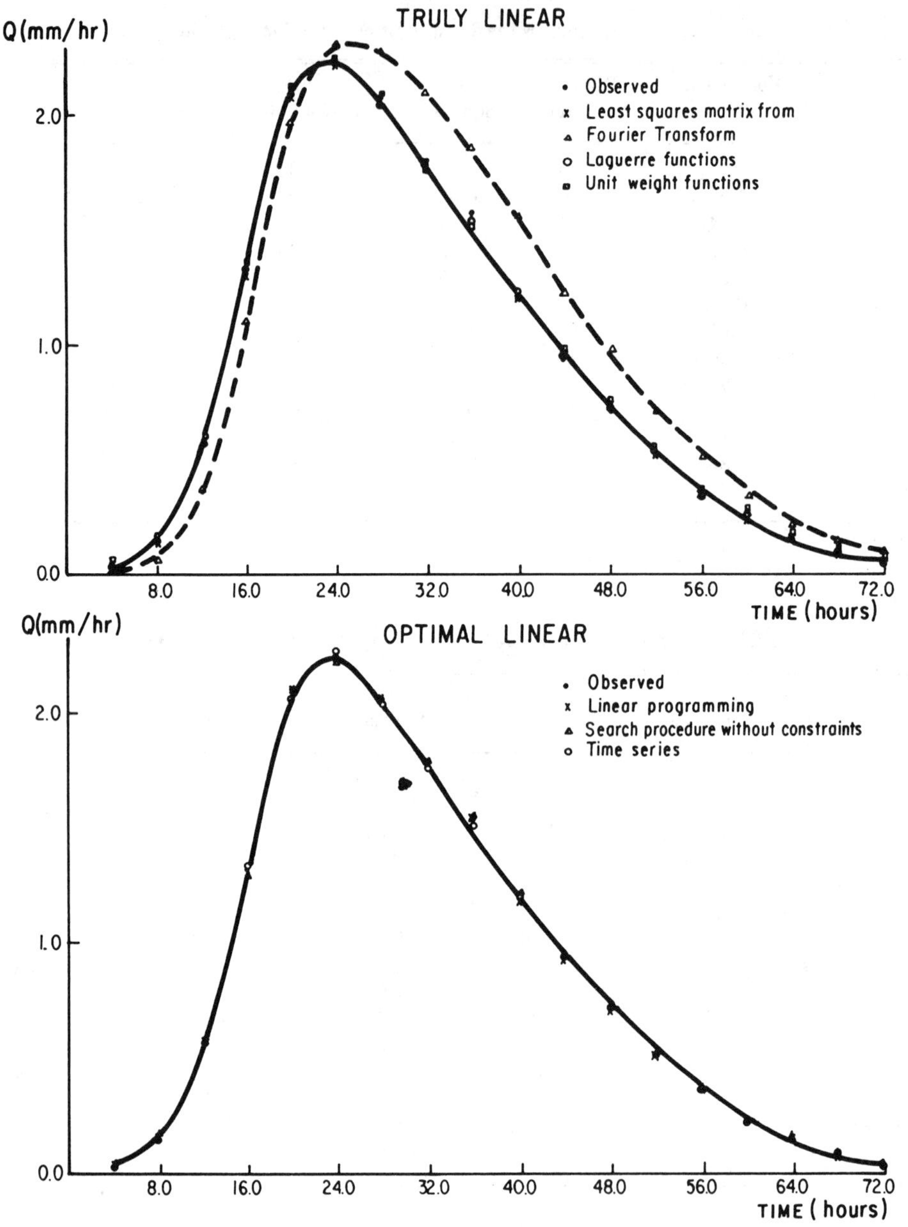

Figure 1. Direct Runoff Hydrograph Predictions.

INTERNATIONAL SYMPOSIUM ON HYDROMETEOROLOGY
JUNE AMERICAN WATER RESOURCES ASSOCIATION 1982

STATISTICAL FLOW FORECASTING MODEL FOR SMALL CATCHMENTS

Soronadi Nnaji[1]

ABSTRACT: The Distributed Lag model is adopted for adaptive flow forecasting within small catchments. The explanatory variables include the current and past values of rainfall intensity obtained, for example, from digitized weather data. Rainfall measurement errors are not considered in this effort in which the dependent variable is the flow rate at a location within the catchment.

Model parameter estimates are obtained by Ordinary Ridge Regression on the first differences of the original variables. The use of first differences reduces serial dependence in the residuals while ridge regression enables one to select a stable set of parameter estimates.

The model is calibrated with observed data. It is validated, in a forecast mode, using the estimated parameters and data sets different from those used in the calibration.

(KEY TERMS: statistics; flow forecasting; small catchments; ridge regression.)

INTRODUCTION

Flash floods cause the loss of hundreds of human lives and millions of dollars of property every year in the United States. The flooding problem is particularly acute in small catchments where there are usually no flood forecasting systems.

Current operational forecasting models combine streamflow simulation and meteorological forecasts to produce flood state forecasts. Stream flow simulation usually involves modeling the abstractive processes which include infiltration, depression storage, and interception. Radar technology provides the most reliable and promising means for effectively providing the meteorological information, primarily rainfall, which is the input into the simulation model.

A real constraint to the use of the above approach in small catchments is that these catchments are usually very poorly instrumented. It becomes necessary to develop models which do not contain explicit representation of the abstractive processes, say. We are motivated towards this end by recent studies (for example Hershfield and Engman, 1978) which showed that the initiation of flash flooding is not significantly sensitive to the abstractive processes. The approach we adopt is a statistical one. In particular we adopt the procedure of Adaptive forecasting which is based on weighted values of current forecast of rainfall and previous observed rainfall and runoff.

THE MODEL

Model Specification

The model we adopt is given below as Equation (1):

$$y_n = \alpha_0 x_n + \alpha_1 x_{n-1} + \alpha_2 y_{n-1} + e_n \quad (1)$$

where y_n is the runoff and x_n is the rainfall for the nth time interval. Thus current flow is a function of the previous flow, the current and previous rainfall. The explanatory variables are assumed to be nonstochastic in this paper though, operationally, x_n is a forecast and so uncertain. By repeated substitutions for the right side (excluding the distrubance term, e_n) y_{n-1}, y_{n-2} ..... can be eliminated from Equation (1) and y_n can be expressed as a function of the first n rainfall rates to give:

$$y_n = \beta_1 x_n + \beta_2 x_{n-1} + \ldots\ldots + \beta_n x_1 + e_n \quad (2)$$

Equation (2) is the distributed lag form of Equation (1) with $\beta_1 = \alpha_0$, $\beta_2 = \alpha_0\alpha_2 + \alpha_1$ and $\beta_i = \beta_{i-1}\alpha_{i-1}$ for $i > 2$ provided it is assumed, as we do, that

$$\alpha_0 + \alpha_1 + \alpha_2 = 1 \quad (3)$$

Because of their lagged nature the explanatory variables are subject to collinearity the degree of which depends on the sampling interval of the variables. Further, in practice observed residuals are used as estimates $\hat{e}_n$ of the disturbance, e_n. Consequently there is also the possibility that the residuals are serially correlated. To minimize this serial correlation we use first differences.

[1] Associate Professor of Civil Engineering, Dept. of Civil Engineering, Clemson University, Clemson, South Carolina 29631.

Given Equation (3) we may eliminate α_0 in Equation (1) and write:

$$x_n - y_n = \alpha_1(x_n - x_{n-1}) + \alpha_2(x - y_{n-1}) + e_n \quad (4)$$

i.e.,

$$z_n = \alpha_1 v_{1,n} + \alpha_2 v_{2,n} + e_n \quad (5)$$

where:

$$v_{1,n} = x_n - x_{n-1}$$

$$v_{2,n} = x_n - y_{n-1}$$

$$z_n = x_n - y_n \quad (6)$$

Thus the explanatory and the dependent variables are all at current time and there is one less variable to consider during parameter estimation.

Estimation of Model Parameters

To address possible collinearity in the explanatory variables we use the method of Ordinary Ridge Regression for model parameter estimation. For the two variable model, Equation (5), the vector of parameters is given by:

$$A = (V'V + \lambda I)^{-1} V'Z \quad (7)$$

where V' is the transpose of the 2 x 2 matrix, V, I is the identity matrix, and Z is the vector of the transformed dependent variable values and λ is the shrinkage factor. Equation (7) reduces to the Ordinary Least Square (OLS) solution when $\lambda = 0$. While the OLS estimator is unbiased, the Ridge estimator is biased and tends to zero as λ tends to ∞. Its primary advantage is that λ can be selected so as to obtain a stable set of parameter estimates.

Now the inverse matrix in Equation (7) is given by:

$$(V'V + \lambda I)^{-1} = \frac{1}{D} \begin{bmatrix} \lambda + \Sigma v_2^2 & -\Sigma v_2 v_1 \\ -\Sigma v_1 v_2 & \lambda + \Sigma v_1^2 \end{bmatrix} \quad (8)$$

where D is the determinant of the matrix $(V'V + \lambda I)$ and we have dropped the time subscript for clarity.

$$D = [(\lambda + \Sigma v_1^2)(\lambda + \Sigma v_2^2) - (\Sigma v_1 v_2)(\Sigma v_2 v_1)] \quad (9)$$

also

$$V'Z = \begin{bmatrix} \Sigma v_1 z \\ \Sigma v_2 z \end{bmatrix} \quad (10)$$

so that

$$\alpha_1 = [(\lambda + \Sigma v_2^2)\, \Sigma v_1 z - \Sigma v_2 v_1\, \Sigma v_2 z] \quad (11)$$

$$\alpha_2 = [(\lambda + \Sigma v_1^2)\, \Sigma v_2 z - \Sigma v_2 v_1\, \Sigma v_1 z] \quad (12)$$

Numerical values for the parameter vector $A = \{a_1, a_2\}$ are obtained in the next section. Note that a_1 and a_2 are the estimate of α_1 and α_2.

Model Calibration

The data used for model calibration and subsequent validation were sampled from the events of June 18, 1942, and October 5, 1941, respectively, over Edwardsville, Illinois, and documented by Holtan and Minshall (1968). Since the data were originally sampled at irregular intervals, each sequence of rainfall/runoff was first smoothed using a spline function interpolatory fit. Rainfall and runoff values were then computed at regular intervals of five minutes. Figure 1 contains the location map for the 52-acre watershed.

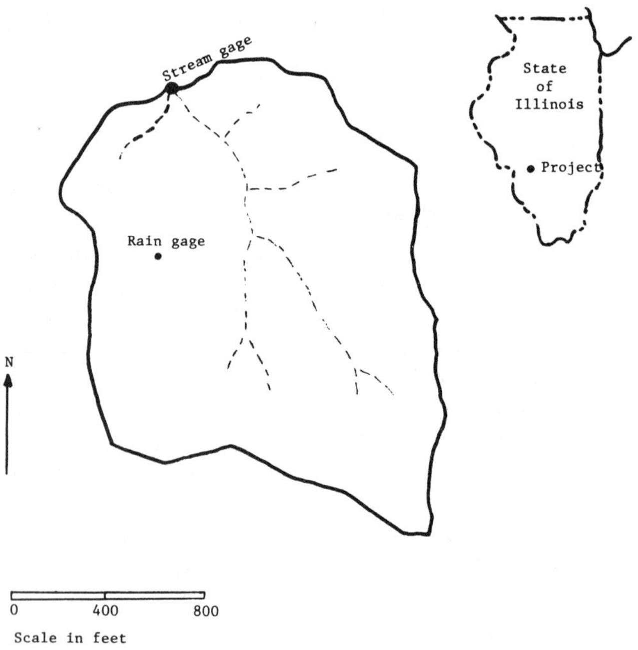

Figure 1. Location Map for Watershed II Near Edwardsville, Illinois.

The parameter estimates with corresponding sample statistics were computed for the June event using a computer program written for the parameter estimation (see Table 1). The statistics computed include standard error and the t-statistic for the parameters α_1 and α_2; the coefficient of determination; Durbin-Watson statistic and the lag one error auto-correlation coefficient. These statistics were computed

TABLE 1. Parameter Estimates and Sample Statistics.

λ	a_1	s_1	τ_1	a_2	s_2	τ_2	R	$\hat{\rho}_1$
0.000	0.028	0.020	1.40	0.934	0.013	71.6	0.94	0.45
0.001	0.056	0.026	2.18	0.880	0.017	50.9	0.92	0.71
0.002	0.074	0.035	2.09	0.833	0.024	33.8	0.88	0.83
0.003	0.087	0.044	1.95	0.793	0.031	25.1	0.84	0.87
0.004	0.095	0.052	1.83	0.756	0.037	20.0	0.81	0.88
0.005	0.101	0.058	1.74	0.723	0.043	16.7	0.77	0.88
0.006	0.105	0.064	1.65	0.694	0.048	14.5	0.74	0.88
0.007	0.107	0.068	1.58	0.666	0.052	12.8	0.71	0.88
0.008	0.109	0.072	1.52	0.642	0.055	11.6	0.69	0.88
0.009	0.110	0.075	1.46	0.618	0.058	10.5	0.66	0.87

NOTES:

s_1 is the standard error of estimate for ith parameter.
τ_1 is the t-statistic for the ith parameter.
R is the square root of the coefficient of determination.
$\hat{\rho}_1$ is the lag-one error autocorrelation coefficient.
Durbin-Watson statistic $\simeq 2(1 - \hat{\rho}_1)$

for λ = 0(0.09) 0.001. Recall that the ridge estimates at λ = 0 correspond to the ordinary least squares estimates.

Table 1 shows that a_1 increases slightly while a_2 decreases with λ. In general both a_1 and a_2 are supposed to decrease with λ. The increase observed in a_1 may be attributed to the constraint, Equation (3), imposed on the parameters. Assuming a significance level of 5 percent, the theoretical t-statistic is 2.09 for a sample size of 21 and two parameters estimated. From the fourth column of Table 1 it is seen that the estimate for α_1 is significant only for λ = 0.001 and 0.002. This may suggest an instability in the estimate a_1 of α_1. In contrast, from the seventh column of the table, the estimate for α_2 is significant for all λ in the range tested. Since the value of R is the highest and the value of $\hat{\rho}_1$ is the lowest for the case of λ = 0.000 we use the values a_1 = 0.028 and a_2 = 0.934 in subsequent analysis. This is reasonable since also the values of a_1 are insignificant for the most part including the case under consideration. Figure 2 shows the hyetograph and the observed runoff hydrograph. It also shows the fitted hydrograph for λ = 0.000. The 95 percent confidence interval on α_1 and α_2 are:

$$-0.01 \leqslant \alpha_1 \leqslant 0.07$$

$$0.91 \leqslant \alpha_2 \leqslant 0.96 \tag{13}$$

The value of the Durbin-Watson is 1.11 given that $\hat{\rho}_1$ = 0.45. The correlation coefficient is found to be significant at a 5 percent level and hence must be considered in further analysis.

The above parameter values are used in the next section for the purpose of validation. The rainfall/runoff data used are those of the event of October 5, 1941.

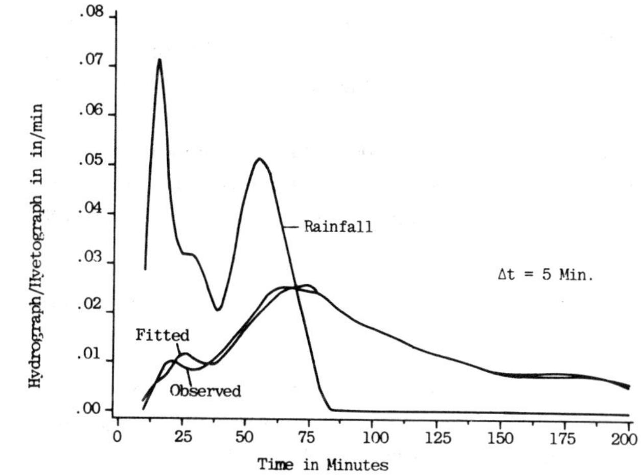

Figure 2. Fitted and Observed Hydrographs (model calibration).

ILLUSTRATIVE EXAMPLE

Forecast Situations

The forecasting process requires the estimates a_1, a_2 and $\hat{\rho}_1$ of α_1, α_2 and ρ_1. It also requires the initial value of stream flow, y_1, and the last residual, which at the beginning, may be assumed to be zero, i.e., $e_1 = 0$.

The forecast equation for the (n + 1)th time interval is

$$\hat{y}_{n+1} = x_{n+1} - a_1(x_{n+1} - x_n) - a_2(x_{n+1} - w_n) + \hat{e}_{n+1} \tag{14}$$

where x_n, x_{n+1} are the current and projected value of rainfall and y_n is the current flow stage. The definition of w_n

and $\hat{e}_{n+1}$ depend on the forecast situation considered. Three situations are examined,

A) $w_n = y_n$

$$\hat{e}_{n+1} = u_{n+1}\hat{\sigma}_e \tag{15}$$

B) $w_n = y_n$

$$e_{n+1} = \hat{\rho}_1(\hat{y}_n - y_n) \tag{16}$$

C) $w_n = \hat{y}_n$

$$\hat{e}_{n+1} = u_{n+1}\hat{\sigma}_e \tag{17}$$

where u_{n+1} is a standardized Normal variate and $\hat{\sigma}_e$ is the standard deviation of error. Situation (C) may be adopted when the value of the current stage, y_n, is not available. Its use, however, may result in a significant error build up in the forecasts. This is borne out in Figure 3 in which we plot the observed and forecast hydrographs for each of the three forecast situations. Figure 3(A) corresponds to forecast situation A.

Quantitatively model performance under each of the three forecast situations is evaluated using the Mean Integral of Time multiplied by Absolute Error (MITAE) index:

$$\text{MITAE} = \frac{1}{N}\sum_{n=1}^{N}(n\,|e_n|) \tag{18}$$

where N is the number of intervals and e_n is the error in the nth interval, i.e., the difference between the forecast flow and the subsequently observed flow. The index has a minimum value of zero which serves as a reference for the purpose of comparing model performance under the three forecast situations.

ANALYSIS AND CONCLUSION

Table 2 contains the calculated values of the MITAE for each situation and time interval. The entries for situations A and C, which used the Normal variate, are the mean of 100 replications. Situation A gives the lowest average MITAE and hence should be the implemented situation for the conditions assumed for the forecast model. While the above MITAE index provides a basis for evaluating the three situations, the ultimate criterion should be one that expresses the forecast errors, e_n in terms of the decision function of the user of the forecasts.

The sensitivity of the sample statistics, in particular R and $\hat{\rho}_1$, to changes in the sampling interval was performed. Such an analysis may serve as a guide in selecting the cost-effective time interval for forecasting. The parameters R and $\hat{\rho}_1$, were observed to decrease with sampling intervals, indicating a loss in predictive capability, as one would expect. The rate of decrease was observed to be less with increasing value of the shrinkage factor λ. Given the forecasting cost as a function of sampling interval it is then possible to make a rational choice of the optimal interval.

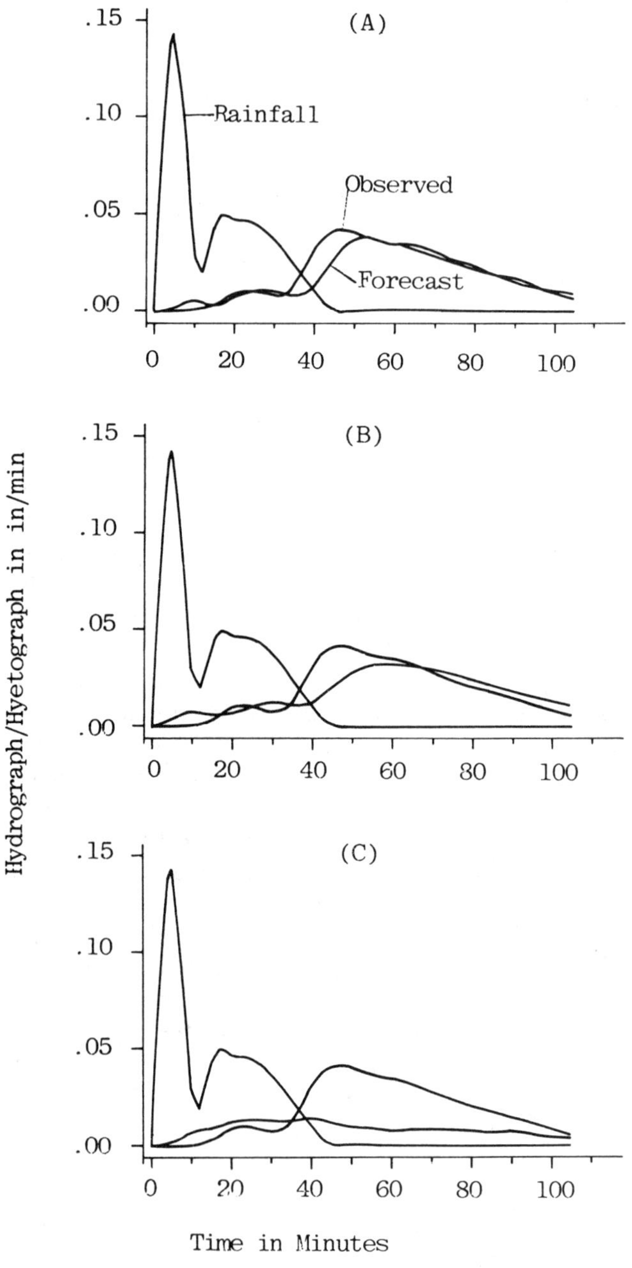

Figure 3. Forecast and Observed Hydrographs (model validation) Δt = 5 min.

In conclusion we observe that no attempts have been made to incorporate some mechanism for updating the values of the parameters during an event. Also for this initial effort we have not incorporated the uncertainty in the form of randomness in the rainfall magnitude and/or errors in rainfall measurements. In spite of the above limitations the proposed model holds promise as a practical tool for forecasting flow stage in small catchments.

TABLE 2. Performance Index for Forecast Situations.

Δ T Min	$w_n = y_n$, $\hat{e}_{n+1} = u_{n+1}\,\sigma_e$ (A)	$w_n = y_n$, $\hat{e}_{n+1} = \hat{\rho}_1(\hat{y}_n - y_n)$ (B)	$w_n = \hat{y}_n$, $\hat{e}_{n+1} = u_{n+1}\,\sigma_e$ (C)
5	0.036	0.060	0.152
10	0.045	0.040	0.062
15	0.046	0.061	0.064
20	0.047	0.049	0.038
25	0.040	0.058	0.056
Avg. MITAE	0.043	0.054	0.074

LITERATURE CITED

Hershfield, D. M. and E. T. Engman, 1978. Some Characteristics of Intense Short-Duration Rainfalls and Associated Runoff. Preprints, Conference on Flash Floods, Hydrometeorological Aspects, Los Angeles, California.

Holtan, H. N. and N. E. Minshall, 1968. Plot Samples of Watershed Hydrology. U.S. Department of Agricultural Research Service, ARS 41-133.

INTERNATIONAL SYMPOSIUM ON HYDROMETEOROLOGY
JUNE AMERICAN WATER RESOURCES ASSOCIATION 1982

A TEST OF RAINFALL-RUNOFF MODELS AS ELEMENTS OF FLOOD FORECASTING MODELS FOR BIG RIVERS

Karl Ludwig[1]

ABSTRACT: Test investigations of different methods for flood forecasting in big rivers are carried out in Germany currently. A relatively small stretch of the Rhine between Karlsruhe and Worms including the Neckar River basin is used as test object for these investigations. As a part of these investigations, different types of rainfall-runoff models are being tested. These shall be connected by flood routing models later to produce system models as a basis of flood forecast in big rivers.
(KEY TERMS: rainfall-runoff model; flood forecast; parameter optimization.)

INTRODUCTION

In order to recommend reasonable methods for flood forecasting in big rivers, several investigations of relevant forecast methods are carried out in the German Federal Republic currently. One of these investigations is a test of different methods of flood forecast by use of river basin models (system models of river basins consisting of rainfall runoff models and flood routing models) on a limited stretch of the Rhine from Maxau (Karlsruhe) to Worms, including the Neckar basin (Figure 1). This investigation is done at the Institute of Water Resources, Hydrology and Hydraulic Engineering in Agriculture of the University in Hannover.

The total project comprises test investigations of rainfall-runoff models and flood routing models, the calibrating of a suitable river basin model and its use for flood forecast, test studies of effects of error correction and on-line calibration methods, and different input data volumes on the forecast effectivity. This paper deals with the first part of the project, the investigation of rainfall runoff models.

AREAL STRUCTURE OF INVESTIGATED FORECAST MODELS

As the investigated methods for flood forecasting are designed for big rivers only, as for instance the Rhine or the Danube in Germany, the desired size of subareas in the forecast river basin model is about 1000 km^2 in order to produce an areal model structure, which is appropriate for this purpose.

The investigation area on Rhine and Neckar between the gages Maxau and Worms has a total area of about 18,600 km^2. It was divided into subareas ranging from 380 to 1500 km^2. The deviations from the wanted size of subareas (1000 km^2) stem from the actual situation in the river basin, especially from the distribution of available gages. Due to this areal model structure, the river basin model contains 24 subareas and 18 channel subreaches. It contains 14 gages and 16 rain gages. The areal structure of the river basin model can be shown quite clearly in the form of a system sketch which is contained in Figure 2.

Though more gaging stations were available the used gages were restricted to the most important ones and the density of rain gages was chosen in a way that one rain gage would fall in one subarea or in a total area of 1000 km^2. This was considered to be a reasonable density for flood forecasts in big rivers under European conditions.

For the test investigation of rainfall runoff models the five gages, which are situated in the head segments of the model structure (Horb, Ebersbach, Pforzheim, Gaildorf, and Doerzbach; see Figures 1 and 2) were used. These are gages, which are not influenced by the flood routing effects in the river basin. They allow an investigation of rainfall runoff models for area sizes, which are of interest in the desired river basin model structure independent of the effects of flood wave transformation in the channels.

GENERAL PROCEDURE FOR THE TEST INVESTIGATION OF RAINFALL-RUNOFF MODELS

For the five gages in the head segments of the river basin model structure the following hydrological submodels were to be tested:

a. Submodels for the calculation of effective rainfall:
   - constant runoff coefficients
   - constant loss rates (PHI-index-method)

[1] Assistant, c/o Institut Für Wasserwirtschaft, Hydrologie und Landwirtschaftlichen Wasserbau, Callinstrabe 32, D-3000 Hannover, Germany.

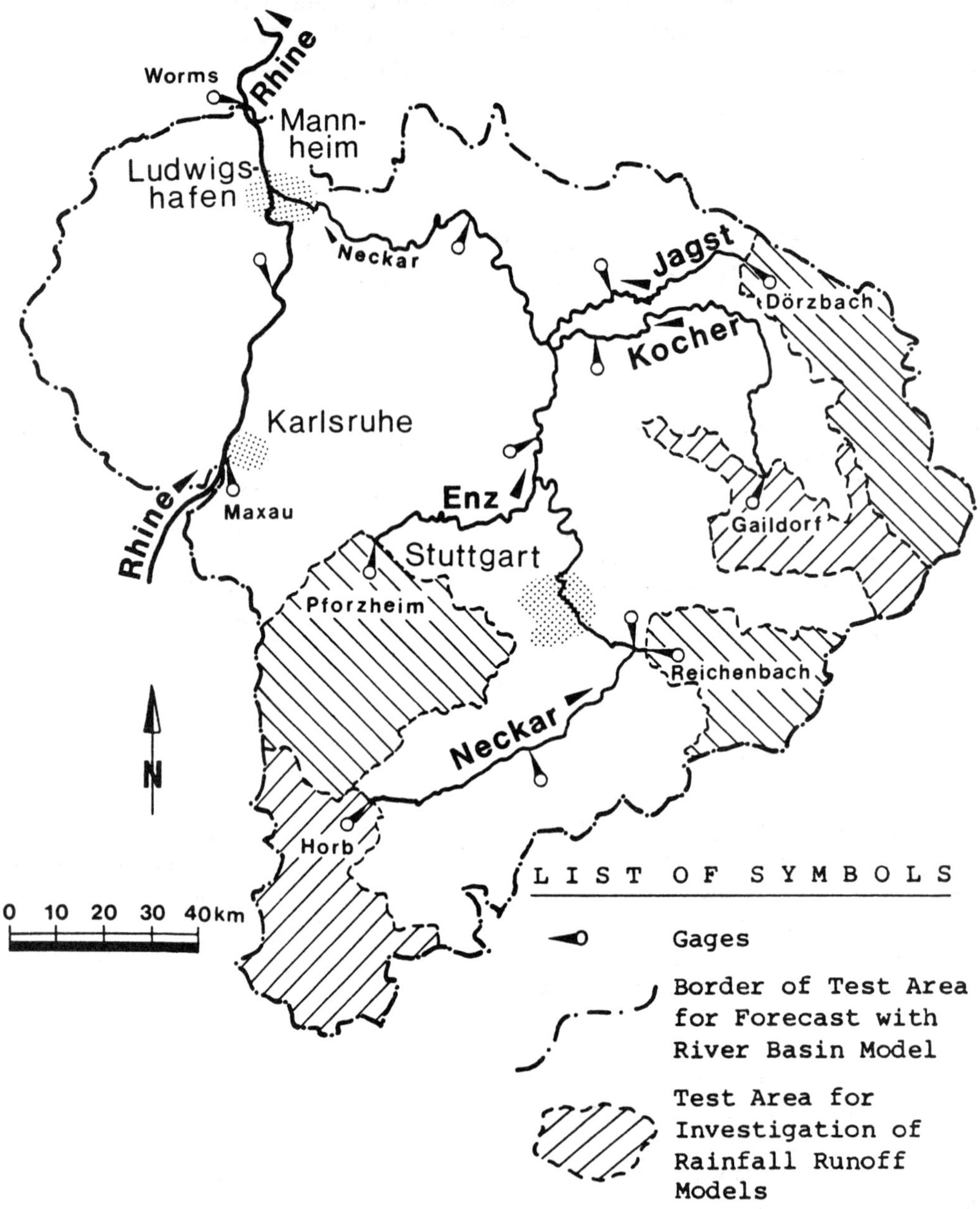

Figure 1. Sketch of the Investigation Area.

- Koehler model (variable runoff coefficients; Koehler, 1971)
- Horton model (variable runoff coefficients; Horton, 1937)

b. Submodels for runoff transformation in subareas:

- linear storage model
- Clark model (Clark, 1945)
- Clark model with simplified time-area diagram
- parallel storage model with constant distribution rate (Ludwig, 1978).

As can be seen from the list of investigated submodels, they are of moderate complexity. Their function is sketched shortly in the following:

Constant Runoff Coefficients: Effective rainfall is a factor (runoff coefficient) times the ordinates of the areal rainfall.

Constant Loss Rates: Effective rainfall is areal rainfall minus a constant loss rate (PHI-index).

Koehler Model: Effective rainfall is a variable factor (runoff coefficient) times the ordinates of areal rainfall. Parameters influencing the variable runoff coefficient are weighting factors. One weighting factor is zero at the start of rainfall and rises linearily to one until a certain time after start of rainfall. It accounts for greater losses at the start period of rainfall. The second weighting factor accounts for the rain intensities; it is the areal rainfall

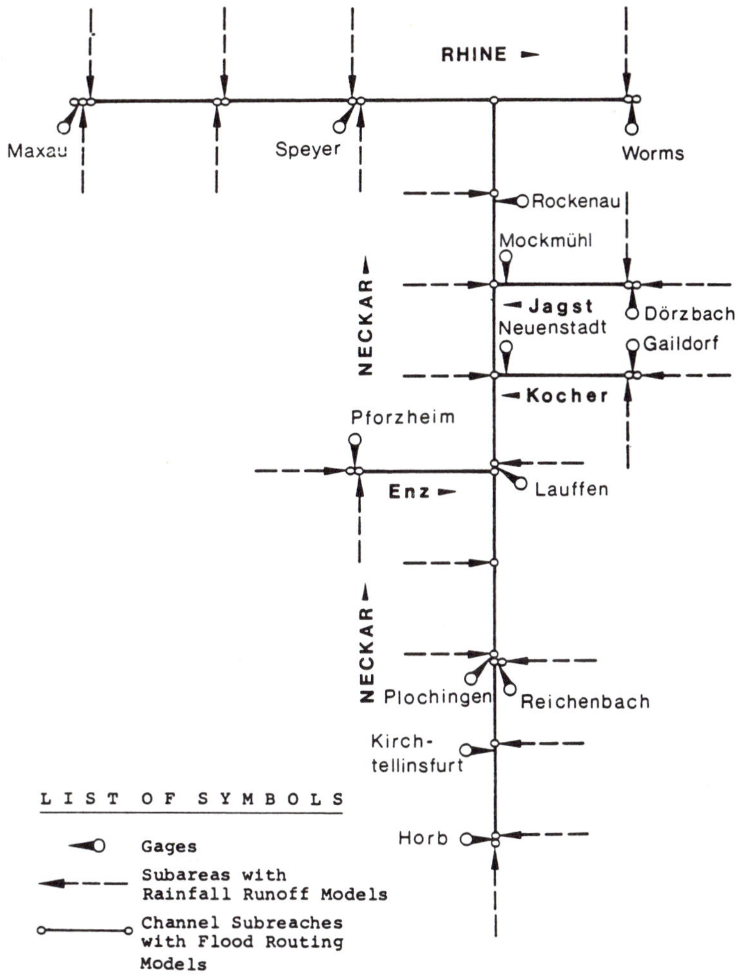

Figure 2. System Sketch of the River Basin Model.

ordinates with an exponent rainging between 1 and 2. Both factors are normalized to produce the total runoff coefficient during an event.

Horton Model: Effective runoff is areal rainfall minus a variable loss rate. Loss rates decrease exponentially in time from an initial value to an end value.

Linear Storage Model: Effective runoff is transformed in the subarea according to the system function of a linear reservoir.

Clark Model: Effective runoff is transformed in the subarea according to a time-area-diagram which represents translation components in the transformation process and a linear storage representing retention components in the subarea.

Simplified Clark Model: The form of the subarea is approximated by a rectangle with equal length. The simplification produces essentially a trapezoidal shape of the time-area-diagram.

Parallel Storage Model: Effective runoff is distributed to a slow (interflow) and quick (direct runoff) storage due to a threshold value (interflow index rate).

USED PROGRAMS

The Program REGMO

Within the program REGMO, a "regional rainfall-runoff model" is being applied. In an analysis part, the flood runoff behavior on several gages within a region with similar runoff characteristics is investigated. Parameters of different rainfall-runoff models are optimized. For the assessment of an advantageous calculation procedure several different rainfall-runoff models are investigated to determine which combination of these is most reliable, simple, and clear.

With the program REGMO, combinations of submodels (a) and (b) have been investigated so that for each runoff event on every gage several different rainfall-runoff models have been calibrated.

The Program FGMOD

FGMOD is a generally structurized program system for the systematic description of runoff processes in river basins. The current program version can be used for analysis, synthesis, and forecast of flood and low-flow runoff processes.

Runoff processes in subareas of river basins are described by rainfall-runoff models. The resulting simulated runoff hydrographs are inflows to channel subreaches, in which runoff transformation is described by flood routing models. Basis for the calculation process is a partition of river basin. This partition depends on the locations of available gages, on (existing or planned) constructions with influences on the runoff processes (reservoirs, essential channel amendments, and so on) and on the configuration of the channel network in river basins.

The program system FGMOD (Ludwig, 1982), was originally designed for the description of runoff processes in river basins for research and planning purposes. It has been modified for the special requirements of flood forecast (Holle, *et al.,* 1981) and will be used later on in this project for the test of flood forecasts for the total area under investigation.

In this part of the investigations, the test of rainfall-runoff models, the program system FGMOD is used only to test parallel storage models in single subareas (Figure 1). Parameters of the parallel storage model are retention parameters for quick and slow storage and the interflow index rate.

The Optimization Routine MARQU

In the programs REGMO and FGMOD, important parameters can be determined by a nonlinear least-square optimization routine (Marquardt, 1963). Figure 3 gives an example for the steps during the optimization of two essential parameters, the Interflow-Index-Rate (A) and a parameter PSI (runoff coefficient) for different start values.

Figure 3 shows the isolines of the target function (in this case a weighted deviation measure) and the positions of parameter sets at different steps of the optimization procedure. The example shows that the optimization routine works very effectively. It has been used satisfactory also for other (single) rainfall-runoff models containing more than 20 parameters.

RESULTS OF THE TESTS OF RAINFALL-RUNOFF MODELS

A visual comparison of the simulation results is not feasible because of the great number of simulations (700 for 10 rainfall-runoff events). So, a deviation measure, the so-called "hydrological deviation" has been used to compare model results. The hydrologic deviation is a mean value of weighted deviations between measurement and simulation expressed in percent of the maximum value. Weights are 1 for runoff zero and 2 for maximum runoff; other weights are interpolated linearly.

A judgment of model efficiency was done by counting the good simulation results, which were expressed by hydrologic deviations below a defined threshold value. This way of model comparison was chosen to prevent, that the comparison was affected by bad simulation results produced by data errors. The numbers of good simulation results of the investigated model combinations are given in Table 1 for two threshold values (10 and 5) for the hydrologic deviation.

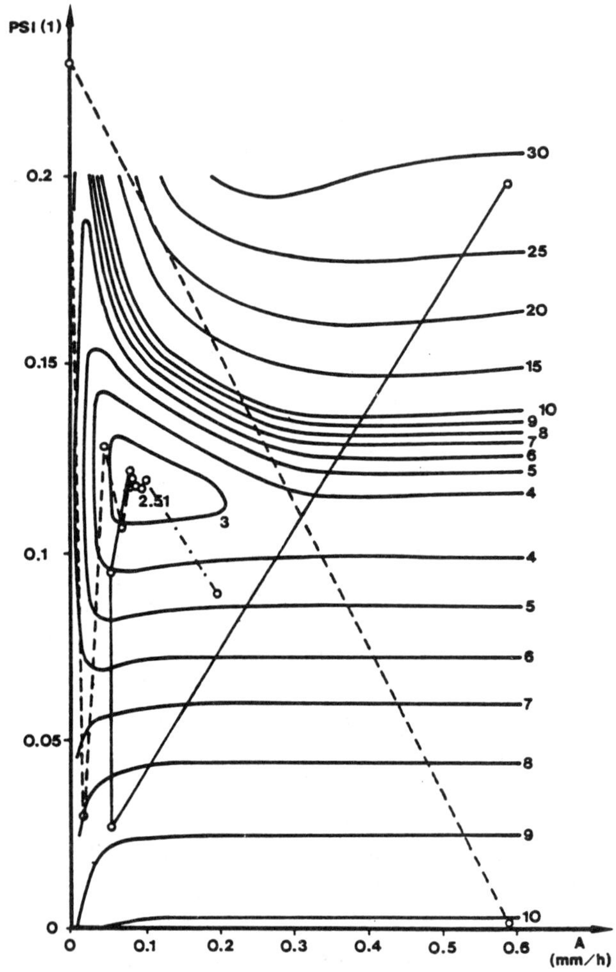

Figure 3. Optimization of Two Model Parameters.

Though several model combinations produce a similar number of good results, it can be seen in Table 1 that best results are produced by the combinations Koehler method – simplified Clark model and constant runoff coefficient – parallel storage model. These model combinations have both a total parameter number of 4. They will both be used in the further investigation of flood forecasts with the river basin model.

TABLE 1. Number of Good Simulation Results for Different Model Combinations
(total number of events on all gages = 45).

Submodels	Horton	PHI-Ind.	Con. Coeff.	Koehler
		DEVIATIONS LESS THAN 10		
Single Linear St.	25	31	32	38
Clark Model	20	22	25	25
Clark, Simplified	28	32	36	40
Parallel Storage	*	34	38	*
		DEVIATIONS LESS THAN 5		
Single Linear St.	13	5	3	15
Clark Model	6	1	3	9
Clark, Simplified	17	12	10	18
Parallel Storage	*	15	17	*

*Combinations not investigated because of high parameter number (6).

LITERATURE CITED

Clark, C. O., 1945. Storage and the Unit Hydrograph. Transactions of the ASCE 110:1419-1446.

Holle, F.-K., P. Leonhardt, and K. Ludwig, 1981. Flood Forecast in River Basins with Regulation Systems. Proceedings of the International Symposium on Real-Time Operation of Hydrosystems, Waterloo, Canada.

Horton, R. E., 1937. Determination of Infiltration Capacity for Large Drainage Basins. Transactions of the American Geophysical Union 8:371.

Koehler, G., 1971. Ermittlung massgebender Abflussdaten fuer kleinere Vorfluter mit Hilfe kurzzeitiger Naturmessungen. Proceedings of the Institut fuer Wasserwirtschaft, Hydrologie, und landwirtschaftlichen Wasserbau, Technical University Hannover, Vol. 23.

Ludwig, K., 1978. Systematische Berechnung von Hochwasserabflussvorgaengen mit Flussgebietsmodellen. Proceedings of the Institut fuer Wasserwirtschaft, Hydrologie, und landwirtschaftlichen Wasserbau, Technical University Hannover, Vol. 45.

Ludwig, K., 1982. The Program System FGMOD for Calculation of Runoff Processes in River Basins. Zeitschrift fuer Kulturtechnik und Flurbereinigung 23(1982):25-37.

Marquardt, D. W., 1963. An Algorithm for Least-Square Estimation of Nonlinear Parameters. Journal of the Society of Industrial Applied Mathematics 11:431.

LOCALIZED QPFs – A STATUS REPORT

H. Michael Mogil[1]

ABSTRACT: During the past five years the Lubbock Weather Service Forecast Office has developed and refined localized QPF procedures for west Texas. Due to the success of their activities, the concept is being studied by other NWS Forecast Offices in Texas, Oklahoma, Arkansas, and Louisiana.

This paper summarizes the status of these projects and future plans to hopefully tie them together into a multi-state QPF program. (KEY TERMS: flash flooding; National Weather Service; Quantitative Precipitation Forecasting (QPF).)

During the 1970s the flash flood problem jumped to the fore-front of weather-related disasters. Major events like the floods in the Big Thompson Canyon (1976); Kansas City, Missouri (1977); Johnstown, Pennsylvania (1977); and Central Texas (1978), coming so close together, were the catalysts which prompted several Federal agencies to accelerate their combative efforts. Some of these efforts included:

- A major flash flood initiative in the central Appalachians (NOAA, 1978). Sophisticated rainfall and river reporting systems, computer links, and close Federal-State cooperation are the cornerstones of this effort.
- An increased flash flood disaster preparedness program. This has included new slide sets, films, and revised publications by the National Weather Service (NWS) and other agencies involved in solving the problem. It has also included the first National Flash Flood Awareness Week, held during the period May 16-23, 1982.
- More flash flood forecast training for National Weather Service forecasters. A special flash flood course (Hollis, 1980) was begun at the NWS Training Center (NWSTTC) in Kansas City in the late 1970s. More than 200 forecasters have attended this comprehensive radar, satellite, hydrometeorology course.
- Expansion of the Heavy Precipitation Branch (HPB) at the National Meteorological Center (NMC) in Suitland, Maryland, and increasing its focus to include heavy rainfall and flash floods.
- Increased emphasis on and skill in producing real-time rainfall estimates from satellite imagery.

Although all of the above are important and are beginning to produce positive results, another effort, begun at the NWS Forecast Office (WSFO) at Lubbock, Texas, during the middle 1970s, is having, perhaps, the greatest impact and explosive growth. This effort, the localized quantitative precipitation forecast (QPF), is what I wish to discuss.

Any effort to improve rainfall forecasting, flash flood, and river flood warnings, and develop a viable flash flood disaster preparedness program must begin with a climatological assessment of the problem. Here the Lubbock NWS forecasters first focused their attention. The hydrometeorological team concentrated on the heavy rain season (May-September), and looked at all cases for each of the four identifiable climatological regions of West Texas. (The reason for the regional breakdown was to account for differences in topography and proximity to the Gulf of Mexico moisture source.) The outcome was a set of maps showing typical mid-level flow patterns favorable for heavy rainfall in each of the four climatological regions.

The Lubbock team also compared observed values of precipitation-favorable meteorological variables (e.g., lifted index, mean relative humidity, precipitable water) to observed rainfall using a linear regression approach. Due to some errors inherent to the system, the team later switched to a 3-variable graph approach. For each of the mid-level flow patterns, for each of the four regions, graphs were prepared relating the atmospheric variables to maximum rainfall and areal coverage. Belville, *et al.* (1978), describes the early Lubbock efforts.

Armed with this information, Lubbock developed a comprehensive decision diagram for preparing a localized QPF for West Texas. Operational forecasts were prepared for the first time in 1976.

The Lubbock staff was not content to simply use the procedures they had developed. Each year they reviewed significant cases, evaluated the overall performance of the procedures, and even modified them by adding new variables (forecast and observed) and by "inflating" the forecasts to compensate for unaccounted biases (Mortimer, *et al.*, 1978).

[1]Deputy Meteorologist-in-Charge, NOAA/National Weather Service Forecast Office, 819 Taylor Street, Room 10A44, Fort Worth, Texas 76102.

For several years the system was used only at Lubbock. Then due to budget cutbacks the number of staff at the Lubbock office was reduced in 1981. As a result, some key members of Lubbock QPF team transferred to other NWS offices. Initially this appeared to be a setback to the entire concept of localized QPFs. However, the "silver lining" quickly appeared and localized QPF projects sprung up at the San Antonio and New Orleans WSFOs. Quite independently, WSFO Fort Worth began looking into developing its own QPF program around the time of the devastating North Texas floods in the fall of 1981.

The next step was to bring these aforementioned offices (and some nearby offices which had not yet begun a QPF program) together. So in early 1982 representatives from the WSFOs at Oklahoma City, Little Rock, San Antonio, New Orleans, and Fort Worth met in Fort Worth for a two day workshop. Offices which had made progress in tackling the QPF problem presented their approaches and their findings. Others listened and learned. Since that meeting Little Rock has made significant progress in locating data sources and in developing its rainfall/flash flood climatology; Oklahoma City is beginning to work on its QPF; Albuquerque did not attend the two day workshop but nevertheless has begun automating the Lubbock system because they have inherited much of Lubbock's former hydrology program; and New Orleans has developed a climatology of significant (i.e., 5 inch or greater) rains and begun applying some of Lubbock's ideas in their operational system. During April 1982, the New Orleans effort resulted in a forecast of 5 to 10 inch maximum rainfalls up to 36 hours in advance. The forecast verified (even if slightly underforecast) when 13 inches of rain fell on Southern Louisiana.

The purpose of the workshop was not to force any one office to follow the lead of another. Rather it was to be a forum for discussion and for sharing information. However, because many of the offices involved in the project acquired people who worked at Lubbock and because of their initial success, it is not surprising that the Lubbock approach (in most aspects) prevails. Nevertheless, differences have begun surfacing. Specifically:

- The New Orleans synoptic climatology is quite different from that developed at Lubbock. This is not surprising, considering Louisiana's proximity to the Gulf of Mexico.
- Fort Worth has noted that vertical motion fields (in particular the 24-hour trajectory model 700 mb net vertical displacement field (Reap, 1972)) may give more reliable and useful information than some of the "inflation" factors developed at Lubbock. This is important, because many of the "inflation" factors were attempts to compensate for seemingly insufficient numerical guidance in predicting vertical motion.
- San Antonio and New Orleans have found that mesoscale and subsynoptic factors (Maddox, *et al.*, 1979 and Mogil, 1978) are crucial for isolating maximum rainfall areas and amounts.
- Lubbock analyzed rainfall amounts and areas using climatological reporting stations several months after the fact (due to data collection and publishing considerations). The River Forecast Center at Forth Worth supplies the local WSFO with daily, state-wide rainfall plots which make post-mortems more timely.

In addition to assisting in forecasting heavy rainfall, development of a climatology has other important benefits. These include indoctrination of new forecasters, isolating anomalies, and having statistics available to support warning/preparedness efforts. WSFO Fort Worth prepared such a climatology based on nearly 10 years of *Storm Data*. Defining a flash flood day as a calendar day in which there was at least one reported flash flood in North Texas, the QPF team at Fort Worth evaluated:

- flash flood frequency
- the seasonal distribution of flash floods
- the diurnal distribution of flash floods
- the geographical distribution of flash floods
- the extensiveness (i.e., number of counties affected by) flash flooding
- the persistence (i.e., number of consecutive days) of flash flooding across North Texas
- the maximum rainfalls associated with the flash flooding
- deaths (number, causes, age categories)
- property losses
- the occurrence of severe weather along with the flash flooding

One anomaly which jumped out immediately was the conspicuous July minimum in flash flood events in North Texas (Figure 1). Meteorologically this can be accounted for since it separates the spring-early summer convective regime from the late summer-early fall tropical cyclone season.

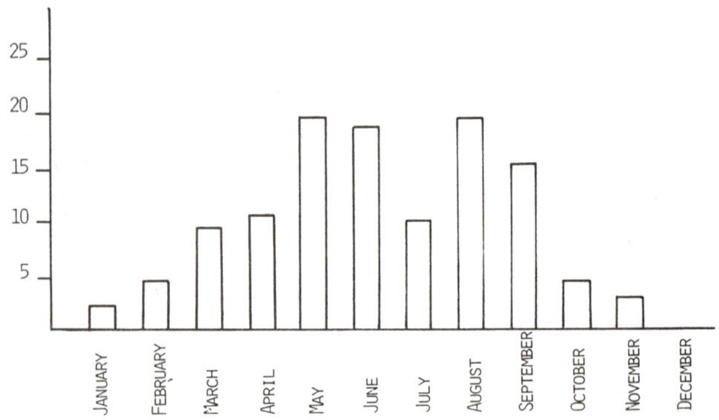

Figure 1. Number of North Texas Flash Flood Days by Month. Information from *Storm Data* 1973-80.

Another anomaly focused on the geographical distribution of flash flood events across North Texas. East Texas, an area with substantially higher annual rainfall, and more frequent heavy rains, had fewer flash flood days than either central or western sections (Figure 2). Population density, media sources, topography, and vegetation could all be operating to produce this apparent anomaly.

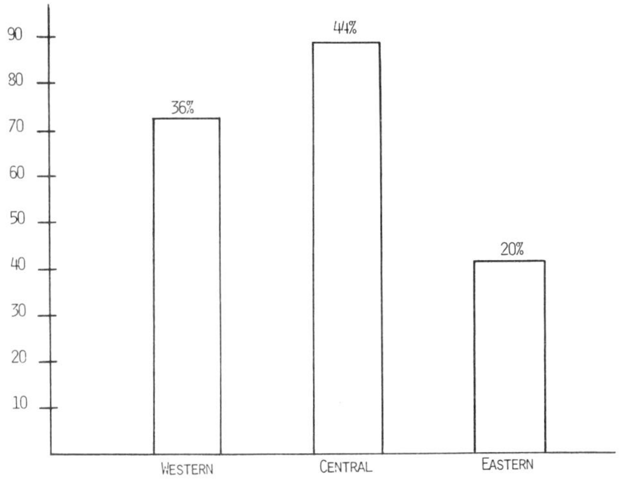

Figure 2. Number of North Texas Flash Flood Days by Region. Information from *Storm Data* 1973-80. Total number of flash flood days does not agree with Figure 1 because a flash flood day could affect more than one region.

Population density is apparent again when flash flood deaths are considered. Of the 34 flash flood-related deaths in North Texas during the period 1973-1980, 30 percent occurred in the Dallas/Fort Worth area.

The seven WSFOs (Albuquerque, Fort Worth, Little Rock, Lubbock, Oklahoma City, New Orleans, and San Antonio) now actively engaged in QPF studies have begun communicating among themselves with increasing frequency, sharing their findings and discussing options, prior to spending considerable time and effort on possibly fruitless approaches. Periodic status reports are also being distributed, not only among the seven offices, but to the River Forecast Centers at Fort Worth, Tulsa, and New Orleans, the National Earth Satellite Service (Washington Headquarters and Kansas City Field Service Station), the Environmental Research Laboratory (Boulder, CO), National Weather Service Headquarters (Washington, D.C.), NWS Southern Region Headquarters (Fort Worth), and the NMC HPB. Information will also be sent to the NWSTTC (Kansas City) for possible inclusion in the flash flood course and for possible use in getting other offices involved in similar efforts.

Obviously the efforts described here are not the only ones underway in the National Weather Service to improve QPFs. However, these efforts are extremely coordinated and have considerable field operations initiative. Hopefully the proper resources (personnel, money, and time) will continue to be available in order to maintain and even expand the effort throughout the National Weather Service.

ACKNOWLEDGMENTS

This paper would never have been written had it not been for the efforts undertaken at WSFO Lubbock years ago. Special thanks must go to the entire QPF team at Lubbock, and especially Billy Crouch, the Meteorologist in Charge for having the foresight to foster the effort. Thanks also go to members of all the QPF teams at the various offices for their efforts at expanding what Lubbock started. Kermit Keeter and Dick Elder of the Fort Worth WSFO prepared the synoptic and general climatologies for North Texas. Susan Beckwith typed the manuscript.

LITERATURE CITED

Belville, J. D., G. A. Johnson, and J. D. Ward, 1978. A Flash Flood Aid – The Limited Area QPF. NOAA Technical Memorandum NWS SR-97, 12 pp.

Hollis, D., 1980. The National Weather Service Flash Flood Forecasting Course. *In:* Proceedings – Second Conference on Flash Floods, American Meteorological Society, Boston, MA, 190-191.

Maddox, R. A., C. F. Chappell, and L. R. Hoxit, 1979. Synoptic and Meso-α Scale Aspects of Flash Flood Events. Bulletin of the American Meteorological Society, 60(2):115-123.

Mogil, H. M., 1978. Mesoscale Analysis – An Important But Overlooked Diagnostic Tool. *In:* Proceedings – Conference of Weather Forecasting and Analysis and Aviation Meteorology, American Meteorological Society, Boston, MA, 58-70.

Mortimer, E., G. A. Johnson, D. Noble, and J. D. Ward, 1980. An Improved Limited Area Quantitative Precipitation Forecast for West Texas. *In:* Proceedings – Second Conference on Flash Floods. American Meteorological Society, Boston, MA, 141-148.

NOAA, 1978. National Flash Flood Program Development Plan FY79-84. Washington, D.C., 47 pp.

Reap, R. M., 1972. An Operational Three-Dimensional Trajectory Model. Journal of Applied Meteorology, 1(8):1193-1202.

INTERNATIONAL SYMPOSIUM ON HYDROMETEOROLOGY
JUNE AMERICAN WATER RESOURCES ASSOCIATION 1982

A SATELLITE TECHNIQUE FOR ESTIMATING RAINFALL FROM FLASH FLOOD PRODUCING THUNDERSTORMS

Roderick A. Scofield[1]

ABSTRACT: Scofield and Oliver developed a technique that gives half-hourly or hourly rainfall estimates from convective systems by using GOES thermal infrared and high resolution visible pictures. The Scofield/Oliver technique is in the form of a decision tree, which a meteorologist uses to determine point or areal rainfall estimates. The technique is applied using imagery displayed with a digital enhancement curve.

During the past year, an experimental modification to the Scofield/Oliver technique was developed for estimating rainfall from warm topped thunderstorms. This paper contains a real-time application of the Scofield/Oliver technique and some thoughts on how to improve the method.

(KEY TERMS: satellites; rainfall estimation; thunderstorms; flooding; mesoscale.)

INTRODUCTION

This paper discusses the use of the Scofield/Oliver Geostationary Operational Environmental Satellite (GOES) technique for estimating convective precipitation from flash flood producing thunderstorms. In the U.S.A. satellite-derived estimates supplement rain gauge observations and radar estimates. In some instances satellite-derived estimates may be the only data available on a world-wide basis. Satellite-derived estimates are the only cost-effective way to obtain precipitation information.

SCOFIELD/OLIVER TECHNIQUE

Scofield and Oliver (1976, 1977, 1978, 1980) have developed a technique which gives half-hourly or hourly rainfall estimates from convective systems by using GOES infrared (IR) and high-resolution visible pictures. The Scofield/Oliver technique is presented in the form of a decision tree, which an analyst uses to determine rainfall estimates. Satellite pictures are analyzed and compared on a half-hourly basis with respect to five different factors to arrive at a numerical rainfall estimate. The estimate is modified according to the available moisture in the storm's environment. Estimates are computed using the following equation:

$$P = (TG + O + M + S)(PW/1.5) \qquad (1)$$

where P is the half-hourly rainfall estimate in inches; TG, a cloud-top temperature and cloud growth factor or divergence aloft factor; O, an overshooting top factor; M, a merger factor; S, a saturated environment factor; and PW/1.5, a precipitable water factor where PW is the observed surface to 500-mb precipitable water for the synoptic-scale environment. At this time, the precipitable water factor is only used to modify estimates for thunderstorms in dry environments or with high bases; estimates are reduced when the observed surface to 500-mb precipitable water is less than or equal to 1 inch. The technique was designed for deep convective systems that occur in tropical air masses with high tropopauses, and it is applied using IR pictures displayed according to a digital enhancement curve. More details on the technique can be found in the previously mentioned references.

APPLICATION OF THE SCOFIELD/OLIVER TECHNIQUE

Currently the National Environmental Satellite Service (NESS) meteorologists compute precipitation estimates and pass them on to the appropriate National Weather Service Forecast Office (WSFO) whenever these estimates are expected to exceed threshold values for flooding as determined by the National Weather Service (NWS) River Forecast Centers (RFC). Receiving the hardcopy prints and computing, summing, and disseminating the estimates takes 45 to 60 minutes. In this section, the results of using the Scofield/Oliver technique in real-time on a convective system are presented.

[1]Meteorologist, NESS Applications Laboratory, National Earth Satellite Service, Applications Laboratory, S/RE3, World Weather Bldg., Room 601, Washington, D.C. 20233.

Georgia-Alabama Rainstorm, April 1, 1981

During April 1, rains of up to 5 inches fell on portions of Georgia and Alabama. The NWS River Forecast Centers provide daily flash flood guidance analyses to the WSFO's. The flash flood guidance analysis prior to the heavy rainfall event over Georgia and Alabama is shown in Figure 1; the values represent the amount of rain needed in a three-hour period to produce flooding. Since each RFC computes their own threshold values, discontinuities in the threshold values can occur between areas of responsibility such as the ones (Figure 1) along the Mississippi-Alabama state line.

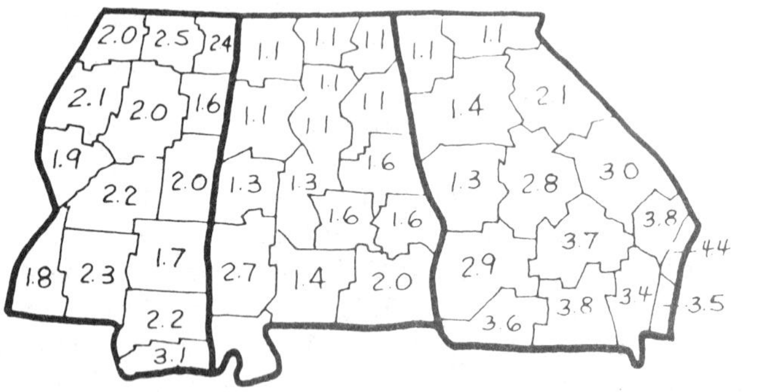

Figure 1. Flash Flood Guidance, 1200 GMT, March 31, 1981.

The IR imagery in Figures 2-5 is enhanced in order to analyze details in the thunderstorm anvils and makes thunderstorm top temperatures more quantitatively determinable. Temperatures associated with each of the contours in the Mb curves are shown in Figure 2: medium gray (–32 to –41°C) represents the warmest tops, white (below –80°C), the coldest. The enhanced IR imagery at 0630 GMT (Figure 2), 0800 GMT (Figure 3), 1000 GMT (Figure 4), and 1200 GMT (Figure 5) shows a persistent area of cold tops located at C. Estimated isohyets were computed and drawn every half-hour and graphically added. In this case, the satellite signatures of cloud-top temperature/cloud growth and stationary storms (saturated environment) were the principal contributors to the estimates. The results of 5½ hours of estimates are shown in Figure 6a; six-hour observed amounts are also plotted in Figure 6b. Both the estimates and the six-hour observed amounts showed that heavy precipitation occurred in western Georgia and eastern Alabama. Where the cold tops persisted in the IR, there were no six-hour observed amounts. However, the more complete 24-hour rainfall observations for this area are shown in Figure 7. In this area of four- to five-inch estimates, flash flood guidance values (Figure 1) indicated that only 1.3 to 1.6 inches of rain were required in three hours to produce a flash flood.

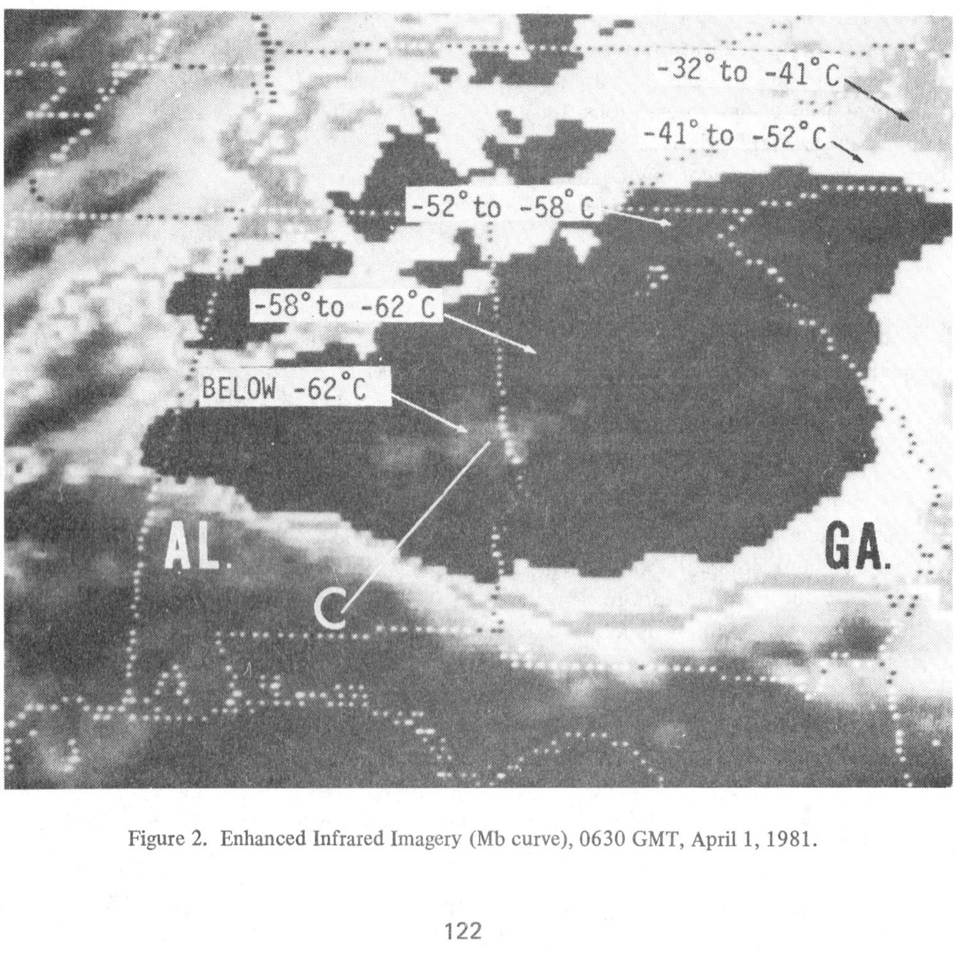

Figure 2. Enhanced Infrared Imagery (Mb curve), 0630 GMT, April 1, 1981.

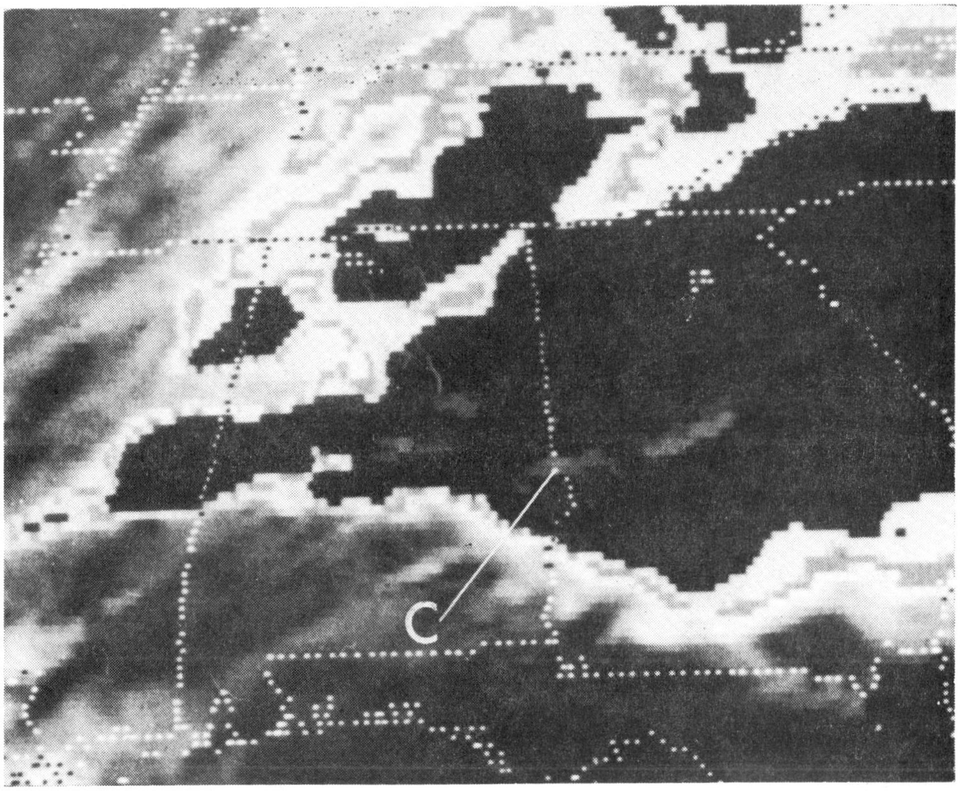

Figure 3. Enhanced Infrared Imagery (Mb curve), 0800 GMT, April 1, 1981.

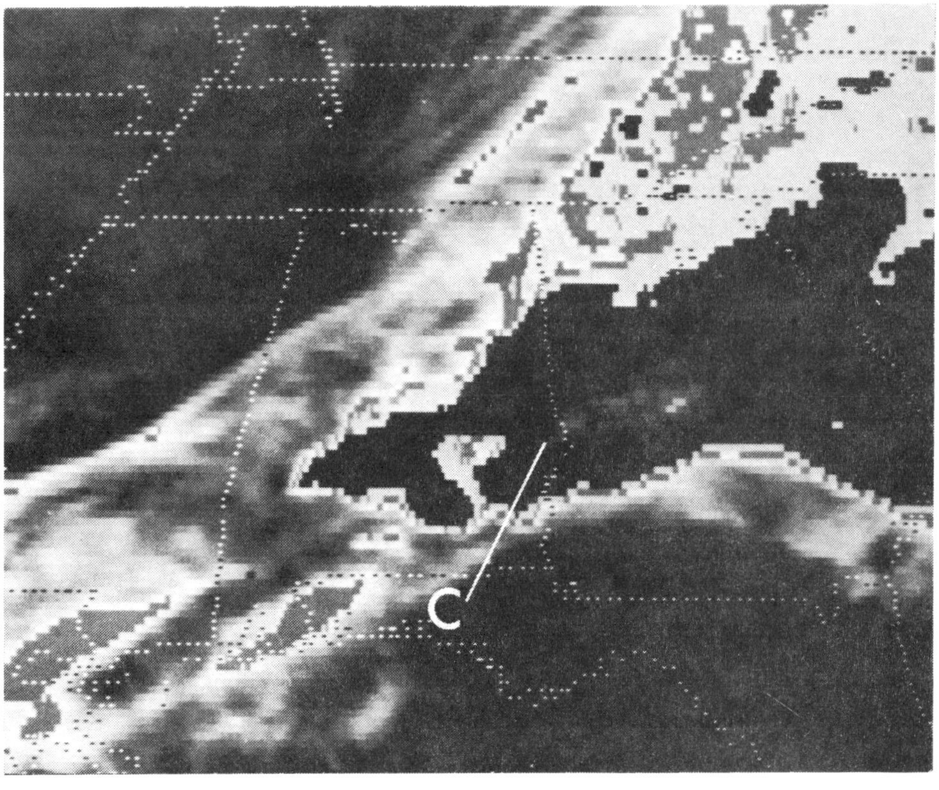

Figure 4. Enhanced Infrared Imagery (Mb curve), 1000 GMT, April 1, 1981.

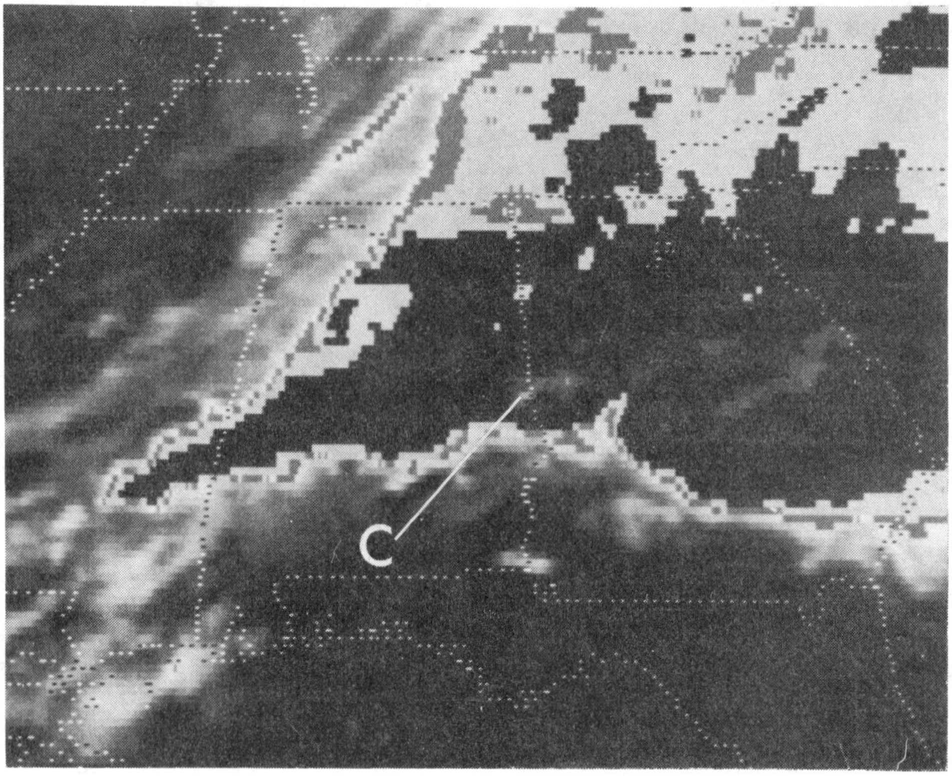

Figure 5. Enhanced Infrared Imagery (Mb curve), 1200 GMT, April 1, 1981.

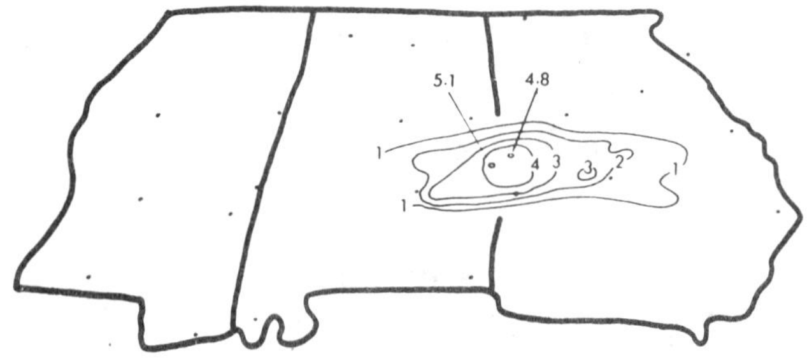

Figure 6a. 0630-1200 GMT Satellite-Derived Rainfall Isohyets in Inches, April 1, 1981.

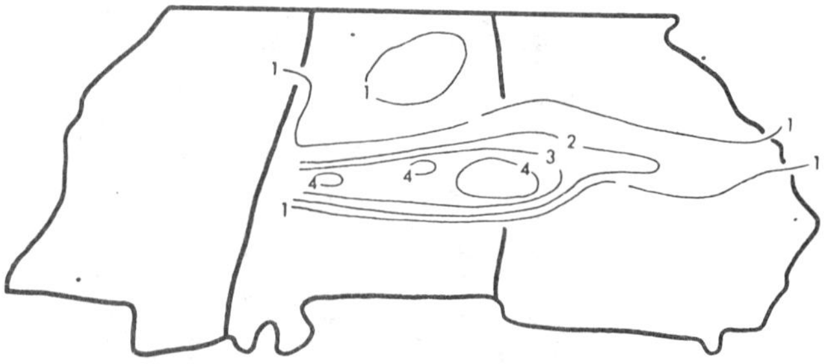

Figure 7. 24-Hour Observed Rainfall Ending at 1200 GMT, April 1, 1981.

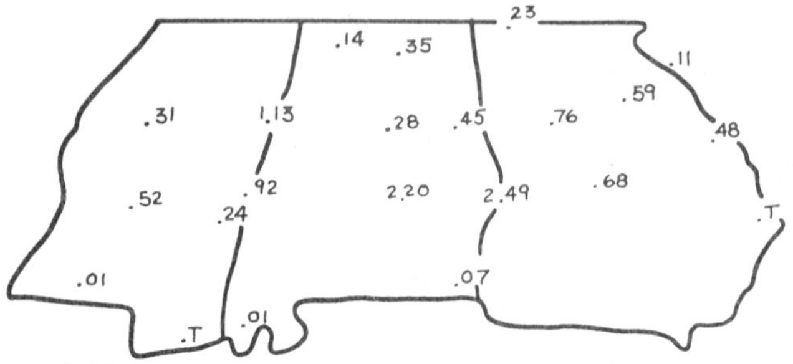

Figure 6b. Six-Hour Observed Rainfall Ending at 1200 GMT, April 1, 1981.

The 24-hour observed amounts agree quite well with the estimates, especially where the cold tops persisted in the IR. Most of the 24-hour rainfall in extreme western Georgia and eastern Alabama fell between 0600-1200 GMT. In this case, the estimates appear to be more representative of the rainfall than the six-hour observed, and the estimates aided the meteorologists in their evaluation of the flash flood potential over Georgia and Alabama. This case is typical of our efforts. From considering only,those convective systems for which the technique was designed, our maximum rainfall estimate for a storm event over a particular county in the U.S.A. is typically within 25 percent of the maximum observed. False alarms are infrequent as we generally underestimate the maximum observed. A systematic verification methodology will be established when the technique is placed on a man-machine interactive computer; this should be accomplished in the very near future.

ESTIMATING RAINFALL FROM WARM TOPS

Currently, there is an experimental warm-top modification to the Scofield/Oliver technique which is designed to more adequately handle heavy rainfall from storms with warm tops (Scofield, *et al.,* 1980; Scofield, 1981). This modification uses the computed temperature at which the convection would stop rising instead of the tropopause temperature. This computed temperature is the best measure of the expected anvil temperature and is used for examining the anvil growth rates. Temperatures equal to or colder than the computed temperature would indicate heavier rainfall rates than warmer ones. The warm-top modification adjusts the estimates computed from the Scofield/Oliver technique by taking into account the computed temperature of the thunderstorm tops. Despite some successes in using this modification, there is still a need to improve our ability to recognize warm-top situations in the conventional data and in the satellite data. A list of features to look for in the conventional and satellite data, which are often associated with warm IR top-heavy rainfall events, is presented in Tables 1 and 2.

TABLE 1. Conventional Data.

1. Computed equilibrium level temperatures warmer than -62°C.
2. Tropopause temperatures warmer than -62°C.
3. A stable layer below the tropopause but above 300 mb.
4. A pronounced trough aloft with some reflection at the surface.
5. A synoptic-scale overrunning situation.

TABLE 2. Satellite Data.

1. Cloud-top temperatures are near or colder than the computed temperatures at which the convection should spread out.
2. A cyclonic circulation is detected in the imagery.
3. Cloud-top temperatures are near or colder than the temperatures of the stable layer located below the tropopause but above 300 mb.
4. Thunderstorms are stationary, growing rapidly, or merging.

FUTURE IMPROVEMENTS TO THE TECHNIQUE

This section presents some thoughts on how to improve the Scofield/Oliver technique.

Thunderstorms in Dry Environments or With High Bases

When thunderstorms occur in air masses much drier than the normal summer air masses of the eastern and southern United States, the computed rainfall is too great unless corrections are made. We have developed a "precipitable water factor" which is a correction scheme to be used when the air is dry. This procedure has been used for the past two years, but on occasion it still gives too heavy rains. Let us examine one of these storms.

Precipitation estimates were computed for a storm system in Texas and Oklahoma shown in Figure 8. The computations based on all the procedures resulted in estimates of nearly 5 inches in southern Oklahoma; only 2 inches were observed. In this case, the precipitable water (Figure 9) was not quite low enough to require this reduction factor and the temperature was unusually warm, giving a relative humidity (Figure 10) of only 50 percent. It appears likely that many of these types of storms occur in the summer in the drier part of the country. Therefore, we are trying to improve the technique by introducing a more adequate reduction factor for dry air. It will take the form of:

$$\frac{(\text{Obs Sfc} - 500 \text{ mb RH}) \text{ X } (\text{Obs Sfc} - 500 \text{ mb PW})}{1.5} \qquad (5)$$

where the factor is only used when the observed precipitable water is 1 inch or less and/or the relative humidity is less than 70 percent. Applying this factor (Equation 2) to the case in Figures 8, 9, and 10, a reduction factor of 0.36 is computed; the factor was obtained from the following ratio:

$$\frac{0.50 \text{ (Sfc 500-mb RH) x } 1.08 \text{ (Sfc} - 500 \text{ mb PW)}}{1.5}$$

$$= \frac{0.50 \text{x} 1.08}{1.5} = 0.36$$

Multiplying the 5-inch estimate by 0.36 reduces the estimate to 1.8 inches; this value is closer to the observed. In addition to modifying the precipitable water factor by the relative humidity derived from rawinsonde data, GOES Visible and IR Spin Scan Radiometer Atmospheric Sounder (VAS) data will be evaluated as to its usefulness in improving estimates from thunderstorms in dry environments or with high bases. As compared to moisture fields derived from rawinsonde data, GOES VAS data will improve the spatial and time resolution of the moisture distribution.

Location of Heavy Thunderstorm Rainfall Using IR Imagery

Locating all areas of heavy thunderstorm precipitation in an IR picture can be a difficult task. Usually the older and bigger a convective system becomes and the stronger

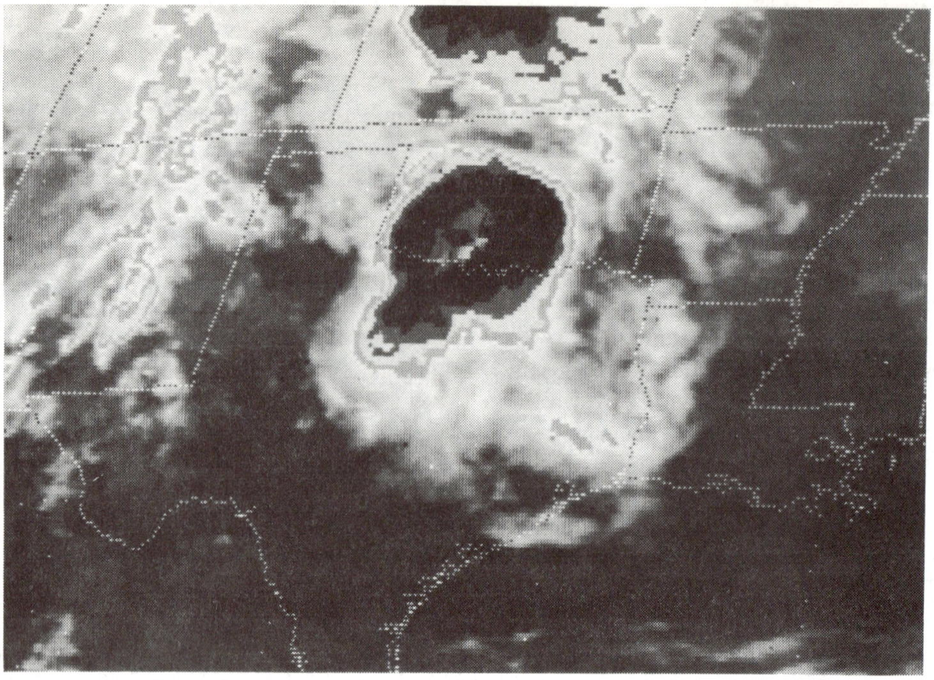

Figure 8. Enhanced Infrared Imagery (Mb curve), 0630 GMT, May 27, 1981.

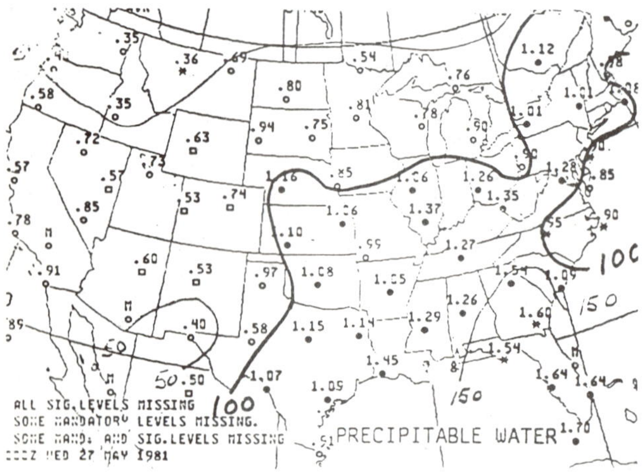

Figure 9. Surface to 500-mb Precipitable Water Analysis, 0000 GMT, May 27, 1981.

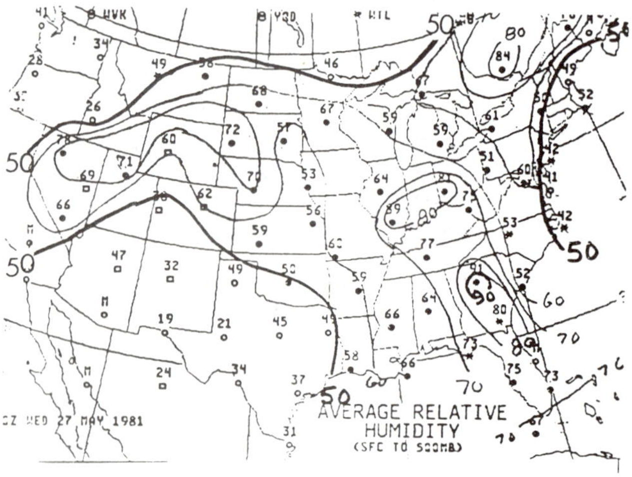

Figure 10. Surface to 500-mb Relative Humidity Analysis, 0000 GMT, May 27, 1981.

the vertical wind shear, the more difficult is the task to analyze the heavy thunderstorm rainfall areas in an IR picture. The convective system in Figure 11 is large and has been in existence for about six hours; vertical wind shear is also present. Heavy rainfall would be expected in the colder top areas located in the active upwind portion of the convective system between A-P. Rainfall analysis is difficult in the downwind portion between P-I (Figure 11). Past experience has shown that the downwind portion is often inactive (little or no rainfall). However, in this case, anvil debris spreading northeastward from the high-top thunderstorms located between A-P could be masking thunderstorms with lower tops in the downwind portion (between P-I). These lower top thunderstorms can produce heavy rainfall. Visible imagery in Figure 12 shows an area of overshooting tops from A-O-T. The radar summary chart (Figure 13) shows that the overshooting tops in the visible portion are associated with a line of thunderstorms from A-O-T.

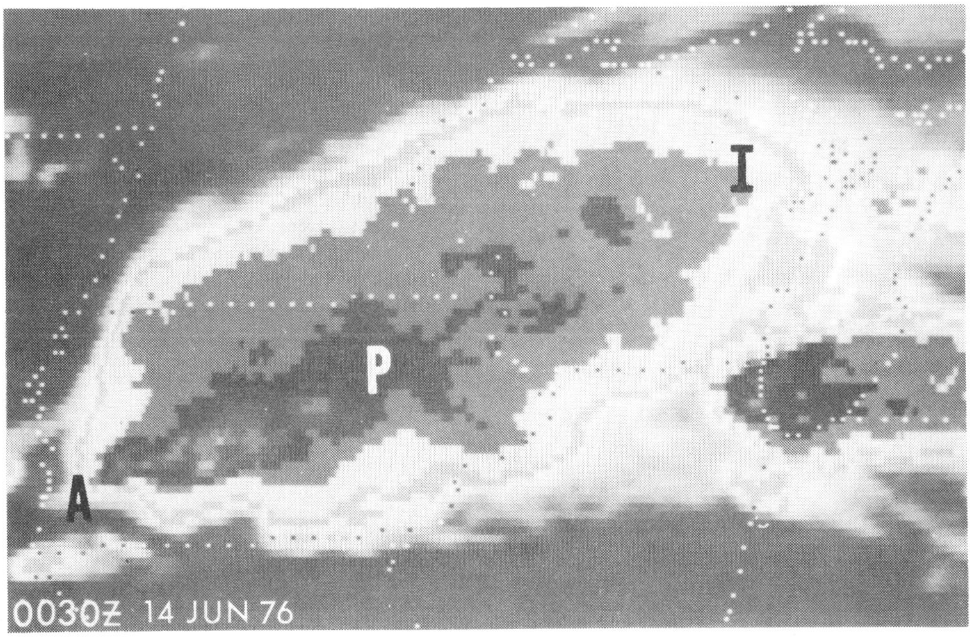

Figure 11. Enhanced Infrared Imagery (Mb curve), 0030 GMT, June 14, 1976.

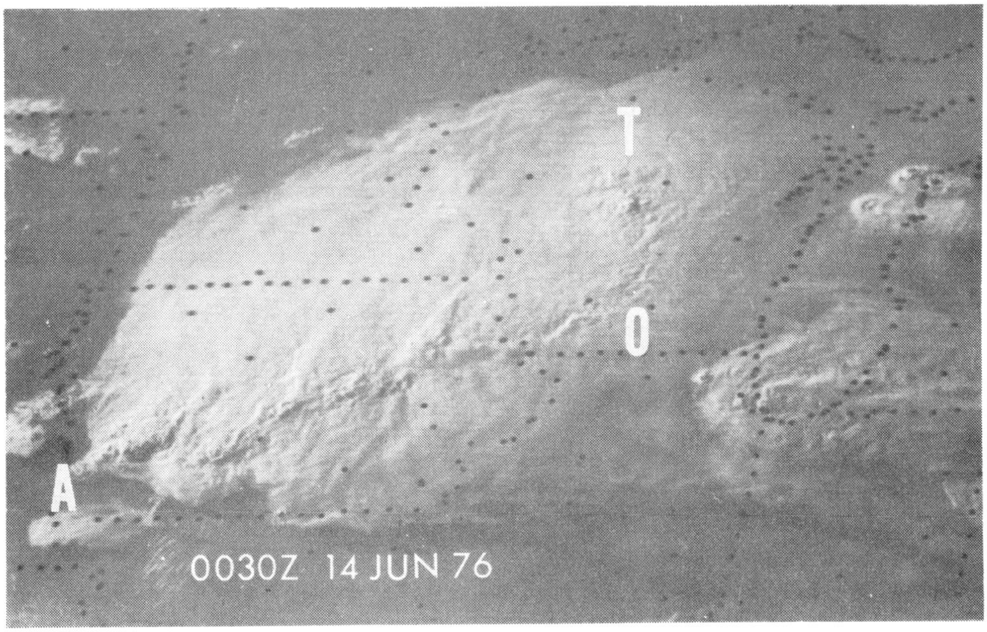

Figure 12. 1-km Visible Imagery, 0030 GMT, June 14, 1976.

The visible image and the radar data verify the existence of active thunderstorms between A-P in the IR picture (Figure 11). However, the visible picture and radar data complement the IR analysis by showing that the thunderstorms also exist in part of the downwind portion of the thunderstorm anvil.

In order to get a better analysis of the location of heavy thunderstorm rainfall in the IR imagery, four efforts need to be continued: (1) developing methods for superimposing radar data on IR pictures, (2) developing methods for superimposing visible data on IR pictures, (3) developing better enhancement curves which utilize the full resolution of the data, and (4) developing methods for using three-minute IR data in the analysis of overshooting tops and other mesoscale features.

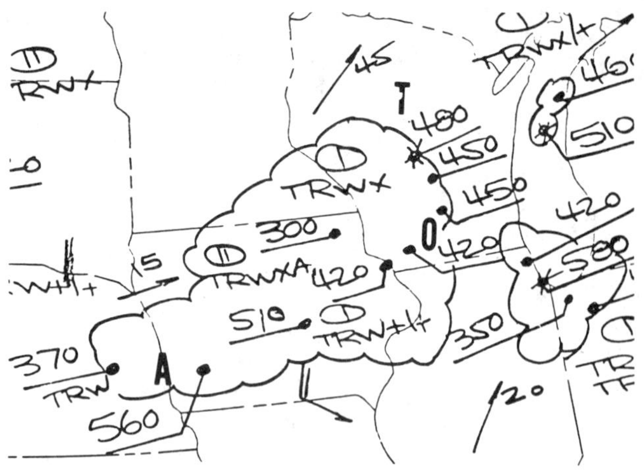

Figure 13. Radar Summary Chart, 0030 GMT, June 14, 1976.

CONCLUSIONS

In this paper, we have shown a real-time application of the Scofield/Oliver technique. This application was a good example of how the technique was used to help meteorologists and hydrologists locate areas of heavy rainfall in western Georgia and eastern Alabama for the issuance of flash flood warnings and to evaluate flood potential. Features to look for in the conventional and satellite data, which are often associated with warm IR top heavy rainfall events, are also presented. Also, we have discussed additional areas where we ,hope to improve the technique. These include estimating rainfall from thunderstorms in dry environments or with high bases and locating areas of heavy thunderstorm rainfall using IR imagery. We will continue to do research and development in these and other areas which will improve the Scofield/Oliver technique for operational use.

ACKNOWLEDGMENTS

The author would like to thank Vincent J. Oliver, NESS Applications Laboratory, for his constructive criticism in the presentation of the manuscript. I also thank John Shadid for the final draft of the figures and illustrations.

LITERATURE CITED

Oliver, V. J. and R. A. Scofield, 1976. Estimation of Rainfall From Satellite Imagery. Proceedings of Conference on Hydrometeorology, April 20-22, 1976, Ft. Worth, Texas, AMS; Proceedings of Sixth Conference on Weather Forecasting and Analysis, May 10-14, 1976, Albany, New York, AMS, pp. 242-245.

Scofield, R. A., 1978. Using Satellite Imagery to Detect and Estimate Rainfall From Flash Flood Producing Thunderstorms. Proceedings of Seventh Conference on Weather Forecasting and Analysis, October 16-19, 1978, Silver Spring, Maryland, AMS, pp. 132-141.

Scofield, R. A., 1981. Satellite-Derived Rainfall Estimates for the Brady's Bend, PA, Flash Flood. Proceedings of Fourth Conference on Hydrometeorology, October 7-9, 1981, Reno, Nevada, AMS, pp. 188-193.

Scofield, R. A. and V. J. Oliver, 1977. A Scheme for Estimating Convective Rainfall From Satellite Imagery. NOAA Technical Memorandum NESS 86, 47 pp.

Scofield, R. A. and V. J. Oliver, 1980. Some Improvements to the Scofield/Oliver Technique. Proceedings of the Second Conference on Flash Floods, March 18-20, 1980, Atlanta, Georgia, AMS, pp. 115-122.

Scofield, R. A., V. J. Oliver, and L. Spayd, 1980. Estimating Rainfall From Thunderstorms with Warm Tops in the Infrared Imagery. Proceedings of Eighth Conference on Weather Forecasting and Analysis, June 10-13, 1980, Denver, Colorado, AMS, pp. 85-91.

INTERNATIONAL SYMPOSIUM ON HYDROMETEOROLOGY
JUNE AMERICAN WATER RESOURCES ASSOCIATION 1982

FORECASTING CONVECTIVE PRECIPITATION BY TRACKING CLOUD TOPS IN GOES IMAGERY

John F. Moses[1]

ABSTRACT: Satellite meteorologists at the National Earth Satellite Service (NESS) have successfully used Geostationary Operational Environmental Satellite (GOES) imagery to locate and estimate half-hourly precipitation accumulations. Interactive processing techniques are presented which help meteorologists provide site-specific estimates within 30 minutes of the observation. Methods are presented which help satellite meteorologists provide a forecast based on tracks of convective cloud tops in GOES imagery.

Experience has shown that the propagation of outflow boundaries is often consistent over several hours when convective activity is supported by a moist, unstable atmosphere. A "normal persistance" method of forecasting was used to extrapolate the position and intensity of the convective activity resulting from an outflow boundary. Automatic precipitation estimates and convective cloud-top tracks were displayed on an interactive computer for an intense rainstorm on July 30, 1979. The speed and direction of thunderstorm outflow boundaries were determined and the point of intersection was predicted. The automatic precipitation estimates and man-machine interactive estimates compared favorably to rainfall reports from the dense automatic rain gauge network operated by the Illinois State Water Survey.
(KEY TERMS: satellites; rainfall estimates; thunderstorms; forecasting; interactive processing.)

INTRODUCTION

Estimates of precipitation provide valuable guidance to meteorologists and hydrologists who must issue flash flood and excessive precipitation watches and warnings. Forecasts of expected rainfall accumulation over several hours must be continually updated to ensure enough lead time for those affected to take appropriate action. To do this, National Weather Service (NWS) flood forecasters must make a consistent and accurate mesoscale interpretation of hourly surface, radar, and satellite observations.

In support of this effort, meteorologists at the National Earth Satellite Service (NESS) in Camp Springs, Maryland, will soon use a new computer system called the Interactive Flash Flood Analyzer (IFFA) to prepare estimates of precipitation every half-hour when convective storms have the potential for producing flash floods. Enhanced Geostationary Operational Environmental Satellite (GOES) images will be received and displayed at half-hourly intervals on the IFFA. Satellite meteorologists can quickly analyze individual convective cloud-top signatures and estimate the maximum half-hour precipitation using guidelines developed by Scofield (1982). The procedure consists of drawing isohyets of half-hour precipitation amounts on the screen of the IFFA terminal. These isohyets are then saved and summed by the computer to provide running total accumulation. The meteorologist will be able to analyze a mesoscale thunderstorm complex in approximately seven minutes and two or three thunderstorm areas in the half-hour period between GOES images (Moses, *et al.,* 1980). This procedure will provide forecasters with the amount of rain that has fallen but will not provide forecast information on expected rainfall during the next 1 to 2 hours.

This paper presents an example of a fully automatic precipitation estimation technique that can help meteorlogists forecast expected accumulation. First, an initial estimate covering a large number of thunderstorm complexes is made automatically by the computer in approximately one minute of computer time. Second, interactive mesoscale analyses and forecasting methods are applied which can help meteorologists forecast the position of the mesohigh boundaries associated with large thunderstorm systems. The future positions of these boundaries along with information about environments of these boundaries are used to estimate the potential for intense convection and heavy precipitation.

AUTOMATIC AND INTERACTIVE RAINFALL ESTIMATES

GOES infrared (IR) images are specially enhanced with repeat gray shades between -62°C and -80°C to show the structure of thunderstorm anvils in 1° increments for display on the video screen. Meteorologists locate the very coldest cloud tops in the active portion of an anvil and estimate a maximum half-hour accumulation for each cloud

[1]Physical Scientist, NESS Applications Laboratory, National Earth Satellite Service, Applications Laboratory, S/RE3, World Weather Bldg., Room 601 (Stop J), Washington, D.C. 20233.

top based on its evolution over the past half hour. The isohyets defining the area of heaviest rainfall (≥0.25 inch per half hour) are drawn. The placement of the isohyets takes into account the apparent motion of the cloud tops over the half-hour period between images. Experience has shown that these isohyets have widths of 15 miles or less (Scofield, 1978).

Measurable displacements have been observed between the cloud-top position and the position of maximum accumulation on the ground (Moses, 1980b). For the half-hour accumulations studied, the maximum rainfall was found to be displaced as much as 12 km from the cloud top. By studying signatures in the cloud-top structure and relative position in the anvil, the meteorologist can locate the areas of heaviest rainfall. In the early, intense stage of convection, a strong updraft will resist tropospheric wind shear and result in the rapid development of very cold tops, with the rainfall accumulating directly under and downwind of the cloud top.

Meteorologists using the interactive rainfall estimation technique on the IFFA will be able to analyze the vertical wind shear in the storm's environment and take into account the displacement of intense rainfall for more correctly positioning isohyets. An automatic rainfall estimation technique must be able to position rainfall under different areas of the cloud top according to measurements of the cloud-top structure and the thunderstorm environment.

In the procedure reported here, meteorologists can select the region of active convection. For cloud tops showing signatures of intense convection, the rainfall is estimated to fall directly under and to the east of the cloud-top area. The centroid of each convective cloud top is detected by the computer and is used as a reference position. The anvil around this reference position is segmented into octants so that the maximum rainfall can be distributed directly under and in the eastern octant of the cloud-top area. The cloud-top temperature, cloud-top shape, and cloud growth are automatically measured and used to estimate a maximum half-hour rainfall accumulation based on guidelines from Scofield and Oliver (Moses, 1980a).

TRACKING OUTFLOW BOUNDARIES

In many situations, the onset of intense convection can be predicted based on the intersection of thunderstorm outflow boundaries. Meteorologists can use GOES imagery to determine the position of thunderstorm outflow or meso-high boundaries which lead to the development of deep convection (Purdom, 1979).

Downdrafts in large mesoscale thunderstorm complexes are known to produce gust fronts. The strongest gust front associated with a thunderstorm complex can usually be found near the leading edge of the strongest IR observed temperature gradient ahead of the coldest tops (Gurka, 1976). Gust fronts and other outflow boundaries represent triggering mechanisms which can initiate the development of deep convection when they interact with an inflow of moist unstable air.

The automatic tracking of thunderstorm cloud tops is accomplished through searches for the coldest cloud-top positions in the current image which are within a fixed radius of the coldest cloud-top positions in the previous image. From the cloud-top positions, cloud-top vectors are produced which represent either (1) motion of the same convective cell, (2) outflow coupling between convective cells, or (3) upper-level winds. Models of the environmental wind profile and of the thunderstorm's motion help meteorologists interpret the cloud-top vectors (Moses, 1981).

Observations of cloud tops in consecutive half-hour images reveal that cloud-top vectors associated with outflows are maintained by convection and are embedded in a moist and unstable atmosphere, the speed and direction of the boundary is often consistent over several hours. By extrapolation, vectors which represent the motion of the outflow boundary and convective activity can be used to successfully predict the location of an intersection of boundaries and the onset of new intense convection.

A FORECASTING EXAMPLE

The rainfall event in Illinois on July 30-31, 1979, illustrates how IFFA could be used to forecast precipitation. At 2100 GMT on July 30, thunderstorms were developing along a cold front which extended from a surface low pressure center over Lake Superior through Wisconsin, eastern Iowa, Kansas, and Oklahoma. Over the next three hours, the storms moved ahead of the front to form a squall line extending from central Illinois through Missouri and Oklahome. The storms were propagating east-southeast in an environment where the winds veered with height and the upper-level winds (15 kts. at 500 mb) were slower than the mid-level winds (35 kts at 700 mb).

The enhanced IR image taken at 2230 GMT on July 30 (Figure 1) shows a cold-top thunderstorm just north of Springfield, Illinois (SPI), and another thunderstorm cluster east of Champaign, Illinois (CMI). A digital array of IR data which encompassed both storms was searched for cloud tops with temperatures equal to or colder than $-69^{\circ}C$. (This temperature is defined by the meteorologist, depending on meteorological conditions.) Two cloud tops in the thunderstorm cluster north of SPI were detected and numbered by the computer. In Figure 1 they are labeled No. 1 and No. 4 with their respective temperatures shown in parenthesis. Two cloud tops were also detected in the storm near CMI; only cloud top No. 2 is shown.

The cloud tops whose computer identification numbers are circled on Figure 2 were detected at 2300 GMT and independently numbered. Cloud top No. 4 found at 2230 GMT in the storm north of SPI was tracked by the

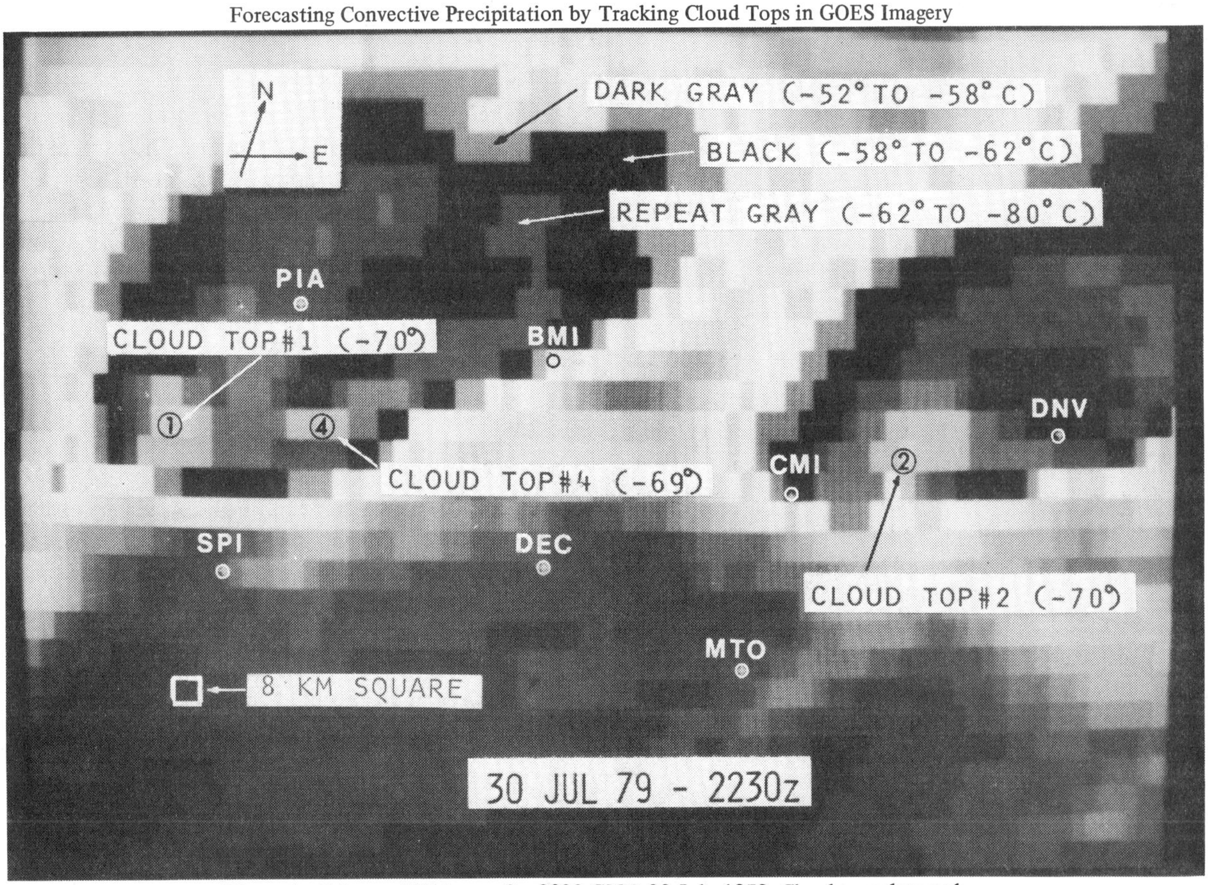

Figure 1. Enhanced IR Image for 2230 GMT, 30 July 1979. Cloud tops detected and numbered by the computer are indicated by circled numbers.

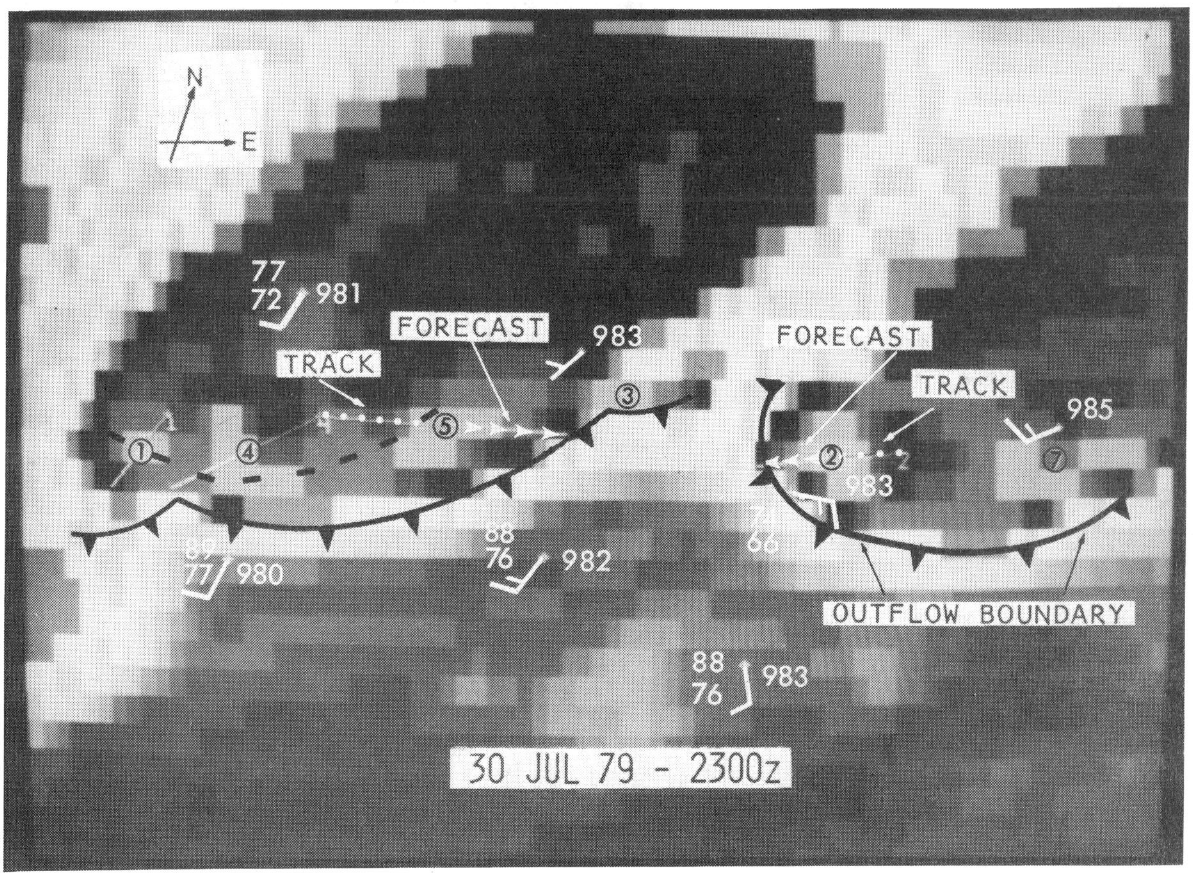

Figure 2. Enhanced IR Image for 2300 GMT, 30 July 1979. Cloud tops detected and numbered by the computer are indicated by circled numbers.

computer to cloud top No. 4 and cloud top No. 5 both found at 2300 GMT. The vector between cloud-top No. 4 and cloud top No. 5 is labeled "TRACK" and the extrapolation of that vector from cloud top No. 5 eastward is labeled "FORECAST." This vector has a heading of 97° and a speed of 14 km per half hour and is interpreted as representing the motion of convective activity resulting from the motion of the outflow boundary.

The outflow boundaries are positioned based on the tight IR temperature gradient ahead of the coldest cloud tops. The dashed line behind the boundary north of SPI and Decator, Illinois (DEC), shows the position of the outflow boundary the previous half hour. This boundary moved between 10 and 15 km to the south and east. The outflow boundary shown near CMI is positioned based on the IR temperature gradient and appears nearly stationary between 2230 and 2300 GMT. If the outflow boundaries near SPI and CMI continue their respective movements toward each other, they are likely to intersect within one hour.

The next IR image taken at 2330 GMT (Figure 3) shows that the outflow boundary of the storm near SPI has moved southeast another 15 km. New cloud-top areas appear along the previous outflow boundary. The thunderstorms in the cluster near CMI are still intense, and a strong, cool air outflow boundary is present. The boundaries are predicted to intersect about 15 km north-northwest of CMI (Figure 3) and result in intense convection and excessive rainfall along the boundaries. Precipitation estimates are computed and used with the predicted position of the outflow boundary for forecasting excessive amounts downstream.

PRECIPITATION ESTIMATES

The automatic half-hour estimates for 2330 GMT in Figure 4 show a maximum of 1.4 inches assigned to the thunderstorms in the thunderstorm cluster near CMI (cloud top No. 3, Figure 3). New growth of the cloud tops in the storm near CMI and western edge of the storm near SPI resulted in increases in the half-hour rainfall estimates. Forecasters can use these estimates as a guide to the intensity of rainfall expected over the next one or two hours during the interaction of outflows.

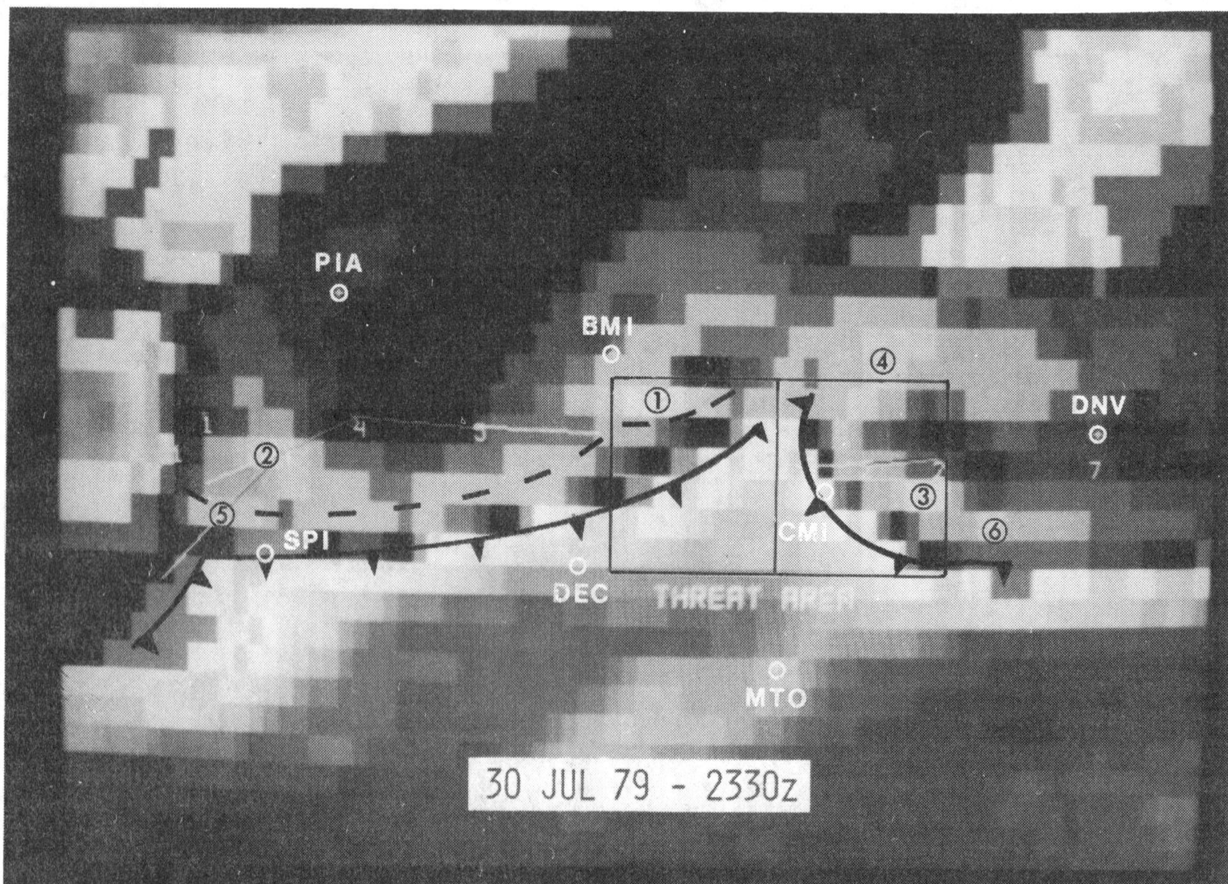

Figure 3. Enhanced IR Image for 2330 GMT, 30 July 1979. Cloud tops detected and numbered by the computer are indicated by circled numbers.

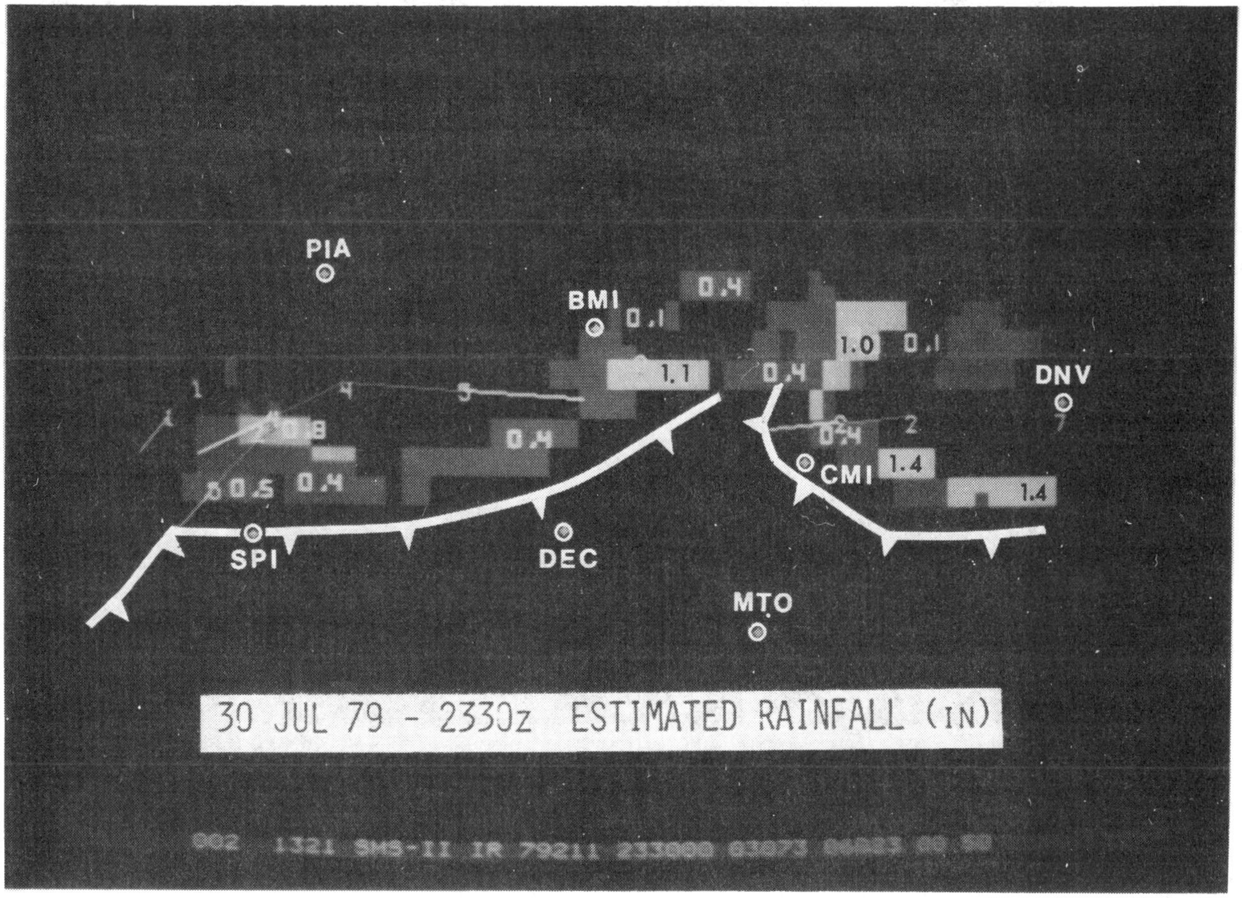

Figure 4. Automatic Satellite Derived Precipitation Estimates for the Half Hour Ending at 2330 GMT, 30 July 1979. The grey tone pattern in Figures 4 and 5 are identical and represent the Satellite derived rainfall amounts. The value of each gray shade is annotated in inches. The local weather service station are Bloomington (BMI), Decatur (DRC), Champaign (CMI), Mattoon (MTO), Danville (DWV), Peoria (PIA), and Springfield (SPI).

For this example, the accuracy of the half-hour estimate is verified through the use of an automatic recording rain gauge network operated by the Illinois State Water Survey. Figure 5 shows the half-hour observed accumulation ending at 2345 GMT for the 260 gauges in the network.

The maximum satellite estimate of 1.1 inches (Figure 4) over the north central part of the network (Figure 5) is derived from measurements of cloud top No. 1 (Figure 3). The observations show a distribution of maximum precipitation oriented in an east-west path under the thunderstorm top. For this intense thunderstorm, the maximum satellite estimate was extended east of the cloud top and corresponds well with the observed distribution.

The total rainfall accumulations using the automatic procedure during the 2½ hour period of outflow interaction are shown in Figure 6 and combined in Figure 7 with the network observations. Estimates using the interactive procedure were made by a satellite mteorologist for the same period and are shown in Figure 8. The precipitation patterns for the automatic and interactive estimates are similar; however, the interactive analysis is considerably less noisy and provides a much more realistic distribution of precipitation.

The two most significant maxima occurred at the northern and eastern edges of the network. The northern maximum of 3.2 inches is attributed to the stationary thunderstorm that remained over the area for approximately 90 minutes. A maximum rainfall of 4.6 inches was observed at the eastern edge of the network and is part of a larger area of maximum accumulations reaching seven inches (storm totals) 12 km southeast of CMI (Changnon and Vogel, 1980). This maximum is attributed to several intense raincells which initiated continuously over the eastern part of the network in the small area where the outflow boundaries intersected. These cells intensified as they moved off the network and merged into the Champaign thunderstorm cluster.

The interaction between the outflow boundaries is believed to have directly initiated these cells and subsequently enhanced the precipitation mechanism in the

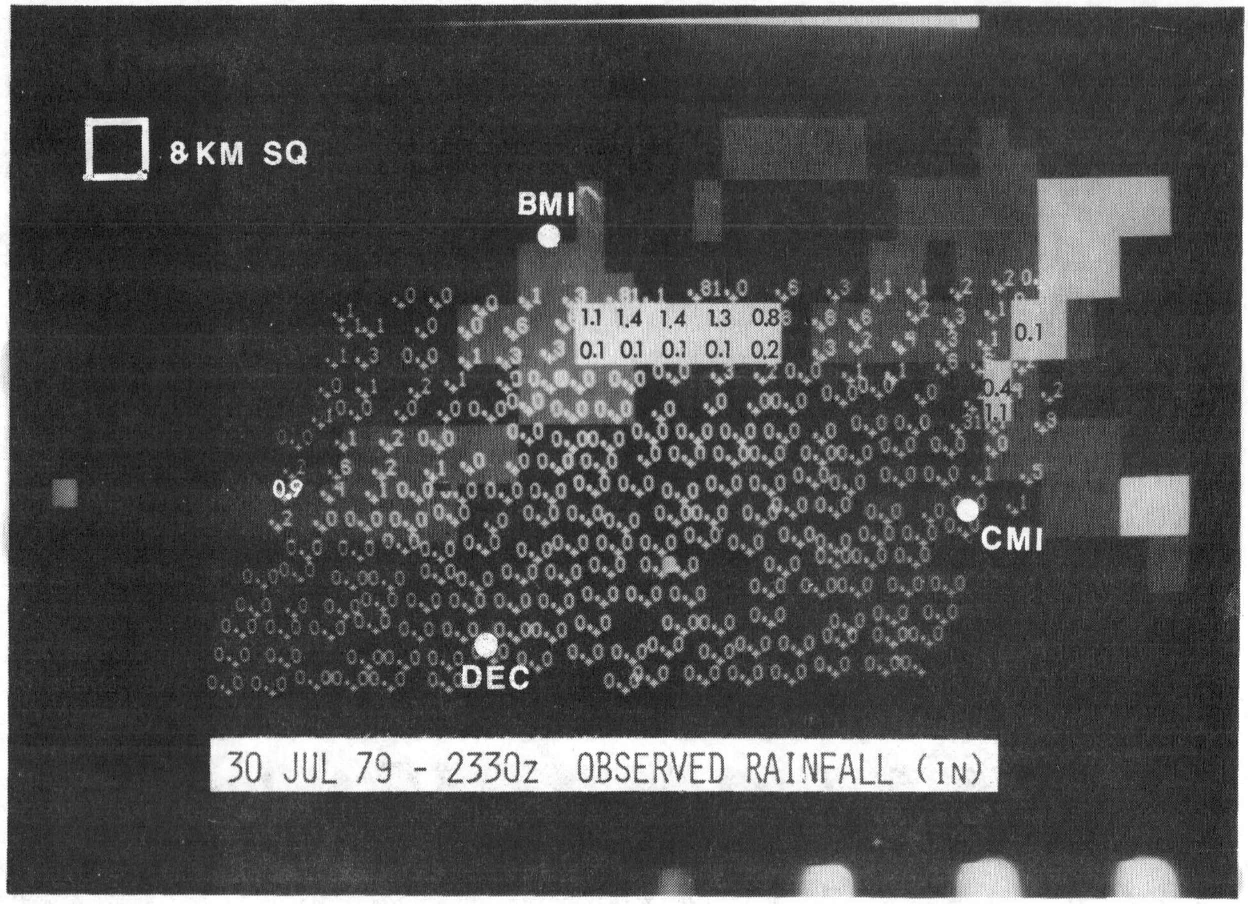

Figure 5. Reports From the Automatic Rain Gauge Network Superimposed Over an Enlarged Portion of the Automatic Precipitation Estimates Shown in Figure 4. Reports every five minutes are summed for the half-hour period ending 2345 GMT, 30 July,1979.

Champaign cluster. *Storm Data and Unusual Weather Phenomena* (1979) reports that hailstones of 1-inch in diameter were observed at about 0100 GMT on July 31 in southeastern Urbana (5 km east of CMI). In addition, the heavy rainfall from these thunderstorms resulted in flash flooding in the Champaign-Urbana area.

The automatic technique underestimated the maximum accumulations during this period because heavy rainfall occurred where the boundaries intersected and thunderstorms merged under cloud-top areas which did not exhibit significant growth or overshooting in the anvil. Consequently, the meteorologist was able to obtain a better distribution of maximum precipitation using the interactive technique.

SUMMARY

Automatic tracking techniques have been used to measure the distance from a thunderstorm top in the current GOES image to the nearest thunderstorm top in the previous half-hour image. Cloud tops continuously regenerated as cells tracked along thunderstorm outflow boundaries. When this process is maintained by the atmosphere, the distance measured can be used by meteorologists to forecast the intersection of outflow boundaries and the potential for excessive rainfall. At about 0000 GMT on July 31, 1979, the intersection of outflow boundaries resulted in the initiation of several intense raincells which merged into existing thunderstorms and caused excessive precipitation and flash flooding in Champaign-Urbana, Illinois.

This fully automatic precipitation estimation technique demonstrates the feasibility of providing an initial field of estimates to the satellite meteorologist who must quickly look for areas of intense convective rainfall that might result in flash flooding. The automatic technique has the unique capability of detecting, segmenting, and tracking individual convective cloud tops so that precipitation can be radially distributed under any part of the cloud according to cloud-top structure and environmental wind conditions.

Meteorologists can use the automatically produced estimates as a first guess, which can be edited, fine-tuned, and adjusted before being used operationally. A preliminary

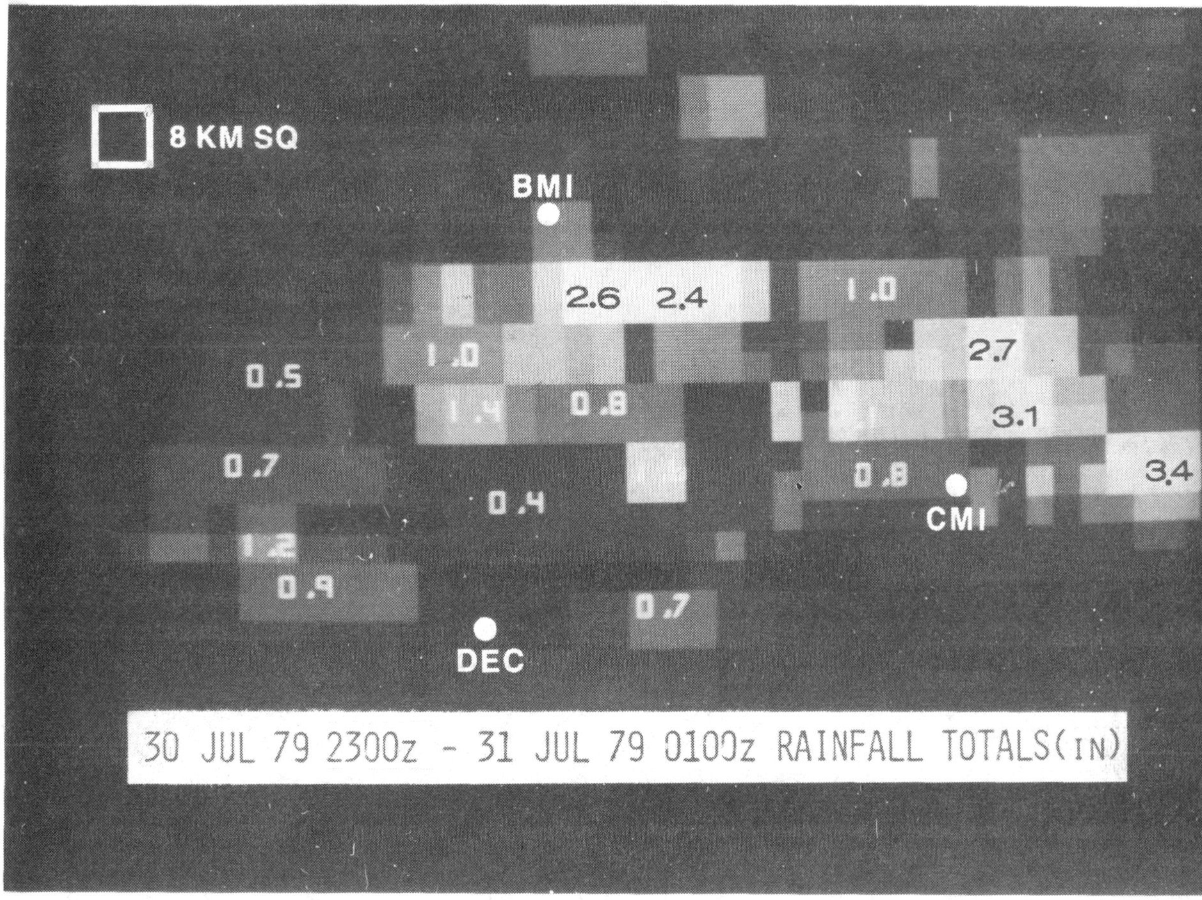

Figure 6. Automatic Precipitation Estimates for the 2½-Hour Period Ending at 0100 GMT, 31 July 1979. The gray tone pattern in Figures 6 and 7 are identical and represent satellite derived rainfall amounts. The value of each gray shade is annotated in inches. Brighter shades refer to larger accumulations. The local weather service stations are Bloomington (BMI), Decator (DRC), and Champaign (CMI).

version was applied to the intense rainfall event which occurred on July 30, 1979, over the dense rain gauge network operated by the Illinois State Water Survey in Central Illinois. For comparison, meteorologists prepared interactive estimates based on the same GOES imagery. Results show that the automatic estimate exhibits the same major features as the interactive estimate but the overall distribution is considerably less realistic. The automatic technique frequently measured less cloud growth than the growth computed subjectively. As a result, the automatic technique produced slightly lower estimates.

LITERATURE CITED

Changnon, S. A. and J. L. Vogel, 1980. The Morphology of an Isolated Severe Rainstorm and Hydroclimatological Characteristics of Such Storms. Second Conference on Flash Floods, March 18-20, 1980, Atlanta, Georgia, AMS, pp. 5-10.

Gurka, J. J., 1976. Satellite and Surface Observations of Strong Wind Zones Accompanying Thunderstorms. Monthly Weather Review 104(12):1484-1493.

Moses, J. F., 1980a. Interactive Techniques for the Estimation of Precipitation From Geostationary Images. Proceedings of the Second Conference8on Flash Floods, March 18-20, 1980, Atlanta, Georgia, AMS, pp. 101-108.

Moses, J. F., 1980b. Numerical Methods for the Analysis of Satellite Rainfall Estimates. Proceedings of the Eighth Conference on Weather Forecasting and Analysis, June 10-13, 1980, Denver, Colorado, AMS, pp. 101-107.

Moses, J. F., 1981. Convective Cloud-Tracking Techniques for Estimating Precipitation From Geostationary Imagery. Proceedings of the Conference on Hydrometeorology, October 7-9, 1981, Reno, Nevada, AMS, pp. 151-158.

Moses, J. F., L. E. Spayd, R. S. Gird, and A. L. Siebers, 1980. Interactive Techniques for Estimating Precipitation From GOES Imagery. Joint Automatic Control Conference, Charlottesville, Virginia, ASME, TP-6D.

National Oceanic and Atmospheric Administration, 1979. Storm Data, Vol. 21, June. National Climatic Center, Asheville, North Carolina, 30 pp.

Purdom, J. F. W., 1979. The Development and Evolution of Deep Convection. Proceedings of the Eleventh Conference on Severe Local Storms, October 2-5, 1979, Kansas City, Missouri, AMS, pp. 143-150.

Scofield, R. A., 1978. Using Satellite Imagery to Detect and Estimate Rainfall From Flash Flood Producing Thunderstorms. Proceedings of the Conference on Weather Forecasting and Analysis and Avaiation Meteorology, Silver Spring, Maryland, AMS, pp. 132-142.

Scofield, R. A., 1983. A Satellite Technique for Estimating Rainfall From Flash Flood Producing Thunderstorms. Proceedings of the International Symposium on Hydrometeorology, Denver, Colorado, AWRA-AMS, pp. 121-128.

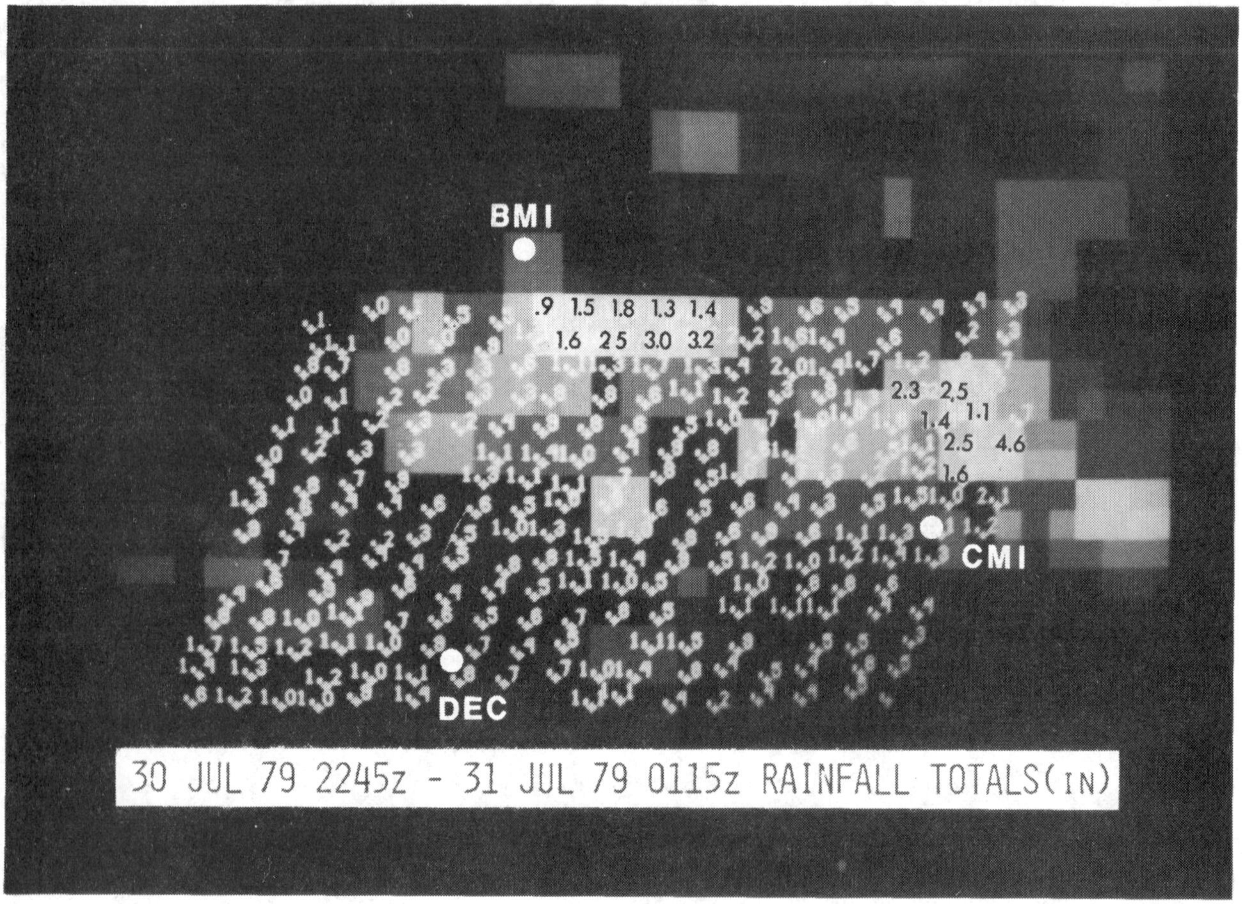

Figure 7. Reports From the Automatic Rain Gauge Network Superimposed Over the Same Area as Figure 6. The gray tone pattern is identical to Figure 6 and represents the satellite derived rainfall with the brighter shades referring to larger estimated accumulations. The local weather service stations are Bloomington (BMI), Decatur (DRC), and Champaign (CMI).

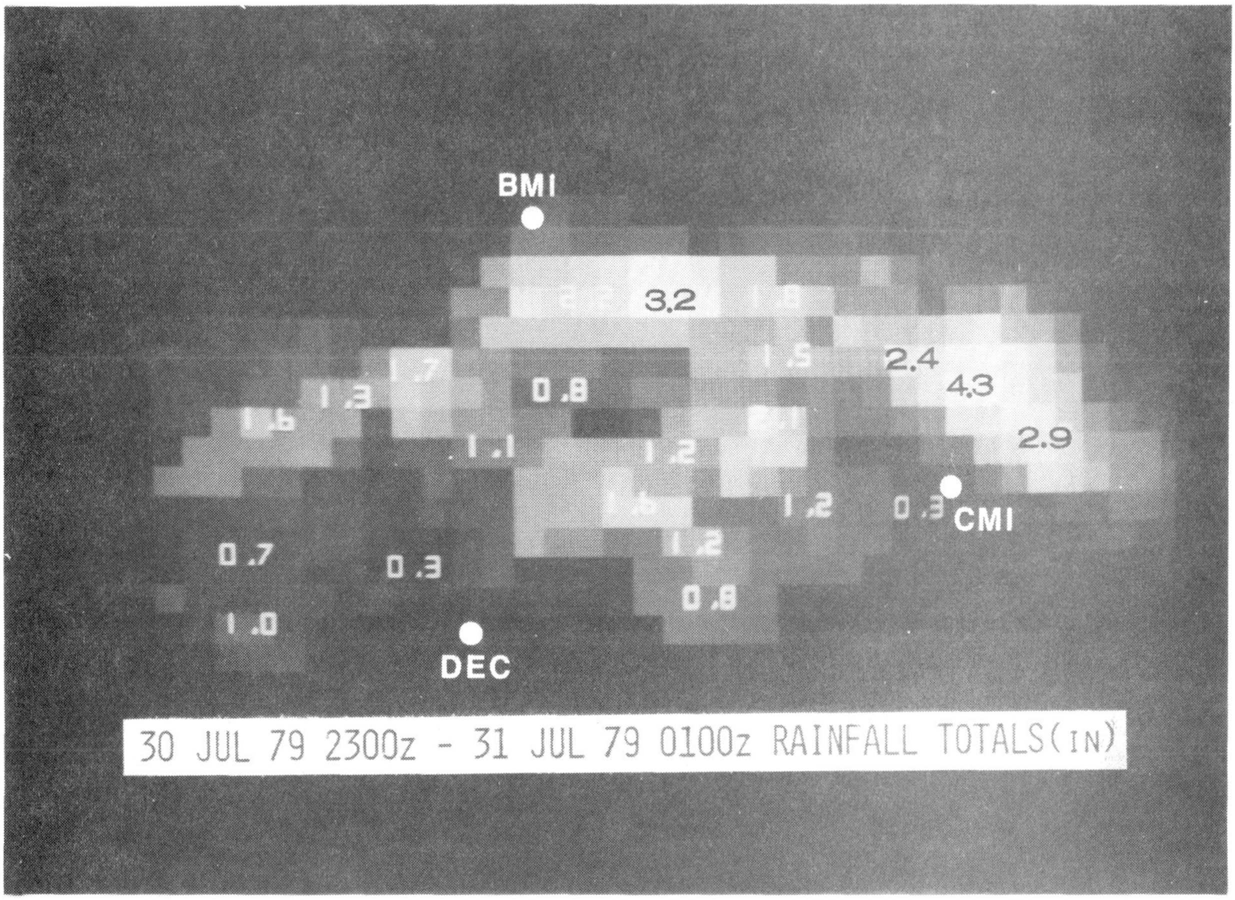

Figure 8. Interactive Estimates for the 2½-Hour Period Ending 0115 GMT, 31 July 1979.
The value annotated for the gray shades are accumulations in inches.

INTERNATIONAL SYMPOSIUM ON HYDROMETEOROLOGY
JUNE AMERICAN WATER RESOURCES ASSOCIATION 1982

ESTIMATING RAINFALL USING SATELLITE IMAGERY FOR WARM-TOP THUNDERSTORMS EMBEDDED IN A SYNOPTIC-SCALE CYCLONIC CIRCULATION

LeRoy E. Spayd, Jr.[1]

ABSTRACT: This paper summarizes several case studies of flash flood producing warm-top thunderstorms embedded in synoptic-scale cyclonic circulations which are easily recognized in the satellite imagery. These synoptic-scale cyclonic circulations are associated with a strong vorticity center at 500 mb and the warm-top thunderstorms occurred north of the 850 mb and surface low centers. The warm-top thunderstorms embedded in these weather systems, once recognized, can be used to make operational rainfall estimates. The estimates are made from consecutive visible and infrared satellite imagery using the Scofield/Oliver convective technique modified to accommodate thunderstorms with tops warmer than -62°C.
(KEY TERMS: satellites; rainfall estimation; warm-top thunderstorms; flash flooding; cyclonic circulation.)

INTRODUCTION

This paper discusses characteristics of flash flood producing warm-top thunderstorms embedded in synoptic-scale cyclonic circulations as observed from Geostationary Operational Environmental Satellite (GOES) data. Examples of these flash flood producing warm-top thunderstorms are shown along with rainfall estimates computed using the Scofield/Oliver convective rainfall estimation technique. A composite of the conventional surface and upper air data observed with these storms is also presented.

SCOFIELD/OLIVER CONVECTIVE TECHNIQUE

Scofield and Oliver (1977, 1980) developed a decision tree technique for estimating precipitation from deep convective systems using visible and enhanced infrared (IR) GOES imagery. The half-hourly or hourly changes in the satellite-observed thunderstorm tops are related to the rainfall rates by the following equation:

$$P = (TG + O + M + S) \times (PW/1.5)$$

where P is the half-hourly rainfall rate in inches; TG, a cloud-top temperature and growth factor; O, an overshooting top factor; M, a cluster or line merger factor; S, a saturated environment factor; and PW, the observed surface to 500-mb precipitable water for the thunderstorm's environment.

The original technique was designed for deep convective systems occurring in tropical air masses with high, cold summer tropopauses (below -62°C). Scofield, *et al.* (1980), proposed a method using conventional upper-air sounding data which modifies the technique to more adequately handle thunderstorms with tops warmer than -62°C. These warm-top thunderstorms occur when the tropopause is lower and warmer than the original technique was designed for or when there is a stable upper atmosphere in which the convective tops do not reach the tropopause. To account for this sub-tropopause stabilization in the environment lapse rate that causes the convection to cap at a warmer level, the warm-top modification to the technique uses the expected temperature of the convective tops as derived from the upper air soundings. To compute this expected temperature from an upper air sounding, a parcel of air at the surface must be lifted along the dry adiabat until the parcel becomes saturated and along the moist adiabat until it becomes colder than the environment. The expected cloud-top temperature represents the point the parcel temperature and the environment temperature are the same. Satellite-observed thunderstorm anvil temperatures that are equal to or colder than the expected cloud-top temperature indicate that the thunderstorms are reaching the capping stable layer and, therefore, should produce more rainfall than the original technique assigned to the thunderstorms.

CURRENT OPERATIONAL SATELLITE RAINFALL ESTIMATES

The Synoptic Analysis Branch (SAB) of the National Earth Satellite Service (NESS) has been using the Scofield/Oliver technique operationally for computing precipitation estimates since 1978. SAB's real-time precipitation estimates are sent to the National Weather Service and are used by the Forecast Offices and River Forecast Centers to assist

[1]Meteorologist, National Earth Satellite Service, Applications Lab, World Weather Bldg., Room 601, Stop J, Washington, D.C. 20233.

them in evaluating a storm's flash-flood potential. The satellite meteorologist monitors the current satellite imagery for known cloud-top signatures associated with heavy rainfall, such as the thunderstorm anvils either expanding or becoming colder with time, and quasi-stationary thunderstorms or thunderstorms regenerating from one location and transversing the same area. To date, the satellite meteorologist's attention has been directed first to thunderstorm tops colder than -62°C since these are most easily identified in operationally produced IR GOES imagery. Heavy rain-producing warm-top thunderstorms often go unnoticed in the operational environment. Since synoptic-scale cyclonic circulations are easily recognized in the satellite imagery, the stationary warm-top thunderstorms embedded in these weather systems can be used to make operational rainfall estimates.

SOME CHARACTERISTICS OF HEAVY RAINFALL EVENTS FROM WARM-TOP THUNDERSTORMS EMBEDDED IN A CYCLONIC CIRCULATION

Data and calculations from five flash flood producing warm-top thunderstorms are shown in Table 1. All of these events did produce flash flooding and were associated with a synoptic-scale cyclonic circulation. Table 1 summarizes these events: Column 1 shows the time, date, and location of the event; column 2, the observed rainfall nearest to where the flood occurred; column 3, the expected temperature of the convective tops as determined by the nearest upper-air sounding in space and time to the event; column 4, the temperature of the thunderstorm anvil as measured by the IR sensor; column 5, the reported tropopause from the upper air soundings; column 6, the temperature of the sub-tropopause stabilization in the environmental lapse rate; and column 7, the maximum height of the convective tops as reported in the hourly MDR radar summaries during the flash flood event. Some brief facts about each flash flood event follows.

Southwestern Indiana, 27 May 1981

During this heavy rainfall event, the satellite meteorologist was analyzing two areas of cold-top thunderstorms: one along the Texas-Oklahoma border and one along the Kansas-Nebraska border as seen on the 0730 GMT, 27 May 1981, image (Figure 1). These thunderstorm systems did not produce any significant flooding because they were moving rapidly and were embedded in a relatively dry environment with surface to 500-mb mean relative humidities of 45 to 60 percent. The flash flooding on this day occurred in southwestern Indiana north of point A. In the enhanced IR imagery, the tops of these quasi-stationary thunderstorms reach into the dark gray range (-52° to -58°C). The cyclonically curved cloud bands extending from B to A and C to A on the image indicate a synoptic-scale cyclonic circulation.

TABLE 1. Data and Calculations From Five Flash Floods Producing Warm-Top Thunderstorms.

Time, Date, Location	Observed Rainfall (unofficial)	Expected Temperature[1]	Anvil Temperature[2]	Tropopause[3]	Sub-Tropopause Lapse-Rate Stabilization[4]	Radar Tops[5] (ft)
27 May 1981 00-12 GMT Southwestern Indiana	5.83" (6"+)	-52°C	Dark Gray (-52° to -58°C)	-52°C	N/A	27,000-34,000
13 July 1979 00-17 GMT Nashville, Indiana	5.3" (10-15")	-42°C	Light Gray (-41° to -52°C)	-68°C	-52°C	N/A
17 August 1981 12-21 GMT Caribou, Maine	6.9"	-48°C	Light Gray (-41° to -52°C)	-48°C	N/A	32,000-37,000
6 May 1981 00-15 GMT Mobile, Alabama	8.0" (16.01")	-34°C	Medium Gray (-32° to -41°C)	-74°C	-32°C	36,000
20 May 1982 21-00 GMT Parkston, South Dakota	(8.5")	-51°C	Dark Gray (-52° to -58°C)	-57°C	-51°C	33,000

[1] Expected temperature of the convective tops computed from the nearest sounding prior to the event.
[2] Coldest thunderstorm anvils as observed in the MB IR enhancement curve.
[3] Reported tropopause temperature from the nearest sounding prior to the event.
[4] Temperature where the sounding lapse rate abruptly becomes more stable than the moist adiabat.
[5] Highest radar tops observed in hourly MDR data during the heavy rainfall period.

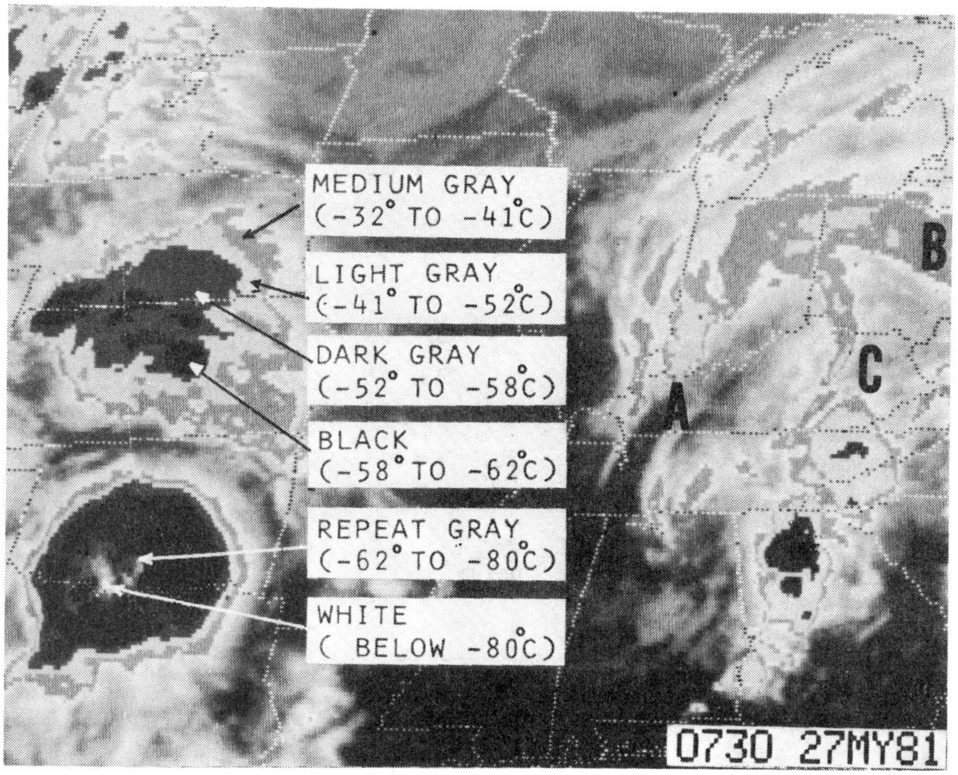

Figure 1. Enhanced IR Images (Mb curve), 0730 GMT, 27 May 1981. The temperature ranges of the Mb curve are annotated and curved cyclonic bands can be seen from B to A and C to A.

Nashville, Indiana, 13 July 1979

This flash flood event was responsible for the deaths of two persons and virtually isolated the town of Nashville, Indiana. Heavy rains, unofficially estimated at 10 to 15 inches, caused damage to three dams and produced the biggest flood in 60 years.

Caribou, Maine, 17 August 1981

Heavy rains totaling nearly 7 inches caused extensive damage to homes, businesses, bridges, roads, and crops in and around Caribou. This rainfall broke the all-time record for the most precipitation in 1 day and in 1 month in Caribou. As seen in Figures 2 and 3, these thunderstorms were embedded in a strong cyclonically curved band from D to D′. While this event was occurring, the operational satellite meteorologist was concerned with tropical storm Dennis, which was dropping 20 to 30 inches of rain over southern Florida. The thunderstorms in southern Florida had tops near $-80^{o}C$ while the thunderstorm tops over Caribou never got colder than $-52^{o}C$. These warm tops were capped by a very low tropopause of $-48^{o}C$ as shown by the Caribou sounding at 1200 GMT, 17 August 1981 (Figure 4).

Mobile, Alabama, 6 May 1981

A sequence of enhanced IR images in Figure 5 shows the thunderstorms which caused the worst flash flooding in Mobile's history. Damage in excess of 35 million dollars was caused by these thunderstorms. In the 0100 GMT image, the small medium-gray area (-32° to $-41^{o}C$) annotated by an F indicates the thunderstorm tops. This area expands with time and moves to the north while new thunderstorms regenerate at the same location. A radar meteorologist would term this type of development as the "train echo" effect. In the atmospheric sounding taken at Boothville, Louisiana, at 0000 GMT, 6 May 1981 (Figure 6), a subtropopause stabilization in the environmental lapse rate begins at $-32^{o}C$. The expected temperature of the convective tops is computed at $-34^{o}C$, which corresponds well to the observed cloud tops between -32° to $-41^{o}C$. The tropopause was reported at $-74^{o}C$. Rainfall estimates computed after the fact yielded 2.95 inches using the basic Scofield/Oliver convective technique and 6.7 inches using the new warm-top modification (Table 2). The 6.7 inches of rain estimated is fairly close to the official 8.0 inches that was observed but is still short of the unofficial reports of 16 inches in the area. However, if this 6.7 inches of rain had been estimated in real time, it would have alerted forecasters that potentially very heavy rainfall was occurring in the Mobile area.

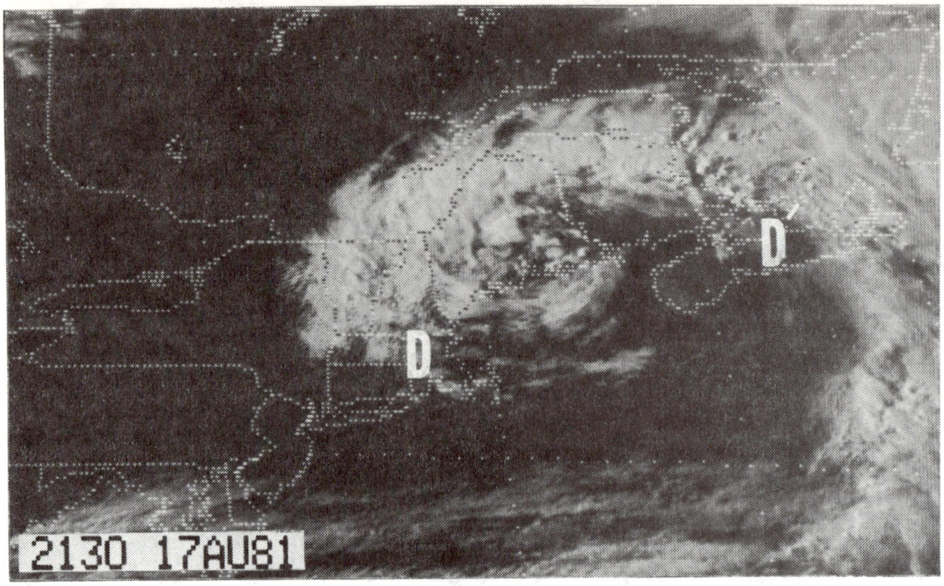

Figure 2. Visible Imagery, 2130 GMT, 17 August 1981. A curved cyclonic band can be seen from D to D′.

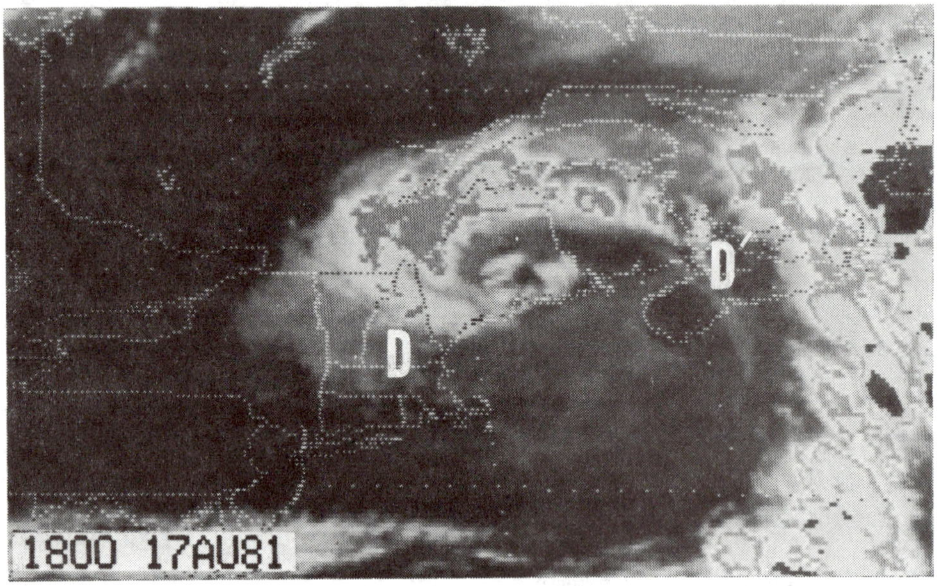

Figure 3. Enhanced IR Imagery, 1800 GMT, 17 August 1981. A curved cyclonic band can be seen from D to D′.

Parkston, South Dakota, 20 May 1982

In this very recent case, an unofficial report of 8.5 inches of rain was observed to fall between 2130 and 2330 GMT. The thunderstorms responsible for this heavy rainfall are indicated by an E on the visible and IR images (Figures 7 and 8). In the IR images, the thunderstorm tops grew and expanded to a temperature about -54°C and overshooting tops and cell mergers were observed. The atmospheric sounding taken at Omaha, Nebraska, at 0000 GMT, 21 May 1982, indicated a sub-tropopause environmental lapse rate change at -51°C and tropopause of -57°C. In this case, the satellite meteorologist was analyzing an area of thunderstorms in the Midwest and Appalachians where cloud-top temperatures were colder than -60°C. After-the-fact rainfall estimates using the warm-top modification to the technique yielded 4.40 inches in a 2½-hour period in the Parkston area.

COMPOSITE OF EVENTS

Atmospheric soundings nearest in space and time to the flash flood producing warm-top thunderstorms were

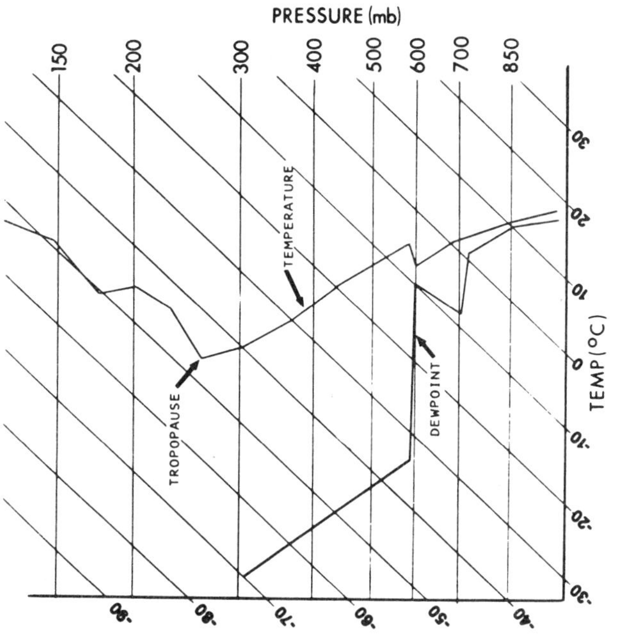

Figure 4. Caribou, Maine, Sounding 1200 GMT, 17 August 1981.

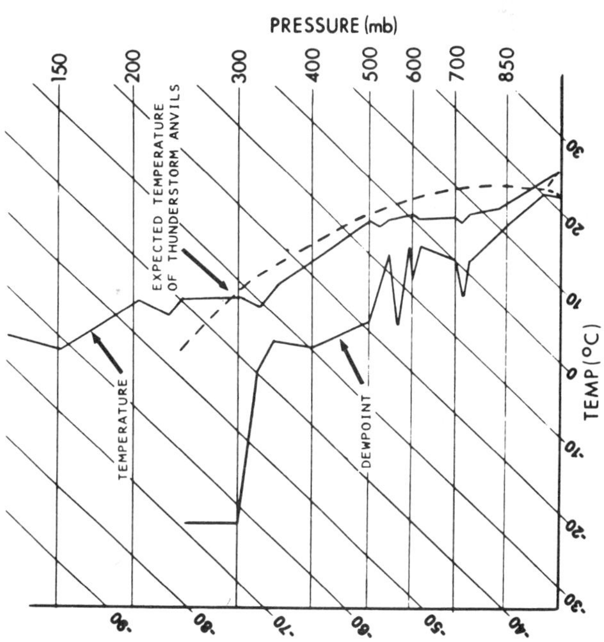

Figure 6. Boothville, Louisiana, Sounding 0000 GMT, 6 May 1981.

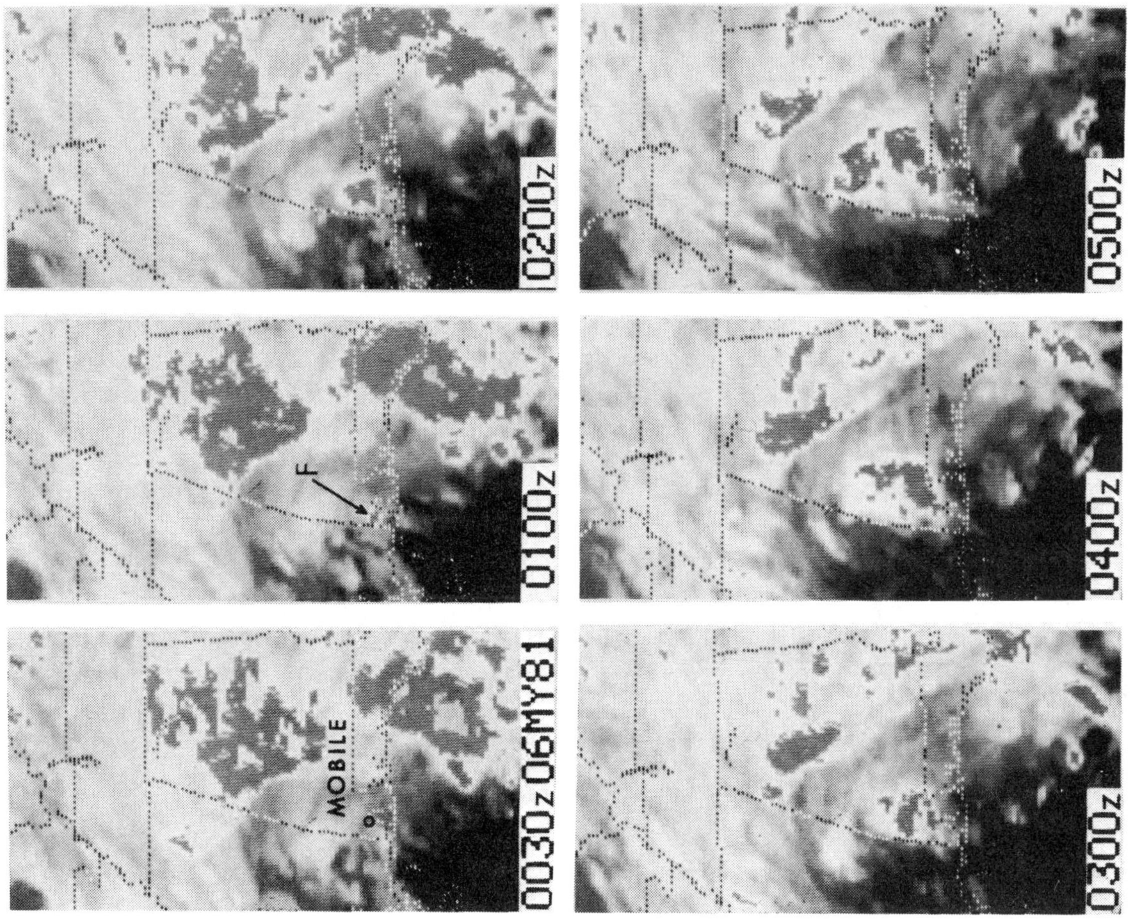

Figure 5. A Sequence of Enhanced IR Imagery (Mb curve) 0030, 0100, 0200, 0300, 0400, and 0500 GMT, 6 May 1981. The warm-top thunderstorms are indicated by an F on the 0100 GMT image.

TABLE 2. Scofield/Oliver Estimates With Warm-Top Modification.

Time (GMT)	S/O* Estimates	S/O* Estimates With Warm-Top Modification	Observed
0030-0100	0.10	0.40	
0100-0200	0.45	1.40	
0200-0300	0.80	1.60	
0300-0400	1.10	2.40	
0400-0500	0.50	0.90	
Total	2.95	6.70	*24-hr* 8.00 *unofficial* 16.01

*Scofield/Oliver Convective Technique.

Figure 7. Visible Imagery, 2230 GMT, 20 May 1982. The flash flood producing thunderstorms are annotated by an E.

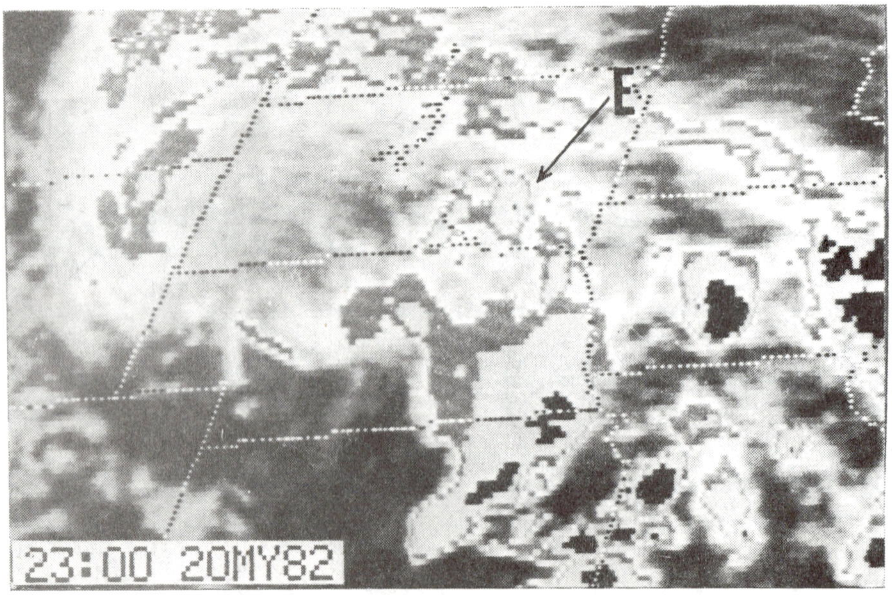

Figure 8. Enhanced IR Imagery, 2300 GMT, 20 May 1982. The flash flood producing thunderstorms are annotated by an E.

compiled. The overall synoptic situation for each event showed distinct similarities between these events. A composite of the 500-mb vorticity analysis, 850 mb, surface, and atmospheric soundings for each of these five flash flood events are shown in Figures 9 to 12. For each of these composites, the boxed area indicates the area where the warm-top thunderstorms occurred. The 500-mb vorticity chart (Figure 9) shows a fairly intense vorticity center cut off from the main jet stream or occurring in the southern part of a split jet stream pattern. The vorticity center is always moving in an easterly or northeasterly direction at a relatively slow speed of about 2° latitude per 12 hours, and the flash-flood area occurs 2 to 3 degrees of latitude to the east, northeast, or north of the vorticity center.

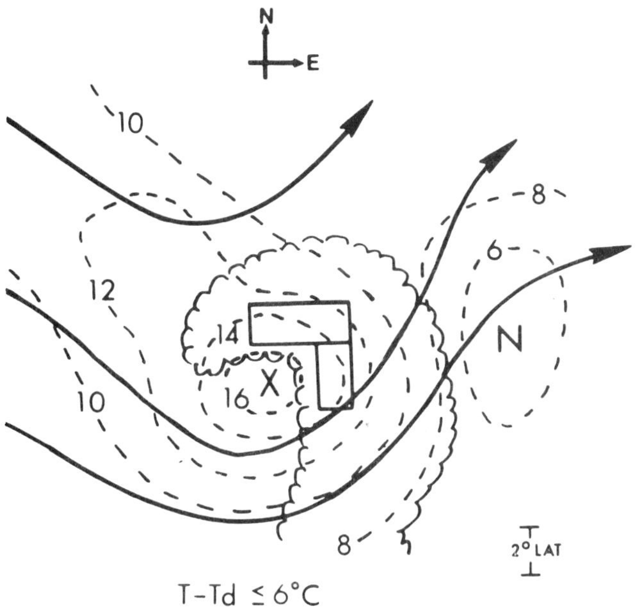

Figure 9. A Composite of the 500-mb Vorticity Chart for the Warm-Top Thunderstorms Embedded in a Synoptic-Scale Cyclonic Circulation. Vorticity isopleths are dashed lines; streamlines are solid lines; and low dewpoint depression areas ($\leqslant 6^{\circ}C$) are scalloped lines.

The warm-top thunderstorms generally formed about 2 to 3 degrees of latitude to the north or northeast of the 850-mb low (Figure 10). An isodrosotherm analysis shows a maximum of moisture wrapping around to the north of the low. As evident in these case studies, the warm-top flash flood area is just to the north and west of the maximum winds at 850 mb; this is different from the flash flood composites produced by Maddox, *et al.* (1979), which had the strongest 850-mb winds pointing into the flash flood area. The warm-top thunderstorms formed about 1° of latitude to the north of the surface low center or surface pressure trough axis in an area of northeasterly winds and average surface temperatures of $68^{\circ}F$. In the surface pressure field, two or three closed 2-mb isobars can be drawn (Figure 11). The events that occurred with the more intense vorticity centers at 500 mb had identifiable surface cold and stationary fronts. When this circulation aloft was weaker, no surface fronts were evident. The average surface to 500-mb mean relative humidity for these events was 83 percent with the precipitable water averaging 1.34 inches. The average winds veered over 180° with height from the surface to 200 mb with wind speeds relatively light at all levels (Figure 12).

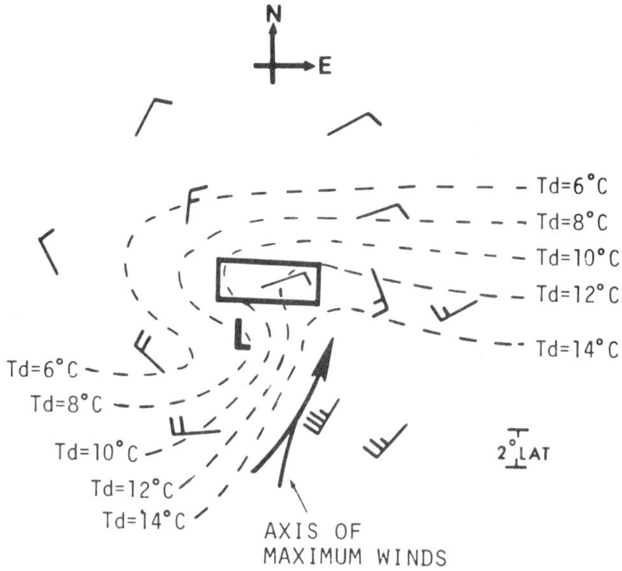

Figure 10. A Composite of the 850-mb Chart for the Warm-Top Thunderstorms Embedded in a Synoptic-Scale Cyclonic Circulation. Wind barbs and dewpoint contours are annotated.

CONCLUSIONS

In this paper, we have shown that one category of flash flood producing warm-top thunderstorms is associated with a synoptic-scale cyclonic circulation at 500 mb, which can be detected from the cloud organization in the satellite imagery. The synoptic-scale cyclonic circulation is usually either quasi-stationary or moving very slowly in an easterly or northeasterly direction. The synoptic characteristics of these storms as revealed by the composite charts and soundings should be helpful to operational meteorologists responsible for detecting these warm-top flash flood producing systems in real time.

ACKNOWLEDGMENTS

The author thanks Roderick Scofield and Ralph Anderson of the NESS Applications Laboratory for their critical reviews of this paper and John Shadid for final drafting of figures.

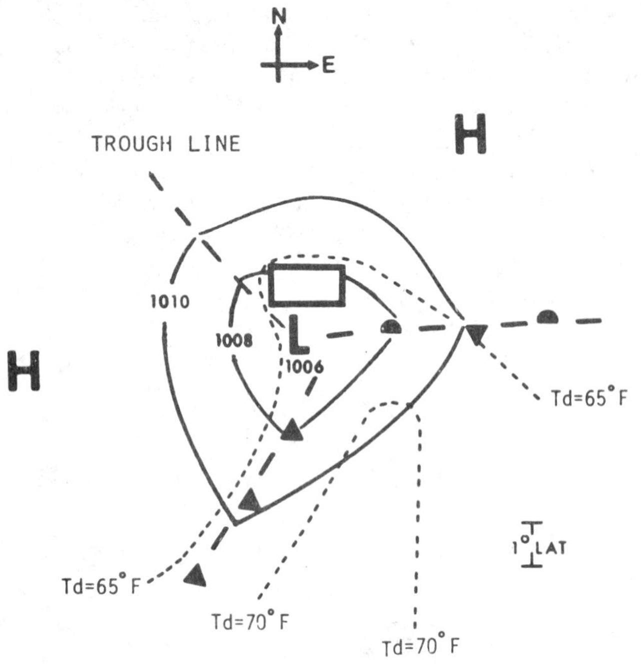

Figure 11. A Composite of the Surface Chart for the Warm-Top Thunderstorms Embedded in a Synoptic-Scale Cyclonic Circulation.

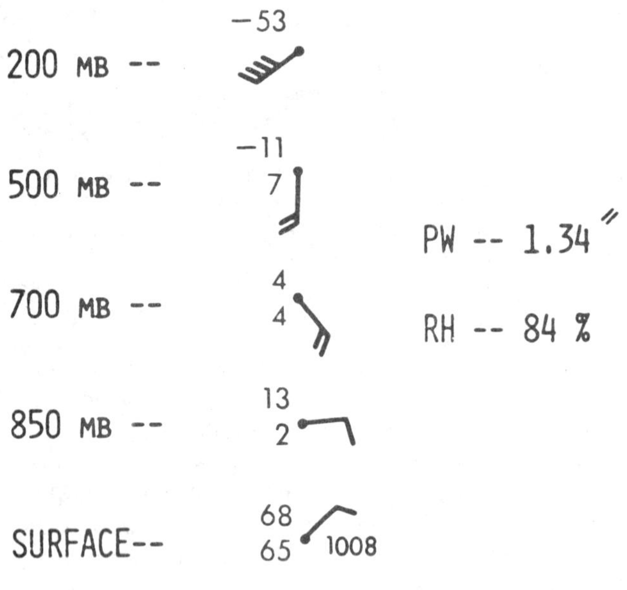

Figure 12. A Composite of the Atmospheric Soundings for the Warm-Top Thunderstorms Embedded in a Synoptic-Scale Cyclonic Circulation. The precipitable water and surface to 500-mb mean relative humidity are annotated.

LITERATURE CITED

Maddox, R. A., C. F. Chappell, and L. R. Hoxit, 1979. Synoptic and Mesoaspects of Flash Flood Events. Bulletin of Am. Meteorological Society 60(2):115-123.

Scofield, R. A. and V. J. Oliver, 1977. A Scheme for Estimating Convective Rainfall From Satellite Imagery. NOAA Technical Memorandum NESS 86, 47 pp.

Scofield, R. A. and V. J. Oliver, 1980. Some Improvements to the Scofield/Oliver Technique. Proceedings of the Second Conference on Flash Floods, March 18-20, 1980, Atlanta, Georgia, AMS, pp. 115-122.

Scofield, R. A., V. J. Oliver, and L. Spayd, 1980. Estimating Rainfall From Thunderstorms With Warm-Tops in the Infrared Imagery. Proceedings of Eighth Conference on Weather Forecasting and Analysis, June 10-13, 1980, Denver, Colorado, AMS, pp. 85-91.

OBJECTIVE ANALYSIS OF RAINFALL DATA FROM DIGITAL RADAR AND RAIN GAGE MEASUREMENTS

Witold F. Krajewski and Kenneth C. Crawford[1]

ABSTRACT: Two devices are used to perform measurements of rainfall: digital radar and rain gages. Each of them gives different types of information. Digital radar data represent a spatial pattern of a storm but the point estimates may be biased through, for example, the reflectivity-rainfall transfer relationship. A standard rain gage optimally sited generally gives a very good estimate of rainfall at a point, but the measurement often is not representative over a larger area. A multivariate regression model has been developed to merge those two kinds of data into an optimal (in a statistical sense) estimate of rainfall. Such optimal estimates can then be used as input to hydrological models. Plans for operational implementation of the regression model is presented as well as results of example data analysis from the National Weather Service Radar at Oklahoma City. Computer requirements are discussed and some aspects of potential future improvements are mentioned. These include the possibility of accounting for the autocorrelation of errors and the inclusion of satellite data.
(KEY TERMS: radar; precipitation; rain gage; objective analysis.)

INTRODUCTION

The Hydrologic Research Laboratory of the National Weather Service (NWS) has many ongoing activities. One research effort is the Hydrologic Rainfall Analysis Project (HRAP), that addresses development of objective techniques for preprocessing, quality controlling, and operationally merging rainfall data from different sources such as multi-radars, rain gages, and satellites (Hydrologic Research Laboratory, 1980; Greene, *et al.*, 1979).

This research deals with the objective analysis of rainfall data from digital radars and standard rain gage stations. It will be shown later that satellite data also can be incorporated after it becomes available "on-line" in required form (digitized).

Our goal is the development of a relatively simple, from a computational viewpoint, method which can be implemented into the operational environment and produce final rainfall analyses to several end users, in particular NWS River Forecast Centers (Figure 1).

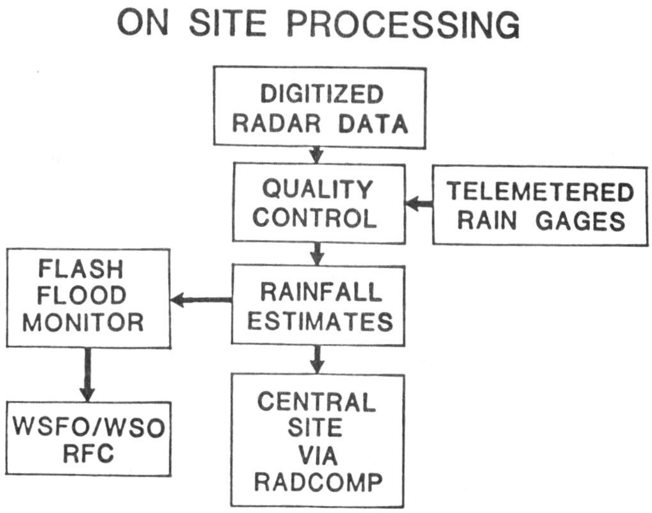

Figure 1. Block Diagram Illustrating Steps Involved in Rainfall Data Processing.

Our portion of producing a uniform grid point rainfall analysis from multiple data sources is illustrated by Figure 2. The grid has been especially designed for HRAP purposes and is fully described by Greene and Hudlow (1982). The rainfall estimates are desired at the grid node locations as a convenience in estimating areal rainfall distributions and other uses as well.

One of the first attempts to solve the problem of merging gage and radar date was made by Brandes (1975) but many consider the best possible results to be provided by optimum interpolation first introduced by Eddy (1963) and Gandin (1963). This technique has been gradually modified, especially for surface rainfall analysis using radar and rain gage data, as examined by Brady (1976), Crawford (1977), Hembree (1980), and others. The characteristics of this methodology and its superiority qualities are given in detail

[1]Respectively, NOAA/NRC Postdoctoral Associate, Hydrologic Research Laboratory, Office of Hydrology National Weather Service, 8060 13th Street, Room 530, Silver Spring, Maryland 20910 (permanent affiliation with the Systems Research Institute of Polish Academy of Sciences); and Deputy Meteorologist in Charge, National Weather Service, New Orleans Area Weather Service Forecast Office, 1120 Old Spanish Trail, Slidell, Louisiana 70458.

by Crawford (1978 and 1979). Using this methodology, rainfall estimates for each grid point are computed from a regression equation of the form:

$$Y_i = X^G_{li} b^G_{li} + \ldots + X^G_{ni} b^G_{ni} + X^R_{li} b^R_{li} + \ldots + X^R_{mi} b^R_{mi}$$

where:

Y_i is final estimate of rainfall at i^{th} grid point;

$X^G_{li}, \ldots, X^G_{ni}$ are n observations of rainfall from gages used for i^{th} grid analysis;

$X^R_{li}, \ldots X^R_{mi}$ are m observations of rainfall from radar bins used for i^{th} grid analysis; and

$b^G_{li}, \ldots, b^G_{ni}$ and $b^R_{li}, \ldots, b^R_{mi}$ are respective regression coefficients.

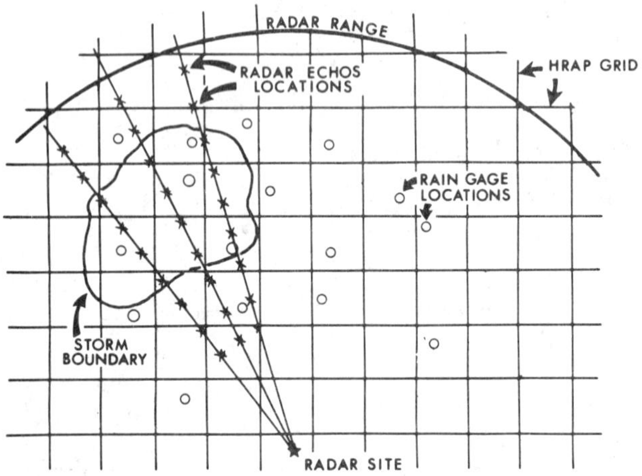

Figure 2. Schematic Illustration of Data Fields Used in Analysis.

If the b coefficients are chosen properly, then the final analysis provides optimal merging of the two data sets with pattern "accuracy" given by radar and point accuracy enhanced by gages.

The algorithm showing the basic concept of the method is described in the following section.

THE METHOD DESCRIPTION

The method applied is a multivariable linear regression,

$$V = Z\beta + \epsilon \qquad (1)$$

where:

V is a predictand matrix;

$Z = (Z_G, Z_R)$ is a matrix of observations consisting of two submatrices containing the observations from gage stations and radar bins, respectively; and

$\beta = (\beta_G, \beta_R)$ is a matrix of respective regression coefficients; and ϵ is a matrix of population errors (i.e., it contains among other features a measure of our model inaccuracies).

The solution for b, the minimum variance estimators of β of Equation (1) is:

$$b = (Z^tZ)^{-1} Z^tV \qquad (2)$$

where Z^tZ is a matrix of covariance between the predictors and Z^tV is a matrix of covariance between the predictand and the predictors. Usually in actual computer regression analysis, the covariance matrix is replaced by a correlation matrix in order to avoid the round-off errors.

The modification in our approach resulted from the fact that the observation matrix V is not known; therefore, the product Z^tV in Equation (2) is not known either. Using the structural similarity of Z^tZ and Z^tV (let us recall that V is also the rainfall estimate but at the location where data were not being collected) and the fact that Z^tZ is a correlation between two variables separated in space, it is possible to estimate Z^tV given the spatial distribution of correlation Z^tZ. In other words, each element of Z^tZ is a value of a function of distance and direction only.

For a bivariate regression, Equation (1) takes the form

$$V = Z_G \beta_G + Z_R \beta_R + \epsilon \qquad (3)$$

and therefore $Z^t Z$ matrix may be presented as

$$Z^t Z = \begin{vmatrix} Z^t_G Z_G & Z^t_G Z_R \\ Z^t_R Z_G & Z^t_R Z_R \end{vmatrix} \qquad (4)$$

which is a symmetric matrix since $(Z^t_R Z_G)^t = Z^t_G Z_R$. Also,

$$Z^t V = \begin{vmatrix} Z^t_G V \\ Z^t_R V \end{vmatrix} \qquad (5)$$

Thus, we see that to perform a regression, we have to model three terms: $Z^t_G Z_G$, $Z^t_G Z_R$, and $Z^t_R Z_R$. The first two terms would be used to estimate $Z_G^t V$ and $Z_R^t V$ (since V should be treated as a gage rather than radar bin). $Z_G^t Z_G$ and $Z_R^t Z_R$ will be called auto-correlation matrices and $Z_G^t Z_R$ will be called cross-correlation matrix. Once the models are available, the matrices Z^tZ and Z^tV can be constructed, b coefficients determined, and the analysis performed.

However, two important questions remain to be answered: how do we model these correlations and how many variables should be taken into regression? To answer the first question, let us explain our construction of raw correlation matrices. Correlations are calculated using each set of paired observations. The atmosphere's anisotropy is incorporated through discrete but variable lag separation distances in both the X and Y directions. The result is that not only station separation distances, but also directional separations are considered. Figure 3 shows an example of three points. More detail is provided in Crawford (1977).

Thus, for the case of using n stations of the same kind, the derived raw auto-correlation matrix is based on a total of n(n−1) station pairs. Of course, constructing the cross-correlation matrix for n_1 stations of the first kind and n_2 of the second, we consider $n_1 \cdot n_2$ pairs. Our next step is to approximate the raw correlation matrix with a two-dimensional function satisfying certain important conditions. More detailed discussion of this matter can be found in Crawford (1977).

In our current study the following function, similar to that used by Brady (1976), Crawford (1977), and then Hembree (1980), is implemented:

$$f(x, y) = \rho \exp \left[- \frac{K}{2(1-\alpha^2)} \right] , \qquad (6)$$

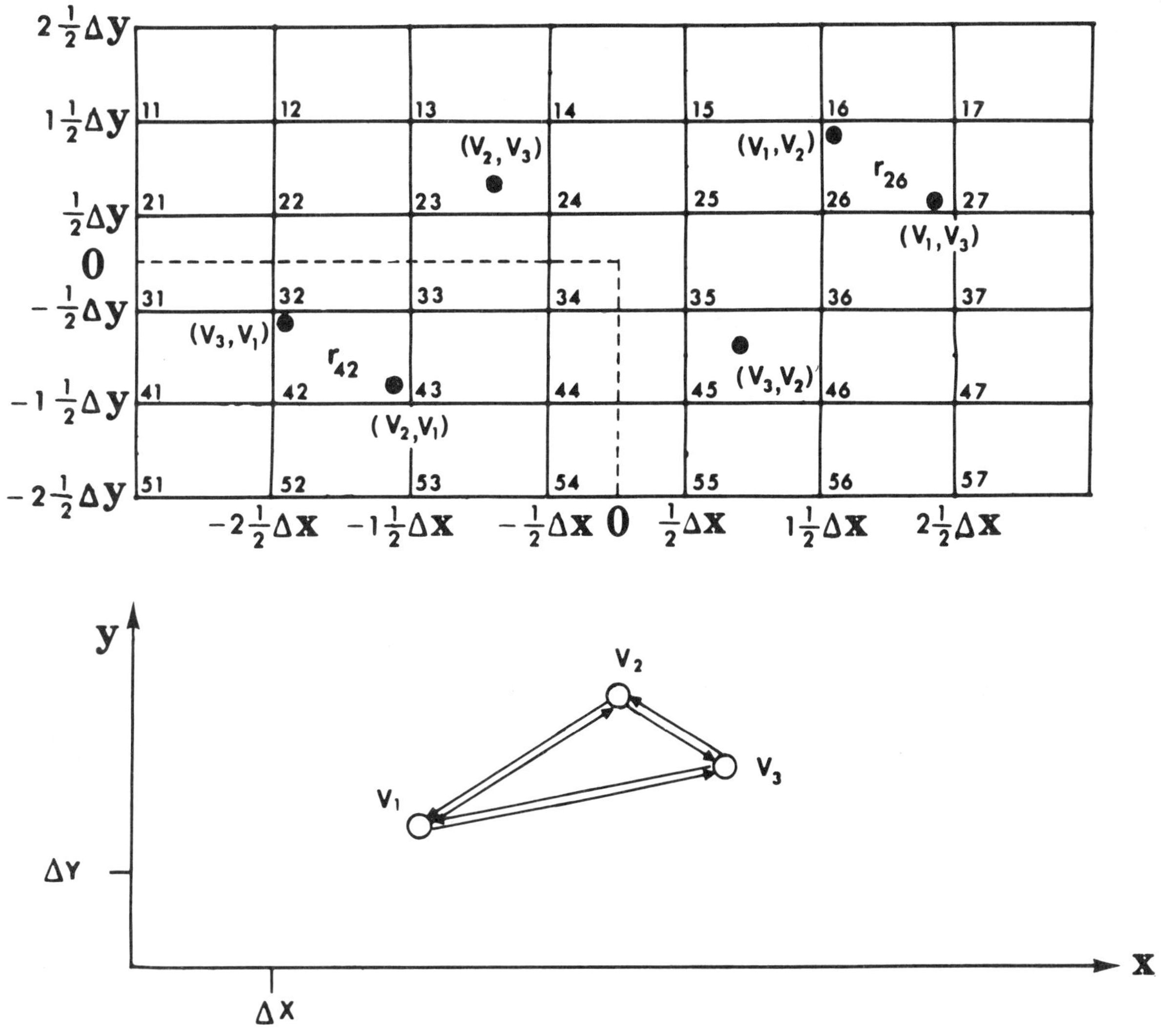

Figure 3. Construction of a Raw Correlation Matrix.

where

$$K = \frac{x^2}{\sigma_x^2} - 2\alpha \frac{xy}{\sigma_x \sigma_y} + \frac{y^2}{\sigma_y^2} \text{ , and} \quad (7)$$

α is an ellipticity parameter in the X-Y plane;

ρ is the lag-zero correlation coefficient; and

σ_x, σ_y are decorrelation distances in X and Y.

Allowing the point of maximum correlation to occur at other than lag zero in X-Y plane, the expression for K takes the form:

$$K = \frac{(x-x_o)^2}{\sigma_x^2} - 2\alpha \frac{(x-x_o)(y-y_o)}{\sigma_x \sigma_y} + \frac{(y-y_o)^2}{\sigma_y^2} \quad (8)$$

where x_o, y_o are now the coordinates of the maximum.

Approximation of the raw covariance matrix using Equation (6) can be obtained by solving the following nonlinear optimization problem:

$$\text{minimize } Q = \sum_{i=1}^{M} N_i \, [r_i - f_i(x,y,x_o,y_o,\alpha,\sigma_x,\sigma_y,\rho)]^2, \quad (9)$$

where

M is the total number of new correlation matrix values over which the function is fit;

N_i is the number of station pairs used to estimate r_i (the raw correlation coefficient);

f_i is a value of the approximation function; and

$x_o, y_o, \alpha, \sigma_x, \sigma_y$, and ρ are parameters to be estimated.

The nonlinear optimization algorithm used was based on the Levenberg-Marquardt method (Brown, *et al.*, 1972).

The second question: "How many variables should be incorporated into regression" can be answered with the help of a step-wise regression algorithm (Draper and Smith, 1966). The set of predictors examined consists of all points (gages and radar bins) located within the "radius of influence" which is fully defined by elipticity parameter α and decorrelation distances σ_X and σ_Y measured in X and Y directions from a grid point under investigation. After selecting possible predictors to be used in Equation (3), the previously derived correlation functions are used to calculate Equation (2). Stepwise regression is performed to then choose the "best" set of predictors.

Finally, the regression coefficients so calculated are used to "filter" the "best" set of predictors to estimate the predictand value at each location. This completes the main part of the analysis. All that remains is an analysis of residual errors and plotting final analysis for an end user.

Some of the computational aspects will be discussed in the following section.

SOME ASPECTS OF THE METHOD APPLICATION

The above described method is being implemented by NWS using data for the Oklahoma region. Data from several radars, including the Oklahoma City radar, and rain gage reports used by the Tulsa River Forecast Center are transmitted to the NWS IBM 360/195 computer in Suitland, Maryland. These digitized data have been previously checked for large errors and thus are ready to be processed. The radar data are full resolution data sets (2 degree azimuthal by 1 nautical mile radial data bins). This gives 180 x 115 = 20,700 possible bins from a single radar. About 700 rain gage reports are available across the region covered by the Oklahoma City radar umbrella. Our grid consists of 101 x 101 nodes with its center located near the radar site. The number of grid points means that for a widespread storm we have to estimate rainfall at 10,201 points. Obviously efficient management of computer resources is vital to our timely analysis. For example, in order to minimize storage, special procedures for rainfall data management have been developed. These include subroutines for bin coordinates computation. To minimize CPU time for each complete analysis, the so-called convex hull (Bentley, *et al.*, 1982) is constructed for each storm, limiting computations only to the region where rainfall actually occurred. Nevertheless, the CPU time consumed for a solution to the total grid network can be considerable (an important factor for using only two-dimensional correlations). Incorporation of vertical correlation and time correlation would: (1) increase the data volume significantly, and (2) increase the computational time for the first phase of our analysis (i.e., construction of the correlation function). The second phase, step-wise regression, and the final analysis would not be affected very much since there is always a possibility of limiting the number of variables in the step-wise regression to be checked.

The inclusion of satellite data into the analysis would expand the covariance matrix to:

$$Z^t Z = \begin{vmatrix} Z_G^t Z_G, & Z_G^t Z_R, & Z_G^t Z_S \\ Z_R^t Z_G, & Z_R^t Z_R, & Z_R^t Z_S \\ Z_S^t Z_G, & Z_S^t Z_R, & Z_S^t Z_S \end{vmatrix}$$

where Z_S is satellite observations matrix. This would necessitate the calculation of three additional correlation matrices.

Figure 4 gives an example for the data obtained from the numerical experiment described by Greene, *et al.* (1980). An advantage for the use of these data is that the true field (original field) is known so the results can be compared against it.

A final point that should be mentioned is that some further investigation of the spatial auto-correlation of errors is needed. Such investigation is planned and will be done in the near future along with other tests of the method.

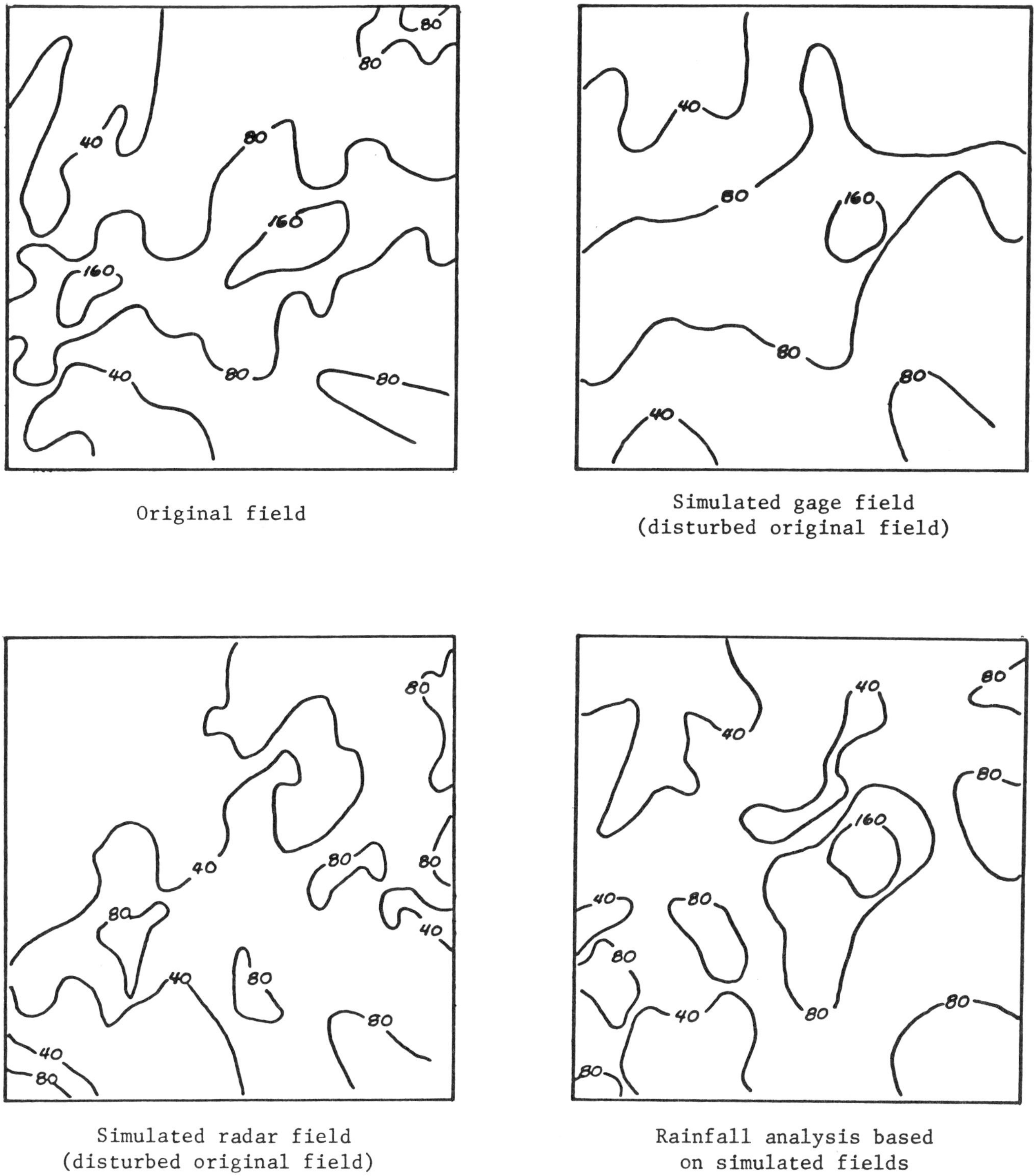

Figure 4. The Results of the Numerical Experiment. The simulated fields were generated out of the original field and then merged using the method described.

ACKNOWLEDGMENTS

This research has been done partly under a National Oceanic and Atmospheric Administration/National Research Council Associateship at the Hydrologic Research Laboratory of the National Weather Service, Silver Spring, Maryland.

LITERATURE CITED

Bentley, J. L., M. G. Faust, and F. P. Preparata, 1982. Approximation Algorithms for Convex Hulls. Communications of the ACM 25(1):64-68.

Brady, P. J., 1976. Multivariate Experimental Design in Meteorology. Dept. of Meteorology, University of Oklahoma, Norman, Oklahoma, 121 pp.

Brandes, E. A., 1975. Optimizing Rainfall Estimates With the Aid of Radar. Journal of Applied Meteorology 14:1339-1345.

Brown, K. M. and J. E. Dennis, 1972. Derivative Free Analogues of the Levenberg-Marquardt and Gauss Algorithms for Nonlinear Least Squares Approximations. Numerische Mathematik 18:289-297.

Crawford, K. C., 1977. The Design of a Multivariate Mesoscale Field Experiment. Dept. of Meteorology, University of Oklahoma, Norman, Oklahoma, pg. 157.

Crawford, K. C., 1978. On the Bivariate Objective Analysis of Surface Rainfall Using Optimum Interpolation. Proceedings of the 18th Radar Meteorology Conference. American Meteorological Society, Boston, Massachusetts, pp. 336-341.

Crawford, K. C., 1979. Considerations for the Design of a Hydrologic Data Network Using Multivariate Sensors. Water Resources Research 15(6):1752-1761.

Draper, N. and H. Smith, 1966. Applied Regression Analysis. John Wiley & Sons, New York, New York.

Eddy, A., 1963. Kinematic Divergence and Large Scale Energy Conservation. Dept. of Meteorology, McGill University, Montreal, Canada.

Gandin, L. S., 1963. Objective Analysis of Meteorological Fields. Leningrad, USSR.

Greene, D. R. and M. D. Hudlow, 1982. Hydrologic Grid Mapping Procedures.

Greene, D. R., M. D. Hudlow, and R. K. Farnsworth, 1979. A Multiple Sensor Rainfall Analysis System. Preprints, Third Conference on Hydrometeorology, Bogota, Columbia, pp. 44-53.

Greene, D. R., M. D. Hudlow, and E. R. Johnson, 1980. A Test of Some Objective Analysis Procedures for Merging Radar and Rain-Gage Data. Preprints, 19th Conference on Radar Meteorology, Miami, Florida. American Meteorological Society, Boston, Massachusetts, pp. 470-478.

Hembree, L. A., 1980. An Investigation of Satellite-Radar Correlation Functions. Report by Amos Eddy, Inc., under Bureau of Reclamation Contract No. 9-07-85-V0024, pg. 192.

Hydrologic Research Laboratory, 1980. Hydrologic Rainfall Analysis Project (HRAP) Activities, NWS, Hydrologic Research Laboratory, Silver Spring, Maryland, 7 pp.

INTERNATIONAL SYMPOSIUM ON HYDROMETEOROLOGY
JUNE AMERICAN WATER RESOURCES ASSOCIATION 1982

ON THE STRUCTURE OF HOMOGENEOUS ANISTROPIC CORRELATION FUNCTIONS FOR REAL-TIME RADAR RAINFALL ESTIMATION

Nicolas R. Dalezios and Nicholas Kouwen[1]

ABSTRACT: Rainfall variability is one of the major problems in hydrology. Conventional raingage data combined with digitized radar reflectivities can provide powerful techniques for monitoring the quantitative rainfall distribution in time and space. An objective analysis technique is performed, which accounts for the analysis of multivariate data sets. In particular, bivariate analyses are discussed that use raingage/radar data sets. This involves the estimation of correlation and cross-correlation functions which model the spatial and temporal structure of a storm. In this work, Gaussian two-dimensional homogeneous anisotropic correlation and cross-correlation models have been derived. Since the application is on real-time surface rainfall estimation, the model focuses on the spatial structure of the phenomenon. Specifically, the dependence of the decorrelation distance upon certain meteorological parameters, which are embedded in the prevailing weather system, is shown to be critical and offers an improvement on the existing correlation models. A Canadian 5 cm weather radar is employed and the analyses are performed on typical spring and summer storms in Southern Ontario. (KEY TERMS: rainfall; radar; multivariate analysis; correlation; isotropy; meteorological parameter.)

INTRODUCTION

Rainfall is a time-space phenomenon and its variability is one of the major problems in hydrology (Greene and Flanders, 1976). In fact, conventional raingage data have not proven "efficient" to describe the areal extent and the time sequence of rainfall. However, weather radar shows good promise in the development of powerful techniques for monitoring the quantitative precipitation distribution in time and space, on a real-time basis, while the events take place.

Until recently, the motive in rainfall measurement by radar was to apply adjustment procedures on the radar-indicated rainfall field to resemble the raingage observations at the ground. The trend now seems to be not simply to adjust the radar measurements, but rather to attempt to better explain the profound discrepancies embodied in the bivariate data sets provided the limitations of the sensors are known. This means that a mathematical description of the spatial structure of the rainfall phenomenon is needed. There are already a number of recent contributions in radar-rainfall measurement (Brady, 1978; Crawford, 1978 and 1980).

This paper belongs to the latter line of thought. The approach followed in this work consists of a bivariate objective analysis. In fact, weights are generated for the bivariate, i.e., radar/raingage, observation data sets in terms of parameter fields using a weighted least-square technique. In this study, a modeled covariance function is used, which is a Gaussian, two-dimensional, anisotropic model. In particular, the correlation parameter is modeled in terms of certain meteorological parameters that characterize the prevailing weather system. These parameters are: the geostrophic vorticity, the thermal wind, the horizontal divergence, the stability index and the cloud height to base ratio. A Canadian 5cm weather radar is employed and the analyses results are presented for two storms in Southern Ontario.

BIVARIATE OBJECTIVE ANALYSIS

Consider the classical linear regression model:

$$Y = X\theta + \epsilon \tag{1}$$

where Y is an nxl predictand vector, X is an nxm predictor matrix, θ is an mxl vector of regression coefficients and ϵ is an nxl noise vector, also called the residual. This is the univariate case.

In the present work a bivariate, i.e., radar/rain-gage, objective analysis is employed. The data used consist of surface rainfall observations from recording raingages and of digitized radar reflectivities. At the preprocessing stage, a 0.5^{o} antenna elevation angle is selected and from each recorded complete radar scan two-dimensional data sets are produced using a polar grid element of 2km in range by 2° in azimuth. Then the radar data, in values of dBz, are transformed to rainfall rates using the Ontario Z-R relationship $Z = 295R^{1.43}$. Moreover, the radar rainfall intensity

[1] Respectively, Research Associate and Associate Professor, Department of Civil Engineering, University of Waterloo, Waterloo, Ontario, Canada.

field is converted from polar to Cartesian coordinates using a 2km x 2km square grid element. If more than one value is contained in each grid element the average is taken. Furthermore, since the selected time step is one hour, time integration is applied to the intensity field over each hour in order to produce hourly radar-rainfall fields. Similarly, hourly raingage-rainfall fields of the rain area are generated. First, a 2km x 2km square grid is superimposed over the target area. Second, a simple method, namely the reciprocal distance weighting technique (Solomon, *et al.*, 1968), interpolates the raingage information to each grid element. On the average, ten recording raingages are used in this study.

The emphasis is on near real-time rainfall estimates for hydrologic purposes. Each hour the bivariate objective analysis may be accomplished by

$$\hat{Y} = X_1\hat{\theta}_1 + X_2\hat{\theta}_2 \qquad (2)$$

where $\hat{Y}$ is an estimate of the "true" surface rainfall Y, X_1 represents the radar-rainfall estimate at a grid element, X_2 refers to the rainfall amount at the same grid element from the raingage field, and $\hat{\theta}_1$ and $\hat{\theta}_2$ are the estimates of the regression weights for the radar and raingage data sets, respectively. The objective analysis is performed iteratively and each hour regression coefficients are estimated, one for each predictor sensor. Furthermore, if anistropy is considered, regression coefficients are estimated for each isotropic sector.

The problem is to estimate the regression coefficients $\hat{\theta}_1$ and $\hat{\theta}_2$ each time. The weighted least-squares analysis is applied to the data sets, and the estimator $\hat{\theta}$ is then:

$$\hat{\theta} = (X^TWX)^{-1}(X^TWY) \qquad (3)$$

where $\hat{\theta}$ is a 2x1 vector, X is an 1x2 matrix containing the correlations of the radar-rainfall field and the raingage field up to lag 1 in space, W is an 1x1 weighting matrix and Y is an 1x1 vector containing a modeled autocorrelation function of the "true" surface rainfall. A fundamental restriction is that W should be aymmetric positive definite matrix to assure a unique nonsingularity. This is performed by letting W become $W = R^{-1}$, where $R = E(\epsilon\epsilon^T)$ is the error covariance matrix with $E(\epsilon)=0$, $R = E(\epsilon\epsilon^T) = \sigma^2 I$, $Var(\epsilon) = R\sigma^2$ and $\epsilon \sim N(0, R\sigma^2)$.

THE COVARIANCE MODEL

To obtain a better understanding of the rainfall phenomenon and its spatial structure it is necessary to simulate the noise or error term of equation (1). In most cases, the covariance function of the rainfall field can be satisfactorily described by an exponential or Gaussian relationship. This fact has been repeatedly stressed by several researchers (Crawford, 1980; Brady, 1978; Greene, Hudlow and Johnson, 1980; Kagan, 1972; Rodriguez-Iturbe and Mejia, 1974). Besides, the so-called exponentially correlated noise is a well-known homogeneous or wide-sense stationary stochastic process.

The correlation of the random field $\epsilon(x,y)$ can be expressed by a modeled correlation function. In this study, the covariance function is given by

$$\rho(r) = \exp(-\alpha(\phi)r^2), \qquad (4)$$

where $r^2 = u^2 + v^2 = (x_1 - x_2)^2 + (y_1 - y_2)^2$ is the square distance between any pair of points in space (x_1, y_1) and (x_2, y_2) and $\alpha(\phi)$ is the correlation parameter. Equation (4) is used to form the 1x1 weighting matrix W of equation (3). The correlation parameter $\alpha(\phi)$ can vary in steps of $\phi=90^o$, thus allowing a 4-subarea anisotropy.

The term X^TWY of equation (3) characterises the covariance domain. Since the "true" surface rainfall Y is unknown, a simple polynomial autocorrelation function is used to model it, namely the triangular autocorrelation function (Ord and Rees, 1979), which is given by

$$\rho(|x - x_i|) = \begin{cases} 1 - \gamma|x - x_i| & \text{for } |x - x_i| \leq 1/\gamma \\ 0 & \text{for } |x - x_i| > 1/\gamma \end{cases} \qquad (5)$$

The value of $\gamma = 0.1$ is assumed in this study.

THE CORRELATION PARAMETER

A multiple linear regression analysis is performed on the correlation parameter $\alpha(\phi)$. The model has the form

$$\alpha(\phi) = b_0 + b_1z_1 + b_2z_2 + b_3z_3 + b_4z_4 + b_5z_5 \qquad (6)$$

where z_1 is the stability index, z_2 is the geostrophic vorticity, z_3 is the cloud top height to base ratio, z_4 is the 500-1000mb thermal wind and z_5 is the horizontal divergence. The regression analysis is applied to the isotropic case, as well as to each sector of the 4-sector anisotropy. The data for the estimation of the five meteorological parameters have been extracted from surface and upper-air weather charts. The correlation parameter is found to vary from 0.10 to 0.55, and its value is kept constant within each storm, since upper-air data are available every 12 hours.

The stability index, also called the Showalter index, is a simple and very important parameter extensively used in synoptic meteorology. The geostrophic vorticity, or more correctly its vertical component, is highly correlated with large-scale storm systems and is estimated at 500mb using finite difference approximations. The cloud top height to base ratio is a very simple, useful and self-explanatory parameter. In fact, if the cloud top is relatively low, it normally means mild and widespread rainfall, whereas if the cloud top is high, this is an indication of convective rainfall of intense and rather localized nature. The thermal wind is the vector difference between the geostrophic wind at the upper and lower levels of a given layer. The significance of the thermal wind lies on its ability to evaluate the "baroclinicity" of the atmosphere and the rain activity of a weather system. In this study, the 500mb-1000mb thermal wind is estimated by the hodograph technique. Finally, the horizontal divergence of the wind field, which is estimated at the surface, is caused by small departures of the wind from the geostrophic approximation. Thus, high accuracy is required in evaluating the wind components, which is a disadvantage of the continuity equation method used in this study.

DISCUSSION

It is undoubtedly necessary to demonstrate the usefulness and the efficiency of the developed bivariate objective analysis. In the present study, an objective comparison of different rainfall analyses is attempted based on a simple statistical validation method. The technique divides the existing raingage stations into a calibrating network and a test network. The calibrating network consists of the recording raingages used in the analyses to estimate point rainfall values at every sampling location of the test network of the non-recording raingages, also called independent test sample. Then, for each storm event, the correlation coefficient r_k is computed between the measured storm total rainfall amounts and the estimated values of the test network. In addition, the Mean Storm Radar Bias (MSRB) or the average G/R ratio of the storm rainfall totals is calculated for the test network.

Two warm season storms are analyzed in Southern Ontario, the storm of April 18, 1975 and the storm of August 16, 1977. Three rainfall analyses are compared: the so-called raingage model based on the reciprocal distance interpolation technique, the bivariate isotropic analysis, and the bivariate four-sector anisotropic analysis. The results are presented in Table 1, and in Figures 1, 2 and 3. Table 1 contains the average G/R ratio of the independent test network, the coefficient of variation c_v of the ratio, and the correlation coefficient r_k for the three rainfall analyses and for the two storm events. Figure 1 contains the system covariance function of a sample hour for the isotrophic case (Figure 1a), as well as for the 4-sector anisotropic case (Figure 1b) with a covariance plot mesh of 0.5km x 0.5km. Figures 2 and 3 show the total storm rainfall patterns for the storms of April 18, 1975 and August 16, 1977, respectively, over the target area of 100 x 70km^2, and for the three rainfall analyses. In particular, Figure 3b shows the rainfall pattern of the raingage model for the storm of April 18, 1975. The contours represent isohyetal lines and the values are in millimeters.

The average G/R ratio or the MSRB ranges, in general, within accepted limits for the three analyses (Table 1). In fact, the MSRB is a measure of the achievement related to

TABLE 1. Comparison of Three Rainfall Analyses.

	Raingage			Isotropic			Anistropic (4)		
	$\overline{G/R}$	c_y (%)	r_k (%)	$\overline{G/R}$	c_y (%)	r_k (%)	$\overline{G/R}$	c_y (%)	r_k (%)
Apr. 18, '75	1.28	18.8	59.95	1.534	25.0	85.65	2.213	47.5	62.98
Aug. 16, '77	1.075	73.1	79.74	0.904	28.0	80.70	1.119	40.1	82.53

the mean and is used to evaluate the accuracy of point rainfall measurements. Similarly, the coefficient of variation c_v of the MSRB ranges within accepted limits except for the raingage model of the storm of August 16, 1977. The coefficient of variation c_v of the MSRB accounts for the relative dispersion about the MSRB and is a measure of the goodness of the areal distribution of the storm. It means, for example, that for the bivariate isotropic analysis of the storm of April 18, 1975, the MSRB is determined with an expected error of 25 percent (Table 1). Furthermore, the correlation coefficient r_k expresses the linear relationship between the measured and the estimated rainfall amounts and results in a measure of the overall correspondence between the two variables.

The results of Table 1 indicate that, in general, the bivariate isotropic analysis performs better than the other methods. Also the rainfall patterns of Figures 2 and 3 show a discontinuity in the anisotropic analysis from sector to sector in some cases. Nevertheless, anisotropy is an important characteristic of rainfall fields. It is felt that the modeled covariance function does not delineate the anisotropic feature of the spatial structure of rainfall to the desired level. The meteorological parameters used in this study are essentially applied to synoptic scale studies. However, the target area is of subsynoptic size. In fact, some parameters have been estimated by selecting a unit length of approximately half the size usually employed. In addition, the information from the upper-air charts has been extracted visually. It is believed that the anisotropic feature of the autocorrelation function deserves further investigation.

CONCLUSIONS

A bivariate objective analysis has been performed on two Southern Ontario warm season storms. The bivariate analysis is in the right direction for a solution to the problem of rainfall measurement by radar. The approach shows the potential for raingage fields to be correlated with radar signals to improve surface rainfall estimates. An important feature of the approach is the regression analysis that models the correlation parameter in terms of certain meteorological parameters, which describe the weather system. Moreover, the near real-time nature of the methodology allows a close look at the behavior of intensifying or dissipating weather systems.

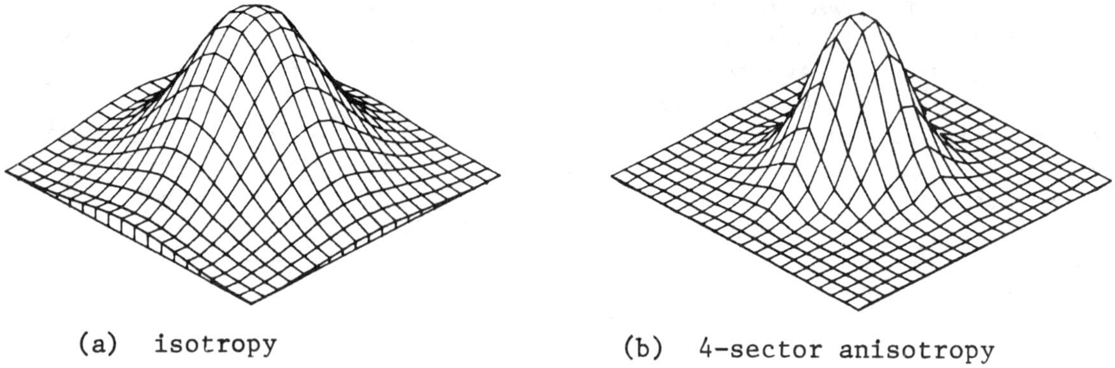

Figure 1. System Covariance Functions for April 18, 1975/19:00 LST.

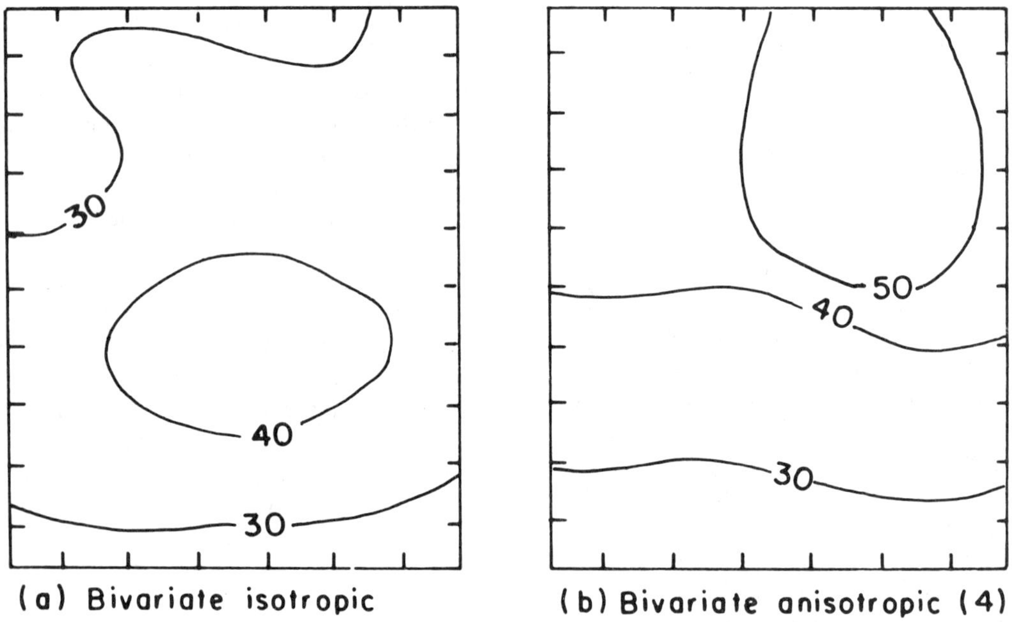

Figure 2. Rainfall Patterns in mm for Storm of April 18, 1975.

ACKNOWLEDGMENTS

This study was partially supported by NSERC Canada and by the Provincial Ontario Government through an Ontario Graduate Scholarship (OGS). The radar data were provided by Environment Canada. In particular, we would like to thank Mr. C. L. Crozier of Environment Canada for his continuous support and his valuable contribution throughout this study.

LITERATURE CITED

Brady, P. J., 1978. Optimal Sampling and Analysis Using Two Variables and Modeled Cross-Covariance Functions. J. Appl. Meteorol., 17(1):12-21.

Crawford, K. C., 1978. On the Bivariate Objective Analysis of Surface Rainfall Using Optimum Interpolation. Reprints, 18th Radar Meteorol. Conf., Atlanta, ASM, pp. 336-341.

Crawford, K. C., 1980. Considerations for the Design of a Hydrology Data Network Using Multivariate Sensors. Water Resour. Res., Vol. 15, 6(12):1752-1762.

Greene, D. R. and A. F. Flanders, 1976. Radar Hydrology – The State of the Art. Reprints, Conf. on Hydrometeorology at Fort Worth, Texas, AMS, pp. 470-479.

Greene, D. R., M. D. Hudlow, and E. R. Johnson, 1980. A Test of Some Objective Analysis Procedures for Merging Radar and Raingage Data. Reprints, 19th Conf. on Radar Meteorol., Miami, Apr. 15-18, AMS, pp. 470-479.

Kagan, R. L., 1972. Precipitation-Statistical Principles. *In:* Casebook on Hydrological Network Design Practice, WMO No. 324.

Ord, K. and M. Rees, 1979. Spatial Processes: Recent Developments with Applications to Hydrology. *In:* The Mathematics of Hydrology and Water Resources, Academic Press, pp. 95-118.

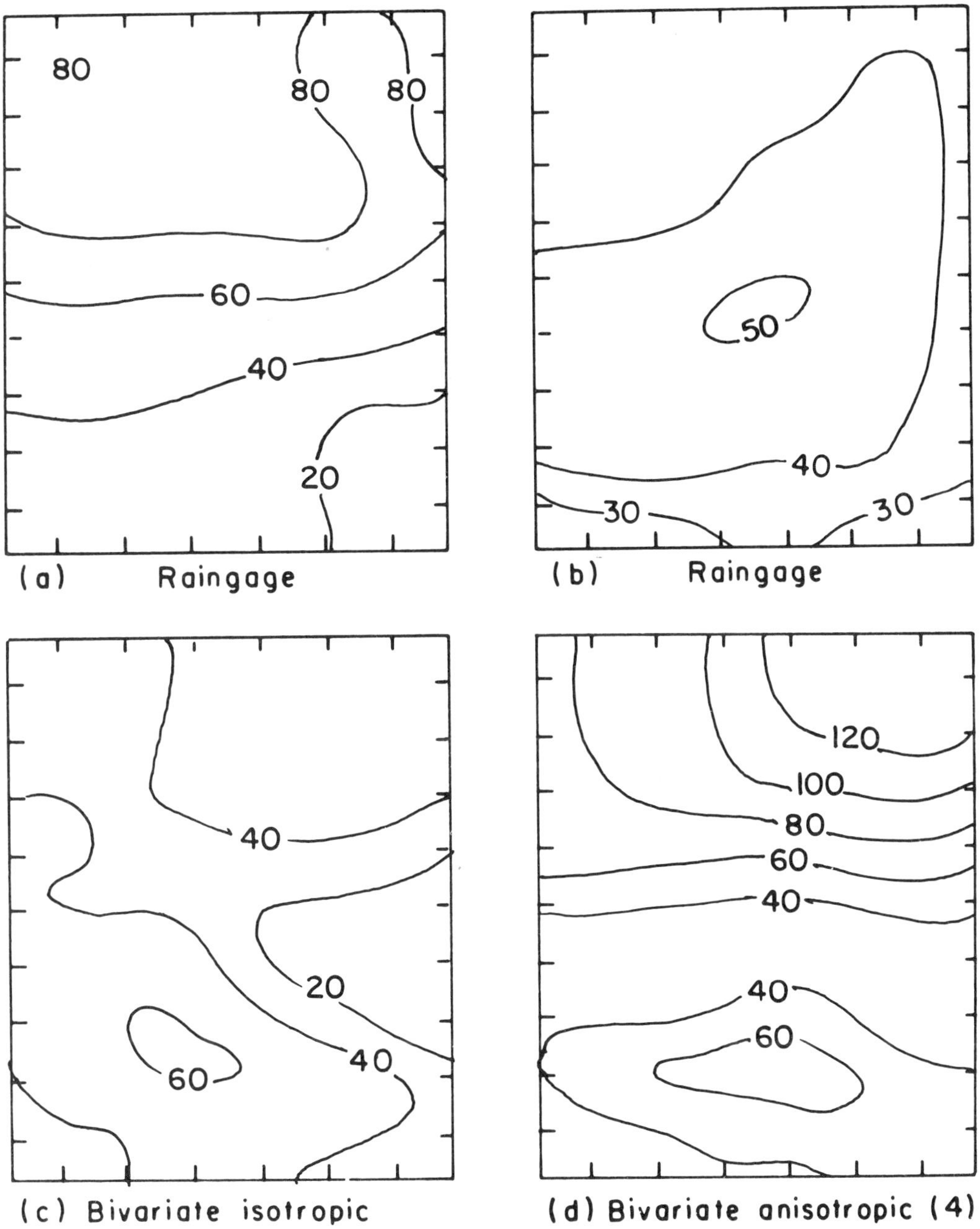

Figure 3. (a), (c), (d) Rainfall Patterns in mm for Storm of August 16, 1977.
(b) Rainfall Pattern in mm for Storm of April 18, 1975.

Rodriguez-Iturbe, I. and J. M. Mejia, 1974. The Design of Rainfall Networks in Time and Space. Water Resour. Res., Vol. 10, 4(8):713-728.

Solomon, S. I. *et al.,* 1968. The Use of a Grid Square System for Computer Estimation of Precipitation, Temperature and Runoff. Water Resour. Res., Vol. 4, 5:919-924.

RAINFALL RELATED PROBLEMS IN OPERATING THE GARD FLOOD WARNING SYSTEM

Charles Obled and Jean-Dominique Creutin[1]

ABSTRACT: The southeastern side of the Massif Central, in France, frequently undergoes very strong rainfall events occurring generally between August and November (total event rainfals in the range 250-500 mm over 2 to 4 days, hourly rainfalls usually over 20 mm/h and may reach 50 to 80 mm/h). A pilot project has been set up in the central part of this region, with a real time data acquisition system and a flood forecasting model. The development of the latter came up against several problems caused by our poor knowledge of the rainfall properties, both spatial and temporal, in this region. These problems are briefly reviewed and discussed. First, some statistical structural properties are given, both for the total rainfall events and for the hourly rainfall fields. Their impact on network density requirements and on uncertainties regarding the estimated average rainfalls in small watersheds is considered. In addition, the role of altitude is considered. However, this variable was eventually deleted from operational use. Some results are also given on sophisticated evaluation of watershed averages, as compared with more common approaches like Thiessen or arithmetic means. Another problem is the lack of reliable short term rainfall forecasts (2 to 6 hours). The usefulness of currently available forecasts (24 h) is considered, as well as the operational use of a synthetic rainfall generator (which takes into account the already observed hourly rainfall), as an aid to the forecaster.
(KEY TERMS: flood warning; precipitation; network design; France.)

DESCRIPTION OF THE SYSTEM

Region and Watersheds Concerned

All the watersheds of the southeastern side of the Massif Central in France are subject to strong rainfall events which occur mainly between August and December (Figure 1). About ten watersheds can be distinguished with areas of about 1000 km^2 or more down to the Rhone or the Mediterranean Sea. Flood events, however, always occur in the upper parts of the watersheds, concern areas of only several hundred km^2.

Within this zone, the Massif Central plateau reaches elevations of between 1000 and 1500 meters in the west and drops eastward down to 200 meters in about 15 or 25 km. In addition to the marked barrier that this presents to the southerly Mediterranean air masses, it causes high river slopes reaching 60/1000 in the upper zone compared with only 2/1000 in the lower plateau region. This explains the very short response time, e.g., 2 to 3 hours for the Vidourle River at Sabatier (200 km^2), which rose from 30 to 500 m3/s in 1½ hours in September 1976, and 8 to 10 hours for the same river at Sommieres (616 km^2), which rose from 80 to 1300 m3/s in 6 hours in October 1958.

The associated rainfall events last an average of 2 to 3 days, giving 200 to 600 mm of precipitation. Over the 1956-76 period, more than 80 significant autumn rainfall events occurred, with about a dozen exceeding 400 mm in terms of event totals. Intensities will be discussed further on but 20 to 50 mm can be given as the usual range with ten-year maximums of 85 mm in one clock hour and 200 mm in three clock hours (September 1976).

Brief Description of the Flood Warning System

Following the 1958 flood disasters, a program involving the construction of retarding dams was begun and, in the early 1970's, it was decided to back this up with a flood warning and forecasting system. For various reasons, this system was restricted to only three watersheds: the Cèze (1130 km^2), the Gard (1855 km^2), and the Vidourle (775 km^2) Rivers.

The system has already been described in detail by Billuart and Tourasse (1980) and includes two main subsystems: (1) a data acquisition and retrieval system and (2) a modeling, simulation, and forecasting system.

Data is radio transmitted to the Nimes Control Center from about 20 telemetered stations via two relay stations. Only tipping bucket rainfall recorders are used in this study although some of the stations are more fully equipped. Due to the short response times already mentioned, the system is scanned automatically every clock hour and may also be activated at any time, as required. These telemetered stations only represent a part of the entire network covering the much larger region concerned by the rain system. In

[1]Respectively, Research Engineer, Institut National Polytechnique, and Attaché de Recherche, CNRS; Institut de Mécanique, Groupe d'Hydrologie, BP53X, 38041 Grenoble-Cedex, France.

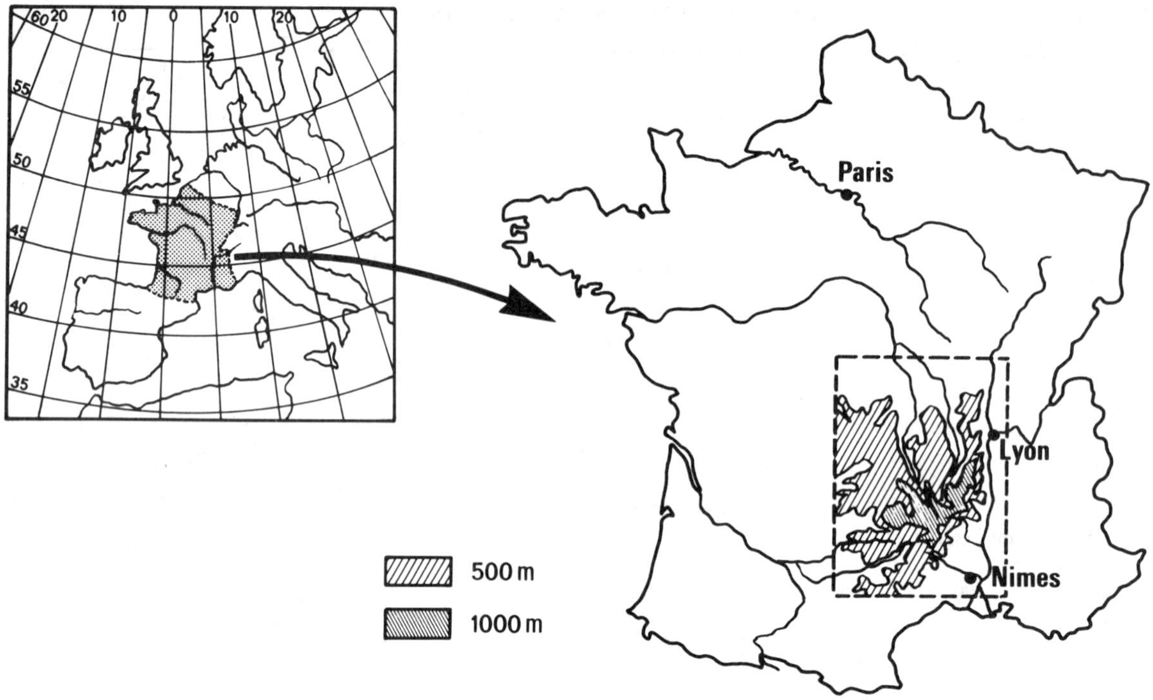

Figure 1. General Location of Study Area.

this way, a network of 58 stations was availab'e for the 1971-79 period (Figure 2).

The modeling and forecasting part of the system is not yet as advanced as the data acquisition part.

Several attempts, based either on distributed or lumped approaches, are currently being developed.

Whichever method is chosen, problems are always encountered with the most critical input, i.e., the definition of the rainfall characteristics.

SPATIAL RAINFALL EVALUATION PROBLEMS

It very rapidly became obvious that the models were extremely sensitive to the input rainfall description and that the high spatial variability of rainfall made it impossible to define a single representative station, even for a small basin. On an hourly basis, this can be revealed by analyzing the climatological range and patterns of the correlations between one station and its neighbors, or by plotting the variograms for the most significant rainfall fields. In addition to a high level of anisotropy, a low correlation range is observed (15 to 30 km) for the two principal directions, one parallel and the other perpendicular to the crest line (Figure 3).

In order to interpolate and plot the isohyets, a sufficiently dense network is therefore required with interstation distances significantly less than the correlation range. This is the case for the complete network (58 stations, with distances between neighboring stations of between 5 and 10 km). For such a network, an extensive comparison of current interpolation methods was made by Creutin and Obled (1980 and 1982).

Among the different methods dealt with, spline function fitting was preferred for practical reasons. The simple methods, such as the Thiessen and arithmetic mean methods, gave significantly poorer results. However, this comparison involved only point interpolation for a few stations selected in order to test the methods.

When integrated over an entire basin, the conclusions are less clear-cut. The methods in fact tend to converge for relatively large basins with relatively dense networks. The Gard River at Anduze can be taken as an example (545 km^2 with 15 stations). The correlation, for about 750 hours of mean rainfalls greater than 1.5 mm, was 0.99 between the Spline and Thiessen methods and the same between the arithmetic mean of the 15 stations and the Spline method.

On the other hand, if the comparison is made using only the telemetered stations, the correlation decreases to 0.92 when using five of them located both inside and outside the watershed, but may drop to 0.55 when using only one in the lower and one in the upper part of this watershed.

This is even more critical for smaller basins since in this case the Thiessen method assigns the entire basin to a single station without taking into account the often highly variable shape of the rainfall surface. Figure 4 gives an example for a 62 km^2 basin (the Galeizon stream at Aube-Morte), where the following are compared:

1. The mean rainfall evaluated by the Spline method using the complete network (taken as reference).
2. The rainfall evaluated from the telemetered network by the Thiessen method, which in this case assigns the entire basin to a single station which is furthermore located outside the basin.
3. The rainfall evaluated by the Spline method using only the telemetered stations, some relatively far from the basin.

Note that (3) is clearly a better estimation than (2), even if rather crude compared to (1). In addition, objective analysis methods, such as Gandin's method or Kriging also give the variance of the error of estimation for the mean rainfall. In the future, it is planned to enter these uncertainties in the rainfall runoff models.

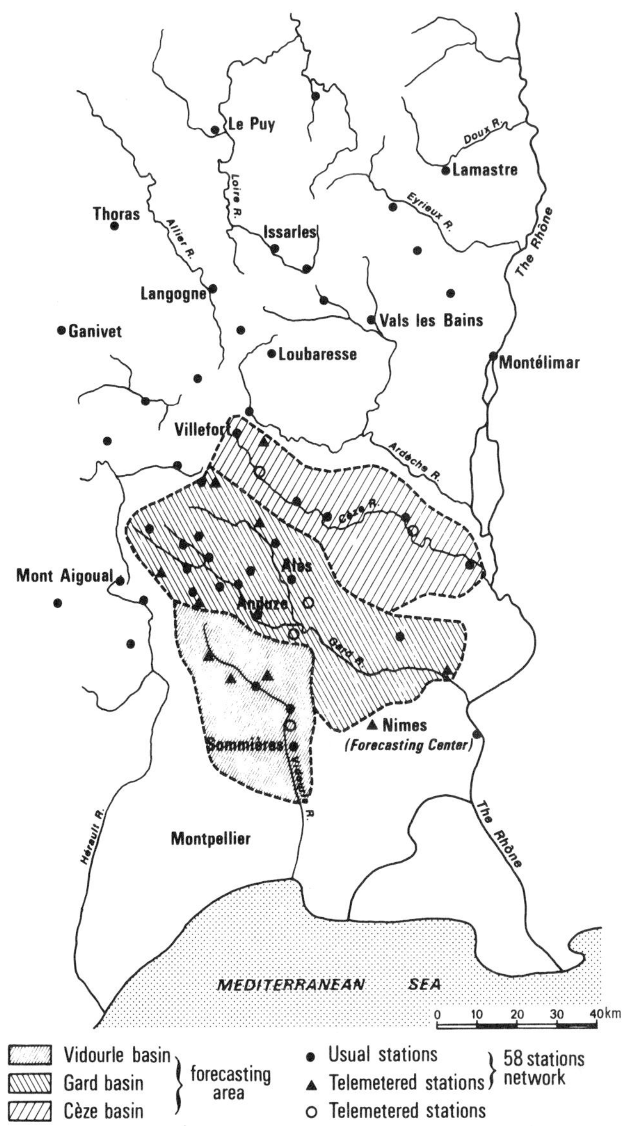

Figure 2. Network Map and Pilot Basins.

THE ROLE OF ELEVATION

In an Interpolation Scheme

It is often surprising that elevations are not included as supplementary coordinates in the interpolation. In actual fact, it is obvious that topographic factors influence the location and the intensity of precipitation. However it is difficult to ascertain whether absolute elevation Z represents an adequate variable.

If there is a strong resemblance between the map of the interannual means and the topographical map, this relationship is much less obvious when considering individual events. Although the total event maximums in this case generally correspond to the topographical ridge, the event is not evenly distributed along it. A relationship may in fact only exist along a transect or a band perpendicular to this ridge.

Correlation tests were carried out for:

Event rainfall ER versus

(1) x, y, x^2, y^2, xy, z

(2) x, y, x^2, y^2, xy

(3) x, y

(4) z

(5) x, y, z

where x, y are coordinates in the plane and z is elevation.

Note first that for 84 events, half the correlation coefficients for (1) are between 0.4 and 0.75 with the other half between 0.75 and 0.9. Comparison between (2) and (3) show that the quadratic terms have nothing to add. Test (4) shows that elevation alone is very poorly correlated with rainfall (between −0.2 and +0.3). On the other hand, the correlation coefficients in (5) increase only slightly with respect to (2) (from 0.0 to 0.07), although the partial correlation coefficients, i.e., r (ER, z /x, y) are nearly all significant and always greater than 0.

These results, concerning events, are basically consistent with those of K. A. Edwards (1972), which were obtained using monthly data. However, when three-hour rainfall fields are considered, corresponding roughly to storm events, the correlation coefficients fall. This is due to the much less smooth structure and the presence of high intensity zones located more often below than on the ridge (see the section on "Rainfall Dynamics"). For these reasons, the z variable was not used in the cartography nor in the calculation of the mean rainfalls.

Climatological Aspects

The effect of elevation was in fact ignored for another reason. The variable z directly explains neither the instantaneous nor the hourly intensity, but in fact the probability of occurrence of precipitation.

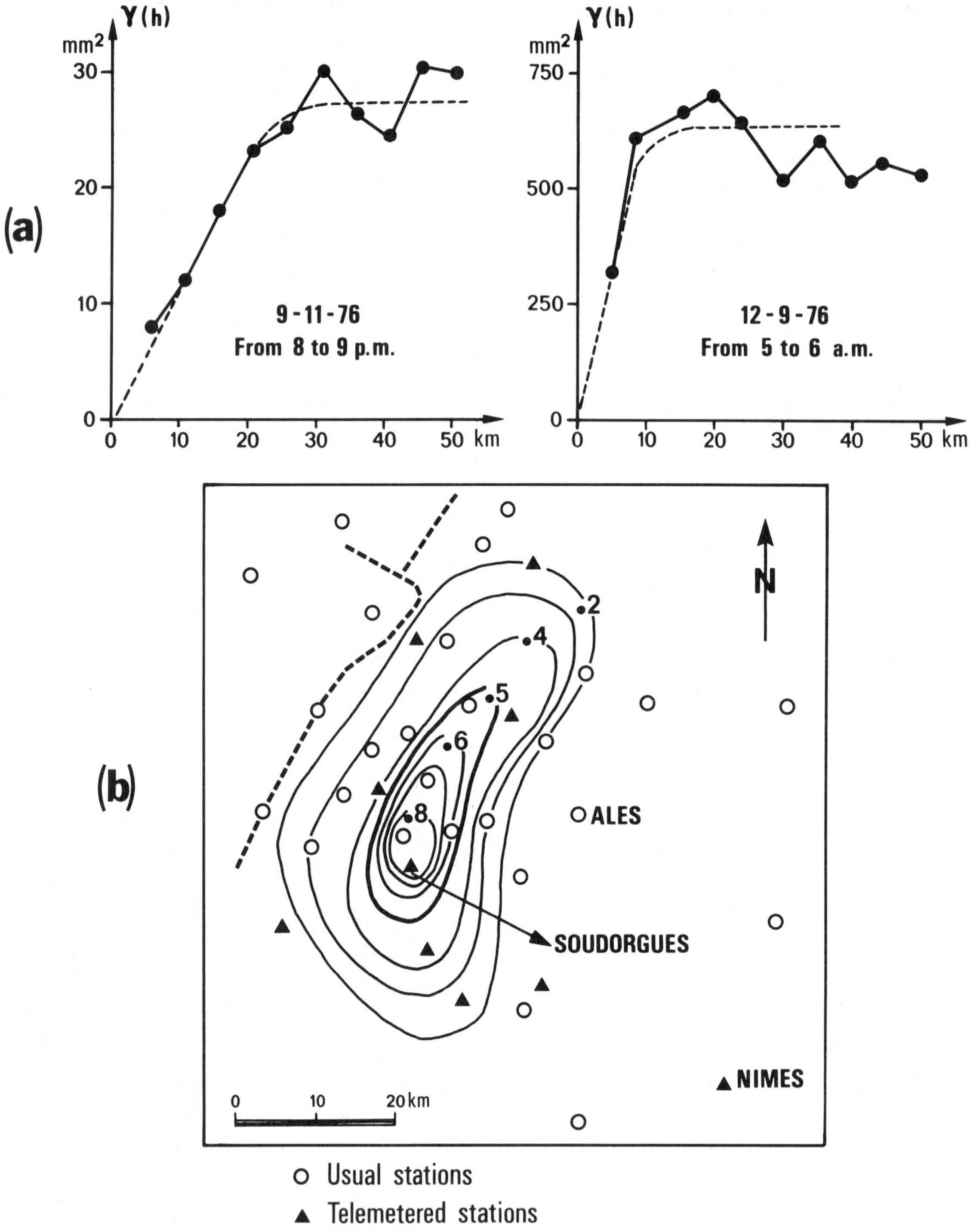

Figure 3. Statistical Structure of Hourly Rainfall Fields: (a) Experimental Variograms of Selected Fields, and (b) Climatic Correlation Pattern Around a Master Station (Soudorgues).

In this way, Mont Aigoual (1565 meters), the highest point in the network: (1) receives on the average (over 20 years), the maximum volume per event; (2) receives the total event maximum for 1 case in 5 (which is otherwise located at another point on the ridge); (3) rarely exceeds 20 mm/hour (on an hourly basis), while the more easterly stations on the rising slope receive 40 to 60 mm/hour during this time; and (4) receives longer duration rainfalls, e.g., for 2500 hours considered, 1350 hours of rainfall at Aigoual, 975 hours at Soudorgues, 710 hours at Anduze,

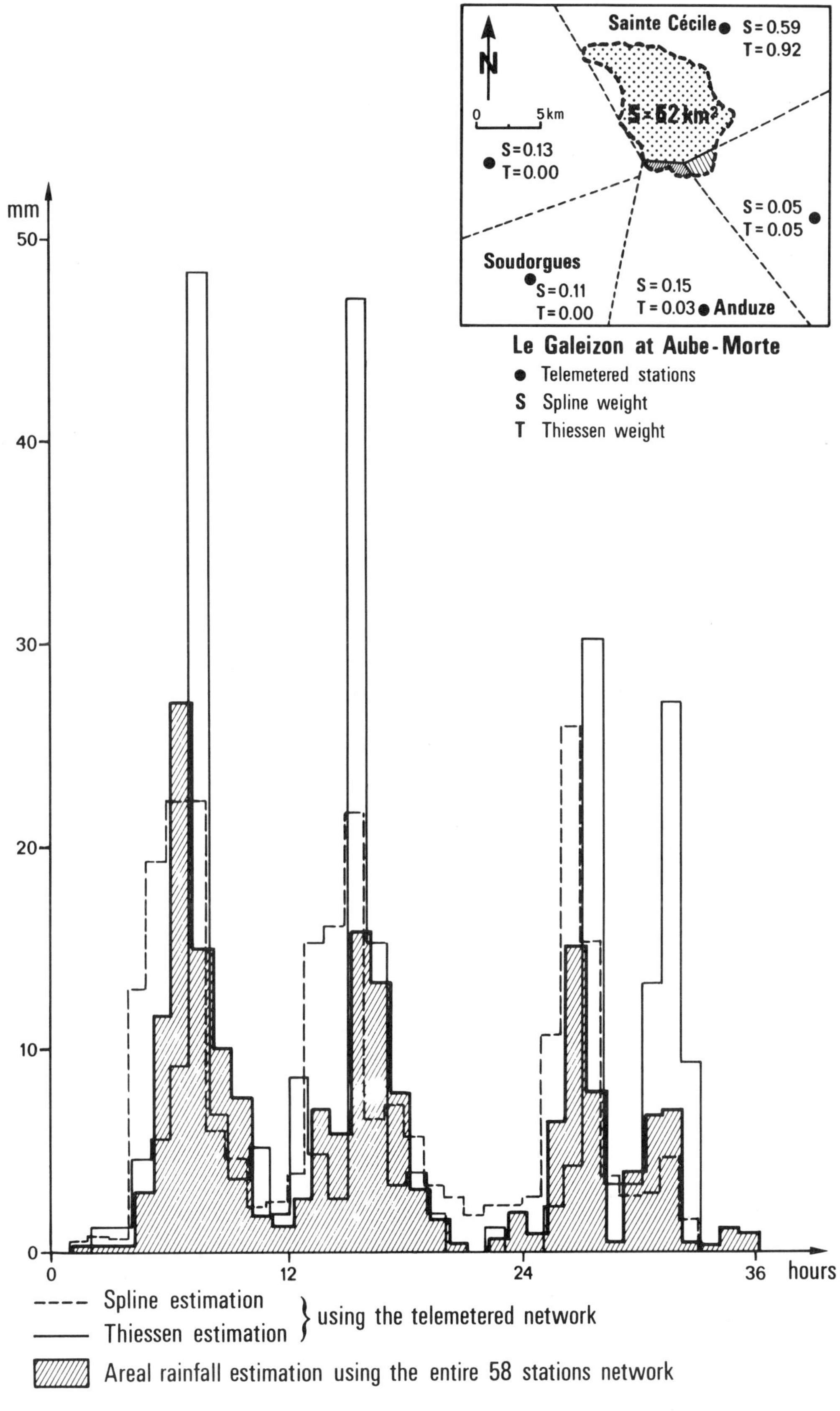

Figure 4. Influence of the Interpolation Scheme on the Basin Areal Rainfall Evaluations.

and 592 hours at Quissac, following a line which is practically a perpendicular transect to the ridge line. (This remains true for the other stations on the ridge.)

On the other hand, the high intensities are clearly shifted towards the bottom of the watersheds, with their maximum along a line parallel to the ridge and located 10 to 20 km downstream (Figure 5). This can be explained by the often unstable nature of the air mass and the appearance of intense convective phenomena as soon as topographical relief is reached. On the other hand, classical orographic effects (forced rise and condensation) tend to explain the higher frequency of rainfall at upper elevations.

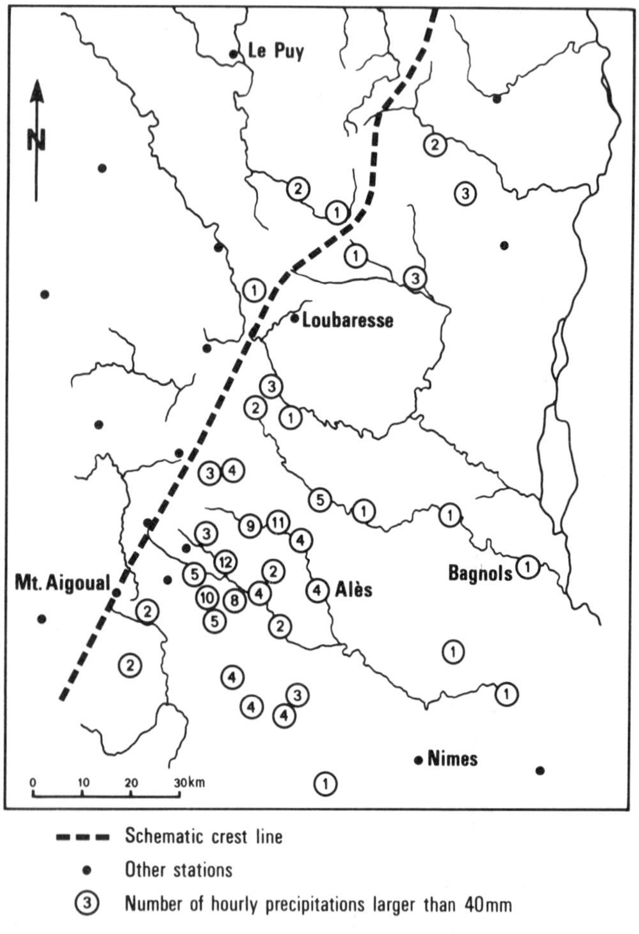

Figure 5. Spatial Pattern of Large Hourly Rainfall Occurrence (among the 28 events selected between 1971 and 1979).

Strangely enough, this fact is often unrecognized by hydrologists who tend to believe in the existence of a precipitation gradient with elevation on all timescales. Note that our results are consistent with those for similar conditions obtained by Fletcher, *et al.* (1981), in small U.S. mountain catchments.

RAINFALL DYNAMICS

To study this aspect, a large number of hourly rainfalls were considered and the interesting sequences were systematically plotted (≈ 500 maps). The following can be observed: (1) rainfall events are first observed, during a preliminary period of variable duration, scattered over the whole region but with low intensities (several mm/hour); (2) next, there is a relatively abrupt transformation into a continuous rainfall band centered in front of the ridge and containing high intensity zones with values up to 80 mm per clock hour; (3) on the whole, this system remains stationary and can last for a very long time (e.g., 24 consecutive hours in October 1977), resulting in the concentration of considerable volumes of precipitation in the upper basins; and (4) finally, the system abruptly disappears toward the east in several hours as the rain stops at upper elevations.

It appears, in fact, that the occurrence of flooding is more a result of rainfall persistence than high intensities and movement from the top to the bottom of the watershed, as sometimes proposed. In one case, the rain system passed the region without stopping and there were no critical rises in streamflows in spite of high intensities. This event is shown on Figure 6, while in the more common events, the band seen on Figure 6c stops moving for several hours.

Complementary meteorological aspects may be found in Tourasse (1980) showing that synoptic scale meteorological features can explain this mesoscale precipitation patterns both in time and space (Figure 7).

FORECASTING AND SIMULATION

The rather special dynamic aspects of these rainfall events, in particular the possibility of cold fronts which remain blocked in one position for varying lengths of time, makes quantitative precipitation forecasting highly uncertain. In addition, even the most sophisticated meteorological models still have a 35 x 35 km mesh size, which is not sufficiently fine to represent the topographical relief. The only operational forecasts available are therefore those provided for the group of watersheds by an elaborate technique involving the determination of analogous situations (Duband, 1980). This method has been used for nearly seven years now but unfortunately only gives daily volumes:

	Forecast Value With Probability of Exceedance		
e.g.: Date	50 Percent	10 Percent	Observed Value
11/09/76	32 mn	83 mn	112 mn
06/11/76	47 mn	105 mn	39 mn

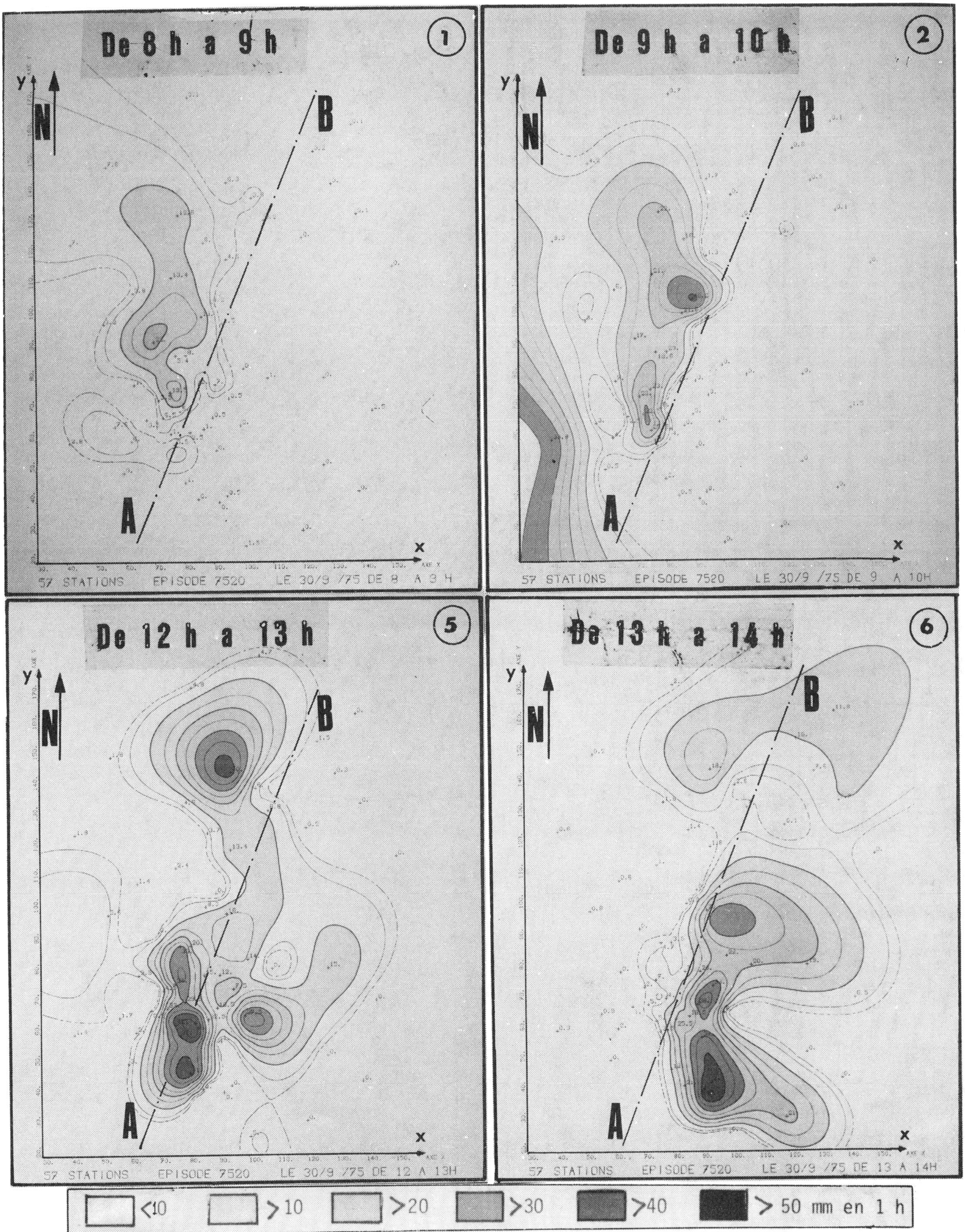

Figure 6. Unusual Rain System Moving Without Blocking During the September 30, 1975, Event.

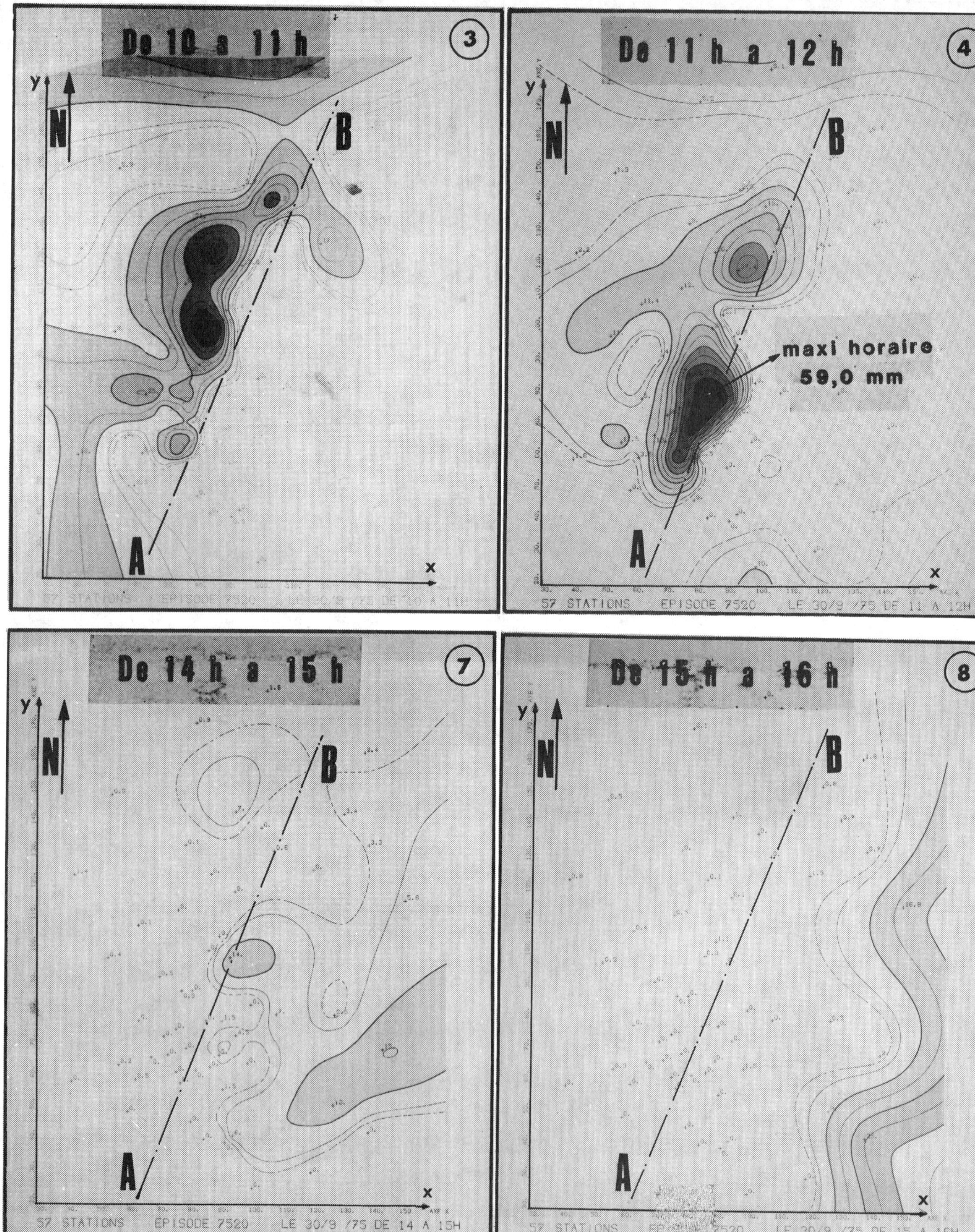

Figure 6. Unusual Rain System Moving Without Blocking During the September 30, 1975 Event (cont'd.).

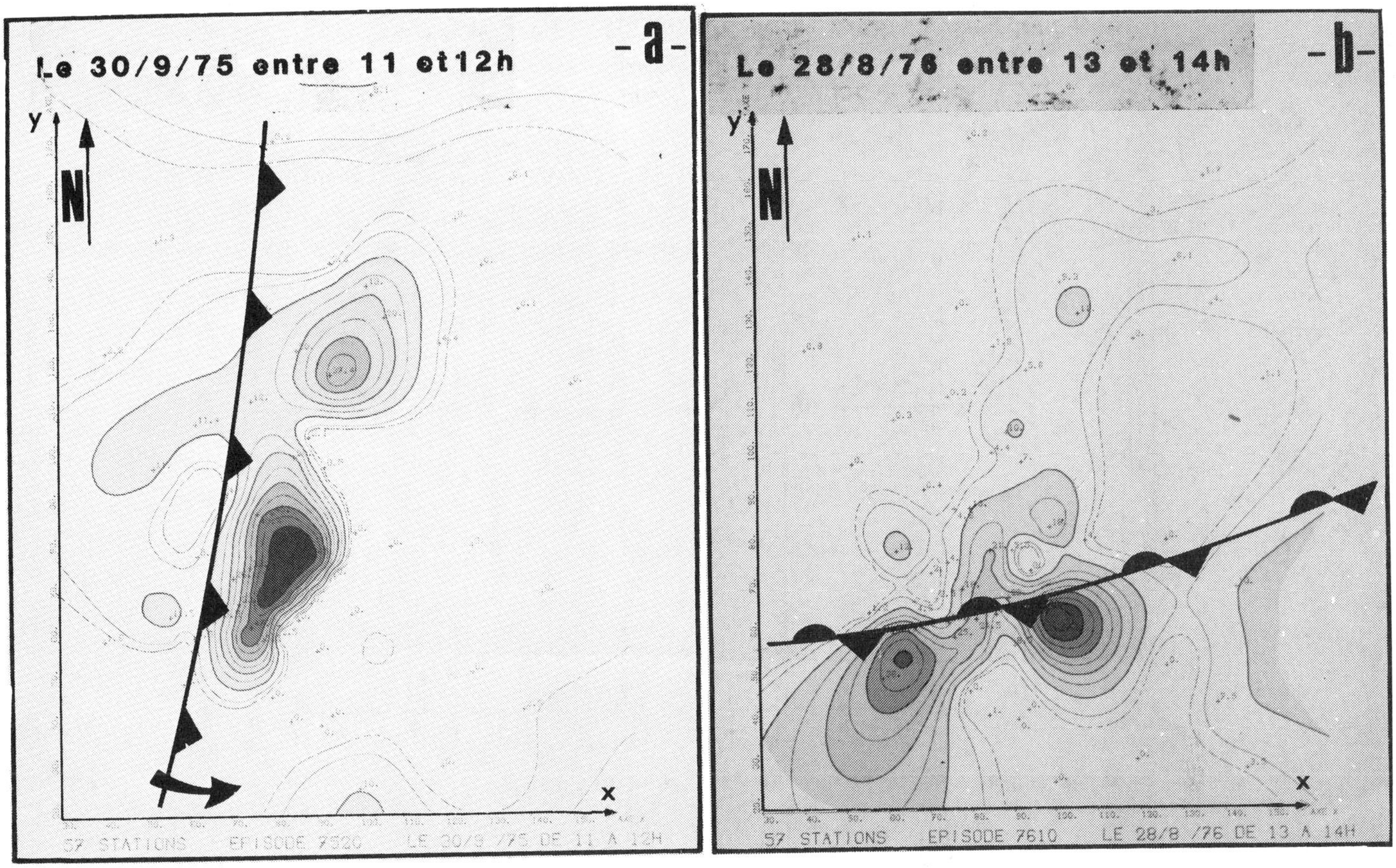

Figure 7. Selected Situations With a Well Identified Front.

Although these forecasts are extremely useful in terms of flood warning, the response time of the watersheds (several hours) makes it impossible to use them for water management based on short time steps.

This problem was already pointed out by Schultz and Klatt (1980) and a more elaborate model (Creutin and Obled, 1980) was suggested. It consisted of a stochastic rainfall generator and was first used to generate separate events at a given station. It has now been adapted to: (1) deal with mean rainfall for a basin and (2) generate the events from hour h taking into account the rainfall already observed. Ten or 20 scenarios are thus generated for the subsequent 6 to 12 hours, bracketing the possible precipitation (Figure 8). These generated rainfalls can be limited by a maximum volume deduced from the 24-hour forecast. If a simple rainfall-streamflow model is available, it is possible to transform these rainfalls into streamflows.

For the time being, this system represents a tool to assist training forecasters who can use it to practise on past flood events. The generated rainfalls are consistent with the observed series except that they do not take into account external factors such as the passage of the cold front, which must be forecast or observed by other means.

CONCLUSION

The aspects dealt with show that the structural characteristics of rainfall, both spatial and temporal, have a major impact on the overall forecasting system, first at the data acquisition level, in terms of the required density of the measurement network, secondly at the rainfall and streamflow modeling level, in terms of their sensitivity to the correct estimation of spatial precipitation, and finally at the operational level, due to the necessity of anticipating rainfall values on a short-term or even hourly basis during the event.

ACKNOWLEDGMENTS

This study has been performed under contract 650-529 from the "Direction Departementale de l'Equipement du Gard."

LITERATURE CITED

Billuart, M. and P. Tourasse, 1980. Telemetering and Simulation. Techniques for a Complete Real Time Flood Forecasting System. Symposium on Flood Forecasting, IAHS – Unesco-WMO, Oxford, 1980, Publ. IAHS No. 129, pp. 3-9.

Creutin, J. D. and Ch. Obled, 1980. Modeling Spatial and Temporal Characteristics on Rainfall as Input to a Flood Forecasting

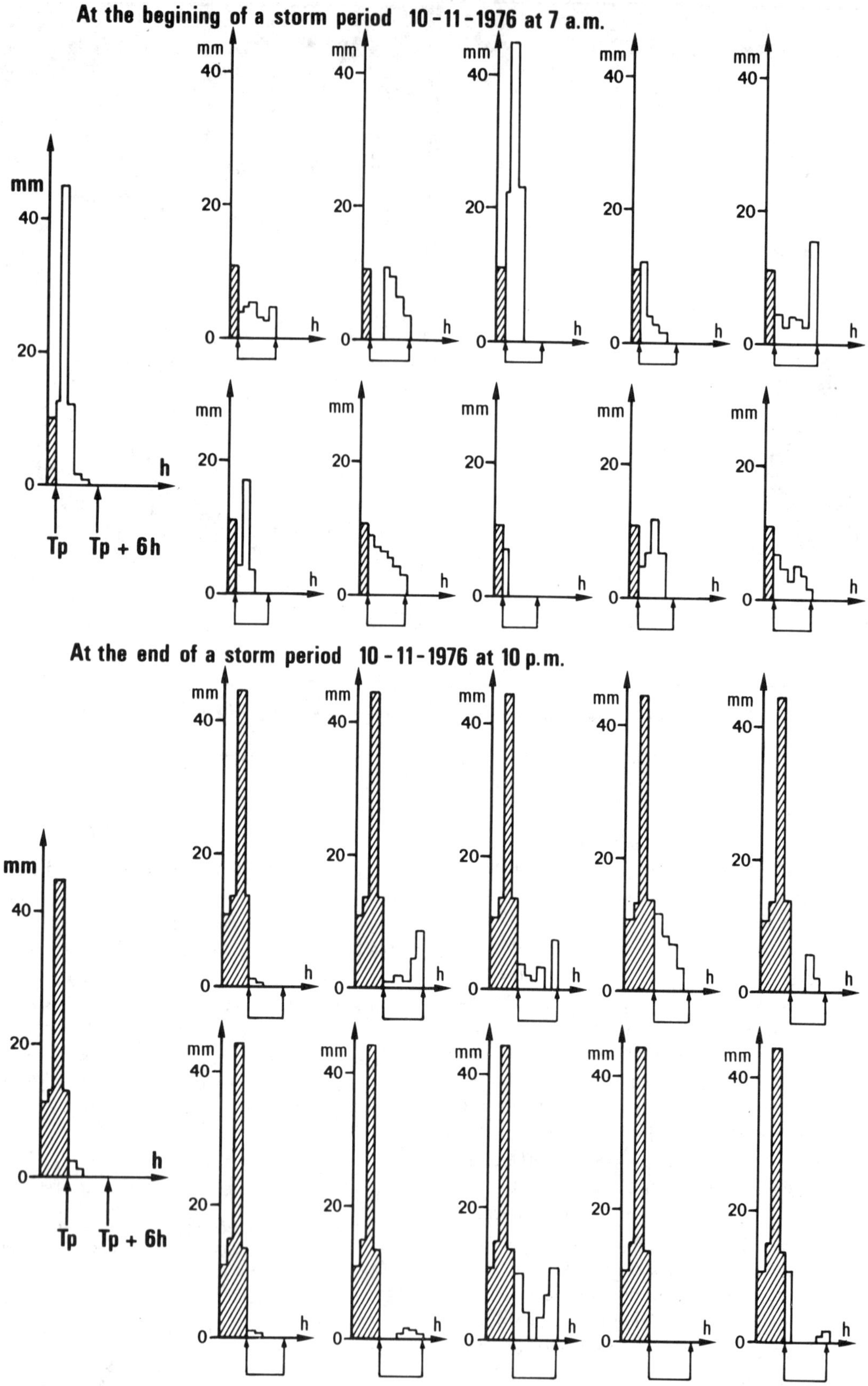

Figure 8. Examples of Scenarios Generated at the Beginning (upper) and the End (lower) of a Storm During the Event of November 11, 1976 (shaded area is already observed rainfall).

Model. Symposium on Flood Forecasting, IAHS – Unesco-WMO, Oxford, 1980, Publ. IAHS No. 129, pp. 41-49.

Creutin, J. D. and Ch. Obled, 1982. Objective Analyses and Mapping Techniques for Rainfall Fields: An Objective Comparison. Water Resources Research 18(2):413-431.

Duband, D, 1980. Dynamic Selection of Analogue Flow Patterns to Produce Quantitative Precipitation Forecasts. WMO Symposium on Probabilistic and Statistical Methods in Weather Forecasting, Nice, September 8-12, 1980, Proceedings, pp. 487-492.

Edwards, K. A., 1972. Estimation Areal Rainfall by Fitting Surfaces to Irregularly Spaced Data. *In:* Distribution of Precipitations in Mountainous Areas. WMO Symposium, Geilo, Norway, 1972, Publ. WMO No. 326, Vol. II, pp. 565-587.

Fletcher, J. E., *et al.*, 1981. Precipitation Characteristics of Summer Storms at Straight Camjou, Barometer Watershed Utah. Research Paper Int 274, Intermountain Forest and Range Experiment Station, Forest Service, USDA, 11 m.

Schultz, G. A. and P. Klatt, 1980. Use of Data From Remote Sensing Sources for Hydrological Forecasting. Research Paper 274, Intermountain Forest and Range Experimental Station, Forest Service, USDA, pp. 75-82.

Tourasse, P., 1980. Analyses spatiales et temporelles de prēcipitations et utilisation opērationnelle dans un systēme de prēvision des crues. These de Docteur-Ingēnieur, Universitē Scientifique et Mēdicale de Grenoble, 190 pp.

GATLINBURG, TENNESSEE, FLASH FLOOD WARNING AND EVACUATION SYSTEM

Corydon W. Bell, Jr.[1]

ABSTRACT: Gatlinburg, Tennessee, is the major gateway to the Great Smoky Mountains National Park. Which together with other area attractions results in a substantial population increase during the summer months. Many of these visitors patronize motels and other businesses located in the West Prong Little Pigeon River floodplain. The river is highly prone to flash flooding. The Gatlinburg flash flood warning system is an automated computer operated rainfall and stream level monitoring system with data analysis capability. It provides a timely warning of the potential for a flash flood. The system consists of a central station; a repeater station; six remote hydrological stations; and a local stream gage. Radio telecommunication is used to address the remote stations and transmit data to the central station. The local stream gage data is transmitted to the central computer over a dedicated phone line. The computer is programmed to take the hydrologic information and utilizing a continually self-adjusting hydrologic model of the watershed forecast flood states in Gatlinburg. The system is being operated by the city personnel and is designed to operate for long periods without attention. The system became operational in December 1981. The evacuation plan has been completed and will be implemented shortly.

(KEY TERMS: flash flood warning systems; floodplain evacuation; telemetered hydrologic information; computer operated; self adjusting hydrologic model; forecasts.)

INTRODUCTION

Gatlinburg, Tennessee, is the major gateway to the Great Smoky Mountains National Park, the most heavily visited in the entire national parks system. The city has a permanent population of about 3,500; however, the park and other area attractions result in a large population increase of up to 25,000 visitors during the summer tourist season. Many of these will patronize motels and other businesses which are located perilously close to the West Prong Little Pigeon River. This places visitors in a potentially dangerous situation, since this stream has all of the characteristics of a stream prone to flash flooding (i.e., fast rise in surface elevation, very high velocities, considerable turbulance, and short duration). A flash flood is extremely dangerous and destructive, offering very little warning time. The situation is worst during the summer months. The combination of the large tourist population and a flash flood threat has the potential to create another Big Thompson Canyon disaster except that the destruction and loss of life would be much greater. The problem is intensified by the fact that the visitors do not appreciate that the mountainous 41.6-square mile watershed above the city can generate flash floods even if it is not raining in Gatlinburg. Various plans for eliminating the flood hazard have been considered. None of these have been found practical or feasible. Until the flood hazard is substantially reduced, a flood forecasting system and evacuation plan are the only way of affording some form of protection. To provide this protection, the Tennessee Valley Authority (TVA), and the city of Gatlinburg have, with assistance from the National Weather Service (NWS), developed a flood warning system and evacuation plan. This system provides warning of flash floods and permits orderly evacuation of the floodplain.

THE FLOOD PROBLEM

Above Gatlinburg the entire basin of the West Prong, including four of its tributaries, Fighting, LeConte, Baskins Creeks, and Roaring Fork, drain the rugged Great Smoky Mountains area. The main stem heads on the crest of the Great Smokies where elevations range from 5,045 feet at Newfound Gap to 6,199 feet on Mount Collins; 6,150 feet on Mount Kephart; and 6,593 feet on Mount LeConte, the highest point on the basin. The watershed drops 5,310 feet in just six airline miles from the crest. LeConte, Baskins Creeks, and Roaring Fork head up on the slopes of Mount LeCounte and fall preciptiously to the city limits. Fighting Creek heads on a long ridge between Mount Collins and Mount Harrison that forms the basin west boundary. Dudley Creek which rises on Mount Winnesoka at elevation 4,316 feet enters the river near the downstream city limits. (Figure 1) Stream channel slopes of 500- to 1,000-feet per mile are common in the upper reaches. The slopes flatten as the streams approach Gatlinburg where the floodplain is

[1] Supervisor of Project Management, Floodplain Management Branch, Tennessee Valley Authority 2B28 Old City Hall, Knoxville, Tennessee 37902.

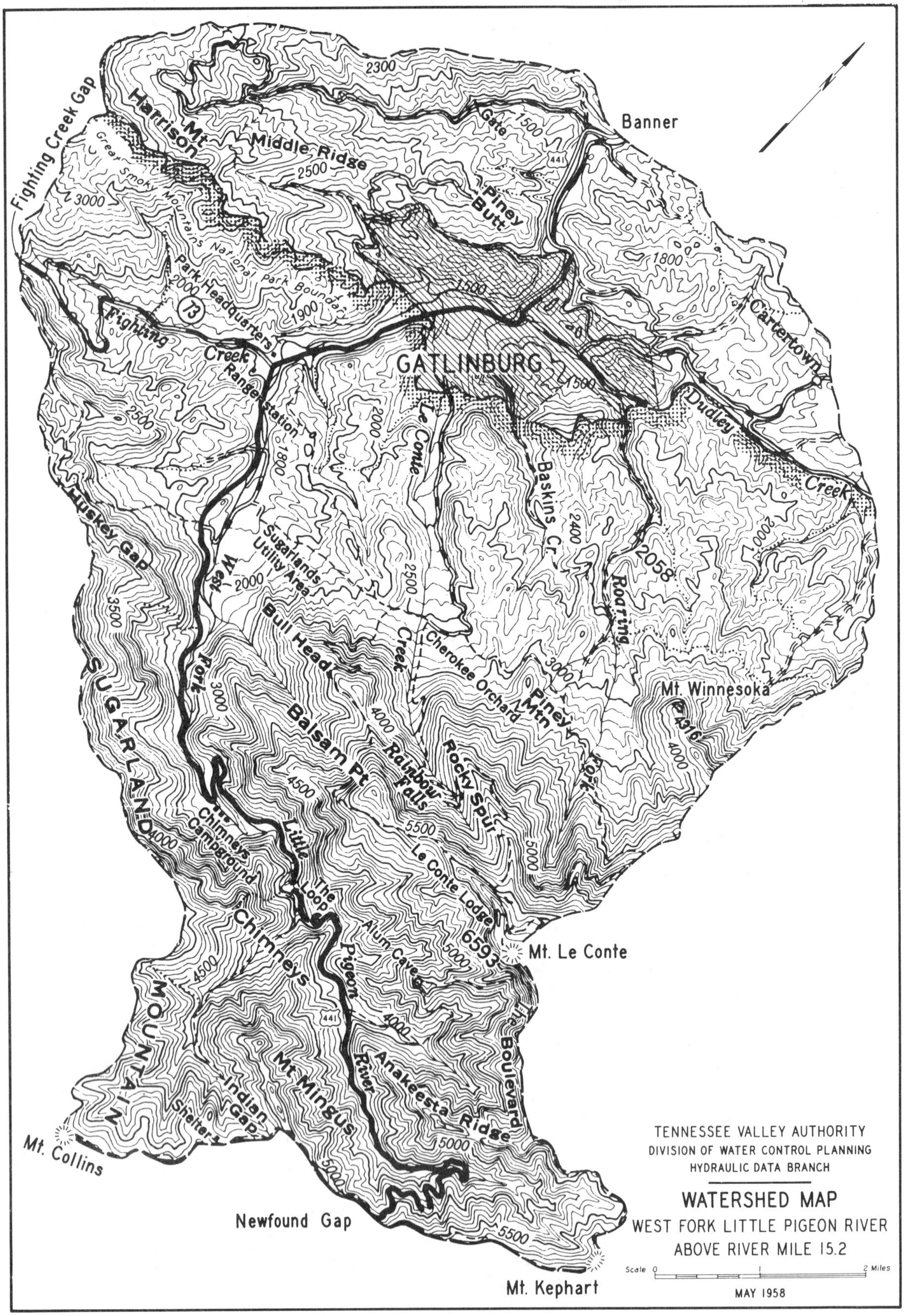

Figure 1. Watershed Map West Prong Little Pigeon River Above Mile 15.2.

quite narrow. Within the city, the average slope of the West Prong is 55-feet per mile.

Rainfall on the watershed ranges from 55- to 90-inches or more with an average annual rainfall of 57.3-inches at Gatlinburg; 82.4-, 86.0-, and 90.5-inches, respectively, at Clingmans Dome, Mount LeConte, and Newfound Gap. There is a considerable potential for intense storms, and with the steep watershed floods can rise rapidly. A one hour storm which produced up to four inches of rainfall on September 1, 1981, resulted in a rise of 10-feet in 15 minutes. Prior to the flood, no rain fell within the city and the only warning was a roaring noise upstream. No lives were lost during this flood. However, flash floods in the immediate vicinity of Gatlinburg have resulted in fatalities. On February 12-13, 1966, a storm produced three to four-inches of rainfall in 27 hours. This, together with saturated conditions from previous melted snow, caused floodwaters to reach near record stages. Major damage occurred to a restaurant and gift shop, six motels, a hotel, one residence, and two city facilities. Many automobiles were damaged or destroyed. Since the flood occurred in February, relatively few people were affected by the floodwaters. Most motel guests in the danger area were evacuated. Those unfamiliar with the rapid rate of rise that occurs with floods on mountain streams waited too long and were trapped.

In its relatively short history as a tourist center, Gatlinburg has been spared severe floods. It is, however, inevitable that large floods in the future will strike the city with disastrous results. At Webb Mountain, only 10 miles from the city, an intense storm in 1938 produced over 11-inches on 27-square miles in three hours and resulted in eight deaths. If this storm occurred today above Gatlinburg, the river would exceed the banks full stage by 7-feet in less than 1½ hours. If it were to center just above the city the warning time would be less than 30 minutes. Floodwater velocity during these floods would reach 11- to 12-feet per second and make the situation even more dangerous.

FLOOD DAMAGE REDUCTION MEASURES

TVA has worked with Gatlinburg for many years to solve the flash flood problem. Flood detention reservoirs in the watershed and park were examined at the feasibility level; these could provide a high degree of protection to the city.

The dams would cause unacceptable environmental damage and otherwise be inconsistent with the purpose of the park. A 6,200-foot tunnel, designed to bypass large floods around the city was investigated. The scheme, while economically feasible, required heavy truck traffic through the city and environmentally unacceptable conditions at the only spoil disposal sites along the river downstream from Gatlinburg. Improvements for the present stream channel to pass large floods and an evacuation scheme, which would have cleared most of the floodway, proved to be economically infeasible and impractible. Thus, the only method that would be both feasible and practical was a flash flood warning system and temporary evacuation plan.

FLOODPLAIN MANAGEMENT PROGRAM

Upon a request by the city, TVA prepared a report in June 1958, concerning the nature and extent of the flood hazard situation. This report, which covered only the West Prong, has been updated and expanded to cover the other streams. In April 1959, the city council passed an ordinance which established emergency procedures to be followed by the city and its inhabitants in the event of a flood. This ordinance provided for early warning of any known high rainfall, or flood events in the watershed; and directed the police to continuously survey the water level within the cities boundaries during wet weather. Provisions were made for warnings to officials at Pigeon Forge and Sevierville downstream. The city also adopted the 500-year flood as its regulatory flood. A water surface rise alarm gage with a horn located in the fire department headquarters, was installed in 1959 by TVA. In 1971 the city entered the emergency phase of the National Flood Insurance Program and is now in the regular program. In 1980 the city requested that its ordinances be revised to make the regulatory flood the 100-year. In the steep valley in which Gatlinburg is located, there is not much difference between the 100- and 500-year flood and TVA agreed to the change.

FLOODPLAIN OCCUPATION

The West Prong floodplain is occupied by the following structures, below the various flood levels shown:

Flood level-year	30	100	500
Hotel and motels	11	22	26
Restaurants	8	15	18
Other commercial	27	63	97
Private residences	11	22	29
Linear feet of public street	8,130	11,060	13,000

PROTECTION THROUGH WARNING AND EVACUATION

By the mid-1970s, it became obvious that there was little prospect of providing significant flood reduction for Gatlinburg through flood modification structures, or other damage reduction measures such as permanent evacuation. This left flood warning and temporary evacuation as the only practical solution. While the city had a flood warning alarm, local officials were hampered by the inability to gather sufficient data to make accurate forecasts and lacked the means of adequately communicating this information to those in the floodplain. Due to the nature of the Gatlinburg flood problem, it was concluded that the traditional flood warning schemes or flood crest forecasts, based upon manually read rain gages, would be inadequate. An environmental analysis report of the proposed flood forecasting and evacuation plan was issued by TVA on May 28, 1976. The report addressed such environmental concerns as locating rain gage structures and stream gages in the park; possible safety hazards to park visitors; visibility of installations; etc. The report showed that the 100-year flood would be 14-feet above stream banks and cause over 200 deaths, and a 500-year flood would rise an additional 2-feet and result in over 500 deaths. These were conservative estimates.

A forecasting scheme was developed by TVA with technical assistance from NWS utilizing state-of-the-art radio telemetry, computer technology, and standard hydrologic flood forecasting methods. An agreement between the city and TVA was executed on June 24, 1976. Under this agreement, the city has: (1) incorporated TVA plans and specifications into bid documents; (2) obtained all necessary construction approvals and permits; (3) upon receipt of bids and concurrence from TVA, awarded contracts to the successful bidders; (4) designated a qualified contract compliance officer; (5) established an operator training program and provided qualified operators on a continuous basis to staff the forecast center and to disseminate warnings; (6) developed an overall action plan in preparation for, and during, flood emergencies, and trained appropriate city employees in their duties and procedures in such events; (7) required that each commercial establishment located wholly or partially within the flood hazard area develop an approved plan for receipt of flood warnings and issuance of specific evacuation instructions; (8) begun providing for a high state-of-maintenance of all equipment associated with the flood warning system; (9) subscribed to appropriate weather advisory services from NWS; (10) provide space and utilities for installation and continuing operators of the forecast center; and (11) continue to enforce zoning ordinances, and land use controls which will prevent unwise use of floodplains. The city will submit semiannual progress and evaluation reports to TVA. TVA has provided: (1) technical assistance and advice to the city in planning and implementation of the project including preparation of plans and specifications; (2) assisted city in developing and maintaining liaison with the National Park Service (NPS), NWS, Civil Defense Agencies, and other federal agencies; and (3) financed the project by reimbursing the city for 90 percent of the cost of obtaining and installing equipment and software up to a maximum of $250,000, and 50 percent of the cities operation and maintenance cost up to a maximum of $15,000 per year for a period of 10 years. Contract A for a computer-based radio telemetry, hydrologic system was awarded to LaBarge, Inc., Tulsa, Oklahoma, for $160,000. Contract B for site preparation including shelter erection and wiring, was awarded to Franks and Nichols, Inc., Knoxville, Tennessee.

THE FLOOD FORECASTING SYSTEM

The Gatlinburg flash flood warning system is an automated, computer-operated rainfall and stream level monitoring system with data analysis capability. It provides a timely warning of the potential for a flash flood. The system consists of a central station; a repeater station; five remote rain gage stations; one remote stream gage station; and a local stream gage. Radio telecommunication through the repeater is used to address the remote stations and transmit data from the remote stations to the central station computer (Figures 2 and 3). Addressing is done through tone coding and data is transmitted using frequency shift keying. The local stream gage data is transmitted to the central computer in an analog form over a dedicated phone line. The computer is programmed to receive the hydrologic information and utilizing a continually self-adjusting hydrologic model of the watershed forecast flood stages in Gatlinburg.

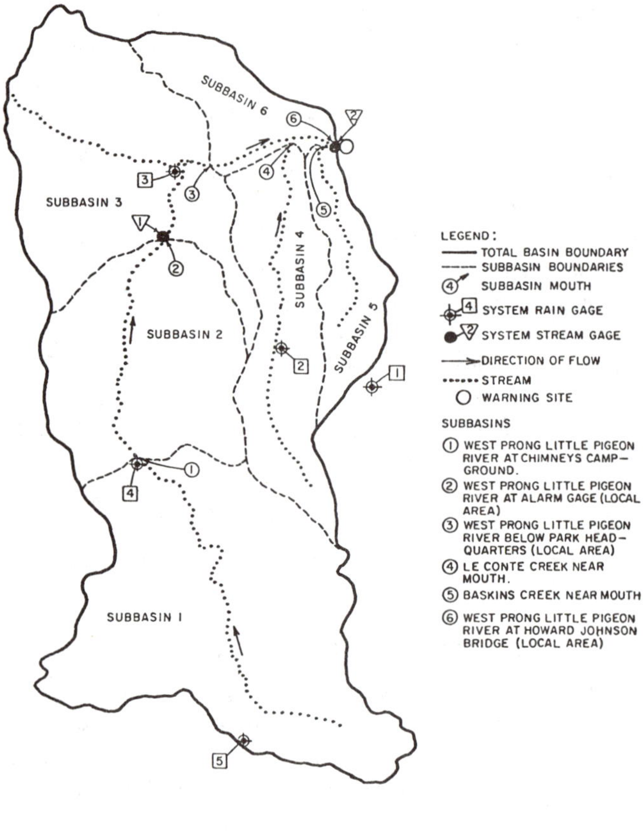

Warning System General Schematic – Six Subbasin System

Figure 2. Map Showing Warning System Equipment Locations. No. 6 is located in the city of Gatlinburg.

The system which is operated by city personnel, must operate with little attention. This dictated that five basic requirements had to be met. These are: minimal interreaction with operating personnel; backup capabilities;

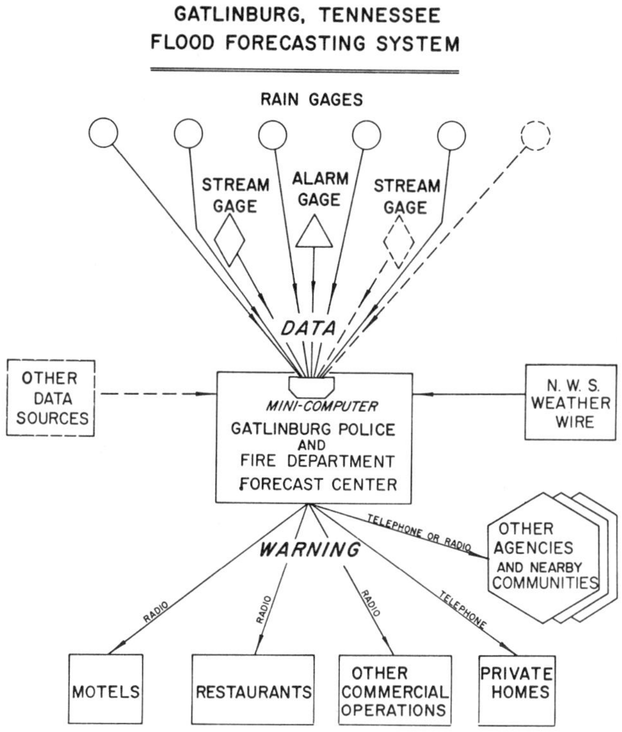

Figure 3. Schematic Diagram of the Gatlinburg, Tennessee, Flood Forecasting System.

Figure 4. Central System Station in the Gatlinburg Fire Department Headquarters. Equipment includes the printer; NOVA 3/12 computer; magnetic tape and disk units; audio visual alarm, system status, and optical display panels; digital clock; radio transceiver and encoder; interface and control logic units; and backup battery. An emergency motor generator set is located elsewhere in the building.

self-diagnostic routine; fail safe operation; and high degree of reliability. The central station equipment (Figure 4) includes a model 8607G NOVA 3/12 minicomputer; model 6021 magnetic tape system; model 6080A DASHER printer; model 8535 NOVA 3 memory management unit; and model 8538 NOVA 3 memory protection unit (all Data General Corporation); model MVP General Electric (GE), transceiver with model 99 encoder; Winchester hard disk unit with flexible disk backup; digital clock with battery backup; LaBarge custom audio-visual alarm, optical display (displaying predicted flood levels and time flood stage will be reached); and status (for displaying abnormal conditions for both the remote and central stations) panels; thumbwheel switches for entering flash flood index constants; Pincore model RS30HH5 emergency alternator; and DB model DB205 antenna. At the repeater station a model DC56YAS66A radio repeater and duplexor and model DB222 antenna are utilized. At the remote stations (Figures 5, 6 and 7), the equipment consists of a LaBarge DARDC data translator unit; Belfort model 5915 rain gage with encoder and servomechanism or Scientific Instrument stream gage (USGS bubble type); model MVP transceiver with encoder; Teledyne 8-watt catalytic thermal generator, and 12-volt automobile battery; LaBarge interface assembly; antenna model DB230; four propane gas bottles; and in case of the stream gage; a nitrogen supply bottle; and circuitry for reporting the station health parameters. The latter consists of four items including the number of full propane gas bottles; status of thermal generator gas flame; battery voltage above or velow 11.5 volts; and rain gage bucket over or under half full. The local stream gage station consists of an existing NWS resistance gage type BE and a dedicated telephone line between the gage and the central station.

SYSTEM OPERATION

Once the program is loaded, the central station minicomputer operates unattended except for the daily flash flood index (FFI) updates. Every ten minutes the computer requests data from each remote station and the local gage. Each station responds by sending two identical messages containing station ID, precipitation or stream level, and "station health" parameters. Health parameters are printed out every six hours and an alarm will sound whenever there is a problem. The central station checks the messages and flags any detected errors, nonresponsive remotes, or any remote in need of servicing. The stream and rainfall data are then analyzed and, depending on their value, time of

Figure 5. Typical Remote Rain Gage Station Shelter. This one is at the National Park Service headquarters. Other stations are similar. The Yagi antenna is used for both transmission and receiving.

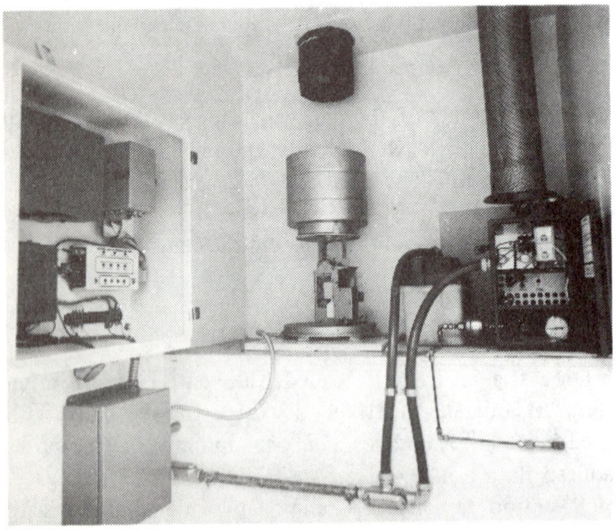

Figure 6. Interior View of a Remote Rain Gage Station Shelter. Equipment, left to right: NEMA enclosure containing remote station health parameter unit, frequency shift key receiver, electronic encoder, interface transmitter, and control logic; rain gage, servomechanism and encoder; lead acid battery; and thermal electric generator. Four propane gas bottles are in the compartment below this equipment.

Figure 7. Interior View of the Remote Stream Gage Station Shelter. Equipment, left to right: NEMA enclosure containing the same equipment as the rain gage station; stream gage gas monitor, encoder, and manometer; battery; thermoelectric generator; and nitrogen gas supply bottle. Propane gas bottles are located in the compartment below this equipment.

day, and history, they will be stored on magnetic tape, printed, discarded, or used to make flash flood and potential flash flood predictions. Other system features include operating program coded in read-only memory; manual interrogation of the remote stations without computer involvement; and overtimer periodically set by software to sound an alarm if there is a computer software or hardware failure; dual transmission for error detection; in the event of a power loss the program will restart automatically upon the return of power or startup of the emergency generator; and with a program failt, when data may be lost, automatic restart is voided and a warning is given to the operator. For ease of writing the program and servicing the hardware, the operating program is written using the Data General Real-Time Operating System (RTOS). Upon startup the operator down loads the program from ready-only memory to random-access memory. This allows the operator to operate the system without having to handle paper tape, magnetic tape, disc packs, etc. This ease of operation and elimination of the need for peripherals with moving parts provides for a more reliable system. The peripherals used – magnetic tape recorder and printer – are for recordkeeping and enhancement and are not essential for successful flash flood prediction.

The minicomputer uses a continually self-adjusting model of the drainage basin to make its forecasts. Rainfall is converted into rainfall excess which is used to compute discharge hydrographs for the subbasins. These are in turn routed and combined to produce a projected hydrograph for the local stream gage. When flooding is eminent the computer will print out a warning, light a control panel warning signal, and sound an audible alarm. When this occurs, appropriate portions of the Gatlinburg evacuation plan will be implemented. The computerized forecast system, except for the input of the FFI, provided daily by the NWS is completely independent of the operator. This minimizes the potential for accidental system impairment. The program in effect provides the onsite services of a trained hydrologist. It is designed with model parameters as variables which can be modified as gaged data from model calibration and experience with the system dictates. Another objective of the system is to provide hydrologic forecasts which enhances operator judgment throughout a flood event. The utility of the system will be enhanced as the operator gains experience in interpreting the data and various forecasts and integrates this with his own evaluation of the event.

INITIAL OPERATIONS

Installation and testing of the system hardware and software was essentially completed in December 1981. On the morning of January 4, 1982, the system alarm warned that a substantial amount of rainfall had occurred above Gatlinburg would cause the West Prong to approach flood stage. This information was used by emergency personnel in answering inquiries from area businessmen regarding the possibility of flooding and the time of crest and the expected height. The initial operation of the Gatlinburg warning system provided emergency adequate time to perform any needed evacuations and proved that Gatlinburg is better prepared to cope with future floods. It is anticipated that the evacuation plan will be implemented shortly.

With over 13,000 linear feet of public street, portions of 26 motels and hotels; 18 restaurants; 29 private residences; and over 97 other commercial establishments at or below the level of the 500-year flood, the evacuation plan for Gatlinburg will not be easy to implement. Besides the magnitude of the problem, there is some resistance. In a public meeting at Boulder, Colorado, to present a flash flood warning plan to the county following the Big Thompson Canyon flood, the NWS flash flood coordinator noted that concern was shown for flood warning, but he believed the general attitude was one of apathy. People with authority to act are not particularly interested in preparing a flood warning plan. During the meeting a number of officials, including two county sheriffs, spoke without mention of the flood warning plan. It was not until Gilbert White spoke that the first mention of the flood warning plan was made.

At Gatlinburg it has been difficult to generate interest and enthusiasm on the part of city officials and property owners. Drafts of the evacuation plan have been reviewed and extensive comments made by TVA during the past several years. Even though it is the city's responsibility to prepare the plan with TVA review and approval, the plan for endangered areas has not been completed, tested, and published. The city does plan to do this before the main part of the tourist season gets underway after July 4. The resistance mentioned comes in the form of commercial enterprise owners disliking the idea of instructions prominently posted at restaurants, hotels, and so forth, detailing to their customers how to proceed once a flood warning has been initiated, yet the customers in any given location need sufficient information to know where to go should that warning occur. The situation is eased somewhat by the fact that safety in the form of higher ground is seldom more than 100 yards away. There are several factors which tend to make the matter more difficult for responsible local officials to effectively implement an evacuation plan. These are: (1) most of the floodplain occupants will be visitors to the area and thus unfamiliar with the situation and their surroundings; (2) floods rise rapidly and at best the system can only give one or two hours warning and in some cases as little as 30 minutes; (3) floods are most likely to occur in the period between the late afternoon and the early morning hours; and (4) most severe floods are likely to occur in late summer when visitation is at its peak. Because of the short warning time and the large numbers of persons likely to be involved, the evacuation plan is designed to protect life and limb and not property. The plan does provide, however, for early dispersal of fire fighting equipment so that all of the community will have protection if floodwaters cut off access.

Gatlinburg evacuation plans are actually a two-part document consisting of the general plan for the city and its officials and a series of annexes which contain individual evacuation procedures for motels, restaurants, hotels, and other commercial and private installations within the hazard area. Essential elements of the plan are: (1) assignment of duties and responsibilities to local officials, public safety forces and others; (2) procedures for dissemination of alarms under various flood conditions; (3) evacuation procedures including routes and safe areas; (4) traffic and crowd control; (5) rescue operations; (6) testing procedures; (7) forecasting center operations; and (8) post flood operations.

Evacuation plans for motels, hotels, restaurants, commercial, and private installations in the hazard area will vary widely in scope and complexity depending upon the degree of hazard exposure and the distance to safety. As noted above, the city of Gatlinburg has a responsibility for development testing and implementation of the evacuation plan. The contract between TVA and the city requires semiannual tests and modification if needed. Because of the ever changing visitor population, it is not possible to include visitors in the test. Testing is limited to local officials, public safety forces, and employees of the various commercial establishments. An outline for a commercial structure evacuation plan would have, but not be limited to, the following elements: (I) Purpose and Scope; (II) Duties and Responsibilities; (III) Alarm; (a) how received; (1) authentication to conformation of receipt; (b) alarm conditions;

(c) alerting procedures by "condition". (IV) Evacuation Procedures; (a) general instructions; (b) route; (c) safe areas; (d) guides. (V) Post Event Instructions. (VI) Methods of Handling Special or Unique Hazards or Conditions. (VII) Readiness and Testing Procedures.

FLOOD FORECASTING/WARNING SYSTEM FOR IBADAN, NIGERIA

Taiwo A. Oyekan and Jerry R. Rogers[1]

ABSTRACT: The city of Ibadan with population exceeding 2 million, is the second largest city in Africa. River Ogunpa and its tributaries transverse the city. Ibadan has been plagued by flooding from torrential rains for years. The last two floods of 1978 and 1980 claimed 40 and over 700 lives, respectively. Additionally, minor flooding frequently occurs during lesser storms. Immediately after the 1980 flood, Ibadan was without drinking water for over 3 weeks due to the damage of Eleiyele dam – the city's major source of drinking water.

Based on previous rainfall records, a typical Stage versus Rainfall Intensity curve is developed. An extensive network of stage recorders and rain gages will also be installed. During heavy storms, adequate forecast and warning will be dispatched once a critical rainfall intensity is attained. A rainfall-runoff model will also be calibrated for more accurate stage predictions. Following the successful calibration of the model, the system will be made available to government officials, radio and television stations for their consideration and eventual implementation of the system.
(KEY TERMS: flood forecasting; flood warning; flood-plain management; rainfall-runoff model; stage; rainfall intensity.)

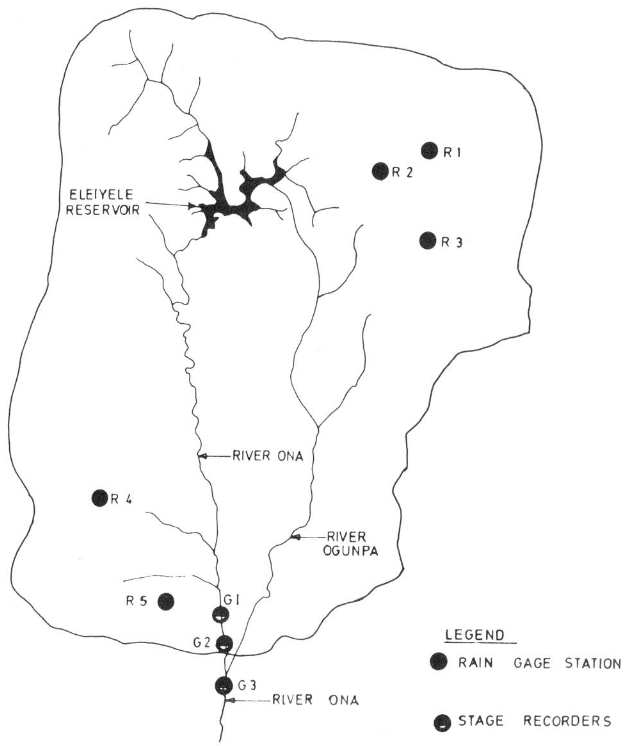

Figure 1. Map of Study Area.

INTRODUCTION

Ibadan, the capital city of Oyo State, Nigeria, with a population exceeding two million spread over an area of approximately 347 mi^2 (900 km^2), is the second largest city in Africa south of the Sahara Desert. The city lies between longitudes 3° 45' and 4° 00' East and Latitude 7° 15' and 7° 30' North. Ibadan is situated on a rolling topography, at altitudes ranging from 459 ft. (140m) to 689 ft. (210m) with isolated ridges and peaks rising to an elevation of 918 ft. (280m) above mean sea level. Figure 1 is a map of the study area.

Rainfall

There are five gauge stations in Ibadan. The approximate locations of the gauges are shown on Figure 1. Table 1 shows typical annual rainfall at these stations from 1952 to 1981.

Brief History of Flooding in Ibadan

Available records indicate that major flooding occurred in Ibadan in June 1951, June 1955, August 1960, August 1963, May 1969, April 1978 and August 1980. Table 2 shows the average amount of rainfall and duration during the major floods.

Documented data on the actual amount of damage caused by the 1951, 1955 and 1960 floods are not

[1]Respectively, Lecturer, Department of Civil Engineering, University of Ibadan, Faculty of Technology, Ibadan, Nigeria; and Associate Professor, Department of Civil Engineering, University of Houston, Houston, Texas 77004.

TABLE 1. Typical Annual Rainfall (1952-1981)

Year	Amount of Rainfall, Inches (mm)
1952	50 (1270)
1953	47 (1193.8)
1954	58 (1473.2)
1955	59 (1498.6)
1956	33 (838.2)
1957	63 (1600.2)
1958	41 (1041.4)
1959	49 (1244.6)
1960	52 (1320.8)
1961	40 (1016)
1962	58 (1473.2)
1963	62 (1574.8)
1964	51 (1295.4)
1965	54 (1371.6)
1966	51 (1295.4)
1967	32 (812.8)
1968	85 (2159)
1969	43 (1092.2)
1970	49 (1244.6)
1971	36 (914.4)
1972	43 (1092.2)
1973	54 (1371.6)
1974	47 (1193.8)
1975	50 (1270)
1976	34 (863.6)
1977	38 (965.2)
1978	62 (1574.8)
1979	69 (1752.6)
1980	76 (1930.4)
1981	60 (1524)

available. However, many of the residents interviewed are of the opinion that it was not as extensive as those recent years. A similar observation has been made by Mogil, *et al.* (1977). It was reported that in the U.S., the average national death toll from flash floods during the 1970's tripled that of the 1940's. Additionally, property damages during the 1970's averaged over 1.5 billion dollars (N 1 billion) annually while that for the 1940's were approximately $125 million dollars (about N82 million) and $375 million dollars (about N250 million) during the 1960's. The 1960 flood in Ibadan was reported to have caused damages in excess of $150,000 U.S. dollars (N100,000) (Maclaren International, 1970).

The 1963 and 1969 floods were not as disastrous as those of 1978 and 1980. Therefore, major emphasis will be on the 1978 and 1980 storms. During the 1978 storm, 4.4 inches (112.7mm) of rain fell within a 4-hour period. Approximately 40 people died during the flooding. Many homes and businesses were inundated. Damages due to flooding were estimated at over $15 million (N10 million).

The August 1980 storm has been described as a 100-year event (Tokun and Adesina, 1981). Over 700 lives were lost as a result of the flooding. It has been described as the worst flood disaster which any city in Nigeria ever experienced. Property damages were in excess of $90 million (N60 million) as a result of a 10.8 inch (275mm) rainfall which occurred in approximately 11 hours. Many bridges and homes were either destroyed or inundated. It was reported that many families were completely eliminated.

Based on the foregoing, it is obvious that a flood forecasting/warning system is needed in Ibadan. However, a flood warning system would be inadequate without public safety and education programs. Mogil, *et al.* (1977), reported that such a program has improved response to flood warnings. Among the reasons that have been given for lack of flood control programs in Ibadan are lack of manpower and funds. Excessive flood damages and other disastrous events could be reduced if a forecasting system were available. Also, the damage to the Eleiyele Dam during the 1980 storm could have been averted if a flood warning or flood watch system were available. Berling (1978) reported on how a project operational alert system played an important role in preventing greater damage and loss of lives at the Big Thompson River Canyon of Colorado. The success of the system came from personal observation of heavy rainfall and measurement of flood flow in the Estes Park area. Based upon observed condition and good judgment, immediate action was taken to place the Project System into an emergency mode of operation. Therefore, the proposed warning system will be useful for flood watch purposes at Eleiyele Dam.

TABLE 2. Major Floods in Ibadan

Date	Amount of Rainfall, Inches (mm)	Duration, Hours
June 1951	3.8 (97.03)	+
June 1955	6.8 (172.7)	5
August 1960	6.76 (168.87)	24
August 1963	7.5 (190.5)	9.5
May 1969	1.0 (25.4)	1.5
April 1978	4.4 (112.7)	4
August 1980	10.8 (274.79)	10.75

+ No available records.

For the reasons stated above the authors suggest a simple and economical method by which floods can be predicted and appropriate warnings issued to residents in the "critical" areas, ahead of time. Cost estimate for such a system has been given (Table 3). Additional work being organized by the authors includes the calibration of a rainfall-runoff model to predict basin runoff. With the aid of such a model, increases in runoff produced at different degrees of urbanization can be measured. Also, the model can be used to evaluate the effectiveness of the "real time"

TABLE 3. Cost Considerations

Description	Quantity	Unit Cost ($)	Total Cost ($)
Initial Capital Costs			
1. Bubbler Type River Level Gage (model 5086) with Self-Reporting Rain Station (model 5054)	5	5,100	25,500
2. Data Receiver	1	2,500	2,500
3. Remote Station Tester	1	3,200	3,200
4. Data Printer	1	2,900	2,900
5. Installation of stations (a) 5 hours/station	5	1,000	5,000
6. Installation of the Central Station (a) 3 hours	1	500	500
Annual Operating Costs			
Maintenance (1 visit/year)		3,000	3,000
Annual Salaries			
3 Central Site operators (8 hours/shift)		9,600	28,800
Unexpected Expenses/year		2,000	2,000
	Total Estimated Cost		**$73,400**

forecasting system proposed in this paper. While there are other more sophisticated forecasting systems, it is believed that the system proposed here will serve to predict floods for several years until more funds can be earmarked for a more advanced forecasting system by either the State or City Government. The cost of the flood warning system will be relatively lower than the cost of other methods of flood mitigation.

FLOOD FORECASTING SYSTEM

The authors believe that it is possible to predict floods in Ibadan and issue warnings via radio and television, to residents along the streams and in other flood-prone areas of the city in time for them to evacuate the flood-threatened areas. In many cases, the warning time may be long enough to permit evacuation of movable property and people. However, the feasibility of a flood forecasting system as a solution to flood problems or as an integral part of a flood control program is dependent on time factor. Since Ibadan is an urbanized area with relatively large streams, hours or even days may elapse before the flood waters begin to approach critical stages in downstream areas. In Ibadan, rarely is the total elapsed time less than two hours and more often, three or four hours has been observed. For example during the August 1980 storm, it took approximately 6 to 7 hours before extreme flooding began in the different parts of the city. During the storm, 11 inches (275mm) of rain fell with an average intensity of 1 inch/hr. (25.4mm/hr.) and a duration of 10.75 hours, hence there was ample time available to issue warnings to residents if a forecasting system were available.

The most important phase in establishing a flood warning system is the development of an effective flood forecasting system. A flood forecasting system must comprise the following: (1) A data collection system which must be capable of obtaining and updating rainfall and river stage data as the event is occurring (this is called "real time" event reporting); (2) The forecasting system must be capable of transforming the information on rainfall and stage data to streamflow at different points, and determine the extent of flooding at points along the stream, where warnings are needed; and (3) The flood forecasting system must have an effective method by which forecasts and warnings can be issued to the public.

Development of the Rainfall-Runoff Model

Extensive and long-term data collection are needed for the successful calibration of the rainfall-runoff model. The model will be used to simulate amount of runoff for intermediate and future levels of urban development in Ibadan. Therefore, the model must simulate the hydrologic conditions as closely as possible in order to be used as a part of the forecasting and warning system.

For comparative purposes, two separate programs will be calibrated. These are the WSP3 Computer Programs written by the U.S. Conservation Service, and the HEC-1 written by the Army Corps of Engineers. Also, the HEC-2 program will be used for flood-plain delineation. These programs will be modified and set up on the IBM 370 computer at the University of Ibadan. (According to Maier (1981), the WSP3 runs in 20 percent of the time needed for HEC-2.) This is an important factor in watershed modeling especially in urbanized areas where times of concentration are relatively low. Among the information which will be fed into the model include the starting water surface elevations, precipitation and antecedent moisture conditions. Water surface elevations, bridge deck and road elevations at critical locations will also be utilized.

INTEGRATION OF THE FORECASTING AND WARNING SYSTEMS

Instrumentation Plan

The proposed network of Liquid Level/Rain Station and the Central Station is shown on Figure 2 as S1 through S5. The extensive network is needed for an adequate and accurate prediction of the magnitude of flooding. Since rainfall is not often uniformly distributed with respect to time and area, five locations are being proposed. The key parameters that will be measured are the amount of rainfall, duration and river stage at each location.

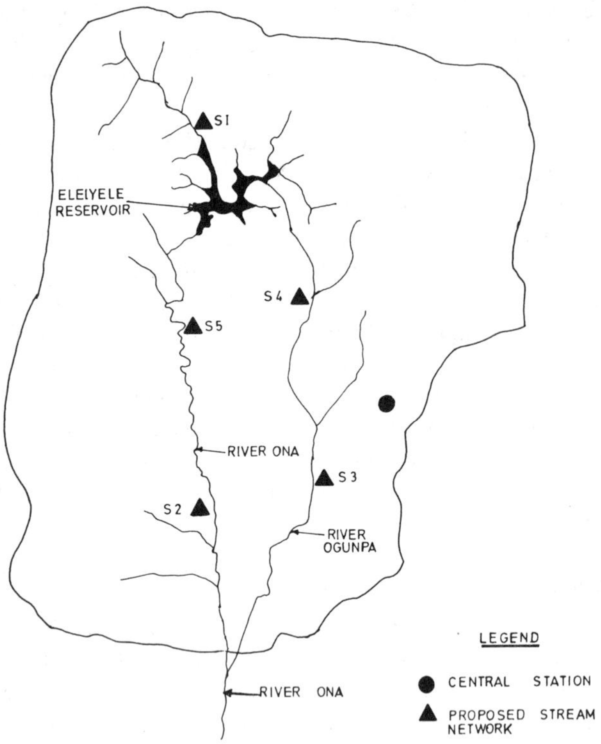

Figure 2. Proposed Stream Network.

Based on the foregoing, stage versus time record of streamflow can be measured at each station. Also, the amount of rainfall can be measured simultaneously. All of the stations will be instrumented with Liquid Level/Rain Station manufactured by the Sierra-Misco, Incorporated. Also, Data Receivers, Data Printers and Remote Station Testers will be used as part of the system. A schematic of the flood warning system is shown on Figure 3. Schematic diagrams of the Liquid Level Station and Self-Reporting Rain Station are shown on Figures 4A and 4B, respectively.

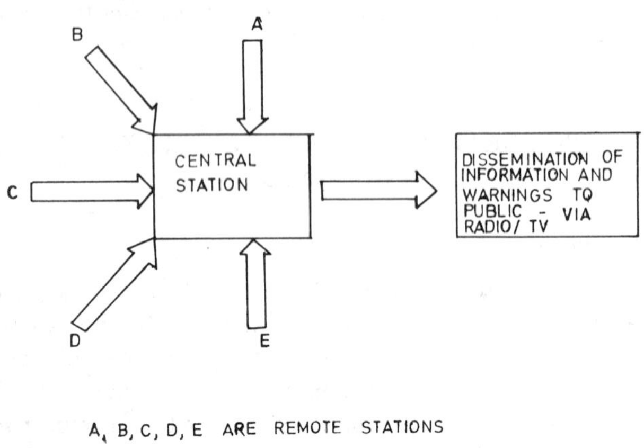

Figure 3. Schematic Diagram of Flood Forecasting/Warning System.

Relationship Between Stage and Rainfall Intensity

A comprehensive river gage network is not currently available for Ibadan. River Stage data is available only for the Ona River on the outskirts of Ibadan, at the three locations shown as G1, G2, G3 on Figure 1, Immediately after the April 1978 and August 1980 floods, the following river levels were recorded at G1 and G2:

River Stage Reading at Gage G1
(Measurement above River bed)

Date	7 a.m.	6 p.m.
April 20, 1978	4.80 ft. (1.46m)	4.80 ft. (1.46m)
April 21, 1978	4.80 ft. (1.46m)	4.80 ft. (1.46m)
August 31, 1980	4.60 ft. (1.4m)	6.80 ft. (2.07m)
September 1, 1980	11.20 ft. (3.41m)	10.40 ft. (3.17m)

River Stage Reading at Gage G2
(Measurement above River bed)

Date	7 a.m.	6 p.m.
April 20, 1978	8.50 ft. (2.59m)	8.50 ft. (2.59m)
April 21-23, 1978	12 ft.+ (3.66m+)	12 ft.+ (3.66m+)
April 24, 1978	11 ft. (3.5m)	11 ft. (3.35m)
August 31, 1980	No gauge readings available.	

No record is available for Station G3.

It is important to note that the gage readings given above as 12 ft.+ (3.66m+) will actually be about 4 to 5 ft. (1.2 to 1.5m) higher. This is because of a concrete impoundment immediately downstream of the gage station. The impoundment was used for construction purposes during the erection of the bridge across the Ona River on the Ibadan-Lagos road in December 1974. However, the impoundment was not removed upon completion of the bridge construction. This created blackwater problems during the 1978 and 1980 floods. Since the maximum reading that could be obtained from the gage is 12 ft. (3.66m), and the gages were overtopped during the floods, depth above river bed during the April 1978 flood has been estimated to be approximately 16 ft. (4.9m).

Previous floods have provided the necessary data utilized in plotting the stage versus rainfall intensity curves shown on Figure 5. Based on the data shown on Table 2, the major flood-producing storms possess average intensities ranging from 0.66 inch/hour (16.8mm/hr.) to 1.1 inches/hr. (27.9mm/hr.). For this reason, a critical range of intensity between 0.5 inch/hr. (12.7mm/hr.) and 1 inch/hr. (25.4 mm/hr.) has been selected for Ibadan. Available rainfall versus duration records, and the time to peak stages (from eye-witness accounts), have been utilized in the development of Stage versus Rainfall Intensity curve of Figure 5.

It should be noted that once the channel geometry changes and the hydraulic capacity is increased, the peak stages will be lesser for the same rainfall intensities shown on Figure 5.

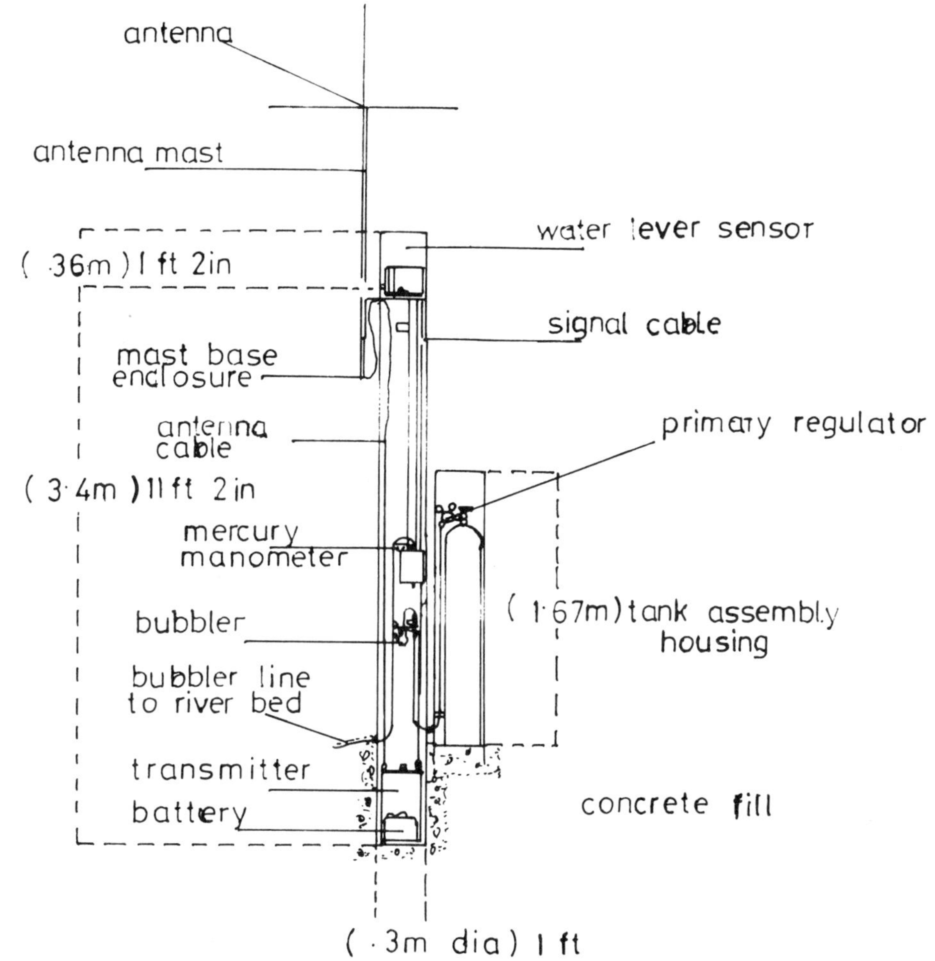

Figure 4A. Schematic of Liquid Level Station (after Sierra Misco Inc., 1981).

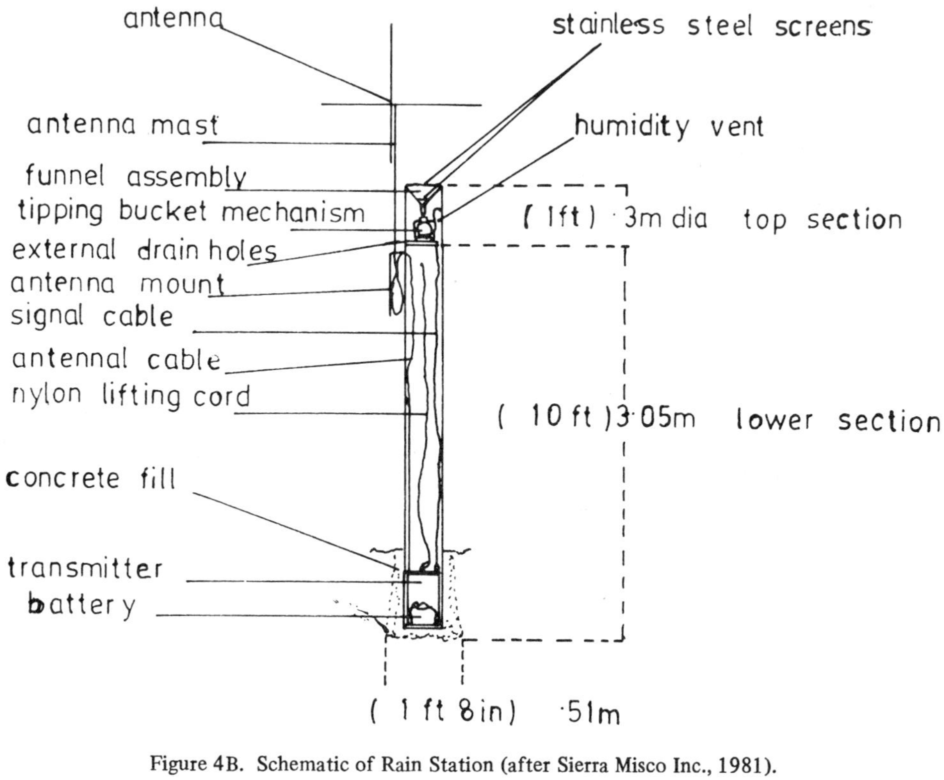

Figure 4B. Schematic of Rain Station (after Sierra Misco Inc., 1981).

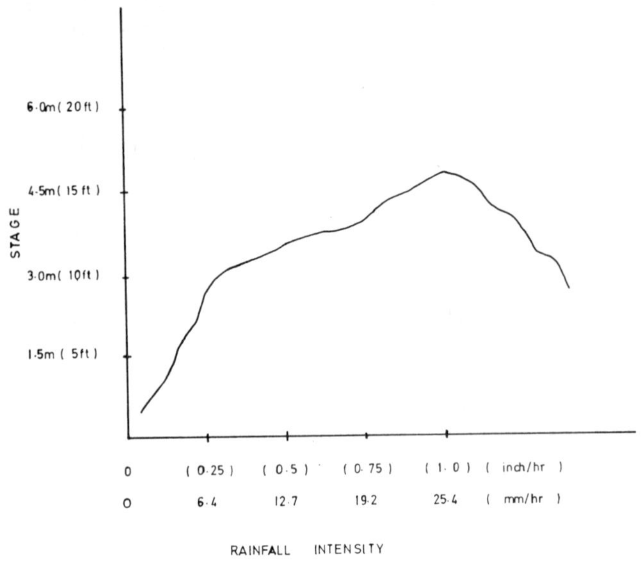

Figure 5. Typical Stage vs. Rainfall Intensity for Ibadan, Nigeria.

Operation of the System

The operation of the Flood Forecasting/Warning System entails both field and office work. The telemetry system has the advantage that it functions properly without being interrogated or supervised. Other advantages include low environmental impact and a design that discourages vandalism, low maintenance costs (only one site visit per year is needed), and the fact that non-availability of solar or AC power at remote sites will not interrupt its operation. The last statement will be an added advantage in Ibadan where there is no constant supply of electricity. During the operation of the Liquid Level Station (Model 5086), the changing head above the bubbler orifice causes a corresponding pressure change which results in a change in the manometer fluid level as the river level changes. Increments in river stage are detected by the Water Level sensor and stored in the Data Transmitter, for transmission to the Central Station. Upon receipt of information at the Central Station, the river stage data will be displayed and printed in the following sequence:

STATION ID / DATA DESCRIPTION / DATA / DATE & TIME

Amount of Rainfall River Level

The Self-Reporting Rain Station (Model 5054) will operate in conjunction with the Liquid Level Recorder. The tipping bucket mechanism generates a digital input signal whenever 1mm (0.04 in.) of rainfall drains through the funnel assembly. The signal is automatically transmitted by the Data Transmitter (Model 5050) to a Receiver (Model 5051) at the Central Station. The transmitted message includes the following:

STATION ID / ACCUMULATED AMOUNT OF RAINFALL

When 4 inches (99mm) of rain has been stored, the next signal automatically resets the accumulator to 00 and the counting process starts over again.

The collected data will be used in the calibration of the rainfall-runoff model previously described. It will be necessary to perform a periodic calibration as data accumulates and urbanization or land use progresses. Based on the peak stage data obtained, starting water surface elevations can be established for use in the computer backwater programs.

As a second requirement for the effective flood forecasting system previously stated, the extent of flooding at points along the stream must be determined. This is called flood-plain mapping. This can be accomplished by use of the HEC-2 computer program of the U.S. Army Corps of Engineers. Since adequate stage data would be available from the accumulated river level data, additional data needed will include channel and bridge cross-sections for input into the program. Once the flood-plain maps are available, extent of flooding will be established and can then be included in the flood warning system. This will constitute a more accurate stage prediction procedure.

Procedures to be Followed at the Central Station

For effective dissemination of information to the public during heavy rainstorms, the authors suggest that the Central Station be located within the premises of the Broadcasting Corporation of Oyo State in Ibadan. Once the critical intensity of 0.5 inch/hr. (12.7mm/hr.) is attained, the Central Station Operator must continuously send out forecast warnings and river levels at gage sites, in form of weather bulletins to residents in flood-proned areas and along the streams and rivers, indicating areas which are likely to be inundated. As an alternative procedure, the operator can monitor the amount of rainfall and river level data being received from the remote stations. As any of the channels approach bank-full conditions, the operator will be provided with data on depths at specific locations and he can send out the flood warnings. The two procedures can be employed conjunctively.

SUMMARY AND CONCLUSIONS

With the increased casualties in Ibadan during the past two floods, the need for a flood forecasting/warning system cannot be over emphasized. The authors have presented a simple, low-cost, low maintenance system for flood warning in Ibadan. A real-time event reporting system is needed so residents can be warned of impending floods ahead of time in that it instantaneously reports meteorological events as they happen. Moreover, the need for expensive interrogation of the system by telephone has been eliminated. This is very advantageous for Ibadan because of the poor telecommunication facilities. The system sends data automatically from remote stations without power supply or solar power. This will prove to be useful in Ibadan because of irregularity in the power supply. An additional feature of the system is that all the equipment is so ruggedly built that it discourages vandalism and can operate effectively under adverse weather conditions.

Since there have not been previous extensive networks for data collection in Ibadan, an extensive data collection system will be implemented. Once the data is collected over a 10-year period, it will then be used for the calibration of a rainfall-runoff model.

The proposed flood forecasting/warning system leaves room for expansion. However, the success of such a system will depend on the response from the citizens. As urbanization progresses, the long-term data collection system and the rainfall-runoff model will be useful as an effective means of predicting floods and river stages during severe rainstorms. While geometry and hydraulic capacity of the streams and rivers will change upon completion of the ongoing channelization scheme and with future flood control programs, it is believed that the system will be an integral part of flood mitigation programs in Ibadan. It will serve the city effectively during major storm events as the amount of rainfall and stage can be measured concurrently. As soon as the model is calibrated, the system will be made available to city and state government officials for their consideration and possible implementation.

ACKNOWLEDGMENTS

The authors wish to express their appreciation to the Department of Geography, University of Ibadan, Forestry Research Institute of Nigeria, Ibadan and Mr. J. A. Oyetunde of the Water Corporation of Oyo State, Ibadan, for making available the data and information necessary for this manuscript.

DEDICATION

This manuscript is dedicated to all the residents of Ibadan, Nigeria, who lost their valuable lives during the fateful floods of April 20, 1978 and August 31, 1980. May their souls rest in perfect peace. Amen.

LITERATURE CITED

Berling, Robert L., 1978. Disaster Response to Flash Floods. J. Water Resources Planning and Management Division, American Society of Civil Engineers 14(1):35-44.

Maclaren International, 1970. Master Plans for Wastes Disposal and Drainage, Ibadan – Immediate Measures Report, World Health Organization, pp. 63-77.

Maier, Nathan D., 1981. The Design of a Flood Forecasting System for Dallas, Texas. Texas Section, American Society of Civil Engineers, Fall Meeting, Houston, Texas, October 24-26, 1981.

Mogil, H. M., *et al.*, 1977. The National Weather Service's Flash Flood Warning and Disaster Preparedness Programs. Second Conference on Hydrometeorology, October 25-27, Toronto, Ontario, Canada, pp. 49-56.

Tokun, A. and M. O. Adesina, 1981. Flood Disaster in the City of Ibadan, Nigeria. Proceedings, International Symposium on Urban Stormwater Management, Lexington, Kentucky, pp. 401-410.

UTILIZATION OF HYDROMETEOROLOGICAL MONITORING SYSTEMS TO ENHANCE RESERVOIR OPERATIONS DURING FLOOD EMERGENCIES

David S. Wilson, Jr. and Daniel H. Phillips[1]

ABSTRACT: The Salt River Project (SRP) is a utility which supplies water and electricity to metropolitan Phoenix, Arizona. Phoenix is located in South Central Arizona on the Salt River near its confluence with the Gila River.

Serious interruption to normal transportation patterns, damages to private and public property and loss of lives are generally associated with flood flows in the Salt River.

SRP operates a system of six water storage reservoirs which regulate the normal flows of the Salt River and its major tributaries, the Verde River and Tonto Creek; these reservoirs were constructed during the period 1910-1946 as conservation facilities. Based on updated hydrometeorology, we know that the dams have inadequate capacities and outlet works to control flood flows such as those experienced in 1978, 1979 and 1980. SRP must rely on a comprehensive emergency reservoir operation procedure to ensure dam safety requirements and minimize the effects of flood flows on the downstream communities until facility improvements can be realized.

The extensive efforts by the Salt River Project and cooperating agencies to monitor weather, watershed, stream flow and reservoir conditions enhance reservoir operation decision making. The paper describes those efforts, the cooperative nature of their maintenance and operation, and the improvements in decision making that can be realized from effective monitoring and forecasting of weather and runoff.
(KEY TERMS: monitoring; forecasting; flood flow; Arizona.)

INTRODUCTION

The Salt River Project (SRP) is a utility which supplies water and electricity to metropolitan Phoenix, Arizona. Phoenix is located in South Central Arizona on the Salt River near its confluence with the Gila River.

The Salt River Project is the nation's oldest multi-purpose reclamation project. It is comprised of the Salt River Valley Water Users' Association and the Salt River Project Agricultural Improvement and Power District.

The District is a political subdivision of the State of Arizona and operates the Project's electrical facilities and provides electricity to over 330,000 residential, commercial, industrial, and agricultural power accounts in Maricopa and portions of Gila and Pinal counties.

The Association is a nonprofit Arizona corporation responsible for the administration of the water rights of its shareholders. As agent for the District, the Association operates and maintains the irrigation transmission and distribution system which provides approximately 1.1 million acre feet of water each year for agricultural, industrial and municipal uses in the Project's 250,000 acre service area. This includes 70 percent of the water used by eight communities having a combined population in excess of one million people. This annual water requirement is supplied from surface water (approximately two-thirds) and ground water.

The surface water supply is controlled by SRP's operation of six water storage reservoirs which regulate the normal flows of the Salt River and its major tributaries, the Verde River and Tonto Creek, draining a watershed area of approximately 12,500 square miles. Watershed elevations range from 1,100 feet to almost 12,000 feet above sea level. The extreme variation in physical terrain produces equally extreme ranges in surface water runoff. For example, in 1977 the combined inflows to the reservoir system totaled only 367,122 acre feet; the combined flows for 1980 exceeded 2,420,000 acre feet; and the average runoff based on 69 years of record is 1,185,000 acre feet.

The reservoirs were constructed during the period 1910-1946 as conservation facilities and have a combined reservoir capacity slightly in excess of 2,000,000 acre feet. Based on recent experience and updated hydrometeorology, we know that the dams have inadequate capacities and outlet works to control major flood flows such as those experienced in 1978, 1979 and 1980. For example, the February/March 1978 storms produced peak combined inflows to the Salt and Verde reservoirs of 229,000 cfs. The December 1978 storm yielded a peak combined inflow of approximately 264,000 cfs. Effective management of the reservoir system during these three events reduced peak combined releases from the lower reservoirs on the Salt and Verde rivers to 125,000 cfs, 120,000 cfs, and 180,000 cfs, respectively.

[1] Respectively, Manager of Water Resource Operations, Salt River Project, P.O. Box 1980, Phoenix, Arizona 85001; and Senior Environmental Analyst, Environmental Services, Salt River Project, P.O. Box 1980, Phoenix, Arizona 85001.

Nevertheless, serious interruption to normal transportation patterns, damage to private and public property and loss of lives are generally associated with such flows in the Salt River through Phoenix. Thus, SRP relies on a comprehensive emergency reservoir operation procedure to ensure dam safety requirements and minimize the effects of flood flows on the downstream communities until facility improvements can be constructed.

A successful flood management operation requires timely implementation of sound decisions and effective communications with affected sectors of the community. With the experience gained from recent major floods, the comprehensive Emergency Reservoir Operations Procedure which SRP has developed, facilitates both decision implementation and communications. The foundation for this procedure rests on the requirement that a steady flow of reliable, timely data is available, and decisions are made based upon evaluation of this data pertaining to weather, watershed condition, river flows, reservoir status, and dam site and downstream limitation considerations. The tools and systems which are utilized can be categorized according to the area of concern just mentioned, i.e., weather, watershed condition, etc.

WEATHER FORECASTING AND MONITORING

The weather is the driving force for all major flood events. However, the difficulties involved in obtaining timely, accurate weather forecasts as well as monitoring real-time occurrences on remote areas of the watershed are major restrictions to optimal reservoir management. SRP meteorologists are responsible for collecting weather forecasts, meteorological data, and watershed weather observations from as many sources as possible. Through their experience in forecasting specifically for the Salt and Verde basins, the information is evaluated and summarized in a forecast tailored for management decision making.

During the primary runoff season of November 1 – May 1, daily contact is maintained with the National Weather Service (NWS) Forecast Office in Phoenix. Access to NWS computer-generated forecast guidance as well as synoptic briefings by NWS forecasters allows for maximum awareness of current conditions and potential weather threats.

A Quantitative Precipitation Forecast/Outlook is issued by SRP meteorologists twice weekly during nonemergency conditions, and lists the expected average watershed precipitation and associated snow level for each 24-hour period for the next nine days. The forecast is intended as the primary vehicle for communicating the long-range potential for severe weather to management.

With the development of a flood emergency or when the potential for excessive rainfall exists, contact with the Phoenix Weather Service Forecast Office is made several times each day. A two-way exchange of information allows SRP to maintain a current knowledge of changing weather conditions, and the NWS to incorporate pertinent information in their public warnings. During these periods, the SRP Quantitative Precipitation Forecast/Outlook is issued daily.

Rapidly changing weather conditions are monitored through the use of GOES (Geostationary Operational Environmental Satellite) satellite imagery. Pictures are received in-house by SRP every half hour as needed.

With the onset of precipitation over the watershed, monitoring and data collection are accomplished by several means. A color radar slave unit to the NWS weather radar is located at SRP. The display is videotaped for later study, and consideration is currently being given to the possibility of digitizing the signal for computer integration.

Precipitation reports are received from both NWS observers and SRP cooperative observers residing on the watershed. In addition, a network of automated precipitation gages provide near real-time observations from remote areas of the watershed. Access to the data is via an in-house terminal capable of interrogating the National Weather Service River Forecast Center DATACOL system in Salt Lake City, Utah.

DATACOL provides comprehensive hydrometeorological data reported by category and location. In addition to streamflow, it provides temperature, precipitation and snow data derived from such systems as SNOTEL, DARDC, GOES and COMSAT. Historic data (since site became operational) is also maintained on file in this system. SRP can interrogate DATACOL using an in-house terminal.

WATERSHED MONITORING AND CONDITION ASSESSMENT

Watershed condition information is derived from manned field operations and automated remote site reporting. SRP personnel conduct both aerial snow flights and ground patrols aimed at assessing snowpack and runoff conditions.

The flights cover the entire 12,500 square mile watershed. Extent of snow coverage, snow depth and condition are carefully mapped. Actual snow depths can be monitored from the air at snow marker locations, and are estimated elsewhere using fences, logs, etc. The accuracy of the data gathered during these flights is dependent on the expertise of the observer. The ground patrols conduct random snow surveys, gage stream flows and collect precipitation data.

Additional data is collected by personnel of the Soil Conservation Service (SCS) and their cooperators in the form of regularly scheduled biweekly snow surveys at specified snow survey sites. This effort is supported financially by SRP.

The efforts of SRP's field personnel are supplemented by those of watershed resident observers. There are 12 volunteer observers who live in or close to various watershed communities. These observers are provided with equipment to collect temperature, precipitation, and snowpack information. Regular monthly reports are compiled and submitted. However, during flood emergency periods, timely reports of changing conditions are made as needed by telephone.

Additional information is received in the form of "snow maps" which are derived from GOES satellite pictures. The snow area from the satellite imagery is transferred to a map of the Salt and Verde watershed, and the percentage of the watershed area covered by snow is determined by planimeter methods. These maps were originally prepared and telecopied to SRP by NOAA analysts. Today, this work is done in-house by SRP's staff.

SRP can interrogate the SCS-managed SNOTEL system daily for precipitation, temperature and snow water content data. This is a system of automatic sensors at 15 strategic locations on or near the watershed. Each site includes a snow pressure pillow, precipitation gage and air temperature sensor.

Utilizing the data available on weather and watershed conditions, an assessment report and executive summary are prepared, which describe the potential for runoff as compared to available space in the reservoirs.

RIVER FLOWS

Telemetered data from seven water stage recording gages on the Salt, Verde and Tonto Creek is received by SRP at hourly intervals. The gages are operated and maintained by the U.S. Geological Survey (USGS) with cooperative funding from SRP. The telemetry system, called METS (Motorola Environmental Telemetry System) is owned and operated by SRP.

Recent development of a State Flood Warning System has resulted in real-time availability of stream gaging data from 24 additional watershed sites. Data from these sites is available every 15 minutes and is received at SRP headquarters directly from the satellite. This data availability is the result of recent acquisition and installation of a "Direct Readout Ground Station." This was accomplished in cooperation with the Arizona Department of Water Resources, the Maricopa County Flood Control District, the U.S. Geological Survey, the U.S. Army Corps of Engineers and the U.S. Bureau of Reclamation. Precipitation gages are currently being added to this system to provide cumulative rainfall data as well as streamflow information.

RESERVOIR STATUS AND OPERATIONS

Each of the six reservoirs has a water level recorder which is linked to SRP's Lake Level Telemetry System. This system telemeters continuous lake level information to SRP's data center and is displayed in digital form. SRP is the sole operator of the system installed on the four Salt River reservoirs. The telemetry equipment for the two Verde River reservoirs is linked to instruments operated and maintained in cooperation with the USGS. Manual readings of water levels at each reservoir are available and can be radioed or telephoned to the data center or Emergency Operations Center as a backup should the telemetry system fail during flood emergency operations.

Hourly reservoir status reports are prepared by computer to show river flows into the reservoirs, individual reservoir contents, available volume, and changes in content for the reporting period. The reports also indicate reservoir releases, and identify ungaged inflows to the reservoirs from local runoff and precipitation directly on the reservoir surface.

RUNOFF FORECASTS

The vast quantity of data collected on weather, watershed, river flow and reservoir conditions must be analyzed and used to predict future runoff and to effectively manage the reservoirs to determine optimum reservoir release rates and schedules during emergency flooding periods. A number of forecasting techniques and tools are utilized for this purpose.

Using an SRP developed method, volumetric estimates of runoff are developed by estimating the amount of water which could be produced by the subwatersheds of the Salt and Verde for predicted storm conditions. Information for each subbasin includes area by elevation zone and delivery efficiency for rain on bare ground, rain on snow, and rapid snow melt with no precipitation. This information is maintained in reference tables. This method of runoff prediction is currently being computerized.

The NWS Antecedent Precipitation Index Model (API) is a short-term flood forecasting model operated by the Salt Lake City River Forecast Center. Antecedent precipitation, time of year, current storm precipitation, the Quantitative Precipitation Forecast/Outlook and storm duration are correlated by computer to predict runoff in inches. The model projects a runoff hydrograph that denotes flow rates at six-hour intervals for gaging stations on the Verde River, Salt River and Tonto Creek immediately upstream of the reservoirs. Model runs are made by NWS personnel at SRP's request, and results are transmitted by telephone.

The NWS Extended Streamflow Prediction Model (ESP) is another weather service computer model which is new to SRP in 1982. This model projects long-term, seasonal volumes of runoff and can be of value to SRP in routine reservoir operations, as well as projecting extended runoff following major flood events. Two other extended streamflow models are being utilized operationally by SRP, and will routinely be compared with the ESP model for evaluation purposes.

The first of these is the SCS model which has been in use since 1968. The SCS model utilized regression equations to express a relationship between certain specific measurable watershed conditions and runoff volume between the forecast date and the end of the runoff season (May 31). Forecasts are made every two weeks between January 1 and April 1 of each year. The measured conditions which are required as input are derived from 14 SNOTEL and snow survey sites, 8 precipitation stations, 3 stream gages and records of past precipitation. This model was developed and is jointly operated by the SCS and SRP.

The second long-term forecast model which SRP utilizes was developed by Wendell Tangborn of Seattle, Washington. SRP has had access to various versions of this model since 1978. However, in the past two years, working as a consultant to SRP, Mr. Tangborn has refined the model for purposes of making intermediate to long-term runoff forecasts for the Salt and Verde basins. The model sees runoff as a function of basin water storage at the time of the forecast, and of precipitation during the forecast period. The outstanding features of this model as compared to others SRP uses, is the small (by comparison) amount of input data required and SRP's in-house possession with access limited only by periods of downtime for the computer used and the availability of daily data.

To evaluate alternative reservoir operation schemes, the river flows forecasted by the various methods just described are routed through the reservoir system using the U.S. Corps of Engineers' HEC-5R program. Because it is quite complex, requires a large amount of computer storage capability, and is being continuously updated by the Corps, the program is maintained on selected commercial computer facilities. SRP is linked to one of those facilities.

COMMUNICATIONS AND ORGANIZATION

In its 79-year history, the Salt River Project has frequently had to manage its water resources to cope with drought, but comparatively few major flood events have had to be managed. The first SRP flood release was made in 1913 and peaked at 3,700 cfs. Peak releases prior to 1978 have been 79,000 cfs (1916); 46,000 cfs (1919); 48,000 cfs (1932); 58,000 csf (1938); 67,000 cfs (1965); and 53,000 cfs (1966).

The potential for flooding due to unusual or extreme watershed conditions, such as an exceptionally large snowpack, can be identified well in advance of the actual flood event, should it occur. However, the recent floods of 1978-1980 were caused primarily by warm, moist tropical storms that developed extensive precipitation with little advance warning. Limited advance warning increases the importance of real-time data collection, functional organization, staffing and communications for emergency reservoir operations.

The current Emergency Reservoir Operating Procedure is the refined product of SRP's recent flood management experience. It identifies procedures and responsibilities for planning emergency reservoir operations, providing timely internal and external communications, and the actual operation of SRP reservoirs and related facilities during flood emergencies. It identifies within the company's organizational framework which areas of management and what staff are required to conduct the functional activities necessary during flood operations, and it establishes for flood emergency periods, a temporary Table of Organization which consists of an Emergency Operations Executive, supported by a Project Reservoir Operations Planning Committee (PROP), an Emergency Operations Center (EOC), and a Communications Liaison.

The PROP Committee is responsible for planning reservoir operations under normal and emergency conditions. During the latter situation, this includes evaluating information pertaining to conditions with potential for destructive flooding and preparing recommendations for emergency reservoir operation, identifying the degree of flood emergency and the development or revision of operating criteria. Committee membership includes a representative for the various functional and expertise areas required.

The EOC is a Center activated during periods of flood emergency to provide communications, control and coordination of reservoir operations. The Center is staffed around the clock with a Shift Manager in charge at all times. The Center is charged with operating in accordance with the emergency PROP plan as approved by the Emergency Operations Executive. However, Center personnel must also react to unanticipated conditions which require deviations from the range of operations intended by the approved plan. The Center staff maintains a complete record of available data and operational activities.

Communications Liaison is responsible for coordination of communications with key personnel in other agencies and jurisdictions outside of SRP. Examples include the City of Phoenix, Maricopa County Civil Defense, Arizona Division of Emergency Services, and the U.S. Bureau of Reclamation. Coordination for briefing meetings is also performed. These meetings, intended for internal personnel as well as external contacts, provide a forum for updating current weather, watershed, river flow and reservoir conditions; explanation of current operating plans; and solicitation of information on other related matters, such as the downstream impacts of current and planned reservoir releases.

CONCLUSIONS

No operation this complex can withstand the scrutiny of hindsight evaluation and appear flawless. However, use of the technology, systems and procedures described have resulted in an impressive operational record. Releases below the reservoirs ranged from 45-63 percent of what flowed into the reservoirs during the floods of 1978-1980. Needed improvements have been identified, and to the extent feasible, incorporated into SRP's Emergency Reservoir Operations. As future needs and improvements are identified, they will be folded into future Procedure modifications. As facility improvements can be realized, they will further reduce the threat of destructive flooding in the Phoenix metropolitan area. A series of cooperative efforts, which include federal, state and county agencies, as well as the Salt River Project, other private organizations and citizens groups, are presently underway to evaluate facility alternatives for flood control on the Salt and Verde rivers.

THE HYDROMETEOROLOGY OF A MAJOR FLOOD IN SOUTH AFRICA

W. A. Nieman[1]

ABSTRACT: The climate of Southern Africa south of 22°S and west of 25°E is semiarid to arid, with an annual precipitation of less than 500 mm. Summer rains are convectional and very localized, causing minor flash floods that peak rapidly and recede quickly. On occasion, however, a westerly wave with exceptional amplitude may cut off a cold low from its source region, resulting in pronounced meridional flow. A blocking high over the Indian Ocean can then cause confluence of warm northerly and easterly air streams, and as cyclonic vorticity in the lower troposphere increases, maximum divergence at all levels will produce heavy rainfall to the east of the cutoff low. This happened in January 1982 over the central Karroo. Runoff from the almost circular catchment of the upper Buffels River converged on a pocket floodplain where the town of Laingsburg is situated upstream from the narrow gap in a ridge through which the flood, peaking at an estimated 6,000 $m^3.s^{-1}$, had to flow. Streets running parallel to the river front acted as subsidiary channels, cutting off escape routes to higher land. More than 100 lives (out of a total population of 3,200) were lost and 300 of the town's 367 houses and business premises were severely damaged.
(KEY TERMS: hydrometeorology; floods; South Africa.)

PROLOGUE

When it started to rain on Friday, January 23, 1981, at and around the town of Laingsburg in the "Murderer's Karroo," it was most welcome. The last "good rain" in this part of the country fell in February 1976 when a small part of the town was flooded to some extent after a single thunderstorm in part of the 3,200 km^2 large catchment of the Buffels River upstream of the town. These summer convectional rains are very localized due to convectional instability developing in a thermal. Like all flash floods of arid and semiarid Southern Africa, the flood peak is rapidly reached and the flood water recedes quickly: after a day or two ephemeral rivers hardly show a trickle on the surface and the winds soon dry the freshly deposited sediment to dust.

There was something unusual about the rain this time. Laingsburg, about 230 km from Cape Town on the main highway to Pretoria, is situated between the folded coastal mountain ranges and the rim of the Great Escarpment of the interior plateau (Figure 1). Half of its annual precipitation of 110 mm is received in summer, mostly in the form of brief convectional showers; the rest of the meager precipitation is received in winter when a cold front, associated with an extra-tropical depression moving eastwards south of the subcontinent, on occasion penetrates inland causing cold, drizzly weather and light snow on the mountains to the south and west of Laingsburg (Schulze, 1965; Weather Bureau, 1965). On January 23, summertime in the Southern Hemisphere, it was a heavy drizzle and not showers from a thermal low falling at Laingsburg. Nevertheless, both the townsfolk, sweltering in the dusty 35°C heat, and the sheep farmers, worried about pasture for their flocks with winter only a few months away, welcomed the rain.

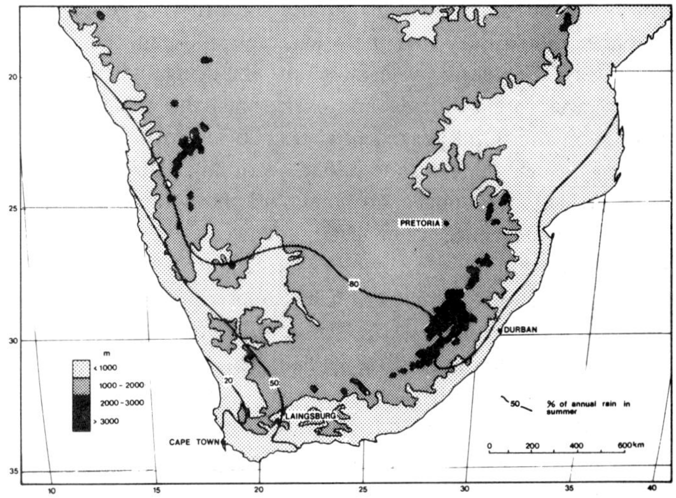

Figure 1. Southern Africa: Topography and Seasonal Precipitation.

The next morning at 08h00 5.5 mm of rain was measured about 30 km northwest of the town where an official rain gauging station of the South African Weather Bureau is located. The parched earth soaked up every drop of the intermittent drizzle. During the day and following night, the skies stayed overcast and heavier showers, still uncommonly

[1] Lecturer, Dept. of Geography, Faculty of Arts, University of Stellenbosch, Stellenbosch 7600, South Africa.

cold, produced 36.5 mm of rain. A large number of people from the district and from town congregated at the church in Laingsburg on Sunday, January 25, 1981, to give thanks both for the rain and because the town would be celebrating its first century in 1981. The Buffels River, rising about 85 km almost due north from the town, and its tributaries were again running because the catchment by now was saturated. When the church service started at 10 o'clock, the water was only 1 meter deep in the river where the national road crosses it about 500 m to the west of the church building. An hour later the water level had risen another half a meter and the farmers were eager to get home before the dirt roads might become inundated where tributaries are crossed by causeways. By now heavy showers poured from the grey-black clouds overhead and the townsfolk hurried home to the snugness of their customary Sunday lunches. Could this be the "heavy, persistent rain" forecasted the previous day by the South African Weather Bureau "over the eastern, central and Southern parts of the Republic?" (Weather Bureau, 1981, 14h00 SAST, 1981-01-24 and 25). At noon, despite the cold showery rain, quite a number of people, including travelers on the national road, had gathered on the bridge deck, about 5 m above the sandy river bed, to watch the swollen river racing past and starting to overflow its banks. An unknown number of persons stayed on the bridge to watch the flood and were soon cut off from escaping to dry ground since the bridge deck is higher than almost half the town. By 3:00 p.m. the last of at least 17 people were swept off the bridge when the water level rose suddenly and overtopped the bridge. At 5:45 p.m. the flood peaked at about 7.6 m above the river bed. Although there was no flood recording gauge in the river, eyewitness reports and numerous photographs showing various locations, water levels, and even the time of the clock on the church spire, made reconstruction of the flood sequence and water levels possib le (Roberts and Alexander, 1982). Three hundred of the town's 367 houses and business premises were severely damaged, a large number being swept from their foundations when, at its peak, about 3,700 $m^3.s^{-1}$ water raced through the town. Out of a total population of about 3,200, more than 90 inhabitants, including 28 senior citizens residing in an old age home on the river's bank, were missing, presumed drowned, the next day. Damage to property and communcations was in excess of R10 million, not counting the damage to farmland along the course of the river where another 12 to 15 people probably drowned, some when two farm houses also collapsed after being flooded.

GEOMORPHOLOGY OF THE DRAINAGE BASIN

Geological History

Sedimentary rock of the Karroo Super Group covers practically all of the upper drainage basin of the Buffels River (Figure 2). The sediments were deposited in a large inland lake during the Late Carboniferous and Early Jurassic geological periods, i.e., 300 Ma to 170 Ma B.P. (Truswell, 1977). In the initial stages continental glaciers deposited the Dwyka tillites presently outcropping to the west and south of Laingsburg, situated at the mouth of the drainage basin (Stratten, 1968). Northward drainage then deposited sediments, originating from the rocks of the southerly situated Cape Super Group, covering the tillites (Ryan, 1968) to form the Ecca and Beaufort sandstone, shale and mudstone of the Karroo Super Group. After lifting and folding, the present river system has been eroding the landscape during the past 140 Ma (Nieman, *et al.,* 1981).

Contemporary Landscape

The shales and mudstones weather easily and sheetwash from the summer showers remove the regolith, resulting in shallow soils on the hillslopes and pediments. Although grass covered more than a century ago (Acocks, 1975), presently the landscape around Laingsburg is covered with sparse, low growing shrubs, offering little protection against fluvial erosion. The sediment load of the rivers are therefore high. Estimates of sediment deposited in the town of Laingsburg by the January 1981 flood run as high as 200,000 m^3, up to 3 m high in many places (Roberts and Alexander, 1981, p. 20). Trees are limited to the sandy river beds. The northern rim of the almost circular catchment basin forms part of the Great Escarpment of the subcontinent and reaches more than 1,700 m above sea level. However, the basin itself is almost plainlike, gently undulating, with east-west ridges, not inhibiting the dendritic drainage pattern (Figure 3). The town, 652 m above sea level, is situated on a pocket floodplain between two such ridges. When precipitation occurs simultaneously over the whole circular catchment, as was the case in January 1981, flood water converges on the town's location.

The narrow gap in the ridge downstream of the town (Figure 4), and the confluence of a major tributary of the Buffels River almost at right angles from the west opposite the southern end of Laingsburg, makes the pocket floodplain extremely vulnerable to flooding. In June 1925 a flood peak discharged 1,326 $m^3.s^{-1}$ downstream of the gap, and in April 1934, December 1937, August 1939, and February 1952 floods in excess of 500 $m^3.s^{-1}$ were recorded in the Buffels River by the Directorate of Water Affairs operating a flow gauging station prior to the erection of a dam about 10 km downstream of Laingsburg. The estimated (Roberts and Alexander, 1982, p. 18) flood peak of 3,700 $m^3.s^{-1}$ upstream of the town in January 1981 was aggravated to more than 6,000 $m^3.s^{-1}$ beyond the confluence of the Wilgehout and Bobbejaans River. The solid earth embankments of the railway bridge, just downstream of the confluence at a narrow part of the river channel before entering the gap in the ridge, further restricts the flow of flood water.

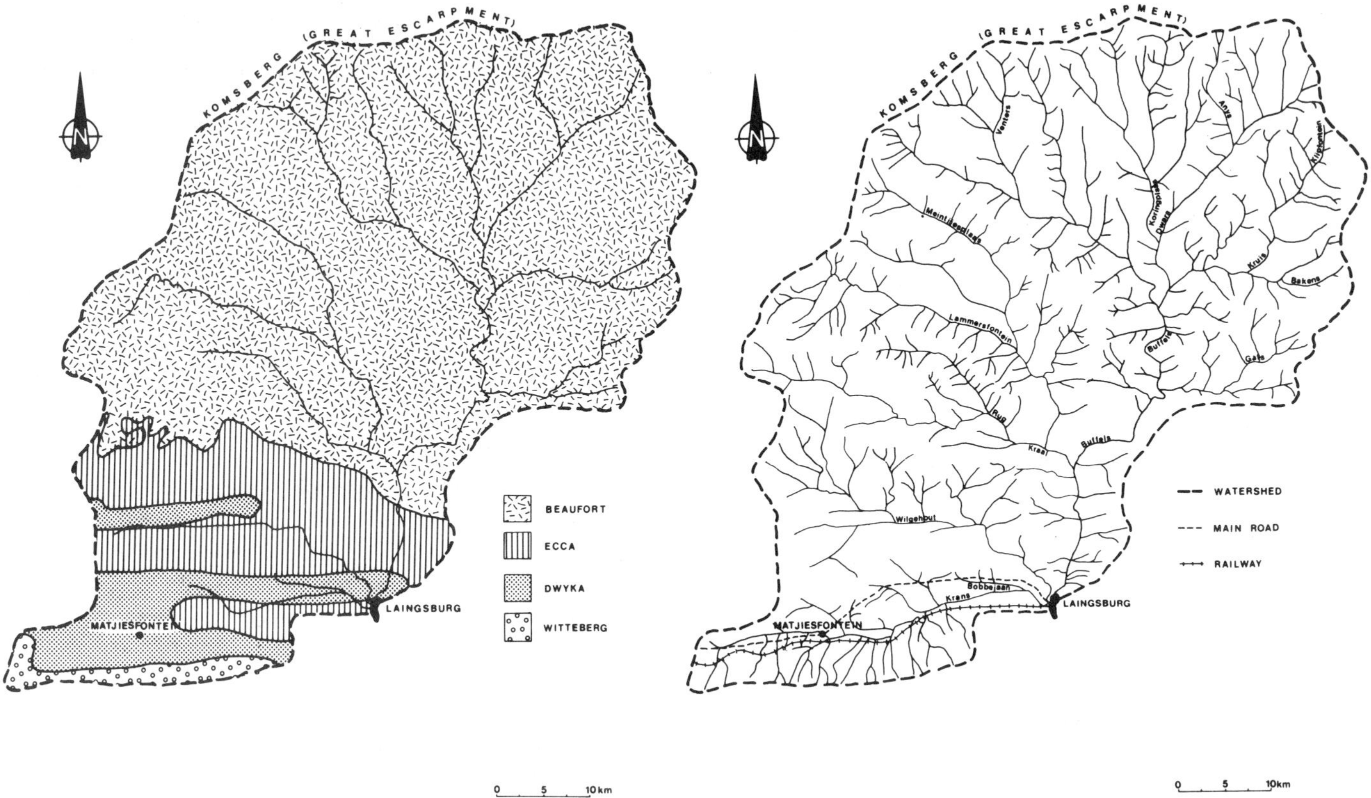

(a) (b)

Figure 2. Buffels River Catchment: (a) Geology, and (b) Drainage.

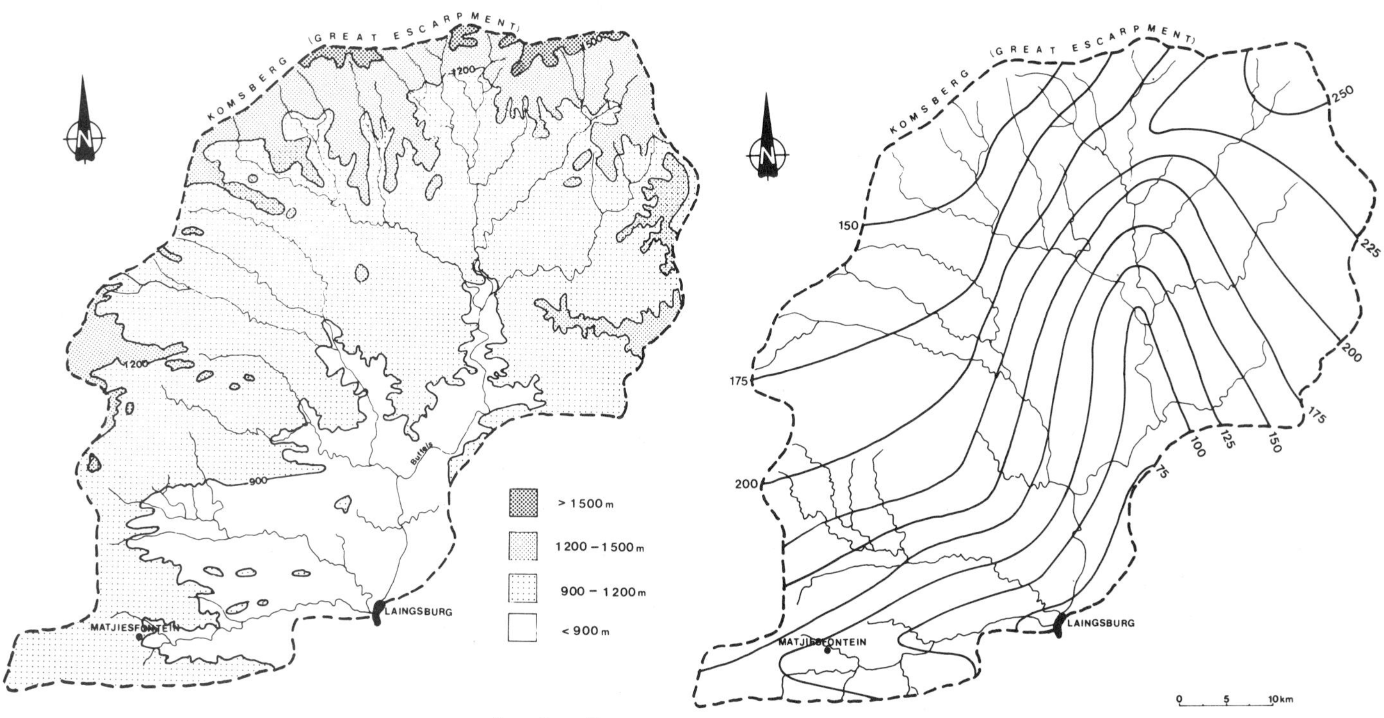

(a) (b)

Figure 3. Buffels River Catchment: (a) Topography, and (b) Rainfall for January 23-25, 1981 (latter after Van Zyl, 1981).

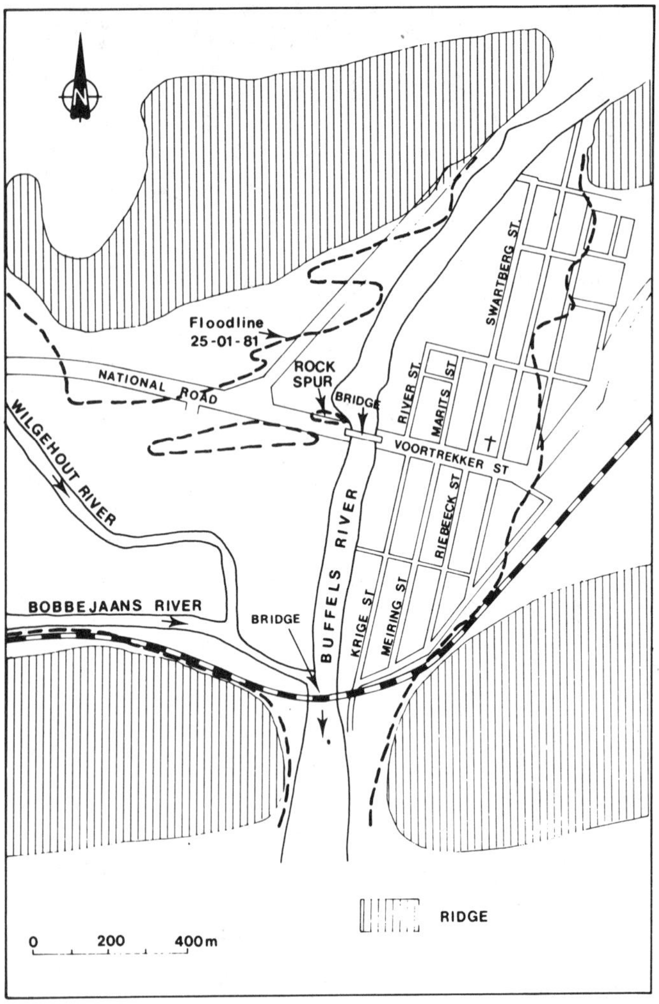

Figure 4. Laingsburg and Environment (after Nieman, *et al.*, 1981; Roberts and Alexander, 1982; Van Zyl, 1981).

The cultural landscape contributed in more than one way to the severity of the flooding disaster of January 1981. The road bridge next to a rock outcrop on the west bank of the Buffels River effectively blocked the flooded river, especially after four of the six 15 m wide clear spans of the bridge were choked by debris, including trees uprooted by the raging torrent. Maximum stream velocity at the constriction was estimated at about 8 $m.s^{-1}$ and eventually critical velocity was exceeded causing a hydraulic jump and a high standing wave upstream of the bridge (Alexander and Roberts, 1981; Roberts and Alexander, 1982). When the river overflowed its normal channel it straightened its course and the streets nearest the river and parallel to it became subsidiary river channels preventing residents to cross the streets to safety at higher levels, even when the water was less than a meter deep (UCT News, 1981). "The total scour volume in the town was 55,000 m^3" (Alexander and Roberts, 1982, p. 19).

HYDROMETEOROLOGY

The first recorded flood in Southern Africa occurred on July 23, 1952, only three months after Jan van Riebeeck established a refreshment station at the Cape of Good Hope for the Dutch East Indies Company. The settlement's garden was washed away by 'the greatest flood on earth,' according to the despondent commander's diary.

The general aridity of Southern Africa can be ascribed to atmospheric subsidence due to the subcontinent's (Figure 5) location in the high subtropics and lower middle latitudes of the Southern Hemisphere. Here a relatively narrow continent is enclosed on three sides by the wide expanses of the South Atlantic, South Indian, and Antarctic Oceans where the meteorological conditions are regulated by the text-booklike anticyclonic nature of the air circulation both on the surface and in the upper air (Figure 6). Wavelike disturbances in the easterly flow over the northern part of the subcontinent establish poleward troughs of the intertropical convergence zone and may cause heavy rains and floods over the northern and eastern parts of the country. Occasionally heavy rains may also result from tropical cyclones moving southwards along the eastern African coast. Waves in the upper air westerlies are associated with the eastward passage of extra-tropical cyclones south of the continent; the accompanying cold fronts in winter brushes the southwestern tip of the continent and the precipitation is enhanced by orographic lifting.

Cutoff Lows

A westerly wave with exceptional amplitude may cut off a cold low from its source region causing pronounced meridional flow (Bowen, 1976, p. 63). With only two upper air observation stations in the southern oceans, *viz.* Gough (40°S, 9°W) and Marion (47°S, 38°E), it is difficult to monitor cold air advection over Southern Africa.

In an appendix to an annual report to both Houses of the Southern African Parliament, a description was given of the floods caused by ". . . heavy downfall of cold rains" (Union of South Africa, 1918, p. 136) in the eastern Karroo and along the south coast in May 1916. Although the very primitive weather maps included in the report do not show upper air contours, one may deduce from the detailed descriptions of surface weather conditions that a cutoff low and blocking high were responsible for the excessive rain and resulting floods. As late as 1961 South African meteorologists were still ". . . tempted to explain rainfall in terms of topography alone" (Triegaardt, 1961, p. 113), but a cutoff low in the upper air over the west coast was found to be the synoptic cause for the floods in the western and central Karroo, well north of Laingsburg.

The Laingsburg Flood

The prolonged drought in the years before the weekend of January 24-25, 1981, dessicated the surface of the Buffels

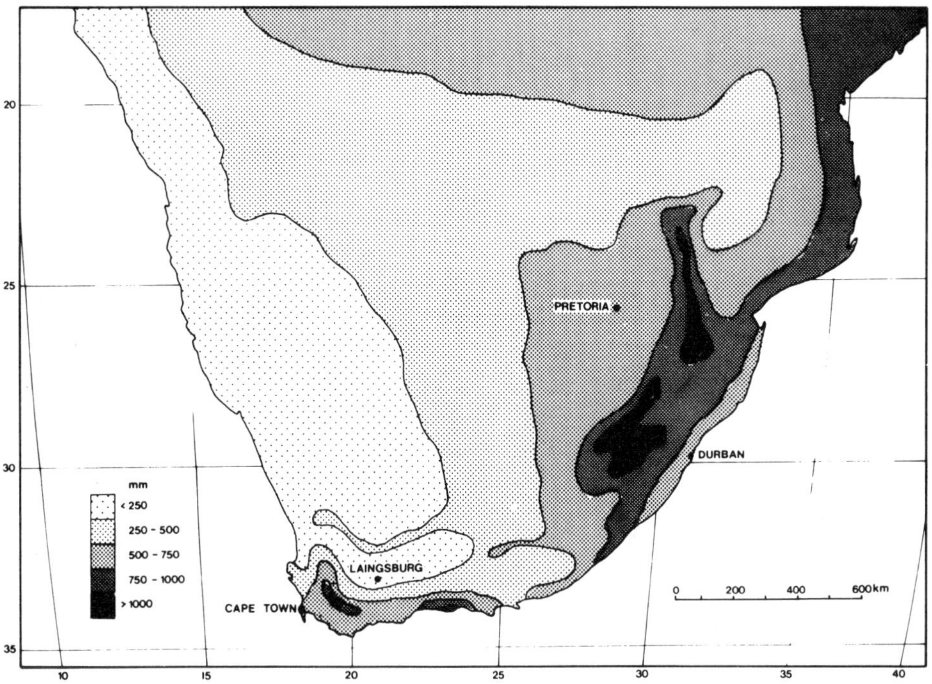

Figure 5. Mean Annual Precipitation of South Africa (after Schulze, 1965).

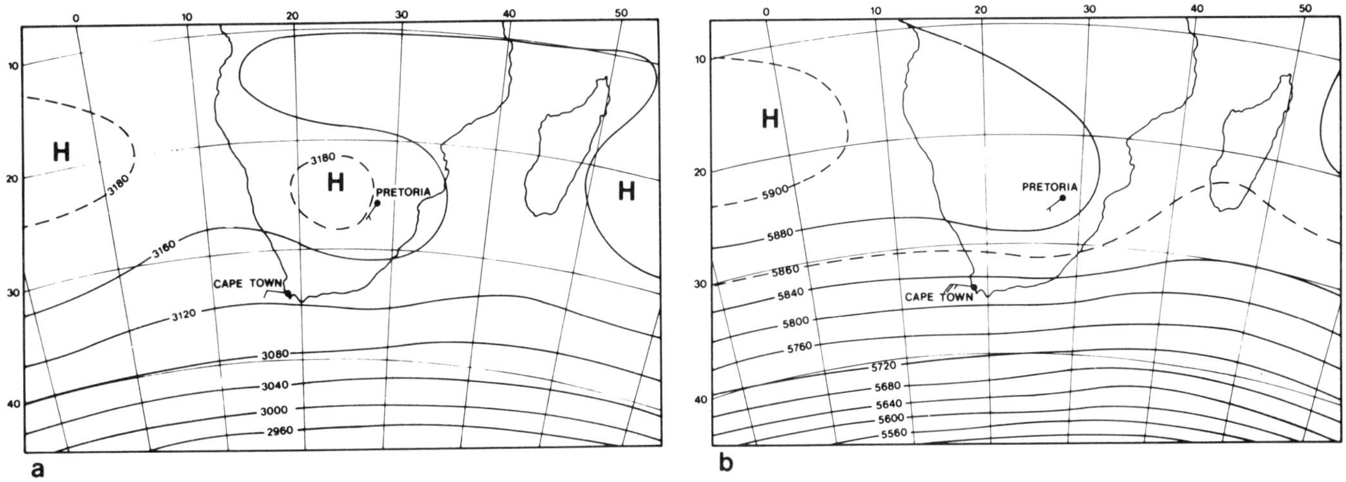

Figure 6. Mean Contours for January: (a) 700 mb Pressure Level, and (b) 500 mb Pressure Level (after Hayward and Steyn, 1967).

River catchment and reduced the natural vegetation drastically. A more common scattered thunderstorm somewhere over the 3,000 km^2 basin, could, at most, have caused a local but minor flood. The prolonged and widespread rain, however, created a dramatic runoff situation, even though the peak intensity of the rain over a period of three hours early on the afternoon of January 25, was most likely less than 60 $mm.h^{-1}$ (Estié, 1981).

The first signs of a disturbance in the upper air westerlies were observed on January 22, 1981. A weak trough of low pressure on the surface (850 mb pressure level for the interior plateau of the subcontinent) well to the east of Laingsburg persisted for most part of the preceeding week; this was associated with scattered thundershowers to the east of the trough axis. Light rain fell along the south coast as a weak cold front passed by eastwards, with the depression center moving along at about 45°S. Upper air temperatures were slightly above normal (3°C at 700 mb, –10°C at 500 mb, –35°C at 300 mb), and the air was dry (dew point temperature on average 40°C lower than the air temperatures). On January 23 the South Indian Ocean high pressure system moved eastwards enabling an unusually strong South Atlantic high pressure system to push an extra-tropical depression towards Marion Island. Moist air then advected over the southern part of the country behind a cold front. The next day a ridge of high pressure progressed eastwards south of the country and fairly heavy rain fell along the southern coastal mountains as the south-westerly winds backed to south-easterly. Over the adjacent interior, despite a surface low, precipitation remained light and intermittent over the Laingsburg area.

In the upper air the amplitude of the westerly wave increased and the surface trough deepened as it narrowed. On Sunday, January 25, a pool of cold air was cut off in a cyclonic movement over the western part of the country (Figure 7). Cyclonic cells were present on all pressure levels up to 200 mb; temperatures were on average 7°C lower than the previous day, and the air was fairly moist (dew point depression averaging 6°C). South of the country and the cyclonic 'funnel' the high pressure ridge was firmly anchored between two anticyclones. Moist tropical air was sucked into the 'funnel' from the north and maritime air converged from the east over the warm Mosambique-Agulhas ocean current off the east and south coast of South Africa. The steep pressure gradient to the southeast of the subcontinent produced easterly winds averaging 50 knots, known colloquially as a "black south-easter" because orographic lifting produces heavy rain over the southern coastal regions and adjacent interior. (Normally the south-easterly wind, or "Cape Doctor," is associated only with light rain on the windward side of the mountains as the stable air of the anticyclone is forced to rise.) The confluence of the northerly and easterly air streams and the orographic lifting increased the instability of the air and as the cyclonic vorticity in the lower troposphere increased, ". . . maximum divergence at all levels took place and heavy rainfall occurred over much of the southern and south-western areas of the Cape Province" (Estiē, 1981, p. 28).

The next day saw the collapse of the steep pressure gradient as the high pressure cell southeast of the country was entrained eastwards by the upper air westerlies; the cutoff low, deprived of its energy source, moved south-westwards and filled rapidly. An almost clear sky allowed the sun to shine down on the havock below.

EPILOGUE

Although the intensity of the precipitation from the cutoff low was not very high, the rain fell on three consecutive days over the whole upper basin of the Buffels River – and adjacent territory– saturating the surface before the highest intensity was reached. Flood water therefore converged almost simultaneously on Laingsburg from all directions in the circular basin.

Usually cutoff lows tend to develop on the leeside of the higher eastern escarpment of South Africa causing flood-producing rains over the south-eastern coast. To predict the exact location of a cutoff low, is still impossible. At present the South African Weather Bureau is experimenting with a "Primitive Equation Model" able to forecast only 12 hours ahead to some degree of accuracy.

Every catchment basin in the country, even in the arid parts, is prone to flooding, but how often? Most towns were established long before modern technology and knowledge were available, at a time when proximity to water was the main locational factor. Also it was much cheaper to build on the relatively flat and fertile floodplain in the hostile natural environment. And a rectangular street plan with streets parallel to the river front was the natural thing to do. Many lives could have been saved in the past if our ancestors were aware of the dangers of a river; if only the streets were angled down towards the river preventing the street to become a subsidiary channel and providing an escape route to higher ground for the inhabitants.

ACKNOWLEDGMENTS

The weather office at D. F. Malan Airport, Cape Town, shared the synoptic maps and charts. Bärbel Haldenwang and Heinrich Ellmann did the artwork; Hetta Engelbrecht typed the manuscript. Many thanks to them and to encouraging friends and colleagues, especially Tertius Harmse, initially co-author.

LITERATURE CITED

Acocks, J. P. H., 1975. Veld Types of South Africa. Memoirs Botanical Survey South Africa, 40, Botanical Res. Inst., Dept. Agric. Techn. Services, Rep. South Africa, Govt. Printer, Pretoria.

Alexander, W. J. R. and C. P. R. Roberts, 1981. Report on Buffels River Flood: 25th January, 1981. Dept. Water Affairs, Forestry and Environmental Conservation, Pretoria.

Bowen, D. Q. (Editor), 1976. A Concise Physical Geography. Hulton Educ. Publ., Amersham.

Estiē, K. E., 1981. The Laingsburg Flood Disaster of 25 January 1981. Weather Bureau News Letter, 383, pp. 19-32.

Hayward, L. Q. and E. E. Steyn, 1967. Aeronautical Climatological Summaries. W.B. 31, Weather Bureau, Dept. Transport, Rep. South Africa, Pretoria.

Nieman, W. A., J. T. Harmse, and H. W. A. Ellmann, 1981. Natuurkundige faktore wat aanleiding gegee het tot die Laingsburgvloedramp: implikasies vir beplanning. Unit for Futures Research, Occ. Paper 81/2, Bureau for Economic Research, Univ. Stellenbosch.

Roberts, C. P. R. and W. J. R. Alexander, 1982. Lessons Learnt from the 1981 Laingsburg Flood. Civ. Engr. S. Afr., 24(1):17-27.

Ryan, P. J., 1968. Stratigraphy of the Ecca Series and Lowermost Beds (Permian) in the Great Karroo Basin of South Africa. Unpublished Ph.D. Thesis, Univ. Witwatersrand, Johannesburg.

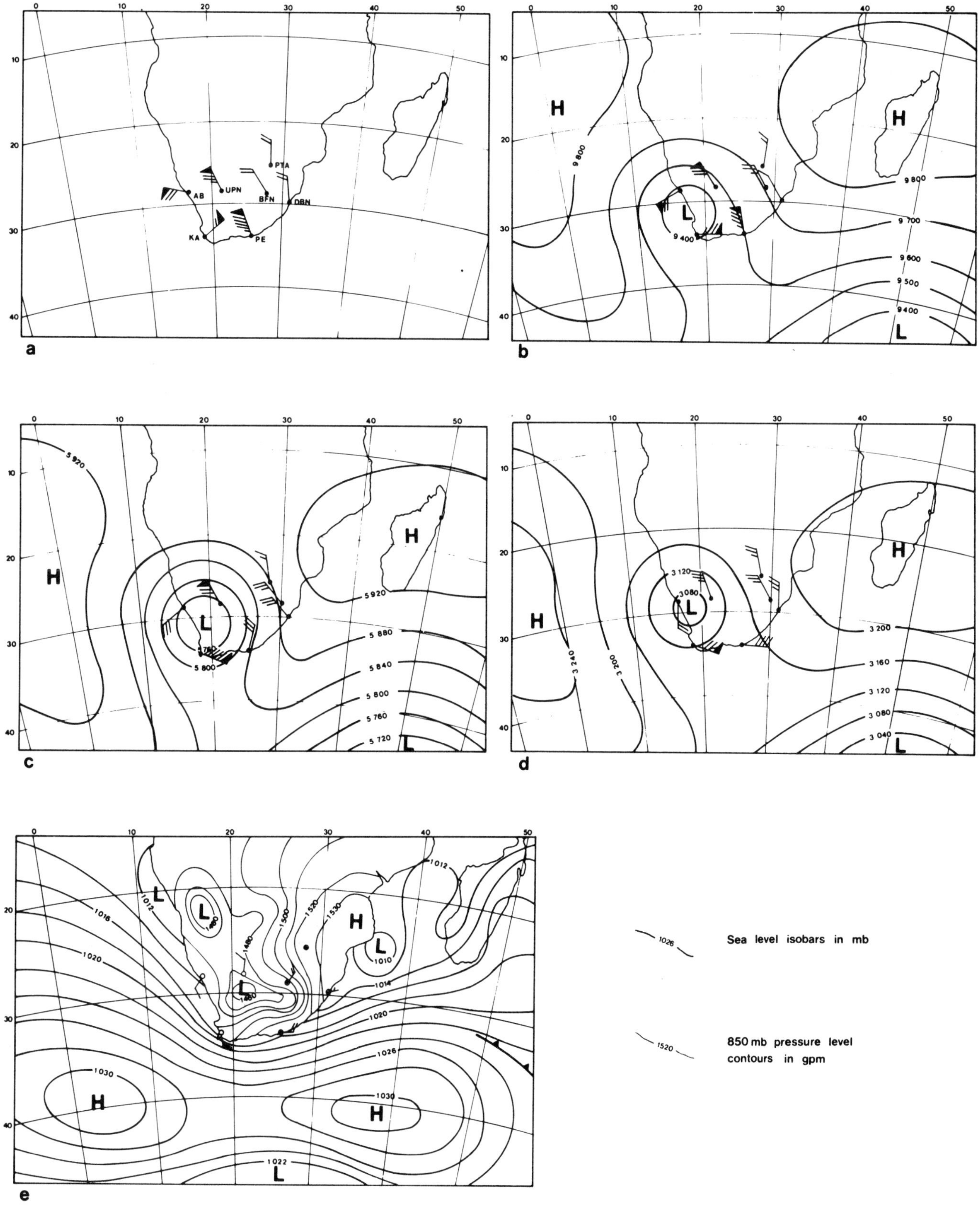

Figure 7. Synoptic Weather Pattern at 14h00 SAST, January 25, 1981: (a) 200 mb Pressure Level, (b) 300 mb Pressure Level, (c) 500 mb Pressure Level, (d) 700 mb Pressure Level, (e) Surface (after original charts from D. F. Malan Airport weather office).

Schulze, B. R., 1965. Climate of South Africa, 8: General Survey. W. B. 28, Weather Bureau, Dept. Transport, Rep. South Africa. Govt. Printer, Pretoria.

Stratten, T., 1968. The Dwyka Glaciation and Its Relationship to the Pre-Karroo Surface. Unpublished Ph.D. Thesis, Univ. Witwatersrand, Johannesburg.

Triegaardt, D. O., 1961. Flood Rains in the Karroo, South Africa. Notos 10(1/4):113-118.

Truswell, J. F., 1977. The Geological Evolution of South Africa. Purnell, Cape Town.

UCT News, 1981. Laingsburg Tragedy – What Happened. S. Afr. Digest, May 1, 12 (abstract of report in Univ. Cape Town newsletter on special meeting of S. A. Inst. Civ. Engr. addressed by Prof. D. Sparks, Univ. Cape Town, and Prof. A. Coetzee, Univ. Stellenbosch).

Union of South Africa, 1918. Report of the Director of Irrigation for the Period 1st April, 1917, to 31st March, 1918. U.G. 49-18. Cape Times Ltd., Cape Town.

Van Zyl, J. A., 1981. A Geomorphological Perspective on the Laingsburg Flood Disaster of 25 January 1981. Unpublished paper read at Conf. Soc. Geography, July 1981, Johannesburg.

Weather Bureau, 1965. Climate of South Africa, 9: Average Monthly Rainfall. W.B. 29, Dept. Transport, Rep. South Africa. Govt. Printer, Pretoria.

Weather Bureau, 1981. Daily Weather Bulletin for the Month January 1981. Dept. Transport, Pretoria.

OPERATIONAL REMOTE SENSING UNDER EXTREME CONDITIONS

James D. Belville and Bill J. Crouch[1]

ABSTRACT: This paper describes the Rio Grande, satellite-interrogated Automatic Hydrologic Observing System flood warning network operated jointly by the International Boundary and Water Commission (IB&WO) and the National Weather Service. Since its establishment in 1976, this system has provided timely and accurate hydrologic information on flow into a lake under the joint control of the United States and Mexico.

In November, 1978, the satellite-interrogated system successfully detected a major Rio Grande Flood under the most extreme conditions which could be expected. Water rose to a level never before observed on this particular river, completely submerging several components of the system within the gage house. However, the system continued to work flawlessly providing the critically needed information.

Also included in the paper is information describing the network design and rationale behind it. A detailed description of the system and its components is also provided.
(KEY TERMS: flood warning; network design; remote sensing; satellites; Rio Grande.)

INTRODUCTION

There are many areas in the United States where heavy rains in remote areas have severe consequences on urban areas or hydrologic structures downstream. The mountainous sections of the country are particularly vulnerable to this type of natural hazard. In remote areas where population is sparse and reliable communication systems nonexistent; the only dependable method of data collection for use in operational hydrology is remote sensing. A method now being used by the National Weather Service (NWS) is the Automatic Hydrologic Observing System which is Satellite interrogated (AHOS/S).

This paper describes the AHOS/S flood warning system, which is interrogated by the Geostationary Operational Environmental Satellite (GOES), for the Devils River, Pecos River, and Rio Grande River. These are all tributaries of Lake Amistad, located northwest of Del Rio, Texas. (See Figure 1.) Also described is how the system functioned under the extreme flood conditions in November 1978.

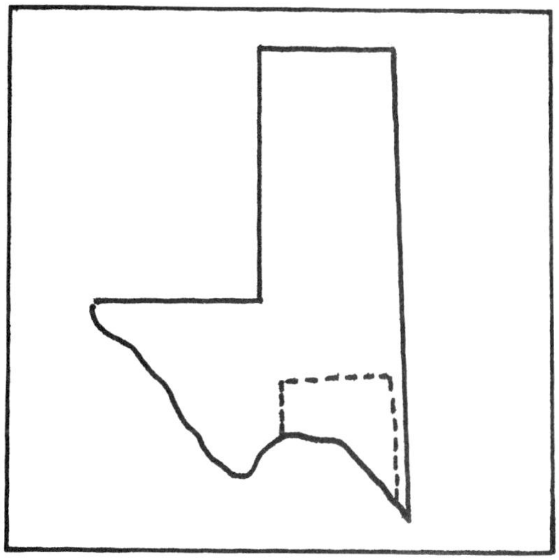

Figure 1. Location of AHOS/S Network with Respect to West Texas as Indicated by Dashed Area.

HYDROLOGIC ASPECTS OF LAKE AMISTAD

Lake Amistad is a large reservoir owned jointly by the governments of both the United States and Mexico. Its primary purposes include water supply and flood control; but facilities to generate power are also available. The operation of Lake Amistad and other structures on the Rio Grande (from El Paso downstream) is by joint agreement between the two governments. This necessitates rapid acquisition of rainfall information and forecast preparation for an efficient operation.

The watersheds of the Pecos, Rio Grande and Devils Rivers have a history of being prolific runoff producers. This is apparent from Table 1 which shows a few selected flows from these tributaries. Runoff volumes have exceeded 1,000 cubic feet per second per square mile over sub-basins within these watersheds several times since the turn of the

[1]Respectively, Meteorologist, National Weather Service, 1120 Old Spanish Trail, Slidell, Louisiana 70458; and Meteorologist-in-Charge, National Weather Service, 830 NE Loop 410, North Crown Building, Suite 300, San Antonio, Texas 78209.

TABLE 1. Selected Peak Flows for the Tributaries of Lake Amistad.

Station/River	Peak Flow (cfs)	Date
Langtry/Pecos	948,000	June 1954
Langtry/Pecos	750,000	September 1974
Foster/Rio Grande	200,000	November 1978
Pafford/Devils	139,000	August 1971

century. Although Lake Amistad has a total capacity of 5 million acre feet, only about 1.5 million is available for flood storage (IB&WC, 1971). Past watershed storms, within 100 miles of the lake, have produced as much as 1.8 million acre feet of runoff in as little as 24 hours. Severe hydrologic problems occur in this area on a regular basis. Therefore, sufficient hydrologic data must be available for safe operation of the reservoir.

Another problem encountered is data acquisition within these watersheds. The lack of population precludes development of a standard rainfall-river reporting network. This was the primary reason for the development of a satellite interrogated system.

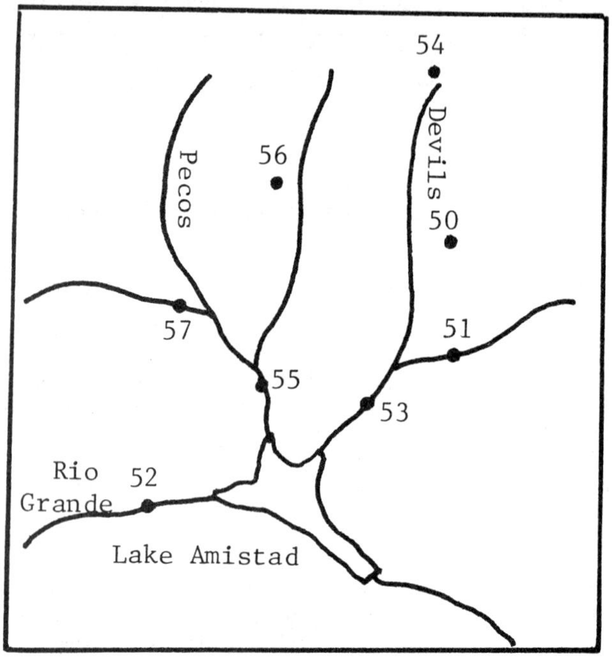

Figure 2. AHOS/S Network in Southwest Texas. Numbers refer to locations as indicated in Table 2.

THE AHOS/S SYSTEM

The AHOS/S flood warning system was established in 1976 as a joint project of the NWS and International Boundary and Water Commission (IB&WC). The network was established to provide the optimal lead time and yet detect runoff from the major portions of the watershed above the reservoir. Figure 2 shows the network configuration as implemented. A list of stations (by name and number) is shown in Table 2. Flood crest travel times from the various warning stations to the lake vary from 6 to 18 hours depending on the particular site. There are eight separate AHOS/S sites of which five are rainfall-river and three are rainfall only.

An AHOS/S installation is made up of several components listed and described below. Using a system as shown in Figure 3 makes it adaptable to almost any location.

1. Power Supply – Consists of two 6-volt motorcycle batteries connected to three panels of solar cells. This provides enough power for 24-hour operation.
2. DARDC – Device for Automatic Remote Data Collection. This device is made by the LaBarge Company. Its purpose is to collect and store the rainfall and river data in digital form.
3. Transmitter – This component is made by Magnavox and has a power output of 10 watts. It is placed in a pressurized canister for protection from the elements. It transmits on a frequency of 402 megahertz and receives on a frequency of 468 megahertz.

TABLE 2. Listing of AHOS/S Sites in Lake Amistad Networks. (ST) self-timed; (T) interrogated; (R/R) rainfall river; (R) rainfall only.

Location	Call-Up	Type	Data
Baggett	TX50	ST	R
Loma Alta	TX51	I	R/R
Dryden 14S	TX52	I	R/R
Juno 5S	TX53	I	R/R
Bissett	TX54	ST	R
Pandale	TX55	I	R/R
Cox	TX56	ST	R
Chandler	TX57	ST	R/R

4. Antenna – This is a helix antenna with a protective covering and has a beam width of about 30 degrees. At the river-rainfall stations it is the only component outside the instrument shelter.

AN EXAMPLE OF AHOS/S OPERATION UNDER EXTREME CONDITIONS

Seldom does a flood warning system encounter an extreme hydrologic event within a year or two after its development. This is primarily because a flood warning system is generally implemented shortly after a large event occurs and weaknesses in the system in use at the time of the flood are discovered. The probability of two large flood

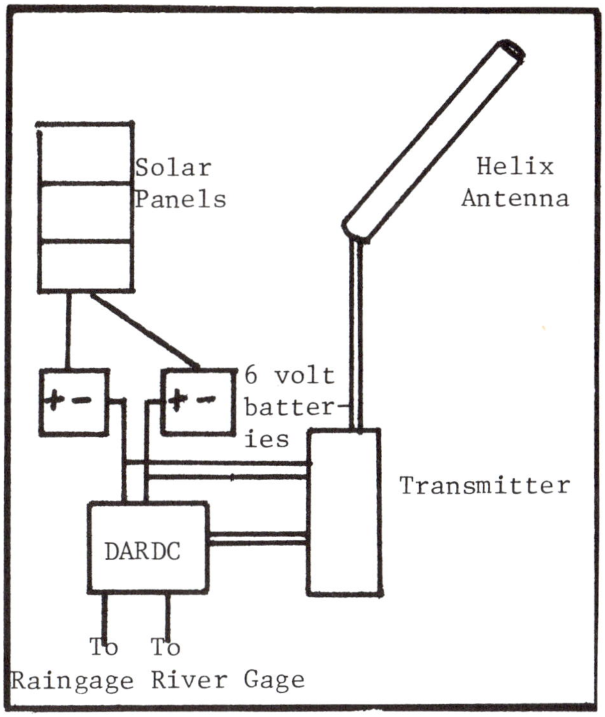

Figure 3. Block Diagram of a Typical Component AHOS/S Installation.

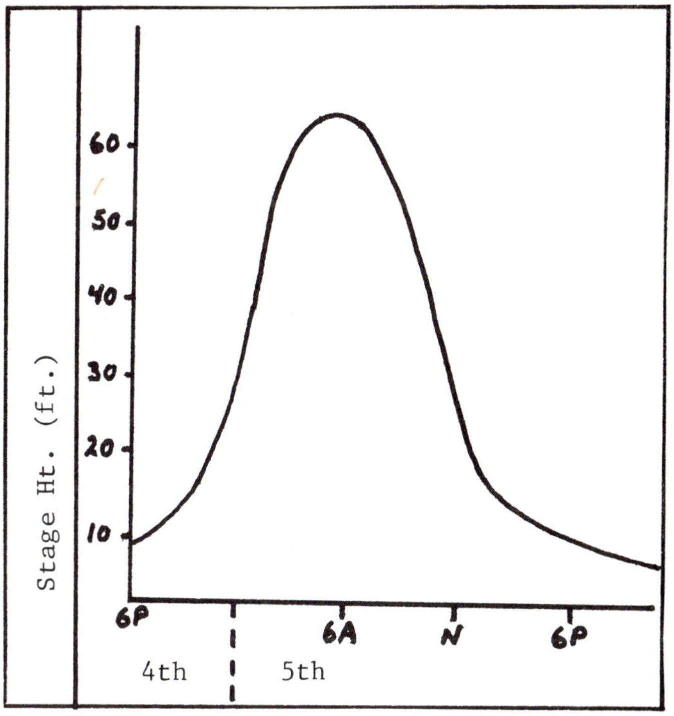

Figure 4. Hydrograph of November 4-5, 1978, on the Rio Grande as Observed by the AHOS/S Site at Dryden 14S. Dot indicates where special interrogations began.

events occurring within a few years of each other is statistically quite low. However, an extreme flood event occurred on the Rio Grande in November 1978, providing the ultimate test for the river-rainfall AHOS/S site on the Rio Grande (south of Dryden, Texas).

During the afternoon and evening of November 4, 1978, a rainstorm occurred on the Rio Grande watershed approximately 40 miles west of the AHOS/S location. This was beyond the range of NWS radar and was warm top convection such that rainfall estimations from satellite imagery (Scofield, *et al.,* 1977) were of little value. It was estimated that approximately 20 inches of precipitation fell over a 1,000 square mile area during a 6-12 hour period beginning during the afternoon of the 4th. The Rio Grande rose from 7.57 feet at 7 p.m. on the 4th to 61.5 feet at 5 a.m. on the 5th. The rate of rise was about 20 feet an hour between 1 a.m. and 3 a.m. (see Figure 4). Special interrogations of the AHOS/S system were begun at midnight on the 5th by the NWS at Lubbock, Texas. This provided a complete picture of the inflow hydrograph which was used in the operation of Lake Amistad.

Even though the flood peak observed was the largest ever recorded on this section of the Rio Grande, the volume of water amounted to only about 30 percent of the total amount of flood storage available. However, had the flood hydrograph not been available, a large reservoir release would have been made immediately based on the rate of rise at the inflow point of the reservoir. This would have caused downstream damage of well over one million dollars based on losses due to releases during previous floods. In essence, the entire AHOS/S system was paid for many times over in terms of property saved during this one flood event.

Some interesting aspects concerning the durability of the equipment were discovered during the flood. The peak stage of 61.5 feet flooded the instrument shelter with water rising to the bottom of the instrument shelf and completely submerging the transmitter canister. Yet, the system responded flawlessly during the entire episode. This success story cannot only be credited to the equipment, but also to the engineering during the implementation of the network.

DISCUSSION

It can be readily seen that an AHOS/S is reliable and cost effective when the system is properly designed. Many advances in the field of remote sensing have been made since the inception of this network. One such improvement was the development of a system that responds to discrete events rather than on a timed basis. However, the purpose of all such systems is the same; to provide reliable data from remote areas.

The total cost of one AHOS/S site was approximately $5,200 including installation but not including rainfall or river gauges. The component system is easily installed in a river gauging shelter or mounted on a tower, as was the case at the rainfall only stations.

There are two types of systems used in this particular network; these are: (1) interrogatable and (2) self-timed. The difference between the two is that the self-timed system reports at regular 6-hour intervals but cannot be interrogated at intermediate times. The interrogatable system can be interrogated hourly, if the need arises. Of the eight installations within the Southwest Texas network, four of the five river-rainfall stations are interrogatable. The three rainfall only stations and remaining river-rainfall station are self-timed. The locations of the self-timed system are such that they are not considered critical of the success of the overall network.

Regular interrogations of the network are made by the National Environmental Satellite Service (NESS) at 6-hourly intervals. The scheduled times are 00Z, 06Z, 12Z, and 18Z. The data is transmitted to the NWS Offices at both San Antonio and Lubbock via teletype circuit.

A federal government study into increased use of automation for the collection of hydrologic data is currently underway and the results should be forthcoming shortly. It is expected that the study will recommend greatly increased use of satellite interrogated systems within the federal hydrologic services, whether pertaining to flash floods or data acquisition. As it stands at the present time, the use of satellite interrogated systems has already gained wide acceptance in several agencies other than the NWS.

ACKNOWLEDGMENTS

First of all, this paper would not have been possible without the excellent work in development and implementation of the AHOS/S network by the NWS Southern Region Facilities Engineering Branch. Thanks are also due Dr. Robert A. Clark, NWS Headquarters; David T. Smith, NWS Southern Region Headquarters; and Scott A. Bottorff, NWS Southern Region Headquarters, for providing valuable information used within this paper. Last but not least, the presentation of this paper would not have been possible had it not been for the skillful typing of the manuscript by Ms. Shirley LeBlanc of NWS Slidell, LA.

LITERATURE CITED

International Boundary and Water Commission, 1971. Flow on the Rio Grande and related data. Water Bulletin No. 47, El Paso, Texas, 100 pp.

Scofield, R. A. and V. J. Oliver, 1977. A scheme for estimating convective rainfall from satellite imagery. NOAA Tech. Memo. NESS-86, 47 pp.

Section 3

WATER SUPPLY AND WATER RESOURCES

THE IMPORTANCE OF STREAMFLOW FORECASTING TO WATER MANAGEMENT WITH INSTREAM FLOW CONSIDERATIONS

Robert T. Milhous[1]

ABSTRACT: The management of water with instream flow considerations is improved by the forecasting of future water availability. In many existing water management schemes, it is the instream flows that are sacrificed to cover the uncertainties in streamflow forecasting and to assure a continuous supply of water for out-of-stream uses. Unfortunately, much of the value received from the instream flows is destroyed by short periods of very low flows. This paper explores the use of streamflow forecasts to reduce the possibility of very low flows and the allocation of available water prior to a drought.
(KEY TERMS: forecasting; water management; streamflow.)

INTRODUCTION

The purpose of this paper is to present some ideas on how streamflow forecasting can help increase benefits from the instream use of water without reducing benefits derived from other uses of water. The instream flow issue will be discussed in this section followed by a section on instream flow worth assessment and, finally, a section on the use of streamflow forecasting.

One of the major conflict zones in water resources management is the allocation of water between out-of-stream and instream uses; in addition, conflict also exists among various instream uses. An approach is available to evaluate the value of water instream for a fishery or recreational use and to help make trade-offs between various instream flow uses. This approach is the instream flow incremental methodology developed by the Instream Flow Group of the U.S. Fish and Wildlife Service.

The basic assumption is that some relationship exists between the quantity of flow (as influenced by water management policy) and the fish habitat or recreational space produced by that flow. In the case of the fishery, little information on the relationship between flow history and fish production is available; hence, fisheries biologists have approached instream flow management as one of providing habitat for fish, and assume that the establishment of a potential carrying capacity for fish means fish will exist. Managers of recreational resources have generally assumed that water would be in a stream when needed therefore managing the recreational use and the constructed facilities, such as boat ramps and camp grounds, without reference to water management.

Two basic aspects of instream flows related to a fishery are: 1) the type of fishery; and 2) the management of the fishery. The type of fishery refers to the species composition of the fish: for instance, a trout-mountain whitefish fishery, a small-mouth bass fishery, a steelhead-chinook salmon fishery, and so forth. The management of the fishery refers to the objective of the fish management agency. Two management extremes for an inland fishery are a put-and-take fishery and a fishery to be maintained entirely by natural reproduction and growth. For any stream fishery, the management alternatives are directly related to the life cycle of the specific species. For instance, steelhead rear in a stream up to two years; in contrast, sockeye salmon leave soon after hatching and rear in freshwater lakes.

The type of fishery and the management of the fishery are directly related to the water supply and the management of the water resources. Some information is available on the relationship between water management and fisheries management, but a logical approach to the interaction of the two management functions has yet to be formulated either in theory or in practice. Three early collections of information are the assessment of instream flow methodologies developed by Stalnaker and Arnette (1976), a thesis by Miller (1976), and the Proceedings of the Symposium and Specialty Conference in Instream Flow Needs held in Boise, Idaho, during May 1976 (Osborn and Allman, 1976). A recent review of fishery flow issues and methods was written by Loar and Sale (1981) together with a paper soon to be available from the Instream Flow Group which will go a long way in formulating a combined approach (Bovee, 1982). The relationship between flow and recreational value of the flow instream depends on the specific recreational activity.

The following sections will discuss techniques which can be used to develop the physical habitat versus discharge relationship and the use of the relationship to select among

[1]Hydraulic Engineer, Instream Flow and Aquatic Systems Group, U.S. Fish and Wildlife Service, Fort Collins, Colorado 80523.

alternate management policies given the forecasted water supply.

THE PHYSICAL HABITAT SIMULATION SYSTEM

An approach to determining the relationship between the quantity of water instream and the value of fisheries habitat has been developed by the Instream Flow and Aquatic Systems Group of the U.S. Fish and Wildlife Service. This approach is referred to as the instream flow incremental methodology and relies heavily on a computer simulation system called the Physical Habitat Simulation (PHABSIM) System. The PHABSIM system is used to determine the fisheries habitat available as a function of the discharge in a stream. Instream needs, identified in an analysis using PHABSIM, can be used in the allocation of water. The PHABSIM system is based on the assumption that:

$$WUA(Q_i) = \int^A f(v,d,s,c)da$$

where WUA is the usable habitat area in the stream at the flow Q_i; v is the velocity at a specific location in the stream (da) at flow Q_i; d is the depth at location da and flow Q_i; s is an index to the bed material (substrate); c is an index to the type of cover present, such as logs, overhangs, and overhanging bush; and da is a small area in the stream. Substrate and cover area are assumed to remain constant, rather than change as a function of discharge. The function f is the criteria which relates the physical parameters (v, d, s, and c) to the habitat suitability for a species of fish or to the usability of the stream for recreation.

There are two basic subsets of the PHABSIM system – a subset used for hydraulic simulation and a subset for physical habitat simulation. These are shown in Figure 1. The purpose of hydraulic simulation is to simulate the velocities and depths needed in the physical habitat simulation if only a limited amount of velocity and depth data are collected. This limited data is used to calibrate the hydraulic model which is then used to simulate the velocities and depths at unmeasured flows. If the user of PHABSIM chooses to collect sufficient velocity and depth sets to define the physical habitat versus flow relationships, then the hydraulic simulation step may be eliminated and the data entered directly into the habitat simulation subset.

The data which the user of PHABSIM must supply are the definition of the channel characteristics, velocity and depth sets to calibrate the hydraulic simulation program, or to go directly to the physical habitat simulation program, and the criteria defining the worth of the habitat versus physical parameter (velocity, depth, etc.) relationship for each species of fish of interest for a given stream.

When the PHABSIM system is used in water management, the assumption is made that the only impact of alternative water management proposals will be on the fishery or on the space available for recreation is through changes in the physical habitat. Where this assumption is not valid, additional biological analyses of the stream will need to be made to supplement PHABSIM.

Use of the PHABSIM system requires the collection of data on stream hydraulics and channel characteristics and the analysis of the data using a library of computer programs. The Instream Flow Group has had considerable success in training biologists, engineers, and others in the use of the system. The system has been used in the planning and management of water resources projects throughout the United States and Canada and is under consideration for use in New Zealand.

The PHABSIM system for simulating the physical habitat for aquatic organisms has and is being used for water management activities at many locations. Training in the use of the simulation system has resulted in many people becoming proficient in the use of a computer in general and PHABSIM, specifically, for instream flow analysis. The best results from instream flow negotiations are obtained when the PHABSIM system is used to simulate physical habitat and a water system model is used to simulate the water system, with the results of both models used to determine the instream flow. See Milhous (1982) for additional information on the combined use of PHABSIM and water resource models.

An example of the habitat discharge relationship as developed using PHABSIM is given in Figure 2. PHABSIM can be used for other types of analysis but the relationship needed for use with streamflow forecasts is the habitat discharge relationship. Additional information on PHABSIM is given by Milhous (1979) and Stalnaker (1980).

THE USE OF STREAMFLOW FORECASTING

Streamflow forecasting fits into the management picture by allowing the water manager and the manager of the habitat use to select operation criteria for the system given the water supply and the relationship between the habitat usability and streamflow. At this point it is not proven that more benefits may be derived from instream uses nor are examples of the use of forecasting available for presentation. What follows are some ideas and concepts that could yield an increase in benefits to society by increasing benefits from instream flows.

Streamflow forecasting is important in two ways – the management of the fishery can change with the management of the water and as a function of either high or low yield to the stream.

If, as a result of streamflow forecasting activities, some idea of the total flow is available, and a functional relationship between streamflow and physical habitat exists, then the physical habitat can be maximized given the water supply and the use of the instream resources managed. As an example, suppose the forecasted streamflow in the fall is for a streamflow that would not support the fish population in the stream in the spring; then the fisheries manager could

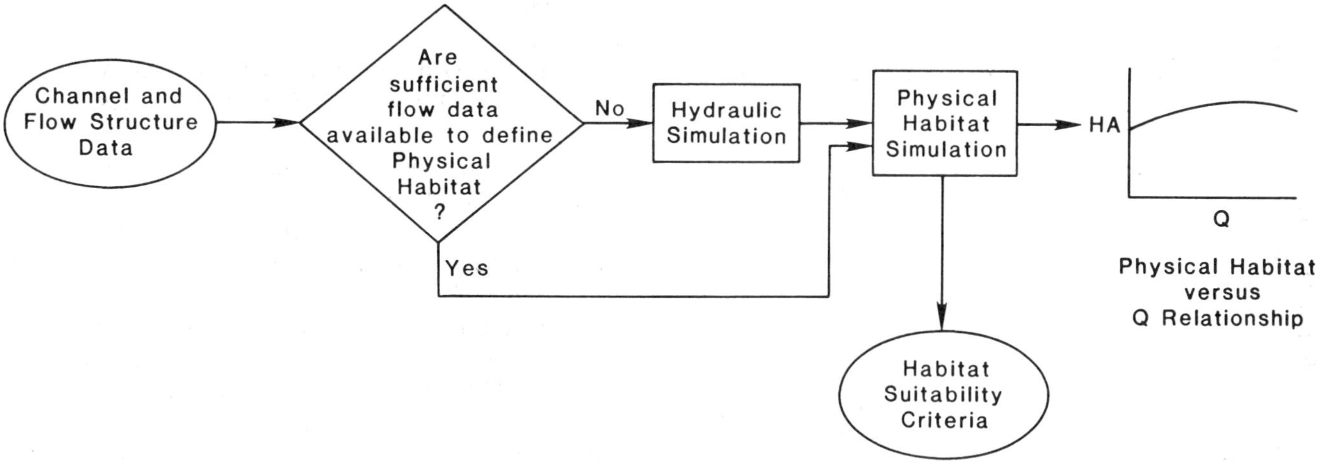

Figure 1. Basic Element of the Physical Habitat Simulation System (PHABSIM).

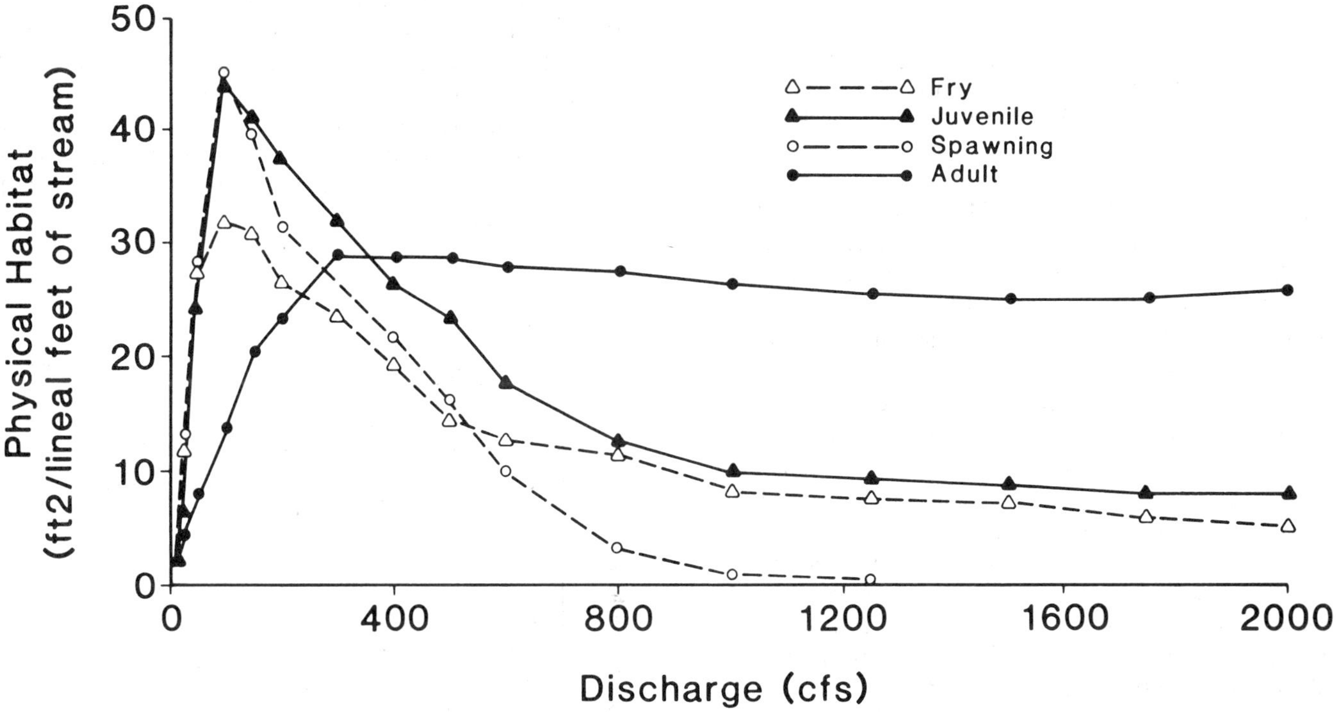

Figure 2. Physical Habitat Versus Streamflow Relationships for Rainbow Trout in the North Fork of the Snoqualmie River, Washington.

allow the excess fish to be harvested. Another example is the case where a reservoir can store water and use some of the water to produce instream benefits. Suppose a situation exists where the "normal" case is as shown in Figure 3 and a severe drought is forecasted, then the fishery manager, in conjunction with the water manager, may negotiate a drought condition flow, as shown in Figure 3, that will use less than 25% of the normal condition quantity of water but produce much more than 25% of the benefits.

The use of physical habitat versus discharge relationship can also be shown by using Figure 2 as an example. From Figure 2 we see that the most useful set of streamflows are between 100 and 300 cfs. Suppose the agreed upon minimum flow for a reservoir upstream was 200 cfs. If the

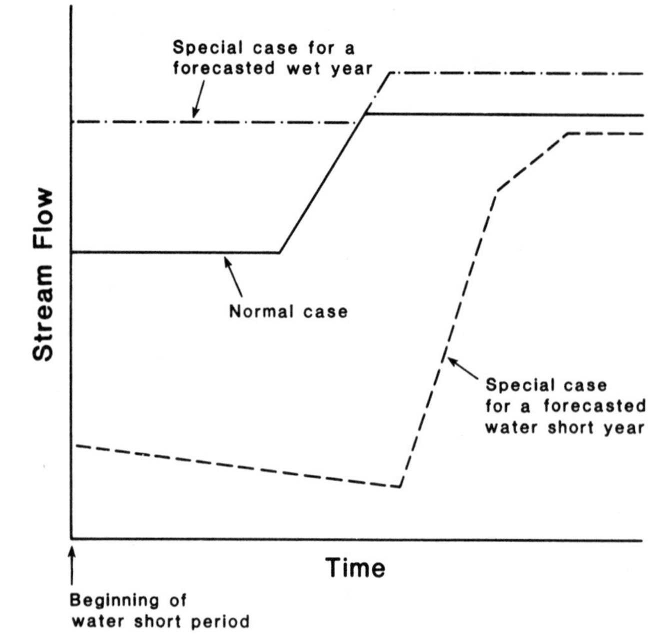

Figure 3. An Example of the Use of Forecasts to Select a Release Schedule for a Reservoir That Will Maximize Streamflow Levels Given a Forecasted Water Supply.

available flow is 100 cfs from May to October a release pattern could be selected that would minimize the impact on the fishery by releasing 150 cfs from May through July; then 50 cfs from August through October. This pattern would yield about the same physical habitat for fry, juveniles, and spawning but significantly reduce the available adult habitat for three months. Another alternative would be 100 cfs through the whole period but this would stress the adults for a much longer period of time. The choice between these, and other, alternatives depends on the knowledge of the fisheries biologist.

The selection of the flow available during a drought period is also assisted by use of Figure 2. In this case, the physical habitat would be very severely stressed by any flows below about 40 cfs. Hence, the 40 cfs figure is just about an absolute minimum for instream flow. For the refill period it is useful to know that flows in excess of about 300 cfs are of no value to any life stage.

If the water use versus benefit function is also known for all water uses it is possible to trade-off uses so as to maximize net-benefits to society prior to any forecasted low flow periods by reducing the water uses early.

Another interesting element of forecasting is the possibility of using long range forecasting to modify operation criteria developed using past records. A criteria used in operation studies of reservoirs in the mid-continent is to base the water supply on the level of the reservoir; the concept is:

$$\text{QAVA} = \text{QDESIRED} \text{ ; IF (STOR} \geqslant \text{STORA)}$$
$$\text{QAVA} < \text{QDEISRED} \text{ ; IF (STOR} < \text{STORA)}$$

where QAVA is the water available for use, QDESIRED is the water supply desired, STOR is the present level of storage in a reservoir, and STORA is some level of storage at which the water use is reduced.

If the predicted reservoir inflow during certain months would be adequate to maintain a substantial rate of use, then both the instream and diversionary use could be higher than would be the case in the absence of a forecast. If the forecast was for low streamflow the use would be lower than without a forecast during the early part of the drought but would not be zero later in the drought. Often the first use to be reduced in times of water shortage is the instream flow use (Nelson, *et al.,* 1976). Hence, forecasts could be more important to instream uses than to diversionary uses.

An example of the concepts discussed above on the impact of forecasting on releases for instream uses is shown in Figure 4. If it is known, because of a reliable and believable forecast, that the changes will be moderate, then the case is shown in 4A, if severer than 4B. In each case the instream uses are improved if the best physical habitat possible is maintained given the water supply. This can be done only if the functional relationship between physical habitat and streamflow is known.

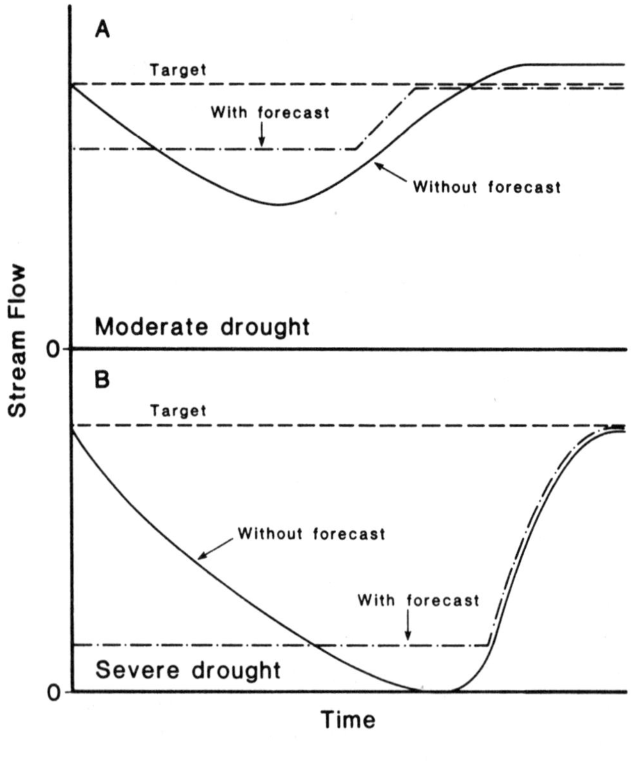

Figure 4. The Impact of Streamflow Forecasting on Releases from a Reservoir for Instream Uses.

To explain how the release rates shown in Figure 4 was derived, the habitat function given in Figure 2 is used. Suppose the target flow is 300 cfs based on Figure 2 and we have the case of a forecasted moderate drought. In the absence of a forecast, the water released for instream uses would be reduced as the storage in the reservoir was reduced. If the forecast is that the drought will be moderate, the reservoir could be drawn down lower than without the forecast. In this case a biologist and system operator could agree on an instream flow of 200 cfs because the reduction would reduce the physical habitat by about 20% for adults while increasing the habitat for the other life stages and at the same time not injure other uses. The reduction of adult habitat may reduce the adult population but also increase the survivability of juveniles and fry. Consequently the impact on the following year's population may be slight.

In the case of a severe drought, the release to the river could be zero during some time period which, for some species, could essentially destroy the fishery for a number of years until the reach is recolonized. If a forecast suggests very low water availability, then the fishery manager may accept a flow of 40 cfs as an absolute minimum but would prefer 100 cfs. The object here would be to have some population of all life stages survive through the drought. In conjunction with the reduced flow the fisheries manager could restrict fishing. In addition, other flow patterns are possible depending upon the judgment of both the fishery and water managers.

In the case where the forecasted supply was less than 40 cfs, then the flows could be maintained at the optimum for harvesting and then made zero when the volume of water available for instream use was exhausted.

SUMMARY

By being able to model physical habitat and the water resources system, and to forecast the water supply, the water manager and fisheries can negotiate a water resources plan for the future which will maximize the benefits derived from the water supply available. A use which can be significantly improved by forecasting is the instream uses of water.

LITERATURE CITED

Bovee, K. D. 1982. A Guide to Stream Habitat Analysis Using the Instream Flow Incremental Methodology. Information Paper No. 12, Cooperative Instream Flow Service Group, U.S. Fish and Wildlife Service. Unpublished Manuscript.

Loar, J. M. and M. J. Sale, 1981. Analysis of Environmental Issues Related to Small-Scale Hydroelectric Development. V. Instream Flow Needs for Fishery Resources. Oak Ridge, Tennessee: Oak Ridge National Laboratory.

Milhous, R. T., 1979. "The PHABSIM System for Instream Flow Studies." *In:* Proceedings: 1979 Summer Computer Simulation Conference held in Toronto, Canada, July 1979, pp. 440-446.

Milhous, R. T., 1982. "The Application of the Physical Habitat Simulation System to Natural Resources Management." *In:* Proceedings: 1982 Summer Computer Simulation Conference held in Denver, Colorado, July 1982.

Miller, J. W., 1976. The Effects of Minimum and Peak Cedar River Streamflows on Fish Production and Water Supply. M.S. Thesis, University of Washington.

Nelson, W., G. Horak, M. Lewis, and J. Colt. 1976. Assessment of Effects of Altered Stream Flow Characteristics on Fish and Wildlife. Part A: Rocky Mountains and Pacific Northwest. Enviro Control, Inc., Rockville, Maryland.

Osborn, J. F. and C. H. Allman (editors). 1976. Proceedings of the Symposium and Specialty Conference on Instream Flow Needs, Boise, Idaho. Published by American Fisheries Society.

Stalnaker, C. B., 1980. "The Use of Habitat Structure Preferenda for Establishing Flow Regimes Necessary for Maintenance of Fish Habitat." *In:* The Ecology of Regulated Streams, J. V. Ward and J. A. Stanford (Editors). Plenum Publishing Corporation, pp. 321-337.

Stalnaker, C. B. and J. L. Arnette (editors). 1976. Methodologies for the Determination of Stream Resource Flow Requirements: An Assessment. Office of Biological Services, U.S. Fish and Wildlife Service.

INTERNATIONAL SYMPOSIUM ON HYDROMETEOROLOGY
JUNE AMERICAN WATER RESOURCES ASSOCIATION 1982

FORECASTING SNOWMELT RUNOFF

Guillermo Q. Tabios III and Jose D. Salas[1]

ABSTRACT: Various forecasting techniques for snowmelt runoff have been proposed in the literature. Most of them are of the multiple regression type where the total seasonal runoff is a function of several variables such as antecedent precipitation, snow-course data, previous seasonal runoff and temperature. Among those noted in particular are the forecasting models developed by Ford (1959), subsequently improved by the USBR (1968) and that proposed by Tangborn and Rasmussen (1976).

This paper discusses some limitations of the forecasting techniques based on multiple regression analysis. In an attempt to improve such techniques, areal averaging by spatial interpolation and recursive model parameter estimation via Kalman filtering are introduced. The original and modified forecasting schemes were compared. The results showed that some improvements can be obtained by the suggested modifications.

(KEY TERMS: forecasting; runoff; snowmelt.)

INTRODUCTION

Water supply obtained from snowmelt has a large economic value for irrigation, hydropower and other uses of water. Snowmelt runoff forecasting in tandem with appropriate reservoir operation permits proper utilization of such water supplies. Various techniques have been proposed in the literature to forecast the total spring and summer runoff which is a function of several variables such as antecedent precipitation, snowpack accumulation, previous seasonal runoff and temperature. Most forecasting schemes available in practice are of multiple regression type. Among those noted in particular is the one proposed by Ford (1959) of the U.S. Bureau of Reclamation (USBR) in which a multiple regression model is used to forecast snowmelt runoff considering the antecedent precipitation, snow-water equivalent or runoff season precipitation as the independent variables. Since the choice of gaging stations affects the reliability of forecasts, the USBR (1968) introduced a procedure where the stations are weighted on the basis of the best multiple correlation coefficient. Likewise, the California Department of Water Resources utilizes a multiple regression model to forecast snowmelt runoff as a function of antecedent precipitation, snowpack accumulation index and previous year seasonal runoff.

Tangborn and Rasmussen (1976) derived a snowmelt runoff forecasting model assuming that snowmelt runoff depends on the amount of snow stored in the basin. Tangborn (1980) extended this model considering the snow ablation energy and runoff water accumulation during a preforecast season and the snow ablation obtained from reliable forecasts of precipitation and temperature. The dependence of snowmelt runoff to the foregoing variables is established using a stage-wise multiple linear regression.

This paper discusses the limitations involved in utilizing forecasting schemes based on regression models. In particular, the USBR and Tangborn-Rasmussen models are analyzed and attempts are made to improve the forecast schemes by introducing the techniques of spatial interpolation and Kalman filtering. Some applications based on the original and modified schemes are presented.

REVIEW OF THE USBR AND TANGBORN-RASMUSSEN MODELS

The USBR model (Ford, 1959) forecasts total seasonal snowmelt runoff using multiple regression. The model is written as:

$$y = a_0 + a_1x_1 + a_2x_2 + a_3x_3 + ... + a_nx_n \qquad (1)$$

where y = total seasonal runoff say, Apr.-Jul. runoff; x_1 = snow water equivalent averaged from various snow course stations, say Feb. 1 snow; x_2 = seasonal precipitation, say Oct.-Jan. precipitation, x_3, ..., x_n = other independent variables; and a_0, a_1, ..., a_n = regression coefficients. The choice of the independent variables in the model is determined based on the highest correlation coefficient.

The x_i's in Eq. (1) are arithmetic averages based on measurements made at the available gaging stations in the watershed. A station weighting procedure, called successive increment method, was later introduced (USBR, 1968) in

[1]Respectively, Graduate Student, Department of Civil Engineering, Colorado State University, Fort Collins, Colorado 80523; and Associate Professor of Civil Engineering, Colorado State University, Fort Collins, Colorado 80523.

order to find weights for each gaging station which gives the best multiple correlation coefficient. Thus, the weighted independent variable x_i of Eq. (1) is given by:

$$x_i = \sum_{j=1}^{n} w_{ij} x'_{ij} / \sum_{j=1}^{n} w_{ij} \quad (2)$$

where w_{ij} = weight of station j for the variable i, x'_{ij} = data of station j for the variable i, and n = number of stations for the variable i. Note that when w_{ij} = 1 for all j, x_i is simply the arithmetic average from all stations. In the successive increment method, the initial solution is the average solution. Then starting with the first station of the first independent variable, its initial weight of 1.0 is decremented or incremented by some fraction say 0.01. Then, the correlation coefficient is computed and compared to the previous one. The process is repeated until the best correlation coefficient is obtained and all stations and variables are exhausted. The minimum station weight is limited to zero while there is no limit to the maximum. However, the maximum station weight is generally 4.0 (USBR, 1968).

The use of the successive increment method which is based on finding the best multiple correlation coefficient is computationally inefficient and arbitrary. First of all, different forecast model parameters and forecast reliability may be obtained when the order of the data input is changed. Secondly, the choice of the best model on the basis of the multiple correlation coefficient does not assure the best forecast which the model is intended for. This is because the correlation coefficient is only a good indicator as far as the historical data is concerned, but not for the future. Lastly, depending on the number of stations available and variables considered, the multiple regression computations may be repeated a large number of times before the best weights are obtained.

Tangborn and Rasmussen (1976) suggested a forecast model based on the annual water balance of the basin considering runoff, precipitation, evapotranspiration and basin water storage. Since the basin primarily accumulates water from winter precipitation and a portion of this goes to winter runoff, the difference of these two approximates the spring storage which eventually becomes the summer runoff. Furthermore, it is assumed that the winter and summer evapotranspiration, the contribution of summer precipitation to runoff and groundwater and snow carry-over effects are the same every year. With the above arguments, the summer runoff forecast R_s^* is given by:

$$R_s^* = a_s(p_w + p_t) + b_s - R_w - R_t \quad (3)$$

where p_w, p_t = winter and spring precipitation, R_w, R_t = winter and spring runoff and a_s, b_s = model parameters obtained by linear regression.

Due to seasonal variations in basin storage calculations the forecast errors resulting from Eq. (3) are rather large. To improve the forecast, Tangborn and Rasmussen introduced a test season (spring) forecast equation. They observed that the errors resulting from Eq. (3) and the test season forecast equation are highly correlated. Thus, their improved summer runoff forecast model is:

$$R_s^{**} = a' p_w + b' P_t + c'(R_w + R_t) + d' \quad (4)$$

where $a' = a_s - ca_t$, $b' = a_s$, $c' = c - 1$, $d' = b_s - cb_t$, a_t, b_t = regression coefficients for the test season, and c = coefficient relating the spring and summer errors. Observe that Eq. (4) is a regression type model similar to the USBR model of Eq. (1).

In both forecasting schemes reviewed above, the dependence of snowmelt runoff to factors such as antecedent precipitation, snow data, etc., are established using linear regression. However, the use of linear regression has some serious drawbacks (Szollosi-Nagy, 1976): linear regression can only be applied to strictly linear and time invariant systems with stationary hydrologic processes; the observations used in the forecast model are considered free from measurement errors, which is seldom realized with current sampling capabilities; and the processing mode of the method requires large computer storage and time as new observations become available.

SUGGESTED MODIFICATIONS

In an attempt to improve the foregoing forecasting schemes, the techniques of spatial interpolation and Kalman filtering are introduced, herein. First, observe that the forecast equations are implicitly a function of the basin water storage, which in turn depends on the catchment characteristics of the basin and areal characteristics of the input hydrologic variable. In watersheds with irregular topography, inadequate coverage of precipitation stations and highly erratic precipitation or snowpack accumulation patterns, a realistic spatial representation of the various variables involved in the forecast model is desirable. For this purpose, three techniques of areal averaging are tried, namely: the Thiessen method, multiquadric interpolation (Hardy, 1971) and optimal interpolation (Belousov, *et al.*, 1971 and Rodriguez-Iturbe, *et al.*, 1972).

The areal average of a process $\bar{h}$ is obtained by discretizing a watershed area into m grids, such that:

$$\bar{h} = \frac{1}{m} \sum_{k=1}^{m} h(k) \quad (5)$$

where h(k) = point estimate at any grid k. The point estimate h(k), with rectangular coordinates x_k and y_k, can be adequately represented by a weighted linear combination of the observed values h_j of station j with coordinates x_j and y_j as:

$$h(k) = \sum_{j=1}^{n} w_j(k) h_j \qquad (6)$$

where $w_j(k)$ = weight of station j for grid k, and n = number of stations.

In the Thiessen method, the weight of the station nearest to point k is equal to 1.0 and the rest of the weights are equal to 0. Using multiquadric interpolation, h(k) is obtained by using the weights:

$$w_j(k) = \sum_{i=1}^{n} \gamma_{ij}\, d_i(k) \text{ for } j = 1, 2, \ldots, n \qquad (7)$$

where $d_i(k)$ is the distance from any point k to station i and γ_{ij} is an element of a matrix which is the inverse of a matrix with distance elements d_{ij}, i,j=1,2,...,n.

Optimal interpolation is based on minimizing the mean-square error of Eq. (6) with respect to the weights. The technique requires defining a correlation function $\rho[d]$ as a function of distance. Under the assumptions of homogeneity of the mean and variance and isotropicity of $\rho[d]$, it can be shown that the weights are obtained by solving:

$$\sum_{i=1}^{n} w_i(k)\, \rho[d_{ij}] + \lambda(k) = \rho[d_i(k)] \quad ; j=1,2,\ldots,n \qquad (8)$$

$$\sum_{i=1}^{n} w_i(k) = 1$$

where $\lambda(k)$ = Lagrange multiplier. Various forms of spatial correlations functions are available (Yevjevich and Karplus, 1973). The reciprocal and exponential models are used in this study. In addition, instead of $\rho[d_i(k)]$, the areal-station correlation $\bar{\rho}_i$ estimated by:

$$\bar{\rho}_i = \frac{1}{m} \sum_{k=1}^{m} \rho[d_i(k)] \quad ; i=1,2,\ldots,n \qquad (9)$$

may be used in Eq. (8) which will give set of n station-basin weights. The use of these weights in Eqs. (2) or (6) yields the areal average of the process.

The technique of Kalman filtering has some advantages over linear regression. As mentioned earlier, the batch mode data processing of linear regression is a limitation. Kalman filtering on the other hand, is computationally- and storage-wise more efficient because of its recursive form of data processing. In addition, it has the advantage of adaptability to non-linear or time-varying processes in contrast to linear regression which is limited to linear and time-invariant processes. Two Kalman filtering algorithms are tried here: the standard Kalman filter (Kalman, 1960) and the adaptive filter (Myers and Tapley, 1976). The Kalman filter formulation is such that the model parameters of Eq. (1), (3) or (4) compose the so-called state vector. Also, the elements of the measurement matrix are the vector of independent variables. In the standard Kalman filter the model parameters are constant in time. Using the adaptive filter, on the other hand, the parameters follow a random walk model. The reader is referred to the original papers for more details.

APPLICATIONS

The USBR model is used to demonstrate the advantages of the suggested areal averaging techniques and the Tangborn-Rasmussen model is used to demonstrate the advantage of Kalman filtering. In the first case, the Apr.-Jul. inflow to the Green Mountain Reservoir in Colorado was used as the forecast variable of interest. Data of seasonal inflows, snow water equivalent and precipitation for the period 1962-1978 were obtained from the USBR, Loveland, Colorado. The following independent variables were considered: the Feb. 1 snow water equivalent, previous year Feb.-June precipitation, Oct.-Jan. precipitation and previous year Nov. inflow. Several forecast models were fitted using the first independent variable, then the first and the second, the first and the third and so on.

The various models and areal averaging techniques were assessed by comparing the one-step ahead forecasts. The linear regression models are treated in a recursive manner. No forecast is made during the first 10 years of records. Thus, starting the forecast on the 11th year, the previous 10 years of data is used. Then the 12th year forecast is based on 11 years, and so on until the 17th year. Thus, 7 data points were used for comparison. Table 1 shows the root-mean square errors (RMSE) between the observed and forecasted values for 6 areal averaging techniques (1 through 6) and 4 (A through D) combinations of independent variables. For the set A methods 3 through 6 have smaller RMSE than 1 and 2 where the multiquadric technique has the lowest RMSE. Similar result is obtained for the set B. However, comparing among sets A and B the addition of the Feb.-June precipitation increase the RMS. On the other hand, the set C yields slightly smaller RMSE in most of the cases. Finally, the set D of 3 variables generally worsens the forecasts as in set B due to the addition of Feb.-June precipitation. Generally, the methods 3 through 6 have smaller RMSE than the others.

The choice of the forecast equation in the USBR method is based on the best correlation coefficient. For the data described above, the correlation coefficients using sets A, B and D have increased in this order for both, the averaged and the weighted solutions. Consequently, the "best" model would correspond to set D which includes Feb.-June precipitation. But, results in this study showed that in general the inclusion of Feb.-June precipitation increased the RMSE. This indicates that the highest correlation coefficient does not assure that a model yields the best forecast. It should be mentioned that Kalman filtering could be

TABLE 1. Root-mean Square Errors Between the Forecasted and Observed Values

Areal Averaging Methods	Set of Independent Variable	(A) Feb. 1 Snow	(B) Feb. 1 Snow Feb.-June Prec.	(C) Feb. 1 Snow Oct.-Jan. Prec.	(D) Feb. 1 Snow Feb.-June Prec., Oct.-Jan. Prec.
1 Average Solution		24.6	24.7	25.6	26.6
2 USBR Weighted Solution		21.5	22.2	–	21.1
3 Thiessen Method		19.7	22.0	20.6	22.7
4 Multiquadric Interpolation		19.2	21.1	18.9	21.1
5 Optimal* Interpolation		19.6	21.4	19.3	21.4
6 Optimal* Weighted Solution		19.6	21.4	19.4	21.4

*Spatial correlation functions:

Feb. 1 Snow: $\rho(d) = \exp(-d/233.60)$
Feb.-June Prec.: $\rho(d) = (1+d/32.68)^{-1}$
Oct.-Jan. Prec.: $\rho(d) = \exp(-d/159.944)$

used instead of the recursive linear regression as was done here by considering the parameters of Eqs. (1), (3) or (4) as the so-called state vector and the measurement variance equal to 1.0.

The Tangborn and Rasmussen model and the runoff data of Thunder Creek, Washington, are used to assess the use of the Kalman filter. Only one precipitation gaging station is used in this case. The available runoff and precipitation are accumulated into seasonal totals where the winter season is from October to March, the spring test season is April and the forecast season is May. The data consist of 44 water years (1931-1974) and both the standard Kalman filter (SKF) and adaptive limited memory filter (ALMF) are tried. The forecast equations (3) and (4) are used for one year ahead forecast. They are compared in a cumulative basis for the periods 11th to 20th years, 11th to 30th and 11th to 44th years. Table 2 shows the RMSE between the observed and forecast values. In using Eq. (3), the SKF with a measurement variance of 1.0 is essentially the same as the recursive linear regression. In this case the ALMF is better than the SKF except for the 11th to 44th years. On the other hand, in using Eq. (4) the ALMF yields better forecast values in all cases.

TABLE 2. Root-mean Square Errors Between the Observed and Forecasted Values

	Equation 3	Equation 4
11th to 20th years		
Linear Regression	649.36	326.56
SKF	649.35	326.56
ALMF	562.49	248.97
11th to 30th years		
Linear Regression	531.82	284.30
SKF	531.81	284.30
ALMF	526.50	273.40
11th to 44th years		
Linear Regression	561.01	270.10
SKF	561.00	270.10
ALMF	590.38	266.31

SKF – standard Kalman filter
ALMF – adaptive limited memory filter

SUMMARY AND CONCLUSIONS

Various snowmelt runoff forecast models have been developed in practice. Most of them are multiple linear regression models. In general, the input variables to these models when based on several stations are obtained by arithmetic averaging or some arbitrary weighting method. As such, they may not necessarily incorporate the areal characteristics or spatial variability of the process. As demonstrated in this study, the use of areal averaging either by Thiessen method, multiquadric or optimal interpolation techniques and the use of Kalman filtering estimation techniques appears promising in reducing forecast errors thus more reliable forecasts.

ACKNOWLEDGMENTS

The support of the U.S.-Spain Project: Conjuctive Water Uses of Complex Surface and Groundwater Systems and the NSF Grant CEE-8110782 is gratefully acknowledged.

LITERATURE CITED

Belousov, S. L., L. S. Gandin, and S. A. Mashkovich, 1971. Computer Processing of Meteorological Data. Israel Program for Scientific Translation, Jerusalem, 210 pp.

Ford, P. M., 1959. Multiple Correlation in Forecasting Seasonal Runoff. Engineering Monographs No. 2, USBR, Denver, Colorado, June, 41 pp.

Hardy, R. L., 1971. Multiquadric Equations of Topography and Other Irregular Surfaces. Journal of Geophysical Research, 76 pp., 1905-1915.

Kalman, R. E., 1960. A New Approach to Linear Filtering and Prediction Problems. Journal Basic Engineering 82D, March, pp. 35-45.

Lee, P. S., P. P. Lynn, and E. M. Shaw, 1974. Comparison of Multiquadric Surfaces for the Estimation of Areal Rainfall. Hydrol. Sci. Bull. XIX(3), September, pp. 303-317.

Myers, K. A. and B. A. Tapley, 1976. Adaptive Sequential Estimation with Unknown Noise Statistics. IEEE Transaction on Automatic Control, August, pp. 520-523.

Rodriquez-Iturbe, I., E. H. Vanmarcke, and J. C. Schaake, 1972. Problems of Analytical Methods in Hydrologic Data Collection. *In:* Proceedings International Symposium on Uncertainties in Hydrologic and Water Resource Systems. University of Arizona, Tucson, Arizona, Volume II, December 11-14, pp. 4.8-1 to 4.8-22.

Szollosi-Nagy, A., 1976. An adaptive Identification and Prediction Algorithm for the Real-Time Forecasting of Hydrological Time Series. Hydrol. Sci. Bull. XXI(1), March, pp. 163-1976.

Tangborn, W. V., 1980. A Model to Forecast Short-Term Snowmelt Runoff Using Synoptic Observations of Streamflow, Temperature and Precipitation. Water Resources Research, Vol. 16, No. 4, August, pp. 778-786.

Tangborn, W. V. and L. A. Rasmussen, 1976. Hydrology of the North Cascades Region, Washington. 2. A Proposed Hydrometeorological Streamflow Prediction Method. Water Resources Research, Vol. 12, No. 2, April, pp. 203-216.

USBR, 1968. Western Division Water Supply Forecasts, Electronic Computer Program, Region 7, Loveland, Colorado, June.

Yevjevich, V. and A. K. Karplus, 1973. Area-time Structure of the Monthly Precipitation Process. Hydrol. Paper No. 64, Colorado State University, Fort Collins, Colorado, August, 45 p.

FORECASTING DAILY PRECIPITATION OCCURRENCES WITH MARKOV CHAIN OF SEASONAL ORDER

Maria Mimikou[1]

ABSTRACT: Forecasting daily precipitation occurrences is very important for planning and management of water resources. It is shown that the proper selection on a seasonal basis of the order of a Markov chain significantly improves its efficiency for forecasting daily precipitation occurrences. The average seasonal storm duration, being the average seasonal ratio of the number of wet days to the number of storm events and containing all meteorological and geographical information found by other researchers as well to influence the seasonally varying order of conditional dependence of daily precipitation occurrences, is proposed as a criterion for determining the proper seasonal Markov order. This thesis, though needing further investigation, is significantly supported in this paper. For seven precipitation stations in Greece, the forecasting error is minimized with chains of varying order, which, on monthly basis, is the closest positive integer to the average storm duration in days reduced by one. There is a prevalence of orders higher than 1 in wet and of orders 1 or 0 in dry periods. The appearance at each station of the period with orders higher than 1 is delayed and its duration decreases going eastwards and with decreasing altitude.
(KEY TERMS: forecasting; daily precipitation occurrences; Markov chain, seasonal order.)

INTRODUCTION

Precipitation forecasting is very important for planning and management of water resources and has many practical applications in engineering and agriculture.

The majority of hydrological methods for precipitation modeling are trying to represent the generating mechanism of the physical process. The models are basically mathematical descriptions of the nature of precipitation and of the structure of its sample time sequence under study. Since the stochastic nature of precipitation is generally accepted, stochastic models are mainly used to describe the process. Two characteristic features of the process are usually distinguished: precipitation occurrences and precipitation depths or intensities, which are treated either separately or together. The present research work is dealing with the precipitation occurrence processes only and, more specifically, with modeling and forecasting daily precipitation occurrences.

The approaches that have been used in developing stochastic models for daily precipitation occurrences have usually subdivided time into wet periods (storm, wet days) and dry periods (interstorm period, dry days). Since Gabriel and Neumann (1962) proposed that the sequence of wet and dry days be represented by a Markov chain, the latter has been one of the most widely and commonly used models for representing daily precipitation occurrence processes (Todorovic and Woolhiser, 1976; Haan, *et al.*, 1976; Buishand, 1978; Waymire and Gupta, 1981).

In this paper some problems for properly modeling point daily precipitation occurrences with Markov chain, concerning the seasonality of the order of the chain, are studied. The purpose is the improvement of the chain's efficiency for forecasting daily precipitation occurrences.

The seasonality of the Markov order of conditional dependence of daily precipitation occurrences and its dependence on the meteorological conditions and geographical location of the precipitation measuring station have been already observed (Chin, 1977). The average seasonal storm duration, being the average seasonal ratio of the number of wet days to the number of storm events and influenced by the precipitation station's meteorology and geography, is herein proposed as an explicit and general criterion for determining the seasonal order of conditional dependence of daily precipitation occurrences and hence the proper Markov order for representing it. For seven precipitation stations in Northwestern and Western Greece the error of forecasting daily precipitation occurrences with Markov chain is minimized with chains of seasonally varying order; their proper monthly Markov order is found to be well related to the average monthly storm duration. The influence of the geographical and topographical location of the precipitation measuring station on Markov order's seasonal behavior is also studied.

CRITERION FOR DETERMINING MARKOV SEASONAL ORDER

The seasonality of the order of conditional dependence of daily precipitation occurrences, and hence of the proper

[1] Lect., Civil Engineering Department, Technical University of Athens, Athens, Greece.

Markov order for representing it, has been already observed and studied (Chin, 1977). In this paper, the problem of selecting the proper seasonal order of a Markov chain used for forecasting daily precipitation occurrences, with the purpose of improving the chain's forecasting efficiency, is discussed. The search of a physical independent variable able to serve as a criterion for explicitly determining the seasonal Markov order and the proper seasonal discretization of time, such that all seasonal changes of the true order of conditional dependence of the physical phenomenon are detected, are the main aspects of the problem. The answers to these are based as well on the observation and conclusions of previous research works (Chin, 1977).

Criterion for Markov Order Determination

Chin (1977) analyzing daily precipitation records for 25 years at more than 100 stations in the conterminous United States observed that "the order of conditional dependence of daily precipitation occurrences depends on the season and geographical location. There exists a prevalence of first-order conditional dependence in summer and higher order conditional dependence in winter." An explanation based on meteorology is proposed and the conclusion is that the proper Markov order for representing the conditional dependence of daily precipitation occurrences has to be determined either by data analysis or by referring to available results of regional type from previous studies and cannot be assumed *a priori.* Based on these observations, an attempt is made in this paper to define an explicit and general criterion sensible to seasonal changes and also influenced by the location of the precipitation measuring station in order to simplify the determination of the seasonal order of conditional dependence of daily precipitation occurrences and therefore of the Markov order for properly representing it. As such the average over a time period storm duration in days is proposed. A storm event is an uninterrupted sequence of wet (rainy) days and the average storm duration is defined here as the ratio of the number of wet days to the number of storm events of a certain time period. Storm duration and frequency are strongly influenced by the meteorological conditions and geographical location of the precipitation measuring station, which were found by Chin (1977) to influence as well the order of conditional dependence of daily precipitation occurrences. For instance, the migratory cyclones especially prevailing in winter months have a broad characteristic length scale and a characteristic life cycle time of several days. This means that at any station the precipitation occurrences associated with cyclone passage would most likely last several days too (long storm durations), exhibiting a conditional dependence with order higher than one. In the summer months much of the precipitation comes from local thundershowers which have a characteristic time from a few minutes (individual cells) to a few hours (cell clusters). This may account for the prevalence of relatively short storm durations and of low (including zero) order of conditional dependence. The location of the station influences also the average storm duration, e.g., the existence of a neighboring lake would increase the atmospheric moisture and would affect the frequency and duration of the storm events and the conditional dependence of the precipitation occurrences in the vicinity. Therefore, the average over a time period storm duration, as defined here, must be related to the order of conditional dependence of daily precipitation occurrences during the same period.

The time period over which the storm duration is averaged must be selected in a way that all seasonal changes of the prevailing meteorological conditions are detected. Consequently a monthly interval seems to be appropriate. The period of a whole season could also be used in cases where the intraseasonal changes of the monthly average storm durations are not significant. Additionally, if the location of the station remains the same, one can expect no significant yearly changes in the order of conditional dependence of the physical phenomenon; therefore, the average storm duration for each month or season can be also averaged over the years of the precipitation record to form the average monthly or seasonal storm duration.

The validity of the average seasonal storm duration as a criterion for determining the seasonal order of conditional dependence of daily precipitation occurrences and hence the proper Markov order for representing it, is significantly supported in the following chapter. Trying to improve Markov chain's efficiency for daily precipitation occurrences forecasting for seven precipitation measuring stations in Greece, it is found that the minimization of the error in forecasting is achieved with chains of seasonally varying order and that a certain empirical relationship exists between the latter and the average seasonal storm duration. The limitations imposed by the relatively limited number of stations and area of application of the study and by the decision procedure followed for selecting the proper Markov order are implying that additional research work is needed for studying the extension to different climatic regimes, geographical locations, Markov chain's applications, etc., of the seasonal relationship between the proposed criterion and the proper Markov order of conditional dependence of daily precipitation occurrences.

Definitions

A Markov chain is a sequence of discrete random variables and is said to be of order r, if r is the smallest positive integer such that for all n the following equation relating the conditional probabilities is satisfied (Chin, 1977):

$$P(X_n/X_{n-1}, X_{n-2}, \ldots, X_{n-r}, X_{n-r-1}) =$$

$$= P(X_n/X_{n-1}, X_{n-2}, \ldots, X_{n-r}), \quad r = 0,1,\ldots \qquad (1)$$

For daily sequences X_n is representing a precipitation occurrence at day n and is said to be equal to 1: $X_n = 1$, if the day is wet (rainfall depth above a threshold) and equal to 0: $X_n = 0$, if the day is dry. In the present work the monthly discretization of time for estimating the average storm duration is adopted. Thus, for a N-years record, p=1,2,. . . , N, the sum of X_n over a month t, t=1, . . . , 12, is the sum of wet days of the month:

$$(\sum_n X_n)_{p,t} = W_{p,t}, \; n=1, 2, 3, \ldots \tag{2}$$

where p is denoting the year of the record. Dividing $W_{p,t}$ by the number k of the individual storms (groups of consequent wet days bounded by dry ones) of the month, one can obtain the average storm duration for the month $d_{p,t}$ as:

$$\frac{W_{p,t}}{k} = d_{p,t} \tag{3}$$

In the cases that a storm is extended to two consecutive months, it is considered to belong to the month in which the majority of daily precipitation occurrences of the storm happened. Averaging $d_{p,t}$ over the N years of the record the average monthly storm duration d_t is obtained as follows:

$$\frac{1}{N} \sum_{p=1}^{N} d_{p,t} = d_t \tag{4}$$

Therefore, for any daily precipitation record there are 12 (or less if another discretization of time is decided) d_t values estimated from the data of the Markov chain's calibration period. These values, as shown in the following, are related to the proper monthly (seasonal) Markov order and hence, can be used for its determination.

FORECASTING WITH MARKOV CHAINS OF PROPER SEASONAL ORDER

The Markov chain's use for forecasting daily precipitation occurrences, the improvement of its efficiency by using proper monthly orders and the establishment of an empirical relationship between the latter and the average monthly storm durations are discussed herein. Daily precipitation data from seven stations widely dispersed in Northwestern and Western Greece are used. Details about the stations (longitude, latitude, altitude, operation) and the precipitation records (time length) are given in Table 1, whereas the location of the stations is shown in Figure 1. A day was defined as wet, if the water equivalent of precipitation recorded on that day was at least 0.25 mm; otherwise it was defined as a dry day.

Forecasting with Markov Chains

Markov chains of four competing orders, r=1 . . . 4, have been calibrated for all seven daily records with time lengths N given in Table 1. All chains were found to be stationary on a monthly basis, which means that the transition probabilities:

$$P(X_n = a_r / X_{n-1} = a_{r-1}, \ldots, X_{n-r} = a_o)$$
$$= P_{a_o, \ldots, a_{r-1}, a_r} = P_a, \tag{5}$$

with $a_o, \ldots, a_r$ the states of the chain defined here to be either 0 (dry day) or 1 (wet day), were independent of n within month t. The stationarity was verified by checking, for all months of each precipitation occurrences record, the standard normal variate z (Spiegel, 1975):

$$z = \frac{P_a^{(t)} - P_a^{(n)}}{\sqrt{P_a^{(t)}(1-P_a^{(t)})(\frac{1}{N_1} + \frac{1}{N_2})}} \tag{6}$$

where $P_a^{(t)}$ the transition probabilities of month t, $P_a^{(n)}$ the transition probabilities of each day n within month t for the unified sample of all N years of the record, estimated according to formulas given in statistical books (Haan, 1977), and N_1, N_2 the sizes of the samples with which the two types of transition probabilities were estimated correspondingly. At the 5 percent significance level the stationarity was accepted for: $|z| \leq 1.96$. These chains of order r ranging between 1 and 4 and stationary on a monthly basis were then used to forecast daily precipitation occurrences one day ahead during the calibration periods for all seven stations. Data generation from Markov chains requires the knowledge of initial state and the transition probability matrix (Haan, 1977). To forecast the state at day n of a month t with a chain of order r a random number is selected between 0 and 1; if this random number is less than or equal to the monthly transition probability for having the day n dry, i.e.: $P_{a_o, \ldots, a_{r-1},0}^{(t)}$, day n is said to be dry; if it is between $P_{a_o, \ldots, a_{r-1},0}^{(t)}$ and 1, day n is said to be wet, for a_o, a_1, . . zero or one.

Selection of the Proper Seasonal Markov Order

The decision procedure for selecting the proper monthly Markov order r_t is based on the minimization of the mean monthly forecasting error ϵ_t percent; it is defined as the ratio of the number of days wrongly forecasted (a wet day forecasted as dry and vice versa) to the total number of days of month t for the unified sample of all N years of each record. Thus, for each station the proper monthly

TABLE 1. Precipitation Measuring Stations.

Station	Catchment	Longitude	Latitude	Altitude (m.a.s.l.)	Operated Since	Record Length (N years)
Pades	Aoos r.	20°55′	40°03′	1170	1967	13
Pramada	Aracthos r.	21°06′	39°32′	817	1963	17
Mikra Gotista	Aracthos r.	21°02′	39°41′	760	1959	20
Damaskinia	Aliakmon r.	21°11′	40°20′	990	1964	16
Anemorachi	Aracthos r.	21°05′	39°19′	390	1965	14
Chalara	Aliakmon r.	21°14′	40°39′	880	1962	17
Eani	Aliakmon r.	21°49′	40°10′	480	1962	18

Figure 1. Location of the Precipitation Measuring Stations.

Markov order r_t is selected among the other competing orders such that the error ϵ_t percent is the minimum. The whole procedure has been implemented in a computer code. The r_t values minimizing the errors for all stations are related to the corresponding average monthly storm durations and are found to be equal to:

$$r_t = D_t - 1, \ r_t \geqslant 0 \qquad (7)$$

where D_t is the closest integer to the average monthly storm duration d_t estimated by Equation (4). In other words, the proper monthly Markov order is the closest positive integer to the average monthly storm duration in days reduced by one. The relationship between r_t and D_t becomes apparent from Table 2. From this table one can see that all Markov orders greater than 1 (in the frame traced inside the table) are concentrated in months between October and April, namely in the wet period, a fact observed by Chin (1977) as well. The period with orders higher than 1 is not the same for all stations; its appearance, relatively to the beginning of the hydrological year, is delayed and its duration decreases going eastwards and with decreasing altitude. Thus, this period at Pramada begins earlier and lasts more than the corresponding periods at Mikra Gotista and Anemorachi, even though the former station lies east of both others; this is due to the altitude of Pramada station which is higher than the altitude of Mikra Gotista and Anemorachi stations. A similar situation is observed between Damaskinia and Anemorachi stations. It is noteworthy that Eani station, located east of all other stations and with relatively low altitude, has a Markov order equal to one in all months except in August where the order is zero, i.e., in this month the precipitation occurrences exhibit no conditional dependence. The latter is also observed in summer months for another two stations with rather short periods with orders higher than 1, namely for the Anemorachi and Chalara stations.

As an example, the monthly errors of forecasting daily precipitation occurrences at Pades station and the improvement of chain's forecasting efficiency by using the proper seasonal Markov order are presented herein. In Figure 2 the monthly ϵ_t percent values are plotted for all Markov orders $r = 1 \ldots 4$, kept constant throughout the year. From this figure it becomes apparent that in the wet period lasting from October to April the chain of order 2 gives the minimum monthly errors, whereas for the other months, May through September, this happens with Markov order equal to 1. The average over the former period minimum mean monthly error is 36 percent, whereas the corresponding value for the latter period is 27 percent. In order to emphasize the improvement of the Markov chain's forecasting efficiency by using the proper seasonal order, it must be

TABLE 2. Relationship Between D_t and r_t in Days.

Month	Pades		Pramada		Mikra Gotista		Damaskinia		Anemorachi		Chalara		Eani	
	D_t	r_t	D_t	r_t	D_t	r_t	D_t	r_t	D_t	r_t	D_t	r_t	D_t	r_t
October	3	2	3	2	2	1	2	1	2	1	2	1	2	1
November	3	2	3	2	3	2	3	2	3	2	2	1	2	1
December	3	2	3	2	3	2	3	2	3	2	2	1	2	1
January	3	2	3	2	3	2	3	2	3	2	3	2	2	1
February	3	2	4	3	3	2	3	2	3	2	3	2	2	1
March	3	2	4	3	3	2	3	2	2	1	2	1	2	1
April	3	2	3	2	3	2	3	2	2	1	2	1	2	1
May	2	1	2	1	2	1	2	1	2	1	2	1	2	1
June	2	1	2	1	2	1	2	1	1	0	2	1	2	1
July	2	1	2	1	2	1	2	1	1	0	1	0	2	1
August	2	1	2	1	2	1	2	1	1	0	1	0	1	0
September	2	1	2	1	2	1	2	1	2	1	2	1	2	1

noticed that for the October through April period the average mean monthly error would be 40 percent, if the order 1 was used as constant throughout the year, whereas for the May through September period the corresponding error value would be 31 percent if the order 2 was consistently used throughout the year. Similarly, significant improvement of the Markov chain's forecasting efficiency by using the proper seasonal Markov order has been observed for all studied stations.

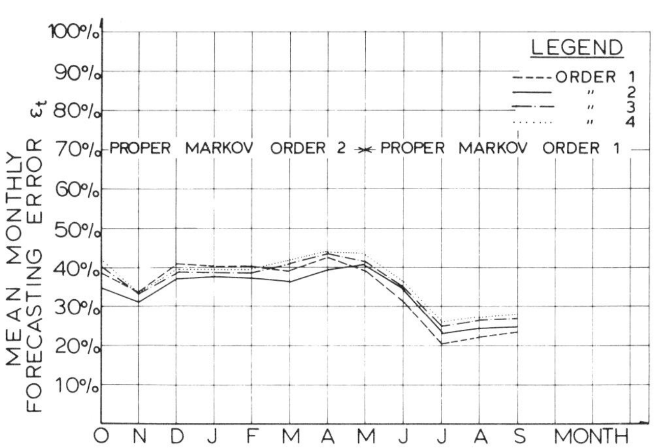

Figure 2. Selection of the Proper Seasonal Markov Order, Pades Station.

CONCLUSIONS

The conclusions drawn from this research are the following:

1. The proper selection on a seasonal basis of the order of a Markov chain significantly improves its efficiency for forecasting daily precipitation occurrences.

2. The average seasonal storm duration, being the average seasonal ratio of the number of wet days to the number of storm events and influenced by the precipitation station's meteorology and geography, is proposed as a criterion for determining the proper seasonal Markov order. For seven precipitation stations in Greece the Markov chain's forecasting error is minimized with chains of seasonally varying order and their proper monthly Markov order is the closest positive integer to the average monthly storm duration in days reduced by one. Additional research work is needed for studying the extension to different climatic regimes, geographical locations, Markov chain's applications, etc., of the seasonal relationship between the proposed criterion and the proper Markov order of conditional dependence of daily precipitation occurrences.

3. There is a prevalence of Markov orders higher than 1 in wet periods and of orders 1 or 0 in dry periods. The appearance at each station of the period with orders higher than 1, relatively to the beginning of the hydrological year, is delayed and its duration decreases going eastwards and with decreasing altitude.

ACKNOWLEDGMENTS

The author wishes to thank the Environmental Research Center of the Greek Public Power Corporation for providing the daily precipitation data.

LITERATURE CITED

Buishand, T. A., 1978. Some Remarks on the Use of Daily Rainfall Models. Journal of Hydrology 36:295-308.

Chin, E. H., 1977. Modeling Daily Precipitation Occurrence Process with Markov Chain. Water Resources Research 13(6):949-956.

Gabriel, K. R. and J. Neumann, 1962. A Markov Chain Model for Daily Rainfall Occurrences at Tel Aviv. Roy. Met. Soc. Quart. Journal 88:90-95.

Haan, C. T., 1977. Statistical Methods in Hydrology. Iowa State Univ. Press, Ames, Iowa.

Haan, C. T., D. M. Allen, and J. O. Street, 1976. A Markov Chain Model for Daily Rainfall. Water Resources Research 12(3):443-449.

Spiegel, M. R., 1975. Theory and Problems of Probability and Statistics. Shaums Outline Series, McGraw-Hill, New York, New York.

Todorovic, P. and D. A. Woolhiser, 1976. Stochastic Structure of the Local Pattern of Precipitation. *In:* Stochastic Approaches to Water Resources II, H. W. Shen (Editor). Fort Collins, Colorado, pp. 15.1-15.37.

Waymire, E. and V. K. Gupta, 1981. The Mathematical Structure of Rainfall Representations, 1. A Review of the Stochastic Rainfall Models. Water Resources Research 17(5):1261-1272.

THE ECONOMIC IMPACT OF CLIMATE ON WATER RESOURCES MANAGEMENT

Amos Eddy[1]

ABSTRACT: The economic well being of our country has become so critically dependent upon its water resources that expected variability in rainfall patterns could impose stresses leading to major social and economic dislocations. Summertime convective storms deliver most of the HIGH PLAINS' surface water supply both directly to crops and indirectly by way of stockponds, reservoirs, and streams to all water requiring sectors of the economy. What is the expectation and the variability of this non-depletable resource? The problem of providing a user-oriented climatology of these convective storms has two interlinked aspects. The first is to find out exactly how the user-activities are climate sensitive and the second is to obtain an adequate sample of data to make the use of the required climatology cost effective. From a scientific point of view, no optimal sampling system can be designed nor can the "best" data processing (climatology) be accomplished until the users' value systems are known.

We are analyzing the Oklahoma daily rainfall, streamflow (discharge) and lake level historical data sets to assess surface water availability; and are examining the agriculture, municipal and industrial sectors as users which turn water into goods, services and hence dollars. One of our results illustrates the variability of the Gross State Product of Oklahoma as a function of the rainfall variability across the state.

(KEY TERMS: climates; economic impact; water management.)

INTRODUCTION

Summer convective storms deliver most of the High Plains surface water supply both directly to crops and indirectly by way of stockponds, reservoirs, and streams to all water-dependent sectors of the economy. A university consortium, of which the Oklahoma Climatological Survey is a member, is making an effort to quantify the manner in which climatic variability leads to economic variability. This paper reports on one aspect of this problem and is focused on the impact on the State of Oklahoma of space-time-intensity distributions of rainfall which are partitioned (i) as soil moisture recharge, (ii) runoff, and (iii) direct use as a component of evapotranspiration.

Figure 1 shows that in the eastern part of Oklahoma the main water usage is by the municipal and industrial sectors and the source of this water is surface streams and lakes. In the west the main water user is agriculture and the present sources are mainly aquifers. Since much of this aquifer water is non renewable, the economy in this sector of the state is rapidly becoming unstable. Calculations indicate that we will likely have a couple of decades to prepare to get along with surface water supplies in the Northwest Water Planning Region. (This is roughly Climate Division 1.)

DISCUSSION

Dr. W. S. Cooter (1981) assembled the water use coefficients shown in Table 1 from a variety of sources, and these are considered to be roughly applicable to the present day Oklahoma scene. It is easy to see that agriculture absorbs most water per dollar of output. In the western part of the state this is associated with (i) winter wheat, (ii) sorghum, (iii) cotton, and (iv) livestock. Winter wheat demands water in the fall and the spring while the other crops use most water in August. Livestock require water all year around, but a summer supply is vital.

W. S. Cooter (1981) performed a rather straightforward simulation which turned soil moisture variability into variability in the Gross State Product. He used only grain crops and a simple State Input-Output transaction matrix to obtain the multiplier effect. Figures 2 and 3 show the general results (note that the mean values of both distributions have been removed). Both distributions are skewed, but a range of almost a half billion dollars in the GSP solely as an EXPECTED climate impact is enough to make the State economists cogitate a little.

In order to refine our knowledge of the variability in surface water, Chris Olsen (1982) has explored some of the western catchment basins in some detail. Figure 4 shows schematically the ten main river basins in the state. We will discuss in detail one sub-basin in the North Fork of the Red River. This river has a reservoir, Lake Altus, with a surface area of 4,100 acres and a mean storage of 73,096 acre feet (capacity = 134,600 acre feet). The river basin area is 113,920 acres.

[1]Director, Oklahoma Climatological Survey, 710 Asp, Suite 8, The University of Oklahoma, Norman, Oklahoma 73019.

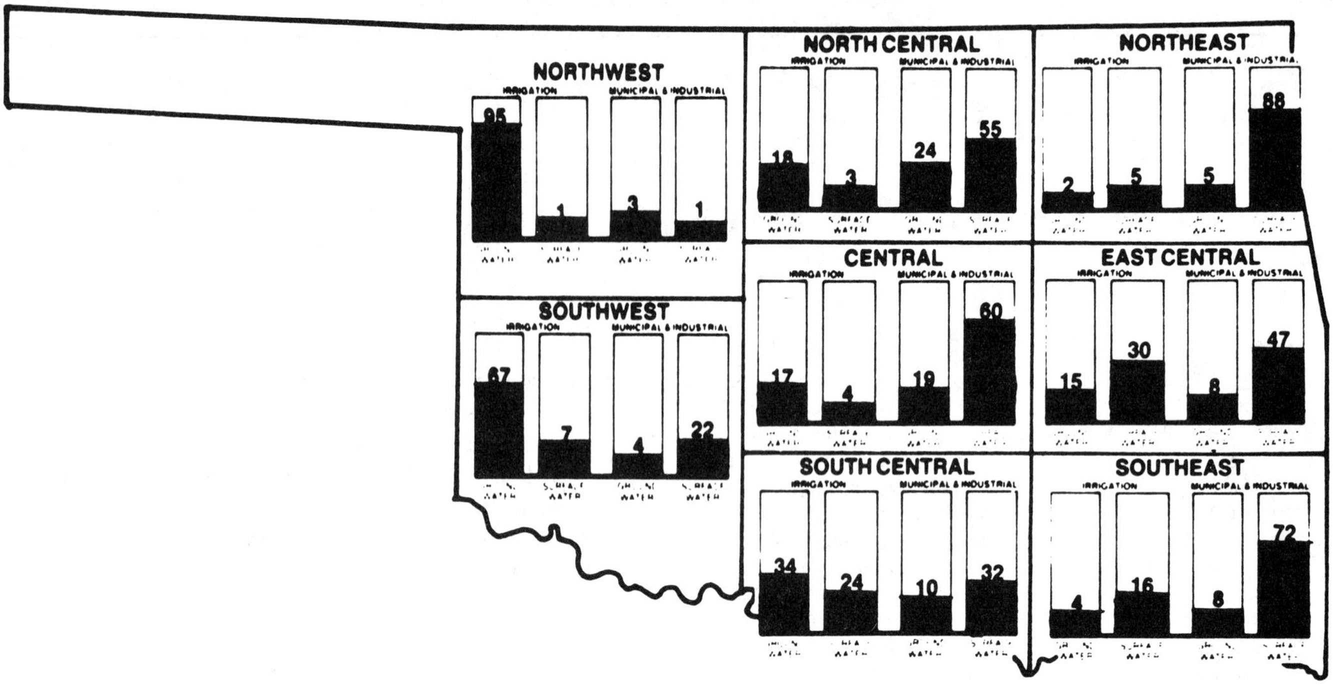

PERCENTAGE BREAKDOWN OF WATER USE BY PLANNING REGION (circa 1978)

After Oklahoma Water Resources Board.

Figure 1. Spatial Distribution of Water Use by Sector and by Source.

TABLE 1. Water Use Coefficients in the Oklahoma Area (Acre Feet/$1,000,000 Output) After Cooter (1981).

Sector	Coefficient
Livestock	300
Dryland farming	20
Irrigated farming	17,000
Agricultural processing	16
Manufacturing	27
Transportation, communication and utilities	460
Real estate, finance, insurance	13
Services	13
Wholesale and retail	8
Mining	16

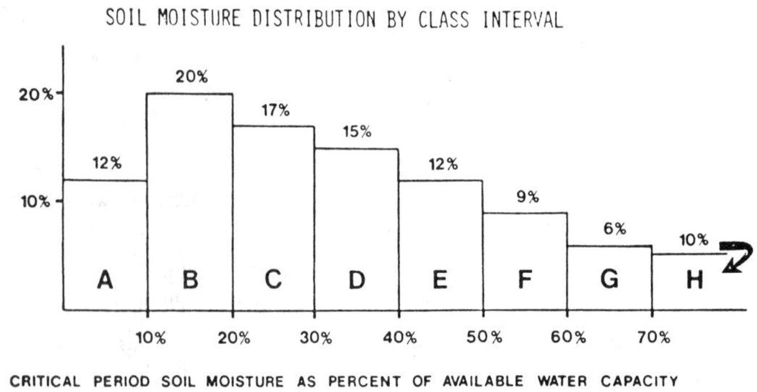

Figure 2. Soil Moisture Distribution by Class Interval.

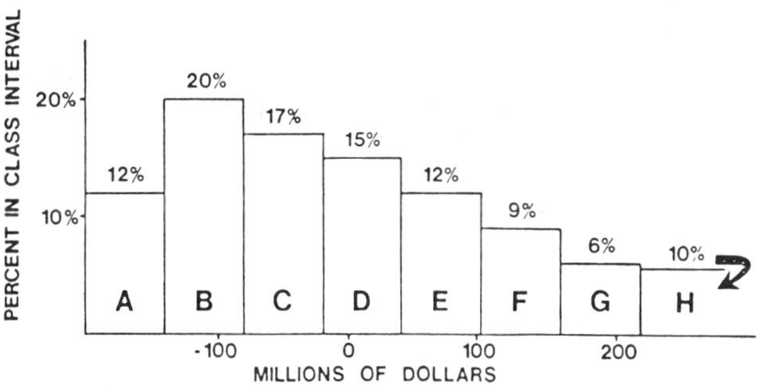

Figure 3. Distribution of Gross State Product Changes by Class Interval.

Olsen defines the concepts of storage and demand in total acre feet for the basin as follows:

Storage

$$S = SM + LC + CC \tag{1}$$

where

S = storage in the basin
SM = soil moisture (calculated)
LC = lake contents (measured)
CC = channel contents

and

$$CC = SC + CS \tag{2}$$

where

SC = stream contents (calculated)
CS = channel supply via ground water (calculated)

Demand

$$D = ET + SE + LE + CL \tag{3}$$

where

D = demand from basin
ET = evapotranspiration (calculated)
LE = lake evaporation (calculated)
CL = channel loss to ground water (calculated)

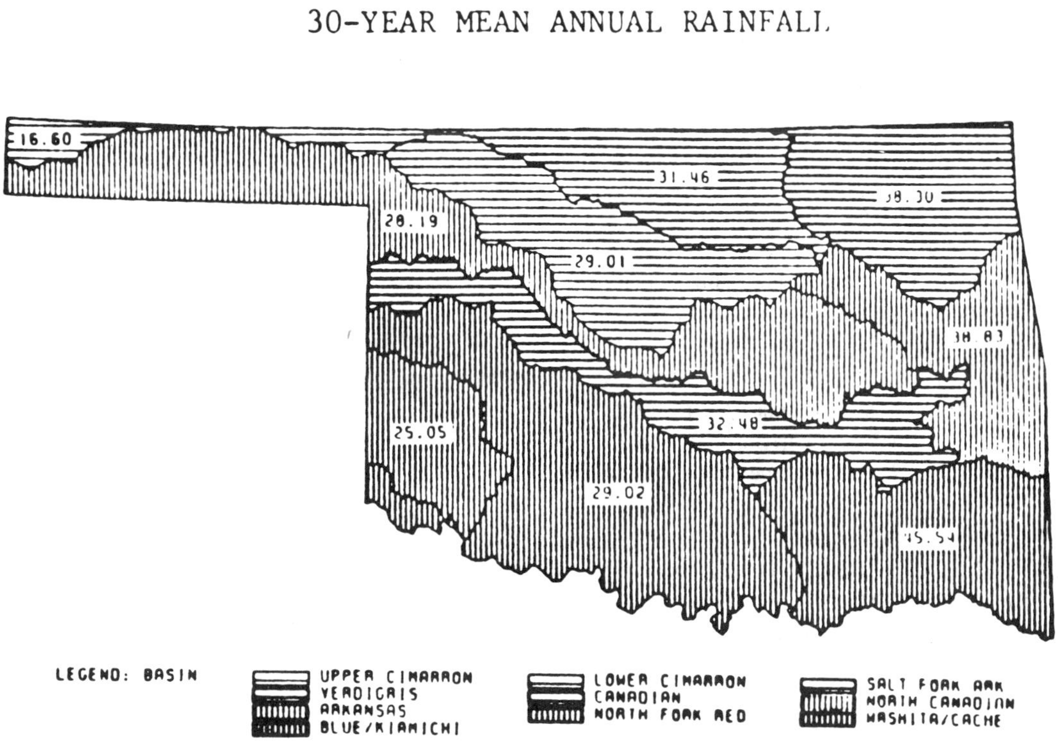

Figure 4. The Major River Basins in Oklahoma with 30 Year (1951-1980) Mean Annual Precipitation Shown.

and

$$SE = (PET - ET)*A \text{ (calculated)} \quad (4)$$

where

PET = potential evapotranspiration over the stream alluvium (calculated)
ET = evapotranspiration over the stream alluvium (calculated)
A = area of the alluvium

and

$$CL = -CS \quad (5)$$

Olsen also highlights the direct use of rainfall for evapotranspiration (as calculated by his hydrological accounting system algorithm) as follows:

$$DELTA = P - R - RO \quad (6)$$

where

DELTA = direct use of rain by evapotranspiration
R = recharge of soil moisture (calculated)
RO = runoff (calculated)

Figure 5 shows three of the above components for the basin under discussion. (Figures 5, 6, 7, and 8 are all 30-year weekly averages which have been smoothed using a 3-point linear filter: weights are .25, .5, .25). It can be seen that moisture constrains growth (defined as evapotranspiration) from week 21 to week 41 on the average in southwestern Oklahoma. The figure emphasizes the fact that the greater part of the dryland crops' growth during this period is dependent directly on precipitation.

Figure 6 shows plots of the standard accounting system variables. It is of interest that the spring rainfall peak leads to runoff while the fall peak does not. Thus, stockponds and reservoirs must be replenished in the springtime according to this analysis. It is known from other analyses that the rainfall inthe autumn is more intense than that in the spring, so the runoff-soil moisture partitioning used in the current model might need some adjustment to produce more runoff – with attendant economic impact implications.

Figure 7 shows STORAGE and DEMAND curves plotted on the same scale for the basin. The 30-year weekly means are shown with the empirical 75% envelope of values in each case. Of considerable interest is the cross section through the two distributions at week 32. About 12% of the years show the storage in the basin with values lower than 88% of the years of demand. In other words, Olsen has quantified the limits to water-dependent economic activity in the area under the current use and storage strategies.

Figure 8 underscores this concept by using the storage and demand values for 1980 as an example. Here, the dry, hot summer depleted the normal storage to the point where the vegetation drew almost entirely on the October rainfall for sustenance – storage would have been completely inadequate to the demand.

SUMMARY

The economic stability in the western great plains has been shown to be heavily water dependent. As the aquifer water becomes economically unavailable, the natural precipitation will play a critical role. A quantitative knowledge of climate variability is a requirement for the development of the appropriate management strategies which will be required to stabilize the economy of the area toward the latter part of this century.

LITERATURE CITED

Cooter, W. S., 1981. The Economic Impact of Climate on Water Management in Oklahoma, Volume VI, Oklahoma Climatological Survey, pp. 57-81.

Olsen, C. C., 1982. The Impact of Climatic Variability on Surface Water Supply in Oklahoma, M.S. Thesis in School of Meteorology, University of Oklahoma.

Oklahoma Climatological Survey, 1981. The Economic Impact of Climate through Water Supply and Demand, Volume V, Oklahoma Climatological Survey, 24 pp.

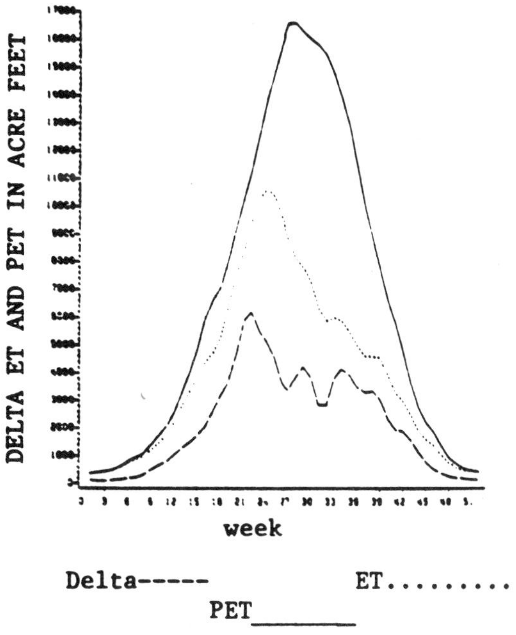

Figure 5. DELTA, ET and PET VS TIME.
Long term weekly means for Subbasin B23 of the North Fork Red River.

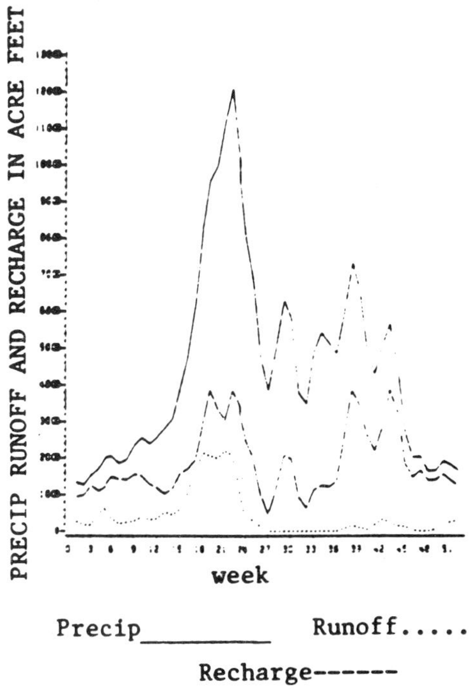

Figure 6. PRECIP, RUNOFF and RECHARGE VS TIME.
Long term weekly means for Subbasin B23 of the North Fork Red River.

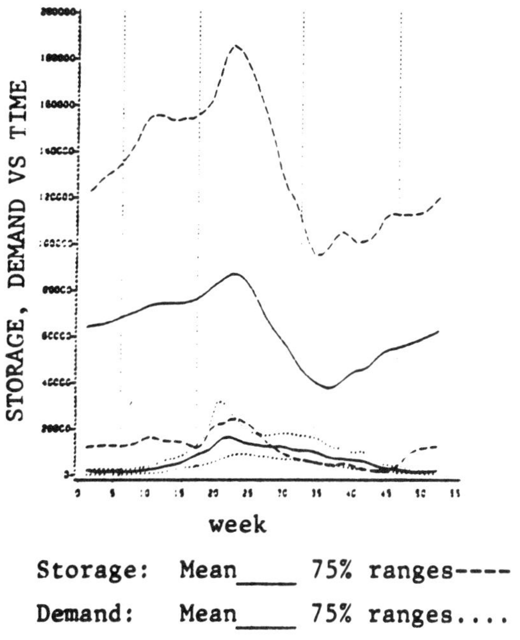

Figure 7. STORAGE, DEMAND VS TIME.
Long term weekly means and 75 percent ranges for Subbasin B23 of the North Fork Red River.

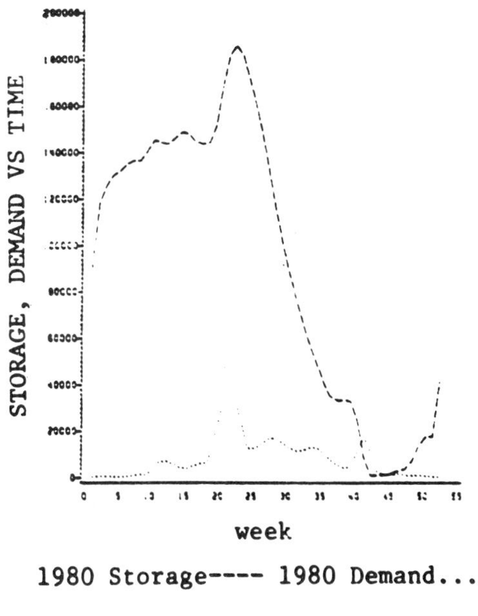

Figure 8. STORAGE, DEMAND VS TIME.
1980 data only for Subbasin B23 of the North Fork Red River.

INTERNATIONAL SYMPOSIUM ON HYDROMETEOROLOGY
JUNE AMERICAN WATER RESOURCES ASSOCIATION 1982

POSSIBLE USES OF LONG-RANGE WEATHER OUTLOOKS IN WATER RESOURCES

Stanley A. Changnon, Jr.[1]

ABSTRACT: A series of world climate abberations that began in the 1970's led to considerable concern over climate and its possible changes (new trends and greater fluctuations). A part of this concern developed because the world, with ever increasing population, has become more sensitive to droughts, floods, and other extremes of weather that comprise climate. This concern led to the recognition that long-range weather outlooks, referred to as those for conditions 1 to 60 months ahead, had enormous utility in the management of water, agricultural, and energy resources. The need for such forecasts has led to intensive research of physical and statistical methods. Those methods available now produce predictions with accuracies promising utility for some water management endeavors. The greatest success so far has come from the statistically based methods. This has been possible because of relatively long climatic records and large computers able to handle large volumes of data and complex statistical analyses. Major areas of application of climate outlooks in water resources include: 1) operations of water management systems (river basins, reservoirs, urban water treatment systems, groundwater recharge, etc.); 2) scheduling of water supply activities (irrigation, structural renovations, etc.); and 3) anticipation of extremes. The needs for predictions range from weeks to years ahead. For example, various reservoir operations can be benefitted by 2-week rain predictions and others by predictions of rain trends over the next 2 years. Predictions correct to the nearest 6 cm are needed in scheduling reservoir releases, but instream flow needs require only a prediction of the relative future normalcy (above, near, or below normal). This paper addresses the potential applications in water resources of outlooks.
(KEY TERMS: climates; weather patterns; forecasting; long-term planning.)

INTRODUCTION

Water and climate are pervasive conditions affecting life processes and most human activities. Water supplies are sensitive to most climate variations, be they associated with man's activities or just with natural fluctuations. Central to this complex climate-water interaction issue is the prediction of the future state of the weather for weeks, seasons, and many years ahead. It appears obvious that such "climate outlooks" can be useful in a myriad of water resource activities. The general areas of application of climate predictions in water resources include: 1) operations of water management systems, 2) scheduling of water supply activities, 3) anticipation of extremes, and 4) planning of major facilities.

This paper is not a result of a long-term study; rather it attempts to review and identify the potential needs and applications of climate predictions to water issues. In a qualitative sense, I also attempted to address the future periods for which predictions are desired, and the general level of predictive accuracy believed useful. As a summary, this review identifies the major problems surrounding the use of climate predictions in water resources, and concludes with some areas of needed research.

NEEDS AND APPLICATIONS OF CLIMATE OUTLOOKS

The application of long-range precipitation outlooks in water resources may seem obvious and the major problem is building an all-inclusive list of uses. These applications embrace two major areas: water quantity, and water quality. Water resource applications, as will be shown, span a wide variety of time scales (from weeks up to 25 years), and span a wide variety of space scales ranging from a square kilometer to the continental scales.

The recognized applications of long-range precipitation outlooks, those defined as being for periods for two weeks up to 25 years, appear in Table 1. There are a large number of operational applications. Probably the most benefit from predictions would come from operating systems involving multiple uses (recreation, irrigation, floods, and water supply). Balancing multiple water requirements using a system approach that includes seasonal climatic differences in precipitation (snow melt, dry periods, or floods) would be very beneficial. Another area of high utility concerns urban runoff management with its conflicting problems of rapid drainage of heavy rainfall, while simultaneously addressing national regulations to maintain quality of drainage waters which requires treatment before release of flood waters.

[1] Chief, Illinois State Water Survey, Box 5050, Station A, Champaign, Illinois 61820.

TABLE 1. Areas of Application of Climate Predictions (2 Weeks to 25 Years) in Water Resources.

A. Operations of water management systems
   1. River basins (multiple control locales)
      a. Snow melt (amount and time) = water supply and flood potential
      b. Water quality – water supply and urban and rural wastes, dilution capacity, and repelling salinity
      c. Transportation adjustments
      d. Hydroelectric power generation
      e. Stormwater management (use of flood control space)
      f. Ice formation and breakup
      g. Irrigation
      h. Recreation
      i. Multiple uses – balancing complex needs
   2. Individual reservoirs
      a. Single or multiple-purpose reservoir operations
      b. Maintain instream flow needs
      c. Short vs long range management options
   3. Urban water treatment systems
      a. Storage and treatment scheduling
      b. Street cleaning
   4. Interbasin transfers or releases
   5. Recharge of groundwater aquifers
   6. Combined groundwater-surface water supply systems
   7. Addressing regulations and their enforcements
   8. Natural lakes
      a. Ice formation and breakup
      b. Navigation
      c. Recreation
      d. Regulation

B. Scheduling water supply activities
   1. Long-term data collection and scientific investigations (streamflow, sediment transport, etc.)
   2. Irrigation
   3. Planned weather modification use and operations
   4. Structural renovations and/or construction
      a. Dams and channel repair
      b. Silt removal, eutrophication modification
      c. Levees
      d. Installation of best management practices
      e. Maintenance of urban quality and supply systems
   5. Farming in bottomlands

C. Anticipation of extremes
   1. Begin and end of wet periods
   2. Begin and end of droughts

D. Planning of major facilities
   1. Design of waste treatment and storm water facilities
   2. Structures to halt or alter river courses
   3. Development of water supply sources
   4. Prediction of sustained yield of aquifers

The scheduling of water supply activities, and particularly means of applying extra water (irrigation and weather modification), certainly would benefit from use of long-term outlooks. The anticipation of extremes is essentially a cross-cutting issue that affects operational scheduling and a host of other water activities. Anticipation of the beginning and ending of droughts, or wet periods, as well as their general magnitude, would be of enormous value.

The final general area of application is identified as planning of major facilities with multi-year life times. Obviously the types of predictions desired are for quite long-term, 10 to 25-year future conditions. In general, past planning input has tended to be either the period of record or normal 30-year precipitation.

A few examples of climate prediction applications are in order. A 2-week precipitation forecast is of great value in reservoir operation for flood control. It could be used to lower the water level of the reservoir to absorb about 1/2 of the expected inflow from a storm, assuming 50 percent accuracy of the predicted rainfall amount. Flow releases can be moderate and extended over a 2-week period. The prediction period duration will depend on the size of the reservoir; a week for a small reservoir; up to a month for a large reservoir.

Longer-term predictions are needed for operating reservoirs to meet instream flow needs. Low flows downstream can be increased by reducing high flow releases for above average inflows if precipitation predictions are available for one to six months in advance and if they indicate below average precipitation. The reliability of the predictions can be useful even if reliability decreases within increasing time. Knowledge of both shorter term (2 to 4 weeks) and the longer term predictions can be utilized in reservoir operations to maximize the overall benefits.

In summary, these illustrations and the applications listed in Table 1 point to the wide potential utility of long-range climate predictions. They vary graphically and they vary seasonally, and they vary from lead time needed as well as the period being predicted. Also, the level of accuracy varies according to the applications. The following sections address these issues.

PREDICTIVE PERIODS

In addressing the temporal aspects of long-term climate outlooks, it is important to recognize and separate the lead time vs the period of predictions. The periods covered by outlooks are those weeks, months, seasons, or years for which the prediction is made. The lead time refers to the period of time before the outlook period begins. Major emphases here is on the relationship between the applications of precipitation predictions and periods of predictions. It is recognized that lead times, those prior to the predicted period, differ to satisfy various design and scheduling operations. Outlooks for future periods of 1 up to 12 months in length would be very helpful in water supply forecasting. It is not so important to hit the right day or week of a storm, but to predict a monthly value that is reasonably close. Predictions of the general magnitude of precipitation for one or two years in the future would have a great impact on the manner in which reservoirs storing water for irrigation are operated in California (McCullough, 1981).

Table 2 presents periods of predictions, for various selected applications of predictions. These are sorted into five classes and the same applications appear under more than one period. Reservoir operations benefit by predictions for 2- to 4-week periods (such as for flood control information); also for periods of 2- to 6-month periods in the future (reservoir operations for maintaining suitable downstream flows), and also for 1- to 2-year future periods (yield management of single-purpose reservoirs in times of drought). A primary finding revealed in Table 2 is that there are needs for predictions of precipitation for a variety of periods, but most occur between 2 weeks and 10 years.

DESIRED PREDICTIVE ACCURACIES

Not much is known about the level of accuracy of precipitation predictions needed to satisfy water resource applications. Since their use has been limited, little thought has been given to this issue. McCullough (1981) indicates that 3-class (terciles) predictions (wet, near normal, and dry) have value. Most hydrologists have indicated, in thinking about precipitation predictions, that they want them "converted" to actual runoff quantities which assumes a capability for predicting a specific precipitation value.

Several research hydrologists were asked to consider some of the major applications (Table 1) and to answer the question of whether precipitation amounts in terciles, or with a specific value being correct to ± 6 to 9 cm, were useful. Results are shown in Table 3. Certain reservoir operations of a longer term nature can be satisfied by precipitation predictions in the 3-class levels, whereas others of the shorter term nature need specific outlooks with error bars of a few centimeters. Many of the applications of the long-range predictions could be satisfied with the 3-class system.

PROBLEMS

This review of long-term precipitation outlooks and their potential applications to water resources reveals a considerable number of applications. Research to develop precipitation outlooks has been emphasized in recent years, and some moderate accuracy exists. A recent meeting (May 1981) of water resource experts and the NWS staff who issue climate predictions revealed that there is a general lack of awareness of the current level of accuracy, and certain major problems exist in the area of utility and development of the precipitation predictions (CAC, 1981).

TABLE 2. Periods Covered by Predictions Desired for Various Applications.

2 to 4 Week Periods	2 to 6 Month Periods	1 to 2 Year Periods	5 to 10 Year Periods	10 to 25 Year Periods
Reservoir operation	Reservoir operation	Reservoir operation	Water supply source develop	Water supply source develop
Reservoir release rates	Instream flow needs	Crop variety selection	Sustained yield from aquifers	
Irrigation requirements	Navigation	Water supply from existing source	Power generation	
Waste treatment plant loadings	Floodplain farming	Plan for stream flow and sediment data collection	Design of waste treatment plants	
Storm water management	Water supply from existing source	Flood mitigation plans	New storm water facilities	
Water quality variation	Storm water management	Construction of new levees	Planning to halt stream course changes	
Flood emergency plans	Water quality variation	Crop variety plantings		
City street cleaning and sweeping	Maintenance of levees	Planning to prevent the deterioration of quality BMP installation		
Water system repairs construction	Snow removal and clearing plans			
	Effects on water quality			
	Hydroelectric scheduling			
	Recreation usage of rivers and reservoir levels			

TABLE 3. General Level of Accuracy Desired in Climate (Seasonal and Annual) Predictions of Precipitation.

Specific Value with Accuracy Limits of 3 to 6 cm	Precipitation Predicted Within Above, Near or Below Normal Terciles
Reservoir operation (short period)	Reservior operation (long period)
Reservoir release rates	Instream flow needs Water supply
Flood emergencies	Navigation Stream flow and sediment Data collection Floods and flood mitigation
Storm water management	Snow removal Crop variety selection Levee construction and maintenance
Irrigation	New storm water facilities BMP applications, aquifer yield, power generation Stream course changes Water quality

A major problem, when one reviews the situation, is a lack of awareness by the atmospheric scientists of the specific needs of hydrologists, and in turn, a lack of awareness of the hydrologists about the emerging capabilities of long-range predictions. Interactions should help in improving the awareness needed by both groups.

Utility of the outlooks by hydrologists depends on their perceptions about the credibility of the predictions. Long-range outlooks represent an area of emerging capabilities and part of the credibility problem relates to the fact that many hydrologists are not aware of how outlooks are calculated. Again, an informational-educational effort is needed.

An important problem with most existing outlooks of precipitation is the lack of expression of the uncertainty levels or error ranges around the predicted value. Most hydrologists have engineering backggrounds and are able to deal with the statistical uncertainty. Hence, expressions of precipitation predictions as probabilities can be understood by hydrologists.

Another problem relating to long-term precipitation predictions relates to the size of area being predicted for. Applications desired by hydrologists can vary widely, from very small basins to very large basins. These may or may not match skills available in precipitation predictions. Some seasonal precipitation predictions are 60 percent accurate for large areas but are no better than chance for small areas. Hence most long-term precipitation predictions relate to very large sized areas and represent regional means.

Another problem area relates to concepts of how to mix or combine long-range outlooks with short-range (hours to days) weather forecasts. Anyone can be confused by their differences, and this interrelationship needs to be studied and explained better by atmospheric scientists. A seasonal forecast for near normal rainfall can have wet periods lasting several days.

Another problem relates to the lead times currently used in most long-range predictions. Many water resource applications have lead time requirements greater than currently used. Water resource experts believe that to be most useful, seasonal predictions of precipitation should be updated monthly and issued with 9-month lead times, as opposed to issuance on the first day of the season.

SOME RESEARCH NEEDS

The above problems point to certain research needs beyond a need to have more accurate predictions. Some research needed relates to studies of how long-range predictions are expressed. The probabilities in space and time, of the predicted magnitude of precipitation need to be defined and expressed in predictions. Secondly, prediction possibilities for weather extremes and the general variability need to be studied. Predictive research also needs to address the occurrences of persistent homogenous (prolonged wet or dry) periods.

It is apparent from these problems, and the research needs, that a major effort of education and interaction is needed between hydrologists and atmospheric scientists. This will help optimize the research and alert each group of the needs and capabilities of each group.

LITERATURE CITED

Changnon, S. A., 1981. Overview of utility of long-range predictions in water resource areas. CAC Users Conference, Gettysburg, PA, 6 pp.

Climate Analysis Center, 1981. CAC Users Conference. Final Report. National Weather Service, Washington, DC, 38 pp.

Hudlow, M., R. Farnsworth, and D. Greene, 1981. Hydrologic forecasting requirements for precipitation data from space measurements. Proceedings Conference on Precipitation Measurements from Space, NASA, Goddard Flight Center, Greenbelt, MD, 8 pp.

McCullough, C. A., 1981. Short-term climate predictions for water resources management. CAC Users Conference Proceedings, Gettysburg, PA, 12 pp.

Schaake, J. C., 1976. Use of mathematical models for hydrologic forecasting in the National Weather Service. Proceedings EPA Conference on Modeling and Simulation, Symposium on Hydrologic Forecasting, Cincinnati, OH, 6 pp.

EXPERIMENTAL WATER SUPPLY FORECASTS BASED ON SEASONAL PRECIPITATION FORECASTS

Robert H. Zettlemoyer[1]

ABSTRACT: In California, a water supply forecast made early in the water year will, by necessity, have a broad probability range associated with the forecast. This wide range is a result of the uncertainty of future precipitation at the time of the forecast. Operational water supply forecasts have always utilized historic precipitation data as input to estimate the possible range of future runoff. With the advent of seasonal precipitation forecasts, there could be an opportunity to reduce the uncertainty of future runoff by incorporating demonstrated seasonal skills into a water supply forecasting procedure.

The California Department of Water Resources has been cooperating with the Scripps Institution of Oceanography for several years in an effort to evaluate seasonal precipitation forecasts. Utilizing data provided by Scripps, techniques were developed to prepare water supply forecasts based on the winter outlook available each December 1. Hindcasts have been made for the last five winters and results show that reductions in probability ranges may be possible.

(KEY TERMS: water supply forecasts; seasonal precipitation forecasts; probability; hindcasts; runoff; rule curve.)

INTRODUCTION

In the United States, precipitation forecasts for 30 to 90 days into the future have been available from several sources in recent years. These forecasts are not "storm specific", but rather only attempt to estimate the total precipitation for the forecast period. Different forecasters use different formats, but each attempts to do better than climatology or persistence. At the Lawrence Livermore Laboratory, an evaluation of seasonal forecasts was made for the years 1974 through 1980 and it was found that "precipitation forecasts produced by Gilman, Namias, and Barnett were significantly better than forecasts of climatological normal . . . . and better than forecasts that the precipitation anomaly from the previous season would persist." (Knox and Orphan, 1981) In general, levels of skills averaged over all seasons, are small for seasonal forecasts.

In California, the State Department of Water Resources has been supporting efforts by Dr. Namias at the Scripps Institution of Oceanography to improve seasonal precipitation forecasts. Although our level of support has been curtailed in recent years, we have continued to review the various aspects of long-range precipitation forecasting. The goal, of course, is to improve the accuracy of water supply forecasts and to optimize water resource operations in the state. Over the years, as knowledge of the weather mechanisms improve, we believe that skills in seasonal precipitation forecasting will improve and eventually it will be possible to make some operational decisions based on these forecasts.

In this study, an attempt is made to translate the winter seasonal precipitation forecasts prepared by Dr. Namias into water supply forecasts. The basin selected was the catchment above Shasta Reservoir, the major supply reservoir of the Central Valley Project. Hindcasts were prepared for the winters of 1975-76 through 1981-82. Water supply hindcasts of water year runoff were developed from each of the December 1 precipitation forecasts and also from the historic range of precipitation.

ANALYSIS

Water supply forecasts are issued in a probabilistic format as a result of two sources of error (ignoring measurement errors). These sources of error are: (1) future precipitation, and (2) an inaccurate transfer function. A water year forecast made for the Lake Shasta Basin on December 1st will be almost totally dominated by the first source of error and so a relatively simple transfer function (multiple regression) was used for this study. More sophisticated transfer functions would have produced very similar results.

Figure 1 shows the seasonal precipitation forecasts and verifications prepared by Dr. Namias for six of the seven winters used in this study (1982 was not available at press time). The forecast format consists of a three category system – light, moderate, or heavy, with each category having an equal probability of happening based on historical

[1]Senior Engineer, California Department of Water Resources, 1416 9th Street, Sacramento, California 95814.

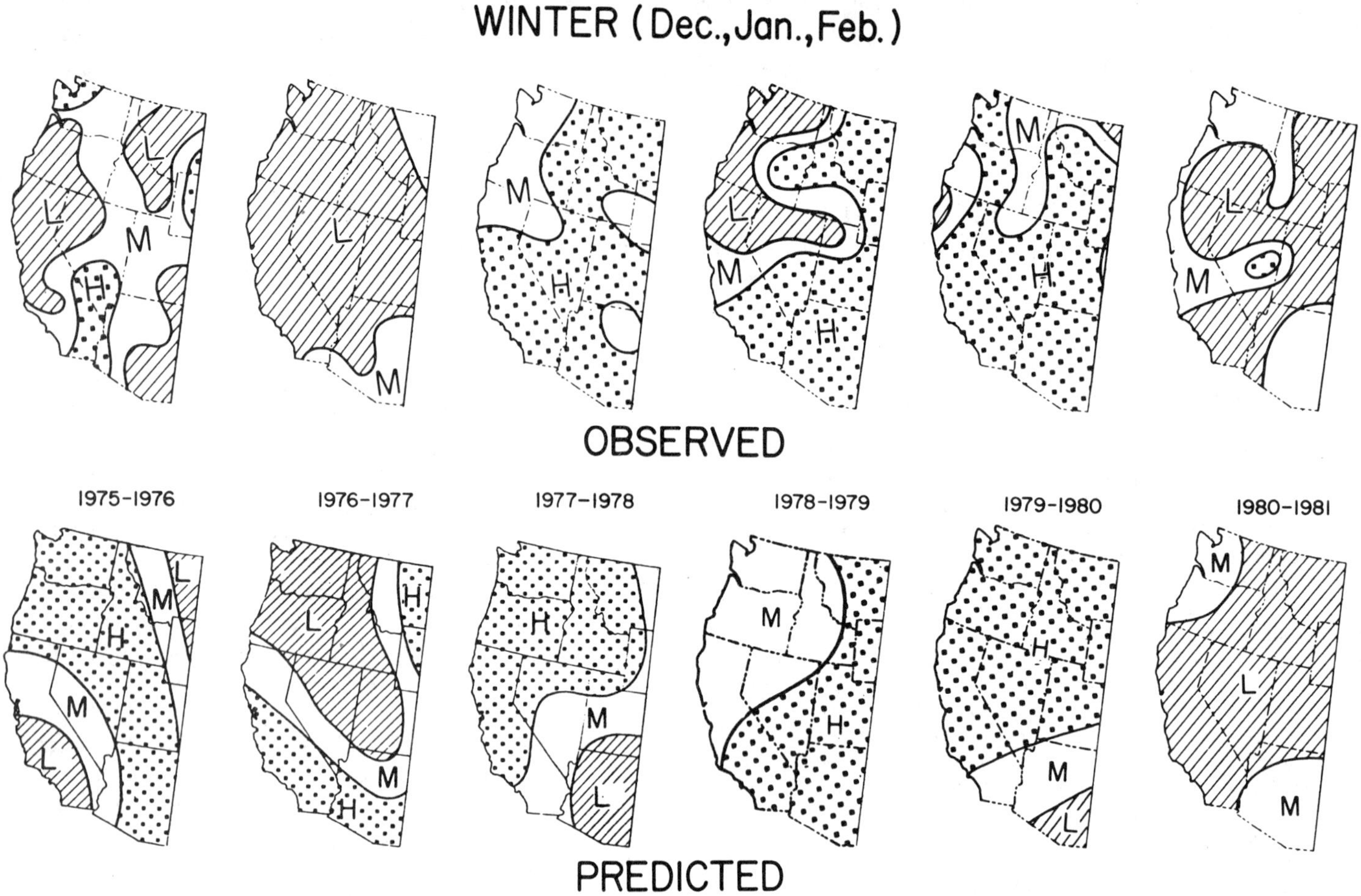

Figure 1. Predicted and Observed Winter Precipitation by Namias.

records. Notice that a forecast of moderate would have a 33-1/3 percent chance of verifying; therefore, if a forecaster were to always forecast moderate his resulting skill score would be zero. A skill score of 0.39 was computed for the California winter season (1976 through 1982), based on three major water producing areas in the state. These areas are defined as: Northern – Shasta and Oroville Basins; Central – Yuba through Stanislaus Basins; and Southern – San Joaquin through Kern Basins. Due to the small sample size and the fact that the data groups are not independent, the true skills could be significantly different than computed. It is also recognized that this skill score is higher than that achieved by Dr. Namias for the total United States winter season.

The major challenge encountered in utilizing seasonal precipitation forecasts for water supply forecasts is to develop a probability density function (PDF) that represents the precipitation forecast and incorporates the forecast skill. In other words, a technique is needed to more precisely define the general category forecast derived from the maps. The PDF for historical precipitation is readily available and serves as the starting point for developing the forecast precipitation PDF (it represents climatology and zero skill). For the Lake Shasta PDF, eight stations within the basin were averaged to represent basin mean precipitation. Sixty-one years of monthly data were available. See Figure 2. For the winter season in this basin upper tercile fell at 107 percent of average and lower tercile was 78 percent of average.

Next we asked the question: What does the PDF look like for a perfect forecast? In theory, a perfect forecast under this type of forecasting system means only that the forecast category verifies within that category every time. Therefore, we can redraw the historic PDF by splitting it into three equal parts and we have the resulting PDF's for each category if perfectly forecast. The remaining step is to redraw each of the perfect PDF's based on the measured skill score. This was done by graphical interpolation between the zero skill PDF and the perfect skill PDF recognizing that a 0.39 skill score means that 59 percent of all forecasts fall within the correct category. The light category is illustrated in Figure 3.

In the final step, water supply forecasts were prepared for the past seven years. As mentioned earlier, a relatively

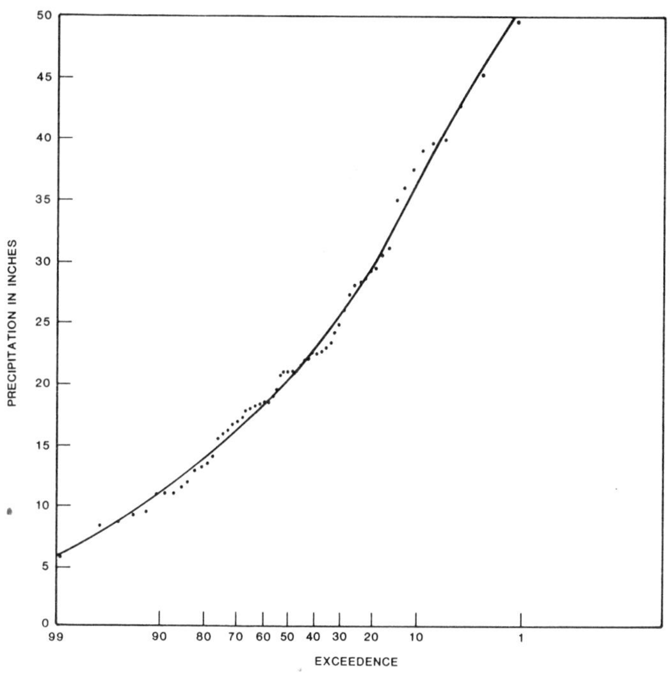

Figure 2. PDF of Shasta Basin Precipitation for the Winter Season.

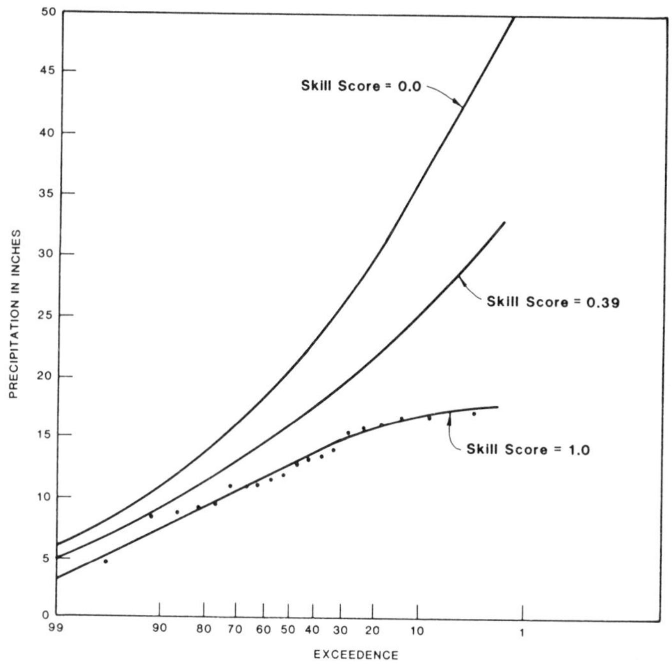

Figure 3. Assumed PDF for a Winter Season Light Category Precipitation Forecast.

simple transfer function was used. Results are shown in Figure 4. Shown are the standard 10, 50, and 90 percent exceedence forecasts for historic precipitation and also for the Scripps seasonal projections. Overall, there is a 25 percent reduction in the 80 percent probability range of the water supply forecasts using the Scripps forecasts as opposed to climatology. Due to the skewed distribution of the precipitation, the reduction is greater when the moderate category is forecast and smaller when the heavy category is forecast. In two of the forecasts, the category boundaries passed through the basin and in those years an interpolation was made. Notice that when the median category is forecast the median projection is the same as the median projection based on historic precipitation and only the 80 percent probability range is reduced.

One of the possible applications for the use of seasonal forecasts is with rule curves such as the one developed for the California State Water Project (SWP) after the two year drought of water years 1976 and 1977. The rule curve relates contractors' deliveries to the available supply from the four major rivers of the Sacramento Valley. A new curve is prepared each year. The initial SWP rule curve is shown in Figure 5. As the season progresses, water supply forecasts are updated and contractors can evaluate the probabilities of deliveries by the project. On December 1, 1977, the Figure 5 rule curve when combined with the normal water supply forecasts would have produced an estimate of 22 percent probability that deficiencies would be necessary in calendar year 1978. Using the procedures described above based on the seasonal projections of Figure 1, the statement would change to about an 11 percent chance of deficiencies.

SUMMARY

With the increasing availability of seasonal precipitation forecasts, there could be an opportunity at certain times and places to reduce the uncertainty of future runoff by incorporating demonstrated skills into water supply forecasts. One such procedure has been demonstrated using the winter forecasts of Dr. Namias. Based on a limited sample, a 25 percent reduction in the 80 percent probability range of a water supply forecast in the Shasta Basin made on December 1 was computed. Potential application requires a careful analysis of all possible consequences.

LITERATURE CITED

California Department of Water Resources. Annual Rule Curve Reports to Contracts.

Knox, J. B. and R. C. Orphan (Editors), 1981. Program Report for Fiscal Year 1980. Atmospheric and Geophysical Sciences Division of the Physics Department, Lawrence Livermore National Laboratory.

Scripps Institution of Oceanography. Annual Reports to the California Department of Water Resources.

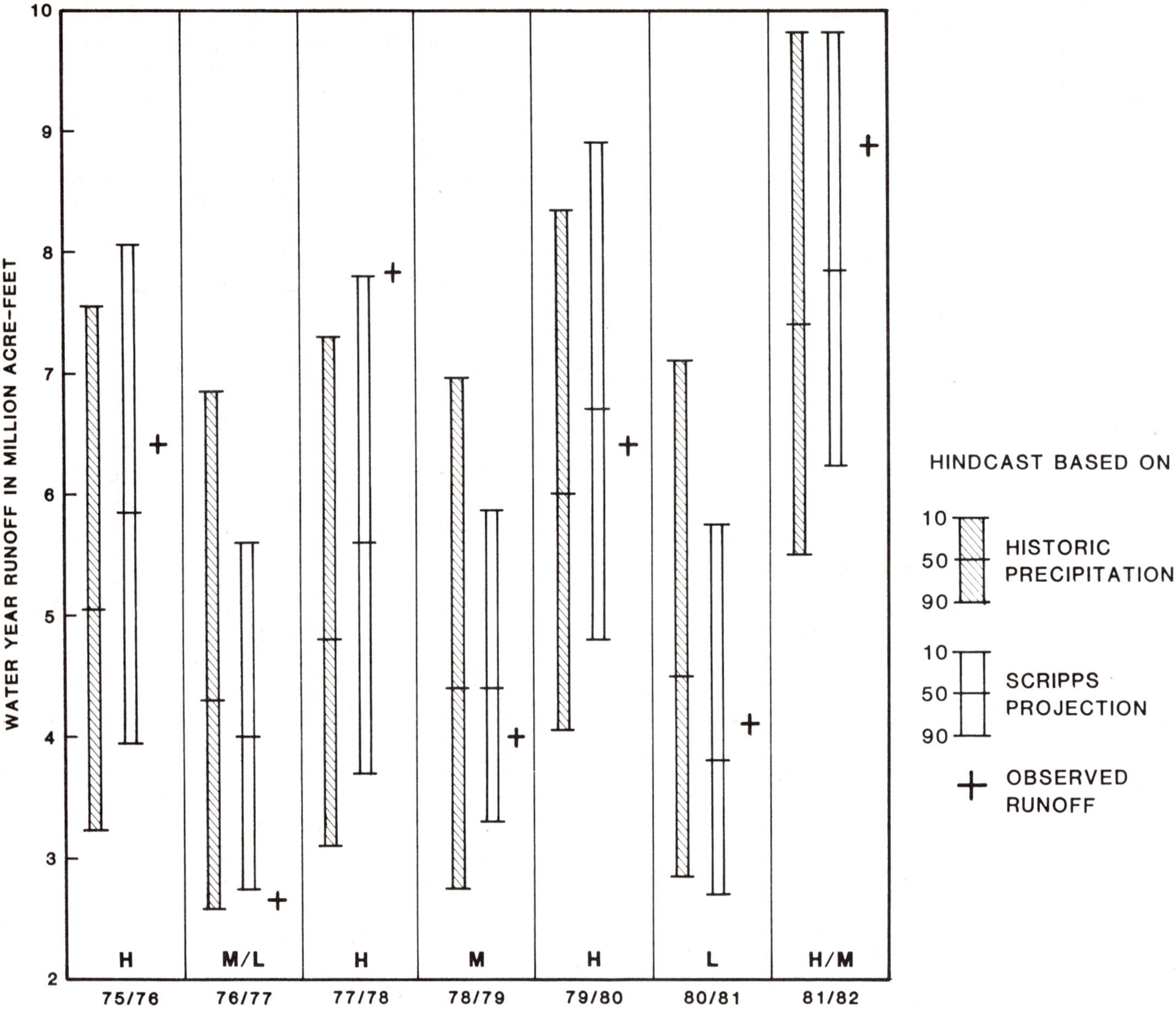

Figure 4. Hindcast Water Supply Forecasts for Lake Shasta Inflow Using Historic Precipitation and Scripps Winter Precipitation Forecasts.

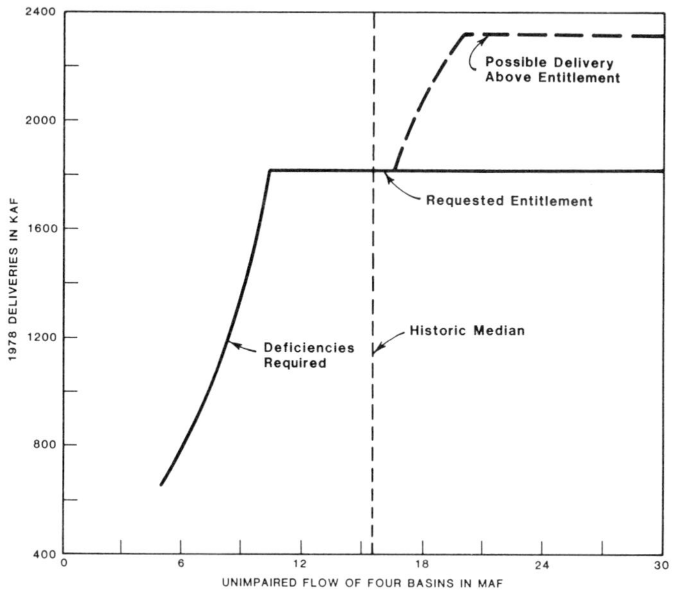

Figure 5. State Water Project Calendar Year 1978 Contractor Deliveries Related to Water Year 1978 Hydrology.

ON THE USE OF WEATHER TYPES FOR MEDIUM RANGE RUNOFF FORECASTING

Adriaan A. van de Griend[1]

ABSTRACT: As a basis for describing the dynamic character of the climate of the European Alpine environment, the Swiss meteorologist Prof. Dr. Max Schüepp developed a classification system of weather types. On the basis of this calssification a selection of hydrometeorologically relevant classes of weather types has been made as a reference basis to investigate the relations between circulation types and runoff phenomena of an Alpine glaciated catchment. To bridge the gap between methods for short range (up to a few days) and long range forecasting of runoff (a few weeks up to months), the usefulness of weather types has been evaluated for purposes of medium range runoff forecasting. Therefore a weather type based model has been developed for forecasting mean runoff for medium range periods of 10 days, to link on to the forecasts of 1000 mb and 500 mb weather charts ten days in advance by the "European Centre for Medium Range Weather Forecasts (ECMWF)" in Reading, England. By doing this an impulse is given to a new form of integration between the disciplines of *hydrology* and *meteorology*, leading to a weather type based approach to runoff forecasting. The model is based on a multiple regression analysis between the dependent variable runoff and a group of independent variables. The latter are split into *initial variables*, related to catchment conditions known at the date of forecast and *future variables* which can be extracted directly from forecast weather charts (*direct method*) or determined by translation of forecast weather charts into weather types (*translation method*). The combination of *direct method* variables and *translation method* variables lead to a *combined method*. The analyses have been performed for the Alpine catchment of the river Ahr (Northern Italy), which has a pluvio-nivo-glacial runoff regime. The analyses are based on the 23-year period 1953-1975.
(KEY TERMS: forecasting; climates; Alpine; runoff.)

INTRODUCTION

Alpine catchments are of considerable importance for hydro-power generation as a result of periodical storage of snow and ice and related nivo-glacial regime components in the discharge. They also form potential water yielding areas for water supply purposes, especially important for irrigation and water supply in many arid regions such as in the Andean countries, the deserts of Soviet and Chinese Central Asia, and in Afganistan, Pakistan, and India (see e.g., Meier and Roots, 1981). Meadium range forecasts of river runoff are important for optimal hydrologic management of these catchments.

As a result of storage and travel time effects, hydrological systems have the property to delay the input of rainfall and snowmelt during runoff generation. This offers the possibility of estimating future runoff without using precipitation and temperature forecasts. The use of this delay plays an important role in current runoff forecasting praxis, where the lag spectrum of the catchment dictates the maximum length of the forecast period.

The natural lag spectrum of high Alpine catchments generally restricts the forecast period to one or two days. Runoff forecasting over periods exceeding this natural lag spectrum are hampered by generally inaccurate precipitation and temperature forecasts especially for the Alpine environments.

As sufficiently accurate forecasts of precipitation and temperature several days in advance are not to be anticipated in the near future, emphasis in this investigation was put (a) on a better insight into the relationships between weather types and runoff phenomena of an Alpine catchment, and (b) on the use of forecast weather types for medium range runoff forecasting. After a short introduction into the regime character of the study area and the problem of weather type classification, the former two topics will be discussed in more detail.

STUDY AREA

The River Ahr Catchment

This investigation concerns the upstream part of the Alpine catchment of the river Ahr (Northern Italy), with a runoff gauging station at Steinhaus (Figure 1). The catchment is situated immediately on the southern side of the Central Alpine ridge of the Zillertal Alps. It covers 148 km^2 and elevation ranges from 1035 m to 3499 m a.s.l. Hydrological and meteorological data from the 23-year period 1953-1975 were used for the investigation.

[1]Member of Scientific Board, Institute of Earth Sciences, Free University Amsterdam, De Boelelaan 1085, 1081 HV Amsterdam, Netherlands.

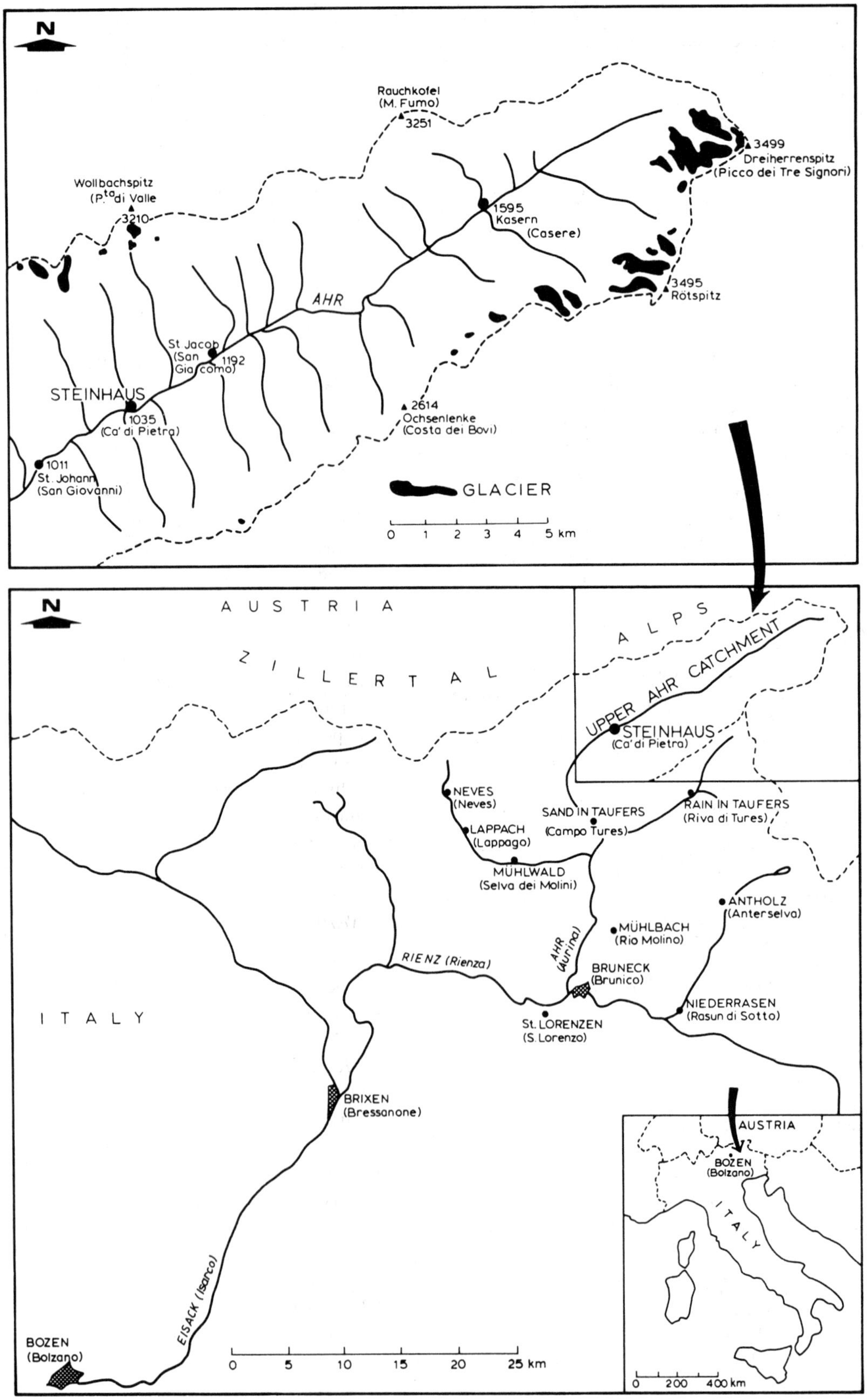

Figure 1. Location Map of the Study Area (the River Ahr Catchment, Northern Italy).

Regime Characteristics of Study Area

The runoff regime of the study area is characterized by Alpine climate and Alpine relief. Dependent on elevation and local climate the following three regimes can be distinguished: (1) *Nival regime*, consisting of runoff from water stored in the snowpack and direct snow melt; (2) *Glacial regime*, consisting of runoff from glaciers; and (3) *Pluvial regime*, consisting of runoff from rainfall.

The three regimes play a combined role throughout almost the entire year. However, their magnitudes and partial contributions to the total regime are related to the time of year, while the transition from one regime to another is gradual. The variations of the (partial) correlation coefficients (Figure 2) between runoff and monthly mean temperature (A) and between runoff and monthly precipitation (B) reflect very well the seasonally dependent nature of the different regimes in the Alpine Ahr catchment.

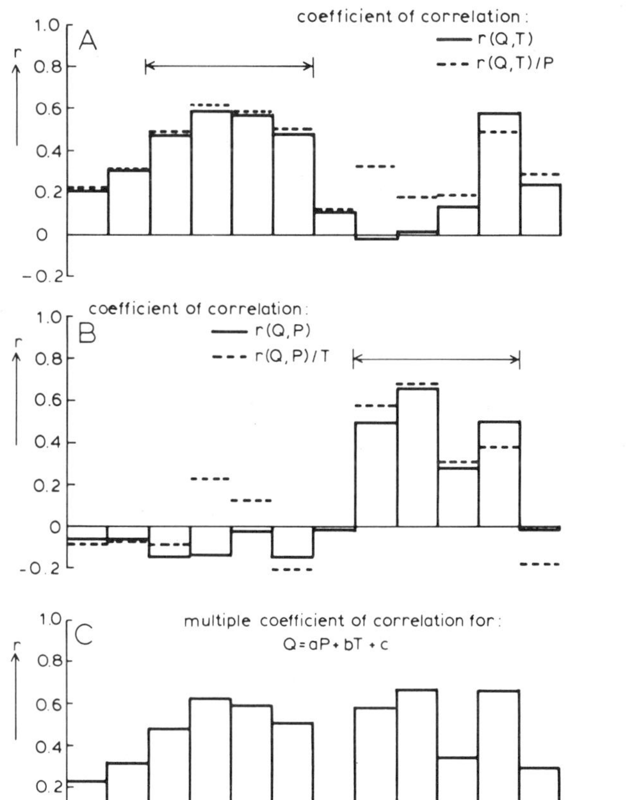

Figure 2. Coefficients of Correlation Between Monthly Runoff (Steinhaus), Monthly Mean Daily Temperature (Antholz), and Monthly Precipitation (St. Jakob), based on the Period 1953-1975: (A) Between Runoff (Q) and Temperature (T) (solid line), Controlled for Precipitation (P) (broken line); (B) Between Q and P (solid line), Controlled for T (broken line); and (C) Multiple Correlation Between Q (dependent variable) and P and T (solid line).

Both the nival and glacial regimes sustain a continuously high level of discharge throughout the summer and so prevent the river from low levels during warm and dry summer periods. Figure 3 shows a plot of 23 annual hydrographs of daily runoff, bounded by a "*minimal discharge curve*." Although the melt regimes are responsible for the bulk of yearly runoff, the pluvial regime is responsible for extreme floods usually occurring in late summer (August and September).

WEATHER TYPES AND HYDROLOGY

Current State of Knowledge

At this moment the field of research covering relationships between circulation types and hydrologic phenomena can be regarded as an unexplored field in hydrology. Only a few have worked in this field (Wehry, 1966; Tröschl, 1967; Schwarzl, 1965, 1971, 1972; De Bruin and Schuurmans, 1978; De Bruin, *et al.*, 1978). However, they investigated primarily relationships between meteorological parameters and circulation types, rather than relationships between circulation types and river runoff. Application of statistical relationships between atmospheric circulation and runoff phenomena of large rivers during spring, with special reference to ice breakup has been reported from Russian experience by Apollov, *et al.* (1970).

Only by systematic investigations can a better insight be achieved into the relationships between circulation types and hydrologic phenomena, which requires a strong integration between both disciplines of hydrology and meteorology.

Weather Type Classification

For Europe several classification systems of weather types have been proposed and developed, each having its own aim and area of application. The most important are those by Hess and Brezowsky (1969, 1977) known as *GROßWETTERLAGEN*, Schüepp (1959, 1968), Kirchhofer (1974), and Kruizinga (1978). Kruizinga classified 500 mb patterns over Europe based on principal components and cluster analysis as earlier proposed by Kuipers (1970). For an overview of existing classification systems see Barry and Perry (1973), and Van de Griend (1981).

Several climatologists investigated the correlation between meteorological parameters in the Alpine region and weather types in order to come to an evaluation of systems for which a catalog of weather types exists. From this evaluation it was concluded (Van de Griend, 1981) that Schüepp's classification results in the best correlation. For the current investigation Schüepp's system has been applied. It relates to a regionally small scale (radius of two degrees of latitude, i.e., 222 km) and is based on a distinction in advective circulation at 500 mb and 1000 mb levels, convective circulation and an accessory distinction

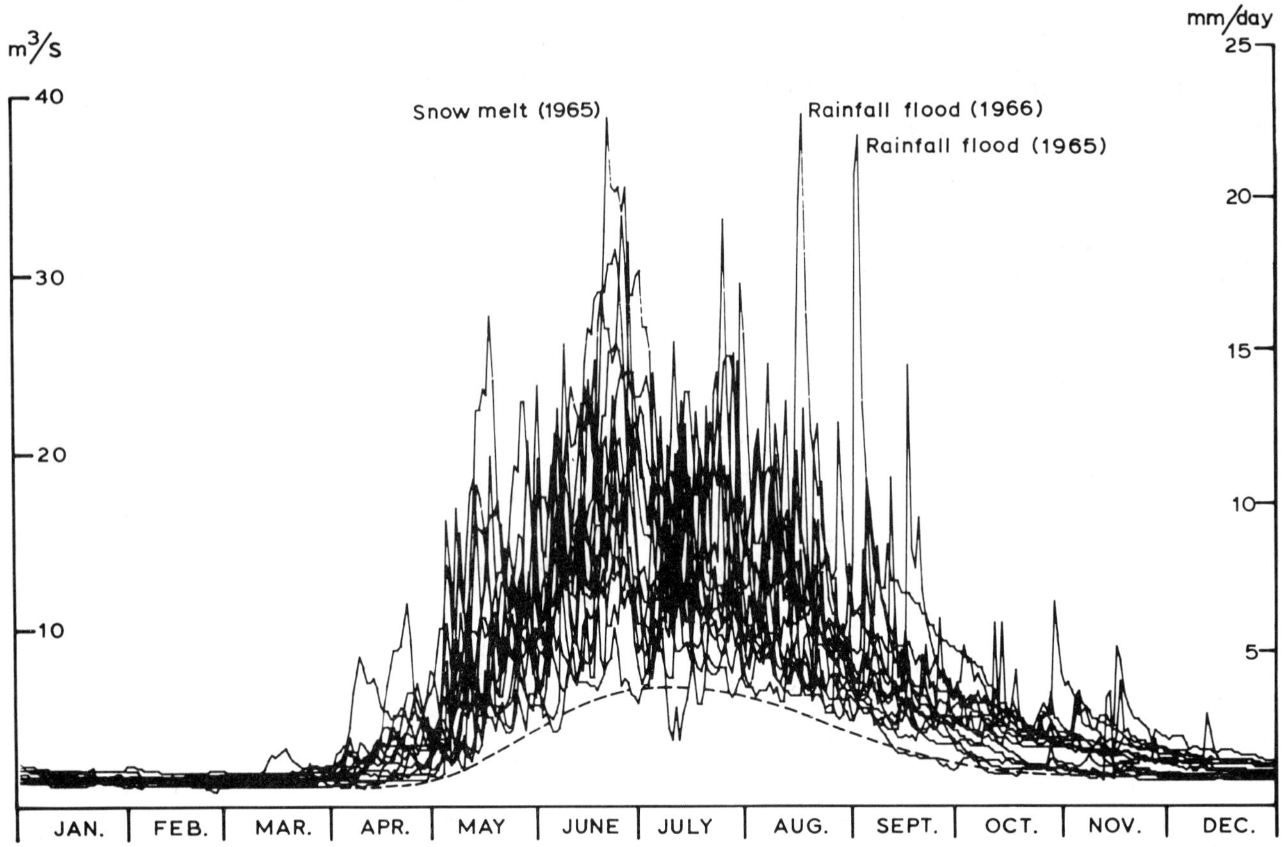

Figure 3. Plot of 23 Annual Hydrographs of Daily Discharge of the River Ahr at Steinhaus (period 1953-1975).

between cyclonic, indifferent and anticyclonic circulation (Schüepp, 1968), leading to 360 theoretical weather situations. However, during the period 1953-1975 only 80 (or 22 percent) occurred on 95 percent of the days and only 32 (or 9 percent) on 75 percent of the days.

Formation of Classes of Weather Types

For statistical analysis it becomes necessary to reduce this total number of weather types to a restricted number of classes which are regarded homogeneous with respect to their influence on hydrology. Otherwise the individual populations would become too small to allow a statistical approach. On the basis of Schüepp's system eight hydrometeorologically relevant classes have been defined based on: (a) a primary distinction on source areas of air masses, and (b) a subsequent division of cyclonicity (cyclonic, indifferent, and anticyclonic circulation). The boundaries between the classes are based on observed temperature departures from normal pentad values (5 days) and on the average rainfall totals per weather type day in the catchment. The scheme is presented in Table 1. A complete outline of the reduction strategy can be found in Van de Griend (1981).

ANALYSIS OF WEATHER TYPE HYDROLOGIC PHENOMENA

The Approach

In order to obtain insight into relationships between runoff phenomena and weather types the following analyses have been performed for each class of weather types and for each month of the year:

1. Analysis of monthly mean daily runoff;
2. Analysis of variance of daily runoff;
3. Mean day-to-day changes of runoff;
4. Coefficients of variation of day-to-day changes of runoff; and
5. Trend analysis of daily runoff during periods of the same class of weather types; all day-to-day changes were analyzed (a) as a function of the time elapsed since the start of that particular class and (b) as a function of the magnitude of runoff.

Monthly Mean Daily Runoff

Starting from the basic concept that the weather types are the driving force for the hydrologic systems, one would expect characteristic populations of daily runoff values and

TABLE 1. Scheme of Hydrometeorologically Relevant Classes of Weather Types, Based on a Primary Distinction on Air Masses and an Additional Distinction on Cyclonicity. The temperature departures from normal pentad (5 days) values (ΔT) and the probability of a day with precipitation >1 mm (P) refer to the seasonal mean (winter, spring, summer, and autumn).

Air Masses	Cyclonicity														
	c	Cyclonic (−)				i	Indifferent (·)				a	Anticyclonic (+)			
I	I^c	−N up to −E				I^i	.N up to .E				I^a	+NW up to +E			
Advection of mainly continental Polar air (cP) and Arctic air (A)	ΔT :	·	−	−	−	ΔT :	−	−	−	−	ΔT :	−	·	·	·
	P :	26%	34%	57%	41%	P :	8%	15%	39%	12%	p :	1%	5%	14%	4%
II	II^c	−W and −NW −L				II^i	.W and .NW .L								
Advection of mainly maritime Polar air (mP) and convective types	ΔT :	·	−	−	−	ΔT :	·	·	·	·					
	P :	40%	45%	67%	47%	P :	16%	27%	43%	25%					
III	III^c	−SE up to −SW				III^i	.SE up to .SW .F, −F				III^a	+SE up to +W +F, all H			
Advection of mainly Tropical air (mT and cT) and convective types	ΔT :	+	+	·	+	ΔT :	+	+	+	+	ΔT :	·	+	+	+
	P :	49%	56%	69%	56%	P :	14%	24%	49%	26%	P :	7%	8%	14%	5%

Explanation of Signs:

P = Percentage of days with precipitation >1 mm.

ΔT = Temperature departure from normal pentad values (^oC)

+ $\Delta T \geqslant +1\ ^oC$

· $-1^oC < \Delta T < +1\ ^oC$

− $\Delta T \leqslant -1\ ^oC$

Advective Types: N up to NW.

Convective Types: H = High pressure situation
F = Flat pressure situation
L = Low pressure situation

characteristic sequences of runoff for different classes of weather types.

To illustrate this a selection has been made of (1) *anticyclonic NW/E-circulation* (I^a), (2) *anticyclonic SE/W-circulation* (III^a), and (3) the combined cyclonic classes II^c and III^c (*cyclonic SE/NW-circulation*). The indifferent classes, which form hybrid types between anticyclonic and cyclonic types, have been omitted in the following qualitative comparisons.

Figures 4 and 5 show the differences in monthly mean daily precipitation and temperature for the selected classes. These differences are much less pronounced when comparing the mean runoff curves for these classes as shown in Figure 6. The small differences in mean runoff even suggest that the populations of runoff occurrences within each class might even not be significantly different, which is confirmed by analysis of variance (Van de Griend, 1981; pp. 3.4/3.6).

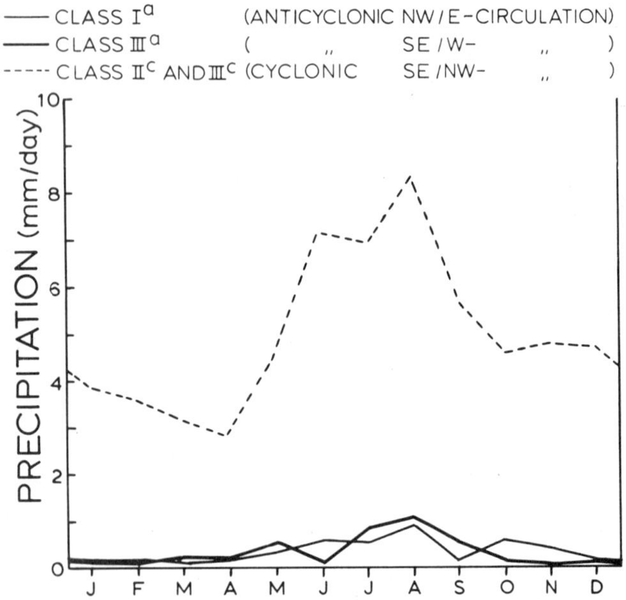

Figure 4. Monthly Mean Daily Precipitation at St. Jakob (1192 m a.s.l.) in the Ahr Valley, for Different Classes of Weather Types (period 1953-1975).

Comparison of Runoff-Variability and Day-to-Day Changes

The different character of runoff during the selected classes, however, appears very well from Figure 7 showing the coefficients of variation (variability) of daily discharge changes for each month. The figure shows the higher variability during late summer for cyclonic SE/NW circulation. The different character is still better expressed by the mean day-to-day changes given in Figure 8.

Although during anticyclonic NW/E-circulation (I^a) Δq shows negative values throughout the year (discharge decreases on average) and Δq shows positive values throughout the year for cyclonic SE/NW-circulation (II^c and III^c) (discharge increases on the average), the differences in mean runoff for both classes are relatively small (Figure 6). Runoff during the weather types preceding anticyclonic NW/E-circulation (I^a) obviously is responsible for this high rate of runoff rather than the particular weather type class itself. Antecedent conditions apparently have extreme influence on the magnitude of runoff for the different classes, which were found to have average durations of about four days.

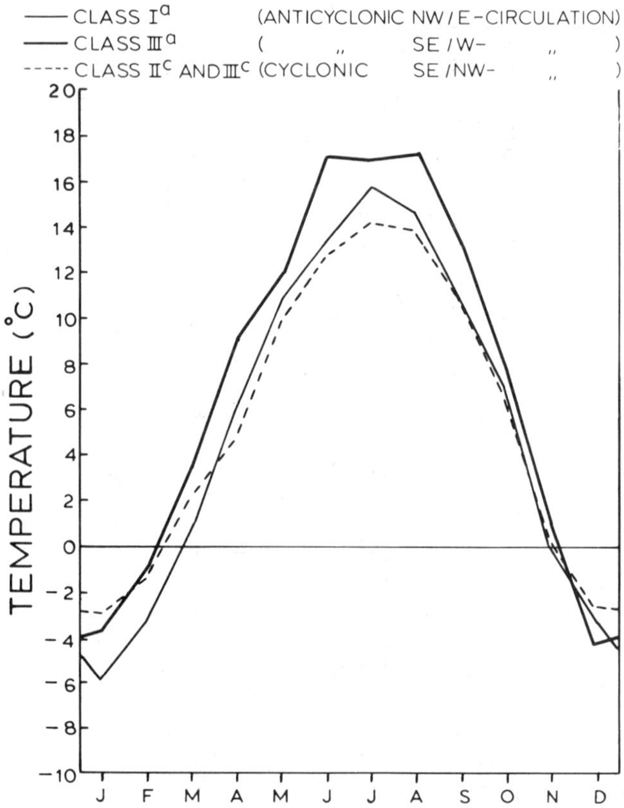

Figure 5. Monthly Mean Daily Temperature at Antholz (1236 m a.s.l.) in an Adjacent Valley of the Area of Investigation, for Different Classes of Weather Types (period 1953-1975).

The Influence of Antecedent Conditions on Runoff During the Classes

The role of antecedent conditions for runoff during the different classes of weather types is probably best illustrated from the analysis of variance applied to daily runoff departures from normal pentad values, within each class of weather types.

If a distinction is made between *internal* and *external variances*, then the internal variance is expected to be primarily governed by the combined effect of weather type and hydrologic system conditions. The external variance, on the contrary, is expected to be primarily governed by factors not related to the particular weather type and may in principle be the consequence of the following two factors:

a. The characteristic times of the systems contributing to runoff at the start of a particular weather type are so large that the corresponding active hydrologic systems are unable to eliminate this influence during the lifetime of that class, in order to reach a runoff level which is characteristic for that weather type and the active systems.

b. The antecedent hydrologic conditions vary from year to year to such a degree that even if the above mentioned characteristic runoff is reached, its magnitude is still governed by the hydrologic conditions (physical conditions of snowcover, etc.)

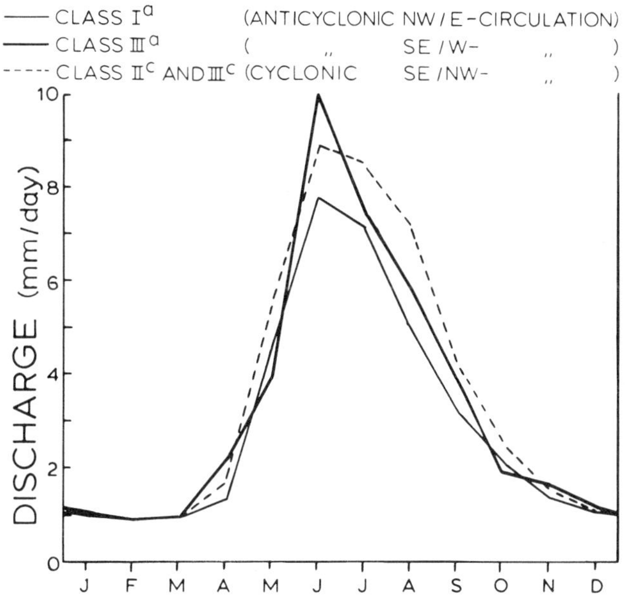

Figure 6. Monthly Mean Daily Discharge of the River Ahr at Steinhaus for Different Classes of Weather Types (period 1953-1975).

From this analysis of variance it has been concluded (Van de Griend, 1981, pp. 3.11/3.12) that the above factors (a and b) are the main factors responsible for the absolute magnitude of runoff. In other words, the magnitude of runoff during a particular weather type – with an average duration of about four days – is largely governed by *autocorrelation* (related to the characteristic times of the passive systems active during the preceding class) and by *antecedent conditions* (probably related to the extent of snowcover and physical properties of the snowpack and glacier ice), rather than the weather type itself.

Trend Analysis

The influence of antecedent conditions on the character of runoff is expected to decrease proportionally as the duration of a class of weather types increases, and thus runoff is expected to be governed in a progressive measure by the corresponding active hydrologic systems.

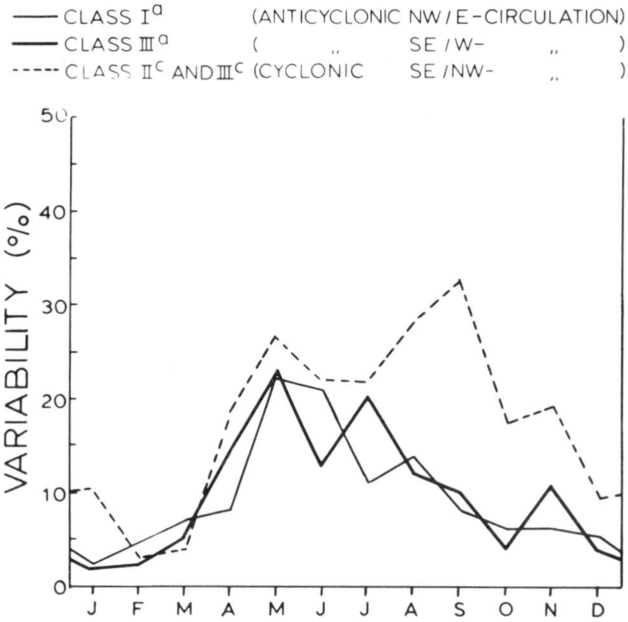

Figure 7. Monthly Mean Variability of Daily Discharge Changes of the River Ahr at Steinhaus, for Different Classes of Weather Types (expressed as a fraction of monthly mean daily discharge, without reference to classes of weather types; period 1953-1975).

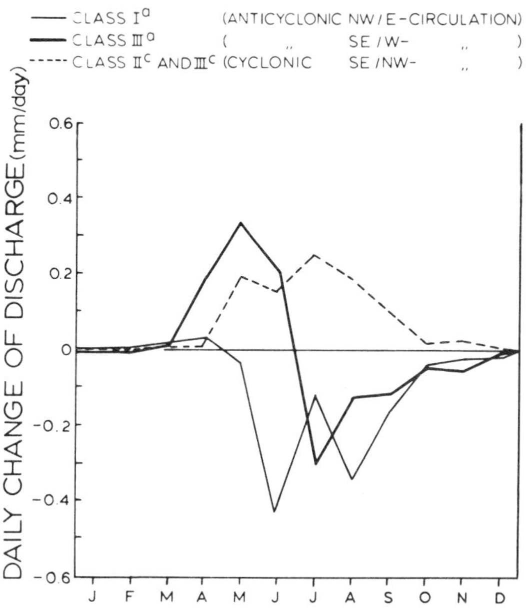

Figure 8. Monthly Mean of Daily Discharge Changes Δq of the River Ahr at Steinhaus, for Different Classes of Weather Types (period 1953-1975).

In order to investigate the influence of antecedent conditions on the character of runoff during periods of the same class, all day-to-day changes were analyzed (a) as a function of the time elapsed since the start of that particular class and (b) as a function of the magnitude of runoff.

This trend analysis has been performed for all days ($\Delta q = q_n - q_{n-1}$), for n = 1, . . . , 6 where q_n is the mean runoff during the nth day of the sequence. Figure 9 gives an example for the anticyclonic SE/W-circulation in July for n=1 and shows the tendency of runoff to increase at low runoff levels and to decrease at high levels. By means of linear regression the average day-to-day change has been estimated at three levels, i.e., for $\bar{q}_{n-1}$, $(\bar{q}_{n-1} + s_q)$ and $(\bar{q}_{n-1} - s_q)$, where $\bar{q}_{n-1}$ and s_q are, respectively, the mean runoff and the standard deviation of daily runoff on day n−1.

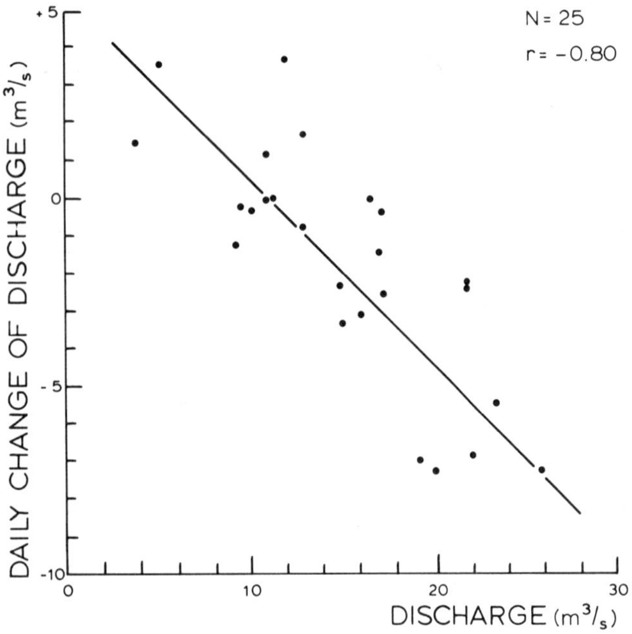

Figure 9. The Relation Between the Change of Discharge Per Day (vertical scale) and the Mean Discharge of the Preceding Day (horizontal scale) in July During Anticyclonic SE/W-Circulation (class III[a]).

Figure 10 gives in this manner the average trend of daily runoff during periods of the same class. The irregularities in this figure result from the reduction of the occurrences with increasing duration. The same analysis has been performed for the mean daily temperature shown in Figure 11.

Interpretation

The mean course of temperature for both anticyclonic classes I[a] and III[a] (Figure 11) shows an average increase on all levels ($\bar{T}$ and $\bar{T} \pm s_T$) as expected for these classes. Daily runoff (Figure 10) does not demonstrate this continuous increase on all levels, but on the contrary displays a distinct dependency on the time elapsed since the start of these particular classes. The expected increase of runoff does not become manifest until after a couple of days and is generally preceded by a period of decreasing runoff.

Class III[a] in particular shows a characteristic course which generally occurs during all months from May until August and in which the following division can be made:

0 → 1st day: High runoff values show a tendency to decrease, while low runoff values show a tendency to increase. Runoff decreases on average. Here we deal with a situation where runoff is a function of antecedent conditions governed by the impact of preceding weather types. Class III[a] was found to be preceded by cyclonic classes in 48.2 percent of the cases and by indifferent classes in 37.8 percent of the cases. During cyclonic and, to a lesser extent, during indifferent circulation, the melting process is tempered, the soil moisture and ground water systems are recharged, and the pluvial runoff system is activated. The initial decrease is therefore explained by the decrease in delayed runoff from the storages of the pluvial runoff system. The nival and glacial systems, active during anticyclonic circulation, are initially still too weak to generate enough melt water to compensate for the decrease of delayed runoff. Only when initial runoff is relatively low, at the $\bar{q} - s_q$ level, the nival and glacial systems are able to generate enough meltwater to increase river runoff at the start of the weather class.

1st → 2nd day: Continuation of the above described processes.

2nd → 3rd day: From the 2nd to the 3rd day we are confronted with a turning point in the average course of runoff. This turning point, which is most clearly expressed in the figure of July, points to the change of contributing systems. The delayed runoff of the pluvial system still continues to decrease, but the nival system generates enough melt water to increase river runoff at all three levels.

3rd → . . . . : During the 3rd and following days the nival (and glacial) system will tend to create a stable level runoff until the melting process is again interrupted by a change in weather conditions corresponding with the succeeding weather class.

The difference in effect between the anticyclonic NW/E-circulation (I[a]) and anticyclonic SE/W-circulation (III[a]), which appeared very clearly from the mean day-to-day changes (Figure 8), is also demonstrated quite distinctly by this trend analysis. Although the above described mechanism also occurs in principle during anticyclonic NW/E-circulation (e.g., in May and July), the effect of the melt generating process is much smaller compared with anticyclonic SE/W-circulation. During August, class I[a] shows a continuous decrease at all three levels while class III[a] still shows an increase after three days.

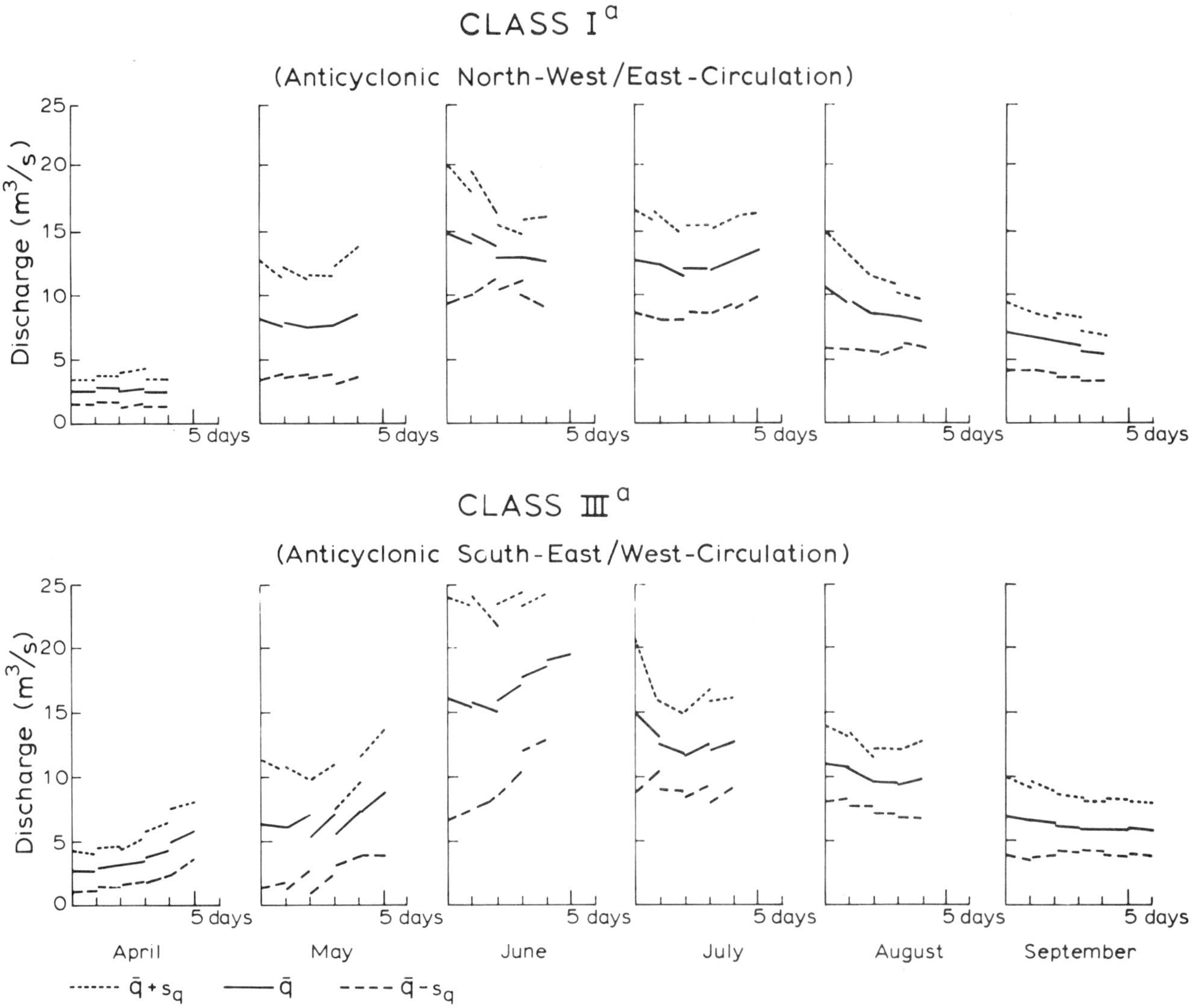

Figure 10. The Average Course of Runoff at Steinhaus (1953-1975) During Five Successive Days, at Three Levels.

Application of the above analysis to the indifferent and cyclonic classes showed that no systematic day-to-day course exists during those weather classes as for anticyclonic classes. This is mainly caused by the capricious, less systematic course of precipitation and by the fast reaction of the pluvial system to precipitation.

Conclusions

On the basis of the above analysis the following conclusions are drawn:

1. The large differences in mean daily precipitation and mean daily temperature between the different classes as shown in Figures 4 and 5 are much less pronounced in the mean runoff curves of Figure 6. This feature can be explained by a number of factors which are related to the complex character and variability of Alpine hydrologic systems, both in space (hypsometric features) and in time (seasonal and year-to-year changes) on the one hand and by the succession of weather types on the other. These main factors are:

– The distinguished hydrologic subsystems bring about an attenuation and a translation of the input from snow and glacier melt and precipitation, such that the runoff depends on the preceding runoff according to the characteristic decay times of the system (autocorrelation). The dependence on preceding runoff appears very clearly from the mean day-to-day discharge changes (Figure 8), in combination with the mean daily runoff (Figure 6).

– The antecedent hydrologic conditions are subject to large year-to-year changes, especially during the months with snow and glacier melt. Differences in snowcover extent and physical characteristics (albedo, snow density, condition of ripening, etc.) cause large variations in discharge which, in combination with the above mentioned autocorrelation, blur the distinctive mean runoff.

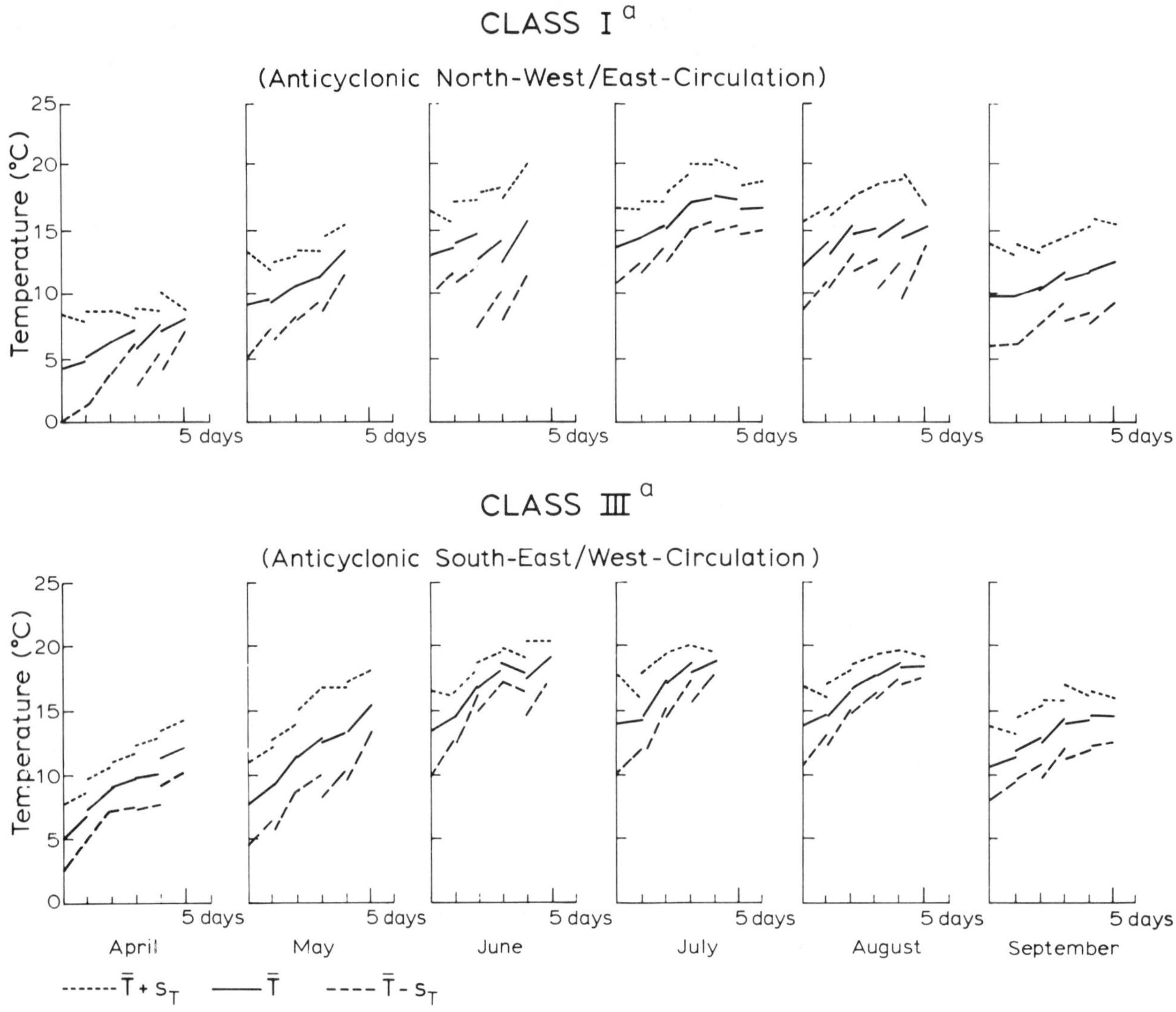

Figure 11. The Average Course of Temperature at Antholz (1236 m a.s.l., 1953-1975) During Five Successive Days at Three Levels.

– Anticyclonic SE/W-circulation (III[a]) and cyclonic SE/NW-circulation (II[c] and III[c]) have the same influence on runoff during the melting season in nivo-glacial discharge areas, in the sense that they are both associated with increasing runoff (Figure 8).

2. From the comparison of statistical phenomena of runoff such as the mean day-to-day changes (Figure 8) and the coefficient of variation (Figure 7) it may be concluded that the character of runoff during the different classes is different.

3. The magnitude of runoff during a particular weather type is to a large degree governed by:

a. Autocorrelation, related to the characteristic decay times of the active systems during the preceding weather types; and

b. By antecedent conditions, probably mainly related to the extent and physical properties of the snowcover.

4. In order to describe the response of the catchment in relation to different weather types, it is necessary:

a. To have knowledge relating to the characteristic decay times of the subsystems; and

b. To define a measure (or measures) which incorporate antecedent conditions.

WEATHER TYPES AND MEDIUM RANGE RUNOFF FORECASTING

During the last few decades great progress has been achieved in the prediction of air circulation several days in advance by means of multilayer numerical models (see e.g., Kletter, *et al.*, 1957; Manabe, *et al.*, 1974; Arpe, *et al.*, 1976). For western Europe, prediction of 1000 mb and 500 mb charts ten days in advance is the aim of the "European Centre for Medium Range Weather Forecasts (ECMWF)" in Reading, England (see "Annual Reports of the ECMWF").

In order to evaluate the usefulness of forecast weather types for purposes of medium range runoff forecasting, aimed at optimum use of forecast weather charts by the ECMWF, a multiple regression analysis has been performed between the dependent variable *10-day mean runoff* and a group of independent variables, related to antecedent catchment conditions and elements of forecast weather charts and weather types.

As the primary aim is to evaluate the applicability of forecast weather charts, a clear distinction has been made between: (a) forecasting weather charts as the field of mteorology, and (b) the use of forecast weather charts for medium range runoff forecasting. Therefore use was made of correct 100 mb and 500 mb charts, constructed after the events.

Selection of Independent Variables (Predictors)

Table 2 shows a complete list of *a priori* selected independent variables, split into *initial variables*, related to catchment conditions known at the date of forecast and *future variables* which can be either extracted directly from forecast weather charts (*direct method*) or determined by translation of forecast weather charts into weather types (*translation method*). *Direct method variables* are retrieved from Schüepp's catalogue and do not automatically represent the most decisive parameters (e.g., the vorticity which is not included as a parameter in Schüepp's calendar would represent probably a relevant parameter for precipitation potentiality).

In the following only the most important variables of Table 2 will be described shortly. For a complete description see Van de Griend (1981).

Initial Variables

Factors to Represent Initial Conditions. Initial or antecedent conditions, in principle show both temporal and spatial variations and are known to belong to the major influences responsible for the dynamic character of the catchment system operation. In order to represent to some extent antecedent conditions, the following variables have been selected:

API = Antecedent Precipitation Index

ATI = Antecedent Temperature Index

which assume an exponential decrease of the influences of precipitation and temperature in time, according to

$$API = \sum_{t=1}^{n} P_t K_1^t \quad \text{and} \quad ATI = \sum_{t=1}^{n} T_t K_2^t$$

TABLE 2. Summary of Dependent and *a Priori* Defined Independent Variables, Selected for Application in Multiple Linear Regression Analysis.

				Independent Variables			
					Future Variables		
Variable X_i	Identification	Dimension	Description	Initial Variables	Translation Method	Direct Method	Dependent Variable
i=1	API	mm	Antecedent Precipitation Index	X			
2	ATI	^oC	Antecedent Temperature Index	X			
3	I_s	mm/^oC/day	Snowcover Index	X			
4	Q_o	mm	Runoff during the day of forecast	X			
5	Q_{-10}	mm/day	Runoff during 10 days preceding the forecast	X			
6	ΔQ	mm	Persistence factor	X			
7	I_T	^oC	Mean future 10-day Temperature Index		X		
8	I_P	mm/day	Mean future 10-day Precipitation Index		X		
9	$I_s \cdot I_T$	mm/day			X		
10	RT_{10}	geop. dam – 500	Mean future 10-day Relative Topography			X	
11	AT_{500}	geop. dam – 500	Mean future 10-day 500 mb-level			X	
12	AT_{1000}	geop. dam	Mean future 10-day 1000 mb-level			X	
13	$I_s \cdot RT_{10}$					X	
14	Q_{10}	mm/day	Mean future 10-day runoff				X

where P_t, T_t represent precipitation (mm) and mean daily temperature (°C), respectively, on a day t before the calculation date and K_1, K_2 are constants (chosen 0.9 resp. 0.8). Further:

$$Q_0 = \text{Runoff at the date of forecast (mm)},$$

to represent autocorrelation phenomena.

The Snow Cover Index. The most important factors involved in snow and glacier melt runoff are related to physical conditions of the snow cover and glacier ice (see e.g., U.S. Army Corps of Engineers, 1966). As for most Alpine regions, sufficient data on these factors are not available for our study area. Therefore all initial conditions with respect to the snowpack and glaciers are condensed into one single index, denoted *Snow Cover Index* I_s and defined as the generated meltwater per degree of maximum daily temperature which can drain freely from the snowpack and glaciers. As this index is time dependent, it has been estimated integrally from the water balance of the 10-day period preceding the date of forecast. The water balance equation for the fluid cycle is given as:

$$R + (\text{Snowmelt} + \text{Glaciermelt}) = Q + E + \Delta S$$

where R = Areal rainfall during 10 days (mm); Q = Runoff during 10 days (mm); E = Evapotranspiration losses during 10 days (mm); and ΔS = Change in directly drainable storages in the catchment during the 10-day balance period (mm).

By setting the effective precipitation for the fluid cycle ($P_{eff} = R - E$) equal to $k.P$ (where k = reduction factor, representing the fraction of total precipitation P that will either runoff directly or will be stored directly in fluid form to leave the catchment as delayed runoff), then the total melt water created during the balance period is estimated from:

$$\text{Snowmelt} + \text{Glaciermelt} = Q - k.P + \Delta S$$

The change in drainable storage (ΔS) is approximated from

$$\Delta S \approx j\,(q_e - q_b)$$

where j is the characteristic decay time (days) of an assumed linear storage and q_e, q_b are the discharge (mm/day) at the end and beginning of the 10-day balance period. The Snow Cover Index then follows from

$$I_s = \frac{\text{Snowmelt} + \text{Glaciermelt}}{\overline{T}_m} = \frac{Q - k.P + j\,(q_e - q_b)}{\overline{T}_m}$$

where $\overline{T}_m$ is the average max. daily temperature (°C) during the 10-day balance period. Values of the characteristic times (j) and the reduction factors (k) have been approximated from mean monthly values, estimated for each month of the 23-year period by means of a mathematical optimization model (Van de Griend, 1981).

Future Variables

Here, future variables refer to known future weather conditions and can be accounted for over the period for which forecast weather charts exist. All future variables in Table 2 can be extracted from forecast weather charts. They have been divided into the following two groups which have been subjected to regression analysis separately.

A. Variables which can be determined from forecast weather charts by translation into weather types (Translation Method).

Temperature Index (I_T). The impact of temperature during the coming 10-day period is estimated by given weather types and corresponding monthly mean temperature departures from normal pentad values (ΔT). These temperature departures have been computed for the distinguished classes for each month from data of the 23-year period. The *Temperature Index* is defined as mean future 10-day temperature expectation, based on forecast classes of weather types and computed according to:

$$I_T = 1/10 \sum_{i=1}^{10} (\Delta T_i + \tau')$$

where, ΔT_i = estimated temperature departure (°C) of day i corresponding to the weather type class and time of year, and τ' = normal pentad temperature (°C) corresponding to day i.

Precipitation Index (I_P). The estimated impact of precipitation is based on the long term (1953-1975) monthly mean daily precipitation for the different classes. The *Precipitation Index,* defined as mean future 10-day expectation of precipitation, is computed according to:

$$I_P = 1/10 \sum_{i=1}^{10} P_i$$

where P_i = expected precipitation on day i based on the forecast weather type class and long term average precipitation for that particular class during the corresponding month.

B. Variables which can be derived directly from forecast 1000 mb and 500 mb charts (Direct Method).

RT_{10} = *Relative Topography*, i.e., 500/1000 mb-thickness in the Alpine region, as an average for the coming 10 days (geop. dam – 500).

AT_{1000} = *Absolute Topography*, i.e., altitude of 1000 mb-level in the Alpine region as an average for the coming 10 days (geop. dam).

On the basis of a combination of *Snow Cover Index* (I_S) and *Temperature Index* (I_T), respectively *Rel. Topography* (RT_{10}) the following derived future variables were defined:

$I_s \cdot I_T$ and $I_s \cdot RT_{10}$

Application of Step-Wise Regression. As an example of step-wise regression, Table 3 shows the dependent variable Q_{10} as a linear function of an increasing number of independent variables for the month of June. The corresponding correlation matrix is given in the upper part of the table. For each introduced variable the criterion is used, that the standard error of estimate expressed as a percentage of the mean of the dependent variable be reduced by at least 1 percent. The significance of the correlation coefficient and the regression coefficients are tested with Fischer's F-test (p=0.01) and the t-distribution (p=0.05), respectively.

Results of Regression Analyses. Regression analyses have been performed for the months May to September.

TABLE 3. Correlation Matrix for the Month of June (period 1953-1975) and Results of Step-Wise Regression.

CORRELATION MATRIX FOR THE MONTH OF JUNE (PERIOD 1953-1975)

	API	ATI	I_s	Q_o	Q_{-10}	ΔQ	I_T	I_P	$I_s \cdot I_T$	RT_{10}	AT_{500}	AT_{1000}	$I_s \cdot RT_{10}$	Q_{10}
API	1.00	−0.33	−0.10	−0.20	−0.19	−0.03	0.13	−0.17	−0.06	0.18	0.18	0.14	−0.05	−0.15
ATI		1.00	0.41	0.65	0.64	−0.01	0.21	−0.04	0.45	0.04	0.07	0.10	0.41	0.23
I_s			1.00	0.82	0.91	−0.04	0.06	−0.21	0.97	0.05	0.14	0.26	0.97	0.63
Q_o				1.00	0.79	0.22	0.11	−0.14	0.82	0.17	0.23	0.27	0.84	0.70
Q_{-10}					1.00	−0.15	0.08	−0.14	0.89	−0.02	0.06	0.19	0.86	0.44
ΔQ						1.00	0.03	−0.03	−0.03	0.15	0.14	0.08	0.00	0.22
I_T							1.00	−0.66	0.29	0.03	0.63	0.55	0.17	0.15
I_P								1.00	−0.35	−0.50	−0.61	−0.63	−0.30	−0.16
$I_s \cdot I_T$									1.00	0.18	0.28	0.37	0.97	0.64
RT_{10}										1.00	0.94	0.58	0.28	0.49
AT_{500}											1.00	0.82	0.35	0.50
AT_{1000}												1.00	0.37	0.39
$I_s \cdot RT_{10}$													1.00	0.73
Q_{10}														1.00

RESULTS OF STEP-WISE REGRESSION (DEPENDENT VARIABLE = Q_{10}; NUMBER OF CASES = 440)

	Intercept	$I_s \cdot RT_{10}$	RT_{10}	I_P	Q_o	ATI	API	AT_{1000}	S.E.E.	t^{c*}	F^c	r	S.E.E.**	$D(=r^2)$
	0.30	0.316							2.17	21.0	442	0.728	23.2	0.530
	−13.30	0.284	0.236						1.95	8.0	291	0.794	20.8	0.630
	−21.44	0.301	0.326	0.548					1.81	6.7	232	0.826	19.3	0.682
	−21.21	0.195	0.338	0.482	0.243				1.73	5.1	192	0.842	18.5	0.709
	−13.96	0.127	0.332	0.435	0.551	−0.119			1.56	5.5	213	0.874	16.7	0.764
Selected →	−13.39	0.127	0.351	0.399	0.539	−0.129	0.056		1.45	5.8	206	0.893	15.5	0.797
	−14.08	0.125	0.325	0.498	0.524	−0.127	0.056	0.145	1.42	2.8	181	0.896	15.2	0.803

r = 0.893
D = 0.797
$\bar{Q}_{10}$ = 9.36 mm/day
S.E.E. = 1.45 mm/day
S.E.E. (% of mean Q_{10}) = 15.5%

Lowest t^c = 5.8
t^t (5%) = 1.6
F^c = 206
F^t(1%) = 3.4
* = lowest value
** = % of $\bar{Q}_{10}$

The remaining months (October to April) have not been subjected to regression analyses because of the rather stable character of runoff during these months. Each of the analyses has been produced both according to *translation method* and *direct method*. In addition, a step-wise regression has been performed on the combined set of *translation* and *direct method variables* (denoted *combined method*), in order to find the optimum regression equation. Results are presented in Table 4. In Table 6 the regression equations have been presented with independent variables in sequential order of decreasing explanation of the variance of the dependent variables.

TABLE 4. Results of Regression Analyses for the Prediction of 10-Day Mean Runoff (Q_{10}) According to the *Translation Method* (A), the *Direct Method* (B), and the *Combined Method* (C).

			Regression Coefficients								
						Future Variables					
			Initial Variables			Translation Method			Direct Method		
Month	Method	Intercept	API	ATI	Q_o	I_T	I_P	$I_s \cdot I_T$	RT_{10}	AT_{1000}	$I_s \cdot RT_{10}$
May	A	−2.82	–	–	+0.303	+0.623	−0.233	+0.214	–	–	–
	B	−10.03	–	−0.032	+0.286	–	–	–	+0.274	–	+0.069
	C	−10.03	–	−0.032	+0.286	–	–	–	+0.274	–	+0.069
June	A	+2.36	–	–	+0.428	–	–	+0.421	–	–	–
	B	−6.84	–	−0.114	+0.293	–	–	–	+0.260	–	+0.161
	C	−13.39	−0.056	−0.129	+0.593	–	+0.399	–	+0.351	–	+0.127
July	A	+0.47	−0.033	–	+0.282	–	+0.298	+0.718	–	–	–
	B	+1.06	−0.021	–	–	–	–	–	–	–	+0.277
	C	−0.03	−0.022	–	–	–	+0.300	–	–	–	+0.281
August	A	+0.69	−0.013	–	+0.416	–	+0.385	+0.267	–	–	–
	B	+6.89	–	–	+0.369	–	–	–	–	−0.323	+0.037
	C	+1.46	−0.012	–	+0.352	–	+0.377	–	–	–	+0.042
September	A	−0.14	–	+0.016	+0.317	–	+0.182	+0.465	–	–	–
	B	+0.77	–	+0.018	+0.327	–	–	–	–	−0.037	+0.086
	C	+0.07	–	+0.013	+0.349	–	+0.157	–	–	–	+0.077

Statistics

Month	Method	r	D (= r^2)	S.E.E.	$\overline{Q_{10}}$ (mm/day)	S.E.E. (% of $\overline{Q_{10}}$)	Lowest* t	F^c	F^t (1%)
May	A	0.788	0.621	1.27	6.22	20.4	2.5	102	3.3
	B	→0.871	0.759	1.02	6.22	16.4	2.8	199	3.3
	C	0.871	0.759	1.02	6.22	16.4	2.8	199	3.3
June	A	0.707	0.500	2.28	9.36	24.4	3.8	183	4.6
	B	0.824	0.679	1.82	9.36	19.4	5.8	223	3.3
	C	→0.893	0.797	1.45	9.36	15.5	5.8	206	2.8
July	A	0.823	0.677	1.19	7.37	16.1	5.7	194	3.3
	B	0.854	0.729	1.09	7.37	14.8	3.4	415	4.6
	C	→0.876	0.767	1.01	7.37	13.7	5.7	382	3.8
August	A	0.840	0.706	1.03	5.26	19.6	3.6	174	3.3
	B	0.843	0.711	1.02	5.26	19.4	3.7	241	3.8
	C	→0.863	0.745	0.96	5.26	18.3	3.4	215	3.3
September	A	→0.878	0.771	0.48	3.08	15.6	4.8	292	3.3
	B	0.861	0.741	0.51	3.08	16.7	4.5	250	3.3
	C	0.874	0.764	0.49	3.08	15.9	3.8	281	3.3

*$t^t_{(5\%)} = 1.6$.

Conclusions. Table 4 shows that in all months except September the *direct method* leads to better results than the *translation method* (compare the values of r, $D(=r^2)$, and S.E.E.). This is explained from the results reported in Table 5, which concerns a comparison between Temperature Index I_T (*translation method*) and Rel. Topography RT_{10} (*direct method*) as a measure to represent heat input by comparing their correlations with observed 10-day mean maximum temperatures (Tm_{10}). From this table it can be seen that except for September the coefficients of correlation between Tm_{10} and RT_{10} exceed those between Tm_{10} and I_T.

TABLE 5. The Coefficients of Correlation for Pr_{10} (*mean 10-day precipitation*) and Tm_{10} (*mean 10-day maximum temperature*), with Parameters AT_{1000} and RT_{10} Used for Their Representation (period 1953-1975).

	May	June	July	August	September
			Temperature		
$r(Tm_{10}/RT_{10})$	0.88	0.88	0.90	0.87	0.53
$r(Tm_{10}/I_T)$	0.40	0.68	0.66	0.62	0.56
			Precipitation		
$r(Pr_{10}/I_P)$	0.39	0.57	0.55	0.75	0.68
$r(Pr_{10}/AT_{1000})$	−0.16	−0.40	−0.41	−0.70	−0.30

From Table 5 it also appears that the Precipitation Index I_P is a better index to represent mean 10-day precipitation than AT_{1000}. This explains why the *combined method* leads to the best regression equations with both RT_{10} (*direct method*) and I_P (*translation method*), except for May and September where the differences are minor.

The coefficients of multiple correlation (r, Table 4) for the *combined method* vary between 0.87 and 0.89 with the explained variance (r^2) between 76 percent and 80 percent. The S.E.E. values expressed as a fraction of the mean Q_{10} vary between 18.3 percent and 13.7 percent.

From the sequential order of variables in the regression equations shown in Table 6 follows the change of relative importance of the variables as a function of the time of year. During the months May to July, $I_s \cdot I_T$ (*translation method*) and $I_s \cdot RT_{10}$ (*direct method*) appear as the first independent variables in the respective equations and thus explain the largest part of the variance of Q_{10}. The importance of the variables, representing the heat input effect on the nivo glacial system, is in accordance with the major contribution of the nivo glacial system to runoff during these months. The decreasing importance of the nivo glacial system in late summer and autumn is reflected in the occurrence of $I_s \cdot I_T$ and $I_s \cdot RT_{10}$ as the second or even the third variable after I_P and AT_{1000} in the equations computed for August and September.

As the above analyses are based on correct 1000 mb and 500 mb charts, constructed after the events, it is clear that the accuracy of estimation of the independent variables from forecast weather charts finally determines the practical meaning for medium range runoff forecasting. So far, no comparative analysis has been performed of the accuracy of estimation of *direct method variables* and *translation method variables* from real forecast weather charts. However, the error of estimating the correct *class of weather types* out of eight classes is expected to be smaller than the error of estimating RT_{10} as a continuous variable.

Although the above analyses already indicate the potentialities of weather types for medium range runoff forecasting, as a means for translation of weather charts to meteorological impact on the hydrological system, the results might probably be improved by regrouping weather types into classes on the basis of still improved regional analyses of precipitation potentialities and temperature departures during the weather types.

ACKNOWLEDGMENTS

The author wishes to thank Prof. Dr. G. B. Engelen (Free University, Amsterdam, the Netherlands), Prof. Dr. M. Schüepp (Meteorological Institute, Zurich, Switzerland), Prof. Dr. F. Fliri (Universität Innsbruck, Austria), Dott. Ing. P. Valentini and Mr. H. Mur (Hydrological Service of Bolzano, Italy), and the many others whose support made this study possible.

This investigation represents one phase of research of the Institute of Earth Sciences, Department of Hydrology, Free University, Amsterdam, the Netherlands.

LITERATURE CITED

Apollov, B. A., G. P. Kalinin, and V. D. Komarov, 1970. Hydrological Forecasting. Israel Program for Scientific Translations, Jerusalem, 338 pp.

Arpe, K., L. Bengtsson, A. Hollingsworth, and Z. Janjič, 1976. A Case Study of a 10-Day Prediction. ECMWF Technical Report No. 1, 105 pp.

Barry, R. G. and A. H. Perry, 1973. Synoptic Climatology. Methods and Applications. Methuen & Co. Ltd., London, England, pp. 94-175.

De Bruin, H. A. R., H. C. Bijvoet, and H. J. Krijnen, 1978. Betrachtungen über das Hochwasser von Februar und März 1970. *In:* Hydrologische Monographie: Das Rheingebiet. Intern. Kommission für die Hydrologie des Rheingebietes, pp. 161-170.

De Bruin, H. A. R. and C. J. E. Schuurmans, 1978. Betrachtungen über die Niedrigwasserperioden von 1959 und 1964. *In:* Hydrologische Monographie: Das Rheingebiet. Intern. Kommission für die Hydrologie des Rheingebietes, pp. 191-203.

Hess, P. and H. Brezowsky, 1969. Katalog der Großwetterlagen Europas. Berichte des Deutschen Wetterdienstes 113, Offenbach a M., 56 pp.

Hess, P. and H. Brezowsky, 1977. Katalog der Großwetterlagen Europas 1881-1976. Berichte des Deutschen Wetterdienstes 113, Band 15, Offenbach a M., 3. verbesserte und ergänzte Auflage.

Kirchhofer, W., 1974. Classification of European 500 mb Patterns. Arbeitsberichte der Schweizerischen Meteorologischen Zentralanstalt, No. 43, 16 pp.

Kletter, L., 1957. Ergebnisse einer Versuchsreihe mit vorausberechneten Höhenkarten über mittelfristige Zeiträume. Arch. Met. Geoph. & Biokl. A-10:144-160.

Kruizinga, S., 1978. Objectieve classificatie van dagelijkse 500 mbar patronen. Koninklijk Nederlands Meteorologisch Instituut. W.R. 78-8, 11 pp.

Kuipers, W. J. A., 1970. An Experiment on Numerical Classification of Scalar Fields. Idöjáràs 74:296-306.

Manabe, S., J. L. Holloway, and M. J. Spelman, 1974. GFDL Global 9-Level Atmospheric Model. GARP Publication Series No. 14: 7-27.

Meier, M. F. and E. F. Roots, 1981. Snow and Ice, Climate and Water Supply. Paper presented on the Intern. Conference on Hydrology and the Scientific Bases for the Rational Management of Water Resources. Paris, France, 18-27 August 1981, 8 pp.

Schüepp, M., 1959. Die Klassifikation der Witterungslagen. Geogr. Pura e Applicata 55:242-248.

Schüepp, M., 1968. Kalender der Wetter- und Witterungslagen von 1955 bis 1967. Veröffentlichungen der Schweizerischen Meteor. Zentralanstalt Nr. 11, 43 pp.

Schwarzl, S., 1965. Die Häufigkeit von Vb-Lagen. Carinthia II, Sonderheft 24:101-106.

Schwarzl, S., 1971. Charakteristische Hochwasserwetterlagen im Alpenraum. Interpraevent 1971, Villach, September 1971.

Schwarzl, S., 1972. Die Vb-Lagen am 21/22 Nov. 1970, eine charakteristische Hochwasserlage im Bereich der Ostalpen. Arch. Met. Geoph. & Biokl., Ser. A. 21:75-93.

Tröschl, H., 1967. Die neuerliche Niederschlags- und Hochwasserkatastrophe im Österreichischen Südalpengebiet, vom 3-5 November 1966. Wetter und Leben, Heft 1-2, Nr. 19:1-12.

Van de Griend, A. A., 1981. A Weather-Type Hydrologic Approach to Runoff Phenomena. Ph.D-thesis, Free University, Amsterdam, the Netherlands, 212 pp.

TABLE 6. Regression Equations for Forecasting Mean 10-Day Runoff According to the *Translation Method* (A), the *Direct Method* (B), and the *Combined Method* (C), for the Months May to September.

Month	Method		Regression Equation
May	A	:	$Q_{10} = -2.82 + 0.214\ (I_s \cdot I_T) + 0.303\ Q_o + 0.623\ I_T - 0.233\ I_p$
	B	:	$Q_{10} = -10.03 + 0.069\ (I_s \cdot RT_{10}) + 0.274\ RT_{10} + 0.286\ Q_o - 0.032\ ATI$
	C	:	Result not improved as compared with B
June	A	:	$Q_{10} = +2.36 + 0.421\ (I_s \cdot I_T) + 0.428\ Q_o$
	B	:	$Q_{10} = -6.84 + 0.161\ (I_s \cdot RT_{10}) + 0.260\ RT_{10} + 0.293\ Q_o - 0.114\ ATI$
	C	:	$Q_{10} = -13.39 + 0.127\ (I_s \cdot RT_{10}) + 0.351\ RT_{10} + 0.399\ I_p + 0.539\ Q_o - 0.129\ ATI - 0.056\ API$
July	A	:	$Q_{10} = +0.47 + 0.718\ (I_s \cdot I_T) + 0.282\ Q_o - 0.033\ API + 0.298\ I_p$
	B	:	$Q_{10} = +1.06 + 0.277\ (I_s \cdot RT_{10}) - 0.021\ API$
	C	:	$Q_{10} = -0.03 + 0.281\ (I_s \cdot RT_{10}) + 0.300\ I_p - 0.022\ API$
August	A	:	$Q_{10} = +0.69 + 0.416\ Q_o + 0.385\ I_p + 0.267\ (I_s \cdot I_T) - 0.013\ API$
	B	:	$Q_{10} = +6.89 + 0.369\ Q_o - 0.323\ AT_{1000} + 0.037\ (I_s \cdot RT_{10})$
	C	:	$Q_{10} = +1.46 + 0.352\ Q_o + 0.377\ I_p + 0.042\ (I_s \cdot RT_{10}) - 0.012\ API$
September	A	:	$Q_{10} = -0.14 + 0.317\ Q_o + 0.182\ I_p + 0.465\ (I_s \cdot I_T) + 0.016\ ATI$
	B	:	$Q_{10} = +0.77 + 0.327\ Q_o + 0.086\ (I_s \cdot RT_{10}) + 0.018\ ATI - 0.037\ AT_{1000}$
	C	:	$Q_{10} = +0.07 + 0.349\ Q_o + 0.077\ (I_s \cdot RT_{10}) + 0.157\ I_p + 0.013\ ATI$

METHODOLOGY FOR DEVELOPING OPTIMAL WATER STORAGE STRATEGIES

Darrell G. Fontane and John W. Labadie[1]

ABSTRACT: Intensifying demands for water in many areas of the western United States, combined with limited opportunities for constructing new water storage projects, have underscored the need for basin-wide, integrated approaches to water development strategies. A computerized model called SCREEN is presented which enables water planners to select storage sites and capacities that meet various water demand scenarios at minimum cost, while considering other objectives such as water quality control and instream flow needs. The model uses an interactive conversational data input and editing format which facilitates use by planners with little familiarity with the computer. The model provides intermediate information during the solution process which allows the planner to make adjustments to reflect noneconomic objectives. The model results can be further refined using a combined simulation-optimization model called OPTRES. Several proposed projects in the Yampa River basin in northwestern Colorado are analyzed by SCREEN as a means of demonstrating its capability.
(KEY TERMS: reservoir sizing; river basin planning; simulation; optimization; interactive computer programs.)

INTRODUCTION

Planners and managers are facing an increasingly complex task of allocating and managing water resources amid conflicting objectives. The western United States faces particularly acute problems due to increasing development pressures and limited water supplies. As a specific example, the State of Colorado is experiencing rapidly increasing in-state water demands, particularly demands for water for coal and oil shale energy development. The State of Colorado has vast resources of low-sulfur coal and oil shale which are an important source of energy to the nation as a whole. As discussed by Morse (1980) and the Missouri River Basin Commission (1979), these resources probably will be developed. This projected development could potentially occur on a large scale and over a short time frame.

Since the demands for water vary over time and space, it is often difficult to meet all requirements, especially under constraints of variable hydrology. Reservoir storage is an effective means of reducing the impacts of hydrologic variability and providing a consistent supply of water to meet varying demands. However, the construction of reservoir storage has become increasingly difficult for a variety of institutional, political, financial, and environmental reasons. Since it appears that the number of future reservoir projects may be limited, those reservoirs that are constructed should represent an optimal, integrated configuration that will meet projected water demands at minimum cost and help maximize total beneficial use of developed water resources.

This paper describes a methodology for use by water resources planners and managers to help select optimal water storage strategies in a multiobjective setting. These storage strategies are based primarily on providing water at minimal total reservoir construction costs, but also indirectly allow consideration of other objectives such as water quality control, instream uses, etc. These strategies are developed by considering all potential and existing reservoir projects in an integrated fashion. The methodology is based upon a series of computer models, including a simulation model to determine the reservoir storage and yield relationship at a given site, a dynamic programming screening model to determine optimum locations and sizes for reservoirs in a basin, and a combined river basin simulation model and nonlinear optimization technique to refine the optimum storage configuration of the screening model and to provide for detailed analysis of the system. The methodology is designed to cover a wide range of planning levels, from preliminary to detailed. It is also designed for ease of use by water resources planners and managers with minimal knowledge of computer languages, such as FORTRAN. Various components of the methodology employ interactive conversational coding. The use of interactive conversational coding allows the user to look at a variety of alternatives and compare results simply by responding to a series of questions about the projects. It also facilitates his ability to go back and change certain aspects of the system being analyzed and note the effects. The methodology is

[1]Respectively, Assistant Professor, Department of Civil Engineering, Colorado State University, Room A327 Engineering Research Center, Fort Collins, Colorado 80523; and Professor, Department of Civil Engineering, Colorado State University, Room A329 Engineering Research Center, Fort Collins, Colorado 80523.

applied to the Yampa River basin in northwestern Colorado. A large number of storage projects have been proposed for this basin, which make it an attractive case study.

FIRM YIELD SIMULATION MODEL

A simulation model called WATPOW was developed to determine the relationship between reservoir firm yield and reservoir storage for all potential reservoir sites. The firm yield is defined as the annual amount of water which can be delivered with certainty (according to a predetermined monthly demand schedule) based upon a given historical or synthetic hydrologic sequence. The model allows monthly varying yields to meet prescribed time variant demands, for a specified total annual yield. The simulation model is based upon an approach proposed by Hall and Dracup (1970). The model output consists of a table of discrete storage values and the associated values of firm yield. Firm yield values range from the minimum to the maximum available for the site. This information is combined with the estimated costs of reservoir storage at each site to yield a cost as a function of firm yield relationship.

SCREENING MODEL

The screening model, called SCREEN, was developed to determine optimal storage configurations (location and size) for a system of reservoirs based upon satisfying firm demands for water supply at a minimal total construction cost. The screening procedure is based on a dynamic programming formulation developed by Hall (1980) and is described in detail by Fontane, Hall and Labadie (1980).

The basic concept of the screening model is to decompose the original problem of multiple reservoirs into a series of small problems each involving only two reservoirs. The essential characteristic of the procedure is the combination via dynamic programming of any two reservoirs, into a single equivalent reservoir which may provide firm yield in some amount at a minimum cost. Appropriate constraints are applied in developing the combination depending upon whether the reservoirs are in series or in parallel. This equivalent reservoir is then combined with another reservoir to find the optimal combination of the three, and the process is continued, as illustrated in Figure 1, until all reservoirs have been combined into a single equivalent reservoir. This final equivalent reservoir will provide the minimum total cost required to produce any discrete quantity of firm yield up to the maximum firm yield available for the entire system. Additionally, for each discrete quantity of firm yield, the optimal combination of reservoirs and their respective sizes which will provide that yield at minimum cost can be determined.

The results of the screening procedure illustrate the advantage of the dynamic programming optimization technique employed. In one run of the screening model a

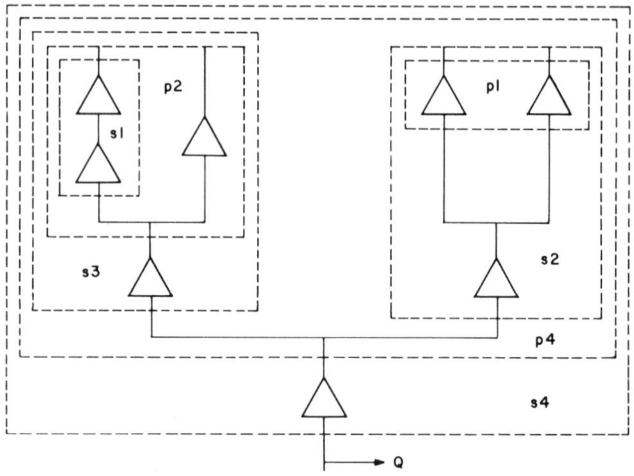

Figure 1. Illustrative Sequence of Subsystem Combinations for Reservoir Sites in Both Series and Parallel.

range of solutions is developed for various levels of total firm yield to be provided by the reservoirs considered. While this is advantageous, what is more significant is the fact that the sensitivity of the individual reservoir sizes to various demand levels can be evaluated. Considering the uncertainty of projecting water demands for the future (Colorado Department of Natural Resources, 1979), this sensitivity information can be extremely useful. In addition, since the model automatically provides a whole family of intermediate solutions for various portions of the basin, various constraints on yield can be introduced where desired to reflect water quality management requirements, instream uses, etc.

The SCREEN model employs interactive conversational coding for ease of use. A separate program, also interactive and conversational, is used to develop the input data for the SCREEN model. This program questions the user to determine the number of reservoirs in the analysis, the firm yield vs. cost information for each reservoir, the order of analysis, and the type of reservoir combination; parallel or series. The output of this program is the required input data file for the SCREEN model in the appropriate format.

The SCREEN model reads the input data file and determines the reservoirs in the first combination (the most upstream projects) and their relationship (parallel or series). Depending upon the type of relationship, the appropriate dynamic programming analysis is employed. The results of the analysis are then displayed. The output consists of a tabular printout which displays a range of values of total firm demand to be satisfied, the optimal firm yield contribution of each reservoir in the analysis, and the total minimized construction costs.

The SCREEN model next questions the user to determine whether the user wishes to proceed to the next reservoir combination. The user therefore has the option of interrupting the analysis after each combination of reservoirs. For example, the user may wish to specify that a reservoir not selected for construction by the screening procedure, should be in fact constructed to some minimum size for reasons other than economic minimization of construction costs. If the user is satisfied with the results of the reservoir combination, the program then proceeds to the next combination. This entire procedure, considering two reservoirs at a time, is repeated until all reservoirs in the system have been analyzed.

COMBINED SIMULATION-OPTIMIZATION MODEL

The output of the dynamic programming screening model will provide the optimal locations and sizes of reservoirs required to meet various levels of projected demand. As the planning process becomes more detailed, however, the optimal storage configuration of the screening model may need to be refined through the use of a simulation model. Simulation models, in contrast to the screening model, can provide a more detailed description of the physical and operational characteristics of the reservoir-river basin system.

For this methodology, a combined simulation-optimization model called OPTRES was developed by linking an existing river basin simulation model, the U.S. Army Corps of Engineers HEC-3 (1974) Model, with a nonlinear optimization technique. The development of the OPTRES model is described in detail by Fontane and Labadie (1980). The HEC-3 simulation model is used to simulate the river-reservoir system for a given storage strategy (such as that obtained from the screening model). The performance of any given system configuration is measured by a user specified function. Generally this function would measure total reservoir costs and the ability of the system to satisfy various target objectives, such as satisfying demands at various locations in the system. The optimization technique then determines refined values for the storage to be provided at each reservoir site and the system is simulated with these refined storage values. This process continues until the function measuring system performance is minimized or maximized as appropriate.

An interactive conversational code was written to develop the input data for the OPTRES model. The code provides instructions on the type of data required for the OPTRES model, questions the user to determine the required input information, and creates the input data file in the appropriate format. This conversational code can be used either to create a new input data file or to modify an existing data file.

DEMONSTRATION STUDY

The developed methodology has been tested on the Yampa River Basin in northwest Colorado. The Yampa River Basin is relatively undeveloped both in terms of population and water resources. There is potential for future development in the basin, particularly due to the availability of coal resources. Recent studies by the U.S. Geological Survey (1979) and the Federal Energy Regulatory Commission (1981) have identified approximately 30 proposed reservoir projects for the basin.

The SCREEN model was applied to a subset of five reservoirs (Table 1) in the Yampa Basin. The configuration of these reservoirs is shown in Figure 2. The relationship of storage and firm yield at each site was developed using the WATPOW simulation model and 50 years (1927 to 1976) of historical flow data. Approximate construction cost relationships for various levels of storage were estimated for each site.

TABLE 1. Test Configuration Reservoir Projects.

Number	Project Name	Site Number
1	Yamcola	01
2	Pleasant Valley	05
3	Dunckley	10
4	California Park	11
5	Thornburg	15

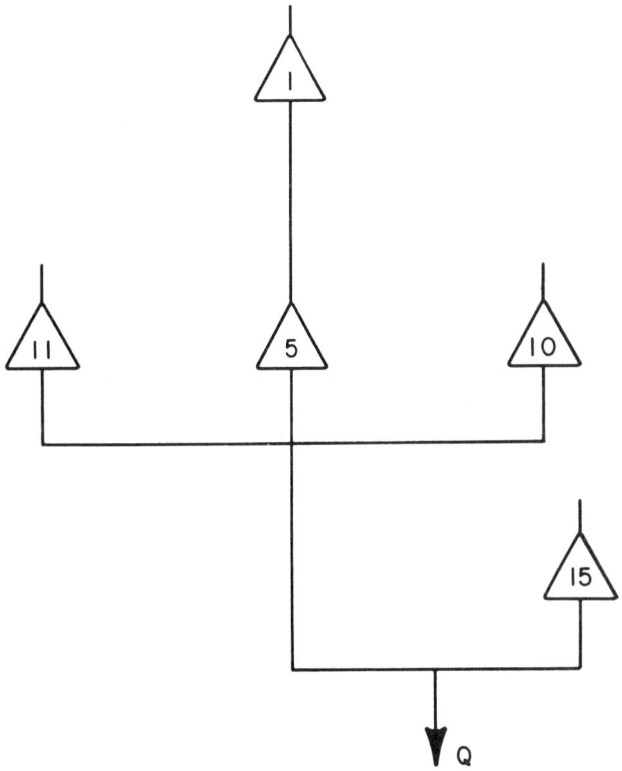

Figure 2. Test Reservoir Configuration for the Screening Procedure.

Three scenarios are presented to illustrate the types of analyses which might be conducted with the screening procedure. These are not presented as actual proposals for construction priority, but only for demonstration purposes. The first scenario is called Case 1: Baseline and assumes that no restrictions are placed on the amount of hydrologically available yield to be provided at any specific location in the test configuration. This condition assumes that the demands for water are, in effect, downstream of the most downstream reservoir. The results of the screening procedure are shown in Table 2, and indicate that two of the reservoirs (Dunckley, Site 10, and Thornburg, Site 15) are not present in the optimal solutions and would not be selected for construction. The Yamcola (Site 1) and California Park (Site 11) Reservoirs are shown to be relatively insensitive to the level of demand. Essentially, these reservoirs are required to produce a limited range of firm yield values once the firm demand reaches a specified level. The projects could then be built to a size to produce the required firm yield and those sizes would be optimum for any future increases in demands. The final reservoir, Pleasant Valley (Site 5), is shown to be highly sensitive to the level of total annual demand. This reservoir must be sized carefully to avoid foreclosure of a future opportunity. If the reservoir is sized too small and future demands increase, the additional demand would have to be provided by the other sites at a higher total cost. This condition is presented as the second scenario called Case 2: Maximum Yield Limited. For this scenario the maximum yield at Pleasant Valley Reservoir (Site 5) was limited to 85,000 ac.-ft. The results of the screening procedure are shown in Table 3 and indicate the additional costs incurred if the future demand requirement increases beyond the level which Pleasant Valley Reservoir is restricted to supply.

The final scenario is called Case 3: Specified Minimum Yields. This scenario represents a condition where the user may wish to specify that certain reservoirs be constructed, other than those selected in Case 1, to provide a specified minimum level of firm yield at certain locations in the basin. For this example, Yamcola Reservoir (Site 1) was constrained to provide a minimum firm yield of 10,000 ac.-ft. (Yamcola Reservoir was actually constructed in 1981 to a size that will approximately satisfy this condition), and California Park (Site 11) was constrained to provide a minimum firm yield of 15,000 ac.-ft. The results of the screening procedure shown in Table 4, show that additional costs are incurred as compared to Case 1. However, these additional costs are only incurred for a certain range of total firm demands.

SUMMARY

The methodology presented has been designed to provide water resources planners and managers with tools for use in evaluating optimal storage strategies. The scenarios discussed illustrate the types of preliminary planning analyses which might be done with the SCREEN model. More detailed planning could refine the screening model output through the use of the OPTRES model. It is hoped this methodology will be useful for Colorado in providing guidance and planning for the development of the water resources of the state.

ACKNOWLEDGMENTS

This study is partially supported by the Colorado Water Resources Research Institute, through funds from the Legislative Council of the State of Colorado. The Institute Director is Dr. Norman A. Evans, whose support we gratefully acknowledge.

LITERATURE CITED

Colorado Department of Natural Resources. The Availability of Water for Oil Shale and Coal Gasification Development in the Upper Colorado River Basin-Summary Report, Public Review Draft Report, Denver, Colorado, October, 1979.

Federal Energy Regulatory Commission. Yampa River Basin: Water Resources Appraisal for Hydroelectric Licensing, San Francisco Regional Office, San Francisco, California, September, 1981.

Fontane, D. G. and J. W. Labadie. Development of Methodologies for Determining Optimal Water Storage Strategies, Technical Report No. 20, Colorado Water Resources Research Institute, Colorado State University, Fort Collins, Colorado, September, 1980.

Fontane, D. G., W. A. Hall, and J. W. Labadie. Optimal Integration of Energy and Water Development Plans, paper presented at the Operations Research Society of America/The Institute for Management Sciences Meeting, November 10-12, 1980, Colorado Springs, Colorado.

Hall, W. A. and J. A. Dracup. Water Resources Systems Engineering, McGraw-Hill, 1970.

Hall, W. A. Multiple Reservoir Systems for Water Supply Planning, Lecture presented at the Application of Dynamic Programming to Water Resources Management Short Course, June 23-27, 1980, Colorado State University, Fort Collins, Colorado.

Missouri River Basin Commission. Forecasts for Western Coal/Energy Development, Resource and Land Investigations Program, U.S. Geological Survey, Reston, Virginia, January, 1979.

Morse, J. G. Energy Resources in Colorado: Coal, Oil, Shale, and Uranium, Westview Press, Boulder, Colorado, 1979.

Steele, T. D., *et al.* The Yampa River Basin, Colorado and Wyoming-A Review to Expanded Coal-Resource Development and Its Impact on Regional Water Resources, Water Resources Investigations 78-126, U.S. Geological Survey, Lakewood, Colorado, September, 1979.

U.S. Army Corps of Engineers. HEC-3 Reservoir System Analysis for Conservation – User's Manual, Hydrologic Engineering Center, Davis, California, July, 1974.

TABLE 2. Output of the Screening Model for Case 1: Baseline.

Level of Annual Firm Demand (ac.-ft.)	Required Firm Yield (ac.-ft.) at Reservoir Sites					Minimized Total System Costs (Million $)
	1 Site # 01	2 Site # 05	3 Site # 10	4 Site # 11	5 Site # 15	
20,000	0	20,000	0	0	0	9.7
40,000	0	40,000	0	0	0	12.7
60,000	0	60,000	0	0	0	14.9
80,000	0	80,000	0	0	0	16.1
100,000	15,000	85,000	0	0	0	27.9
120,000	15,000	85,000	0	20,000	0	44.5
140,000	20,000	120,000	0	0	0	67.8
160,000	20,000	120,000	0	20,000	0	86.0
180,000	20,000	130,000	0	20,000	10,000	117.0

TABLE 3. Output of the Screening Model for Case 2: Modified-Maximum Yield Limited at Pleasant Valley Reservoir.

Level of Annual Firm Demand (ac.-ft.)	Required Firm Yield (ac.-ft.) at Reservoir Sites					Minimized Total System Costs (Million $)
	1 Site # 01	2 Site # 05	3 Site # 10	4 Site # 11	5 Site # 15	
85,000	0	85,000	0	0	0	16.2
100,000	15,000	85,000	0	0	0	27.9
120,000	15,000	85,000	0	20,000	0	44.5
140,000	20,000	85,000	5,000	20,000	10,000	76.4

TABLE 4. Output of the Screening Model for Case 3: Specified Minimum Yields.

Level of Annual Firm Demand (ac.-ft.)	Required Firm Yield (ac.-ft.) at Reservoir Sites					Minimized Total System Costs (Million $)
	1 Site # 01	2 Site # 05	3 Site # 10	4 Site # 11	5 Site # 15	
25,000	10,000	0	0	15,000	0	25.0
40,000	20,000	0	0	20,000	0	31.8
60,000	10,000	35,000	0	15,000	0	36.9
80,000	10,000	55,000	0	15,000	0	39.4
100,000	10,000	75,000	0	15,000	0	40.9
120,000	15,000	85,000	0	20,000	0	44.5
140,000	15,000	105,000	0	20,000	0	70.7
160,000	20,000	120,000	0	20,000	0	86.0
180,000	20,000	130,000	0	20,000	10,000	117.0

ECONOMIC GAINS FROM PROBABILISTIC TEMPERATURE FORECASTS

Mark G. Alexandridis and Roman Krzysztofowicz[1]

ABSTRACT: Probabilistic and categorical daily temperature forecasts are evaluated in the context of power load scheduling by an electric utility. Two decision procedures are considered: an optimal one which accounts for forecast uncertainty and a suboptimal one which does not. Probabilistic forecasts are found superior to categorical forecasts because they: (1) result in higher economic gains for both decision procedures, and (2) apprise users that forecasts are inherently uncertain. This latter attribute may impel the sub-optimal users to refine their decision procedures and, consequently, reap still higher economic gains.
(KEY TERMS: decision theory; power load scheduling; forecast evaluation; probabilistic forecasts.)

INTRODUCTION

It is widely held in meteorologic literature that probabilistic forecasts are more valuable than categorical forecasts and these are more valuable than climatological forecasts (Thompson, 1952; Glahn, 1964; Winkler and Murphy, 1979), where value is measured by the difference between the expected loss incurred by the decision maker in the absence of a forecast and with a forecast. This hypothesis implicitly assumes that the forecasts are used optimally by the decision maker, i.e., he accounts for the intrinsic uncertainty in the forecasts. This is seldom done in practice. More conventional, nonoptimal, decision procedures do not take forecast uncertainty into account but utilize the forecasts as if they were error-free.

This paper reports the results of an *ex ante* evaluation of categorical and probabilistic temperature forecasts for both optimal and nonoptimal decision procedures. The error statistics of the forecasts are taken from experiments conducted by Murphy and Winkler (1975) at the National Weather Service Office in Denver. These experiments were designed to test the ability of the forecasters to quantify the uncertainty in their forecasts. By all statistical measures, the categorical forecasts were quite accurate, and probabilistic forecasts were remarkably well calibrated. From the viewpoint of a decision maker, the key questions are: (1) how to optimally use categorical or probabilistic forecasts, and (2) what economic gains are to be expected from such forecasts?

A Bayesian decision methodology for the optimal use of forecasts in static decision problems is described elsewhere (Alexandridis and Krzysztofowicz, 1982). This paper discusses the expected gains. These gains are evaluated in the context of the daily average power load scheduling by an electric utility.

FORECASTING SCHEMES AND DECISION PROCEDURES

Decision Problem

Every day the operator of an electric utility makes a decision a as to the average power to be generated on the following day. Associated with the decision a is an economic loss incurred when the actual power demanded is ω:

$$\ell(\omega,a) = \begin{cases} c_o(\omega - a)^2 & \text{if } \omega \leqslant a, \\ c_u(\omega - a)^2 & \text{if } \omega \geqslant a. \end{cases}$$

The utility suffers losses for both generating more or less power than that which is actually demanded.

The forecasting of short-term (less than 24 hours) power load is typically bifurcated into weather sensitive and insensitive components (Gupta and Yamada, 1972; Corpening, *et al.*, 1973). Only the former component is of concern herein. Furthermore, it is assumed that given perfect knowledge of tomorrow's actual average temperature θ, one knows precisely the average power load demanded ω. This relationship, $\omega = \psi(\theta)$, is defined by:

[1] Respectively, Research Assistant and Assistant Professor, Massachusetts Institute of Technology, Department of Civil Engineering, Cambridge, Massachusetts 02139.

$$\omega = \begin{cases} AM & \text{if } \theta < t_a \text{ or } t_d < \theta \\ B + a_h(\theta - t_b) & \text{if } t_a \leqslant \theta < t_b \\ B & \text{if } t_b \leqslant \theta \leqslant t_c \\ B + a_c(\theta - t_c) & \text{if } t_c < \theta \leqslant t_d \end{cases}$$

where: AM – maximum load [MW]; B – weather insensitive daily average base load [MW]; a_h, a_c – heating and cooling coefficients [MW/^{O}F]; t_a,t_b,t_c,t_d – reference temperatures [^{O}F].

Naive and Perfect Forecasts

Naive and perfect forecasts are considered first to establish an upper bound on the value of any real-time forecast. A naive, or climatological, forecast is simply the probability distribution of θ estimated on the basis of a climatological record. Let RN denote the minimum expected loss (risk) under this distribution:

$$RN = \min_a E[\ell(\psi(\theta),a)]$$

If perfect forecasts of θ were available, then the risk would be:

$$RF = E[\min_a \ell(\psi(\theta),a)]$$

Categorical Forecasts

The decision maker is supplied, on a daily basis, with a categorical (point) forecast t of θ. This forecast is uncertain. To optimally account for the possible forecast error, the forecast has to be processed through Bayes theorem (Alexandridis and Krzysztofowicz, 1982), and the optimal decision should minimize the expected loss under the posterior distribution of the state θ given the forecast t. Let RC denote the associated risk.

More often than not, decision makers are unaware of the normative guides to optimal integration of probabilistic information and decision making (Slovic, 1981). They ignore uncertainty or employ simple, suboptimal, heuristics to synthesize probabilistic information before it is integrated with the decision consequences. We consider a suboptimal decision procedure which disregards uncertainty, or error, in the forecast. That is, the decision maker instead of minimizing the posterior expected loss, minimizes the loss as if the forecast were perfect. Let the associated risk be $\overline{RC}$. Clearly $\overline{RC} \geqslant RC$.

Probabilistic Forecasts

In their experiment, Murphy and Winkler (1975) tested two probability elicitation techniques: fixed-width, variable-probability, and variable-width, fixed-probability. The latter technique proved to be more reliable, and its results are evaluated herein. The forecasts are constructed by having the forecaster first assess the median t of the state of nature θ, and then assess the widths t_{50} and t_{75} of the intervals within which he expects the state θ to fall with probabilities .50 and .75. If the credible intervals t_{50} and t_{75} are centered about the median t, then the vector $\underline{\mu} = (t,t_{50},t_{75})$ uniquely defines five points on the distribution function of θ. This distribution encodes the forecaster's uncertainty about the true value of θ. The smaller his uncertainty, the narrower the credible intervals, and, consequently, the more precise forecast. (Precision is defined as the inverse of the variance.)

A measure of the reliability of probabilistic forecasts is their calibration: the forecasted frequency of θ should match the observed frequency. Again, to account for imperfect calibration, the probabilistic forecast $\underline{\mu}$ should be processed through Bayes theorem, and the optimal decision should minimize the expected loss under the posterior distribution of θ given the forecast $\underline{\mu}$. Let RP denote the associated risk. If a decision maker presumes perfect calibration and minimizes the expected loss under the forecast distribution of θ, then his risk $\overline{RP}$ is larger than RP.

The risks defined above are summarized in Table 1. One can prove (Alexandridis and Krzysztofowicz, 1982) that the following orders hold: $RF \leqslant RP \leqslant RC \leqslant RN$, $RC \leqslant \overline{RC}$, $RP \leqslant \overline{RP}$.

PERFORMANCE MEASURES

A forecasting scheme coupled with a decision procedure forms a system. Four systems, defined in Table 2, are evaluated. The expected gains in a given system depend upon both the quality of the forecasts and the degree of optimality of the decisions. To capture this interplay, three measures of performance are employed.

Values. The values are defined in Table 2. The potential value (PV) represents an upper bound on the value of any forecast for a given decision problem. The optimal value (OV) represents the value of a given real-time forecast when it is used in an optimal decision procedure. The actual value (AV) represents the value of a given real-time forecast when it is used in a given decision procedure; if this procedure is optimal, then obviously AV = OV. Generally, $AV \leqslant OV \leqslant PV$.

Efficiencies. The generic definitions of the forecast (FE), decision (DE), and total (TE) efficiencies are given by:

$$FE = \frac{OV}{PV}; DE = \frac{AV}{OV}; TE = \frac{AV}{PV} = FE \cdot DE$$

FE is an indicator as to how skillful the actual forecasting scheme is relative to a perfect scheme. DE is a measure of how inept the given decision procedure is when compared to an optimal one. TE is a composite metric of the actual total system performance relative to potential performance.

TABLE 1. Risks of Forecasting Schemes and Decision Procedures.

Forecast	Decision: Optimal	Nonoptimal
Perfect	RF	
Probabilistic	RP	$\overline{RP}$
Categorical	RC	$\overline{RC}$
Naive (Climatological)	RN	

TABLE 2. Values of the Forecast-Decision Systems.

Values	System: Categorical F: Optimal D 1	Nonoptimal D 2	Probabilistic F: Optimal D 3	Nonoptimal D 4
potential PV	RN – RF			
optimal OV	RN – RC		RN – RP	
actual AV	RN – RC	RN – $\overline{RC}$	RN – RP	RN – $\overline{RP}$

F – Forecasting scheme
D – Decision procedure

Expected opportunity losses. The generic definitions of the forecast (FOL), decision (DOL), and total (TOL) expected opportunity losses are:

$$FOL = PV - OV \ ; DOL = OV - AV \ ;$$
$$TOL = PV - AV = FOL + DOL$$

FOL represents the value of replacing the existing forecasting scheme by a perfect one. DOL is the value of replacing the actual decision procedure by an optimal one. The value of supplanting the current forecast-decision system by one in which the forecasts are perfect and decision procedure optimal is given by TOL.

CASE STUDY

Input Data

The cost coefficients were: $c_o = 10$ and $c_u = 20$ [\$/MW^2]. The parameters of the load operator were: AM – B = 300 [MW]; $a_h = -10$, $a_c = 15$ [MW/^oF]; $t_a = 20$, $t_b = 50$, $t_c = 70$, $t_d = 90$ [^oF]. The climatological distribution of θ was forged from nine years of daily average temperature data for the month of April. The statistics of the forecast errors and calibration were those reported by Murphy and Winkler (1975, p. 4 and 7). The evaluation results are shown in Table 3.

TABLE 3. Comparative Evaluation of Four Forecast-Decision Systems.

Performance Measure	System: Categorical F: Optimal D 1	Nonoptimal D 2	Probabilistic F: Optimal D 3	Nonoptimal D 4
Expected Daily Loss [10^3 \$/day]	RC = 12.3	$\overline{RC}$ = 22.3	RP = 11.7	$\overline{RP}$ = 16.6
Values [10^3 \$/day]				
potential PV	38.9	38.9	38.9	38.9
optimal OV	26.6	26.6	27.2	27.2
actual AV	26.6	16.6	27.2	22.3
Efficiencies				
forecast FE	0.68	0.68	0.70	0.70
decision DE	1.00	0.62	1.00	0.82
total TE	0.68	0.43	0.70	0.57
Expected Opportunity Losses [10^3 \$/day]				
forecast FOL	12.3	12.3	11.7	11.7
decision DOL	0	10.0	0	4.9
total TOL	12.3	22.3	11.7	16.6

RN = 38.9 [10^3 \$/day]
RF = 0 [10^3 \$/day]
F – Forecasting scheme
D – Decision procedure

Results

1. When both categorical and probabilistic forecasts are used optimally, their values are AV(1) = 26.6 and AV(3) = 27.2 [10^3 \$/day]. If the forecasts were perfect, they would be worth PV = 38.9 [10^3 \$/day], or roughly 14.2 million dollars per year. The forecast expected opportunity loss, FOL, constitutes an upper bound on the economically justified investments towards improving the quality of the forecasts. Thus, a decision maker already receiving probabilistic forecasts should spend no more than FOL(3) = 11.7 [10^3 \$/day] for a prospective forecast enhancement.

2. High efficiencies, FE(1) = 68% and FE(3) = 70%, indicate that both types of forecasts are of good quality and are much more valuable than naive forecasts which are based solely on a climatological record (efficiency of a naive forecast is zero).

3. Higher efficiency of probabilistic forecasts, by FE(3) – FE(1) = 2%, indicates that the forecasters were able to quantify uncertainty in their forecasts with a degree of reliability that has economic impact. The expected gain from using probabilistic forecasts instead of categorical forecasts is AV(3) – AV(1) = .6 [10^3 \$/day], or about

240 thousand dollars annually. Considering the fact that the issuance of subjective probabilistic forecasts does not require any new hardware but only additional skill and effort on the part of a forecaster, the expected net gain would be positive (even after subtracting an extra remuneration for a forecaster).

4. When both categorical and probabilistic forecasts are used nonoptimally, their values are AV(2) = 16.6 and AV(4) = 22.3 [10^3 \$/day]. The efficiencies of the nonoptimal decision procedures are DE(2) = 62% and DE(4) = 82%, while the expected opportunity losses incurred by the decision makers due to the nonoptimal use of the forecasts are DOL(2) = 10.0 and DOL(4) = 4.9 [10^3 \$/day], or 3.6 and 1.8 million dollars per year. The decision makers could economically justify investments up to these amounts for improving their decision procedures.

5. The value of the optimally used categorical forecasts, AV(1), is larger than the value of the nonoptimally used probabilistic forecasts, AV(4). This belies the prevailing sentiment that more information is necessarily better than less. Obviously, this is only true when the information is processed and used optimally.

6. When both types of forecasts are used nonoptimally, then probabilistic forecasts are TE(4) – TE(2) = 14% more valuable than categorical forecasts. The corresponding expected gain from using probabilistic instead of categorical forecasts is AV(4) – AV(2) = 5.7 [10^3 \$/day], or roughly 2.1 million dollars annually. That is, even if the decision maker does not account for the possible imperfect calibration of probabilistic forecasts, he is much better-off using probabilistic forecasts than using categorical forecasts under the (false) presumption that they are error-free.

CONCLUSIONS

1. Probabilistic forecasts are more valuable than categorical forecasts when used in the same (optimal or suboptimal) decision procedure.

2. The potential gains from probabilistic forecasts are relatively greater for suboptimal decision procedures (which ignore forecast uncertainty) than for the optimal ones (which account for forecast uncertainty).

3. In view of the mounting evidence compiled by behavioral scientists that people tend to ignore uncertainties, particularly those not directly apparent or pointed out explicitly, probabilistic forecasts should be considered superior to categorical forecasts because they do convey, explicitly and quantitatively, the information about uncertainty, and because this information can be transformed into tangible gains by economically motivated decision makers even if the decision procedures they employ are suboptimal.

ACKNOWLEDGMENTS

This material is based upon work supported by the National Science Foundation under Grant No. CEE-8107204.

LITERATURE CITED

Alexandridis, M. G. and R. Krzysztofowicz, 1982. Value of Categorical and Probabilistic Temperature Forecasts for Scheduling of Power Generation. Technical Report No. 282, Ralph M. Parsons Laboratory, Massachusetts Institute of Technology, Cambridge, Massachusetts, May.

Corpening, S. L., N. D. Reppen, and R. J. Ringlee, 1973. Experience with Weather Sensitive Load Models for Short and Long Term Forecasting. IEEE Trans., PAS-92(6):1966-1972.

Glahn, H. R., 1964. The Use of Decision Theory in Meteorology. Monthly Weather Review, 92(2):383-388.

Gupta, P. C. and K. Yamada, 1972. Adaptive Short Term Forecasting of Hourly Loads Using Weather Information. IEEE Trans., PAS-91(5):2085-2094.

Murphy, A. H. and R. L. Winkler, 1975. The Use of Credible Intervals in Temperature Forecasting: Some Experimental Results. IIASA, Rm-75-70, Laxenburg, Austria.

Slovic, P., 1981. Toward Understanding and Improving Decisions. Proceedings, International Symposium on Real-Time Operation of Hydrosystems, University of Waterloo, Waterloo, Ontario, Canada, June, pp. 426-450.

Thompson, J. C., 1952. On the Operational Deficiencies in Categorical Weather Forecasts. Bulletin of the American Meteorological Society, 33(6):223-226.

Winkler, R. L. and A. H. Murphy, 1979. The Value of Weather Forecasts in the Cost-Loss Ratio Situation: An Ex Ante Approach. Proceedings, Sixth Conference on Probability and Statistics, American Meteorological Society, pp. 134-138.

TRENDS AND EXTREMES OF LAKE ERIE WATER SUPPLIES

Frank H. Quinn[1]

ABSTRACT: Lake Erie is the fourth largest of the Great Lakes, with a surface area of 25,700 km^2 and a land basin area of 61,100 km^2. Its water is derived from the upper Great Lakes and its own net basin supply consisting of precipitation, runoff, and lake evaporation. In this study, a data base consisting of monthly time series of air temperature, precipitation, runoff, and lake evaporation is used to examine trends and extremes in the net basin supply. This averages about 62 cm, with extremes ranging from 0.0 cm in 1934 to 145 cm in 1950. Of particular note is an apparent major change in the supplies occurring during the late 1930's. The average annual basin supplies for 1940-79 were approximately 50 percent greater than for 1900-39. A monthly climatic water balance model was developed and calibrated to analyze the impact of changes in air temperature and precipitation regimes on the basin runoff. Particularly emphasized in the study are the conditions causing an increase in water supplies during the present regime, the record low Lake Erie water levels of 1964, and the extreme high levels of 1952 and the early 1970's. Future applications of the model are also discussed.
(KEY TERMS: Great Lakes; Lake Erie; water balance; water supply.)

INTRODUCTION

Lake Erie (Figure 1) is the fourth largest of the Great Lakes, with a surface area of 25,700 km^2 and a land basin area of 61,100 km^2. Throughout the 1970's and continuing into the 1980's, the lake has received excessive water supplies, resulting in extremely high lake levels. During late 1972 and 1973, record monthly mean water levels were set for all 12 months. Since 1966 the annual mean lake level has continually exceeded the 1900-75 long term average. Major low supply periods occurred during the mid-1930's and the mid-1960's, with record low monthly mean lake levels set in the mid-1930's.

This study was undertaken to analyze the reasons for the current high supply regime and to put it in historical perspective. The water supply for Lake Erie is derived from the upper Great Lakes and its own net basin supply, consisting of precipitation, runoff, and lake evaporation. The upstream water transported through the Detroit River averages about 640 cm on the lake, while the net basin supply averages about 62 cm. The outflow from the lake through the Niagara River averages about 690 cm. With relatively stable flow relationships between the inflow and outflow channels, major changes in the net basin supply can result in major changes in lake levels. In this study, a data base consisting of monthly time series of air temperature, precipitation, runoff, and lake evaporation is used to examine trends and extremes in the net basin supply. The runoff time series prior to 1937 is generated from a monthly climatic water balance model based on the Thornwaite procedure.

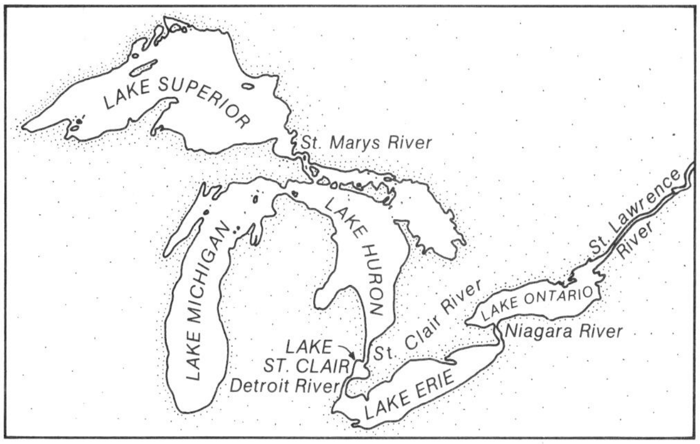

Figure 1. The Great Lakes.

BASIC DATA

Long-term time series for Lake Erie data going back to 1900 consist of recorded air temperatures from Buffalo, New York, Cleveland and Toledo, Ohio, and London, Ontario, computed overbasin and overlake precipitation, Niagara and Detroit River flows, and changes in lake storage. The 1900-30 precipitation values were taken from the National Ocean Survey (1980), while the 1931-1979 data were recently computed by Quinn and Norton (1982) using the grid-square technique. The Niagara and Detroit River flows are from the Coordinating Committee (1976, 1982) and the changes in lake storage obtained from Quinn and Derecki

[1]Head, Lake Hydrology Group, Great Lakes Environmental Research Laboratory/NOAA, 2300 Washtenaw Avenue, Ann Arbor, Michigan 48104.

(1976). Data available from 1937 to present consists of lake runoff (Derecki, 1976; Quinn and Kelley, 1982) and monthly lake evaporation values (Derecki, 1976, as modified; Derecki, 1981).

LAKE ERIE NET BASIN SUPPLIES

The net basin supply to Lake Erie is defined as

$$NBS = P + R - E \qquad (1)$$

where:

NBS = the net basin supply,

P = the precipitation,

R = the runoff into the lake, and

E = the evaporation from the lake surface.

Until recently, the net basin supplies were calculated indirectly from the inflow and outflow through the Detroit and Niagara Rivers, respectively, and from the monthly rates of change in lake storage. The complete hydrologic balance for Lake Erie can be expressed as

$$P + R + Q_D = E + Q_N + D_W \pm \Delta S \qquad (2)$$

where:

Q_D = the Detroit River flow,

Q_N = the Niagara River flow,

D_W = the diversion through the Welland Canal, and

ΔS = the rate of change of lake storage.

Equations (1) and (2) can be combined to yield

$$NBS = Q_N - Q_D + D_W \pm \Delta S \qquad (3)$$

The latter equation has been used until recent times because of the availability of the required data. The major disadvantage is that, with an average annual NBS of 62 cm on the lake, even small errors in the Detroit or Niagara River flows, averaging about 650 cm, can cause major errors. As an example, a 5 percent error in either of the flow rates result in an approximate error of 50 percent in the NBS.

With Equation (1) the magnitudes of each of the components are approximately equal, yielding equivalent errors in the NBS for errors in either the precipitation evaporation or runoff. However, lake evaporation and runoff data are only available subsequent to 1937, limiting the usefulness of Equation (1) to the last 43 years.

The annual time series of NBS calculated from Equation (1) is shown in Figure 2. A five-year binomially derived weighted moving average with weights of 0.06, 0.25, 0.38, 0.25, and 0.06 from Panofsky and Brier (1968) was used to filter the data. This eliminates short period oscillations in the time series. Figure 2 shows the existence of two distinct regimes, a low regime lasting from 1900 to approximately 1939 and a high regime lasting from 1940 to 1979. The latter regime indicates a 48 percent increase in the net basin supply, with a corresponding 9 percent increase in precipitation. The figure also shows two major high supply periods, the early 1950's and the present time. These regimes also correspond to changes in the Great Lakes precipitation regime as reported by Quinn (1981). Record high lake levels were set on Lake Erie during 1972 and 1973. Major low supply periods occurred during the mid-1960's, the mid-1930's, and around 1918. A comparison of the NBS computed from Equation (1) for 1937-79 shows a mean of 98 cm, with a standard deviation of 31 cm. The corresponding values from Equation (3) are a mean of 73 cm and a standard deviation of 32 cm. The difference of 25 cm in the mean is equivalent to about 3.5 percent of the Detroit or Niagara River flow.

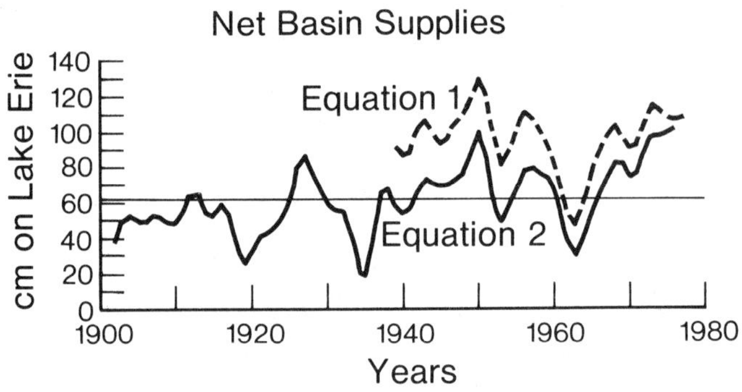

Figure 2. Net Basin Supplies.

LAKE ERIE WATER BALANCE MODEL

Since the change in precipitation between the two regimes was approximately 9 percent, the remainder of the apparent increase in NBS must be accounted for by either a major increase in runoff, a large decrease in evaporation, or both. Since runoff data are not available prior to 1937, monthly climatic water balance models based upon the Thornwaite procedure (Thornwaite and Mather, 1957) were developed and calibrated for both the United States and Canadian land basins. If successful, the models can be used to compute the annual runoff prior to 1937 and to analyze the impact of changes in air temperature and precipitation regimes on the basin runoff.

Sanderson (1966) demonstrated the application of the Thornwaite procedure in a climatic water balance study of

Lake Erie for 1958-63. The models for the United States and Canadian basins were calibrated with the 1960-79 data and verified over 1937-59. The combined models had a rms error of 5 cm (7 percent) in the runoff for 1937-79 and a correlation coefficient between model and measured annual runoff in excess of 92 percent. The input precipitation and air temperature data used to drive the model, computed runoff, potential evapotranspiration, and combined precipitation and runoff are shown in Figure 3.

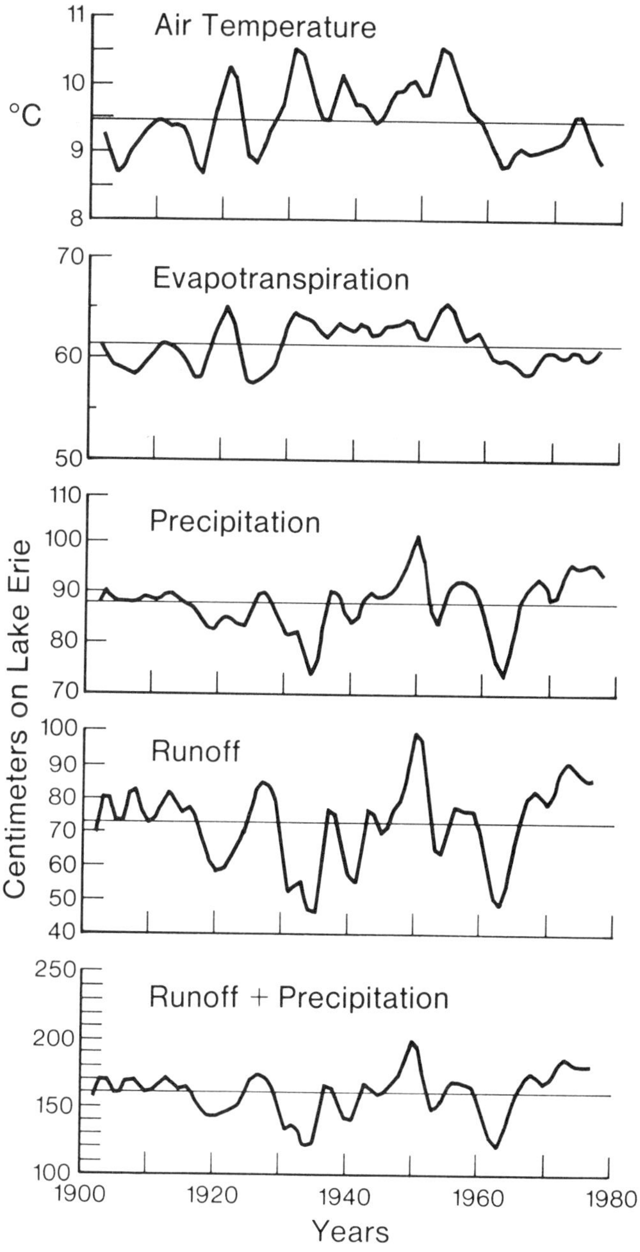

Figure 3. Hydrologic Variables.

The air temperature time series, comprised of the mean of the Buffalo, Cleveland, and Toledo data, indicates three separate regimes. A relatively warm regime between the late 1920's and late 1950's was preceded and followed by relatively cold regimes. The difference in annual average temperature between the high and low periods is approximately 1°C. The model indicates that a temperature decrease of 1°C increases the runoff by approximately 12 percent owing to decreased evapotranspiration. This results in an increase of about 9 percent in the NBS. However, the cooler air temperature also appears to result in increased evaporation from the lake surface, thus partially compensating for the increased runoff. The combined precipitation and runoff time series indicates that the last 11 years have constituted one of themost prolonged wet periods since 1880 and that there is no indication of an early change to dryer conditions.

Figure 3 also indicates above average combined precipitation and runoff during the first 15 years of record, thus casting doubt on the computed values of the NBS.

CONCLUSIONS

The primary cause of the recent high supplies since 1966 is an increase in precipitation averaging 6 percent over the long-term average. This has resulted in a 16 percent increase in runoff. Also contributing to increased runoff has been a 0.3°C decrease in the average air temperature for this period. Tests with the model have indicated that, in general, a 0.5°C decrease in temperature will increase runoff by 6 percent. These increases have been offset by an apparent increase in lake evaporation produced by the cooler temperatures and perhaps a different circulation pattern. By contrast, the net basin supply, as computed from the connecting channel flows, shows an increase in the net basin supply of 42 percent over the long-term average. This would require a long-term average evaporation of 86 cm, approximately 20 percent above the 1966-79 average. This seems highly unlikely. Thus the long-term average NBS is probably in error and the extremely low supplies in the early years of the century are probably more apparent than real. The period 1900-10 had the same average annual air temperature as the recent 1966-79 period. Assuming the same evaporation in both periods yields a NBS of about 89 cm for the early period compared with 105 cm in the latter period, an increase of 18 percent. The corresponding values computed using the connecting channel flows are 45 cm and 88 cm, respectively, an increase of 96 percent. These differences in the NBS are extremely important because the values from 1900 to date are used in the development of the Great Lakes regulation plans and water resource studies.

The low supplies in 1936 were due to 7 years of below average precipitation, while lower supplies in 1962-64 resulted from 5 years of below average precipitation. In the case of the 1930's, the drought was also aggrevated by 4 years of above average air temperatures.

The outlook for the future is for continued high supplies to the Lake Erie basin. The current high precipitation regime, in place since about 1940, is continuing, coupled with an ongoing cooling trend, which results in less evapotranspiration and high runoff.

This study also demonstrates the usefulness of the climatic water balance model for large drainage basins with desired temporal scales of 1 month and greater. The model provides the capability to fill in runoff data gaps and also to access quantitatively the impact of climatic changes and trends. Thus, it is amenable to both simulation and forecast operations.

LITERATURE CITED

Coordinating Committee on Great Lakes Basic Hydraulic and Hydrologic Data, 1976. Lake Erie Outflow 1806-1964 with Addendum 1965-1975. Report, Coordinating Committee on Great Lakes Basic Hydraulic and Hydrologic Data, Chicago, Illinois, and Cornwall, Ontario, June, 28 pp., 15 plates, 16 tables.

Coordinating Committee on Great Lakes Basic Hydraulic and Hydrologic Data, 1982. Lakes Michigan-Huron Outflows, St. Clair and Detroit Rivers, 1900-1978. Coordinating Committee on Great Lakes Basic Hydraulic and Hydrologic Data, Chicago, Illinois, and Cornwall, Ontario (in press).

Derecki, J. A., 1976. Evaporation from Lake Erie. NOAA Technical Report ERL 342-GLERL 3. National Technical Information Service, Springfield, Virginia, 89 pp.

Derecki, J. A., 1981. Stability Effects on Great Lakes Evaporation. Journal of Great Lakes Research 7(4):357-362.

National Ocean Survey, 1980. Great Lakes Precipitation Bulletin, Report. National Oceanic and Atmospheric Administration, National Ocean Survey, Rockville, Maryland.

Panofsky, H. A. and G. W. Brier, 1968. Some Applications of Statistics to Meteorology Report. Pennsylvania State University, University Park, Pennsylvania, 224 pp.

Quinn, F. H., 1981. Secular Changes in Annual and Seasonal Great Lakes Precipitation, 1854-1979, and Their Implications for Great Lakes Water Resource Studies. Water Resources Research 17: 1619-1624.

Quinn, F. H. and J. A. Derecki, 1976. Lake Erie Beginning-of-Month Water Levels and Monthly Rates of Change of Storage. NOAA Technical Report ERL 364-GLERL 9. National Technical Information Service, Springfield, Virginia, 37 pp.

Quinn, F. H. and R. N. Kelley, 1982. Great Lakes Hydrologic Data Base. Unpublished Data Base, Great Lakes Environmental Research Laboratory, Ann Arbor, Michigan.

Quinn, F. H. and D. C. Norton, 1982. Great Lakes Precipitation by Months. NOAA Data Report (in press).

Sanderson, M., 1966. A Climatic Water Balance of the Lake Erie Basin, 1958-1963. *In:* Publications in Climatology, Vol. XIX, No. 1. C. W. Thornwaite Associates, Laboratory of Climatology, Elmer, New Jersey, 87 pp.

Thornwaite, C. W. and J. R. Mather, 1957. Instructions and Tables for Computing Potential Evapotranspiration and the Water Balance. Drexel Institute of Technology, Laboratory of Climatology, Centerton, New Jersey, 243 pp.

Section 4

LONG-TERM PLANNING FOR WATER QUALITY

DURATION OF STREAM TEMPERATURE INCREASES FOLLOWING FOREST CUTTING IN THE SOUTHERN APPALACHIAN MOUNTAINS

Lloyd W. Swift, Jr.[1]

ABSTRACT: Cutting timber along small streams in the southern Appalachian Mountains increases water temperature. Although the effect is assumed to be short-lived because of rapid regrowth of dense vegetation, timber harvest in streamside shade zones on National Forest land is carefully regulated to minimize temperature changes and maintain trout habitat. Clearcutting all vegetation over 2.5 cm DBH from a 59.6-ha south-facing watershed in western North Carolina allowed both the magnitude and duration of water temperature increases to be studied. About 958 m^2 of stream were exposed. Daily maximum temperatures at the downstream margin of the cutting were increased an average of 3.3°C the first two summers after cutting. The increases declined in the next three summers to 1.2°C. Daily minimums were increased about 1.3°C both winter and summer, but only in the first year. The daily range of water temperatures (maximum minus minimum) was increased during all five summers. A method for predicting water temperature changes was tested and found to overestimate the summer increases.
(KEY TERMS: water temperature; stream temperature; thermal pollution; forest cutting.)

INTRODUCTION

Wooldridge and Stern (1979) and Patton (1973) have reviewed the literature and summarized the impacts of forest cutting upon stream temperature. Removal of forest cover from cold-water streams has repeatedly been shown to increase summer water temperatures. Only a few studies have reported changes in other seasons, changes in minimum temperature and the daily temperature range or studied how long temperature effects last. The duration and magnitude of temperature increases are limited by rapid regrowth of forest vegetation typical for the mountains of the eastern United States. For example, summer maximum water temperatures were less under an 8-year-old regrowing hardwood forest in western North Carolina than under an uncut stand (Swift and Messer, 1971). Even in the first year after cutting, buffer strips or streamside shade zones reduce or eliminate temperature increases (Swift and Baker, 1973; Burton and Likens, 1973; Brown and Krygier, 1970).

This study describes changes, after clearcutting, of both minimum and maximum water temperatures in all seasons and during five years of forest regrowth.

THE EXPERIMENT

A cable logging experiment on a 59.6-ha watershed at the Coweeta Hydrologic Laboratory began in spring 1977. The climate of this south-facing slope in the Appalachian Mountains of North Carolina is characterized by high precipitation (1,770 mm/year) and moderate temperatures (12.8°C annual mean). The elevation of the watershed ranges from 720 m at the V-notch weir to 1,065 m. The total length of flowing stream is 2,052 m, 55 percent of which is in the central channel. The maximum stream width at a pool 36 m above the weir is 2.4 m.

The vegetation on this experimental basin had been undisturbed since a woodland grazing demonstration 28 years earlier. Before logging, the basal area was 25.3 m^2/ha on 3,058 stems/ha > 2.5 cm (Boring, *et al.*, 1981). During logging, essentially all living vegetation was cut. Clearcutting adjacent to all branches of the stream was completed by late August 1977. Debris was removed from the lower two-thirds of the main stream channel as a condition of the timber sale contract. All other logging slash was left where it fell. Approximately 38 percent of the 1,535 m^2 of stream surface was shaded, mainly by dead logging debris bridging the narrow channel. Minor areas of stream covered by overhanging banks or road culverts were included in the shaded percent but topographic shading was judged to be negligible at midday.

Regeneration of vegetation was mainly stump and root sprouts although approximately 34 percent of the first year total biomass was in herbaceous plants and vines (Boring, *et al.*, 1981). Herbaceous plants are not a significant part of the mature forest but play an important role in providing the initial shade cover of a regrowing forest. Most woody vegetation was distributed as scattered sprout clumps. In fall 1977, the year of cutting, the leaf biomass of streamside vegetation was only 16 percent of that for the precut forest. At the end of the second and third growing seasons (1978 and 1979), the streamside leaf biomass had reached 60 and 78 percent of the precut amount and by the end of

[1] Research Meteorologist, Coweeta Hydrologic Laboratory, Southeastern Forest Experiment Station, Forest Service, USDA, Route 1, Box 216, Otto, North Carolina 28763.

the 1981 growing season, only about 100 m of the wider stream channel above the weir remained unshaded.

Temperature recorders were installed just above the weir of the logged watershed and on an adjacent uncut stream. Measurements began three years before the forest was cut and are continuing. The measurement sites have similar upstream drainage areas and stream lengths. Data were recorded by remote bulb recording thermographs with 7-day charts. The sensing bulbs were immersed in flowing water and shaded from direct sunlight. The chart trace was checked for accuracy with a thermometer observation each time a chart was changed. Air temperature was recorded at climatic stations within and adjacent to the logged watershed. Daily minimum and maximum temperatures were picked from the charts to the nearest 0.6°C. Temperatures of the streams were similar before logging. The daily difference (experimental minus control) between maximums for the two streams and the daily difference between minimums were calculated for the period of record. Shifts in magnitude of between-stream differences for either maximum or minimum temperatures are taken to be measures of changes in the temperature of the experimental stream. Monthly means of these daily differences were used in this analysis.

RESULTS

During the 34 months before cutting, both the maximum and minimum temperatures were always warmer in the experimental stream and temperature differences between streams varied seasonally (Figure 1). Small differences of less than 1°C occurred in late summer and fall of 1974 and 1975, but temperatures were 1.0° to 2.4° warmer in the experimental stream in spring 1975 and spring and summer 1976. Year-to-year fluctuations in air temperature were small and could not explain variations in stream temperature between the three precutting years. The average monthly difference across the precut period was +0.9°C for the daily minimum and +1.5°C for the daily maximum (standard error < 0.18°C).

After the forest was cut, monthly differences between maximum water temperatures were consistently larger suggesting that maximum temperatures were increased in all seasons for the duration of the study. Differences between minimum temperatures increased during the first two years and then declined.

The largest differences in maximum temperatures occurred in June 1978 and was 5.4°C greater than the mean difference of precut months. The average increase over pretreatment maximum temperatures was 3.3°C during the first 17 months after cutting. The peak increase in minimum temperature difference occurred immediately after cutting in August 1977; the average increase was 1.3°C for the 17-month period. Both minimum and maximum temperatures of the stream were also raised after cutting.

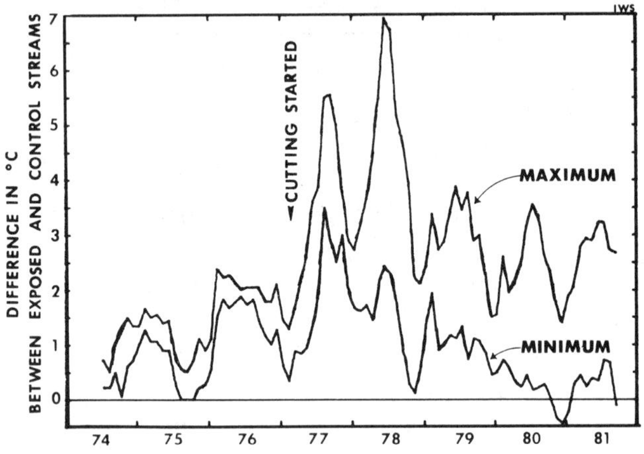

Figure 1. Monthly Mean Differences for Minimum and Maximum Water Temperatures Between the Experimental and Control Streams.

After the 1979 growing season, both differences were less. The difference between monthly minimums declined to an average of 0.4°C below pretreatment levels for the last three years, whereas, maximum differences averaged 1.2°C above pretreatment levels.

The sustained increase in maximum water temperature differences, in contrast to the declining minimums (Figure 1), suggests that the daily range of water temperature was dramatically and consistently increased after cutting. The temperature recordings confirm that the range was increased by 3° to 4°C in summer and fall 1978. Smaller increases in range occurred thereafter in each season.

Brown (1970) proposed an equation which allows the maximum change in water temperature to be estimated if area of exposed stream, flow rate, and net radiation for the exposed stream can be measured or estimated. The equation can be adjusted for topographic and vegetative shading and for heat storage in the streambed (Forest Service, 1980). Estimated increases were calculated and compared to observed maximum water temperatures from the set of 40 days in August to September 1977 and June to July 1978 that had the greatest observed increases. The unshaded area of stream was used and the other corrections were unneeded. Estimated increases were 13° to 18°C for days when 3° to 6°C rises were measured.

DISCUSSION

The temperature increases reported for this stream exceed North Carolina water quality standards which require that the temperature of mountain streams may not be increased 2.8°C above the natural water temperature, and trout water temperatures may not be increased by more than 0.5°C nor exceed 20°C (North Carolina Administrative Code, Title 15, Chapter 2B). National Forests are managed to reduce the heating of mountain streams and protect trout habitat; clearcutting is limited to considerably smaller areas than this study site and buffer or shade strips are left next to the wider streams. Thus, the temperature record of this experiment is an example of an extreme case where a large area of stream is exposed on a south slope.

In the first two years after clearcutting, temperatures in all seasons were raised. Both daily minimums and maximums were warmer, the daily means increased, and the daily range expanded. By the third year, leaf biomass in the streamside area had reached 78 percent of that for mature forests but some increase in stream temperature was still apparent five years after cutting. Prolonged influence may be due to the type and stature of the regrowing vegetation. Three years after cutting, only 38 percent of the streamside leaf biomass was produced by woody plants. The young vegetation was too short to completely shade the wider portions of the stream channel, but vines and tall herbaceous plants did shade much of the middle and upper reaches of the stream.

Temperature increases developed late in the first summer, in August through October. The largest increases in daily maximum water temperatures occurred in June-July 1978, the second summer after cutting began. The largest increases would have come during the first year if all the streamside forest had been cut in the spring. Instead, only the midportion of the stream network (39 percent of the unshaded water surface) was cut by May. Forty-seven percent of the unshaded water surface was in the lower portion of the watershed. Until the vegetation around this lower 487 m of the stream was cut in August 1977, the water heated in the upstream area was probably cooled by shading and dilution before reaching the temperature measurement point, a phenomenon observed elsewhere by Swift and Baker (1973). Midsummer maximum temperatures remained well above those in the control stream through 1981 and probably will continue above normal for several more seasons. A declining trend of about 0.3°C per year is suggested by data from 1979 through 1981, although more rapid temperature recovery may occur as streamside forest regrows to shade the lower stream.

Winter minimums were increased only during the first and part of the second seasons. By the third and fourth winters, minimum water temperatures were below those experienced before the forest was cut. This finding is not unusual. Lee and Samuel (1976) and Reinhart, *et al.* (1963), report 2°C reductions of winter minimum stream temperatures and Burton and Likens (1973) found lowered minimums even in October. A possible explanation for the reduced winter minimums in the later years of this study would be the alteration of the energy balance between the water surface and streamside vegetation. Because the solar altitude is low in winter, young vegetation can shade the stream and moderate the heating of the water. In addition, longwave radiation toward the night sky is not blocked by short vegetation, thus, the water is cooled more than when a forest canopy is present to intercept and hold part of the energy. This explanation was not confirmed experimentally at Coweeta.

Estimated stream temperature increases were two to five times greater than observed increases. The differences between estimated and measured temperature changes could be due to stream length and the time a parcel of water is exposed to heating. The model assumes a channel length of less than 610 m, whereas, the main channel of this stream is nearly twice that length. The duration of exposure in this long channel could be longer than the two hours of the day receiving the peak solar radiation used in the calculations. However, doubling the exposure time to obtain a lower average radiation gives only a minor reduction of estimated temperature. Hewlett and Fortson (1982) also found a large difference, but in the opposite direction, between the measured and estimated temperature increases after clearing a Georgia Piedmont watershed. Brown's equation is a simple representation of several complex energy exchange processes. These studies indicate a need for further development of the stream temperature equation before it can be applied in the southeastern United States.

LITERATURE CITED

Boring, L. R., C. D. Monk, and W. T. Swank, 1981. Early regeneration of a clear-cut southern Appalachian forest. Ecology 62: 1244-1253.

Brown, G. W., 1970. Predicting the effect of clear-cutting on stream temperature. Journal of Soil and Water Conservation 25:11-13.

Brown, G. W. and J. T. Krygier, 1970. Effects of clearcutting on stream temperature. Water Resources Research 6:1133-1139.

Burton, T. M. and G. E. Likens, 1973. The effect of strip-cutting on stream temperatures in the Hubbard Brook Experiment Forest, New Hampshire. BioScience 23:433-435.

Forest Service, 1980. Temperature, Chapter VII. *In* An approach to water resources evaluation of non-point silvicultural sources. U.S. Environmental Protection Agency, EPA-600/8-80-012, pp. VII.1-VII.30.

Hewlett, J. D. and J. C. Fortson, 1982. Stream temperature under an inadequate buffer strip in the southeast Piedmont. Water Resources Bulletin 18:983-988.

Lee, R. and D. E. Samuel, 1976. Some thermal and biological effects of forest cutting in West Virginia. Journal of Environmental Quality 5:362-366.

Patton, D. R., 1973. A literature review of timber harvesting effects on stream temperatures: research needs for the Southwest. USDA, Forest Service, Rocky Mountain Forest and Range Experiment Station, Research Note RM-249, Fort Collins, Colorado, 4 pp.

Reinhart, K. G., A. R. Eschner, and G. R. Trimble, Jr., 1963. Effect on streamflow of four forest practices in the mountains of West Virginia, USDA, Forest Service, Northeastern Forest Experiment Station, Research Paper NE-1, Upper Darby, Pennsylvania, 79 pp.

Swift, L. W., Jr. and S. E. Baker, 1973. Lower water temperatures within a streamside buffer strip. USDA, Forest Service, Southeastern Forest Experiment Station, Research Note SE-193, Asheville, North Carolina, 7 pp.

Swift, L. W., Jr. and J. B. Messer, 1971. Forest cuttings raise temperatures of small streams in the Southern Appalachians. Journal of Soil and Water Conservation 26:111-116.

Wooldridge, D. D. and D. Stern, 1979. Relationships of silvicultural activities and thermally sensitive forest streams. Report DOE 79-5a-5, College of Forest Resources, University of Washington, Seattle, Washington, 90 pp.

INTERNATIONAL SYMPOSIUM ON HYDROMETEOROLOGY
JUNE AMERICAN WATER RESOURCES ASSOCIATION 1982

DYNAMIC OPTIMAL LAKE REGULATION FOR QUANTITY AND QUALITY OBJECTIVES

Bruce Loftis, John W. Labadie, and Darrell G. Fontane[1]

ABSTRACT: Presented herein is a computer modeling procedure for determining optimal operating strategies in a system of lakes subject to quality and quantity constraints. Several types of objectives can be considered in the general strategy with the one selected for this study being to minimize adverse water quality impacts. Integrated operation of lakes involves an interplay between port selection in multi-level intake structures, water quality parameters, operational time increments, the number and types of projects in a system, and project purposes. The complex lake operation problem can be decomposed into more limited subproblems, with iterative strategies used to achieve an overall solution. An optimization phase is used to suggest possible operation strategies followed by a simulation phase which evaluates each alternative suggested by the optimization phase. A case study is presented illustrating the complexity of lake operation and the applicability of the solution procedure.
(KEY TERMS: simulation; optimization; lake operation; water quality; trade-off.)

INTRODUCTION

As a result of increasing public awareness and recent state and federal legislation, water quality considerations in the operation of water resource systems are assuming a significant priority. Multilevel intake structures are the primary method of controlling the quality of releases from stratified lakes. These structures allow the release of water from various vertical strata in the lake. Typically, lakes are operated to enhance a downstream fishery or to maintain preproject downstream temperature characteristics. However, a wide range of operating strategies can be envisioned to satisfy different water quality needs.

The purpose of this paper is to present an approach for determining operation strategies in a system of multipurpose lakes subject to quality and quantity constraints.

Operation of a system of lakes is fundamentally a matter of complex interactions between port selection policies, water quality criteria, operational scope and detail in time and space, project purposes, and the types of projects. Operation with consideration of both water quality and water quantity adds an additional dimension of complexity in operational problems, requiring careful consideration of these tradeoffs.

In a density stratified lake, concentrations of water quality constituents vary with elevation. If a multilevel intake structure exists for a particular project, then tradeoffs between ports must be evaluated to determine which ports should be opened and what flow rate should pass through each open port in order to meet downstream water quality requirements.

Another type of tradeoff in lake management is that between water quality constituents. Often, operation to meet goals for several water quality parameters results in conflict. An obvious example is a downstream water quality requirement for water with a low temperature and a high dissolved oxygen (D.O.) content to promote good fish habitat. In the summer, however, water at the top of a lake has a high temperature and a high D.O. content, while water at the bottom of the lake is normally low in temperature and low in D.O. A decision to withdraw surface water achieves the D.O. requirement, but does not meet the temperature requirement. Withdrawal of bottom water achieves the temperature objective, but does not meet the D.O. objective. It is possible to operate to meet both objectives as closely as possible. But the complexity of the problem increases drastically as the number of parameters increases.

A third type of tradeoff involves water quality changes over time. It is possible to operate a lake optimally on a day-to-day basis with no anticipation of future conditions. Such operation may result in meeting the objectives early in the stratification period and deviating from the objectives late in the stratification period. However, if future conditions are known or can be estimated, then optimal operation can smooth out daily deviations to achieve an overall lower level of deviation. The principle behind this dynamic-optimal operation strategy is that a large number of small deviations may have a less adverse impact on the ecosystem than a smaller number of larger deviations.

A fourth type of tradeoff requires making total release decisions from a lake in order to claim water quality benefits, which may be in conflict with benefits or demands

[1] Respectively, Instructor, Rm. A323; Associate Professor, Rm. 331; Instructor, Rm. A327; Colorado State University, Dept. of Civil Engineering, Engineering Research Center, Fort Collins, Colorado 80523.

from other project purposes such as flood control or hydropower production. During a major flood, it is apparent that operation of a lake for flood protection has the highest priority during normal hydrologic events; however, it is possible to consider tradeoffs between water quality and other project purposes. For example, a peaking power operation could require a sufficiently large release flow rate such that the zone of withdrawal reaches the bottom. In this case, anoxic water would be released downstream or bottom sediments disturbed. Reducing the peaking power generated could improve the release water quality.

A final type of tradeoff relates to operation of a system of lakes and connecting streams to achieve overall water quality and quantity objectives at different locations throughout the system.

In related work, Kaplan (1974) used nonlinear optimization in conjunction with a lake simulation model to develop optimal port operation strategies that would optimize a scalar water quality index encompassing several parameters. This model does not include water quantity regulation in conjunction with water quality decisions. Martin (1973) developed a comprehensive water quantity-quality operational model for a river basin. This model fails to include the effects of lake stratification and selective withdrawal strategies. Fontane, *et al.* (1981), developed a fully dynamic optimizing model for water quality regulation of reservoir releases, but did not include the associated water quantity regulation.

In dealing with the complexity of operation of a multilake system for both quality and quantity objectives, a modeling procedure is needed to decompose this complex problem into a number of smaller more tractable problems and then recombine them together into an overall optimum strategy with respect whatever operational objectives are given priority. Such a procedure is proposed and developed subsequently, along with some initial, limited computational experience with a case study in western Maryland. The problem associated with the case study is outlined first in order to set the stage for development of the decomposition algorithm. However, the procedure developed is a general one that is applicable to many other reservoir systems besides the one considered here.

DESCRIPTION OF CASE STUDY

Presentation of a procedure for determining optimal lake operation stretegies for a system of lakes is facilitated by selection of a realistic case study. The Bloomington-Savage system in western Maryland is selected for this purpose. Bloomington Lake is located on the north branch of the Potomac River (NBPR) a few miles upstream of its confluence with the Savage River. Savage River Lake is located just upstream of the confluence with the NBPR. The Bloomington Lake watershed is heavily mined with both deep and strip mines. The resulting water quality is poor throughout the watershed. The flows into Bloomington Lake are characterized by sulphuric acid, heavy metals, high dissolved solids, and copious precipitates. The acid mine drainage has a devastating effect on aquatic life in the NBPR and in Bloomington Lake. The Savage River Lake watershed has little mining activity and the quality of the water is excellent. A location map for the system is shown in Figure 1.

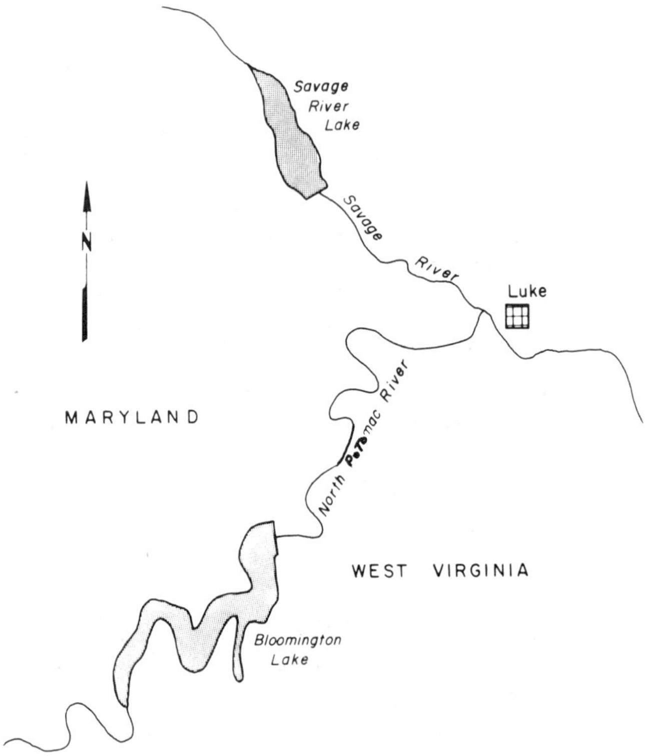

Figure 1. Location Map.

The control objective for this system of two lakes is to blend releases from Bloomington and Savage River Lakes to provide the best downstream water quality. Thus, the water in Savage River Lake is the resource to be allocated optimally over time. A minimum flow requirement and the overall water quality objective is provided by a paper mill located at the confluence of the two rivers at Lake Maryland.

The paper mill has a requirement for 100 cfs of water from the river for every day throughout the year. This serves as a minimum flow constraint at the confluence of the two rivers. The paper mill also has a water quality requirement that the acid concentration at the confluence remain constant throughout the year. A less important water quality requirement is that the temperature at the confluence remains at or near the natural stream temperature for each day of the year.

The control problem is to determine operational strategies for each day of the year for each of the lakes such that the downstream acid concentration is constant from day to day. Constraints on the operational decisions include a minimum release from each of the lakes and a minimum total release from both of the lakes for each day. Also, a pool elevation-versus-day guide curve is provided for each lake. For Bloomington and Savage River Lakes, substantial flexibility is available to deviate from the guide curve, but the operation decisions must be such that the general shape of the curve is maintained.

SOLUTION PROCEDURE

The problem to be solved is the determination of operation strategies for each lake for each day in order to produce the *best* downstream water quality, subject to quantity constraints and the hydrodynamic characteristics of each lake. The full problem could be formulated as one comprehensive mathematical optimization problem, but the problem is more logically attacked by decomposition into independent subproblems. The first subproblem is an optimization phase to compute possible release strategies for each lake and the second is a simulation phase to evaluate the performance of each alternative suggested by the optimization module in a more detailed fashion.

The general solution procedure is as follows:

1. *Identify an initial possible operating strategy for each lake.* This is accomplished by considering each lake independently and computing those release flow rates such that end-of-day pool elevation exactly matches the elevation day guide curve for each day for each lake. Thus, for a one-year period of interest, 365 release flow rates are computed for each lake. These flow rates serve as an initial trajectory for the simulation phase. It might be noted that water quality considerations play no role in the selection of initial release trajectories. Improvement can be made with respect to water quality objectives by modifying the initial release flows in some systemmatic manner.

2. *Determine the release quality characteristics.* The quality characteristics of the release flows are computed with a lake water quality simulation model that maintains heat, mass, and quality balances throughout a simulation period.

3. *Optimize release decisions to meet downstream water quality objectives.* Assume the release quality characteristics from the simulation module are invariant with release flow rate. Then an optimization module can be used to generate operation strategies that provide *best* downstream water quality conditions. The assumption that release quality characteristics are independent of flow rate is not strictly true, but is valid in the context of an iterative solution.

Subsequent steps in the solution procedure involve iteration between the simulation and optimization phases of the problem until some measure of convergence is reached. The simulation phase produces release water quality concentrations from each lake. The optimization phase uses the quality information to produce a release flow strategy for each lake for each day. The release flow strategy can then be used as input to the simulation module to update the release quality characteristics. Convergence is expected since water quality characteristics will not tend to change dramatically under discrete changes in water quantity decisions.

Decomposition is a powerful tool that allows solution of large scale problems by combining the solutions of smaller independent problems. Decomposition is a logical approach for the lake system regulation problem under consideration because two phases of the solution, simulation and optimization, can be identified such that the only interaction between the phases is that the output from one is used as input to the other.

WATER QUALITY SIMULATION

The WESTEX Lake Simulation Model, developed at the U.S. Army Engineer Waterways Experiment Station, provides a procedure for examining the thermal energy imposed on a lake. These energy balances and lake hydrodynamic phenomena are used to dynamically simulate vertical water quality profiles and release concentrations. The model is primarily a heat budget model. That is, temperature is the fundamental water quality constituent under consideration. However, the model is capable of performing a simple routing of conservative water quality constituents. Certain nonconservative water quality constituents can be simulated if the nonconservative aspects can be isolated and described, and if there are no complex interactions between constituents.

The model includes computational procedures for simulating heat transfer at the air-water interface, advective heat transfer of inflow and outflow, and the internal dispersion of thermal energy. The model is conceptually based on one dimensional division of the lake into discrete horizontal layers of uniform thickness (see Figure 2).

Ports are selected for each simulation day in order to satisfy downstream water quality objectives. A linear programming algorithm is embedded in the simulation model for this purpose. The physical constraints of operation of a multilevel intake structure are included in the linear program so that realistic port strategies are selected.

WATER QUANTITY ALLOCATION

The water quantity allocation phase of the lake regulation problem involves the determination of the flow to be released from each lake of a system of lakes in order to most nearly meet water quality requirements at downstream control points. In a real-time operation mode, decisions often must be made on a daily basis. If information about the future state of the system is available, then this information can be included in the decision-making process. Presumably, decisions based on information for an extended time frame will be better in the long run than decisions based solely on short term information, assuming the long term information is not greatly inaccurate.

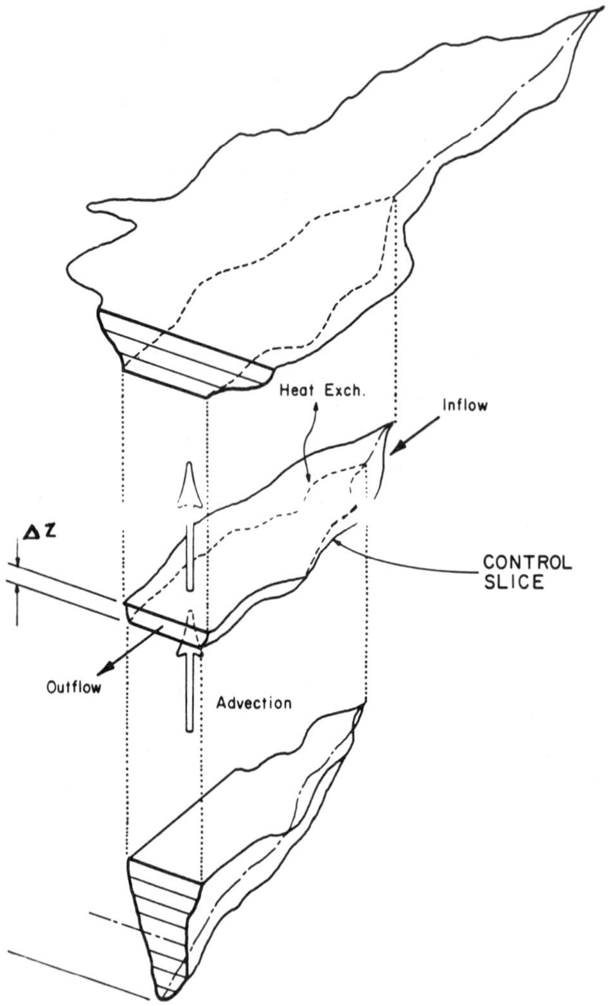

Figure 2. Discretization of a Stratified Lake.

An important assumption for the quantity allocation problem is that the release concentration for each water quality constituent for each lake for each day is known and is independent of the release flow rates. This assumption is valid because of the iterative nature of the overall solution approach; that is, iterations between a simulation model to produce release quality characteristics and an optimization model to produce release flows.

The objective for the quantity allocation problem is to determine release flows from each lake such that water quality concentrations at downstream control points are most nearly met. The downstream quality characteristics are flow-weighted averages of the release characteristics from each lake. For a two-lake problem with temperature and one quality constituent under consideration, the downstream quality characteristics are computed as follows:

$$T_D = \frac{T_1 q_1 + T_2 q_2}{q_1 + q_2} \; ; \; C_D = \frac{C_1 q_1 + C_2 q_2}{q_1 + q_2} \qquad (1)$$

where T_D and C_D are the mixed temperature and quality concentrations at the downstream control point, T_1, T_2, C_1, C_2 are the release temperatures and quality concentrations from lake-1 and lake-2. The flow rates q_1 and q_2 are the releases from the two lakes and are the decision variables for the problem.

The objective function is a scalar measure of how well a particular decision strategy meets the downstream target quality concentrations throughout the period of interest.

$$A = \sum_{N=1}^{365} \{W_T[T_{D,N} - T_N^*]^2 + W_C[C_{D,N} - C_N^*]^2\} \qquad (2)$$

where N is an index for days, W_T and W_C are relative weights, T_D and C_D are computed downstream quality characteristics, and T^* and C^* are specified quality *targets* at the control point.

The constraint set for the quantity allocation problem takes the form of constraints on release flow and constraints on pool elevation. For every time step and each lake, there are constraints of the form

$$q_1 \geqslant q_{1,min} \; ; q_2 \geqslant q_{2,min} \qquad (3)$$

For the two parallel lake problems, there may be constraints of the form

$$q_1 + q_2 \geqslant Q_{min} \qquad (4)$$

for each time step, where q_1 and q_2 are the release flows from lake-1 and lake-2, respectively, and Q_{min} is the minimum flow requirement at the downstream control point.

A pool elevation guide curve for a lake often has the shape of the curve in Figure 3.

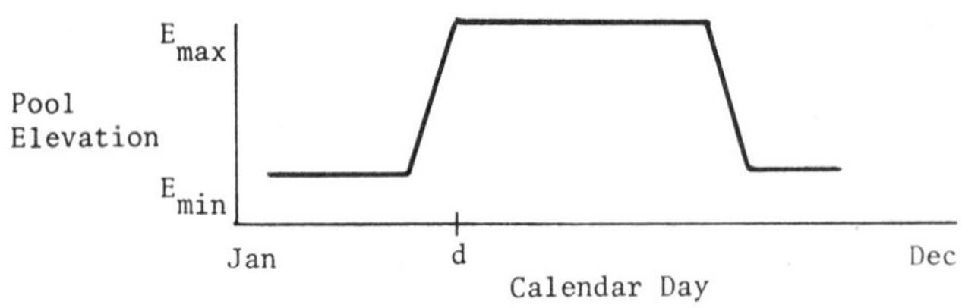

Figure 3. Example Guide Curve for a Lake.

This guide curve can be included in the constraint set by defining constraints at a few of the points on the curve. For example, to assure that the lake is filled in the summer, the problem could be constrained such that the pool elevation must be above E_{max} on day d. Corresponding to E_{max} is a volume V_{max} which is a function of the geometry of the lake. The volume of the lake for any day d can be computed from a mass balance of inflow and outflow volumes on previous days.

$$V_d = V_i + \sum_{t=i}^{d} I_t - \sum_{t=i}^{d} q_t \pm \text{(other net gains, losses)} \quad (5)$$

where V_d is the lake volume on day d, V_i is the initial lake volume on day i, t is an index for time steps; i_t is the inflow volume at time step t, q_t is the outflow volume at time step t. Thus a constraint of the form

$$V_d \geqslant V_{max} \quad (6)$$

can be written in terms of known information and a linear function of the decision variables.

COMPUTATIONAL EXPERIENCE

The optimization problem can be formulated in terms of a nonlinear programming problem with a nonlinear objective function and linear constraints. All of the water quality considerations are embodied in the objective function and all of the water quantity considerations are contained in the constraint set. A gradient projection technique is used since it is especially adaptable to linearly constrained optimization problems.

Preliminary results are shown in Figure 4. These results indicate release acid concentrations computed by the simulation model for a year with average meteorologic and hydrologic conditions. Operation of the model to meet a downstream temperature objective is represented as iteration-1. Iteration-2 represents operation of the model to achieve a downstream quality objective. Iteration-2 indicates release quality concentrations that are much closer to the objective quality than release temperatures from the two simulations would show a similar pattern. These results illustrate the tradeoffs between parameters required for operation of lake system. Work is continuing relative to interfacing the optimization model and the simulation model. The combined iterative procedure will produce results that include optimal quantity allocation in time and will provide a methodology for analyzing the various quality-quantity tradeoffs.

ACKNOWLEDGMENTS

The authors gratefully acknowledge Mr. Pete Juhle and the U.S. Army Engineer District, Baltimore, for support of the research effort presented herein.

LITERATURE CITED

Fontane, D. G., J. W. Labadie, and Bruce Loftis, 1981. Optimal Control of Reservoir Discharge Quality Through Selective Withdrawal. Water Resources Research 17(6):1594-1604.

Kaplan, E., 1974. Reservoir Optimization for Water Quality Control. Ph.D. Dissertation, University of Pennsylvania, Philadelphia, Pennsylvania.

Loftis, Bruce, 1980. WESTEX-A Reservoir Heat Budget Model. Draft Report, U.S. Army Engineer Waterways Experiment Station, Vicksburg, Mississippi.

Martin, Q. W., *et al.*, 1973. Instrumentation for Engineering Management of a Multi-Purpose River Basin System. Center for Research in Water Resources, The University of Texas at Austin.

Poore, A. B. and B. Loftis, 1982. Water Quality Optimization by Selective Withdrawal. Draft Report, U.S. Army Engineer Waterways Experiment Station, Vicksburg, Mississippi.

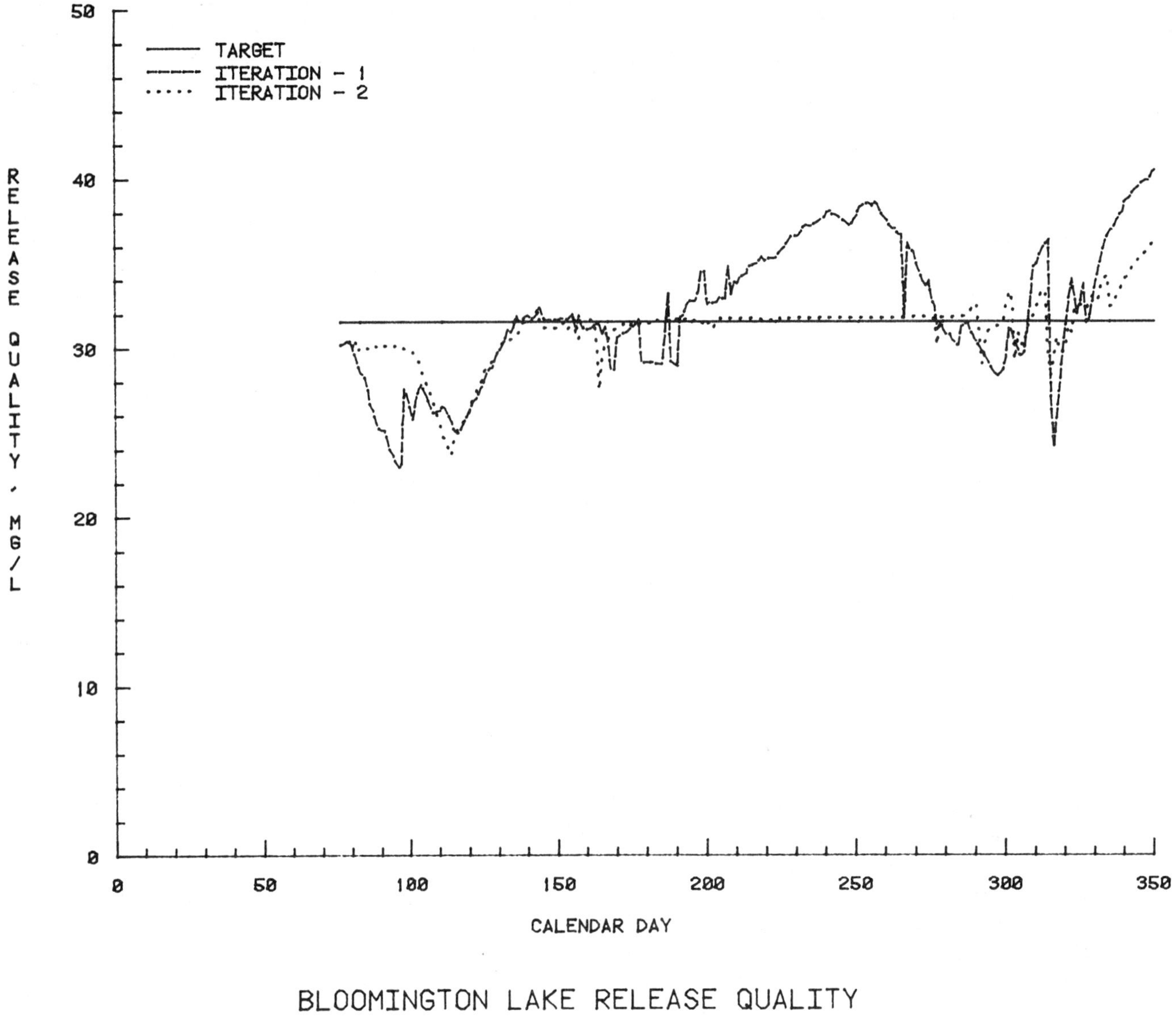

Figure 4. Simulation Results.

ENVIRONMENTAL CHANGES IN THE BLACK SEA ESTUARINE REGIONS

David Tolmazin[1]

ABSTRACT: Retention of river water in huge storage lakes built on the rivers entering the Black Sea have severely affected the hydrology and chemistry of the estuarine regions (the Northwestern Black Sea and Azov Sea) and have caused a substantial decline in fisheries. The introduction of large scale irrigation programs have contributed to further deterioration of ecological conditions (hypoxia and mass mortality of benthic forms). Several proposed rehabilitation schemes included use of special structures to reduce water exchange between the estuaries and the sea, complete partition of the estuaries from the sea by dams, and interbasin water transfer from the lower reaches of the Danube River. However, the Soviet leadership has not yet committed itself to any remedial measures.

(KEY TERMS: estuarine ecosystems; water management and control; water pollution; rehabilitation of estuaries; Black Sea coastal regions.)

INTRODUCTION

Estuaries are said to be at the bottom of the ecological ladder in the riverine environment and are, therefore, very sensitive to transformations of the hydrographic network feeding them. These complex river-estuary relations have become extremely strained in the southern USSR. Nationwide fresh water management and control during the past 30 years have led to the distinct deterioration of the Black Sea estuarine regions and shrinking of the inland seas.

On the whole, the USSR is not a water short country. It ranks first in terms of fresh water supply (about 4700 km^3/year). However, the geographical distribution of fresh water resources is not in close accord with the national needs. The most heavily populated areas (European part, Caucasus, and central Asia), which comprise approximately 80 percent of the country's population, 98 percent of its arable land, and which produce up to 80 percent of all industrial output, contain only 24 percent of the total fresh water flow (Babich, *et al.,* 1980). The situation is even worse on the northern slope of the Black Sea basin. It represents a relatively modest 4.1 percent of the territory but serves as a home for 29 percent of the USSR's population (Vendrov, 1979). The rivers draining into the Black Sea – the Dnepr, Dnestr, Don and Kuban – flow through areas of inadequate moisture and zones of intensive economic activities.

The major unfavorable changes in ecological conditions in the Black Sea followed the erratic Soviet strategy to regulate river flows. Starting in the 1950s, all southern plain rivers in the USSR were used mainly for hydraulic energy production and industrial water supply. This resulted in an accumulation of river water in a number of huge water reservoirs. In the mid-1960s the chief goal of the Soviet economy became the creation of self-sufficient agriculture. A number of heavily funded irrigation programs were implemented with the construction of a network of open channels for interbasin water transfers. The substantial reduction in the amount and quality of fresh water, due to extensive withdrawals from the rivers, and marked increase in water pollution threatens to inhibit further economic expansion. The most distinct adverse effects are felt in estuarine regions and the lower reaches of the rivers of the Black, Caspian, and Aral Sea. This paper analyzes the environmental problems of the Black Sea.

Decline in Fish Production

Over the last 25 years the fishing industry of the Black and Azov Seas has suffered a disastrous decline. Both sea basins were previously populated by a large stock of valuable fishes. However, in the early 1960s the water quality began to deteriorate (Krotov, 1976). The first fish to suffer was the Black Sea mackerel. The whole standing stock of some 15,000 to 20,000 tons had disappeared by 1967, never to be recovered. At the time there appeared to be no obvious reason for the disaster. The experts thought it was due to the rapid reproduction of predators, bluefish, and bonito, and concluded that the Black Sea mackerel would be replaced by a large type of scad (Table 1). However, the latter also vanished, followed by the bonito and bluefish.

In the Azov Sea, the changes were even more destructive (Bronfman, *et al.,* 1979). Until recently this shallow basin (average depth 10 m) ranked first among other basins of the world ocean in terms of fish production (about 8 tons/ sq. km). Due to low salinity (10.5 ppt), the brackish water inhabitants could utilize the pelagic and benthic food resources of the entire basin. During three decades, the fish

[1] Associate Professor, Department of Marine Sciences, University of Connecticut, Groton, Connecticut 06340.

TABLE 1. Overall Stocks and Prices of Individual Species in the Black Sea.

Species	Overall Stock* in the 1950s	Overall Stock* Presently	Annual Catch* in 1970-74	Overall Price**
		FISHES		
Anchovy	800-1200	1000-1200	130	28.0
Scad	200	Disappeared After 1967	– –	– –
Scad (small type)	50-100	100-600	22	15.2
Black Sea Mackerel	15-20	Disappeared Before 1967	– –	– –
Sprat	150-175	490	0.9 (1974)	0.19
Bonito	400	Unknown	– –	– –
Other Fishes	200	200	8.3	5.2
		MOLLUSCS		
Mussels and Oysters	1000-1200	600-800(?)	0.2	0.5
		ALGAE		
Phillophora Rubens	9000	9000	1.5-2.0(?)	– –
Cystoseira Barbata	2000	2000	– –	– –

*Thousand tons.
**Million rubles.

catch declined dramatically. Moreover, those fishes caught were less desirable species. Presently, the species of high commercial value constitute only 7 to 8 percent of the total catch (Table 2).

TABLE 2. Annual Catches in the Azov Basin.

	Years				
	1936-1939 1949-1952	1956	1963-1966	1970	1974
Thousand Tons	250-300	95	150-170	120	98
Percent Valuable Fishes	70-80	30	10-15	12	7-8

Valuable Fishes: Sturgeon, Giant Sturgeon (Beluga), Pike-Perch, Bream, Roach, Herring.
Fishes of Little Value: Anchovy, Sprat, Gobies.

Economic considerations present the same problem. The present overall value of raw fish production in the Black and Azov Seas is estimated at 130 million rubles per year. Food products manufactured from this raw material cost approximately 520 to 650 million rubles. If the present deterioration in the environment continues, the annual fish losses (including the destruction of mussel and oyster fields and algae) will be about 150 million rubles a year (Krotov, 1976).

The next step in the deterioration of estuarine regions was marked by severe hypoxia and anoxia (lack of dissolved oxygen) in the Northwestern Black Sea, close to the river mouths (Tolmazin, 1977). The mussel fields and bottom fauna between the Danube and Dnestr estuaries and south of the Dnestr estuary (an area of over 350,000 ha) were destroyed during the summers of 1973 to 1975 (Figure 1).

Transformation of Drainage System

The piscatorial disasters coincided with the introduction of schemes to regulate the water runoff in major rivers. The postwar Soviet water economy, particularly in the southern regions, was based on the conception of retaining as much fresh water as possible within the land. The principal "instruments" of such control were to be water storage lakes. It seemed at the time that storage lakes served all aspects of progress. They were primarily designed to provide cheap power from nonpolluting hydroelectric stations and guarantee stable and safe depths for river navigation. They were to serve as reserve water masses for irrigation as well as for municipal and industrial water supply. Some storage lakes were to be used to process various pollutants coming from nearby towns and industries. And finally, they were to be used for controlled fish farming and recreation.

Six large storage lakes and several smaller ones on the Dnepr (with an overall volume of 43.6 cu km), storage basins on the Don, Kuban, and Dnester Rivers, all ensured an enormous water reserve inland. The result was unrestricted use of fresh water by industry. On the debit side, the water returned into the rivers insufficiently treated, sometimes highly enriched with toxic chemicals (Gusev, 1967). A further decline in the quality and amount of water resources was caused by irrigation programs to boost agricultural production in the Ukraine and Northern Caucasus. Water was brought to arid areas via complex irrigation systems. They included wide open canals (e.g., the 340 km long North Crimea Canal), widespread narrow canals and ditches, locks and pumping stations. No measures were taken to prevent extensive seepage into the ground, as well as evaporation from the open surfaces of canals and water reservoirs. The efficiency of the existing irrigation network did not exceed 35-40 percent. The underground water table was raised; as a consequence, a sizeable portion of arable acreage has become marshy. Some parts have been affected by an outcrop of deep saline water, causing salt patches on arable land. About 50-55 percent of irrigated water, saturated with agricultural chemicals, returns into the river, contributing to water pollution.

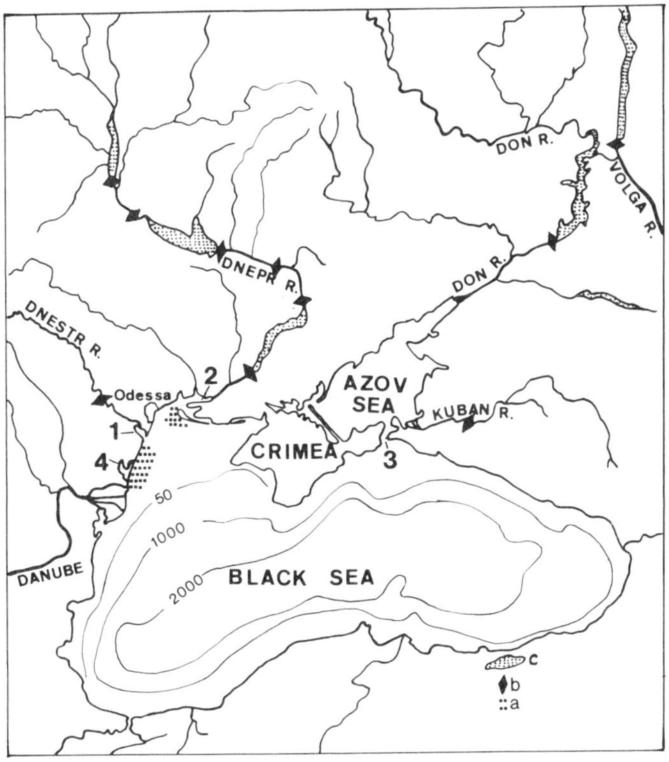

Figure 1. The Black Sea and Major Tributaries (depth in meters) – 1. Dnestr Estuary; 2. Dnepr Estuary; 3. Kerch Strait; 4. Lagoon Sasyk; a. Hydropower Stations; b. Area of Hypoxia and Mass Mortality; c. Storage Lakes.

The total area of irrigated land in 1980 reached nearly 2 million ha (Golovko, 1981). Plans to irrigate 7.8 million ha in the Ukraine are being considered. This will substantially increase irretrievable water withdrawals.

The annual flow of the Dnepr and Dnestr Rivers has fallen by 17 percent to 25 percent (Babich, *et al.*, 1980). The amount of water flowing into the Azov Sea decreased 23 percent. By the year 2000, the demand for fresh water from various water users is expected to double. The available water reserves in the Black Sea basin and in the southern part of the country will be inadequate for continued growth of agriculture and industry in the drier regions. To accommodate all future water needs of the national economy, the long disputed diversion of river flow from the northern slope of the country will be necessary (Gustafson, 1980). In this program, the Black Sea basin will receive only a small part of clean northern water. The Danube River is regarded as a most likely source of water to the drought prone northern slope of the Black Sea. However, even in this case it would be difficult, if not impossible, to step the described deterioration of the estuarine regions.

Causes of Environmental Changes in the Coastal Areas

The Black Sea and its tributaries have reacted so strongly to industrial impact because of special oceanographic conditions. Excess of precipitation and river runoff over evaporation caused a sharp density discontinuity layer between the upper (75 m deep) low saline (17.5 ppt) and deeper, more saline (up to 22.1 ppt) water. This layer inhibits vertical mixing and the depths of the Black Sea, poisoned by hydrogen sulfide, are essentially lifeless. However, the distribution of life in the upper layer is patchy.

The bulk of marine life is concentrated near the coast, over the sea shelf. The most developed shelf is in the northwest, where three major rivers – the Danube, the Dnepr, and Dnestr (which, between them, provide 75-80 percent of river runoff) – enter the sea. Another river affected area is the Azov Sea. The rivers deliver life sustaining substances, including nutrients, organic materials, phosphates, nitrates to the nearshore regions. Plankton blooms at the river mouths, providing food for fish and molluscs. In the past, the upper reaches of the Dnepr and Dnestr estuaries and the Azov Sea were flooded each spring and fall by fresh water. The Don and Kuban Rivers had 40-50 thousand sq. km of marches which were very productive spawning and nursery grounds.

Clearly, neither the Black Sea nor the Azov Sea are today as productive as they once were. The cause seems to be the restrictions placed on the rivers – restrictions that siphoned off vital fresh water for other developments. The zones of low salinity habitat diminished substantially. The average salinity of the Azov Sea increased from 10.5 ppt in 1951 to 13.5 ppt in 1975 (Bronfman, *et al.*, 1979).

The water areas with conditions favorable for anadromous and semi-anadromous fishes have decreased substantially. For instance, intensive water withdrawals from the Don River reduced these areas for the two major Azov Sea species – the pike-perch and the European bream, from 25 and 12 sq. km, respectively, to 6 and 4 sq. km. In the Black Sea these two fishes became rarities.

The brackish water habitat has also been severely damaged by the salt wedges, which now penetrates far upstream into the Dnepr, Dnestr, and even the Danube Rivers. Increased stability of the water column caused reduced mixing, which, in turn, inhibits oxygen supply to the sea bed. Rapid oxidation of abundant organic outwash from the irrigated lands facilitates the above mentioned effects of hypoxia and mass mortalities in the coastal sea.

Projects for Rehabilitation of the Estuaries

There are several projects to curb further deterioration of the estuarine environment in spite of increasing fresh water deficit. One project suggests a decreased water exchange between the estuaries and the sea (Lagutin and Tolmazin, 1965; Tolmazin, 1974; Rozengurt, 1974) by the use of dams of various geometry. For instance, model predictions show that if the cross-section of the Kerch Strait is reduced by four times by a dam with a long (4 km) channel, 10-12 km^3 can be taken from the rivers entering the Asov Sea without a noticeable decrease in its salinity. This remedial measure would be efficient only for limited irrevocable consumption of fresh water.

Another well justified project (Osmer, 1973) called for complete partition of the Azov Sea which should be brought under man's full control. A rather sophisticated ecologic-economic model was developed (Gorstko, 1975) to simulate physical, chemical, and biological processes in this future lake. The current fields and salinity distributions, based upon numerical modeling, may adequately represent real oceanographic situations in the future. However, rather subtle chemical and biological interactions remain obscure. However, rather subtle chemical and biological interactions remain obscure. This project was part of the official Scheme of Complex Utilization and Protection of the Azov Sea (completed in 1971) but was never realized. This was due, perhaps, to the fact that the environmental funds in the 1972-1976 Five Year Plan were channeled into the expensive reclamation of the so-called non-black-earth zone (the wet and arid regions around Moscow and the northwest.

Another project concerns the transfer of Danube water through a number of shallow lagoons into the Dnestr and Dnepr estuaries (Ponomarenko, 1980). This diversion scheme envisages four various routes to transfer northward up to 15 km^3 annually (the first phase). The least expensive version of 4 billion rubles capital investment will include construction of several fresh water storage lakes instead of the shallow lagoons (Figure 1), open canals, and closed conduits to channel water over 290 km into the Dnepr River. The scheme will include eight pumping stations and six hydroelectric stations to absorb the energy of the falling water. The proponents of this project claim that the Danube water will arrest the decline of the Dnestr and Dnepr estuaries for at least 30-40 years, will alleviate local pollution problems, and irrigate 8 million acres of arable lands. In the more distant future the waterway from the Danube should be lengthened to bring water to the Azov Sea.

It should be noted that the above projects can hardly solve major ecological problems in the river basin and estuaries. Nationwide water deficits can be prevented by the diversion of the northern rivers. However, many environmentalists suggest the use of extreme caution in further restructuring of the country's drainage system (Vendrov, 1979). Some planners (Babich, *et al.*, 1980) recommend imposing restrictions on industrial and agricultural consumption of water resources by more efficient water protection and the use of advanced nonwater technology. However, even these measures will hardly restore the previous biological productivity of the Black Sea and other basins.

LITERATURE CITED

Babich, B., V. Lozansky, and A. Kuzin, 1980. The Conservation and Rational Utilization of Water Resources as a Major Economic Problem. The Current Digest of the Soviet Press 32(12):2-4.

Bronfman, A. M., V. G. Dubinina, and G. D. Makarova, 1979. Hydrologic Elements of Productivity in the Azov Sea. Pischevaya Promyshlennost, Moscow, USSR, 287 pp. (in Russian).

Golovko, L. P., 1981. Rational Structure of Utilization of Irrigated Land in the Ukraine. Hydrotechnique and Melioration 10:58-60 (in Russian).

Gorstko, A. B., 1975. Descriptive Simulation Model for Forecasting the Condition of a Water Ecosystem. *In:* American-Soviet Symposium on Use of Mathematical Model to Optimize Water Quality Management. U.S. EPA, Gulf Breeze, Florida, pp. 433-441.

Gusev, A. M., 1967. Influence of Waste Waters and Timber-Rafting on Fishery Basins and Damage Inflicted by Pollution to the USSR Fishery. Proceedings of All-Union Research Institute of Lakes and Fishery, Leningrad, 53 pp. (in Russian).

Gustafson, T., 1980. Technology Assessment, Soviet Style. Science 208:1343-1348.

Krotov, A. V., 1976. Economic Assessment of Losses of Fishery in the Black and Azov Sea Basins Due to the River Water Control. *In:* Problems of Sea Economics, the Ukrainean Academy of Sciences, Odessa, pp. 135-141 (in Russian).

Lagutin, B. L. and D. Tolmazin, 1965. Theoretical Solution to the Problem of Artificial Regulation of Water Exchange Through Kerch Strait. Soviet Hydrology, Selected Papers, No. 3, AGU, pp. 302-304.

Osmer, N. A., 1973. On Necessity to Regulate Water Exchange Between the Black and Azov Seas. *In:* Proceedings of the North Caucasian Scientific Center of High Education. Rostov-on-Don, USSR, pp. 25-27 (in Russian).

Ponomarenko, V. C., 1980. Canal Danube-Dnepr. Hydrotechnique and Melioration 11:10-13 (in Russian).

Rozengurt, M. Sh., 1974. Hydrology and Prospects of Restructuring of Natural Resources of the Odessa Limans (lagoons). Naukova Dumka, Kiev, 223 pp. (in Russian).

Tolmazin, D., 1974. Sea Straits. *In:* Oceanology, 1, A. P. Kapitsa (Editor). G. K. Hall and Co., Boston, Massachusetts, pp. 35-55.

Tolmazin, D., 1977. Hydrological and Hydrochemical Structure in the Areas of Hypoxia and Mass Mortality in the Northwestern Black Sea. *In:* Marine Biology, Naukova Dumka, Kiev, USSR, pp. 12-17 (in Russian).

Vendrov, S. L., 1979. The Problems of Transformation of River Networks in the USSR. Hydrometeoizdat, Leningrad, USSR, 207 pp. (in Russian).

APPLICATION OF IFG'S INSTREAM WATER TEMPERATURE MODEL IN THE UPPER COLORADO RIVER

Fred D. Theurer, Kenneth A. Voos, and C. G. Prewitt[1]

ABSTRACT: An instream water temperature model has been developed, verified, validated, and applied by the Instream Flow and Aquatic Systems Group (IFG). Using readily available hydrometeorological data, the model predicts average daily water temperature and diurnal fluctuations within a stream network. Various solution techniques are available ranging from desktop calculator programs designed for simple applications to a sophisticated computer program designed for complex systems, large data volumes, or repetitive applications. The application of the model and possible interpretations of its results are discussed in this paper with respect to two endangered fish species in the Upper Colorado River Basin. Complete documentation for the model and solution techniques is available from IFG (Theurer, 1982).
(KEY TERMS: water temperature; stream flow; modeling; Upper Colorado River.)

INTRODUCTION

The instream water temperature model is designed for general application, using standard hydrologic and hydraulic data, and a few readily available meteorological parameters. Mean daily (24-hour average), minimum nighttime, and maximum daytime water temperatures are predicted for the time periods represented by the hydrometeorological input data. However, only the mean daily water temperatures for normal (arithmetic mean of historical data) June and July meteorology are applied and interpreted in this paper.

The model was used by the U.S. Fish and Wildlife Service (FWS) to simulate historical and current temperature regimes and to predict possible future temperature changes in the Upper Colorado River Basin (UCRB) (Prewitt, *et al.*, 1981). Figure 1 shows the UCRB network. The network contains 15 U.S. Geological Survey (USGS) stream flow gages. Data from ten of these gages provided initial conditions for the physical process portion of the water temperature model. The other five are interior network gages, which were used to validate and calibrate the model.

The instream water temperature model has two principal components: a regression model for smoothing data at locations of recorded water temperatures, and a physical process model for water temperature predictions elsewhere in the network. A validation check of the mean daily water temperature was made on each component, using historical hydrometeorological data (Theurer and Voos, 1982). Input data were summarized by months for the years 1964 through 1977. This paper summarizes the comparison of some mean daily water temperature regimes and temperature requirements for the Colorado squawfish (*Ptychocheilus lucius*) and the humpback chub (*Gila cypha*) in the UCRB.

TEMPERATURE REQUIREMENTS

Water temperature is important for fish survival and reproduction. Temperature regimes may be primarily responsible for the longitudinal distribution of species, production rates, and other population dynamics. It can even affect the ultimate survival of a species.

Perhaps the most notable case documenting temperature influences, and the need to predict temperature regimes, is the effect of Flaming Gorge and Glen Canyon dams on the UCRB fisheries. The Colorado squawfish and humpback chub, resident native fish species in the Green and Colorado Rivers, are now on the Federal endangered species list primarily because of the change in the temperature regimes of their native rivers (Holden and Stalnaker, 1975; Miller, *et al.*, 1982).

Both species evolved in the turbid, warm waters of the mainstem Green and Colorado Rivers. Both prefer warm water, and the squawfish has demonstrated a need for water temperatures possibly as warm as 20°C to initiate spawning (Hammond and Inslee, 1981; Miller, *et al.*, 1982). Various chub spawning studies have been summarized by Miller, *et al.* (1982). They document that a water temperature range between 16-18°C may be suitable for chub spawning. June and July are the UCRB spawning periods for the chub and squawfish, respectively (Miller, *et al.*, 1982).

[1] Respectively, Civil Engineer, Soil Conservation Service; Biological Systems Engineer, OAO Corporation; and Aquatic Systems Analyst, IPA/ Colorado State University; U.S. Fish and Wildlife Service, Instream Flow and Aquatic Systems Group, Drake Creekside Bldg. 1, 2625 Redwing Road, Ft. Collins, Colorado 80526.

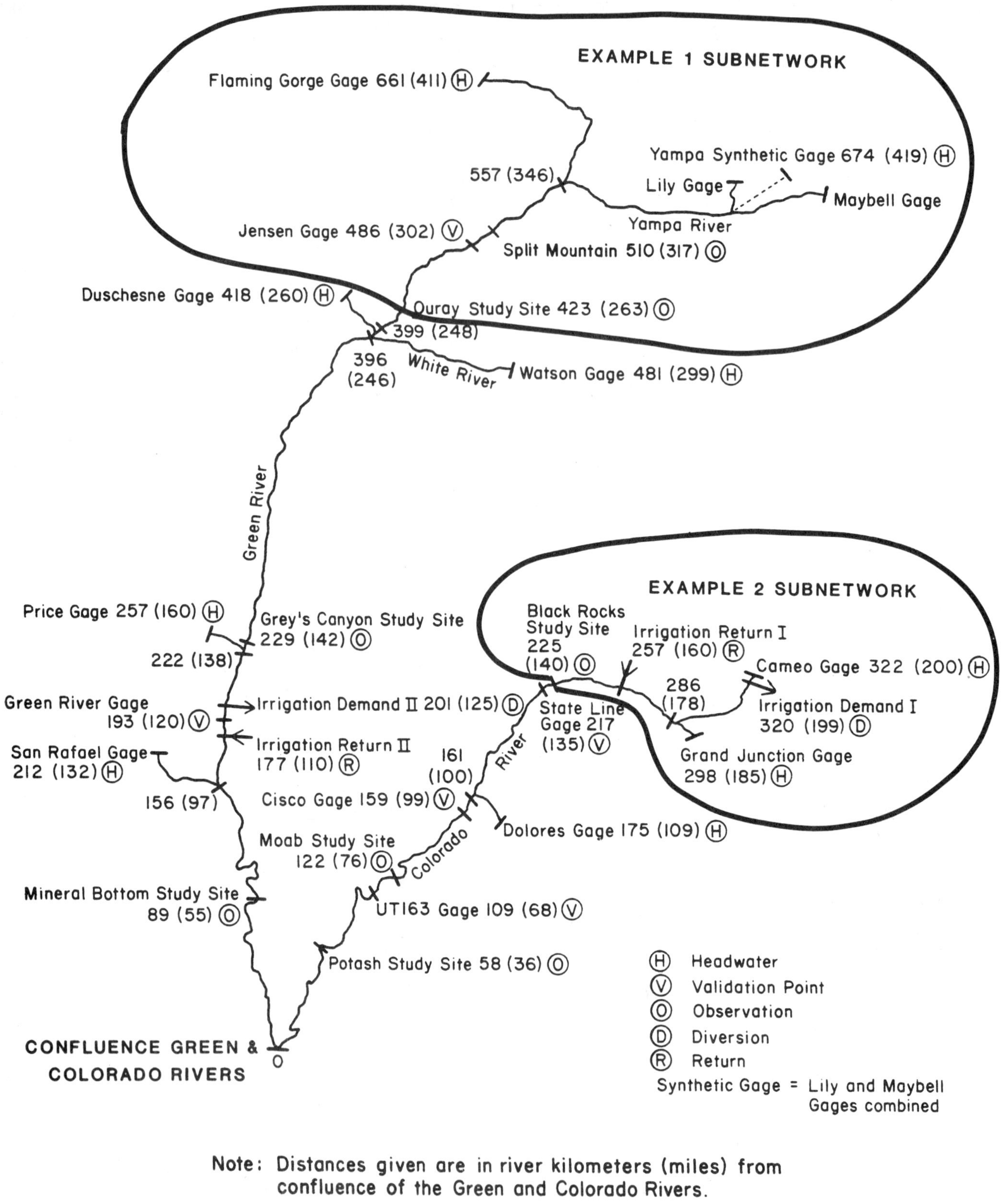

Figure 1. Upper Colorado River Basin Network Showing the Significant Hydrometeorological and Selected Study Site Locations. (Also shown are the subnetworks for examples 1 and 2.)

The Colorado River below Glen Canyon dam previously supported large populations of squawfish and chub; now, none are present (Minckley and Deacon, 1968). Chub, however, have been observed in the tailwaters of the Little Colorado River before these waters mix with the cold mainstem waters.

The hypolimnentic (deep water) release from the reservoir subsequent to its closure provides relatively cold water below Glen Canyon dam. For example, according to published water quality data for the USGS gage at Lees Ferry, which is located 26 kilometers downstream from the dam, the mean July temperature in 1973 was 8.5^{o}C (U.S. Geological Survey, 1973). The cold thermal regime induced by the reservoir persists far downstream because there are no large tributaries to ameliorate these very cold release waters. The resulting lack of successful spawning has eliminated squawfish and chub populations in these cold water reaches (Miller, *et al.,* 1982).

Below Flaming Gorge reservoir, a similar situation exists. No squawfish or chub are currently found in the Green River between the dam and the confluence with the Yampa River, a distance of 104 kilometers. The Yampa is an uncontrolled tributary that contributes more than one-third of the Green River flow. Both fish species still inhabit the Yampa and are also found in the Green below the Yampa confluence (Holden and Crist, 1981). Water temperature is sufficient to explain the absence of squawfish and chub in the Green above the confluence.

To comply with the Endangered Species Act, future developers in the UCRB will be required to ensure that proposed projects maintain necessary water temperatures. Because the instream water temperature model can predict downstream water temperatures under various conditions of initial reservoir release temperatures and discharges, the model could be used to analyze proposed projects and to suggest means of mitigating, or even enhancing, aquatic habitat.

MODEL APPLICATIONS

The impacts of reservoir release temperatures and discharges, or any other form of diversion or augmentation, on downstream water temperatures can be simulated by the model. The effects of otherwise natural flows can be included in the simulation. This paper will present two applications of interest to water resource planners and fishery managers. Two output displays are useful to visualize these applications – a longitudinal temperature profile and a temperature versus site discharge curve. The first example illustrates an application using the longitudinal profile output display. This display locates the river reaches of suitable spawning habitat for squawfish. The second example illustrates an application using the temperature discharge output display. It is used to determine site discharge ranges necessary to maintain suitable temperature for squawfish and chub spawning.

The first example demonstrates downstream temperature changes due to various reservoir release temperatures. Output was obtained for a given reservoir release discharge and a single constant set of hydrometeorological conditions. The simulated water temperatures include the effects of all additional normal flows in the stream network. The water temperatures are plotted versus river location for each release temperature (Figure 2). The plots are called longitudinal temperature profiles.

In this example, only the profile from the Yampa synthetic gage (location of assumed reservoir) down to the Ouray study site are shown. Normal July hydrometeorological conditions were used. Each profile is for the specified reservoir release temperature varying from 4-18°C and one at 19.40^{o}C. The latter profile is the normal July water temperature at the Yampa synthetic gage and, therefore, represents existing normal July conditions. The normal July diurnal fluctuation is less than $\pm$2.0^{o}C at Ouray.

Suitable squawfish spawning temperatures are superimposed on the profiles to clearly see the resulting river reaches of suitable spawning temperature habitat. Also added is substrate suitability. The river reach below Split Mountain contains a sand substrate which is unsuitable for spawning (Prewitt, *et al.,* 1981).

This example shows that assumed reservoir release temperatures between 4 and 18°C would not provide the minimal spawning temperature threshold for squawfish at any suitable downstream location between the assumed reservoir (Yampa synthetic gage) and Split Mountain. Only releases approaching existing conditions (19.40^{o}C) begin to ensure proper spawning temperatures in some of the remaining known spawning reaches between these two locations. The Green River between the Yampa-Green confluence and Split Mountain provides only marginal spawning habitat (Joseph, *et al.,* 1977).

A validation check was made at interior gages within the network. Measured water temperatures (smoothed by the respective gage's regression model) were compared to values that were independently predicted by the physical process model. The difference between the predicted and measured values is the validation check. The result of the validation check at Jensen gage, which is located between the Yampa-Green confluence and the Ouray study site, was +0.52^{o}C for the normal July hydrometeorological conditions. Obviously, this prediction is the uncalibrated value; the model could easily be calibrated to completely eliminate the difference.

The above analysis was made using historical data for the operation of Flaming Gorge. The historical normal July temperature release is 6.51^{o}C. Recently, a variable level selective withdrawl operating scheme has been implemented at Flaming Gorge. The new normal July release temperature is 11.11^{o}C Wegner, *et al.,* 1981). This results in a Green River water temperature of 19.8^{o}C immediately below its confluence with the Yampa for existing normal July hydrometeorological conditions. Therefore, the Green River

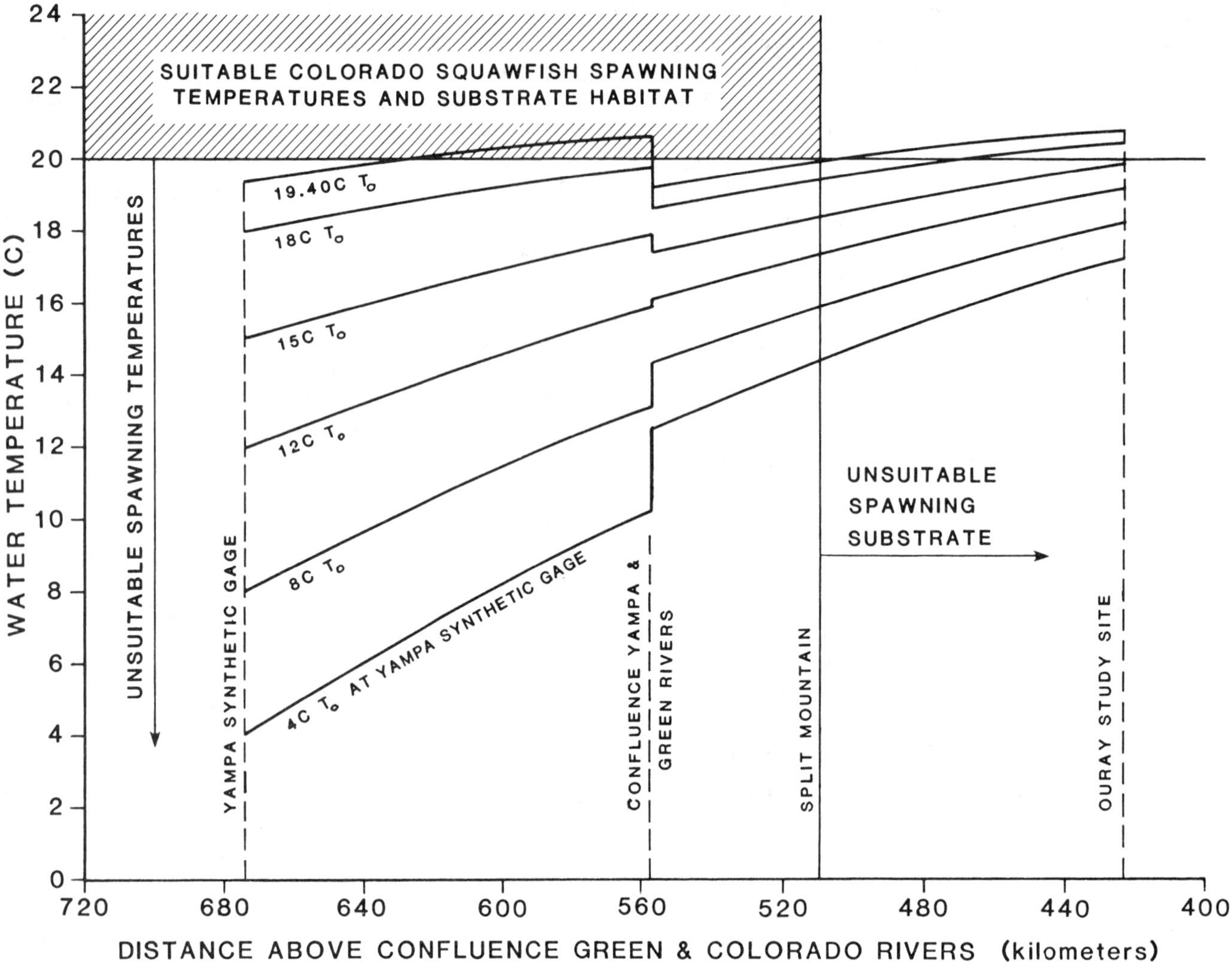

Figure 2. Longitudinal Temperature Profiles from the Yampa Synthetic Gage to the Ouray Study Site for Nornal July Hydrometeorological Conditions. The mean July flow at the Yampa synthetic gage (103 cms) was used as the assumed reservoir release discharge. Normal July discharges were used for the remaining tributaries below this gage. Profiles are shown for various release temperatures ranging from 4-18°C, plus the existing normal water temperature of 19.40°C, at the Yampa synthetic gage.

below this confluence down to Split Mountain will have an improved suitable spawning temperature and substrate habitat for existing conditions.

The second example demonstrates downstream temperature changes at a specified study site due to alteration of flows at a given upstream location. It is done for two water temperature criteria at the upstream location with two constant sets of hydrometeorological conditions. The simulated water temperatures at the site include the impacts of all additional normal flows in the network. These temperatures are plotted versus the discharge at the site (Figure 3). A separate curve is shown for both governing upstream water temperature criteria and each set of hydrometeorological conditions. They are called temperature versus site discharge curves.

In this example, only the temperature versus site discharge curves at the Black Rocks study site – a suspected squawfish and known chub spawning habitat (Miller, *et al.*, 1982) – are shown. Normal June and July hydrometeorological conditions were used. Cameo gage was the given upstream location for assuming various flows ranging from zero discharge to an amount sufficient to produce 700 cms at Black Rocks. Two water temperature criteria at Cameo gage were selected. The first criterion varied the water temperature at Cameo with the assumed discharge; i.e., the assumed discharge was used with the normal monthly meteorology to calculate water temperatures at Cameo. The second criterion assumed that the water temperature at Cameo would always be equal to the current mean monthly water temperature regardless of assumed flow at Cameo; i.e., the water temperature associated with the current mean monthly flow at Cameo. The normal June and July diurnal fluctuations are slightly more than ±2.0°C for both mean monthly discharges at Black Rocks.

Any combination of reservoirs, diversions, or augmentations above the Cameo gage, that would result in either

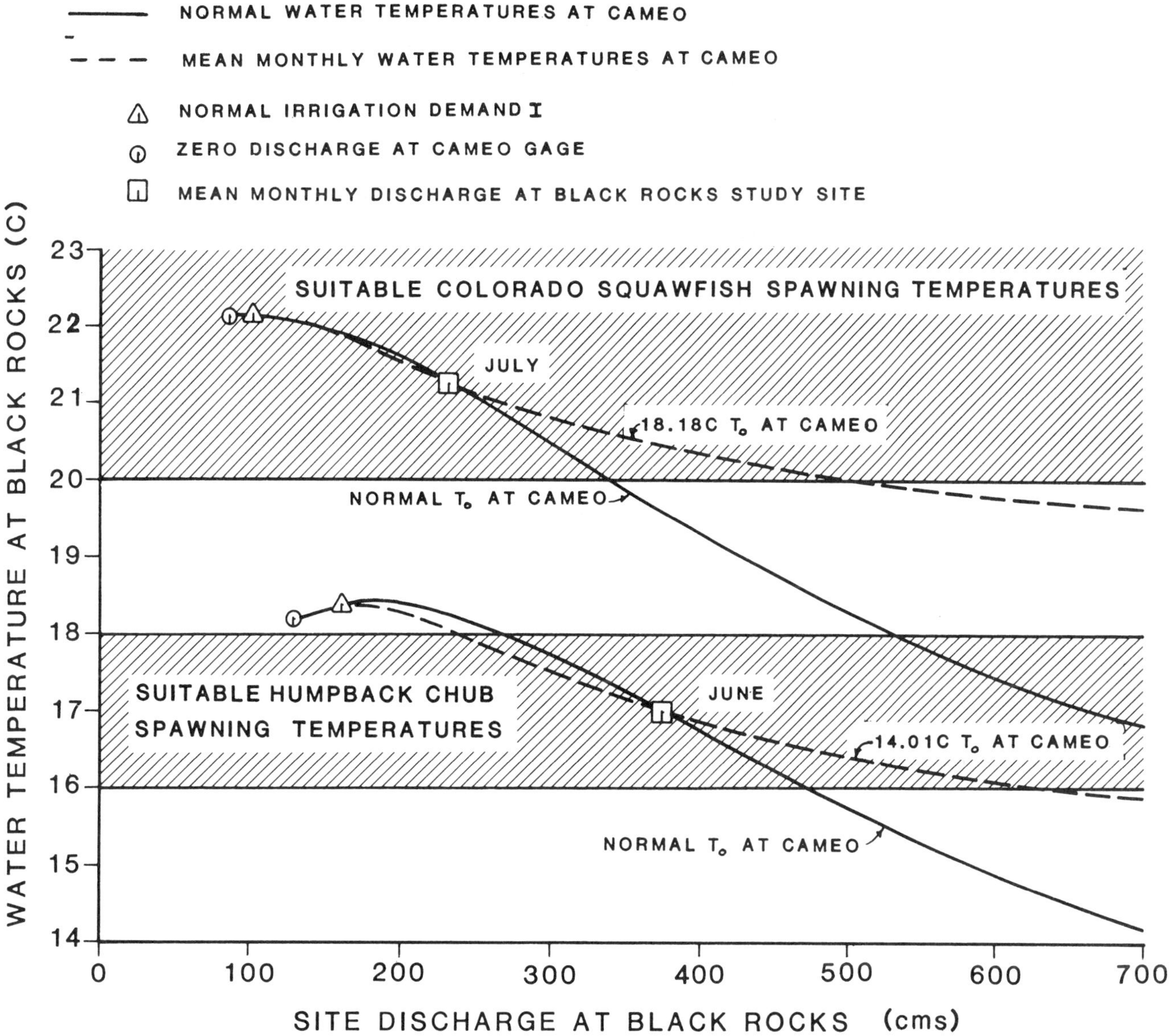

Figure 3. Temperature vs. Site Discharge at Black Rocks Study Site for the Normal June and July Hydrometeorological Conditions, Respectively. Various discharges were assumed at Cameo gage with the remaining tributaries below this gage contributing normal June and July discharges. Two criteria to meet water temperatures at Cameo gage are shown: (1) the normal water temperature at Cameo as a function of assumed discharge and normal monthly meteorological conditions at Cameo gage, and (2) the mean monthly water temperature associated with the normal monthly hydrometeorological conditions at Cameo gage (14.01°C in June and 18.18°C in July).

assumed criterion, are simulated by the temperature versus site discharge curves at Black Rocks. A reservoir at Cameo could be managed according to either water temperature criteria. Water resource management projects upstream from Cameo may be better simulated by the first criterion (normal water temperatures at Cameo) because the stream flow above Cameo would be subject to otherwise natural hydrometeorological conditions.

The suitable squawfish and chub spawning temperatures are superimposed on the temperature discharge plot. Clearly, the spawning temperature requirement for squawfish at Black Rocks would be met for all discharges less than the mean July flow (233 cms) at Black Rocks and for discharges at least up to 350 cms depending upon the applicable water temperature criteria at Cameo. However, the chub spawning temperature could require a Black Rocks flow range in June as narrow as 270 to 480 cms depending upon the applicable water temperature criteria at Cameo. The mean June flow is 375 cms at Black Rocks.

The results of the validation check at the State Line gage, which is located 8 kilometers below Black Rocks, was −0.13°C and −0.18°C for the normal June and July hydrometeorological conditions, respectively.

CONCLUSION

The two examples clearly demonstrate the power of the instream water temperature modeling capability. The model allows the user to predict, with reasonable confidence, downstream temperature impacts of changes in upstream discharges, water temperatures, and even stream geometry regardless of the complexity of the stream network.

This model can be used in numerous applications to analyze projects planned to enhance or suspected to affect the aquatic habitat temperature regime. As knowledge of the temperature requirements of aquatic species expand, more sophisticated use of this model is expected.

LITERATURE CITED

Hammond, R. L. and T. D. Inslee, 1981. Culture of Endangered Colorado River Fishes. U.S. Fish Wildl. Serv. Nat. Fish Hatchery, Willow Beach, Arizona.

Holden, P. B. and L. W. Crist, 1981. Documentation of Changes in the Macroinvertebrate Fish Populations in the Green River Due to Inlet Modification of Flaming Gorge Dam. Bio West, Inc., Logan, Utah, 52 pp.

Holden, P. B. and C. B. Stalnaker, 1975. Distribution and Abundance of Mainstream Fishes of the Middle and Upper Colorado River Basins, 1967-73. Trans. Am. Fish. Soc. 104(2):217-231.

Joseph, Timothy W., James A. Sinning, Robert J. Behnke, and Paul B. Holden, 1977. An Evaluation of the Status, Life History, and Habitat Requirements of Endangered and Threatened Fishes in the Upper Colorado River System. FWS/OBS-77/62, September, U.S. Fish Wildl. Serv., Fort Collins, Colorado, 184 pp.

Miller, W. N., J. J. Valentine, D. L. Archer, H. M. Tyus, R. A. Valdez, and L. Kaeding, 1982. Final Report, Colorado River Fisheries Project. U.S. Fish Wildl. Serv., Salt Lake City, Utah, 36 pp., appendices.

Minckley, W. L. and J. E. Deacon, 1968. Southwestern Fishes and the Enigma of "Endangered Species." Science 159:1424-1432.

Prewitt, Charles G., Kenneth A. Voos, and Terry Waddle, 1981. Completion Report – IFG Study of Physical-Chemical Habitat Conditions on the Green and Colorado Rivers in Colorado and Utah. Coop. Instream Flow Serv. Group, U.S. Fish Wildl. Serv., Fort Collins, Colorado, 44 pp.

Theurer, Fred D., 1982. Instream Water Temperature Model. Instream Flow Inf. Paper 16, Coop. Instream Flow Serv. Group, U.S. Fish Wildl. Serv., Fort Collins, Colorado.

Theurer, Fred D. and Kenneth A. Voos, 1982. IFG's Instream Water Temperature Model Validation. Conference on Water and Energy: Technical and Policy Issues. ASCE Proceedings of the Hydraulics Conference, Pittsburgh, Pennsylvania, and Fort Collins, Colorado, May 23-26 and June 23-27, 1982, 6 pp.

U.S. Geological Survey, 1973. Water Resources Data for Arizona. Part 2. Water Quality Records. U.S. Geol. Surv., Tucson, Arizona, 178 pp.

Wegner, David L., Jerry B. Miller, and Donald R. Brummer, 1981. Salinity and Phosphorus Routing Through the Colorado River Reservoir System. Colorado River Aquatic Habitat Symposium, November, 26 pp.

INTERNATIONAL SYMPOSIUM ON HYDROMETEOROLOGY
JUNE AMERICAN WATER RESOURCES ASSOCIATION 1982

APPLICATION OF STATISTICAL LOW FLOW ANALYSIS AS A BASIS FOR WATER QUALITY PLANNING

Henning Rubach[1]

ABSTRACT: Relations between runoff and water quality are analyzed on a statistical basis. Relevant investigations were made possible because, for the first time, parallel measurements of runoff and water quality parameters are available. Both relations between runoff and water quality and methods for improvement of water quality by low flow augmentation are discussed.
(KEY TERMS: low flow statistics; water quality statistics; low flow augmentation.)

INTRODUCTION

Statistical analysis of low flow periods is a necessary basis for water quality management. Especially the relations between low flow values, their duration and the frequency are of importance for many tasks related with water quality management.

Two practical examples for application of statistical low flow analysis are discussed. The examples refer to the Leine River basin in northern Germany which consists both of mountains up to 1000 m above sea level and very flat areas. Due to the relatively dense population, management of both runoff and water quality is necessary in this basin and is already done to a certain extent by use of existing reservoirs.

In the first example, water quality parameters are described as a function of runoff of defined frequency and may thus be used as information for the necessary amount of low flow augmentation.

In the second example, it is shown what size of storage volumes is necessary for the desired amount of low flow augmentation. The storage volumes depend on the desired minimal runoff value, the frequency and the duration of the augmentation process.

For the relevant calculations, the gage Greene in the Leine is used for assessment of runoff frequency, the gage Foerste, which lies upstream of a big pollution source and the gages Brueggen, Poppenburg and Pattensen, which are situated below this point, were used for analysis of potassium permanganate consumption.

The relatively long series of measurements (10 years) for potassium permanganate consumption and dissolved oxygen are a relatively good basis for the investigation of runoff-quality relations (Figure 1).

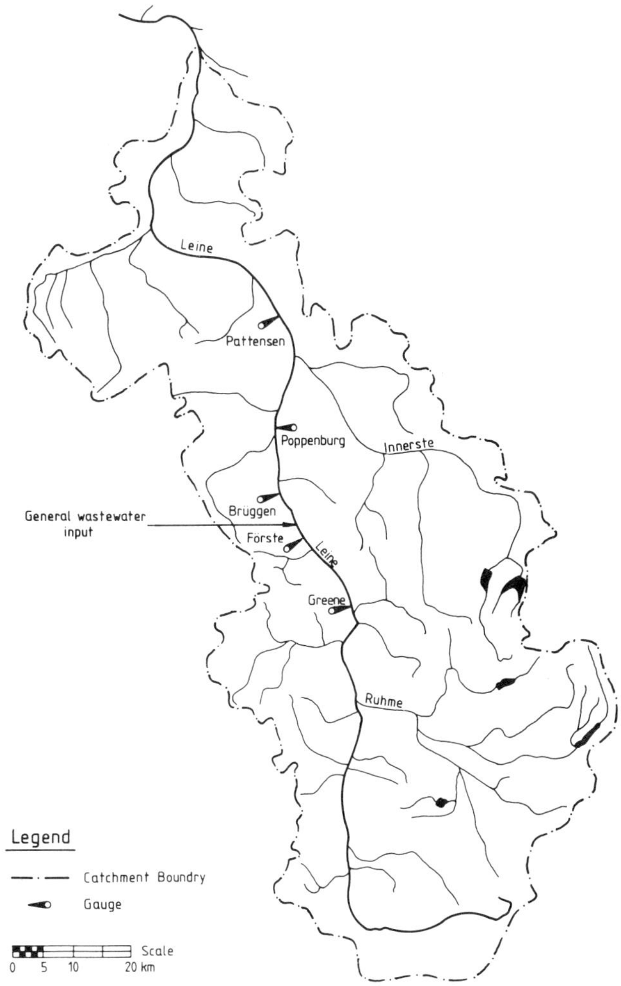

Figure 1. Runoff and Quality Gages in the Leine River Basin.

[1] Assistant, Institut Für Wasserwirtschaft, Hydrologie und Landwirtschaftlichen Wasserbau, Callinstrabe 32, D-3000 Hannover, Germany.

STATISTICAL ANALYSIS OF LOW FLOW

In case of low flow analysis it is often necessary to describe the statistical properties by more than one parameter. Especially knowledge of both minimum runoff and low flow duration is important for many tasks. A further important parameter is the termine of the start of low flow periods.

On the basis of daily runoff values, the mean low flow is calculated for a set of different durations of low flow periods. For each duration step, the yearly lowest value of mean low flow in the period of this duration is used as an element of a series of extreme values. A frequency distribution is fitted to this empirical series of extreme values to extrapolate low flows of defined frequencies.

If this is done for a set of durations, a duration-runoff frequency function can be determined, which gives information about mean low flows for defined durations and frequencies. For instance the mean low flow for a period of 20 days and a return period of 5 years is 11.5 m^3/s.

FREQUENCY ANALYSIS OF WATER QUALITY PARAMETERS

Water quality parameters are analyzed anologous to the method described above. Results are duration-quality frequency functions. As an example for this type of functions, the relations between frequency of potassium permanganate consumption depending on duration is given in Figure 3 for the gage Poppenburg. So for instance, the exceedance of a mean value of 52 mg/L potassium permanganate consumption in 7 days is related to a return period of 2 months or a mean value of 58 mg/L to a return period of 4 months.

As expected, the mean potassium permanganate consumption decreases as the duration increases.

ANALYSIS OF THE CORRELATION WATER QUALITY-RUNOFF

An important question is whether frequency relations of runoff data can be used as an index for frequency relations

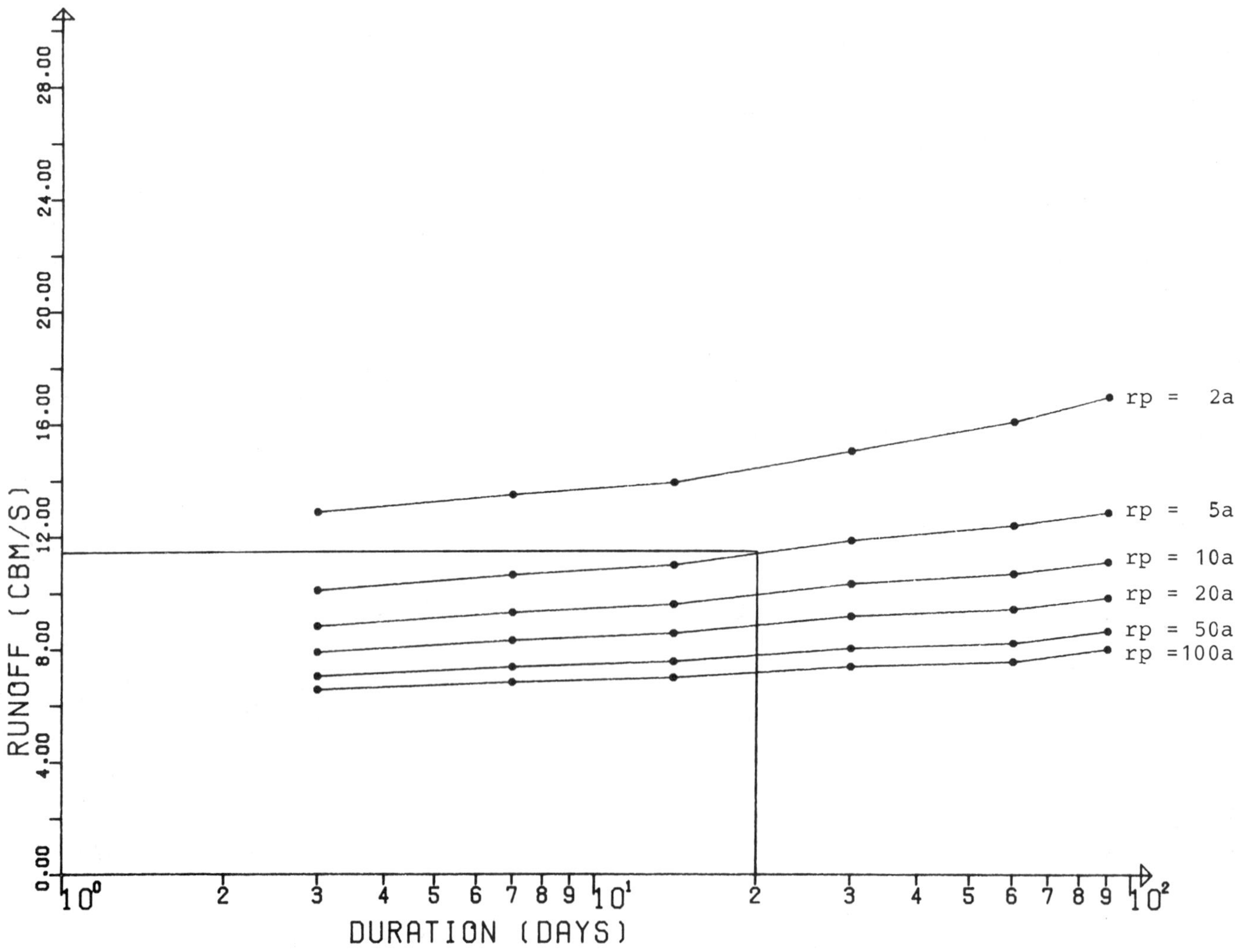

Figure 2. Duration-Runoff-Frequency Function – Greene.

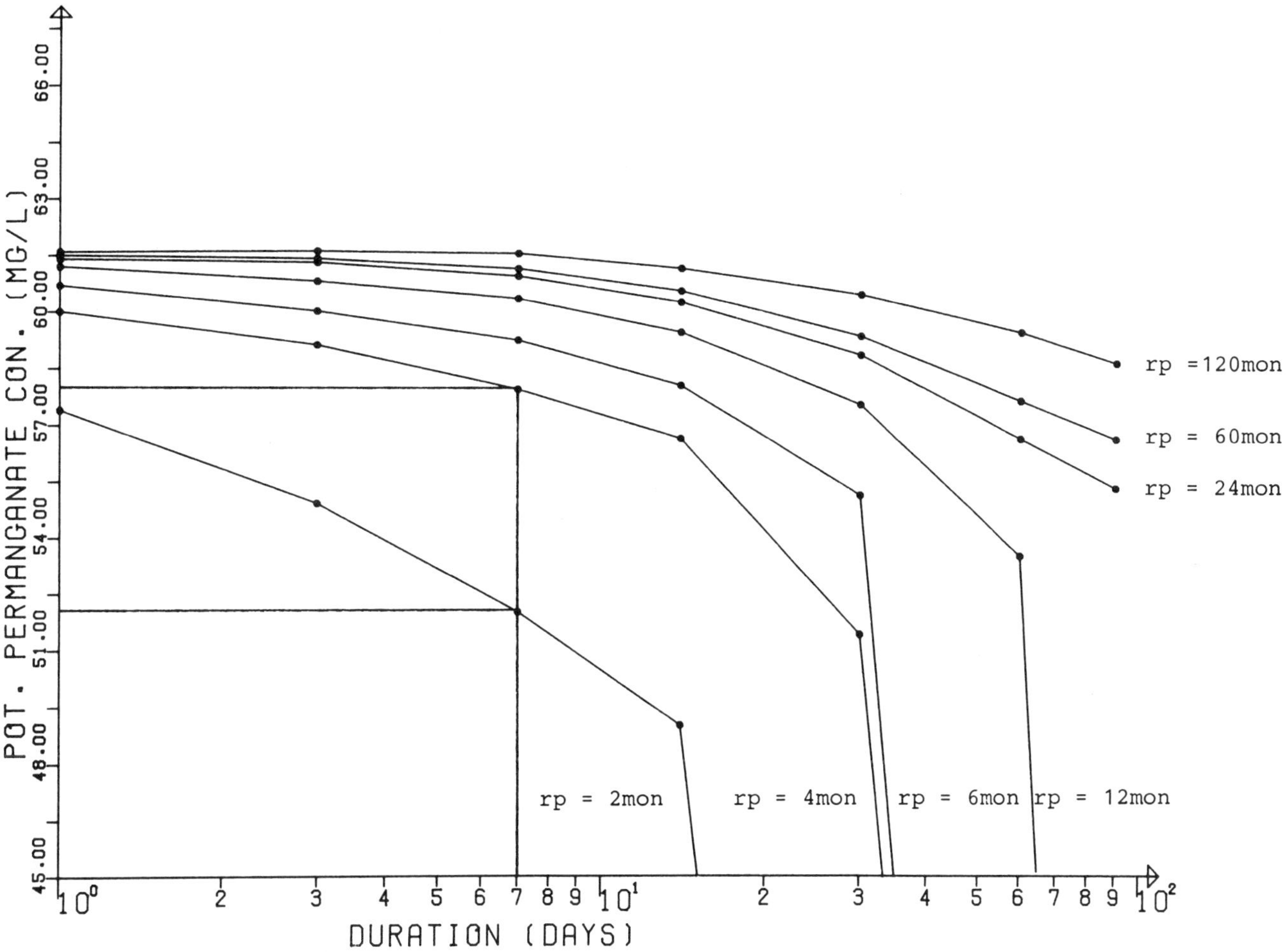

Figure 3. Example for Duration-Quality-Frequency Function – Poppenburg (potassium permanganate consumption).

of water quality parameters. To check this, the (linear) correlation between synchronous values of mean low flow and water quality parameters was tested for all durations. The result was that correlation coefficients between 0.1 and 0.3 indicated that the correlation was very poor. Depending on the sample size, only in 10 percent of the cases correlation coefficients were significantly different from zero on a 95 percent significance level. This means that it is not possible to assess water quality parameters using runoff information only.

Relevant tests had been executed both with potassium permanganate consumption and dissolved oxygen values. The nonexistent correlation between runoff and water quality parameters may be explained by the manyfold influences on water quality.

Nevertheless, it could be found that certain regions of synchronous runoff values and water quality parameters as, for instance, the potassium permanganate consumption (Figures 4, 5, and 6) and for different durations may be defined by enveloping curves in a runoff-quality diagram.

Figures 4 and 5 show these curves for different durations on one gage and Figure 6 for one duration on four gages, which are situated lower on the Leine River. The enveloping curves in Figures 4, 5, and 6 include more than 95 percent of the values.

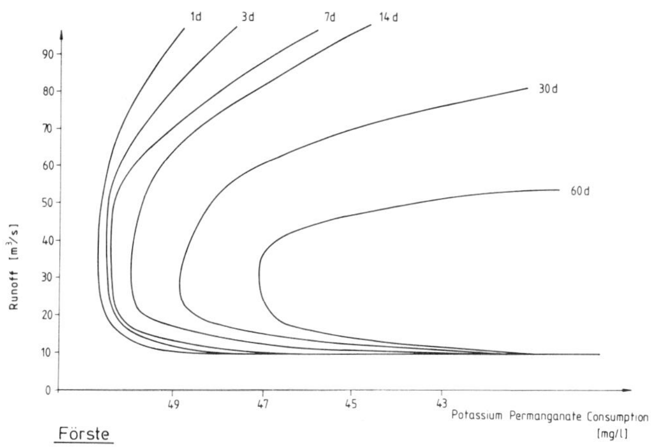

Figure 4. Runoff-Quality Diagram for the Gage Föerste for Different Durations.

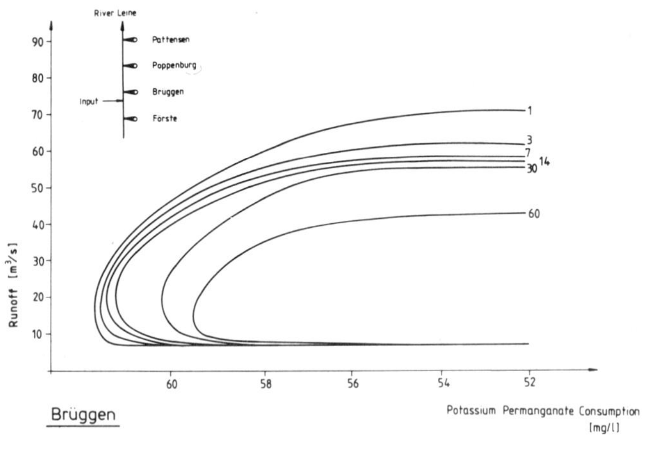

Figure 5. Runoff-Quality Diagram for the Gage Brüeggen for Different Durations.

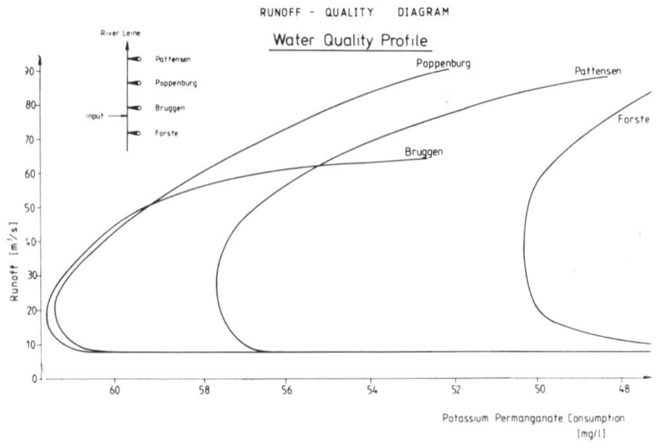

Figure 6. Runoff-Quality Diagram for the Gages Föerste, Brüeggen, Poppenburg, and Pattensen for 7 Days Duration.

An analysis of the runoff-quality diagrams allows the following interpretations:

1. For one gage, the enveloping curves of longer durations are included by the enveloping curves related with shorter durations. This is a consequence of the fact, that in the case of short durations, influences by small and short floods affect quality parameters more than in cases of longer durations. Further, a smoothing of natural variances takes place in case of longer durations and produces a dampening effect.

Figure 5 shows the envelopes for a river reach with a high pollution level situated downstream of a significant pollution source. The distances between different enveloping curves are smaller than in Figure 4. Distances between enveloping curves for shorter durations in Figure 5 are smaller than in Figure 4, as a consequence of the higher pollution level.

2. The regional characteristic of quality parameters may be considered by comparing enveloping curves for one duration (7 days) and different subsequent gages. Above the pollution source, on the gage Foerste, the potassium permanganate consumption is rather small, below 50 mg/L and is not very sensitive to runoff changes.

Below the pollution source, the parameter is on a considerably higher level of more than 60 mg/L. These pollution peaks exist on the gage Brueggen only for runoff values of less than 60 m^3/s. Also at Poppenburg this peak may be observed, but also at higher runoff levels. On the gage Pattensen, the maximal potassium permanganate consumption goes down to 57 mg/L, the enveloping curve is shifted parallel to the relevant curve for Poppenburg.

3. It can be seen that the highest pollution does not occur in times of extreme low flow. This means that self purification effects in general increase if runoff increases, but may improve again at very low runoff states in direction to the extreme low flows. Thus it is possible to define a set of runoff values which are critical with respect to water quality states.

FREQUENCY ANALYSIS OF RUNOFF VALUES WITH DEFINED WATER QUALITY CRITERIA

The frequency of runoff states which are related with a synchronous exceedence of defined water quality criteria may be assessed by a statistical analysis. A set of runoff values, at which a given threshold value of a water quality parameter is exceeded, is used as a sample. If several of these samples related to different threshold values are used and frequency distributions are fitted to these, a runoff-quality frequency function is produced (see Figure 7).

For the parameter potassium permanganate consumption, there are considerable differences between the gage Foerste, which has a relatively low pollution level and the gages Brueggen, Poppenburg, and Pattensen, at which a higher pollution level exists.

As has been discussed above in relation with the runoff-quality diagrams, the differences between runoff of the frequencies 0.1 and 0.9 are greater on the gage Foerste than on the other gages (see gage Brueggen in Figure 7). For the gages with high pollution levels (Brueggen, Poppenburg, and Pattensen) runoff values selected by use of a quality criterion are, for a certain frequency, smaller than original runoff values of the same frequency.

ASSESSMENT OF THRESHOLD VALUES IN CASE OF LOW FLOW AUGMENTATION

By low flow augmentation from reservoirs the frequency of water quality parameters can be changed, but the occurrence of water quality parameters below the desired level cannot be prevented.

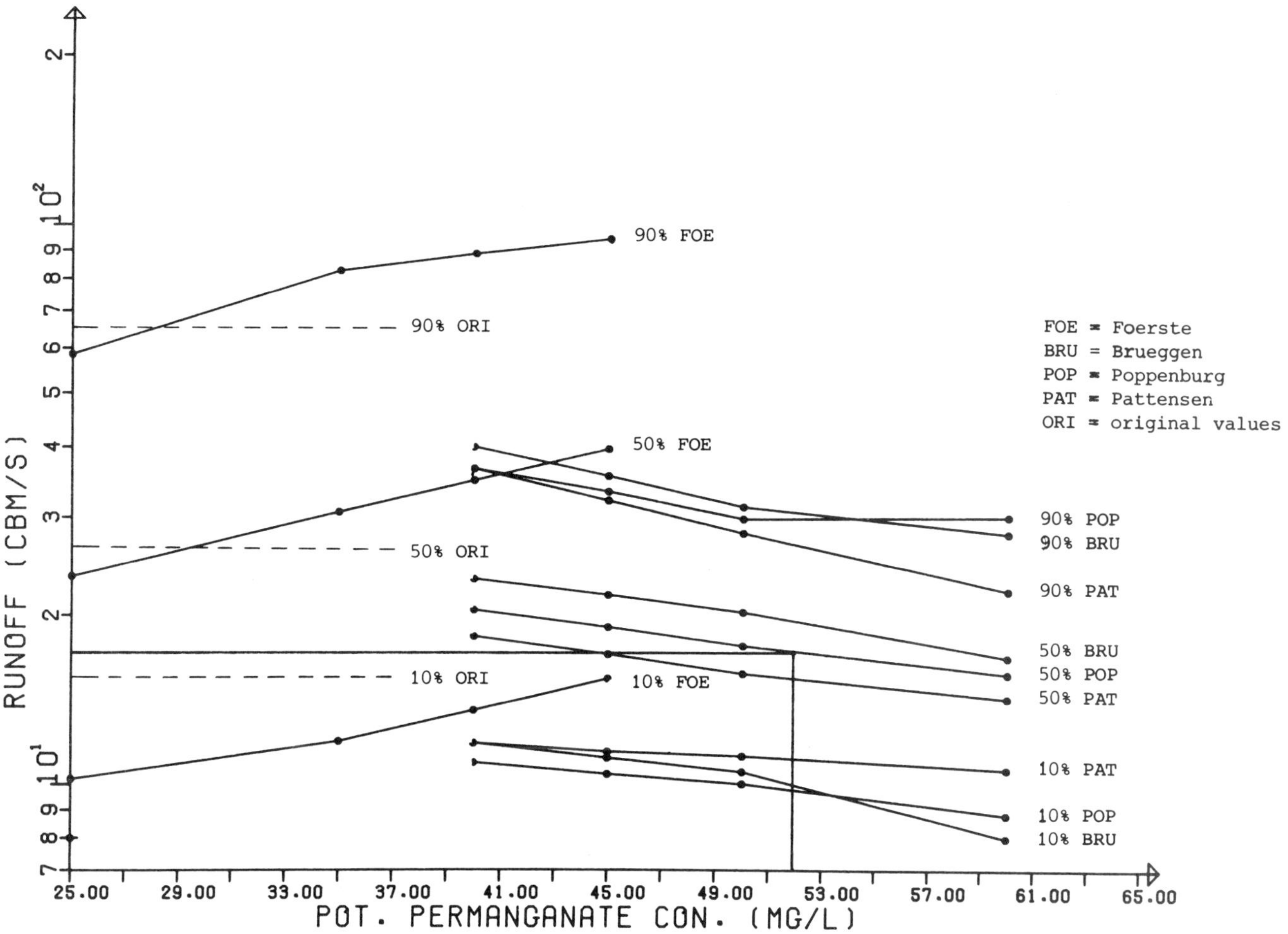

Figure 7. Runoff-Quality-Frequency Function for 7 Days Duration (potassium permanganate consumption).

As an example, on the gage Poppenburg, a value of potassium permanganate consumption of 52 mg/L in a period of 7 days corresponds with a return period of 2 months under present conditions (see Figure 3). It is desired to change the quality in such a way that the same quality conditions occur with a return period of 4 months.

It follows, that in 50 percent of the cases, with unacceptable water quality conditions, must be prevented. As shown in Figure 7, for a potassium permanganate consumption of 52 mg/L, the frequency curve of 50 percent for gage Poppenburg gives a runoff threshold of 17.4 m^3/s. It must be considered that, in this case, the number of events above the 50 percent line increases. For a proper control of the found threshold value, the following considerations must be made:

1. The frequency curves in Figure 7 are shifted in direction of bigger runoff values depending on quality of water which is used for augmentation, and the original runoff level.

2. It may be determined, for instance from a deterministic water quality model, how water quality changes with the augmentation.

ASSESSMENT OF STORAGE VOLUMES FOR LOW FLOW AUGMENTATION

If a proper threshold value for runoff augmentation and an essential water quality improvement are defined as described above, the storage volumes must be assessed, which are necessary to produce this augmentation.

During the analysis of the duration-runoff frequency function (Figure 2) the mean low flow of periods with defined durations has been used, the mean runoff which is necessary for the augmentation can be assessed directly as difference runoff.

The necessary storage volume can be found by multiplying the difference runoff by the duration. If for instance in a period of 7 days a threshold value of runoff of 17.4 m^3/s is to be produced, a volume of $(17.4 - 13.5) \times 7 \times 86400 = 2.4\ 10^6 m^3$ is necessary.

The found storage volume guarantees the augmentation of a 7-day low flow period with a return period of 2 years. It is assumed, in this case, that runoff control is possible in such a way that such a constant runoff threshold can be held.

If the above described procedure is repeated for all durations, a maximal volume can be found and used for the dimensioning of storage basins (Figures 8 and 9).

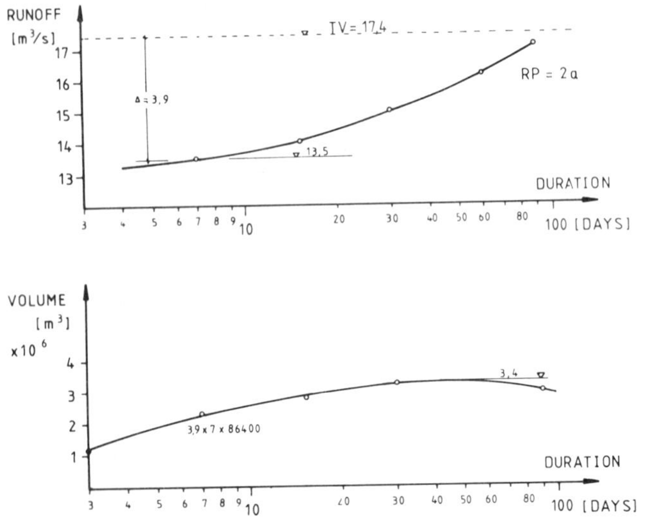

Figure 8. Duration-Runoff-Frequency Function.

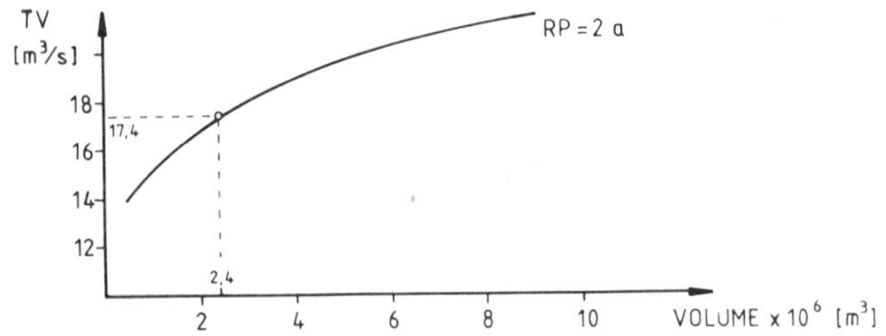

Figure 9. Graph to Determine Necessary Storage Volumes.

The repeating of the procedure for different runoff threshold values and frequencies leads to a function, which shows the necessary storage volumes which depend on threshold value and frequency (Figure 9).

LITERATURE CITED

Computergerechte Wasserwirtschaftliche Rahmenplanung, Mitteilungen aus dem Institut fuer Wasserwirtschaft, Hydrologie und landwirtschaftlichen Wasserbau der Universitaet Hannover, Heft 44, Band 1 und Band 2, 1978.

Gumbel, E. J., 1958. Statistics of Extrems. Columbia University Press, New York, New York.

Lecher, K., H. Lehmann, and K. Ludwig, 1982. Bewirtschaf tungsplaene fuer Fliessgewaesser – Water Management Plans for River Systems. Zeitschrift fuer Kulturtechnik und Flurbereinigung, Heft 1.

Prellberg, D., 1977. Beitrag zur Ermittlung von Niedrigwasserperioden, Mitteilungen des Leichtweiss. Institutes fuer Wasserbau der Technischen Universitaet Braunschweig, Heft 57.

NATURAL ACIDITY OF WATERS IN PODZOLIZED SOILS AND POTENTIAL IMPACTS FROM ACID PRECIPITATION

J. D. Stednick and D. W. Johnson[1]

ABSTRACT: Nutrient movements through sites in southeast Alaska and Washington were documented to determine net changes in chemical composition of precipitation water as it passed through a forest soil and became stream flow. These sites were not subject to acid precipitation (rainfall pH 5.8-7.2), yet soil water was acidified to 4.2 by natural organic acid forming processes in the podzol soils. Organic acids precipitated in the subsoils, allowing a pH increase. Stream water pH ranged from 6.5 to 7.2 indicating a natural buffering capacity that may exceed any additional acid input from acid rain. Precipitation composition was dominated by magnesium, sodium, and chloride due to the proximity of the ocean at the southeast Alaska site. Anionic constituents of the precipitation were dominated by bicarbonate at the Washington site. Soil podzolization processes concurrently increased solution color and iron concentrations in the litter and surface horizons leachates. The anion flux through the soil profile was dominated by chloride and sulfate at the southeast Alaska site, whereas at the Washington site anion flux appeared to be dominated by organic acids. Electroneutrality calculations indicated a cation deficit for the southeast Alaska site.
(KEY TERMS: nutrient cycling; podzolization; soil water; acid precipitation.)

INTRODUCTION

Acid rain is but one of several processes causing H^+ input to forest ecosystems. Carbonic and organic acids can produce H^+ within soils and often account for the majority of cations leached from undisturbed forest soils (Johnson, *et al.,* 1977). Carbonic acid does not normally produce low pH solutions since H^+ rapidly exchanges for cations on soil cation exchange sites, but the more highly buffered organic acids are known to significantly acidify waters, sometimes lowering rain pH from 5.6 to less than 4.5 as it passes through forest canopy and litter layers in alpine or northern ecosystems (Johnson, *et al.,* 1977; Johnson, 1981).

In this paper we describe solution composition and ion fluxes for unpolluted sites in southeastern Alaska and in the subalpine zone of the Washington Cascades. The purpose of this comparison is to examine similarities and differences in natural leaching processes within these two ecosystems which are somewhat similar in climate but quite different in geology and vegetation. Definition of natural leaching processes will enable speculation on potential impacts of acid precipitation.

Sites

The subalpine site is located at an elevation of 1130 m in the Washington Cascades, in the upper drainage of the Cedar River Watershed, at Findley Lake, a 10 ha oligotrophic lake in a cirque with a maximum elevation of 1450 m. Nutrient cycling studies were located on a moraine near the outlet of the lake where the soil is a heavily podzolized Cryandept overlying andesite.

Vegetation at Findley Lake consists of an overstory of silver fir [*Abies amabilis* (Dougl.) Forbes] with mountain hemlock [*Tsuga mertensiana* (Bong.) Carr.] as a secondary species. Understory vegetation is blueberry (*Vaccinium* spp) and beargrass (*Xerophyllum tenax*). Mean annual temperature is 5.4^{o}C, with an annual precipitation mean of 270 cm, most of which is in the form of snow which lasts 9-10 months.

The Alaskan study was conducted in Indian River Valley near Tenakee Springs on Chichagof Island in southeast Alaska. The experimental site (lat. 57° 49′ 50″ N, long. 135° 16′ 00″ W) was at an elevation of 105 m with mountain peaks up to 960 m, forested up to 770 m. The mature coastal Sitka spruce-western hemlock forest [*Picia sitchensis* (Bong.) Carr. and *Tsuga heterophylla* (Raf.) Sarg.] with subordinate vegetation of *Vaccinium ovalifoluim - V. alaskensis* and *Menziesia ferruginea* (Viereck and Dryness, 1980) was 400-600 years old and approximately 70 percent hemlock by volume. Area climate is characteristically humid with mean annual precipitation of 255 cm. The average annual air temperature is 7°C.

The soil at Indian River was identified as belonging to the Ulloa gravelly silt loam series, a Dystric Cryandept and thixotropic soil. These soils typically have thin A2 horizons; thin dark, reddish-brown to dusky red B21 horizons over slightly acid to neutral dark, yellowish-brown to dark brown B2 horizons. Ulloa soils occur on steep side slopes

[1]Respectively, Assistant Professor, Department of Earth Resources, Colorado State University, Fort Collins, Colorado 80523; and Research Staff, Environmental Sciences Division, Oak Ridge National Laboratory, Oak Ridge, Tennessee 37830.

of glaciated valleys. The regolith is weathered colluvium derived from limestone and metamorphosed marble.

METHODOLOGY

At the Findley Lake site, precipitation and throughfall at all sites were collected by funnels and plastic containers, and soil solutions were collected by means of tension lysimeters (Cole, 1968). Vacuum systems were installed at all sites to maintain lysimeter plate tension (10kPa).

Solutions were analyzed for pH and bicarbonate within hours of sample collection. Samples were collected frequently to avoid losses of dissolved CO_2 and pH rise. Bicarbonate was determined by titration with 0.02N H_2SO_4 to pH 4.5. Sulfate was determined on a Technicon Autoanalyzer II. Chloride was determined by specific ion electrode (Orion Model 801). Calcium, magnesium, sodium, and potassium were analyzed by atomic adsorption spectrophotometry (Instrumentation Laboratories spectrophotometer 353). Periodic analyses were made for aluminum, iron, ammonium, and orthophosphate, but concentrations were negligible relative to the other ions. Color was measured as percent transmittance (Bausch and Lomb Spectronic 70) and expressed as platinum color units (alpha-platinum standard).

At the Indian River site, water samples were collected as: (1) precipitation, (2) forest floor leachate, (3) mineral soil leachate (near the bottom of the rooting zone), (4) ground water from a spring, and (5) stream flow. Precipitation samples were collected at three sites. Precipitation inlets were covered with spun fiberglass to prevent sample contamination from windblown litterfall and detritus. Forest floor leachate (organic horizon) was collected by ceramic lysimeter plates (Cole, 1968) installed under the forest floor. A hanging column of water provided a tension of approximately 10kPa and drained into 20 liter carboys buried in the soil. Mineral soil leachates were collected by suction cup lysimeters on 1 meter tubes evacuated to 15kPa and buried from 0.6-1.0 meters deep. Samples were analyzed for pH, electrical conductivity, alkalinity, total-nitrogen, ammonium-nitrogen, nitrate-nitrogen, total-phosphorus and ortho-phosphate, sulfate-sulfur, chloride, silica, calcium, magnesium, sodium, and potassium using standard methods similar to those described above.

RESULTS AND DISCUSSION

The acidification processes naturally occurring in the podzol soils may create acid solutions as low as pH 4.2 (Johnson, *et al.,* 1977) (Table 1). The solutions are buffered as they pass through the solum and emerge as stream flow close to neutral. Soil podzolization processes concurrently increase solution color and ion concentrations in the litter and surface horizons leachates. Soil organic matter production results in proton release and subsequent acidification of the near surface horizons. But more important is the production and mobilization of organic acids. This decrease in pH is due to the leaching of organic acids from these ecosystem components, and the organic acids (mainly fulvic acid) are responsible for the ongoing podzolization processes in soils from this site (Ugolini, *et al.,* 1977a, b; Dawson, *et al.,* 1978). These organic acids with their chelated Fe and Al precipitate in the B2hir horizon (Dawson, *et al.,* 1978) allow a pH rise and bicarbonate dominance of solutions beneath the B2hir (Johnson, *et al.,* 1977).

In contrast, precipitation is only slightly acidified as it passes through the litter layer at the Indian River site (Table 1; throughfall data were not taken at this site). The soils at this site are heavily podzolized and the presence of very highly colored solutions suggests that organic compounds are present. However, pH's are high and bicarbonate plays a major role in buffering forest floor solutions and dominates soil and stream solutions (Table 2). Clearly, the limestone present in soil parent material at this site is exerting a major influence on not only the geochemical components of the ecosystem (spring and stream water chemistry) but also on the biogeochemical components of the ecosystem (forest floor and soil solution).

Findley Lake site generally has an anion deficit at monitored transferred sites (the difference between the sum of cations and sum of anions). This anion deficit has been attributed to the presence of organic acids (Johnson, *et al.,* 1977). In contrast, Indian River exhibited a cation deficit in solutions from the forest floor, mineral soil, and ground water sampling points. Calculated electrical conductivities were less than measured electrical conductivities (Stednick, 1982) and suggests that organic compounds (as evidenced by solution color) may be positively charged.

Potential Impacts of Acid Rain

Assuming anthropogenic sources contaminated the atmosphere such that the precipitation pH reached 4.0 (0.100 meq/L as H^+) and that SO_4^{2-} increased by 0.100 meq/L to account for the H^+ increase, then assume SO_4^{2-} in soil solution and streams increased by 0.200 meq/L due to concentration effects from evapotranspiration. This causes an increase in total anion concentration of 0.200 meq/L in solutions containing no bicarbonate. Since total anions increase, total cations will increase by the same amount to maintain electroneutrality. If H^+ increases in proportion to total cations (Seip, 1980; Johnson, 1981) pH will decrease. In solutions of pH 5 or higher, this will cause a decrease in bicarbonate concentration which will, in turn, cause a decrease in total anions and thus total cations. In short, solutions of pH 5 or above will experience an anion shift from HCO_3^- to SO_4^{2-} and total anions and cations will increase by less than the amount of SO_4^{2-} added (0.200 meq/L). The exact extent of the shift is easily solved by iterative techniques (Johnson, 1975). Solutions

TABLE 1. Some Properties of Natural Waters from Indian River, Alaska, and Findley Lake, Washington.

	pH	HCO_3^- (meq/L)	SO_4^{2-} (meq/L)	Cl^- (meq/L)	Σ Anions (meq/L)	Σ Cations (meq/L)	Color Units
			INDIAN RIVER, ALASKA				
Precipitation	6.5	0.087	0.023	0.021	0.143	0.142	5
Forest Floor	5.5	0.200	0.261	0.274	0.739	0.377	1300
Mineral Soil	6.5	0.985	0.251	0.390	1.632	0.831	110
Spring	6.8	1.337	0.064	0.330	1.736	1.601	60
Stream Water	7.2	0.717	0.067	0.119	0.912	1.120	40
			FINDLEY LAKE, WASHINGTON				
Precipitation	5.6	0.066	0.035	<0.010	0.101	0.106	0
Throughfall	4.3	0.000	0.064	<0.010	0.064	0.116	44
Forest Floor	4.2	0.000	0.044	0.029	0.074	0.165	301
A2 Horizon	4.2	0.000	0.056	0.026	0.065	0.186	306
B2hir Horizon	4.7	0.000	0.049	0.025	0.106	0.133	66
B3 Horizon	5.4	0.090	0.049	0.032	0.171	0.189	92
Lake Outlet	6.2	0.193	0.037	0.042	0.272	0.230	0

TABLE 2. Theoretical Composition of Waters from Indian River, Alaska, and Findley Lake, Washington, if Impacted by Acid Rain. Sulfate precipitation and throughfall are increased by 0.1 meq/L and SO_4^{2-} in soil solution and stream water increased by 0.2 meq/L. Total cations, H^+ and HCO_3^- are then solved for by iteration. See text for details of calculations.

	pH	HCO_3^-	SO_4^{2-}	Σ Anions	Σ Cations	Percent Increase in Cations
			INDIAN RIVER, ALASKA			
Precipitation	4.0	0.000	0.123	0.243	0.242	70
Forest Floor	5.4	0.145	0.461	0.884	0.522	38
Mineral Soil	6.5	0.882	0.451	1.729	1.928	12
Spring	6.8	1.249	0.264	1.546	1.713	7
Stream Water	7.1	0.644	0.267	1.039	1.247	11
			FINDLEY LAKE, WASHINGTON			
Precipitation	4.0	0.000	0.135	0.201	0.206	94
Throughfall	3.9	0.000	0.164	0.174	0.226	95
Forest Floor	3.9	0.000	0.244	0.274	0.365	120
A2 Horizon	3.9	0.000	0.256	0.282	0.386	110
B2hir Horizon	4.3	0.000	0.249	0.274	0.333	150
B3 Horizon	5.1	0.048	0.249	0.330	0.348	84
Lake Outlet	6.0	0.123	0.237	0.402	0.360	57

with a pH of less than 5 would behave differently as the bicarbonate endpoint is about pH 4.5.

Using this simple model, we have calculated the potential effects of acid rain on the anion and total cation composition and pH of solutions from the Indian River and Findley Lake sites. Due to the large influence of limestone on solution compositions at Indian River, acid rain would have only a minor effect on pH, total cations, and total anions in all solutions (except precipitation) according to this model (Table 2). On the other hand, total anions and cations increased by a factor of two or more in the more dilute Findley Lake solutions. Again, pH effects were minor but sufficient to cause substantial reductions in HCO_3^- due to the log scale for pH (i.e., see changes in HCO_3^- in Indian River forest floor, Findley Lake B3 horizon and Lake Outlet).

CONCLUSIONS

In both of the unpolluted site studies here, theoretical considerations suggest that acid rain would have a minimal effect on soil solution and stream water pH. In the Alaskan case, the influence of limestone overwhelms the effects of acid rain, and in the Washington casê solutions are either very acid already or sufficiently concentrated such that pH effects are theoretically minimal. It does seem, however, that cation leaching could increase substantially in the Washington case. The degree to which this would affect soils and solutions in the long run would depend upon the rate of cation weathering from soil minerals, a process that is difficult to quantify and likely to increase with increased H^+ inputs.

ACKNOWLEDGMENTS

Research sponsored by Contract RP-1813-1 with the Electric Power Research Institute for the Office of Health and Environmental Research, U.S. Department of Energy, under Contract W-7405-eng-26 with Union Carbide Corporation. Publication No. 2016, Environmental Sciences Division, Oak Ridge National Laboratory.

LITERATURE CITED

Cole, D. W., 1968. A System for Measuring Conductivity, Acidity, and Rate Flow in a Forest Soil. Water Resour. Res. 4:1127-1136.

Dawson, H. J., F. C. Ugolini, B. F. Hrutfjord, and J. Zachara, 1978. Role of Soluble Organics in the Soil Processes of a Podzol, Central Cascades, Washington. Soil Sci. 126:290-296.

Johnson, D. W., 1975. Processes of Elemental Leaching in Some Tropical, Temperate, Alpine, and Northern Forest Soils: Factors Influencing the Availability and Mobility of Major Leaching Agents. Ph.D. Thesis, University of Washington, Seattle, 169 pp.

Johnson, D. W., 1981. The Natural Acidity of Some Unpolluted Waters in Southeastern Alaska and Potential Impacts of Acid Rain. Water, Air, and Soil Pollut. 16:243-252.

Johnson, D. W., D. W. Cole, S. P. Gessel, M. J. Singer, and R. V. Minden, 1977. Carbonic Acid Leaching in a Tropical, Temperate, Subalpine, and Northern Forest Soil. Arctic and Alpine Res. 9:329-343.

Seip, H. M., 1980. Acidification of Freshwater – Sources and Mechanisms. *In:* Ecological Impacts of Acid Precipitation, D. Drabløs and A. Tollan (Editors). Johs. Grefslie Trykkeri A/S, Mysen, Norway, pp. 358-366.

Stednick, J. D. (in press). Nutrient Movement Through a Sitka Spruce-Western Hemlock Forest in Southeast Alaska.

Ugolini, F. C., H. Dawson, and J. Zachara, 1977a. Direct Evidence of Particle Migration in the Soil Solution of a Podzol. Science 198:603-605.

Ugolini, F. C., R. Minden, H. Dawson, and J. Zachara, 1977b. An Example of Soil Processes in the *Abies amabilis* Zone of Central Cascades, Washington. Soil Sci. 124:291-302.

Viereck, L. A. and C. T. Dyrness, 1980. A Preliminary Classification System for Vegetation of Alaska. USDA Forest Service, Pacific Northwest Forest and Range Exp. Sta. Gen. Tech. Rep. PNW-106, 38 pp.

A NEW LOOK AT FORECASTING NATURAL PROCESSES

G. M. Petrie[1]

ABSTRACT: In an effort to aid the Department of Energy in the identification and characterization of future natural processes which may degrade the performance of a mined repository for high-level nuclear waste, a stochastic/deterministic computer model was developed for application to specific sites. This model was designed to consider long time periods over which relatively slow geologic processes can be of importance. Evaluation of the results of the first analyses suggests that this technique gives useful insight into identifying and characterizing the important processes, and appears to be particularly useful in characterizing synergistic effects. Moreover, analysis of the results suggests that applying the same technique, using a shorter time scale, to water management would be rewarding. Some of the suggested advantages of such a new application of existing technology include: 1) early detection of errors in either the data or conceptual model, 2) quantification of the probabilities for the timing and magnitude of future natural events, 3) simulation of synergistic interactions, 4) quantification of uncertainty through the use of probability density functions in a manner that is relatively easy for non-specialists to understand, 5) exchange of information among technical disciplines and across political boundaries, 6) rapid evaluation of the effects of changes in regulations, and 7) clear separation of objective and subjective inputs.
(KEY TERMS: model; geohydrology; nuclear wastes; forecasting.)

INTRODUCTION

For several years, a group at Pacific Northwest Laboratories has been working to develop methods needed in assessing the behavior of deep underground radioactive waste repositories. Part of this work required developing methods to identify and characterize possible future geologic events and processes which might significantly promote the migration of waste to the biosphere.

Because the methodology will ultimately be used as a tool to inform those involved in the repository licensing process, the methodology had to meet a wide range of criteria. These criteria included:

1. ability to provide a degree of auditability;
2. ability to accommodate objective and subjective input;
3. ability to facilitate parametric and sensitivity studies;
4. ability to assist in describing disruptive sequences and their frequencies;
5. ability to establish limits or initial conditions for input into consequence analysis transport models; and
6. ability to provide the necessary flexibility to accommodate an increasing data base.

While several methods were considered, including Delphi and fault tree analysis, a computer-enhanced methodology that met the above criteria was developed for the following reasons:

1. Significant interactions exist among disruptive phenomena. For example, climate, precipitation, glaciation, sedimentation, and ground-water recharge are all synergistic. Parallel integration of process-response models of the various phenomena accounts for these interactions.
2. Most of the disruptive phenomena are time-dependent; therefore, initiation and consequences cannot always be considered simultaneously. The computer is an effective bookkeeper, keeping track of all the interrelated and time-dependent events and processes. Each submodel updates the system state which is defined by a series of variables describing the current environment (e.g., the distance from the river to the repository). The next submodel interrogates the system variables (which may have been changed by the previous submodels) and reacts to them. Thus, the interrelations and time lags do not have to be explicitly introduced into the system, but will be found by the computer simulation by superposition as it reacts to the variable changes during the step-wise integration.
3. Defining data needs, and then finding and including the necessary data, is an obstacle faced by the geologist. Because data needs are identified during model development (the translation of the conceptual model into computer code) the Geologic Simulation Model (GSM) lessens the danger that important data, logic, or elements might be overlooked. Because the model is computerized, all information must be broken down into a series of simple and logical commands.
4. Data uncertainties dictate reliance on probability distributions rather than on discrete numbers. The simulation model is an effective tool in that the data are collected in

[1] Research Scientist, Battelle – Pacific Northwest Laboratory, 3200 George Washington Way, P. O. Box 999, Richland, Washington 99352.

terms of probability density functions. If the data had to be collected in 'hard and fast' terms, they would be nearly impossible to gather and accept as accurate.

5. The sensitivity of the output to variations in the input must be analyzed for efficient use of time and money. The timely responsiveness to new information is perhaps the most valuable function of the GSM. It is designed to respond to new ideas through changes in the data rather than through "hard wiring in" concepts into the computer code. The methodology explicitly reveals the conceptual model to geoscience users and provides a degree of auditability to the study.

As well as mitigating various technical problems, a computer simulation is a valuable tool for dealing with human-related obstacles, such as possible oversimplification/abstraction of data as they are summarized and disseminated by the scientist. The GSM presents its uncertainties in terms of probabilities. As this material is in relatively compact numerical values, it has less chance of getting lost in the abstraction process.

Any risk assessment methodology must recognize that the storage of nuclear waste is an emotionally charged subject. To be successful, a methodology must present an open, believable package to a large audience. The simulation model lends a degree of objectivity. It is like a machine that solves a differential equation: the user sets up the equation and supplies the boundary conditions. Once the machine is turned on, the user cannot bias the results.

Many of the problems which were found to be mitigated by this approach in dealing with nuclear waste appear to be also common in water resources planning. The remainder of this article will briefly describe the system as it currently exists and then concentrate on describing how the current system might have to be changed in order to meet the needs of water resources management. Lastly, the possible advantages and disadvantages that such a system might offer water resources manage will be considered.

CURRENT SYSTEM

The GSM is designed to simulate evolution of the geologic and hydrologic system of a portion of the Columbia Plateau for a million years. The time frame chosen is long enough to accommodate future requirements for analyses for thousands, tens of thousands, or even hundreds of thousands of years. It is not, however, so long that the model must consider significant changes in detail of global tectonics activities and rates. Elevel submodels (Petrie *et al.*, 1981), each of which address a class of disruptive phenomena, compose the GSM:

1. climate
2. continental glaciation
3. deformation
4. geomorphic events
5. hydrology
6. magmatic events
7. meteorite impact
8. sea-level fluctuations
9. shaft seal failure
10. sub-basalt basement faulting
11. undetected features.

These submodels (Figure 1) address the disruptive natural phenomena that may, alone or in concert, affect the safety of a waste repository located in a layered basalt geologic system.

The GSM is designed to be flexible enough to incorporate objective and subjective input from a variety of geoscience disciplines, and is operated in single-run and Monte Carlo modes. The single-run mode provides for user selection of desired disruptive events and associated rates and magnitudes. The Monte Carlo mode involves pre-established distributions for individual phenomenon occurrences, rates, magnitudes and phase relations. This technique generates a large number of disruptive event sequences which may then be analyzed by geoscientists for plausibility. The ones found plausible are assessed for likelihood by a statistical package, and those which exceed regulatory probability standards (e.g., those of the Nuclear Regulatory Commission) become the event queues for consequence analysis.

Input from geoscientists is critical at all stages of the disruptive sequence selection and consequence analysis. This model is designed to complement the efforts of the geoscientist by providing broad outlines of the long-term geologic and hydrologic changes that might occur in the vicinity of the repository. The geoscientist abstracts this information, identifies patterns and trends, classifies histories, and provides detail where needed. He may also change the input data to test any insights gained in the abstraction process (Figure 2).

In this testing and abstracting process, the need for interactive computer graphics became apparent. The vast amount of data used and generated by the model would otherwise be extremely difficult to handle and interpret. In addition, many geologists are not proficient in the use of computers. It is imperative, however, that the same geologist be involved in peer review of the geologic system conceptualization and conversion of that conceptualization to a computer model.

Figure 3 illustrates the nature of the output of the GSM during the Monte Carlo phase. Shown are the average values (solid line), envelope of maximum and minimum values (dahsed line) and standard deviation (long-dashed, short-dashed line) of the Darcy velocity values (volumetric discharge per unit area assuming a homogeneous, permeable medium) through a hypothetical repository as functions of time for the first 20,000 years of simulated time (Figure 3a) and for the full million years (Figure 3b). The range in values can be attributed to several sources of uncertainty which include:

1. inhomogeneities of the geologic/hydrologic system which precludes accurate characterization at a reasonable

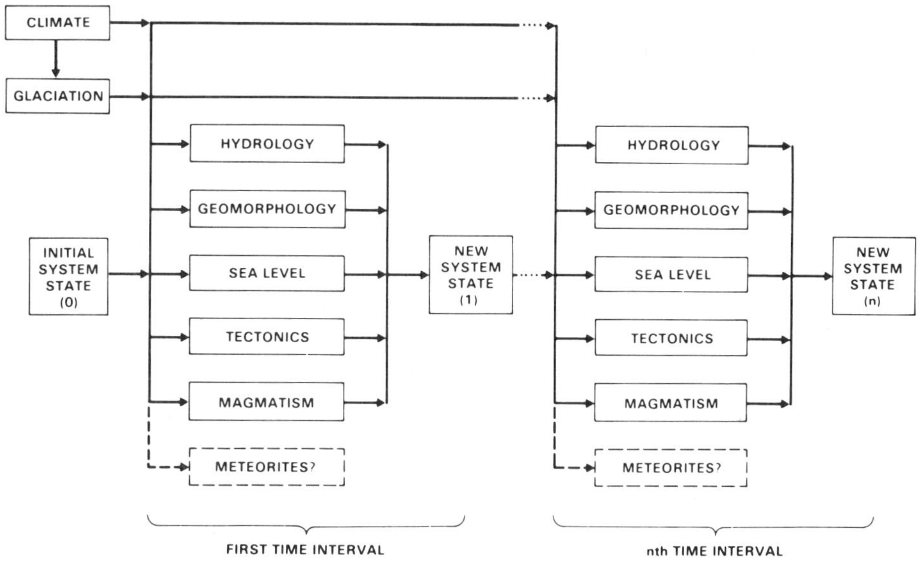

Figure 1. Schematic Diagram of the Operation of the AEGIS Geologic Simulation Model.

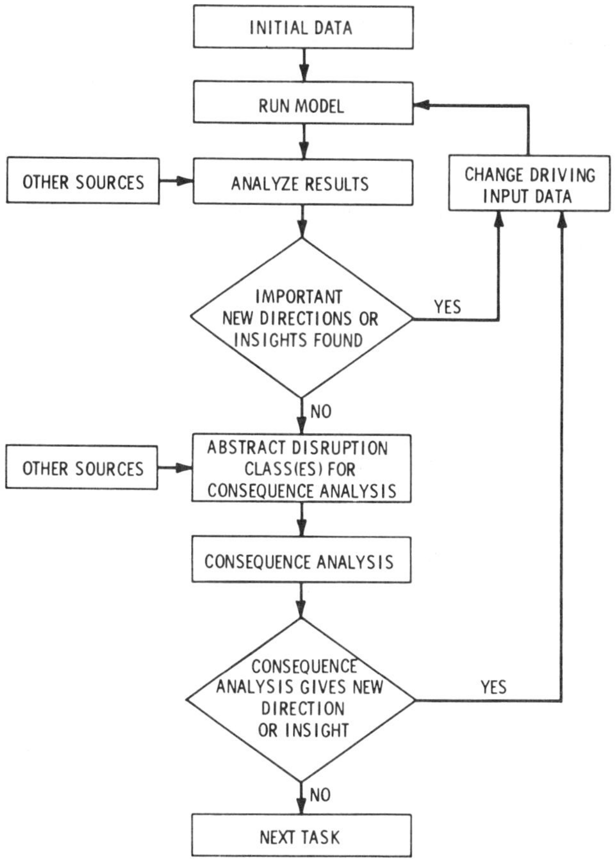

Figure 2. Model Use.

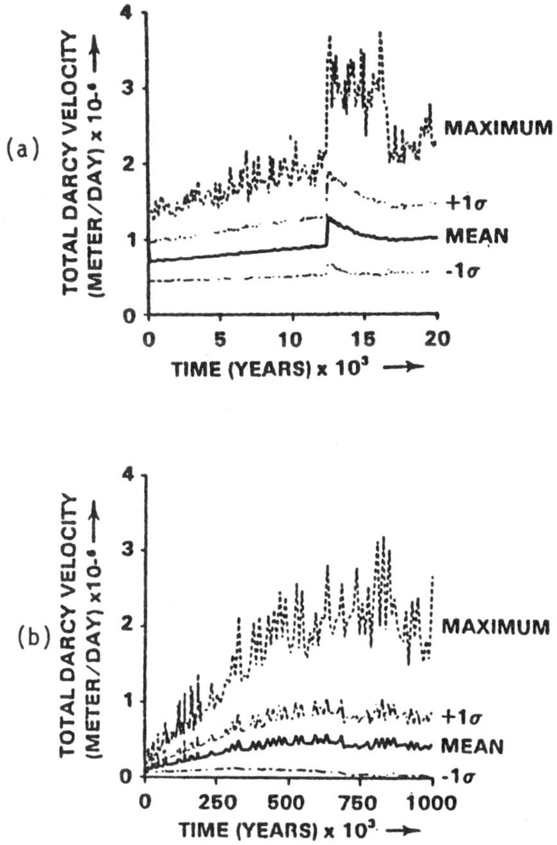

Figure 3. Total Darcy Velocity Through the Repository (a) For the First 20,000 Years, and (b) For the Entire 10^6 Years (Solid Line). Dashed lines are maximum anm and minimum values in 500 simulations. Long-dashed/short-dashed lines show ± one standard deviation (minimum values lie on horizontal axis).

level of data collection and model complexity,

2. lack of knowledge of underlying physics driving some phenomena, and

3. inherently stochastic phenomena.

In spite of the inherent "noise" in the geologic/hydrologic system under analysis, the GSM performed several useful functions. These included:

1. obtaining an estimation for the timing and magnitude of important events,

2. obtaining a better understanding of their interrelations,

3. obtaining an idea of what the driving forces are, and

4. obtaining a reasonable rationale for allocating limited resources in fugure geologic investigations.

For a detailed description of the model results, the reader is referred to Foley, *et al.* (1982).

MODIFICATIONS TO THE CURRENT SYSTEM

Not claiming to be an expert in water resources management, it would perhaps be presumptuous for the author to lay out a detailed plan of attack on how the current model could be modified to meet water resources management needs. Nevertheless, it is obvious that at least three aspects of the GSM would have to be changed. The most obvious change is the large time scale. The current model runs for a million years, with the shortest time steps being 100-yr. for the first 20,000 yr. A Water Resources Model (WRM) should have time steps on the order of a month over a 10-yr. period. Alternatively, for some regions, it may be simpler to partition a year into dry and wet seasons. Our experience suggests that the length of the time step is a trade-off between output resolution, input data quality, conceptualization limitations, run time, and development cost. For instance, the GSM uses a longer time step toward the end of the million year run for the following reasons:

1. the run time costs were reduced,

2. the resolution was not needed (i.e., the first 20,000 yr. were of prime interest), and

3. the uncertainties in both the input data and the basic assumptions used in building the model did not justify short time steps.

With this in mind, it might be most cost effective to use month-long time steps in a WRM since:

1. data appear to be available on a month-to-month basis,

2. month-long time steps allow for a greater resolution than seasonal long time steps, and

3. month-long time steps make the model easier to transfer to other areas with more than just a simple dry and wet season (which implies a savings in development cost).

Another change would be to de-emphasize the geologic events. For example, faulting is a particularly important event in geologic repository risk assessment. If included at all in a WRM, faulting need not be modeled in as much detail.

The last, and perhaps the most difficult modification to convert the GSM to a WRM, would be the increased attention given to man-induced changes. The current strategy with the GSM is to first let the model define the range of possible natural changes in the environment and then speculate on further changes which man might superimpose on these natural changes.

One characteristic of the current system, which experience suggests should not be changed, is the use of "user friendly" computer graphics. Past history in risk assessment suggests that "solving" the scientific problems is only part of the battle. It is important that the assumptions and driving data used in building and running the model be made as clear as possible. If this is not done, then the temptation on the user's part appears to be to:

1. misuse the model,
2. distrust it,
3. fear it, or
4. Ignore it.

Interactive computer graphics have been a great help in dispelling much of the mystery often associated with complicated models and computer simulations in particular.

ADVANTAGES AND DISADVANTAGES

At the outset it should be stated that building a model of useful sophistication and adequate documentation is an expensive and time-consuming process. For a Monte Carlo model, of the type described here, the run time and output analysis cost can also be significant. Nevertheless, there are several advantages that should balance out the cost. For instance, during model development the analysis tends to be forced into a holistic approach. This can be very important since the GSM effort suggests that the whole is indeed greater than the sum of its parts. Not only is the model developer forced to think in terms of interacting processes, as opposed to isolated forces; but specialists are also forced to talk to one another. Furthermore, a WRM would help promote communication across political boundaries since the natural processes being modeled tend to ignore these boundaries. In this light, even the cost of model development may be an advantage since it may force different political entities to cooperate, by sharing development cost, in order to keep the cost at a reasonable level for each party. Another important benefit that occurs during the development process is the early detection of errors. Since the translation of the conceptual model into computer code requires that concepts be reduced into simple, straightforward steps, much less "hand waving" is allowed to pass for valid answers. In addition, during model runs, errors in input data or invalid assumptions can be identified when they cause implausible results. Another advantage is that, when necessary, data can be gathered in terms of probability density functions and a measure of the degree of subjectivity can be assigned to each value.

As alluded to above, the model "accounts for" the synergistic interactions. If experience with the GSM is any indication, the synergistic effects can be important and are not always found with Delphi or fault-tree analysis. The GSM type of model appears to have the best chance of correctly accounting for these effects.

With a WRM, it should be possible to respond to changes in either scientific understanding or political regulations in a relatively fast manner. At the very least, it should be possible to do cost-effective sensitivity analysis.

One final advantage is that the results of the model can be abstracted in terms the general public can use and understand.

CONCLUSIONS

Many of the same basic problems faced in a risk assessment analysis for a nuclear waste repository appear to also be of concern in water management. It may be to the advantage of both fields to exchange information with the goal of developing sophisticated stochastic/deterministic computer models which could be useful in quantifying possible future changes in the environment.

LITERATURE CITED

Foley, M. G., 1982. Geologic Simulation Model for a Hypothetical Repository Site in the Columbia Plateau – Volume 2: Results. PNL-3542-2, Pacific Northwest Laboratory, Richland, Washington (to be published).

Petrie, G. M., *et al.*, 1981. Geologic Simulation Model for a Hypothetical Site in the Columbia Plateau, PNL-3542, Pacific Northwest Laboratory, Richland, Washington.

INTERNATIONAL SYMPOSIUM ON HYDROMETEOROLOGY
JUNE AMERICAN WATER RESOURCES ASSOCIATION 1982

AGRICULTURE EXPANSION PLANNING: INCORPORATING WATER REUSE

Mohamed N. Allam and David H. Marks[1]

ABSTRACT: A mathematical model has been built to guide decisions required for agricultural expansion planning. Of particular interest is the case when some or all of the irrigation sources available have saline water. These decisions generally are quantities and locations of all resource inputs, mixing ratio between different waters, if mixing is possible, irrigation network design, required enlargement in the existing irrigation system, and crop pattern distribution in the new lands. These decisions are based on the maximum net benefit criteria and are carried out in a mathematical optimization framework. A comprehensive study of the use of this model in a large scale planning problem has been done. This case is based upon the proposed agricultural expansion in the Nile Delta and the Sinai in Egypt. The available irrigation sources in these regions are fresh and saline waters from the River Nile, and from the drains of the existing cultivated lands, respectively. Different alternative schemes for irrigating the new lands have been obtained. An economic approach for enabling the decision makers to analyze the different alternatives has been presented.
(KEY TERMS: water reuse; model; irrigation; Egypt; saline water.)

INTRODUCTION

When using low quality water in agricultural practices, various soil and cropping problems are to be expected. The most common problems are salinity, soil permeability and toxicity. The results of numerous studies (U.S. Salinity Lab., 1954; Ayers, 1977; and Mass and Hoffman, 1977) have shown that the main damage to plant growth is usually due to salinity of the irrigation water. An increase in the irrigation water salinity causes a decrease in growth and yield of the plants. Also, the irrigation water requirement has to increase with a leaching fraction which is necessary for keeping a long term salt balance in the root zone. Therefore, in general the saline irrigation water causes an increase in the irrigation water requirement and decrease in the crop yield.

Ayers (1977) and Mass and Hoffman (1977) used the criterion that the crops vary greatly in their salt tolerance and constructed crop tolerance tables. These tables indicate for each crop the maximum allowable salinity without yield reduction and the percent yield decrease per unit salinity increase beyond the defined threshold. These tables give us a wide range for selecting crops according to their tolerance compatible with salinity of available water to obtain maximum possible yield. Goldberg, *et al.* (1971), provided another tool for reducing the irrigation water salinity effects on crops yield by irrigation more frequently using trickly irrigations methods. However, until now, no research had been done to use the trade-off between the crops salt tolerance, water requirement, and economic value as a criterion for crop selection in the context of planning agricultural expansion to new lands.

This work provides a tool for approaching investment in infrastructure, operational strategies and crop selection based on the above trade-off. This has been done through a large scale agricultural expansion planning model in such a way that in addition to the optimum crop pattern distribution it determines the necessary enlargement of the existing irrigation system for supplying the new lands with irrigated water. For each area of the new lands the model gives the most economic conjunctive use of available water sources and the optimum design of the irrigation network. In addition, as will be shown through the case study, the model can provide us with enough information to judge the trade-off between the mixing ratio of fresh to recycled waters and the economic return from the agricultural expansion project.

IRRIGATION WATER SALINITY

A salinity problem due to irrigation water quality occurs if salts from the applied irrigation water accumulate in the crop root zone and reduce the availability of soil water to the crops. To avoid salt accumulation to an excess level, it must be removed in amounts about equal to the salts applied (salt balance concept). To dissolve and remove the salts, adequate water must be applied to allow percolation through the entire root zone (leaching). This can be done during each irrigation but needs to be done only after the salts have

[1]Respectively, Research Associate, Ralph M. Parsons Laboratory, Department of Civil Engineering, Building 48-212, Massachusetts Institute of Technology, Cambridge, Massachusetts 02139; and Professor of Civil Engineering, Ralph M. Parsons Laboratory, Department of Civil Engineering, Building 48-305, Massachusetts Institute of Technology, Cambridge, Massachusetts 02139.

accumulated to near damaging concentrations. In this state, the average soil salinity of the root zone will be closely associated with the quality of the irrigation water applied as well as with the fraction of water moving in the root zone. The crop primarily responds to the average and any increase in water salinity will result in an increase in average soil salinity. Such an increase may have little practical significance, unless the salt content rises sufficiently to affect the crop yield.

The leaching fraction (LR) based on field and laboratory experience (Ayers, 1976) was computed for different irrigation methods. For surface irrigation (including sprinkler),

$$LR = \frac{EC_w}{5EC_e - EC_w} \qquad (1)$$

where:

EC_w = salt concentration of the irrigation water in millimhos at 25°C, and

EC_e = soil salinity which causes a yield reduction of 10 percent or less for a given crop.

Crops vary greatly in their salt tolerance, and therefore the suitability of a water for irrigation will also vary with crops. This gives us a wide choice of crops and expands the usable range of water salinity for irrigation. An evaluation of the relative salt tolerance of agriculture crops had been done by Mass and Hoffman (1977). They provided two essential parameters sufficient for expressing salt tolerance, the maximum allowable salinity without yield reduction; and the percent yield decrease per unit salinity increase beyond the threshold. In computing these parameters they assumed that yield decreases linearly as salt concentration increases above the threshold level as:

$$Y = 100 - B(EC_e - A) \qquad (2)$$

where:

A = salinity threshold, in millimhos per centimeter, and

B = percent yield decrease per unit salinity increase.

EC_e = soil salinity which can be assumed is equal to 1½ times in terms of irrigation, water salinity ($1.5 \times EC_w = EC_e$) (Mass and Hoffman, 1977).

DESCRIPTION OF THE PLANNING MODEL

The maximum net-benefit criterion is used here to determine the above planning decisions. The net-benefits resulting from an agricultural expansion is usually the agricultural production revenues less irrigation water transmission cost, new land development cost and some other costs on farm fuel, like costs of seeds, fertilizer, labor and machinery. Irrigation transmission cost includes excavation, lining, irrigation structures and maintenance costs of the new canals; enlarging cost of existing canals and water lifting cost. Land development cost includes the costs of drainage works, housing, electricity, transportation, communication, land leveling and social services.

This maximization procedure is constrained with nine constraint sets. The first set is to ensure that the outflow from each node of the irrigation network is less than or equal to the inflow to this node. The second set is to keep the sum of the cultivated lands with the various crops within each new area less than or equal to the size of the area itself. The third set consists of sequential planting constraints. Some crops need to be planted before others to ensure enhancement of soil nitrogen. These constraints are formulated in such a way that the desired planting sequence can be achieved.

The following three constraint sets deal with irrigation water salinity and its effect on agricultural production. The first is a set of salt conservation equations to determine the water salinity at each node of the irrigation network. This set is also to determine the salt concentration of mixed water at each mixing station. The second set is to compute the yields of the different crops according to the salinity degree of water supply via equation (2). The third set is to determine the water requirements of the various crops. This is done by adding the required leaching fraction according to irrigation water salinity (equation (1)) to the freshwater requirements of the crops.

The seventh and eighth constraint sets are on the upper and lower capacities of the irrigation canals. These constraints are to keep the flow through a canal less (more) than an upper (lower) limit. The upper limit is the upper capacity of a canal which is one of the model's decisions. This capacity will be equal to the maximum seasonal flow. The lower limit (capacity) can take different values rather than zero for different purposes like navigation, power generation, etc. The lower capacities of the various irrigation canals represent an input to the model. Finally, the last constraint set is to ensure the non-negativity of the planning decisions.

The model deals with the maximization of convex objective functions accounting for economies of scale of water transmission costs. This leads to difficulties in seeking the global optimum solution because many local optimum are likely to exist. There is really no way around this and the solutions produced will be local optimum. The decision variables appear separately in both the objective function and constraint sets. Then the separable programming, such as the one available in MPSX (the IBM LP package) can be used to solve the model for the case study.

CASE STUDY: THE AGRICULTURAL EXPANSION IN THE NILE DELTA AND SINAI

The Nile Delta is naturally divided by the two major branches of the River Nile – Domietta and Rosetta Branches into three main regions. These regions are the Eastern Delta, Middle Delta, and Western Delta. In this paper we are focusing only on the Eastern Delta and Sinai. The agricultural expansion in this region is greater than 1,500,000 feddans (1 feddan = 1.038 acres). This expansion is proposed in twenty-two areas. The irrigation water resources for these new areas are:

1. El-Salam Canal which is now under planning. Its water is fresh water from the Domietta branch of the Nile and drainage water from El-Sarw pump station and Bahr Hadus drain. There are two proposed pump stations on this canal. The first pumping station is to lift the water after being mixed with El Sarw drainage water. The second is at the end of Bahr Hadus to lift the drainage water to be mixed with the canal's water. This canal extends from its intake at Domietta branch to Sinai across the Suez Canal as shown in Figure 1.

2. Ismailia Canal, after some enlargement, is proposed to feed directly some new areas close to it rather than two new canals. These canals are the irrigation Suez Canal for supplying new lands in both west and east of the Suez Gulf, and El-Salhia Canal which is planned to feed new areas in the east and west of the Suez Canal.

3. El-Bahr El-Saghier after an expansion to its cross section.

4. A new canal which takes its water directly from Domietta branch to supply area (20) with irrigation water in the northern part of the Eastern Delta.

5. A drainage water from Bahr El-Bakar pump station, Bahr El-Bakar sub-drain, and Faraskour pump station.

6. Treated sewage water for feeding 100,000 feddans south to Ismalia Canal (area 22).

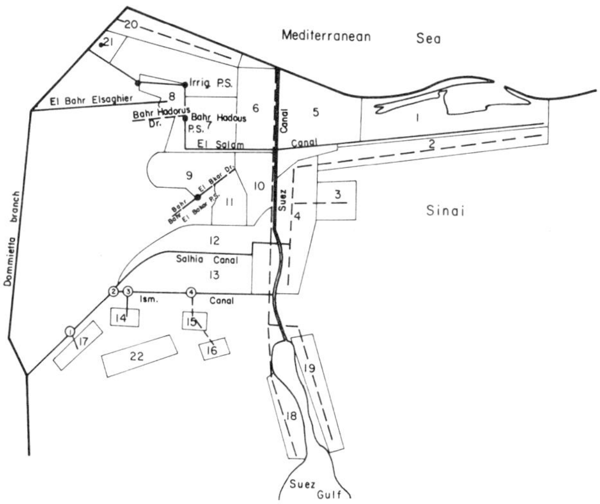

Figure 1. A Schematic Map of the New Lands and the Irrigation Water Sources in the Eastern Delta and Sinai Regions.

A network representation of the irrigation system in the new land is shown in Figure 2.

In Egypt, the agricultural year is divided into two main seasons, winter which starts at the beginning of November, and summer which starts at the beginning of March. The main winter crops are short-season clover, wheat, vegetables, barley, beans, and long-season clover. During the summer, cotton, maize, rice and summer vegetables are grown.

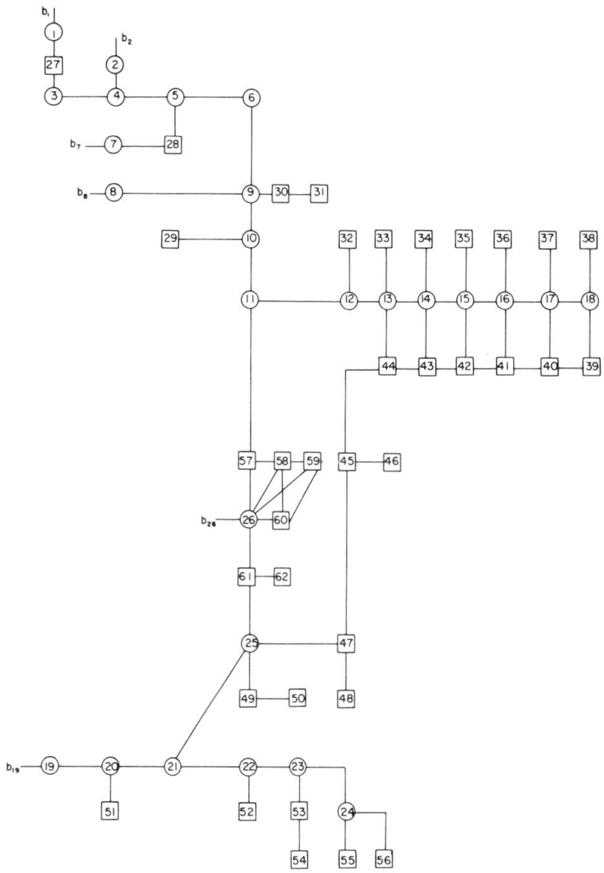

Figure 2. A Network Presentation of the Irrigation System in the Eastern Delta and Sinai Regions.

SOLUTION TO THE CASE STUDY

To apply the mathematical model, some data has to be prepared as an input to the model. Some of this data is collected and the other is assumed, for more details see Allam (1980). The model was applied three times using three different mixing ratios of drainage water to fresh water. These ratios were 2.5, 1.4, and .8 through the winter; and 2.5, 1.3, and .7 through the summer. The separable programming available on MPSX was used for solving the model. The computation time for each run in an IBM 370 was less than a minute for the case study which has more than 500 constraints.

Using first mixing ratios 2.5 and 2.5 during the winter and the summer seasons, respectively, the irrigation network design was obtained. The capacities, cross-sectional areas, seasonal water flows, total costs of the new canals, enlargement cost of old canals, and crop pattern distribution in the new lands had been determined (Allam, 1980).

The following remarks about the crop pattern distribution were observed:

1. Large areas are cultivated with beans which have a higher price and are moderate in their water consumption.
2. During the summer the crops of high water requirements like rice are chosen to be cultivated close to the water sources for minimizing the transmission costs.
3. The crops which have a higher salinity tolerance are cultivated in the areas which have saline irrigation water like the cotton in areas (29), (30), and (31). The sensitive crops to salinity are located in the areas which have fresh water supplies like the beans in area (50).
4. More complicated crop selection was done, when the maize had been chosen in sub-areas (57), (58), (59), and (60) where the irrigation water is very saline. The maize is preferred to cotton because of its higher market price, although the latter has a higher salinity tolerance and lower water requirement. However, as will be shown in the second solution, cotton was preferred to maize whenever we have a shortage in irrigation water.

Two other runs have been done using smaller mixing ratios equal to 1.4 and 1.3 for the second run and 0.8 and 0.7 for the third solution through the winter and summer seasons, respectively. By decreasing the mixing ratios the potential water for irrigation decreases. By looking at the two runs output the following can be noticed:

1. During the winter and as shown in the first solution, large areas are cultivated with beans.
2. The summer vegetables are selected in area (50) instead of the rice in the first solution. This happens because of the reduction in the irrigation water. Because of the same reason the cotton is chosen in the areas (57), (58), (59), and (60) instead of the maize in the first solution.
3. While most of the new lands are cultivated in the winter, the only completely cultivated areas during the summer are (27), (28), (51), (58), (59), (60), (61), and (62) because they are very close to the irrigation water sources. The other areas are either partially cultivated or not cultivated at all. This happens because of the higher water requirements of the summer crops.

The benefit cost ratio (B/C) was used for comparing the different solutions shown in Figure 3. This figure shows that when the mixing ratio increases, the benefit cost ratio decreases until reaching its minimum value 1.18 corresponding to 1.35 mixing ratio. The (B/C) starts to increase again. This happens because when using the lowest mixing ratio, two main canals were selected for supplying the north part of the Sinai with irrigation water. These canals are El-Salam Canal carrying the mixed water and El-Salhia Canal which takes its irrigation water from Ismailia Canal. These areas

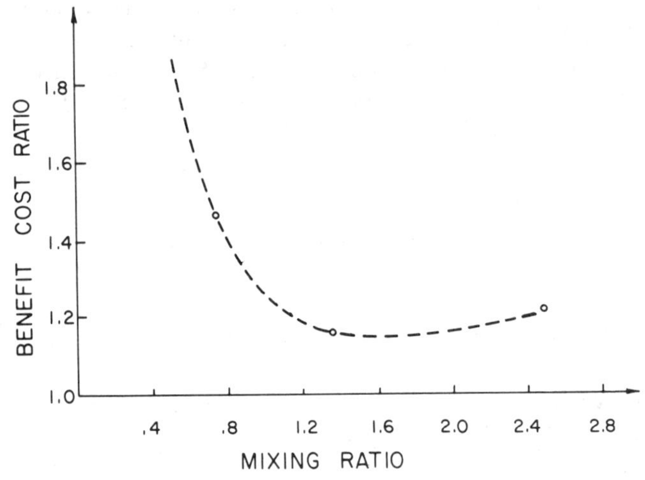

Figure 3. The Effect of Water Mixing on the Benefit Cost Ratio.

which are served with El-Salam Canal are mainly cultivated with beans during the winter, rice and maize during the summer. Then by increasing the mixing ratio to 1.4 and 1.3 during the winter and summer seasons, respectively, the irrigation water increased and we were able to irrigate two more areas, (37) and (38). Therefore the total cost and the gross benefit are increased, but due to the increase of water salanity with its bad effects on both water requirements and yields of the crops we got a lower (B/C) equal to 1.18. With more increase in the mixing ratio it is expected to get a lower (B/C), due to the increase of the water salinity. This was not the case as the excess in irrigation water has enabled us to irrigate all new lands in the north part of the Sinai using only El-Salam Canal where El-Salhia Canal stopped at sub-area (45) and the transmission costs highly decreased. On the other hand this excess in irrigation water gave us a chance to extend a new canal from El-Salam Canal to area (57). This canal has improved the crop pattern in this area from clover and cotton to vegetables, maize, and rice which have higher market prices, and hence the gross benefit increased. Therefore, we got an increase in B/C from 1.18 to 1.21 which still is less than the highest value of 1.46.

In comparison between the various solutions, the third one is the best alternative which has the highest B/C. But in fact we could get higher B/C if we use a lower mixing ratio, or when no mixing at all which will result in leaving more lands without reclamation. Therefore the optimum mixing ratio will be mainly constrained with the size of lands to be reclaimed.

SUMMARY AND CONCLUSIONS

This study has provided a working mathematical model for analyzing and optimizing the conjunctive use of fresh and saline water sources in agricultural practices. This model includes crop choice based on a trade-off between crop's salt tolerance, water requirement, and economic value. As shown in the case study, the model has been applied successfully and leads to interesting insights for this area of Egypt.

ACKNOWLEDGMENTS

This study was sponsored by the MIT Technology Adaptation Program which is funded through a grant from the Agency for International Development.

BIBLIOGRAPHY

Allam, M., 1980. A Planning Model for Irrigated Agricultural Expansion: A Methodology and Case Study of the Nile Delta. S.M. Thesis, Department of Civil Engineering, M.I.T.

Allam, M. and D. H. Marks, 1982a. Scheduling of Agricultural Expansion: Incorporating Drainage Water Reuse. Submitted to Water Resources Research for publication.

Allam, M. and D. H. Marks, 1982b. Scheduling of Agricultural Expansion: Income Redistribution Objectives. Submitted to Water Resources Research for publication.

Ayers, R. S., 1976. Water Quality for Agriculture. Food and Agriculture Organization of the United Nations, Rome.

Ayers, R. S., 1977a. Quality of Water for Irrigation. Journal of the Irrigation and Drainage Division, Proceedings of the American Society of Civil Engineers, Vol. 103(IR2).

Ayers, R. S., 1977b. The Horizontal Expansion and Land Reclamation Policy. Ministry of Irrigation and Land Reclamation, A.R.E., Cairo.

Bradley, Hax and Magnanti, 1977. Applied Mathematical Programming. Addison Wesley, Menlo Park, California.

Douglas, J. L. and R. R. Lee, 1971. Economics of Water Resources Planning. McGraw-Hill, New York, New York.

Douglas, J. L. and R. R. Lee, 1978. National Program in Irrigation and Drainage. Ministry of Irrigation, A.R.E., Cairo.

Goldberg, D., B. Gorant, M. Schueli, Ben-Asher, and M. Rinet, 1971. Increasing the Agricultural Use of Saline Water by Means of Trickle Irrigation. Paper No. 71071 of the Water Resources Bulletin.

Hayward, H. E., 1956. Plant Growth Under Saline Conditions: Arid Zone Research IV. UNESCO.

Hayward, H. E., 1973. Irrigation, Drainage and Salinity. UNESCO/FAO.

Mass, E. V. and G. J. Hoffman, 1977. Crop Salt Tolerance – Current Assessment. Journal of the Irrigation and Drainage Division, Vol. 103(IR2).

U.S. Salinity Laboratory Staff, 1954. Diagnosis and Improvement of Salinity and Alkali Soils. U.S.D.A. Agricultural Handbook, No. 60.

Section 5

DROUGHT AND CLIMATE

INTERNATIONAL SYMPOSIUM ON HYDROMETEOROLOGY
JUNE AMERICAN WATER RESOURCES ASSOCIATION 1982

CLIMATIC VARIABILITY AND HYDROLOGIC PROCESSES: AN ASSESSMENT FOR THE SOUTHWESTERN UNITED STATES

Charles W. Stockton and William R. Boggess[1]

ABSTRACT: Based on the likely upward trend in global temperatures, due to increasing amounts of man-induced atmospheric carbon dioxide, the possible effects of a warmer and drier climate on surface water supplies in the Rio Grande and Colorado River basins has been investigated. Should average annual temperatures increase by 2°C and precipitation decrease by 10 percent, annual runoff from the upper Rio Grande would likely decrease by about 30 percent. Decrease for the Colorado would be about 35 percent in the Upper and 50 percent in the Lower basins, respectively. Since demand is already exceeding or approaching supplies in these regions, any dimunition in water yield would have serious regional and national ramifications.
(KEY TERMS: climatic change; water supplies; Colorado River; Rio Grande River; southwestern United States.)

INTRODUCTION

There is little doubt that the development of the western United States, especially beyond the end of the present century, is tied to the availability of surface water for agriculture, industry, hydropower, and human consumption. Added to these established demands are the so-called "instream uses" which require that sufficient flow be maintained to enhance recreation, scenic and aesthetic values, fish and wildlife, etc. With the increased demands caused by the burgeoning population of the southwest, along with the development of the region's energy resources, it is highly doubtful both the offstream and instream demands can be met by most streams and rivers in the area (Anderson, 1982). As a result, increasing conflict between users and potential users can be expected. Temporary shortages created by droughts will become more critical and, unless mitigated by stored water or groundwater augmentation, may cause hardship through the entire spectrum of users.

In most planning processes related to water resources, climate is rarely considered as a variable. Yet climatologists generally agree that nowhere is climate invariant. Short-term (high frequency) variations that produce above or below normal temperature or pecipitation are familiar since they occur periodically and are documented by instrumented data. Reliable weather records, however, are generally not long enough in most parts of the world to reveal the low frequency (long-term) variations or the even longer trends of which high and low frequency oscillations are a part. The occurrence of such variations have been documented by proxy sources such as tree rings, ice cores, stream stratigraphy, and pollen profiles (Stockton and Boggess, 1980).

In a recent paper, Hansen, *et al.* (1981), point out that the world has had both warming and cooling during the past century. They further show that a sharp warming trend has been underway since the mid-1960's and suggest that the current warming trend is likely to continue due to increases of atmospheric carbon dioxide and trace gases related to the combustion of fossil fuels. The magnitude of changes has been greatest in northern latitudes. For instance, between 1880-1940 the higher latitudes warmed about 0.8°C and then cooled about 0.5°C from 1940-1970. In contrast, low latitudes increased by about 0.3°C during 1880-1930 with little change since then. The same writers state that global temperatures are about as high now as they were in 1940 and that the common misconception that the world is cooling is based on Northern Hemisphere experience to 1970. The effect of continued warming on precipitation is a matter of conjecture but there will undoubtedly be both increases and decreases in various parts of the world.

In an earlier paper (Stockton and Boggess, 1980) we emphasized the need to build both robustness and resilience into water resources planning. As defined by Matalas and Fiering (1977), robustness refers to the intentional planning of management systems that will perform well under a variety of climatic variations. Resilience is a property of robustness that allows a system designed for a particular degree of climatic variation to be modified in response to changes in the magnitude and persistence of those variations.

[1] Respectively, Associate Professor of Dendrochronology and Research Associate; Laboratory of Tree-Ring Research, University of Arizona, Tucson, Arizona 85721.

The Stockton and Boggess, 1980, paper resulted from our evaluation of the resilience of each of the Water Resources Regions of the coterminous United States, done for the U.S. Army Corps of Engineers (Stockton and Boggess, 1979). Specifically the likely effects of two climatic change scenarios were considered: (1) a 2°C increase in annual temperature with a 10 percent decrease in precipitation (warmer and drier); and (2) a 2°C decrease in annual temperature with a 10 percent increase in precipitation (cooler and wetter). Although the techniques used in this study were far from optimal (they represented our opinions based on the best data available), it was apparent that with the warmer and drier scenario most of the western regions would be in real trouble.

Because runoff is a residual process, it is apparent that adverse climatic variability should be more readily detectable in arid regions and possibly will be more severe. These regions simply do not have the resilience necessary to cope with such a change. The likelihood that a warmer and drier climate may be in the offing is certainly not beyond the realm of reason, given the projected effects of increasing levels of atmospheric carbon dioxide on global temperatures.

More recently, with support from the National Science Foundation, we have proceeded with a more detailed study to determine the effects of the warmer and drier climatic scenario on the availability of surface water in each of the western water resources regions. In general, an empirical water balance approach has been used to model the effects of warmer and drier conditions on water yields. Due to limitations on both time and space, only the results for the Upper Colorado, Lower Colorado, and Rio Grande Regions are discussed here. The Colorado and Rio Grande Rivers originate and receive most of their runoff from high mountainous areas and then flow through arid to semiarid country where they supply water for many purposes. The two rivers literally represent the "life-blood" of the southwestern United States. Their waters are already overallocated and any dimunition of flow would have disastrous regional and national effects.

REGIONAL CLIMATIC EFFECTS

In the following sections, numbers for mean annual runoff, reservoir storage, evaporation, withdrawals, consumption, and instream uses are from the U.S. Water Resources Council (1978).

Rio Grande Region

The total drainage area of the Rio Grande is 230,000 sq. mi. but 93,000 sq. mi. is in Mexico. Only 89,000 of the 137,000 sq. mi. in the U.S. contributes runoff to the river as 44,000 sq. mi. drains into closed basins.

The arid climate of the region is characterized by low humidity and erratic rainfall. Average annual precipitation ranges from 30 inches in both the high mountains of the headwaters and the Gulf Coastal Plain to about 8 inches in the central portion. Although regional precipitation averages 86 million acre-feet, its variability, coupled with high evapotranspiration rates and the recurrence of prolonged dry periods, creates major water problems. Temperatures over the region vary from quite harsh in the mountains to very mild in the lower portions of the valleys. Mean annual temperatures vary from 40°F in the higher mountain areas to 74°F in the lower Rio Grande Valley of Texas.

In our earlier study we used empirically derived curves relating annual runoff to mean annual temperature and determined that a scenario 1 (warmer and drier) climatic change would reduce the mean annual runoff from the upper portion of the Rio Grande drainage by about 40 percent. In our more recent investigations we have used a general water balance approach including methods of Thornthwaite and Mather (1955), Penman (1948), Budyko (after Sellers, 1965), and Sellers (1965) to determine the impact of the warmer and drier scenario upon mean annual runoff. To do this, we evaluated the climatic effects upon runoff from smaller basins where we have better data and somewhat greater control. This was necessary because diversions of water from the Rio Grande occurred prior to European settlement and reached significant amounts before systematic streamflow measurements were instituted. Thus, few actual measurements of natural streamflow are available, forcing one to use hydrologic estimates based on the best available data whose reliability may be questionable due to incomplete knowledge. By selecting smaller drainages for closer scrutiny, we hope to minimize these data discrepancies. The three drainages investigated include the Rio Grande above Del Norte, Colorado; the Conejos above Mogote, Colorado; and the Rio Chama above Elvado, New Mexico. In each case, many model forms with varying estimated parameters were utilized to evaluate the effects of a warmer and drier climate. Although the different models produced a diversity of absolute results, the input of the climatic change scenario (warmer and drier) has been remarkably consistent. That is, the 2°C increase in mean annual temperature accompanied by a 10 percent decrease in mean annual precipitation would result in an estimated 30 percent decline in mean annual runoff. This is not inconsistent with the estimates produced earlier from empirical curves.

Several important points must be kept in mind when considering streamflow as a source of water within the Rio Grande Region and the effects of climatic variability: (1) there is no surplus water in the region either for new use or for expanding those already in existence; (2) the states within the region have developed legal institutions for the allocation and administration of water and these are legally enforced; (3) the Rio Grande is an international river, and decisions must be made on this basis; (4) there are unsolved Indian Rights problems; and (5) the generally arid climate requires conservation of existing supplies to meet regional water needs and requirements.

The historical runoff data show significant downward trends throughout the region (Langbein, 1979). This trend is typical for streams within the southern intermountain region and represents a decrease in mean annual flow ranging from 0.6 to as much as 4.0 percent per year decrease from the early 1900's to the present. The present ratio of total reservoir storage to mean annual runoff for the Rio Grande Region is 3.6, the second highest of any region in the nation. Consequently, one is faced with the reality that additional surface water storage would not help to ameliorate water supply problems in the regions should a warmer and drier climate occur. In fact, our studies suggest that the warmer and drier scenario would increase the ratio of total available storage to mean annual runoff to about 14.2, indicating that roughly 14 years of mean annual flow would be required to fill the existing available storage.

The Colorado River Region

For consistency, this region is divided into the Upper and Lower Basin, although most of the water in the lower basin region originates in the Upper Basin.

Upper Colorado River Region. The Upper Colorado Region covers about 113,500 square miles and includes parts of Arizona, Colorado, New Mexico, Utah, and Wyoming. The region encompasses high mountains and plateaus, narrow valleys, and broad expanses of desert. The overall climate of the region is semiarid to arid, with precipitation ranging from more than 40 inches in the higher mountains to about 6 inches in the desert. Mean annual temperature ranges from less than 30°F in the high mountains to more than 50°F in the southcentral part.

The mean annual streamflow from the region is 11.2 maf and the median is the same. The estimate of the mean annual runoff is uncertain. Within the WRC (1978) publication, three estimates are given: 13.0 maf, 14.0 maf, and 15.0 maf. For 1975, fresh water withdrawals amounted to 7.7 maf, of which about 2.4 maf were consumed. By the year 2000, total withdrawals are estimated to be 8.4 maf and consumption to be 3.4 maf. Total streamflow that is necessary for optimal fish and wildlife habitat condition is 8.9 maf per year. Evaporation losses (1975) were estimated to be 0.8 maf and 0.9 were expected for the basin. By the year 2000, these values are expected to be 0.8 maf (little or no change) and 1.2 maf, respectivelyy.

It is estimated that 85 percent of the total runoff from the region results from snowmelt in the higher mountain region with the peak occurring in May and June; the remaining months are low flow months. Flow records from all parts of the region show a significant downward trend from 1897-1976. The trend varies from –0.44 percent to –0.70 percent per year; the average appears to be about –0.60 percent per year.

We estimate the occurrence of a climate change similar to that described by scenario 1 (warmer and drier) would reduce runoff by about 35 percent. Consequently, if the 15.0 maf estimate of current average runoff is accepted, the mean annual flow would be reduced to about 9.75 maf. There would likely be a general decrease in firm reservoir yield, as the mean flow would become more variable and less persistent. Recharge to the major aquifer would decline with the decrease in direct surface runoff. General water usage would increase as would the already high salinity of the river water.

The capacity of the water system to respond would be poor, as presently available water is overappropriated. The current ratio of requirement to supply is 0.84, whereas under the warmer-drier scenario we estimate this ratio would increase to 1.26 or an obvious greater demand than supply. The ratio of total available storage to mean annual runoff would increase from 1.38 to 2.21 should we experience a regional climatic change toward warmer and drier conditions.

Lower Colorado River Basin. The Lower Colorado River Basin has an area of 154,860 square miles located largely in Arizona with lesser portions in New Mexico, Nevada, Utah, and California. Climate varies widely because of considerable differences in altitude, latitude, and the distribution of mountain ranges. Winter rains result from moisture from frontal systems moving in from the Pacific Ocean. Moisture from the Gulf of Mexico is associated with the often violent convective summer thunderstorms. Average annual precipitation in the desert areas of the region may be as low as 2.5 inches or greater than 30 inches on some of the higher mountain peaks; the regional average is less than 10 inches per year. Evaporation depletes more than 95 percent of the precipitation before it reaches the streams. Temperature also varies widely across the region. In January, the values range from 20-25°F in the north to near 55°F in the south. In July these same areas record temperatures averaging close to 55°F and 95°F, respectively.

There are three main sources of water in the region: (1) 3.15 maf of Colorado River water, appropriated under the "Law of the River" and originating in the Upper Colorado River Basin, (2) local runoff estimated to total about 3.12 maf originating within the region, and (3) local ground water resources estimated to be 1 billion acre feet. The flow available from the upper basin is dependent upon the resources available and the operating procedures established by the Colorado River Compact (1922) and under the jurisdiction of the Secretary of the Interior. In 1975, inflows from the upper basin amounted to about 11.2 maf per year. The mean streamflow in the region is estimated to be 1.8 maf with the median being equal to the mean. Although estimates vary somewhat, total water withdrawals in the region in 1975 amounted to nearly 10.0 maf, but this value is expected to decline to about 8.85 maf by the year 2000. Because of excessive evaporation and transpiration, the total annual usable runoff from the region is estimated to be a negative 0.7 maf. Mined ground water is estimated to be

2.7 maf per year. In 1975, a total of 5.15 maf were consumed, 1.3 maf lost to evaporation, and 5.0 maf exported from the region. By the year 2000, the annual consumption is estimated to be 5.28 maf, evaporation 1.4 maf, and exportation 4.4 maf. At these levels, the annual average remaining streamflow will be a negative 1.72 maf, assuming zero ground water mined from storage. In addition, the U.S. Fish and Wildlife Service estimates that a remaining streamflow of 7.68 maf per year is necessary for optimal fish and wildlife habitat conditions.

Should the postulated climate warmer and drier scenario occur, we estimate runoff from the region would be reduced by about 50 percent. The present requirement/supply ratio is 1.18 and we estimate this would reach 2.74 under a warmer and drier scenario. The current ratio of total available storage to mean annual runoff for the region is 12.65 and this would rise to nearly 19.0 should the warmer-drier change occur.

CONCLUSIONS

Although a climatic change of the magnitude we have described may never occur, the implications of any adverse climatic conditions should be obvious. With the existing delicate balance between supply and demand it is highly problematical whether either region could continue at present levels of usage, much less tolerate any expansion, should there be less water available. It should be emphasized that the adverse effects of reduced runoff would not be confined to the basin boundaries. This is especially true of the Colorado where about one-third of the average annual flow is exported from the basin. A major portion of the diversions goes to California and provides water for irrigating the Imperial Valley and to help supply the metropolitan areas of Los Angeles and San Diego. It is essential, therefore, that present supplies be used prudently, with the knowledge that less, not more, water is apt to be available in the future.

ACKNOWLEDGMENTS

This research was supported by National Science Foundation Grant No. ATM 79-24365.

LITERATURE CITED

Anderson, Raymond L., 1982. Conflict Between Establishment of Instream Flows and Other Water Uses on Western Streams. Water Resources Bulletin 18(1):61-65.

Hansen, J., D. Johnson, A. Lacis, S. Lebedoff, P. Lee, D. Rind, and G. Russel, 1981. Climatic Impact of Increasing Atmospheric Carbon Dioxide. Science 213(4511):957-966.

Langbein, W. B., 1979. Annual Water Supplies in the West. U.S. Geological Survey Water Resources Review for January, p. 22.

Matalas, N. C. and M. B. Fiering, 1977. Resource Systems Planning. *In:* Climate, Climatic Change and Water Supply. National Research Council, National Academy of Sciences, Washington, D.C., pp. 99-110.

Penman, H. L., 1948. Natural Evaporation from Open Water, Bare Soil and Grass. Proc. Roy. Soc, London A-193, pp. 120-145.

Sellers, W. D., 1965. Physical Climatology. University of Chicago Press, Chicago, Illinois, 272 pp.

Stockton, C. W. and W. R. Boggess, 1980. Tree Rings: A Proxy Data Source for Hydrologic Forecasting. *In:* Symposium Proceedings, Unified River Basin Management, R. M. North, L. B. Dworsky, and D. J. Allee (Editors). American Water Resources Association, Minneapolis, Minnesota, pp. 609-624.

Stockton, C. W. and W. R. Boggess, 1979. Geohydrological Implications of Climate on Water Resource Development. Contract Rept. DACW 72-78-C-0031, U.S. Corps of Engineers, C. W. Stockton and Associates, Tucson, Arizona.

Thornthwaite, C. W. and J. R. Mather, 1955. The Water Balance. Publications in Climatology 8(1), Drexel Institute of Technology, Centerton, New Jersey.

United States Water Resources Council, 1978. The Nation's Water Resources – The Second National Water Assessment by the U.S. Water Resources Council. U.S. Water Res. Coun., Washington, D.C.

DROUGHT HISTORY IN THE WESTERN GREAT PLAINS FROM TREE RINGS

David M. Meko[1]

ABSTRACT: Drought history in the western Great Plains was reconstructed from 1640 to the present from recently collected tree-ring samples. Seven outlier ponderosa pine sites, comprising more than 100 trees, were sampled in the summer of 1980 in North Dakota, South Dakota, Nebraska, Montana, and Wyoming. Equations to reconstruct annual (September-August) precipitation were estimated by multiple linear regression of divisionally averaged precipitation against indices of ring width. Reconstructions emphasized the long-term significance of the 1930's drought: the years 1936, 1939, and 1934 ranked 1, 4, and 14, respectively, in drought severity out of 338 years. On the other hand, when drought conditions were averaged over periods of 10 or more years, the 1930's drought paled in comparison with major droughts in the 1750's, 1820's, and 1860's. Spectral analysis indicated that the apparent 20-year return period of major drought in modern times is not a feature of the long-term drought history. A strong (95-99 percent significance level) spectral peak was found, however, at near 60 years. Water use planning should consider the possibility of long wavelength precipitation fluctuations that may not be discernible in the relatively short gaged records.
(KEY TERMS: drought; tree rings; Great Plains.)

INTRODUCTION

The most severe droughts of this century in the western Great Plains were characterized by a clustering together of dry years. Annual precipitation in eastern Montana and Wyoming averaged more than 10 percent below the long-term mean in 1930-40 and 1952-62. Precipitation records suggest a rythmic recurrence in drought: the large negative anomalies in the 30's and 50's were counterbalanced by large positive anomalies in 1919-29, 1941-51, and 1963-73 (Perry, 1980).

Drought at the western extreme of the Great Plains is likely to become more important as water demand increases for expected energy development (Taylor, 1978). Water resources planning should consider the possibility of long lasting droughts, cycles in drought, and irregular low frequency variations in precipitation. Gaged precipitation records may be too short to adequately represent these features.

Recent studies have shown that useful information on regional drought history can be obtained from tree-rings (Cook and Jacoby, 1979; Duvick and Blasing, 1981; Meko, *et al.,* 1980; Puckett, 1981). This paper describes reconstructions of drought history in the western Great Plains back to the mid-1600's from recently collected tree-ring data. Reconstructions were analyzed to put the 1930's and 1950's droughts in long-term perspective, and to search for evidence of drought cycles.

DATA

Locations of tree-ring sites are shown in Figure 1, and descriptive information on sites is given in Table 1. Tree-ring samples were collected and processed by standard techniques of the Laboratory of Tree-Ring Research at the University of Arizona (Stokes and Smiley, 1968). Several trees were sampled at each site. Cores were dated and measured to get time series of annual ring width, which were converted to indices by dividing each ring width by a value from a fitted curve that approximated the variation of mean ring width with tree age. Tree-ring indices from all usable cores at a site were averaged to form the site tree-ring index. This index is a measure of annual anomalies in diameter growth adjusted for distortion due to aging of trees.

Climatic data consisted of station and divisional monthly precipitation and temperature. Long station series were obtained from the files of the Tree-Ring Laboratory. Divisional series were obtained from the Climatic Data Center in Ashville, North Carolina. Divisional series covered the years 1931-1977; station series were of varying length, the longest going back to 1878.

TREE RINGS AS INDICATORS OF DROUGHT

Annual diameter growth in drought sensitive trees is reduced by long periods of water stress within the trees (Kramer and Kozlowski, 1960). Since low soil moisture and high evaporative demand favor water stress, likely candidates for correlation with tree-ring indices are precipitation and temperature.

[1]Research Associate, Laboratory of Tree-Ring Research, University of Arizona, Tucson, Arizona 85721.

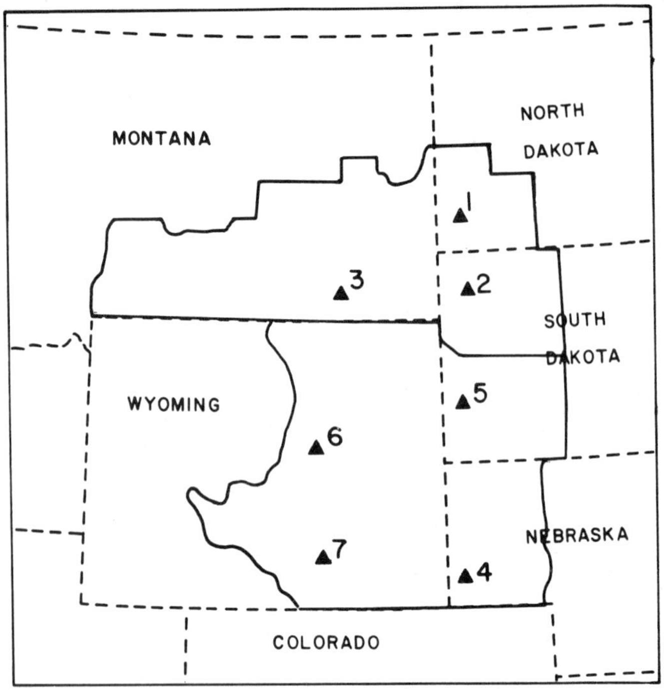

Figure 1. Map Showing Region Boundaries (solid line) and Location of Tree-Ring Sites. Sites numbered as in Table 1.

Inference of seasonal climate anomalies from tree rings is difficult because the set of climatic conditions that may lead to a given ring width is not unique. Timing of precipitation or temperature anomalies relative to phenology of the tree is likely to affect ring width, as are intensity, duration, and time between climatic anomalies. For example, the seasonal temperature and precipitation conditions in years with the four narrowest rings at the Slim Butte, North Dakota, site are shown graphically in Figure 2. In general, two or more seasons were hot and two or more seasons dry in these narrow-ring years. Variability, however, was the rule: winters and summers ranged from hot and dry to cold and wet, and the only characteristic common to all four years was a dry spring.

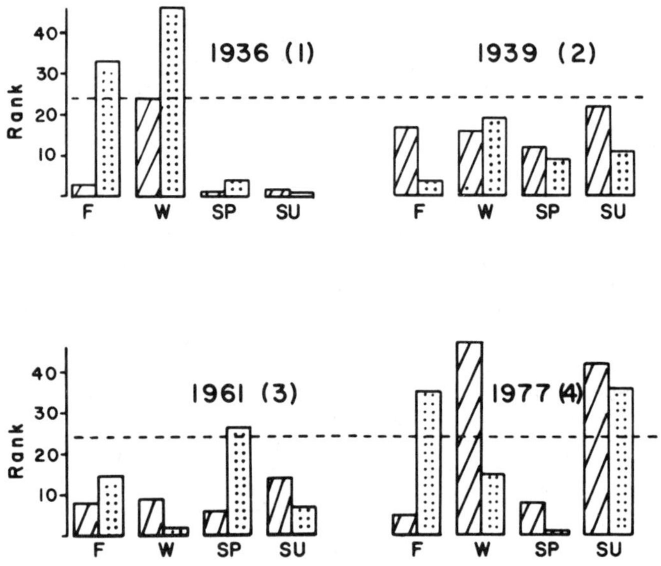

Figure 2. Rank of Seasonal Precipitation (lined bar) and Temperature (stippled bar) in Years with Four Narrowest Rings at Slim Butte, South Dakota (site No. 2 in Figure 1). Climatic data are for Northwestern Division in South Dakota and are ranked from driest or hottest (1) to wettest or Coldest (47). Period for analysis was 1932-1978.

Partial correlation coefficients were used to statistically summarize the relationships of all tree-ring series to divisional-average precipitation and temperature. The partial correlation coefficient measures correlation between two variables, with influence of one or more other variables removed (Panofsky and Brier, 1968). Shown below are

TABLE 1. Listing of Tree-Ring Sites.[1]

Site[2] No.	Region	Name	State	Elevation (ft)	Sample Depth[3] 1900	1800	1700	1600
1	North	Burning Coal	North Dakota	2600	20	18	8	1
2	North	Slim Butte	South Dakota	3500	34	32	9	0
3	North	Otter	Montana	4000	38	34	7	0
4	South	Pumpkin	Nebraska	4800	12	10	4	0
5	South	Custer	South Dakota	5900	38	35	14	1
6	South	Teapot	Wyoming	5300	30	30	12	2
7	South	Rock River	Wyoming	7100	26	24	13	8

[1] All samples were Ponderosa Pine (*Pinus ponderosa* Laws.) collected in the summer of 1980 by T. P. Harlan and D. M. Meko.
[2] Corresponds to numbers on map in Figure 1.
[3] Number of cores dating back to the specified year.

median values of partial correlation coefficients between tree-ring indices and climatic variables (1933-1977), with influence of previous year's growth removed:

	Climatic Variable	
Grouping	**Precipitation**	**Temperature**
September-August	0.65	0.30
January-August	0.57	0.28
March-August	0.56	0.38
June-August	0.38	0.39

Tree-ring indices were usually more highly correlated with precipitation than with temperature. Correlation with precipitation increased with length of period over which precipitation was summed. Of the variables tested, the best candidate for reconstruction was annual (September-August) precipitation. Tree-ring indices obviously integrate precipitation over many months in a complex way, however, and influences of other variables, especially warm season temperature anomalies, are not negligible.

RECONSTRUCTION

Reconstructions of annual (September-August) precipitation were made by multiple linear regression. First, divisional precipitation was regressed on the tree-ring index in a lagged predictor model. Potential predictors for precipitation in year t were tree-ring index in years t−1, t, and t+1. Separate regressions were run for each of the seven tree-ring series. Predictors were entered stepwise until proportion of variance explained (R^2) leveled off. Values of R^2 ranged from 0.13 at site No. 4 to 0.49 at site No. 3; for five of the sites, R^2 was between 0.40 and 0.50.

Long-term tree-ring indices were substituted into the regression equations to generate reconstructions to the mid-1600's of divisional precipitation. Because large scale anomalies were of interest, these single site reconstructions were weighted into regional reconstructions. Sites 1-3 were grouped into a North region, and sites 4-7 into a South region. Divisional precipitation for all divisions within a region (see regional boundaries in Figure 1) were averaged to form regional precipitation, which was the predictand in the regional regression model. Predictors were the single site reconstructions mentioned earlier. Regression coefficients were estimated and used to weigh single site reconstructions into regional reconstructions.

ACCURACY

Reconstructions explained 55 percent of the variance of regional precipitation (1933-1977) in the North region and 56 percent in the South. Standard errors of reconstructions expressed as percent of 1933-1977 normal were 12 percent in the North and 10 percent in the South.

These statistics can be misleading as indicators of accuracy of reconstructions if the climate in the calibration happened to favor unusually high sensitivity of tree growth to rainfall. This possibility was investigated by comparing reconstructions with long station precipitation series for evidence of drop off in agreement from the calibration period (1933-1977) to earlier years (Figure 3). No noticeable drop off was found. In addition, both low and high frequency variations in actual precipitation appear to have been tracked by the reconstructions.

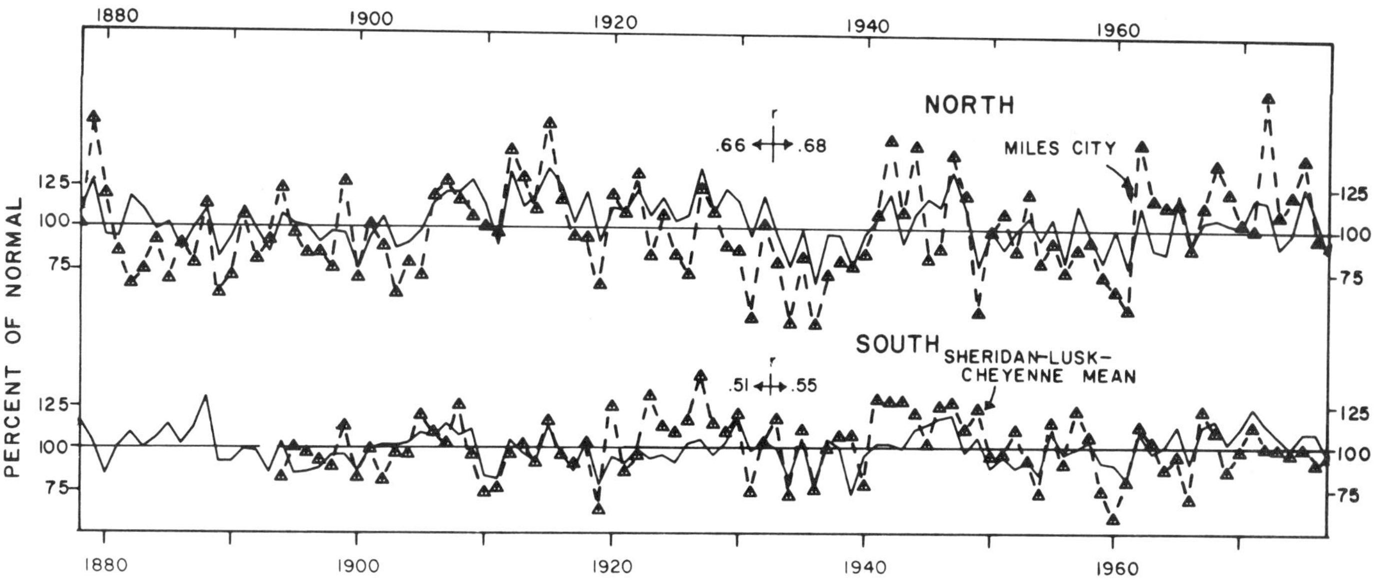

Figure 3. Annual (September-August) Precipitation Series from Selected Stations and Reconstructed Regional Precipitation. Series are plotted as percent of 1933-1977 normal. Correlation coefficients between actual and reconstructed series are given for both the 1933-1977 period and the earlier years.

DISCUSSION

Time series plots of regional reconstructions indicate that periods of drought or wetness in the North region usually coincided with similar periods in the South (Figure 4). The parallelism was especially strong from 1760 to 1900. The mean of the two regional reconstructions is a measure of drought history over the study area as a whole. The droughts of the 1930's and 1950's in the mean series were milder than several earlier droughts. Nine major droughts were identified by nonoverlapping 10-year averages of the mean reconstruction (Figure 5). The relative severity of droughts in the 1750's and 1860's were emphasized: reconstructed annual precipitation averaged 88 percent of the 1933-1977 normal in the driest decades of those droughts, compared with 93 percent of normal in the 1930's.

The 1930's did stand out for single-year drought intensity. The years 1936, 1939, and 1934 ranked 1, 4, and 14 in size of anomaly out of 338 reconstructed years. The driest year – 1961 – of the 1950's drought ranked 12th of 338 years.

PERIODICITY

The long wavelength variations evident in the time series plots in Figure 4 were reflected strikingly in estimated spectral density functions of regional reconstructions (Figure 6). A spectral peak near wavelength 55 years occurred in both regional reconstructions. These peaks were stable under the window closing procedure of Jenkins and Watts (1968): wavelength ranged from 53 to 70 years as number of lags was varied from 1/3 to 1/5 of the sample size (338 years). Significance remained at higher than the 95 percent confidence level throughout window closing.

The 22-year recurrence interval in drought referred to by Perry (1980) was not found in the long-term reconstruction. The later parts of the reconstructions – years covered

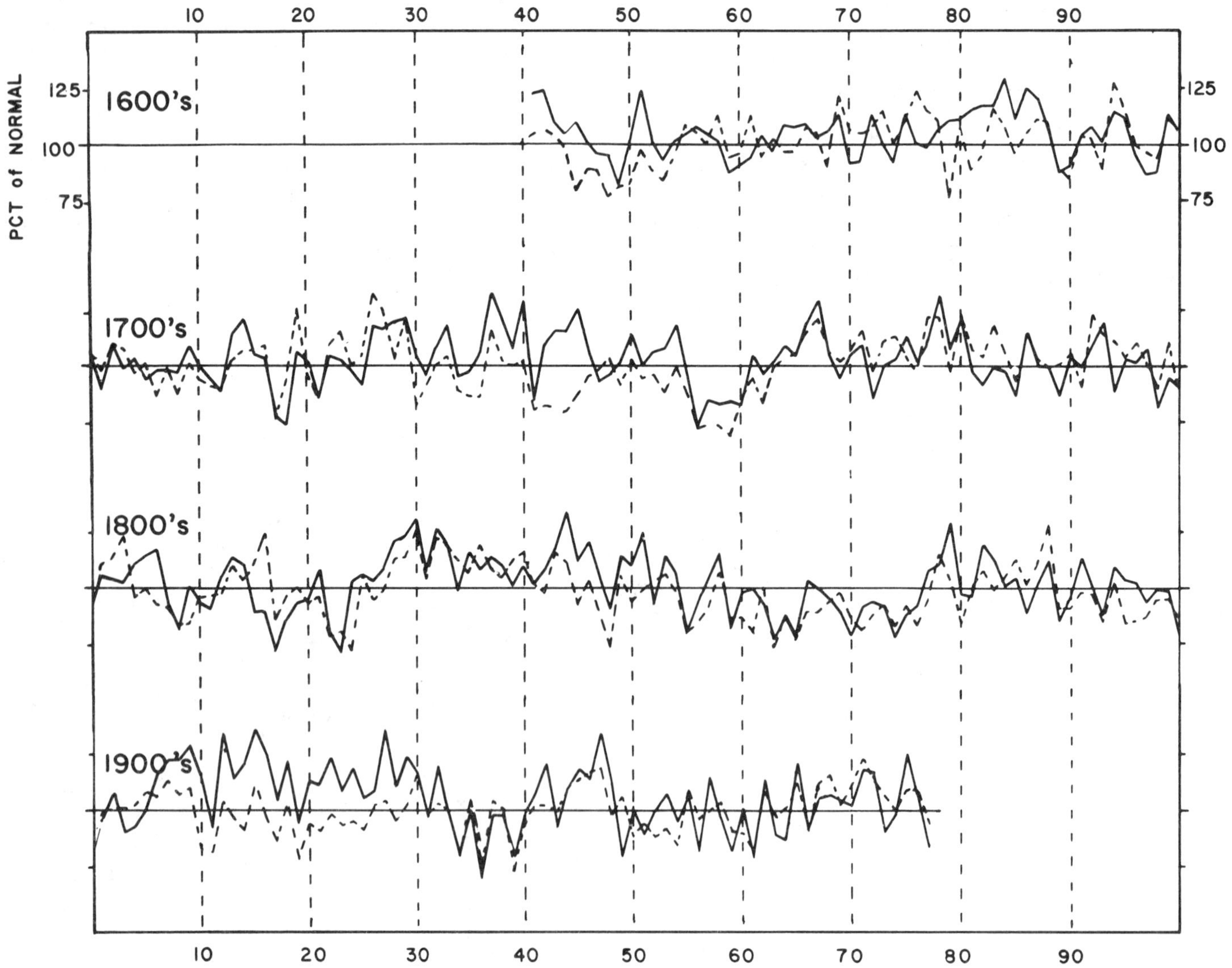

Figure 4. Reconstructed Regional Precipitation as Percent of 1933-1977 Mean. Solid line for North region, dashed for South.

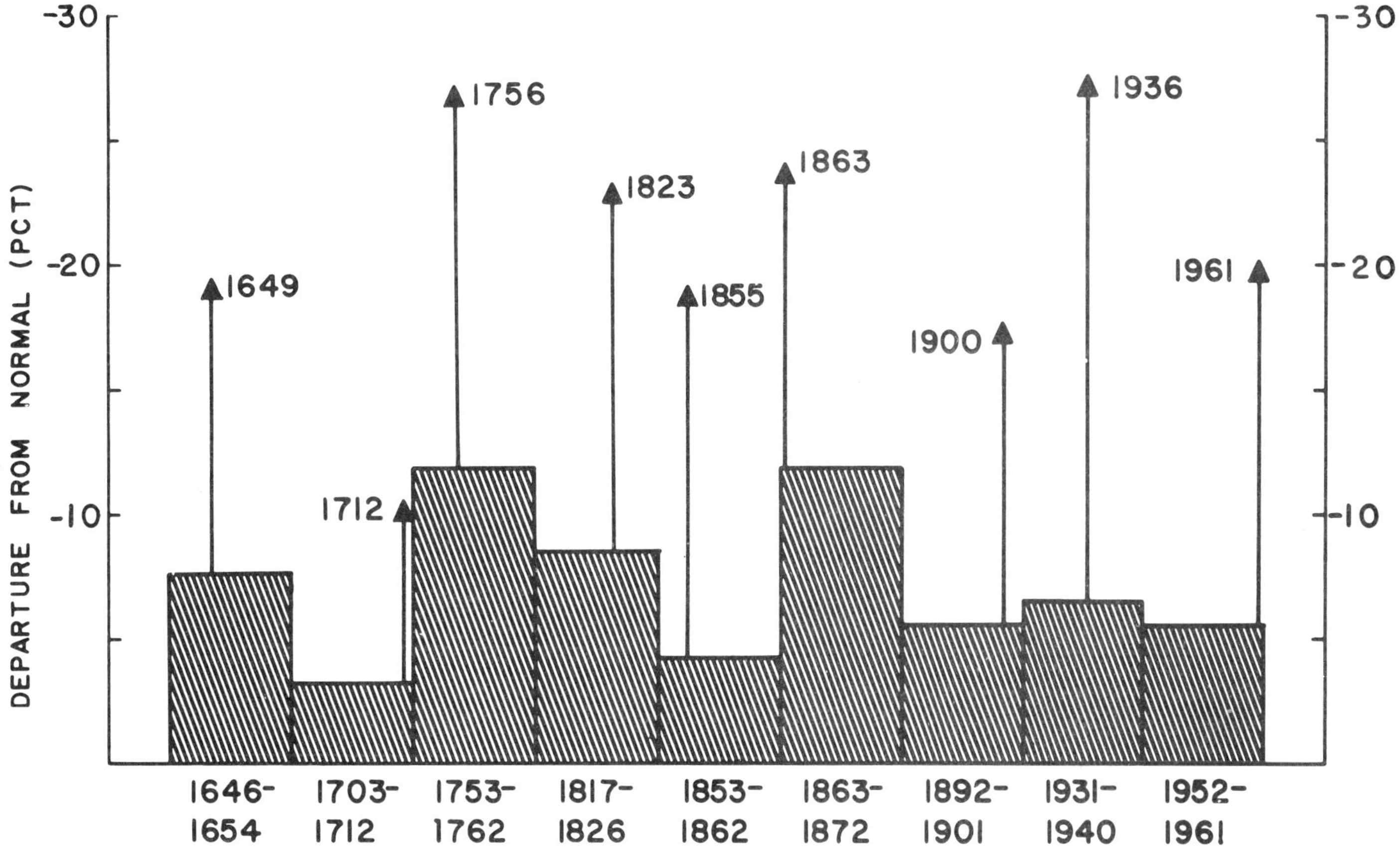

Figure 5. Nine Worst Droughts in Terms of Nonoverlapping 10-Year Average Anomalies in Reconstructed Precipitation. Ordinate is reconstructed precipitation as departure from 1933-1977 normal. Anomaly for direst year within each drought is shown by arrow.

covered by gaged precipitation records – provided the only evidence of drought recurrence in alternating decades.

It may be argued that the spectral peaks in Figure 6 are artifacts of the tree-ring series, and not truly representative of rainfall variations. This is possible, but unlikely. Spectra and time series plots from the two regions are remarkably similar considering regional reconstructions came from two completely different sets of tree-ring data. Also, comparison of regional reconstructions with station precipitation series suggests that long wavelength variations are reflected in reconstructions. Further work is required to quantify the frequency distortion in tree-ring indices.

ACKNOWLEDGMENTS

T. P. Harlan is thanked for his help in collecting the tree-ring samples. This research was supported by National Science Foundation Grant No. ATM 7924365.

LITERATURE CITED

Cooke, Edward R. and Gordon C. Jacoby, Jr., 1979. Evidence for Quasi-Periodic July Drought in the Hudson Valley, New York. Nature 282:390-392.

Duvick, D. N. and T. J. Blasing, 1981. A Dendroclimatic Reconstruction of Annual Precipitation Amounts in Iowa Since 1680. Water Resources Research 17(4):1183-1189.

Jenkins, G. M. and D. G. Watts, 1968. Spectral Analysis and its Applications. Holden-Day, San Francisco, California, 525 pp.

Kramer, P. J. and T. T. Kozlowski, 1960. Physiology of Trees. McGraw Hill, New York, New York, 642 pp.

Meko, D. M., C. W. Stockton, and W. R. Boggess, 1980. A Tree-Ring Reconstruction of Drought in Southern California. Water Resources Bulletin 16(4):594-600.

Panofsky, H. A. and G. W. Brier, 1968. Some Applications of Statistics to Meteorology. The Pennsylvania State University, University Park, Pennsylvania, 224 pp.

Perry, C. A., 1980. Preliminary Analysis of Regional Precipitation Periodicity. U.S. Geological Survey Water Resources Investigations 80-74, 26 pp.

Puckett, L. J., 1981. Dendroclimatic Estimates of a Drought Index for Northern Virginia. U.S. Water-Supply Paper 2080. Washington, D.C., U.S. Government Printing Office, 39 pp.

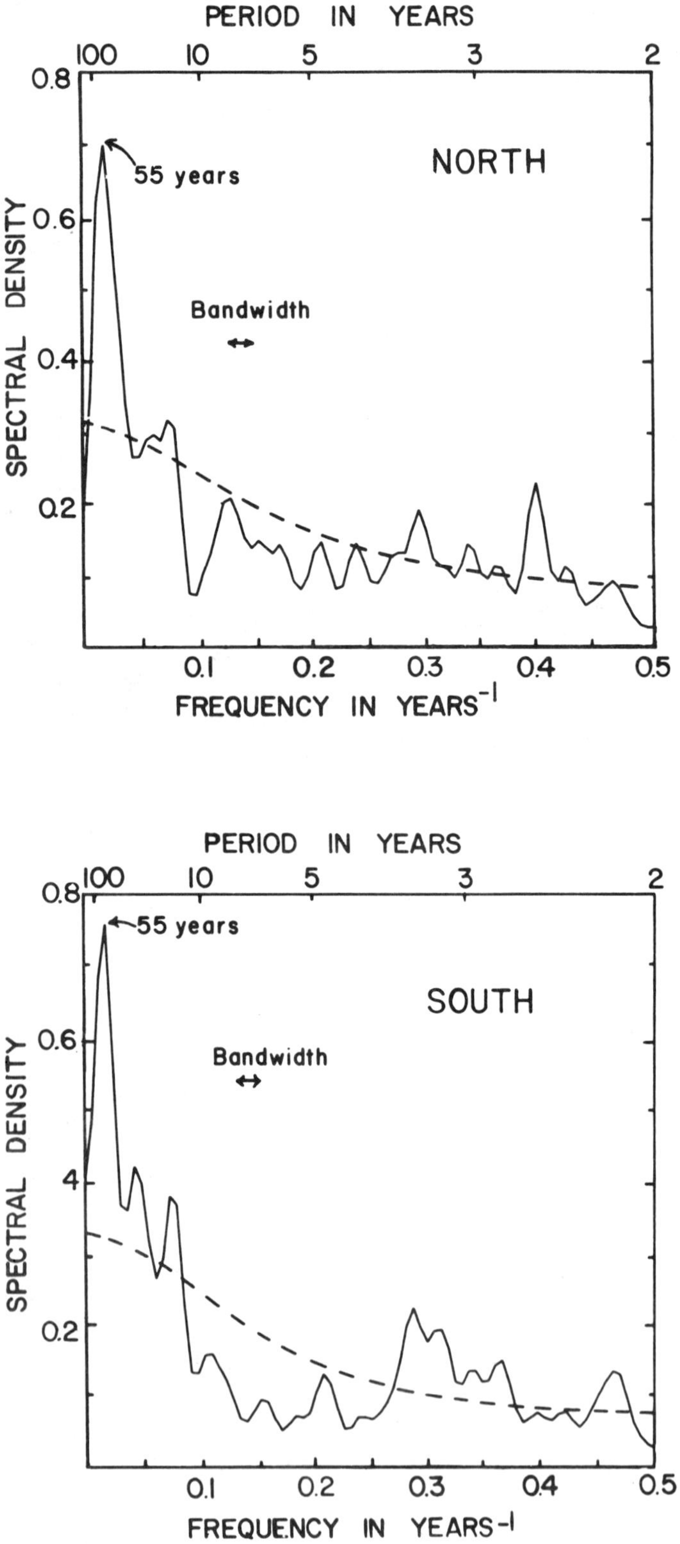

Figure 6. Sample Spectral Density Functions for Regional Reconstructions, 1650-1977. Analysis run with Parzen window, 82 lags. Peaks at 55 years significant at 99 percent level when tested against first order autoregressive null hypothesis (dashed line).

Stokes, M. A. and T. L. Smiley, 1968. An Introduction to Tree-Ring Dating. University of Chicago Press, Chicago, Illinois.

Taylor, O. J., 1978. Summary Appraisals of the Nation's Ground-Water Resources – Missouri Basin Region. U.S. Geological Survey Professional Paper 813-Q, 41 pp.

INTERNATIONAL SYMPOSIUM ON HYDROMETEOROLOGY
JUNE AMERICAN WATER RESOURCES ASSOCIATION 1982

PROBABILITY OF DROUGHT COVERAGE WITH VARIOUS RETURN PERIODS FOR JAPAN

Norio Tase[1]

ABSTRACT: Probabilities of droughts have been analyzed as a point process of rainfall or stream flow. This paper deals with how large droughts with various return periods cover the area. That is, conditional probabilities of drought coverage once given that drought with a certain return period occurs are obtained from data analysis. Data used here are annual precipitation values at 82 stations in Japan for 50 years (1921-1970), which can be assumed to be normally distributed. After standardizing annual precipitation amounts, isolines of return periods of 2, 10, 20, 50, 100, 200, and 500 years are drawn for the 50 years, and drought area, number and area of drought cells are measured. The characteristics of drought occurrence are discussed and probabilities are obtained from frequency distributions.
(KEY TERMS: Japan; drought; drought area; drought cell; probability; return period.)

INTRODUCTION

As reviewed by Şen (1980), the classical approach to the drought problem is the evaluation of the instantaneous smallest value of rainfall or stream flow by means of the theory of extremes. Since this approach does not tell anything about drought duration and drought coverage, the next step was to consider the duration, especially by the theory of runs (Yevjevich, 1967). This approach has been developed well since then.

On the other hand, few investigations on areal coverage of droughts have been carried out. Even a descriptive method of areal characteristics of drought has not been well developed. Pinkayan (1966) studied the probability of occurrence of wet and dry years over the western United States. Gibbs and Maher (1967) analyzed the area extent of past droughts in Australia by classifying the annual rainfall with the decile range, which does not necessarily correspond to the same probability over all stations. Tase (1976) studied area-deficit-intensity characteristics of drought experimentally or by Monte Carlo method over the Great Plains. Further, Tase and Yevjevich (1978) investigated the effects of size and shape of a region on drought coverage and showed an importance of spatial dependency to drought characteristics. Sen (1980) derived theoretical relations or probabilities among the indices developed by Tase (1976) for a random field or a spatially independent case. This study was an important step but Şen's assumption of spatial independency is completely unrealistic.

The purpose of this study is to analyze characteristics of areal coverage of drought with various return periods, with a special reference to relations among areas of drought, area and number of drought cells, and their intensity. This gives an idea of how large of an area we should consider in designing a water resources project. For the first step, annual precipitation over Japan is analyzed in this paper, which can be treated easily in a statistical sense.

ANNUAL PRECIPITATION AND DROUGHT IN JAPAN

Data used here are annual precipitation amounts of 82 stations in Japan for 50 years (1921-1970) shown in Figure 1. Figure 1 also shows the mean annual precipitation which has a range between 1000 and 3000 mm with coefficient of variation of about 0.15 to 0.2. There are great seasonal variations in annual precipitation. On the Pacific Ocean side, the precipitation is concentrated in the rainy season (June and July) and in the typhoon period (September and October), while a severe dry period occurs in winter and a weak dry month occurs in August. On the Japan Sea side, heavy snowfalls abound in winter besides rainfall in the rainy season and typhoon period. Generally, droughts occur mainly due to less rainfall in the rainy season. While a year with less annual precipitation may be a dry or drought year, a drought year does not always have less precipitation. However, a severe drought year generally has much less annual precipitation.

Annual precipitation can be assumed to be normally distributed at most of the stations (Tase, 1983). After standardizing annual precipitation amounts by mean and standard deviation, iso-probability lines of return periods(T) of 2, 10, 20, 50, 100, 200, and 500 years are drawn for the 50 years. The standardized value(z) corresponding to return

[1] Assistant Professor, Institute of Geoscience, The University of Tsukuba, Ibaraki, 305 Japan.

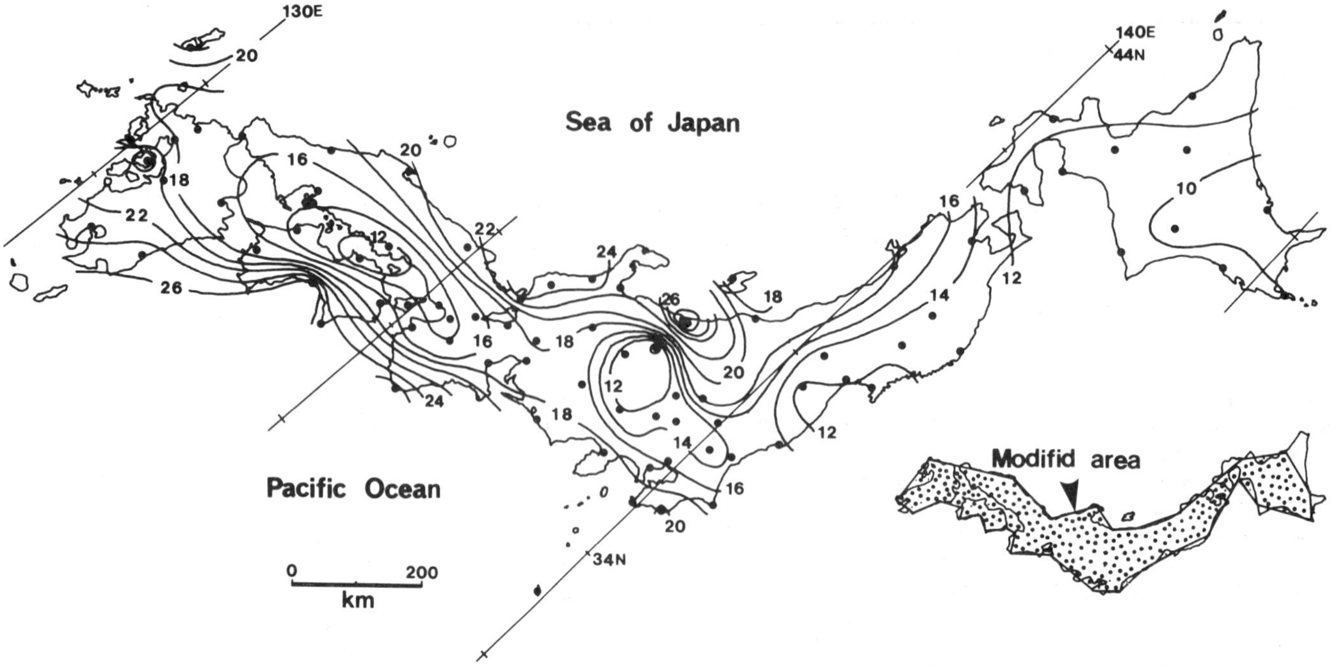

Figure 1. Location of Stations (closed circles), Mean Annual Precipitation (100 mm), and Modified Area for Measurement.

periods of 2, 10, 20, 50, 100, 200, and 500 years are 0.000, −1.282, −1.645, −2.054, −2.327, −2.575, and −2.880, respectively. Here drought area is defined as the total area with values less than z_T corresponding to a given return period over the whole area in each year. The area with a return period of longer than T years is measured for each year. This drought area (A_T) is shown by the percentage to the total area of Japan (403,000 km^2) defined in Figure 1. Since drought area may consist of several isolated drought cells, area and number of the drought cells are also measured for each year. The area of the drought cell (C_T) is also shown by the percentage to the total area of Japan. Maximum intensity (I) in drought area or a drought cell is defined by the maximum values of $|z|$ in drought area or the drought cell.

CHARACTERISTICS OF AREAL COVERAGE OF DROUGHTS

Figure 2 shows secular variations of drought areas of A_2, A_{10}, A_{20}, A_{50}, and A_{100}. According to drought area of A_{10} or A_{20}, there may exist three dry periods and one or two wet periods, but there seem to be no clear dry or wet periods in the series of A_2 except one wet period of 1950's. The severest drought in the last 100 years was drought in 1939, where A_{10}=46.1 percent, A_{50}=15.9 percent, A_{100}= 9.2 percent, A_{200}=2.0 percent, and I=3.074 are the largest in 50 years, though A_2=75.2 percent is the twelfth. The largest drought area of A_2 is 95.9 percent in 1924 and the smallest is 0.4 percent in 1923.

As shown in Figure 3, there seems to be three groups in a relation between drought area of A_2 and maximum intensity. That is, above the maximum intensity of 2.0 and below the drought area of 10 percent, there is no clear relation. In others, there is a positive linear relation. Figures 2 and 3 tell that very severe drought such as 1939 and 1947 does not cover the whole area, but includes large drought area with longer return periods such as $A_{10}>15$ percent, $A_{50}>5$ percent, and $A_{100}>0$.

Figure 4 shows frequency distributions of drought area. Distribution types are exponential except for A_2, which is asymmetric and negatively skew. This is different from that for a spatially independent case by Şen (1980). Drought with T=2 occurred in all 50 years and tended to have relatively large area with mean drought area of 53.4 percent and mode of about 70 percent. Relative frequencies of occurrences of drought seem to be high (84 percent for T=10, 50 percent for T=20, 30 percent for T=50, 18 percent for T=100, and 10 percent for T=200). However, the mean drought areas are smaller than the expected values except for A_2 [A_2=53.4 percent (50 percent), A_{10}=7.6 percent (10 percent), A_{20}=2.6 percent (5 percent), A_{50}=0.7 percent (2 percent), and A_{100}=0.3 percent (1 percent)]. Drought with a return period of 100 years appears once every five years, its average area is 1,130 km^2, and the maximum drought area is 37,000 km^2 which can completely cover the water supply area for the Tokyo Metropolis.

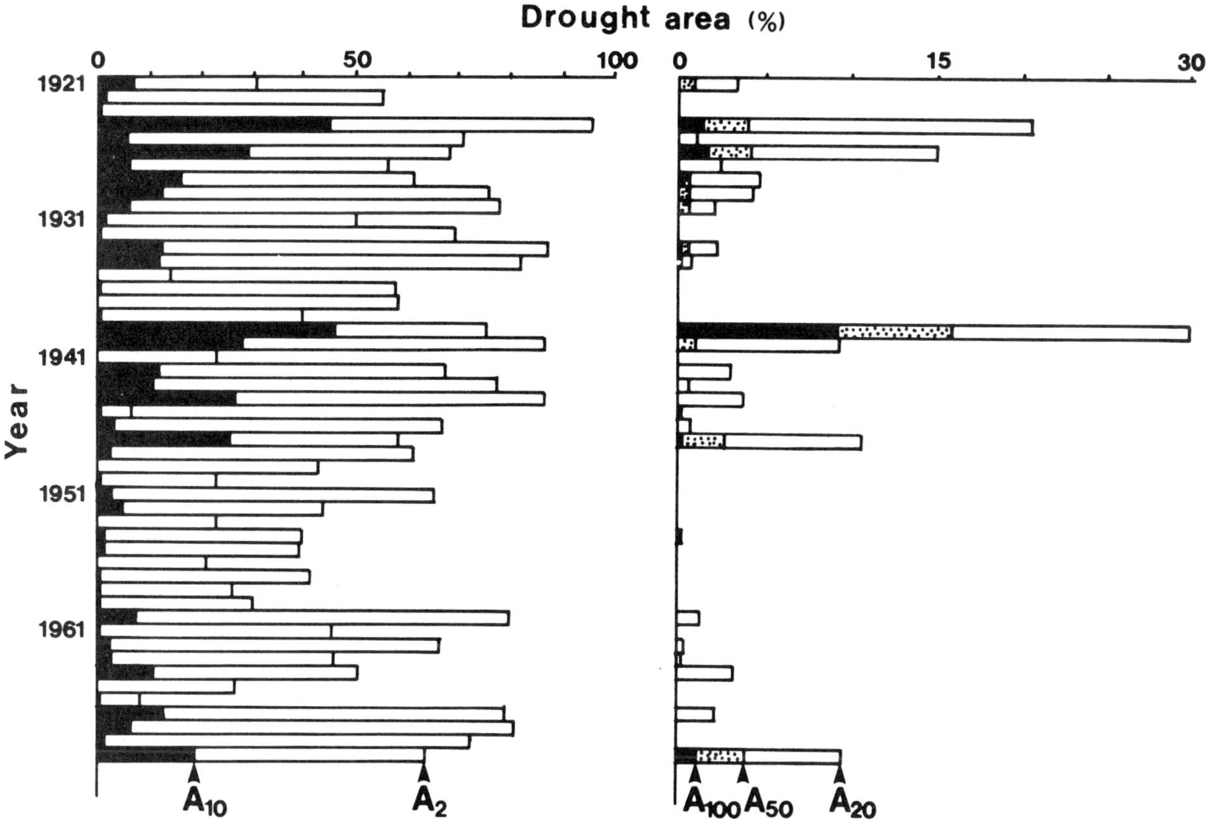

Figure 2. Secular Variations of Drought Areas of A_2, A_{10}, A_{20}, A_{50}, and A_{100}.

Figures 5 and 6 show relations among number (N) and area of drought cell and drought area. Figure 5 indicates that many drought cells with long return periods do not appear simultaneously in a year. Especially, drought area of $T \geqslant 100$ consists of only one drought cell except in 1926. With an increase of the number of drought cells with $T \geqslant 10$, drought area of A_2 increases. For drought cells of C_2, however, drought area of A_2 decreases with an increase of the number. This means that one or two drought cells of C_2 cover either very large or very small areas, and so many drought cells of C_2 cannot bring about severe drought. Drought cells with cells of longer return periods have larger area than those without longer cells as expected.

These facts might indicate that drought area and drought cells with the return periods of less than 10 years and those longer than 10 years have different mechanisms of occurrence.

CONCLUSIONS

This study is rather qualitative because of the short observation periods. However, relative frequencies of droughts with various return periods can be calculated even from the figures, or more sophisticated by fitting of distribution functions. This analysis also gives a crude idea of how large droughts of various return periods cover the area. One other important result is that severe and weak droughts have different mechanisms of occurrence, which may be supported by meteorological facts.

ACKNOWLEDGMENTS

The financial supports for the research leading to this paper were provided by the Ministry of Education and Science under Grant No. 56790166 (1981) and the University of Tsukuba (1980), which are gratefully acknowledged.

LITERATURE CITED

Gibbs, W. J. and J. V. Maher, 1967. Rainfall Deciles as Drought Indicators. Bulletin No. 48, Bureau of Meteorology, Melbourne, 33 pp.

Pinkayan, S., 1966. Conditional Probabilities of Occurrence of Wet and Dry Years Over a Large Continental Area. Hydrology Paper No. 12, Colorado State University, Fort Collins, Colorado, 53 pp.

Şen, Z., 1980. Regional Drought and Flood Frequency Analysis: Theoretical Consideration. Journal of Hydrology 46:265-279.

Tase, N., 1976. Area-Deficit-Intensity Characteristics of Droughts. Hydrology Paper No. 87, Colorado State University, Fort Collins, Colorado, 40 pp.

Tase, N., 1983. Regional Occurrences of Wet and Dry Years in Japan. Annual Report of Institute of Geoscience, The University of Tsukuba, No. 8:37-40.

Tase, N. and V. Yevjevich, 1978. Effects of Size and Shape of a Region on Drought Coverage. Hydrological Sciences Bulletin 23(2):203-212.

Yevjevich, V., 1967. An Objective Approach to Definitions and Investigations of Continental Hydrologic Drought. Hydrology Paper No. 23, Colorado State University, Fort Collins, Colorado, 18 pp.

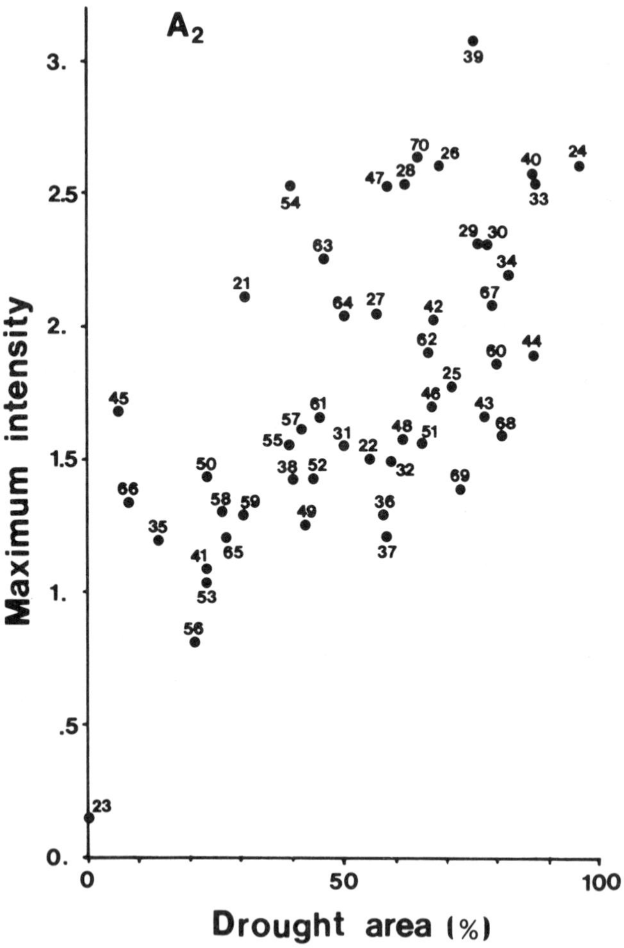

Figure 3. Relation Between Drought Area of A_2 and Maximum Intensity (figures show the last two digits of year).

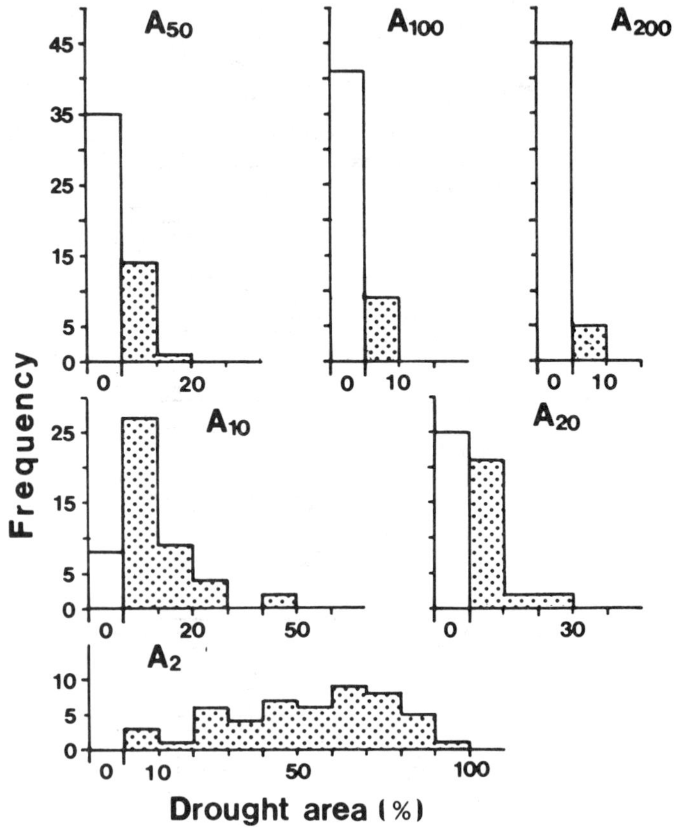

Figure 4. Frequency Distributions of Drought Area.

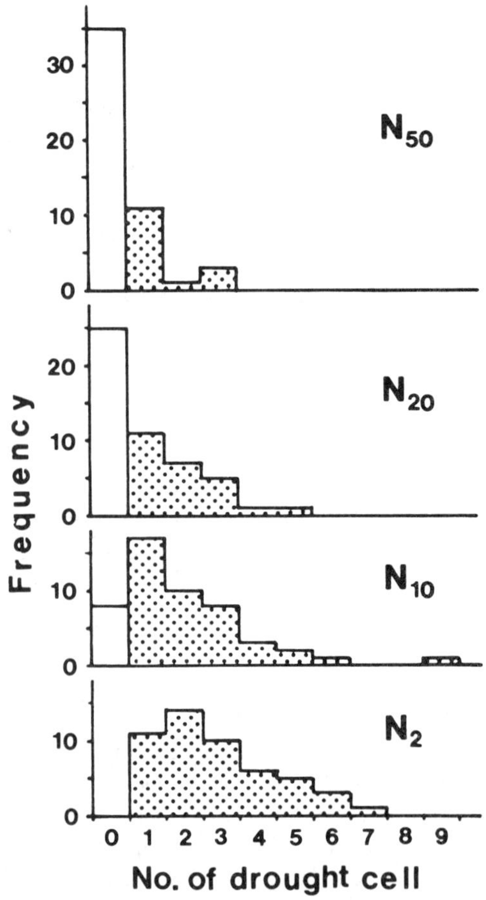

Figure 5. Frequency Distributions of Number of Drought Cells.

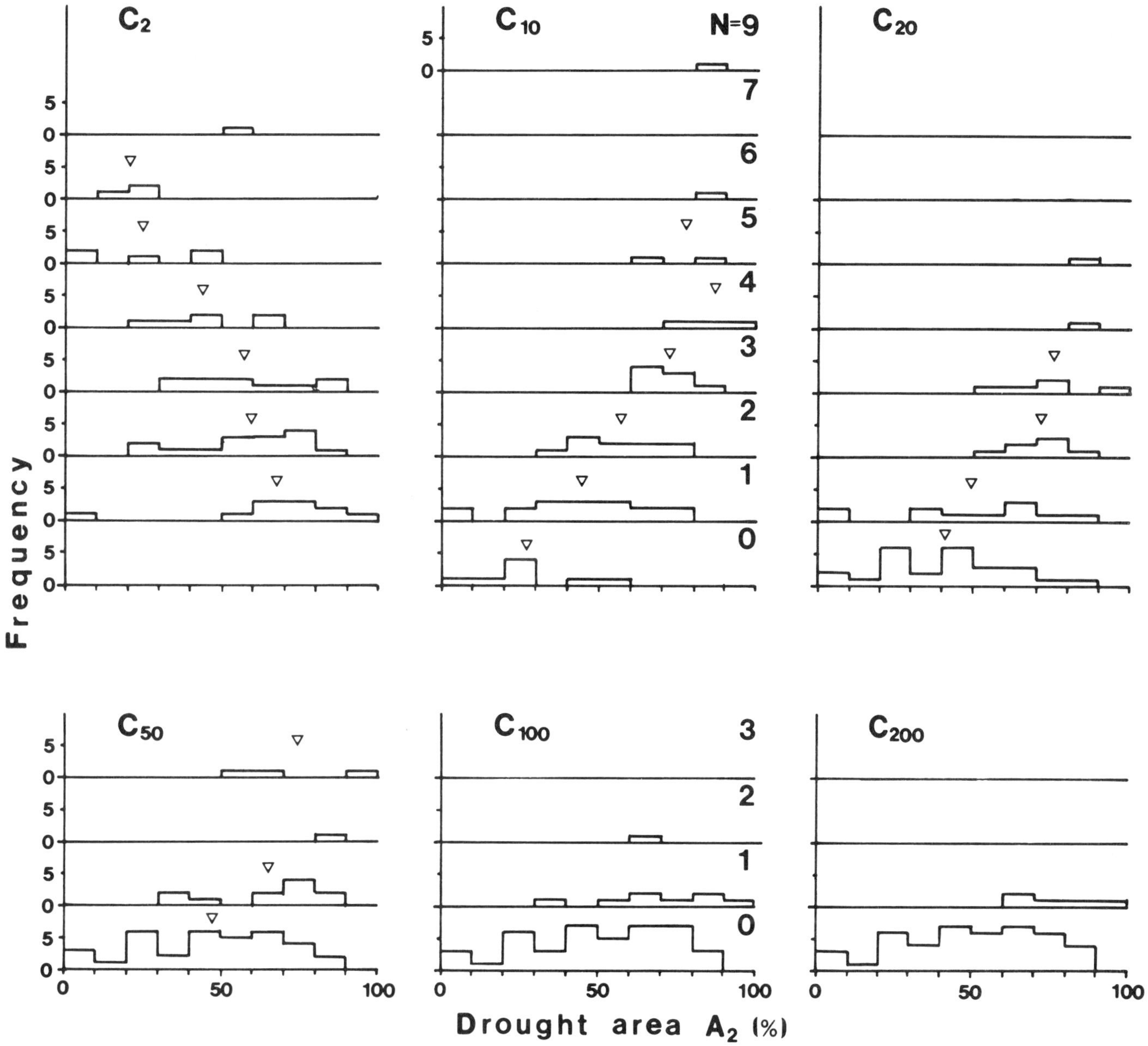

Figure 6. Frequency Distributions of Area of Drought Cells with a Relation to Occurrence Number and Means (open triangles).

INTERNATIONAL SYMPOSIUM ON HYDROMETEOROLOGY
JUNE AMERICAN WATER RESOURCES ASSOCIATION 1982

MEASURING DROUGHT SEVERITY AND ASSESSING IMPACT

Donald A. Wilhite[1]

ABSTRACT: Our sensitivity to variations in climate is increasing in response to world population pressures and diminishing natural resources. The impact of the droughts of the mid-1970s and early 1980s is recent evidence of this sensitivity. In the U.S. annual losses due to drought are estimated to be approximately $1.2 billion. As a result, decision makers at all levels are experiencing an expanding need for accurate, near real-time information about droughts and their effects.

The Palmer Drought Severity Index (PDSI) has been widely used as a general measure of water conditions. This index is based on a hydrologic accounting system which produces an index of zero under normal weather conditions for all climatic regimes. However, the PDSI is not infallible. In fact, the reliability of this index for many applications has become increasingly suspect. This paper will focus on the shortcomings of the PDSI and suggest some alternative approaches to the characterization of drought severity and impact. (KEY TERMS: drought; PDSI, drought index.)

INTRODUCTION

Throughout human history drought has been a threat to man's existence, often altering the course of history itself. Today, amidst growing population pressures, political instability and dwindling natural resources, the role of drought in the disruption of social, economic, and physical systems is of great concern. The Sahelian drought, for example, resulted in the death of as many as 100,000 persons between 1968-1973 (Eckholm and Brown, 1977) and left countless thousands suffering from malnutrition. Drought impacts were even felt in the technologically advanced nations of North America and Western Europe during the mid-1970s. Massive government relief was applied to avert serious problems in water short areas.

Attempts to respond to and, hopefully, mitigate the effects of drought must rely on accurate and timely assessments of the severity and areal extent of the drought. Many of the technological tools required to provide such information are available or within our grasp, e.g., near real-time weather information, operational crop yield models, thermal waveband satellite imagery, etc. Yet, little progress has been made in drought assessment techniques. This paper focuses on difficulties in assessing drought impact, with particular emphasis on: the agricultural sector, specific shortcomings of the Palmer Drought Severity Index (PDSI), and factors to consider when assessing agricultural drought severity.

DROUGHT – DEFINITION AND CHARACTERISTICS

Drought differs from other natural hazards in several ways. First, it is a 'creeping phenomenon,' the effects of which accumulate slowly over a considerable period of time. Second, drought impacts are less obvious than are damages that result from other natural hazards such as floods and hurricanes. Third, the absence of a precise and universally accepted definition of drought adds to the confusion about whether or not a drought exists and, if it does, its severity.

In this paper drought is defined as a climatic excursion involving a shortage of precipitation sufficient to adversely affect crop or range productivity. This definition is purposely general. It is intended to help focus attention on a critical ingredient of a "working" definition – impact. Ideally, some measure of impact should be included in any definition of drought.

Droughts differ in three essential characteristics – intensity, duration, and spatial coverage. Intensity is generally measured by the departure of some climatic index from normal and is closely linked to duration in the determination of impact. Percent of normal precipitation has been, historically, the most commonly used index (see e.g., Hoyt, 1936; Nace, 1965). More recently, Changnon (1980) has defined intensity in terms of the impact of a 12-month precipitation shortage on a wide range of water users in Illinois. He found that, where 12-month precipitation in Illinois is less than 50 percent of normal, virtually all endeavors are affected.

Each drought episode possesses markedly dissimilar spatial characteristics. Compare, for example, the three major droughts of this century. In July 1934, the northern half of the U.S. experienced extreme drought. By contrast the pattern of extreme drought in July 1956 was concentrated primarily in the southwest and in the southern and central Great Plains states. In the mid-1970s the far

[1]Assistant Professor, Center for Agricultural Meteorology and Climatology, University of Nebraska-Lincoln, 241 L. W. Chase Hall, Lincoln, Nebraska 68583.

western, northern Great Plains and upper midwest states were most heavily affected. Although the drought of the mid-1970s equaled these previous droughts in intensity for a small portion of the U.S., neither duration or spatial extent were comparable.

DROUGHT IMPACT

Most Americans are only vaguely familiar with the losses caused by drought. While agricultural losses are most obvious, changes in the Consumer Price Index or Business Index are not.

The U.S. has experienced several widespread droughts since 1976, each producing substantial economic losses. The first occurred in 1976-77 and affected a large portion of the far western U.S., especially California, and sections of the Great Plains and midwest. A recently completed assessment of this drought and the eastern/midwestern freeze (NOAA, 1982) estimated losses at $36.6 billion (expressed in 1980 dollars). These losses were distributed as follows: production, 33 percent; foodstuffs, 26 percent; transporation, 18 percent; sales, 13 percent; and energy, 10 percent.

In addition to the impacts cited above, the 1976-77 drought resulted in a federal expenditure of billions of dollars for relief to drought victims. Some 40 separate programs were administered by 16 federal agencies during the period (GAO, 1979). Aid to water users alone amounted to five billion dollars during 1976-77 (GAO, 1979).

A second drought occurred during the summer of 1980 and was unusual because of its association with a record setting heat wave. Beginning in southwest Texas in mid-June, the heat wave spread northeastward, reaching the Ohio Valley and mid-Atlantic region by mid-July (Anonymous, 1981). Drought was associated with the heat wave in portions of the nation while in the northern plains and upper midwest drought occurred independent of the heat wave. Losses have been placed at $18 billion dollars (NOAA, 1982) and were distributed as follows: production of foodstuffs, 85 percent; energy, 9 percent; and transportion, 6 percent.

PALMER DROUGHT SEVERITY INDEX

It is society's vulnerability to drought that makes it imperative to develop the capability for accurate and timely assessments of drought severity and, ultimately, of drought impact. The Palmer Drought Severity Index (PDSI) (Palmer, 1965) and the Crop Moisture Index (CMI) (Palmer, 1968) are the primary tools currently used by federal agencies to describe prevailing drought conditions. The PDSI is intended to reflect long-term moisture conditions and the CMI short-term conditions. The PDSI and CMI are published monthly and weekly, respectively, in the Weekly Weather and Crop Bulletin.

Palmer's purpose in developing the PDSI was to derive a "general methodology for evaluating the meteorological anomaly (of drought) in terms of an index which permits time and space comparisons of drought severity" (1965). Although it is commonly referred to as a drought index, the PDSI actually is used to evaluate prolonged periods of abnormally wet or dry weather. Index values generally fall within a range of –6 to +6. Values between –2 and +2 represent near normal conditions while values less than –4 or greater than +4 reflect extreme dryness or wetness, respectively. The PDSI integrates antecedent weather conditions over a period of months and, therefore, reacts slowly to deteriorating soil moisture conditions. As a result, each computed monthly value reflects antecedent or accumulated weather conditions over a period of several months.

SHORTCOMINGS OF THE PDSI

Although the PDSI has come under increasing criticism in recent years from members of the scientific community, little of this criticism has found its way into the literature. Changnon (1980) has documented a situation for south central Illinois between early 1976 and early 1977 where precipitation was 65 percent of normal for a 20,800 km^2 area. However, the PDSI indicated near normal conditions in March 1977 when the accumulated precipitation deficit was greatest. Inconsistencies of this type need to be documented and investigated to determine reasons for the discrepancy. These inconsistencies may be associated with certain locations and situations. In this paper I will present other examples of apparent conflicts between the PDSI and actual observed drought severity.

Several explanations are proposed for the lack of consistency between calculated PDSI values and the reality of drought severity. First, plant response was not considered in the derivation of the index, nor does it figure in the calculation of index values. Yet, the PDSI is used to assess the impact of drought on crops. Crops differ in their response to moisture and heat stress. Thus, an agricultural drought index should be crop specific, wherever possible. Second, the PDSI uses the Thornthwaite method of estimating evapotranspiration (ET) in the calculation of index values. The Thornthwaite method is unable to account for sensible heat advection, a major source of the energy that drives the ET process in the Great Plains region. Thus, there is concern that the Thornthwaite method severely underestimates ET in subhumid and semiarid regions (Rosenberg, 1974) and, accordingly, that the Palmer Index tends to underestimate drought severity. Third, the PDSI is calculated only on the basis of climatic divisions. Nebraska is divided into eight such divisions, Arizona into seven, etc. Drought assessment can be difficult at this scale (Wilhite and Neild, 1982).

For the reasons discussed above, the validity of using the PDSI to assess agricultural drought impact is questionable. In addition, differences in land use further complicate the impact assessment issue. Index values cannot be directly linked to impact because each crop responds differently to moisture and heat stress. For example, a PDSI of −3.0 in July may signal substantial reductions in dryland corn yield. Were moisture conditions to improve, corn yield would still be low but soybean yields may turn out near normal. In important agricultural states such as Nebraska, where crop production is diversified according to region, a great deal of supplemental information is necessary before the nature of the impact can be fully understood.

SUMMARY

Drought results in repeated and substantial impacts. It is imperative that tools be developed that allow for accurate and timely assessments of drought severity and impact. Currently, the PDSI is the primary tool used in the U.S. to monitor drought conditions. However, the overall usefulness of the index is subject to question. The index should be tested more critically than has been done heretofore and modified as appropriate. Until that is done, the PDSI should always be used cautiously and reinforced with supplementary information.

LITERATURE CITED

Anonymous, 1981. A $20 Billion Heat Wave. Bulletin of Amer. Meteorol. Society 62:91.

Changnon, S. A., 1980. Removing the Confusion Over Floods and Droughts: The Interface Between Scientists and Policymakers. Water International, June, pp. 10-18.

Eckholm, E. and L. Brown, 1977. Spreading Deserts – The Hand of Man. Worldwatch Paper 13, August, Worldwatch Institute.

General Accounting Office, 1979. Federal Response to the 1976-77 Drought: What Should Be Done Next? Report to the Comptroller General, January 31, Washington, D.C.

Hoyt, J. D., 1936. Droughts of 1930-34. U.S.G.S. Water-Supply Paper 680, USGPO, Washington, D.C.

Nace, R. L., 1965. Drought of the 1950s with Special Reference to the Midcontinent. U.S.G.S. Water-Supply Paper 1804, USGPO, Washington, D.C.

NOAA, 1982. Climate Impact Assessment. U.S. Department of Commerce, EDIS-CEAS, January 25.

Palmer, W. C., 1965. Meteorological Drought. Research Paper 45. U.S. Weather Bureau, Washington, D.C.

Palmer, W. C., 1968. Keeping Track of Crop Moisture Conditions Nationwide: The New Crop Moisture Index. Weatherwise, August, pp. 156-161.

Rosenberg, N. J., 1974. Microclimate: The Biological Environment. John Wiley and Sons, New York, New York.

Wilhite, D. A. and R. E. Neild, 1982. Determining Drought Frequency and Intensity on the Basis of Plant Response: Wild Hay in the Sandhills of Nebraska, U.S.A. Agric. Meteorol. 25:257-265.

INTERNATIONAL SYMPOSIUM ON HYDROMETEOROLOGY
JUNE AMERICAN WATER RESOURCES ASSOCIATION 1982

DEVELOPMENT OF A SURFACE WATER SUPPLY INDEX – A DROUGHT SEVERITY INDICATOR FOR COLORADO

L. E. Dezman, B. A. Shafer, H. D. Simpson, and J. A. Danielson[1]

ABSTRACT: Economy-damaging drought conditions in two of the past four years have prompted formulation of a drought response plan by the Colorado state government. The Surface Water Supply Index (SWSI) was devised for use in conjunction with the Palmer Index to provide an objective indicator of water supply conditions in each major basin in the state. The SWSI was designed to quantify surface water supply available for those sectors of the economy dependent on river and reservoir water. The Palmer Index is used for dry land sectors. The State's drought response plan is activated when SWSI or Palmer values fall below specified levels. Though physically based, the SWSI utilizes historical data and statistical methods to combine present reservoir storage, stream flow and high elevation precipitation (rainfall and snowpack) into a single index number. Commonly used distributions are employed to determine probabilities of non-exceedence for each of the above mentioned variables. Probabilities are then merged using weights representative of the variable's probable impact on surface water supply. SWSI's for each basin are published monthly by the Colorado Division of Water Resources and through the winter months by the cooperator in this project, the USDA Soil Conservation Service, Snow Survey Unit.
(KEY TERMS: drought response plan; water availability; frequency analysis; Palmer Index; climatic variability.)

INTRODUCTION

The droughts of 1976-77 and 1980-81 were the first periods of serious moisture shortage since the early 1950's. State officials became aware that a method of assessing the onset, severity, and termination of drought was needed to efficiently deal with the negative impacts. Although numerous sources of hydrometeorological data are available, an objective technique was lacking which adequately integrated them into a generalized indicator of water availability. The Palmer Index (PI), useful in quantifying drought periods in the eastern plains of Colorado, is seriously flawed when applied to the mountainous western region.

In response to this need, a Surface Water Supply Index (SWSI) was developed which incorporates multiple hydrologic/climatological components into a single, objectively derived index value for each major basin in the state. The index is an integral part of Colorado's Drought Response Plan (DRP) (Lamm, 1981), that outlines the state's organized reaction to identifiable drought conditions.

Colorado's DRP designates ten task forces with responsibilities for various aspects of drought assessment and response. The Water Availability Task Force (WATF) evaluates and disseminates water supply information to other task forces. SWSI and the Palmer Index are used by the WATF in judging whether a drought condition exists and if so, its severity. Thus they "trigger" both activation and deactivation of the DRP.

SWSI focuses on surface water supplies derived from melting snow which accounts for 65 to 85 percent of the annual flow of the state's major streams. Reservoir storage and current precipitation amounts are also taken into account in appraising available or forecasted water supplies. When spring arrives, the moisture represented by the melting snowpack is translated into streamflow and is included in the water supply assessment. SWSI unifies these various components into a single index value useful for management decisions in times of impending or current water shortage. To give managers an overview, an index value is generated monthly for each of seven Colorado basins and presented in map format (Figure 1).

RATIONALE FOR SWSI DEVELOPMENT

Information Available Prior to SWSI

Prior to SWSI, several reports not specifically designed for the purposes of the DRP were used to obtain basic overviews. Among them were the SCS Water Supply Outlook (WSO), reports by Division of Water Resources (DWR), the State Climatologist, the National Weather Service, and the U.S. Geological Survey. These reports, supplemented by expert opinions, often led to dissimilar conclusions.

[1]Respectively, Senior Water Resources Engineer, Colorado Division of Water Resources, 1313 Sherman Street, Room 818, Denver, Colorado 80203; Snow Survey Supervisor, USDA Soil Conservation Service, Snow Survey Unit, P. O. Box 17107, Denver, Colorado 80217; and Assistant State Engineer and State Engineer, Colorado Division of Water Resources, 1313 Sherman Street, Room 818, Denver, Colorado 80203.

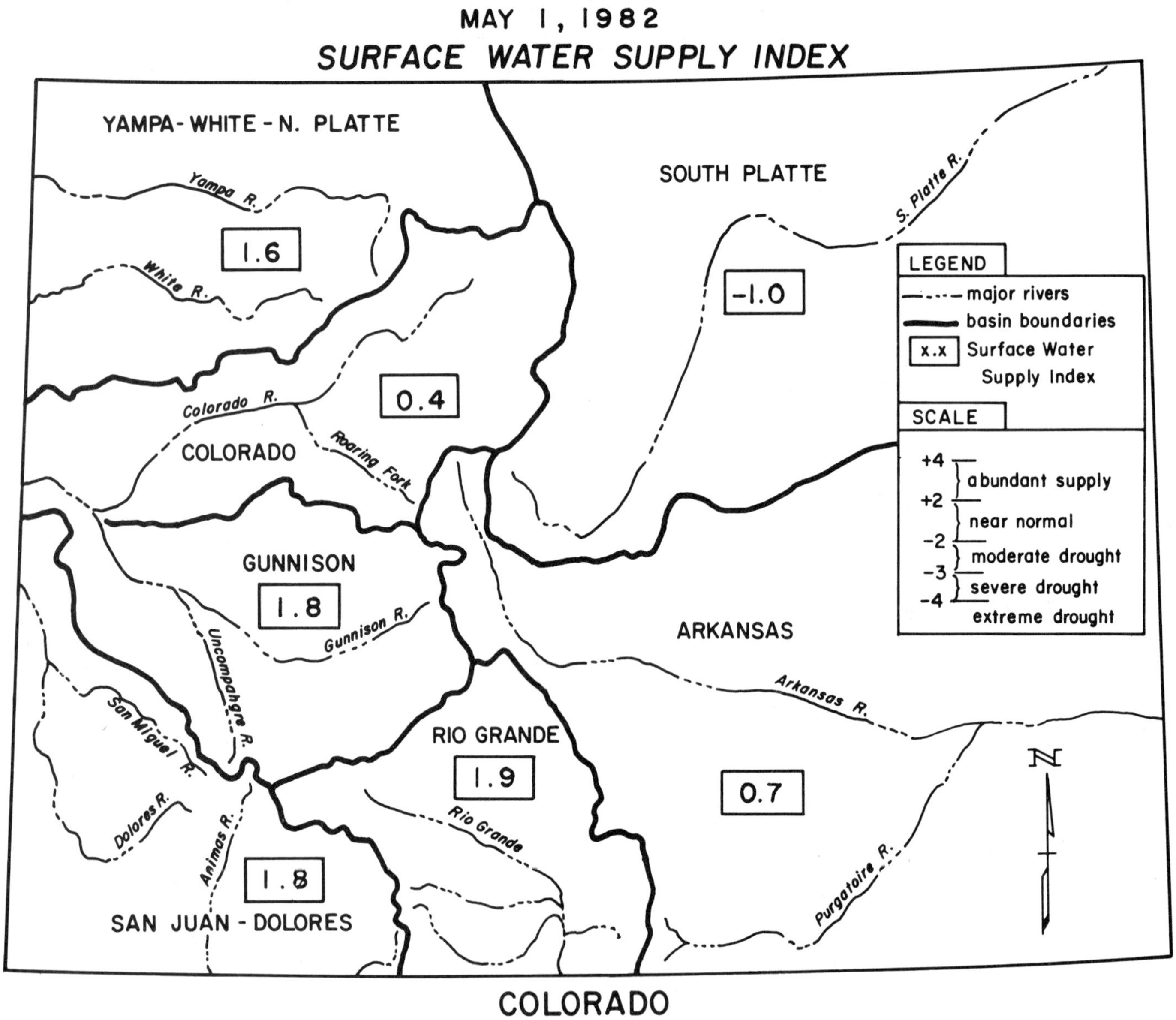

Figure 1. Surface Water Supply Index Map of Colorado.

Climatic and Hydrologic Variability

Colorado's climate exhibits systematic variation both in an east-west and a north-south direction. The east-west variation is primarily caused by the presence of high mountain ranges in the west and relatively low elevation plains in the east. The mountainous western half of the state has a pronounced north-south variability in winter precipitation and streamflow (Doeskin and Shafer, 1981). Normalized April 1 snowpack data as well as April-September streamflow volumes for the Rio Grande and Yampa River basins were found to have significantly different variances. Therefore, it was believed that a procedure that statistically acknowledged the existing hydrologic variation between basins would provide a more realistic appraisal of the probable impacts of extreme events.

Complementary Indices

It was first necessary to clarify the role the SWSI would serve as a complement to the PI. The PI was developed to identify the onset, severity and duration of a drought expressed as a departure of weather conditions from some norm (Palmer, 1965). It is essentially a soil moisture accounting algorithm that ignores snowpack and is calibrated to relatively homogenous hydrologic areas. Consequently, the PI is only valid in Colorado as a general indicator of moisture conditions in low elevation areas which are essentially mountain water independent, i.e., nonirrigated areas.

SWSI, by comparison, was conceptualized as an index that could serve areas that are mountain water dependent (MWD), i.e., areas that rely on snowmelt runoff as their main source of supply. SWSI and PI are mutually complementary indices descriptive of different hydrologic regimes; used together, they more completely and accurately reflect overall moisture conditions in the state.

FORMULATION OF SWSI

General

The SWSI was designed so that it would be an indicator of basinwide water availability for the MWD sector, be predictive, and permit comparison of water supply conditions between basins to assess relative drought severity. Several data with inherent predictive abilities are merged objectively to form basin index numbers using the following steps:

1. Individual monthly station data are summed to obtain a composite reservoir datum, snowpack datum, and precipitation datum. This reduces the number of data, smooths the effect of anomalous events at individual stations, and enhances the applicability of statistical methods to reservoirs.

2. Component data are normalized to one scale – the non-exceedence probability (PN) by using frequency analyses. This integrates the effect of hydrologic variability of each component and permits comparison between components. The PN is superior in some respects to percent of average (as given in many water supply reports) because it uses important additional information inherent in the collective data set, i.e., the dispersion tendency (standard deviation). Percent of average incorporates the station's past history in computation of the mean value (central tendency) but ignores dispersion.

3. PN's for each component are utilized in a basin weighting/scale equation to determine a SWSI for each basin in the state. A winter season SWSI equation is:

$$SWSI = \frac{[(a \times PN_{SP}) + (b \times PN_{PCP}) + (c \times PN_{RS}) - 50]}{12} \qquad (1)$$

where a, b, c are weights for each component and must total one. PN is non-exceedence probability (%) and SP, PCP, and RS refer to snowpack, precipitation and reservoir components, respectively. During summer, streamflow (PN_{SF}) replaces the snowpack component in the equation. Each basin has unique a, b, and c's for winter and summer.

The weighting/scale equation (Eq. 1) does three things: (1) weights each component, (2) centers the SWSI about zero, and (3) compresses the scale. Steps (2) and (3) produce numbers comparable to PI. The following table gives important scale points and their designations.

SWSI	**Designation**
+4	Abundant supply
+2	Near normal
–1	WATF activated
–2	Moderate drought – other DRP task forces activated
–3	Severe drought – agency drought programs activated
–4	Extreme drought

The flowchart shown in Figure 2 depicts the stages of SWSI development and operational application.

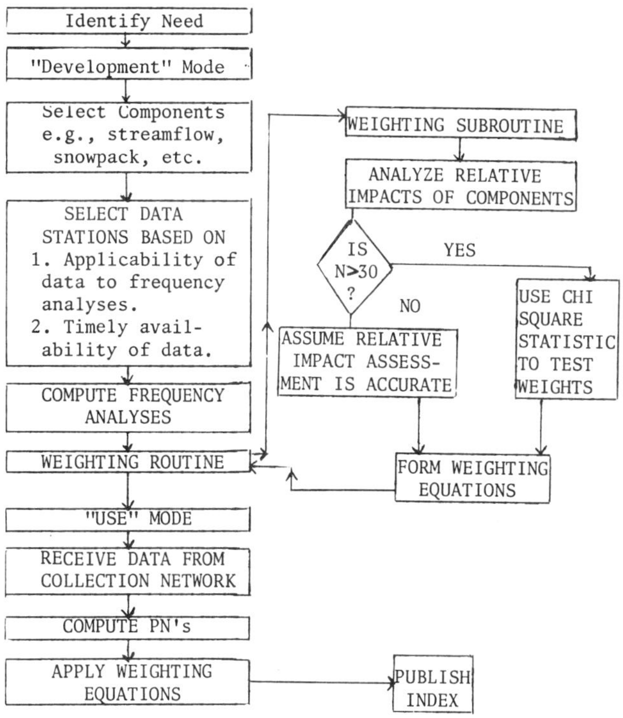

Figure 2. SWSI Development and Operations Flowchart.

Data Requirements

Stations were selected to give representative geographical coverage. Major considerations in staisfying this requirement were finding gages with sufficient length of record to develop frequency analyses that could submit data in a timely manner for actual operations. Streamflow and reservoir stations had to meet the additional requirements of compatible sizes. Some form of cumulative data is used for all components, except summer precipitation, to enhance SWSI's predictive abilities.

Reservoir storage is the only component that is not clearly a random variable due to possible serial correlation. To reduce the dependence of data on previous values (i.e., to system memory), both irrigation and municipal reservoirs were selected and their data summed.

Frequency Analyses

Frequency analyses were done for all the components so that PN could be determined. Components with sufficient length of record were tested to ensure the applicability of frequency analyses. Only reservoir data in the South Platte basin were found to be dependent; these reservoirs can store two and one-half times the average annual streamflow at Kersey.

Component Weightings

The weightings a, b and c in Eq. 1 are based on the component's potential impact on water available in the basin. Annual streamflow and reservoir volumes were compared to determine relative weightings. Expert opinions by the National Weather Service, the Colorado State Climatologist and the SCS yielded initial approximattions of the impact of precipitation and snowpack.

Final weightings were based on the hypothesis that the additive nature of the components cause the SWSI to be normally distributed. The Chi-square statistic was used to optimize goodness of fit for trial component weights.

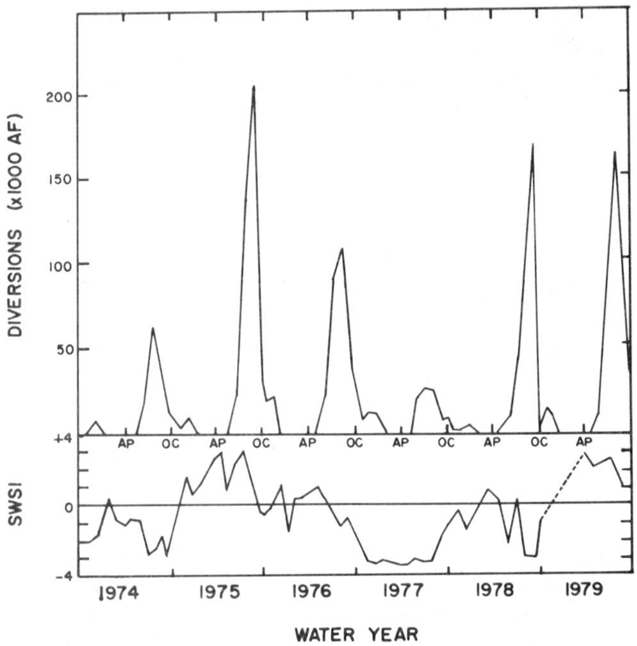

Figure 3. Selected Rio Grande Basin Surface Diversions and SWSI Time Series.

SWSI PERFORMANCE

Rio Grande Basin surface diversions were used to measure SWSI performance. Correlations between monthly Rio Grande north bank diversion data (dependent variable) and SWSI (independent variable) were investigated. The years 1974 through 1979 were used since they include extreme and near-normal events.

July and August diversions, usually the largest of the year, appear increasingly predictable progressing from January 1 to May 1 (Figure 3). Results indicated the August 1 SWSI's were most predictive of September diversions. In summary, winter SWSI's are most useful in predicting peak season, and summer SWSI's in predicting late season water availability.

OPERATION AND CONSUMER RESPONSE

The SWSI map (Figure 1) appears in the monthly Colorado WSO, December through May, and in the DWR newsletter issued monthly for the entire year. Because much of the value of these reports lies in their timeliness, it is important that any new product they contain be producible within the first five days of each month. The SWSI computational procedure is straightforward and entails less than an hour once the requisite data are assembled.

The Water Availability Task Force has used SWSI and PI thresholds according to plan. Response by the professional community concerning both the concept and results through the 1981 drought period have been favorable.

REFINEMENTS

Monthly SWSI's indicate drought intensity, but not explicitly duration. An objective means of quantifying duration would involve a gradient component in addition to snowpack, precipitation, reservoir storage, and streamflow. The gradient component would be based on frequency analyses of monthly changes in the other components. The net effect of this step would be to incorporate duration into each month's SWSI and thereby damp wide monthly variations.

SUMMARY

Colorado's droughts of 1976-77 and 1980-81 were the impetus for the development of a Surface Water Supply Index. SWSI and the Palmer Index are currently the two indices recognized in the Colorado Drought Response Plan as triggering mechanisms to officially monitor drought conditions and activate appropriate impact task forces to deal with its effects. SWSI is meant to serve those areas dependent on surface water supplies originating as snowmelt in the mountains and classified as mountain water dependent.

A single SWSI number is generated monthly and displayed in map form for each major basin in Colorado. The index number is derived from the probability distributions of snow course, precipitation, reservoir, and streamflow data to accurately reflect the state's hydrologic variability. Each of the constituent variables in the index is weighted in proportion to its probable impact on total surface water resources in the basin. The SWSI's produced are thus numerical values that express the current and future availability of water supplies to meet a multitude of competitive demands. The range of SWSI values is from –4.0 (prospective water supplies extremely poor) to +4.0 (prospective water supplies abundant). The SWSI number is a general indicator of basin-wide surface water supply conditions.

LITERATURE CITED

Doeskin, N. J. and B. A. Shafer, 1981. Winter Drought in the Colorado Mountains. American Meteorological Society 2nd Conference on Mountain Meteorology. Steamboat Springs, Colorado, November 1981. pp. 57-58.

Lamm, D. R., 1981. The Colorado Drought Response Plan, State of Colorado, Office of the Governor, Division of Disaster Emergency Services.

Palmer, W. C., 1965. Meteorological Drought. Research Paper No. 45, U. S. Department of Commerce, National Weather Service. 58 pp.

INTERNATIONAL SYMPOSIUM ON HYDROMETEOROLOGY
JUNE AMERICAN WATER RESOURCES ASSOCIATION 1982

TRANSBORDER SALINITY AND DROUGHT MANAGEMENT: U.S./MEXICO UNRESOLVED ISSUES

Roger Lee Eldridge[1]

ABSTRACT: The U.S. government has guaranteed the quantity and the quality of the water delivered at the border to Mexico in the Colorado River. The history and substance of these commitments are briefly reviewed. The unresolved issues of salinity control and drought response are described. Particular attention is given to how we define the salt loading problem, and to the lack of definition of "extraordinary drought." The conclusion is that a Salinity Control and Drought Response Compact is needed among the states of the Colorado River Basin.
(KEY WORDS: salinity control; drought response; Colorado River; U.S./Mexico Treaties; border water.)

Explicit guarantees by the U.S. government to Mexico regarding the salinity of and drought on the Colorado River exist. In the review of the history and substance of these commitments that follows, several unresolved questions facing water managers and users of the Colorado River Basin are discussed.

A. **Salinity**. The most intensively managed river on the continent drains four major sub-basins along the western Rocky Mountains, through the Green, Upper Colorado, San Juan and Yuma rivers. This now mountainous drainage area was once the bed of a prehistoric sea, which as it dried deposited many beds of salty shales. The rivers now washing through those lofted, folded sediments leach out much of that ancient salt. The vast delta created by the river, ironically left dry and desert-like after its watery birth, stretches from above Palm Springs to the present Gulf. It supports marvels of crop production when touched with irrigation. On the American side of this delta is the largest diversion on the whole river, the 2.9 million acre feet serving the 450,000 acres of the Imperial Irrigation District. On the Mexican side, a valley twice as large contains 490,000 irrigated acres and 12,000 farms.

There are two major agreements between the U.S. and Mexico on the quality of river water that the U.S. must deliver at the border. The matter was first addressed on the Rio Grande side in 1906, when Mexico obtained the right to 60,000 acre feet/yr. (74 million cubic meters) at the Juarez delivery point. Only in 1939 did the upper Rio Grande basin states – Colorado, New Mexico and Texas – agree to the delivery responsibilities of each state, after 16 years of negotiations. The second major agreement in 1944 divided the Rio Grande below the Juarez area, and determined also that Mexico is to receive from the Colorado River 1,500,000 acre feet/yr. (1,850,224 cubic meters) at the Arizona-California border. Although the Upper and Lower Basin states are each responsible for delivery of half of this amount according to the treaty, they have not yet agreed about the precise delivery responsibilities of each state and delivery schedules. The states have with various degrees of formality made it clear that they do not intend to let the quantitative or qualitative agreements by the federal government with Mexico interfere with their full utilization of the water appropriated to them by the Colorado River Compacts.

In the early 1970s, Soluble Mineral Content (SMC) of the water reaching the delta passed the lower threshhold of crop tolerances, threatening to ruin the agricultural economy on both sides of the border. Mexico also received the return flow from the Wellton-Mohawk Irrigation District that enters the river below the last American diversion, creating SMC for them of over 1200 mg/l! A strong diplomatic effort led by President Echeverria in 1973 produced Minute 242 of the International Boundary and Waters Commission, establishing water quality standards for the river at the Mexican border as no more than 115 ppm higher than the water diverted for use just north of the border at Imperial Dam. At the time of the signing of the Minute, salinity at Imperial Dam was 850 ppm and at the Mexican border was 980 ppm.

The 1974 Colorado River Basin Salinity Control Act was the U.S. government's follow-up on the treaty with Mexico. It authorized the construction of a desalinization plant at Yuma, Arizona, and four up-stream projects to prevent salts from entering the river, plus studies of twelve more projects. Amendments in 1980 eliminated two of the earlier projects and created a slightly different list of ten for more study.

In the year preceding the water quality agreement with Mexico, the Congress had dealt with domestic water quality problems with the Water Pollution Control Act. This act set

[1] Director, State Supported Organized Research, Colorado Commission on Higher Education, 1550 Lincoln, Room 207, Denver, Colorado 80203.

a goal of removing 2.8 million tons of the 10 million tons of salt in the Colorado River each year, and set numerical salinity standards for the river at 879 ppm at Imperial Dam. Then in 1977, the Clean Water Act did more than all its predecessors to establish water management methods. Water quality management planning (Section 208), reviewing cycles (Section 303), point source discharge controls (Section 402), and dredge and fill controls (Section 404) have become major administrative tools for states, councils of government and counties for handling water quality issues.

The major import of all this for our present topic is the growing federalization of the management of the river. The U.S. government has become the guarantor of water quality at the Mexican border, as well as the manager and financier of border area and up-stream construction projects necessary for reducing SMC levels to meet treaty commitments. The seven basin states, in an impressive show of unity, applied their political muscle during the writing of the 1974 Salinity Control Act, to protect their rights to develop all the water apportioned to them, and to start the federal government paying for removing salt from the river. As more water is used upstream for irrigation, salinity in the river increases. It was clever of the states, or at least it appeared so in 1974, to get federal taxes to pay for taking the salt out while they were putting it in. We are discovering, however, that federally mandated standards and management practices remain in place, while the federal paymaster is leaving the scene.

In the meantime, the group created by the basin states to collaborate in the administration of the Act and to defend state interests, the Salinity Control Forum, reports (1981):

> There are many entities and levels of government concerned with the salinity of the Colorado River. However, only the federal government is involved in all the major Basin-wide aspects of the salinity problem, and a solution is only possible in a Basin-wide context . . . . Federal lands, including Indian reservation lands, are the source of most of the naturally occurring salts in the river. Accordingly, it is believed that the federal government is the appropriate unit of government to finance the salinity control projects and to be allocated a major share of costs.

But as the same document reports in another section, the Bureau of Land Management reported in 1977 that the salt load contributed by surface runoff from public domain lands in the upper basin states was 8 percent of total salt loading. Runoff from private, state and Indian lands and all other federal lands contributed another 7 percent. Thus, federal land sources cannot be "the source of most of the naturally occurring salts in the river." It should also be noted that all six of the construction projects now underway upstream for blocking salt sources are on private, not public lands. The Bureau of Reclamation in fact had problems at both Paradox Valley and Meeker Dome getting title to private lands so that government salinity control projects could be built! Grazing practices in the watersheds can contribute to salt loading, but grazing is a private, not a federal enterprise.

One basis for claiming federal tax dollars for offsetting the effects of local commercial ventures is the erroneous claim that federal lands contribute most of the natural salt loading of the river. Another basis is the equally erroneous claim that natural salt loading itself is the cause of the salinity problem. The undeniable popularity of blaming natural sources for the salinity problem is the desire to capitalize on the American habit of holding the federal government responsible for helping to restore damages done by nature. A big enough storm becomes a "natural disaster area," defined by governors flying around the scene in helicopters. So long as natural forces can be blamed, subsidies from the federal treasury can be claimed.

But as the following illustration shows, the real causes of salinity damages are artificial salt loads from irrigation, and municipal and industrial developments. As in good prior appropriation doctrine, nature was there first with respect to salt sources, and with respect to the natural tolerances of plants for salt. The key around which to organize our inventory of salt in the river is the 800 ppm threshold of plant damage. Natural salt loading of the Colorado River consists of about 400 ppm, and scientifically speaking is not responsible for degradation of drinking water or for crop damage. In fact, before crop damage begins, the river has absorbed almost two-thirds of man-made salt loads. It is true that the river drains a basin full of salt laden shales, but it is still not valid to blame nature for the entire salinity problem of the Colorado River. The river in its natural state would pose no salinity problem at all. But we have not left the river in its natural state; we have developed and used its water in order to live in the basin – and to export water entirely out of the basin. It is our use of the river that creates the salinity problem, not nature! This point is made clear when we relate salt sources to natural salt tolerances, as illustrated below. How we define a problem very often determines how we perceive the best means of solving it. The easiest way to keep salt out of the Colorado River may be to block natural sources, but a statement of the easiest solution should not be confused with a statement of the causes of the problem, unless we are lobbying for federal funds!

The task of the public policy maker is to draw a line between public spending for public benefits, and public spending for private interests, and it is not an easy task. The U.S. salinity control program must deal sooner or later with the fact that public spending to reduce natural salt loading on the river gives amnesty to earlier private developers of water, so that they do not have to pay for downstream damages for which their developments are responsible. Over the long range, federal desalinization subsidies create a free ticket for future private water developments who can add to the top of the salt load to the degree that federal projects

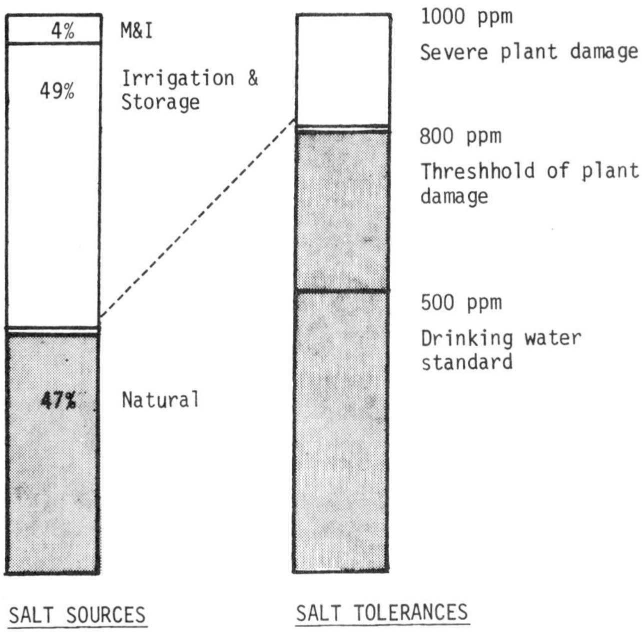

have subtracted from the bottom. That is an attractive prospect from the viewpoint of the previously quoted Salinity Control Forum. Whether or not it is a realistic prospect has become a critical question in 1982. Federal subsidies of all kinds are progressively harder to pry out of Washington, and the relatively easy point source construction solutions are few and will soon be behind us, while salinity levels at Imperial Dam could climb past 1000 ppm by 2000 because of the high rates of private water development in the southwestern United States. The basin states in 1974 ratified basin-wide management of the river in the process of securing federal funds to solve their problem. That money is drying up but the management structure will remain, along with the commitments to Mexico.

The unresolved issues are what the responsibilities of the Upper Basin states for helping to meet water quality obligations to Mexico, and who is going to pay for salinity control projects on the Colorado River. Given the new enthusiasm of many western state politicians for the new federalism, with its hostility toward federal subsidies for special interest groups, it will be interesting to see if the Salinity Control Forum can achieve the dream of every special interest group – to avoid being perceived as a special interest while at the same time keeping its special federal subsidy.

B. **Drought.** The most intensively managed river on the continent contains, depending on which series of years are used to calculate the average flow, 12 to 16 million acre feet of water. The 1922 Compact that divided the river between the Upper and Lower Basins relied on Reclamation Service figures of 16 million acre feet. Since then, however, the average flow has been only 13.9 million acre feet. Because the Lower Basin states got a guarantee, not of half the flow, but of 7.5 million acre feet a year, the Upper Basin states just have to share what is left. By 1949 when the Upper Basin states divided their portion among themselves, they had learned not to use volumes of water in the rules, so they share by percentages: Colorado (51.75 percent), Utah (23 percent), Wyoming (14 percent), and New Mexico (11.25 percent).

We should note that the Compact also expresses the amount of water due to the Lower Basin from the Upper Basin in terms of 75 million acre feet within a ten year period. This technically would allow the Upper Basin to tote up excess deliveries in wet years as an excuse to withhold water during two or three years of a ten year averaging period. We perhaps could anticipate the Upper and Lower Basins selecting different ten year periods for averaging purposes. We also might anticipate downstream municipalities exerting considerable political pressure for minimum guaranteed annual water supplies, despite the averaging formula in the Compact.

The Upper Basin states are not now using all the water to which they are entitled, even using the lowest average flow figures. Therefore, the issue has yet to be joined as to what will happen when active uses of the river's water exceeds the supply. The entire flow of the river has long since been legally assigned to potential users, and the time is approaching when all these users will take their shares. Our apportioning system has created legal shares to the river that are larger than the amount of water that flows in the river most of the time; it is an over-apportioned river. Quiet but intense maneuvering by the states to protect their apportionments influences almost every other feature of the western political landscape. It is into this intensely political context that a drought will fall when it comes.

The Upper and Lower Basins are equally responsible for protecting the 1.5 million acre feet/yr. that Mexico expects under the 1944 Treaty. In the case of extraordinary drought in the United States, the treaty allows reductions in the flow to Mexico in the same proportion as reductions of consumptive use in the United States. The concept of proportionate reductions (although undoubtedly difficult to administer in the field) presents no conceptual difficulties. The unresolved issue is what constitutes an extraordinary drought. As observed by Cesar Sepulveda (1978), one of Mexico's foremost water treaty scholars:

> The Treaty of 1944 failed to specify whether the drought could occur in the total region served by a river system or only in a portion of it, and also did not define the intensity nor duration of the drought.

We should add that we do not have any agreed formulas for calculating how transmission losses or stored reserves on the U.S. side should fit into the proportionate reductions, or how supplemental groundwater pumping in the basin and at the border are to fit in.

If we had agreements with Mexico on such formulas, would we be in a position adequately to administer the quotas and limits on our side? Institutional response capacities to drought are very spotty in the Southwest; only California and Colorado have prepared drought emergency plans. Opinion among water managers is often sanguine on this matter; many worthy citizens believe that the senior water rights system will automatically take care of adjustments required by extraordinary drought, while others point to the emergency powers of the state executive branch that can override normally vested rights. It is doubtful whether a senior water right devoted to agricultural uses could withstand the political pressures from large, thirsty urban populations in time of drought. A major question is whether temporary re-allocation procedures in time of drought will be orderly or chaotic. Even in the few states where drought emergency plans exist, there are insufficient manpower, budget reserves, or information and communication systems to handle water emergencies (Lamm, 1981). There are no basin-wide management authorities to implement and enforce water emergency measures.

As population concentrations increase in the semi-arid portions of the Southwest, the vulnerability of large numbers of people to even small downturns in water supply in the Colorado River becomes a more and more important policy issue. The unresolved questions are the ambiguity in the treaty language with Mexico that governs how the two countries will share water shortages, and the lack of institutional preparedness for coping with drought conditions in the Colorado River Basin. On these, as well as the salinity control issues, our work is far from done at both the state and regional levels. If past history is any guide, it will take some crisis to stimulate the basin states into formulating the next great compact on the Colorado River – the Salinity Control and Drought Response Compact.

LITERATURE CITED

Colorado River Basin Salinity Control Forum, 1981. Water Quality Standards for Salinity, Colorado River System, Proposed Report on the 1981 Review. Salt Lake City, Utah.

Lamm, Richard, 1981. The Colorado Drought Response Plan. Denver, Colorado.

Sepulveda, Cesar, 1978. Instituciones Para la Solucion de Problemas de Superficie Entre Mexico y Los Estados Unidos. 18 Natural Resources Journal, 140.

INTERNATIONAL SYMPOSIUM ON HYDROMETEOROLOGY
JUNE AMERICAN WATER RESOURCES ASSOCIATION 1982

THE USE OF HYDROMETEOROLOGICAL DATA IN DROUGHT MANAGEMENT: POTOMAC RIVER BASIN CASE STUDY

James A. Smith, Daniel P. Sheer, and John C. Schaake, Jr.[1]

ABSTRACT: Development and implementation of a drought management system for the Potomac River basin are discussed in this paper. The drought management system is comprised of hydrologic models which are used for streamflow forecasting and models of the regional water supply system. It is shown that the hydrologic models, especially the Sacramento model, are particularly well suited for representing streamflow characteristics in the Potomac River basin. An important component of the streamflow forecast system is the Extended Streamflow Prediction (ESP) procedure, which is used to produce long-range water supply forecasts. The streamflow forecast system and water supply models are linked to form a drought management system which is used for allocating water supplies in the Washington, D.C., metropolitan area during drought periods.
(KEY TERMS: streamflow forecasting; drought management.)

INTRODUCTION

Three topics pertaining to the development and implementation of a drought management system for the Potomac River basin are discussed in this paper. First of all the physical characteristics of low streamflow in the Potomac River basin are discussed. Spatial variation of low streamflow is shown to be closely related to geologic features of the basin and a map of hydrologic provinces is presented. Our second topic concerns the hydrologic models that are used for streamflow forecasting. The main components of the streamflow forecast system are the Sacramento Soil Moisture Accounting Model and the Extended Streamflow Prediction (ESP) procedure, both of which are components of the National Weather Service River Forecast System (NWSRFS). The relations between the physical characteristics of low streamflow in the Potomac River basin and the form of the hydrologic models are discussed. Our final topic concerns linking the streamflow forecast system with a model of the Potomac River basin water supply system. This combined system is used for allocating water supplies during drought periods.

HYDROGEOLOGY OF LOW STREAMFLOW IN THE POTOMAC RIVER BASIN

Streamflow data for water years 1950-1979 from gages within the upper Potomac River basin along with geologic maps of the states of Maryland, Virginia, West Virginia, and Pennsylvania were used in developing the map of hydrologic provinces presented in Figure 1. Streamflow data for the upper Potomac River basin are summarized in Table 1. Following Olmsted and Hely (1963) the statistic Q_{90}/Q_A is used as the primary measure of low flow regimen, where Q_A represents the mean daily flow (in cfs/drainage area in square miles) and Q_{90} represents the daily flow that is exceeded with probability 0.9.

The central features in the relationship between low flow and geology can be summarized as follows:

1) Regions underlain by shale are the poorest sources of baseflow in the basin.

2) The Piedmont metasedimentary rocks have uniformly high baseflow.

3) The metavolcanic rocks of the Piedmont and Blue Ridge have uniformly poor baseflow.

4) In the folded sedimentary rocks of the Valley and Ridge, the presence of carbonate rocks is necessary for high baseflow conditions.

5) The region with highest baseflow in the Potomac River basin is located in the eastern portion of the Great Valley at the western slope of the Blue Ridge (Antietam Creek and the South Fork of the Shenandoah River).

6) Flat-lying sedimentary rocks of the Allegheny Plateau have poor baseflow even when carbonate rocks are present.

The physical basis for the sharp contrasts between hydrologic provinces is best illustrated in the Piedmont of Virginia, Maryland, and Pennsylvania. Figure 2a shows flow duration curves for three streams in the Virginia Piedmont. Bull Run and Cedar Run drain Triassic shale; Difficult Run,

[1]Respectively, Hydrologist and Director, Interstate Commission on the Potomac River Basin Cooperative Water Supply Operations on the Potomac River (CO-OP), 1055 First Street, Rockville, Maryland 20850; and Chief, Hydrologic Services Division, Office of Hydrology, National Weather Service, Silver Spring, Maryland 20910.

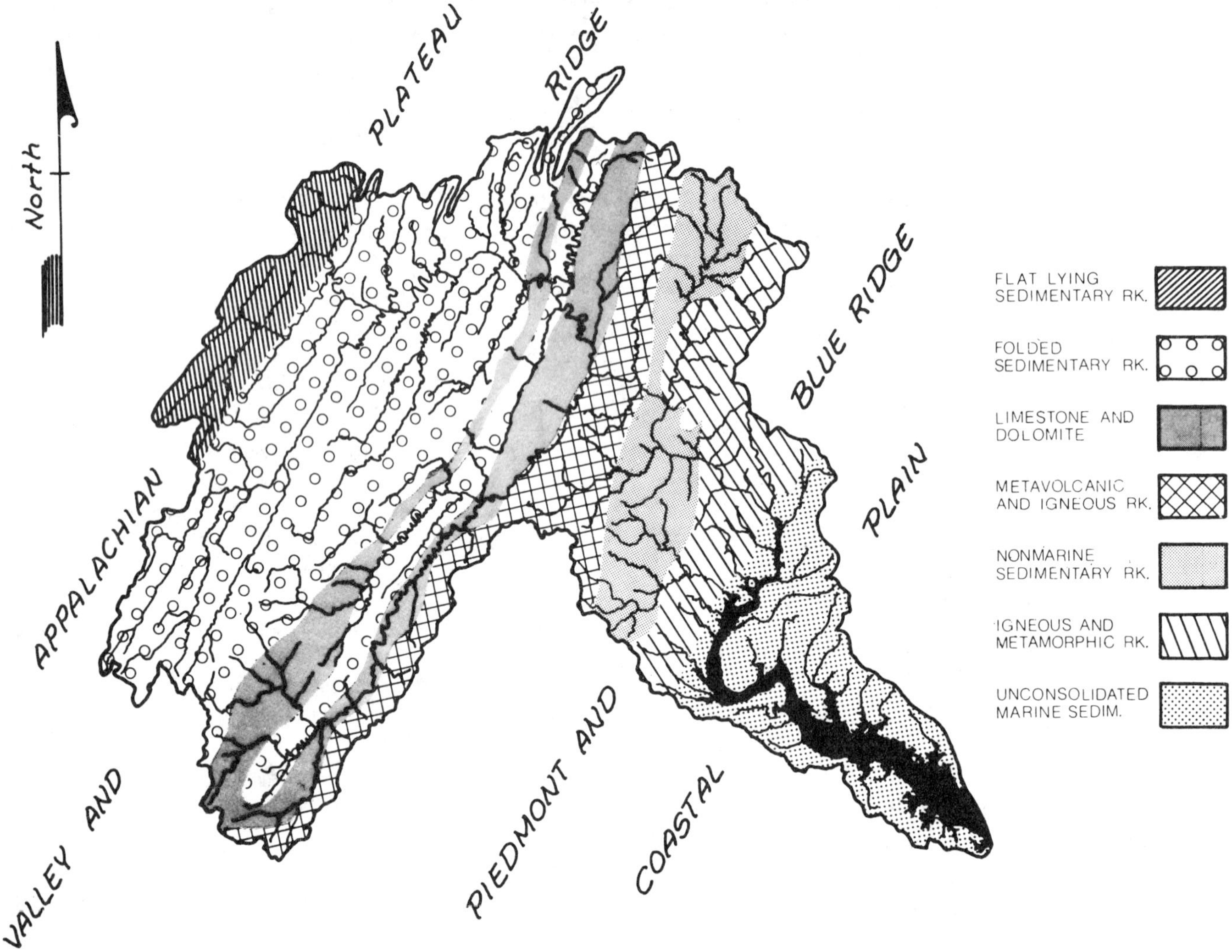

Figure 1. Hydrologic Provinces of the Potomac River Basin.

which borders Bull Run, drains Paleozoic metasedimentary rocks. Figure 2b shows flow duration curves for Monocacy River, Senaca Creek, and Patuxent River. Monocacy River, which drains Triassic shales in Pennsylvania, has flow duration values nearly identical to those of Bull Run and Cedar Run, while Seneca Creek and Patuxent River, which drain Paleozoic metasedimentary rocks in Maryland, have flow duration values similar to Difficult Run.

The contrasts in flow properties between shales and crystalline rocks of the Piedmont are most readily explained in terms of soil moisture storage and permeability properties. The Piedmont crystalline rocks are covered by a thick mantle of saprolite. Nutter and Otton (1966) report that depth to bedrock averages 45 feet in the crystalline Piedmont of Maryland. Nutter and Otton also note that a region of high permeability occurs at the base of the saprolite. Conversely, soils forming on the Triassic shales are thin (Roberts, 1928) and, due to the high clay content, have low permeability.

Storage and permeability are also the key features in producing the region of high baseflow at the western border of the Blue Ridge. Surficial deposits on the west slope of the Blue Ridge in Virginia were reported by Hack (1965) to range from 100-300 feet. Similar values were reported for the corresponding region in the Antietam Creek basin to Maryland (Nutter, 1976). This region of high groundwater storage is underlain by intensely folded carbonate rocks. Joint planes in the carbonate ocks, which are especially numerous due to the intense folding, have been enlarged by solution to produce the most extensive caves in the Potomac River basin (Nutter, 1976). Storage in the carbonate rocks is hydraulically connected with stream channels largely through joint planes. This region is thus similar to the crystalline Piedmont in that it has high ground water storage capacity and zones of high permeability.

Thin soils provide the primary explanation for low baseflow conditions in the metavolcanic rocks of the Blue Ridge

TABLE 1. Summary Statistics for Streams in the Potomac River Basin.

	DA (sq. mi.)	Q_A (cfsm)	Q_{90}/Q_A
1) Flat-Lying Sedimentary Rocks			
North Branch	225	2.00	0.10
Abram Creek	47	1.62	0.06
Crabtree Creek	17	1.72	0.08
Savage River	49	1.52	0.05
2) Folded Sedimentary Rocks			
A) Shales and Sandstones			
Patterson Creek	219	0.78	0.05
Back Creek	243	0.85	0.06
South Fork South Branch	102	0.98	0.07
N.F. Shenandoal River	210	0.93	0.02
Passage Creek	88	0.76	0.07
B) Carbonate Rocks			
South Branch	182	0.88	0.18
Opequon Creek	272	0.84	0.24
Antietam Creek	281	1.04	0.33
Middle River	375	0.81	0.26
South River	127	1.09	0.22
S.F. Shenandoah River	1642	0.93	0.25
3) Metavolcanic Rocks			
Catoctin Creek	67	1.13	0.07
Goose Creek	332	0.97	0.06
Owens Creek	6	1.61	0.07
Hunting Creek	18	1.45	0.10
4) Metasedimentary and Igneous Rocks			
Seneca Creek	101	1.06	0.24
Patuxent River	35	1.14	0.22
Difficult Run	58	1.06	0.21
Big Pipe Creek	102	0.96	0.21
5) Nonmarine Sedimentary Rocks			
Bull Run	148	1.08	0.02
Broad Run	51	1.03	0.08
Cedar Run	93	0.95	0.02
Monocacy River	173	1.08	0.04

and Piedmont, the noncarbonate folded sedimentary rocks of the Valley and Ridge, and the flat-lying sedimentary rocks of the Allegheny Plateau (see Trainer and Watkins, 1975). Thin soils form on metavolcanic rocks due, in large part, to the lack of quartz in the parent material. The shales which predominate in the noncarbonate regions of the Valley and Ridge and Allegheny Plateau have similar hydrologic properties to the shales of the Piedmont described above. The contrast in hydrologic properties between flat-lying carbonate rocks and folded carbonate rocks is, in our opinion, related to the relative importance of bedding (high in flat-lying carbonates) and joint planes (high in folded carbonate ocks) in determining storage and permeability characteristics. A different perspective on this problem is presented by White (1977).

The relations between low streamflow and geology outlined above are not peculiar to the Potomac River basin. Table 2 summarizes streamflow data at selected sites in the Central and Southern Appalachians. It should be noted that two of the classifications listed in Table 1 are missing from Table 2. Metavolcanic rocks of the Central and Southern Appalachians, similar to those exposed along the Blue Ridge in the Potomac River basin, are restricted to a small area of southwestern Virginia. Nonmarine sedimentary rocks are prominent in the Piedmont of New Jersey, Delaware, Pennsylvania, and Virginia. Unfortunately, historical streamflow records that are unaffected by regulation are scarce. A detailed discussion of the relations between low streamflow and geology in the Central and Southern Appalachians is in preparation.

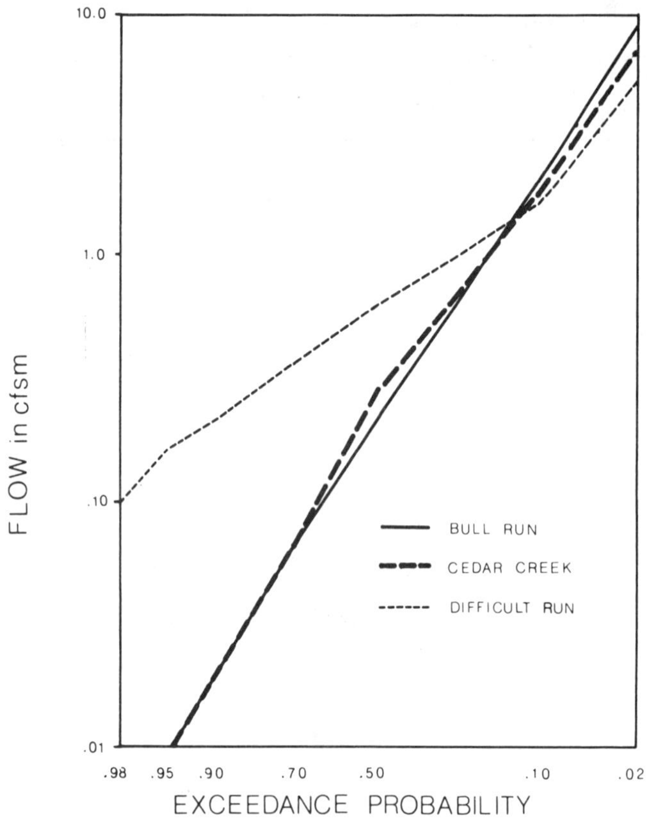

Figure 2a. Flow Duration Curves for Piedmont Streams – Virginia.

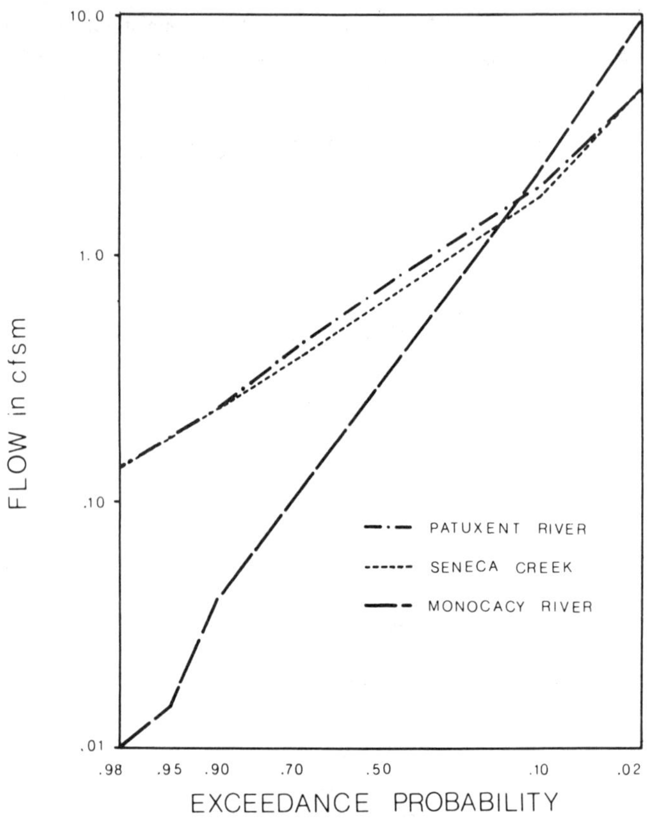

Figure 2b. Flow Duration Curves for Piedmont Streams – Maryland, Pennsylvania.

DESCRIPTION OF THE STREAMFLOW FORECASTING SYSTEM

Low summer streamflow in the Potomac River is the product of low summer rainfall combined with dry antecedent soil moisture conditions. This is illustrated by comparing summer streamflow records from 1951 and 1966 (Figure 3). During the period July-September nearly identical rainfall totals were recorded in 1951 and 1966, yet in 1966 (following three *years* of abnormally low rainfall) Potomac River streamflow at Point of Rocks reached a daily minimum of 547 cfs while in 1951 (following a wet Fall and Winter) the minimum daily flow was 1170 cfs. Streamflow forecasting, used as a drought management tool, must consider information concerning soil moisture conditions as well as uncertainty in future precipitation. The approach taken for streamflow forecasting in the Potomac River basin relies on a conceptual hydrologic model, the Sacramento Soil Moisture Accounting Model, which is a component of the National Weather Service River Forecast System (NWSRFS). The Sacramento model can accurately reproduce historical streamflow using historical meteorological data as input. The model uses parameters which are surrogates for the physical structure of the drainage basin. It provides the forecasting flexibility required for water supply management.

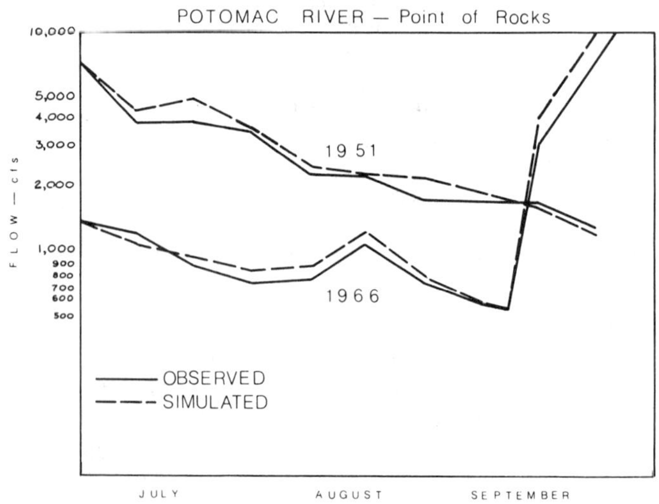

Figure 3. Potomac River Streamflow – 1951 and 1966.

The Sacramento model, illustrated in Figure 4, provides a conceptual accounting of the precipitation input to a drainage basin. The model represents the storage and movement of water beneath the surface, the transmission of ground water to stream channels, and the evaporation and transpiration of water from the soil and stream channel.

TABLE 2. Summary Statistics for Central and Southern Appalachian Rivers.

	DA (sq. mi.)	Q_A (cfsm)	Q_{90}/Q_A	Drainage
1) Flat-Lying Sedimentary Rocks				
East Branch Delaware River	163	1.92	0.09	Delaware River
Pine Creek	944	1.45	0.07	Susquehanna River
Youghiogheny River	295	2.23	0.07	Monongahela River
Allegheny River	550	1.71	0.08	Allegheny River
Greenbrier River	540	1.66	0.06	New River
S.F. Kentucky River	722	1.60	0.03	Kentucky River
Cumberland River	374	1.95	0.07	Tennessee River
2) Folded Sedimentary Rocks				
A) Shales and Sandstones				
Jordan Creek	76	1.50	0.09	Delaware River
Bald Eagle Creek	44	1.72	0.09	Susquehanna River
Aughwick Creek	205	1.25	0.06	Susquehanna River
Back Creek	134	1.37	0.06	James River
Calfpasture River	144	1.12	0.04	James River
B) Carbonate Rocks				
Little Lehigh River	81	1.21	0.37	Delaware River
Yellow Breeches Creek	216	1.34	0.38	Susquehanna River
Conodoquinet Creek	470	1.22	0.20	Susquehanna River
Little Juniata River	220	1.75	0.22	Susquehanna River
Kerrs Creek	35	0.95	0.21	James River
Reed Creek	247	1.13	0.27	New River
S.F. Holston River	301	1.50	0.23	Tennessee River
Nolichucky River	805	1.76	0.30	Tennessee River
3) Metasedimentary and Igneous Rocks				
Chester Creek	61	1.48	0.29	Delaware River
Deer Creek	94	1.36	0.34	Susquehanna River
South Branch Patapsco River	64	1.14	0.25	Patapsco River
North Mayo River	108	1.24	0.36	Roanoke River
S.F. New River	207	2.13	0.40	New River
French Broad River	68	3.59	0.38	Tennessee River
South Yadkin River	306	1.15	0.35	Pee Dee River
Chattooga River	207	3.28	0.37	Savannah River

Soil moisture storage is represented in the model by upper zone and lower zone storages. Each of these zones is further subdivided into tension water and free water storages. Traditional components of hydrograph separation can be interpreted as drainage from specific storages in the model. Output from the lower zone free water storages can be interpreted as baseflow. Output from upper zone free water into the stream channel can be interpreted as interflow. Precipitation in excess of the storage and drainage capacity of the upper zone, which is transmitted directly to the stream channel, can be interpreted as overland flow. A detailed description of the Sacramento model can be found in Burnash and Ferral (1972) and Peck (1976).

The main parameters of the Sacramento model are maximum storage capacities and drainage rates. It will be recalled that a conclusion of the previous section was that spatial variability of low streamflow in the Potomac River basin could be explained largely in terms of ground water storage capacity and permeability. The form of the Sacramento model is thus attractive in terms of its consistency with physical features of the Potomac River basin.

The Potomac River basin was segmented into 24 subregions for which the Sacramento model was calibrated. A hydrologic routing model, lag/K (Linsley, *et al.*, 1972), is used to route the outflow from one basin to downstream segments. Segmentation of the basin was based in part on the location of streamflow gages. Another goal of segmentation was to provide regions of homogeneous hydrologic properties. This was done by utilizing the results discussed in the section concerning hydrologic provinces. As an aside, we note that an important potential application of a hydrologic province map such as the one presented in Figure 1 is obtaining "good" initial estimates for Sacramento model parameters for uncalibrated areas. For example, the results presented in Figure 1 and Tables 1 and 2 (along with the geologic map of Pennsylvania) suggest that the calibrated

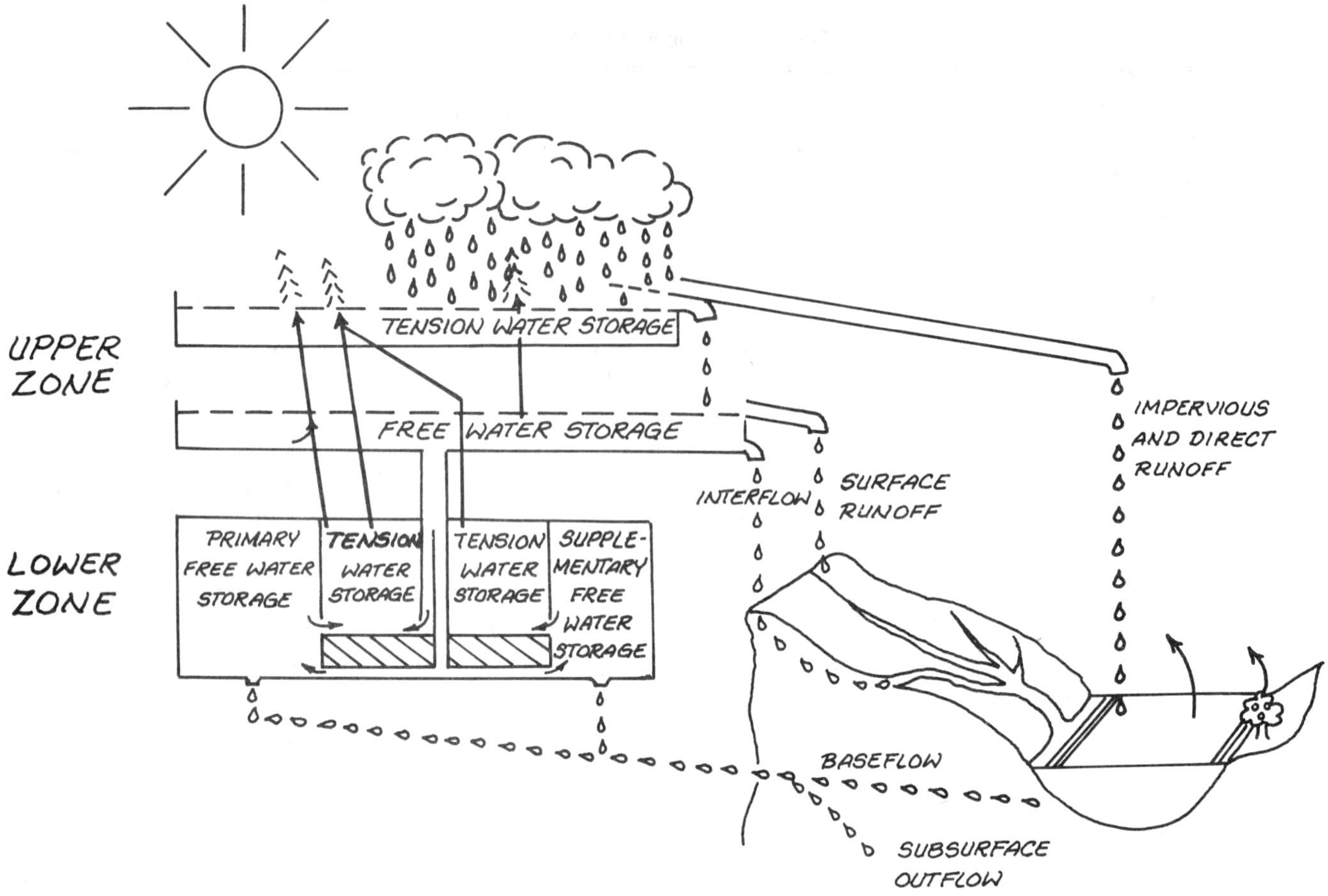

Figure 4. Illustration of the Sacramento Model.

parameters for Antietam Creek in the Potomac River basin would be good initial estimates for the parameters of Yellow Breeches Creek in the Susquehanna River basin and Little Lehigh Creek in the Delaware River basin.

The procedure which is used to convert the hydrologic models discussed above into a forecast system is the Extended Streamflow Prediction (ESP) component of NWSRFS. The ESP procedure, as applied to the Potomac River basin, incorporates current information pertaining to soil moisture conditions by using "current" storages in the Sacramento model as the starting point in a simulation of streamflow (thus storages in the Sacramento model must be regularly updated with observed precipitation). In simulating future streamflow the ESP procedure uses historical precipitation data.

Table 3 illustrates the main features of the procedure. This table is produced from an ESP forecast of minimum mean daily flow of the Potomac River for the period May 1, 1982 to October 1, 1982. Twenty-four Years (1951-1974) of historical precipitation data (restricted to the period May 1-October 1) are used to simulate 24 streamflow sequences. The minimum values for each year are listed in Table 3 under "Conditional Simulation." Each simulation begins with the same Sacramento soil moisture storages, namely the storages of May 1, 1982. Comparison of the simulated values with the observed values indicates that the risk of streamflow falling to the levels of the mid-1960s in 1982 is small.

APPLICATION OF THE FORECAST SYSTEM TO DROUGHT MANAGEMENT

The principal sources of water for the Washington Metropolitan Area (WMA) are the Potomac River, two local reservoir systems, and two upstream reservoirs (for which the travel time to the WMA is approximately five days). Palmer, *et al.* (1982), concluded that the WMA water needs could be met (beyond the year 2000) without constructing new reservoirs if available resources were jointly operated. The efficiency of joint operation of WMA water supplies is due in part to the regional variability of supplies. For example, the two local reservoir system, although separated by a short distance, have very different supply characteristics due to the fact that one is underlain by Triassic shales, which are flashy and have very low baseflow, while the other is underlain by high baseflow metasedimentary rocks. One of the conclusions reported by Palmer, *et al.* (1982), was that streamflow forecasting should play a central role in drought management.

TABLE 3. Sample ESP Output From Forecast of Minimum Daily Flow of Potomac River – May 1, 1982 to October 1, 1982.

ESP Forecast: Minimum Daily Flow for Potomac River – May 1, 1982 to October 1, 1982		
Water Year	**Conditional Simulation in CFS**	**Observed in CFS**
1951	1290	1310
1952	1823	2110
1953	1385	1200
1954	1569	1080
1955	2196	1460
1956	1806	1860
1957	1121	806
1958	1420	1810
1959	1208	787
1960	1871	1660
1961	1403	1310
1962	1173	1050
1963	982	880
1964	974	665
1965	1125	734
1966	890	547
1967	1928	1640
1968	1621	1010
1969	1854	885
1970	1405	1480
1971	2183	1810
1972	1989	2160
1973	1640	2270
1974	1560	1740
Mean	1517	1344
Standard Deviation	373	498

As indicated in the previous section, the forecast system that has been developed for drought management in the Potomac River basin is based on the ESP procedure. Operating rules for reservoirs, however, are not based directly on probabilistic forecasts of minimum daily flow (such as shown in Table 3) but rather on the probability of meeting demands and refilling reservoirs. The transition from an ESP forecast involving flow related variables such as minimum daily flow or total flow volume to an ESP forecast of meeting demands is accomplished by linking ESP to a model which simulates the water supply system of the basin. This is done by using the simulated streamflow sequences of ESP as input to the water supply model. Thus for the forecast period used in Table 3, 24 sequences of May 1-October 1 streamflow for five sites (Potomac River and inflow to four reservoirs) are used as input to a water supply model producing, for example, 24 values of total shortages. The 24 values of shortages are then used to produce a probability distribution of shortages for the period May 1, 1982 to October 1, 1982.

The CO-OP model, which simulates daily water supply operations in the Potomac River basin, is used in conjunction with ESP to form a drought management system. CO-OP has as its predecessor the PRISM model developed by Palmer (Palmer, *et al.*, 1981). PRISM has been extensively used as a water supply planning tool. The CO-OP model serves both as a planning tool and, through its link with ESP, as an operational model.

SUMMARY AND CONCLUSIONS

Low streamflow regimen in the Potomac River basin is shown to be strongly associated with geologic features of the basin. The sharp hydrologic contrasts between regions draining shale, metasedimentary rocks, and carbonate rocks are not peculiar to the Potomac River basin, but are generally valid for the Central and Southern Appalachians.

The Sacramento Soil Moisture Accounting Model is the central component of the streamflow forecasting system developed for the Potomac River basin. The Sacramento model is especially appealing in that its parameters have physical interpretations which are consistent with the results discussed in the section on "Hydrogeology of Low Streamflow in the Potomac River Basin." As an extension of this observation, we suggest that hydrologic province maps such as presented in Figure 1 may be useful in obtaining good initial estimates of Sacramento model parameters for uncalibrated areas from the calibrated Potomac River basin parameters.

Our final topic concerns application of the streamflow forecasting system to water supply management. The CO-OP model simulates, on a daily time step, the principal water supply and demand features of the Washington Metropolitan Area. The CO-OP model is linked with the ESP procedure to produce a forecasting tool for water supply management during drought periods. In particular the system produces probabilities of water shortage and refilling reservoirs.

LITERATURE CITED

Burnash, R. J. C. and R. L. Ferral, 1972. Generalized Hydrologic Modeling, A Key to Drought Analysis. Second Int. Symp. in Hydrol., Ft. Collins, Colorado.

Hack, J. T., 1965. Geomorphology of the Shenandoah Valley, Virginia, and Origin of the Residual Ore Deposits. U.S. Geological Survey Prof. Paper 484, 84 pp.

Hely, A. G. and F. H. Olmsted, 1963. Some Relations Between Streamflow Characteristics and the Environment in the Delaware River Region. U.S. Geol. Survey Prof. Paper 417-B.

Linsley, R. K., M. A. Kohler, and J. L. H. Paulhus. Applied Hydrology, McGraw Hill, New York, New York.

Nutter, L. J., 1973. Hydrogeology of the Carbonate Rocks, Frederick and Hagerstown Valleys. Maryland Geol. Survey Rept. Inv. 19.

Nutter, L. J. and E. G. Otton, 1969. Groundwater Occurrence in the Maryland Piedmont. Maryland Geol. Survey Rept. Inv. 10.

Palmer, R. N., J. A. Smith, J. L. Cohon, and C. S. ReVelle, 1982. Reservoir Management in the Potomac River Basin. Jour. of Water Resources, Planning and Management Div. – ASCE 108(WR1):47-66.

Peck, E. L., 1976. Catchment Modeling and Initial Parameter Estimation for the National Weather Service River Forecast System. NOAA Technical Memorandum NWS HYDRO-31.

Robert, J. K., 1922. The Triassic of Northern Virginia. Ph.D. Dissertation, The Johns Hopkins University, Baltimore, Maryland.

Trainer, F. W. and F. A. Watkins, 1975. Geohydrologic Reconnaissance of the Upper Potomac River Basin. U.S. Geol. Survey Water-Supply Paper 2035.

White, E. L., 1977. Sustained Flow in Small Appalachian Watersheds Underlain by Carbonate Rocks. Journal of Hydrology 32: 71-86.

INTERNATIONAL SYMPOSIUM ON HYDROMETEOROLOGY
JUNE AMERICAN WATER RESOURCES ASSOCIATION 1982

SPATIAL CHARACTERISTICS OF SEVERE DROUGHT IN THE SACRAMENTO RIVER BASIN

M. L. Shelton[1]

ABSTRACT: Precipitation in California during water year 1977 was the lowest in 138 years of record. Hydrologic effects of the drought were especially significant in the Sacramento River basin because this watershed contains about one-third of the reservoirs in California and about one-half of the total intrastate reservoir storage capacity. A severe drought in 1924 stimulated construction of many of the water storage and transmission facilities that exist in the basin today. The severity of the 1924 and 1977 droughts is compared by analysis of cumulative water deficiencies derived from parameters representing natural moisture flux processes and calculated using Thornthwaite's climatic water budget. Isopleth maps of cumulative moisture deficiencies for 1924 and 1977 facilitate inclusion of a spatial factor in quantifying drought severity. Area weighting of cumulative moisture deficiencies reveals that runoff during the longer duration 1977 drought was 14.6 km^3 less than during 1924 and moisture availability during typically dry months was 5 hm^3 less during the 1977 drought.
(KEY TERMS: drought; Sacramento River; California; climatic water budget; runoff.)

INTRODUCTION

Water year 1977 was the driest year in 138 years of record in California, and the drought pervaded all but the arid southeastern portion of the state. In the Sacramento River basin, precipitation was only 30 percent of normal and unimpaired runoff was 20 percent of normal. The record setting dryness in 1977 followed the fourth driest year on record, and the severity of the 1977 drought in the Sacramento River basin impinged upon water availability statewide because water from the basin is exported to urban and agricultural users in the San Joaquin Valley and Southern California.

The Sacramento River basin (Figure 1) covers 70,800 km^2 and produces approximately 26.2 km^3 of average annual runoff. A severe drought in 1924 was one of the stimuli for constructing many of the water storage and transmission facilities present in the watershed today, and these structures represent about one-third of the reservoirs in California and about one-half of the total intrastate reservoir storage capacity. Water management planning requires quantitative information concerning severe droughts, but hydrologic comparisons of the 1924 and 1977 droughts are inconclusive due to changes in population, land use, and water use practices. This study employs climatic water budget analysis to define spatial and temporal moisture deficit variations and to quantify the cumulative moisture deficiency during the severe droughts of 1924 and 1977.

A WATER BUDGET PERSPECTIVE OF DROUGHT

Quantifying moisture deficiencies in natural moisture flux processes emphasizes stress in biologic and hydrologic systems as a measure of drought. A basic assumption underlying this analysis is that plant life and hydrologic processes are adjusted to the long-term mean relationship between moisture supply and demand. Thus, drought can be expressed meaningfully by deviations in moisture relationships from those expected for the locale, and the severity of drought can be expressed by the cumulative moisture deficiency (Dracup, *et al.*, 1980). The climatic water budget (Thornthwaite, 1948) offers an appropriate methodology to quantify deviations in natural moisture supply and demand relationships and for analyzing spatial and temporal variations in the deviations.

In the water budget approach, a deficit occurs when precipitation and available soil moisture are unable to satisfy potential evapotranspiration (PE). Since deficit (D) is stated in relation to PE, D/PE expresses the moisture deficiency quantitatively, and the effect of water deficit relative to available energy is standardized from place to place. As a monthly index, D/PE represents the degree of stress imposed upon growing plants by prevailing energy and moisture conditions. The index incorporates the effects of both soil moisture storage and PE. In the Sacramento River basin, monthly D/PE indicates the level of water deficiency for plants during the warm season when rainfall is typically small. However, some level of moisture deficiency is expected in this environment during the summer and the assessment of drought requires quantifying the

[1]Associate Professor, Department of Geography, University of California, Davis, California 95616.

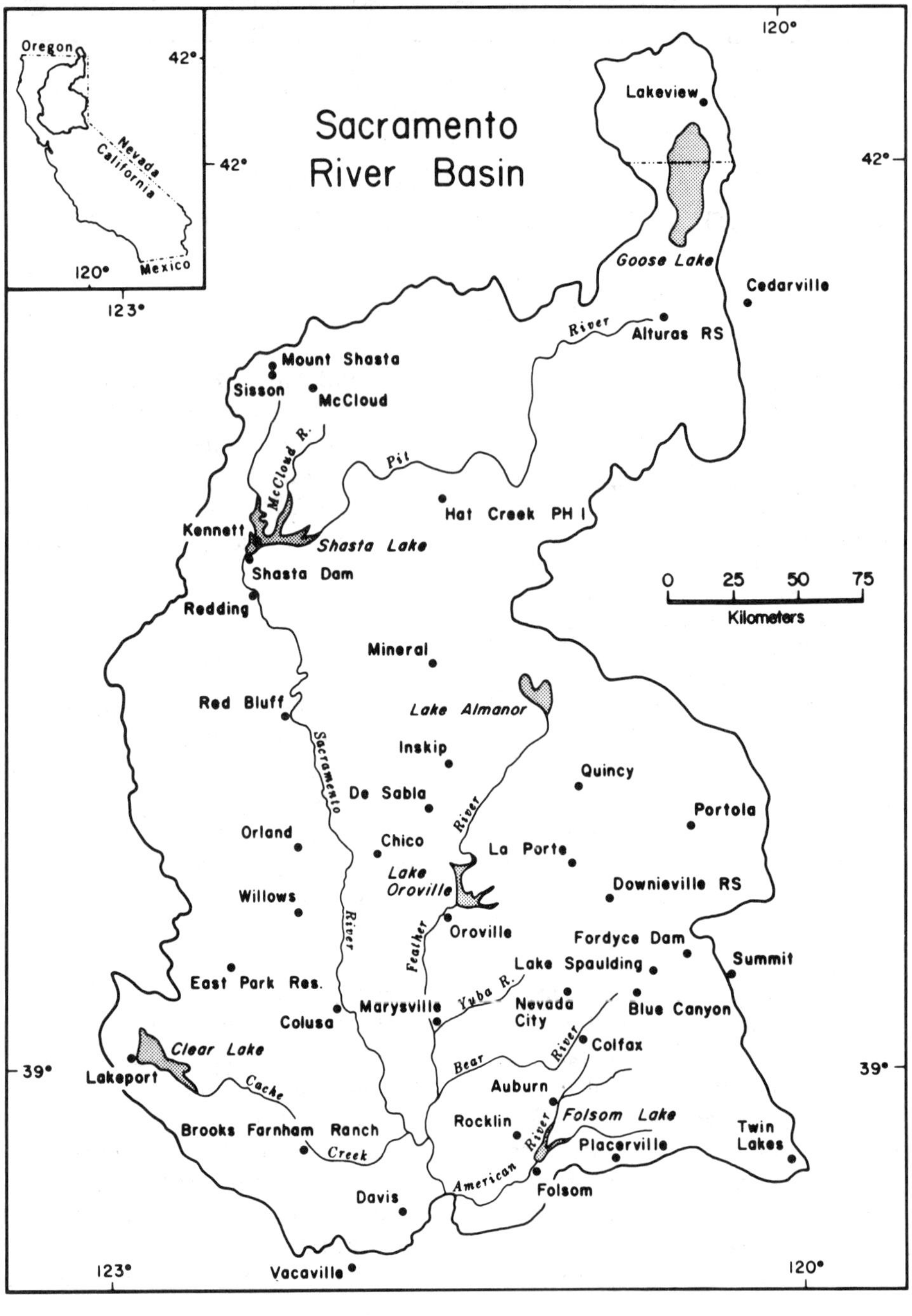

Figure 1. Study Area and Weather Stations Used for the Climatic Water Budget Analysis.

deviation between the actual and expected deficiency.

During wet months, moisture deficiencies are manifested by reduced runoff and limited groundwater recharge which produce stress in hydrologic systems. Surplus (S) from the climatic water budget procedure is a surrogate parameter for moisture allocated to groundwater and surface runoff. Surplus from the long-term mean monthly water budget indicates the proportion of precipitation needed to sustain hydrologic processes at typical levels. It is assumed that stress in the hydrologic system will occur when the moisture contribution falls below some threshold level which maintains lakes, rivers, and groundwater reservoirs at near normal levels.

Since a moisture deficiency can occur in either a wet month or a dry month, quantifying the cumulative moisture deficiency requires more than a summation of D/PE. Precipitation may be less than expected during a normally wet month but not so limited as to produce a deficit as defined by the climatic water budget. Nevertheless, a moisture deficiency may occur in hydrologic systems and some expression other than D/PE is required to characterize the deficiency. Surplus quantified by the climatic water budget methodology serves this purpose because it is a function of the coincidence of precipitation and PE which represent natural moisture supply and demand.

QUANTIFYING THE MOISTURE DEFICIENCY

The moisture deficiency for a given month important in defining drought is a function of the expected occurrence of deficit or surplus in that month based on the long-term mean relationship between energy and moisture. Therefore, quantifying the moisture deficiency requires computing an average monthly water budget for the station and identifying the expected occurrence of deficit or surplus. Since by definition a specific month cannot have both a deficit and surplus, the moisture deficiency for a given month is determined by employing either a dry month expression of moisture deficiency based on expected plant stress or a wet month expression based on expected stress in the runoff process.

The dry-month moisture deficit denotes the extent to which the calculated deficit deviates from the deficit that would be expected under typical energy and moisture loadings. The dry-month deficit (DD_m) attributed to drought is derived by solving:

$$DD_m = AMD_m - (PE_m \cdot D_{\bar{m}}/PE_{\bar{m}}) \tag{1}$$

where AMD_m is the computed water budget deficit for the month, PE_m is the computed energy demand for moisture, $D_{\bar{m}}$ is the water balance deficit for the month computed from long-term monthly records, and $PE_{\bar{m}}$ is the energy demand for the month computed from long-term records. $D_{\bar{m}}$ cannot be used directly as a measure of the expected moisture deficiency during a drought because $PE_{\bar{m}}$ is unlikely to be a good estimator of PE_m when sparse precipitation is accompanied by enhanced energy loadings.

During a normally wet month, drought causes depressed levels of surplus and stress in groundwater and surface-water systems. Quantifying the moisture deficiency requires estimating the extent to which surplus deviates from the moisture necessary for maintaining normal levels of streamflow and groundwater storage. Moisture deficiencies during a wet month (WD_m) are estimated by solving:

$$WD_m = C_b S_{\bar{m}} - S_m \tag{2}$$

where C_b is a basin specific coefficient for drought related runoff, $S_{\bar{m}}$ is the water budget surplus computed from long-term monthly records, and S_m is the computed water budget surplus for the month. The coefficient C_b for the Sacramento River basin is 0.8438 which is the ratio of the third decile of runoff to median runoff. It is reasoned that the median runoff is sustained by surplus calculated using the long-term monthly water budget and the third decile runoff results from diminished surplus occurring during a drought.

The cumulative moisture deficiency (CMD) calculated for the duration of the drought is represented by:

$$CMD = \sum_1^n DD_m + \sum_1^n WD_m \tag{3}$$

where DD_m and WD_m are defined in equations 1 and 2, and n is the number of dry months or wet months with a moisture deficit as defined by equations 1 and 2.

CHARACTERIZING SEVERE DROUGHTS

The 1924 and 1977 droughts in the Sacramento River basin are analyzed using climatic water budgets for 33 stations computed for the years 1922-25 and 1975-78 (Figure 1). A computer program (Willmott, 1977) employing evapotranspiration adjustments for California (Shelton, 1978) facilitates the water budget calculations. The quantities corresponding to the appropriate water budget parameters are substituted in equations 1 and 2 for calculating the moisture deficiency. The number of stations used in the analysis is limited by the availability of climatic data for the 1924 drought, but 26 of the 33 stations are used for both events. Water budgets are computed using a representative soil moisture capacity of 114 mm for all stations.

Soil moisture storage is the key parameter used for developing rules for identifying the beginning and the end of drought. These rules are subjective and they represent a first approximation for addressing this problem. The beginning of drought is attributed to the first month in which soil moisture storage is reduced below one-half of capacity without recharge above this level in the following month. The rationale for this selection is that stress in many biologic systems would begin when soil moisture is reduced to this amount. Drought is considered to end when a month has surplus of at least 90 percent of monthly surplus based on a long-term mean conditions and moisture storage in the following month is at capacity. The rationale for the selection of these conditions is that such moisture availability would be unlikely to contribute to stress in either biologic or hydrologic systems.

Using these criteria, the duration of the 1924 drought in the Sacramento River basin is from May 1923 to November 1924. However, drought persisted at a limited number of stations until February 1925. The moisture deficiency during the core 19 months of drought is less than 25 cm in the lower valley from Marysville to south of Davis (Figure 2). The deficiency is over 100 cm near Blue Canyon and to the northwest in the Sierra Nevada. The general area around Shasta Lake experienced a deficiency of 100 cm also. About 50 percent of the basin has a moisture deficiency of 50 cm or less while only 10 percent has a moisture deficiency of over 100 cm.

The 1977 drought in the Sacramento River basin begins as early as May 1975 at many stations and lasts until

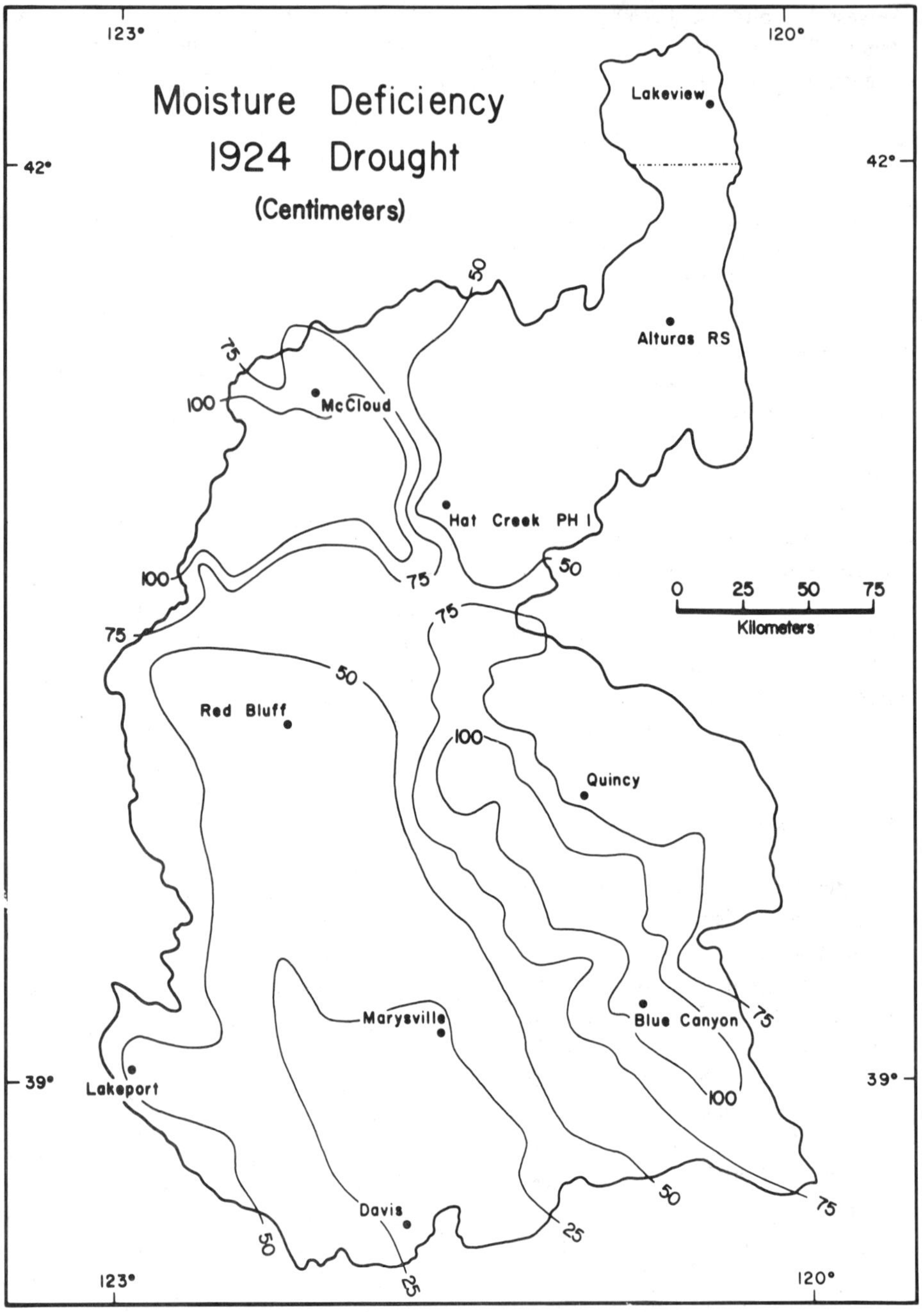

Figure 2. Cumulative Moisture Deficiency for the 1924 Drought.

November 1977. During these 31 months, a moisture deficiency of less than 25 cm is limited to a narrow fringe of the basin northeast of Alturas (Figure 3). In the Sierra, the moisture deficiency exceeds 150 cm around Blue Canyon and west of Quincy. Approximately 24 percent of the basin has a moisture deficiency of 50 cm or less and about 26 percent has a deficiency of over 100 cm.

The duration of the 1977 drought exceeds that of the 1924 drought by about 12 months. However, the severity as indicated by the CMD is more difficult to assess quantitatively because the spatial characteristics of the deficiencies must be considered. The spatial pattern of the CMD derived from solution of equation 3 for each of the study stations and illustrated in Figures 2 and 3 is used to calculate an area weighted cumulative moisture deficiency (AWCMD) for the basin. In this way, spatial variations in moisture conditions are placed in a prominent role in calculating drought severity.

Planimetering the area between the isopleths in Figures 2 and 3 reveals that the AWCMD for the entire basin during the 1924 drought is equivalent to 58 cm of moisture. The comparable figure for the 1977 drought is 79 cm or an

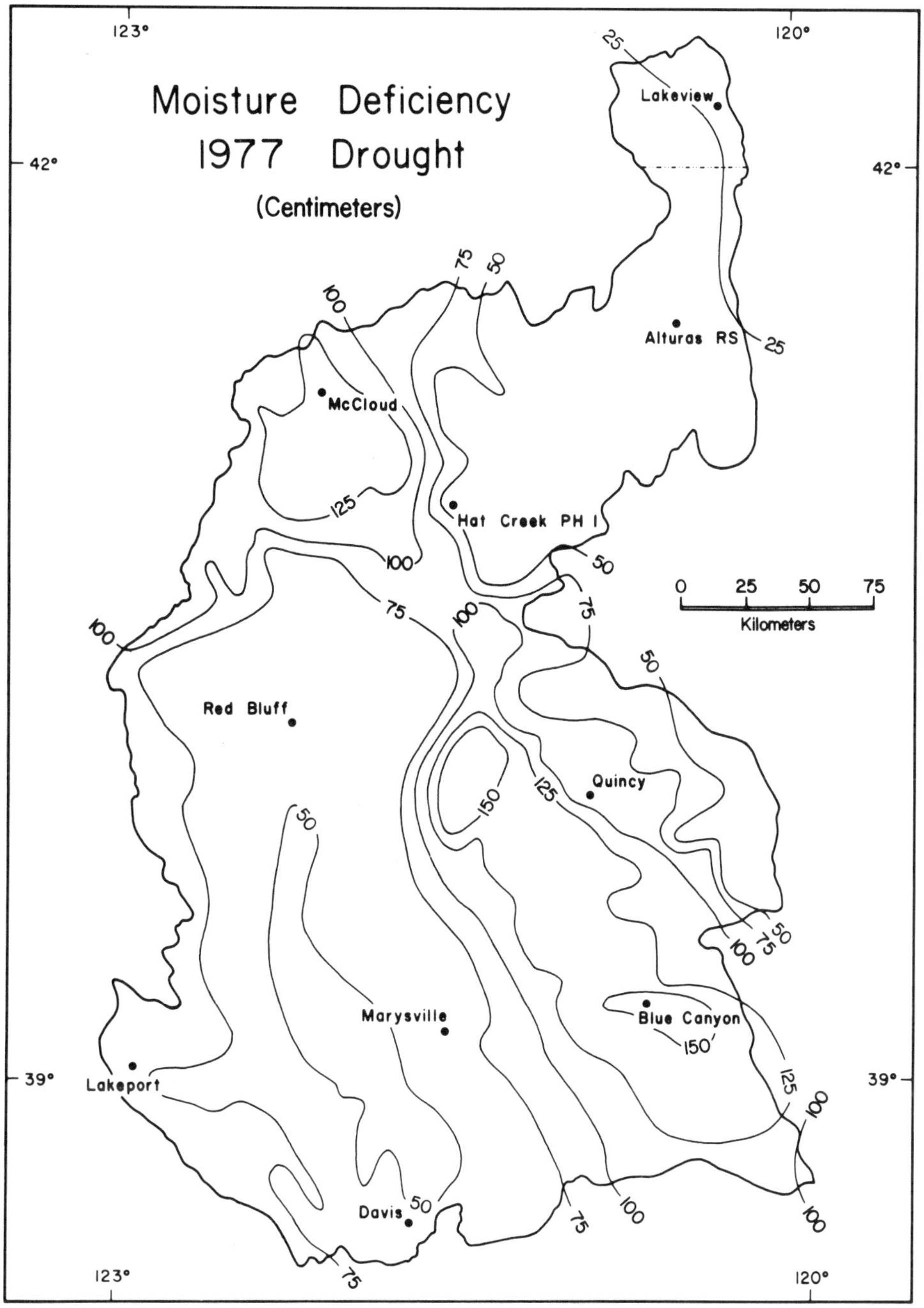

Figure 3. Cumulative Moisture Deficiency for the 1977 Drought.

AWCMD during the 1977 drought that is 21 cm greater than the deficiency during the 1924 drought. The 36 percent increase in the AWCMD during the 1977 drought can be interpreted as a measure of the greater severity of this drought. For the Sacramento River basin as a whole, the additional 21 cm of moisture deficiency during the 1977 drought is equivalent to about 14.6 km^3 of runoff or an amount equal to approximately 56 percent of average runoff for one year.

It should be noted that the average monthly AWCMD is slightly larger for the 1924 drought than for the 1977 drought. However, the average monthly deficiency for the 1977 drought is expected to be smaller. Most California stations receive meager amounts of precipitation during the months of May to October and moisture deficiencies are calculated for these months using equation 1. Elevated energy loadings during dry months are the major determinant of deviations between calculated moisture deficits and expected deficits, but the magnitude of the increase in available energy during a dry-season month is restricted to a much narrower range than the magnitude of precipitation variations during a wet-season month. This means that the contribution of a dry-season month to the total CMD in equation 3 is likely to be smaller than the contribution by a wet-season month experiencing a reduced precipitation input and a moisture deficiency defined by equation 2. Since the 1977 drought spans three summer dry seasons while the 1924 drought includes two summers, the 1977 drought is expected to have a slightly smaller average monthly moisture deficiency.

The moisture deficiency pattern for both the 1924 and 1977 droughts (Figures 2 and 3) reflects important spatial characteristics regulated by precipitation in the basin. Deficiencies at lower elevations in the basin must be smaller than those at high elevations because the valley floor typically receives less precipitation which places an upper limit on the magnitude of the moisture deficiency. The deficiencies northeast of Hat Creek Power House No. 1 are smaller for the same reason. The dominant influence of areas receiving greater precipitation is clearly revealed by the seasonal contribution of moisture deficiency during the 1977 drought. Wet-season deficiency, or the summation of wet months in equation 3, strongly dictates the general pattern of the total deficit except in the low precipitation zones of the basin. The area weighted basin value for the wet-season deficiency is 56 cm or about 70 percent of the total deficiency.

The longer duration of the 1977 drought produces significant temporal differences that distinguish it from the 1924 drought. The 1924 drought is confined to one winter, or wet season, and this limits the total moisture deficiency accumulation. During the early stages of the drought in 1923, only small deficiencies accumulated throughout the normally dry summer months. As a result, the moisture deficiency for water year 1924 is a reasonably close approximation of the magnitude and pattern of the moisture deficiency for the entire drought. Since the 1977 drought includes two winters, the moisture deficiency computed for water year 1977 is a poor approximation of the cumulative moisture deficit or the severity of the 1977 drought. In general, the 1977 water year moisture deficiency represents about one-half or less of the CMD for all months during the 1977 drought.

COMPARING DROUGHT SEVERITY

Natural flow calculations for streams in the Sacramento River basin (CDWR, 1980) suggest that the 1924 drought was less severe than the 1977 drought because runoff in water year 1924 was about 1.1 km^3 greater. However, the climatic water budget analysis indicates that water year 1977 is an inadequate representation of the 1977 drought.

Expressing drought in terms of the volume of moisture deficiency derived from planimetering spatial variations in deficits provides a quantitative basis for comparing the severity of the two droughts that is independent of other temporal limits or data summations. The AWCMD for the 1977 drought is equivalent to 55.4 km^3 while the total deficiency for the 1924 drought is equivalent to 40.8 km^3. This volumetric expression of drought severity emphasizes the hydrologic implications of the two droughts. In addition, it provides a quantitative measure of drought for water management purposes that incorporates the temporal characteristics of the two droughts more realistically than either water year runoff or the average monthly deficiency.

Equation 2 defines wet-month deficiencies during the 1977 drought that, in total, are equivalent to approximately 38.6 km^3 of runoff or nearly 1.6 times greater than the wet-month deficiencies of 24.5 km^3 of equivalent runoff during the 1924 drought. Dry-month deficiencies defined by equation 1 are equivalent to 16.3 km^3 during 1924 and 16.8 km^3 for the 1977 drought. Although the dry-month deficiencies are less disparate, they indicate the occurrence of an elevated need for moisture equivalent to 5 hm^3 during the 1977 drought. This amount of water was needed during the dry months of 1975-77 to maintain levels of moisture use in the biologic system comparable to the natural moisture demand during the shorter 1924 drought.

Volumetric expression of the total moisture deficiency indicates that water year calculations of natural flow for the Sacramento River basin understate the greater severity of the 1977 drought by one order of magnitude. The climatic water budget analysis demonstrates the importance of incorporating spatial variations in moisture deficits when quantifying deficiencies as a measure of drought severity.

LITERATURE CITED

California Department of Water Resources, 1980. California Central Valley Natural Flow Data. The Resources Agency, Department of Water Resources, Division of Planning, Sacramento, California, 77 pp.

Dracup, J. A., K. S. Lee, and E. G. Paulson, Jr., 1980. On the Definition of Droughts. Water Resources Research 16(2):297-302.

Shelton, M. L., 1978. Calibrations for Computing Thornthwaite's Potential Evapotranspiration in California. Professional Geographer, 30(4):389-396.

Thornthwaite, C. W., 1948. An Approach Toward a Rational Classification of Climate. The Geographical Review, 38(1):55-94.

Willmott, C. J., 1977. WATBUG: A FORTRAN IV Algorithm for Calculating the Climatic Water Budget. Publications in Climatology, 30(2):1-55.

DETERMINING DROUGHT EFFECTS ON SURFACE WATER BODIES NEAR A NEW WELLFIELD

Charles Palmer, Jr.[1]

ABSTRACT: The start-up of the Cross Bar Ranch Wellfield in central Pasco County, Florida, was coincident with a major state-wide drought. During this drought many shallow surface water bodies experienced low water levels or dry conditions. If these low water levels were significantly due to wellfield operations, mitigating actions were to be taken by the West Coast Regional Water Supply Authority under the Special Conditions of the permit to operate the wellfield. An intensive hydrologic analysis of these surface water bodies was undertaken. Central to this analysi were historical data from aerial photographs. This, combined with classical hydrologic analysis, demonstrated that the observed low water levels were drought related and not caused by the Water Year 1981 Cross Bar Ranch pumpage of 12.9 MGD.
(KEY TERMS: drought; aerial photography; surface water bodies.)

INTRODUCTION

The West Coast Regional Water Supply Authority (the Authority) is an intergovernment agency formed to develop in a technically, economically, and environmentally sound manner public raw water supplies for the tri-county area of the Tampa Bay region of Florida (Figure 1). Thus, the Authority functions as a wholesaler of water to the governments within the area it serves. Its Governing Board is made up of one representative each of the five major governments of the tri-county area: Hillsborough, Pasco, and Pinellas Counties and the Cities of Tampa and St. Petersburg (Figure 1).

Water use in Florida is regulated under Chapter 373 of the Florida Statutes which stipulates the requirements under which a permit must be obtained to use water in Florida. Basically, the permittee must demonstrate that the proposed use is reasonable, beneficial, and in the public interest; and will not adversely affect the resource.

The permitting agency is the Southwest Florida Water Management District (SWFWMD), one of five such districts under the State Department of Environmental Regulation.

On February 6, 1980, SWFWMD granted the Authority a Consumptive Use Permit to operate the Cross Bar Ranch Wellfield (Figure 1).

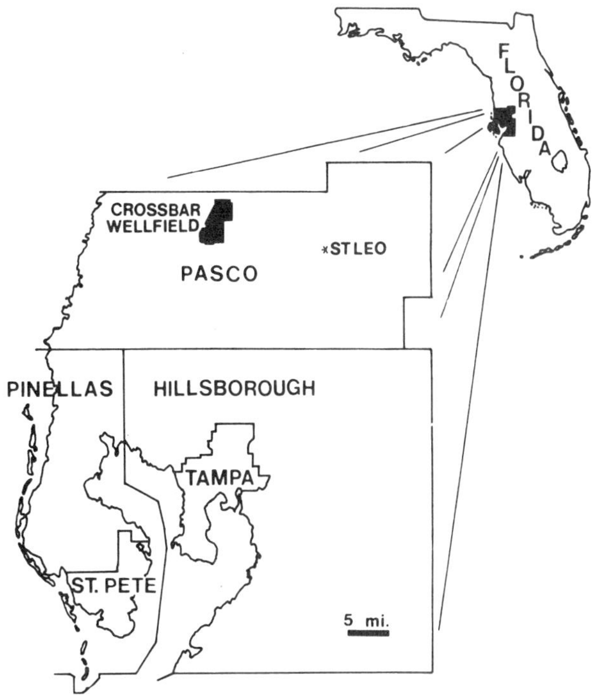

Figure 1. Authority Location Map – Cross Bar Wellfield.

The start-up of the Cross Bar Ranch Wellfield (Cross Bar) in central Pasco County, Florida, was coincident with a major statewide drought. This wellfield is a major water source to meet demands in Pinellas County, Florida, and is permitted to produce 30 MGD average – 45 MGD peak day. However, during Water Year 1981 (October 1, 1980, to September 30, 1981) its average day production was 12.9 MGD.

During Water Year 1981, many shallow surface water bodies experienced low water levels or dry conditions. If these low water levels were significantly due to wellfield operations, mitigating actions were to be taken by the

[1]Water Resource Engineer, West Coast Regional Water Supply Authority, Suite 121, 2280 U.S. 19 North, Clearwater, Florida 33515.

Authority under the Special Conditions of the permit to operate the wellfield.

DROUGHT

All rainfall data, unless otherwise noted, is from the NOAA weather station at St. Leo, Florida (Figure 1).

Both Water Year 1980 and Water Year 1981 (October 1 to September 30) were drought years throughout the entire state of Florida. Near crisis conditions were reached in many parts of the state as a result of two consecutive years of less than normal rainfall. Florida has a wet season (normal rainfall, 32.33 inches) that extends from June through September and a dry season (normal rainfall, 24.26 inches) that extends from October through May. Thus, the total annual normal rainfall is 56.59 inches. In Water Year 1980 the total rainfall was 41.96 inches making it the eighth driest water year in 86 years of record at St. Leo. Water Year 1981 had a total rainfall of 48.29 inches making it the 22nd driest water year on record. Together the total rainfall for water years 1980-1981 was 90.25 inches, making them the sixth driest two-year period on record and the driest since Water Years 1961-1962.

If the wet season is not sufficient to recharge the hydrologic system and the following spring is dry, water levels fall to less than normal levels. This is especially noted in shallow (less than 3 feet deep) surface water bodies in and around Cross Bar.

This was the case in the spring of 1981. The cumulative deficit entering Water Year 1981 was 14.63 inches. By the end of May 1981, the deficit was 23.55 inches.

HISTORIC SURFACE WATER BODY BEHAVIOR FROM AERIAL PHOTOGRAPHS

For surface water bodies, long-term records of historic water levels near Cross Bar do not exist. Even if such records did exist, the huge number of surface water bodies would result in incomplete data. Each pond must be defined in terms of its own site specific character such as leakance, depth, drainage, etc. Data from one surface water body does not, in general, apply to another surface water body. Thus, some means had to be developed to evaluate wellfield effects upon surface water bodies. A simple declaration that a major drought was in progress was not sufficient. A common reply to such a declaration was, "I have lived here x years. We have had dry spells before and the lake has never been this low. It did not start going down until you turned that wellfield on." Fortunately, historic aerial photography combined with rainfall data from the NOAA weather station at St. Leo proved to be sufficient to show that any wellfield effect was overshadowed by the drought.

West-central Florida has a good history of air photo missions. Although these missions were flown for purposes other than to document high and low surface water levels, enough were flown on or near times of interest to provide area-wide information of the behavior of the surface water systems.

From these photographs it could be shown that low or nearly dry surface water bodies correlated nicely with periods of low recorded rainfall. In addition, otherevents often blamed on wellfields, such as sinkholes, can be assessed and correlated with climatic conditions. To illustrate such use of aerial photographs, the Pasco Lake-Jumping Gully watershed will be used as an example (Figure 2).

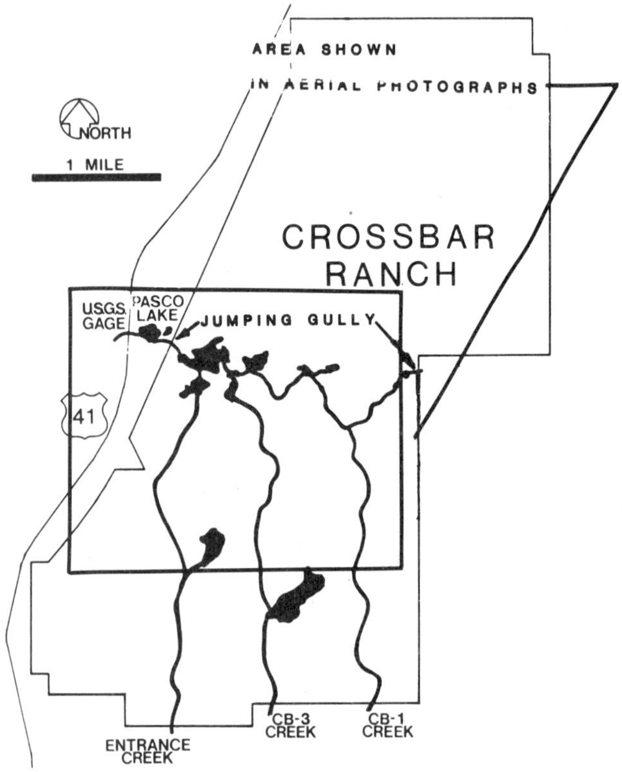

Figure 2. Cross Bar Wellfield Map With Jumping Gully and Its Tributaries.

PASCO LAKE-JUMPING GULLY WATERSHED

Jumping Gully is a small creek that begins on the eastern boundary of Cross Bar in a large marsh. It is fed by three creeks that enter from the southern edge of Cross Bar and flow north through a series of marshes. These creeks are named "Entrance Creek," "CB-3 Creek," and "CB-1 Creek" for identification. Jumping Gully flows west through a series of marshes into Pasco Lake just outside the western boundary of Cross Bar (Figure 2). It is this flow that fills and maintains Pasco Lake.

Pasco Lake began as a small marsh pond. Between 1957 and 1965 the pond was deepened and enlarged by dredging; the removed fill beging used to raise the adjacent lands to provide home building sites. A weir was installed at the outlet of the pond to retain water within the larger, deeper area. The present water surface ware when full is approximately 10 acres.

The USGS has collected stage-discharge data on Jumping Gully as it leaves Pasco Lake since May 1964. This gage is on the U.S. 41 Highway bridge, 100 feet downstream of the Pasco Lake weir. In June 1977, the USGS also began collecting water level data on Pasco Lake.

HISTORIC AERIAL PHOTOGRAPHY

March 23, 1957 – USDA's Agricultural Stabilization and Conservation Service, Frame CTT-IT-31 (Figure 3)

This photography was flown just before the end of a three-year drought that was called the worst on record (Pride and Crooks, 1962). For comparison, the total rainfall for Water Years 1980 and 1981 was 90.25 inches vs. the Water Years' 1955 and 1956 total of 85.19 inches. Thus, this is an excellent benchmark year of low water levels.

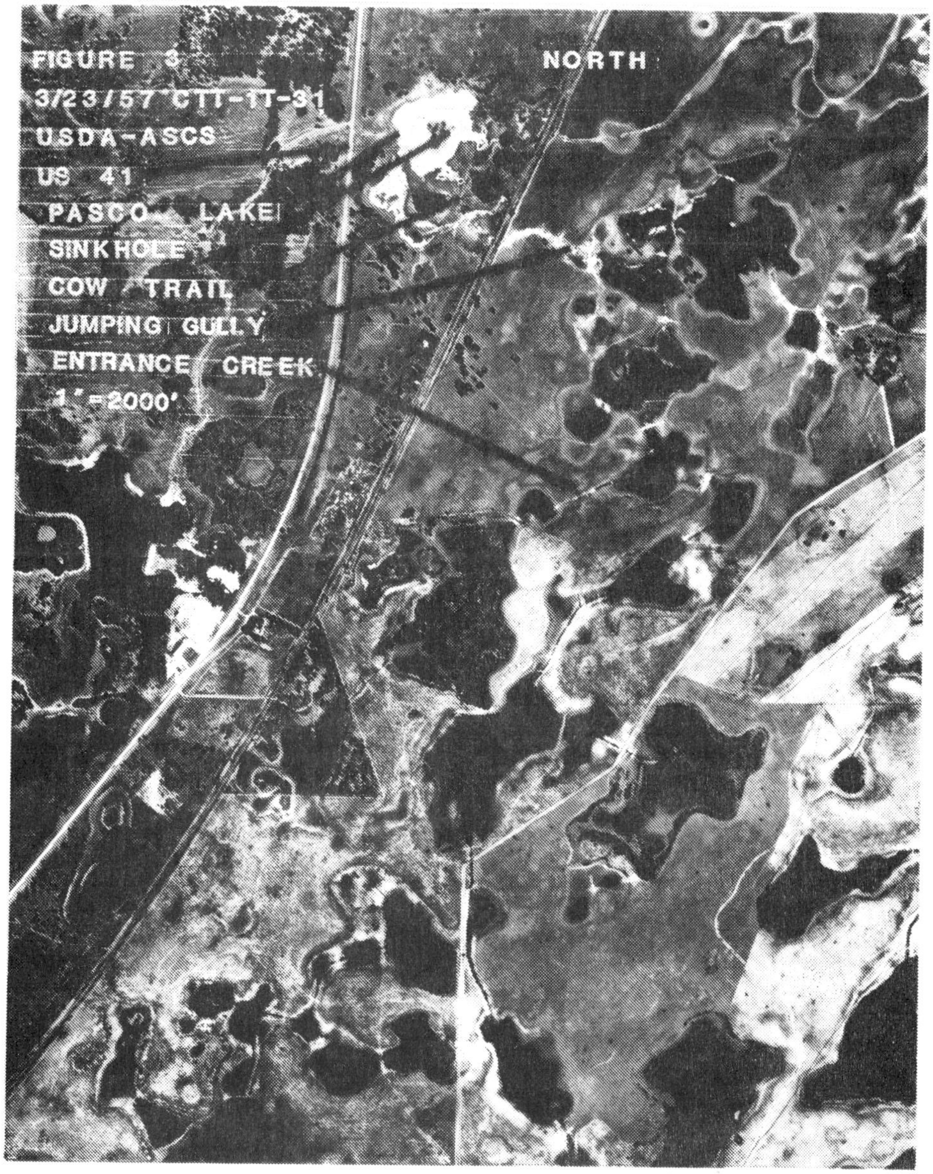

Figure 3. USDA-ASCS 3/23/57 Aerial Photograph of Pasco Lake Area.

Figure 3 shows the northern end of Entrance Creek, Jumping Gully, Pasco Lake, and numerous small ponds and marshes. Other frames of this coverage are very similar. From this photography the following is apparent: ponds, marshes, and creeks are all dry or nearly so throughout the Cross Bar area.

Pasco Lake is shown before the dredging and enlargement. Several things about Pasco Lake should be noted: it is dry and had been for some time since cattle trails in the southeast quadrant of the lake are well established; its size and shape, and the small dark spot next to the larger dark spot in the northeast quadrant of the lake which is a small sinkhole. The small sinkhole was of great interest. It was alleged by residents on the lake that Cross Bar had caused this sinkhole to form in November 1980 and drain the lake in an extremely short period of time.

When this photography was compared to the end of May 1981, the conditions on and near Cross Bar were found to be analogous.

December 10, 1972 – Mark Hurd Aerial Surveys, Frame IRC-CUJ 1236 (Figure 4A)/February 17, 1973 – Florida Department of Transportation, Frame PD 1223-10-13 (Figure 4B)

This photography aided in answering the question of residents around Pasco Lake as to when Pasco Lake could be expected to again have water in it. Pasco Lake is a leaky impoundment that is solely dependent on streamflow to maintain it. Therefore, the obvious answer to the question posed by the residents was that it would not fill until the Jumping Gully watershed was replenished by sufficient rainfall.

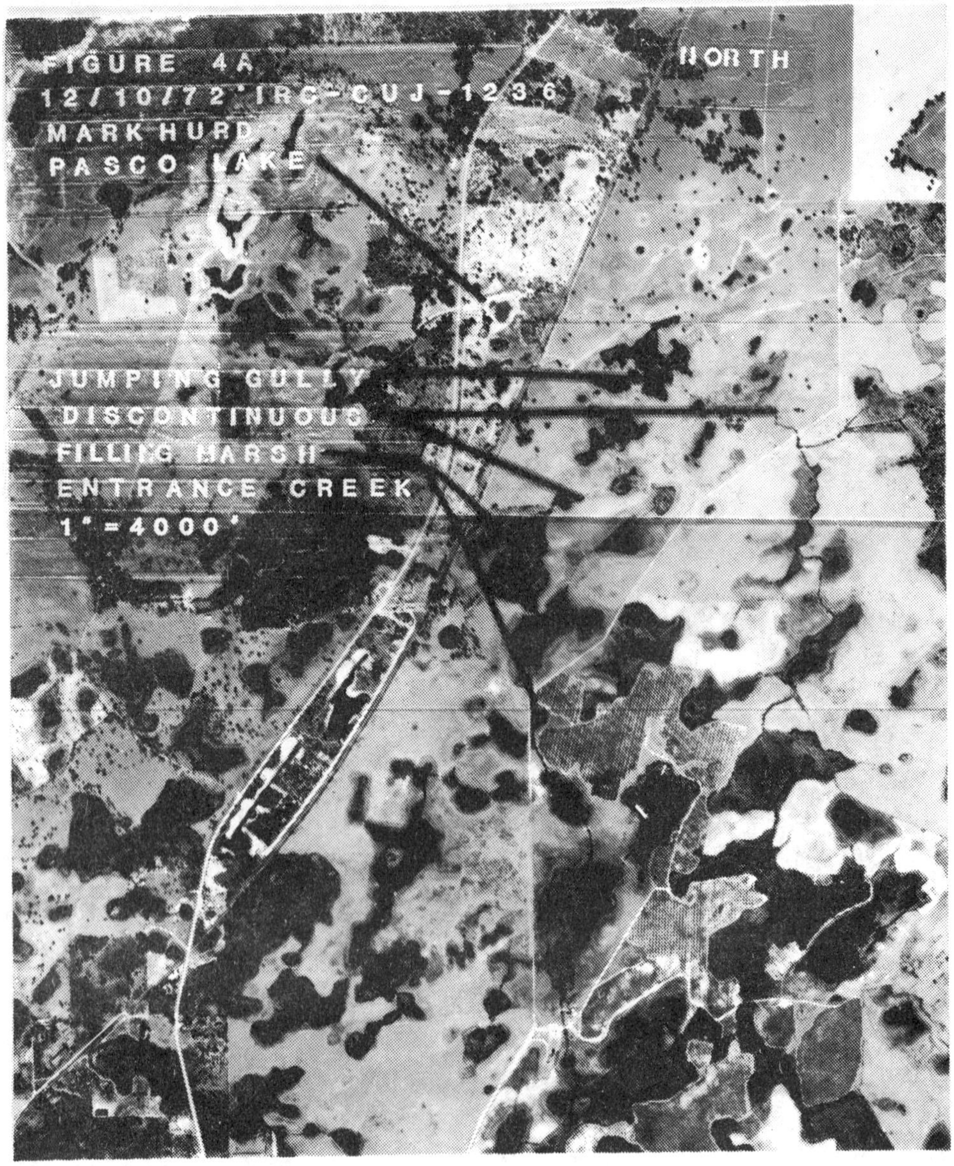

Figure 4A. Mark Hurd 12/10/72 Aerial Photograph of Pasco Lake Area.

Figure 4B. FDOT 2/17/73 Aerial Photograph of Pasco Lake Area.

Given that Pasco Lake needed streamflow, how much rain would be needed to fill the watershed and cause flow in Jumping Gully? These two photographs gave data to address that question.

Figure 4A shows Pasco Lake nearly dry with no flow in Jumping Gully. However, the watershed is filling as indicated by the nearly full marshes in the southern end of Entrance Creek. As we move north the marshes become less and less full and Entrance Creek becomes discontinuous before it reaches Jumping Gully. Each marsh in the Entrance Creek system must fill and overflow for water to move down the watershed to Jumping Gully and Pasco Lake. This condition of the watershed was the result of 32.64 inches of rainfall in the preceeding six months. Conditions shown in this photograph are similar to October 1981 after the summer rains.

Figure 4B shows Pasco Lake nearly full, the entire Jumping Gully system flowing, the marshes nearly full throughout the watershed, and all tributary creeks flowing. Between November 1, 1972, and February 1, 1973, 15.25 inches of rain fell. This was enough rain to trigger the system which was on the verge of flowing in the headwaters to Entrance and CB-3 Creeks to a flowing Jumping Gully and a full lake. This photograph shows conditions similar to February 1982.

Based on studies of these photographs it was estimated that due to the state of the watershed at the end of September 1981, approximately 15 inches of rain over the fall and winter would be needed to refill Pasco Lake. This turned out to be a remarkably good estimate. October 1981 through February 1982 rainfall measured at Cross Bar was

15.69 inches (Pasco Lake gage) with Pasco Lake beginning to refill on the 25th of January 1982.

CLASSICAL HYDROLOGIC ANALYSES

As a result of a large outpouring of citizen complaints and concerns, the Pasco Lake-Jumping Gully watershed was intensively studied by the Authority and the staff of SWFWMD. The Authority's consulting hydrogeologists, Leggette, Brashears and Graham, Inc., made a major contribution to the study. In order to address the citizen's concerns, these staffs reviewed and generated a massive amount of data and numerous analyses of the history, nature, and behavior of Pasco Lake.

CONCLUSIONS

The purpose of these studies was to determine the hydrologic cause of low surface water levels observed in and around Cross Bar in 1980 and 1981. The results of these studies was best summed up by Mr. Harry Oleson of Leggette, Brashears and Graham, Inc.: "Based on the large amount of data, and a detailed review of analyses of these data, we have concluded that there has been no significant nor adverse impact on Pasco Lake nor the entire surface-water system resulting from the wellfield operation" (Oleson, October 14, 1981, Summary of the Hydrologic Conditions on the Cross Bar Ranch Wellfield, with Special Emphasis on Pasco Lake; Report to the Governing Board of the West Coast Regional Water Supply Authority).

The historical aerial photographs and the data showed clearly that the hydrologic cause of low surface water levels was a direct result of the drought of 1980-1981.

ACKNOWLEDGMENTS

The author extends thanks to the following sources of aerial photography: Florida Department of Transportation, Tallahassee, Florida; Mark Hurd Aerial Surveys, Minneapolis, Minnesota; and the U.S. Department of Agriculture-Agricultural Stabilization and Conservation Service, Salt Lake City, Utah.

Understanding the drought of 1980-1981 required a massive effort by the staffs of the West Coast Regional Water Supply Authority, Clearwater, Florida; Southwest Florida Water Management District, Brooksville, Florida; and Leggette, Brashears and Graham, Inc., Tampa, Florida; with special thanks to Mr. Bob Evans, SWFWMD, and Mr. Harry Oleson, LBG.

LITERATURE CITED

Martin, R. J. (Editor), 1912 (Revised 1933). Climate Survey of the United States, Section 105 – Southern Florida. Superintendent of Documents, USGPO, Washington, D.C.

NOAA, 1953. Climate Summary of the United States, Supplement for 1931 Through 1952, Number 11-6 (Florida). National Oceanic and Atmospheric Administration, Asheville, North Carolina.

NOAA, 1952-1982 Annual. Climatological Data, Florida. National Oceanic and Atmospheric Administration, Asheville, North Carolina.

Pride, R. W. and J. W. Crooks, 1962. The Drought of 1954-56, Its Effect on Florida's Surface Water Resources. Report of Investigations No. 26, United States Geological Survey – Florida Geological Survey, Tallahassee, Florida.

USGS, 1977-1980 Annual. Water Resources Data for Florida, Surface Water – Southwest Florida. United States Geological Survey, Tallahassee, Florida.

INTERNATIONAL SYMPOSIUM ON HYDROMETEOROLOGY
JUNE AMERICAN WATER RESOURCES ASSOCIATION 1982

SCALE PROBLEMS IN THE SYNTHESIS OF MULTI-SITE PRECIPITATION

K. Malcolm Leytham[1]

ABSTRACT: Consideration of the qualitative relationships between precipitation and atmospheric circulation over the west coast of North America suggests that inter-station precipitation relationships are nonlinear with higher cross correlations during drought than during normal or wet periods. This indicates that current multi-site stochastic models, which all assume linear inter-station relationships, may underestimate the areal extent of severe drought. The presence of nonlinear inter-station relationships is confirmed by analysis of monthly precipitation data. Evaluation of the performance of a simple multi-site stochastic model shows that current methods may seriously distort spatial drought characteristics in the synthesis of data at widely separated sites (separation > 1000 km).
(KEY TERMS: stochastic hydrology; drought; multi-site precipitation.)

INTRODUCTION

The development of stochastic techniques for the design and operation of water resource systems has been the subject of considerable research in the last twenty years. A detailed review of past work is given by Lawrence and Kottegoda (1977). Until recently most work has been directed toward the synthesis of hydrometeorologic data at single sites. Great effort has been expended in studying the serial correlation structure of streamflow and precipitation data, but the spatial structure of such data has received little attention, especially for monthly or seasonal data covering large areas. Techniques for the synthesis of multi-site data have been developed largely as extensions of the methods used in single-site synthesis. Despite the importance of droughts in system design and operation, the ability of conventional models to reproduce the spatial structure of widespread drought has also received little attention.

This paper addresses problems of synthesizing monthly precipitation at widely separated sites on the west coast of North America, with particular emphasis on maintaining the frequency of widespread below normal precipitation depths observed in the historic record. The paper is comprised of two parts. In the first part the spatial structure of monthly rainfall is analyzed and the gross features of the spatial structure are related in a qualitative sense to features of the regional atmospheric circulation. In the second part of the paper, a conventional multi-site model is evaluated with respect to its ability to maintain the observed spatial structure of monthly precipitation data. A more complete treatment is given by Leytham (1982).

DATA ANALYSIS

Precipitation on the west coast of North America is almost exclusively associated with frontal systems generated over the Pacific Ocean. The frontal systems are associated with the jet stream and are steered to the west coast by the jet stream. The track of the jet stream is thus significant in identifying those areas receiving rainfall.

Drought on the west coast arises when the jet stream follows a persistent anomalous track. For example, during the winter of 1976-1977 the jet stream followed a persistent northerly track which brought wetter and warmer conditions than normal to the Gulf of Alaska while severe drought prevailed from southern British Columbia to southern California (Edmon, 1980). In this case, drought extended over 2000 km along the west coast. It is clear that drought affects much larger areas than do frontal systems and one might therefore expect higher cross correlations during drought than during wet or normal periods, particularly for sites separated by large distances (> 1000 km). This suggests that inter-station precipitation relationships are nonlinear for widely separated sites. The cross correlation structure of monthly rainfall was investigated using data from 12 sites along the west coast, shown in Figure 1. The variation of the January cross correlation with distance from selected stations is shown in Figure 2. Plots for other winter months are similar.

The cross correlation for the monthly data shows a consistent pattern of variation with distance. The most notable feature of Figure 2 is the consistent negative correlation at large distances. Negative correlations may again be explained by considering the large scale circulation patterns. For example, very dry conditions at Eureka are generally

[1]Consultant, 9538 44th Avenue NE, Seattle, Washington 98115. Formerly Research Assistant, Department of Civil Engineering, University of Washington, Seattle, Washington 98195.

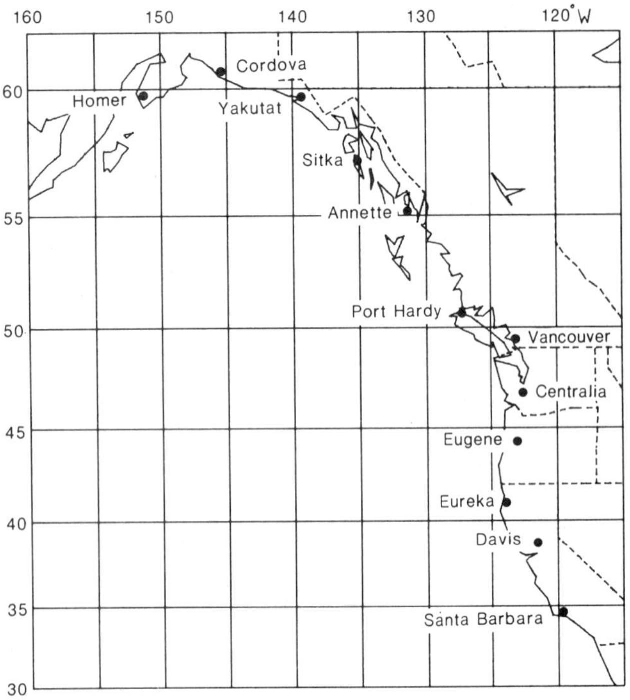

Figure 1. Location of Precipitation Stations.

associated with a persistent northerly jet stream track and, hence, unusually wet conditions in the Gulf of Alaska.

Although negative cross correlations at large distances are physically reasonable, as pointed out earlier the inter-station relationships may have a nonlinear structure and the simple cross correlation may not be an appropriate measure of dependence. This problem is illustrated in scatterplots for the January data shown in Figure 3.

Figure 3a shows a scatterplot of January data for Port Hardy and Eureka (separation 1300 km). The cross correlation coefficient, 0.1, is greatly influenced by widespread precipitation shortfalls in January 1949, 1963 and 1977 which appear as a cluster of points in the bottom left of the plot. Elimination of these points would give a slight negative correlation. A similar feature, though not as pronounced, is apparent in the scatterplot for Vancouver vs. Eureka (separation 1000 km) in Figure 3b. As the separation between stations decreases these nonlinearities become less obvious. At separations less than about 700 km the assumption of linearity seems reasonable.

It is evident in Figures 3a and 3b that the scale of meteorologic phenomenona is important in determining the nature of inter-station relationships. Frontal systems on the west coast may affect stations up to 500 km apart whereas drought may affect stations up to 2000 km apart. In Figure

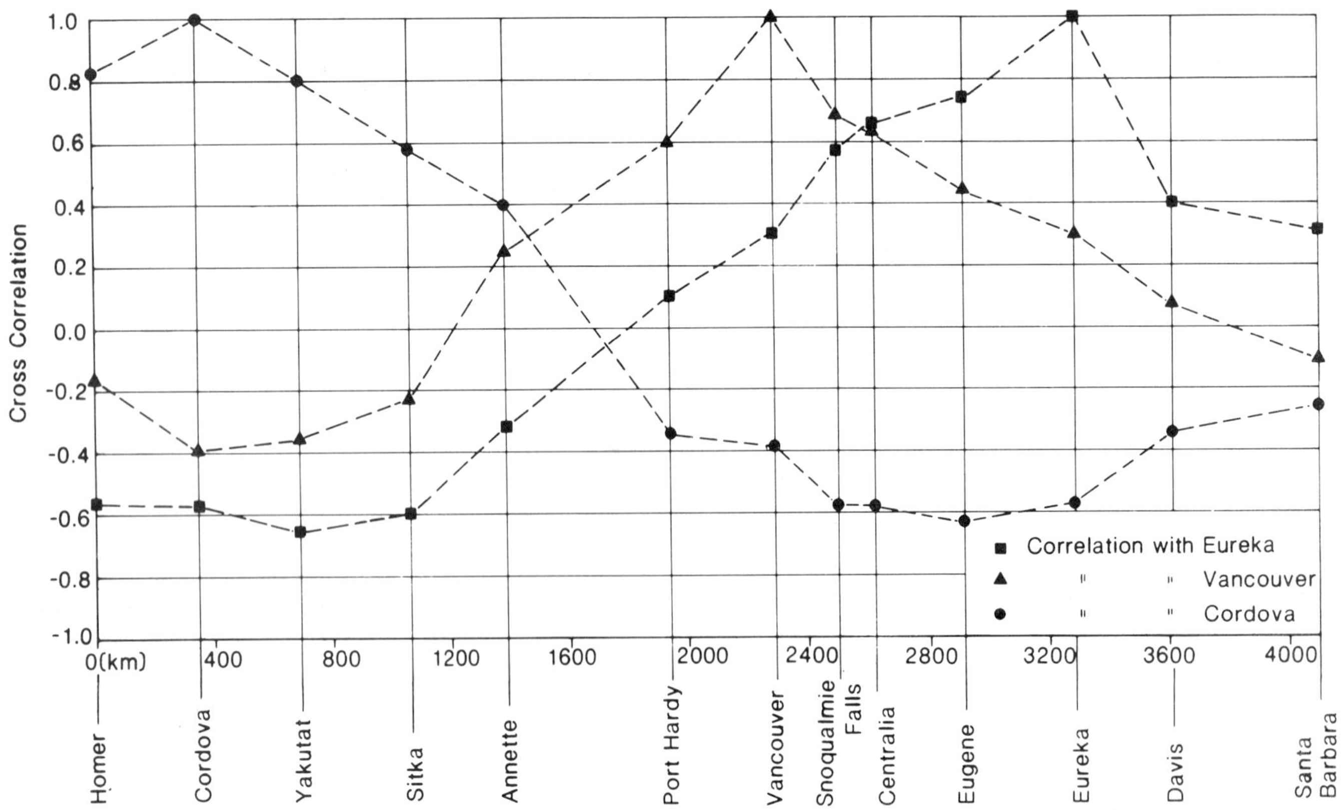

Figure 2. Variation of January Cross Correlation With Distance.

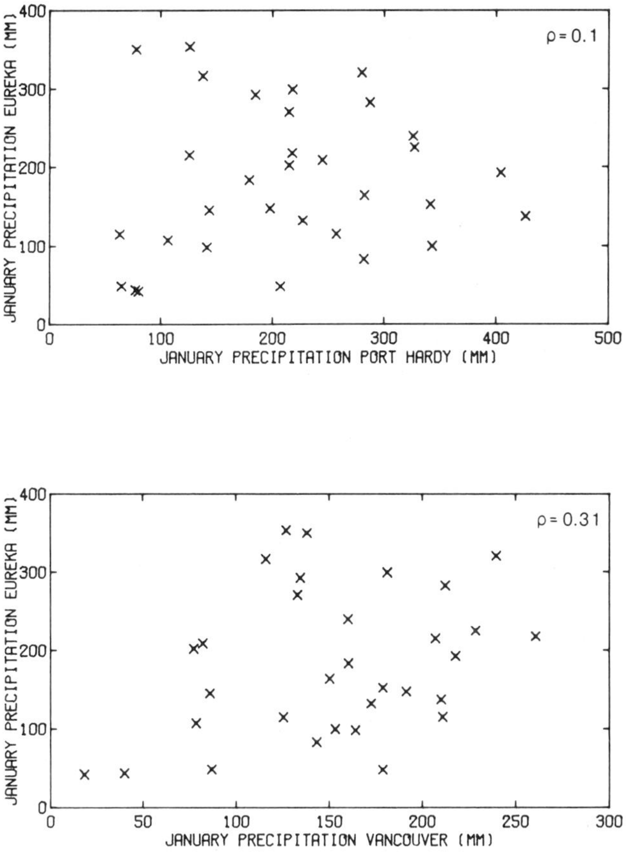

Figure 3. Scatterplots for January Precipitation Data.
a) Port Hardy vs. Eureka.
b) Vancouver vs. Eureka.

3a (station separation 1300 km) rainfall depths during normal or wet conditions at either station are essentially uncorrelated at the monthly interval. This is because: (1) station separation is large in comparison to the size of frontal systems, (2) frontal systems pass through the region in three to five days thus the rainfall in one month may be made up of contributions from a number of storms, and (3) apparently random fluctuations in the position of the jet stream track can move frontal systems several hundred kilometers north or south of the mean monthly position without greatly affecting the predominant pattern of circulation. In contrast, severe drought often extends from central California to southern Alaska; thus drought conditions may occur simultaneously at both Eureka and Port Hardy.

The scatterplots shown here demonstrate nonlinearities in the inter-station rainfall relationships, which can be attributed qualitatively to features of the large scale atmospheric circulation. The significance of these nonlinearities in stochastic hydrology is explored in the next section.

EVALUATION OF CONVENTIONAL MULTI-SITE MODELS

All current multi-site stochastic models use simple linear cross correlations to express inter-station dependence. Such models are commonly based on the assumption that, after suitable transformations, the multi-site data can be regarded as samples from a multivariate normal distribution. This assumption is, however, generally an act of faith made, not because the data have been shown to come from a multivariate normal distribution, but becuase of the attractive practical computational features of the distribution. Of particular importance from the operational point of view is the fact tht the marginal distributions are normal.

The usual approach to modeling has been first to fit a simple distribution, such as a normal or lognormal, to the data from each site. The choice of distribution is dictated by the requirement that application of a simple transformation should permit the transformed single-site data to be treated as samples from normal distributions. These normal distributions are of course the marginal distributions of the transformed multi-site data. Multi-site data in the transformed domain can hence be obtained by sampling from a multivariate normal distribution having the fitted marginal distributions and with an appropriate covariance matrix and serial correlation structure. The primary concern in most modeling efforts to date has been to ensure that the theoretical marginal distributions fit the data adequately. The validity of assuming joint normality has received little or no attention.

The performance of a conventional multi-site model was evaluated using the January monthly precipitation record of 32 observations (1947-1978) from Eureka and Port Hardy. The location of these sites was shown in Figure 1. January data alone were used to simplify the evaluation and also because the primary concern in the study is in winter precipitation. Most major water resource facilities on the west coast rely heavily on snowmelt from winter precipitation and thus deficits in winter precipitation are of principal interest. Although a dry January does not necessarily imply a deficient winter snowpack, the January data can be used as an index of winter conditions. The basic statistics for the January data at Port Hardy and Eureka are shown in Table 1, and a scatterplot of the data was shown in Figure 3a.

The serial-correlations for the January data are small. Moreover, our principal interest is in the spatial characteristics of the synthetic series in mathematical expectation. Thus the January data, after suitable transformation, will be assumed to be temporally independent samples from a bivariate normal population.

Cumulative distribution functions for the Port Hardy and Eureka data are plotted as January precipitation divided by mean January precipitation on normal probability paper in Figure 4 along with fitted three parameter log normal (LN3) distributions. The LN3 distribution was used because it is well understood and flexible, and is related by a

TABLE 1. Basic Statistics for January Precipitation Data at Port Hardy and Eureka (1947-1978).

	Port Hardy	Eureka
Mean (mm)	212.70	182.80
Standard deviation (mm)	98.00	91.70
Skew	0.27	0.23
Serial correlation (lag 1)	0.18	0.19
Cross correlation	0.10	

TABLE 2. LN3 Paramters for January Precipitation Data at Port Hardy and Eureka.

Station	Transformation	Mean μ_y	St. Dev. σ_y	Shift a
Port Hardy	y = log(x − a)	7.784	0.0436	−2187
Eureka	y = log(x − a)	7.314	0.0676	−1317

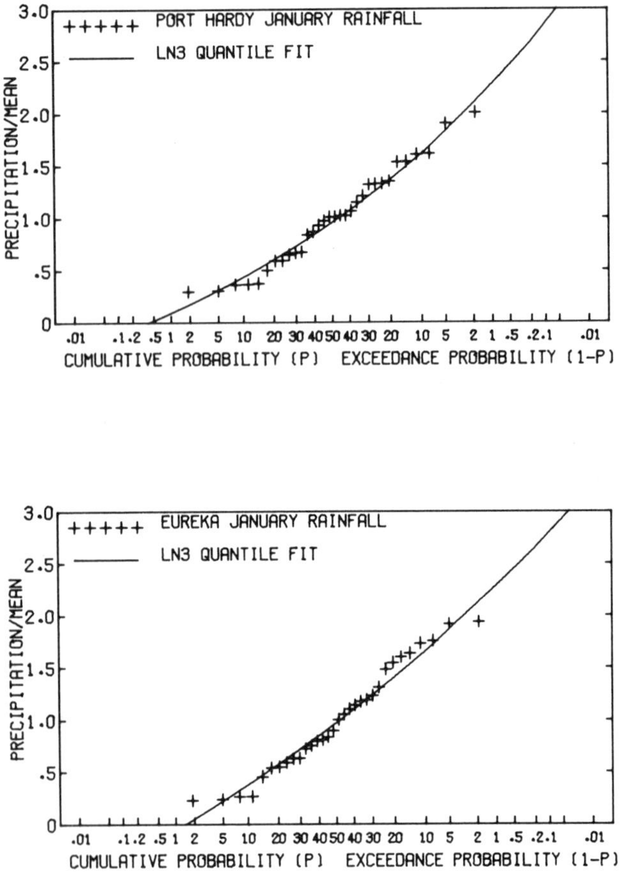

Figure 4. CDF's for January Precipitation Data.
a) Port Hardy.
b) Eureka.

simple transformation to the normal distribution. The LN3 were fitted by the mixed maximum likelihood/quantile method suggested by Stedinger (1980). The parameters of the fitted distributions are given in Table 2. The fit of alternative distributions was not investigated in detail. However minor variations in the shapes of the fitted distributions would have little effect on the synthesis of joint low events and would not affect the results presented here.

The LN3 distributions give a reasonable fit to the data at both sites. Synthetic data in the log domain can thus be generated by independent sampling from a bivariate normal distribution with means and standard deviations from Table 2 and with a cross correlation coefficient of 0.1 estimated by moments from the raw data. The correction to the cross correlation to account for the log transformation (Mejia and Rodriguez-Iturbe, 1974) was found to be negligible. The scheme adopted for sampling from the bivariate normal distribution was similar to that given by Matalas (1967). Application of the inverse transform (x = exp(y) + a) returns the synthetic data to the natural domain.

The procedure used for evaluating the multi-site characteristics of the above model was through Monte Carlo simulation. For Port Hardy and Eureka 3200 years of synthetic January data were generated. This corresponds to one hundred 32-year sequences (the historic record was 32 years in length). The 10, 15, 20,...,50 percent quantiles for the two synthetic sequences were determined and then joint occurrences were counted in which both sites had rainfall less than or equal to their respective 10, 15, 20,..., 50 percent quantiles. These counts were divided by 100 to give an estimate of the expected number of joint occurrences in a period of 32 years. These data and the corresponding data from the 32 year historic record are shown in Figure 5.

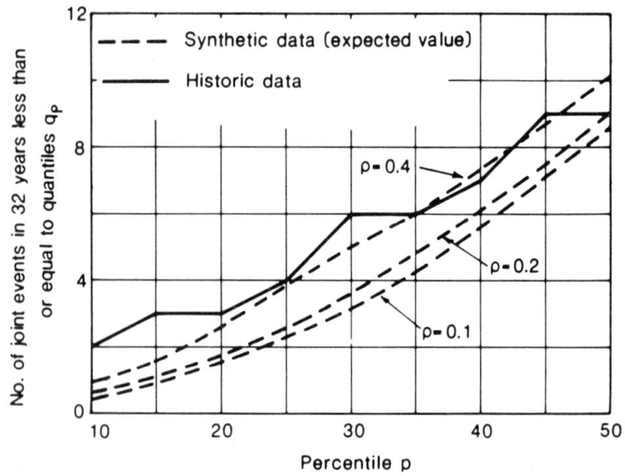

Figure 5. Occurrences of Joint Low Events at Port Hardy and Eureka for Historic and Synthetic January Data.

To indicate the distribution of synthetic joint low events, the actual numbers of joint events in the one hundred 32-year periods comprising the synthetic record are given in Table 3 for various quantile levels. For example, the table shows that of the one hundred 32-year periods in the synthetic sequence there were 60 periods with no events in which rainfall at both sites was concurrently below the 10 percent quantile and 37 periods in which there was one such event.

TABLE 3. Occurrences of Joint Low Events in January Synthetic Record for Port Hardy and Eureka.

Percentile	Number of 32-Year Synthetic Sequences Having n Joint Events Less Than or Equal to Quantiles q_p									
p	n	0	1	2	3	4	5	6	7	8
10		60	37	3						
15		36	41	20	2	1				
20	18	18	34	30	13	3	2			
25		10	22	24	26	12	2	3	0	1

The sensitivity of the result in Figure 5 to changes in the cross correlation coefficient was investigated by repeating the above experiment with values of the cross correlation of 0.2 and 0.4. The results of these experiments are also shown in Figure 5. For comparative purposes Figure 6 shows a similar plot for joint occurrences where both sites had rainfall greater than their respective 10, 15, 20,..., 50 percent quantiles.

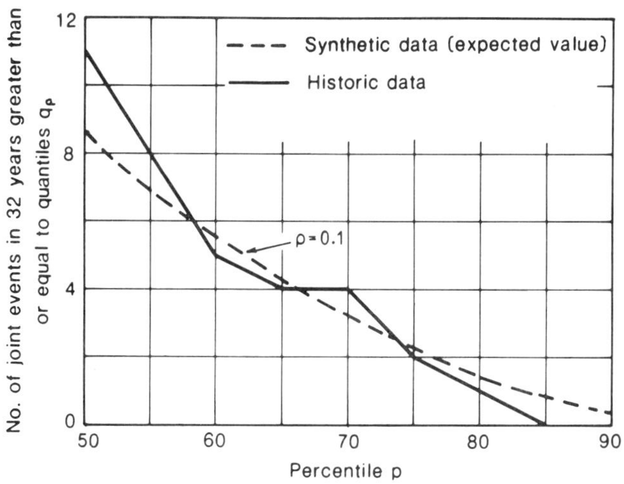

Figure 6. Occurrences of Joint High Events at Port Hardy and Eureka for Historic and Synthetic January Data.

Figure 5 and Table 3 demonstrate, as expected, that the model used cannot produce droughts as wide in areal extent as those found in the historic record from Port Hardy and Eureka. Of particular concern is the fact that few of the synthetic 32 year periods contained more joint low events than were found in the historic record. An alternative modeling approach based on sampling from a mixture of two multivariate normal distributions has been investigated by Leytham (1982) who shows mixture distributions to be capable of maintaining both the marginal distributions and cross properties of the Port Hardy and Eureka data.

CONCLUSIONS

The statistical significance of the above results is difficult to determine in view of the low power of statistical tests, the short record available to us, and possible errors in fitting the marginal distributions. However, in expectation the synthetic data should replicate reasonably closely the joint characteristics of the historic data. The model clearly fails to meet this goal. A useful measure of the lack of fit is the return period of severe joint events. The historic data indicates that joint events less than the 10 percent quantile have a return period of approximately 16 years. The synthetic data in contrast indicates a return period of 74 years. This is a major discrepancy and could clearly have an important effect on projects with inter-connected components separated by large distances (> 1000 km). This of course raises the question of the practical significance of the result. Water and hydro power transfers currently take place between northern and southern California involving distances in excess of 1000 km. Hydropower transfers along the Pacific Intertie between Washington and southern California involve distances in excess of 1800 km. Similar hydropower transfers between southern Alaska and the lower states of the U.S.A. will become feasible in the future. The results presented here are of considerable importance in the design or operation of such projects.

ACKNOWLEDGMENTS

I would like to thank Steve Burges for helpful comments on this paper. The work described is part of a larger study on the analysis and synthesis of multi-site drought sequences funded by the Office of Water Research and Technology (Project Nos. A-107-WASH and A-115-WASH) through the Washington State Water Research Center.

LITERATURE CITED

Edmon, H. J., 1980. A Study of the General Circulation Over the Northern Hemisphere During the Winter of 1976-1977 and 1977-1978. Monthly Weather Review, 108:1538-1553.

Lawrance, A. J. and N. T. Kottegoda, 1977. Stochastic Modeling of Riverflow Time Series. Journal of the Royal Statistical Society, Ser. A(104):1-47.

Leytham, K. M., 1982. Physical Considerations in the Analysis and Synthesis of Hydrologic Sequences. Technical Report No. 76, Harris Hydraulics Laboratory, Department of Civil Engineering, University of Washington, Seattle, Washington.

Matalas, N. C., 1967. Mathematical Assessment of Synthetic Hydrology. Water Resources Research, 3(4):937-945.

Mejia, J. M. and I. Rodriguez-Iturbe, 1974. Correlation Links Between Normal and Log-normal Processes. Water Resources Research, 10(4):689-690.

Stedinger, J. R., 1980. Fitting Log-normal Distributions to Hydrologic Data. Water Resources Research, 16(3):481-490.

INTERNATIONAL SYMPOSIUM ON HYDROMETEOROLOGY
JUNE AMERICAN WATER RESOURCES ASSOCIATION 1982

INFLUENCES OF WEATHER PATTERNS ON WETLAND FLOODINGS

Flora C. Wang and P. D. Scarlatos[1]

ABSTRACT: The coastal wetland of Louisiana is a depressional watershed characterized by large areas of swamps and marshes and slow natural drainage through bayous and channels. Elevations are low, generally ranging from a maximum of about 30 feet on natural levee terrain to sea level shoreward into coastal marshes. The climatic data of coastal Louisiana have been analyzed and grouped into eight weather types. The synoptic classes are related directly to variations in selected environmental parameters, thereby providing a framework for assessing climate's contribution to the functioning of a natural system. Western Terrebonne Parish, Louisiana, an extensive marsh floodplain, is located at the east of Atchafalaya Bay. A levee was constructed to limit flooding in the area east of the floodway. The area is subject not only to backwater flooding from the Lower Atchafalaya River, but also to tidal flooding from the Gulf and headwater flooding from local rainfall. This paper examines the relationships between weather patterns and wetland floodings as related to water levels, velocities, and stage characteristics of floodplain.
(KEY TERMS: weather patterns; wetland floodings; levee systems; water level and velocity; computer simulation.)

INTRODUCTION

The Atchafalaya River Basin, located in south-central Louisiana, is a flood plain for Mississippi and Red River floodwaters. It extends inland from the Gulf of Mexico 120 miles to the confluence of the Mississippi and Red Rivers, as shown in Figure 1 (after Van Beek, *et al.*, 1979). Protection levees were constructed east and west of and parallel to the Atchafalaya River course at a distance of 15 miles apart. To protect the area east of Morgan City from backwater flooding, the 13-mile Avoca Island Levee was also built south of Morgan City, as displayed in Figure 1.

The active development of a new delta in the Atchafalaya Bay has resulted in the elongation of the river's course and has raised the stage at the end of the existing Avoca Island levee. To compensate for this raise, the U.S. Army Corps of Engineers (1981) proposed a 14,000-foot extension of the existing levee to continue backwater-flooding protection for the area east of the lower floodway.

Concerns have been expressed that the extension of the existing levee would cut off the normal amount of freshwater and nourishing sediment presently discharged by the river into the backwater area, which in turn could result in further accelerating the deterioration of the marsh. Many feel that the extension should be delayed until its effects can be evaluated.

The purpose of this paper is to investigate by computer simulation the relationships between climatology and wetland flooding. Our objectives are to study the influence of prevailing weather patterns on current velocities in bayous and canals and the stage characteristics of wetlands.

WESTERN TERREBONNE PARISH MARSH

As shown in Figure 1, our study area is designated as the western Terrebonne Parish Marsh. A drainage basin located east of Atchafalaya Bay and south of Morgan City, the area is primarily a low-lying marsh that is at most a few feet above mean sea level (Wang and Gosselink, 1982). These marshlands are drained by numerous meandering bayous and have undergone some changes over the last 20 years. The changes are associated with both natural and man-made activities (CEI, 1980) such as the break-up of marshes by floods and storms, the subsidence due to the compaction of sediments and the rise of mean sea level, the replacement of marshland by oil and gas access canals, and the impoundment resulting from spoil disposals.

Floodings in the marsh area comes from three sources. Backwater flooding is caused by high river stages in the Lower Atchafalaya River. Headwater flooding is produced by local heavy rainfall or rainfall in the upstream drainage basin. And tidal flooding is generated by wind-induced tides in the marshes and bays. As the Corps of Engineers' report (COE, 1981) stated, the backwater flooding has always occurred in combination with wind-tide and rainfall-runoff floodings.

AVOCA ISLAND LEVEE AND EXTENSION

The existing Avoca Island levee was built to limit the

[1]Respectively, Associate Professor of Department of Marine Sciences; and Post-Doctoral Research Associate of Coastal Ecology Laboratory, Center for Wetland Resources, Louisiana State University, Baton Rouge, Louisiana 70803-7503.

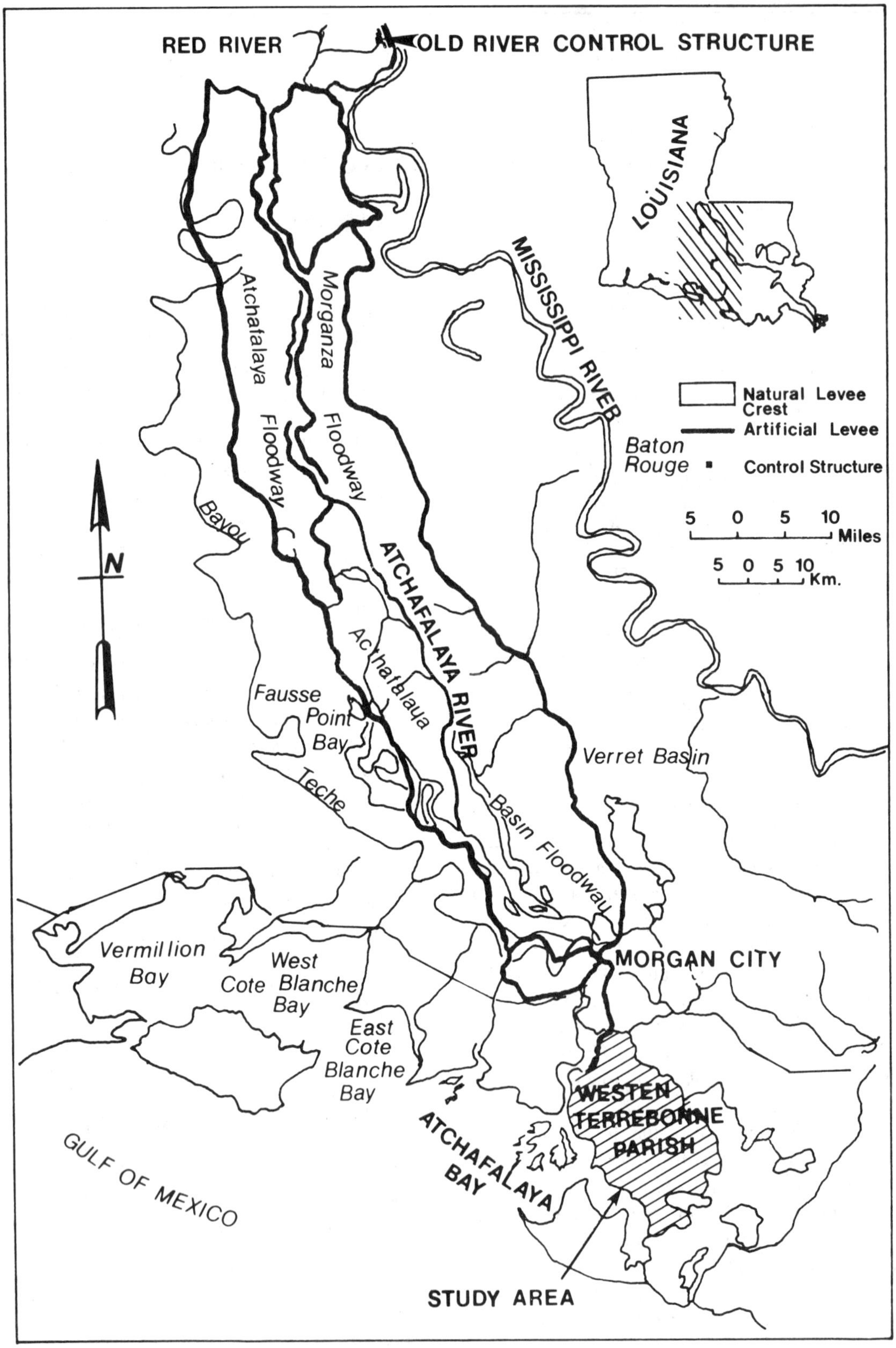

Figure 1. Atchafalaya River Basin and Its Levee Systems (After Van Beek, *et al.,* 1979).

flood stage east of Morgan City to the same levels as those that occurred in the 1945 flood. To achieve the projected flood stages, about 13 miles of levee was constructed south of Morgan City, as delineated in Figure 1, and was aligned parallel to the Atchafalaya River.

The growth of the Atchafalaya delta is a recent event in the history of the Mississippi River delta systems. The new detla has resulted from the steady capture of Mississippi River flow by the hydraulically more efficient Atchafalaya River. As river sediments have filled in Atchafalaya Bay, the stage at the end of the existing Avoca Island levee has risen by about 2 feet (COE, 1981). To counterbalance this increase, a 14,000-foot levee extension is proposed (COE, 1981) to contain backwater flooding in the area east of the floodway.

The protected area east of Morgan City is of great significance to Louisiana's regional economy, as it is the home base for the manufacture of products used in exploration and production of oil and gas in the Gulf region. On the other hand, the marshland immediately adjacent to the proposed levee extension, an enclosed area of about 160 square miles (Figure 1), represents a large, high-quality fish and wildlife habitat, long used by local fishermen, trappers, and others. This conflict of interests has sharpened the focus on the effects of levee extension on marsh flooding.

SYNOPTIC CLIMATES IN COASTAL LOUISIANA

A synoptic climatic system based on the National Weather Service daily weather and observational data taken at the first-order weather station has been developed by Muller (1977). Eight weather types are classified and are organized into a synoptic framework for the regional climate. The weather types are related directly to the physical parameters of weathers such as air temperature, dew point temperature, relative humidity, wind speed and direction, and cloud cover. This synoptic approach is from the local perspective and the regional generalization. However, the categories pinpoint an orderly system of regular winds produced by regular atmospheric conditions.

Figure 2 shows the eight synoptic weather types in coastal Louisiana derived from the observational data taken every third hour at New Orleans, Louisiana, in 1971-74 (Wax, *et al.,* 1978). The eight weather types are listed as: (1) Frontal Overrunning (FOR), (2) Frontal Gulf Return (FGR), (3) Continential High (CH), (4) Coastal Return (CR), (5) Gulf High (GH), (6) Gulf Return (GR), (7) Pacific High (PH), and (8) Gulf Tropical Disturbance (GTD). The meteorological properties associated with each weather type are discussed in detail elsewhere (Muller, 1977).

Precipitation is closely related to the weather types. Muller (1977) showed that the frontal activities (FOR and FGR) accounted for more than two-thirds of coastal Louisiana's rainfall, but were present only one-third of the time during the 1971-1974 period. It is these two weather types that prevail during the Louisiana winter and spring rainy seasons.

THE SWMM COMPUTER MODEL

Many watershed models have been developed since 1960 due to the advent of digital computer. The EPA Storm Water Management Model (SWMM) is used in this study to investigate the influence of climate on wetland's response. The SWMM (EPA, 1975) is a comprehensive mathematical model. The model is structured into six independent computer programs or blocks.

The RECEIVE block is used in this paper to estimate current velocity and water stage during wetland floodings induced by the weather activity. Physical abstractions of western Terrebonne drainage basin is the first step in utilizing the RECEIVE model. The basin is divided into three types of hydraulic elements: bayou and canal, floodplain, marsh and swamp. Bayous and canals serve as conduits for water flow. Marshes and swamps serve as water storage areas. Floodplains adjacent to bayous and canals are characterized by wide and shallow channels with high roughness to simulate sluggish overland sheet flow.

Figure 3a delineates the physical abstraction and Figure 3b shows the schematic junction-channel network for western Terrebonne Parish Marsh. In the figure, the numbers pertain to the names of junctions and channels for which input parameters and output results are reported in the text.

The dynamic behavior of water in the network is influenced by forces that are described in the equation of motion and the continuity equation (EPA, 1971). These equations are applied to each hydraulic element and solved simultaneously to produce a temporal history of the water characteristics of wetlands. The basic steps of computational procedure in the RECEIVE block have been described in detail elsewhere (EPA, 1975; Hopkinson and Day, 1980).

RESULTS OF SIMULATION

The coarse network of western Terrebonne marsh was composed of 16 discrete junctions and 30 channels. Floodplains (marshlands adjacent to bayous) were simulated as shallow and wide channels designated by dash lines and parallel to bayous. Bayous and canals were represented by solid lines (Figure 3b). The physical dimensions of junctions (surface areas and width) and channel (length and width) were measured directly from the map. The Manning's roughness coefficients for various water depths were inferred from measurements of flow over sawgrass in the Florida Everglades (COE, 1956). The depth of water was taken as positive downward from the low low-water level (LLW).

The boundary conditions imposed to the system are tidal fluctuations for junctions #3, 4, 5, 6, 7, 8, 9, and 10

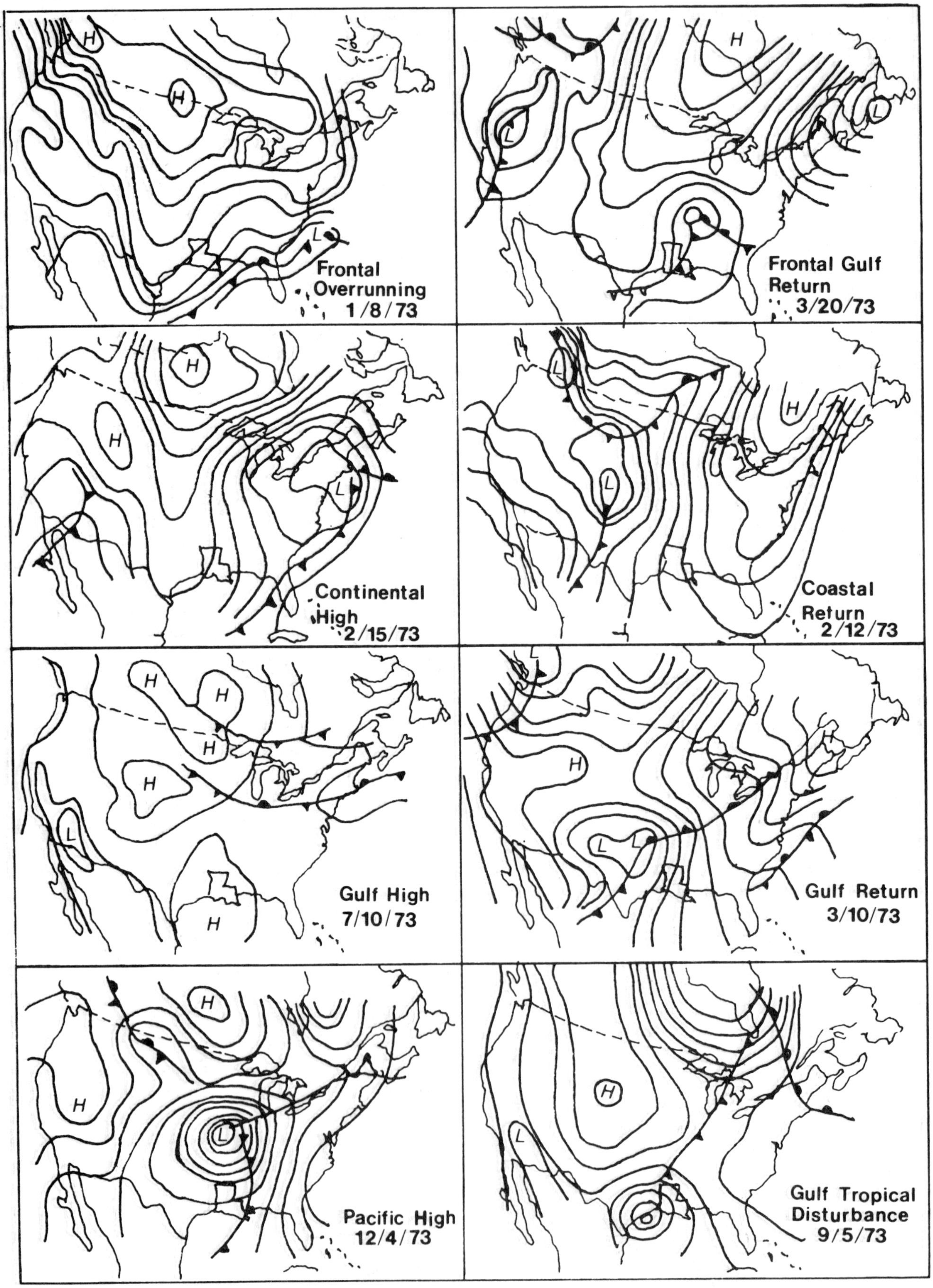

Figure 2. Synoptic Weather Types in Coastal Louisiana (Wax, *et al.,* 1978).

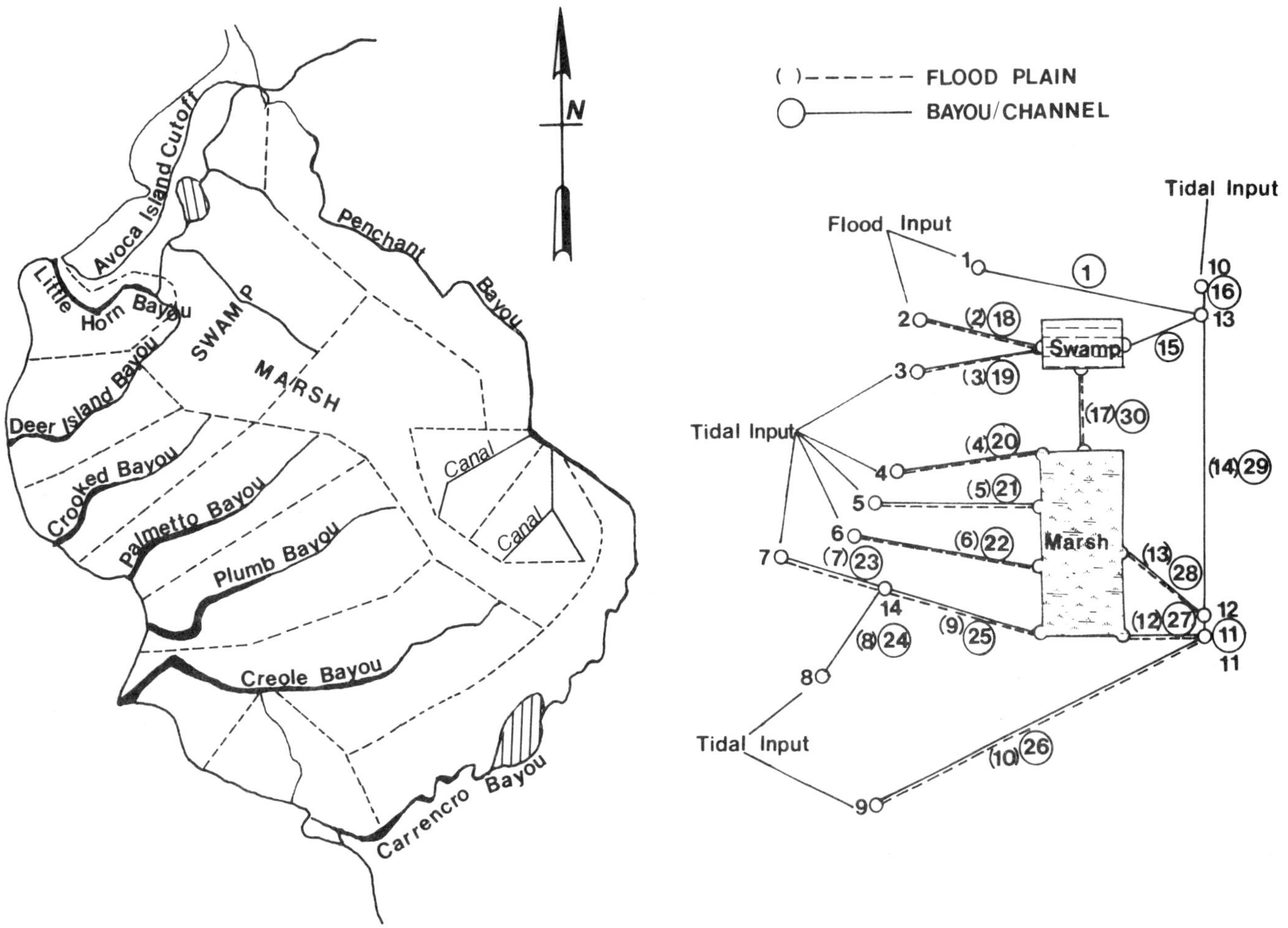

Figure 3. Junction-Channel Network for Western Terrebonne Parish Marsh.
(a) Physical Abstraction. (b) Schematic Network.

and flooding discharge for junctions #1 and 2. Wind stresses generated by various weather patterns were superimposed throughout the whole network. Table 1 lists the 10 cases that were simulated. In the table, two seasons are broadly defined, with "winter" spanning the period from late fall to early spring, and "spring covering the rest of the year. Various values of weather types, wind (speed and direction), river discharge, and tidal range are assumed to be representative of prototype data.

The initial water velocities along the various channels were taken as zero. However, each simulation was run for three tidal periods so that a quasi-steady state was reached. Figures 4a and 4b exhibit the water stages for the two major components of floodplain, the marshland (#16) and swampland (#15), respectively. The run numbers (R1 to R10) in these figures indicate the same simulation cases as listed in Table 1 for various weather conditions. The current velocities in Plumb Bayou (channel 22), the channel adjacent to the bay; and Penchant Bayou (channel 29), the inland bayou, are displayed accordingly in Figures 5a and 5b. In these figures, the positive velocities are referred to water flow from bayous to marshes or swamps; and the negative velocities are those flow from marshes or swamps to bayous.

CONCLUSION AND DISCUSSION

The behavior of the western Terrebonne Marsh under various hydrodynamic and weather conditions was simulated by the SWMM computer model. The results obtained are referred to the case where the whole network is under the low low-water level, so that the dynamic interaction of the various elements of the systems, such as bayous, canals, marshes, and swamps, can be properly evaluated.

Because of a lack of field data *in situ,* regional information and general experience are relied upon. The analysis presented here may be subject to change as new information becomes available. The following conclusions are drawn, based on the assumptions and analyses we made in this paper.

TABLE 1. Prototype Data on Weather Patterns, Tidal Ranges, and River Discharges for Simulation.

Run Number	Weather Type	Wind Speed (Mile/Hour)	Wind Direction	River Discharge (cfs)	Tidal Range (ft.)
R1 (winter)	FOR	12	North	200,000	0.70
R2 (winter)	FOR	12	North	345,000	1.30
R3 (winter)	FGR	10	South	345,000	1.30
R4 (winter)	FGR	8	South	380,000	1.40
R5 (winter)	CH	8	North	120,000	0.90
R6 (summer)	GH	8	North	120,000	1.40
R7 (summer)	GR	8	South	120,000	1.40
R8 (summer)	CH	8	North	280,000	1.40
R9 (summer)	CR	8	South	280,000	1.40
R10 (summer)	CR	8	South	120,000	0.90

Data source: Wax, *et al.* (1978); Bayley and Day (1981).
Note: 1 ft. = 0.305m; 1 mile = 1.609 km; 1 cfs = 0.028 cms.

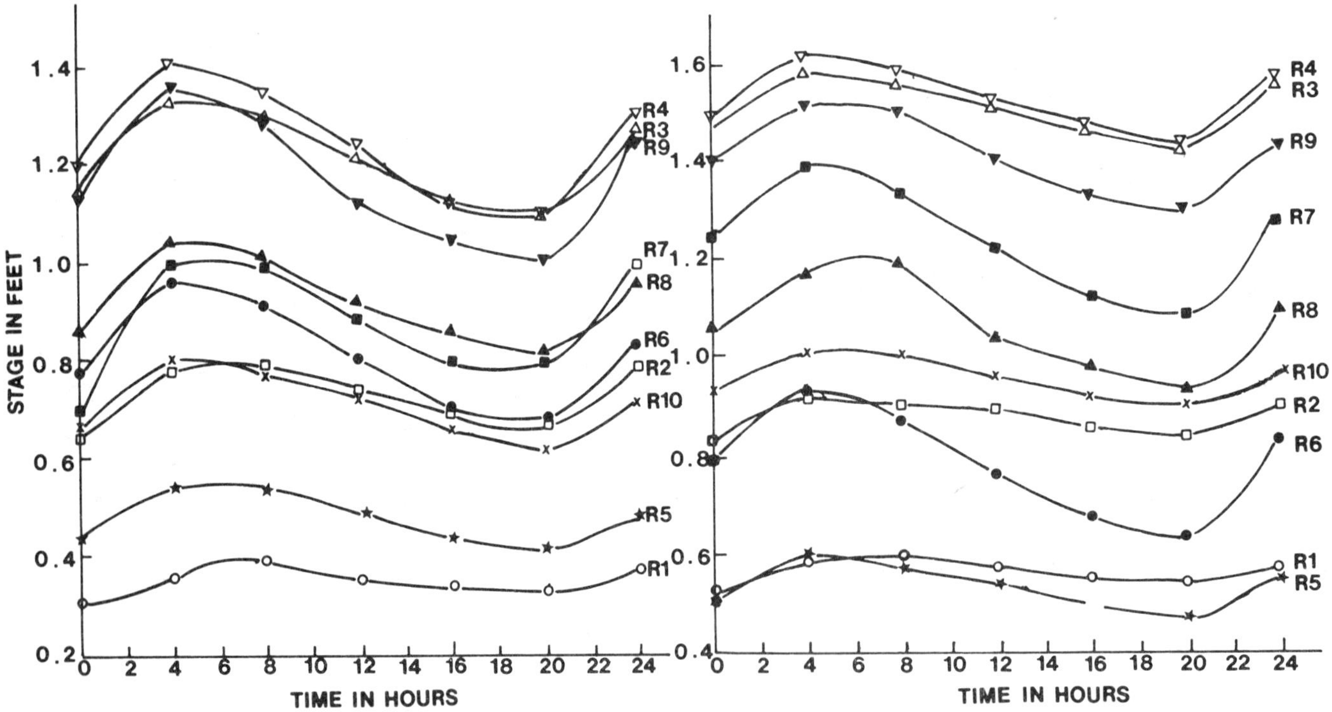

Figure 4. Simulated Water Stages for Western Terrebonne Parish Marsh.
(a) Marshland (Junction 16). (b) Swampland (Junction 15).

1. Southerly winds, higher river discharges, and high tidal ranges cause a rise in the water stages of marshes and swamps.

2. Under the same climatic conditions, the water stages are higher in the swamps than in the marshes.

3. The greater oscillatory behavior of the marshes shows that tides influence marshes more than swamps.

4. Bayous adjacent to the bay have predominately tidally induced currents, while bayous further inland are less affected by tides.

5. Flooding is more predominant in inland bayous and their adjacent wetland than over areas close to the bay.

6. Winds seem to have less effect on the bayou currents than on the marsh and swamp water stages.

ACKNOWLEDGMENTS

The research work reported in this paper is supported in part by the Louisiana Sea Grant College Program, a part of the National Sea Grant College Program administrated by NOAA, U.S. Department

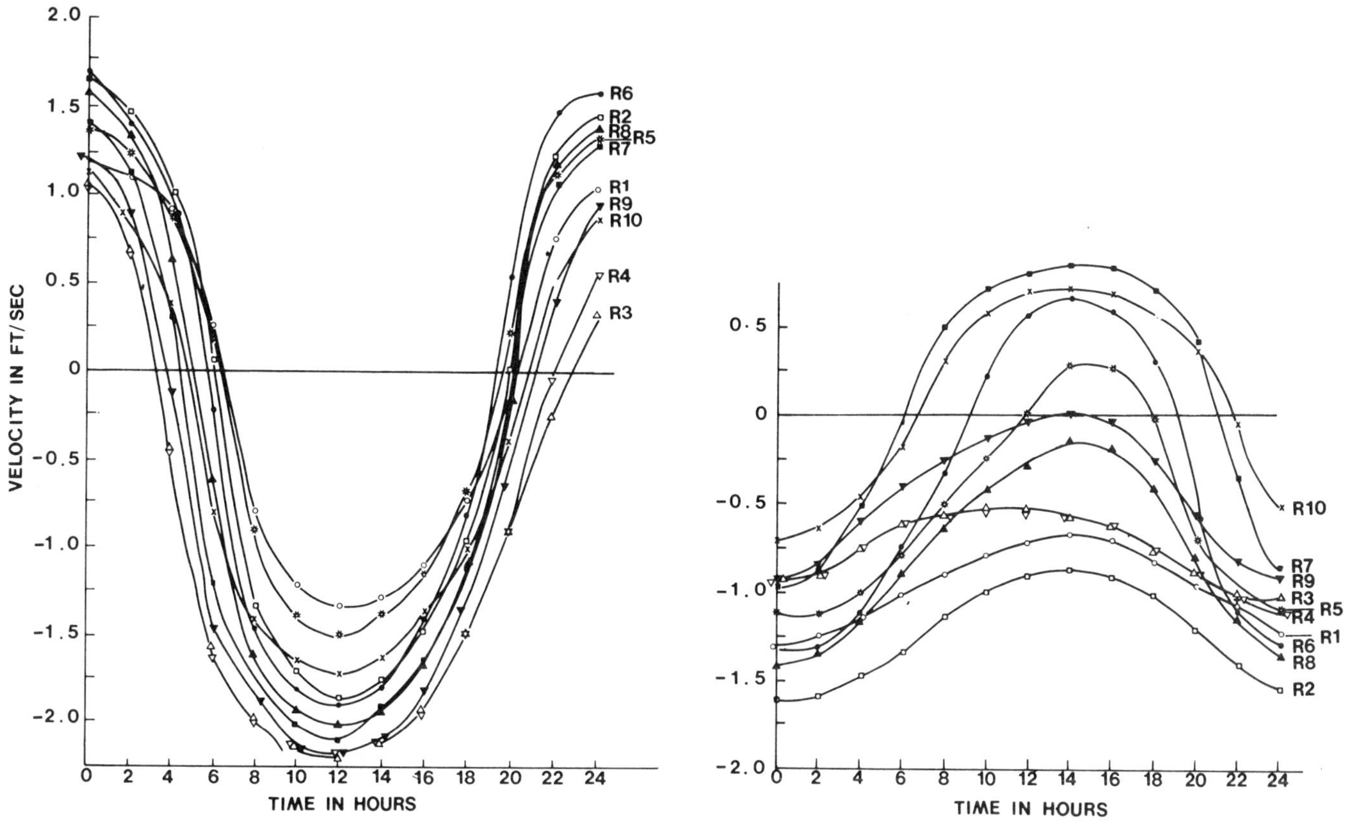

Figure 5. Simulated Current Velocities for Western Terrebonne Parish Marsh.
(a) Plumb Bayou (Channel 22). (b) Penchant Bayou (Channel 29).

of Commerce. Louisiana State University Coastal Ecology Laboratory Publication No. LSU-CEL-83-7.

LITERATURE CITED

Bayley, S. E. and J. W. Day, 1981. Nutrient Dynamics and Habitat Characterization of the Atchafalaya Bay System, Project R/MPE-9, Louisiana Sea Grant Program, Vol. II, 1981-82, pp. 297-314.

Coastal Environments, Inc., 1980. Environmental Characterization of Terrebonne Parish: 1955-1978. Final Report for Terrebonne Parish Policy Jury, Coastal Environments, Inc., Baton Rouge, Louisiana.

Corps of Engineers, 1956. Everglades Gaging Program, Progress Report, Everglades Area, Florida. U.S. Army Corps of Engineers, Jacksonville District, Florida, Report No. 6.

Corps of Engineers, 1981. A Feasibility Report: Environmental Impack Statement on the Atchafalaya Basin Floodway System, Louisiana. Vol. 1 – Main Report and Environmental Impact Statement; Vol. 3 – Appendix C: Engineering Investigations, U.S. Army Corps of Engineers, New Orleans District.

Environmental Protection Agency, 1971. Storm Water Management Model. U.S. EPA Water Pollution Control Research Series 11024DOC07/71.

Environmental Protection Agency, 1975. Storm Water Management Model User's Manual, Version II. Environmental Protection Technology Series, EPA-670/2-75-017, National Environmental Research Center, Office of Research and Development, USEPA, Cincinnati, Ohio.

Hopkinson, C. S., Jr., and J. W. Day, Jr., 1980. Modeling Hydrological Eutrophication in a Louisiana Swamp Forest Ecosystem. Environmental Management, 4(4):325-335.

Muller, R. A., 1977. A Synoptic Climatology for Environmental Baseline Analysis: New Orleans. Journal of Applied Meteorology, 16(1):20-33.

Van Beek, J. L., A. L. Harman, C. L. Wax, and K. M. Wicker, 1979. Operation of the Old River Control Project, Atchafalaya Basin: An Evaluation from a Multiuse Management Standpoint, Environmental Monitoring and Support Laboratory, U.S. EPA, Las Vegas, Nevada.

Wang, F. C. and L. Gosselink, 1982. Issues of Levee Extension and Wetland Flooding. *In:* Proceedings of Water and Energy, Technical and Policy Issues, American Society of Civil Engineers, Fort Collins, Colorado.

Wax, C. L., M. J. Borengasser, and R. A. Muller, 1978. Barataria Basin: Synoptic Weather Types and Environmental Responses. Sea Grant Publication No. LSU-T-78-001, Center for Wetland Resources, Louisiana State University, Baton Rouge, Louisiana, 60 pp.

Section 6

WEATHER MODIFICATION

ON THE EVALUATION OF OPERATIONAL CLOUD SEEDING PROJECTS

C.-F. Hsu and S. A. Changnon, Jr.[1]

ABSTRACT: There is a growing demand to "properly" perform statistical-physical evaluations of operational weather modification projects. Controversies generated by reanalyses of the past operational projects have been increasing. We believe that useful evaluations can be accomplished with "careful" target-control analyses (temporal and spatial). Extensive simulation studies have been performed to investigate the statistical powers of a collection of evaluation techniques for various modification scenarios, and high power technique for each scenario has been identified.

A few past operational projects also have been evaluated to test the more powerful techniques found in the simulation studies. Key issues affecting statistical evaluations have been identified and innovative methods to obtain scientific information (from operational projects) concerning the seeding effects also have been investigated. (KEY TERMS: weather modification; cloud seeding; evaluation; precipitation; rain enhancement.)

INTRODUCTION

Since the historical experiments of Langmuir's CIRRUS project (1948), it has long been recognized that modifying weather by cloud seeding is a viable alternative in supplementing deficient natural precipitation, hence a useful tool in water resources planning (Haragan, 1974; Howell, 1977; Changnon, 1980a). The potential beneficiaries of increased precipitation covered a wide spectrum of users including agriculture, power supply, water supply, irrigation, forest fire suppression, recreation, and so on. Rain falling at the "right" moment becomes critical to certain agricultural activities, and the "random" occurrence of natural precipitation rendered a great number of operational cloud seeding projects being done as an answer to this problem. The existence and growth of those operational, or commercial weather modification projects are generally based on a premise that weather modification is effective, and consequently their evaluation has not always been addressed properly or believably. However, our knowledge of the effectiveness of weather modification up to the present time is still a big "unknown." More than 30 years of scientific research on the weather modification did not produce enough convincing evidences that we have successfully modified the weather (Changnon, 1980b). The blame for this inconclusive endeavor includes poor project design, poor conduct of the project, bad draw in randomization, poor data collection procedures, reanalyses of data, poor planning and leadership, lack of coordination, and most critically, insufficient funding both at the federal and state levels (Changnon, 1980b). In spite of these drawbacks, many operational cloud seeding projects have been conducted over the years covering a vast portion of the U.S. (Figure 1).

The recognition that operational cloud seeding projects can provide vast amounts of useful information, and that their evaluation has potential for gaining scientific knowledge, was described in detail by Changnon, *et al.* (1979). The recommendation that considerable attention be given to the evaluation of selected operational projects was made by the President's Weather Modification Advisory Board (1978a) and its Statistical Task Force (1978b). Another reason for the need for evaluation of operational cloud seeding projects comes from the operational community's need for standardization, development, and adoption of more stringent operational criteria and instrumentation to provide better evaluations. The third reason arose from the user group wanting to know quantitatively the cost-effectiveness of their paid seeding operations.

The complexity of the evaluation of operational cloud seeding projects is sufficiently difficult to require a sizeable scientific effort. A multi-year project now in progress (Operational Seeding Evaluation Techniques) is aimed at this difficult problem of evaluating operational weather modification projects. Its primary goal is to devise innovative approaches and to study and assess statistical-physical techniques for use in the evaluation of weather modification operations. Additional objectives of the project include the evaluation of selected operational projects to test the techniques developed and the recommendation of the criteria and the planning needed for future operational projects to make them easier to evaluate.

TECHNIQUE DEVELOPMENT AND SIMULATION STUDIES

After an extensive literature search (Hsu, 1981), a collection of statistical-physical techniques to evaluate weather

[1] Respectively, Meteorological Statistician and Chief, Illinois State Water Survey, 605 E. Springfield, Champaign, Illinois 61820.

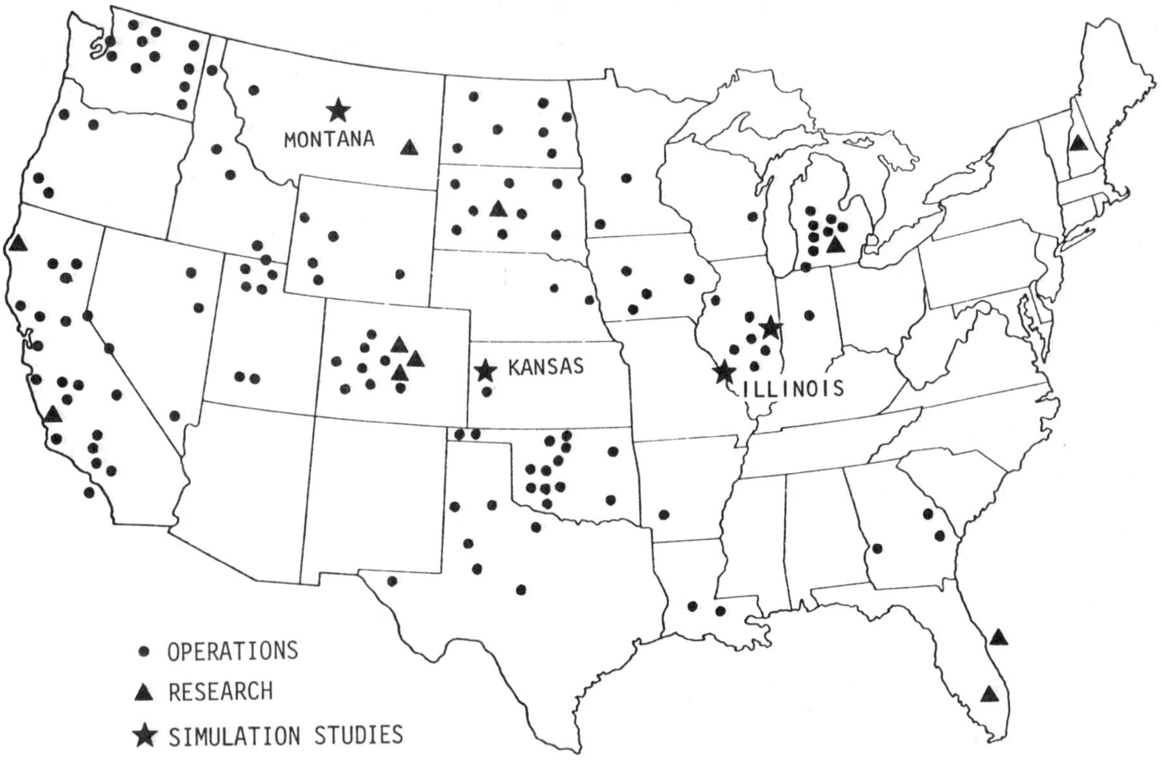

Figure 1. Locations of Research and Operational Weather Modification Projects in the United States, 1973-1977. (Two operational projects were in south-central Alaska and no project in Hawaii, from WMAB, 1978a).

modification projects was compared primarily through extensive simulation testing by superimposing assumed seeding effects upon natural precipitation distributions (Hsu, *et al.*, 1981a). Statistical power (Hsu, 1979) was the main index used for the comparison.

Five data sets from four areas were selected for simulations. They included monthly (May-September) and seasonal rains in both southwestern Kansas and east central Illinois, annual hail (insurance) lost-cost values in central Montana, and 48-hour and storm total rains in southwestern Illinois (Figure 1). Findings from the simulation studies revealed the following:

(1) In the Kansas rain simulation the principal component regression retaining the first component (PCR[1]) was one of the most powerful techniques for various summer months and target control designs.

(2) In the east-central Illinois (ILL-EC) rainfall study, PCR[1] and double ratios (DR) were generally the most powerful, and the two regressions (2Reg) were the next powerful. The results of the ILL-EC simulation, when compared with those of the Kansas simulation, indicated that PCR[1] had high powers in both simulations in every month except June, when only the double ratio had high power (in the averaged target simulation).

(3) In the Montana hail suppression simulation, for a larger target the principal component regression with three components (PCR[3]) was the most powerful in both three-seeded and six-seeded years simulations. For smaller targets PCR[3] worked well only in the three-seeded year study, while DR was most powerful in the six-seeded year study, followed closely by PCR[3] and the sum of rank power test (SRP). SRP had high powers in the three-seeded year study only when the assumed seeding effect was large.

(4) In the Illinois storm simulation, results from using the constant seeding induced increases revealed that MR and PCR[1] were the most powerful techniques. SRP, MR, or PCR[1] were the more powerful techniques when the varying seeding effect models were assumed (Hsu, *et al.*, 1981a).

(5) In the Illinois 48-hour simulation, MR was the most powerful technique in all cases; PCR[1] was a close second. The double ratio had high powers in some cases.

EVALUATION OF OPERATIONAL PROJECTS

Past seeding projects of the commercial type were reviewed to determine which were most suitable for testing the statistical-physical techniques developed under the project. Suitability was based upon location (climatically), length of project, goal of seeding (rain enhancement and/or hail suppression), and adequacy and availability of data. Rain enhancement projects selected for testing included several small scale airborne seeding projects operated in

Illinois within the past seven years, and one project using ground based generators in northwestern Oklahoma. For testing the techniques developed in the Montana hail study, a hail suppression project carried out in the Texas Panhandle during 1970-1976 was selected. In addition, a combined hail suppression and rain enhancement project during the warm seasons of 1975-1979 (the Muddy Road Project) in southwestern Kansas has been selected for evaluation. This relatively large scale project encompassed 15 target counties.

For the five small scale rain enhancement projects in Illinois (Changnon and Hsu, 1981), results generally reflected mixed outcomes. Two of the projects indicated increased rainfall in the target rainfall and/or radar echoes (39 percent in 1976, nonsignificant; 39 percent in 1979, significant at 10 percent level). (For 1979 seeding results, see Figure 2 as well as Hsu and Changnon, 1981.) One year (1978) indicated a 29 percent rain decrease (nonsignificant). The target echo results were also mixed. In all instances, the one-year (usually one or two months) projects were too short, regardless of the apparent increases or decreases of rainfall or echo in the target areas, to draw conclusions that have any statistical or physical significance when taken alone.

A large-scale hail suppression and rain enhancement project, the Muddy Road Project in southwestern Kansas, was evaluated using both monthly rain and annual crop hail loss cost data. Additional evaluation using daily and storm rains is in progress. The evaluation of the hail suppression efforts indicated that there was a reduction of annual hail loss cost values during the 1975-1979 seeded period compared to the 1948-1971 period; however, only the reduction in the east subtarget was significant at the 10 percent significance level (Figure 3). This example also demonstrated that the principal component regression is a better technique for evaluating hail suppression than the multiple regression (Hsu, *et al.,* 1981a). On the other hand, the statistical evaluation on the monthly and seasonal rainfall observations indicated that there was a nonsignificant reduction of rainfall in the target area during the seeded period.

A seven-year (1970-1976) operational project in the Texas Panhandle which dealt with hail suppression was also selected for evaluation. A technique of rotated factor analysis was applied to a 12-county data set for the period of 1947-1976. Seven factors explained 91 percent of the total hail loss cost variance. From the loading matrix it was shown that the target counties, Hale and Lamb, were both heavily loaded on one of the factors, Factor 4. Furthermore, both Welch's t-test and the Mann-Whitney test showed that only Factor 2 (south controls) and Factor 4 (target) displayed decreases which were significant at the 5 percent significance level between the historical and seeded factor scores. The indication was that the two target counties together showed a significant change of hail loss cost between the historical and seeded period, and except for part of southern counties, most control counties did not show any significant decrease of loss cost values between the two periods.

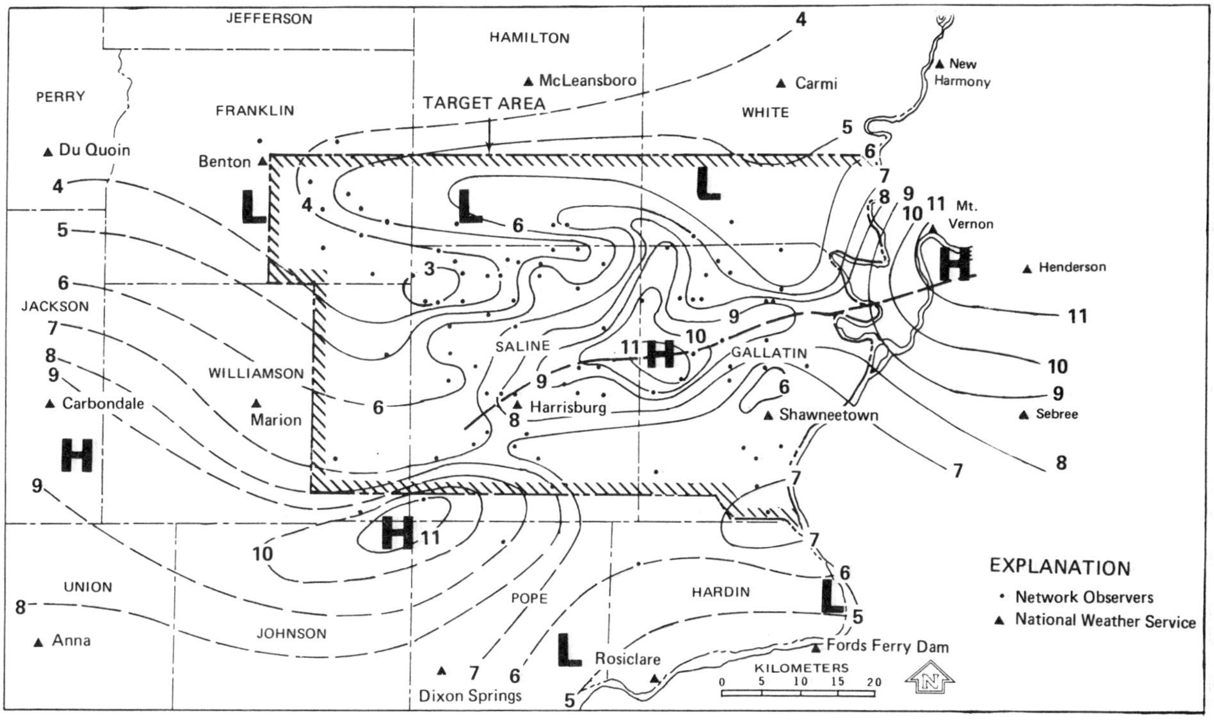

Figure 2. Total Rainfall (inches) in 1979 Operational Period of Weather Modification, South-Eastern Illinois.

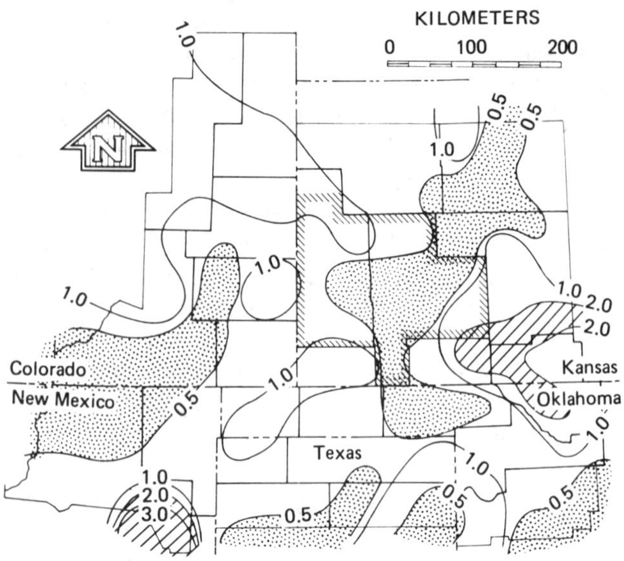

Figure 3. Ratio of Hail Lost Cost, 1975-1979 Average to 1948-1971 Average, Muddy Road Project.

USE OF COVARIATES

The use of surface meteorological covariates to aid evaluation has long been advocated. Gabriel (1981) gave an illustration on the usefulness of covariates. We had collected a number of appropriate meteorological covariates (Achtemeier, 1981; Westcott, 1979), and then screened them by correlating with rainfall variables observed in an Illinois raingage network (METROMEX). The selected covariates were used in the technique comparison simulation. Findings indicated thatthe use of meteorological covariates improved the powers of evaluation techniques. A two-stage screening of the covariates was found to yield higher power; in addition, the use of upwind controls further ensured an even higher power. If the degree of spuriousness in the variables is unknown, then the use of principal component derived covariates is recommended over point covariates.

HISTORICAL COMPARISON

The validity of using the historical comparison approach to evaluate weather modification projects was studied. A number of statistical tests were employed and the resulting distributions of P-values were compared by using a number of pairs of long term National Weather Service stations in Illinois (Hsu, *et al.,* 1981a). From these distributions of P-values, no striking evidence of departures from uniformity was apparent. At most, there was a very vague suggestion of a tendency to smaller P-values then expected from uniformity. On the other hand, a high significance for too large P-values with excessive dispersion was discovered. Five vs. five-year comparisons suggested that the standard tests were slightly radical, whereas 10 vs. 10-year comparisons yielded apparently conservative tests. At best, we may conclude that the standard statistical techniques of comparing operational with historical precipitations are not blatantly invalid.

OPERATIONAL CRITERIA

A report by Huff and Changnon (1980) provided guidance for achieving effective, reliable evaluation of seeding results and, consequently, established credibility in these evaluations. It also provided scientific information leading to better understanding and greater skills in future weather modification operations. Four tasks were discussed – design, selection of seeding criteria, conduct of seeding mission, and collection and recording of data. A number of key issues and recommendations are presented at the end of the report. They included personnel qualifications required for operations; seeding criteria; requirements for operations for different types of seeding; needs for radar and other instrumentation; and requirements for detailed documentation of operational activities.

RADOMIZATION APPROACH TO FUTURE PIGGYBACK PROJECTS

The addition of scientific measurements to operational weather modification projects has been referred to as 'piggyback' science (Weather Modification Advisory Board, 1978b). The topic of piggyback projects is relatively recent, and because of their future potential as a viable means of conducting scientific experiments in conjunction with commercial operations, efforts were exerted to obtain a better understanding of the feasibility of piggyback experiments (Gabriel and Changnon, 1982). Some form of randomization was to be employed in this approach. Inventive ways to incorporate randomized piggyback research were studied. Two types of piggyback projects appear scientifically possible and acceptable to the users of operational weather modification. The first type utilizes partial randomization on some (rainfall) occasions before, during, or after the designated operational seeding period. The weather and climate conditions before or after would have to be very similar to the designated seeding period. Also, minimal randomization was to be undertaken, if at all possible, during the period of user need for modification.

The second type of potential piggyback research would employ randomization only during seeding operations, with sufficient frequency (percentage of seedable situations) to provide adequate statistical data for reliable evaluations. For instance, different seeding agents or different seeding rates could be compared. Combination of both the first and second type piggyback approaches is also feasible. Two examples of possible piggyback experiments and outlines of the experimental designs were explicitly given by Gabriel and Changnon (1982).

THE SYNOPTIC APPROACH

Another major effort of the project involves the evaluation of the Muddy Road Project (1975-1979) and the northwestern Oklahoma ground based project (1972-1976) employing daily and storm-by-storm synoptic weather information. For each seeded occasion, storm motion(s) and synoptic type(s) were assigned by looking at radar echo characteristics and surface weather charts obtained from the National Weather Service. We have looked at a total of approximately 150 days from the Muddy Road Project, and about 150 days (50 percent randomly selected days) from the Oklahoma project. The majority of the storms occurring over the Muddy Road Project area belonged to squall zone (23.4 percent), airmass (22.8 percent), cold front (21.0 percent), or stationary front (15.6 percent); while those occurring over the Oklahoma project belonged to airmass (33.7 percent), cold front (20.4 percent), or squall zone (17.3 percent). More than half of the storms occurring in both project areas came from the west.

KEY ISSUES IN THE EVALUATION

The fundamental principle of any evaluation of seeding effect is that of "comparison" – be they temporal, spatial, or both (Hsu, *et al.,* 1981b). Factors relevant to evaluation include choice of response variables; sampling unit; project type – rain, snow, or hail; period used for data formation; target control design; sample size – seeded and unseeded; covariates; areal sizes – target and controls; assumed seeding effect – additive, multiplicative, or varying; and statistical techniques employed. Rerandomization testing was recommended for the actual evaluation of operational cloud seeding projects (Gabriel, 1979; Hsu and Changnon, 1981; Hsu, *et al.,* 1981a). Valid comparisons can be obtained only if one avoids certain obvious biases, advertent or inadvertent. Elements essential to better evaluations are: using uniformly defined observations for seeded and nonseeded occasions; using uniform method in measuring response variables; selecting target control design and statistical technique *a priori*; adhering rigorously to a preordained protocol; documenting as completely as possible seeding operation such as seeding criteria used, methods of collecting and recording data, seeding agent amount and timing, instrumentation, seeding rate, and so on.

ACKNOWLEDGMENTS

This work was supported by the grants ATM 79-05007 and ATM 81-07027 from the National Science Foundation. The opinions and conclusions expressed here are those of the authors and do not necessarily reflect views of NSF. Appreciation is extended to Julia Chen, who provided computer programming assistance, and to Bob Scott, who performed the synoptic analyses of the Muddy Road and Oklahoma projects.

LITERATURE CITED

Achtemerier, G. L., 1981. Relationships Between Rainfall and Meteorological Surface Variables: Applications to Midwestern Operational Weather Modification. Final Report, NSF ATM79-05007, Illinois State Water Survey, Champaign, Illinois, 69 pp.

Changnon, S. A., Jr., 1980a. Is Your WEather Being Modified? J. Irrig. and Drainage Div., Proc. of Amer. Soc. of Civil Eng. 106: 37-48.

Changnon, S. A., Jr., 1980b. The Rationale for Future WEather Modification Program. Bull. Amer. Meteor. Soc. 61(6):546-551.

Changnon, S. A., Jr. and C. F. Hsu, 1981. Evaluation of Illinois Weather Modification Projects of 1976-1980: A Summary. Circular 148, Illinois State Water Survey, Champaign, Illinois, 31 pp.

Changnon, S. A., Jr., F. A. Huff, and C. F. Hsu, 1979. On the Need to Evaluate Operational WEather Modification Projects. Bull. Amer. Meteor. Soc., 60(7):770-774.

Gabriel, K. R., 1979. Re-randomization Tests on Weather Experiments. Sixth Conf. on Planned and Inadv. Wea. Mod., Amer. Meteor. Soc., Urbana, Illinois, 10-13 Oct., pp. J5-J6.

Gabriel, K. R., 1981. On the Roles of Physicists and Statisticians in Weather Modification Experimentation. Bull. Amer. Meteor. Soc. 62(1):62-69.

Gabriel, K. R. and S. A. Changnon, Jr., 1982. Piggyback Weather Experiments: Superimposing Randomized Treatment Comparisons. J. Wea. Mod. 14(1):7-10.

Haragan, D. R., 1974. Precipitation Augmentation-Problems and Progress. Water Resources Bulletin 10(3):547-554.

Howell, W. E., 1977. Environmental Impacts of Precipitation Management: Results and Inferences from Project Skywater. Bull. Amer. Meteor. Soc. 58(6):488-501.

Hsu, C. F., 1979. Two Methods of Computing Statistical Powers with Application to Weather Modification. Proc. Statist. Comp. Sec., Amer. Statist. Assoc., Washington, D.C. 243-246.

Hsu, C. F., 1981. An Annotated Bibliography on the Evaluation and Statistical Issues of Weather Modification. Contract Report 265, Illinois State Water Survey, Champaign, Illinois, 93 pp.

Hsu, C. F. and S. A. Changnon, Jr., 1981. Assessment of Summer 1979 Weather Modification Effort in Southeastern Illinois. J. Wea. Mod. 13(1):132-140.

Hsu, C. F., S. A. Changnon, Jr., F. A. Huff, and K. R. Gabriel, 1981a. The Assessment of Statistical-Physical Techniques for the Evaluation of Weather Modification Operations. Contract Report 286, Illinois State Water Survey, Champaign, Illinois, 90 pp.

Hsu, C. F., K. R. Gabriel, and S. A. Changnon, Jr., 1981b. Statistical Techniques and Key Issues for the Evaluation of Operational Weather Modification. J. Wea. Mod. 13(1):195-199.

Huff, F. A. and S. A. Changnon, Jr., 1980. Criteria for Weather Modification Operations and Effective Evaluation. Contract Report, NSF ATM 79-05007, Illinois State Water Survey, Champaign, Illinois, 22 pp.

Langmuir, I., 1948. The Production of Rain by a Chain Reaction in Cumulus Clouds at Temperatures Above Freezing. J. Meteor. 5(5):175-192.

Weather Modification Advisory Board, 1978a. The Management of Weather Resources. Volume I, "Proposals for a National Policy and Program." Report to the Secretary of Commerce, Dept. of Commerce, Washington, D.C., 229 pp.

Weather Modification Advisory Board, 1978b. The Management of Weather Resources. Volume II, "The Role of Statistics in Weather Resources Management." Report of the Statistical Task Force, Dept. of Commerce, Washington, D.C.

Westcott, N., 1979. Annotated Bibliography of Predictor Variables for Weather Modification Applications. Technical Report, Illinois State Water Survey, Champaign, Illinois, 117 pp.

A FEDERAL/STATE COOPERATIVE WEATHER MODIFICATION PROGRAM

Merlin C. Williams[1]

ABSTRACT: In 1980 NOAA initiated a federally sponsored program of research in weather modification in cooperation with two state sponsored operational programs. The general purpose of this research was to investigate the means for evaluation of large, continuous operational programs, and responded to both expressed needs by state groups and recommendations of scientific study groups. The operational programs selected were sponsored by the states of North Dakota and Utah and represent modification attempts for summer, convective precipitation augmentation, and winter snowpack augmentation. The approach to the program involved extensive design activities utilizing contractually obtained expertise for a broad sector of the field. The design activities were followed by field investigations in 1980 and 1981. Specific goals in the early stages of the program focused on definition of modification potentials and verification of appropriate hypotheses. Preliminary results are available from both of these years. Renewed funding support for activities in the two states provides the basis for continuation in 1982. This program provides an example of cost-effective research achieved by cooperative efforts with on-going operations in a technical area with potential for enhancing water supply.
(KEY TERMS: weather modification; North Dakota; Utah; Federal program.)

INTRODUCTION

Weather modification research in the United States has proceeded through a series of multiagency approaches over the past 30 plus years. These programs have also proceeded through a number of phases and philosophies ranging from emphasis on cloud physics through mathematical modeling to statistical testing. These efforts have been distinguished by a lack of definitive results that are acceptable to the scientific community with one or two exceptions. The first of these is the demonstration of the ability to effect single clouds by a number of investigators (Woodley, *et al.*, 1977; Hess, 1974). The second exception is represented by the results of the Israeli project that was conducted over a long period (Hess, 1974). It is generally agreed that the results of this project represent the most conclusive proof to date. However, definite results for areawide effects from research programs in the United States have remained elusive.

Throughout the period of research during the past 30 years, operational programs have represented a continuous background of activity in the field. These programs have been distinguished by two features:

1. The level of activity in terms of funding representing projects has remained relatively constant despite the fact that the geographical locations of these projects have changed.
2. There has been a continuous increase in knowledge, data base, and experience associated with these projects that provides a certain degree of continuity for the entire field.

These two facts provide the basis for a conclusion that the potential benefits from successful weather modification are recognized and the continuing demand represented by projects is an expression of need for a useable technology.

BACKGROUND

A number of recommendations have been made by national panels and boards going back to 1957 (U.S. Advisory Committee on Weather Modification, 1957; National Academy of Sciences, 1966) to make use of operational programs as a base for efforts to increase understanding. These recommendations have varied from proposing direct support to independent analysis of the data routinely collected by such projects. In part, the recommendations have been based on the constancy of operational activities and the desirability of achieving cost-effectiveness by sharing equipment and activities costs.

The most recent recommendation for federal/state efforts was made by the Weather Modification Advisory Board (WMAB), commonly called the Cleveland Committee, for its chairman in 1978 (Weather Modification Advisory Board, 1978). Of the large number of recommendations made by this Board, the most pertinent points from the standpoint of this discussion were:

1. Basic research to increase understanding of the cloud systems involved was a primary need.
2. Federal aid should be provided to State or local weather modification projects that present an opportunity for increasing understanding.

[1] Director, Weather Modifiation Program Office, Environmental Research Laboratories/NOAA, 325 Broadway, Boulder, Colorado 80303.

3. The development and operation by a new Federal unit of selected joint use facilities for weather modification research is needed.

While these recommendations have not and cannot be implemented completely because of budgetary constraints, the desirability of joint, cooperative activities is clearly indicated.

The efforts of the Weather Modification Program Office (WMPO) in developing joint efforts go back to 1976 when a study of perceived needs of operational programs was conducted (Dennis, 1977). This study was based in part on discussions with a number of planners and managers in states where weather modification operations were in progress or contemplated. The results of the study indicated that two needs were felt by these officials:

1. Development of evaluation techniques for operational projects is of primary importance.
2. Research to improve treatment technology is needed to provide cost-effective operations.

The study revealed that a majority of the people contacted placed these needs in the priority given, and felt federal support should be provided for this purpose.

Based on the recommendations of the WMAB and the WMPO study, planning was undertaken in 1978 for a suitable federal/state cooperative program. Planning was done in concert with key people throughout the field of weather modification. Funding to initiate a program was provided by Congress in fiscal year 1980.

GOALS

The goals developed for the program were based on the various recommendations, and took cognizance of the expressions of need discussed above. The primary goal was to conduct research necessary for the development of effective evaluation techniques for operational programs. A secondary goal of contributing to the basic research function of developing understanding resulted from the fact that the evaluation research includes many research activities that are common to a number of weather modification research efforts.

It was recognized at the onset that achievement of these goals was a long term process requiring a minimum of five years. It was also recognized that the goals established represented a first phase of a program encompassing much broader needs. Thus, the goals of the current program are not intended to limit the scope of a federal/state cooperative program. Rather, they contribute to a more comprehensive program needed to address the entire range of problems facing the development of a cost-effective operational technology.

APPROACH

The first step in the development of a long term program was to conduct a technical evaluation of the status of operational programs and provide a sound design for needed research. The basic philosophy employed was to seek inputs from as broad a sector of the weather modification scientific community as possible. A contract was awarded to the Department of Atmospheric Science (DAS) at Colorado State University to serve as coordinator for achieving broad inputs. Through their efforts, scientists, operators, and managers representing federal, state, university, and private groups were brought together to develop the design. These participants were assigned to task groups (Design Task Groups) on the basis of their expertise. The resulting task group reports were integrated into recommendations and a final report (Grant, *et al.,* 1981) by the Department of Atmospheric Sciences.

Initiation of a particular program in a state was based on proposals that reflected the comprehensive design produced by the Department of Atmospheric Sciences. Such proposals developed by state managers and operators are assumed to reflect local needs and capabilities, and be guided by the technical design of a larger peer group. The proposals in the past have included subcontracts by appropriate groups as well as direct involvement of the state agencies.

General program management and administration is provided by the Weather Modification Program Office (WMPO) of the Environmental Research Laboratories of the National Oceanographic and Atmospheric Administration (NOAA). Proposals received by this office are distributed for review of their technical content by outside scientists and staff members. Subsequently, the scientific and funding aspects of the programs are negotiated with the states to ensure soundness and fit with the overall program of weather modification in NOAA. Contractual arrangements are used as a vehicle for conducting the activities and contract oversight, monitoring and technical guidance are provided by WMPO.

PROGRAM ACTIVITIES
(Summer Convective)

Initial conditions for consideration of programs as a part of a cooperative effort included the requirements that they be on-going, have some historical background, and have significant funding support from state and local governmental levels. Two states with long term operational programs were selected – North Dakota and Utah. From an overall, national standpoint, these selections provided geographical differences and different climatic regimes typical of a broad range of problems in the field of weather modification. One summer convective precipitation regime and one winter snowfall regime were represented.

The operations program in North Dakota is concerned with both rainfall augmentation and hail suppression from convective systems. Evaluation of this program by the Design Task Groups in 1980 indicated that inadequate information was available to evaluate the applicability of the dynamic seeding hypothesis (Woodley, 1977) being used. They recommended that initial activities concentrate on verification of the appropriate hypothesis before more specific studies of evaluation techniques for seeding effects be undertaken.

First year efforts during the summer of 1980 began a radar cloud climatology for use in identifying general features in the dynamic seeding hypothesis. A second aspect to the program considered the effectiveness of the operational procedures used in delivering seeding materials to potentially suitable clouds. The cloud climatology activities addressed the long term question of the proportion of cloud systems in North Dakota that could be expected to react dynamically to seeding, while the second activities addressed the question of differences in amounts of seeding materials used in North Dakota and for other projects using the dynamic seeding hypothesis. Preliminary results from analysis of the radar observations indicate that the operational procedures used are being effective insofar as the observations permit interpretation.

Activities in the summer of 1981 considered more specific questions related to the assumed hypothesis. A six-week period representing the climatological maximum for suitable conditions was selected in June and July for field observations. The questions addressed during this period involved:

1. The applicability of the dynamic seeding hypothesis to clouds in North Dakota as determined by model predictions of growth versus actual in-cloud and remotely sensed observations of growth and precipitation.
2. Differences in microphysical characteristics of seeded and unseeded clouds.
3. Difference in microphysical characteristics of clouds resulting from different methods of seeding.

Observations were provided by radar, radiosonde, instrumented aircraft, and surface sensors.

Approximately 10 to 12 test case days were obtained during the field operations. Detailed analyses are underway for six of these test cases considered to be most suitable. These analyses consider such features as the relationship between cloud size and precipitation (both observed and predicted by the one-dimensional cloud model), time and location of precipitation formation in seeded and unseeded clouds, and liquid water versus ice concentrations in seeded and unseeded clouds. Preliminary results provide indications of differences, but point out the need for greater sample size.

Plans for the summer of 1982 provide for increasing the data base for the studies initiated in 1981. The same types of equipment will be used during a one-month field program, beginning in June. Analyses of the various observing programs will be conducted during the following year.

PROGRAM ACTIVITIES
(Winter Orographic)

A large scale operational program has been conducted by the State of Utah for the purpose of increasing snowfall over mountainous areas. This is primarily a case of dealing with a managed water resource, since the intent is to increase runoff into reservoirs for subsequent irrigation use. In this program the Design Task Group felt that an adequate hypothesis existed and evaluation studies could be undertaken at a more advanced stage than in North Dakota. The approach recommended consisted of attempting to detect signatures in both particle trajectories and water budget under seeded conditions.

The initial effort was implemented in the winter of 1980-81 utilizing a combination of surface, remote sensing, and airborne equipment to attempt to establish the signatures produced by the ground based seeding employed in the operational program. The field observational program was conducted during the February to March period of 1981. Specific questions addressed included:

1. Is there a detectable change in cloud liquid water and precipitation under seeded conditions?
2. Is there a change in cloud particle trajectories under seeded conditions?
3. Can seeding materials be detected in appropriate locations in storm systems using the existing operational techniques?
4. What statistical tests and procedures are best suited to evaluation of the program?

Despite a relatively dry season in 1981, five test case storms were observed during the field program.

Measurements were made with a microwave radiometer (liquid water), K-band radar (cloud depth), 3-cm radar (cloud characteristics), radiosonde (state parameters), instrumented aircraft (microphysical parameters), nuclei counter (seeding materials), and a variety of surface sensing and sampling equipment. Preliminary analyses have been completed and a final report is available. The preliminary results indicate that the observational program was generally adequate for addressing the questions posed. However, a larger data base is required for definitive answers.

Plans for a 1982 program provide for concentration on identifying changes in liquid water contents in clouds resulting from seeding with an array of observing systems, similar to those used in 1981; however, less emphasis will be placed on the instrumented aircraft, particle tracing, and statistical studies and more emphasis will be placed on seed and no seed identification. A two-month field program during the 1982-83 winter season is anticipated.

FUTURE PLANS

It is the policy of NOAA/Environmental Research Laboratories to respond to Congressional direction and support for such programs, and will continue to conduct them as long as this support is provided. Additional states have expressed interest in having similar programs, but no specific proposals have been developed at this time. Planning is underway to provide a comprehensive ERL plan for weather modification such that the federal/state cooperative efforts complement and extend the basic research efforts underway in that organization.

LITERATURE CITED

Dennis, A. S. and A. Gagin, 1977. Recommendations for Future Research in Weather Modification. Weather Modification Program Office ERL/NOAA, Boulder, Colorado.

Grant, L. O., P. Mielke, Jr., and T. Vonderltar, 1981. A Program for Federal/State/Local Cooperative Weather Modification Research. Department of Atmospheric Science, Colorado State University, Ft. Collins, Colorado (draft).

Hess, W. N., 1974. Weather and Climate Modification. John Wiley and Sons, New York, New York, pp. 229-273.

National Academy of Sciences, 1966. Weather and Climate Modification Problems and Prospects. Vol. I: Summary and Recommendations; Vol. 2: Research and Development. Final Report of the Panel on Weather and Climate Modification, Committee on Atmospheric Sciences, NAS-NRC Publication 1350.

U.S. Advisory Committee on Weather Control, 1957. Final Report, Vols. 1 and 2.

Weather Modification Advisory Board, 1978. The Management of Weather Resources, Vol. I. Report to the Secretary of Commerce.

Woodley, W. L., J. Simpson, R. Biondini, and J. Jordan, 1977. NOAA's Florida Area Cumulus Experiment, Rainfall Results 1970-1976. Preprints, Sixth Conference on Planned and Inadvertent Weather Modification, Champaign-Urbana, Illinois, October 10-13, 1977, American Meteorology Society, Boston, Massachusetts, pp. 206-273.

CLIMATOLOGICAL BACKGROUND STUDIES FOR OPERATIONAL WEATHER MODIFICATION

Amos Eddy[1]

ABSTRACT: An effective weather modification program requires information concerning the space-time distributions of seeding opportunities in the area of interest. The successful evaluation of such programs also requires information concerning the expected values of rainfall characteristics. The Oklahoma Climatological Survey is producing such information using: a) Manually Digitized Radar data to infer cloud climatologies as a function of time of day and week of the year, b) Daily COOP data sets with a space resolution of about 270 square miles per station to infer surface rainfall, runoff, and soil moisture distribution characteristics, and c) Rawinsonde data transformed by the Hirsch one-dimensional Great Plains Cloud model to infer the climatology of atmospheric structure. With these climatologies in hand, one can assess the time, place and frequency of occurrence of seeding opportunities under stated physical hypotheses concerning precipitation enhancement (or reduction); furthermore, he can estimate the time required to obtain a data sample adequate to evaluate the project for given statistical hypotheses. The marginal value of additional observation systems can also be estimated.
(KEY TERMS: weather modification;

INTRODUCTION

A successful State weather modification program must be based on the premise that there is a high probability of an associated net gain to that State over a given period of time. Consequently, before such a program is undertaken, knowledge must be accumulated concerning:

1. a. user benefits and user losses which could be associated with both increased and decreased rainfall,
   b. present rainfall – user product interactions, and
   c. details of the rainfall climatology relevant to these users;
2. a. quantifiable cloud physics hypotheses concerning atmospheric conditions considered to be associated with rainfall increases, and rainfall decreases,
   b. explicit seeding procedures to be associated with the cloud physics hypotheses, and
   c. details of the rainfall climatology relevant to these hypotheses;
3. a. explicit statistical hypotheses which can lead to a verification of the effects of the cloud seeding, and
   b. an experimental design specifying the observation sampling procedure to be used in order to accomplish the statistical hypothesis testing with the required degree of confidence and within the given time period.

The Oklahoma Climatologial Survey has been working toward these goals, and this paper is in the nature of a progress report on the effort.

DISCUSSION

Our users are mainly agriculture, municipalities and industry. The consumptive use of water by these sectors is discussed in more detail in the companion paper (by Eddy, OCS) in these same proceedings. Here we state simply that an appropriate rainfall climatology must be resolved in space, in time, and in intensity in order to assess user-impact.

Figure 1 shows the rainfall in the panhandle is less than optimal to support the sorghum/feed lot industry which thrives there at present by using aquifer water. In fact over the western half of the State, the timing of the rainfall and its variability are both critical to agriculture.

Figure 2 shows that this variability can have dramatic features. Weather modification which might have produced (or even been associated with) this storm would likely have been considered undesirable even though the effort might have been to decrease the rain. Figure 3 shows the gage network, the climate divisions and the location of the Manually Digitized Radar (MDR) box selected for illustration (see Meisner and Eddy, 1981).

Figure 4 implies that (at least in the late 1970's) the western part of the panhandle received most of its rainfall around suppertime in the summer. The appearance of clouds about 25 percent of the time in this season agrees with the raingage results shown in Table 1, where the 30-year summertime average percent no-rain station dats is given as 74.84 percent.

[1]Director, Oklahoma Climatological Survey, University of Oklahoma, 710 Asp, Suite 8, Norman, Oklahoma 73019. (Supported under Contract Number 0-07-81-V0144 from BUREC to Oklahoma Water Resources Board to OCS.)

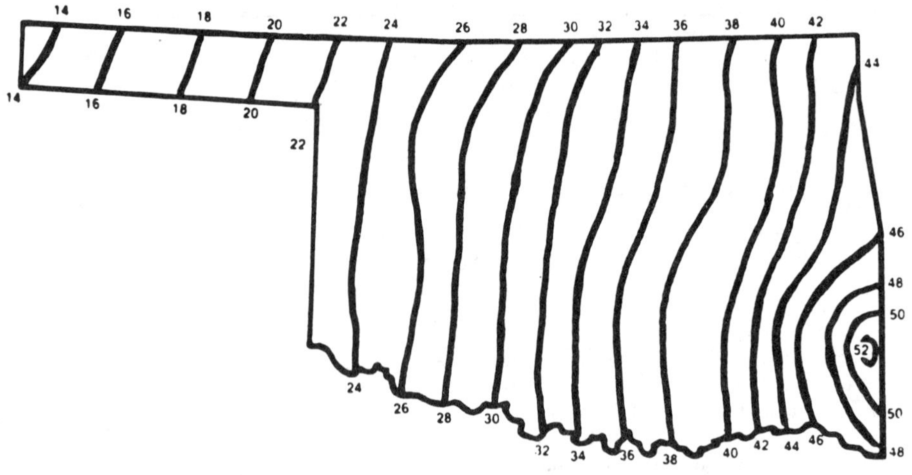

Figure 1. The Average Annual Precipitation of Oklahoma Differs Greatly from West to East.

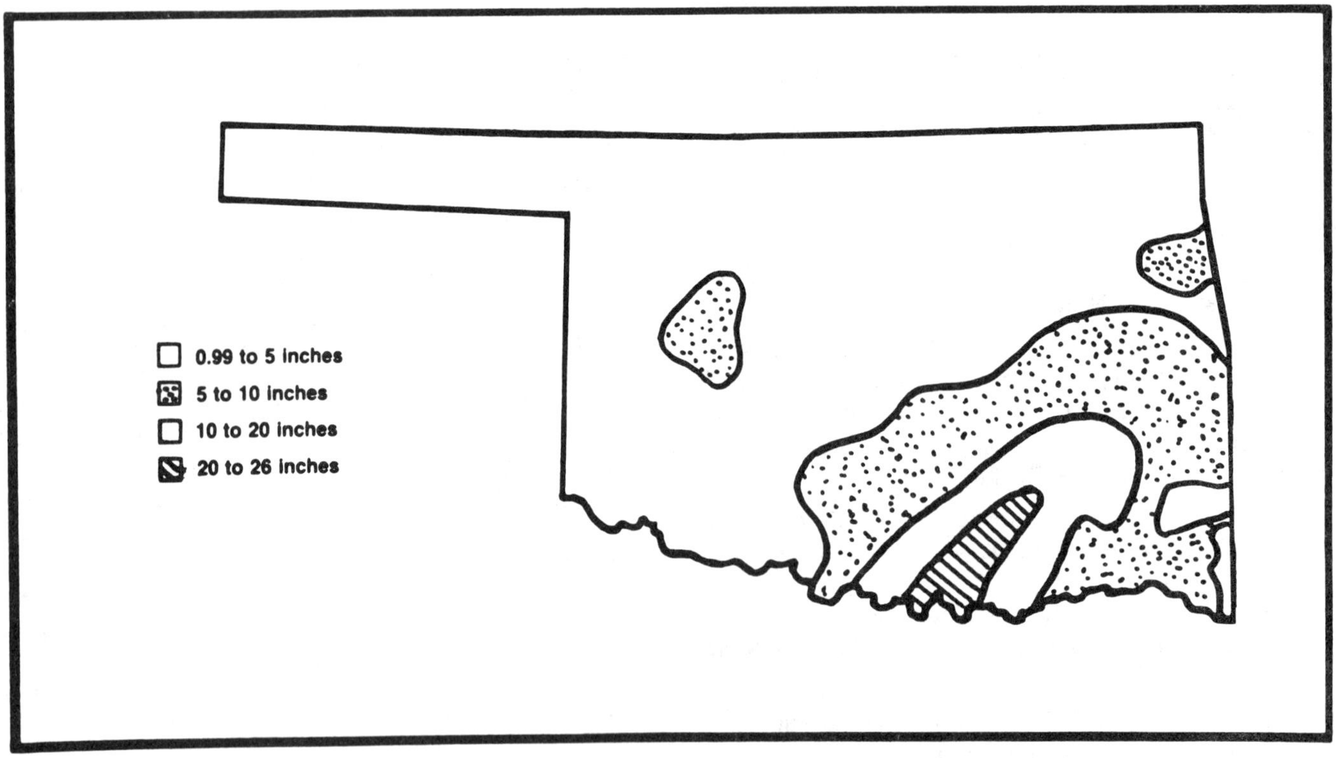

Figure 2. Over 20 Inches of Rain Fell in Some Parts of Oklahoma During the One-Week Period from October 12 to October 18, 1981. (This natural event was not the consequence of cloud seeding.) (From Pflaum and Eddy, 1982)

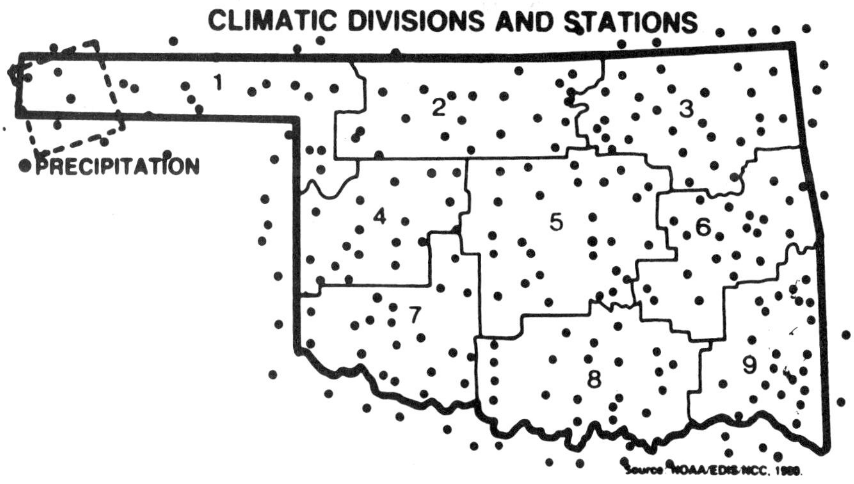

Figure 3. Dashed Line Box in Panhandle Shows Location of MDR Echoes of Figures 4 and 5.

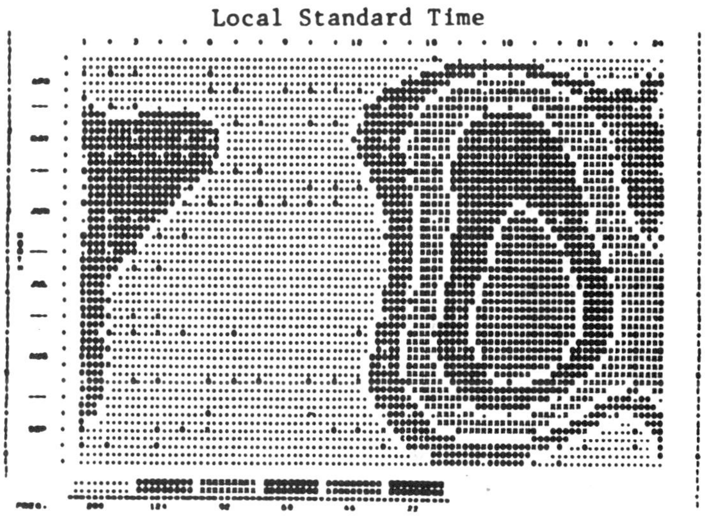

Figure 4. MDR Ranges 1-9 (Light to Moderate Rain) 1974-1980 Averages. Darkest area shows frequencies from 25-30 percent of possible occurrences. For location see Figure 3.

Figure 5 shows the MDR observations depiction of the abnormally dry summer which followed an abnormally wet spring in 1980.

Table 1 is a very small sample of the space-time-intensity variability analyses done for the State. One can note that CD1 is drier than the State average both because stations show rain less often (21.4 percent of the time vs. 24.1 percent of the time) and the intensity of the rainfall is less. (Statewide only 19.9 percent of the rainfall is logged as a trace, whereas in CD1 26.8 percent of the rain is a trace.) The very heavy rains are also very infrequent in the panhandle. Traces of rainfall in warm weather are often ideal fungus (rust) producing conditions.

Table 1 also shows analyses for the driest five years out of the 30 analyzed to indicate the rainfall regimes which one might expect when the State really needs rainfall enhancement. Note here that these five years were dry both because there were 3-4 percent fewer rainy station days and because the rainfall was less intense when it did rain. Although the former condition cannot likely be alleviated by cloud seeding, the latter possibly can.

We have established using two completely distinct data sources that the driest of years in Oklahoma usually have rain producing clouds in the critical seasons of these years.

The next question concerns the possibility of increasing the rainfall from these clouds.

Firstly, if one follows the widespread concept that increases may be possible for daily rainfall intensities of in the range from .1 inch to 1.0 inch, then Table 1 shows the frequencies of occurrence of station rainfall in these categories. Table 2 shows some of the characteristics of the 1980 dry-year clouds versus the 1976-1980 year averages as deduced by the Great Plains Cloud Model (GPCM) of Hirsch (1971). Note that in the dry year the cloud bases were about 1.6^{o}C colder and about 1300 feet higher than "normal". This implies less moisture than normal available for precipitation. One also sees that the GPCM seeding mechanism translated into only about 6 percent less potential for growth in the dry year than in the "normal" year (5-year average).

Cloud physicists are also interested in the multivariate climatology of cloud base temperature, cloud top temperature and maximum vertical motion (Todd and Howell, 1981).

Tables 3 show such climatologies for the period April-September for both 1976-1980 and 1980. Tables 4 show similar climatologies for the three summer months only of the set given in Tables 3.

TABLE 1. Relative Frequency Distributions of Daily Rainfall Intensities for Oklahoma Comparing the 1951-1980 Statistics with Those of the Driest Five Years of the Period (1952, 1954, 1956, 1963, 1966).
Summer (SU) = weeks 25-35 inclusive. Spring (SP) = weeks 10-22 inclusive. AN = annual.

	Percent of Daily Station Reports Showing No Rain	Percent of Rainy Station Days in Categories Shown								Total Number Reports
		Trace	.01"	.01" to .1"	.1" to .5"	.5" to 1.0"	1.0" to 2.0"	2.0" to 4.0"	Over 4.0"	
STATE										
Dry 5 Years										
Su	78.6									
An	79.7	23.0	3.0	23.3	29.9	12.4	6.5	1.4	.1	398,893
Sp	74.98									
30 Years										
Su	76.08									
An	75.90	19.9	2.8	22.5	30.2	13.4	8.2	2.3	.2	2,265,186
Sp	70.31									
CD 1										
Dry 5 Years										
Su	74.84									
An	82.0	30.1	4.0	25.7	28.3	7.9	3.2	.4	0	28,044
Sp	78.39									
30 Years										
Su	73.80									
An	78.6	26.8	3.9	25.4	29.1	9.2	4.3	.8	0	148,608
Sp	74.35									

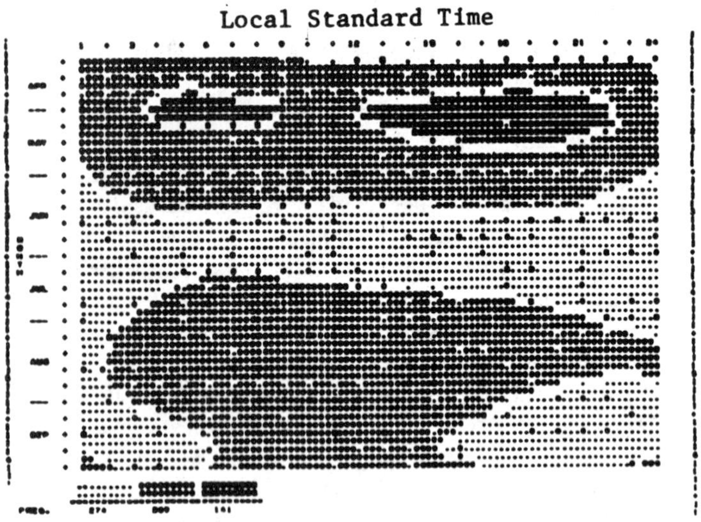

Figure 5. MDR Ranges 1-9 Giving Departure of 1980 Frequencies from 7-Year Means of Figure 3. Lightest areas show negative departures of over 5 percent. Darkest area indicates a wetter than normal spring.

Although the details are interesting, one can summarize the salient feature as follows.

During the summer season of the dry year, the GPCM analysis of the Oklahoma City R/S showed 51 days out of 92 (55 percent) to have cloud base temperatures greater than 10°C, cloud top temperatures less than −45°C, and maximum vertical motions of over 20 m/s. The 5-year average for the period gave 191 days out of 460 (42 percent) to be in this category. If this domain [CBT>10°C; CTT<−45°C; W_{MAX}>20 m/s] could be logically associated with potential seeding success or failure, then the summer of 1980 in Central Oklahoma was 25 percent more susceptible than normal to such hypothesized expectations from cloud seeding. It must be stressed that this example is used for illustrative purposes only, in order to show the potential use of one-dimensional cloud models in an operational program. The cloud physics associated with this model-implied stratification are the subject of current close examination. Note that Table 4d shows that the 1980 summer had colder than average cloud base temperatures.

The Evaluation Model

Eddy (1981) discussed an appropriate model for weather modification response variables, in this case the response variable could be daily rainfall. The model makes use of the well known analysis of covariance (ANCOVA).

TABLE 2. Some Summary Statistics for OKC 1200z R/S Analysis via GPCM.
Dynamic lifting potential = vertical growth of cloud under the GPCM seeding hypothesis.

	1976-1980				1980			
	Low	Mean	High	N	Low	Mean	High	N
Cloud Base Temperature	-16°C	10.03^{o}C	22.1^{o}C	1685	-15°C	8.4^{o}C	20.8^{o}C	321
Cloud Base Height	.36 km	2.66 km	6.78 km	1685	.4 km	3.06 km	5.72 km	321
Dynamic Lifting Potential	−.4 km	.34 km	7.2 km	1268	0 km	.32 km	5.3 km	227

$$(R - \mu_R) = A\tau + (X - \mu_X)\beta + \epsilon \, . \quad (1)$$

Response Variable	Non-Homogeneities	Covariates	Noise
Radar Reflectivities	Separate Populations		
	Seeding Treatments		

We have looked to cloud physicists, mesoscale and synoptic scale meteorologists, and weather modification experts to provide knowledge regarding the set of τ and X variables we should consider. This has led to the climatology outlined above and is being used to examine the data from a variety of projects.

SUMMARY

Admittedly, the above climatology is still incomplete. Two glaring omissions are: i) mass and moisture convergence in the ambient atmosphere climatologies, and ii) nuclei characteristics climatologies. Nonetheless, we feel we are on the right track in providing linkages between the user interests and the scientists, and between non-seeded climatological expectations and those anticipated from cloud seeding.

LITERATURE CITED

Hirsch, J., 1971. Computer Modeling of Cumulus Clouds During Project Cloud Catcher, Institute of Atmospheric Science, South Dakota School of Mines Report 71-7, 61 pp. NTIS #PB281028/AS.

Pflaum, J. C. and Amos Eddy, 1982. A Course of Action for Oklahoma. Volume 3 of the Oklahoma Climatological Survey Weather Modification Board by Amos Eddy Inc., 95 pp.

Todd, C. and W. Howell, 1981. A Seminar Series. Volume 1 of the Oklahoma Climatological Survey Weather Modification Series, pp. 100-150.

Meisner, Bernard N. and Amos Eddy, 1981. A Cloud Climatology Using Manually Digitized Radar Echoes Over Oklahoma. Volume 2 of the Oklahoma Climatological Survey Weather Modification Series, 240 pp.

TABLE 3. Frequencies of Occurrence of Days with OKC 1200 z R/S Analyses in Given Categories (via GPCM) Six Months: April through September.

Cloud Base Temperature (columns, boundaries 0°C, 5°C, 10°C, 15°C, 20°C); Cloud Top Temperature (rows, boundaries -60°C, -45°C, -30°C, -15°C)

Cloud Top Temperature	0°C	5°C	10°C	15°C	20°C	
	0	4	2	19	5	30
-60°C	5	10	15	26	10	66
-45°C	6	6	8	5	0	25
-30°C	18	15	9	3	2	47
-15°C	14	23	29	44	30	140
	43	58	63	97	47	308

Table 3a: 1976-1980 [April-September] $W_{MAX}<20$ m/s

Cloud Base Temperature

Cloud Top Temperature	0°C	5°C	10°C	15°C	20°C	
	1	7	34	108	108	258
-60°C	1	3	13	28	21	66
-45°C	0	0	0	4	3	7
-30°C	0	0	1	2	1	4
-15°C	0	0	0	1	0	1
	2	10	48	143	133	336

Table 3c: 1976-1980 [April-September] $W_{MAX}>20$ m/s

Clous Base Temperature

Cloud Top Temperature	0°C	5°C	10°C	15°C	20°C	
	0	2	0	6	0	8
-60°C	4	3	5	5	0	17
-45°C	0	2	4	0	0	6
-30°C	6	2	2	0	0	10
-15°C	5	7	7	9	3	31
	15	16	18	20	3	72

Table 3b: 1980 [April-September] $W_{MAX}<20$ m/s

Cloud Base Temperature

Cloud Top Temperature	0°C	5°C	10°C	15°C	20°C	
	1	1	9	43	15	69
-60°C	0	1	2	9	5	17
-45°C	0	0	0	1	1	2
-30°C	0	0	0	1	0	1
-15°C	0	0	0	0	0	0
	1	2	11	54	21	89

Table 3d: 1980 [April-September] $W_{MAX}>20$ m/s

TABLE 4. Frequencies of Occurrence of Days with OKC 1200 z R/S Analyses in Given Categories (via GPCM) Dry Season: June, July August.

Table 4a: 1976-1980 dry season W_{MAX}<20 m/s

Cloud Top Temperature \ Cloud Base Temperature	< 0°C	0°C–5°C	5°C–10°C	10°C–15°C	15°C–20°C	> 20°C
< -60°C	0	1	0	9	4	14
-60°C to -45°C	0	3	7	10	9	29
-45°C to -30°C	0	3	3	3	0	9
-30°C to -15°C	0	2	2	1	2	7
> -15°C	0	4	9	22	21	56
Total	0	13	21	45	36	115

Table 4b: 1980 dry season W_{MAX} <20 m/s

Cloud Top Temperature \ Cloud Base Temperature	< 0°C	0°C–5°C	5°C–10°C	10°C–15°C	15°C–20°C	> 20°C
< -60°C	0	0	0	2	0	2
-60°C to -45°C	0	2	3	3	0	8
-45°C to -30°C	0	1	1	0	0	2
-30°C to -15°C	0	1	1	0	0	2
> -15°C	0	1	3	4	3	11
Total	0	5	8	9	3	25

Table 4c: 1976-1980 dry season W_{MAX}>20 m/s

Cloud Top Temperature \ Cloud Base Temperature	< 0°C	0°C–5°C	5°C–10°C	10°C–15°C	15°C–20°C	> 20°C
< -60°C	0	0	16	72	81	169
-60°C to -45°C	0	0	4	22	16	42
-45°C to -30°C	0	0	0	2	2	4
-30°C to -15°C	0	0	1	1	1	3
> -15°C	0	0	0	1	0	1
Total	0	0	21	98	100	219

Table 4d: 1980 dry season W_{MAX}>20 m/s

Cloud Top Temperature \ Cloud Base Temperature	< 0°C	0°C–5°C	5°C–10°C	10°C–15°C	15°C–20°C	> 20°C
< -60°C	0	0	6	30	9	45
-60°C to -45°C	0	0	1	9	3	13
-45°C to -30°C	0	0	0	0	0	0
-30°C to -15°C	0	0	0	0	0	0
> -15°C	0	0	0	0	0	0
Total	0	0	7	39	12	58

JUNE INTERNATIONAL SYMPOSIUM ON HYDROMETEOROLOGY
AMERICAN WATER RESOURCES ASSOCIATION 1982

EFFECTS OF WEATHER MODIFICATION ON WATER SUPPLY FROM RESERVOIRS IN UTAH

David S. Bowles and Meg Frantz[1]

ABSTRACT: As part of Southwest Drought Research Program, the effects of a proposed standby cloud seeding program on water supply from reservoirs in Utah were estimated. Reservoir inflows for four systems of reservoirs were estimated using a procedure based on the National Weather Service (NWS) water supply forecasting procedures. Typically, these procedures comprise a multiple regression of seasonal runoff volume on rainfall, snowwater equivalent, and runoff in the previous year. It was necessary to adapt the basic NWS procedure to improve its applicability for predicting reservoir inflows in drought years. Daily hydrographs of reservoir inflow were obtained by backrouting recorded outflow hydrographs through reservoirs using records of reservoir contents and evaporation estimates. At reservoirs where outflows are not recorded, inflow hydrographs were developed using hydrograph transfer techniques. Reservoir inflow volumes and hydrographs were obtained for two standby cloud seeding designs for the drought years of 1934 and 1977. With the help of river commissioners and other knowledgeable individuals, the ownership of increases in water supply from each reservoir was determined according to the water rights system. This information was utilized as the basis for a benefit/cost evaluation of alternative proposed standby cloud seeding designs. The paper describes both the procedures for estimating water supply under various cloud seeding designs and the results in terms of estimates of water supply increases and ownership of these increases.
(KEY TERMS: hydrology; reservoirs; water rights; weather modification.)

INTRODUCTION

Background

The Southwest Drought Research Program is a cooperative effort between the U.S. Bureau of Reclamation and State agencies in Arizona, Colorado, Kansas, New Mexico, Oklahoma, Texas and Utah. Under the program, multidisciplinary research is being performed leading toward a regional capability for applying weather modification to lessen the impacts of drought. The State of Utah chose to use program funds to prepare a weather modification contingency plan under the direction of the Utah Division of Water Resources. North American Weather Consultants (NAWC) developed standby cloudseeding designs and estimated their effectiveness at augmenting precipitation. Increased precipitation was converted into greater water availability on range and dry land forms and as increased reservoir inflow by the authors, who were associated with Utah Water Research Laboratory. Economists from Utah State University evaluated the benefits of weather modification in terms of the value of increased agricultural production. An additional aspect of the Utah study was a study of drought climatology by the Utah State Climatologist.

This paper presents the hydrology component of the Utah part of the Southwest Drought Research Program. Specifically, it describes the estimation of the amount and ownership of the additional usable water from several representative Utah reservoirs that would be expected based on predictions of the effectiveness of cloudseeding. We refer to such water as "developed water". These estimates were made because they were needed as an input to the evaluation of economic benefits to agriculture resulting from the use of developed water, and also to better understand the fate of water developed from cloud seeding.

The variability in duration, intensity, geographic extent and responses to drought are characteristics which make it difficult to define a generalized "design drought", as one might define a design flood. Therefore, the meteorologic and hydrologic conditions of two historic years, 1934 and 1977, were studied as examples of the last year of a multi-year drought and a single-year drought, respectively. The effects on precipitation and snowpack of two alternative emergency seeding programs, as estimated by NAWC, were superimposed on the natural hydrology of these drought years to obtain the basis for the estimation of water availability under seeded conditions. The two emergency program designs, which are described fully by NAWC (1982), represent a conservative and an optimistic estimate of the percent increase over the amount of natural precipitation. They are referred to as the "M1" and "M2" designs, respectively.

In Utah, the quantity of water available for irrigation is a function of both the physical hydrology and the water

[1]Respectively, Branch Manager, Law Engineering Testing Company, 181 Inverness Drive, West, Suite 100, Englewood, Colorado 80112; and Graduate Research Assistant, Utah Water Research Laboratory, UMC 82, College of Engineering, Utah State University, Logan, Utah 84322.

rights allocation system. For instance, infiltration characteristics, snowpack ablation factors, groundwater flow patterns, and reservoir evaporation rates all affect the timing and quantity of water available in a reservoir. But the quantity of water available to agricultural water users is also determined by their stockholdings in reservoirs and by the seniority of their water rights. The problem, then, was to determine for the meteorologic and hydrologic conditions in 1934 and 1977, how much water was available to agriculture in those years, and to estimate the increases in that quantity which would result based on the predicted effectiveness of each proposed standby cloudseeding programs. Since the goal of the overall project was to predict how cloudseeding might be used to mitigate drought impacts in the 1980's, current water resources facilities and operating policies were assumed. Thus, in this study, runoff resulting from the 1934 and 1977 droughts was analyzed within the context of modern water resources development.

Objectives

Four "reservoir-buffered" river basins were selected for study. Objectives with respect to these systems were two-fold: 1) to determine how the quantity of water available for reservoir storage would be affected under weather modification, and 2) to determine the ownership of the developed water in accordance with established water rights and other institutional considerations.

Existing water supply forecast models were adapted to predict reservoir inflow from observed precipitation. These models were applied to the drought years of 1934 and 1977, yielding an estimate of inflows without cloudseeding. The model was then rerun to estimate inflows that would be expected if each seeding program design had been implemented under 1934 and 1977 conditions. Once hydrographs of reservoir inflows for the "no seeding" and the 'seeding" cases had been developed, the quantity of developed water was allocated according to whatever water rights currently pertain in each basin. Interviews were held with the river commissioners and other knowledgeable individuals in each area to learn the water rights and rules governing reservoir operations and the allocation of natural flow. A more detailed discussion of the procedures and results of this study is presented in Frantz and Bowles (1982).

STUDY RESERVOIRS

Since project resources did not permit the study of all Utah reservoirs, a sample of four reservoirs or reservoir systems was selected for use as indicators of the changes in water supply resulting from cloudseeding performed in various Utah climate divisions. These reservoir systems were selected on the basis of their size, importance as an irrigation water supply, and representativeness of various climate divisions. The availability of hydrometeorological data in and around each basin was also taken into consideration in the selection process.

Two single-reservoir systems, Ashley Creek in eastern Utah and the Price River drainage in central Utah, were selected together with two multiple-reservoir systems, the Weber River in northeastern Utah and the Sevier River in southcentral Utah. These four systems comprised a total of ten reservoirs.

METHODOLOGY FOR PREDICTING EFFECTS OF WEATHER MODIFICATION ON INFLOW VOLUMES

Methods Investigated

Several methods for predicting the effects of weather modification on inflow volumes were investigated. A parametric hydrology model was considered to be the most desirable approach but was eliminated from consideration because the level of effort required to properly calibrate and validate such a model for four river basins was beyond the scope of the project. The development of new regression procedures for predicting runoff occurring as reservoir inflow from precipitation and snowpack data was attempted. Ultimately, existing National Weather Service (NWS) water supply forecasting procedures were adapted to predict reservoir inflow volumes with and without cloudseeding.

National Weather Service Method

Seasonal water supply forecasting models have been developed for most Utah rivers by the NWS Colorado River Forecast Center, Salt Lake City. Personnel at the Center willingly made their work available for use in the study. The models are multiple linear regressions of the following form:

$$Q_t = b_1 + b_1 P_f + b_2 P_w + b_3 P_s + b_4 S_{we} + b_5 Q_{t-1} \quad (1)$$

in which:

Q_t = seasonal flow volume
b_i = regression coefficient, i = 0,5
P_f = fall precipitation
P_w = winter precipitation
P_s = spring precipitation
S_{we} = April 1 snowpack water equivalent
Q_{t-1} = previous year flow volume

Generally, the runoff season used by NWS is April 1 through September 30.

Procedure for Applying National Weather Service Method

The NWS precipitation-runoff relationships are available for predicting upstream flow or inflow to nearly all the selected reservoirs. Although the NWS method was developed for real-time forecasting of water supply volumes, it was applied in this study for off-line prediction of reservoir inflow volumes. It was first supposed that historic values of the independent variables for 1934 and 1977 could be input to the regression models to obtain estimates of inflow without cloudseeding; next the historic values of the independent variables would be increased by factors representing the precipitation and snowpack increases induced by cloudseeding to provide estimates of inflow with cloudseeding. The ratio of these estimates would then be computed and used to adjust the observed runoff, resulting in predictions of reservoir inflow under cloudseeding for each drought year. Two problems complicated this procedure:

1) Since the State of Utah operated a seeding program in 1977, historic values of precipitation and streamflow would first have to be reduced to represent natural conditions as a base from which to estimate inflows under the proposed cloudseeding programs.

2) The NWS models, which were developed using 20 to 30 years of data prior to 1970, do not appear to be good predictors of inflow volumes in extremely dry years, and in some cases, yield negative estimates of runoff volumes.

Clearly, the second problem was the more serious of the two since whatever approach was used to resolve the second problem could be operated in reverse in order to resolve the first problem. Resolution of the second problem was achieved by plotting predicted increases in inflow volumes for several dry years against the observed inflow volumes and then extrapolating a trend line to the observed inflows for 1934 and 1977. A sample of the 10 to 12 driest years of record, excluding those for which the NWS models yield a negative prediction, was used to prepare such plots for each reservoir inflow. Runoff was estimated using historic and augmented values of the independent variables and the difference between these two estimates was calculated as a prediction of the increase in inflow volume due to cloudseeding. Separate calculations and plots were prepared for the 1977 operational seeding program and for the two proposed seeding designs, M1 and M2. The trend lines were generally found to be either linear or parabolic in form and in most cases were forced to pass through the origin. Examples of these plots are presented in section on estimates of reservoir inflow.

Table 1 contains the NAWC estimates of the augmented seasonal precipitation expressed as a percentage of observed precipitation for each study basin and for the M1, M2 and 1977 seeding programs. Monthly precipitation input to the NWS equations was increased in accordance with these percentages. Snow-course terms were augmented by the percentage corresponding to the January-to-April period.

TECHNIQUE FOR ESTIMATING TIMING OF RESERVOIR INFLOWS

Water allocation rules in Utah are usually based on daily flows, and therefore daily inflow hydrographs were required for each reservoir. In general, reservoir inflow can be estimated from gaged outflow using the following mass balance equation:

$$I = O \pm \Delta S + E \qquad (2)$$

in which:

I = reservoir inflow
O = gaged reservoir outflow
ΔS = change in reservoir contents
E = reservoir evaporation loss

Pan evaporation data do not exist at most of the study reservoir sites. Therefore, it was necessary to estimate pan evaporation using a procedure developed by Richardson (1982).

Storage contents are recorded only bimonthly or monthly for most study reservoirs. Therefore, Equation 2 was solved on a monthly basis only. A nearby station, preferably upstream but in some cases a well-correlated gage in a different basin, was selected to characterize the distribution of daily flows. Specifically, the ratio of reservoir inflow to the nearby station flow was computed for each month. Ordinates of the nearby station hydrograph were multiplied by this ratio to provide an estimated daily reservoir inflow hydrograph. In order to avoid abrupt jumps in estimated daily flows at changes between calendar months, ratios in successive months were linearly interpolated to provide a ratio which changed smoothly from day to day.

To distribute the increase in inflow resulting from cloudseeding, it was assumed that cloudseeding would have negligible effects on the timing of runoff. Each daily flow rate from the beginning of the snowmelt season through the end of the water year was multiplied by the ratio of augmented seasonal inflow volume based on the procedure described in the section on procedure for applying NWS method to the observed seasonal inflow volume. For 1934, this resulted in two new hydrographs, one for the M1 design and another for the M2 design. For 1977, an additional hydrograph was generated, representing streamflows which would have occurred in the absence of the 1977 cloudseeding program.

SUMMARY OF METHOD FOR ESTIMATING HYDROGRAPH OF RESERVOIR INFLOWS AND CLOUDSEEDING

In summary, the procedure described in the preceeding two sections can be condensed into five steps: 1) compute

TABLE 1. Estimated Percent Increase in Precipitation Under Three Cloudseeding Programs (NAWC, 1982).

	Cloud Seeding Program					
	M1		M2		1977	
Basin	Percent	Period of Application	Percent	Period of Application	Percent	Period of Application
Weber River	10	Dec. 15 – Mar. 31	10 20	Nov. 1 – Dec. 31 Jan. 1 – May 31	15	Feb. 1 – May 31
Price River	10	Dec. 15 – Mar. 31	10 20	Nov. 1 – Dec. 31 Jan. 1 – May 31	15	Jan. 1 – May 31
Ashley Creek	[1]	[1]	5 10	Nov. 1 – Dec. 31 Jan. 1 – May 31	10	Mar. 1 – May 31
Sevier River	10	Dec. 15 – Mar. 31	10 20	Nov. 1 – Dec. 31 Jan. 1 – May 31	15	Jan. 1 – May 31

[1] M1 program is not applicable to Ashley Creek Basin.

seasonal runoff estimates under various seeding designs for a sample of dry years; 2) plot the predicted changes against the observed seasonal volumes and fit a curve; 3) enter the fitted curve with observed 1934 or 1977 seasonal volume and read the volume of the newly developed reservoir inflow water; 4) construct historic daily hydrograph by backrouting through the reservoir on a monthly basis, and distributing the flows based on those recorded at a well-correlated nearby gage; and 5) multiply ordinates of the hydrograph obtained in Step 4 by the ratio of the inflow volume with seeding to the inflow volume without seeding based on the results of Step 3. Several reservoirs did not conform precisely to each step outlined above. The special considerations in these cases are described in Frantz and Bowles (1982).

ESTIMATES OF RESERVOIR INFLOW

Table 2 summarizes the predictions of increases in seasonal inflows at each reservoir under the three seeding program designs (i.e., M1, M2, and 1977). Although most of the regression of developed water on observed reservoir inflow equations are parabolic or linear in form, the data were most satisfactorily fitted with hand-drawn curves in some cases. Examples of parabolic, linear, and hand-drawn curves are shown in Figures 1a through 1c, respectively.

It is interesting to note the similarity in inflow volumes under the M1 design and the 1977 program (see Table 2). The difference in the two seeding designs is a later startup date for the 1977 program, and a greater estimated percent increase in precipitation than for the M1 design (see Table 1).

Table 3 presents the percent increases in inflow, based on the non-seeded seasonal runoff, for each seeding design and for each of the study years. Clearly, the Weber and Price River Basins of the northern Wasatch Mountains are the most responsive to weather modification. This would be expected, given the higher elevation of the catchment areas contributing to these reservoirs, and therefore, the greater influence of snowpack on runoff and the relatively smaller losses at higher elevations. Also larger proportions of these drainage areas are subject to the orographic storms to which seeding technology is suited.

The comparatively small increases for Ashley Creek inflows are explained by the smaller proportion of high elevation, high yield area in its catchment area, and by other physical and meteorological factors which make the area less conducive to successful weather modification. These are discussed in more detail in NAWC (1982).

The Sevier River models indicate moderate gains in flow compared to those in the northern mountains. This is partly due to the diversions and consumptive use above the locations at which water supply volumes are predicted. Water generated in the mountainous boundaries of the basin may never reach Piute or Sevier Bridge Reservoirs. Another factor is the relative importance of groundwater and return flow in the Sevier Basin. Some benefits from seeding in the Sevier Basin would not be expected to be seen until years following the year of seeding, but these later year influences were not explored in this study.

ALLOCATION OF DEVELOPED WATER

General Procedure

Knowledge of the physical effects of cloudseeding on streamflow need to be converted into economic benefits if the feasibility of weather modification is to be assessed and the value of such a program justified. Such an economic evaluation, in turn, requires that one knows which users are entitled to use the developed water. Although many authors have attempted to estimate the quantity of water that might be induced by cloudseeding, few if any have analyzed its ownership. For the purposes of the economic model utilized in this project, developed water was divided

TABLE 2. Observed and Predicted Seasonal Reservoir Inflow Volumes (1000's ac. ft.).

Reservoir and Seeding Design	1934 Observed[1]	1934 Seeding	1977 Observed	1977 No Seeding	1977 Seeding
Weber River Basin					
Weber River at Oakley					
(Wanship Reservoir Inflow)	39.54		42.86		
M1		45.64			41.82
M2		55.37			50.87
1977				36.12	
Pineview Reservoir	20.59		21.17		
M1		28.36			21.10
M2		37.90			28.35
1977				15.13	
East Canyon Reservoir	4.80		7.83		
M1		7.50			8.79
M2		11.25			13.06
1977				5.70	
Echo Reservoir	9.91		16.99		
M1		13.52			17.39
M2		18.45			23.58
1977				12.59	
Lost Creek Reservoir	2.57		5.02		
M1		3.51			5.14
M2		4.78			6.97
1977				3.72	
Price River Basin					
Scofield Reservoir	9.77		7.14		
M1		15.77			5.65
M2		19.57			8.47
1977				3.83	
Ashley Creek Basin					
Ashley Creek near Vernal[2]					
(Steinaker Reservoir diversion)	24.87		23.66		
M2		28.97			25.34
1977				21.60	
Sevier River Basin					
Sevier River at Hatch					
(Upper Sevier increment)	122.72		108.61		
M1		136.22			100.91
M2		160.15			118.64
1977				90.91	
Piute Reservoir	68.91		65.90		
M1		70.04			64.65
M2					66.50
1977		73.51		64.32	
Sevier River at Gunnison	29.6		49.2		
M1		32.09			46.97
M2		36.20			52.95
1977				43.33	

[1] Observed flows in 1934 are identical to no seeding design.
[2] No M1 design is proposed for Ashley Creek.

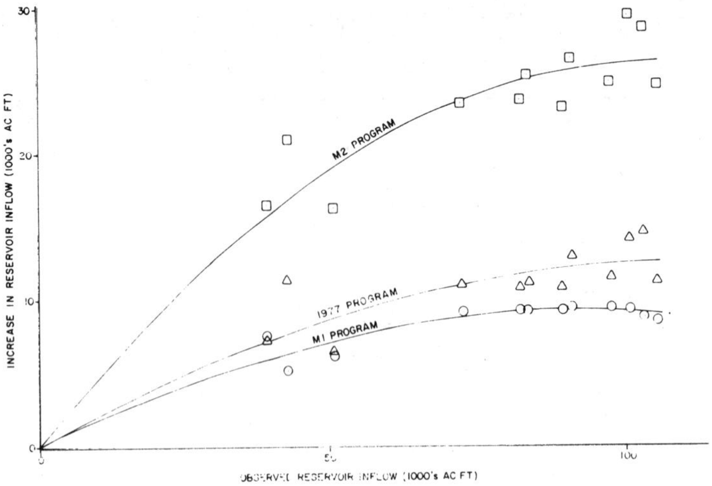

a. Parabolic Curves-Wanship Reservoir (April-September)

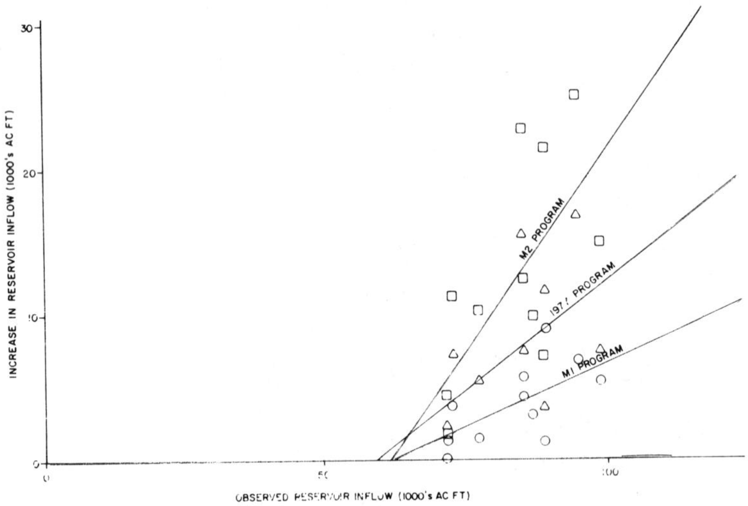

b. Linear Curves-Piute Reservoir (April-September)

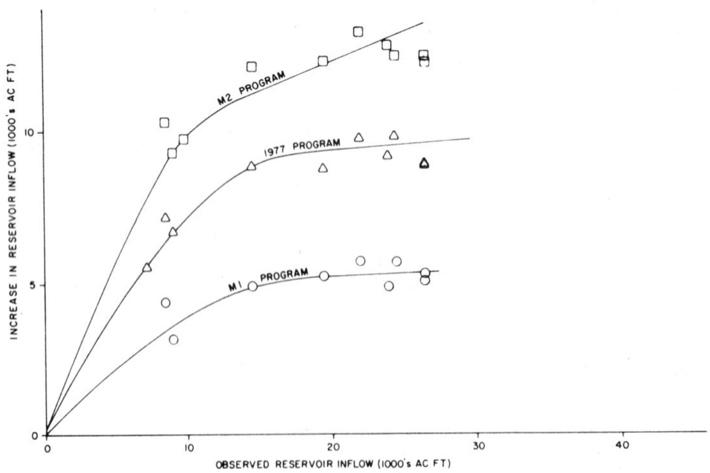

c. Hand Drawn Curves-Scofield Reservoir (April-September)

Figure 1. Increase in Reservoir Inflow vs. Observed Reservoir Inflow for M1, M2, and 1977 Programs.

TABLE 3. Percent Increases in April-September Reservoir Inflows.

Reservoir	1934 M1	1934 M2	1977 M1	1977 M2
Weber River Basin				
East Canyon Reservoir	54	129	56	134
Pineview Reservoir	37	84	40	87
Wanship Reservoir	15	40	16	41
Price River Basin				
Scofield Reservoir	61	100	48	121
Ashley Creek Basin[1]				
Ashley Creek near Vernal	–	17	–	17
Sevier River Basin				
Sevier River at Hatch	11	31	11	31
Piute Reservoir	1.6	6.7	0.5	3.4
Sigurd Gunnison	8.4	22	8.4	22

[1] M1 program is not applicable in Ashley Creek Basin.

both by climate division and by use sector (i.e., agricultural and non-agricultural).

Space permits only a general discussion of the procedure selected to determine ownership of developed water, along with its legal justification. The interested reader is referred to Frantz and Bowles (1982) for a detailed description of basin-specific rules and other water allocation policies for each reservoir system for the study years 1977 and 1934.

The allocation of developed water poses some interesting legal questions. Particularly where privately-sponsored projects are undertaken, it would seem to be desirable to separate developed water from the natural supply, and to permit the sponsor to make use of water which results from his investment. Utah statute, however, treats pragmatically the obvious difficulties in proving the precise volume of water creditable to seeding, and resolving conflicting claims of cloud seeders who believe that their areas of influence have overlapped. The Utah Weather Modification Act (Utah Code Annotated 1953, 73-15-4) states that developed rainfall ". . . shall be considered as part of Utah's basic water supply the same as all natural precipitation . . . , and all statutory provisions that apply to water from natural precipitation shall also apply to water derived from cloudseeding." Thus it is clear that hydrographs of augmented reservoir inflow obtained in this study should be allocated under the established water allocation rules which apply under non-seeded conditions.

Utah water rights, although administered by the state and based on doctrines accepted throughout the Western U.S., are locally variable. The rules reflect the historical development of a basin, its physical and geographical features, and the tradition and desires of the local users. In each basin, it was necessary to interview individuals familiar with the daily procedures of allocating water. Usually this was the river commissioner, but help was also received from officers and employees of water users groups. Despite variations in the details, the general approach taken was to examine river commissioners' records of the actual allocation in 1977, which revealed the limits of the historical supply, and areas of short-fall. The changes with respect to the historical water supply were then allocated. The amount of water use recorded in the river commissioners' reports was also noted. Basin-wide use was then adjusted by the allocation of developed water. Since the same kinds of records did not exist for 1934, the predicted supply for the observed or non-seeded conditions was allocated using the same relative distribution of water among users as occurred in 1977.

Allocation Results

An interesting finding from the study of the allocation of developed water is that in most basins during extreme drought years, natural flow users are allocated more developed water than are reservoir stockholders (see Table 4). The allocation for 1934 in the Weber basin may be anomalous and may be largely the result of assumptions made because of the lack of information for that year. Since user demand in the Weber River basin for 1934 was not known, it was assumed that all flow without seeding would be allocated as natural flow and that the developed water during April, May and half of June would be stored. A more realistic approach would require that demand be represented as a function of the supply, but a dynamic procedure of this type was beyond the scope of the project.

TABLE 4. Percent of Developed Water Which is Allocated to Reservoir Storage.

Basin	1934 M1	1934 M2	1977 M1	1977 M2
Weber	77%	77%	26%	32%
Scofield	46%	52%	28%	28%
Ashley Creek	1	16%	1	7%
Sevier	0%	0%	0%	0%

[1] M1 program is not applicable to Ashley Creek Basin.

SUMMARY AND CONCLUSIONS

Ten reservoirs in four Utah river basins were selected based on their size, importance for irrigation supply, and representativeness of Utah's climate divisions. National Weather Service seasonal volume forecast proecedures were adapted so that they could be used to estimate the volume change in runoff which might be expected under two different cloudseeding programs under the 1934 and 1977 conditions. The adaptation involved fitting a curve to plots of predicted volume increases versus observed flows for approximately ten low flow years. Hydrographs of reservoir inflows in 1934 and 1977 were generated by backrouting through the reservoir on a monthly basis, and using a nearby daily station to distribute the reservoir inflows on a daily basis. Ordinates of the daily hydrograph from the beginning of the snowmelt season until the end of September were multiplied by the ratio of the seasonal volume as predicted with cloudseeding to observed seasonal volume. In this manner, reservoir inflow hydrographs for two cloudseeding design programs, and without seeding, were obtained for the two study years. The water supplies represented by the alternative hydrographs were allocated in accordance with current water allocation, based on water rights, in order to determine the ownership of the developed water.

The estimated increases in streamflow resulting from assumed increases in precipitation in different basins vary from 3 percent in the Sevier River basin to 150 percent in the Weber River basin for the M2 program. The small increase in the Sevier River basin is explained by upstream diversion of developed water and the carryover of developed water as groundwater. The larger increases in the Wasatch range are attributable to the deeper snow accumulation and to lower consumptive losses in these northern drainage basins. The difference in the percent increase in streamflow between the two study years is very small, with the percentage for 1934 being slightly smaller in most cases.

Although there have been many attempts to estimate the quantity of water that might be induced by cloudseeding, few if any have analyzed its ownership. In this study, which examined the extreme drought years of 1934 and 1977, streamflow was so low that most water which might have been developed through cloudseeding would be allocated to natural flow users rather than to reservoir storage. In each of the four basins studied, the developed water supply would be divided among a group of users in proportion to their right within a class of rights. Exactly which users benefit from cloudseeding depends on the unique water rights system within a basin, which reflects the history of settlement, industrialization, and peculiar characteristics in the resolution of local conflicts.

ACKNOWLEDGMENTS

Work on which this paper is based was jointly funded by the U.S. Bureau of Reclamation and the Utah Division of Water Resources as part of the southwest Drought Research Program.

LITERATURE CITED

Frantz, M. and D. S. Bowles, 1982. Potential Effects of Weather Modification on Water Supply. Chapter 2 in Draft Utah Report on Southwest Drought Research Project. 75 pp.

NAWC, 1982. Chapters 5 and 7 in Draft Utah Report on Southwest Drought Research Project.

Richardson, E. A. Appendix A in Draft Utah Report on Southwest Drought Research Project.

Utah Code Annotated, 1953. S 73-15-4.

ECONOMIC VIABILITY OF WEATHER MODIFICATION WATER PROJECTS

Lewis Moore[1]

ABSTRACT: Weather modification demonstration and operational development programs can provide economically viable water projects for the 1980's. Despite the recent high interest rates and austere fiscal policy, weather modification programs are becoming more attractive because of their high projected benefit-to-cost ratios and comparatively small capital costs. While more conventional water projects are receiving increasing scrutiny and criticism on cost recovery, applied cloud seeding projects offer prompt payback periods even if interest is computed at the current prime rate.

Demonstration of operational weather modification techniques, as in the Bureau of Reclamation's proposed Colorado River Augmentation Demonstration Program, offers a variety of substantial benefits through enhancing the effectiveness of existing water projects. This program to confirm and demonstrate the capability of cloud seeding to increase the flow of the Colorado River is expected to establish a technology with a benefit-to-cost ratio of 7:1 to 10:1. By producing an annual average of 1.6 million acre-feet (MAF) in the Colorado River Basin and 0.5 MAF in adjacent basins, operational weather modification could generate an additional $30 million in additional hydropower, create $41 million benefits in salinity reduction, and $39 million in new water supplies.

An integral part of this as well as any other weather modification demonstration program is a continuing economic feasibility assessment to provide updated projections of program viability. This continuing assessment must keep the program objectives focused on the goal of augmented water supplies while continually evaluating the probable environment for the technology being demonstrated. Essentially this feasibility assessment will be a high-level management tool to refocus or even suspend the program should the projected benefits begin to fade in view of new data.
(KEY TERMS: weather modification; Colorado River Basin; economic feasibility.)

Weather Modification water projects today offer attractive possibilities for investment by both public and private interests, provided sufficient management control can be exercised to ensure probable returns outweigh development and operational costs. The Bureau of Reclamation's proposed Colorado River Enhanced Snowpack Test (CREST) to demonstrate the potential of cloud seeding in augmenting the Colorado's flow appears to offer an excellent investment for the private sector. Bureau officials are optimistic that CREST will receive the required financial support because of the high benefit-cost (B/C) ratio to projections, the limited risk involved and the opportunity to segment the program into discrete phases requiring decisions on whether to proceed.

The high B/C ratio projected by many weather modification programs is particularly important in recent periods of double-digit interest rates and corresponding fiscal restraint. High, prompt, and tangible benefits are necessary to help justify new starts or even continuation of ongoing projects since the increased cost of capital can drive water projects out of the feasible range for investment. While many water projects may have a B/C ratio not greatly exceeding unity, CREST with a projected B/C in the range of 10:1, exhibits a much greater degree of resilience to high interest rates.

In addition to its potentially high return, CREST would have comparatively little capital cost and much of the outlay necessary would be spread over the duration of the program. Complimenting the moderate initial cost of the demonstration is the ability to stop and start weather modification operations with little or no residual equipment, facilities, or environmental disruption.

CLOUD SEEDING POTENTIAL IN THE COLORADO RIVER BASIN

The Colorado River Basin offers excellent possibilities for weather modification to enhance streamflow and offset a growing water shortage. The Colorado's long term average annual flow of approximately 15 million acre-feet (MAF) is over-committed between the Upper and Lower Basins even before deducting the 1.5 MAF presently obligated to Mexico each year. Continued development of the Lower Basin and extensive, energy-related expansion in the Upper Basin are expected to result in serious and continuing water shortages before the end of the 20th century.

Three-fourths of the Colorado's flow originates in less than an eighth of the Basin area. In these high-yield, alpine areas above 9000 feet in elevation, there appear to be good possibilities for significantly enhancing snowpack through

[1]Staff Assistant for Atmospheric Resources Management, Bureau of Reclamation, Office of Liaison-Engineering and Research, U.S. Department of the Interior, Washington, D.C. 20240.

wintertime cloud seeding. The Bureau has proposed to demonstrate cloud seeding potential over these areas, 87 percent of which are public lands, to thoroughly test the possibilities of augmenting the water supply.

To test operational cloud seeding procedures and to quantify the results, CREST would conduct an initial three-year phase of site selection, studies, and equipment installation and subsequent five-year randomized cloud seeding demonstrations in two of the Colorado's subbasins with transferability investigation for other areas of the basin. Studies to develop new instrumentation to improve the ability to recognize cloud seeding opportunities and to improve seeding techniques would be conducted concurrently. The eight-year program would cost $68.4 million (in 1981 dollars) and could be totally financed through adding a fraction of a mill per kilowatt hour to the Bureau's operation and maintenance (O&M) charge for hydropower generated at its Colorado River dams.

The Bureau estimates the augmentation potential of weather modification at 1.3 MAF in the Upper Basin, 0.3 MAF on the Mogollon Rim of the Lower Basin, and 0.5 MAF in adjoining river basins as a result of seeding in the Colorado drainage. This estimate, prepared by North American Weather Consultants, is the most conservative of the three projections made for the Bureau. At the other extreme, a 1974 report by Stanford Research Institute estimated an average annual augmentation potential of 2.3 MAF in the Upper Basin (Bureau, 1980).

The Bureau has projected benefits from the North American estimate using its Colorado River Simulation System Model (Lease, 1981). In 1981 dollars, it forecast $80 million in Upper Colorado River Basin benefits from the following sources: $29 million from increased hydroelectric capacity based on a coal replacement of 20 mills per kilowatt-hour (kWh), $41 million from salinity reduction using an average value of $450,000 for each milligram per liter reduction in dissolved solids at Imperial Dam, and $10 million from water supplies to reduce deficits and for new uses.

Adding benefits could be achieved from seeding in the Lower Basin as well as in "spill-over" effects in adjacent river basins. The Bureau used a value of $50 per acre-foot of water on Arizona's Mogollon Rim and $25 per acre-foot in adjoining basins to calculate that additional benefits of over $29 million would accrue to areas outside the Upper Basin. These estimates of value appear to be comparable with other projections of benefit for augmented flow and when combined with Upper Basin benefits would be expected to produce a B/C of approximately 10:1 in an operational program (Bureau, 1982).

NEED FOR A DEMONSTRATION OF WEATHER MODIFICATION IN THE COLORADO RIVER BASIN

Demonstration of cloud seeding effects is absolutely essential before large-scale sustained operations use of weather modification is practiced in the Colorado River Basin. The Weather Modification Advisory Board (WMAB), convened by the Secretary of Commerce in 1976, reported that snowpack enhancement in the West:

> "is the most advanced, and closest to significant, broad scale operational use of all cloud seeding possibilities. A successful confirmatory experiment, including, of course, provision for assessing the seeding results must be completed before large-scale seeding of winter orographic storms is considered to be an acceptable tool available to water-resource decision makers." (WMAB, 1978)

One of the most obvious reasons for the demonstration program in the Colorado River Basin is the need to quantify the augmentation potential. The Interior Department which is charged to investigate augmentation techniques first must verify that benefits from seeding convincingly outweigh costs. Further, administrators must have proof that seeding certain clouds in a prescribed matter and under specific conditions produces predictable results and without large social or environmental costs. Finally, adjudication of the claims to augmented Colorado River water hinge on the demonstration of how and in what quantity these water resources can be enhanced.

The Administration endorses immediate implementation of CREST provided the Basin states agree to funding the demonstration program through power sales from the Bureau's Colorado River hydroelectric facilities or through any alternative method the states might want to use to reimburse program cost. The Interior Department's rationale for this "user pay" financing is that it has already invested over $50 million of public funds for the research and development of winter orographic cloud seeding techniques; consequently, the Administration contends that private sector beneficiaries ought to pay for the final steps in making this technology operational. When operational readiness is demonstrated, authorization by Congress would permit a basin-wide, self-supporting cloud seeding program for augmentation of the Colorado River.

While it is desirable that all beneficiaries contribute equally to the cost of CREST based on some pro-rata formula of benefit from water, power, reduced salinity and other amenities resulting from augmented flows, practical public policy would seem to dictate that hydropower revenues alone should be the source of funding. This is so because hydropower is by far the principal vendible benefit to be realized in augmenting the Colorado. It is consumed primarily by those who would also gain most from the more intangible benefits of streamflow augmentation and its sale offers the most expeditious means of cost recovery through existing operations and maintenance charges (Coldiron).

Inasmuch as increased hydropower offers an immediate, tangible benefit to the Colorado Basin and the adjacent allottees of hydropower, other potential revenues are much more indefinite and disputed. In the case of new water, there seem to be conflicting claims and responsibilities. While the Secretary of the Interior is charged to study means of augmenting the flow to satisfy the requirements of the 1944 Mexican Water Treaty, the Upper and Lower Basin States are feuding over which states now have the responsibility to supply the Mexican water. Also, some of the Basin States claim all water produced from weather modification within their borders. These uncertainties of who owns the water, like the questionable possibility of charging for improved water quality, seem to dictate that the technology of cloud seeding and its eventual consequences in terms of water rights and quality be demonstrated and subsequently adjudicated. In the interim, most of the new water from weather modification would be flowing through Colorado River dams and generating more power.

With the proposed user pay financing of CREST comes the need to assure the power consumers that weather modification technology is a viable investment. Assessments of economic viability will be made prior to starting each major phase of CREST which should be able to assure investors that the venture is a good one. In the event seeding potentials or the interim results are not favorable as compared with alternative uses of scarce capital, the program can be terminated and losses minimized.

WHAT IS MORE WATER REALLY WORTH IN THE COLORADO RIVER BASIN

The value of water in the Colorado Basin obviously depends on who you are, where you are, and what you consume. While the chief benefits from operational cloud seeding are more water, less salinity and greater hydropower potential, increasing the power output of the Colorado offers the most immediate, tangible opportunity for program cost recovery in the Colorado River Basin. Hence from the standpoint of an individual or a firm, the direct hydropower benefit would be the most appropriate on which to base the project's value.

Added hydropower, especially when it is consumed as peaking power or replaces oil or gas-fired generation, is the most versatile and valuable form of electrical energy. More water behind the Colorado River dams with operational weather modification means a potential 1.5 billion killowatt hours that could be put to optimum use. This assumes that required streamflow, base power generation, and other operational requirements can be covered by natural flow of the Colorado and that additional storage would be available for use with augmented flows.

The value of the extra increment of electrical energy that could be generated with augmented streamflow is critical to the economic justification of CREST and that value varies greatly depending on who is estimating and how the power is to be consumed. The Bureau's projection of 20 mills per kilowatt-hour or the approximate present cost of coal-fired, base load generation is realistic if the additional power is used to offset coal-fired thermal generation. The figure of 20 mills is also realistic given the present Federal pricing policy which limits the Bureau and the Western Area Power Administration to the recovery of operational expenses and certain project costs in delivering Colorado River hydropower. (DOE order RA 6120.2, September 20, 1979). The current contracts are delivering that energy at an average wholesale price of about 10 mills per kilowatt-hour.

Preference consumers with firm commitments of this power source are getting a bargain proportional to the effectiveness of their use of it. If these consumers could make near optimum use of additional hydropower available with weather modification, the actual value might be several multiples of the Bureau's 20 mills projection.

The following diagrams present various scenarios for CREST benefits. The net present values of CREST-added hydropower has been projected at a range of discount rates using the values of 20 and 50 mills. The former figure is what the Bureau is estimating it would recover; the latter would seem to be a fairly conservative estimate of value for the consumer. In view of other studies, and the need to justify consumer financing of CREST, it is important to emphasize just how good an investment weather modification in the Colorado Basin ought to be for the power consumers.

The future benefits of CREST consist of those realized during the demonstration program as well as during subsequent operations. A net vendible benefit of about 200 million extra kWh of hydropower with the production of 200,000 acre-feet of additional water is expected during each of the last five years of the eight-year CREST program. Subsequently, 1.3 million acre-feet of augmentation is expected with the advent of basinwide operations. The value of CREST will be enhanced by the minimization to the interval between the demonstration program and the start of operational seeding. Two scenarios have been projected in computing the present value of CREST, one with basinwide seeding starting 10 years after the start of CREST and the other 20 years later. The former assumes only a year's delay for evaluation and authorization of operation while the latter assumes an 11-year delay.

In Figure 1, future costs of CREST projected by the Bureau are reduced to present values and deducted from expected total benefits in curves A and B. In C and D, calculations for the two extremes of estimated operational cost define a range of present value for the prompt application, hydropower benefit only contingency. In Figure 2, present value was calculated assuming a future value of 50 mills/kWh for both delayed and prompt applications to test CREST sensitivity to an increased power value.

While these projections would seem to be attractive to the investor, there is a need for an agreement providing that if present allottees of Colorado River hydropower pay for

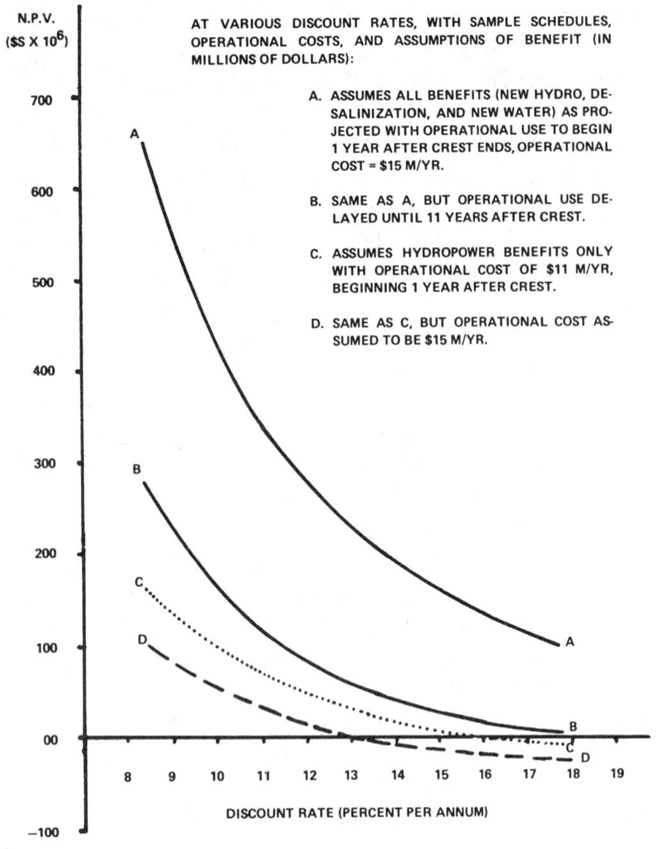

Figure 1. Colorado River Enhanced Snowpack Test (CREST) Net Present Value.

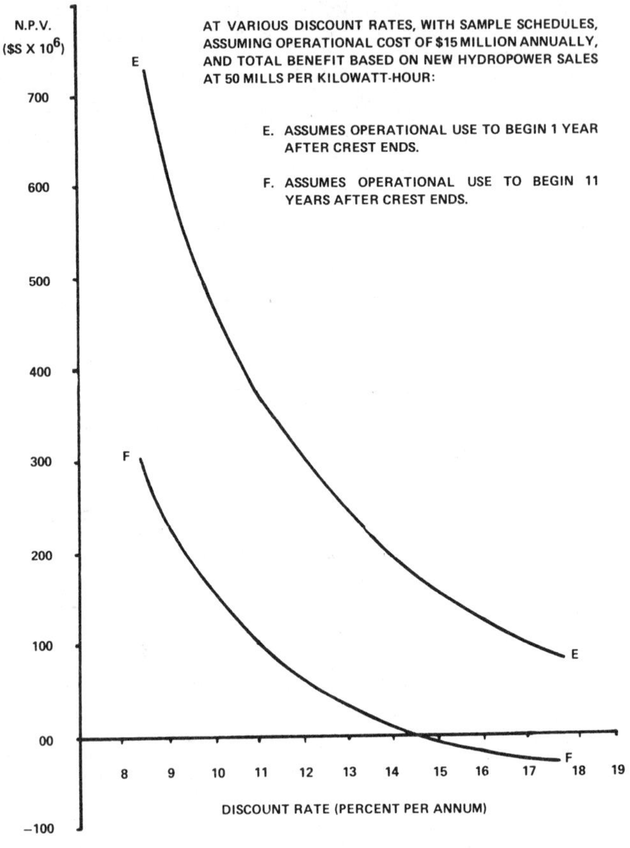

Figure 2. Colorado River Enhanced Snowpack Test (CREST) Net Present Value.

the CREST demonstration, they will receive a proportional share of benefits from new power generated. Such an agreement would also need to address the future pricing policy to be followed and define the future subsidy preference customers could expect. Under existing contracts it is not clear how additional power would be allotted or that CREST contributions would necessarily be its beneficiaries.

CONCLUSION

The continuing demand for hydroelectric energy as well as the availability of quality water is critical to the economy of the Colorado River Basin. This is the driving force behind the Bureau of Reclamation proposal to demonstrate weather modification in augmenting the Colorado's flow.

Demonstration of real benefits and operational effectiveness is required before widespread use of weather modification can be justified and issues of water rights settled. The Administration has proposed that this demonstration be financed through the sales of hydropower from the Colorado River.

The value of new hydropower to the consumers would appear to justify their investment in CREST even assuming double digit discount rates or some delay in getting an operational program authorized. What is needed now is an agreement between the Federal Government and Colorado River hydropower consumers specifying how additional power would be allotted and under what pricing policy.

CREST provides an example of the economic resiliency of weather modification projects in periods of exceptionally high interest rates. The critical assumption here is that continuing assessments of physical, legal, environmental and political factors are made and the probability of achieving the expected seeding result is realistically updated.

LITERATURE CITED

Bureau of Reclamation, Division of Atmospheric Resources Research, March 31, 1982. Briefing Document for Colorado River Enhanced Snowpack Test, Briefing presented to interested Members of Congress, Washington, D.C.

Bureau of Reclamation, Office of Atmospheric Resources Management, April 1980. A Proposal for Precipitation Management in the Colorado River Basin, Denver, Colorado.

Coldiron, William A., February 12, 1982. Weather Modification Demonstration Program Financing Through Colorado River Hydropower Revenues. Interior Department Solicitor's Opinion, Washington, D.C.

Hurley, Patrick A., October 18, 1967. Augmenting Upper Colorado River Basin Water Supply by Weather Modification. A paper prepared for the Annual Meeting of the American Society of Civil Engineers, New York.

Lease, John C., November 16-18, 1981. Simulated Impact of Weather Modification on the Colorado River. A paper prepared for the Symposium of the Aquatic Resources Management of the Colorado River Ecosystem, Las Vegas, Nevada.

Weather Modification Advisory Board, June 30, 1978. The Management of Weather Resources. Vol. 1, Final Report of the Weather Modification Advisory Board, Washington: U.S. Department of Commerce.

INTERNATIONAL SYMPOSIUM ON HYDROMETEOROLOGY
JUNE AMERICAN WATER RESOURCES ASSOCIATION 1982

STATE PERSPECTIVE ON WEATHER MODIFICATION ACTIVITIES

Barbara C. Welles[1]

ABSTRACT: This paper discusses regulatory provisions, controversies surrounding the use of weather modification and significant advances obtainable from broad communication, utilizing the experiences of Colorado State government. The Colorado mountains provide appropriate topography for winter snowpack augmentation. Interest in use of the technology is accelerated by commercial projects funded by recreation and water users and a major demonstration project proposed by the Bureau of Reclamation. Scientific and human dimension questions raised by the public and important to regulatory agencies include quantification of snowpack increases, downwind effects, and long range environmental consequences. Until understanding of the physical cloud processes is more precise from studies now possible with advanced weather-measuring equipment, the social and economic costs of weather modification can be addressed by analyzing the differences in expenses between light and heavier nonaugmented snowstorms. Research and continued development of the technology can be encouraged by combining regulation and discussion of controversial concerns with a wide variety of interest groups.
(KEY TERMS: weather modification; cloud seeding; snowpack augmentation; regulation; human dimension controveries and costs; commercial projects and research projects.)

As interest in weather and the ability to modify it increases, what are the responsibilities of the states or countries where the cloud seeding activity takes place? This paper addresses the roles of regulation, public information, and research in providing conditions in which weather modification can develop in a timely, productive manner.

REGULATION

In the United States there are federal requirements for reporting cloud seeding activity, but the actual regulation has been left to the individual states. Two-thirds of the states in the nation have passed laws controlling weather modification within their boundaries.

As in many states, regulation is important in Colorado. Because of its mountain topography, the state occupies a unique position in the United States as a location for application of winter orographic cloud seeding. If winter seeding to augment snowpack proves as successful as advocates believe it can be and if pressures to augment the flow of the Colorado River continue, the State of Colorado may be heavily impacted. Colorado includes over 70 percent of the seedable terrain in the Upper Colorado River Basin. That same terrain provides the setting for an enviable tourist industry. To assure adequate snowpack privately financed cloud seeding projects are conducted annually to improve snow conditions at ski areas and to increase spring runoff for water users.

The remarks in this paper relate to winter snowpack augmentation because of the current interest in its application in Colorado both for commercial projects and for the Bureau of Reclamation's Colorado River Enhanced Snow Test (CREST). The purpose of CREST is to augment the flow of the Colorado River to meet United States water oblications to Mexico and the growing demand for water in the basin.

STATUTORY PROVISIONS

The current Colorado Weather Modification Act, passed in 1972, directs responsibility for regulation and support of research activities to the executive director of the Colorado Department of Natural Resources.

Typical of many state weather modification statutes, the Colorado law prescribes a state advisory committee appointed by the Governor and requires licenses and permits before a specific project may be conducted. Penalty for seeding without a license or permit in Colorado is a class V felony.

Colorado, like many other states, requires the following conditions prior to issuance of a permit for a specific project:

- The project offers promise of providing economic benefit.
- If a commercial project, it is scientifically and technologically feasible.
- If a scientific or research project, it is designed to expand the knowledge and technology of weather modification.

[1] Director, Colorado Weather Modification Program, Colorado State Natural Resources Department, 1313 Sherman Street, Room 718, Denver, Colorado 80203.

- It does not involve a high degree of risk of substantial harm to land, people, health, safety, property, or the environment.
- It will not adversely affect another project.
- It is designed to minimize risk and maximize scientific gain or economic benefits to residents of the area or the state.
- It is designed to include adequate safeguards to prevent substantial damage to land, water rights, people, health, safety, or the environment.

In addition to meeting the above criteria to obtain a permit, project operators must conduct operations under additional safeguard conditions prescribed for each project.

The statute recognizes the potential economic value afforded by seeding and encourages operations, research, and development. However, state funding in most years has been minimal. With the exception of two wintertime drought relief programs, seeding has been limited to locally funded commercial projects and federally funded research.

CONTROVERSY

In spite of restrictive measures applied to cloud seeding projects, public reaction is almost always controversial. Commercial projects are funded by those believing in the potential of the technology and those who are likely to receive economic benefit. Their expectations of increased snowfall are based on federally sponsored research in the Colorado rockies in the 60s and early 70s. The studies indicate a 10 to 20 percent increase in annual snowfall for seeding conducted under specified conditions. Although unsure of precise results, sponsors believe the likelihood of benefit is so much greater than program cost that it should be tried. Opponents, however, are concerned that the benefits may not exceed the costs.

The agency responsible for seeding must address the controversy which is rarely totally absent in cloud seeding discussions. Major interest in the subject and its application frequently rises and falls with the weather. Calls for seeding wax in times of drought and wane when moisture is ample. In dry years, fewer seedable clouds are available as targets than during average or above average winters. Limited seeding is advocated by some on a regular basis in all but very wet years to maintain a higher average water supply.

Frequently legislator's constituencies include both proponents and opponents of cloud seeding and, as is to be expected in such cases, legislators often avoid taking sides. Despite limited seedable opportunities in times of drought, more legislative support can be expected in dryer years.

PROPONENTS

Interests and concerns of those who support seeding and those who question the use of the technology must be considered. Commercial project sponsors who fund winter seeding in Colorado include ski areas, summer recreation interests, e.g., river runners, small ditch companies, large municipal water providers, conservancy districts, towns and counties benefiting from tourism, resorts, and commercial operations.

SKEPTICS AND OPPONENTS

Frequently leading the opposition are residents of high mountain communities who normally experience considerable annual natural snowfall. Others are residents of communities where ranching is an important economic sector and those who believe man should leave nature well enough alone.

County commissioners from some high mountain communities, while supporting a need for additional research to determine seeding effects, worry that their counties incur additional heavy snow removal costs as a result of cloud seeding programs. They seek financial assistance in meeting those additional costs.

Those engaged in ranching question seeding on several grounds. Many believe snow will be increased on their pastures, rather than in the higher areas targeted for snow increase. Some fear programs allowed to begin in early November will make their range grass less accessible to the cattle and thus create need for earlier supplemental feeding. In the spring, ranchers worry about increasing snowfall during calving time.

PUBLIC INFORMATION

Public information and citizen understanding developed through grass roots participation and education are essential if widespread cloud seeding is to be politically acceptable. It is Colorado policy to keep the public well informed of all proposed cloud seeding or seeding related activities and to encourage public participation and interest.

If questions and concerns raised by the public, particularly in the areas of specific projects, are met in a straight forward and timely fashion, chances are improved that commercial application and research activity to determine important answers will be able to continue.

Means of informing the public used effectively in Colorado include public hearings on permit applications, state and local project advisory committees, news media releases and a public official open-door policy. Local control provisions on project start-up dates dependent upon conditions in a specific area and meetings to discuss cloud seeding in geographic areas of unique concern are also used on occasion.

WHO "OWNS" THE WATER

Many legal questions surround "developed" water and are raised as additional issues for study. Water developed for seeding for a long time to come likely will remain subject to the laws of the river as opposed to being claimed, stored, and put to use by the individual or entity paying for the seeding. Ownership of water from snowpack augmentation will depend upon interpretation by the courts of water law after sufficient verification of water produced is available for its use to be claimed.

RESEARCH AND DEVELOPMENT

From a state perspective, the government agency accountable for cloud seeding should encourage research and development of the technology and study of the socio-economic, and environmental effects of the use of that technology in the area of the agency's governance. Answers to the questions about seeding effects raised by the public are essential for effective regulation. Questions are raised by the scientific community, by those downstream and downwind, and by those who would reap the benefit of increased runoff from augmented snowpack.

TECHNOLOGICAL ADVANCES

Significant technological advances in weather-measuring equipment including C-band radars, radiosondes and dual wave length radiometers, have occurred since the Climax and Colorado River Basin Pilot Project (CRBPP) studies were completed. This potential for more precise research findings is a catalyst for new studies of physical cloud processes relative to seeding and to the human dimension aspects.

STUDY NEEDS

- From a state perspective, identification of the location and amount of increased snow resulting from ground based generators or aircraft seeding is important to maintain the confidence of the public in the potential of the technology. This information will also encourage use of the technology over a sustained period adequate to assess long term results.
- To date snowfall increases anticipated from snowpack augmentation have been estimated seasonally. Knowledge of per-storm increases will play an important role in advancing the state of the art in cost/benefit analysis of cloud seeding and developing mitigation procedures.
- The delivery capability for seeding materials and the downwind effects are important to a regulatory agency and a subject of frequent interest to citizens.
- While ecology projects indicate negligible short-term effects, the need for long term assessment is raised often by the public.
- Work toward developing an equitable process to compensate those who incur additional costs as a result of increased snowfall or runoff resulting from specific cloud seeding projects is necessary to address citizen concerns.
- At the same time research proceeds on the physical processes of winter cloud seeding, it is important to remove some of the mystery from use of this new technology by analyzing the human dimensions of natural snowfall. More precise understanding of the costs and benefits of natural snowfall will aid understanding the costs and benefits, whether real or imaginary, of cloud seeding in the future.
- Until more is known about the quantities of snow resulting from snowpack augmentation and the per storm delivery, social and economic human dimension aspects should be addressed through study of effects in below, average, and above average snowpack years. The timing of snowfall and the length of time snow remains on the ground will also affect the findings.

Perceived and actual impacts must be addressed to differentiate the realities and the myths in the questions asked about cloud seeding. Myths create difficult political problems for state and local officials.

COMMUNICATION

The mission of an agency responsible for cloud seeding is aided by interaction with as wide as possible a spectrum of the available weather modification community. The spectrum can include those whose livelihoods result from the study or utilization of the technology, such as representatives of Federal mission agencies, state agencies responsible for weather modification and institutions of higher education. Other ends of the spectrum include private weather modification firms and manufacturers of cloud seeding supplies, water interest and user groups, county and municipal government, and private citizens interested in research and commercial application of cloud seeding.

SUMMARY

In summary, a state perspective on weather modification, from Colorado as an example, emphasizes the importance of public information and encouraging research in conjunction with regulatory procedures. The three functions together enable a governing agency to best meet its responsibility to the citizens it represents.

INTERNATIONAL SYMPOSIUM ON HYDROMETEOROLOGY
JUNE AMERICAN WATER RESOURCES ASSOCIATION 1982

RESULTS OF THE SHAKEDOWN SEASON FOR THE FIRST PRECIPITATION AUGMENTATION EXPERIMENT OF THE SIERRA COOPERATIVE PILOT PROJECT

Larry Vardiman[1]

ABSTRACT: The Sierra Cooperative Pilot Project (SCPP) conducted a shakedown of an exploratory precipitation augmentation experiment in the winter of 1981-82. Equipment and procedures were tested for suitability in the experiment. Dry ice was released from an aircraft in post-frontal convection over the central Sierra Nevada. Seeding was conducted over three-hour time blocks along 20-mile legs parallel to the mountain. Seeding effects were observed by aircraft, radar, and precipitation gages.

Results of the shakedown season are described in terms of the seeding responses observed, the procedures developed, and the equipment utilized in the experiment. The modified experimental design, to be implemented in the winter of 1982-83, is described with estimates of the expected seeding effects.

(KEY TERMS: weather modification; precipitation enhancement; snowpack augmentation; cloud seeding.)

INTRODUCTION

The SCPP (Sierra Cooperative Pilot Project) conducted the shakedown of an exploratory precipitation augmentation experiment in the winter of 1981-82 based on a design developed over four previous years of experimentation, L. Vardiman and J. D. Marwitz (1981) and Vardiman (1982). Equipment and procedures were tested for suitability in such an experiment over the central Sierra Nevada. Dry ice seeding was conducted by aircraft in post-frontal convection over three-hour time blocks along 20-mile legs parallel to the mountain barrier. Seeding effects were observed by aircraft, radar, and precipitation gages. Based upon the results of this shakedown season the experimental design will be modified and hopefully a multi-year experiment will be initiated in the winter of 1982-83.

OPERATIONS

Ten experimental units were obtained during the shakedown season. Table 1 shows the dates and times of the experimental units. The experimental units were classified into three categories of seeding potential – A, B, and C. The greater the amount of supercooled liquid water in a cloud and the fewer the ice crystals, the greater the seeding potential. The ratio of liquid water content (LWC) to ice crystal concentration (ICC) in units of micrograms/crystal was used to classify the clouds. Category A had the highest seeding potential, exceeding 100 micrograms of LWC per crystal over 10 percent or more of a measured portion of an experimental region. Category B had the next highest, exceeding 10 micrograms per crystal and category C was the lowest, exceeding 1 microgram per crystal.

Unfortunately, the three most seedable occurrences of experimental conditions were not seeded because of suspensions due to flood threat or severe weather. Seeding was conducted in both B and C conditions and yielded detectable seeding signatures, at least in aircraft measurements.

TABLE 1. 1982 SCPP-1 Experimental Units

Date	Time (Z)	Seedability* Category	Seed/ No Seed**
Jan 18	2006-2306	C	S
Jan 19	2058-2358	C	S
Jan 21	1858-2158	A	NS
Jan 28	2014-2314	B	R(NS)
Mar 02	2228-0128	C	R(S)
Mar 03	2001-2301	B	R(S)
Mar 15	2007-2307	B	R(S)
Mar 16	2201-0101	C	R(NS)
Mar 18	1945-2245	A	NS
Mar 19	2020-2320	A	NS

*General Cloud Seedability (LWC/ICC is the ratio of liquid water content to the ice crystal concentration in the experimental region).
A: LWC/ICC ⩾ 100 μgm/crystal
B: LWC/ICC ⩾ 10 μgm/crystal
C: LWC/ICC ⩾ 1 μgm/crystal

**Seeding Mode
S: Seeded
NS: Not Seeded
R: Seeding Decision Randomized

[1]Christian Heritage College, 2100 Greenfield Drive, El Cajon, California 92021.

Of the ten experimental units observed, only five were randomized. The first two cases were deliberately seeded to shakedown the procedures while seeding was known to be occurring. The January 21, March 18, and March 19 cases were conducted under suspension so no seeding was possible.

MEASUREMENTS

Seeding effects were measured with three basic systems: 1) a microphysics aircraft, 2) radar, and 3) precipitation instrumentation at the ground. Sufficient observations were not obtained during 1981-82 to permit conclusive statements about seeding effects. Two to four years are required to obtain enough cases to detect the seeding signal above the natural variability. The purpose of the 1981-82 season was to shakedown the field and analysis procedures.

Although the quality of observations from prior calibration years is less than that in 1981-82, inclusion of that data allows some comparison of seeded and non-seeded cases.

Microphysics Aircraft

The microphysics aircraft observed many variables as it penetrated seeded and non-seeded clouds, (J. D. Marwitz and R. E. Stewart, 1981). Figure 1a and 1b show the average change in time of one of the primary response variables measured by the aircraft. Figure 1a shows the average ice crystal concentration measured with a Particle Measuring System 2D-C optical imaging probe in four seeded and 16 non-seeded A-type cells.

It is evident that the seeded cells had considerably more ice crystals than the non-seeded cells. In fact, after the first few minutes the seeded cells exceeded the non-seeded cells by over an order of magnitude. Figure 1b shows the same variable as Figure 1a but for C-type clouds. In this case the non-seeded cells show more ice crystals than the seeded cells.

Radar Analysis

The radar analysis for the 1981-82 season has not been completed as of this paper. However, one of the main analysis procedures, shown in Figure 2, is the percent coverage of radar reflectivity within boxes downwind of the seeding line (A. W. Huggins, 1981). Figure 2 shows boxes centered at 30 and 60 minutes after seeding for the 14 February 1979 case. The boxes are moved downwind at the speed of the cell motion and the sides of the boxes diverge to allow for dispersion of seeding material laterally. The coverage greater than 20 decibels of radar reflectivity (DBZ) at 30 and 60 minutes after seeding is 23.3 percent and 57.3 percent, respectively. Similar boxes are evaluated each 30 minutes after seeding and averaged together over the experimental unit. Percentage coverage greater than a fixed DBZ level is expected to be greater for seeded cases than for non-seeded cases.

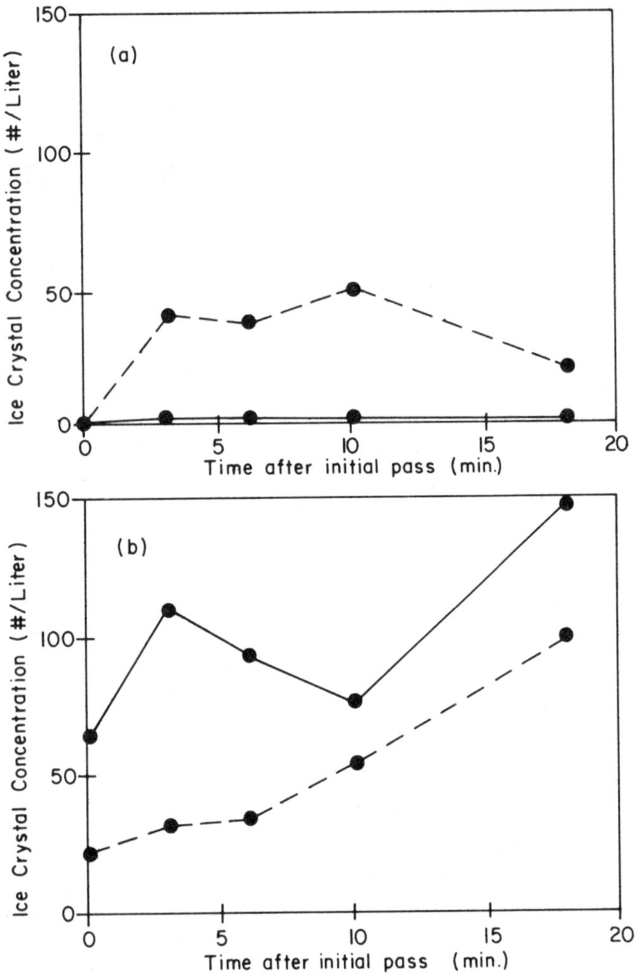

Figure 1. Average Ice Crystal Concentration Versus Time. (a) A-type cells. (b) C-type cells.

Precipitation Instrumentation

Gages – A network of approximately 40 Belfort weighing bucket rain gages was distributed over the American and Truckee-Tahoe River Basins. These gages were organized in 20 km wide regions parallel to the crest. By overlaying the expected seeding plume for each case over the gage regions, the contribution of seeding by elevation zone may be estimated. The average intensity and probability of exceeding fixed precipitation rates were calculated for each region within the seeding plume (Solak, *et. al.,* 1981).

Table 2 shows the average intensity and precipitation probability for the region 20-40 km upwind of the crest for the 10 cases observed in 1981-82. Although the average intensity is greater for the seeded cases than for the non-seeded cases, the average probability of exceeding 0 mm/hr is less. The random noise for such a small sample is too great

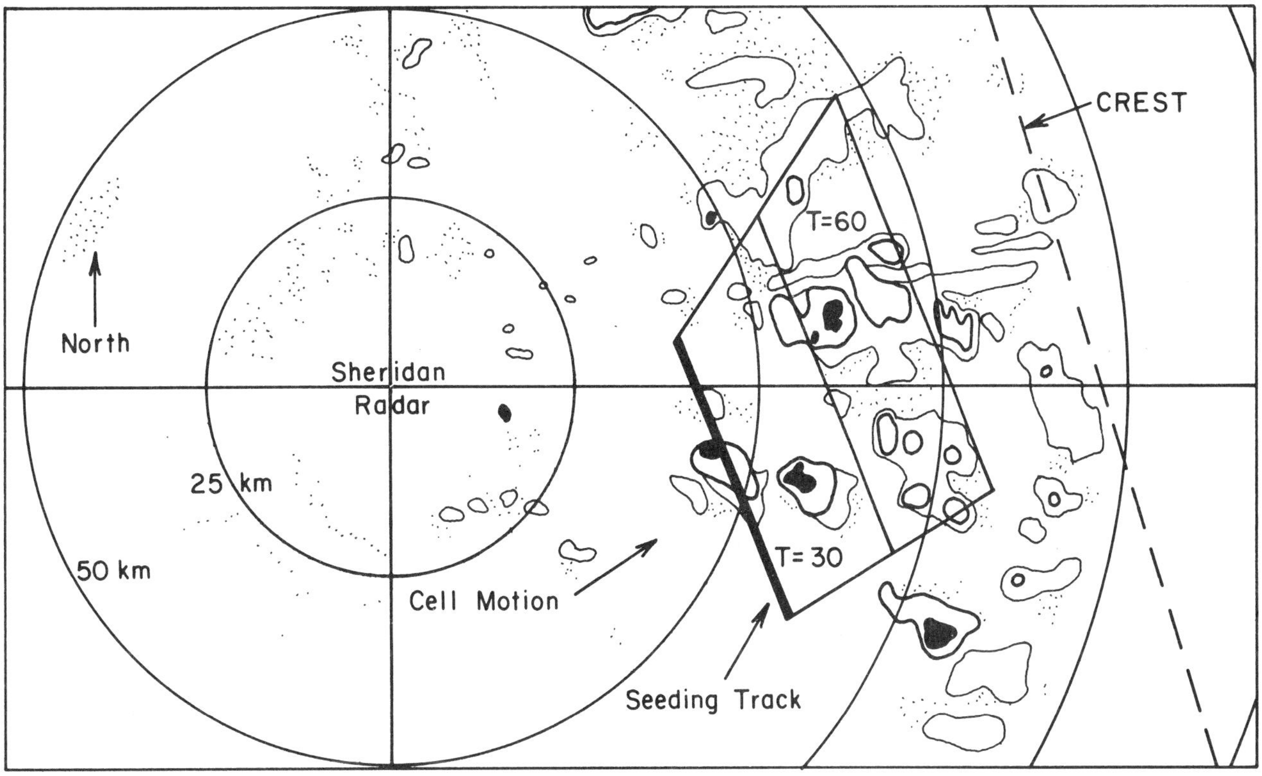

Figure 2. PPI Plot of Reflectivity on 14 February 1979 at 2044 PST. Speckles > 10 DBZ, Light Contours > DBZ, Heavy Contours > 30 DBZ, Solid > 40 DBZ.

to permit any conclusions. Estimates of the sample size needed to determine if seeding is producing a consistent effects range from 40 cases to over 100.

Ground Microphysics – A PMS 2D-C probe has been adapted for use on the SCPP to measure changes in ice crystal concentration and size distributions at the ground. This instrument has great promise for measuring seeding effects which may not be possible with standard precipitation gages.

Figure 3 shows the ice crystal concentration as a function of time when two CO_2 seeding lines were released about 40 km upwind of the instrument. The lines were released about 10 minutes apart, perpendicular to the wind, and about 40 minutes upwind of the probe. The cloud type was a uniform, orographic nature rather than the cellular convective type of SCPP-1. The relatively uniform background concentration for this type makes the two spikes at 2009 and 2021 Pacific Standard Time produced by seeding easily detectable.

CONCLUSION

The 1981-82 SCPP-1 shakedown season adequately tested the instrumentation and procedures for the experimental design of an exploratory experiment on cellular convection. With some modification to the experimental procedures, it is concluded that a multi-year experiment starting in 1982-83 may be possible. Before a final decision can be made to proceed, statistical evaluation of the period of time needed to detect effects and a benefit cost study must be completed.

LITERATURE CITED

Huggins, A. W., 1981. Classification and Distribution of Radar Echoes for the SCPP: Eighth Conference on Inadvertent and Planned Weather Modification, Reno, Nevada, pp. 36-37.

Marwitz, J. D. and R. E. Stewart, 1981. Some Seeding Signatures in Sierra Storms. J. Appl. Meteor., Vol. 20, No. 20, No. 10, pp. 1129-1144.

Solak, M. R., Allan, T. Henderson, M. Henderson, S. Pinion, E. Pinion, S. Barnes, H. Duckering, 1981. SCPP Data Collection and Analysis for the Period 1 July 1980 through 31 August 1981. Interim Progress Report No. 3, Atmospherics, Inc., WSBR Contrack No. 9-07-85-Voo, 160 pp.

Vardiman, L., 1982. The Design of SCPP-1. A Randomized Precipitation Augmentation Experiment on Winter Cellular Convection in the Central Sierra Nevada. Available from: Bureau of Reclamation, Division of Atmospheric Resources Research, D-1200, P.O. Box 25007, Denver, CO 80225.

Vardiman, L. and J. D. Marwitz, 1981. A Scenario for an Exploratory Seeding Experiment on Post-Frontal Convection in the Sierra Cooperative Pilot Project. Eighth Conference on Inadvertent and Planned Weather Modification. Reno, Nevada, pp. 82-83.

TABLE 2. Precipitation Intensity and Probability.

Date	Intensity (^{mm}/hr.)	Probability (P > 0 ^{mm}/hr.)
	Seeded Cases	
18 Jan	.90	1.00
19 Jan	.34	.61
02 Mar	.36	.80
03 Mar	.17	.35
15 Mar	.00	.00
	$\bar{I}$ = .35	$\bar{P}$ = .55
	Non-Seeded Cases	
21 Jan	.36	.66
28 Jan	.01	.21
16 Mar	.02	.50
18 Mar	.45	1.00
19 Mar	.48	.67
	$\bar{I}$ = .26	$\bar{P}$ = .61

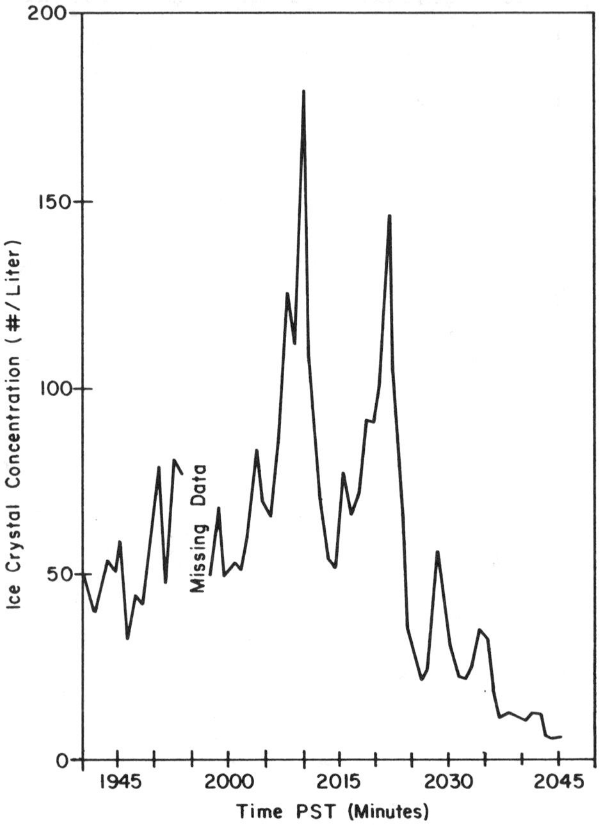

Figure 3. Ice Crystal Concentration vs. Time for Ground-Based PMS-2DC Probe – 26 February 1982.

JUNE INTERNATIONAL SYMPOSIUM ON HYDROMETEOROLOGY
AMERICAN WATER RESOURCES ASSOCIATION 1982

COUNTER-SEEDING AND HOW TO AVOID IT

Clement J. Todd and Wallace E. Howell[1]

ABSTRACT: If clouds are overseeded, the intention of increasing the precipitation may be defeated and decreases may result. The hypothesis is put forward that ice-phase overseeding may convert supercooled cloudwater into ice particles that cannot be collected by precipitation embryos. In summer cumulus, overseeding may freeze drizzle drops into ice pellets that bounce away from collisons with embryos. In winter clouds, overseeding may wipe out the supercooled water that is necessary for maintaining the high vapor pressure essential for sticky collisions among ice crystals, thus preventing them from growing into snowflakes large enough to fall to the ground. The new generation of airborne cloud-physics instruments is looked to for measurements by which this hypothesis may be tested.
(KEY TERMS: cloud seeding; overseeding; cloud physics; precipitation management; precipitation decrease.)

INTRODUCTION

For over 30 years there has been concern that it is possible to overseed clouds. There have been angry public controversies, in Pennsylvania, Maryland, Texas, Colorado, New Mexico, North Dakota, as well as elsewhere over whether cloud seeding done to increase rain or suppress hail has decreased precipitation. The principal investigator for Project Whitetop, a research project conducted on summer cumulus clouds in Missouri from 1960 to 1965, is convinced that the seeding there resulted in a decrease in precipitation (Braham, 1979). The Colorado River Pilot Project, 1970-75, produced an indeterminate result. Reexamination by Todd and Howell (1980) indicates this was due to seeding producing an almost equal mixture of increases and decreases. Many experts are concerned that there must be care not to decrease precipitation when the intention is to increase it. But generally comments are vague about the conditions for which seeding is expected to decrease precipitation.

The writers have worked to develop a system of hypotheses that defines the various conditions for which seeding should be expected to increase, have no appreciable effect, or decrease precipitation. We have reevaluated some 20 projects (work in progress) to see if we could find evidence of increases and decreases in precipitation supporting these hypotheses. Most of the projects showed evidence of increases most of the time. Seven of them showed decreases some of the time. The evidence of decreases ranged from strongly significant to weakly suggestive, in what seems to agree with the hypotheses.

We describe briefly how we think seeding can produce decreases, and how the mechanism differs between summer and winter clouds. Then we make suggestions how to learn to minimize the decreases and maximize the increases.

THE HYPOTHESES

Much operational ice-phase seeding has been done under the hypothesis that seeding would cause earlier development of more precipitation embryos or collectors and that this would make precipitation develop more efficiently. There has been concern, though, that making too many precipitation collectors would overseed. The overabundant collectors would compete with each other so that fewer would grow to be effective precipitation particles.

We did a careful reexamination of the theory of precipitation formation (Todd and Howell, 1982), and found that some very important aspects were generally overlooked. Surely collectors are important, but during most of the seeding time, nature was producing some precipitation, so there were some natural collectors. For example, we estimate that out of every hundred seeded hours in the Colorado River Pilot Project when precipitation was falling in the target, there were about 85 hours when precipitation was likewise falling so far upwind that it must have come from an unseeded part of the cloud. For many of the times that there were seeding responses, we calculated that there was not time for silver iodide to have made effective collectors. So we asked ourselves if seeding on these occasions made the collectables more collectable or less so (precipitation increases or decreases). The collectables may be cloud droplets, ice crystals, snow flakes, aggregates, or liquid or frozen drizzle. The more we looked at this, the

[1]Respectively, Consultant, 34 South Lookout Mt. Cir., Golden, Colorado 80401; and Senior Scientist, Bureau of Reclamation, E&R Center, Attention 1240, P.O. Box 25007, Denver, Colorado 80225.

more sense it seemed to make. So we worked on constructing a system of hypotheses focusing on what seeding could do to the collectables.

The details of the hypotheses are complex, and are presented in the citation above, and are expected to appear in the soon to be published Handbook in Applied Meteorology. The essence of the hypotheses we present separately for summer and winter clouds.

SUMMER CUMULUS

In summer cumulus, coalescence often turns most of the cloud water into drizzle 0.5 to 2 mm in diameter by the time it reaches –5 C to –20 C. Without seeding, at typical updraft speeds, the drizzle remains unfrozen and easily collectable. However, moderate ice-phase seeding can freeze practically all the drizzle in a minute or two near the warm end of the effective nucleation zone. Likewise, any larger drop acting as a collector would quickly freeze. We are concerned, then, that seeding in this situation will render the now-frozen drizzle uncollectable during normal collisions and hence greatly slow down the growth of precipitation. Some of the uncollectable frozen drizzle might continue to grow as collectors but in fast updrafts their likely fate is to be carried to great heights and sheared away from the active cloud to fall through dry air and be lost to evaporation.

If the drops are much larger than 3 mm in diameter when they are frozen, they will carry much of their water to the ground even if they do not grow any more and no matter how high and dry their trajectory. If they are appreciably smaller than 0.5 mm in diameter, they sweep less volume per minute, and it takes much heavier seeding to freeze them all rapidly at the warm end of the ice-nucleating range. So the same seeding rate will freeze a lesser proportion of them by the time they reach the same level in the cloud. They are smaller and have less energetic collisions, and will stick to the collectors if enough supercooled water is present to keep the collectors sticky. Thus our hypothesis suggests the a critical range for coalesence-growth when it is vulnerable to ice-phase sending. This vulnerable range is a function of the cloud base temperature, cloud top temperature, updraft speed, the wind shear, and both the cloud nuclei and giant condensation nuclei concentration in the inflow air.

By observing enough cases one should be able to verify empirically and statistically what conditions of seeding rates, cloud base temperature, etc., make clouds less effective precipitators. It would be more effective, though, to fly into the critical regions of the clouds in question and make direct observations of the progress in precipitation formation, to see if seeding was speeding it up or interfering with it.

Hygroscopic seeding, according to this hypothesis, can change the effect of simultaneous ice-phase seeding on a cloud. If the hygroscopic seeding causes drizzle to replace cloud droplets in the –5 C to –20 C zone, then it will make the cloud vulnerable to precipitation decrease by ice-phase seeding. If, on the other hand, the hygroscopic seeding causes drizzle to form and be collected into raindrops before they ascent too high, then ice-phase seeding may freeze particles that are already large enough to be sure of becoming collectors and reaching the ground.

These ideas lead to a short list of "don'ts". DON'T put ice-phase seeding in the part of a cumulus where supercooled drizzle is being collected, lest being frozen it becomes uncollectable, to be carried aloft, and lost. If ice-phase seeding is being done for the purpose of producing collectors intended to be active when the ascent rate is too rapid for drizzle to form naturally, DON'T add hygroscopic seeding that would convert cloud water to drizzle in the –5 C to –20 C zone. How is the weather modifier to know when these conditions prevail? That we discuss in a later paragraph.

WINTER CLOUDS

Winter clouds generally have colder bases so the coalesence precipitation processes are hardly started by the time cloud parcels pass through the zone where ice-phase seeding is effective. Ice-phase precipitation development by diffusional growth plays the critical role. Our detailed review of the cloud physics has convinced us that here the most frequent effect of seeding is to make the collectables more collectable so that naturally provided ice collectors can grow more effectively. But sometimes seeding has the opposite effect and makes the collectables less collectable.

Just one to three minutes after ice-nucleation an ice crystal grows to have a diameter ten times as large as the supercooled cloud drops that it robbed for its diffusional growth. By virtue of this, collectors falling through the same amount of collectable water collide with two to four times as much water substance if it is in ice crystals rather than in supercooled cloud drops. Of two identical collectors, the one falling through these ice crystals will grow two to four times as fast as the one falling through the same concentration of water substance in the form of diffusion-grown cloud droplets, if the ice crystals are caught on collision. Furthermore, the collector collecting ice crystals is less likely to shed water, as may happen to the water-collector if it cannot dispose fast enough of its heat of fusion.

Conditions for sticking versus bouncing-off have been investigated in the laboratory by Hosler, *et. al.* (1957), and Hallgren and Hosier (1960). Further research is needed, however, especially under actual cloud conditions, to give us a firmer basis for our hypotheses.

We have reason to believe, though, that low energy impact collisions of ice to ice will lead to collection whenever the crystal surfaces are wet. We expect them to be wet enough for sticking: (1) if the cloud is mixed water and ice, (2) if the cloud is all ice in rising air warmer than about −20 C, or (3) all ice in sinking air warmer than −3 C. We do not expect sticking on contact in an all-ice cloud colder than −20 C, or in any but the warmest descending all-ice clouds. Electrification must play a role, too. Electric charges that promote collision and sticking are generally strong in mixed water-and-ice clouds.

How fast a cloud turns to ice depends upon the number of active ice nuclei and the temperature at which the nucleated crystals are growing, but it also depends upon the ice multiplication rate, which is seldom well known. There is uncertainty as to where seeding will slow or stop precipitation development in winter clouds. It looks to us as though it will be when the updraft pushes appreciably above −20 C but passes through the fast crystal growth region between −12 C and −20 C slowly enough for the crystals to consume all of the supercooled cloud water before cloud parcel moves into the much slower diffusional growth region colder than −20 C. Probably about two to five meters per second is the critical speed range. Rising faster than this, the cloud will pass the fast growth region before the cloud water is consumed so there will still be a water/ice cloud colder than −20 C in this region where the resulting particles will be too small and will be lost. Rising slower than this the crystals will have the time they need to aggregate into precipitation before they reach the no-sticking region.

According to our hypothesis, there should be care in seeding for a target downwind of a mountain crest where the air is apt to be descending. If all of the cloud is turned to ice, aggregation may be stopped unless the temperature is warmer than −3 C.

Otherwise conditions for decreasing precipitation are most apt to occur as the trough passes. Here the storm gets deeper and the updrafts get stronger, and the natural aggregation zone is apt to be pushed substantially higher than −20 C into the region where seeding may stop aggregation. Seeding that starts before the trough passes is apt to flood the area with nuclei that are sucked up as the trough passes and causes precipitation loss.

MEASUREMENTS

We need measurement. The most effective measurements, for our purposes, are from cloud physics aircraft. They are expensive but if used judiciously in sampling, ought to be cost effective for commercial projects. About two weeks at a commercial project with sampling generally in the critical precipitation formation zones should reveal a lot. If this could be done on each of a large number of commercial projects, it would help each project learn when seeding can be productive and when to avoid it or when to do it differently. If this information would be summarized and generalized it would in addition provide much information needed for advancing the technology. In a short time it would provide a large number of observations from very diverse cases in a manner that big science at big projects would not.

There are about six very effective cloud physics aircraft available now owned by the universities, private companies, and the government. These could do a tremendous job of sampling the operational projects throughout the country and around the world. Such sampling would increase the cost per project by maybe 50 or 100 percent, but if the apparent value of the projects is 10 to 40 times their cost, and if this improved the performance even moderately, it would be worth it. After all doesn't major learning in every field, medicine, computers, and consumer products come from the application? And isn't the trend to get at the heart of the problem with the best instrumentation even though it is expensive?

LITERATURE CITED

Braham, R. R., 1979. Field Experimentation in Weather Modification. Journal of the American Statistical Association, 74(365): 57-104.

Hallgren, R. E. and C. L. Hosler, 1960. Preliminary Results on the Aggregation of Ice Crystals, Physics of Precipitation, Geophysical Monograph No. 5, Am. Geophysical Union, Washington, pp. 257-263.

Hosler, C. L., D. C. Jensen, and L. Goldshalk, 1957. On the Aggregation of Ice Crystals to Form Snow, Journal of Meteorology, 14:415-420.

Todd, C. J. and W. E. Howell, 1980. General and Special Hypotheses for Winter Orographic Cloud Seeding. The Journal of Weather Modification, 12(1):1-15.

Todd, C. J. and W. E. Howell, 1982. A System of Hypotheses for Precipitation Management Today. The Journal of Weather Modification, 14(1):61-74.

INTERNATIONAL ASPECTS OF ACID DEPOSITION

Ellis B. Cowling[1]

ABSTRACT: In recent decades, "acid rain" has become a dominant feature of man induced change in the chemical climate of the Earth. Wherever fossil fuels are burned, metal ores are smelted, and materials are processed on a large scale, various gaseous, aerosol, and particulate waste products are released into the atmosphere. These substances and their reaction products are dispersed by meteorological processes and deposited on vegetation, soils, and surface waters, often at great distances from the source of emissions. Concern has arisen in many countries about effects on forests, fish, crops, water quality, materials, and human health. The result is a growing concern about international exchange of air pollutants, wet and dry deposition of strong acids and other acidifying substances, as well as the associated deposition and mobilization of toxic metals and the leaching of nutrient substances.

These concerns developed originally among the Scandinavian countries, Great Britain and Germany, and other industrial nations of central Europe. More recently they have also developed between Canada and the United States. International cooperation between atmospheric scientists and biologists and among leaders of forestry, agricultural, water resource, industrial, and governmental organizations is resulting in a general concensus about both the phenomena involved and about strategies for the management of acid deposition.

(KEY TERMS: acid precipitation; Europe; North America.)

INTRODUCTION

During the past two decades, scientific research and public debate about acid deposition and its consequences in aquatic and terrestrial ecosystems have developed in many countries around the world. The most urgent concerns have been in western Europe and eastern North America. Here, regions of great sensitivity to atmospheric deposition of acidic and acidifying substances exist within a few hundred kilometers of areas of intensive industrial and urban development. Sweden, Norway, and Canada have been especially worried — each country has large numbers of lakes, streams, and rivers in which populations of sport and commercial fish have declined over a period of years. Great Britain, Germany, and other nations of central Europe and the United States have become progressively more involved as acid deposition effects have become more apparent in both their own and in neighboring countries.

Beginning about 1850, but most notable since the late 1960's, researches have been conducted in many countries on the manifold relationships and interactions among the following: combustion of fossil fuels, smelting of metals, emissions of sulfur and nitrogen oxides, transport and dispersion of air pollutants over both short and long distances, chemical transformations in the atmosphere, wet and dry deposition processes, acidity of rain and snow, deposition and mobilization of nutrient substances and toxic metal ions, changes in the chemistry of surface waters, alteration of ecosystem processes, increases and decreases in the productivity and stability of aquatic and terrestrial ecosystems, accelerated corrosion of engineering and structural materials, management of acid deposition precursors, and strategies and tactics for amelioration of acid deposition effects in various regions of the world.

The purpose of this brief paper is to introduce certain international perspectives on the scientific research and public debates about acid deposition and its effects. Primary attention will be given to European and North American experiences since they are best known; but the concepts and ideas developed here should be understood to be relevant in any part of the world in which regions sensitive to acid deposition occur within the 2- to 5-day average travel distance of air masses from regions of intense industrialization.

EARLY AWARENESS OF THE ACID DEPOSITION PHENOMENON

Since the industrial revolution began in England, it should be no surprise that the first scientific studies of acid deposition and its effects were made in Great Britain. An English chemist named Robert Angus Smith was the first to discover the phenomenon in the middle of the 19th century. In 1852, Smith (1852) published a detailed report on the chemistry of rain in and around the city of Manchester, England. In this remarkably early account, Smith called attention to the changes in precipitation chemistry as one

[1]Associate Dean for Research, School of Forest Resources, North Carolina State University, Raleigh, North Carolina 27650.

moves from the middle of a polluted city to its surrounding countryside: "We may therefore find three kinds of air, – that with carbonate of ammonia in the fields at a distance, – that with sulphate of ammonia in the suburbs, – and that with sulphuric acid, or acid sulphate, in the town." Smith also pointed out that the sulfuric acid in city air caused the colors of textiles to fade and metals to corrode.

Twenty years later, in an extraordinary book entitled "Air and Rain: The Beginnings of a Chemical Climatology," Smith (1872) first used the term "acid rain" and enunciated many of the principal ideas that are part of our present understanding. On the basis of detailed studies in England, Scotland, and Germany, Smith demonstrated that precipitation chemistry is influenced by such regional factors as combustion of coal, decomposition of organic matter, wind trajectories, proximity to the sea, and the amount and frequency of rain or snow. Smith proposed detailed procedures for the proper collection and chemical analysis of precipitation. He also noted acid rain damage to plants and materials and commented on the atmospheric deposition of arsenic, copper, and other metals in industrial regions.

Unfortunately, however, Smith's pioneering and prophetic book apparently has been overlooked by essentially every subsequent investigator of the acid deposition phenomenon. Eville Gorham (1981) developed the first detailed analysis of Smith's early work for a report by the National Academy of Sciences.

MODERN AWARENESS OF ACID DEPOSITION

Contemporary concern about acid deposition and its effects originated in three seemingly unrelated fields of science: limnology (the study of lakes), agriculture, and atmospheric chemistry.

Progress of Limnology

The relationship between rain or snow and the water in streams and lakes was obvious even to prehistoric man. But the connection between changes in the chemistry of rain and snow and alterations in the chemistry of lake and stream waters remained obscure until the middle of the 20th century.

In a long series of papers beginning in 1955, Eville Gorham (1955, 1957, 1958a-d, 1961, 1965), now Professor of Ecology at the University of Minnesota, built the major foundations for our present understanding of the causes of acid deposition and its impact on aquatic ecosystems. On the basis of research both in England and in Canada, Gorham and his colleagues demonstrated the following principles:

1) much of the acidity in precipitation near industrial regions can be attributed to atmospheric emissions produced by the combustion of fossil fuels;

2) progressive losses of alkalinity in surface waters and increases in the acidity of bog waters can be traced to the atmospheric deposition of acid substances in precipitation;

3) much of the free acidity in soils receiving acid precipitation is due to sulfuric acid;

4) the incidence of bronchitis in man can be correlated with the acidity of precipitation; and

5) fumigation with sulfur dioxide and resultant acid rain contribute to the deterioration of vegetation, soils, and lake water quality around metal smelters.

Thus, by the mid-1950's and early 1960's, Gorham was the second scientist to establish a major part of our present understanding of the sources and environmental consequences of acid deposition. But his pioneering researches, like those of Smith a century before, were met by a thundering silence from both the scientific community and the public at large. One plausible explanation is that Gorham's work, being highly interdisciplinary, was published in a diverse array of scientific journals. In any event, lack of recognition of Gorham's work resulted in a further lag in both scientific and public awareness of acid deposition.

Progress in the Agricultural Sciences

The importance of the atmosphere as a source of nutrients for the growth and development of plants was first recognized by Robert Hooke in 1687 (see also Gorham, 1965). From 1855 to 1916, scientists at the Rothamstead Experiment Station in England also demonstrated the relationship between nutrients in air and the growth of various crop plants (Way, 1855).

In the mid-1940's an imaginative Swedish soil scientist, Hans Egnér, developed a systematic way to investigate the fertilization of crops by nutrients from the atmosphere. Working from his laboratory at the Agricultural College near Uppsala, Egnér created the first 20th century network for the collection and chemical analysis of precipitation. A large number of sampling buckets were set out at experimental farms all over Sweden, and the major chemical constituents in what we now call bulk deposition (rain, snow, and dust fall) were measured on a regular monthly basis. The acidity of precipitation was one of several chemical parameters that were measured. Other agricultural scientists gradually expanded this network – first to Norway, Denmark, and Finland and later to most of western and central Europe. This network came to be called the European Air Chemistry Network and it provided the first large scale and long term data on the changing chemistry of precipitation and its importance for agriculture and forestry (Emanuelsson, *et al.*, 1954; Egnér, *et al.*, 1955). In 1956, the International Meteorological Institute in Stockholm assumed responsibility for further coordination of the network. In 1957, as part of the International Geophysical Year, it was further expanded to include Poland and much of the Soviet Union. In marked contrast to monitoring efforts elsewhere in the world, the original European Network has remained in continuing operation with more than 100 collection stations for nearly three decades.

Progress in the Atmospheric Sciences

Under the imaginative leadership of Christian Junge in Germany and Carl Gustav Rossby and Erik Eriksson in Sweden, the science of atmospheric chemistry began in Europe and later spread to North America (NAS, 1975). Junge, Rossby, and Eriksson were convinced that atmospheric processes were efficient mechanisms for the long distance as well as the short distance dispersal of many different substances. The data from Egnér's precipitation chemistry network provided the means to test various hypotheses about the trajectories of air masses, turbulent dispersal processes, and atmospheric scavenging and deposition processes (Rossby and Egnér, 1955). These observations led Eriksson to enunciate a general theory to describe the biogeochemical circulation of matter on the earth (Eriksson, 1952, 1959, 1960).

The transport and deposition of water by atmospheric processes has been well known since the earliest observations of evaporation, cloud movement, and rain. But the notion that atmospheric transport and deposition were major means for dispersal and chemical transformation of many other substances was still only a working hypothesis in the early 1950's. Junge, Rossby, and Eriksson championed these then-novel ideas and initiated various experimental tests of their hypotheses. Junge worked for a period of time in the United States (Junge and Werby, 1958). Rossby and Eriksson sponsored a series of European conferences on atmospheric chemistry and dispersal processes and attracted the interest of scientists in many other fields of inquiry including biology, forestry, agriculture, meteorology, and medicine (Eriksson, 1954). Both Eville Gorham and Svante Odén were among the students at these conferences, but they never met personally until 1975.

Integrated Studies of Limnology, Agriculture, and Atmospheric Chemistry

The first major integration of knowledge about acid deposition in these three fields of science was achieved by Svante Odén, a soil scientist at the Agricultural College near Uppsala in Sweden. In 1960, as a young colleague of Egnér, Rossby, and Eriksson, Odén started a Scandinavian network to measure surface water chemistry. When data from this network were combined with those from the European Air Chemistry Network, a series of general trends and relationships began to emerge and were published by Odén in two different media – Stockholm's prestigious newspaper *Dagens Nyheter* (Odén, 1967) and in an Ecology Committee Bulletin (Odén, 1968). The newspaper report outlining Odén's ideas about an "insidious chemical war" among the nations of Europe captured the attention of the press who began the process of public education about acid deposition in Europe. In much the same way, the Ecology Committee Bulletin stimulated scientific interest in both the acid deposition phenomenon and its ecological effects. Odén analyses of air mass trajectories and temporal and geographical changes in precipitation chemistry clearly showed that:

1) acid deposition was a large scale regional phenomenon in much of Europe with well defined source and sink regions;
2) both precipitation and surface waters were becoming more acidic;
3) long distance (100-2000 km) transport of sulfur and nitrogen containing air pollutants was taking place among the various nations of Europe;
4) there were marked seasonal trends in deposition of major ions and acidity; and
5) long term trends in acidity of precipitation could be detected in many countries of Europe.

Odén also postulated that the long term ecological consequences of acid precipitation would include the following:

1) changes in surface water chemistry,
2) decline of fish populations,
3) leaching of toxic metals from soils into surface waters,
4) decreased forest growth,
5) increased plant diseases, and
6) accelerated damage to materials.

These conclusions and hypotheses led to a veritable storm of scientific and public debate about acid deposition and its effects. Suddenly, limnological, agricultural, and atmospheric scientists began to argue and debate with each other about Odén's unconventional ideas and his general theory of atmospheric influences. Multidisciplinary discussions and international conferences ensued all over Europe and around the world as scientist after scientist was inspired (or provoked!) into designing experimental tests to prove or to disprove Odén's audacious ideas.

The Swedish government responded to the growing public and scientific controversies by initiating an inquiry which culminated in Sweden's Case Study for the United States Conference on the Human Environment – "Air Pollution Across National Boundaries: The Impact of Sulfur in Air and Precipitation" (Bolin, *et al.*, 1972).

The important ideas in both the Ecology Committee Bulletin and the Swedish Case Study were debated intensively all over Europe. Two major scientific initiatives followed in short order. The first occurred in 1972 when three organizations in Norway joined together to establish the so called SNSF Project – The Norwegian Interdisciplinary Research Programme, "Acid Precipitation – Effects on Forest and Fish." The annual budget for the SNSF Project was about 10,000,000 Norwegian Kroner (U.S. $2,000,000) per year from 1972 to 1980. This huge project had two comprehensive goals:

1) to establish as precisely as possible, the effects of acid precipitation on forests and fresh water fish; and

2) to investigate the effects of air pollutants on soils, vegetation, and water to the extent required to support the primary objective.

The project produced a steady stream of technical and scientific reports on various aspects of acid deposition and its effects. The SNSF project also sponsored two major international scientific conferences – one at Telemark, Norway, in June 1976 and the second at Sandefjord, Norway, in March 1980. Braekke (1976) and AMBIO (1976) published major reports in connection with the conference at Telemark. The conference at Sandefjord was designed to provide a forum for evaluation not only of recent research within the SNSF Project, but also the large amount of research currently being done elsewhere in the world (Drabløs and Tollan, 1980). A final report and bibliography from the SNSF Project was published at the end of 1980 (Overrein, *et al.,* 1980; Tollan, 1980).

The second major scientific initiative was an international study of the long range transport and deposition of atmospheric sulfur in eastern and western Europe. It was initiated by the Organization for Economic Cooperation and Development (OECD) in 1973-75. Brunjulf Ottar (1976) of the Norwegian Institute for Air Research provided leadership for this study. The OECD findings, published in 1977, showed that the area receiving highly acid deposition included almost all of northwestern Europe. The findings confirmed the ideas that air pollutants are transported over both short and long distances and showed that air quality in each European country is measurably affected by emissions from all other European countries (OECD, 1977).

In October 1977, the Economic Commission for Europe (ECE) originated the Cooperative Programme for Monitoring and Evaluating the Long-Range Transmission of Air Pollutants in Europe, which is now in operation. The Economic Commission has nearly completed formulating a multinational convention governing nations' responsibilities for combating the long range transboundary transport of air pollutants.

In June 1982, the Swedish government sponsored the 1982 Stockholm Conference on Acidification of the Environment. This major international conference consisted of two scientific Expert Meetings followed by a Ministerial Conference attended by representatives from 28 other countries in Europe and North America. The conclusions from the Expert Meeting and the preconference document, "Acidification Today and Tomorrow" (Swedish Ministry of Agriculture and Environment, 1982) provided the scientific foundation for a major new advance in international understanding and commitments about: 1) the nature and magnitude of biological effects, and 2) the ECE Convention and other approaches that could be used to achieve significant decreases in emissions of acid deposition precursors, especially sulfur dioxide. A prominant role in the Ministerial Conference was played by representatives of various central European countries who acknowledged, some for the first time, that control of long distance transport at its source was both necessary and desirable, and that the magnitude of desirable decreases in emissions would be of the order of 50 percent of 1980-81 emissions.

SCIENTIFIC AND PUBLIC AWARENESS IN NORTH AMERICA

Concern about acid deposition and its ecological effects in North America developed first in Canada and then later in the United States. Initial interests were focused on the effects of sulfur dioxide exposure and associated wet and dry deposition of acids and heavy metals in the vicinity of metal smelting and sintering operations (Katz, 1939; Gordon and Gorham, 1963), especially those near Sudbury, Ontario (Gorham and Gordon, 1960; Hutchinson and Whitby, 1974). During the early 1970's, interest spread to other parts of Canada as declining fish populations were discovered in more and more lakes, streams, and rivers in southern Ontario and Nova Scotia remote from local sources of atmospheric sulfur (Beamish and Harvey, 1972).

The first detailed research about precipitation chemistry in the United States was completed by MacIntrye and Young (1923). Emphasis was given to the importance of airborne nutrients for the growth of crops. This work was followed by the work of Junge and other atmospheric scientists during the 1950's (Junge and Werby, 1958). The first regional monitoring network for precipitation chemistry was maintained by a group of State Agricultural Experiment Station scientists from 1953-55 (Jordan, *et al.,* 1959). The first national monitoring program was established under the auspices of the Air Pollution Program within the Public Health Service Laboratory at Cincinnati, Ohio. The data for 1960-66 was summarized by Lodge, *et al.* (1968). As has been the case for all such studies in North America, however, these early programs were redirected and/or terminated so that no continuing records are available of long term trends in precipitation chemistry.

Scientific and public interest in acid deposition and its ecological consequences were stimulated in North America by Svante Odén with a series of 14 lectures at various institutions in the United States during the fall of 1971, and also by Torsten Ahl and Odén at the 19th International Limnological Congress in Winnipeg, Manitoba, in 1974. A series of publications by Gene Likens, Charles Cogbill, James Galloway, Carl Schofield, and others, provided further stimulus (Likens, *et al.,* 1972; Cogbill and Likens, 1974; Galloway and Likens, 1976; Likens, 1976; Schofield, 1976; Galloway, *et al.,* 1978; Likens, *et al.,* 1979). Experimental studies of various biological effects of acid deposition were initiated at Cornell, North Carolina State, and other universities. David Shriner's (1978) dissertation demonstrated both direct injury to vegetation and various indirect effects through pathogens and parasites. Carl Schofield's (1976) research on the extinction of fish populations in the Adirondack Mountains was especially alarming.

Growing awareness of the important impacts of acid deposition on fish populations and potential effects on forests led the U.S. Forest Service to sponsor the First International Symposium on Acid Precipitation and the Forest Ecosystem in May 1975. The Proceedings of this Symposium and the associated Workshop Reports were published by Dochinger and Seliga (1976a, b). At Congressional hearings in July 1975, Cowling (1976) testified about the inadequacy of research on acid deposition in the United States. Specifically, he said that the lack of a coordinated research program on ecological effects and the absence of a stable monitoring network were the primary causes of the profound ignorance in North America about acid deposition and its effects.

Both Canada and the United States have recently started long term programs for the chemical analysis of acid deposition. The Canadian Network for Sampling Precipitation began in 1976; the National Atmospheric Deposition Program was started in the United States in 1978 (Galloway and Cowling, 1978). By early 1982, some 50 sampling stations were operating in Canada and 100 in the United States.

The data from these programs (Figure 1) show that the area receiving acid precipitation now embraces about two-thirds of the total land area of North America. Although sulfuric acid has been found to be the dominant source of atmospheric acidity both here and in Europe, nitric acid accounts for almost one-third, and the fraction is rising.

In 1978, the governments of Canada and the United States established a Bilateral Research Consultation Group on the Long-Range Transport of Air Pollutants to coordinate the exchange of scientific information on acid deposition. This group documented the transboundary flow of air pollutants in eastern North America, and showed that about 11 times more oxides of nitrogen and two to four times more sulfur oxides are transported from the United States to Canada than the reverse (Altshuller and McBean, 1979, 1980).

During the late fall of 1978, the U.S. Congress passed a resolution calling for bilateral discussions with Canada to preserve and protect mutual air resources. A joint statement on the issues was prepared in July 1979, and on August 5, 1980, the governments of Canada and the United States signed a Memorandum of Intent to develop a bilateral agreement on transboundary air pollution including "the already serious problem on acid rain." To provide a suitable technical and scientific foundation for the formulation of such an agreement, five bilateral work groups were established to prepare scientific reports on the following specific aspects of the problem:

1) impact assessment;
2) atmospheric modeling;
3) strategies development and implementation;
4) emissions, costs, and engineering assessments; and
5) legal, institutional arrangements, and drafting.

Preliminary drafts of the first four of these reports were completed in early 1982 and are presently being reviewed in preparation for their submission to the Canadian Department of External Affairs and the United States Department of State.

FUTURE INTERNATIONAL RESEARCH AND NEGOTIATIONS

In May 1981 the following steps leading to the development of a management plan for acid deposition and its effects were submitted in testimony before a subcommittee of the United States Senate Committee on Environment and Public Works:

(1) Determine the geographical location of sensitive natural resources and materials (lakes, streams, crop lands, forests, man-made structures, etc.) which are now being affected by acid deposition or may be affected in the future.

(2) Identify those resources which society wishes to protect.

(3) Conduct a comparative analysis of various strategies by which the protection desired by society might be achieved. Alternative strategies would include the following:

(a) decrease the amount of emissions leading to acid deposition;
(b) use ameliorative treatments such as liming of lakes or application of protective coatings to materials; or
(c) develop acid tolerant forms of life and structural materials.

(4) Assuming that decreased emissions is the alternative of choice, as suggested in the recent Atmosphere-Biosphere Report of the National Academy of Sciences (1981), determine the amount by which acid deposition will need to be decreased in order to avoid undesired effects on the resources selected for protection.

(5) Determine where and by how much emissions of acid deposition percursors will have to be decreased in order to achieve the extent of decrease in deposition that will be required in the regions selected for protection.

(6) Evaluate specific management strategies and tactics by which each selected source region could do its part to achieve the total emissions decrease that would be necessary. Possible methods would include the following:

(a) conserve fossil fuels;
(b) alter the mix of energy sources to include a higher proportion from nonpolluting sources such as solar or possibly nuclear;

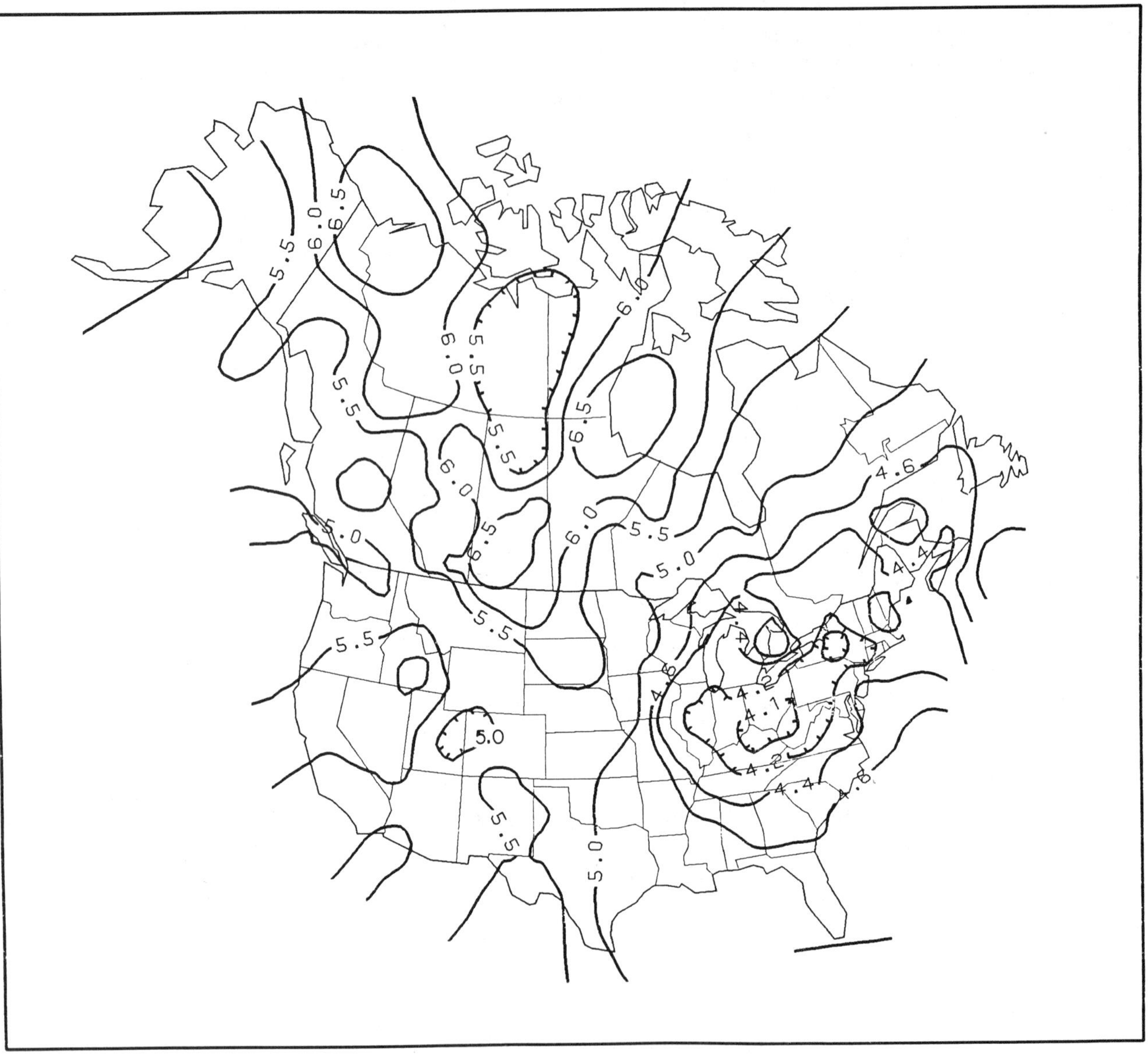

Figure 1. Average Annual pH of Precipitation in Canada and the United States During 1980. The data shown on this map were developed by the Canadian Network for Sampling Precipitation (CANSAP) in Canada and the National Atmospheric Deposition Program (NADP) in the United States.

(c) use of some combination of fuel cleaning, combustion modification, flue-gas treatment, etc.

(d) alter regulations to include existing as well as new sources, encourage "least-emissions-dispatching," permit the purchase and exchange of "pollution rights," etc.

A major part of the requirements for step 1 were provided by the data shown in Figure 2. These data were developed originally for an earlier House of Representatives hearing on acid rain.

A major increment of progress toward steps 3 and 4 was achieved in the Atmosphere-Biosphere Report of the National Academy of Sciences (NAS, 1981). Page 181 of that report reads as follows:

"Of the options presently available only the control of emissions of sulfur and nitrogen oxides can significantly reduce the rate of deterioration of sensitive freshwater ecosystems. It is desirable to have precipitation with pH values no lower than 4.6 to 4.7 throughout such areas, the value at which rates of degradation are detectable by current survey methods, as mentioned above. In the most seriously affected areas (average precipitation pH 4.1 to 4.2), this would mean a reduction of 50% in deposited hydrogen ions."

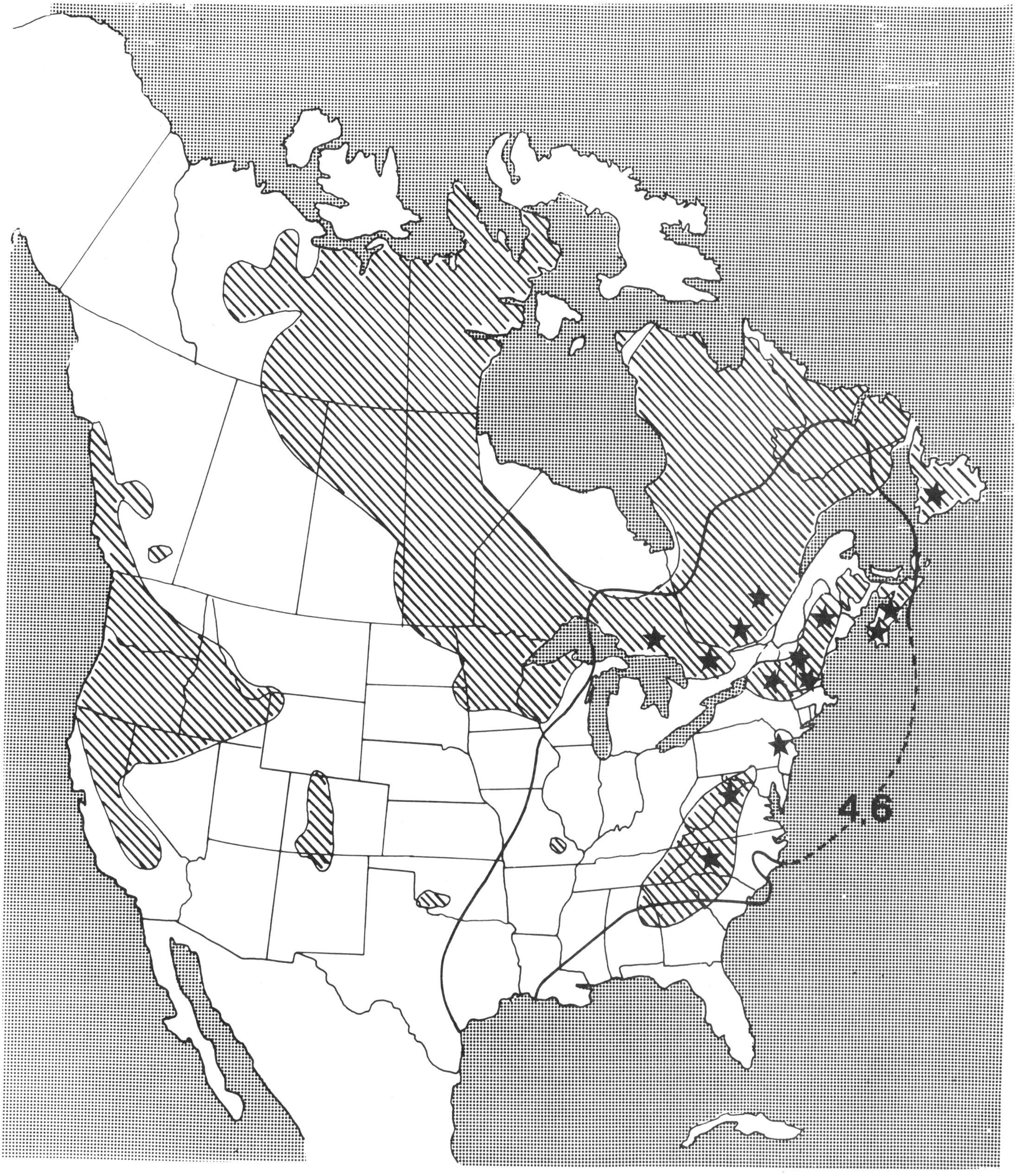

Figure 2. Regions of Canada and the United States Where Lakes and Streams Have Been Acidified by Acid Deposition. The isopleth line defines the area receiving a precipitation weighted annual average pH of precipitation below 4.6 during 1979. These data were provided by the CANSAP and NADP monitoring programs. The shaded area outlines regions containing lakes and streams that are sensitive to acidification by acid deposition because most surface waters within this region have alkalinity values below 50 milliequivalents per liter. The stars indicate where lakes and streams have been acidified by acid deposition from the atmosphere. (Used with permission of J. N. Galloway.)

In June 1982 the Canadian-United States Memorandum of Intent Working Group on Impact Assessment drafted the following statement which also defines a "biologically acceptable" target loading:

"Based on limited empirical observations available to the working group which are supported by theory and modeling, it is concluded that a loading rate of less than 20 kg/ha/year of sulfate in precipitation would protect aquatic resources currently observed to be experiencing acidification, pH depression and biological alteration. The extent of additional aquatic resources that would be protected or not protected by loadings less than 20 kg/ha/year are not known at this time. Further scientific evidence must be considered when available and the loading target revised accordingly."

Also in June 1982, the Swedish Ministry of Agriculture and Environment (1982) submitted the following statement for consideration at the 1982 Stockholm Conference on Acidification of the Environment:

"Given the knowledge we now have, there are several ways of formulating what might be an 'acceptable' level of deposition, that is to say, how much the sensitive ecosystems can stand without suffering cumulative damage. All these bases of calculation give similar results. A sulphate deposition of about 0.5 grams of sulphur per square metre per year could be tolerated without entailing any risk of large-scale acidification damage. If we wish to prevent the acidification of even the most susceptible lakes and water sources, the sulfate deposition will have to be reduced to not more than 0.3 grams of sulphur per square meter per year. It should be pointed out, however, that every measure which reduces the deposition is positive and that there is no threshold value that has to be achieved before one can discern a positive effect on the environment."

These related recommendations of "biologically acceptable" target loadings were developed by three separate groups of scientists. If they are generally accepted, as it now appears that they may be, the stage is set in both North America and in Europe for step 5 in the above list of steps leading toward management of acid deposition in both continents. The critical question then becomes:

"Where and by how much will it be necessary to decrease emissions of acid-deposition precursors so as to achieve the decrease in deposition that is recommended?"

CONCLUSIONS

Acid deposition has become a major environmental issue in both Europe and North America during the past two decades. Much has been learned already but much more remains to be learned about various aspects of the problem. The challenge for all of us as scientists, in Europe and in North America, is to satisfy the public need for information by developing a more complete understanding of the atmospheric processes, soil transformations, vegetational changes, alterations in water quality, effects on materials, and physiological influences of acid precipitation.

The challenge for industrial and political leaders in North America and Europe alike continues as always – to make decisions about complex issues under conditions of substantial uncertainty. The uncertainties involved in management of acid deposition and its effects are analogous with those that surrounded the debates about the role of phosphorus in the eutrophication of Lake Ontario and Lake Erie in the early 1970's. The experts then could not agree. Some said a 40 percent decrease in loading would be needed; others said 75 percent; still others said nitrogen was to blame, not phosphorus. Environmentalists warned of "possible irreversible harm." Industry said more research was needed. Finally, a political decision was reached – a plan should be developed. More debates were held. A theoretical model was used to predict that a 50 percent decrease in phosphorus loading might be sufficient and a management plan was finally implemented. After some time the lakes began to improve. The theoretical estimate was too low but with some further adjustments the plan worked and the lakes are now on the road to recovery.

After a decision is made to proceed with preliminary evaluation of policies to protect sensitive regions from acid deposition, it probably will take at least five years to debate the scientific and management issues sufficiently to formulate an appropriate management plan. Also, at least another five years may well be required to achieve an actual reduction in emissions. During the period of debate, new biological, atmospheric chemical, and meteorological insights will be developed which will help refine whatever management strategies may be recommended. Therefore, it seems prudent that the nations in North America and Europe press on earnestly with the debates now underway recognizing that:

1) research is continuing,
2) a decision to extend present policies is one among many possible alternatives, and
3) being good international neighbors requires a timely search for mutually acceptable understandings and courses of action.

ACKNOWLEDGMENTS

Although the research described in this article has been funded in part by the USEPA through a Cooperative Agreement with North Carolina State University, it has not been subjected to EPA's policy review and therefore does not necessarily reflect the views of the Agency. Portions of this paper are adopted from earlier, more detailed reports on the history of acid precipitation (Cowling, 1981, 1982) and are incorporated in this paper with permission from the SNSF Project in Norway and from Environmental Science and Technology (1982, Vol. 16, pp. 110A-123A, American Chemical Society).

LITERATURE CITED

Altshuller, A. P. and G. A. McBean, Co-Chairmen, 1979. The LRTAP Problem in North America: A Preliminary Overview Prepared by the United States-Canada Bilateral Research Consultation Group on the Long-Range Transport of Air Pollutants. Atmospheric Environmental Service, Downsview, Ontario, 48 pp.

AMBIO, 1976. Report of the International Conference on the Effects of Acid Precipitation in Telemark, Norway. AMBIO 5:200-252.

Beamish, R. J. and H. H. Harvey, 1972. Acidification of the LoChoche Mountain Lakes, Ontario and Resulting Fish Mortalities. J. Fisheries Research Board of Canada 29:1131-1143.

Bolin, B., *et al.,* 1972. The Impact on the Environment of Sulfur in Air and Precipitation. *In:* Sweden's Case Study for the United Nations Conference on Human Environment: Air Pollution Across National Boundaries. Norstadt and Sons, Stockholm, 97 pp.

Braekke, F. H. (Editor), 1976. Impact of Acid Precipitation on Forest and Freshwater Ecosystem in Norway. Res. Rep. 6/76, SNSF Project, Ås, Norway, 111 pp.

Cogbill, C. V. and G. E. Likens, 1974. Acid Precipitation in the Northeastern United States. Water Resources Research 10:1133-1137.

Cowling, E. B., 1976. Testimony on Research and Development Relating to Sulfates in the Atmosphere, pp. 398-440. *In:* Research and Development Relating to Sulfates in the Atmosphere – Hearings before the Subcommittee on the Environment and the Atmosphere of the Committee on Science and Technology. U.S. House of Representatives, 94th Congress, First Session, Washington, D.C., 1029 pp.

Cowling, E. B., 1980. An Historical Resume of Progress in Scientific and Public Understanding of Acid Precipitation and its Biological Consequences. Research Report No. 18, SNSF Project, Oslo, Norway, 29 pp.

Cowling, E. B., 1982. Acid Precipitation in Historical Perspective. Environ. Sci. Technol. 16:110A-123A.

Dochinger, L. S. and T. A. Seliga, 1976a. Proceedings First International Symposium on Acid Precipitation and the Forest Ecosystem. USDA Forest Service, Gen. Tech. Rept. NE-23, Northeast. For. Exp. Sta., Upper Darby, Pennsylvania, USA, 1079 pp.

Dochinger, L. S. and T. A. Seliga, 1976b. Workshop Report on Acid Precipitation and the Forest Ecosystem. USDA Forest Service, Gen. Tech. Rept. NE-25, Northeast. For. Expt. Sta., Upper Darby, Pennsylvania, USA, 18 pp.

Drablφs, D. and A. Tollan (Editors), 1980. Ecological Impact of Acid Precipitation. Proc. Intl. Conf. SNSF Project, Oslo, Norway.

Egnér, H., G. Brodin, and O. Johansson, 1955. Sampling Technique and Chemical Examination of Air and Precipitation. Kungl. Lantbrukhogskolan Ann. Uppsala, Sweden, 22:369-410.

Emanuelsson, A., E. Riksson, and H. Egnér, 1954. Composition of Atmospheric Precipitation in Sweden. Tellus 6:261-267.

Eriksson, E., 1952. Composition of Atmospheric Precipitation. I. Nitrogen Compounds. Tellus 4:215-232; II. Sulfur, Chloride, Iodine Compounds. Bibliography, Tellus 4:280-303.

Eriksson, E., 1954. Report of an Informal Conference in Atmospheric Chemistry Held at the Meteorological Institute, University of Stockholm, 24-26 May 1954. Tellus 6:302-307.

Eriksson, E., 1959. The Yearly Circulation of Chloride and Sulfur in Nature: Meteorological, Geochemical and Pedological Implications. Part I: Tellus 11:375-403.

Galloway, J. N. and E. B. Cowling, 1978. The Effec ts of Precipitation of Aquatic and Terrestrial Ecosystems: A Proposed Precipitation Chemistry Network. J. Air Pollut. Control Assoc. 38: 228-235.

Galloway, J. N. and G. E. Likens, 1976. Calibration of Collection Procedures for the Determination of Precipitation Chemistry. Water, Air, and Soil Pollut. 6:241-258.

Galloway, J. N., E. B. Cowling, E. Gorham, and W. W. McFee, 1978. A National Program for Assessing the Problem of Atmospheric Deposition (acid rain). Nat. Atm. Deposition Prog., Nat. Res. Ecol. Lab., Fort Collins, Colorado, USA, 97 pp.

Gordon, A. G. and E. Gorham, 1963. Ecological Aspects of Air Pollution from an Iron Sintering Plant at Wawa, Ontario. Can. J. Bot. 41:1063-1078.

Gorham, E., 1955. On the Acidity and Salinity of Rain. Geochim. Cosmochim. Acta 7:231-239.

Gorham, E., 1957. The Ionic Composition of Snow Lowland Lake Waters from Cheshire, England. Limnol. and Oceanogr. 2:22-27.

Gorham, E., 1958a. Atmospheric Pollution by Hydrochloric Acid. Quart. J. Royal Meteorol. Soc. 84:274-276.

Gorham, E., 1958b. The Influence and Importance of Daily Weather Conditions in the Supply of Chloride, Sulphate, and Other Ions to Fresh Waters from Atmospheric Precipitation. Phil. Trans. Royal Soc. London, Ser. B. 247:147-178.

Gorham, E., 1958c. Free Acid in British Soils. Nature 181:106.

Gorham, E., 1958d. Bronchitis and the Acidity of Urban Precipitation. Lancet ii:691.

Gorham, E., 1961. Factors Influencing Supply of Major Ions to Inland Waters, with Special Reference to the Atmosphere. Geol. Soc. Amer. Bull. 72:795-840.

Gorham, E., 1965. Thomas Brotherton, Robert Hook, and Some Neglected Experiments in the Plant Physiology During the Late Seventeenth Century. BioScience 15:412.

Gorham, E., 1981. *In:* Atmosphere-Biosphere Internations Toward a Better Understanding of the Ecological Consequences of Fossil Fuel Combustion. National Academy of Sciences, Washington, D.C., pp. 9-21.

Gorham, E. and A. G. Gordon, 1960. The Influence of Smelter Fumes Upon the Chemical Composition of Lake Waters Near Sudbury, Ontario and Upon the Surrounding Vegetation. Can. J. Bot. 38:477-487.

Hutchinson, T. C. and L. M. Whitby, 1974. Heavy Metal Pollution in the Sudbury Mining and Smelting Region of Canada. 1. Soil and Vegetation Contamination by Nickel, Copper, and Other Metals. Environ. Conserv. 1(2):123.

Jordan, H. V., C. E. Bardsley, L. E. Ensminger, and J. A. Lutz, 1959. Sulfur Content of Rain Water and Atmosphere in Southern States as Related to Crop Needs. U.S. Dept. Agr. Tech. Bull. No. 1196, 16 pp.

Junge, G. E. and R. Werby, 1958. The Concentration of Chloride, Sodium, Potassium, Calcium, and Sulfate in Rain Water Over the United States. J. Meteorology 15:417-425.

Katz, M. (Editor), 1939. Effect of Sulphur Dioxide on Vegetation. National Res. Council of Canada. Publ. No. 815, Ottawa, 447 pp.

Likens, G. E. 1976. Acid Precipitation. Chem. Eng. News 54(48): 29-44.

Likens, G. E., F. H. Bormann, and N. M. Johnson, 1972. Acid Rain. Environment 14:33-40.

Likens, G. E., R. F. Wright, J. N. Galloway, and T. J. Butler, 1979. Acid Rain. Sci. Am. 241(4):43-51.

Lodge, J. P., J. B. Pate, W. Basbergill, G. S. Swanson, K. C. Hill, E. Lorange, and A. L. Lazrus, 1968. Chemistry of United States Precipitation. Final Report on the National Precipitation Sampling Network. National Center for Atmospheric Research, Boulder, Colorado, 66 pp.

MacIntyre, W. H. and I. B. Young, 1923. Sulfur, Calcium, Magnesium, and Potassium Content and Reaction of Rainfall of Different Points in Tennessee. Soil Sci. 15:205-227.

National Academy of Sciences, 1975. Atmospheric Chemistry: Problems and Scope. Com. Atm. Sci., Nat. Res. Council, Washington, D.C., 130 pp.

National Academy of Sciences, 1981. Atmosphere-Biosphere Interactions; Toward a Better Understanding of the Ecological

Consequences of Fossil-Fuel Combustion. Com. Atm. Bios. Nat. Acad. Press, Washington, D.C., 263 pp.

Odén, S., 1967. Dagens Nyheter. October 24, 1967.

Odén, S., 1968. The Acidification of Air and Precipitation and Its Consequences in the Natural Environment. Ecology Committee Bull. No. 1. Swedish National Science Research Council, Stockholm, Translation Consultants, Ltd., Arlington, Virginia, USA, 117 pp.

Organization for Economic Cooperation and Development, 1977. The OECD Programme on Long Range Transport of Air Pollutants. OECD, Paris, France.

Ottar, B., 1976. Monitoring Long-Range Transport of Air Pollutants: The OECD Study. AMBIO 5:203-206.

Overrein, L. N., H. M. Seip, and A. Tollan, 1980. Acid Precipitation – Effects on Forests and Fish. Final Report to the SNSF Project, 1972-80. SNSF Research Report No. 19, SNSF Project, Oslo, Norway.

Rossby, C. G. and H. Egnér, 1955. On the Chemical Climate and Its Variation with the Atmospheric Circulation Pattern. Tellus 7: 118-133.

Schofield, C. L., 1976. Effects of Acid Precipitation on Fish. AMBIO 5:228-230.

Shriner, D. S., 1976. Effects of Simulated Rain Acidified with Sulfuric Acid on Host-Parasite Interactions, pp. 919-925. *In:* Proc. First Int. Symp. on Acid Precipitation and the Forest Ecosystem, L. S. Dochinger and T. A. Seliga (Editors). USDA For. Serv. Gen. Tech. Rep. NE-23, Upper Darby, Pennsylvania.

Smith, R. A., 1852. On the Air and Rain of Manchester. Mem. Lit. Phil. Soc. Manchester, Series 2, 10:207-217.

Smith, R. A., 1872. Air and Rain: The Beginnings of Chemical Climatology. Longmans, Green, London, 600 pp.

Swedish Ministry of Agriculture and Environment, 1982. Acidification Today and Tomorrow. Ministry of Agriculture, Stockholm, Sweden, 232 pp.

Tollan, A., 1980. Annotated Bibliography 1974-80. SNSF Project, Oslo, Norway.

Way, T., 1855. The Atmosphere as a Source of Nitrogen to Plants. Roy. Agr. Soc. Jour. 16:249-267 (see also 17:123-162 and 618-621).

INTERNATIONAL SYMPOSIUM ON HYDROMETEOROLOGY
JUNE AMERICAN WATER RESOURCES ASSOCIATION 1982

THE U.S. NATIONAL ACID PRECIPITATION ASSESSMENT PROGRAM

Chris Bernabo and Raymond Herrmann[1]

ABSTRACT: The National Acid Precipitation Assessment Program (NAPAP) was mandated by Congress (P.L. 96-294) to provide a better understanding of the causes, effects, and options for managing "acid rain." The U.S. Interagency Task Force on Acid Precipitation is charged with planning and implementing the National Program. The Task Force integrates and coordinates the diverse federal efforts to establish a firmer scientific basis for formulating energy and environmental policy. Twelve federal agencies participate in the Task Force, which is jointly chaired by USDA, EPA, and NOAA. A 10-year National Plan was issued in June 1982, and it provides a general description of the priorities, objectives, focus, and resources of the program. Nine technical "task groups" are responsible for developing detailed operational plans for research categories including: natural sources, man-made sources, atmospheric processes, deposition monitoring, aquatic impacts, terrestrial impacts, materials effects, control technology, and assessments and policy analysis. The Task Force also coordinates the federal program with the research and monitoring activities of states, universities, private sector groups, and other nations. In conducting the National Program, the Task Force is utilizing various cooperative assessment activities and interdisciplinary analyses of complex regional, national, and international data sets such as those generated by NADP, ILWAS, SNSF, and CANSAP.
(KEY TERMS: acid precipitation; United States; federal program.)

INTRODUCTION

The National Acid Precipitation Assessment Program coordinates federally funded research on the causes and effects of acid deposition. This effort links the research activities of 12 federal agencies and focuses their work on the timely development of a firmer scientific basis for decision making. Federally funded acid precipitation research has increased from $11 million in FY 1980 to $18 million for FY 1982. The President's budget for FY 1983 requests $22 million for acid rain research.

The National Program involves comprehensive research efforts encompassing studies of man-made and natural sources, atmospheric processes, deposition monitoring, aquatic and terrestrial impacts, effects on materials, examination of control strategies, and assessment of policy options. The National Program coordinates the federal efforts with the research and monitoring activities of the states, private sector groups, and other nations and is responsible for providing annual progress reports to the President and Congress describing the status of research and its implications.

STRATEGY OF THE ACID PRECIPITATION PROGRAM

The Acid Precipitation Act of 1980 (Title VII of the Energy Security Act of 1980 – P.L. 96-294) established the Interagency Task Force on Acid Precipitation (the Task Force) to develop and implement a comprehensive National Acid Precipitation Assessment Program. The purpose of the National Program is to increase our understanding of the causes and effects of acid precipitation.

The Task Force has a National Plan to broadly outline the ten-year research program mandated by the Act. This general document is supplemented by a detailed operational research plan for managing the federal research program. The Task Force also issues Annual Reports describing: research progress; the current state of knowledge about acid precipitation and its implications; and future research needs. The first Annual Report to the President and the Congress was issued in January 1982.

The National Program employs a general strategy that includes:

- Building upon previous efforts to develop a federal acid rain program. The former Acid Rain Coordinating Committee (ARCC) has been reconstituted to form the statutory Task Force; the ARCC draft plan was refined and served as a foundation for developing the current plan.
- Using existing scientific knowledge for timely assessments and, where appropriate, policy recommendations. Currently available data and information from the U.S. and other nations will be critically analyzed and applied where possible.
- Conducting research to develop more knowledge. The research emphasis is on activities that contribute most effectively to establishing a firmer scientific basis for decision making.

[1]Respectively, Executive Director, National Acid Precipitation Assessment Program, 722 Jackson Place, N.W., Washington, D.C. 20006; and Chief, Water Resources, Field Support Lab, National Park Service, 107 C Natural Resources, Colorado State Univ., Fort Collins, Colorado 80523.

• Establishing a long term National Trends Network (NTN). The development and maintenance of a well designed acid precipitation (wet and dry) monitoring network is essential for documenting and understanding acid precipitation and its effects. Existing monitoring efforts, such as those of the National Atmospheric Deposition Program (NADP), have been incorporated into the National Trends Network protocol.

• Evaluating information and the policy implications. Information generated by the National Acid Precipitation Assessment Program and other research efforts in this country and abroad, are expected to contribute significantly to our knowledge. The information produced will be synthesized periodically, subjected to scientific peer review, published, and interpreted to guide decisions on future research.

• The Task Force reports annually to the Congress, the President, and the nation on the research program's progress and the implications of our expanding knowledge about acid rain.

The Task Force works with all those concerned to ensure that the federal activities are conducted as part of a comprehensive, well coordinated, and long term National Program. The National Program will utilize and coordinate the expertise available in federal and state agencies, universities, industry, private contractors, and research institutions.

Proposed Research

Research is proposed in nine basic categories where more information about acid precipitation is urgently needed. The research tasks described in the National Plan each focus on a specific area and generally involve the coordinated participation of several agencies. The number and extent of the proposed research tasks eventually carried out will depend on the availability of funds in future years.

The research is being focused to answer a number of broad questions such as:

• What is the quantitative relationship between emissions of precursors and deposition of acids?

• How do chemical and physical atmospheric processes control acid deposition?

• What are the interactions between acid rain and other pollutants affecting ecosystems?

• What is the extent of damaged or sensitive aquatic ecosystems in the U.S.?

• What is the relative contribution of local versus distant sources of acidic materials?

• How great is the potential for acid deposition damaging forests, crops, and soils?

• What are the most cost effective ways to manage acid deposition?

Numerous more specific questions about acid rain are being addressed. As an example within the water resources area, several important issues have been identified.

Impacts on water resources of the U.S. are considered under two topics: First, the chemical alteration of water quality, including ground water, drinking water supplies, streams, and lakes; and secondly, the effects on the species and populations that make up biologically productive components of aquatic ecosystems.

Studies comparing recent water quality data with those of 25 or 50 years ago show that lakes and streams in certain remote regions now have lower pH and alkalinity, and higher sulfate ion ($SO_4^=$) and aluminum concentrations. Great differences are observed to exist from one watershed to the next, leading to the conclusion that the surrounding bedrock, soils, and vegetation have a great influence over the observed effects. Thus, understanding water quality effects from acid deposition requires a broad set of information about the watershed.

Principal questions about water quality alterations relate to both resources and human health impacts which could result from deposition of the ions H^+, $SO_4^=$, NO_3^-, and NH_4^+, and include the following:

1. What are the trends of sulfate, nitrate, ammonium, hydrogen ion, and alkalinity concentrations in shallow ground water and in streams and lakes? Can a monitoring network be established to document these water quality parameters, taking into consideration the need to understand the effects of local adjacent soils, vegetation, and land use differences? Are there unique water monitoring programs in the acid deposition regions that already provide quality assured records on recent trends? Can quality assurance procedures be extended from the atmospheric deposition program to water quality effects monitoring, including measurements of watershed influences?

2. Since nitrogen is frequently a limiting nutrient in aquatic systems (more so than sulfur), what is the fate of nitric acid and ammonium ion deposition in watersheds? Under what conditions might nitric acid and ammonium ion inputs function and contribute to surface water acidity?

3. What factors in the surrounding soils, bedrock geology, and patterns of weather and climate make lakes or streams vulnerable to (or tolerant of) acid deposition? What is the relative contribution to lakes and streams or hydrogen ions from precipitation, runoff, and subsurface flows? Can existing data or field responses and information on processes be used to develop predictive models of watershed and water quality response;

4. Have pH and heavy metal concentrations and recalcitrant organics such as DDT and PCB's in surface and ground sources of drinking water changed over time? How extensive are the effects of acid deposition on septic field microorganics and decomposition processes? How many home, livestock, and municipal drinking water systems are affected by changes in water chemistry, and how large are the populations they serve? Where are these water systems?

5. What methods do we have for treating drinking water to reduce its content of acidity or heavy metals? Can the

acidity of drinking water systems be safely adjusted by liming or other treatment?

Organization and Implementation

The Interagency Task Force on Acid Precipitation is jointly chaired by the Department of Agriculture (DOA), the Environmental Protection Agency (EPA), and the National Oceanic and Atmospheric Administration (NOAA). The other participating federal entities are: the Departments of Interior (DOI), Health and Human Services (HHS), Commerce (DOC), Energy (DOE), State (DOS); the National Aeronautics and Space Administration (NASA); the Council on Environmental Quality (CEQ); the National Science Foundation (NSF); and the Tennessee Valley Authority (TVA). The Task Force also includes four Presidential appointees and the Directors of the Argonne National Laboratory, Brookhaven National Laboratory, Oak Ridge National Laboratory, and the Pacific Northwest National Laboratory.

The responsibilities of the Task Force are to:

- Develop and update the National Acid Precipitation Assessment Plan;
- Oversee and implement a ten-year comprehensive research program that coordinates and focuses the acid rain activities of the federal agencies (no regulatory authority is vested in the Task Force);
- Maintain an inventory of federally funded acid precipitation research projects;
- Develop an annual interagency budget for the federal program;
- Provide Annual Reports on the program's progress and policy implications;
- Encourage productive interaction between the federal program and private sector, academic, state and local governmental and international activities; and
- Obtain nonfederal input to the planning, review, and program activities.

The Task Force meets at least three times a year to finalize budgets, establish objectives, set priorities, approve plans and reports, and conduct program reviews. All federal acid precipitation research and assessment activities are coordinated and integrated by the Task Force to form a cohesive national program.

Individual agencies' projects are carried out in the context of the National Program defined by the current ten-year Plan (as updated and revised by the Annual Reports) and with oversight by the Task Force. Technical Task Groups for each of the nine research categories oversee and facilitate detailed planning of activities in their assigned areas. The Research Coordination Council integrates the outputs of the Task Groups as well as develops and updates the detailed operational research plan for the National Program. A Program Coordination Office serves as the focal point for planning and managing the interagency federal effort. The Office is managed by the Task Force's Executive Director and is responsible for disseminating information and providing liaison with nonfederal activities, other nations, and the public. The Task Force maintains a Task Group on international activities. This group is chaired by the Department of State and assists the Task Force in ensuring that the U.S. National Acid Precipitation Assessment Program is effectively coordinated with relevant international activities.

The role of the Task Force in planning the interagency budget for the National Program is an important aspect of the federal effort. The Task Force develops a coordinated interagency budget for the National Acid Precipitation Assessment Program. By working together through the Task Force, the agencies have established a research program that focuses on addressing national needs while building on the research interests and capabilities of the individual agencies. The strong interagency planning process eliminates undesirable duplication and avoids crucial omissions in the National Program.

The Task Force provides the mechanism for establishing the research goals for the National Acid Precipitation Assessment Program, identifying the projects needed to meet those goals, and deciding which agencies are best suited to conduct the necessary work. The result is a comprehensive program of interlocking projects, with each agency contributing to specific aspects of the overall national effort.

INTERNATIONAL SYMPOSIUM ON HYDROMETEOROLOGY
JUNE AMERICAN WATER RESOURCES ASSOCIATION 1982

GAS PHASE AND PRECIPITATION ACIDITIES IN THE COLORADO MOUNTAINS

B. J. Huebert, R. B. Norton, M. J. Bollinger, D. D. Parrish, C. Hahn, Y. A. Bush, P. C. Murphy, F. C. Fehsenfeld, and D. L. Albritton[1]

ABSTRACT: For the past few years, both the gas phase concentrations of nitric acid, its precursors, and nitrate aerosols and the precipitation concentrations of nitrate were measured at Niwot Ridge, a remote area field site located at 3 km elevation in the Colorado mountains west of the Denver metropolitan area. The measurements were made using a variety of techniques: filter collection, ion chromatographic analysis, direct pH and conductivity determinations, and chemiluminescence detection. An extensive wind speed/wind direction network, coupled with trajectory analyses, provided meteorological support. The prevailing winds at the site are from the west, although occasionally the wind is from the east, across the metropolitan area in that direction. Correlations between the wind direction and the acidity levels, both in the air and in the preparation, show that, despite the fact that east winds are atypical, the acidic components accompanying these winds are the likely major source of the relatively high acid deposition that occurs at the site. The observed nitric acid/precursor correlations are consistent with the current picture of the transformation chemistry. During the winter of 1980/81, it was shown that snow scavenges nitric acid and nitrate very efficiently.
(KEY TERMS: acid precipitation; Colorado; wind direction; snow; scavenging; chemical reaction; nitric acid; nitrate.)

INTRODUCTION

Acid rain is one of today's most widely publicized environmental controversies. The issue was raised by the discovery in the 1950's that high acidity rainfall was occurring in northern and western Europe. Later it was recognized that a similar situation exists in the northeastern United States and the neighboring parts of Canada, and, in recent years, acid precipitation has been documented at numerous other locations in the United States and around the world. In particular, Lewis and Grant (1980a) have recently reported that the mean pH of rainfall at the Continental Divide in Colorado has decreased from 5.4 to 4.7 between May 1975 and May 1978. They concluded that this change was the result of an increasing deposition of nitric acid, HNO_3, which is an oxidation product of the anthropogenically produced nitric oxide molecule, NO, in the atmosphere. While these valuable observations have drawn attention to yet another region affected by acid precipitation, Lewis and Grant (1980a), Kelly and Stedman (1980), and Lewis and Grant (1980b) have pointed out that these pH data also raise a puzzling question. Since the prevailing winds at the Continental Divide are from the west and the only major metropolitan areas are to the east, where is the source of the NO that produces the HNO_3?

We have addressed that question at a remote area field site near the Continental Divide where the gas phase concentrations of nitric acid, its odd-nitrogen precursors, and nitrate aerosols were measured. An extensive wind speed/wind direction network, coupled with trajectory analyses, provided meteorological support. Since much of the precipitation at this site is snow, the measurements were made not only during clear periods, but also during snow storms. The pH and anion content of the snow were determined. The goals of the study were to examine the acidity levels of the air, the likely origins of the acidity, the scavenging efficiency of snow storms, and the atmospheric chemistry of nitric acid. The results obtained are summarized here.

EXPERIMENTAL DETAILS

The sampling site is located on Niwot Ridge at an elevation of 3,000 m (10,000 ft), which, as shown in Figure 1, is about 1,500 m (5,000 ft) above and 30 km (50 miles) northwest of the Denver metropolitan area. No major urban areas lie to the west of the site for hundreds of miles. Niwot Ridge is within a biopreserve maintained by the Mountain Research Station of the University of Colorado. Vehicular traffic, and other human activities, are kept at a minimum in the immediate vicinity of the site.

Our research van is located in a clearing in a pine and spruce forest at a point just below timberline. Figure 2

[1]Respectively, Huebert, Department of Chemistry, Colorado College, Colorado Springs, Colorado 80903; Others, Aeronomy Laboratory, National Oceanic and Atmospheric Administration, Boulder, Colorado 80303. (Bollinger also Department of Chemistry, University of Colorado, Boulder, Colorado; Parrish also Metropolitan State College, Denver, Colorado, and Cooperative Institute for Research in Environmental Sciences, University of Colorado/NOAA, Boulder, Colorado; Hahn, Murphy, and Fehsenfeld also Cooperative Institute for Research in Environmental Sciences, University of Colorado/NOAA, Boulder, Colorado.)

gives a cutaway view that shows the instruments. Nitric oxide and nitrogen dioxide, denoted collectively here as NO_x ($\equiv NO + NO_2$), are measured by two differently configured chemiluminescence/photolysis instruments, NO_x[A] and NO_x[B], of our own design (McFarland, *et al.*, 1979; Kley, *et al.*, 1981) that sample continuously through inlets atop the van approximately 4 m above the ground. The detection limit for a 10 s data average is 2 pptv (parts per trillion by volume) for NO and 10 pptv for NO_2. By intercomparing the two instruments as one is modified temporarily, extensive potential artifact tests have been made. The instruments are routinely calibrated by passing standard mixtures through the inlets a few times each day. We estimate the uncertainty in a clean air measurement to be less than 30 percent.

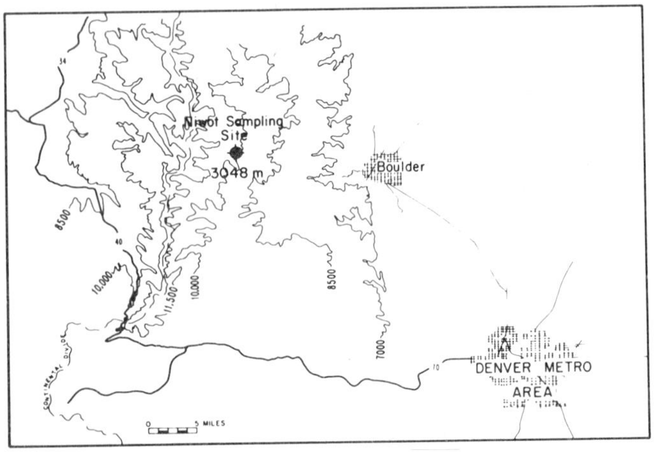

Figure 1. Location of the Niwot Ridge Sampling Site. The metropolitan areas of Boulder and Denver, Colorado, are shown hatched. The major highways are the heavy solid lines. Contour lines of elevation (in ft) are the light solid lines. The Continental Divide is the broken line.

Nitric acid vapor and aerosol nitrate, NO_3^-, were measured using a filter collection technique (Huebert and Lazrus, 1980; Huebert, 1980) in which nitrate particulates and nitric acid are sequentially collected from air drawn through a series of two filters. The sampling time is about an hour, depending on nitric acid levels. The filter bank is located on the top of the van, shown on the right hand side in Figure 2. After exposure, the filters are sealed in noncontaminating plastic bags, the captured material later extracted, and the extract analyzed for nitrate by ion chromatography. The detection limit for a two-hour sample is about 1 pptv for HNO_3 and 2 pptm (parts per trillion by mass in air) of equivalent nitrate for NO_3^-. At loadings well above the detection limit, the uncertainty in the absolute measurement of HNO_3 is less than 30 percent, based on our accuracy tests (Goldan, *et al.*, 1983).

Other instruments in or near the van shown in Figure 2 make ancillary measurements: (a) the concentrations of ozone, O_3; carbon monoxide, CO; and heavy hydrocarbons, HHC's; (b) the intensity of solar UV flux; and (c) temperature, humidity, wind speed, and wind direction, MET DATA. The signals from all of the instruments are recorded on two microprocessors: the meteorological data on PROFS (Prototype Regional Observations and Forecast Service of NOAA) and the remainder on the GAMS (General Atmospheric Measurement System) developed and built in our laboratory. Precipitation samples are collected in noncontaminating plastic bags by an on-site researcher during the course of the event, sealed immediately, and analyzed for pH and anions by ion chromatography, within a day or two afterwards.

RESULTS

The instruments described above have made atmospheric measurements at Niwot Ridge for approximately three

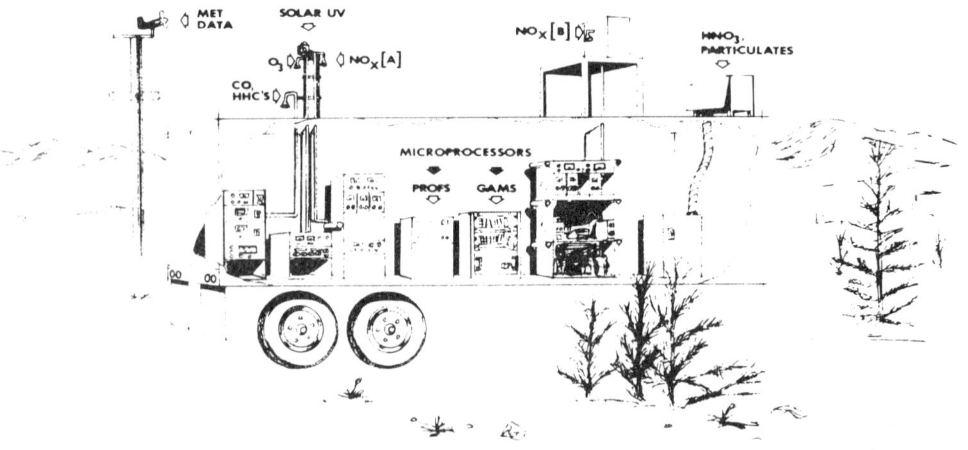

Figure 2. Cutaway Sketch of the Research Van. Shown are the instruments for measuring HNO_3, particulates, NO_x, solar uv, O_3, CO, heavy hydrocarbons, and the meteorological data. PROFS and GAMS denote two microprocessors for instrument control and data recording.

years. The focus has been on annual and seasonal cycles, the snowstorms of the winter of 1980/81, clean air episodes, the role of hydrocarbons in the nitrogen chemistry, and the photolysis of NO_2. Here we emphasize the acid related results. The others are summarized elsewhere (Bollinger, *et al.,* 1982, Huebert, *et al.,* 1982; Roberts, *et al.,* 1982; Parrish, *et al.,* 1983). In the following discussion, the clear air results are described first, the snowstorm results second. The former address the atmospheric acidity levels at the site; the latter address the mechanisms whereby these acidic compounds are deposited on the surface.

Clean Air Studies

NO_x is a remarkably sensitive indicator of air that has been altered by man's activities. Although there are natural sources of NO (microbial soil activity, lightning, stratospheric intrusions, and oxidation of ammonia – Liu, *et al.,* 1980), man's fossil fuel and biomass burnings are the major source of NO in urban areas. NO is rapidly converted to NO_2 via the reaction with ozone, O_3,

$$NO + O_3 \rightarrow NO_2 + O_2, \quad (1)$$

and NO_2 is photodissociated back to NO by solar ultraviolet radiation, $h\nu$,

$$h\nu + NO_2 \rightarrow NO + O. \quad (2)$$

Thus, the NO emission flux, the ozone levels, and solar intensity establish an ambient NO_x level. The contrast between these levels in pristine air and in polluted air is dramatic. In the equatorial Pacific region, McFarland, *et al.* (1979), found midday NO levels of 5 pptv, the lowest ever measured. At Niwot Ridge, we occasionally observed such clean levels, but only under special meteorological conditions. Specifically, when inversions isolate the site from metropolitan areas in vicinity and the air masses reaching our site come from the northwest over the sparsely populated areas of Montana and Wyoming, the NO_x levels observed at Niwot Ridge can be as low as 14 pptv (Bollinger, *et al.,* 1982). In contrast, when the wind is from the east, over the heavily populated Denver metropolitan area, the NO_x levels can be as high as 30 ppbv (parts per billion by volume), an increase over pristine air levels by three orders of magnitude.

This dependence of NO_x levels on wind direction is shown graphically by the example in Figure 3, which gives the wind direction, NO_x concentrations, and HNO_3 concentrations as a function of time on 1 March 1979, a clear, cool day. The morning hours represent the predominant wind conditions in the Continental Divide area, namely, a wind from the west or northwest (Barry, 1972, 1973). The air masses sampled during the early morning hours had passed over the relatively unpopulated region to the west of the Divide. Figure 3 shows that the NO_x concentrations in these air masses were low, less than 500 pptv. However, the prevailing westerlies were interrupted by east winds for about 5 hours in the early afternoon. This is another characteristic feature of the front range climatology: occasional "upslopes" in the early afternoon, particularly in the summer months. During these episodes, the polluted air of the Denver metropolitan area is convected to the site by the rising, warmed air mass. As Figure 3 shows, the arrival of the polluted air is signaled by a sharp increase in the NO_x levels, which reach a maximum of 10 ppbv at midafternoon.

NIWOT RIDGE, I MARCH, 1979
-8 to 0 C, 0 to 90% Clear

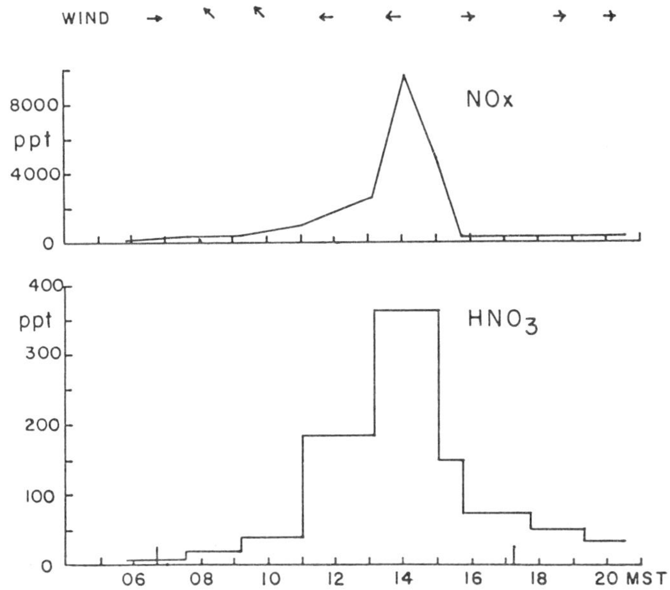

Figure 3. HNO_3 and NO_x Concentrations and Wind Direction Measured at Niwot Ridge on 1 March 1979. Wind from the West: →, etc.

Accompanying the high NO_x levels are high HNO_3 levels, which increase from typical clean air concentrations of about 40 pptv to a peak of 350 pptv. This positive correlation between east winds and large NO_x and HNO_3 concentrations is typical of our observations, which now span three years and include all seasons and a variety of meteorological conditions, as well as the more limited observations by Kelly and Stedman (1980) and Kelly, *et al.* (1980), at the same site. As an example, for a two-month period in the winter of 1979, the wind direction/NO_x/HNO_3 inventory shows (a) that about 70 percent of the time the wind was steady from the west and average concentrations were 250 and 35 pptv for NO_x and HNO_3, respectively; (b) that about 15 percent of the time the wind was steady from the east and the average NO_x and HNO_3 concentrations were

2000 and 240 pptv; and (3) about 15 percent of the time the winds were light and variable. Thus, although the wind is from the east only a small fraction of the time, the almost order of magnitude greater NO_x and HNO_3 concentrations associated with these winds imply that at least 60 percent of the total concentration of these species during the entire observing period came from sources to the east. This is a conservative percentage since it does not include concentrations observed during light and variable wind conditions, which do include some short duration and weak east wind conditions.

As a further indication of the strong correlation between NO_x, and hence anthropogenic influence, and HNO_3 in the Colorado Mountains, Figure 4 shows the observed levels of the sum of the HNO_3 and NO_3^- mixing ratios (i.e., a measure of the "air acidity") as a function of the simultaneously observed NO_2 levels. The data were taken in the winter of 1980 (triangles) and the summer of 1981 (squares) and the larger symbols denote 24-hour samples and the smaller symbols denote 12-hour samples (Huebert, *et al.*, 1982). Over about two decades of variation, the concentration of acidic compounds in the air at Niwot Ridge is strongly correlated with that of NO_2.

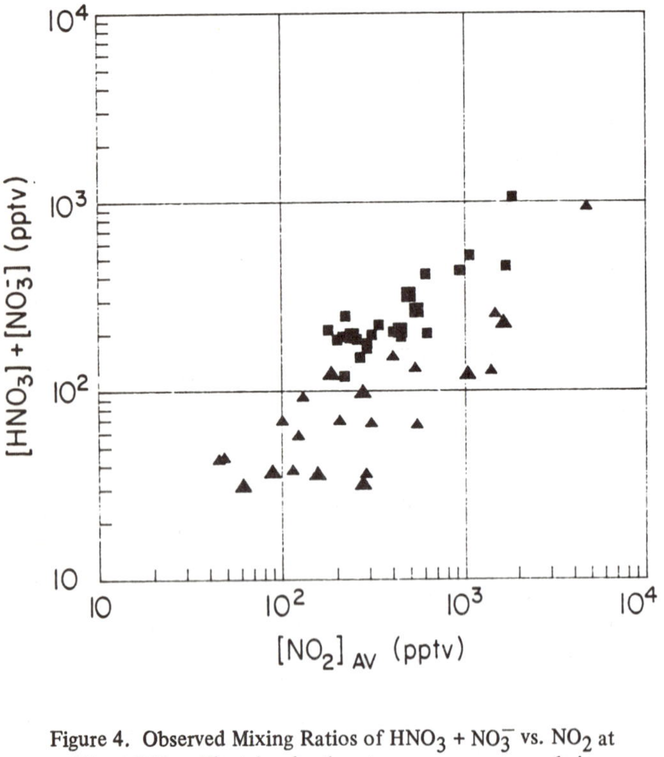

Figure 4. Observed Mixing Ratios of $HNO_3 + NO_3^-$ vs. NO_2 at Niwot Ridge. The triangles denote measurements made in December 1980, and the squares denote those made in July-September 1981. The small symbols correspond to 12-hour averages and the large symbols correspond to 24-hour averages.

Figure 4 also gives some insight into the atmospheric processes that form the acidic compounds from the NO_2 precursor. First, the observed positive correlation between $HNO_3 + NO_3^-$ and NO_2 confirms the qualitative theoretical picture of the chemistry that is involved, namely, that NO_2 is the precursor to both HNO_3 and NO_3^-. The atmospheric chemical processes involved are complex, but a major process is thought to be

$$OH + NO_2 \xrightarrow{M} HNO_3 \quad (3)$$

and

$$HNO_3 \xrightarrow{Het} NO_3^- , \quad (4)$$

where M denotes a three body reaction and Het denotes a gas-to-particle (heterogeneous) conversion. Thus, via the hydroxyl radical, OH, NO_2 is converted to HNO_3, which leads to NO_3^-.

Figure 4 shows an additional feature of this chemistry. Namely, at a given NO_2 mixing ratio, the summer levels of $HNO_3 + NO_3^-$ (squares) tend to be higher than the winter levels (triangles). This may be due to the higher diurnally averaged OH levels that are expected in the summer; hence, more rapid conversion of NO_2 to HNO_3 and thence to NO_3^- (Logan, *et al.*, 1981). Since simultaneous NO_2, HNO_3, and NO_3^- measurements over a wide range of concentrations have not been made before, the data in Figure 4 are the first to reveal this seasonal variation of the $(HNO_3 + NO_3^-)/NO_2$ ratio.

Snow Storm Studies

It is clear from the foregoing observations that man-made combustion emissions originating from the urban areas from the east of Niwot Ridge are the likely major contributor to the acidic compounds in the air over this mountain site. Various mechanisms can deposit these compounds on the surface, where they then can potentially increase the acidic loading on the mountain ecosystem, as is implied by the observations of Lewis and Grant (1980a). Dry deposition can occur in the absence of rain or snow. Unfortunately, very little is known about deposition velocities of HNO_3 and NO_3^-, because of the great difficulties in making such measurements (Hicks, 1980). Indeed, there are no estimates in mountainous terrain. Therefore, one cannot presently assess the efficiency with which dry deposition mechanisms transport the observed acidic compounds from the air to the ground. However, wet deposition processes (snow, rain, hail, etc.) play at least as an important role in the total deposition. We have examined the efficiency of the washout of acidic compounds by snow, since this is the major form of precipitation at the site (Barry, 1972, 1973).

We investigated the acidic characteristics of the snow storms of the winter of 1980/81. The first of these occurred on 12 November 1980. Figure 5 shows the temporal behavior of NO_x and HNO_3 on the days 11-14 November. The shaded area represents the observed HNO_3 levels (left-hand scale) and the broken curve represents the observed NO_x levels (right-hand scale). Prior to the onset of the snowfall, the winds were from the west and the HNO_3 and NO_x correspond to moderately clean air. About noon on 12 November, the winds reversed and an upslope brought the storm and urban air to Niwot Ridge, a pattern that is the typical mountain snow storm. The NO_x levels jumped dramatically when this occurred. However, the HNO_3 levels declined steadily as the snowfall continued, reaching the detection limit after a few hours. NO_x, which is not very soluble, was relatively unchanged during the snowfall. Before the storm the HNO_3/NO_x ratio was about 1/2; during the storm it dropped to about 1/800. Falling snow is clearly a very efficient scavenger of HNO_3.

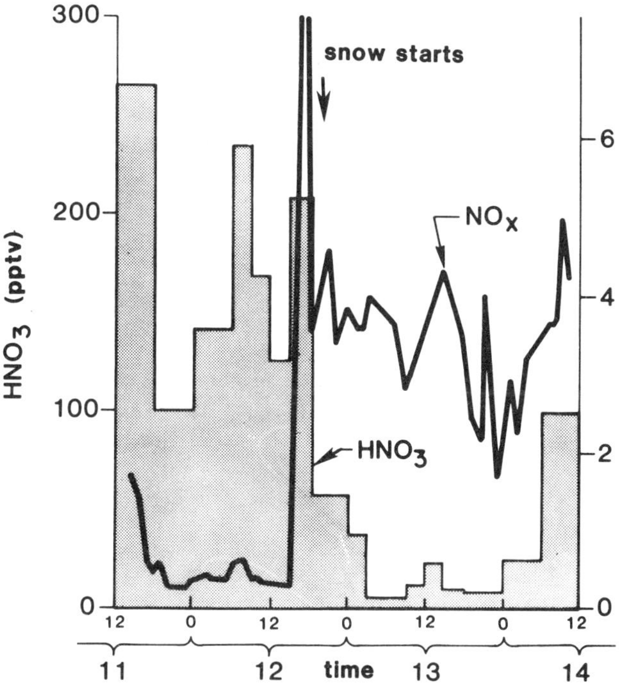

Figure 5. Mixing Ratios of HNO_3 and NO_x Observed Before, During, and After the Snow Storm on 12 November 1980.

During the course of the winter, the storms were examined in greater chemical, precipitation, and meteorological detail. Long range air trajectory calculations (Heffter, 1980), as well as a local, 150-km-square grid of meteorological instruments (PROFS), were used to identify the path over which the storm traveled to reach our observing station. Upslopes accompanied most snowstorms, as is typical, and hence brought relatively polluted local air. An example of the local air trajectories for such a "dirty" snow storm is shown in Figure 6 for 1100 MST on 2 March 1981. Denver is at the center of the figure and the filled circles are the meteorological stations. Our research site is at the station WRD on the left center. The pH level in the snow sample that we collected during this storm was 4.3. Following this snowfall was a snowfall event occurring in air masses that had come from the west. The local trajectories for 0200 MST on 3 March 1981 are shown in Figure 7. The snow samples collected from this storm had a pH of typically 5.3, near the normal, clean air level of 5.6. Thus, acidity seen in the collected precipitation reflects the same correlation with air mass origin that the data in Figure 3 show for the atmospheric acidity levels.

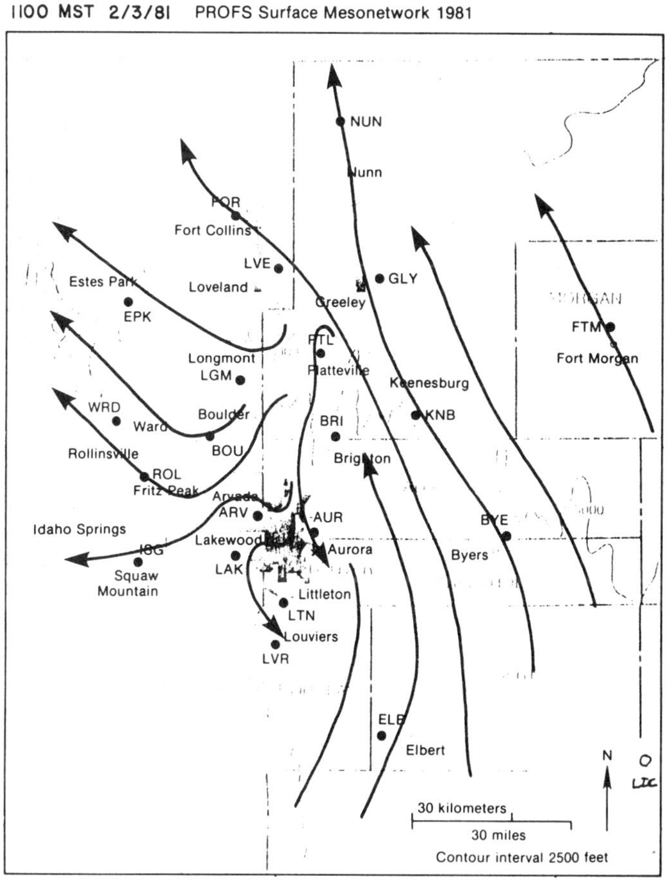

Figure 6. Computed Surface Level Air Trajectories in the Denver, Colorado, Vicinity at 1100 MST on 2 March 1981. The solid symbols are the meteorological stations of the NOAA Prototype Regional Observation and Forecast Service (PROFS), on whose data the trajectory calculations were based.

The details of the scavenging efficiencies of snow were examined by sequential sampling of the changing air concentrations of the acidic compounds, the latter subjected to subsequent pH and anion analyses. These data show the correlated reduction of airborne acidity and the increase in sampled precipitation acidity during the early part of the snow storm. These measurements were used (Huebert, *et al.*, in preparation, 1983) to determine washout ratios, W, for $HNO_3 + NO_3^-$, defined as

$$W = k\rho / [HNO_3 + NO_3^-] \qquad (5)$$

where k is the nitrate mixing ratio in the precipitation (μg/g), ρ is the air density (g/m^3), and [ ] denotes the HNO_3 and NO_3^- concentration in the unscavenged air ($\mu g/m^3$). Typical values obtained were 5000. Although such values are required in model calculations of acidic deposition, only a few have been reported: SO_2, nitrate, and sulfate. Our data for HNO_3 shows that the process is at least, if not more so, as efficient for HNO_3 removal by snow as it is for the three aforementioned species removed by rain.

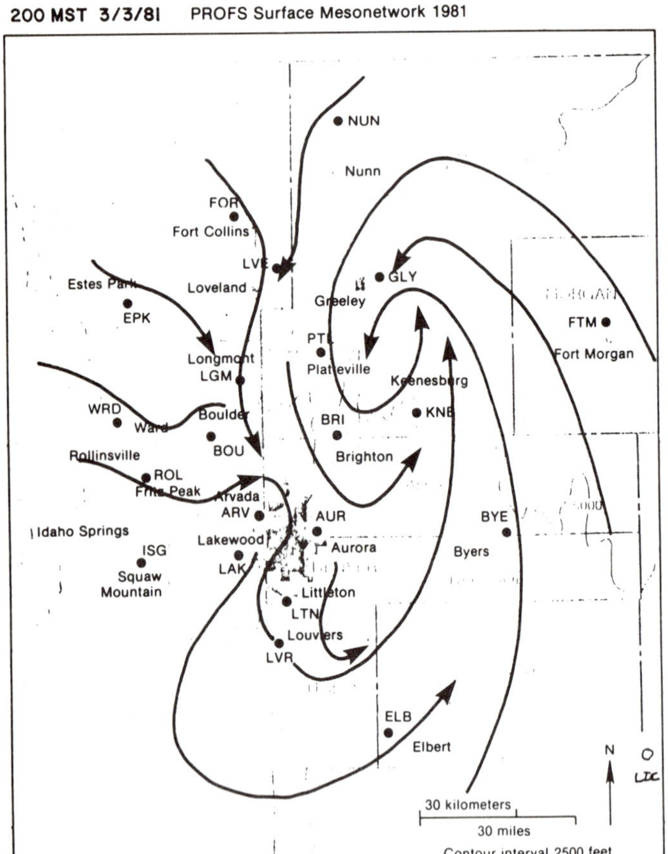

Figure 7. Computer Surface Level Air Trajectories in the Denver, Colorado, Vicinity at 0200 MST on 3 March 1981. The solid symbols are the meteorological stations of the NOAA Prototype Regional Observation and Forecast Service (PROFS), on whose data the trajectory calculations were based.

CONCLUSIONS

As Lewis and Grant (1980a) have noted earlier, acid precipitation is occurring in the Colorado mountains. Our relatively few event pH measurements of falling snow are in the range of values observed in their long term collector sampling program. What we have added in this regard is an identification of the likely major source, namely, the urban areas to the east, despite the fact that the prevailing winds are from the west. Furthermore, the measurements of the air concentrations of the acidic materials and their precursors, and the seasonal variations therein, are qualitatively consistent with the chemical transformations believed to be involved.

ACKNOWLEDGMENTS

We acknowledge the support of the staff of the Mountain Research Station in the course of these measurements. We also have profited from useful discussions and help from J. M. Miller, W. M. Lewis, M. C. Grant, and R. E. Sievers, and many of our colleagues in the Aeronomy Laboratory.

LITERATURE CITED

Barry, R. G., 1972. Climatic Environment of the East Slope of the Colorado Front Range. Inst. Arct. Alp. Res. Occas. Pap. No. 3.

Barry, R. G., 1973. A Climatological Transect on the East Slope of the Front Range, Colorado. Art. Alp. Res. 5:89-100.

Bollinger, M. J., D. D. Parrish, C. Hahn, D. L. Albritton, and F. C. Fehsenfeld, 1982. NO_x Measurements in Clean Continental Air. *In:* Composition of the Nonurban Troposphere. 2nd Symposium. Williamsburg, Virginia, 25-28 May, 1982, American Meteorological Society, Boston, Massachusetts.

Goldan, P. D., W. C. Kuster, D. L. Albritton, F. C. Fehsenfeld, P. S. Connell, R. B. Norton, and B. J. Huebert, 1983. Calibration and Tests of the Filter-Collection Method for Measuring Clean-Air Ambient Levels of Nitric Acid. Atmos. Environ. (in press).

Heffter, J. L., 1980. Air Resources Laboratories Atmospheric Transport and Dispersion Model. NOAA Tech. Memo, ERL-ARL-81.

Hicks, B. B., 1980. On the Dry Deposition of Acid Particles to Natural Surfaces. *In:* Polluted Rain, T. Y. Toribara, M. W. Miller, and P. E. Morrow (Editors). Plenum Press, New York, New York, pp. 327-339.

Huebert, B. J., 1980. Nitric Acid and Aerosol Nitrate Measurements in the Equatorial Pacific Region. Geophys. Res. Lett. 7:325-328.

Huebert, B. J. and A. L. Lazrus, 1980. Tropospheric Gas Phase and Particulate Nitrate Measurements. J. Geophys. Res. 85:7337-7344.

Huebert, B. J., R. B. Norton, M. J. Bollinger, D. D. Parrish, P. C. Murphy, D. L. Albritton, and F. C. Fehsenfeld, 1982. HNO_3 and NO_3^- Particulates Measured at a Remote Rural Site. *In:* Composition of the Nonurban Troposphere. 2nd Symposium, Williamsburg, Virginia, 25-28 May 1982, American Meteorological Society, Boston, Massachusetts.

Kelly, T. J. and D. H. Stedman, 1980. Effect of Urban Sources on Acid Precipitation in the Western United States. Science 210: 1043.

Kelly, T. J., D. H. Stedman, J. A. Ritter, and R. H. Harvey, 1980. Measurements of Oxides of Nitrogen and Nitric Acid in Clean Air. J. Geophys. Res. 85:7417-7425.

Kley, D., J. W. Drummond, M. McFarland, and S. C. Liu, 1981. Tropospheric Profiles of NO_x. J. Geophys. Res. 86:3153-3161.

Lewis, W. M. and M. C. Grant, 1980a. Acid Precipitation in the Western United States. Science 207:176-177.
Lewis, W. M. and M. C. Grant, 1980b. Effect of Urban Sources on Acid Precipitation in the Western United States. Science 210: 1043.
Liu, S. C., D. Kley, M. McFarland, J. D. Mahlman, and H. Levy II, 1980. On the Origin of Tropospheric Ozone. J. Geophys. Res. 85:7546-7552.
Logan, J. A., M. J. Prather, S. C. Wofsy, and M. B. McElroy, 1981. Tropospheric Chemistry: A Global Perspective. J. Geophys. Res. 86:7210-7254.
McFarland, M., D. Kley, A. L. Schmeltekopf, and R. H. Winkler, 1979. Nitric Oxide Measurements in the Equatorial Pacific Region. Geophys. Res. Lett. 6:605-608.
Parrish, D. D., P. C. Murphy, D. L. Albritton, and F. C. Fehsenfeld, 1983. The Measurement of the Photodissociation Rate of NO_2 in the Atmosphere. Atmos. Environ. (in press).
Roberts, J. M., P. D. Goldan, D. L. Albritton, F. C. Fehsenfeld, and R. E. Sievers, 1982. Measurements of Atmospheric Hydrocarbons at a Remote Rural Site. *In:* Composition of the Nonurban Troposphere. 2nd Symposium, Williamsburg, Virginia, 25-28 May 1982, American Meteorological Society, Boston, Massachusetts.

SPATIAL AND TEMPORAL TRENDS IN THE CHEMISTRY OF ATMOSPHERIC DEPOSITION IN NEW ENGLAND

Denis W. Hanson and Stephen A. Norton[1]

ABSTRACT: We have evaluated atmospheric deposition related geochemical changes in New England and Quebec by means of (1) transect studies of soil chemistry parallel to pH and metal deposition gradients; and (2) chemical analysis of lake sediments with the following results: (a) Mn, Ca, Mg, K, Zn, Fe, and Pb concentrations and pH in organic litter of high altitude spruce-fir forests relate to the present pH gradients for precipitation and inferred gradients of heavy metal loading; (b) atmospheric deposition of heavy metals (including Pb, Zn, and Cu) began increasing in the 19th century in New England and may be leveling out in the last 5-10 years. Accelerated mechanical erosion may mask increases in concentration of these elements in lake sediments but deposition rate calculations indicate as much as a 12X increase in sediment accumulation rate of Pb due to atmospheric loading. Relative increases depend on increased loading, background values, and total deposition rate. Inferred acidification of precipitation and surface waters has resulted in increased leaching of Zn, Ca, Mg, Mn, and Al from soils and sediments with inferred concurrent increases in concentrations of these elements in surface waters. Results (a) and (b) imply that increasing atmospheric deposition of acids and metals has strongly influenced forest soils, sediment, and surface water chemistry.
(KEY TERMS: acidic precipitation; heavy metals; sediment; soils; deposition rates; air pollution.)

INTRODUCTION

Precipitation in eastern North America is generally recognized as being acidic and increasingly so; the acidity is due to anthropogenic emissions of SO_X and NO_X gases, precursors of H_2SO_4 and HNO_3. However, the time frame of the acidification is not well known and is of critical importance in evaluating the urgency of the "acid rain problem." Furthermore, it is not known if the area affected is expanding with time. Associated with atmospheric emissions of acid precursors are metal pollutants, including Pb, Zn, and Cu. Little is known of their distribution, both in space and time. Also, the linkage between metal emissions and the deposition rates of these metals (as well as for H^+) is unclear and probably changing with time due to different source terms, changing stack heights, changing fuels, the advent of various emission controls, and other factors.

Historic synoptic data for precipitation chemistry (H^+, $SO_4^=$, NO_3^-) are sparse prior to the advent of the National Atmospheric Deposition Program. Data for metals are nearly nonexistent. Consequently, we have employed indirect environmental measurements to assess both modern and historic patterns for atmospheric deposition of acids and metals.

Regional studies of acidic precipitation induced changes in terrestrial ecosystems have not been undertaken due to the relative complexity of terrestrial ecosystems, the lack of preacidic precipitation baseline data, and the long time scale of observable changes. In northeastern North America, forests grow on young, generally infertile soils in which most exchangeable nutrients are in the surface organic layer. Increased cation leaching by acidic precipitation, or slowed decomposition due to increasing soil acidity and heavy metal accumulation may affect ecosystem productivity. Concurrently, the leached cations will alter downstream water chemistry. We used a regional soil sampling transect to study acidic precipitation induced changes in forest soils.

In eastern North America, precipitation pH isopleths (and probably heavy metal deposition isopleths) are probably expanding. The pH isopleths in New England and Atlantic Canada form a northeast trending gradient controlled by regional storm track patterns. A transect on this pH gradient is effectively a time study of precipitation pH induced changes. For example, if the pH of precipitation in Vermont has decreased from 4.5 to 4.0 in the last 30 years, sampling and comparing a Vermont soil site with a similar soil site in Maine now receiving precipitation of pH 4.5 will indicate the direction of changes which occurred in the Vermont soil due to increasingly acidic precipitation in that 30-year period. Changes in element accumulation and depletion can be assessed. At the same time, chemical analysis of recent forest litter gives a synoptic view of precipitation related changes.

Lake sediments integrate and record the chemistry of many processes within the watershed as well as airshed changes. For example, TiO_2 in lake sediment is a relatively immobile soil component which other elements may be compared to for relative rates of chemical and mechanical erosion. Other elements which are not mobile, concentrated

[1]Geological Sciences, University of Maine at Orono, Orono, Maine 04469.

in biological material, or subject to diagenetic mobilization (e.g., Cr) should co-vary with Ti. All these immobile elements may vary reciprocally with variations in organic matter (including diatom frustules) as well as autochthonous inorganic matter. Some elements (e.g., Na) may vary vertically in a sediment core due to the changing proportions of allochthonous inorganic matter and other sediment components or due to variations in the degree of chemical weathering of the soil precursors of sediment (Davis and Norton, 1978). Other elements are partially controlled by organic content of the sediment. For example, Ca is readily leached from inorganic and organic soils but may form as much as 25% of the ash from the organic matter. Thus, sediments high in organic matter may have relatively high Ca. Elements such as Fe and Mn may be mobilized upward by diagenetic processes as the sediment accumulates.

With all these factors determining bulk composition of the sediment, unambiguous interpretation of time dependent changes in sediment chemistry is difficult. However, in lakes which have relatively undisturbed drainage basins, many of these confounding variables have been relatively constant while atmospheric chemistry has changed. Thus, meaningful interpretation of chemical changes recorded in the sediment is possible.

METHODS

We sampled subalpine conifer forest soil litter from 14 mountains in northern New England and eastern Canada (Figure 1). The pH of precipitation presently ranges from about 4.0 (site 1) to 4.6 (site 14) (Semonin, *et al.,* 1981). All sites are located on noncalcareous granitic or pelitic bedrock.

The subalpine conifer forest was selected for study because: (1) it is present on higher peaks throughout the region; (2) it is a structurally simple ecosystem dominated by balsam fir (*Abies balsamea*) (Reiners and Lang, 1979); (3) anthropogenic effects are restricted to airborne pollution; and (4) both wet and dry pollutant inputs are greater than at lower elevations due to increased wet deposition, and to cloud droplet and aerosol impaction on the dense foliage because of frequent cloud and fog cover and high winds (Schlesinger and Reiners, 1974; Reiners, *et al.,* 1975). These factors suggested that soils of subalpine conifer forests might be the first component of the terrestrial ecosystem to exhibit acidic precipitation impact.

Soil litter samples (15 from each site, Figure 1) were air dried at 50°C, ground in a Wiley mill to 40 mesh, or ground in a mortar and pestle, and then digested in HNO_3 and $HClO_4$ (Hanson, 1980). The filtered solution was analyzed by atomic absorption spectrophotometry (Perkin-Elmer 703). Soil pH was measured in a 1:10 soil to 0.01M $CaCl_2$ slurry.

Sediment cores were obtained with a stationary piston corer (Davis and Doyle, 1969) from the profundal region of two lakes in Maine (Figure 1). Mountain Pond (Rangeley) has the following characteristics: Summer pH = 5.8, 2.6 ha, 837 m.a.s.l., z_{max} = 3.5 m. Fish River Lake has the following characteristics: Summer pH = 7.5, 1070 ha, 217 m.a.s.l., $z_{max} \cong 18$ m. The sediments have been dated by Cs^{137} and Pb^{210} (Johnston, 1980). Chronology has been established with the constant rate of supply (CRS) model (Appelby, 1978). Sediment was placed in solution by the method of Buckley and Cranston (1971). Chemical analyses were performed by flame atomic absorption spectrophotometry (all elements except Pb) and by graphite furnace atomic absorption spectrophotometry for Pb. Deposition rates are based on the CRS chronology in the upper part of each core but extrapolated to ages greater than 100 years by using the constant input concentrations model (CIC) age-depth relationships.

Two lake's sediment records are discussed because of a contrasting history of mechanical stability in the watershed, particularly in approximately the last 100 years. Accelerated erosions of soils and deposition of sediment due to logging, road construction, etc., confuses the interpretation of the chemical record preserved in the sediment (Davis and Norton, 1978). The relatively constant concentration of TiO_2 in the sediment in Mountain Pond and Fish River Lake suggests a lack of serious disturbance in the watershed over the last 200 + years. However, the Fish River watershed has undergone intense logging in recent time (~ 25 years ago), and the Mountain Pond drainage basin was logged sometime prior to 50 years ago.

RESULTS

Soil Transect

Chemical analyses indicate that cation leaching and heavy metal accumulation gradients exist which follow the precipitation pH gradient. Leaching of Mn (Figure 2) and Ca from litter increases as precipitation acidity increases. Increased K, Mg (Figure 3), and Zn (Figure 4) leaching appears to occur only at the Vermont sites, the part of the sampling transect where precipitation acidity is greatest. This suggests that K, Mg, and Zn leaching rates increase only when some threshold precipitation pH (ca. 4.0) is reached. The order of relative susceptibility to increased leaching due to increased precipitation acidity is Mn $>$ Ca $>$ Mg $\geqslant$ K $\geqslant$ Zn. However, Zn leaching is probably greater than Figure 4 indicates because Zn loading is greatest at the Vermont sites (Galloway, *et al.,* 1982), even though litter Zn concentrations are lower. Sodium and Cd possibly show leaching patterns similar to those of K, Mg, and Zn.

Iron and Pb (Figure 5) in litter increase southwesterly. Concentrations are higher where precipitation acidity is higher. The Fe and Pb gradients are at least partially caused by relative enrichment of Fe and Pb in litter as more mobile cations and compounds are leached. Relative enrichment

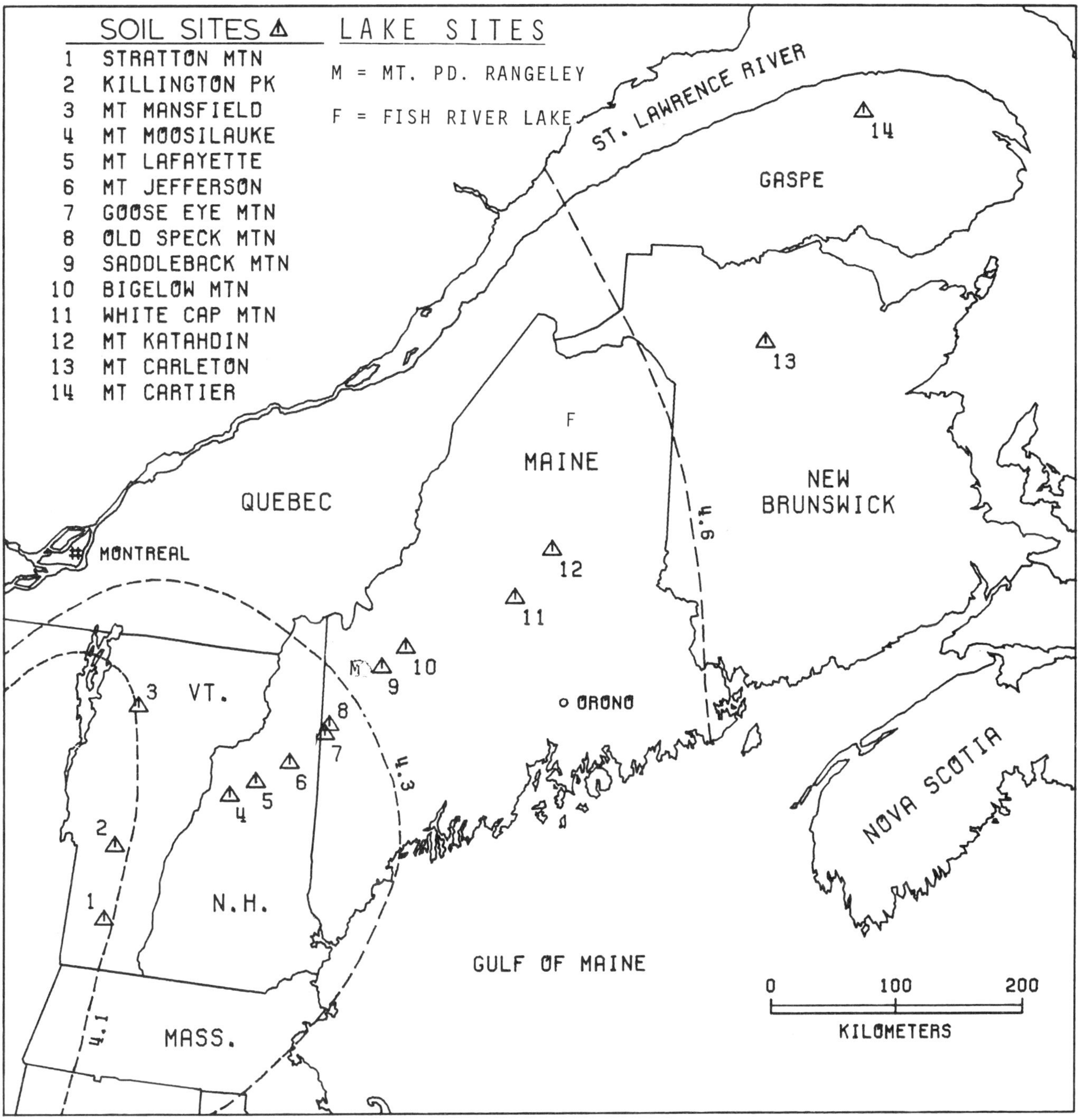

Figure 1. Location of Soil Transect Sites and the Two Study Lakes. Approximate pH isopleths (4.1, 4.3, and 4.6) are based on Semonin, *et al.* (1981).

is greatest at sites receiving precipitation of greater acidity. In addition, a large part of the Pb and Fe concentration gradient in litter is due to an atmospheric Pb and Fe deposition gradient and accumulation which follows the regional precipitation-Ph gradient (Galloway, *et al.*, 1982). The regional Pb concentration pattern indicates that much of the accumulated Pb is not local in origin, even though motor vehicle and population densities in parts of the region are high. Lead concentrations are highest in soil L and F layers (Hanson, 1980), indicating that Pb accumulation is a recent, continuing phenomenon. Yearly Pb input rates (> 100 mg m^{-2} y^{-1}) to subalpine fir forest ecosystems are much greater than those at remote lowland areas (Chow and Johnstone, 1965) and are comparable with some urban area Pb input rates.

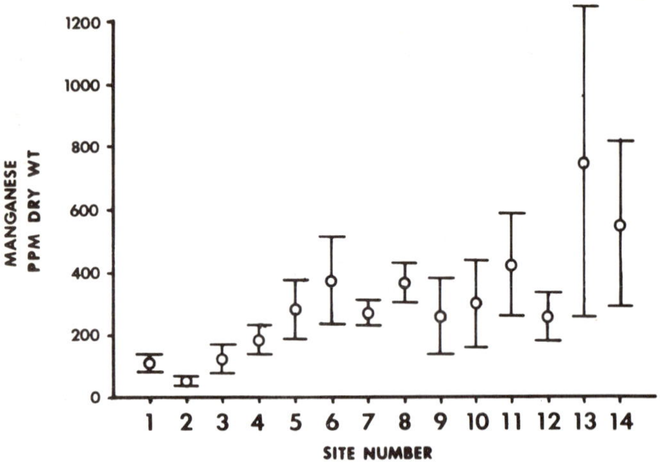

Figure 2. Manganese Concentrations of Organic Soils From Southern Vermont (1) to the Gaspe' Peninsula, Quebec (14). The error bar is one standard deviation.

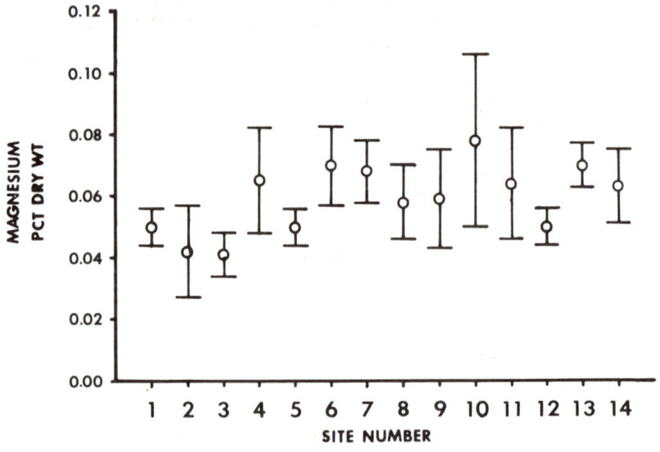

Figure 3. Magnesium Concentrations of Organic Soils From Southern Vermont (1) to the Gaspe' Peninsula, Quebec (14). The error bar is one standard deviation.

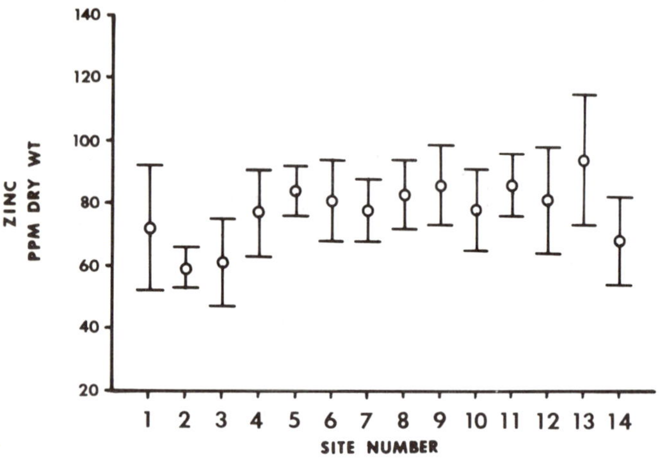

Figure 4. Zinc Concentrations of Organic Soils From Southern Vermont (1) to the Gaspe' Peninsula, Quebec (14). The error bar is one standard deviation.

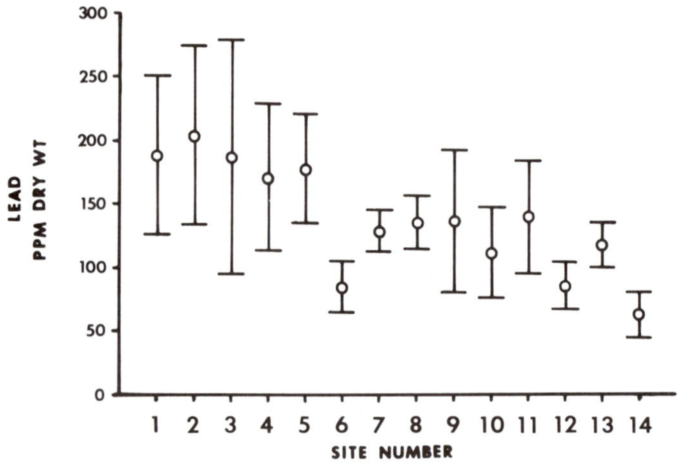

Figure 5. Lead Concentrations of Organic Soils From Southern Vermont (1) to the Gaspe' Peninsula, Quebec (14). The error bar is one standard deviation.

Soil litter also shows a pH gradient across the transect (Figure 6). Litter acidity generally increases as precipitation acidity increases. The soil litter pH pattern suggests that K, Mg, and Zn leaching increases at lower pH's than those necessary for increased Mn and Ca leaching.

Cluster analysis of soil litter chemistry data orders the mountain sites, with one exception, according to their position along the precipitation pH gradient (Figure 7). This implies that precipitation-pH, and associated heavy metal deposition, control soil litter chemistry in subalpine fir forests. Other possible sources of regional concentration gradients are probably not the cause of the concentration gradients determined.

Sediment Chemistry

The chemistry of the Fish River Lake sediment core (Figure 8) suggests steady state conditions except for the pronounced increase in total Pb (~ 17 cm), an indistinct increase in Zn, and an increase in Cu at about 10 cm. The presence of a sharp inflexion in the Pb^{210} curve at ~ 11 cm suggests, however, a sharp increase in the sedimentation rate.

These two observations indicate nonsteady state conditions for Zn, Pb, and Cu deposition as well as an independent increase in total sediment deposition per unit time. Figures 9a and 9b for Pb and K_2O reinforce this interpretation. The Pb deposition rate increases from about 0.75

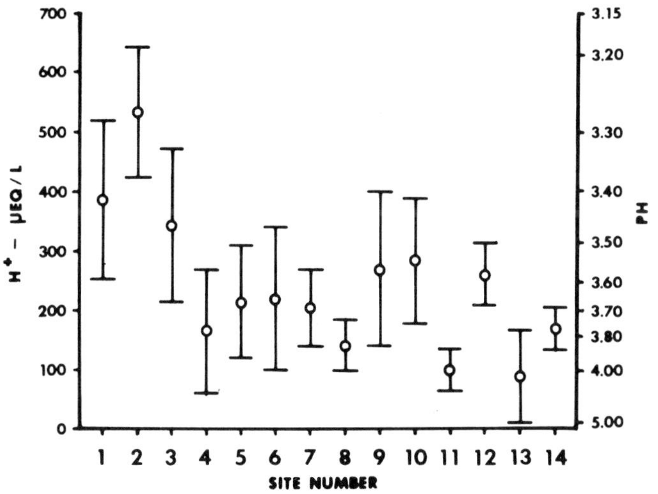

Figure 6. Soil Litter pH From Southern Vermont (1) to the Gaspe' Peninsula, Quebec (14). The error bar is one standard deviation.

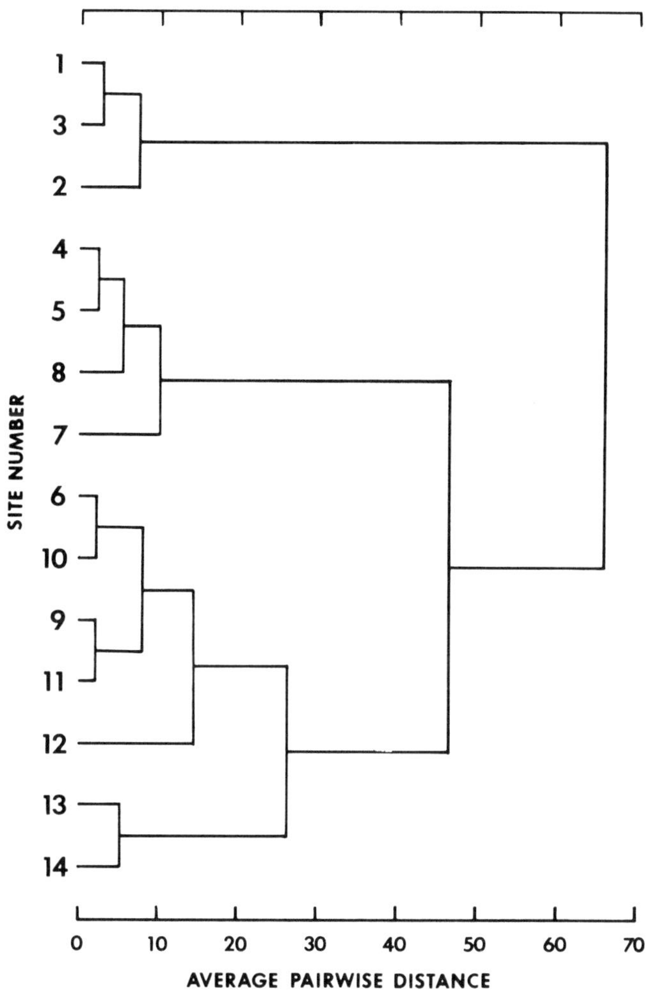

Figure 7. Cluster Analysis of 14 Soil Sites From Southern Vermont (1) to the Gaspe' Peninsula (14) Using Concentrations of Ca, Mg, K, Na, Al, Fe, Mn, Pb, Zn, and Cd.

μg/cm^2/y to 3 μg/cm^2/y during 1880 to 1978. K_2O does not increase concurrently. The K_2O deposition rate (along with all other major rock forming elements such as Na, Ca, Mg, Fe, Mn, Al, and Si) increases approximately 70 percent after 1950. This period coincides with major logging and land disturbance in the drainage basin. Major elements are dominantly derived from within the drainage basin and all major element deposition rates increase by approximately this amount. However, this increase accounts for only a fraction of the increase for Pb, Zn, and Cu. The latter increases are attributed to increased deposition from the atmosphere. The relative increases for Pb, Zn, and Cu seen in the cores are a function of background concentrations, atmospheric loading rates, and deposition rates of inorganic detritus.

The chemistry of the Mountain Pond sediment core (Figure 10) also suggests relatively steady state conditions and a concurrent increase in Pb and Zn, with the additional observation that Ca and especially Zn decrease near the surface. However, the deposition rate curves (Figure 11a and 11b) for Pb and K_2O suggest that steady state does not prevail. The deposition rate of K_2O increases from 8 to about 19 μg/cm^2/y (a 2.5 x increase) over the last 75 years. The Pb curve starts rising at about 1875 and increases 14X; only 2.5X of the increase can be linked with increased mechanical erosion, the rest must be atmospheric Pb. Similar trends exist for Zn and Cu.

Comparison of deposition rates (for the same element) between lakes reveals that Fish River Lake is receiving sediment at the coring site at 10 to 100X the rate of Mountain Pond. This is caused primarily by sediment focusing processes.

SUMMARY

1. Precipitation-pH and associated heavy metal deposition strongly influence soil litter chemistry in subalpine fir forests.

2. Lead and Fe accumulate in subalpine soils relative to Zn, Mn, Ca, Mg, and K in areas receiving the most acidic precipitation.

3. Zn, Mn, Ca, and to a lesser extent Mg and K are preferentially leached from subalpine soils receiving acidic precipitation.

4. Increased deposition rates for Pb, Zn, and Cu start about 1880 and increase to the present (Pb) or nearly to the present (Zn and Cu). Zn may be leached from soil detritus and sediment during progressive acidification of the lake ecosystem.

5. Increased mechanical erosion, commonly out of phase with increased atmospheric deposition of Pb, Zn, and Cu, also accelerates deposition of these elements.

6. Pb^{210} dating is necessary to accurately assess sediment chemistry, and particularly to compare deposition rates within and between lakes.

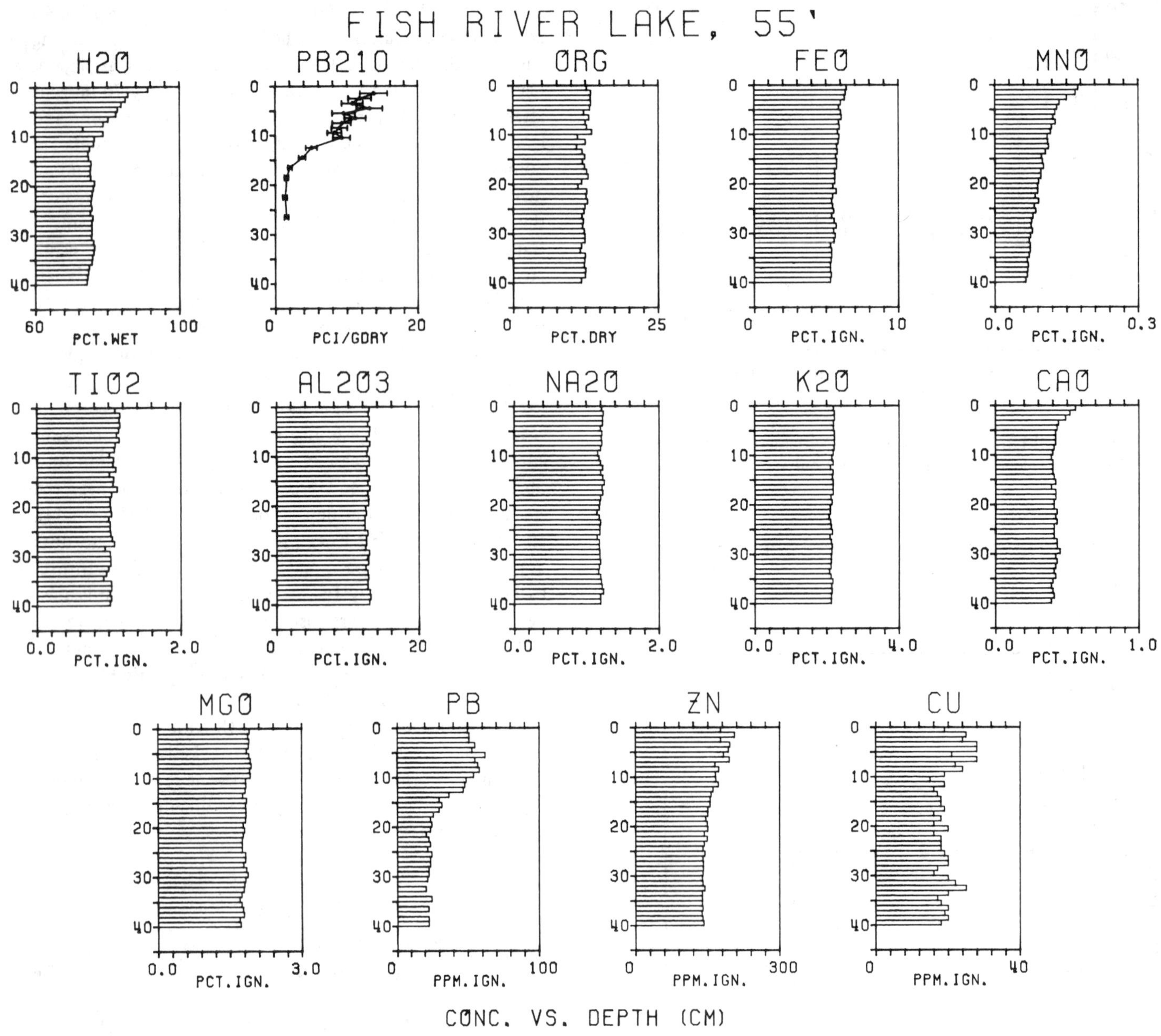

Figure 8. Chemistry of the Lake Sediment Core From Fish River Lake, Maine. Units are weight percent or ppm of ignited sediment.

7. If metal deposition from the atmosphere can be used as an approximate surrogate for deposition of acid, acid precipitation started at least 100 years ago and has been increasing in an exponential fashion. Presently, it is the most acidic it has been for the last 100 years.

ACKNOWLEDGMENTS

Portions of this data (soils) were published in "Water, Air and Soil Pollution," Sediment analyses for Mountain Pond (Rangeley) were supported by the U.S. National Science Foundation (Grant DEB-78-10641 to R. B. Davis and S. A. Norton). The Fish River Lake data were gathered with support from the Superior Mining Company, employees of which helped in the field work. We were greatly assisted in the field and laboratory by Geneva Blake, Elaine Hall, and Marilyn Morrison.

LITERATURE CITED

Appleby, P. G. and F. Oldfield, 1978. The Calculation of Lead-210 Dates Assuming a Constant Rate of Supply of Unsupported ^{210}Pb to the Sediment. Catena 5:1-8.

Buckley, D. E. and R. E. Cranston, 1971. Atomic Absorption Analyses of 18 Elements From a Single Decomposition of Aluminosilicate. Chem. Geol. 7:273-284.

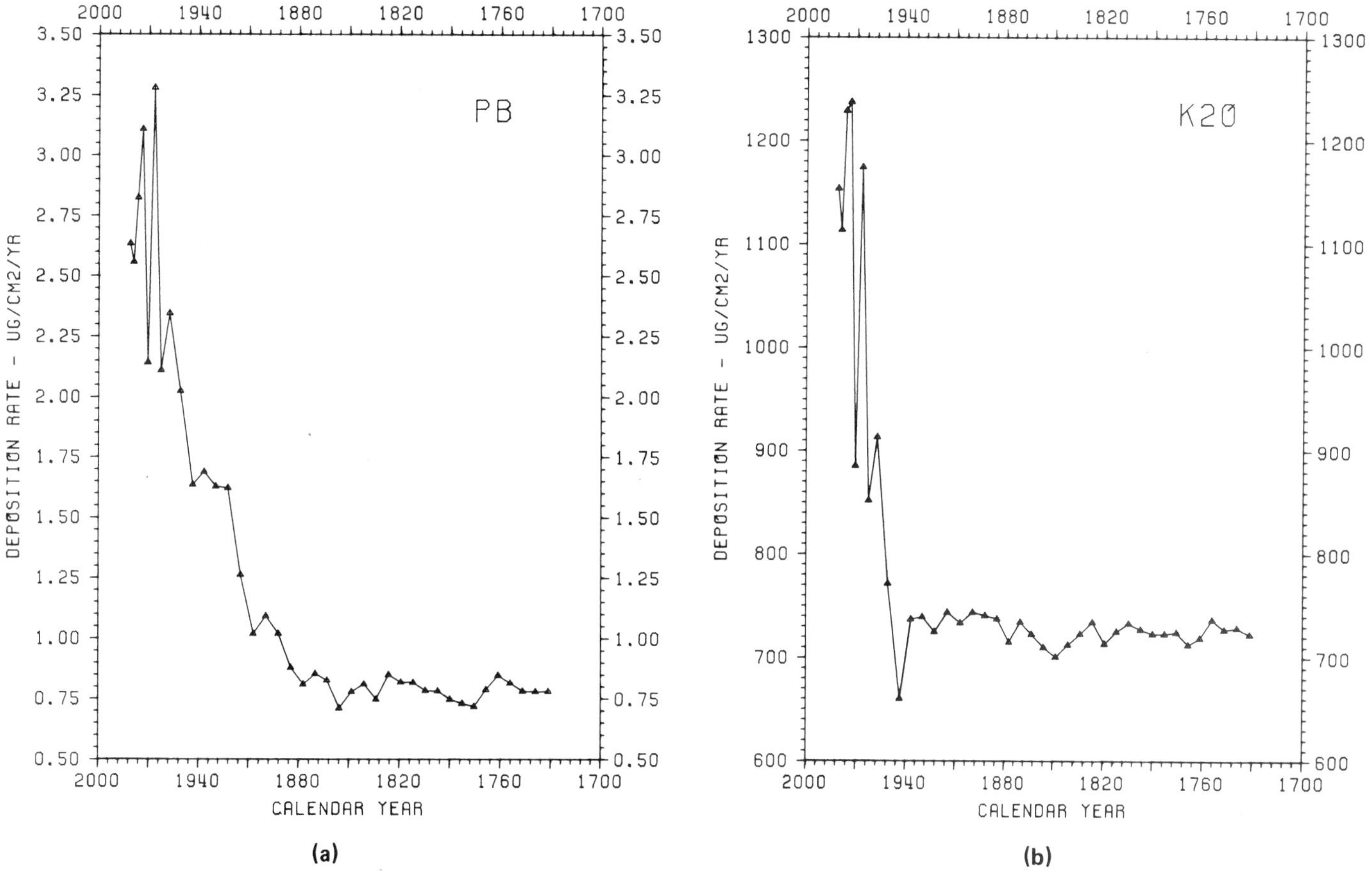

Figure 9. (a) Deposition Rate of Pb for Fish River Lake, Maine; and (b) Deposition Rate of K_2O for Fish River Lake, Maine.

Chow, T. J. and M. S. Johnstone, 1965. Lead Isotopes in Gasoline and Aerosols of Los Angeles Basin, California. Science 147:502-503.

Davis, R. B. and R. W. Doyle, 1969. A Piston Corer for Upper Sediment in Lakes. Limnol. Oceanogr. 14:643-648.

Davis, R. B. and S. A. Norton, 1978. Paleolimnologic Studies of Human Impact on Lakes in the United States, With Emphasis on Recent Research in New England. Pol. Arch. Hydrobiol. 25:99-115.

Galloway, J. N., J. D. Thornton, S. A. Norton, and H. L. Volchok, 1982. Trace Metals in Atmospheric Deposition: A Review and Assessment. Atmos. Environ. 16:1677-1700.

Hanson, D. W., 1980. Acidic Precipitation-Induced Chemical Changes in Subalpine Fir Forest Organic Soil Layers. Unpubl. Ms. Thesis, Univ. Maine-Orono, 90 pp.

Johnston, S. E., 1980. A Comparison of Dating Methods in Laminated Lake Sediments in Maine. Unpubl. Ms. Thesis, Univ. Maine-Orono, 79 pp.

Reiners, W. A. and G. E. Lang, 1979. Vegetational Patterns and Processes in the Balsam Fir Zone, White Mountains, New Hampshire. Ecology 60:403-417.

Reiners, W. A., R. H. Marks, and P. D. Vitousek, 1975. Heavy Metals in Subalpine and Alpine Soils of New Hampshire. Oikos 26:264-275.

Schlesinger, W. H. and W. A. Reiners, 1974. Deposition of Water and Cations on Artificial Foliar Collectors in Fir Krummholz of New England Mountains. Ecology 55:378-386.

Semonin, R. G., V. C. Bowersox, D. F. Gatz, M. E. Peden, and G. J. Stensland, 1981. Study of Atmospheric Pollution Scavenging. Ill. Inst. Nat. Resourc., 19th Progress Report.

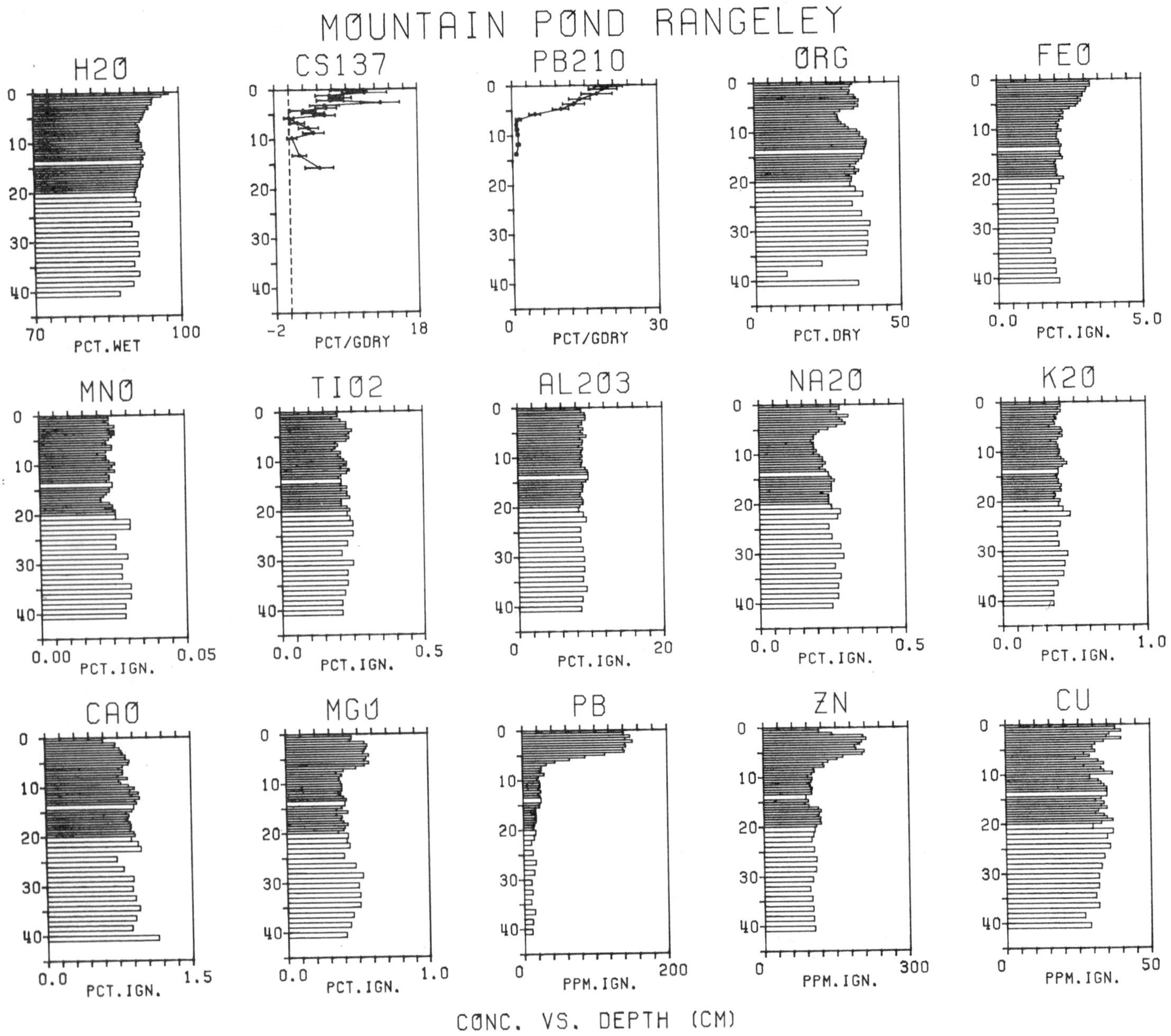

Figure 10. Chemistry of the Lake Sediment Core From Mountain Pond (Rangeley), Maine. Units are weight percent or ppm of ignited sediment.

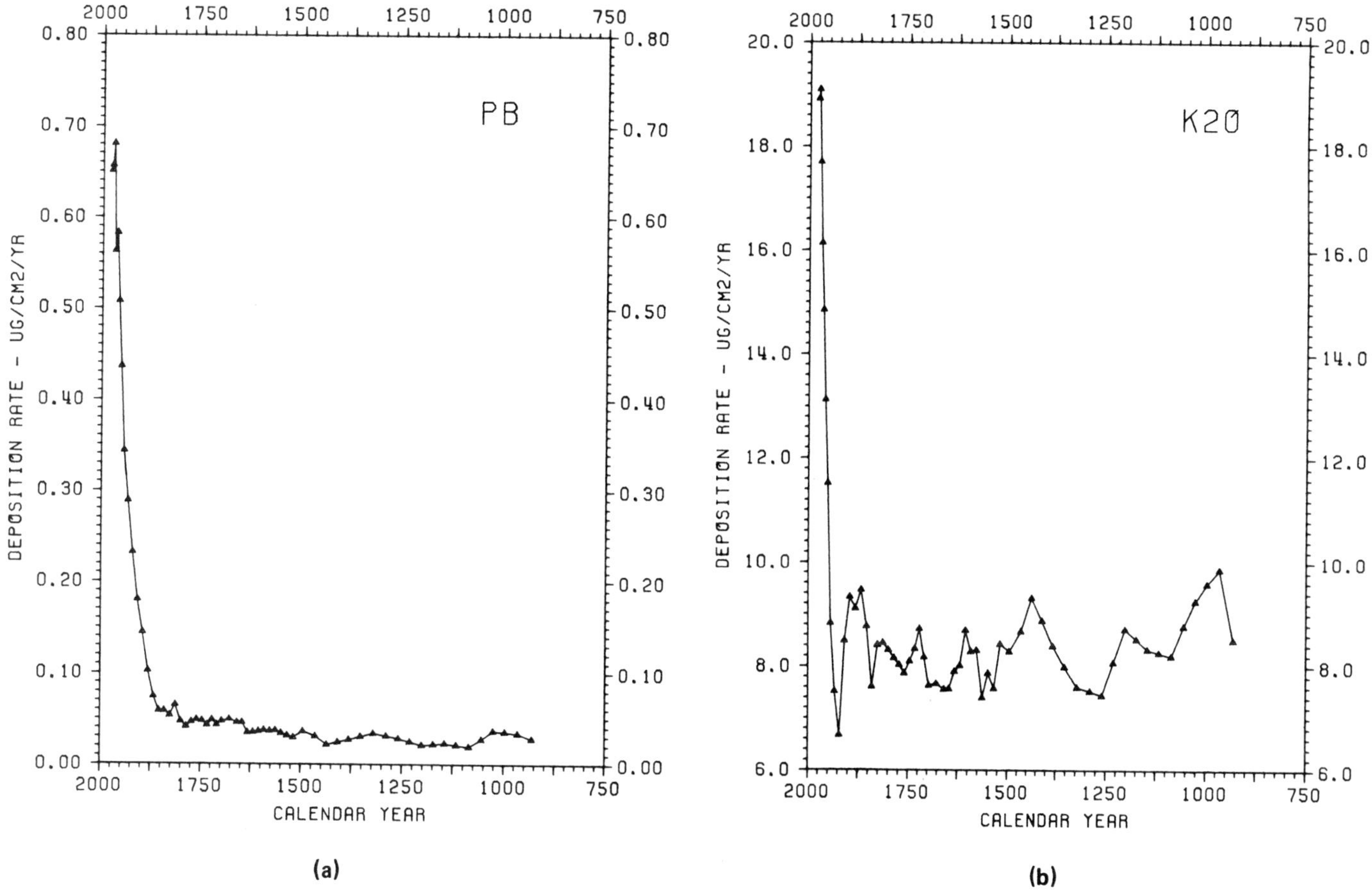

Figure 11. (a) Deposition Rate of Pb for Mountain Pond (Rangeley), Maine; and (b) Deposition Rate of K_2O for Mountain Pond (Rangeley), Maine.

INTERNATIONAL SYMPOSIUM ON HYDROMETEOROLOGY
JUNE AMERICAN WATER RESOURCES ASSOCIATION 1982

RAPID FLUCTUATIONS IN STREAMFLOW pH AND ASSOCIATED WATER QUALITY PARAMETERS DURING A STORMFLOW EVENT

Edward S. Corbett and James A. Lynch[1]

ABSTRACT: Three aspects of hydrometeorological significance – the quality of precipitation, the watershed response of converting precipitation into streamflow, and the resulting change in streamflow quality – were studied on a forested experimental watershed in central Pennsylvania. Changes in streamflow quality over time due to a 4.38-inch rainstorm are discussed. Hydrograph analysis indicated 23.6 percent of the rainfall input was discharged as stormflow. Streamflow pH declined from 7.32 to 4.95 during the storm, and H+ concentration increased over 200 times. Alkalinity decreased from 480 to 20 μeq/L, while acidity increased from −328 to 220 μeq/L during the stormflow period. Seven days after rainfall ended and during which time no additional rain occurred, alkalinity had only recovered to 73 μeq/L.

(KEY TERMS: acid precipitation; streamflow pH; acidity; hydrogen ion concentration; alkalinity; storm hydrograph response.)

INTRODUCTION

Although the effects of long-term acidification on the chemistry and biota of lakes has received much attention, the impact of short-term depressions of streamflow pH, such as those occurring during snowmelt or individual stormflows, is less well known. Many investigators believe the acidification results from increased inputs of hydrogen ion from atmospheric sources. Other potential sources of hydrogen ion include changes in land use, weathering and oxidation of reduced sulfur materials, or generation of organic acids from organic litter decomposition (Bisogni and Driscoll, 1979).

Acidification of surface waters may have serious consequences for the aquatic biota inhabiting these systems. Of particular significance are episodic events, such as snowmelt and initial rains after drought, which cause rapid changes in water chemistry. When these events coincide with vulnerable life stages of aquatic organisms in the affected waters, they can be particularly devastating (Cowling and Linthurst, 1981).

Sudden drops in pH cause severe physiological stress that may produce fish kills at pH levels above those normally toxic to adult fish. Resistance to such shocks is species dependent (Wright, *et al.*, 1976). Trojnar (1977) found that the survival of brook trout fry, especially those incubated in nonacidic water, decreased significantly when exposed to waters at pH less than 5.0. Sudden drops in pH are often observed in the early spring when the pollutants accumulated in the snowpack are concentrated in the meltwater during the first phases of melting (Hagen and Langeland, 1973; Hultberg, 1976; Johannessen and Henriksen, 1978). Oligotrophic lakes and streams with low buffering capacity are especially susceptible to pH drops. Survival and emergency of many aquatic insects are reduced by short exposures to low pH waters (Bell, 1971). Several studies attribute fish mortality to the increased aluminum concentrations that are associated with decreases in lake or streamwater pH. An extensive review of how the aquatic ecosystem may be impacted by acid precipitation is presented by Haines (1981).

Fluctuations in Streamwater pH

Although changes in lake and stream pH during the snowmelt runoff period have been sampled in several studies, there still remains a lack of quantitative data on rapid changes in streamwater pH due to individual storm events. Even where watersheds have considerable buffering capability, the buffering reactions may not be fast enough to prevent a decline in stream pH during snowmelt or other major storm events (Scheider, *et al.*, 1979; Jeffries, *et al.*, 1979).

Jeffries, *et al.* (1979), compared the pH of several streams in mid-March before spring runoff began to pH measured at peak runoff in mid-April. An increase in H+ concentration at peak runoff was observed in all streams. Small streams generally had greater changes in terms of increased H+ concentrations. However, even in large rivers the pH declined about 0.2 to 0.4 pH units, accompanied by a reduction in alkalinity.

Following a 2-cm rainstorm in July, Scheider, *et al.* (1979), found that the H+ content of a stream draining one

[1] Respectively, Principal Hydrologist, Northeastern Forest Experiment Station, U.S. Forest Service, 104 Academic Projects Bldg., University Park, Pennsylvania 16802; and Associate Professor of Forest Hydrology, School of Forest Resources, The Pennsylvania State University, 311 Forest Resources Laboratory, University Park, Pennsylvania 16802.

of the subwatersheds of Red Chalk Lake in Southern Ontario increased by a factor of 10. The H+ content began to increase 8 hours after the start of the 4-hour storm and peaked 3 hours later. Two hours after the storm peak, the H+ content began to approach prestorm levels.

THE STUDY

Study Site

A study to quantify and characterize the changes in streamwater pH and associated water quality parameters due to single storm events was conducted on the Leading Ridge Experimental Watersheds located in the Ridge and Valley Province of central Pennsylvania. They are representative of a substantial portion of Pennsylvania forest land, much of which lies within established municipal watershed boundaries.

The vegetative cover consists of an uneven aged forest of oak, hickory, and maple. Most of the soils are residual, having developed in place through the weathering of the underlying strata. Soils on the lower slopes are primarily silt and stony loams that are well drained and have a high moisture holding capacity. The middle and upper slopes are classified as cobbly and stony loams, which are also well drained. Rough, stony land covers the ridge top. The soil mantle generally ranges in depth from 3 to 5 feet.

Three major geologic formations underlie the watersheds. Most of the lower slope is underlain by the Rose Hill shale formation. Castanea sandstone underlies the upper part of the lower slope, middle slope, and most of the upper slope. Tuscarora quartzite underlies the ridge top.

Measurement Procedures

A modified broad crested Trenton weir with a sharp crested, 90-degree v-notch in the center was used to measure stream discharge. Changes in streamflow were monitored with a FW-1 water level recorder. A continuous recording pH meter (Analytical Measurements Model 30) monitored changes in streamflow pH. Water samples were taken by both grab sampling and by battery powered sequential water samplers to verify the recorder results and for laboratory analysis of water quality parameters. Precipitation samples were collected with an Aerochem Metrics Model 301 automatic wet/dry precipitation collector located at the research site. Alkalinity and acidity were analyzed with a Fisher Titrimeter II system, pH with an Orion model 901 ionalyzer, and specific conductance on a Hach model 16300 conductivity meter. Concentrations of H+ were calculated directly from the pH data. (The use of trade, firm, or corporation names in this paper is for the information and convenience of the reader. Such use does not constitute an official endorsement or approval by the U.S. Department of Agriculture or the Forest Service of any product or service to the exclusion of others that may be suitable.)

RESULTS

Rainfall Input

This paper presents results of a single rainstorm on the completely forested 303-acre Leading Ridge Watershed. The precipitation event was a 4.38-inch rainstorm on October 25-28, 1981, with a duration of 57 hours. Hourly rain intensities ranged from 0.01 inch/hour to 1.32 inches/hour. Average rainfall pH was 4.38 with the most highly acidic rain occurring in the first half of the storm. The rainstorm was followed by a rainless period through November 5, 1981. Chemical data on the storm are given in Table 1.

TABLE 1. Concentration and Deposition of Selected Chemical Components for Precipitation Event of October 25-28, 1981.*

Component	Concentration μeq/L	Deposition (mg/m^2)
$SO_4^{=}$	20.4	109.03
NO_3^-	11.3	77.88
NH_4^+	11.1	22.25
H+	41.7	4.45

*Based on concentration data from the National Atmospheric Deposition Project (NADP/NC-141).

Hydrologic Response

For the 18-day period before this rainstorm, a total of 0.67 inch of rain had fallen on the watershed. Antecedent flow rate at the start of the storm was 0.10 csm (ft^3/sec/mi^2). The stormflow separation procedure of Hewlett and Hibbert (1967) was used to calculate total stormflow volume, which was 1.03 inches or 23.6 percent of the rainfall input. Stormflow duration was 83.8 hours. Quickflow made up 81.2 percent of the total stormflow volume, while 18.8 percent came off as delayed flow after subtracting baseflow volumes.

Streamflow responded only slightly to the first 14 hours of rainfall, which totaled 0.26 inch. The next 12 hours of rainfall (0.37 inch) produced a gradual increase in streamflow from 0.10 to 0.34 csm. The next 30 hours of the storm (3.69 inches) produced rapid increases in streamflow with a peak flow rate of 30.18 csm occurring at 2400 hours on October 27. Rainfall stopped at 0100 hours on October 28. The stormflow recession ended at 1749 hours on October 29.

Changes in pH and H+ Concentration

The changes in streamwater quality as a result of the storm event are shown in Table 2. Streamwater pH did not change during the first 14 hours of rainfall. For the next 12 hours, streamflow pH declined from 7.32 to 7.07. This corresponded to the period of gradually increasing streamflow. Concentration of H+ rose from 0.048 to 0.08 μeq/L.

Streamflow pH dropped sharply over the next 30 hours, especially during two particular intervals. During the first interval, pH changed from 6.92 to 5.73 over a 7-hour period with an input of 1.16 inches of rainfall. Stream discharge increased from 0.68 to 7.51 csm while H+ increased from 0.12 to 1.86 μeq/L during this time span. Stream discharge then leveled off at 9 to 10 csm for a period of 11 hours during which an additional 0.75 inch of rain was recorded. Streamflow pH remained fairly constant at 5.74.

The second interval occurred when a sharp burst of rainfall, 1.32 inches in 1 hour, caused streamflow pH to decrease from 5.74 to 4.95 during a 2-hour period. This decrease took place as the rate of stormflow runoff began to increase rapidly to the maximum peak flow rate for the storm. Concentration of H+ during this period reached a level of 11.22 μeq/L.

Over the next 8 hours, streamflow pH rebounded to 5.56 and then began a slow recovery reaching 6.23 at the end of the storm runoff event (1749 hours October 29) when streamflow discharge receded to 3.90 csm. By the end of the month, streamflow pH was 6.53 at a flow rate of 1.43 csm. On November 3, before the start of the next rainfall event, streamflow pH had recovered to 6.62, still 0.70 pH unit below its initial pH value. Concentrations of H+ decreased rapidly to 2.75 μeq/L 8 hours after reaching its maximum value and then began a slow decline reaching 0.30 μeq/L at the end of October.

Changes in Alkalinity and Acidity

Alkalinity decreased (480 to 80 μeq/L) and acidity increased (−328 to 80 μeq/L) steadily as streamflow rates increased in response to the first 2.5 inches of rainfall (Table 2). This corresponds to the period through the first interval of rapid streamflow pH decline. Alkalinity then decreased slowly to 20 μeq/L through the end of the storm runoff event.

Acidity, however, increased sharply from 80 to 220 μeq/L, corresponding to the second interval of rapid streamflow pH decline. Thus, the stream could be classified as acidic through most of the remainder of the stormflow event. Acidity had declined to 4 μeq/L shortly before the storm runoff event ended. Alkalinity remained depressed for some time following the storm event. As of November 3, alkalinity had recovered to only 73 μeq/L and specific conductance to 35 μmhos/cm from its prestorm level of 92 μmhos/cm.

DISCUSSION

The abrupt change in streamflow pH due to a single rainstorm event indicates the potential for creating stressful conditions within aquatic habitats. The change in streamflow pH, 7.32 to 4.95, or 2.37 pH units is substantial. The H+ concentration in the stream was increased by a factor of 233 within a 1½-day period.

Corbett, *et al.* (1975), have shown that up to 74 percent of the rainfall reaching the soil surface can be discharged as rapid subsurface stormflow when soil moisture is near field capacity. This rapid movement of acid precipitation through macropore pathways in the soil may help to explain why some lakes and streams may be subject to pulses of acidification even though the soils around them possess a considerable buffering capacity.

As only 23.6 percent of the rainfall input was converted to stormflow during this storm event, the potential may exist for even greater changes in streamflow quality. And with alkalinity remaining depressed after the storm, the stream would be more vulnerable to significant inputs of acid precipitation for some time. It remains to be determined exactly how soil and vegetation cover alter the inputs of acid precipitation.

TABLE 2. Streamwater Quality Fluctuations on Leading Ridge Watershed 1 During the Stormflow of October 26-29, 1981.

Variable	At Start of Stormflow Event, October 26	During Peak Flow Period, October 27	At End of Stormflow Event, October 29	Units
pH	7.32	4.95	6.23	pH
H+	0.05	11.22	0.59	μeq/L
Acidity	−328	220	4	μeq/L
Alkalinity	480	20	56	μeq/L
Specific Conductance	92	40	32	μmhos/cm

LITERATURE CITED

Bell, H. L., 1971. Effect of Low pH on the Survival and Emergence of Aquatic Insects. Water Res. 5:313-319.

Bisogni, Jr., J. J. and C. T. Driscoll, Jr., 1979. Characterization of the pH Buffering Systems in Dilute Adirondak Surface Waters. Res. Prj. Tech. Completion Rept. A-073-NY, NTIS, PB80-113426, 134 pp.

Corbett, E. S., W. E. Sopper, and J. A. Lynch, 1975. Watershed Response to Partial Area Applications of Simulated Rainfall. Proc. Int. Symp. on Hydrol. Characteristics of River Basins, Publ. No. 117, Int. Assoc. Hydrol. Sci., pp. 63-73.

Cowling, E. B. and R. A. Linthurst, 1981. The Acid Precipitation Phenomenon and Its Ecological Consequences. Bio-Science 31: 649-654.

Hagen, O. and A. Langeland, 1973. Polluted Snow in Southern Norway and the Effect of the Meltwater on Freshwater and Aquatic Organisms. Environ. Poll. 5:45-47.

Haines, T. A., 1981. Acidic Precipitation and Its Consequences for Aquatic Ecosystems: A Review. Trans. Am. Fish. Soc. 110:669-707.

Hewlett, J. D. and A. R. Hibbert, 1967. Factors Affecting the Response of Small Watersheds to Precipitation in Humid Areas. Int. Symp. on Forest Hydrology, Pergamon Press, New York, New York, pp. 275-290.

Hornbeck, J. W., G. E. Likens, and J. S. Eaton, 1976. Seasonal Variation in Acidity of Precipitation and the Implications for Forest-Stream Ecosystems. USDA For. Serv. Gen. Tech. Rep. NE-23, pp. 597-609.

Hultberg, H., 1976. Thermally Stratified Acid Water in Late Winter – A Key Factor Inducing Self-Accelerating Processes Which Increase Acidification. USDA For. Serv. Gen. Tech. Rep. NE-23, pp. 503-517.

Jeffries, D. S., C. M. Cox, and P. J. Dillon, 1979. Depression of pH in Lakes and Streams in Central Ontario During Snowmelt. J. Fish. Res. Board Can. 36:640-646.

Johannessen, M. and A. Henriksen, 1978. Chemistry of Snow Melt-water: Changes in Concentration During Melting. Water Resour. Res. 14:615-619.

Scheider, W. A., D. S. Jeffries, and P. J. Dillon, 1979. Effects of Acidic Precipitation on Precambrian Freshwaters in Southern Ontario. J. Great Lakes Res. 5:34-51.

Trojnar, J. R., 1977. Egg Hatchability and Tolerance of Brook Trout (*Salvelinus fontinalis*) Fry at Low pH. J. Fish. Res. Board Can. 34:574-579.

Wright, R. F., T. Dale, E. T. Gjessing, G. R. Hendrey, A. Henriksen, M. Johannessen, and I. P. Muniz, 1976. Impact of Acid Precipitation on Freshwater Ecosystems on Norway. USDA For. Serv. Gen. Tech. Rep. NE-23, pp. 459-476.

INTERNATIONAL SYMPOSIUM ON HYDROMETEOROLOGY
JUNE AMERICAN WATER RESOURCES ASSOCIATION 1982

ACID PRECIPITATION AND BUFFER CAPACITY OF LAKES IN THE SIERRA NEVADA, CALIFORNIA

John M. Melack, John L. Stoddard, and Daniel R. Dawson[1]

ABSTRACT: The east central Sierra Nevada received acid precipitation (pH 3.7 to 4.9) during convective storms interspersed through the dry season of 1981. Sulfuric acid contributed about twice the acidity of nitric acid. In contrast, late autumn, winter, and early spring snow (1981-1982) ranged in pH from 5.2 to 6.1 (mean 5.7) and had low ammonium, nitrate, and sulfate concentrations. As of 1981 most of the alpine lakes of the Sierra Nevada remain very weakly buffered, bicarbonate lakes that receive a small loading of acid precipitation and a large annual input of snowmelt uncontaminated by strong acids. If the acidity of the precipitation increases, the pH of the lakes will decrease rapidly because the lakes and their basins have extremely low buffer capacity especially when runoff peaks during snow melt.
(KEY TERMS: acid precipitation; Sierra Nevada; lake alkalinity.)

Acid rain and snow fall on California's coast, inland valleys, and Sierra Nevada. Acid precipitation occurs in the Los Angeles basin (Liljestrand and Morgan, 1981), the San Francisco Bay area (McColl and Bush, 1978), and in the Central Valley (McColl, 1980). Within the Sierra Nevada, the Tahoe basin (McColl, 1980) and the western slope of Sequoia National Park (NADP station, Thomas Nichols, Sequoia-Kings Canyon National Parks, personal communication, 1982) are known to receive acid precipitation. As such evidence accumulates, concern for the alpine waters of the Sierra Nevada is justifiably growing. Lakes and streams in granitic basins of other regions of the world have been the first habitats to show the impacts of acid precipitation (e.g., Almer, *et al.*, 1974; Schofield, 1976).

The objectives of this paper are twofold. First, we will present the first evidence that acid precipitation occurs in the alpine region and eastern flank of the Sierra Nevada. Second, we will evaluate the potential impact of this precipitation on the alpine lakes. The data we present include the first reliable measurements of the buffer capacity of alpine lakes that span the length of the Sierra Nevada crest and include lakes lying on all the major rock types.

METHODS

Precipitation was collected at two (winter) or three (summer) sites in the east central Sierra Nevada (Figure 1). The primary site was located at 3000 m ASL on Mammoth Mountain; additional sites were located at 2150 m ASL in Long Valley (Sierra Nevada Aquatic Research Laboratory, SNARL), and at 2450 m ASL about four kilometers southeast of the Mammoth Mountain site (Valentine Reserve). Collections began in June 1981 and have continued through the winter and spring of 1982.

Rain was sampled on an event basis during the summer at the Mammoth Mountain site in an Aerochemetrics wet and dry fall collector (Galloway and Likens, 1978) located 6 m above the ground. Seven weekly accumulations of dry fall were collected by rinsing the dry bucket with 100 to 300 ml of distilled water. At SNARL and Valentine Reserve, plastic buckets designed for use with the Aerochemetrics collector were placed in areas away from roads, buildings, or trees and opened during rain events only. The collectors and linear polyethylene sample bottles used to store the samples were cleaned with 10 percent HCl and rinsed at least six times with deionized, distilled water before use. Duplicate plastic, wedge shaped rain gages (Huff, 1955) were used at each site.

Snow samples were obtained on an event basis at Mammouth Mountain and SNARL. Integrated snow samples were collected from a calibrated snow board that was lifted to the top of the snow pack after each event. The snow was melted in a water bath at room temperature before analysis. Snow water equivalences were measured daily on Mammouth Mountain by collecting a column of snow from a snow board and weighing the snow contained in a known volume (U.S. Weather Bureau 8-inch precipitation gauge and scale calibrated in inches of water). In mid-April, when the snow depth was maximal, a snow pit was sampled at intervals from top to bottom at the Mammoth Mountain site.

[1] Respectively, Associate Professor and Graduate Research Assistant, Dept. of Biological Sciences; and Resident Reserve Manager, Marine Science Institute; University of California, Santa Barbara, California 93106.

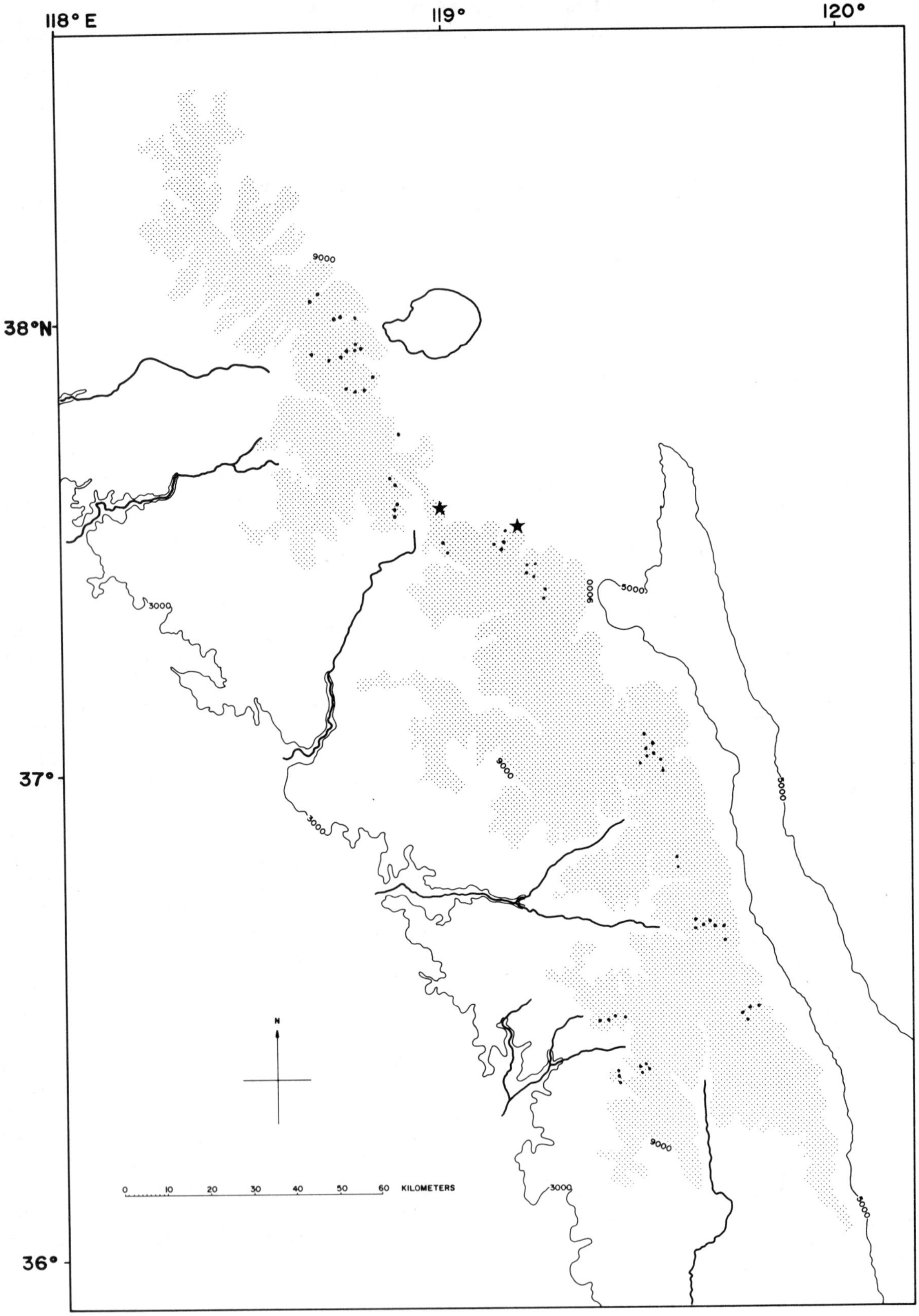

Figure 1. Location of Lakes Sampled in the Summer of 1981 (●) and of Precipitation Collectors Sampled Year Round (★). The Central Valley is on the left and the Owens Valley on the right. Shaded area lies along the Sierra Nevada crest and is over 9000 ft (ca. 3000 m) in elevation.

Hydrogen ion activity (pH) was measured immediately after collection of the precipitation samples with a null balancing meter (Perkin Elmer Coleman 37A) equipped with a low ionic strength electrode (Jena Glass Works; Sargent Welch S-30072-15). The electrode was calibrated with pH 7 or 4 buffers, washed twice for ten minutes with stirred, distilled water, and then placed in an unstirred aliquot of sample for 15 minutes. This aliquot was then discarded and the pH determination made on a fresh, unstirred sample. Occasional calibration with dilute solutions of strong acid verified the validity of this procedure (Galloway, *et al.*, 1979).

Major cation and anion analyses were performed on samples filtered through Gelman A/E glass fiber filters. Ammonium was determined immediately by the indophenol blue method (Strickland and Parsons, 1972). Samples for nitrate were stored frozen and those for other solutes were stored at $4^{o}C$. Nitrate was determined as nitrite after hydrazine reduction (Kamphake, *et al.*, 1967; McColl, 1980). Sulfate was measured with the barium chloranilate method of Shafer (1967) adapted to liquid samples; the lower limit of sensitivity was ca. 5 μ eq 1^{-1}. Chloride was determined by the mercury thiocyanate method of Florence and Farrar (1971); 1 μ eq 1^{-1} could be detected. Cation analyses (calcium, magnesium, sodium, and potassium) were performed on an atomic absorption spectrophotometer (Varian Techtron Model AA6). Samples for calcium and magnesium were spiked with lanthanum chloride; a nitrous oxide-acetylene flame was used for calcium and an air-acetylene flame for the other cations.

Alpine lakes were usually sampled at a depth of 20 cm near the center of the lake from an inflatable boat. The linear polyethylene bottles used for sampling were washed with 10 percent HCl, rinsed four times with deionized, distilled water, and three times with the lake water. pH measurements were made immediately after sampling with a digital pH meter (Markson Model 95) and combination electrode designed for low ionic strength waters (Graphic Controls PHE 52539). The electrode was calibrated daily with pH 7 buffer and thoroughly rinsed after calibration. In a laboratory comparison results were similar to those obtained with the meter and electrode used to measure precipitation pH. Alkalinity was measured within eight hours of sampling by the Gran titration procedure (Stumm and Morgan, 1972; Talling, 1973); 0.01 N HCl was dispensed with a micrometer buret. Major solutes were determined on filtered samples with the same methods as described for precipitation.

RESULTS

Chemical composition of precipitation collected in the east central Sierra Nevada is summarized in Table 1 as volume weighted means. The period with convective, summer rain storms is listed separately from the seasons with snow derived primarily from frontal storms because of the considerable differences between these two periods. Summer rain on Mammoth Mountain had high ammonium, nitrate, and sulfate concentrations and a mean pH of 4.6; pH ranged from 4.4 to 5.3. Sulfuric acid contributed about twice the acidity of nitric acid. The pH of seven summer rains at SNARL ranged from 3.7 to 4.9; six rains at Valentine Reserve ranged from 4.2 to 5.3. Rain fell infrequently during the summer as brief showers; a total of 4 cm of rain fell. In early October a combination of wet snow and rain with a pH of 4.5 and totalling 3 cm of water fell on Mammoth Mountain. In contrast, late autumn, winter, and early spring snow on Mammoth Mountain ranged in pH from 5.2 to 6.1 (mean 5.7) and had low ammonium, nitrate, and sulfate concentrations; similar snow fell at SNARL. The water equivalence of the total precipitation from mid-October to mid-April was 142 cm; the 14 samples collected from the end of November to mid-April constituted 80 percent of the precipitation. Snowfalls between mid-October and the end of November totalled 48 cm of water and were not sampled on an event basis but were included in the snow sampled from a pit.

A snow pit on Mammoth Mountain was sampled on 22 April 1982 before significant snow melt had occurred. These collections represent an integrated sample of the total deposition of solutes since snow accumulation began in mid-October. The average concentrations (μ eq 1^{-1}) of solutes are as follows: NH_4, 1.3; Na, 3.9; K, 0.8; Ca, 2.4; Mg, 0.3; NO_3, 1.3; SO_4, <7; Cl, 2. The integrated pH was 5.9. These values are quite similar to those in Table 1.

Areal atmosphere inputs of summer wet and dry deposition and as snowfall for a portion of 1981-1982 are listed in Table 2. The limited extent of these data preclude detailed evaluation but permit some general observations. With the exception of H^+, cation and anion loading by dry fall during the summer is similar to that by rain. Although considerably more precipitation fell as snow, the areal inputs of ammonium are greater during the summer, and areal inputs for the other solutes are disproportionately due to wet and dry deposition in the summer. Furthermore, the Mammoth Mountain area had about 1.6 times more precipitation than is usual during the winter of 1981-82 (California Cooperative Snow Surveys, 1982). Within the general area that includes Yosemite, Mammoth Mountain, and the Rock Creek basin, the long term, average water content of the snow pack on 1 April among eight routine stations at about 3000 m ASL ranges from 39 cm to 107 cm and averages 75 cm (California Department of Water Resources, 1982). The long term average at Mammoth Mountain is 80 cm.

A survey of lakes that spanned the length of the Sierra Nevada crest was conducted in June, July, August, and September 1981 (Figure 1). Seventy lakes were visited; of these, 23 were sampled two to four times. Eighty percent of the lakes lie above 3000 m in elevation and many were headwaters. The lakes range in surface area from about 0.3 ha to 160 ha; watershed surface areas range from 15 ha to

TABLE 1. Chemical Composition of Precipitation, East Central Sierra Nevada.

Location/Dates	Volume Weighted Mean – μ eq 1^{-1}								
	pH	NH_4	Na	K	Ca	Mg	NO_3	SO_4	Cl
MAMMOTH MOUNTAIN									
30 June 1981 – 3 October 1981	4.6	25.0	11.3	2.7	10.4	3.0	17.0	36	12.0
30 November 1981 – 14 April 1982	5.7	0.3	4.0	0.6	1.5	0.3	1.3	<6	4.1
30 June 1981 – 14 April 1982	5.5	2.4	4.3	0.7	1.8	0.4	2.8	<7	4.4
SNARL									
29 November 1981 – 11 April 1982	5.7	0.5	3.4	0.8	3.0	0.5	2.2	<7	3.0

TABLE 2. Precipitation Loading at Mammoth Mountain, California.

Type/Dates	m eq m^{-2}								
	H	NH_4	Na	K	Ca	Mg	NO_3	SO_4	Cl
DRY DEPOSITION 30 June 1981 – 3 October 1981	0.1	0.6	0.5	0.2	1.0	0.2	–	0.8	0.4
WET DEPOSITION 30 June 1981 – 3 October 1981	1.0	1.0	0.5	0.1	0.4	0.1	0.7	1.4	0.5
SNOW DEPOSITION 30 November 1981 – 14 April 1982	2.8	0.4	5.7	0.9	2.1	0.4	1.8	8.5	5.8

more than 1000 ha. Most of the bedrock in the basins consists of granitic rocks, but basins were selected whose bedrocks represent a range of weatherability from calcareous metamorphic rocks (limestone and dolomite) to orthoclase granites (quartz monzonite and alaskite). Vegetation is very sparse or lacking in the majority of the watersheds, but some basins had white bark pine (*Pinus albicaulis*) and sedge dominated turf near the lakes.

Buffer capacity of the water in the lakes of the Sierra Nevada is extremely low (Figure 2). Seventy percent of the lakes have summer alkalinities below 90 μ eq 1^{-1}, and the few with alkalinities above 210 μ eq 1^{-1} are either located below 3000 m ASL or in basins with rocks that weather more easily than granites. The major cation in the Sierran lakes is calcium, and in most lakes bicarbonate is the major anion; a few lakes had as much or more sulfate or chloride as bicarbonate. About 80 percent of lakes have sulfate concentrations less than 40 μ eq 1^{-1}, and about 90 percent of lakes have nitrate values less than 4 μ eq 1^{-1}. Summer pH values in surface waters are usually between 6 and 8 (Figure 3). The lakes with pHs above 8 often had abundant aquatic plants; as the plants use CO_2 for photosynthesis, they can readily raise the pH in these dilute waters.

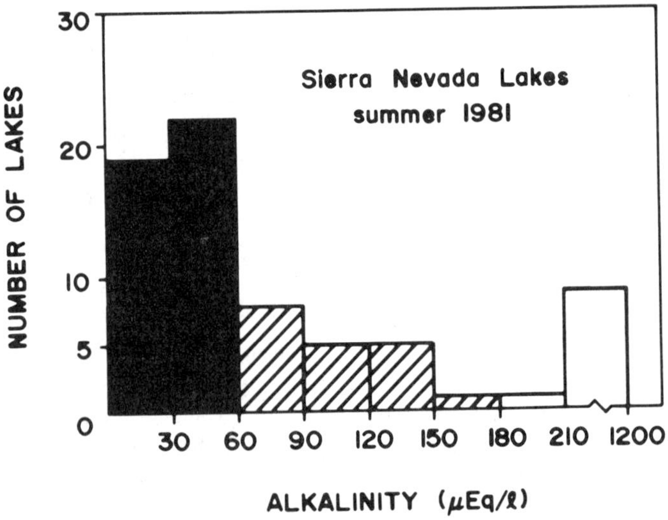

Figure 2. Alkalinity of 70 Sierra Nevada Lakes Measured in Summer of 1981. Black section of histogram represents lakes extremely sensitive to acid inputs; cross-hatched section represents lakes very sensitive to acid input.

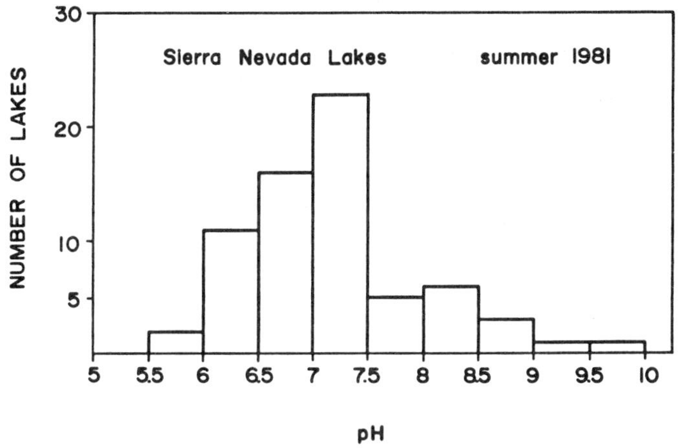

Figure 3. Midsummer pH Values for 68 Sierra Nevada Lakes.

DISCUSSION

While the east central Sierra Nevada do receive some acid precipitation (pH <5), over 90 percent of the precipitation falls as snow with a pH near 5.7, the expected acidity of a solution of atmospheric CO_2 and water. The acid precipitation is associated with occasional convective storms during the prolonged dry season. The strong acid contaminants most likely come from urban and agricultural sources within California (Lawson and Wendt, 1982). In contrast, the majority of the precipitation in the Sierra Nevada is associated with large frontal systems that sweep across California from the Pacific Ocean. Precipitation from such storms in sparsely populated coastal regions is not contaminated with strong acids. For example, Kennedy, *et al.* (1979), reported an average pH of 5.3 and average nitrate and sulfate concentrations of 1.6 μ eq 1^{-1} and <2.7 μ eq 1^{-1}, respectively, in the Mattole River basin of northern California. However, these same storm systems pass over coastal and inland cities and agricultural lands that are potential sources of SO_x and NO_x. The pH of precipitation in these areas is usually less than 5 (McColl, 1980). Apparently, at least in 1982, these potential contaminants were not present in the east central Sierra Nevada in sufficient quantity to produce acid snow.

Few additional analyses of the chemical composition of precipitation in the Sierra Nevada are available, and no event sampling above tree line other than that reported here for Mammoth Mountain has been accomplished. Feth, *et al.* (1964), collected snow along major east-west highways in the northern Sierra during the winters of 1958-59 and 1959-60. They judged their data to indicate that no detectable contamination from activities of man influenced the chemical composition of the snow. Their laboratory determinations of pH ranged from 4.2 to 8.3 with a median of 5.8, but these values were considered by Feth, *et al.* (1964), not to represent the true pH of freshly melted snow. Brown and Skau (1975) collected snow from snow boards four times at 26 locations in the east central Sierra Nevada from January to April 1975. Snow fall was low to moderate (1.5 to 38 cm of water), and chemical concentrations varied little among these locations. Major anion and cation concentrations were similar to those reported here; pH ranged from 5.3 to 5.9. McColl (1980) reported a volume weighted pH of 5.2 to Tahoe City, but the general applicability of this value may be questioned becuase of the temperature inversions and large number of automobiles in the Tahoe basin (Cahill, *et al.*, 1978). Lawson and Wendt (1982) calculated a volume weighted mean pH of 5.0 for a full year of samples collected at the NADP site in Sequoia National Park. During the winter the pH of snow at this location is usually above 5.3 (Thomas Nichols, Sequoia-Kings Canyon National Parks, personal communication, 1982).

The buffer capacity, pH, and major solutes of the lakes of the Sierra Nevada have been measured infrequently and in few of the thousands of lakes. Moreover, very seldom have the analyses of pH and alkalinity been performed with sufficient attention to the rapid changes that occur in these dilute, weakly buffered waters. For example, Bradford, *et al.* (1968), surveyed 170 alpine lakes in the summer of 1965 but made their pH measurements about two weeks after collection of the samples. During a repeat sampling of many of these same lakes in the summer of 1981, Bradford, *et al.* (1981), measured pH within about 4 hours of collection and reported a pH range of 5.6 to 6.9. Neither of these surveys included alkalinity determinations. Tonnessen and Harte (1982) sampled 26 subalpine lakes on the western slopes of the Sierra Nevada during the summers of 1980 and 1981 and reported a pH range of 6.0 to 8.5 and that about 40 percent of the lakes had alkalinities below 100 μ eq 1^{-1}. Because there are no reliable and representative measurements of alkalinity in high altitude lakes prior to those reported here and those of Tonnessen and Harte, it is not possible to recognize if a change through the years in the alkalinity such as that reported by Lewis (1982) for the Colorado Rockies is occurring in the Sierra Nevada.

To predict the impact or potential impact of the atmospheric deposition of acidity on a lake requires measurements of the areal loading to the drainage basin and subsequently to the lake as modified by buffering within the basin and the lake. In the Sierra Nevada most of the precipitation accumulates as snow and then more or less abruptly enters the lake during snow melt. Furthermore, laboratory and field examinations of snow melt chemistry (e.g., Colbeck, 1981; Jefferies, *et al.,* 1979) indicate that solutes are concentrated in the first stages of runoff.

Among the alpine lakes sampled in our survey the ratio of the surface area of the basin to the volume of the lake varied from 3 to 200. As a hypothetical exercise, if a range of snow pack depths (0.8 to 1.6 m of water) with different acidities (pHs 5.1 to 5.4) is applied to this range of basins, an estimate of the impact of spring snowmelt on the pH of the lakes can be made. The results of such a simulation for Sierra Nevada lakes indicate that most of the lakes will attain the pH of the runoff water. The model assumes that 80 percent of the solutes enter the runoff in the first 50 percent of the snowmelt, a snow alkalinity of 2 μ eq 1^{-1}, no buffering of the runoff before it reaches the lake, no thermal stratification in the lakes, and no contribution from the sediments to the alkalinity of the lake during the brief period of snowmelt. If a greater proportion of the solutes leave the snow earlier in the melt or if the lake is stratified and thus the runoff is added to a smaller volume of lake water, the acidic loading would be greater. These two possibilities are likely. The major uncertainty in this model is the extent of buffering and contributions of additional alkalinity from weathering in the basin and from the lake sediments. In the alpine basins of the Sierra Nevada thick snow packs with ice layers interspersed will tend to route some portion of the runoff through the snow itself, and the underlying ground is largely bare granites with little soil. These factors would reduce the extent of the buffering in the basin. However, Feth, *et al.* (1964), emphasize that melt waters rapidly increase in solute concentration once they contact the lithosphere. The lake sediments can contribute to the alaklinity of the overlying water as demonstrated in laboratory microcosms by Tonnessen and Harte (1982). However, during the short period with large inflows of snow melt into a lake that may be stratified, the influence of the sediments is likely to be low.

As of 1981 most of the alpine lakes of the Sierra Nevada remain very weakly buffered, bicarbonate lakes that receive a small amount of acid precipitation and a large annual input of snowmelt uncontaminated by strong acids. If the acidity of the precipitation increases, the pH of the lakes will decrease because the lakes and their basins have extremely low buffer capacity especially when the runoff peaks during snowmelt.

ACKNOWLEDGMENTS

We thank Cliff Ochs for dedicated field and laboratory assistance, James Sickman for help with chemical analyses, and Danny Marks for helpful discussion. We appreciate the cooperation of David McCoy and Mammoth Mountain Ski Area, the assistance and water equivalence data provided by the U.S. Forest Service, Mammoth Range District, Inyo National Forest, the facilities provided by the Sierra Nevada Aquatic Research Laboratory, and the research permission and help offered by Sequoia-Kings Canyon and Yosemite National Parks. The research leading to this report was supported by the Office of Water Research and Technology, USDI, under the Annual Cooperative Program of Public Law 95-467, and by the University of California Water Resources Center, as part of Office of Water Research and Technology Project No. A-080-CAL and Water Resources Center Project UCAL-WRC-W-590. Contents of this publication do not necessarily reflect the views and policies of the Office of Water Research and Technology, U.S. Department of the Interior, nor does mention of trade names or commercial products constitute their endorsement or recommendation for use by the U.S. Government.

LITERATURE CITED

Almer, B., W. Dickson, C. Ekstrom, E. Hörnström, and U. Miller, 1974. Effects of Acidification on Swedish Lakes. Ambio 3:330-336.

Bradford, G. R., F. L. Bair, and V. Hunsaker, 1968. Trace Metal and Major Element Content of 170 High Sierra Lakes in California. Limnol. Oceanogr. 13:526-530.

Bradford, G. R., A. L. Page, and I. R. Straughan, 1981. Are Sierra Lakes Becoming Acid? Calif. Agr. 35:6-7.

Brown, J. C. and C. M. Skau, 1975. Chemical Composition of Snow in the East Central Sierra Nevada. Cooperative Report Series Publ. No. AG-1, Renewable Natural Resources Div., Univ. of Nevada, Reno, Nevada.

Cahill, T. A., L. L. Ashbaugh, J. B. Barone, and P. J. Feeny, 1978. Spatial Distribution of Primary Automotive Pollutants at Lake Tahoe. *In:* Proc. Conf. Sierra Nevada Meteorology, American Meteorological Society, pp. 29-33.

California Cooperative Snow Surveys, 1982. Water Conditions in California. Report 4, May 1, 1982, Bulletin No. 120-82, 16 pp.

California Department of Water Resources, 1982. California Snow Survey Measurement Schedule. State of California, 43 pp.

Colbeck, S. C., 1981. A Simulation of the Enrichment of Atmospheric Pollutants in Snow Cover Runoff. Water Resour. Res. 17: 1383-1388.

Feth, J. H., S. M. Rogers, and G. E. Roberson, 1964. Chemical Composition of Snow in the Northern Sierra Nevada and Other Areas. U.S. Geol. Surv. Water Supply Paper 1535-J, 39 pp.

Florence, T. M. and Y. J. Farrar, 1976. Spectrophotometric Determination of Chloride at Parts-Per-Billion Level by the Mercury (II) Thiocyanate Method. Anal. Chim. Acta 54:373-377.

Galloway, J. N. and G. E. Likens, 1976. Calibration of Collection Procedures for the Determination of Precipitation Chemistry. Water, Air and Soil Pollut. 6:241-258.

Galloway, J. N., B. J. Cosby, Jr., and G. E. Likens, 1979. Acid Precipitation: Measurement of pH and Acidity. Limnol. Oceanogr. 24:1161-1165.

Huff, F. A., 1955. Comparison Between Standard and Small Orifice Rain Gauges. Trans. Amer. Geophy. Union 37:686-694.

Jefferies, D. S., C. M. Cox, and P. J. Dillion, 1979. Depression of pH in Lakes and Streams in Central Ontario During Snowmelt. J. Fish. Res. Bd. Can. 36:640-646.

Kamphake, L., S. A. Hannak, and J. M. Cohen, 1967. Automated Analyses for Nitrate by Hydrazine Reduction. Water Res. 1:206-211.

Kennedy, V. C., G. W. Zellweger, and R. J. Avanzino, 1979. Variation of Rain Chemistry During Storms at Two Sites in Northern California. Water Resour. Res. 15:687-702.

Lawson, D. R. and J. G. Wendt, 1982. Acid Deposition in California. Proc. Soc. Automotive Engineers Sym. (in press).

Lewis, W. M., Jr., 1982. Changes in pH and Buffer Capacity of Lakes in the Colorado Rockies. Limnol. Oceanogr. 17:167-172.

Liljestrand, H. M. and J. J. Morgan, 1981. Spatial Variation of Acid Precipitation in Southern California. Envir. Sci. Tech. 15:333-338.

McColl, J. G., 1980. A Survey of Acid Precipitation in Northern California. Final Report of Calif. Air Resources Board Cont. A7-139-30.

McColl, J. G. and D. S. Bush, 1978. Precipitation and Throughfall Chemistry in the San Francisco Bay Area. J. Environ. Qual. 7: 352-357.

Schafer, H. N. S., 1967. An Improved Spectrophotometric Method for the Determination of Sulfate With Barium Chloranilate as Applied to Coal Ash and Related Materials. Anal. Chem. 39: 1719-1726.

Schofield, C. L., 1976. Acid Precipitation – Effects on Fish. Ambio 5:228-230.

Strickland, J. D. H. and T. R. Parsons, 1972. A Practical Handbook of Seawater Analysis. Fish Res. Bd. Can. Bull. 167, 310 pp.

Stumm, W. and J. J. Morgan, 1972. Aquatic Chemistry. John Wiley and Sons, New York, New York, 583 pp.

Talling, J. F., 1973. The Application of Some Electrochemical Methods to the Measurement of Photosynthesis and Respiration in Freshwaters. Freshwat. Biol. 3:335-362.

Tonnessen, K. and J. Harte, 1982. Acid Rain and Ecological Damage: Implications of Sierra Nevada Lake Studies. Public Affairs Report (in press).

ACID PRECIPITATION: THE IMPACT ON TWO HEADWATER STREAMS IN SHENANDOAH NATIONAL PARK, VIRGINIA

Paul W. Shaffer and James N. Galloway[1]

ABSTRACT: The Shenandoah Watershed Acidification Study (SWAS) was initiated in October 1979; the purpose of this study is to evaluate the present and future impact of acid precipitation on soil and stream water chemistry in the Shenandoah National Park. The study focuses on two small watersheds in the park. Weekly precipitation and stream water chemistry sampling has been carried out since November 1979; results for the first two years are presented. Temporal data shows dilute but stable water chemistry except for a modest summer increase in bicarbonate and associated base cations in one of the study streams. Sulfate is the dominant anion, magnesium and potassium the dominant cations. Input-output budgets show that many major cations and anions are accumulating in the watersheds, including H^+, NH_4^+, NO_3^-, $SO_4^=$, and Ca^{2+}, while there is net export of K^+, Mg^{2+}, HCO_3^-, and silica. Sodium and chloride show net export during periods of high stream discharge, and accumulation during low flow; long term input and output are probably about equal. Stability of stream water chemistry appears to be controlled primarily by sulfate adsorption, bicarbonate production, and cation exchange equilibria in the soil. Because sulfate adsorption capacity is finite and because there is net export of base cations from soil, current stability is achieved at the cost of long term degradation of the soil system.
(KEY TERMS: acid precipitation; stream water chemistry; small watershed; input-output budgets; temporal trends; Shenandoah National Park.)

INTRODUCTION

During the past decade, acid precipitation has become recognized as a major environmental problem in North America. Available data indicates that the area receiving acidic rainfall has grown to include almost all of the eastern United States, and that the acidity of rainfall has increased at most locations (Cogbill and Likens, 1974; Likens, *et al.*, 1979). In the northeastern United States (and southeastern Canada and Scandinavia), effects of acid precipitation have been widely observed as declines in pH and alkalinity of lakes, leading to mobilization of aluminum and some heavy metals and also to loss of fish populations (Dochinger and Seliga, 1976; Schofield, 1976; Drabløs and Tollan, 1980; Hendrey, *et al.*, 1980; Overrein, *et al.*, 1980). Recent data also suggests that acid precipitation has resulted in changes in stream water chemistry in Pennsylvania and North Carolina (Arnold, *et al.*, 1980; Burns, *et al.*, 1981).

Research efforts have focused primarily on areas in the Northeastern U.S., Canada, and southern Norway and Sweden, areas which share a fairly limited set of environmental conditions. Lakes have been studied much more than streams, all of these areas have cold climates where a significant fraction of annual acid input occurs as a pulse during snowmelt, bedrock is typically granitic, and soils are usually spodosols formed from glacial till. Previous studies (Galloway and Cowling, 1978. Hendrey, *et al.,* 1980) have suggested that streams in large areas in the southeastern U.S. are vulnerable to acidification, but there have been few systematic water quality studies in these areas, so actual effects remain largely unknown.

The Shenandoah National Park is located in the Blue Ridge Mountains of central Virginia and is characterized by conditions unlike those outlined above. Surface waters are limited to small headwater streams, formation of a substantial snowpack is rare; bedrock is variable and includes orthoquartzite, shale, granite, and metabasalt, and soils are typically thin and rocky (usually ultisols) formed from underlying rock. Systematic precipitation and stream water chemistry data were lacking until the onset of our research, so the possible extent of stream acidification was previously unknown.

The Shenandoah Watershed Acidification Study (SWAS) was initiated in October 1979; the purpose of this project is to assess the present and potential future impact of acid precipitation on soil and stream system in the Shenandoah National park. The project was designed as a two-phase research program, with Stage I as a "black box" assessment of inputs and outputs from two small calibrated watersheds, looking at temporal changes in stream water chemistry, annual input/output budgets, and stream water chemistry changes associated with rainfall events. This period has also been used to collect background data on hydrology and on soil and bedrock chemistry and mineralogy. Stage II, now

[1]Respectively, Graduate Research Assistant and Associate Professor, Dept. of Environmental Sciences, Clark Hall, University of Virginia, Charlottesville, Virginia 22903.

beginning, is a process-oriented phase in which we are developing and testing hypotheses to clarify the processes regulating solution chemistry. This paper presents two years of Stage I data and summarizes our current understanding of chemical behavior of the SWAS watersheds.

SITE DESCRIPTION

The SWAS study area is located in the Blue Ridge physiographic province of Virginia near the southern end of the Shenandoah National Park (Figure 1). Two watersheds are included in the study, Deep Run and White Oak Run. Deep Run watershed (306 ha) is a narrow elongate valley, with the stream flowing over bedrock along much of the stream channel. Hillslopes are steep, soils are thin and rocky, and tributaries are limited to a few ephemeral streams. White Oak Run (515 ha) has one major perennial tributary; hillslopes are once again steep and soils thin and rocky, but a well developed alluvial plain occurs along the lowest mile of the stream.

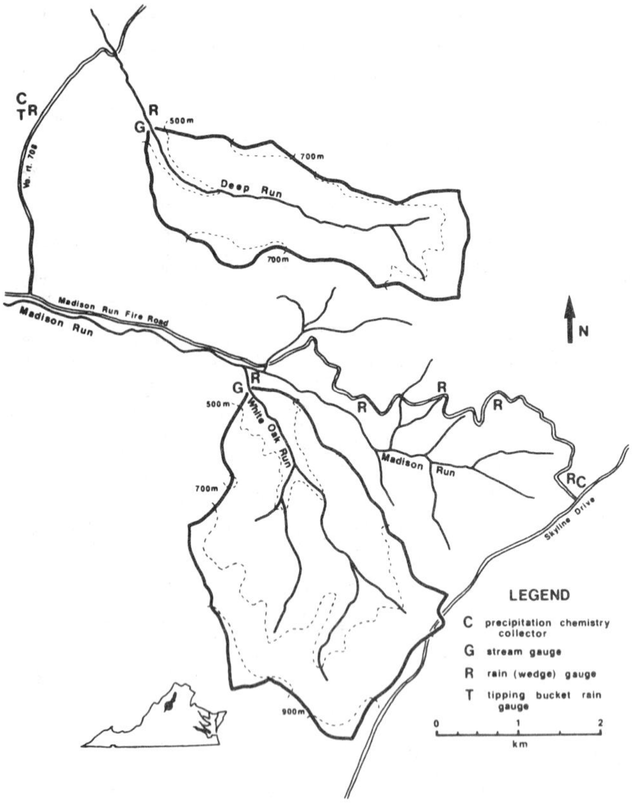

Figure 1. Map of the SWAS Study Area, Showing the Location of Deep Run and White Oak Run Watersheds and Field Monitoring Equipment. Watershed boundaries are indicated by the heavy lines. Inset map at lower left indicates the location of the study area in the southern end of the Shenandoah National Park.

Both watersheds are covered with second growth forest 50-80 years old; mixed chestnut and white oak cover most of the watersheds with scrub pine on drier south facing slopes. Both watersheds are underlain by early Cambrian quartzites and shales (Weverton, Hampton, and Erwin formations); quartz and sericite are the dominant minerals in bedrock with minor orthoclase (Schwab, 1970, 1971; Gathright, 1976). Soils are formed from underlying bedrock and are generally similar throughout the two watersheds. A thin A horizon (0-5 cm in depth) overlies a B horizon which grades into saprolite at 60-100 cm. Color and texture are uniform through the B horizon, although there are chemical and mineralogical changes with depth. Soils are silty or silty clay loams, and are classified as hapludults, dystrochrepts, and udorthents (Hockman, *et al.,* 1979; Garrahan, T., 1979, USDA Soil Conservation Service, Rockingham County, Virginia, personal communication).

Soil chemistry reflects the aged, highly weathered soils in Deep Run and White Oak Run; soils are strongly acidic and base saturation averages 2 percent. Soil mineralogy is dominated by quartz and dioctahedral mica (sericite) at the base of the profile, with the mica weathering to a 10-14Å interlayer structure and finally to a 14Å "dioctahedral vermiculite" at the top of the B horizon. In the A horizon, mica has collapsed to 10.5Å. Minor kaolinite is present throughout the column, probably formed from orthoclase in bedrock. Typical soil chemistry and mineralogical profiles are given in Figure 2.

METHODS AND MATERIALS

Routine field sampling of Deep Run, White Oak Run, and precipitation began in November 1979 and has continued on a weekly basis since that time. Precipitation input is measured using a network of wedge gages and a tipping bucket gage (Weathermeasure P-501 with P521 event recorder) is used to determine the time and intensity of input. Stream flow is monitored using punched tape stage recorders (Fisher-Porter Model ADR 1542); gages were installed in pools at natural bedrock control sites. Gage installation and maintenance has been performed by personnel from the U.S. Geological Survey.

Precipitation samples for chemical analysis are collected using a wet/dry collector (Aero-Chem Metrics) and/or funnel collectors (Likens, *et al.,* 1977) located at two sites adjacent to the watersheds. Stream water field measurements include pH, temperature, and specific conductance, and replicate samples are collected in plastic bottles for laboratory analysis. All sample collectors and containers are acid washed, rinsed in deionized water, and individually checked before use to avoid contamination. Because suspended solids in stream water are negligibly low ($<$ 0.2 mg/L except during storm periods), samples are not routinely filtered. Samples are stored on ice immediately following collection, then refrigerated at $4^{o}C$ until analysis. During storm events,

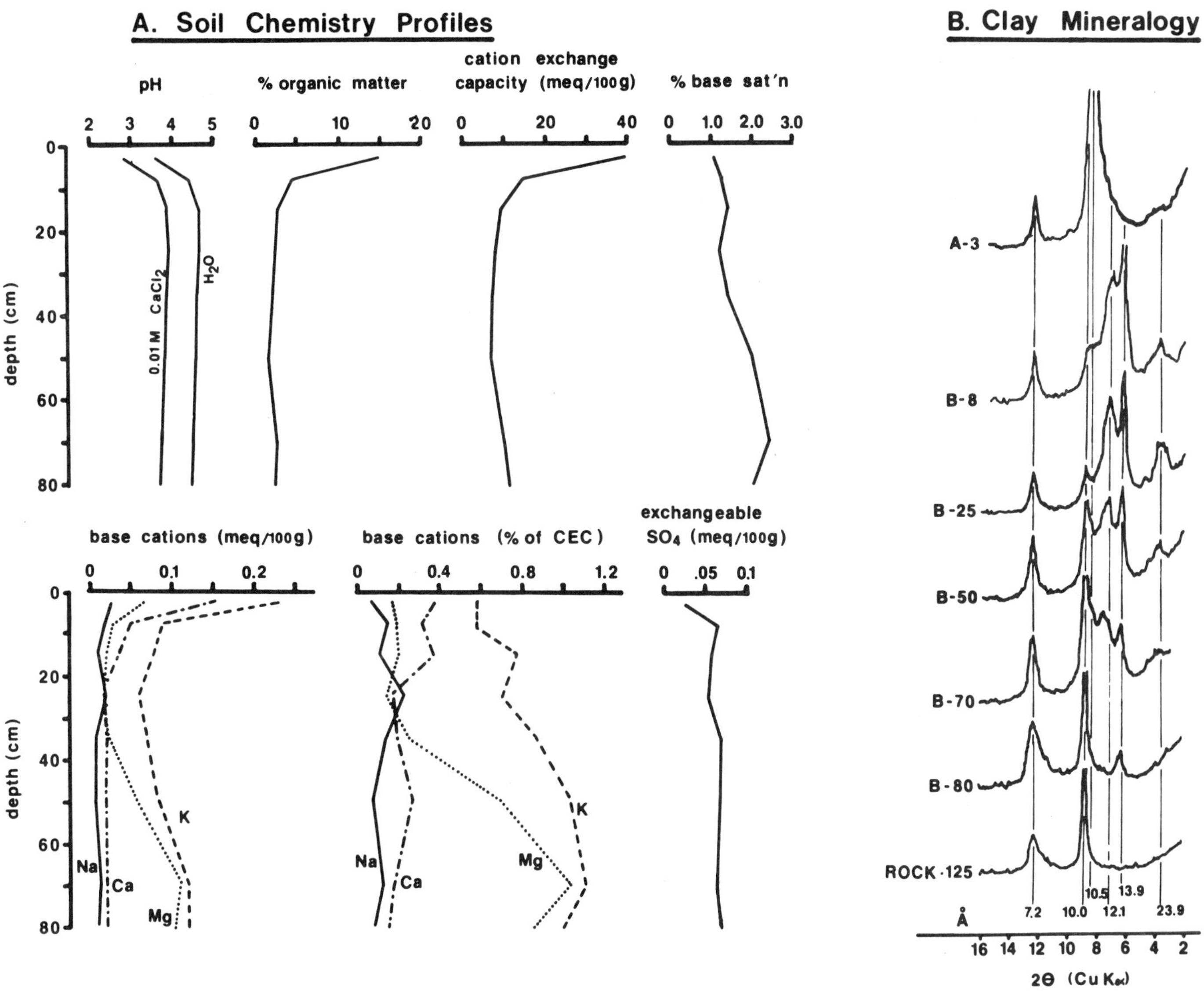

Figure 2. Typical Soil Chemical and Clay Mineralogy Profiles in the White Oak Run Watershed (Site WH07 is shown). pH was measured using a 1:1 suspension of soil and deionized water or 0.01M $CaCl_2$, other methods as described in text. In the clay profiles, the dominant feature is the expension of 10A mica to a 10-14A interlayer and then to a 14A "dioctahedral vermiculite" at the top of the B horizon, then an abrupt collapse to 10.5A in the A horizon.

samples are collected using automatic water samplers (S4040, Manning Environmental Corp.) activated manually or by stage sensing devices.

Laboratory analyses include measurement of pH and conductivity, and alkalinity is measured using a two end-point (pH 4.5, 4.2) potentiometric titration (APHA, 1971). Major anions (NO_3, Cl, SO_4), ammonium and silica are measured using standard procedures adapted for Technicon II autoanalyzer (Galloway and Likens, 1976). Major cations (Ca, Mg, K, Na) are analyzed by flame atomic adsorption using an Instrumentation Laboratories 751 spectrophotometer.

Soil samples were collected from about 50 sites in the two watersheds during 1979-80, and nine profiles were collected in November 1981 for more detailed chemical and mineralogical analysis; samples were collected from hand dug pits, air dried, and sieved to remove coarse (>2 mm) material prior to analysis. Analyses were performed using methods from Black (1965), except that exchangeable cations were measured by the method of Hajek, *et al.* (1972), and exchangeable sulfate measurements were made by the method of Ensminger (1954) using K_2HPO_4 as the extractant. Soil and bedrock mineralogy were determined by x-ray diffraction, using random and preferred orientation slides for clay mineral identification of the $\leqslant 2\ \mu m$ fraction.

Samples for clay analysis were pretreated to remove organic matter using hydrogen peroxide and to remove free iron oxides using sodium dithionite (Kunze, 1965).

Chemical input budgets were calculated using average rainfall depth over each watershed (data are nearly equal for the two watersheds, White Oak Run input fluxes are reported) coupled with chemistry data for individual storm events. Output budgets were calculated using weekly chemical data coupled with daily discharge data for the preceding week. All flux data are reported as equivalents per hectare except silica, which is reported as moles per ha.

RESULTS AND DISCUSSION

Temporal stream water chemistry trends are shown for White Oak Run and Deep Run in Figures 3 and 4, and volume weighted average concentrations of major ions in the two streams and in precipitation are given in Table 1.

Precipitation chemistry is comparable to that reported elsewhere in the eastern United States (Anon., 1981; Likens, *et al.*, 1977; Parker, 1980, M.S. Thesis, Dept. of Environmental Sciences, University of Virginia). Hydrogen and sulfate are the dominant species, each accounting for about 60 percent of total cation or anion equivalents. Nitrate

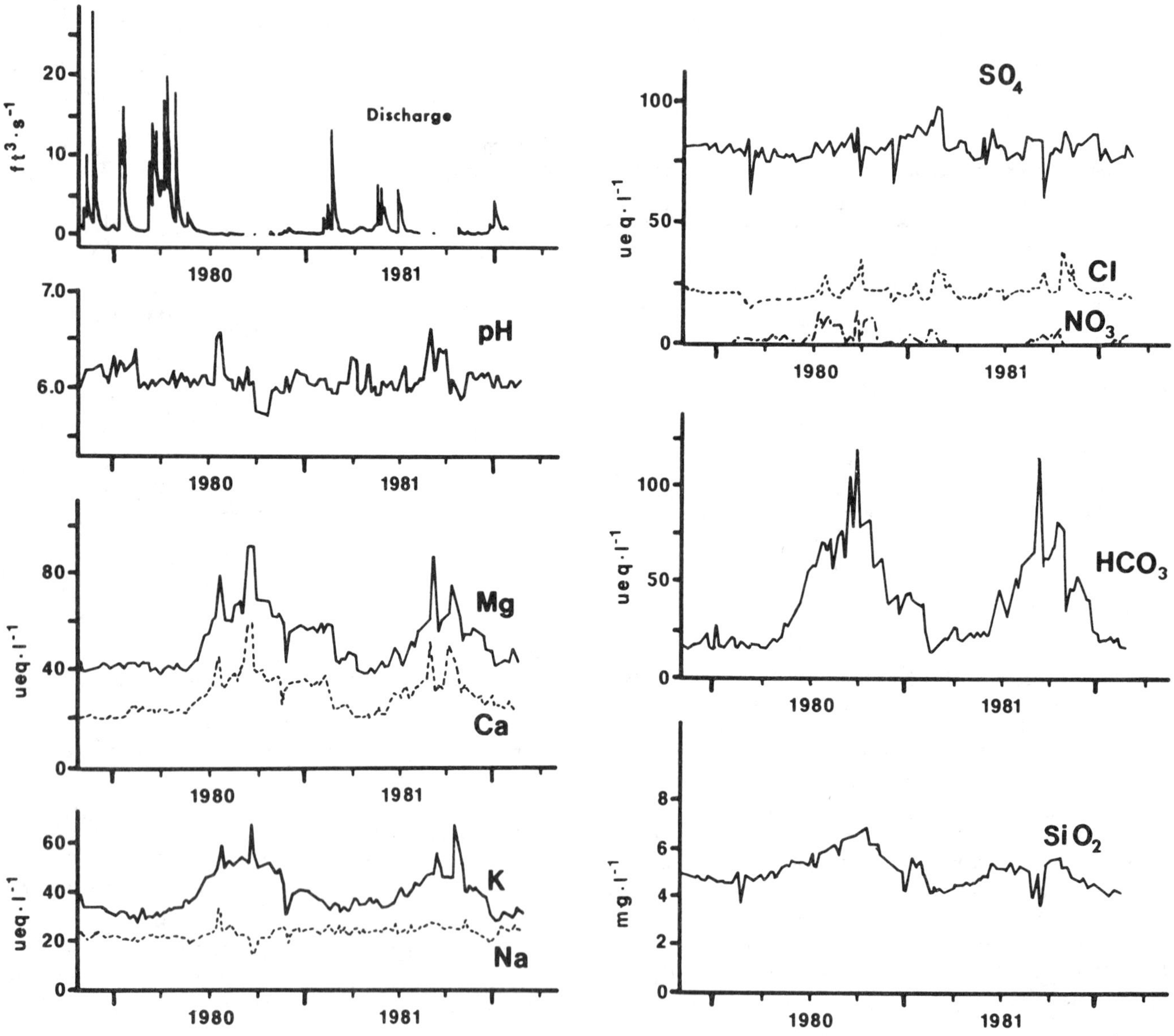

Figure 3. Temporal Chemical Trends in White Oak Run From November 1, 1979, to Mid-February 1982. The discharge curve is discontinuous during periods when the stage was below the stilling pipe; at about 0.01 cfs.

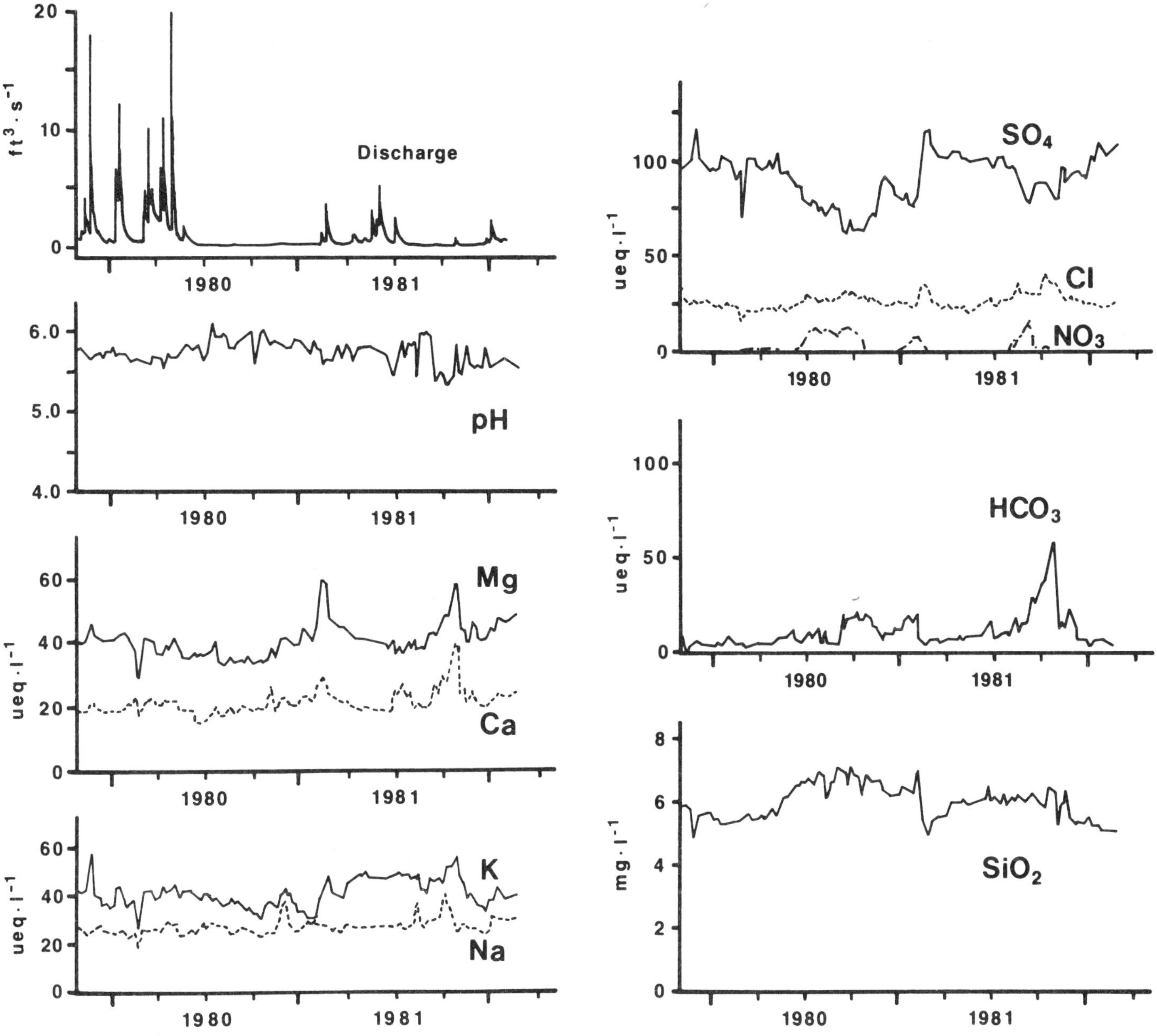

Figure 4. Temporal Chemical Trends in Deep Run During the Period November 1, 1979, to Mid-Frebruary 1982.

accounts for 27 percent of anion equivalents and chloride the remainder; the nitrate-sulfate ratio is 0.43. Base cations accounts for 27 percent of total cations, and ammonia 14 percent. Storm-to-storm precipitation chemistry is highly variable, seasonal trends have not been determined for our data.

In White Oak Run, pH has remained near 6.0 throughout the study period, with no evidence of change in response to changes in season or discharge. Sulfate, the major anion in solution, and chloride show little change during the study period, while nitrate levels remain low year-round. Bicarbonate shows a pronounced seasonal increase during the summer and fall, presumably associated with seasonal increases in soil CO_2 concentration. Base cations, with the exception of sodium, undergo a corresponding increase during the summer months. Levels of calcium, magnesium, and potassium all increase to about 1.6 times winter levels, while sodium levels are almost unchanged. Magnesium and potassium are the dominant cations in solution, reflecting their presence in bedrock, while the net source of calcium and sodium is precipitation. Silica levels show only small variation; increasing in concentration during the summer and fall months in both 1980 and 1981.

TABLE 1. Volume Weighted Mean Concentration of Major Ions and Silica in Precipitation and Stream Water in 1980-1981.

		Precipitation		White Oak Run		Deep Run	
		mg/ℓ	μeq/ℓ	mg/ℓ	μeq/ℓ	mg/ℓ	μeq/ℓ
Ca	1980	0.30		0.49		0.41	
	1981	0.20		0.53		0.42	
	2 Year Mean	0.25	12.6	0.50	24.9	0.41	20.5
Mg	1980	0.064		0.51		0.49	
	1981	0.043		0.55		0.51	
	2 Year Mean	0.054	4.4	0.52	42.6	0.50	40.8
K	1980	0.086		1.29		1.58	
	1981	0.075		1.39		1.79	
	2 Year Mean	0.081	2.1	1.31	33.6	1.63	41.7
Na	1980	0.13		0.49		0.60	
	1981	0.15		0.57		0.63	
	2 Year Mean	0.14	6.0	0.51	22.1	0.60	26.3
NH_4	1980	0.28		0.009		0.008	
	1981	0.18		0.003		0.003	
	2 Year Mean	0.23	13.1	0.007	0.4	0.007	0.4
H	1980	– –		– –			
	1981	– –		– –			
	2 Year Mean	– –	53.7 (pH 4.27)	– –	0.9 (pH 6.06)		2.2 (pH 5.66)
NO_3	1980	1.72		0.11		0.08	
	1981	1.30		0.05		0.03	
	2 Year Mean	1.52	24.9	0.09	1.5	0.06	1.0
Cl	1980	0.36		0.67		0.77	
	1981	0.34		0.83		0.90	
	2 Year Mean	0.35	9.8	0.71	20.1	0.80	22.7
SO_4	1980	3.16		3.76		4.67	
	1981	2.40		3.98		4.85	
	2 Year Mean	2.79	58.1	3.82	79.6	4.72	98.3
HCO_3	1980	– –		1.14		0.39	
	1981	– –		1.40		0.61	
	2 Year Mean	– –		1.21	19.8	0.45	7.3
SiO_2	1980	0.043	– –	4.70		5.53	
	1981	0.015	– –	4.68		5.87	
	2 Year Mean	0.029	– –	4.69	– –	5.62	– –
Σ Cations	2 Year Ave.		91.9	– –	124.4	– –	131.9
Σ Anions	2 Year Ave.		92.8	– –	121.1	– –	129.3
				cfs		cfs	
H_2O (discharge)	1980			1.76		1.04	
	1981			0.65		0.36	
	2 Year Mean			1.21		0.70	

Deep Run chemistry is in many respects similar to that of White Oak Run. pH is lower at 5.7, but once again shows little variation. Sulfate is again the major anion, accounting for nearly 75 percent of total anion equivalents. Unlike White Oak Run, sulfate levels in Deep Run show some variability over time; the cause of this variability is unclear at at this time. Bicarbonate levels in Deep Run are lower and generally stable at about 5 μeq/L, except for a period during autumn 1981 at the end of a prolonged drought. As in White Oak Run, magnesium and potassium are the most abundant cations, but seasonal variability is much less pronounced than in White Oak Run. Silica levels increase in the summer in Deep Run, although the variation is not as well defined as in White Oak Run. Annual mean concentrations show little difference for 1980 and 1981, despite the large differences in discharge.

From April 1980 through 1981, aluminum chemistry was monitored in White Oak Run; aluminum and major ion data for the period was processed using the chemical equilibrium model WATEQ (Truesdell and Jones, 1974) to determine whether major ion concentrations were controlled by mineral saturation. These data (a typical set of calculations is provided in Table 2) show that White Oak Run is saturated with respect to quartz, kaolinite, and a variety of aluminum oxyhydroxide minerals, slightly undersaturated for muscovite (K-mica), and greatly undersaturated for all other relevant mineral phases. Such data suggests that saturation controls may play a role for silica and aluminum in the SWAS watersheds, but are unimportant for other major ions in this study. Major ion concentrations are somewhat different for the two SWAS watersheds, but show a high degree of stability, suggesting very efficient control. Determination of the nature of controls will be the major objective of SWAS Stage II.

TABLE 2. Summary of Typical WATEQ Chemical Equilibrium Data for White Oak Run. Data from May 2, 1980, is presented. A saturation index of 1.0 indicates saturation, >1.0 supersaturation, <1.0 undersaturation.

Mineral	Ion Activity Product	K_T	Saturation Index (IAP/K_T)
Adularia	3.97×10^{-25}	2.51×10^{-22}	1.58×10^{-3}
Amorphous $Al(OH)_3$	4.87×10^{-34}	9.04×10^{-33}	5.39×10^{-2}
Aragonite	5.60×10^{-15}	7.65×10^{-9}	7.31×10^{-7}
Boehmite	4.87×10^{-34}	1.56×10^{-34}	3.13×10^{0}
Brucite	2.30×10^{-22}	3.64×10^{-12}	6.30×10^{-11}
Calcite	5.60×10^{-16}	3.87×10^{-9}	1.45×10^{-6}
Dolmite	5.71×10^{-29}	1.81×10^{-17}	3.16×10^{-12}
Gibbsite	4.87×10^{-34}	4.23×10^{-35}	1.15×10^{1}
Gypsum	4.11×10^{-10}	2.52×10^{-5}	1.63×10^{-5}
Halite	3.79×10^{-10}	3.56×10^{1}	1.06×10^{-11}
Halloysite	3.94×10^{-36}	4.86×10^{-35}	8.10×10^{-2}
Illite	1.20×10^{-43}	7.29×10^{-43}	1.64×10^{-1}
Kaolinite	3.94×10^{-36}	2.80×10^{-39}	1.41×10^{3}
K-Mica	2.46×10^{-52}	4.39×10^{-52}	5.61×10^{-1}
Pyrophyllite	2.50×10^{-44}	3.72×10^{-43}	6.74×10^{-2}
Quartz	7.97×10^{-5}	6.13×10^{-5}	1.30×10^{0}
Amorphous SiO_2	7.97×10^{-5}	6.83×10^{-4}	1.17×10^{-1}

Input/output budget data is presented in Figure 5 and summarized in Table 3. Given the stability of stream water chemical composition, hydrologic fluxes play a major role in determining chemical fluxes, so water flux data is included. Precipitation fluxes during 1980 and 1981 were nearly equal, but their timing was quite different. The fall and winter of 1979-80 were quite wet, resulting in high runoff and consequently in high chemical output fluxes. Conversely, 1981 was preceded by a very dry fall, and precipitation was light during winter and early spring, but high during summer months; thus water was removed from the watershed by evapotranspiration rather than streamflow. Total runoff was 31 cm from each stream in 1980, only 11 cm in 1981.

As stated above, chemical fluxes are strongly influenced by hydrologic fluxes; to a first approximation input rates are constant over the study period while output fluxes are directly proportional to stream discharge and thus very irregular over time. Output fluxes were high in early 1980, low during winter and spring 1981, and negligible during summer and fall of both years. Input and output net fluxes are thus a reflection not only of biogeochemical processes, but also hydrologic events.

Input/output budgets suggest several types of ionic behavior in the SWAS watersheds and are indicative of processes regulating stream water chemistry. Major ions and silica can be described by four classes of behavior.

1. Input > > output (H^+, NH_4^+, NO_3^-). Despite the acidic soils and poorly buffered stream water in the SWAS watersheds, less than one percent of hydrogen input is lost via stream flow. Ion exchange and mineral weathering are the primary sinks for H^+. Nitrogen species are also almost entirely retained in the watersheds, with biological uptake the presumed sink.

2. Input > output ($SO_4^=$, Ca^{2+}). Sulfate adsorption has been widely observed in highly weathered soils, although acidified lake/stream systems are usually characterized by approximately equal sulfate inputs and outputs. Sulfate adsorption is believed to be responsible for sulfate accumulation in the SWAS watersheds and is probably acting to control sulfate concentrations in stream water. Because sulfate is already the dominant anion in solution into both SWAS streams, sulfate adsorption is acting to reduce total ionic strength of stream water and thus reducing the impact of acid precipitation in these watersheds. Evaluating sulfate adsorption processes and sulfate adsorption capacity will be a major focus of research during Stage II of SWAS.

Calcium accumulation in the SWAS watersheds is substantial (~ 60 percent of input for 1980-1981) and is primarily a reflection of very calcium deficient bedrock. Exchangeable calcium levels in the soil are extremely low, and calcium is probably accumulating by ion exchange and biological uptake.

3. Input/output relationship variable (Cl^-, Na^+). Precipitation is the major long term source of both these ions to the watershed, probably the sole significant source.

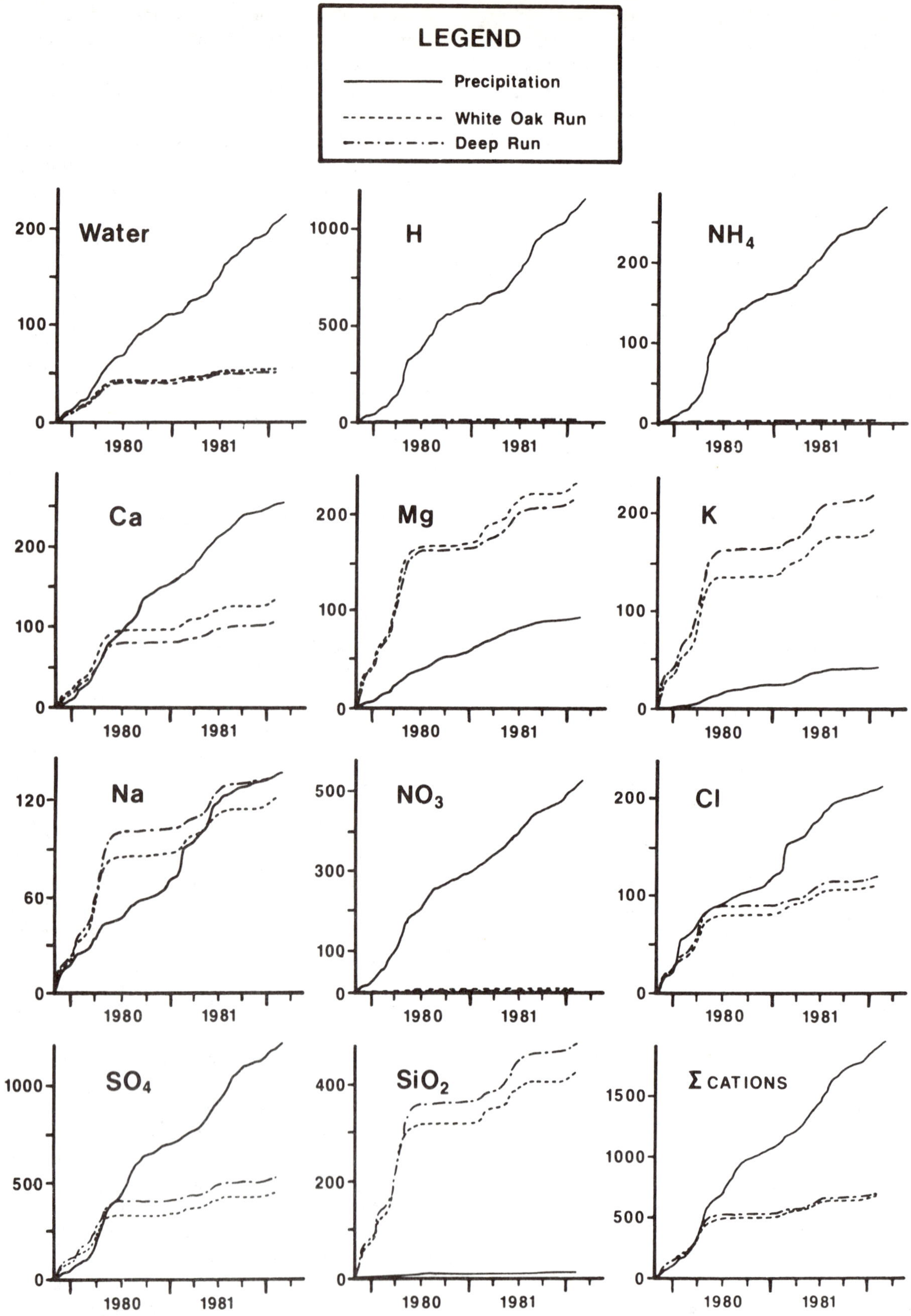

Figure 5. Cumulative Input and Output Flux Curves for Precipitation and Stream Water for the SWAS Watersheds from November 1, 1979, to January 31, 1982. Note the close relationship between water flux from the watersheds and chemical output fluxes. All data are expressed as eq·ha^{-1} except for water (cm) and silica (moles·ha^{-1}).

TABLE 3. Summary of Annual Input and Output Fluxes for Precipitation and Streams in the SWAS Watersheds in 1980-1981. Units are eq·ha^{-1} except for water (cm) and silica (moles·ha^{-1}). Positive net flux values represent accumulation in the watershed.

		PPT	White Oak Run		Deep Run	
		Input	Output	Net Flux	Output	Net Flux
H_2O	1980	98.7	31.5	– –	30.9	– –
	1981	93.2	11.6	– –	10.6	– –
	Mean	96.0	21.6	– –	20.7	– –
Ca	1980	147.0	76.0	+71	63.0	+84
	1981	95.0	31.0	+64	22.0	+73
	Mean	121.0	54.0	+67	43.0	+78
Mg	1980	52.0	131.0	−79	125.0	−73
	1981	33.0	52.0	−19	45.0	−12
	Mean	42.0	92.0	−50	85.0	−43
K	1980	22.0	104.0	−82	125.0	−103
	1981	18.0	41.0	−23	49.0	−31
	Mean	20.0	72.0	−52	87.0	−67
Na	1980	54.0	67.0	−13	80.0	−26
	1981	63.0	29.0	+34	29.0	+34
	Mean	58.0	48.0	−10	55.0	+3
NH_4	1980	156.0	1.6	+154	1.4	+155
	1981	94.0	0.2	+94	0.2	+94
	Mean	125.0	0.9	+124	0.8	+124
H^+	1980	574.0	2.6	+571	6.9	+567
	1981	468.0	1.1	+467	2.3	+466
	Mean	521.0	1.9	+519	4.6	+516
NO_3	1980	279.0	5.6	+273	3.8	+275
	1981	198.0	0.9	+197	0.5	+197
	Mean	239.0	3.2	+236	2.1	+237
Cl	1980	100.0	60.0	+40	67.0	+33
	1981	89.0	27.0	62	27.0	+62
	Mean	95.0	44.0	51	47.0	+48
SO_4	1980	649.0	247.0	+402	301.0	+348
	1981	466.0	96.0	+370	107.0	+359
	Mean	557.0	172.0	+385	204.0	+353
SiO_2	1980	7.0	246.0	−239	285.0	−278
	1981	2.4	90.0	−88	104.0,	−102
	Mean	4.7	168.0	−163	194.0	−189
Total Cations	1980	1005.0	382.0	+623	401.0	+604
	1981	771.0	154.0	+619	147.0	+624
	Mean	888.0	268.0	+620	274.0	+614
Total Anions	1980	1028.0	371.0	+657	391.0	+637
	1981	753.0	151.0	+602	145.0	+608
	Mean	891.0	261.0	+630	268.0	+623

Neither sodium nor chloride is strongly retained by exchange or adsorption, and neither is accumulated to a large degree in biological material. Consequently, both are probably controlled primarily by hydrologic processes; under high discharge conditions, there is net export from the watershed and at low flow, net accumulation. Long term input and output are expected to be nearly equal.

4. Output > input (Mg^{2+}, K^+, HCO_3^-, SiO_2). Net losses for the SWAS watersheds occur only for the few species for which inputs are small and an internal source exists. Magnesium, potassium, and silica are solubilized during mineral weathering. Magnesium and potassium are the predominant cation species in bedrock (along with iron and aluminum which are virtually insoluble at pH $\geqslant$ 5), and are also the dominant exchangeable cations in soil. Bicarbonate forms from dissociation of carbonic acid and subsequent weathering or exchange of H^+ for a base cation. Mineral weathering is the long term source of these cations to solution, but the short term source may be either ion exchange or weathering. Identification of the short term source(s) of base cations to solution and the relevant control mechanisms will be another topic of SWAS Stage II.

As is true for several key ions, total cation and anion input in substantially larger than total output, indicating that the overall biogeochemical process ongoing in the watersheds is one of accumulation. Three sets of factors interact to cause net accumulation; first and most important is the very inert bedrock, which weathers very slowly and yields little to solution. Second, biological accretion is important as a sink for nitrogen and perhaps calcium. Finally, adsorption is an efficient (although finite) sink for sulfate and some counterion, probably hydrogen.

It is important to keep net flux data in perspective relative to hydrologic processes in the SWAS watersheds. Chemical inputs and outputs are proportional to precipitation and stream flow, thus the precipitation to stream flow relationship can have a major impact on net flux of certain ions. For those ions for which input is either much greater (H^+, HN_4^+, NO_3^-) or much smaller (K^+, Mg^{++}, HCO_3^-, SiO_2) than stream flow export, hydrology is of little significance except in determining the magnitude of the net flux. For sodium and chloride, however, long term net fluxes are expected to be small, and the sign of the net flux will change from month-to-month or year-to-year depending on precipitation-discharge ratios. Calcium and sulfate output fluxes in the SWAS watersheds are about equal to inputs during periods of high discharge (fall 1979-early spring, 1980, Figure 5), but much smaller during low flow seasons; thus for the SWAS watersheds we anticipate continuing accumulation of these two ions.

The stability and fairly high pH of the SWAS streams is probably the most surprising aspect of this research. Bedrock is very unreactive, and soils are thin, highly acidic, and nearly devoid of base cations. In spite of these facts, pH remains constant at 5.7 and 6.0 in Deep Run and White Oak Run, respectively, and major ion chemistry is very stable on an annual basis and over large changes in discharge. Storm related chemical changes are generally minor in nature, and of very short duration – a few hours to a day or two. The most pronounced changes are associated with heavy thundershowers during summer low flow, when pH reduction may be 0.5 units and alkalinity reduced by about 50 percent, but recovery from these storms is complete within 12 hours.

Stability in these watersheds appears to be related to several processes. Sulfate adsorption is probably the single most important process, because it reduces sulfate concentrations by at least 50 percent (based on flux data) and thus has a large impact on total anion concentration. Based on the undersaturation of mineral phases and the simultaneous changes in the concentration of three base cations (Ca, Mg, K), it is suggested that cation exchange is the primary control on cation concentrations. Adsorption and exchange controls are also supported by the rapid recovery of stream water chemistry following storm events.

In light of these probable controls, there is significant concern about the long term stability of stream water chemistry in the watersheds. Sulfate adsorption capacity of soils is finite; as these sites become saturated, sulfate output will rise until input and output are equal, resulting in a major increase in total anion concentration. Simultaneously, there is currently a net export of base cations from the watersheds; at current loss rates and in the absence of significant primary weathering, we estimate depletion of the soil base cation pool within 50-200 years. As sulfate output increases, this time period will shorten accordingly.

In summary, both stream systems show a high degree of chemical stability, but it is maintained at the expense of long term degradation of the soil buffer systems. As cation reservoirs are depleted, there will be an inevitable increase in hydrogen (and aluminum and heavy metal) concentrations in solution, leading to degradation of associated forest and aquatic ecosystems. The time frame over which significant acidification will occur is the subject of ongoing research; current efforts are focused on determination of sulfate adsorption capacity and primary weathering rates in the watersheds. Available data suggests that if current sulfate input levels to SWAS watersheds are maintained or increased over the coming decades, the inevitable consequences will be significant acidification and consequent degradation of the SWAS ecosystem.

ACKNOWLEDGMENTS

The authors wish to express appreciation for generous financial support provided by the National Park Service. We also wish to acknowledge the cooperation and assistance of personnel from the Shenandoah National Park, and thank landowners adjacent to the park for allowing access to study sites.

LITERATURE CITED

American Public Health Association, 1981. Standard Methods for the Examination of Water and Wastewater. 13th Edition, APHA, Washington, D.C.

Anonymous, 1981. The MAP3S Precipitation Chemistry Network: Fourth Periodic Summary Report (1980). Report to US EPA prepared by Pacific Northwest Laboratory, Report No. PNL-4144.

Arnold, D. E., R. W. Light, and V. J. Dymond, 1980. Probable Effects of Acid Precipitation on Pennsylvania Waters. EPA Ecological Research Series EPA-600/3-80-012-20.

Black, C. A. (Editor), 1965. Methods of Soil Analysis, Part 2. American Society of Agronomy, Madison, Wisconsin, 639 pp.

Burns, D. A., J. N. Galloway, and G. R. Hendrey, 1981. Acidification of Surface Waters in Two Areas of the Eastern United States. Water, Air and Soil Pollution 16:277-285.

Cogbill, C. V. and G. E. Likens, 1974. Acid Precipitation in the Northeastern United States. Water Resources Research 10(6): 1133-1137.

Dochinger, L. S. and T. A. Seliga (Editors), 1976. Proceedings of the First International Symposium on Acid Precipitation and the Forest Ecosystem. USDA Forest Service Technical Report NE-23, 1074 pp.

Drabløs, D. and A. Tollan (Editors), 1980. Ecological Impact of Acid Precipitation. Proceedings of an International Conference, Sandefjord, Norway, March 11-14, 1980, 379 pp.

Ensminger, L. E., 1954. Some Factors Affecting the Adsorption of Sulfate by Alabama Soils. Soil Science Society of America Proceedings 18:259-264.

Galloway, J. N. and E. B. Cowling, 1978. The Effects of Precipitation on Aquatic and Terrestrial Ecosystems: A Proposed Precipitation Chemistry Network. Air Pollution Control Association Journal 28(3):229-235.

Galloway, J. N. and G. E. Likens, 1976. Calibration of Collection Procedures for the Determination of Precipitation Chemistry. Water, Air and Soil Pollution 6:241-258.

Gathright, T. M. II, 1976. Geology of the Shenandoah National Park, Virginia. Virginia Division of Mineral Resources Bulletin No. 86, Charlottesville, Virginia, 93 pp.

Hajek, B. F., F. Adams, and J. R. Cope, Jr., 1972. Rapid Determination of Exchangeable Bases, Acidity and Base Saturation for Soil Characterization. Soil Science Society of American Proceedings 36:436-438.

Hendrey, G. R., J. N. Galloway, S. A. Norton, C. L. Schofield, P. W. Shaffer, and D. A. Burns, 1980. Geological and Hydrochemical Sensitivity of the Eastern United States to Acid Precipitation. EPA Ecological Research Series EPA-600/3-80-024, 100 pp.

Hockman, J. R., J. C. McKinney, T. R. Burruss, D. Jones, R. E. Modesitt, L. G. Manhurt, and W. R. Waite, Jr., 1979. Soil Survey of Augusta County, Virginia, 248 pp.

Kunze, G. W., 1965. Pretreatment for Mineralogical Analysis. *In:* Methods of Soil Analysis, Part 2, C. A. Black (Editor). American Society for Agronomy, Madison, Wisconsin, pp. 568-577.

Likens, G. E., F. H. Bormann, R. S. Pierce, J. S. Eaton, and N. M. Johnson, 1977. Biogeochemistry of a Forested Ecosystem. Springer-Verlag, New York, New York.

Likens, G. E., R. F. Wright, J. N. Galloway, and T. J. Butler, 1979. Acid Rain. Scientific American 241:43-51.

Overrein, L. N., H. M. Seip, and A. Tollan (Editors), 1980. Acid Precipitation – Effects on Forest and Fish. Final Report of SNAF Project, Oslo, Norway.

Schofield, C. S., 1976. Acid Precipitation: Effects on Fish. Ambio 5:228-230.

Schwab, F. L., 1970. Origin of the Antietam Formation (Late Precambrian-Lower Cambrian), Central Virginia. Journal of Sedimentary Petrology 40(1):354-366.

Schwab, F. L., 1971. Harpers Formation, Central Virginia: A Sedimentary Model. Journal of Sedimentary Petrology 41(1):139-149.

Truesdell, A. H. and B. F. Jones, 1974. WATEQ, a Computer Program for Calculating Chemical Equilibria of Natural Waters. Journal of Research, U.S. Geological Survey 2:233-248.

THE IMPACT OF ATMOSPHERIC DEPOSITION ON THE WATER QUALITY OF EVERGLADES NATIONAL PARK

Mark D. Flora and Peter C. Rosendahl[1]

ABSTRACT: A preliminary investigation of the relationship between surface loading and atmospheric deposition indicates that precipitation is responsible for a significant percentage of the chemical constituents entering the unique fresh water marsh system of Everglades National Park. From June 1980 through August 1981, wetfall contributed more than 95 percent of the ammonium and nitrate and 83 percent of the orthophosphate entering the 1000 km^2 Shark River Slough, the major fresh water drainage of Everglades National Park. Conversely, surface water delivery is responsible for more than 95 percent of the calcium, sodium, magnesium, and chloride; and more than 85 percent of the sulfate input into the slough. The mean volume weighted pH of precipitation during this 14-month period was 4.9, which while acidic, is not sufficient to alter the pH of the highly buffered surface water of the slough, which has a mean alkalinity of 182 mg L^{-1} (as $CaCO_3$).
(KEY TERMS: acid precipitation; water quality; Everglades; fresh water marsh; National Atmospheric Deposition Program.)

INTRODUCTION

Everglades National Park, encompassing more than 6000 km^2 of the southern tip of Florida, was established by Congress in 1947 to protect remaining wetland areas that once covered most of Florida south of Lake Okeechobee. A predominant feature of the Everglades wetland system is the fresh water slough, a shallow (less than 1.5 m), seasonally dynamic, slow flowing marsh system. The 1000 km^2 Shark River Slough is the largest fresh water flow system in Everglades National Park. It provides critical habitat for several threatened or endangered species, serves as an area of water recharge for the important Biscayne Aquifer, and functions as a major source of fresh water input into Everglades estuarine zone. Because of the importance of this aquatic system, considerable effort has been expended to characterize and preserve its excellent water quality (Rosendahl and Rose, 1979; Flora and Rosendahl, 1981, 1982).

Over the past decade, much interest has been generated concerning the magnitude, extent, and environmental effects of acid precipitation and atmospheric deposition. Brezonik, *et al.* (1980), demonstrate that acid precipitation is not limited to northeastern North America, but exists as far south as Florida. Crisman, *et al.* (1980), in assessing the impact of acid precipitation on poorly buffered lakes located along the central ridge of Florida documents significant adverse effects on both water quality and the biota which can be attributed to acidification.

The purpose of this paper include: 1) the quantification of the initial 14 months of acid precipitation and atmospheric deposition collected as part of the National Atmospheric Deposition Program, 2) the determination of the relationship between atmospheric deposition and surface water inflow as loading mechanisms, and 3) a preliminary evaluation of the Shark River Slough's susceptibility to acidification or other water quality modification as a result of atmospheric deposition.

METHODOLOGY

Data concerning the chemical quality of precipitation throughout southern Florida prior to the late 1970's are scarce. The first comprehensive study designed specifically to quantify the chemistry of wetfall and bulk precipitation throughout Florida was begun by the University of Florida in the north-central portion of the state in 1976 (Hendry and Brezonik, 1980), and expanded throughout the state in 1977 (Hendry, *et al.*, 1980). Several stations included in this network were located in southern Florida, including Miami (urban), Corkscrew Swamp Sanctuary (cypress swamp), and Bahia Honda State Park (marine). In early 1980, an additional wetfall station located at the Tamiami Ranger Station along the northern boundary of Everglades National Park (Figure 1) was incorporated as part of this network's final year of operation.

In June 1980, with the establishment of a wetfall/dryfall station at the South Florida Research Center, Everglades National Park became part of the long term national monitoring network as a member of the National Atmospheric Deposition Program (N.A.D.P). In June 1981, additional wetfall/dryfall stations were established at several of the

[1] Respectively, Hydrologist and Research Hydrologist, Everglades National Park, South Florida Research Center, P.O. Box 279, Homestead, Florida 33030.

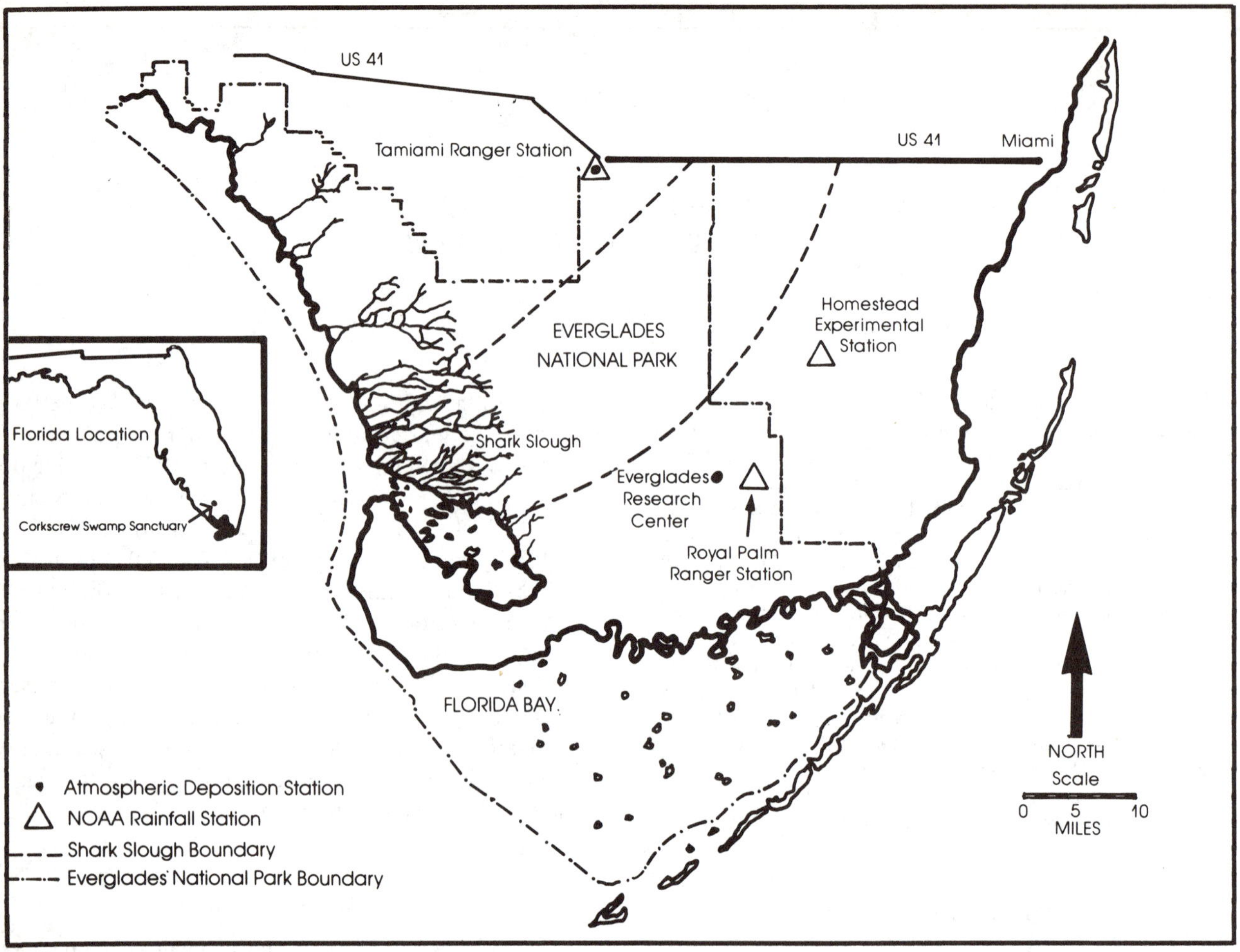

Figure 1. Map of South Florida Atmospheric Deposition and Precipitation Monitoring Sites.

terminated University of Florida locations in southern Florida, including the Corkscrew Swamp Sanctuary and the Tamiami Ranger Station, by Environmental Science and Engineering under contract to the Florida Electric Power Coordinating Group. While the results reported in this paper necessarily incorporate published data from these other studies, care is taken to utilize only those data that were collected and analyzed by equivalent methodology.

Wetfall/dryfall samples were collected at all sites by means of an Aerochem Metrics Model 301 wetfall/dryfall sampler. Wetfall sample buckets were changed weekly with field specific conductance and pH being determined at the site by using a Beckman Model RC-16C conductivity bridge and either Altex or Corning pH meters equipped with semi-micro electrodes. Buckets are then shipped to central laboratory facilities (University of Florida, Illinois State Water Survey, Enviornmental Science and Engineering Laboratories, respectively) for analysis. The laboratory determinations of pH, specific conductance, ionic composition, and nutrient concentration are accomplished using equivalent methods (Hendry, 1977; Peden, *et al.*, 1979; ESE, 1982), with care taken in each laboratory to assure adequate quality control.

Ionic concentrations are determined using flame atomic emission and ICAP for major cations and Technicon Autoanalyzer methodology for the determination of chloride and sulfate, and the major nutrients including ammonium, nitrate, and phosphate.

Surface water chemical analysis was completed by the U.S. Geological Survey state laboratory in Ocala, Florida, utilizing standard USGS analytical techniques (USGS, 1979; APHA, 1979).

Monthly rainfall volumes are calculated utilizing the Thiessen polygon weighted mean averaging technique and monthly rainfall data collected at three National Oceanographic and Atmospheric Administration rainfall stations located at the Tamiami Ranger Station, the Royal Palm Ranger Station, and the Homestead Agricultural Experiment Station.

Monthly surface water delivery to the Shark River Slough is provided by the U.S. Army Corps of Engineers, based upon gate opening height and upstream/downstream head differential at the four surface water delivery structures located along the northern boundary of Everglades National Park.

These data are then utilized to determine monthly chemical loading rates to the system from atmospheric and surface inflow sources in terms of both total loading (metric tons) and deposition per unit area (g m^{-2} mo^{-1}).

RESULTS

Atmospheric Deposition

Monthly volume weighted mean precipitation chemistry is available from the Everglades Research Center N.A.D.P. site from June 1980 through August 1981. Additionally, areawide comparisons of wetfall chemistry can be made for two 10-11 week periods from June-August 1980 and from July-September 1981 by utilizing concurrent data from stations of various networks located at the Everglades Research Center, Tamiami Ranger Station, and Corkscrew Swamp Sanctuary.

Table 1 depicts monthly volume weighted mean wetfall chemistry at the Everglades N.A.D.P. site. Volume weighted mean wetfall chemistry during this period was 4.9, with monthly volume weighted pH ranging from 4.5 to 5.7. Monthly volume weighted mean concentrations were also determined for the major cations, anions, and nutrients at the Everglades site.

Of the major cations, calcium, magnesium, and potassium concentrations were approximately half of those reported for wetfall in north-central Florida (Gainesville) in 1976-1977 (Hendry and Brezonik, 1980), while sodium concentrations were similar at both sites. Significant differences (p=0.05) were also noted for the anions, with sulfate at the Everglades site (1.05 mg L^{-1}) being approximately 50 percent of that reported for Gainesville (2.05 mg L^{-1}). However, considerable monthly variability occurred with the mean monthly sulfate concentration ranging from 0.33 mg L^{-1} to 2.70 mg L^{-1}. Seasonal trends of lower pH and higher sulfate concentrations reported in the summer months for northern Florida (Brezonik, *et al.,* 1980) were not noted at the Everglades site.

The reason for the differences in seasonal trends between north-central Florida and subtropical southern Florida may be related to both southern Florida's climatological patterns and the proximity of the Everglades site to oceanic influence. Rainfall patterns in subtropical southern Florida are dominated during the wet season (summer) by localized convective showers influenced by daily sea breezes. During the dry season (winter) synoptic scale processes, such as frontal systems are more common. In addition, this study was affected in August 1981 by a subtropical cyclonic depression during which 42.3 cm of rainfall occurred at the Everglades site during one 48-hour period, effectively diluting the volume weighted monthly concentration for August.

Chloride concentrations at the Everglades site (0.77 mg L^{-1}) were slightly less than those reported at the Gainesville site but nitrate concentrations were significantly higher in the Everglades (0.63 mg L^{-1}) than those in north-central Florida (0.19 mgL^{-1}).

TABLE 1. Monthly Volume Weighted Mean Wetfall Chemistry at the South Florida Research Center, Everglades National Park, Florida.

Month	Rainfall (cm)	pH	Sp. Cond. (mhos)	Ca (mgL^{-1})	Mg (mgL^{-1})	K (mgL^{-1})	Na (mgL^{-1})	NH_4-N (mgL^{-1})	NO_3-N (mgL^{-1})	Cl (mgL^{-1})	SO_4 (mgL^{-1})	PO_4 (mgL^{-1})
6-80	33.5	4.57	14.4	0.28	0.08	0.02	0.65	0.06	1.41	0.45	1.75	0.003
7-80	27.8	4.49	19.3	0.16	0.06	0.03	0.48	0.11	1.46	0.64	1.58	0.003
8-80	21.7	5.21	9.3	0.23	0.06	0.15	0.32	0.16	0.65	0.93	1.31	0.083
9-80	18.9	5.41	6.3	0.28	0.03	0.04	0.21	0.03	0.07	0.46	0.35	0.007
10-80	9.9	5.19	11.9	0.39	0.10	0.18	0.52	0.03	0.50	0.83	1.35	0.014
11-80	16.6	NA	NA	NA	NA	NA	NA	NA	NA	NA	NA	NA
12-80	3.1	5.83	21.6	0.32	0.35	0.13	NA	0.02	0.02	4.59	1.40	0.003
1-81	1.1	NA	NA	NA	NA	NA	NA	NA	NA	NA	NA	NA
2-81	13.1	NA	13.9	0.53	0.08	0.11	0.39	0.58	0.09	0.68	0.48	0.003
3-81	4.7	4.75	30.2	0.54	0.18	0.12	1.20	0.27	1.45	2.21	2.7	0.017
4-81	0.3	NA	NA	NA	NA	NA	NA	NA	NA	NA	NA	NA
5-81	10.2	5.30	16.3	0.23	0.09	0.05	0.61	0.75	0.96	1.11	2.20	0.003
6-81	9.6	4.96	17.3	0.18	0.11	0.05	0.36	0.15	0.65	1.49	1.14	0.003
7-81	16.4	5.81	12.7	0.23	0.13	0.30	0.43	0.02	0.02	0.84	0.33	0.013
8-81	51.0	5.42	5.0	0.03	0.05	0.02	0.31	0.02	0.14	0.53	0.37	0.003
Volume Weighted Average		4.91	12.0	0.217	0.076	0.076	0.429	0.137	0.634	0.772	1.05	0.013

The concentration of nutrients in rainfall and their relative contribution to the overall nutrient inputs into the aquatic systems are of interest in nutrient loading studies. In this study, the wetfall concentrations of ammonium (0.14 mgL^{-1}) and soluble reactive phosphorus (0.013 mg L^{-1}) were similar to those reported for much of Florida.

An areawide comparison of volume weighted wetfall chemistry throughout southern Florida (Table 2) depicts an areal similarity in the chemical composition throughout the nonurban areas of southern Florida. pH values were similar at all stations, but lower during the 1980 (4.67-4.71) comparative period than during the similar period in 1981 (4.97-5.53). This difference may reflect the unusual precipitation patterns in 1981 with drought conditions during most of the year being interrupted by an unusually large rainfall event associated with a subtropical depression occurring during late August. An areal distribution was noted, however, for both sodium and chloride which decreased from east to west with increasing distance from the oceanic influence. The other major cations (calcium, magnesium, and potassium), anions (sulfate), and nutrients (ammonium, nitrate, and phosphate) were found in similar concentrations at each site.

Surface Water Delivery

The long term water quality characteristics of the surface water in the Shark River Slough have been summarized by Flora and Rosendahl (1981, 1982b). Data presented in Table 3 demonstrate that these are chemically well buffered waters with a pH range of from 7.0 to 8.3 and a mean total alkalinity of 182 mg L^{-1} (as $CaCO_3$). The dominant cations, calcium (70 mg L^{-1}) and sodium (50 mg L^{-1}) result from the regional lithology, as does bicarbonate, which serves as the predominant anion.

The soluble nutrient fraction of these waters are very low, reflecting the efficient uptake mechanisms for available nutrients found throughout southern Florida's fresh water marsh system.

Loading Characteristics

The overall quality of water in the Shark River Slough is dependent upon the quality and quantity of the water delivered to the slough. It is therefore important to determine the relative loading to the system from each of the major inputs.

During the period analyzed here, precipitation accounted for 83.9 percent and surface water delivery for 16.1 percent of the water entering the slough. The nutrient concentrations of the wetfall were in slight excess of those measured in the surface water, while the conc entrations of major cations and anions in the surface water were typically 20 to 300 times those found in the wetfall.

Table 4 and 5 document the monthly loads of ionic and nutrient constituents delivered to the slough as either surface water inflow or wetfall deposition. While dryfall deposition data are not yet available for this site, Hendry and Brezonik (1980) report that in north-central Florida wetfall

TABLE 2. Volume Weighted Mean Wetfall Chemistry at Three South Florida Stations from Mid-June Through August 1980 (10 weeks), and from early July Through September 1981 (11 weeks).

	Rainfall (cm)	Lab (pH)	Field (pH)	Sp. Cond. (mhos)	Ca (mgL^{-1})	Mg (mgL^{-1})	K (mgL^{-1})	Na (mgL^{-1})	NH_4-N (mgL^{-1})	NO_3-N (mgL^{-1})	PO_4-P (mgL^{-1})	Cl (mgL^{-1})	SO_4 (mgL^{-1})
MID-JUNE THROUGH AUGUST 1980													
Corkscrew Audubon Sanctuary[1]	71.7	4.67	NA	14.3	0.12	0.02	0.01	0.19	0.07	0.18	NA	0.34	1.06
Tamiami Ranger Station[1]	31.8	4.71	NA	16.3	0.41	0.13	0.09	0.27	0.09	0.22	NA	0.61	1.23
Everglades Research Center[2]	59.2	4.68	NA	14.8	0.19	0.06	0.08	0.41	0.13	NA	0.017	0.56	1.46
EARLY JULY THROUGH SEPTEMBER 1981													
Corkscrew Audubon Sanctuary[3]	68.3	4.97	4.82	NA	0.14	0.05	0.02	0.19	0.03	0.13	0.009	0.42	0.62
Tamiami Ranger Station[3]	61.1	5.19	5.12	NA	0.19	0.05	0.05	0.22	0.14	0.16	0.009	0.50	0.75
Everglades Research Center[4]	79.0	5.53	5.22	8.9	0.09	0.08	0.10	0.49	0.03	0.20	0.021	0.62	0.44

[1]Data from University of Florida network (personal communication with C. D. Hendry).
[2]Data from N.A.D.P. network (N.A.D.P., 1981a, 1981b).
[3]Data from E.S.E. (1982).
[4]Data from N.A.D.P. (unpublished).

TABLE 3. Surface Water Chemistry and Actual Monthly Delivery to the Shark River Slough at S-12D.

Month	Surface Water Inflow ($m^3 x 10^6$)	pH	Alkalinity (mgL^{-1})	Sp. Cond. (mhos)	Ca (mgL^{-1})	Mg (mgL^{-1})	K (mgL^{-1})	Na (mgL^{-1})	Cl (mgL^{-1})	SO_4 (mgL^{-1})	NH_4–N (mgL^{-1})	NO_3–N (mgL^{-1})	TKN (mgL^{-1})	PO_4–P (mgL^{-1})	T-P (mgL^{-1})
7-80	1.4	7.9	132	400	47	19	1.3	24	44	7.1	0.11	0.03	1.23	0.014	0.014
8-80	18.0														
9-80	95.8		200	795	55	18	4.7	81	110	34.0	0.03	0.02	1.20	0.010	0.014
10-80	83.1	7.1	208	800	65	17	4.0	55	99	33.0	0.01	0.21	2.08	0.014	0.036
11-80	74.9														
12-80	40.8														
1-81	23.6	7.0	156	530	58	9	0.9	28	51	7.7	0.08	0.04	0.59	0.010	0.020
2-81	11.0														
3-81	4.1														
4-81	1.6	7.3	228	900	92	16	3.0	64	75	13.0	0.19	0.26	1.94	0.003	0.070
5-81	2.8														
6-81	6.0														
7-81	9.1	8.3	169	780	75	14	1.4	38	73	24.9	0.04	0.01	1.84	0.010	0.010
8-81	15.0	7.8		820	100	16	2.6	62	93	30.0	0.02	0.00	1.30	0.010	0.010
Mean	27.7	7.4	182	718	70	16	2.6	50	78	21.4	0.07	0.08	1.45	0.010	0.025

NOTE: Data ssembled from U.S. Geological Survey and U.S. Army Corps of Engineers Sampling Programs for delivery water to Everglades National Park.

TABLE 4. Monthly Nutrient Loading (metric tons) from Atmospheric Wetfall and Surface Sources to the Shark River Slough.

Month	Rain ($m^3 x 10^6$)	Surface Inflow ($m^3 x 10^6$)	NH_4–N		NO_3–N		PO_4–P	
			Rain	Surface	Rain	Surface	Rain	Surface
7-80	275.1	1.4	30.1	0.2	382.9	0.1	0.5	0.0
8-80	215.2	18.0	34.3	2.0	139.9	0.5	17.8	2.5
9-80	187.0	95.8	5.6	2.9	13.0	1.9	1.2	1.0
10-80	97.9	83.1	3.0	0.8	48.9	17.5	1.3	1.2
11-80	164.1	74.9	NA	0.7	NA	15.7	NA	1.0
12-80	30.7	40.8	0.6	3.3	0.6	1.6	0.1	0.4
1-81	11.1	23.6	NA	1.9	NA	0.9	NA	0.2
2-81	129.9	11.0	75.3	0.9	11.7	0.4	NA	0.1
3-81	46.6	4.1	12.4	0.8	67.5	1.1	0.8	0.0
4-81	3.0	1.6	NA	0.3	NA	0.4	NA	0.0
5-81	100.9	2.8	75.5	0.5	75.6	0.7	0.3	0.0
6-81	94.9	6.0	17.6	0.2	61.6	0.1	0.3	0.1
7-81	163.5	9.1	3.2	0.4	3.2	0.1	2.1	0.1
8-81	505.5	15.0	10.1	0.3	70.8	0.0	1.5	0.1
$\bar{x}$ Load/Month (metric tons)			24.3	1.1	79.6	2.9	2.4	0.5
Percent of Load	83.9	16.1	95.6	4.4	96.5	3.5	82.8	17.2
$\bar{x}$ Load g/m^2–mo			0.025	0.001	0.080	0.003	0.002	0.0005

is the primary source of deposition for sulfate, ammonium, nitrate, magnesium, and potassium, while wetfall and dryfall are of comparable importance in the deposition of sodium, calcium, and chloride.

Atmospheric deposition is found to be more important than surface water delivery in the loading of nutrients into the slough. Wetfall contributes 95 percent of the ammonium, 96 percent of the nitrate, and 83 percent of the phosphate annually entering the system (Table 4). Surface water delivery constitutes the predominant loading mechanism for the major ionic constituents entering the slough. Total ionic loading varies greatly from month to month as a function of surface water delivery through the water delivery structures. Surface water delivery accounts for 98

TABLE 5. Monthly Ionic Constituent Loading (metric tons) from Atmospheric Wetfall and Surface Water Sources, Shark River Slough.

			Ca		Na		K		Mg		Cl		SO_4	
Month	Rain	Surface Inflow	Rain	Surface	Rain	Surface	Rain	Surface	Rain	Surface	Rain	Surface	Rain	Surface
7-80	275.1	1.4	42	66	119	34	6	2	11	27	117	62	288	10
8-80	215.2	18.0	50	846	84	432	26	23	10	342	162	792	228	127
9-80	187.0	95.8	52	5269	23	7759	4	450	3	1724	51	10538	38	3257
10-80	97.9	83.1	38	5401	31	4571	11	332	6	1413	49	8227	80	2742
11-80	164.1	74.9	NA	4869	NA	4120	NA	300	NA	1273	NA	7415	NA	2472
12-80	30.7	40.8	10	2366	NA	1142	3	37	7	367	95	2081	29	314
1-81	11.1	23.6	NA	1368	NA	661	NA	21	NA	212	NA	1204	NA	181
2-81	129.9	11.0	69	638	32	308	9	10	7	99	55	561	39	85
3-81	46.6	4.1	25	377	50	262	5	12	8	66	92	307	112	53
4-81	3.0	1.6	NA	147	NA	102	NA	5	NA	26	NA	120	NA	21
5-81	100.9	2.8	23	258	40	179	3	8	6	45	74	210	146	36
6-81	94.9	6.0	17	450	38	228	5	8	12	84	156	204	119	149
7-81	163.5	9.1	38	683	85	346	59	13	26	127	167	438	65	227
8-81	505.5	15.0	15	1500	124	930	8	39	8	240	212	1395	148	450
$\bar{x}$ Load/mo (metric tons)			34	1731	63	1505	12	90	9	432	112	2396	117	723
% of Load	83.9	16.1	1.9	98.1	4.0	96.0	11.7	88.3	2.0	98.0	4.4	95.6	13.9	86.1
$\bar{x}$ Load h/m^2-mo			0.034	1.745	0.064	1.518	0.012	0.091	0.009	0.437	0.113	2.42	0.118	0.729

NOTES: Precipitation and surface inflow in units of $m^3 \times 10^6$.
Ionic constituent loads in metric tons (unless indicated).

percent of the calcium, 96 percent of the sodium, 88 percent of the potassium, 98 percent of the magnesium, 95 percent of the chloride, and 86 percent of the sulfate entering the slough from June 1980 through August 1981.

These results demonstrate, then, that the total ionic and nutrient loading to the system are directly influenced by the quantitative relationship between the amount of surface water delivered to the slough and the amount of atmospheric deposition, and that changes in the water management/hydrological regime are likely to influence the chemical loading to the system.

CONCLUSIONS

The negative impacts of atmospheric deposition on poorly buffered surface waters are well documented for areas of Scandinavia and northeastern North America where a large number of lakes have already been found to be devoid of fish, impoverished in the numbers and diversity of phytoplankton and zooplankton, and to be changing towards conditions of low aquatic productivity (Glass, *et al.*, 1980) . The impact of atmospheric deposition on the aquatic resources of Florida are not well documented though Crisman, *et al.* (1980), report decreased pH and increased sulfate concentrations in a number of poorly buffered oligotrophic lakes located in a sandy soil regime along the central highlands of Florida. In addition to changes in water quality, Crisman, *et al.* (1980), report depressed Chlorophyll *a* concentrations, a species shift from blue-green to green algae, and a reduction in both the numbers and species diversity of the zooplankton in slightly acidified Florida lakes (pH=4.98).

The effects of atmospheric deposition on an aquatic system is a function of both the quantities of acid forming substances being deposited on a watershed, and the assimulative capacity of the watershed and aquatic system to neutralize them. Wright (1976) defines a threshold level at which acid precipitation begins to affect Swedish lakes as an annual volume weighted precipitation pH of 4.6. Glass, *et al.* (1981), further define a mean danger threshold for lake acidification based upon the relationship between pH and alkalinity in northern Minnesota lakes. Lakes with an alkalinity of less than 15 mg L^{-1} (as $CaCO_3$) are highly susceptible to acidification while those with an alkalinity of between 15 and 30 mg L^{-1} are potentially susceptible.

The subthreshold volume weighted mean pH of wetfall into the Shark River Slough (pH=4.9) coupled with the high buffering capacity of the surface water indicate that the water of the slough are not threatened by acidification.

However, Galloway and Likens (1979) and others demonstrate that heavy metal deposition including Ag, Au, Cr, Cu, Pb, Sb, V, and Zn increase on a regional basis as atmospheric emissions from fossil fuel combustion increases. No data concerning the deposition of heavy metals are available for southern Florida.

Atmospheric deposition is reported earlier to be the predominant mechanism for nutrient loading to the slough, contributing approximately 90 percent of the soluble reactive nutrients. The effect of this loading is not apparent on the overall surface water chemistry, where soluble nutrient concentrations are naturally low, probably due to both rapid nutrient uptake by biological communities and immobilization by the sediments.

Because of its dilute nature, precipitation contributes only 2 percent to 11 percent of the ionic constituent load to the slough. Flora and Rosendahl (1982b) show that overall patterns of specific conductance and ionic composition throughout the slough are determined primarily by the ionic composition of the surface water delivery to the slough, though modified seasonally by rainfall and evaporation patterns.

LITERATURE CITED

American Public Health Association, 1979. Standard Methods for the Examination of Water and Wastewater, 14th Edition. Washington, D.C., 1193 pp.

Brezonik, P. L., E. S. Edgerton, and C. D. Hendry, 1980. Acid Precipitation and Sulfate Deposition in Florida. Science 208:1027-1029.

Crisman, T. L., R. L. Schulze, P. L. Brezonik, and S. A. Bloom, 1980. Acid Precipitation: The Biotic Response in Florida Lakes. Proc. Int. Conf. Ecol. Impact Acid Precip., SNSF Project, Norway.

Environmental Science and Engineering, 1982. Florida Acid Deposition Study, Monitoring Program – Phase I, Summary Report. Gainesville, Florida.

Flora, M. D. and P. C. Rosendahl, 1981. Specific Conductance and Ionic Characteristics of the Shark River Slough, Everglades National Park, Florida. National Park Service, South Florida Research Center, Homestead, Florida, 55 pp.

Flora, M. D. and P. C. Rosendahl, 1982a. The Response of Specific Conductance to Environmental Conditions in Everglades National Park. Water, Air and Soil Pollution 17:51-59.

Flora, M. D. and P. C. Rosendahl, 1982b. An Analysis of Surface Water Nutrient Concentrations in the Shark River Slough, 1972-1980. National Park Service, South Florida Research Center, Homestead, Florida, 48 pp.

Galloway, J. N. and G. E. Likens, 1979. Atmospheric Enhancement of Metal Deposition in Adirondack Lake Sediments. Limnology and Oceanography 24(3):427-433.

Glass, N. R., G. E. Glass, and P. J. Rennie, 1980. Effects of Acid Precipitation in North America. Environment International 4: 443-452.

Glass, G. E., T. G. Brydges, and O. L. Loucks, 1981. Impact Assessment of Airborne Acidic Deposition on the Aquatic Environment of the United States and Canada. U.S. Environmental Protection Agency, EPA-600/10-81-000, Washington, D.C., 152 pp.

Hendry, C. D., 1977. The Chemistry of Precipitation in Northcentral Florida. M.S. Thesis, College of Engineering, University of Florida, Gainesville, Florida, 141 pp.

Hendry, C. D. and P. L. Brezonik, 1980. Chemistry of Precipitation at Gainesville, Florida. Environmental Science and Technology 14(7):843-849.

Hendry, C. D., P. L. Brezonik, and E. S. Edgerton, 1980. Acid Precipitation in Florida (U.S.A.): Results of a Statewide Monitoring Network. Proc. Int. Conf. Ecol. Impact Acid Precip., SNSF Project, Norway.

National Atmospheric Deposition Program, 1981a. NADP Data Report – Precipitation Chemistry, Second Quarter 1980. Natural Resource Ecology Laboratory, Colorado State University, Fort Collins, Colorado, 281 pp.

National Atmospheric Deposition Program, 1981b. NADP Data Report – Precipitation Chemistry, Third Quarter 1980. Natural Resource Ecology Laboratory, Colorado State University, Fort Collins, Colorado, 314 pp.

Peden, M. E., L. M. Skowron, and F. F. McGurk, 1979. Precipitation Sample Handling, Analysis and Storage Procedures. Report No. 4, Contract EY-76-S-02-1199, U.S. Department of Energy, Pollutant Characterization and Safety Research Division, Office of Health and Environmental Research, Washington, D.C.

Rosendahl, P. C. and P. W. Rose, 1979. Water Quality Standards: Everglades National Park. Environmental Management 3(6):483-491.

U.S. Geological Survey, 1979. Methods for the Determination of Inorganic Substances in Water and Fluvial Sediments. Techniques for Water Resources Investigations of the U.S. Geological Survey, U.S. Government Printing Office, Washington, D.C., 626 pp.

Wright, R. F., 1976. Historical Changes in the pH of 128 Lakes in Southern Norway and 130 Lakes in Southern Sweden Over the Period 1973-1976. SNSF Project TN-34-77 NISK, Ås, Norway.

RED SPRUCE DIEBACK IN VERMONT AND NEW HAMPSHIRE: IS ACID PRECIPITATION A CONTRIBUTING STRESS?

Arthur H. Johnson, Deborah G. Lord, and Thomas G. Siccama[1]

ABSTRACT: A dieback of red spruce (*Picea rubens*) has been quantitatively documented in the high elevation boreal forest of Vermont, and observed in New Hampshire and New York. No primary pathogens or obvious abiotic factors appear to be responsible for the dieback. The boreal forest of the Green and White Mountain peaks is above cloud base for extended periods (800-1600 hours per year), and cloud moisture is very acid (average pH approximately 3.7). A study of foliar, root, and soil chemistry in healthy and declining stands in Vermont and New Hampshire leaves open the possibility of acid rain induced stress due to effects on foliage or, less likely, effects due to high levels of root aluminum.
(KEY TERMS: acid precipitation; New England; forests.)

INTRODUCTION

Controlled experiments have indicated that acid rain or mist can adversely affect foliage (i.e., Wood and Bormann, 1975; Evans and Curry, 1979; Scherbatskoy and Klein, Univ. of Vermont, unpublished manuscript). Other studies suggest the possibility of detrimental effects on plants through unfavorable effects on soils (i.e., Tamm and Wiklander, 1980; Abrahamsen, 1980; Ulrich, 1981a, b). Surveys of tree growth in Scandinavia (Jonsson and Sundberg, 1972; Strand, 1980) and in the U.S. (Johnson, *et al.*, 1981) documented or suggested recent, apparently abnormal declines in growth rate, but there is no convincing evidence that acid deposition is a contributing stress in those cases.

Very substantial mortality of red spruce in the northern Green Mountains of Vermont has been documented by Siccama, *et al.* (1982). There was a 50 percent decrease in density and basal area between 1965 and 1980 on the four mountains sampled. Trees in all size classes were affected. The cause is presently unknown, but is not likely due to successional dynamics. Roman and Raynal (1980) showed lower than expected growth in red spruce cored in the Adirondack Mountains of New York. Analysis using temperature and moisture data showed that the effect was not climatically controlled. The spruce dieback has been noted in New Hampshire by U.S. Forest Service personnel (unpub. data). In 1968, J. S. Hadley (unpub.) studied declining spruce in the Green Mountains and observed the presence of secondary fungal pathogens, particularly *Fomes pini* (red ring rot) and *Armillaria mellea* (shoestring root rot). These pathogens normally infect weakened trees, and Hadley judged that the drought of the early 1960's was the triggering stress. However, trees have continued to die in the intervening 14 years, and tree rings from the spruce in the boreal forest of Vermont and New Hampshire do not record negative effects on growth in 1962-65.

Figure 1 shows precipitation and cloud data for the three forest zones on Camels Hump Mountain (Huntington, Vermont). The northern hardwood forest (NHW) occurs at elevations up to about 730 m. Spruce is a minor component, and sugar maple (*Acer saccharum*), yellow birch (*Betula alleghaniensis*), and american beech (*Fagus grandifolia*) are dominant. The boreal forest zone (BOR) occurs at elevations above approximately 850 meters and red spruce, balsam fir (*Abies balsamea*), and white birch (*Betula papyrifera* var. *cordifolia*) are dominant. The intervening transition forest (TRNS, 730-850 m) is a mixture of the boreal and northern hardwood forest species. Spruce in the boreal zone are above cloud base for 800 to 1600 hours per year Siccama and co-workers (Yale Univ., unpublished data) collected cloud moisture in the White Mountains during the growing season of 1981. Preliminary results indicate that the average cloud moisture pH was approximately 3.7 or 0.5 pH lower than rainfall values collected at the same stations. Similar cloud moisture pH was measured in the Green Mountains during 1981 (R. Klein, Univ. of Vermont, personal communication). The high elevation boreal forest of New England is frequently exposed to very acid moisture. Few other remote forest ecosystems anywhere are exposed to such acid moisture for such extended periods. Accordingly, we studied chemical characteristics of red spruce roots and foliage at Camels Hump to see if any "signal" resulted which might indicate nutritional stress induced by acid precipitation.

[1] Respectively, Associate Professor and Graduate Assistant, University of Pennsylvania, Dept. of Geology D4, Philadelphia, Pennsylvania 19104; and Staff Ecologist, Yale School of Forestry and Environmental Studies, Yale University, New Haven, Connecticut 06511.

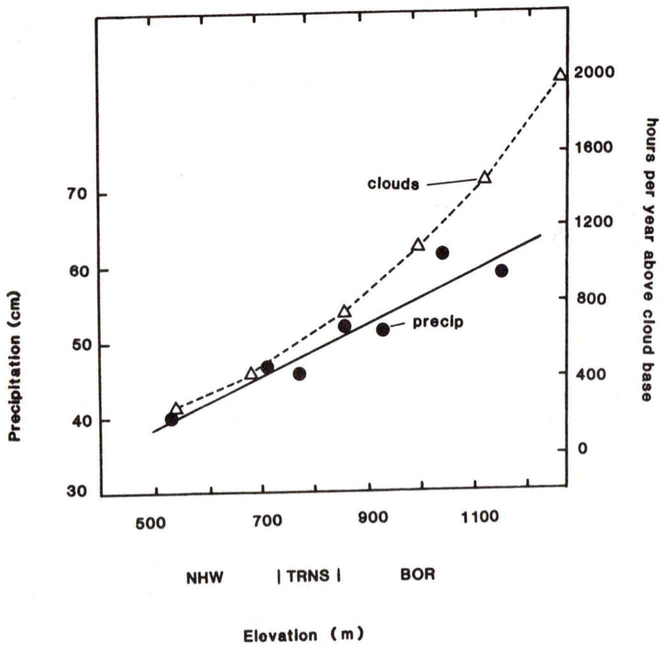

Figure 1. Precipitation and Cloud Data for Camels Hump Mountain, Vermont. Precipitation is growing season moisture collected by Siccama (1974). Cloud base data from Burlington, Vermont airport (Siccama, 1974).

METHODS

In June 1981 we sampled red spruce at three elevations on Camels Hump (500, 800, and 1000 m). We sampled trees which were < 2 cm dbh and 20-30 years old, and trees which were in the 2-9 cm size class which were up to 120 years old. Trees were judged to be in good condition if they had needles in the crown and buds. Trees were judged to be in poor condition if they lacked needles in the crown. Such trees had few or no buds. As a "control" site, we sampled spruce in a healthy stand at 800 m in the Hubbard Brook Experimental Forest (West Thornton, New Hampshire). The stand sampled was comprised of healthy spruce and fir, but it should be noted that there is spruce mortality in other parts of the Hubbard Brook forest. Red spruce is shallow rooted, so roots of < 2 cm dbh trees were generally confined to the forest floor 0 horizons. Trees were excavated, and roots and branches collected using plastic gloves. In the laboratory, the plant tissue was washed in double deionized water and dried at 80°C. Roots were separated into < 2 and > 2 mm classes, ground, then dried to constant weight at 80°C. Tissue samples were dry ashed and digested in HNO_3 as described by Andresen, *et al.* (1980). Analysis of the extracts was carried out by plasma emission spectroscopy for Ca, Mg, K, P, Mn, Cu, Zn, Al, Pb, Cd, Ni, Cr, and Mo. Total N and total S were determined at the University of Georgia Plant and Soil Testing Laboratory, Athens, Georgia, on dry tissue samples.

RESULTS AND DISCUSSION

Soil pH in the 0 horizons (which is the root zone for the < 2 cm dbh trees) did not change between 1965 and 1980 (Figure 2). This is attributable to the relatively large, natural, internal production of H^+ compared to the deposition of H^+ in precipitation. Fine root (i.e., "feeder" roots, < 2 mm diameter) chemistry is summarized in Table 1, and significant differences among groups are shown in Table 2. Owing to the rather large difference in root Al between control and declining stands, we sampled healthy looking spruce in several other areas to assess root aluminum concentrations (Table 3). Particular interest in aluminum was prompted by the work of Cronan and Schofield (1979) which showed the mobilization of Al in forest soils, and the findings of Ulrich (1981a, b) which suggest a link between norway spruce (*Picea abies*) decline and high concentrations of Al in soil solution. Our data suggest that roots of trees in good condition can have high Al concentrations (> 2000 ppm), and trees in poor condition can have low (< 500 ppm) Al concentrations in fine roots. In light of the fact that spruce growth continued to decline during a period when there was no change in soil pH, and given the lack of significant differences in root composition among the groups of trees, there is no real evidence that acid precipitation affected red spruce through a soil pathway. The evidence does not support the contention that Al accumulation in root tissue is an important link between acid precipitation and spruce mortality.

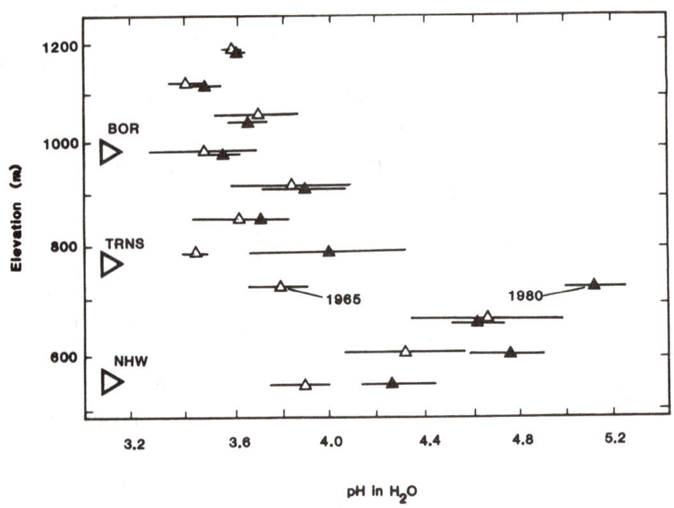

Figure 2. Soil pH (0 horizon) measured in H_2O (1:1 by weight). 1965 data are from Siccama (1974) and three samples per site were measured. In 1980, five samples were measured. Darts show approximate locations were spruce were sampled.

Results of the foliar analyses are shown in Tables 4 and 5. Foliar N and S are directly related to elevation (Figure 3)

TABLE 1. Composition of Fine Roots (*Picea rubens*).

		Percent						ppm								
Forest Type	**Condition**	**N**	**S**	**P**	**K**	**Ca**	**Mg**	**Mn**	**Cu**	**Zn**	**Al**	**Pb**	**Cd**	**Ni**	**Cr**	**Mo**
								CAMELS HUMP / 2-9 cm dbh								
TRNS (n=10)	Poor	1.1	0.09	0.11	0.27	0.19	0.07	280	15	290	4800	65	7	7	6	14
	Good	1.1	0.08	0.10	0.24	0.24	0.06	280	11	220	3100	82	5	7	5	<5
BOR (n=20)	Poor	0.91	0.07	0.08	0.25	0.22	0.06	480	8	180	1300	47	<5	<5	<5	<5
	Good	0.98	0.08	0.10	0.25	0.22	0.07	200	7	170	1700	48	<5	<5	<5	<5
								CAMELS HUMP / <2 cm dbh								
NHW (n=10)	Poor	0.61	0.07	0.07	0.30	0.50	0.10	350	9	170	3500	25	<5	12	<5	<5
	Good	0.72	0.06	0.07	0.31	0.48	0.14	250	12	190	2800	30	<5	13	<5	<5
TRNS (n=20)	Poor	1.0	0.7	0.9	0.31	0.29	0.08	350	9	220	1300	45	6	6	<5	<5
	Good	1.1	0.8	0.11	0.36	0.27	0.09	280	9	260	1200	50	6	6	<5	<5
BOR (n=20)	Poor	n.d.	n.d.	0.07	0.24	0.29	0.08	170	9	210	490	60	5	<5	<5	<5
	Good	n.d.	n.d.	0.10	0.27	0.24	0.08	230	11	190	1400	65	<5	<5	<5	<5
Hubbard Brook – no obvious mortality (n=11)		0.75	0.06	0.08	0.42	0.25	0.08	190	7	98	410	31	<5	<5	<5	<5

and foliar S and S:N ratios are higher in trees which were in poor condition (Figures 3 and 4). In the absence of details on precipitation chemistry and plant response, we might speculate that the elevational trends in tissue composition result from increased HNO_3 and H_2SO_3 deposition at higher elevations. The ratio of S to N is generally a better measure of foliar sulfar status than S alone, since plant proteins have a nearly constant S:N ratio of 0.069 (weight/weight). S in excess of that ratio is generally present as SO_4. It is interesting to note that the only significant difference between declining and healthy trees was the excess foliar sulfur in declining individuals. As we do not know how the excess S was delivered to the foliage (i.e., absorption by foliage, uptake by roots), speculation about the cause and effect of this excess sulfur is not warranted.

TABLE 2. Comparisons Among Root Compositions.

Significant differences in root composition between Camels Hump and Hubbard Brook:

1. More K at Hubbard Brook (1.5x)
2. More Zn and Pb at Camels Hump (2x)
3. More Al at Camels Hump (5x)

Significant differences in root composition between healthy and declining trees at Camels Hump:

None

TABLE 3. Al Concentrations in Fine Roots of Red Spruce in the <2 cm Diameter Class.

Location	Mortality*	Al (ppm)	n	Overall Mean
Camels Hump NWH	Yes	3130 ± 760	10	
Camels Hump TRNS	Yes	1260 ± 202	20	1513 ± 225
Camels Hump BOR	Yes	957 ± 356	20	
Hubbard Brook TRNS	No	410 ± 228	11	
Hubbard Brook Road Cut	No	2144 ± 208	10	
Lincoln, Vermont Old Field	No	2063 ± 176	10	1168 ± 167
Bristol, Vermont Undisturbed	No	821 ± 62	8	
Manchester, Vermont				
Old Field (A)	No	762 ± 131	4	
Old Field (B)	No	729 ± 104	6	
Old Field (C)	No	497 ± 64	5	

*Refers to mortality in the <2 cm size class. Canopy trees were declining or dead in the Manchester, Vermont, stands and were infected with *Armillaria*.

There has been little work published on the effects of acid rain or mist on red spruce, and meaningful comparisons with other species are difficult. S levels in declining trees at Camels Hump are in the range associated with physiological stress in scots pine (*Pinus sylvestris*) (Huttunen, *et al.*, 1980, 1981) and mesophyll cell damage in norway

TABLE 4. Composition of Foliage (*Picea rubens*).

		Percent						ppm								
Forest Type	Condition	N	S	P	K	Ca	Mg	Mn	Cu	Zn	Al	Pb	Cd	Ni	Cr	Mo
CAMELS HUMP / 2-9 cm dbh																
TRNS (n=10)	Poor	1.1	0.10	0.12	0.48	0.22	0.05	347	3	15	60	3	<3	<3	<3	<3
	Good	1.3	0.09	0.19	0.56	0.29	0.08	565	6	23	39	<3	<3	<3	<3	<3
BOR (n=20)	Poor	1.0	0.12	0.12	0.46	0.31	0.08	500	<3	23	35	3	<3	3	<3	<3
	Good	1.2	0.10	0.16	0.48	0.23	0.08	400	<3	19	40	<3	<3	4	<3	<3
CAMELS HUMP / <2 cm dbh																
NHW (n=10)	Poor	0.89	0.08	0.10	0.50	0.30	0.06	220	<3	23	52	<3	<3	3	<3	5
	Good	0.89	0.07	0.10	0.48	0.29	0.06	250	3	17	43	<3	<3	3	<3	<3
TRNS (n=20)	Poor	1.0	0.10	0.10	0.45	0.28	0.05	420	<3	19	28	<3	<3	<3	<3	<3
	Good	1.0	0.08	0.10	0.43	0.26	0.05	410	<3	20	32	<3	<3	<3	<3	<3
BOR (n=20)	Poor	1.1	0.12	0.10	0.40	0.44	0.09	400	<3	30	32	3	<3	<3	<3	<3
	Good	1.1	0.10	0.10	0.39	0.34	0.07	340	3	26	63	<3	<3	<3	<3	<3
Hubbard Brook – no obvious mortality (n=11)		0.99	0.08	0.09	0.56	0.32	0.06	250	4	21	23	<3	<3	<3	<3	<3

spruce when excess S is the result of SO_2 pollution. On the other hand, when S in excess of protein requirements was delivered by artificial acid rain, norway spruce and scots pine accumulated the same levels observed at Camels Hump without apparent adverse effects (Tveite, 1980). SO_2 measurements at Hubbard Brook, New Hampshire, and Whiteface Mountain, New York, indicate that growing season concentrations are about 2.5 $\mu g\ m^{-3}$ which is about an order of magnitude less than the levels thought to cause damage through long term exposure (Smith, 1981).

TABLE 5. Comparisons Among Foliar Compositions.

Significant differences in foliar composition between Camels Hump and Hubbard Brook:

1. More K at Hubbard Brook
2. More S at Camels Hump
3. More Al at Camels Hump

Significant differences in foliar composition between healthy and declining trees at Camels Hump:

1. Higher S in declining trees
2. Higher S:N ratio in declining trees

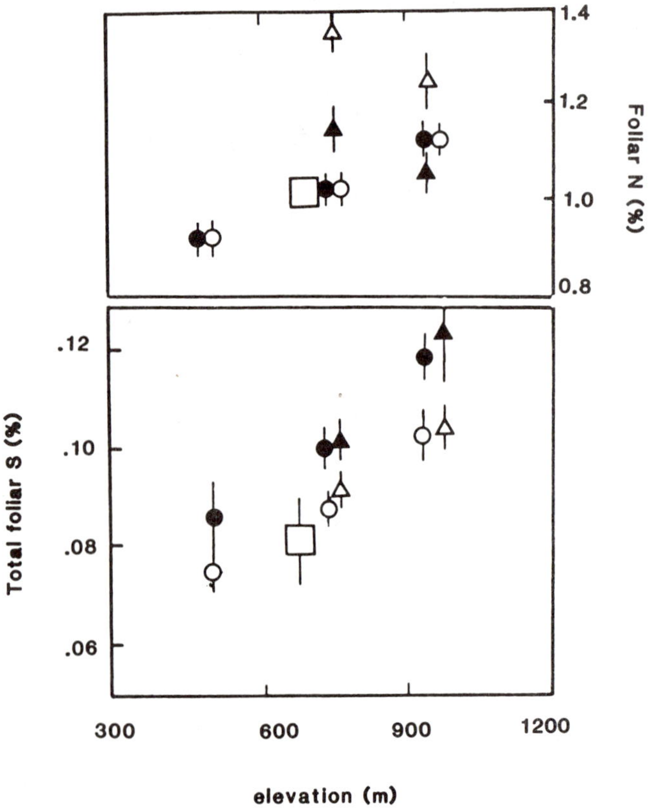

Figure 3. Foliar S and N in Red Spruce at Camels Hump (triangles and circles) and Hubbard Brook (squares). Open symbols are trees in good condition. Solid symbols are trees in poor condition. Circles are <2 cm trees, triangles are 2-9 cm trees.

SUMMARY

We regard these data as inconclusive, but a useful first step in defining directions for future research. A better understanding of soil plant nutritional relationships is needed, and data from controlled experiments would be helpful in trying to explain the differences in Al, S, and N.

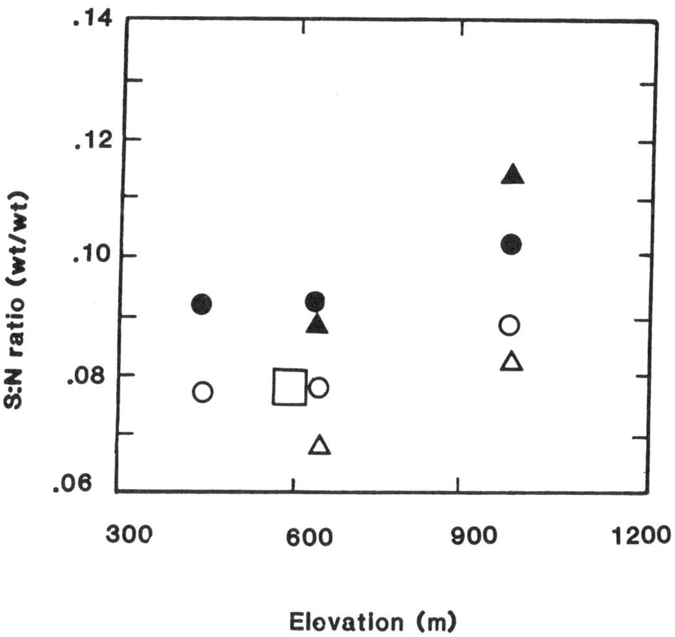

Figure 4. S:N Ratio in Red Spruce Foliage at Camels Hump and Hubbard Brook. Symbols are the same as in Figure 3.

No conclusions regarding the presence or absence of acid precipitation induced stress are warranted, however the data presented suggest that the possibility of a biogeochemical link between acid precipitation and spruce dieback exists. It is important to pursue other mechanisms which could be important stresses. Biotic agents such as primary root pathogens have not been thoroughly studied. Stress from oxidant pollutants alone or in combination with acid precipitation remains uninvestigated.

ACKNOWLEDGMENTS

This work was supported by funds from the U.S. EPA and the Andrew W. Mellon Foundation.

LITERATURE CITED

Abrahamsen, G., 1980. Acid Precipitation, Plant Nutrients and Forest Growth. *In:* Ecological Impact of Acid Precipitation. Proceedings of an International Conference, Sandefjord, Norway, March 1980, D. Drabløs and A. Tollan (Editors), pp. 58-63.

Andresen, A. M., A. H. Johnson, and T. G. Siccama, 1980. Levels of Lead, Copper, and Zinc in the Forest Floor of the Northeastern U.S. J. Environ. Qual. 9(2):293-297.

Cronan, C. S. and C. L. Schofield, 1979. Aluminum Leaching Response to Acid Precipitation: Effects on High Elevation Watersheds in the Northeast. Science (204):304-306.

Evans, L. S. and T. M. Curry, 1979. Differential Response of Plant Foliage to Simulated Acid Rain. Am. J. Bot. 66:953-962.

Huttenen, S., K. Laine, and T. Pakonen, 1980. Air Pollutants as Additional Stress Factors on Conifers Under Northern Conditions. *In:* Ecological Impact of Acid Precipitation. Proceedings of an International Conference, Sandefjord, Norway, March 1980, D. Drabløs and A. Tollan (Editors), pp. 168-169.

Huttunen, S., L. Karenlampi, and K. Kolari, 1981. Changes in Osmotic Potential and Some Related Physiological Variables in Needles of Polluted Norway Spruce (*Picea Abies*). Ann. Bot. Fennici 18:63-71.

Johnson, A. H., T. G. Siccama, D. Wang, R. S. Turner, and T. H. Barringer, 1981. Recent Changes in Patterns of Tree Growth Rate in the New Jersey Pinelands: A Possible Effect of Acid Rain. J. Environ. Qual. 10(4):427-430.

Jonsson, B. and R. Sundberg, 1972. Has Acification by Atmospheric Pollution Caused a Growth Reduction in Swedish Forests? Institute for Skogsproduktion, Report 20, 48 pp.

Roman, J. R. and D. J. Raynal, 1980. Effects of Acid Precipitation on Vegetation. *In:* Actual and Potential Effects of Acid Precipitation on a Forest Ecosystem in the Adirondack Mountains. New York State Energy Research and Development Authority, Report ERDA 80-28, pp. 4-1–4-63.

Siccama, T. G., 1974. Vegetation, Soil and Climate on the Green Mountains of Vermont. Ecol. Monographs 44:325-349.

Siccama, T. G., M. Bliss, and H. W. Vogelmann, 1982. Decline of Red Spruce in the Green Mountains of Vermont. Bull. Torrey Bot. Club (in press).

Smith, W. H., 1981. Air Pollution and Forests. Springer Verlag, New York, New York.

Strand, L., 1980. The Effect of Acid Precipitation on Tree Growth. *In:* Ecological Impact of Acid Precipitation. Proceedings of an International Conference, Sandefjord, Norway, March 1980, D. Drabløs and A. Tollan (Editors), pp. 64-65.

Tamm, C. O. and G. Wiklander, 1980. Effects of Artificial Acidification with Sulphuric Acid on the Growth in Scots Pine Forest. *In:* Ecological Impact of Acid Precipitation. Proceedings of an International Conference, Sandefjord, Norway, March 1980, D. Drabløs and A. Tollan (Editors), pp. 188-189.

Tveite, B., 1980. Foliar Nutrient Concentrations in Field Experiments. *In:* Ecological Effects of Acid Precipitation. Proceedings of an International Conference, Sandefjord, Norway, March 1980, D. Drabløs and A. Tollan (Editors), pp. 204-205.

Ulrich, B., 1981a. Die Walder in Mitteleuropa: Messergebnisse ihrer Umweltbelastung, Theorie ihrer Gefahrdung, Prognose ihrer Entwicklung. Library Environment Canada Translation No. 00ENV TR-2038, 11 pp.

Ulrich, B., 1981b. Bodenchemische und Umwelt-Aspekte der Stabilitat von Waldokosystemen. Library Environment Canada Translation No. 00ENV TR-2046, 13 pp.

Wood, T. and F. H. Bormann, 1975. Increases in Foliar Leaching Caused by Acification with an Artificial Mist. Ambio 4(2):169-173.

EFFECTS OF AMBIENT CONCENTRATIONS OF AIR POLLUTANTS ON VEGETATION INDIGENOUS TO THE BLUE RIDGE MOUNTAINS OF VIRGINIA

John M. Skelly, Boris I. Chevone, and Yaw-Shing Yang[1]

ABSTRACT: Vegetation indigenous to the Shenandoah National Park (SNP) and Blue Ridge Mountains of Virginia has been exposed to phytotoxic concentrations of oxidant air pollution, primarily ozone (O_3), over the past several years. Exposure of vegetation to episodic O_3 incursions has resulted in visible symptom expression on several native plant species. Longer term, low dosage exposures typically recorded throughout the summer oxidant season have been demonstrated to reduce foliar biomass production of native forbes, grasses, and sedges via an air pollutant exclusion chamber system in the Big Meadows area of the SNP. Seedling height growth of several forest tree species was also adversely affected by ambient oxidant concentrations. Monitoring data for O_3 and SO_2 at SNP from 1979-1981 have demonstrated that O_3 was present in eight-hour (1100-1800 EST) average concentrations of 0.023 to 0.059 ppm O_3 during April through September; hourly average SO_2 concentrations rarely exceeded 0.006 ppm. Analysis of weekly rainfall samples has also been conducted since November of 1978 with the lowest pH rainfall recorded of 3.39. The average pH of all weekly samples is about 4.35. No effects to vegetation in the Blue Ridge Mountains have been ascribed to acidified rainfall.
(KEY TERMS: ozone; photochemical oxidants; acid precipitation; open top chambers; foliar biomass.)

INTRODUCTION

Ambient ozone (O_3) concentrations currently occurring during the summer months in much of the Eastern United States are often of sufficient magnitude to result in visible foliar injury on sensitive plants. Since O_3 was initially identified as the cause of weather fleck in tobacco (Heggestad and Middleton, 1959), numerous studies have been conducted, primarily under laboratory conditions, on the phytotoxic effects of O_3 to various plant species. In most of these studies, macroscopic foliar symptoms were used as an index of plant sensitivty. However, recent research has shown that plant growth and productivity can be adversely affected by O_3 exposure in the absence of any visible symptoms (Bennett and Runeckles, 1977; MacLean, *et al.*, 1977; Houston and Dochinger, 1977; Kress and Skelly, 1981). In spite of such recent advances, little is known about the response of plants in their native habitat to ambient pollutant concentrations.

Mesoscale transport of O_3 and its photochemical precursors into remote areas distant from urban and industrial sources has been well documented (Miller, *et al.*, 1972; Cleveland and Kleiner, 1975; Hayes and Skelly, 1977; Husar, *et al.*, 1977). Episodic O_3 concentrations, with peak hourly averages greater than 0.08 ppm, have been reported in forested areas in the midwestern United States and in the Blue Ridge Mountains of Virginia (Wolff, *et al.*, 1977; Yang and Chevone, 1982). Lioy and Sampson (1979) characterized 40 O_3 episodes occurring over the midwest and northeastern United States as originating from air masses entering the U.S. over Minnesota, stagnating at various places over the area of study, and having average durations of seven days. Ozone concentrations during these episodes ranged from 0.10 to 0.20 ppm.

The atmospheric deposition of acidic substances to terrestrial and aquatic ecosystems is currently under considerable investigation. Acidic aerosols and gases may be present in the same air masses containing high concentrations of O_3 and these pollutants can be transported long distances concurrently. Recent studies have suggested that rainfall acidity and sulfate and nitrate concentrations have been increasing, most predominantly in the northeastern U.S., during the last 15 years (Cogbill and Likens, 1974; Galloway, *et al.*, 1976; Galloway and Likens, 1981; Likens and Butler, 1981). However, a comprehensive precipitation sampling network, encompassing the eastern U.S. and utilizing standardized collection and analytical procedures, has only been operational since 1978 (Galloway and Cowling, 1978). The data base from this network is currently insufficient to establish any trends in the acidity or chemical composition of precipitation.

This report describes results of a three-year monitoring program characterizing ambient O_3 concentrations and the chemistry of weekly precipitation samples in the Blue Ridge Mountains of Virginia. The effects of ambient O_3 concentrations on foliar biomass production of native forbes and

[1]Respectively, Professor and Head, Dept. of Plant Pathology, The Pennsylvania State University, 211 Buckhout Laboratory, University Park, Pennsylvania 16802; and Assistant Professors, Dept. of Plant Pathology and Physiology, Virginia Polytechnic Institute and State University, Blacksburg, Virginia 24061.

grasses and seedling height growth of several forest tree species during this period are also presented.

MATERIALS AND METHODS

Ambient O_3 concentrations were monitored continuously at three sites in the Blue Ridge Mountains of Virginia from 1979 through 1981. Ambient sulfur dioxide (SO_2) concentrations were also monitored, but on a variable basis. The monitoring sites were located at Big Meadows in the Shenandoah National Park (SNP) (elev. 1037 m MSL), Rocky Knob in the Blue Ridge Parkway (elev. 945 m MSL), and the Horton Research Center, Salt Pond Mountain in Giles County (elev. 964 m MSL). The Rocky Knob monitoring site was located approximately 260 km south-southeast of Big Meadows and the Horton Research Center approximately 230 km southwest of Big Meadows.

Ozone concentrations were monitored with a Bendix Model 8002 chemiluminescent O_3 analyzer (Bendix Corporation, Lewisburg, West Virginia). The O_3 monitor was calibrated at 2- to 4-week intervals using a Bendix Model 8852 Dynamic Calibration Unit and a Dasibi UV spectrophotometric O_3 analyzer as a primary calibration standard. Sulfur dioxide concentrations were monitored using a TECO Model 43 pulsed fluorescent SO_2 analyzer (Thermo Electron Corporation, Hopkington, Massachusetts). A National Bureau of Standards traceable source of bottled SO_2 was used as a primary calibration standard for the SO_2 analyzer. Ozone and SO_2 concentrations were recorded continuously with a strip chart recorder and expressed as hourly concentrations.

Precipitation samples were collected on a weekly basis from November 1978 through September 1981 at the Horton Research Center using an Aerochemetrics wet-dry sampling unit (Aerochem Metrics, Inc., Miami, Florida). The pH and conductivity of each sample was measured in the field and the samples were then sent to the National Atmospheric Deposition Program's (NADP) Central Analytical Laboratory (CAL) in Illinois for chemical analysis. Ionic concentrations of Ca, K, Mg, Na, NH_4, Cl, NO_3, SO_4, and PO_4 were determined at CAL and sample pH and conductivity were remeasured.

Within a two-hectare meadow community located at Big Meadows, twelve 3 m diameter plots were established 3 m apart in May 1979 to study the effects of ambient O_3 concentrations on growth of forest tree seedlings and foliar biomass production of other native vegetation. Treatments were randomized over the area and four plots each received charcoal filtered air or nonfiltered air via blowers attached to open top chambers (Heagle, *et al.*, 1973); four additional plots were left undisturbed and received ambient air. Chambers operated continuously during the growing season in 1979, 1980, and 1981.

In May of 1979, year old, wild type seedlings of tulip poplar (*Liriodendron tulipifera* L.), sweet gum (*Liquidambar styraciflua* L.), green ash (*Fraxinus pennsylvanica* Marsh.), black locust (*Robinia pseudocacia* L.), table mountain pine (*Pinus pungens* Lamb), eastern white pine (*Pinus strobus* L.), Virginia pine (*Pinus virginana* Mill.), and eastern hemlock (*Tsuga canadensis* L.) were planted in each of the 12 chambers (2 seedlings/plot). Seedlings were observed for foliar symptoms at monthly intervals during each growing season. Height measurements to terminal bud were taken at the time of planting and in early September of each year.

Native ground vegetation growing within the 12 plots was harvested by clipping to a height of 1.25 cm, twice each growing season. Harvests occurred on August 14 and October 9, in 1979, June 23 and September 15 in 1980, and June 23 and September 9 in 1981. Thee clipped biomass was bagged and dried to constant weight.

Standard analysis of variance techniques were used to determine differences in seedling height growth and foliar biomass production in charcoal filtered, nonfiltered, and ambient air plots.

RESULTS

A typical oxidant season occurred each year usually beginning in mid-April and ending in early October (Table 1). During the oxidant season, monthly hourly average O_3 concentrations were generally 0.035 to 0.060 ppm and peak hourly concentrations ranged from 0.080 to 0.120 ppm. During the summer months, O_3 episodes frequently occurred having durations of 1 to 3 days and peak hourly concentrations often exceeding 0.09 ppm. These episodes were primarily regional in nature as evidenced by the generally high O_3 concentrations at each of the three monitoring sites (Table 2).

TABLE 1. Monthly 8-Hour Average (1100-1800 EST), Peak 1-Hour Average, and Cumulative Seasonal Ozone Concentrations (ppm) Monitored at Big Meadows, Shenandoah National Park, During 1979-1981.

	1979		1980		1981	
Month	Average	Peak	Average	Peak	Average	Peak
January	0.041	0.06	0.020	0.03	0.033	0.06
February	– –*	– –	0.024	0.04	0.030	0.06
March	0.062	0.10	0.035	0.04	0.028	0.05
April	0.055	0.10	0.042	0.07	0.043	0.07
May	0.052	0.08	0.048	0.10	0.038	0.08
June	0.055	0.08	0.059	0.09	0.023	0.06
July	0.047	0.09	0.058	0.09	0.037	0.10
August	0.054	0.07	0.051	0.10	0.030	0.06
September	0.046	0.09	0.046	0.09	0.037	0.06
October	0.042	0.08	0.033	0.08	0.044	0.09
November	0.039	0.07	0.035	0.06	0.055	0.08
December	0.028	0.05	0.041	0.06	– –*	– –
Total Ozone Dosage Between 1 April–30 September (ppm/hr)	73.38		74.21		50.51	

*Data not available.

TABLE 2. Peak Hourly Ozone Concentrations (ppm) Recorded at Three Monitoring Stations in Western Virginia During Ozone Episodes in the Spring and Summer of 1979, 1980, and 1981.

Date	Monitoring Site		
	Big Meadows	Rocky Knob	Horton Center
1979, April 11-12	0.100	0.128	– –*
1979, June 5-6	0.082	0.112	– –
1979, September 12	0.095	0.079	0.072
1980, May 29-30	0.100	– –	0.069
1980, June 13-15	0.093	0.088	0.063
1980, July 14-15	0.089	0.080	– –
1980, July 31	0.084	0.049	0.043
1980, August 29-30	0.102	0.113	0.129
1981, May 24	0.084	0.054	0.094
1981, June 22-23	– –	0.106	0.073
1981, June 29-30	– –	0.090	0.072
1981, July 8-11	0.096	0.129	0.114

*Data not available.

The monthly average and peak hourly average O_3 concentrations monitored at Big Meadows for 1100-1800 EST during 1979 through 1981 are presented in Table 1. The cumulative O_3 dosage during the plant growing season was 73.4, 74.2, and 50.5 ppm-hr for 1979, 1980, and 1981, respectively. Ozone concentrations exceeding 0.06 ppm occurred over extended periods of time each year.

Ambient SO_2 concentrations monitored at Big Meadows were usually below instrument detection levels. Sulfur dioxide, therefore, was considered insignificant relative to potential adverse effects on plant productivity.

The pH of weekly precipitation samples varied considerably, but was generally acidic, and ranged from 7.40 to 3.39. Mean ionic concentrations in samples for each year are presented in Table 3. Mean sulfate and nitrate concentrations increased and mean pH decreased during the three-year monitoring period. The average pH of all samples was about 4.35.

TABLE 3. Mean Ionic Concentrations (mg/L)[1] in Wet Deposition Collected Weekly at the Horton Research Center, Giles County, Virginia, from November 1978 to September 1981.

Ionic Species	1978-79[2]	1980	1981[3]
Ca	0.40	0.72	1.93
Mg	0.07	0.11	0.31
K	0.13	0.10	0.39
Na	1.12	0.71	0.34
NH_4	0.43	0.37	1.04
NO_3	1.27	2.41	4.41
Cl	0.32	0.48	0.59
SO_4	3.62	4.73	8.46
PO_4	0.19	0.00	0.15
pH	4.56	4.30	4.18

[1]Nonvolume weighted means.
[2]Includes November and December 1978.
[3]Through September.

Correlation analysis of hydrogen ion concentration and sulfate or nitrate concentration yielded coefficients ranging from 0.31 to 0.60 and 0.38 to 0.86, respectively, over the three years. Sulfate was most highly correlated with ammonium, with coefficients greater than 0.89.

The mean concentration of all ionic species varied with pH of the sample (Table 4). Maximum concentrations occurred at pHs less than 4.00 and greater than 6.50. Minimum concentrations occurred between pH 4.50 and 5.50.

TABLE 4. Mean Ionic Composition of Wet Deposition Collected Weekly at Horton Research Center, Giles County, Virginia, from November 1978 to September 1981 as a Function of pH.

Species*	pH				
	<4.00	4.01-4.50	4.51-5.50	5.51-6.50	<6.50
Ca	142.1	29.9	16.2	56.1	182.9
Mg	42.2	8.7	5.3	13.2	30.6
K	17.9	3.2	1.9	2.7	21.0
Na	25.0	14.3	15.6	65.6	276.7
NH_4	114.5	24.4	7.7	21.7	151.3
NO_3	121.5	35.9	18.4	30.9	108.8
Cl	25.1	10.4	6.9	13.3	47.9
SO_4	376.6	99.0	47.1	75.7	261.2
PO_4	8.7	1.2	0.1	0.4	53.6
n	13	47	53	17	6

*Concentrations expressed as hydrogen ion microequivalents.

Foliar biomass production of native meadow vegetation at Big Meadows in the charcoal filtered chambers, nonfiltered chambers, and open plots for each harvest and each year, 1979 through 1981, is presented in Table 5. Biomass production was greatest in the filtered chambers each year. Total dry weight of vegetation during the study period was 40.3, 31.9, and 25.9 kg, respectively, in the filtered chambers, nonfiltered chambers, and ambient plots and differed significantly ($P \leqslant 0.05$) among treatments.

Analysis of variance by harvest dates and years demonstrated a significant difference in biomass weight among treatments ($P \leqslant 0.01$) and among years ($P \leqslant 0.01$). A significant interaction ($P \leqslant 0.01$) between year and harvest date also occurred.

Visible foliar injury occurred on several plant species in the ambient air and nonfiltered chamber plots. Tulip poplar, green ash, black locust, and common milkweek exhibited purple stipple on the upper leaf surfaces. A red pigmentation occurred on the foliage of sweetgum, blackberry, and wild strawberry, and eastern white pine showed needle tip burn and chlorotic mottle. These symptoms are typical of O_3 induced injury on these plants.

Annual height growth of tree seedlings was, in general, least in 1979 when seedlings were becoming established and increased each year thereafter. Mean annual growth rates in all tree species, except black locust, in 1980 and 1981 were greatest in those chambers receiving charcoal filtered air.

TABLE 5. Dry Weight (gm) of the Above Ground Biomass Collections from 1979-1981 for Charcoal Filtered Chambers, Nonfiltered Chambers, and Open Plots at Big Meadows, Shenandoah National Park. Biomass collections were made two times per summer.

	1st Harvest	2nd Harvest	1st Harvest	2nd Harvest	1st Harvest	2nd Harvest	1st Harvest	2nd Harvest	TOTAL
FILTERED CHAMBERS									
	No. 1		No. 8		No. 10		No. 12		
1979	1280	651	1808	363	1641	523	2534	433	9,233
1980	2337	3542	2017	2173	1690	2528	2255	3453	19,995
1981	2982	934	1733	937	1458	1017	1757	1208	11,126
TOTAL	10,826		9,031		8,857		11,640		40,354
NONFILTERED CHAMBERS									
	No. 2		No. 5		No. 9		No. 11		
1979	706	113	759	501	1636	402	1836	387	6,340
1980	1002	1759	1280	4062	1738	2572	1408	2262	16,083
1981	755	609	2108	1472	1650	618	1265	974	9,451
TOTAL	4,944		10,182		8,616		8,132		31,874
OPEN PLOTS									
	No. 3		No. 4		No. 6		No. 7		
1979	706	100	847	217	785	354	790	187	3,986
1980	809	1554	1299	2666	1754	2779	1111	2148	14,120
1981	1113	530	1332	581	2088	800	867	499	7,810
TOTAL	4,812		6,942		8,560		5,602		25,916

Mean annual percent increase in height of conifer seedlings in charcoal filtered chambers compared with open plots was 11 percent for eastern white pine, 42 percent for Virginia pine, and 65 percent for table mountain pine in 1980 and 1981. Although these increases are substantial, analyses of variance indicated no significant difference ($P \leq 0.05$) between treatments because of large variation between replicates within treatments.

DISCUSSION

This study and several other reports (Skelly, *et al.*, 1979; Skelly, 1980; Duchelle, *et al.*, 1982; Yang and Chevone, 1982) indicate that ozone is the major phytotoxic air pollutant present in the Blue Ridge Mountains of Virginia and the SNP. While SO_2 does occur in the ambient air in remote mountain forests, concentrations are well below threshold doses necessary to injure even sensitive plant species. Ozone episodes occurred during the summer oxidant season each year lasting 1-3 days with peak hourly concentrations of 0.08-0.12 ppm. These results are in agreement with Wolff, *et al.* (1977), although episodic peak concentrations were somewhat lower in this current study.

Acidic precipitation also occurs frequently in the forested mountains of Virginia. Although a decrease in mean pH of precipitation samples and an increase in mean sulfate and nitrate concentrations occurred during the three-year study period, the data base in insufficient to establish meaningful trends at this time. The variation in mean concentrations of ionic species as a function of sample pH has also been reported for rain samples collected from central Minnesota (Pratt, *et al.*, 1982). While the chemical and physical parameters contributing to this distribution are not known, such a distribution will affect any typical correlation analysis involving the hydrogen ion. No effects to vegetation have as yet been attributed to acidic precipitation.

Ambient concentration of O_3 in the SNP were of sufficient magnitude to cause a reduction in foliar biomass production in a high meadow community. While some plant species showed visible symptoms induced by O_3, foliar biomass loss occurred primarily without the appearance of such symptoms. These results are consistent with those of many other researchers who have found adverse plant effects that could not be explained upon the basis of symptom expression (Treshow, *et al.*, 1967; Stone and Skelly, 1974; Houston and Dochinger, 1977). The repression of seedling height growth by ambient O_3 concentrations, while not statistically significant, suggest that growth reductions in seedling trees (10 percent/year minimally in O_3 sensitive tree species) may currently be occurring in remote forested areas in Virginia.

Both O_3 and acidic precipitation are currently impacting forest vegetation in the Blue Ridge Mountains of Virginia. At present, only ambient O_3 concentrations have been shown to reduce growth and productivity of native plant species. However, the potential exists for interactive effects of acidic precipitation and O_3, and additional research is required to determine if such effects may be occurring.

ACKNOWLEDGMENTS

The authors wish to thank the U.S. Department of Interior, National Park Service, Division of Air Quality for supporting portions of this research through Cooperative Agreement No. CX001-9-0011.

LITERATURE CITED

Bennett, J. P. and V. C. Runeckles, 1977. Effects of Low Levels of Ozone on Growth of Crimson Clover and Annual Ryegrass. Crop Sci. 17:443-445.

Cleveland, W. S. and B. Kleiner, 1975. Transport of Photochemical Air Pollution from Camden-Philadelphia Urban Complex. Envir. Sci. and Tech. 9:869-872.

Cogbill, C. V. and C. E. Likens, 1974. Acid Precipitation in the Northeastern United States. Water Resour. Res. 10:1133-1137.

Duchelle, S. F., J. M. Skelly, and B. I. Chevone, 1982. Oxidant Effects on Forest Tree Seedlings in the Appalachian Mountains. J. Soil, Water, Air Pollut. 16:363-373.

Galloway, J. N. and E. B. Cowling, 1978. The Effects of Precipitation on Aquatic and Terrestrial Ecosystems – A Proposed Precipitation Chemistry Network. JAPCA 28:229-235.

Galloway, J. N. and G. E. Likens, 1981. Acid Precipitation. The Importance of Nitric Acid. Atmos. Environ. 15:1081-1085.

Galloway, J. N., G. E. Likens, and E. S. Edgerton, 1976. Acid Precipitation in Northeastern United States: pH and Acidity. Science 194:722-724.

Hayes, E. M. and J. M. Skelly, 1977. Transport of Ozone from the Northeast U.S. into Virginia and Its Effects on Eastern White Pines. Plant Dis. Rep. 61:778-782.

Heagle, A. S., D. E. Body, and W. W. Heck, 1973. An Open-Top Field Chamber to Assess the Impact of Air Pollution on Plants. J. Environ. Qual. 2:365-368.

Heggestad, H. E. and J. T. Middleton, 1959. Ozone in High Concentrations as Cause of Tobacco Leaf Injury. Science 129:208-210.

Houston, D. B. and L. S. Dochinger, 1977. Effects of Ambient Air Pollution on Cone, Seed, and Pollen Characteristics in Eastern White and Red Pines. Environ. Pollut. 12:1-5.

Husar, R. P., D. E. Patterson, D. C. Paley, and N. V. Gillani, 1977. Ozone in Hazy Air Masses. *In:* Proceedings of the International Conference on Photochemical Oxidant Pollution and Its Control, B. Dimitroades (Editor). USEPA-600/3-77-001a, Res. Triangle, North Carolina, pp. 275-282.

Kress, L. W. and J. M. Skelly, 1982. Response of Several Eastern Forest Tree Species to Chronic Doses of Ozone and Nitrogen Dioxide. Plant Disease 66:1149-1152.

Likens, G. E. and T. J. Butler, 1982. Recent Acidification of Precipitation in North America. Atmos. Environ. 15:1103-1109.

Lioy, P. J. and P. J. Samson, 1979. Ozone Concentration Patterns Observed During the 1976-1977 Long Range Transport Study. Environ. Int'l. 2:77-83.

MacLean, D. C., R. E. Schneider, and D. C. McCune, 1977. Effect of Chronic Exposure to Gaseous Fluoride on Yield of Field-Grown Bean and Tomato Plants. J. Amer. Soc. Hort. Sci. 102: 279-299.

Miller, P. L., M. H. McCulchan, and H. P. Milligan, 1972. Oxidant Air Pollution in the Central Valley, Sierra Nevada Foothills, and Mineral King Valley of California. Atmos. Environ. 6:623-633.

Pratt, G. C., M. Coscio, D. W. Gardner, B. I. Chevone, and S. V. Krupa, 1982. An Analysis of the Chemical Properties of Rain in Minnesota. Atmos. Environ. (in press).

Skelly, J. M., 1980. Photochemical Oxidant Impact on Mediterranean and Temperate Forest Ecosystems: Real and Potential Effects. *In:* Proceedings Symposia on Effects of Air Pollutants on Mediterranean and Temperate Forest Ecosystems. Riverside, California, pp. 38-51.

Skelly, J. M., S. F. Duchelle, and L. W. Kress, 1979. Impact of Photochemical Oxidant Air Pollution on Eastern White Pine in the Shenandoah, Blue Ridge Parkway, and Great Smoky Mountains National Parks. *In:* Proc. II Conf. on Sci. Res. in Nat'l. Parks, San Francisco, California.

Stone, L. L. and J. M. Skelly, 1974. The Growth of Two Forest Tree Species Adjacent to a Periodic Source of Air Pollution. Phytopathology 64:773-778.

Treshow, J., F. K. Anderson, and F. Harner, 1967. Responses of Douglas Fir to Elevated Atmospheric Fluorides. For. Sci. 13: 114-120.

Wolff, T. T., P. J. Lioy, G. D. Wight, R. E. Meyers, and R. J. Cederwall, 1977. An Investigation of Long-Range Transport of Ozone Across the Midwestern and Eastern United States. Atmos. Environ. 11:797-802.

Yang, Y. S. and B. I. Chevone, 1982. Characterization of Ambient Ozone in the Blue Ridge Mountains of Virginia (Abstr.). Phytopathology 72:712.

INTERNATIONAL SYMPOSIUM ON HYDROMETEOROLOGY
JUNE AMERICAN WATER RESOURCES ASSOCIATION 1982

POTENTIAL FOR ACID PRECIPITATION DAMAGE TO LAKES OF THE SIERRA NEVADA, CALIFORNIA

Kathy A. Tonnessen and John Harte[1]

ABSTRACT: Acid precipitation has been measured in many areas of California, including the Sierra Nevada. This region is characterized by high elevation, granite based lakes which may be sensitive to acid inputs. Possible damage to these aquatic systems due to continued acid deposition was investigated during a program of field monitoring and laboratory microcosm experiments. During 1979-1981 Sierra lake water samples were analyzed for pH, alkalinity, and concentrations of major and minor cations. Microcosm systems, established in a controlled laboratory environment using lake water and sediments, were treated with nitric acid and then analyzed for changes in chemical and biological variables.
(KEY TERMS: acid precipitation; aquatic effects; Sierra Nevada lakes; microcosm experiments.)

INTRODUCTION

Acid precipitation in the western United States is now being measured and monitoring networks are being set up to determine the nature and extent of its occurrence. In California two recent studies by McColl (1980) and Morgan and Liljestrand (1980) have documented instances of acidic deposition. It would appear that in the State the nitrate (NO_3^-) component often outweighs the sulfate ($SO_4^=$) contribution in rainfall.

One of the areas of California potentially sensitive to acidic deposition is the Sierra Nevada, located along the eastern boundary. A report on sensitive areas in North America by Galloway and Cowling (1978) identifies the Sierra as a region characterized by poorly buffered soils and granite based lakes. The subalpine and alpine lakes in this region share many of the characteristics of lakes adversely affected by acid deposition in other parts of the U.S. and the world (Cowling, 1980).

We chose to study selected subalpine lakes of the western slope of the Sierra to establish baseline water quality which would allow for the identification of chemical and biological changes due to acidic deposition. We then attempted to simulate the ecosystem stress of increased acidic deposition, particularly in the form of snowmelt (Johannessen and Henriksen, 1978), on these systems by performing microcosm experiments in the laboratory. In these experiments we were particularly concerned with recording changes in concentrations of micronutrients (iron, manganese) and trace elements (aluminum, cadmium, copper, lead, zinc) which might be leached from lake sediments with increasing acidification. This phenomenon is particularly important to study in light of findings by Cronan and Schofield (1979) on the importance of aluminum leaching in the Northeast which has led to toxic effects on biota in Adirondack lakes.

METHODS

Between 1979-81 we sampled 35 lakes in the Sierra Nevada, most located in subalpine basins between 1600-2600 m. Two lower elevation reservoirs were included to allow for comparison of water quality parameters. Each lakes was sampled once, at three stations in the lake: shore; midlake, at the surface; and midlake, at secchi disc depth. Samples were collected and analyzed for pH (Sargent-Welch model PBX pH meter), alkalinity (by Gran titration), temperature, transparency, major and minor cations (Ca, Mg, Al, Cd, Cu, Pb, Mn, Fe, Zn) (by graphite furnace atomic absorption spectrophotometry, Perkin-Elmer) and biological populations (phytoplankton and zooplankton). Trace analysis was performed on filtered (0.45 μm Nucleopore) and unfiltered samples. Phytoplankton samples were collected using a 5-mL pipette and counted with a Wild microscope using Utermohl inverted microscope techniques. Zooplankton samples were collected using a 64 μm mesh sieve and species populations were estimated from microscope counts.

Three sets of microcosms were set up in a light and temperature controlled facility at Lawrence Berkeley Laboratory. Whole water samples and sediments from two Sierra lakes: Tenaya Lake (2480 m) and Mosquito Lake (2440 m) and one Bay Area reservoir, Briones (70 m), were collected and used to fill 18-liter Nalgene tanks. These

[1]Respectively, Research Associate and Professor, Energy and Resources Group, Energy and Environment Division, 70-110A, Lawrence Berkeley Lab, University of California, Berkeley, California 94720.

microcosms were sampled weekly and the samples analyzed for pH, alkalinity/acidity, cations (Ca, Mg, Al, Cd, Cu, Fe, Pb, Mn, Zn) (total and dissolved) and populations of phytoplankton and zooplankton. The same analytical techniques were used in both field and lab.

RESULTS

Sierra lakes sampled were in the circumneutral pH range (6.0-8.0), with low alkalinities (29.0-200.0 μ eq/L). Table 1 presents data on water quality for some of these subalpine lakes. Calcium and magnesium levels were low, indicative of low alkalinity systems. The trace element analysis indicated the presence of detectable (> 1.0 μ g/L) levels of aluminum, iron, and manganese. Figure 1 shows a plot of pH vs. alkalinity for 24 Sierra lakes and the two comparison reservoirs.

Of the three sets of microcosm experiments run, the Tenaya Lake experiment will be discussed here. For additional data from these experiments we refer you to Tonnessen and Harte (1980) and Tonnessen (1982).

During the Tenaya Lake experiment three replicate tanks with sediment and three without sediment were maintained as controls at a system pH of 6.3-6.5 for a 7-week period (see Figure 2). On day 1 sufficient nitric acid was added to the three treatments with sediment and three treatments without sediment to bring the system pH down to 4.0. Following this "acidification" there was little change (or buffering) in system pH. The treatments with sediment increased to a pH of 4.2 and then remained stable. The acidification of these treatments was accompanied by small increases in dissolved Ca, Mg, Mn, and Fe levels. Dissolved aluminum concentrations in the treatments with sediments showed the largest increases (from 18.0 to 68.0 μg/L) (see Figure 3). Smaller increases in aluminum in the treatments without sediment reflect leaching from suspended particles in the water column.

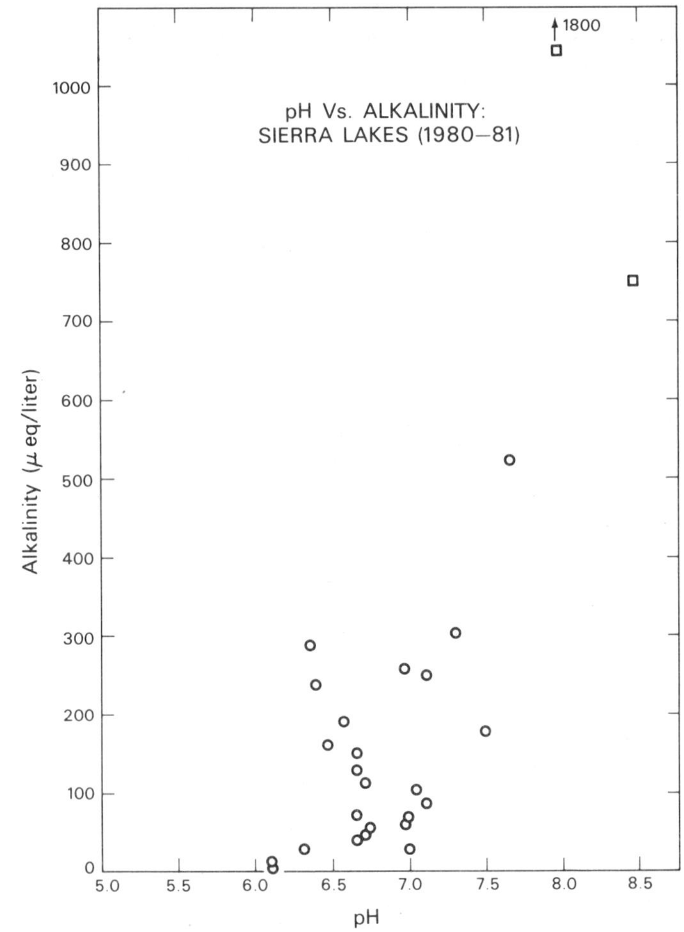

Figure 1. pH vs. Alkalinity for 26 California Lakes: ○ indicates subalpine lakes, □ represents lower elevation reservoirs.

Among the zooplankton counted in this experiment, the largest number belonged to the group *Rotifera.* In the control systems this population increased from eight individuals/liter at week 3 to 195 individuals/liter at the end of

TABLE 1. Water Quality of Ten Lakes of the Sierra Nevada.

	Altitude (m)	pH	Alkalinity (μeq/L)	Ca (mg/L)	Mg (mg/L)	Al (μg/L)	Fe (μg/L)	Mn (μg/L)
Woods Lake	2510	6.5	178.0	1.90	0.55	11.4	16.0	3.4
Tenaya Lake	2480	6.3	29.0	0.30	0.03	21.2	4.6	4.5
Mosquito Lake	2440	6.9	62.0	0.35	0.09	18.2	50.6	4.7
Kirkwood Lake	2340	7.1	104.0	0.76	0.24	5.4	29.5	0
Angora Lake	2270	6.6	41.8	0.20	0.06	30.9	12.9	1.3
Echo Lake	2260	6.7	56.0	0.35	0.09	4.9	2.2	3.4
Huntington Lake	2120	6.6	150.0	0.58	0.12	6.1	7.3	2.7
Haven Lake	2050	6.7	87.0	0.30	0.36	26.4	3.5	1.1
Forebay Lake	2000	6.6	128.0	0.60	0.15	14.8	5.0	10.4
Fuller Lake	1630	6.6	192.0	1.50	0.42	12.7	19.2	12.0

seven weeks. The acid treated systems supported fewer numbers of rotifers, with the largest population being recorded at week 7 in the treatments with sediment (60 individuals/liter).

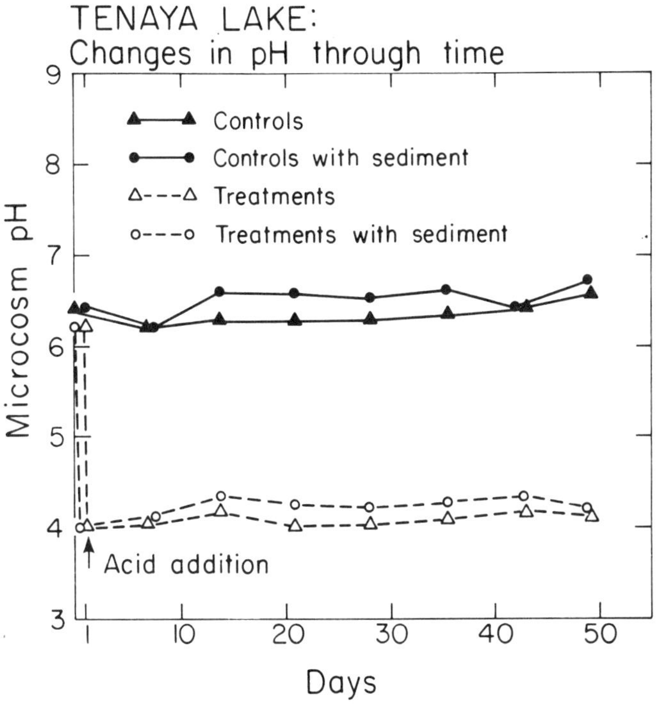

Figure 2. Changes in Microcosm pH Through Time in the Tenaya Lake Experiment. Nitric acid was added to treatments at day 1.

Phytoplankton groups responded variably to the acid stress. In general, planktonic chlorophyte populations remained relatively unaffected by the acid treatment, while chrysophytes (particularly diatoms) were adversely affected. Cryptomonads increased significantly in the acidified systems without sediment. In all treatments a mat of filamentous green algae (including *Mougeotia* sp.) formed on the bottom of the tanks.

DISCUSSION

The field data we collected in the Sierra Nevada indicate the sensitivity of these subalpine lakes to acid deposition. The thin, poorly buffered soils, small watershed areas, granitic basins, and low alkalinity waters of this region indicate the potential susceptibility of these aquatic systems to acid stress. Baseline monitoring of pH, alkalinity, principal anions and cations, and trace elements, such as aluminum, needs to be continued in both the subalpine and alpine systems of the Sierra to detect changes due to increasing acidification. Evidence from other parts of the world point to the early spring snowmelt period as the time of greatest vulnerability. Sampling should be carried out at this time.

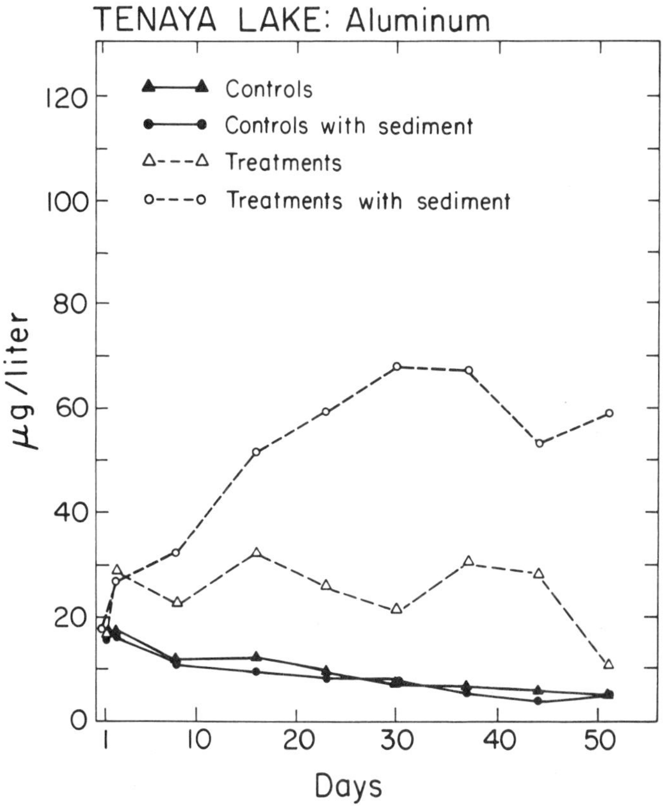

Figure 3. Changes in Concentrations of Dissolved Aluminum in Microcosms During the Tenaya Lake Experiment.

Possible impacts of this acidification have been simulated in our microcosm experiment. By manipulating these laboratory systems we have attempted to predict what effects might be detected as acidification proceeds and what processes might be altered as a result of this stress. The experiments we have performed indicate the importance of littoral sediment leaching of aluminum, manganese, and iron. A progressive loss of alkalinity coupled with increases in levels of these elements in the water column of Sierra lakes might be an early sign of progressive watershed acidification.

Among the plankton the identification of "indicator" species would be useful in detecting system acidification. Yan and Stokes (1978) made use of experimental enclosures in Canadian lakes and demonstrated certain systematic phytoplankton changes with increasing acidification. In a number of *in situ* experiments, including the whole lake acidification by Schindler (1980) in the Experimental Lakes Area of Canada, the development of an algal mat (including *Mougeotia* sp.) was characteristic of accelerating acidification in lakes.

Certain of these same systematic population changes following acid stress were also observed during our microcosm experiments. Those groups whose growth rates appeared to be stimulated by the addition of acid included dinoflagellates and cryptomonads. The diatoms experienced population decreases in the treatments. In one set of microcosms an algal mat developed, a result similar to that observed in other experiments and field observations.

These results suggest the validity of using microcosm systems to investigate changes in lake geochemical processes and to identify indicator species in sensitive systems which are currently being subjected to acid stress.

ACKNOWLEDGMENTS

This study was supported in part by grants from the California Policy Seminar and the Water Resources Center of the State of California.

LITERATURE CITED

Cowling, E. B., 1980. Acid Rain: An Emerging Ecological Issue. *In:* Issues and Current Studies. National Research Council, Washington, D.C., pp. 78-87.

Cronan, C. S. and C. L. Schofield, 1979. Aluminum Leaching Response to Acid Precipitation: Effects on High-Elevation Watersheds in the Northeast. Science 204(2):304-306.

Galloway, J. N. and E. B. Cowling, 1978. The Effects of Precipitation on Aquatic and Terrestrial Ecosystems: A Proposed Precipitation Chemistry Network. J. Air Pollut. Control Assoc. 28: 229-235.

Johannessen, M. and A. Henriksen, 1978. Chemistry of Snow Melt Water: Changes in Concentration During Melting. Water Resource Research 14:615-619.

McColl, J., 1980. A Survey of Acid Precipitation in Northern California. Final Report to the California Air Resources Board (A7-149-30), Sacramento, California.

Morgan, J. J. and H. M. Liljestrand, 1980. Measurement and Interpretation of Acid Rainfall in the Los Angeles Basin. Final Report to the California Air Resources Board (A7-110-30), Sacramento, California.

Schindler, D. W., 1980. Experimental Acidification of a Whole Lake: A Test of the Oligotrophication Hypothesis. *In:* Ecological Impact of Acid Precipitation, D. Drabløs and A. Tollan (Editors). Proc. Int. Conf., SNSF, Sandefjord, Norway, pp. 370-374.

Tonnessen, K., 1983. Potential for Aquatic Ecosystem Acidification in the Sierra Nevada, California. *In:* Acid Precipitation, Aquatic Effects, G. Hendrey (Editor). Ann Arbor Science Publisher Inc., Ann Arbor, Michigan (in press).

Tonnessen, K. and J. Harte, 1980. The Potential for Acid-Precipitation Damage to Aquatic Ecosystems of the Sierra Nevada, California, U.S.A. *In:* Ecological Impact of Acid Precipitation, D. Drabløs and A. Tollan (Editors). Proc. Int. Conf., SNSF, Sandefjord, Norway, p. 338.

Yan, N. D. and P. M. Stokes, 1978. Phytoplankton of an Acidic Lake, and Its Response to Experimental Alteration of pH. Environmental Conservation 5:93-100.

VARIATION IN ECOSYSTEM SENSITIVITY AND RESPONSE TO ANTHROPOGENIC ATMOSPHERIC INPUTS, UPPER GREAT LAKES REGION

J. Robert Stottlemyer[1]

ABSTRACT: Precipitation throughout the Upper Great Lakes Region is considered to be acidic. Average annual values range from pH 4.4 in lower Michigan to 5.2 in western Minnesota. In the central portion of the region the comparative values for average annual pH has fallen from 6 to 4.6 since 1955. Baseline data are few, but it appears the regional loading of acidity and associated contaminants is increasing. The Lake Superior basin typifies regional ecological diversity with northern hardwood forest and sedimentary bedrock on the south shore and boreal coniferous forests on Precambrian substrates on the north shore. Isle Royale and basin ecosystem studies show that Lake Superior significantly modifies precipitation inputs temporally and spatially. Often over 50 percent of the annual input is snow. Snowfall can increase 3x with a rise of 100m above the lake. This results in the heaviest load of atmospheric contaminants in the more sensitive low order watersheds. Hardwood forest canopies seasonally neutralize acid precipitation. Boreal conifer canopies have minimal effect. Glacial till quality appears more important than bedrock composition in determining whether acid precipitation is neutralized by carbonates or basic aluminum salts. The latter case can result in the release of soluble Al. The vulnerability of lakes and streams is known to be dependent on the nature and degree of watershed buffering. Watersheds with northern hardwoods generally have extensive glacial till, and this vegetation/soil interaction suggests that such ecosystems are much more resistent to the direct impacts of acid precipitation than are boreal ecosystems.
(KEY TERMS: acid rain; Great Lakes Region; ecosystems.)

INTRODUCTION

The objectives of this paper are to discuss the variation in ecosystem response to present inputs of atmospheric contaminants especially acid precipitation in the Upper Great Lakes region, and to demonstrate the value of using the ecosystem approach in such an assessment. The results presented and discussed below summarize pertinent parts of the first three years' of Lake Superior basin baseline watershed ecosystem studies conducted in Isle Royale National Park, the Keweenaw Peninsula, and in Pictured Rocks National Lakeshore (Figure 1).

Ecosystems have complex budgets for input and output of energy and nutrients; these are partitioned among living and nonliving components which in turn are interacting with one another. To date most research on natural systems has emphasized the study of selected ecosystem components. Relatively little attention has been devoted to component interaction. It is increasingly evident that knowledge of ecosystem interactions is essential to understand system functioning and response to anthropogenic impacts. The study of mineral nutrient cycling in watershed ecosystems of varying character is an effective approach for assessing ecosystem response to human impacts such as atmospheric deposition. The sensitivity of surface aquatic resources to inputs such as acid rain is closely related to the bedrock geology of the region, the texture and buffering capacity of the soil, and the distribution of precipitation.

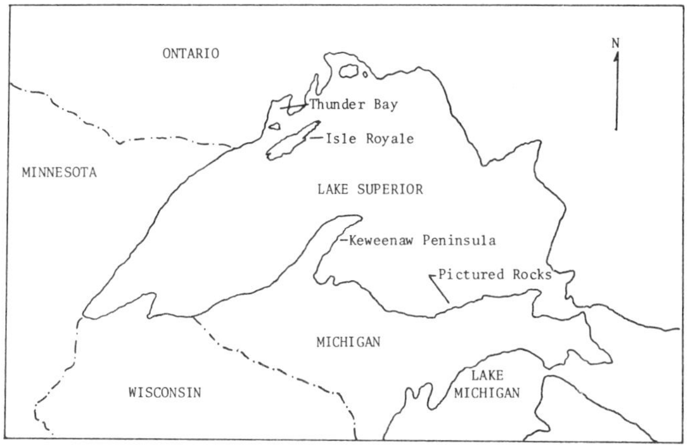

Figure 1. Lake Superior Basin and Vicinity.

Ecosystems of the Upper Great Lakes region are particularly complex due to a combination of factors. Bedrock geology is comprised of the erosion resistent Canadian shield which extends south to the Keweenaw Peninsula on Lake Superior's south shore. Further south the bedrock is

[1] Adj. Professor, Biological Sciences, Great Lakes Resource Studies Unit, Dept. Biological Sciences, Michigan Technological University, Houghton, Michigan 49931.

primarily Cambrian with more recent sedimentary deposits. The region has been extensively glaciated as recently as 10,000 years ago. Ancient beachlines, areas of thin glacial till and outwash, and moraines are often superimposed resulting in very different soil conditions over short distances. Stream and lake water chemistry reflect this soil diversity with soft and hard water streams often in close proximity. Because of the very thin glacial till, sand, and the outcropping of weather resistent bedrock, northeastern Minnesota, northwestern Wisconsin, and portions of Michigan's Upper Peninsula are considered to be very sensitive to present levels of anthropogenic atmospheric deposition (Galloway and Cowling, 1978).

The region is a major ecotonal zone between the Northern Hardwood forest which predominantes along the south shore of Lake Superior and the Boreal Coniferous forest found on Isle Royale and along Lake Superior's north shore. Due to its size Lake Superior never truly freezes over. This moderates cold winter air masses from the northwest, but results in high snowfall to the south with annual averages between 600 and 700 cms. One-half the annual precipitation can be held in the snowpack by middle March (Stottlemyer, 1982).

Present sources of atmospheric contamination are principally from the south and east (Hidy, 1982; Dohrenwend, *et al.,* 1980). There is some evidence that local sources such as Duluth, north central Minnesota, and Thunder Bay, Ontario, could also be significant contributors.

SITE DESCRIPTION AND METHODS

The bedrock of Isle Royale is primarily Precambrian metamorphosed volcanics. Except for the southwestern third of the island the soils are generally thin and poorly buffered. On the southwestern portion extensive alkaline till deposits affect the water chemistry of study watersheds such as Washington Creek and its tributaries. Edaphic and climatic factors also allow northern hardwood species such as sugar maple to prevail locally. The remainder of the island including the Moskey study watershed is dominated by boreal coniferous forest and moderately to poorly buffered substrates. The Keweenaw study site has a substrate very similar to Washington Creek, but its vegetation is young northern hardwoods. The Little Beaver drainage at Pictured Rocks has a Cambrian sedimentary substrate with local deposits of limestone and dolomite, and its vegetation cover is 80-year old northern hardwood forest. The research has centered on the use of small undisturbed gaged watersheds to determine how water chemistry is altered by precipitation, soil, and dominant vegetation. Much effort has been devoted to examining precipitation alteration by dominant vegetation species because most of the region is forested. There is a considerable nutrient and contaminant load in the snowpack. Thus, snow accumulation and quality has also been studied at all three sites for three winters. Field methods are outlined in Stottlemyer (1982). All water chemistry, except for aluminum, is done at Michigan Technological University or on Isle Royale using ion chromatography.

RESULTS

The amount and distribution of precipitation varies widely due to the influence of large lakes. The mean annual precipitation amount for Thunder Bay, Ontario, is 67.8 cm (17.7 cm), for Isle Royale 75.9 (9.6), the Keweenaw Peninsula 87.7 (12.4), and Pictured Rocks 77.6 (14.2). Standard deviations are shown in parentheses and the sampling period is 1967-1980. The variation results from the prevailing westerly and northwesterly winds across Lake Superior and the precipitation "shadow" of the Keweenaw Peninsula (see Figure 1). The amount of rainfall does not appear to increase with elevation above Lake Superior. Forty to 50 percent of the annual precipitation is snowfall. Snowfall amounts increase markedly with minor elevation increases above Lake Superior (Table 1).

TABLE 1. Mean Moisture Content (cm) of January-February Snowpack on Northwest Slope Adjacent Lake Superior (1979-1982). Standard deviation in parentheses.

Station	Elevation			
	190 m	225 m	260 m	295 m*
Isle Royale	10.7 (2.6)	13.3 (2.8)	11.9 (2.6)	14.1 (3.6)
Keweenaw Penin.	13.8 (7.6)	15.7 (5.2)	18.5 (4.5)	24.8 (5.4)
Pictured Rocks	11.8 (3.3)	14.9 (4.2)	15.8 (4.4)	19.0 (4.3)

*Lake Superior's Mean Elevation is about 183 m.

There is no significant variation in precipitation quality with increasing elevation for the stations sampled (n=350+). Precipitation throughout the region is acidic (Table 2). The longest sampling record is on Isle Royale (May 1979-present), and a regression of H^+ concentration with time shows a gradual but statistically insignificant increase.

The chemistry of incident precipitation is considerably modified as it passes through the forest canopy (Table 3). The H^+ concentration of incident precipitation is reduced by 73 percent as it passes through a sugar maple canopy, by 50 percent when passing through white cedar, and by 37 percent in white spruce throughfall (stemflow not included here).

The reaction of two first order Isle Royale streams to precipitation inputs is summarized in Tables 4 and 5. Moskey is about 1 km in length, and WATR 1.5 km. The sampling stations are spaced along the stream length with increasing numbers downstream, and the sampling period for these data is April 1981 to March 1982. The Moskey watershed contains very scattered shallow glacial till, and is

TABLE 2. Weighted Mean Chemical Composition (mg/L) of Incident Precipitation from May 1979 to March 1982 for Selected Stations in Lake Superior Basin.

Station	pH	Ca	Mg	Na	K	NH_4	NO_3	PO_4	SO_4	Cl
Isle Royale	4.39	0.26	0.08	0.15	0.15	– –	2.62	0.035	1.94	0.29
Isle Royale*	6.30	1.61	0.44	0.72	0.31	0.40	0.33	0.021	3.02	3.99
Pictured Rocks**	4.74	0.84	0.35	0.09	0.26	0.42	3.13	0.043	3.41	0.11

*Historical data collected May 1973-October 1974 at same station as present.
**Sample period is from November 1980 to present.

vegetated by boreal forest species. WATR has continuous and deep glacial till deposits and is vegetated mostly by northern hardwood species. Both drainages have precambrian metamorphosed volcanic substrates.

TABLE 3. Weighted Average Annual Chemical Composition (mg/L) of Incident Precipitation and Throughfall Under Different Species, March 1981-1982.

		Throughfall			Net Change		
	Open	Maple	Cedar	Spruce	Maple	Cedar	Spruce
Ca	0.26	6.00	5.54	3.61	5.74	5.28	3.35
Mg	0.08	4.59	2.38	1.73	4.51	2.30	1.65
Na	0.15	0.36	0.57	0.54	0.21	0.21	0.42
K	0.15	20.74	7.78	10.44	20.59	7.63	10.29
NH_4	– –	0.49	0.86	0.49	– –	– –	– –
NO_3	2.62	3.67	8.51	2.24	1.05	5.89	−0.38
PO_4	0.035	0.019	0.278	0.031	−0.016	0.243	−0.004
SO_4	1.94	13.13	13.63	13.17	11.19	11.69	11.23
Cl	0.29	1.94	2.84	4.04	1.65	2.55	3.75
pH	4.39	4.96	4.69	4.59	0.57	0.30	0.20

TABLE 4. Mean Chemical Composition (mg/L) with Increasing Distance Downstream for Low-Moderate Buffered Stream (Moskey), Isle Royale, Michigan (1981).

Substance	Station 1	Station 2	Station 3
Ca	8.35(1.67)*	8.87(2.53)	11.6(3.00)
Mg	4.40(0.98)	4.67(1.43)	5.64(1.35)
Na	1.51(0.24)	1.55(0.43)	1.92(0.46)
K	0.09(0.09)	0.20(0.20)	0.19(0.06)
NH_4	0.03(0.06)	0.08(0.08)	0.008(0.02)
NO_3	0.75(0.61)	0.00	0.23(0.47)
SO_4**	3.85(5.64)	2.86(2.76)	4.70(2.27)
Cl	0.22(0.39)	0.05(0.03)	0.12(0.09)
pH	6.11	6.34	6.41
Alk	24(14)	29(11)	44(17)

*Standard deviation in parentheses.
**No PO_4 found in stream samples.

DISCUSSION

Most of the above analyses are confined to events of the past year. This eliminates yearly variation in precipitation inputs which can alter stream chemistry (Lewis and Grant, 1979). Last year was one of the driest on record followed by a very turbulent winter which increased the variance in station precipitation quantity over that measured previously. The increase in snowfall with elevation is a significant ecological factor. It results in nearly a 2x increase in loading of pollutants in the uppermost reaches of Superior basin watersheds which are also the most sensitive to such loading. Precipitation throughout the region is acidic with average annual values ranging from about pH 4.4 in lower Michigan to 5.2 in western Minnesota (NADP, 1981). In the central portion of the region it appears possible the pH has declined from around pH 6 to pH 4.6 since 1955 (Cogbill and Likens, 1976). Unfortunately, historical data are few. Those collected in Isle Royale during 1973-74 (Table 2) as part of a Western Superior Basin Study are probably of limited value since the cation and Cl values indicate contamination probably by blowing dust. The 1973-74 levels of NO_3 are much below those observed today which is consistent with the general increase in NO_3 observed in the midwest. But the SO_4 level is more in line with that observed today. An earlier analysis (Stottlemyer, 1982) of SO_4 levels in water quality data collected by the U.S. Geological Survey since 1967 for Washington Creek on Isle Royale reveals little change over this 15-year period. Since SO_4 does not appear to be retained by the ecosystem (Tables 4 and 5) in this region, stream flow levels can indicate the nature of past atmospheric inputs. Cation levels in Isle Royale precipitation today are a little above those levels observed in New England (Likens, *et al.*, 1977). This could be the result of slightly higher airborne dust levels from the western plains or aerosols from Lake Superior.

The alteration of incident precipitation chemistry by the forest canopy over a period of one year reveals the relative importance of hardwood species and conifers in reducing the H^+ load to the forest floor and stream. The loss of hydrogen is made up principally by increased K and Ca in throughfall. The effectiveness of maple in this alteration process is accentuated by the fact that almost 90 percent of incident precipitation occurs as throughfall while spruce

TABLE 5. Mean Chemical Composition (mg/L) with Increasing Distance Downstream for Moderately Buffered Stream (WATR), Isle Royale, Michigan (1981).*

Substance	Station 1	Station 2	Station 3	Station 4	Station 5	Station 6
Ca	6.76(3.7)	7.52(4.0)	8.53(5.1)	7.42(5.9)	8.00(5.1)	11.94(8.5)
Mg	5.55(2.8)	6.75(2.7)	7.65(2.7)	6.62(3.1)	6.75(2.1)	7.84(2.9)
Na	1.95(0.5)	2.17(0.4)	2.74(0.6)	2.53(1.1)	2.88(0.7)	2.55(0.7)
K	0.40(0.3)	0.34(0.2)	0.46(0.2)	0.33(0.2)	0.39(0.2)	0.61(0.2)
NH_4	– – – –	0.07(0.1)	0.08(0.2)	0.01(0.01)	0.04(0.07)	0.01(0.04)
NO_3	0.10(0.11)	0.05(0.05)	0.09(0.1)	0.33(0.3)	0.22(0.2)	0.92(0.8)
PO_4	0.013(0.03)	0.03(0.06)	0.01(0.03)	0.006(0.01)	0.006(0.01)	0.04(0.1)
SO_4	4.23(4.1)	2.18(1.8)	2.98(2.3)	– – – –	4.72(2.5)	4.98(1.6)
Cl	0.16(0.2)	0.16(0.13)	0.13(0.13)	0.17(0.15)	0.16(0.15)	0.28(0.2)
pH	6.04	6.24	6.47	6.85	6.84	7.07
Alk	28(23)	51(12)	63(17)	67(18)	65(16)	72(21)

*Station numbers increase downstream in Tables 4 and 5.

and cedar throughfall averages about 65 percent of incident precipitation. The overall importance of this nutrient loss from the canopy is not known. These results suggest, however, that the occurrence of northern hardwood forest reduces the direct impact of present levels of atmospheric acidity on stream environments when compared to the boreal forest where conifers such as white spruce are more abundant.

Soil characteristics of the two watersheds accentuate their vegetation differences. Moskey [mean alkalinity 44(17) at Station 3], dominated by boreal forest, has a surprisingly similar water chemistry to WATR at its upper station, but water chemistry and neutralization are little changed once in the stream. WATR [mean alkalinity 72(22) at Station 6], dominated by maple, shows much higher levels of K and Mg likely due to canopy leaching and the presence of alkaline till. pH increases markedly downstream. With such pH values neutralization of precipitation by aluminum salts appears doubtful. Spring snowmelt has reduced pH values up to 1.2 units (from mean), and such "shocks" at such a critical time could have a long term impact. The Keweenaw and Pictured Rocks watersheds under study also show a significant spring reduction in pH, but overall are better buffered than either of the Isle Royale watersheds.

Ecosystems have many mechanisms to conserve essential mineral plant nutrients. An indication of this is the general inverse relationship between nutrient concentration and stream flow from undisturbed watersheds. At lower concentrations a linear model has generally been found to best describe the relationship. The closer the slope of the regression to zero or a positive value the greater probability there are outside inputs or there is some disturbance or imbalance within the ecosystem. The Moskey drainage shows a significant decline in total cation concentration with increased stream flow except for K^+ and H^+ which show a slight increase (slope not statistically significant). K losses are probably due to increased leaching by excess H^+. NO_3 and SO_4 both show an insignificant decline with increased discharge possibly reflecting some inability of the ecosystem to use these atmospheric inputs. WATR shows a similar anion response but a much different cation relationship. Total cation concentration decreases more dramatically with increasing discharge, and both K and H^+ decrease with increased discharge. This indicates that WATR has a greater capacity to effectively absorb the present acid loading without increased nutrient loss from the ecosystem. This probably indicates a vegetation-soil interaction. However, until these data are converted to budgets and the sampling period extended little can be concluded.

The above watersheds are two of six under study. Higher buffered systems have been found in Minnesota (Wagner and Lemire, 1976) and at Pictured Rocks (Stottlemyer, 1982). Likewise two far more sensitive systems have been found in northwest Wisconsin (Tom Scheffy, personal communication), by this investigator on the Keweenaw Peninsula, and near Sault Ste. Marie, Ontario (Kelso, *et al.*, 1982). Generalizations regarding ecosystem sensitivity to the direct and indirect impacts of anthropogenic atmospheric inputs will be difficult in this glaciated region because of the geological variation. The interaction of precipitation amount, vegetation type, and the presence or absence of alkaline glacial tills result in a complex mosaic with very sensitive ecosystems intermixed with more resistent ecosystems.

LITERATURE CITED

Cogbill, C. V. and G. E. Likens, 1974. Acid Precipitation in the Northeastern United States. Water Resources Res. 10:1133-1137.

Dohrenwend, R. E., S. G. Shetron, J. R. Stottlemyer, and R. J. Olszewski, 1980. Acid Precipitation in the Keweenaw Peninsula of Michigan's Upper Peninsula. *In:* Ecological Impact of Acid Precipitation, D. Drabløs and A. Tallan (Editors), pp. 106-107.

Galloway, J. N. and E. B. Cowling, 1978. The Effects of Precipitation on Aquatic and Terrestrial Ecosystems: A Proposed Precipitation Chemistry Network. J. Air Pollut. Control Assoc. 28: 229-235.

Hidy, G. M., 1982. Large-Scale Pollution Events Over the Northeastern United States. Abstracts of Papers of the 148th National Meeting, American Association for the Advancement of Science, p. 59.

Kelso, J. R., R. J. Love, J. H. Lipsit, and R. Dermott, 1981. Whole Lake Response: Ecological Effects. *In:* The Effects of Acid Precipitation on Ecological Systems in the Upper Great Lakes, F. D'Itri (Editor). Ann Arbor Science (in press).

Lewis, W. M., Jr. and M. C. Grant, 1979. Relationships Between Stream Discharge and Yield of Dissolved Substances from a Colorado Mountain Watershed. Soil Science 128:353-363.

Likens, G. E., F. H. Bormann, R. S. Pierce, J. S. Eaton, and N. M. Johnson, 1977. Biogeochemistry of a Forested Ecosystem. Springer-Verlag, New York, New York, 146 pp.

NADP, 1981. National Atmospheric Deposition Program Data Report – Precipitation Chemistry. Natural Resource Ecology Laboratory, Colorado State University, Fort Collins, Colorato, Vol. III, No. 3.

Stottlemyer, J. R., 1982. Ecosystem Evaluation of Anthropogenic Atmospheric Inputs. *In:* Symposium on External Threats to Ecosystems of the National Parks. American Association for the Advancement of Science Annual Meeting, January 3-8, 1982, Washington, D.C.

Wagner, D. M. and R. S. Lemire, 1976. Water Quality of Sixteen Minnesota Rivers Tributary to Lake Superior. J. Great Lakes Res. 2:111-123.

Section 8

PROBABLE MAXIMUM PRECIPITATION AND FREQUENCY FOR DESIGN

SITE SPECIFIC PROBABLE MAXIMUM PRECIPITATION ESTIMATES, UPPER SOUTH PLATTE RIVER BASIN, COLORADO

J. T. Riedel, B. H. Wang, and J. L. Diebel[1]

ABSTRACT: Probable maximum floods (PMF) at Antero, Elevenmile and Cheesman Dams on the headwaters of the South Platte River in Colorado were estimated through a site specific hydrometeorological and hydrologic study. The study included estimation of probable maximum precipitation (PMP) by storm transposition and maximization and a determination of the rainfall-runoff relationship by the dimensionless graph-lag curve method. Drainage sites of interest are between approximately 200 and 2000 mi.

Major rainfalls in the South Platte River basin and vicinity show patterns that are closely associated with the terrain features. The storm rainfall history in the region is one of modest rainfall in the mountainous areas and very severe rainfalls in the areas where the plains meet the foothills. Due to the great difference in climate within a short distance, major efforts were needed to determine which storms should be transposed and how these storms should be adjusted in estimating PMP. This paper presents, in detail, the procedure and results of PMP analysis.
(KEY TERMS: probable maximum precipitation; Colorado; Platte River.)

INTRODUCTION

Probable maximum precipitation (PMP) may be defined as the greatest depth of rainfall for a given duration that the atmosphere can produce over a selected basin. Meteorologists have not been able to determine entirely which parameters, and the values of them in combination, will result in the greatest rainfall. Therefore, it is customary to refer to PMP values as estimates.

Details of the procedures and methods used in determining PMP are described in the Hydrometeorological Report (HMR) series of the National Weather Service (NWS). A publication (World Meteorological Organization, 1973) also summarizes the various methods. Basically, PMP estimates are obtained either directly or indirectly by transposition and moisture maximization of recorded major storm rainfalls.

PMP estimates often are taken from regional maps on which isolines of "generalized" PMP are defined. Such estimates result from detailed studies. For the basin under discussion, no detailed generalized PMP estimates are available although adequate studies have covered areas near the basin (Hansen, *et al.,* 1977; Schreiner and Riedel, 1978). The unique topographic characteristics of the basins also require special consideration. Therefore, a site specific study was made of three basins (see Figure 1) in which terrain features of the basins were carefully considered.

GENERAL CLIMATOLOGY OF MAJOR RAINSTORMS NEAR THE UPPER SOUTH PLATTE BASIN

The magnitude and isohyetal patterns of major rainstorms in and near the Upper South Platte River Basin above Cheesman (referred to as "the Basin" hereafter) are closely associated with the terrain. The greatest rainfalls are found in the relatively flat region east of approximately the 7,000 ft. msl contour line (mountain front). Point values as great as 24 inches in less than 6 hours (near Cherry Creek, May 30, 1935) have been measured east of this line. This extreme event was imbedded in a general storm lasting over two days and covering thousands of square miles.

For the east facing slopes, the data show relatively smaller rains because winds at higher elevations usually become westerly, and therefore prevent little additional orographic intensification. The Big Thompson Canyon storm of July 31, 1976 which gave over 12 inches in four hours is one of the most intense storms known on the east slopes.

Smaller rains are found to the lee side of the eastern facing barriers. Extreme point rainfalls are possible for relatively short durations, drawing needed atmospheric moisture from the locality. Longer-duration rainfalls of large areal coverage caused by orographic lifting also occur on the eastward facing slopes and "spillover" into the lee areas.

[1] Respectively, Consulting Hydrometeorologist, Route 2, Box 62, Huddleston, Virginia 24106; Associate and Principal Hydrologist, Harza Engineering Company, 150 South Wacker Drive, Chicago, Illinois 60606; and Engineering Manager, Denver Water Department, 1600 West 12th Avenue, Denver, Colorado 80254.

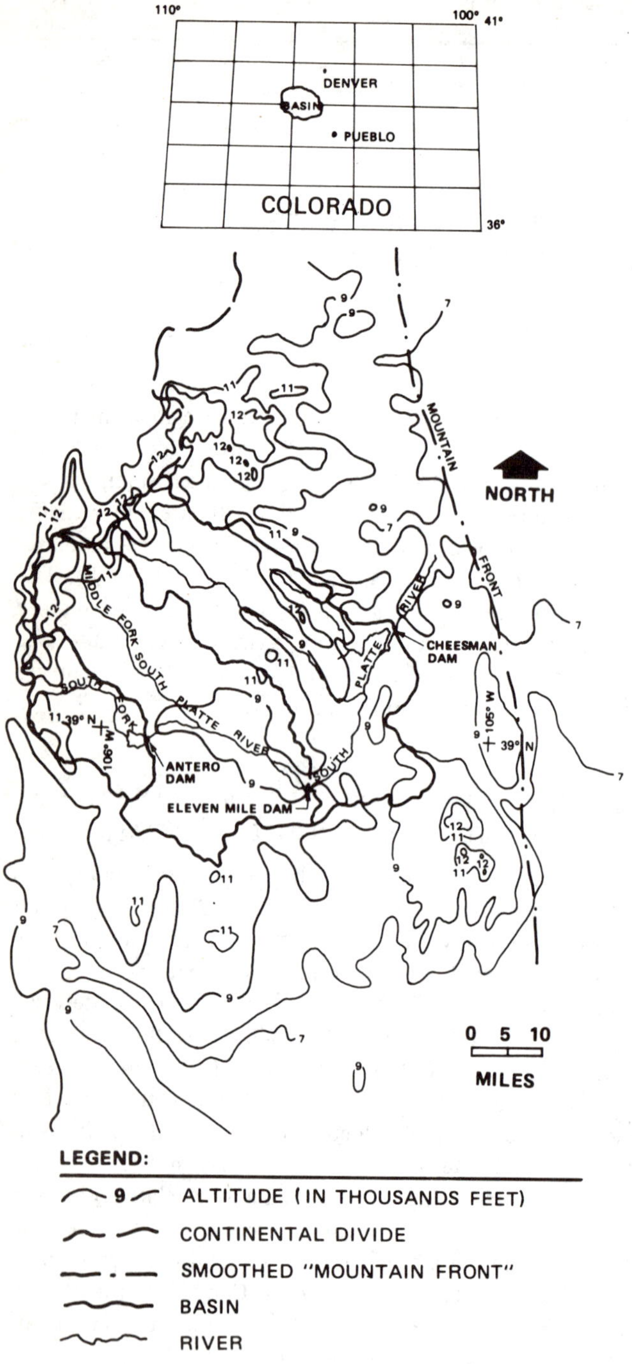

Figure 1. Basin Outlines and General Topography.

Most extreme point rainfalls in the mountainous region west of the Continental Divide are local both in time and space. This contrasts with extreme rainfalls east of the mountain front where they occur in large area and long-duration rain. For the region between the Continental Divide and the 105th meridian, it appears that there should be a gradual change from point PMP of a local storm near the Divide to the point PMP "imbedded" in a general storm near the mountain front.

ESTIMATED PMP

Data Sources

A publication (Shipe and Riedel, 1976) lists greatest known areal rainfalls covering standard area sizes and durations. An update of that publication was the primary data source for this study.

Cheesman Basin

For this 1750 mi^2 basin, a storm lasting at least two days would cause critical flood. A major study effort was therefore devoted to locating extreme rainfalls of similar duration that had occurred in similar topographic settings somewhere between the mountain front and the Continental Divide. The June 6-8, 1964 storm centered near Gibson Dam, Montana, was found to be the most severe with a topographic setting similar to that of the Basin. The storm isohyetal pattern is shown in Figure 2. The portion of the pattern west of the Continental Divide is of major interest to the present study.

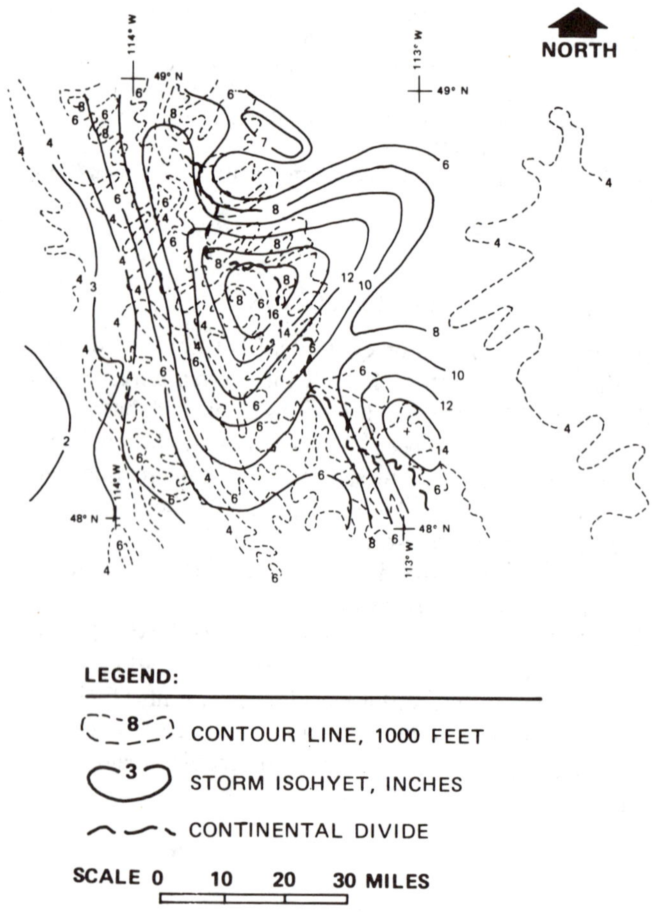

Figure 2. Isohyets of June 6-8, 1964, Montana Storm.

In summary, meteorological factors important to this storm are: a) high moisture inflow from the Gulf of Mexico, b) an upper air low pressure center to the southwest, c) a surface low pressure with a weak frontal system approaching from the west, and d) instability for convergence rain.

This and other severe storms of large areal coverage centered along the eastern slopes in Colorado are listed in Table 1. By and large the features of the June 1964 storm were also found in the east slope storms of Colorado. The June 1964 Montana storm was therefore transposed to the Basin as shown in Figure 3.

TABLE 1. General Storms In or Near the Cheesman Basin and the Montana Storm.

Date	Rain Center	Maximum Depth ($2000\ mi^2$, 48 hrs)	Approximate Distance from Center of Cheesman Basin (mi)
May 29-31, 1894	Ward District, Colorado	5.9	70
April 14-16, 1921	Frys Ranch, Colorado	4.4	60
July 19-24, 1929	Cheesman, Colorado	1.8	20
April 19-22, 1933	Westcliffe, Colorado	4.0	80
June 6-8, 1964	Gibson Dam, Montana	9.0*	700

*For a $2000\ mi^2$ area west of Continental Divide.

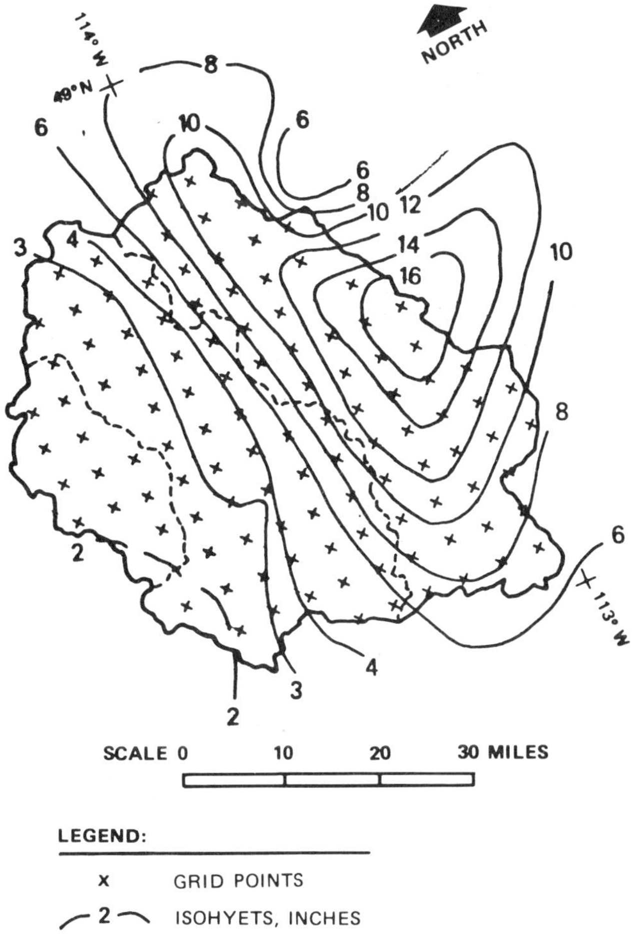

Figure 3. 1964 Montana Storm Transposed to Cheesman Basin.

The topography where the storm occurred was compared with that in and near the Basin. The northeast ridge of the Basin is higher (10,000 ft msl compared to about 7,000 ft msl) and more continuous. On the other hand, the length of the eastward facing slope is considerably longer than that in Montana. This tends to decrease the average rainfall on the slope.

We first increased the storm rainfall for maximum moisture, using a maximization factor of 1.67. We then adjusted the maximized rainfall by multiplying it by the ratio of the 100-yr 24-hr rainfall (Miller, *et al.*, 1973) in the transposed location to the corresponding value where the storm occurred. This was done on a 5-mi grid shown on Figure 3. We assumed that the 100-yr 24-hr rainfall reflects many factors important to heavy rain, such as length, height, orientation and degree of slopes as well as opportunity for high moisture. The resulting moisture maximized and transposed isohyetal pattern is shown in Figure 4. This PMP was distributed with time by a composite depth-duration relation based on the 1964 storm and generalized PMP to the east and west. Table 2 gives the PMP values.

Antero Basin

While a six-hour thunderstorm (local) PMP for this 192 mi^2 basin (see Figure 1) is most critical, a two-day general storm PMP was also estimated for comparison.

Local PMP. Because observation data was sparse in the Basin, rainfall extremes were collected from other locations considered comparable. The selected extremes are listed in Table 3.

The variation in rainfall with elevation was investigated. First, the extremes of the storms were plotted against elevation and the one-hour duration interpolated. Other plots were made after all values were normalized to a one-hour duration by: (a) a depth-duration relation of local PMP west of the Continental Divide (HMR 49), and (b) the depth-duration relation of the Big Thompson Canyon rainfall. Additional guidance was drawn from maximum one-hour and six-hour rainfalls of records (1943-1972) for 48 recorder stations west of the Continental Divide in Colorado. These values showed no consistent variation with elevation up to about 7,000 ft. Above this elevation, the averaged data indicated a decrease.

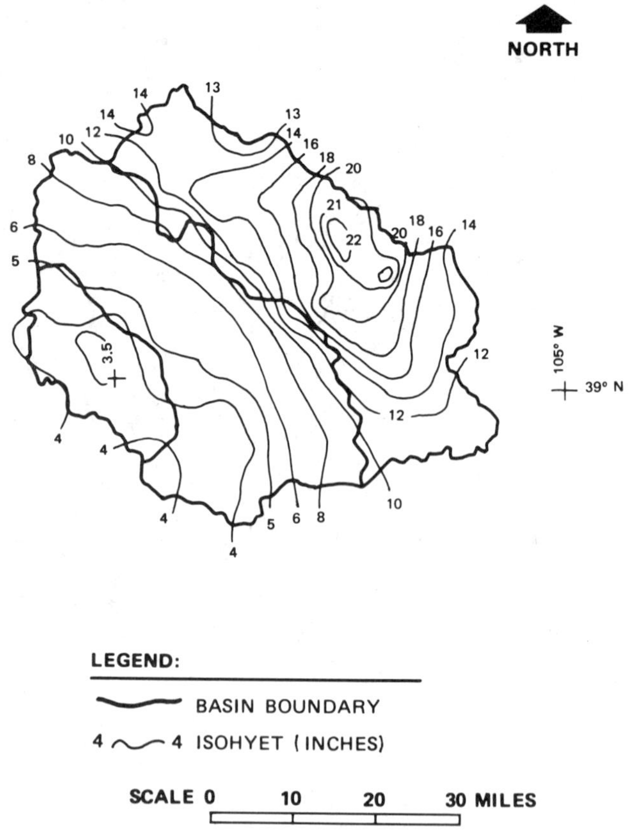

Figure 4. Smoothed Isohyetal Map.

Figure 5 shows the elevation relations found and variation of atmospheric moisture with elevation. We used the latter to adjust each of the selected storms by the ratio of the moisture above 9,000 ft (the elevation of the Basin) to that above the elevation where the storm occurred. Table 3 shows the adjustments and the resulting rainfall depths. The observed and adjusted data are plotted on Figure 6 along with the adopted envelopment used to define the local point PMP. The point values from Figure 6 were reduced to average values over the Antero Basin by the depth-area relations of local PMP west of the Continental Divide (HMR 49).

General Storm PMP. Transposition of the June 1964 storm, critical for the Cheesman Basin, gives very low values over the Antero Basin. Therefore, the storms in or near the Upper South Platte Basin in Table 1, were evaluated. Two stood out above the others, the May 29-31, 1894 and the April 19-22, 1933 storms.

The center of the 1948 storm was not considered transposible to the Antero Basin. The average rainfall observed over the Antero Basin was 3.05 inches. Moisture maximization resulted in an adjustment factor of 2.46. We used a maximum value of 2.0 because too high a moisture adjustment may not be compatible with optimum values of other parameters important to extreme rainfall. We thus arrived at a basin average depth of 6.1 (3.05 x 2) inches for the storm

TABLE 2. PMP Estimates for Cheesman, Eleven-Mile and Anteror Basin of the South Platte River, Colorado.

Basin	Duration (Hours)											
	6	12	18	24	30	36	42	48	54	60	66	72
	Depth (in.)											
A. General Storm PMP												
Cheesman (1750 mi^2)	3.4	5.6	7.3	8.4	9.1	9.5	9.8	10.0	10.2	10.4	10.5	10.6
Eleven-mile (971 mi^2)	2.4	4.0	5.2	6.0	6.5	6.7	6.9	7.1	7.3	7.4	7.5	7.6
Antero (192 mi^2)	3.0	4.7	5.9	6.6	7.0	7.3	7.5	7.7	7.8	8.0	8.1	8.2

Basin	Duration (Hours)								
	1/4	1/2	3/4	1	2	3	4	5	6
	Depth (in.)								
Antero (192 mi^2)	1.0	1.8	2.5	3.1	4.6	5.7	6.5	7.2	7.8
Eleven-mile (200 mi^2)	0.9	1.6	2.2	2.7	4.0	5.0	5.7	6.3	6.9

TABLE 3. Most Extreme Point Rainfalls Known Considered Transposable to Antero and Eleven-Mile Basins

Date	Location Town, State	Elevation (1000 ft)	Rain Depth (in.)	Duration (hr.)	Storm T_D (°F)	Maximum[1] T_D (°F)	Distance[2] Direction	Adjustment[3] (%)	Adj. Rainfall (in.)
1. 8/17/1922	Virsylvia, New Mexico	7.6	7.5	4	71[4]	77	–	126	9.4
2. 8/29/1935	Las Cruces New Mexico	3.9	7.2 9.0	3 6	71	77	40 SSE	84	6.0 7.6
3. 7/27/1937	Leadville, Colorado	10.0	4.3 4.4	¾ 1	70[4]	75	–	159	6.8 6.9
4. 7/12/1976	Sweetwater (nr), Colo.	7.0	12.5	2½	71	77	38 SSE	118	14.7
5. 7/12/1976	Big Thompson C., Colorado	8.0	4.5 6.7 9.5 11.3 12.5	½ 1 2 3 4	72	76	150 SSE	115	5.2 7.7 10.9 13.0 14.4
6. 8/19/1978	White Sands, New Mexico	4.0	10	4	76	78	650 SE	68	6.8
7. 9/10/1938	Masonville (nr), Colo.	5.4	7	½	70	76	500 S	97	6.8

NOTES:
[1]12-hr persisting 1000-mb dewpoint.
[2]Distance (mi) and direction from storm center for dewpoints.
[3]Ratio of precipitable water (W_P) for max. T_D to W_P for storm T_D. W_P for max. T_D reduced for 9000 ft elev.
[4]Storm dewpoint estimated.

duration of 60 hours. This was adjusted to 5.86 inches for a duration of 48 hours.

The Westcliffe, Colorado, storm of April 19-22, 1933, centered in the valley between the Sangre De Cristo Mountains to the west and the Wet Mountains to the east. Exposure to southerly moisture flow is similar to that of Antero Basin. We transposed this storm center to the Antero Basin by means of the 100-yr 24-hr rainfall. This gave 8.0 inches for the 60-hr storm duration which was adopted as the general storm PMP. A depth-duration relation gives the values shown in Table 2.

Eleven-Mile Basin

Both general and thunderstorm PMPs were estimated ror this 971 mi^2 basin.

Local PMP. The point local PMP derived from the Antero Basin is considered adequate for this basin also. We decided to use an area limit of 200 mi^2 and place the storm critically over the basin. The resulting estimate for basin fit is shown in Table 2.

General Storm PMP. The two storms of May 1894 and April 1933, were treated in a fashion for this basin similar to that for the Antero Basin. The May 1894 storm gave an estimate of 7.14 inches of rainfall for 48 hours. This exceeds the 6.21 inches obtained from the April 1933 storm. The adopted general storm PMP and its time distribution are shown in Table 2.

LITERATURE CITED

Hansen, E. M., F. K. Schwarz, and J. T. Riedel, 1977. Probable Maximum Precipitation Estimates, Colorado River and Great Basin Drainages, HMR No. 49, NWS, U.S. Department of Commerce, Silver Spring, MD., 161 pp.

Miller, J. F., R. H. Frederick, and R. J. Tracey, 1973. Precipitation-Frequency Atlas of the Western United States. Vol. 1: Montana, Vol. III: Colorado, Atlas 2, NWS, NOAA, U.S. Dept. of Commerce, Silver Springs, MD.

Schreiner, L. C. and J. T. Riedel, 1978. Probable Maximum Precipitation Estimates, United States, East of the 105th Meridan HMR No. 51, NWS, NOAA, Silver Springs, MD, 87 pp.

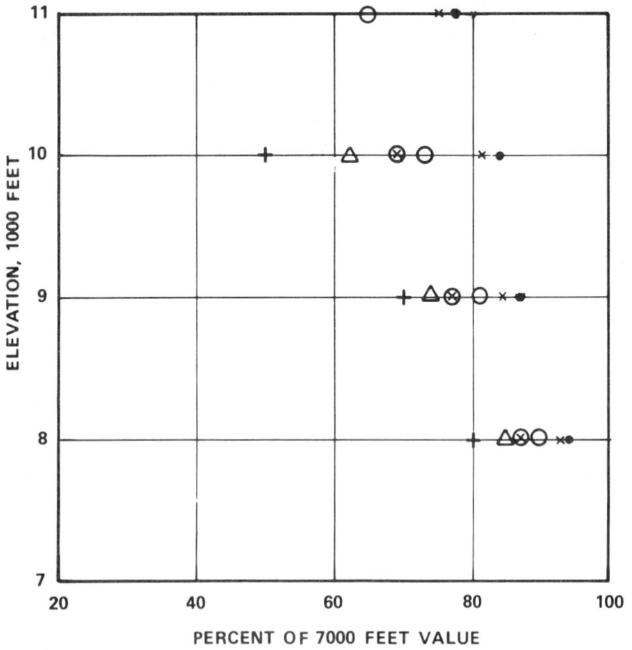

LEGEND:

MAXIMUM RAINFALL AT 48 RECORDER STATIONS WEST OF MOUNTAIN FRONT IN COLORADO

- ● 1-HOUR
- × 6-HOURS
- ○ PRECIPITABLE WATER FOR 1000 mb DEWPOINT OF 77° F

EXTREMES FOR 7 STATIONS (TABLE 3)

- \+ NORMALIZED TO 1-HOUR BY DEPTH-DURATION OF HMR 49
- ⊗ NORMALIZED TO 1-HOUR BY DEPTH-DURATION OF BIG THOMPSON STORM
- △ 1-HOUR DURATION INTERPOLATED

Figure 5. Variation of Station Rainfall With Elevation.

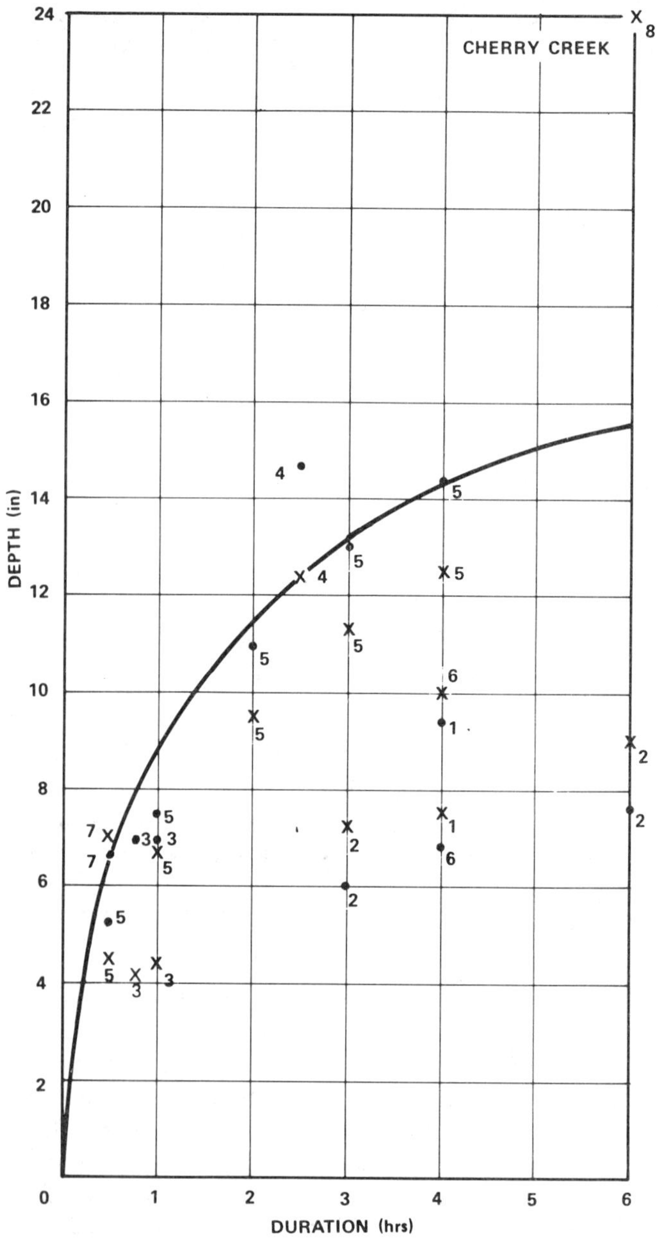

STORM

1. VIRSYLVIA, N. MEXICO
2. LAS CRUCES, N. MEXICO
3. LEADVILLE, COL.
4. SWEETWATER (nr), COL.[a]
5. BIG THOMPSON, COL.
6. WHITE SANDS, N. MEXICO
7. MASONVILLE, COL.[b]
8. EAST OF MTN. FRONT DATA

X. OBSERVED

● TRANSPOSED AND MOISTURE MAXIMIZED

a] DEPTH AND DURATION ESTIMATED

b] STROM OCCURRED AT MOUNTAIN FRONT;

Figure 6. Depth Duration Relation of Point Thunderstorm PMP, Antero and Eleven-Mile Basins.

Shipe, Albert P. and J. T. Riedel, 1976. Greatest known areal storm rainfall depths for the contiguous United States, NOAA Technical Memorandum NWS HYDRO-33, NWS, NOAA, U.S. Dept. of Commerce, Silver Spring, MD, 174 pp.

World Meteorological Organization, 1973. Manual for estimation of probable maximum precipitation, WMO No. 332, Operational Hydrology Report No. 1, Geneva, Switzerland, 190 pp.

REGIONAL HYDROMETEOROLOGICAL ANALYSIS FOR SPILLWAY DESIGN

Melvin G. Schaefer[1]

ABSTRACT: The development or revision of guidelines for the design of spillways for earthfill dams is a problem that is currently being faced by many regulatory agencies. Intuitive in any spillway design guideline is the concept of balancing the risk of under design with the consequences of a spillway failure. Unfortunately, existing guidelines rely heavily on engineering judgment and do not incorporate any actual risk assessment of spillway inadequacy in the decision-making process.

In this paper a new approach to the problem of developing spillway guidelines is presented which utilizes a frequency analysis of extreme precipitation events for assessing risk. The analysis is accomplished from a regional perspective based on the concept that extreme storms can be more reliably analyzed on a regional rather than a point basis. The results of the analysis can be used in a risk based decision-making process for developing spillway design guidelines.

This methodology should be especially attractive to regulatory agencies because it can provide information on the relative safety that a given design level will provide to projects in their region of jurisdiction.

(KEY TERMS: spillway design; hydrometeorology; regional analysis; dam safety; risk assessment.)

INTRODUCTION

The selection of an appropriate spillway design for a major hydraulic structure is a problem frequently encountered by the hydrologist. Normally, the selection of a spillway design level is primarily based upon the guidelines or standards of the governing regulatory agency. As a result, those guidelines and/or standards largely predetermine the "safety" of the final design. Unfortunately, past guidelines have been mostly subjective in the manner in which the concepts of risk or safety have been applied.

With the impetus from the National Dam Safety Program, many regulatory agencies are currently updating or revising their spillway design guidelines. Because the purpose of the design guideline is to provide safety, these new guidelines should be based on some quantitative measure of the safety that they will provide.

It is the intent of this paper to present some new concepts and methodologies which can be used to quantitize the risk associated with the design of spillways. The approach that is taken utilizes a regional frequency analysis of extreme precipitation events to assess the risk of spillway inadeaquacy. Through this approach, regulatory agencies charged with the responsibility of protecting public safety can develop risk based spillway design guidelines which have the capability of providing a reasonably known and predeterminable level of safety.

OVERVIEW

At the present time, most of the spillway design guidelines have a format wherein the design level is expressed as a percentage of the Probable Maximum Flood (PMF) or Probable Maximum Precipitation (PMP). The design level is typically determined based upon the potential hazards that a dam failure would present to downstream areas. The magnitude of the potential hazards are used as the basis for a subjective judgment of the level of design required.

Conceptually, this guideline structure is simple and logical; the greater the hazard, the more stringent the design criteria. However, when viewed from the perspective of a regulatory agency with regional responsibilities (local, state, or Federal) there are several major inadequacies. First, a design level based solely on percentages of PMP or PMF does not contain any quantitative measure of the actual safety of the design. What is the probability that a 40, 50, or 75 percent PMP or PMF level will be exceeded in a given year or during the life of the structure? Certainly, if one is to balance the risk of underdesign with the consequences of failure of the structure, one must have a good handle on the magnitude of the risk involved. Second, a given design level may or may not provide the same level of safety at all locations in the region (Figures 1A and 1B). It is seen in the Midwestern region that the 6-hr., 100-yr precipitation (Hershfield, 1961) is roughly 20 percent of the 6-hr., 10-mi^2 PMP (Schreiner, 1978), and there is not much variability of this ratio across the region. In contrast, the region of Western Washington is extremely variable with the 6-hr., 100-yr. precipitation (Miller, 1973) ranging from 30 percent

[1]Surface Water Hydrologist, Washington State Department of Ecology, Olympia, Washington 98504.

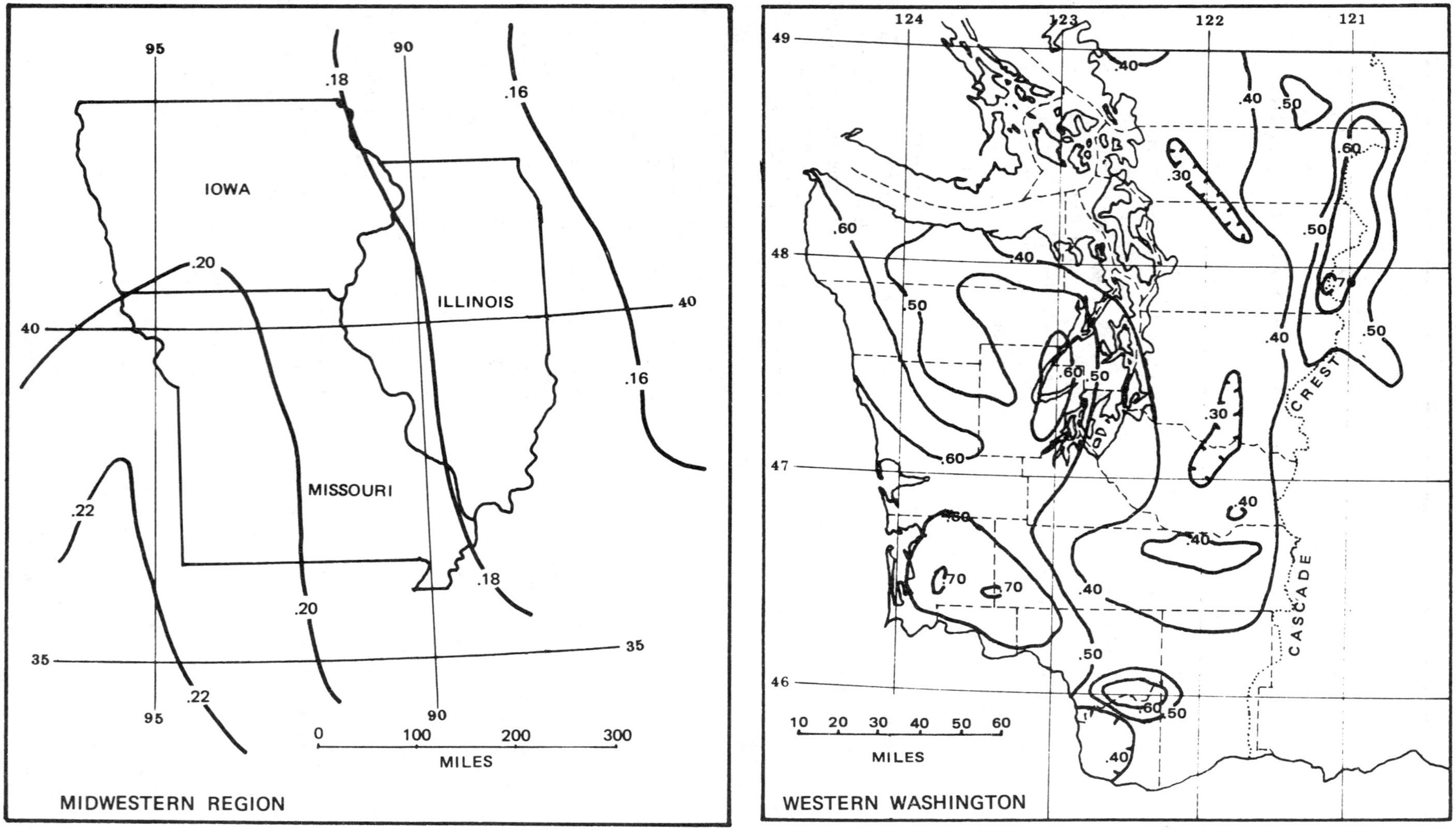

Figures 1A, 1B. Generalized Map – Ratio of 6-Hour, 100-Year Precipitation to 6-Hour 10 mi^2 PMP.

to 70 percent of the 6-hr., 10-mi^2 PMP (Myers, 1966). Certainly any design level expressed as a fixed percentage of PMP or PMF will not provide the same level of safety within these two regions.

The third weakness with the aforementioned guideline format is it does not take into account that the guidelines are being applied to numerous projects. The implication being that the risk of a spillway inadequacy increases as the number of projects and the size of the region increases. This occurs because extreme precipitation events and thus extreme floods occur more frequently on a regional basis than on a point basis.

A solution to the problems raised above can be accomplished by employing frequency analysis methods using a regional perspective – one which analyzes the regional statistical characteristics of extreme precipitation events in order to obtain information about the magnitude and frequency of occurrence of major storm events on both a regional and point basis. Once the frequency characteristics of major storms are determined, a risk based approach for selecting spillway design levels can be developed.

THEORETICAL EXPECTATIONS

The first step in the regional analysis (RA) of extreme storms is to develop a theoretical foundation for the analysis. This is accomplished by developing a conceptual model of a regional precipitation gaging network for use in investigating the regional statistical characteristics of extreme storms. Some of the questions that this model will address are: What probability distribution best describes the annual series data for maximum precipitation events within a region? What are the parameters of the regional distribution? How does the regional probability distribution differ from the distribution used to describe the data from an individual precipitation station? And what confidence can be placed in the results from a regional analysis?

Before proceeding with the development of the model, it is appropriate to first discuss some of the concepts and methodologies which will be used in the analysis.

Regional Concept

For purposes of this paper, a region is defined as any geographical area in which two homogeneity conditions are met. First, there must be climatic homogeneity. All storms in the region being analyzed must have originated from the same storm type, i.e., general storm, thunderstorm, or hurricane. Second, the statistical characteristics of the precipitation data at each station must be homogeneous. This requires that the same probability distribution with a common coefficient of skewness (γ) can be used to describe each station's data. It is not necessary that the exact probability distribution be known prior to the analysis, only that the station data are Identically Distributed (ID).

Method of Analysis

The RA of extreme storms is analogous to the frequency analysis of annual maximum events at a precipitation station. Each year, the most extreme event of the region is selected for inclusion in an annual series data set for frequency analysis. In order to identify the most extreme precipitation events, the station data must first be made compatible with each other. Statistically, this requires that the data at each station be standardized. Using the convention initiated by Chow (1951); for any storm duration to be investigated each station's data (X_i) can be expressed in terms of the station mean ($\bar{x}$), station standard deviation (S_x) and a frequency factor (K_i) as:

$$X_i = \bar{x} + K_i S_x \tag{1}$$

However, after the station data are standardized and have mean = 0 and standard deviation = 1, [0,1], the station data are then comprised of a series of frequency factors. Knowing that all stations are ID, the maximum event in the region in any year is the event having the largest K value. A frequency analysis can then be conducted on the data set comprised of the annual maximum frequency factors.

Conceptual Model

Consider a homogeneous geographical region having a number of precipitation stations, n, which make up a precipitation gaging network. The spatial distribution of the stations within the region is such that the station data can be considered to be independent of each other. All storms being analyzed are of the same type, the station data are identically distributed, an infinite length record is available at each station, and all station data have been standardized to [0,1]. What probability distribution will describe the data that is obtained from selecting the maximum event each year from the n stations?

The process which has just been outlined is not unique. It is nearly identical to the formulation used by Fisher and Tippet (1928), Gumbel (1958) and others which leads to the derivation of the Extreme Value family of distributions. Drawing from their work, it can be shown that if the probability distribution describing the individual station data is Normal, Log-Normal, Exponential, Gamma, or Extreme Value Type 1, then the limiting form of the distribution which describes the maximum events from the region will be the Extreme Value Type 1 distribution (EV).

This is an important outcome because if a region is in fact homogeneous, then there exists a probability distribution (EV) toward which the regional data is converging. It still remains to be determined how quickly the regionalized data will converge to EV and what will be the parameters of the regional distribution.

These aspects of the problem can be investigated by considering several different probability distributions as possible "correct" distributions for describing precipitation data. Referring to Figure 2, a variety of [0,1] probability distributions are depicted which cover the range of distributions which have been used to describe short duration precipitation data at a station.

These initial (station) distributions were analyzed in accordance with the regionalization process, as outlined earlier, using conventional statistical methods as presented by Gumbel (1958) and by Monte Carlo simulation techniques. The results of that analysis are shown in Figures 3, 4, and 5.

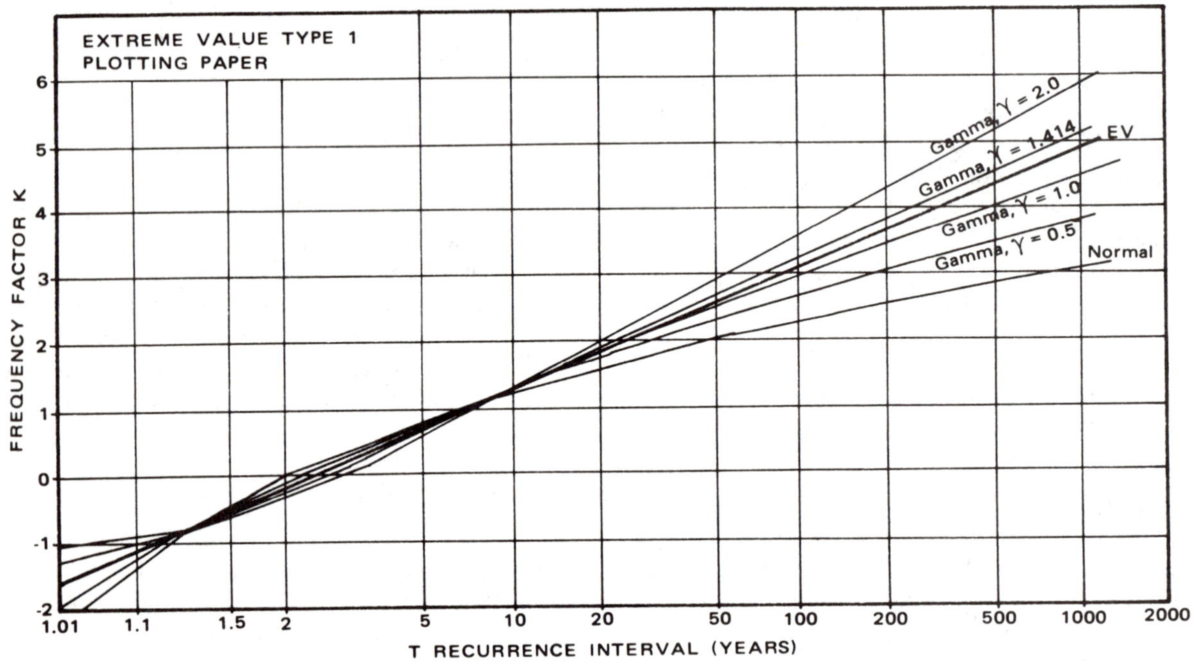

Figure 2. K v. T Relationships for [0,1] Distributions.

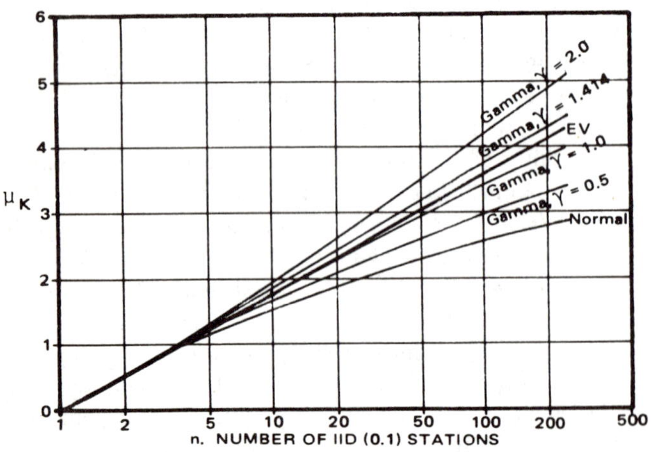

Figure 3. Variation of Regional Mean Value With Number of IID Stations.

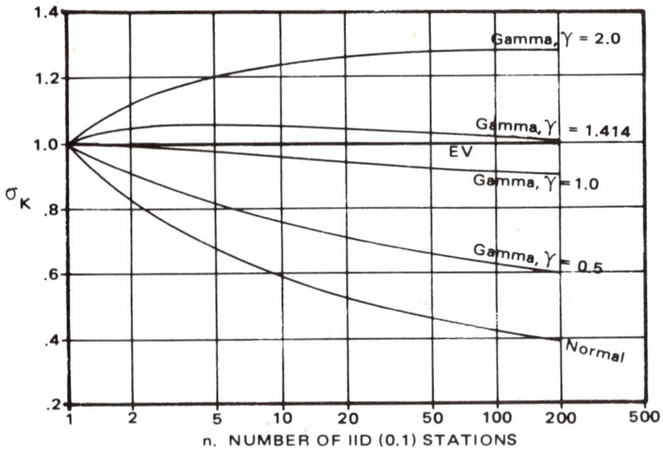

Figure 4. Variation of Regional Standard Deviation With Number of IID Stations.

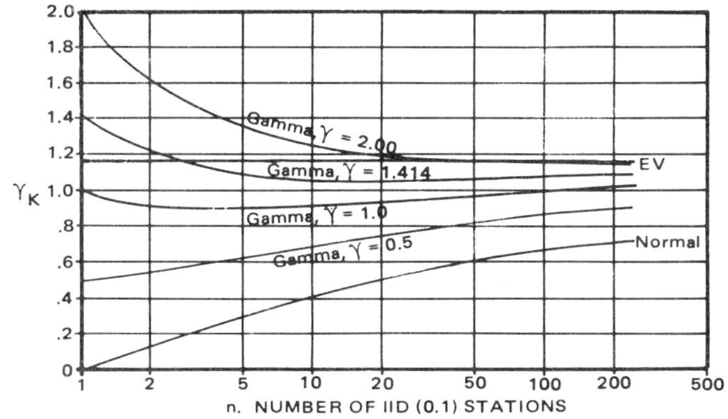

Figure 5. Variation of Regional Skew Coefficient With Number of IID Stations.

It is seen in Figure 3 that as the region increases in size and the number of independent and identically distributed (IID) stations increase, the magnitude of the population mean value (μ_k) of the regional frequency factor data increases. Thus, extreme events occur more frequently when a regional rather than a point perspective is adopted.

The next consideration in analyzing the statistical characteristics of the regionalization process is the variability of the regional data. It is seen in Figure 4 that the regional population standard deviation (σ_k) may increase or decrease in magnitude compared to its original value of unity. Its value is dependent upon the number of IID stations, that is the size of the region and the statistical characteristics of the initial distribution. In general, σ_k is greater than unity for regions where the distribution describing the station data has a skew coefficient which is greater than that for the EV which has $\gamma = 1.14$ and σ_k is less than unity for regions where the skew coefficient of the distribution describing the station data is less than 1.14.

The final aspect of the regionalization process is the convergence of the initial distribution to the EV. It is seen in Figure 5 that the regional population skew coefficient (γ_k) converges from its original value specified by the initial distribution to a value of $\gamma_k = 1.14$ corresponding to the EV distribution.

A more quantitative measure of convergence can be obtained using the Kolmogorov-Smirnov goodness of fit test. This statistical test is not used in the conventional manner, instead the test statistic, D_2, is computed for an increasing number of stations, n, to follow the convergence of the initial distribution to that of the EV. Table 1 can be used as a guide to assist in the determination of convergence. In general, regions having station distributions with skew coefficients greater than that of the EV converge rapidly, while those having skew coefficients less than the EV converge slowly

In summary, the information produced by the conceptual model provides the theoretical expectations for the behavior of the statistical characteristics of a regional analysis.

TABLE 1. Convergence Characteristics of Initial Distributions.

Initial [0,1] Distribution	Initial D_2	No. of Stations to Converge $D_2 \leq .0150$	$D_2 < .0100$
Gamma $\gamma = 2.0$	.127	6	10
Gamma $\gamma = 1.414$	.017	2	10
Ev $\gamma = 1.14$	.000	1	1
Gamma $\gamma = 1.0$	.015	1	65
Gamma $\gamma = 0.5$	.036	175	> 1,000
Normal $\gamma = 0.0$	.077	> 1,000	> 10,000

REGIONAL ANALYSIS FOR WESTERN WASHINGTON

Homogeneity Verification

The cornerstone of the regional approach is the existence of regional homogeneity. For this reason, a nonparametric homogeneity test was developed (Schaefer, 1982) to ascertain the validity of the ID assumption. The test was applied to subareas within Western Washington and to the region itself, and in all cases the ID assumption was found to be valid. The second aspect of regional homogeneity requires that all storms being analyzed are from the same storm type. The climate in Western Washington is a mid-latitude, west coast, marine type, and major storms are usually produced by "synoptic scale cyclones" (Frederick, 1981). Storm events are characteristically long in duration, often lasting several days with extensive areal coverage and moderate intensities. All major storm events are of the General Storm type. Thus, the two regional homogeneity conditions were satisfied

Results of Analysis

A regional frequency analysis was conducted for storm interdurations of 6 hrs. and 24 hrs. for the 26,000 mi^2 region of Western Washington. The method of moments was used to compute the sample statistics ($\bar{x}$, s_x,g) of the station data. The statistics s_x and g were adjusted for sample size based on the bias correction factors presented by Wallis (1974) and the station data were then standardized to [0,1]. (See Schaefer (1982) for a complete description of the methods used in data analysis.) A frequency analysis was conducted on the regional annual series frequency factor data for the period 1941-1979. The background information for the station networks and the results of the analysis are shown in Table 2. The 24-hr. data are graphically displayed in Figure 6, and the results from the 24-hr. RA will be discussed in detail.

TABLE 2. Statistical Results of Regional Analysis for Western Washington.

Storm Inter-Duration	No. of Stations	Station Years of Record	Station Density	Regional Population Estimates		
				$\hat{\mu}_k$	$\hat{\sigma}_k$	$\hat{\gamma}_k$
24 hrs.	106	3,560	1/285 mi^2	2.56	1.00	1.04
6 hrs.	60	1,750	1/580 mi^2	2.43	0.92	0.60

The first item of interest in the 24-hr. data is that the regional mean annual event has a frequency factor value of μ_k = 2.56. This corresponds to a point event with a recurrence interval of approximately 50 years (Figure 2). Further inspection of Figure 6 reveals that extreme events, i.e., large K, while rare for any given station, are quite commonplace for the region. This is an important fact to consider when one has the responsibility for the design, operation, or safety of a large number of hydraulic facilities located throughout a region.

The greatest advantage obtained from the RA is that the results may be used as a type of "tail test" to identify the initial (station) distribution. Referring to Figure 3 with $\hat{\mu}_k$ = 2.56 indicates that the existing 106 station network corresponds to a theoretically independent network of approximately 25 stations. Now, referring to Figure 4 with $\hat{\sigma}_k$ = 1.00 and n = 25 indicates that the stations' distribution was EV. Because one estimate of σ_k is not sufficiently reliable, additional estimates were obtained by conducting RA on subareas within the region and on regional station networks of various densities. The estimates for μ_k and σ_k from these studies were applied to Figures 3 and 4 as previously described and the results supported the hypothesis that $\sigma_k \simeq 1$ and that the stations' distribution was EV or very nearly EV.

One additional "test" which assists in the identification of the stations' distribution was made by computing the average skew coefficient ($\bar{g}$) from the 106 stations. It was found that $\bar{g}$ = 1.09; remembering that γ = 1.14 for the EV, this is strong supporting evidence that the stations' distribution is EV.

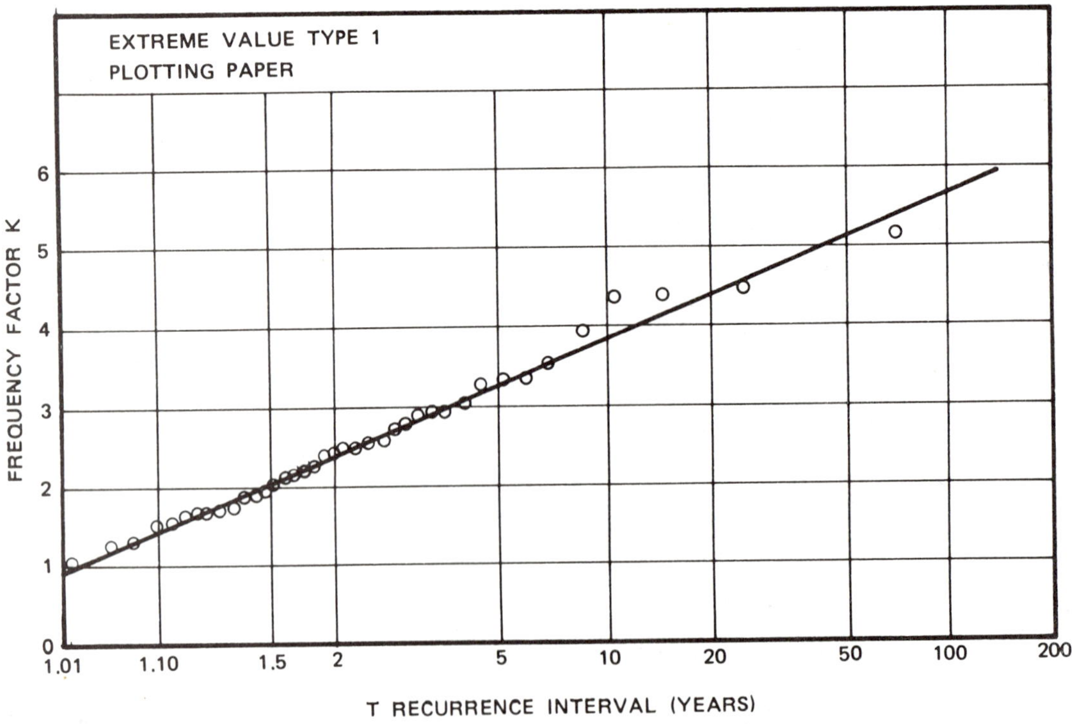

Figure 6. Frequency Plot of Regional 24-Hour Data for Western Washington.

Knowing that the stations' distribution is very close to EV, the regional distribution will have, for all practical purposes, converged to EV (Figure 5, Table 1). Furthermore, knowing the identity of the regional distribution and having reliable estimates of its parameters, much greater confidence can be placed in the prediction of regional events beyond those that have been experienced.

The implications to be drawn from the RA largely depend on one's perspective. For example, if a design level of K = 5 was chosen for a project, this would correspond to an annual exceedance probability of about 1 in a 1,000 (Figure 2) for that site. This design level may be quite acceptable to the dam owner, but in any given year that design level has a chance of 1 in 40 (Figure 6) of being exceeded somewhere within the region. That level of exposure may not be acceptable to the regulatory agency responsible for protecting public safety.

Incorporating Risk Factors

At this point, the results from the RA can be adapted into a risk approach. To be technically correct, risk should be defined as the probability of failure of the structure resulting from inadequate spillway capacity. For guideline purposes, however, it would appear reasonable to define risk as the probability of the design level being exceeded with the tacit assumption that failure would be imminent.

Regional risk would be the joint probability that a storm of a given magnitude occurs and that it hits a project which was designed for something less than the event that has occurred. Computation of the joint probability thus requires information about the current and/or future density of projects in the region and the actual and/or proposed spillway design standards for those projects. The solution of the joint probability can be obtained directly from the RA for various network densities previously discussed. Regional risk computed in this manner is an annual risk, one experienced each year that the projects exist. Computation of the risk (R) during the lives of the projects or during any interval of time could be computed using the Binominal distribution as:

$$R = 1-(1-P)^N \qquad (2)$$

where:

P = Probability of a given design level being exceeded in any year at some project in the region.

N = Number of years of exposure

SPILLWAY DESIGN GUIDELINE STRUCTURE

A spillway design guideline formulated from a regional risk approach employs a design level expressed in terms of the magnitude of the frequency factor K. The storm volume corresponding to this design level can be computed from equation 1 for each storm duration to be investigated as part of the flood analysis for spillway design. The maximum design level would be equivalent to PMP.

The guideline structure through which a design level is selected for a given dam and reservoir project and downstream hazard setting may take any form deemed appropriate by the regulatory agency developing the guidelines. Once a guideline structure is proposed, the results from a RA can be used to determine the risk for each proposed design level. A determination can then be made on whether the guideline will perform as desired.

Implementing the Design Level

The storm duration and temporal distribution of the storm volume are two critical elements in the implementation of the design level that should be part of the guidelines. Several storm durations as well as probabilistic-based hyetographs (Frederick, 1979, 1981; Huff, 1967) should be used in the flood analysis for spillway design. This will ensure that realistic storm hyetographs are used in the flood analysis and will allow the critical storm duration to be determined. This procedure will provide greater assurance that the risk associated with the design level is a good estimate of the true risk of spillway inadequacy.

SUMMARY

A method has been presented for the regional analysis of the magnitude and frequency of occurrence of major storms. It was shown that there are several distinct advantages to be gained from using a regional approach in the analysis of extreme precipitation events. With a regional approach, methods and sufficient data are available to make a reliable identification of the proper probability distribution to be used to describe regional and point (station) annual maximum precipitation data. There is also sufficient data to make reliable estimates of the parameters for the regional probability distribution. Thus, the uncertainties of probability model identification and parameter estimations are minimal in a regional approach and prediction of the extreme precipitation events used in spillway design become much more reliable.

Risk concepts have been presented which can be used to quantitize the risk that a given design level will be exceeded. Through this approach, regulatory agencies can develop spillway design guidelines which will provide a reasonably known and predeterminable level of safety.

LITERATURE CITED

Chow, V. T., 1951. A General Formula for Hydrologic Frequency Analysis. Transactions, American Geophysical Union, Vol. 32: 231-237.

Fisher, R. A. and L. H. C. Tippet, 1928. Limiting Forms of the Frequency Distribution of the Smallest and Largest Member of a Sample. Proceedings, Cambridge Society, Vol. 24:180-190.

Frederick, R. H., 1979. Interduration Precipitation Relations for Storms – Southeast States. NOAA Technical Report NWS 21, National Weather Service, Silver Springs, MD.

Frederick, R. H., J. F. Miller, F. P. Richards, and R. W. Schwerdt, 1981. Interduration Precipitation Relations for Storms – Western United States. NOAA Technical Report NWS 27, National Weather Service, Silver Spring, MD.

Gumbel, E. J., 1958. Statistics of Extremes. Columbia University Press, New York.

Hershfield, D. M., 1961. Rainfall Frequency Atlas of the United States for Duration from 20 Minutes to 24 Hours and Return Periods from 1 to 100 Years. Technical Paper No. 40, Weather Bureau, U.S. Department of Commerce, Washington, D.C.

Huff, F. A., 1967. Time Distribution of Rainfall in Heavy Storms. Water Resources Research 3, No. 4:1007-1019.

Miller, J. F., R. H. Frederick, and R. J. Tracey, 1973. Precipitation – Frequency Atlas of the Western United States. Vol. X, Washington. NOAA Atlas 2, National Weather Service, Silver Spring, MD.

Myers, V. A., J. T. Riedel, F. K. Schwarz, and R. L. Weaver, 1966. Probable Maximum Precipitation, Northwest States. Hydrometeorological Report No. 43. Weather Bureau, Washington, D.C.

Schaefer, M. G., 1982. Regional Analysis of the Magnitude and Frequency of Occurrence of Major Storms in Western Washington. To be published by Washington State Department of Ecology.

Schreiner, L. C. and J. T. Riedel, 1978. Probable Maximum Precipitation Estimates, United States East of the 105th Meridian. Hydrometerological Report No. 51. National Weather Service, Silver Spring, MD.

Wallis, J. R., N. C. Matalas, and J. R. Slack, 1974. Just A Moment. Water Resources Research, Vol. 10, No. 2:211-219.

INTERNATIONAL SYMPOSIUM ON HYDROMETEOROLOGY
JUNE AMERICAN WATER RESOURCES ASSOCIATION 1982

THE 100-YEAR 24-HOUR EVENT: FACT OR FICTION

Jerald S. Fifield[1]

ABSTRACT: For the hydrologic engineer, it is important that peak flows resulting from a 24-hour climatic event be representative of the area under investigation. An important parameter for determining peak flow (and in generating the accompanying hydrograph) from an ungaged watershed is utilizing an accurate time distribution for the design storm. Frederick, *et al.*(1981), developed a method for determining how precipitation is distributed with respect to time for various areas throughout the western United States. By comparing 100-year 24-hour flood hydrographs simulating flows from ungaged watersheds using the Frederick, *et al.* (1981), distribution to the Soil Conservation Service Type I or Type II distribution, it becomes evident that serious hydrologic errors could develop for which the hydrologic engineer would be responsible.
(KEY TERMS: precipitation; flood hydrographs; rainfall distribution.)

PROBLEM

For the hydrologic engineer, regulations that govern the design of sedimentation ponds, development of diversion structures, culvert sizing, and so forth, often require an accurate determination of flood flows from ungaged watersheds resulting from a 100-year 24-hour climatic event. As analyses are undertaken, it becomes evident that the concept of a 24-hour event is not fully understood by the engineer nor the time distribution of precipitation accurately represented. Consequently, results may be in error causing excess expenditures or creating situations where substantial damage may occur if the 100-year 24-hour event becomes a reality.

THE 24-HOUR EVENT

In parts of the nation, a single storm may have a duration lasting 24 hours. However, for the western United States a storm lasting 24 hours is not a common occurrence (except for the Pacific Northwest) and usually not of the type that produces flood flows. Fletcher, *et al.* (1981), observed in their studies of precipitation events for central Utah only 4 percent of the storms recorded had durations longer than 7 hours and rarely produced floods in the study area. However, they found that about 90 percent of the storms had a duration of 4 hours or less.

The U.S. Weather Bureau (1963) indicates that the months having maximum recorded 24-hour rainfall for the western United States are March through November except for areas west of the Sierra Nevada and Cascade Mountains. Review of climatic records further illustrate that maximum recorded point rainfall for the hours of 1, 2, 3, 6, 12, and 24 have a greater occurrence during the months of May through August when thunderstorm activity is more likely to occur. Since thunderstorm activities are usually of short duration, a 24-hour event may consist of more than one climatic event. Hence, a suggested definition for hydrologic engineers to use concerning a 24-hour climatic event during the thunderstorm season may be "the amount of precipitation that has accumulated at a point for 24 hours following the first measurable rainfall."

RAINFALL DISTRIBUTION

Perhaps more important to the hydrologist predicting flood flows from ungaged watersheds is how the distribution of rainfall will occur over the area being studied. A set of widely used distributions were developed by the Soil Conservation Service (SCS) using the U.S. Weather Bureau rainfall frequency atlas, namely Type I and Type II distribution. The Type I distribution is designed for Hawaii, Alaska, and the Coastal side of the Sierra Nevada and Cascade Mountains in California, Oregon, and Washington. The Type II distribution was intended to be used in the remainder of the United States, Puerto Rico, and the Virgin Islands. Figure 1 illustrates these distributions.

Since thunderstorm activities can result in high intensity precipitation for short durations and may create destructive flood flows, one may question whether a Type II distribution is adequate for most of the western United States. To further analyze this problem, Frederick, *et al.* (1981),

[1]President, HydroDynamics Inc., P.O. Box 1327, Parker, Colorado 80134.

evaluated maximum annual precipitation events for durations of 1, 2, 3, 6, 12, and 24 hours. Figure 2 illustrates the study area, boundaries created, and number of station years of data per region evaluated. Their analysis indicated that only in the Pacific Northwest (West) do 70 percent or more of the climatic events result in measurable precipitation in all of the 24 hours of a 24-hour event. When the time duration is reduced to 6 hours, then all regions had storms with measurable precipitation in each hour at least 40 percent of the time. Frederick also indicated that for all sections except the Pacific Northwest (West) the beginning hour of shorter duration storms or rainfall within a longer duration of rainfall or storm usually commenced within the first hour.

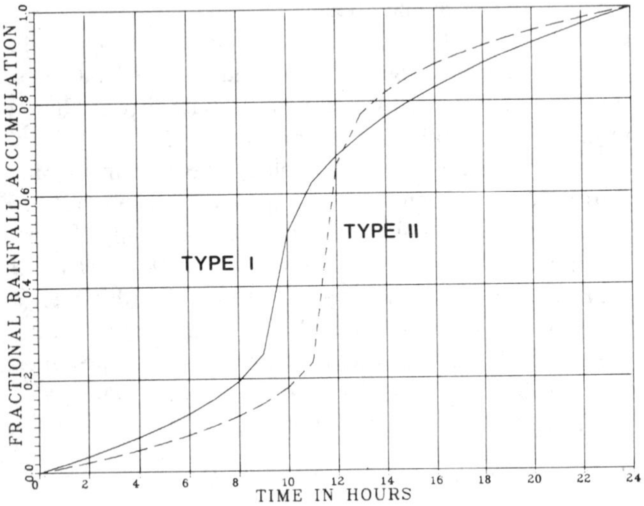

Figure 1. SCS Precipitation Distribution for the 24-Hour Synthetic Storm.

To examine the differences between large and small storms, Frederick ranked data according to magnitude and separated the data sets into quartiles where Quartile I represented the top 25 percent of the storms analyzed with regard to magnitude and Quartile IV representing the lower 25 percent of the storms. A technique was developed by Frederick such that the 100-year 24-hour precipitation event can be distributed over 24 hours. Figure 3 illustrates the 100-year 24-hour distribution for 4.0 inches of precipitation at Gillette, Wyoming, based upon Frederick's method and compared to values presented in the U.S. Weather Bureau NOAA 2 Atlas for the same time periods. Figure 3 indicates that substantial differences exist concerning the distribution of precipitation which in turn have an impact in evaluating hydrologic characteristics of flood flows from ungaged watersheds. Similar distribution differences exist for each of the regions illustrated in Figure 2. Figure 4 compares Fredericks' method to actual conditions observed for 24-hour climatic events of various stations in the Southwest Mountains section and the Great Basin section. Values plotted were chosen to represent the climatic events that may occur when severe thunderstorm activity is prevalent in the area of analysis. It should be noted that the time of year and climatic conditions contribute to the way precipitation is distributed with respect to time.

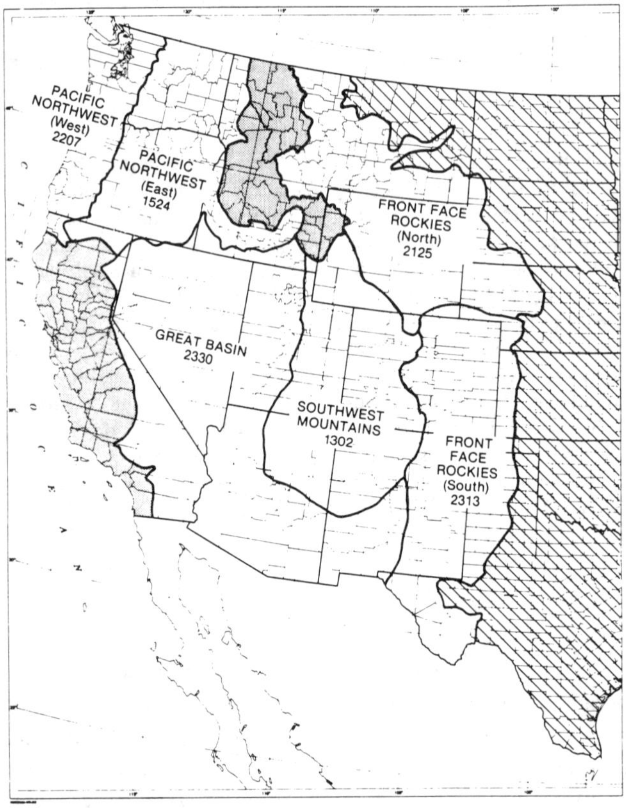

Figure 2. Study Area Showing Station Years of Data Per Region (from Frederick, *et al.*, 1981).

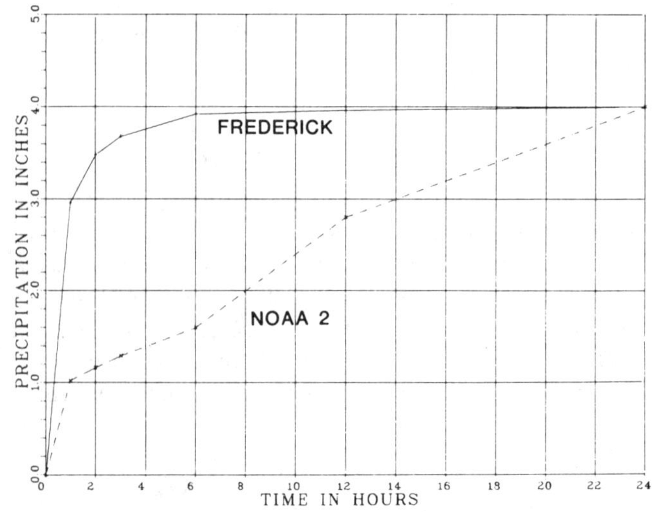

Figure 3. 100-Year 24-Hour Precipitation Distribution for Gillette, Wyoming.

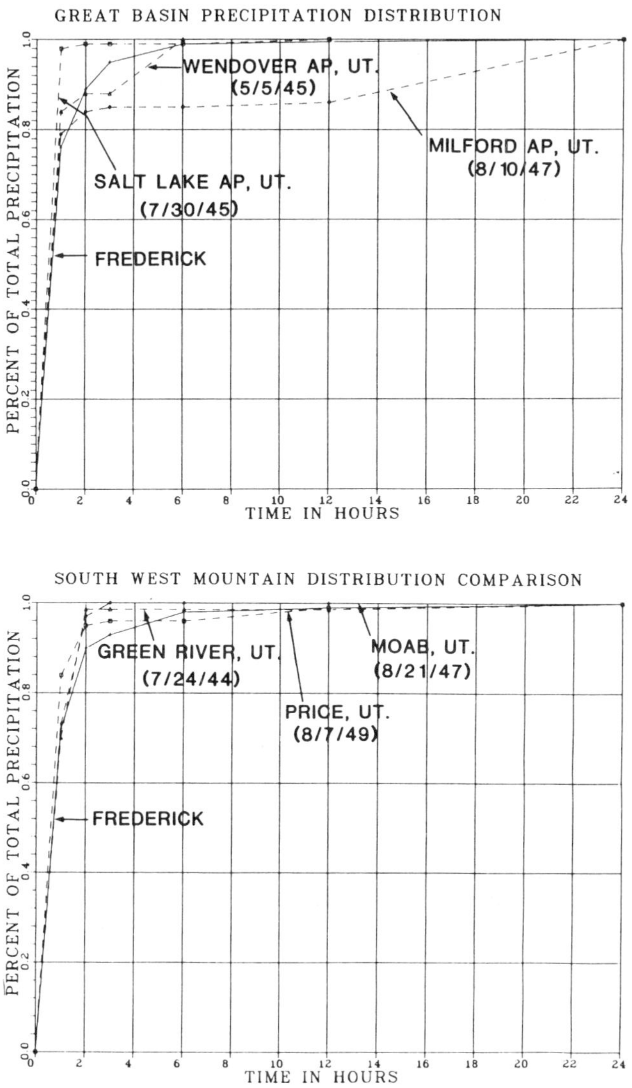

Figure 4. Precipitation Distribution for the Great Basin and Southwest Mountain Regions.

HYDROGRAPH COMPARISONS

To evaluate the technique presented by Frederick, flood flows from ungaged watersheds resulting from 100-year 24-hour events were developed for various regions of the western United States. The SCS Curve Number Method was utilized with a Type I or Type II distribution to create flood hydrographs that were compared to flood hydrographs resulting from 0.90 probability level distribution based upon Fredericks' work and assuming Quartile I storms occur. All watersheds were assumed to be 1 square mile in area, have a CN value of 86, except for the Pacific Northwest (West) where CN = 70, a drainage length of 5,000 ft. and a slope of 10 percent.

Figure 5 illustrates the difference in the hydrographs for Parker, Colorado, which is representative of the Front Face Rockies (South). Notice that the peak flow is approximately twice as large utilizing the Frederick method as compared to the Type II distribution. Figure 6 illustrates a flood hydrograph for Gillette, Wyoming, which is representative of the Front Face Rockies (North). Once again, the peak flows differ in magnitude but not as severe as Parker, Colorado. Within the Pacific Northwest (East) section, the roles are reversed with respect to peak flows depending on the distributions used. Figure 7 illustrates the flood hydrograph for a 100-year 24-hour event in Spokane, Washington, where the Type II distribution yields a higher peak flow than the Frederick method. Finally, when the Pacific Northwest (West) is analyzed (see Figure 8) with regard to the flood flows from 100-year 24-hour event, the technique developed by Frederick illustrates a substantial difference in peak value flows occurs. Figures 5 through 8 illustrate that when determining peak flood flows, if storms do not have time distributions representative of the area under investigation, substantial errors can result. Consequently, it is important that when the hydrologic engineer determines peak flood flows, a time distribution that adequately represents how precipitation falls on an area be used to ensure accuracy.

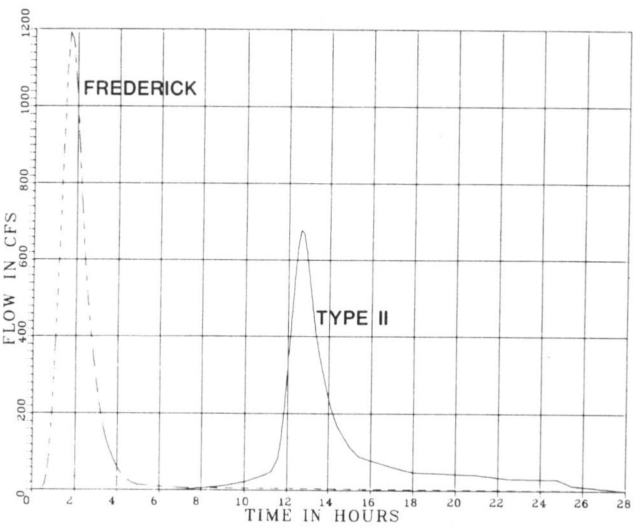

Figure 5. 100-Year 24-Hour Hydrograph for Parker, Colorado.

SUMMARY AND CONCLUSIONS

For the Western United States, the hydrologic engineer must realize that a 24-hour climatic event is the accumulation of precipitation for 24 hours after measurement of initial rainfall. Since severe flood damage from ungaged watersheds can occur during the thunderstorm season, it is essential to adequately represent the time distribution of precipitation for the area being analyzed. A time distribution of precipitation developed by Frederick can better simulate climatic conditions in the Western United States than that represented by the SCS Type I or Type II distribution assuming a 0.90 probability level distribution. As a result, synthetic hydrographs of flood flows can be generated to accurately represent hydrologic characteristics of an area.

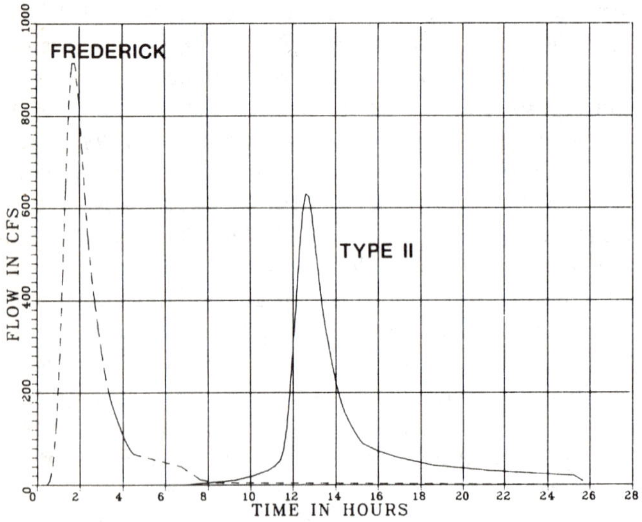

Figure 6. 100-Year 24-Hour Hydrograph for Gillette, Wyoming.

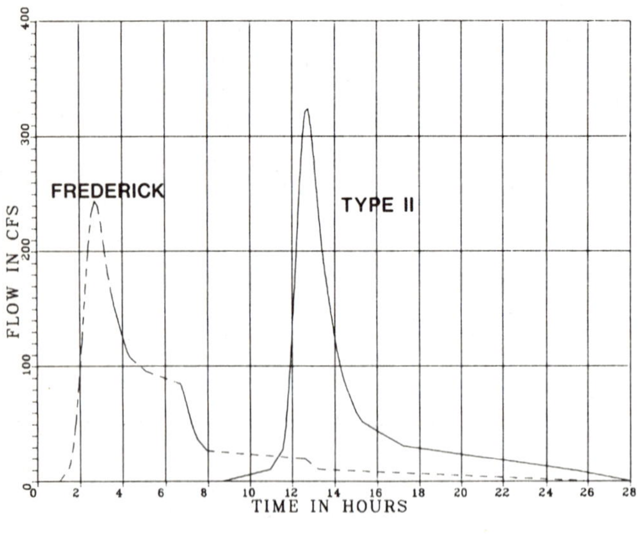

Figure 7. 100-Year 24-Hour Hydrograph for Spokane, Washington.

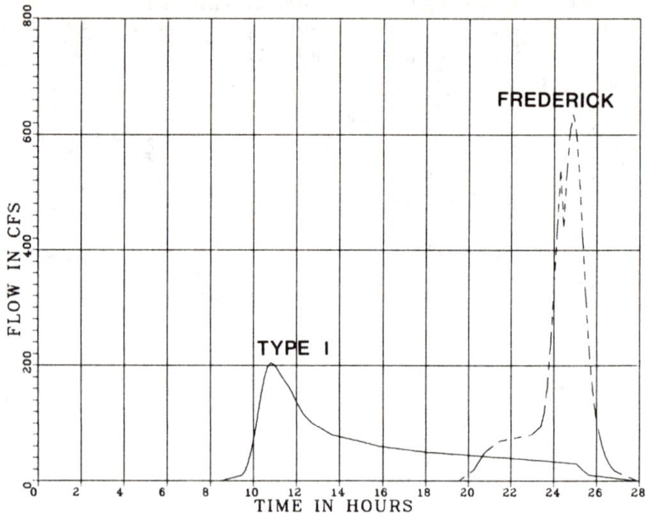

Figure 8. 100-Year 24-Hour Hydrograph for Port Angeles, Washington.

LITERATURE CITED

Fletcher, Joel E., A. L. Huber, E. E. Farner, K. R. McLaughlin, J. Rector, and L. J. Schmidt, 1981. Precipitation Characteristics of Summer Storms at Straight Canyon Barometer Watershed, Utah. U.S. Forest Service, Research Paper INT 274, U.S. Department of Agriculture, Washington, D.C., 11 pp.

Frederick, Ralph H., J. F. Miller, F. P. Richards, and R. W. Schwerdt, 1981. Interduration Precipitation Relations for Storms – Western United States. NOAA Technical Report NWS 27, U.S. Department of Commerce, Washington, D.C., 159 pp.

U.S. Weather Bureau, 1963. Maximum Recorded United States Point Rainfall for 5 Minutes to 24 Hours at 296 First-Order Stations. Technical Paper No. 2, U.S. Department of Commerce, Washington, D.C., 56 pp.

INTERNATIONAL SYMPOSIUM ON HYDROMETEOROLOGY
JUNE AMERICAN WATER RESOURCES ASSOCIATION 1982

FREQUENCY ESTIMATES OF WATER AVAILABLE FOR RUNOFF DUE TO PRECIPITATION AND/OR SNOWMELT

Frank Richards, John Miller, Edward Zurndorfer, and Normalee Foat[1]

ABSTRACT: Design of hydraulic structures requires estimates of the amount of water those structures are expected to withstand. Commonly used sources of this information includes precipitation-frequency atlases. The underlying assumption in the use of precipitation-frequency estimates is that the precipitation is immediately available for runoff. This assumption may be misleading in areas where a significant amount of the annual precipitation falls as snow. In such areas, the water stored in the snowpack as the result of a number of preceding precipitation events can be released over a relatively short period of time. For a given year, spring melt, which may also be accompanied by rain, can yield far more water than any single precipitation event (of the same duration). Water amounts due to precipitation in conjunction with snowmelt are not now routinely determined. The modification of a snow accumulation and ablation model presently used by the National Weather Service permitted the simulation of the build-up and melt-off of snowpacks for selected agricultural areas of the northwestern United States. This adaptation made use of climatological observations of daily precipitation, and maximum and minimum temperatures. Extreme 5- and 15-day totals of predicted water release from the snowpack were determined. These values were compared to the largest 5- and 15-day precipitation amounts for the remainder of the year, and the larger of the two values was used to produce an annual maximum water available for runoff series at nearly 150 stations. Maps of water available for runoff, based on frequency analysis of these series will be presented.
(KEY TERMS: frequency analysis; runoff; precipitation; snowmelt.)

INTRODUCTION

Precipitation-frequency studies are available which depict the amount of precipitation likely to occur for various durations and return periods (e.g., Hershfield, 1961; Miller, 1964; Miller, *et al.*, 1973). Such information is used for planning and design of hydrologic structures and for flood evaluation reports. This use of precipitation-frequency values assumes that the precipitation is immediately available for runoff or infiltration. There are, however, areas where a significant amount of the annual precipitation falls as snow and is accumulated in a snowpack. At some later time this stored water is released for runoff. The release will occur during warm weather and may or may not be accompanied by rainfall. Depending on the local climate, this period might come as one or more thaws during the winter, or the snowpack could continue to increase until spring. In either case, over a period of a few days, the melting snow could release greater amounts of water than any single precipitation event during that year. Since hydrologic structures must be designed to handle runoff no matter what the source, frequency of precipitation alone is inadequate for design purposes where melting snow releases water at a rate exceeding precipitation on a scale significant to the structure.

In this study the amount of water that flows out of the base of a snowpack is termed water available for runoff (WAR). This water may be due to melt alone, or melt accompanied by rain that flows through the snowpack. To account for all possible conditions, WAR also includes the amount of rain that falls if there is no snowpack. Although snowfall, snow on the ground, and water equivalent measurements are made at some stations, no WAR estimates are presently made. In an attempt to quantify WAR amounts, climatic observations of precipitation and temperature were used in conjunction with a snow accumulation and ablation model to calculate estimates of water released from snowpacks.

The area considered in this study encompasses the Palouse region, between the Cascades and the Bitterroot Mountains including parts of Washington, Oregon, and Idaho, and is part of a larger study that also included portions of northern Utah (Richards, *et al.*, 1983). WAR estimates for the agricultural areas of the Snake River Basin in southern Idaho, which lies between these two areas, were previously determined by Frederick and Tracey (1976). While infiltration rates due to varying soil type and moisture content are an important part of the total design problem, this study is limited to the meteorological aspects of snow accumulation and melt.

[1]Meteorologists, Office of Hydrology, NOAA, National Weather Service, W212, 8060 13th Street, Silver Spring, Maryland 20910.

SNOW ACCUMULATION AND ABLATION MODEL

The approach taken in this study to estimate WAR values was to model the buildup and melt-off of the seasonal snowpack using the snow accumulation and ablation model developed by Anderson (1973). This model is an energy balance model in which the relevant physical processes are parameterized on the basis of surface air temperature. The only other critical quantity necessary for our application was precipitation. The operational use of this model requires 6-hr estimates or observations of precipitation and temperature. For the purposes of this study, the present observation network reporting on a 6-hr interval is not sufficient to define the desired spatial variations. For this reason, it was necessary to use a denser network reporting daily observations of precipitation, and maximum and minimum temperatures. The maximum and minimum temperatures were used to estimate representative "daytime" and "nighttime" (i.e., 12-hr) temperatures. The model is calibrated to a given area through the determination of "melt factors" which relate the heat exchange at the air-snow interface to air temperature. The melt factors implicitly contain the effects of long- and short-wave radiation and attempt to model the annual variation of the radiation budget at the earth's surface.

The model was calibrated by determining melt factors at 30 stations where observations of precipitation, snow on the ground, and maximum and minimum air temperatures were available. The calibration process attempted to simulate a time series of water equivalent of snow on the ground that was similar to the observed time series of snow depth, and to have both series reach zero within a few days of each other. The calibration stations were representative of all geographical sections of the study area and included a wide range of elevations. The calibration factors determined at these stations were then plotted and smooth fields were drawn. Factors for all other stations in the region that had precipitation and temperature data were then read from the resulting analysis.

Figure 1 shows an example of model results for the 1954-55 snow season at Elk River, Idaho. The solid curve (left scale) shows the variation of the observed snow on the ground and the dots are the model calculated water equivalent of the snowpack (right scale). The time variation of the model output generally matches the observations, and both the observed values and the model results reached zero on April 19, 1955. The scales used imply a constant snowpack density throughout the snow season. At shorter time scales associated with individual snowfall events, the increases shown by the observed snow on the ground tend to be larger than those for the model-generated water equivalent values because the freshly fallen snow tends to be fluffy and have lower density. As the snow settles, the observed snow on the ground values decrease even when no melt or loss of water content is predicted by the model. The variation of density with time is also evident on a seasonal time scale. Figure 1 shows that the water equivalent values are generally less than the observed snow on the ground values through early March. By late March, during the spring melt, the observed snow on the ground values are consistently lower than the model water equivalent estimates, and reflect a "ripe," higher density snowpack.

The calibration process attempted to optimize the correspondence between observed data and model results over the entire snow season. Because of this, and because of the empirical nature of the parameterizations within the model, a well calibrated station could have notable differences between model results and observations at durations of a day or two. Totals over several days usually appeared to be in considerably better agreement. In general, the calibration process was most successful in matching model results to snow-on-the-ground observations for those years when a given station experienced substantial snowfall.

FREQUENCY ANALYSIS

The annual maximum WAR series based on an October-September water year for each station were fit, using the Gumbel fitting technique, to a Fisher-Tippett Type I frequency distribution (Gumbel, 1958). This distribution has been used in previous precipitation-frequency studies (Hershfield, 1961; Miller, 1964; Miller, *et al.*, 1973) and its adequacy for precipitation is discussed by Hershfield and Kohler (1962). Its use in this study facilitates comparison of results with those studies as well as with the WAR frequency estimates for the Snake River Basin (Frederick and Tracey, 1976). No attempt was made to determine whether the Fisher-Tippett distribution was optimally suited to WAR data.

It is generally accepted that with extreme rainfall events the annual series, made up of the largest event in each year, and the corresponding partial duration series, made up of the largest "N" events from an "N-year" period of record, can differ because two or more large precipitation events can occur in the same year. At stations with substantial snowpacks, extreme snowmelt events, with or without additional precipitation, are likely to occur only once during the melt season, especially for longer duration events. It was concluded that partial-duration adjustments would, in general, not be appropriate in the higher elevations. We did not attempt to determine whether any adjustment should be used for the lower elevations or how the adjustment factor might vary with elevation. Therefore, results in this study are presented in terms of annual series values.

An example of the resulting analyses for 5-day WAR estimates at the 2-yr return period is shown in Figure 2. The detail in this field is not based solely on the station data. Guidance to interpolation between stations was provided by developing relations between the frequency estimates and meteorological and topographic quantities at the

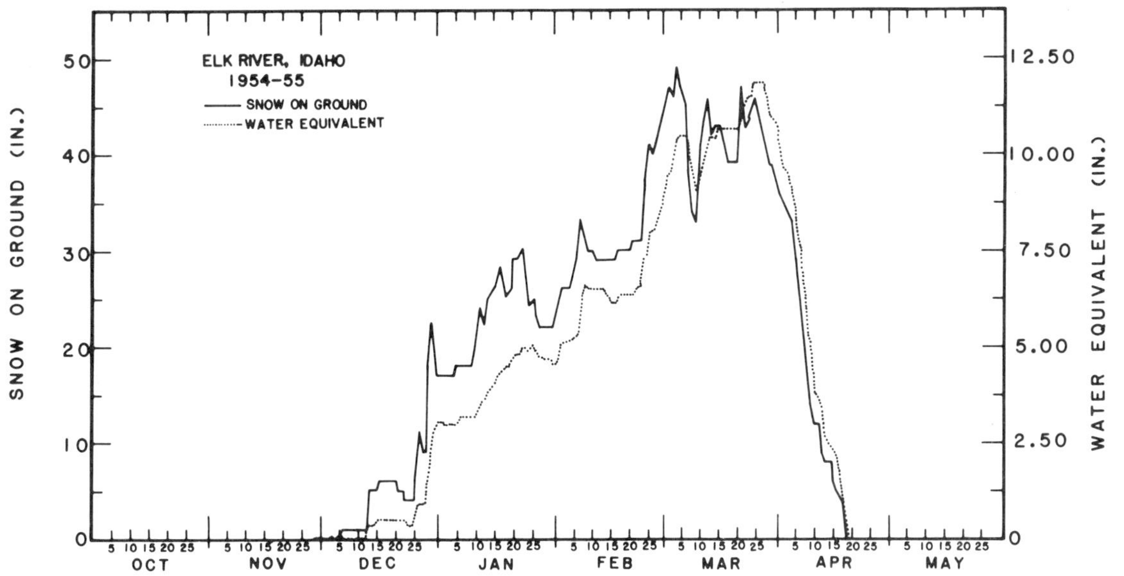

Figure 1. Time Series of Observed Snow on the Ground and Model-Generated Water Equivalent of Snow on the Ground for the 1954-55 Snow Season at Elk River, Idaho. The solid line and the left scale are snow on the ground and the dotted line and right-hand scale are water equivalent.

stations. The resulting linear equations were applied on a 5-min latitude/longitude grid to provide estimates between the stations. The approach is similar to that used by Miller, *et al.* (1973), in their development of precipitation-frequency maps for the same area.

DISCUSSION

The spatial variation of the 2-yr, 5-day WAR map (Figure 2) is typical of both the 2-yr maps for longer and shorter durations, and the 100-yr maps at all durations. There is a broad minimum in the center of the figure associated with the basin formed by the Cascades to the west and the Bitterroot Mountains to the east. The variation shown is generally related to the topography (see Figure 3 for elevations contours), with the larger values associated with the higher elevations and localized minima usually associated with valleys. In a few areas, minima occurred in the lee of major topographic barriers. A prominent example is the broad minimum just to the west of the point where the Washington, Oregon, and Idaho state lines meet. The WAR maximum to the west along the crest of the Blue Mountains suggests that the moisture supply is depleted when westerly flow, forced up the Blue Mountains, condenses and results in considerable precipitation and higher WAR amounts. Large valleys such as the Yakima Valley and the John Day River Valley are evident in the analysis. Although the WAR data indicated lower values in some of the smaller valleys, the resolution of the data was not sufficient to determine whether all such valleys were consistently associated with localized minima. Valleys, especially those in the higher elevations, may have been too limited in size to affect the larger-scale meteorological flow that produced high WAR values over broad mountain areas. An enhanced snowpack due to accumulation of snow blowing into the smaller valleys may also have increased the snowpack above that due to local precipitation. Because of these considerations, we chose not to include such small-scale detail in the analysis.

WAR values from this study were compared to precipitation-frequency estimates from NOAA Atlas 2 (Miller, *et al.*, 1973). This comparison indicated that the WAR values were generally about 10 percent larger than the 2-yr, 24-hr precipitation amounts, with some differences associated with higher elevations as large as 25 percent. The modest differences result, in part, from the fact that short-duration snowmelt processes generally cannot make water available as fast as extreme short duration rainfall rates. Hence, short duration precipitation events are more likely to produce the maximum annual WAR event than snowmelt, although these precipitation events may be enhanced by snowmelt. Also, if a given precipitation amount associated with the 24-hr annual maximum precipitation amount fell as snow, it would not necessarily be immediately available for runoff, leading to a lower annual maximum WAR amount.

Differences between precipitation and WAR frequency estimates at longer durations could be expected to be larger because the time scale associated with extreme snowmelt events is probably longer than 24 hr. To evaluate these differences we determined 2-yr, 5-day precipitation amounts using an approach similar to that used by Miller (1964). Miller estimated 2-yr, 2- to 10-day precipitation-frequency

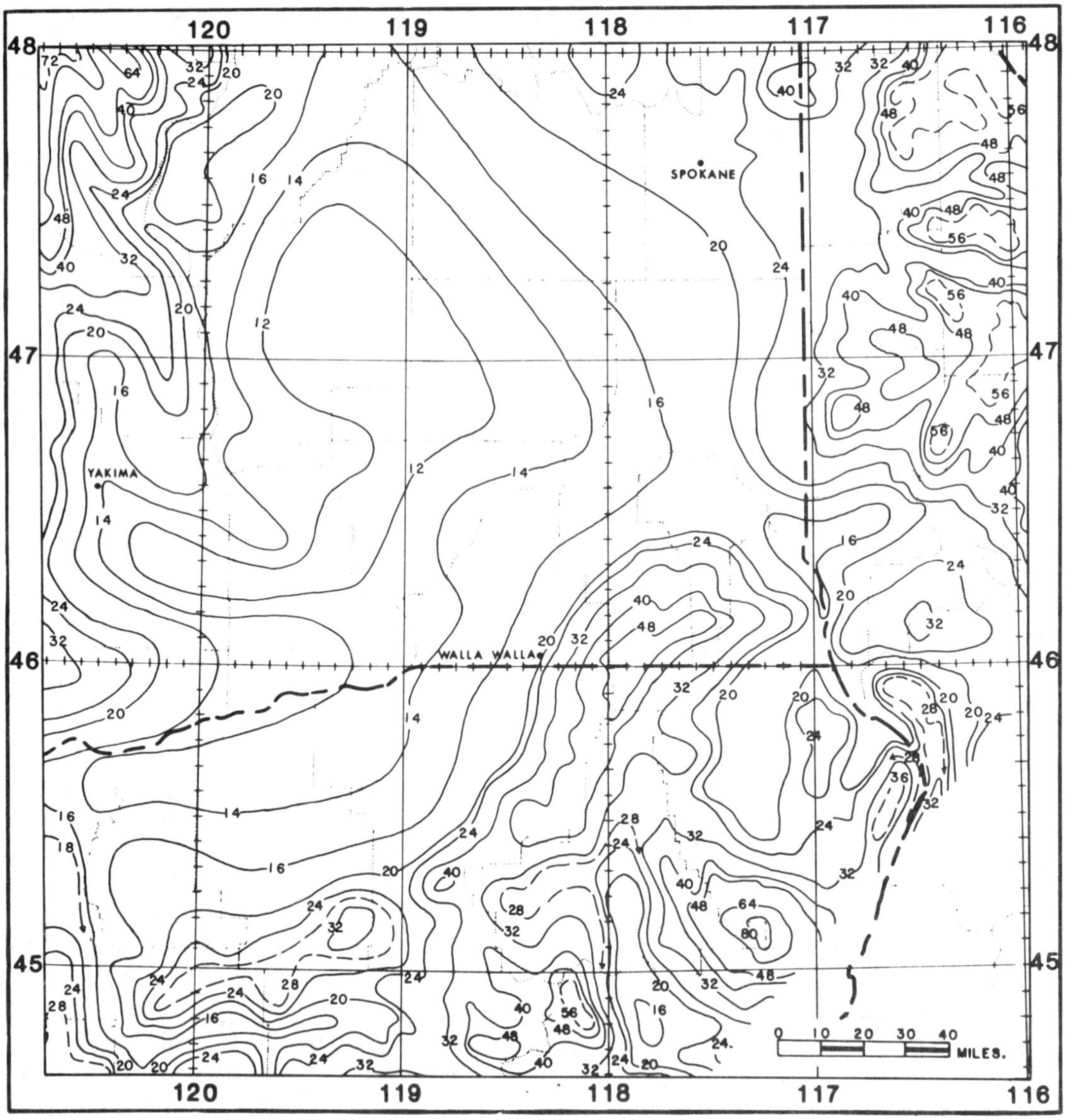

Figure 2. Two-Year, 5-Day Water Available for Runoff Amounts.

values by using a set of index relationships based on latitude, and both the 2-yr, 1- and 24-hr precipitation-frequency values. Miller's results were not used for comparison with the WAR values because NOAA Atlas 2 provided revised 2-yr, 1- and 24-hr values. We applied the method outlined by Miller to the revised 2-yr, 1- and 24-hr values to obtain "improved" 2-yr, 5-day precipitation-frequency values.

The results of the comparison between these precipitation-frequency values and the corresponding 2-yr, 5-day WAR values are presented in Figure 3. While the general pattern is similar to the differences between the 2-yr, 24-hr NOAA Atlas 2 and WAR frequency values (not shown), the magnitude of the differences is considerably larger. This figure highlights the relationship of the differences to elevation, which is also shown. Smaller differences are generally associated with major valleys and in the broad basin between the Cascades and the Bitterroots. The largest differences are associated with the higher elevations where deeper snowpacks generally occur. Some localized values associated with the 7,000 foot elevation contours had WAR values twice as large as the precipitation estimates. Other local maxima, as well as localized minimz associated with small valleys also occur but are not shown by the analysis. The main purpose of this figure was to focus on the broad-scale differences between the precipitation and WAR estimates.

A final comparison was made to isolate effects due entirely to the physical processes embodied in the snowmelt model. At 93 stations within the area shown in Figure 3, 1- to 15-day annual maximum precipitation series were determined, and frequency values estimated. The stations were stratified into three groups: stations located in those

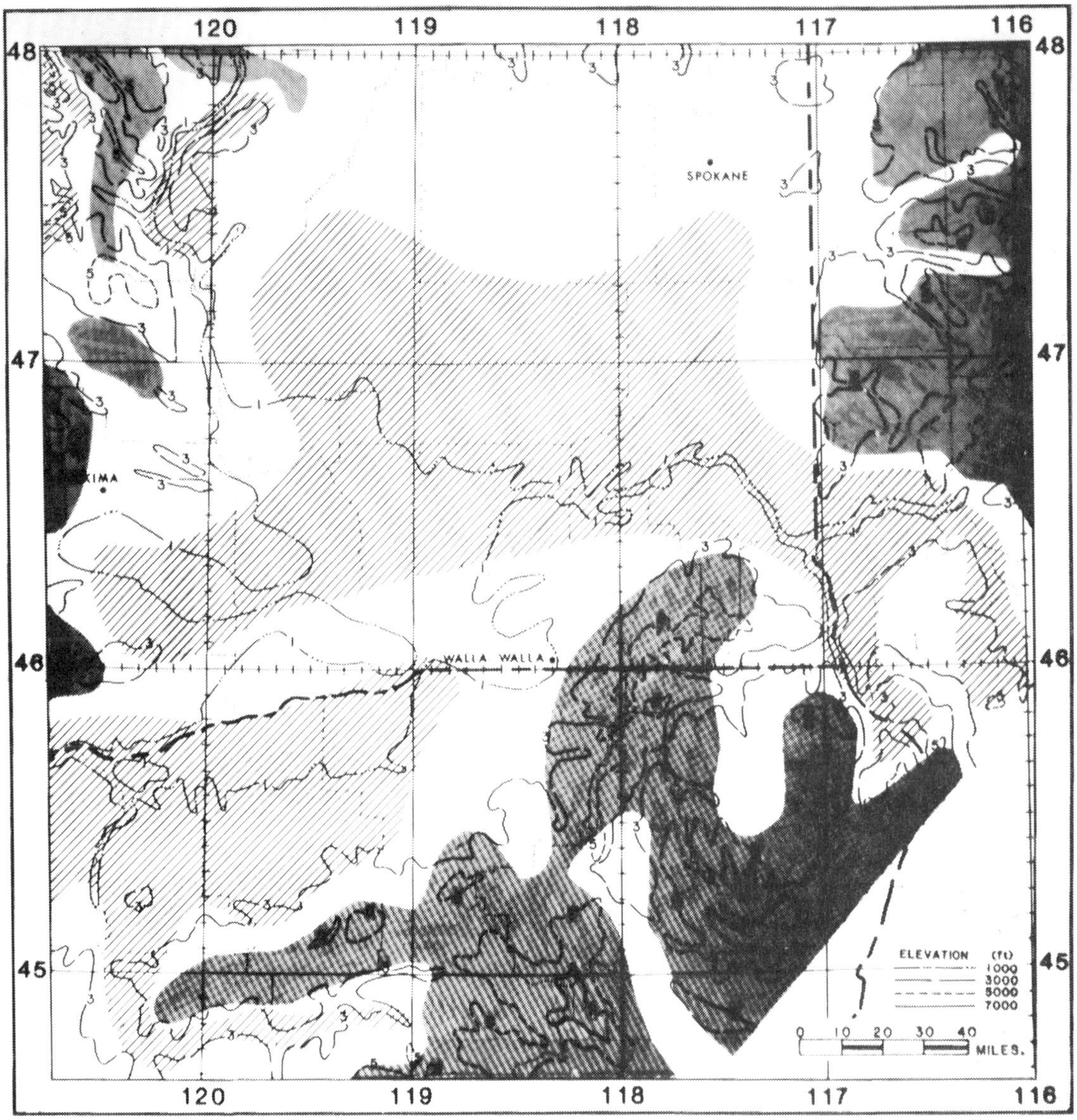

Figure 3. Smoothed Elevation Field for the Study Area and Regions Where the 2-Yr, 5-Day WAR Amounts are Larger Than the Corresponding 2-Yr, 5-Day Precipitation Amounts, (1) By Less Than 25 Percent (Hatched Areas), (2) Between 25 and 50 Percent, and (3) More Than 50 Percent (Shaded Areas).

regions of Figure 3 where the precipitation and WAR estimates were: (1) less than 25 percent (hatched areas), (2) between 25 percent and 50 percent, and (3) greater than 50 percent (shaded areas). At all stations within each group, the 2-yr, 1- to 15-day precipitation and WAR values were normalized by the 2-yr, 15-day WAR value. Similarly, all 100-yr values (both precipitation and WAR) were normalized by the 100-yr, 15-day WAR value. The normalized values for all stations within each group were averaged and the results plotted as a function of duration, as shown in Figure 4. At both the 2- and 100-yr return periods the WAR values exceed the precipitation values at all durations in all three groups. The differences increase with duration, and the differences for the 100-yr values are consistently larger than those for the 2-yr values for all three groups.

SUMMARY AND CONCLUSIONS

Climatic precipitation and temperature data were used as input to a snow accumulation and ablation model that yielded annual maximum WAR series. These data were fit to a Fisher-Tippett Type I frequency distribution and the resulting values were used to produce maps of WAR-frequency estimates. The broad features of the spatial variation of the WAR field were qualitatively similar to previously published precipitation-frequency maps at the shorter durations, with the WAR values generally about 10 percent larger for the 24-hr duration. As the duration increased, the WAR values became progressively larger than corresponding precipitation values. The increase was greatest in higher elevations where snowpack accumulation and spring snowmelt were considerable. At durations in excess of five days,

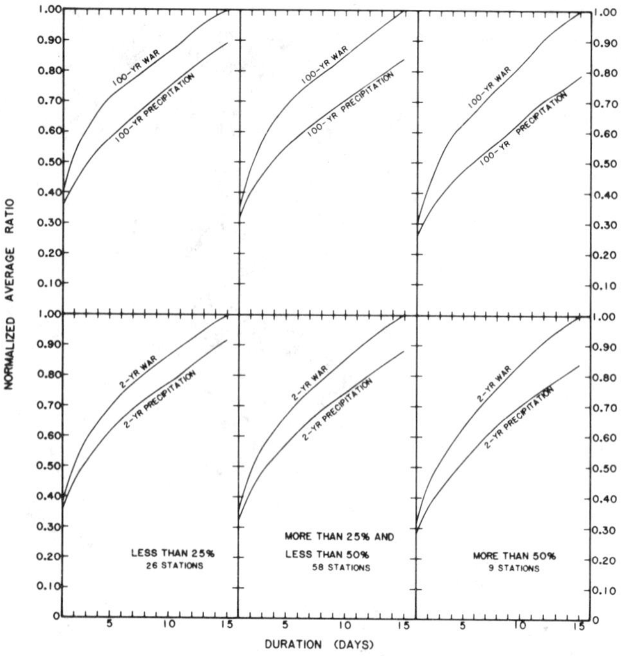

Figure 4. Relationship Between WAR and Precipitation as a Function of Duration. See text for details.

there were broad higher elevation regions where the WAR-frequency values were more than 50 percent greater than corresponding precipitation-frequency values.

ACKNOWLEDGMENTS

Sponsorship and financial support for this project was provided by the Soil Conservation Service, U.S. Department of Agriculture.

LITERATURE CITED

Anderson, E. A., 1973. National Weather Service River Forecast System – Snow Accumulation and Ablation Model. NOAA Tech Memo. NWS HYDRO-17, National Weather Service, Silver Spring, MD.

Fredrick, R. H. and R. J. Tracey, 1976. Water Available for Runoff for 4 to 15 Days Duration in the Snake River Basin in Idaho. NOAA Tech Memo. NWS HYDRO-29, National Weather Service, Silver Spring, MD.

Gumbel, E. J., 1958. Statistics of Extremes. Columbia University Press, New York, NY.

Hershfield, D. M., 1961. Rainfall Frequency Atlas of the United States for Durations from 30 Minutes to 24 Hours and Return Periods from 1 to 100 Years. U.S. Weather Bureau Tech. Paper No. 40, Washington, D.C.

Hershfield, D. M. and M. A. Kohler, 1962. An Empirical Comparison of the Predictive Value of Three Extreme-Value Procedures. J. Geophys. Res., 67(4):1535-1542.

Miller, J. F., 1964. Two- to Ten-Day Precipitation for Return Periods of 2 to 100 Years in the Contiguous United States. U.S. Weather Bureau Tech. Paper No. 49, Washington, D.C.

Miller, J. F., R. H. Frederick, and R. J. Tracey, 1973. Precipitation Frequency Atlas of the Western United States. NOAA Atlas 2, Vols. I-XI, National Weather Service, Silver Spring, MD.

Richards, F., J. F. Miller, E. A. Zurndorfer, and N. S. Foat, 1983. Water Available for Runoff for Durations up to 15 days for Selected Agricultural Regions in the Northwest United States. NOAA Tech. Report, in preparation.

FLOOD FREQUENCY ANALYSIS WITH DETECTION AND MODIFICATION OF OUTLIERS

Krishan P. Singh and Masahiro Nakashima[1]

ABSTRACT: An outlier in a set of data is an observation which appears to be inconsistent with the remainder of that set of data. In flood frequency analysis, such an outlier significantly affects or distorts the distribution and hence the design flood estimates. Therefore, it is desirable to detect and modify outliers before conducting flood frequency analysis. Test statistics have been developed by extensive Monte Carlo experiments for detecting and modifying outliers in an observed flood series. Methods of transforming flood series into quasi-normal series were investigated, and power transformation was found to be the most suitable. A new flood frequency methodology has been developed to: (1) transform a flood series into a quasi-normal series; (2) perform statistical tests on the series to detect outliers at various probability levels; (3) modify outliers, if any, and transform the modified series back to a flood series; and (4) estimate design floods for various recurrence intervals using normal, log-Pearson type III, and mixed (mixture of two normals) distributions. The flood frequency analysis with the mixed distribution, coupled with objective detection and modification of outliers, provided very satisfactory results for about 200 flood series analyzed.
(KEY TERMS: flood frequency; outliers; statistical tests; power transformation; two distribution; Monte Carlo methods.)

INTRODUCTION

Prerequisites for deriving satisfactory estimates of design floods are the objective detection and modification of outliers/inliers at desired levels of significance and a versatile technique of flood frequency analysis. An outlier in a set of data is defined as an observation or subset of observations which appear to be inconsistent with the remainder of that set of data (Barnett and Lewis, 1978). The inconsistency can be interpreted as the observation being either significantly higher at the high end and lower at the low end (or lower at the high end and higher at the low end) than the value indicated by the rest of the data; the observation is termed as an outlier or inlier, respectively. A literature search did not show any statistical tests for checking outliers (at higher than 5 percent level) and inliers at any level of their occurrence. Tests are also not available for multiple outliers and inliers. These tests were developed from normally-distributed random sequences: the normal distribution has only two parameters – mean μ and variance σ^2.

The observed flood distribution needs to be transformed to a series distributed as $N(\mu, \sigma^2)$ for application of statistical tests to detect outliers and inliers. Three transformations were tested extensively to determine the most suitable transformation. A preliminary analysis of 200 observed flood series showed that they did not conform satisfactorily to the conventionally used distributions such as normal, lognormal, and log-Pearson type III. These distributions were fitted much better by the two-distribution method (Singh and Sinclair, 1972). A new flood frequency methodology was developed to detect and modify any outliers and inliers at various probability levels as well as to compute various frequency floods with power transformation, log-Pearson type III, and two-distribution methods.

STATISTICAL TESTS FOR OUTLIERS AND INLIERS

Four methods or algorithms for generating normally distributed random numbers were tested extensively regarding their suitability, stability, and effectiveness in generating such numbers. The polar method developed by Box, Muller, and Marsaglia (Knuth, 1969) was found better than the other three methods in generating random numbers with statistical attributes closely resembling samples drawn from a population distributed as N(0, 1). The other three methods tested were a method of Box and Muller (1958), inverse normal function, and central limit theorem.

Departure (a test statistic for detecting outliers/inliers) is defined here as the standard normal deviate corresponding to the plotting position of the high or low point of the series under consideration, minus the sample standard deviate for that point. The plotting position (Cunnane, 1978) is obtained from the expression, $p = (m-\alpha)/(n + 1-2\alpha)$, in which α is a shift parameter, m is the rank,

[1]Respectively, Principal Hydrologist, Illinois State Water Survey, 605 East Springfield, Champaign, Illinois 61820; and Graduate Research Assistant, Illinois State Water Survey, 605 East Springfield, Champaign, Illinois 61820.

and n is the sample size. One thousand samples were generated for each size n of 10, 15, 20, 25, 30, 40, 50, 60, 75, and 100, and these were ranked in an ascending order of magnitude. The values of the five lowest and five highest deviates from 1000 samples of size n were stored in 10 series, each series of 1000 corresponding to a high or low value. Each of these series was standardized and departures were determined. The proper value of α was obtained by generating departure series with α = 0.00, 0.25, 0.32, 0.38, 0.43, and 0.50. The α value which makes the mean of the computed departure series equal to zero was used. This value varied from 0.40 to 0.50 depending on the sample size and the rank of the outlier or inlier. One thousand samples each of the 10 sizes were generated 40 times to obtain the departure distributions of outliers and inliers at both high and low ends of the flood spectrum. These departure distributions are shown in Figure 1; the numbers 1 to 5 represent the rank (lowest event is 1, second lowest is 2, and so on; highest event is 1, second highest is 2, and so on). The departures for a single outlier at the 1 percent and 5 percent level check exactly with those given by Barnett and Lewis (1978) for the range of n from 20 to 100.

METHODS OF NORMALIZING DATA

Three methods of transforming an observed flood series to a series distributed as $N(\mu,\sigma^2)$, where μ is the mean and σ^2 is the variance, were tested extensively on flood series observed at 28 gaging stations in Illinois. The methods are: power transformation, Wilson-Hilferty (1931) transformation, and 3-parameter lognormal transformation. The power transformation was judged the best for converting an observed flood series to a quasi-normally distributed series. The pertinent reasons are: 1) power transformed series are more stable and consistent even when some outliers/inliers are present, and 2) overall results obtained from the flood series analyzed with the power transformation are much better than from the Wilson-Hilferty transformation. The 3-parameter lognormal transformation is unsuitable for general use.

Box and Cox (1964) suggested power transformation for normalizing a data set:

$$y_i = \frac{Q_i^\lambda - 1}{\lambda}, \lambda \neq 0; y_i = \log Q_i, \lambda = 0$$

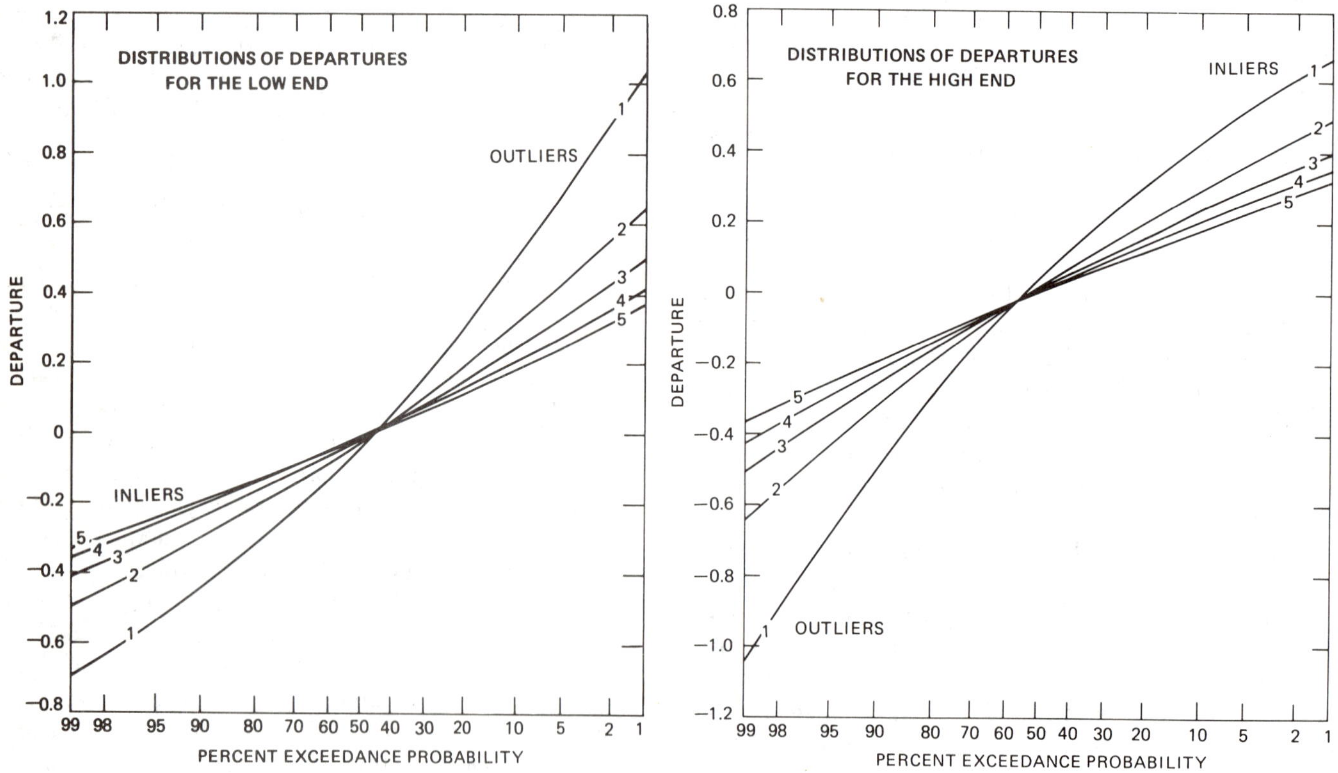

Figure 1. Distributions of Departures for the Low and High End.

in which Q_i is the annual flood from a sample of size n, λ is a constant of transformation, and i = 1, 2, . . ., n. Maximum log-likelihood (ML) estimator of λ, when $\lambda \neq 0$, can be obtained (Singh, 1980) from

$$L_{max}(\lambda) = -0.5 \log \sigma_y^2(\lambda) + \log J(\lambda;Q)$$

and

$$\log J(\lambda;Q) = (\lambda-1) \sum_{i=1}^{n} \log Q_i$$

A computer algorithm was developed for computing $L_{max}(\lambda)$.

The maximum log-likelihood estimators of λ were used in the flood frequency methodology and the computer program. The y_i transformed series has a skew very close to zero but the kurtosis may not equal 3 as for a normal distribution. Symmetric distributions with kurtosis less than 3 have flatter tops and those with kurtosis greater than 3 have sharper tops than a normal distribution. Correction factors were developed for adjusting the standard deviates corresponding to various recurrence intervals for the kurtosis computed for the y_i series.

NEW FLOOD FREQUENCY METHODOLOGY

A new flood frequency methodology has been developed and computerized. It objectively detects the outliers and inliers at various probability levels and modifies them if necessary. The computer program prints 2- to 1000-year floods from power-transformed series as a symmetric distribution, both with and without kurtosis correction; from log-Pearson type III distribution, both with sample skew and weighted skew; and from the mixed distribution, for levels of 0, 1, 2, 3, 4, 5, and 6. Levels 1 through 6 correspond to outlier-inlier probability pairs of 0.01, 0.99; 0.05, 0.95; 0.10, 0.90; 0.20, 0.80; 0.30, 0.70; and 0.40, 0.60. The level 0 involves no detection or modification of outliers/inliers. The relevant information on statistics of the three methods and the given and modified values of outliers/inliers are also printed at all the levels.

The mixed distribution concept considers the observed floods (or their logarithms) to belong to two populations with means μ_1 and μ_2, variances of σ_1^2 and σ_2^2, and relative weights of α and $1-\alpha$. The mixed distribution method developed from various studies is based on the following equations:

$$p_1\{x\} = \frac{1}{\sigma_1\sqrt{2\pi}} \int_{-\infty}^{x} \exp - [\frac{(x'-\mu_1)^2}{2\sigma_1^2}]\, dx'$$

$$p_1\{x\} = \frac{1}{\sigma_2\sqrt{2\pi}} \int_{-\infty}^{x} \exp - [\frac{(x'-\mu_2)^2}{2\sigma_2^2}]\, dx'$$

$$p\{x\} = \alpha p_1\{x\} + (1-\alpha)\, p_2\{x\}\,; 0 \leqslant \alpha \leqslant 1$$

in which p is the probability of being equal to or less than x, and x = log Q. The distribution parameters were obtained with a nonlinear programming algorithm (Singh and Nakashima, 1981). The asymmetry of the power-transformed or log-transformed flood series, as evidenced by the 3rd and/or the 5th moment being significantly different from zero, is accommodated easily by the mixed distribution concept.

SOME EXAMPLES

The new flood frequency methodology is applied to the observed annual flood series for the Sangamon River at Oakford, Illinois (USGS No. 05 583000, drainage area 5093 square miles, and n = 62 years). The number of high or low floods considered for outlier/inlier detection and modification is given by [n/10] with a maximum of 5. The five lowest and five highest observed floods are shown under level 0 in Table 1. The observed highest flood is about 2.2 times the second highest flood. According to Singh (1980), the flood of 123,000 cfs which occurred on May 20, 1943, was caused by a 50-year storm, covering most of the basin, over very wet antecedent soil moisture conditions, giving a runoff factor of about 2.2 times that for the next four high floods caused by 10- to 25-year storms. The detection and modification of any outliers and inliers from one level to the other is shown in Table 1. For the low floods, the lowest flood L1 is detected as an inlier at levels 5 and 6 and the 5th lowest flood L5 is detected as an outlier at levels 4 through 6. For the high floods, the H1 is detected as an outlier at level 2 (corresponding to 0.05 probability level) through 6 (corresponding to 0.40 probability level). The other four high floods are detected as inliers starting from level 1 or 2 through 6. An evaluation of flood frequency analyses for 200 flood series has indicated that level 5 is sufficient for satisfactory elimination of the effects of outliers and inliers on the estimated floods.

The 100-year floods with: 1) y's as normally distributed, 2) log-Pearson type III or LP3 distribution, and 3) mixed distribution are given in Table 1. The weighted skew for LP3 is obtained with the regional skew of –0.4. The relevant statistics for y's and log Q's are:

TABLE 1. Sample Computer Output with the New Methodology

Station No. 05 583000 Sangamon River at Oxford
Drainage Area 5093 Sq. Mi. – Years of Record 62 (1910-1979)

Method		Level No. 0	1	2	3	4	5	6
					100-Year Flood in cfs			
Normal, y's								
With kt = 3.0		82,029	82,498	78,988	75,264	72,347	70,357	69,059
With sample kt		89,513	89,814	83,010	76,541	71,464	68,673	67,080
Log Transform								
LP3, sample skew		78,851	79,420	77,082	74,442	72,444	70,037	67,484
LP3, weighted skew		82,331	82,803	81,409	79,623	78,185	77,061	76,063
Mixed distrib., MD		89,279	89,772	82,699	75,229	69,725	69,535	70,023
Type	**No.**				**Observed and Modified Floods in cfs**			
Low	1*	3,480	3,480	3,480	3,480	3,480	2,927	2,353
	2*	3,800	3,800	3,800	3,800	3,800	3,800	3,800
	3*	4,630	4,630	4,630	4,630	4,630	4,630	4,748
	4*	5,670	5,670	5,670	5,670	5,670	5,670	5,673
	5*	5,960	5,960	5,960	5,960	6,142	6,448	6,654
High	5*	44,700	44,700	46,285	46,909	47,461	47,649	47,927
	4*	45,800	46,403	48,920	49,564	50,127	50,322	50,591
	3*	46,300	49,331	52,305	52,933	53,542	53,703	53,999
	2*	55,900	55,900	56,753	57,476	58,223	58,451	58,810
	1*	123,000	123,000	106,954	92,964	81,678	75,145	70,771
					Flood in cfs for Recurrence Intervals (Years)			
Method	**Level**	**2**	**10**	**25**	**50**	**100**	**500**	**1000**
y's, kt=3.0	0	21,738	47,712	61,422	71,717	82,029	106,247	116,843
sample kt		21,738	46,495	62,345	75,439	89,513	126,090	143,606
LP3, sample skew		21,857	48,001	60,903	70,090	78,851	97,730	105,279
weighted skew		21,649	48,439	62,293	72,434	82,331	104,456	113,630
MD, mixed dist.		23,411	43,791	57,244	71,571	89,279	140,073	166,295
y's, kt=3.0	5	22,484	45,139	55,687	63,169	70,357	86,256	92,855
sample kt		22,484	45,404	55,380	62,230	68,673	82,580	88,212
LP3, sample skew		22,271	46,392	56,882	63,817	70,037	82,195	86,602
weighted skew		21,807	47,424	59,895	68,707	77,061	94,894	101,959
MD, mixed dist.		23,451	44,336	54,384	61,915	69,535	88,063	96,621

*High and low floods considered for outlier detection and modification, kt is kurtosis.

	Level 0			Level 5		
	Skew	**Kurtosis**	**5th**	**Skew**	**Kurtosis**	**5th**
y's	0.018	3.820	3.900	−0.055	2.792	0.416
log Q's	−0.562	3.421	−3.027	−0.764	3.317	−5.653

The asymmetry as represented by the absolute value of the 5th moment not being close to zero is rather small for the y's at level 5. Very small values of skews for y's underscore the utility of the power transformation.

The 100-year floods (with the exception of LP3 with weighted skew) are very close to each other at level 5 but they are close to each other only with the normal y's with kurtosis correction and the mixed distribution, MD, at level 0. The relative values of 2- to 1000-year floods with all the methods are shown in Table 1 for levels 0 and 5. Observed and modified floods and the fitted mixed distribution curve (level or window 5) for the Sangamon River near Oakford are shown in Figure 2.

Another example is the Sangamon River near Oakley, Illinois (USGS No. 05 572500, drainage area 774 square miles, and n = 27 years). The highest observed flood of 16,000 cfs is detected as an inlier at level or window 4. It is modified to 17,766 cfs in window 4 and to 20,085 cfs in

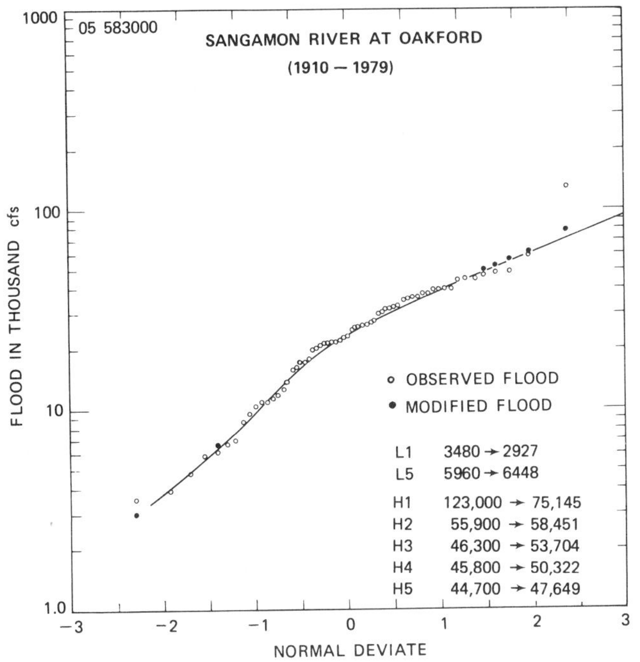

Figure 2. Observed and Modified Floods and the Fitted Mixed Distribution Curve (Window 5) for the Sangamon River at Oakford.

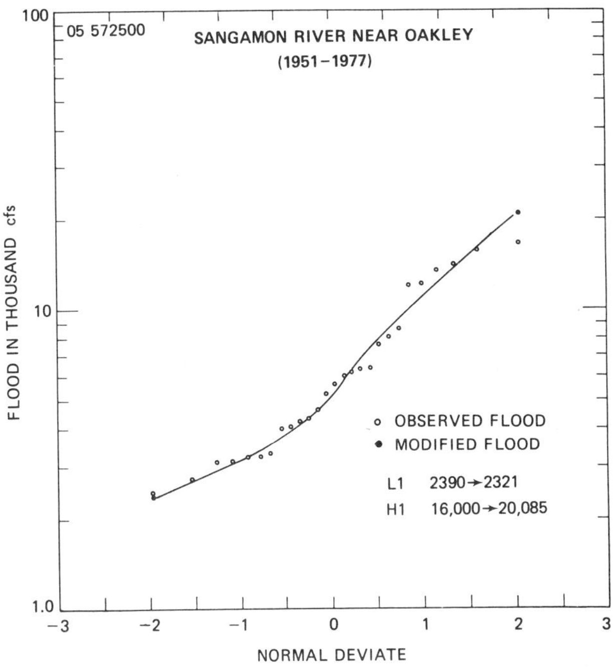

Figure 3. Observed and Modified Floods and the Fitted Mixed Distribution Curve (Window 5) for the Sangamon River Near Oakley.

window 5. The lowest flood of 2390 cfs is detected as an inlier at level 5 and modified to 2321 cfs. The observed and modified floods and the fitted mixed distribution curve are shown in Figure 3. The 100-, 500-, and 1000-year floods with the MD method for levels 0 and 5 are:

	Floods, cfs		
	100-Year	**500-Year**	**1000-Year**
Level 0	21,888	28,984	32,290
Level 5	24,557	33,731	38,116

The detection and modification of a significant high inlier increases flood estimates, whereas the detection and modification of a significant high outlier decreases flood estimates. The flood estimates derived with the MD and level or window 5 exhibited satisfactory regional relations.

CONCLUSIONS

Satisfactory tests for detecting single and multiple outliers and inliers in the two extreme tails of a suitably transformed flood series at various levels of significance, have been developed. Extensive testing of various transformations for converting observed flood series to quasi-normal series indicates the power transformation to be the best for use. A new methodology for flood frequency has been conceptualized, developed, and computerized. It detects and modifies outliers/inliers at different levels and computes various frequency floods at these levels with normal y's, log-Pearson type III distribution, and the mixed distribution method. The kurtosis correction with the power transformation method is reasonably valid if the 3rd and higher odd moments are close to zero. However, the mixed distribution method covers a wide range of asymmetries and kurtosis values.

LITERATURE CITED

Barnett, V. and T. Lewis, 1978. Outliers in Statistical Data. John Wiley and Sons, New York, 365 pp.

Box, G. E. P. and D. R. Cox, 1964. An Analysis of Transformation. Journal of Royal Statistical Society, B26:211-252.

Box, G. E. P. and E. M. Muller, 1958. A Note on the Generation of Random Normal Deviates. Annals of Mathematical Statistics, 29:610-611.

Cunnane, J., 1978. Unbiased Plotting Position – A Review. Journal of Hydrology, 37:205-222.

Knuth, D. E., 1969. The Art of Computer Programming. Addison-Wesley, Reading, Massachusetts, 105-113.

Singh, K. P. and R. A. Sinclair, 1972. Two-Distribution Method for Flood Frequency Analysis. Journal of Hydraulics Division, ASCE, 98-HY1:29-44.

Singh, K. P., 1980. Flood Frequency Analysis by Power Transformation – Discussion. Journal of Hydraulics Division, ASCE, 106-HY3:462-465.

Singh, K. P. and M. Nakashima, 1981. A New Methodology for Flood Frequency Analysis with Objective Detection and Modification of Outliers/Inliers. Illinois State Water Survey Contract Report 272, 145 pp.

Wilson, E. B. and M. M. Hilferty, 1931. Distribution of Chi Squares. Proceedings of National Academy of Sciences, 17(12):684-688.

INTERNATIONAL SYMPOSIUM ON HYDROMETEOROLOGY
JUNE AMERICAN WATER RESOURCES ASSOCIATION 1982

OVERVIEW APPLICATION OF HMR 51-52 AND SITE-SPECIFIC STORM TRANSPOSITION AND MAXIMIZATION TECHNIQUE

Donald T. Jensen[1]

ABSTRACT: A 1,208-square-mile basin and a 638-square mile basin above Box Butte Dam, Nebraska, are used as test cases for computing probable maximum precipitation (PMP). Noting new definitions for storm and basin PMP, Hydrometeorological Report numbers 51 and 52 (HMR 51-52) procedures are applied to the test basins. Traditional storm transposition and maximization assumptions are also considered. Nine critical storms from the Box Butte Dam region are transposed to the test basins and are maximized for moisture charge. Depth-durational aspects of the storms are presented. Comparisons of PMP estimates are made for the test cases using HMR 51-52 and tradiitional transposition and maximization techniques. Results show most differences in PMP estimates are within the reasonable range of measurable error.
(KEY TERMS: probable maximum precipitation; storm transposition.)

INTRODUCTION

In the United States and in many areas in the world, the concept of probable maximum precipitation (PMP) is used in water resources planning and in evaluating the adequacy of existing structures. Specifically, PMP is necessary in the development of the maximum probable flood (MPF) and the determination of peak discharge and maximum volume.

In order to evaluate the MPF for a specific location, it is necessary to determine temporal and spatial distribution of PMP. This may be accomplished either by using the method of storm transposition and moisture maximization or by using generalized procedures formalized in the Hydrometeorological Report Series published by the National Weather Service (NWS). The purpose of this paper is to compare basin PMP estimates derived using the two methods for site-specific locations on the Great Plains of the United States.

For the United States east of the 105th meridian, PMP has been defined by the Bureau of Reclamation (Bureau), Corps of Engineers (Corps) and NWS as ". . . theoretically the greatest depth of precipitation for a given duration that is physically possible over a given size storm area at a particular geographical location at a certain time of the year" (Hansen, Schreiner and Miller, 1982). Precipitation estimates meeting these definition requirements may be obtained following procedures established in Hydrometeorological Report No. 51 (HMR 51) (Schreiner and Riedel, 1978). Comparisons showing observed precipitation values and calculated PMP estimates have been published by Riedel and Schreiner (1980). Seasonal variations of PMP estimates for the area may be obtained using methodologies found in Hydrometeorological Report No. 53 (HMR 53) (Ho and Riedel, 1980).

Bureau meteorologists and hydrologists have traditionally developed site-specific PMP estimates for particular basins using standard transposition and moisture maximization techniques. During the past four years, the Bureau has been very involved with the Corps and NWS in developing methodologies, procedures and criteria which would aid in meeting the needs of the major Federal construction agencies. Through cooperation, a definition of storm and basin PMP has evolved. Storm PMP is noted above and basin PMP may be described as ". . . theoretically the greatest depth of precipitation for a given duration that is physically possible over a particular drainage basin at a certain time of the year."

The definition of basin PMP is slightly different than storm PMP and implies how severely an actual PMP storm might affect a specific river basin. Because this is what flood hydrologists need to compute the MPF for a particular dam, methods were developed to evaluate site specific criteria from the generalized criteria. These methods are presented in the forthcoming Hydrometeorological Report No. 52 (HMR 52) (Hansen, Schreiner and Miller, 1982) which will be published in the early fall of 1982.

TEST BASIN DESCRIPTION

Box Butte Dam is located about 26 miles north of Alliance in northwestern Nebraska on the Niobrara River. The basin drains an area of 1208 square miles and elevations range from about 4000 feet at the dam to over 5500 feet on

[1]Meteorologist, Bureau of Reclamation Engineering and Research Center, P. O. Box 25007, Denver Federal Center, Denver, Colorado 80225.

the western rim of the basin. For hydrologic considerations, the basin was divided into two subbasins. Subbasin I (638 sq. miles) consists of the drainage area immediately above Box Butte Dam and Subbasin II (570 sq. miles) represents the headwaters of the Niobrara River (Figure 1).

The climate of the Box Butte drainage basin is similar to other areas on the Great Plains in the Central United States with relatively little rainfall, low humidity, hot summers and cold winters. Although there is great variation in rainfall from year to year, mean annual precipitation for the basin is about 16 inches. Over 75 percent of the precipitation falls in the months from April through September with the period of May and June receiving the heaviest rainfall.

Severe storms in the basin usually result from either moist air from the Gulf of Mexico interacting with frontal systems in the region or from decadent hurricanes. Because of the influence of the Rocky Mountain barrier to the west, no significant rain from the Pacific Ocean reaches the basin. Historic severe storms in the region include:

Greeley, Nebraska storm of June 4-7, 1896
Warrick, Montana storm of June 6-8, 1906
Springbrook, Montana storm of June 17-21, 1921
Savageton, Wyoming storm of September 27-October 1, 1923
Boyden, Iowa storm of September 17-18, 1926
Burlington, Colorado storm of May 30-31, 1935
Grant Township, Nebraska storm of June 3-4, 1940
Stanton, Nebraska storm of June 10-13, 1944
Holt, Missouri storm of June 22, 1947
Council Grove, Kansas storm of July 9-13, 1951
Ritter, Iowa storm of June 7, 1953
Glen Ullin, North Dakota storm of June 25, 1966
Greeley, Nebraska storm of August 12-13, 1966
Harlan, Iowa storm of September 10-12, 1972

STORM COMPUTATIONS USING HMR 51 AND HMR 52

To test the adequacy of HMR 51 and HMR 52, many studies and comparisons have been made by Bureau, Corps and NWS hydrologists and meteorologists. As an example, for the Box Butte River Basin, Bureau meteorologists compared basin PMP (HMR 51 and HMR 52) estimates with values derived using Bureau methods of site-specific analysis.

Basin PMP estimates were derived using HMR 51 and a 1981 draft of HMR 52. All calculations were done "by hand." Since that time, computer programs have been developed by the Bureau, Corps and NWS to handle the tedious and extensive calculations required to derive basin PMP.

To determine generalized storm PMP estimates for the basins, index values for the geographic location and size were read from HMR 51 for 6, 12, 24, 48, and 72 hours.

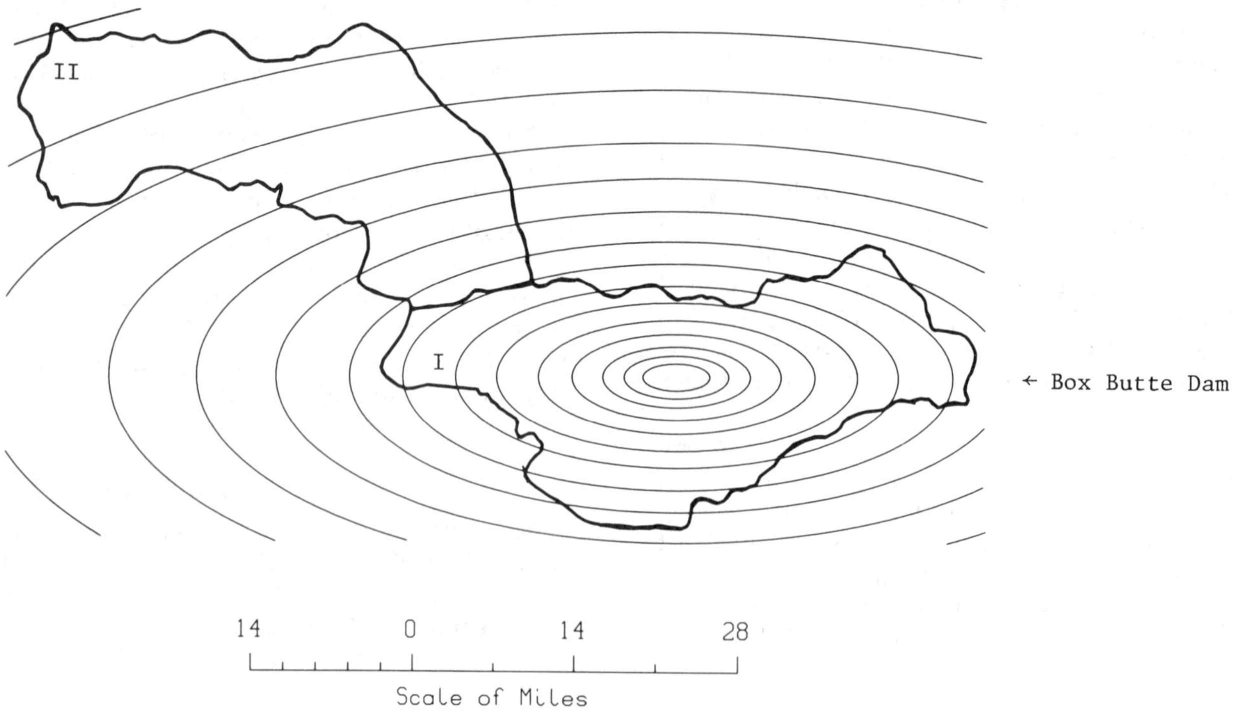

Figure 1. HMR 52 Ellipse Applied to Box Butte Dam and Drainage Basin, Nebraska.

The depth and area values were plotted on semilogarithmic paper for each duration and smooth curves drawn through the points. Values for the basin area were read from the smooth curves and then plotted as depth-duration on linear graph paper and the points smoothly enveloped.

In order to apply the HMR 51 estimates to the drainage basin, the accepted 2.5:1 elliptical pattern from HMR 52 was placed over Subbasin I outline in the most critical position. Basin area between each isohyet was planimetered. Multiplying incremental areas by durational isohyetal label values from HMR 52 nomograms resulted in an average volume of precipitation. The average volume was divided by area to calculate average rainfall depth for each subbasin. A reduction was made for the difference between the orientation of the elliptical pattern over the basin and the orientation of observed historic storms in the region. The same procedures were followed for various storm areas and for storm durations of 6 to 72 hours to determine the storm area that produced the most critical hydrologic event.

STORM CALCULATIONS USING TRANSPOSITION AND MAXIMIZATION METHODS

The Bureau method of determining PMP using transposition and moisture maximization is based on the assumption that the historical record of storms within the basin is not sufficient to define PMP. In order to find a sufficiently severe storm set for the basin, isohyetal patterns of storms within the same homogeneous meteorological region are assumed to have the potential to occur in the basin in the most critical position. However, orientation of the storm pattern may be limited by meteorological considerations. It is also assumed that the most efficient meteorological mechanism exists in the historic storm, so that an optimal amount of precipitable water in the atmosphere would occur as rainfall over the basin. These assumptions are also true for HMR 51 and HMR 52.

Following these assumptions, storm isohyetal patterns were transposed over the Box Butte River Basin and fitted over Subbasin I in the most critical position. Basin average precipitation was determined by calculating the volume of rain that would have fallen within the basin.

Timing of the precipitation was determined by using rain gage records from recording stations within the storm area. Spacial distribution for timing was calculated using Thiessen polygons which provided the weighting factor for each of the temporal distributions developed from the mass curves of rainfall. Rainfall depths for hourly time increments for the total storm were determined for each mass curve. These rainfall depths were multiplied by the weighting factor for the Thiessen polygons which they represent. Values for each Thiessen polygon and for each time increment were then summed with values for all polygons for each time increment. The result was the weighted incremental rainfall for the basin.

Basin average precipitation from the transposition of the isohyetal pattern was divided by the total storm precipitation amount derived in the Thiessen polygon procedure. All time incremental amounts were then multiplied by the resulting ratio, yielding representative transposed precipitation amounts for the specific time increments.

In harmony with the assumption that the atmosphere will be charged with an optimal amount of precipitable water during a PMP event, moisture charge was computed for historical storms using storm and maximum observed dewpoint indices. The storm was maximized in place for moisture charge and then transposed to the drainage basin under study and adjusted for elevation and barriers to the inflow moisture. Barrier and elevation adjustments were made to take into account reduced moisture in the atmosphere which occurs with increased ground surface elevation. The observed storm rainfall amounts transposed to the new location were multiplied by the ratio of the maximum moisture available at the observed location to the maximum moisture available at the transposed location.

In order to envelop and make comparative plots of the important transposed and maximized storms, storm rainfall depths must be evaluated for comparable time increments. This was accomplished by plotting on a depth-duration graph, the greatest depth for the shortest time increment for each storm. The greatest total depth for two adjacent time increments was then computed and plotted. The process continued until storm depths for the total storm had been computed and plotted with other storm depths. Highest values for each time increment were enveloped and the enveloping curve was taken as PMS for Box Butte Drainage Basin (Flood Hydrology, 1977).

COMPARISONS

Comparisons of PMP for the two procedures for the total 1,208-square-mile basin and the 638-square-mile area of Subbasin I are shown in Table 1 for some durations. For Subbasin I, the differences between the two methods were less than 14 percent. As the comparative table shows for the hydrologically critical short durations, most differences were much less than 10 percent. Bureau meteorologists consider such differences within the reasonable range of measurable error.

CONCLUSIONS

Figures 2 and 3 show the envelopment of storms transposed to the drainages above Box Butte Dam, Nebraska. Comparison of these values with those derived using HMR 51 and HMR 52 are shown in Table 1. These comparisons show there is little significant difference between values computed using the two methods.

Normally, the generalized method using HMR 51 and HMR 52 is preferred over the site-specific transposition and

TABLE 1. Comparisons of HMR 51/52 and Traditional Method for Computing PMP.

	Subbasin I (638 mi^2)		Total Basin Including Subbasins I and II (1,208 mi^2)	
Duration (hours)	HMR 51/52 Precipitation Center Above Box Butte Dam in Subbasin I (inches)	Traditional Method-Precipitation Centers Above Box Butte Dam in Subbasin I (inches)	HMR 51/52 Precipitation Center Above Box Butte Dam in Subbasin I (inches)	Traditional Method-Precipitation Centers Above Box Butte Dam in Subbasin I (inches)
1	4.64	4.75	2.82	3.28
2	7.42	7.66	4.51	5.83
3	9.28	8.95	5.64	6.54
4	10.62	10.23	6.45	7.26
5	11.70	10.73	7.11	7.97
6	12.54	11.23	7.62	8.96
12	14.71	12.96	9.05	10.09
18	15.97	13.98	9.91	11.04
24	16.92	14.69	10.56	11.43
36	18.33	16.11	11.53	11.94
48	19.31	17.51	12.20	12.44
72	20.46	17.88	12.98	12.85

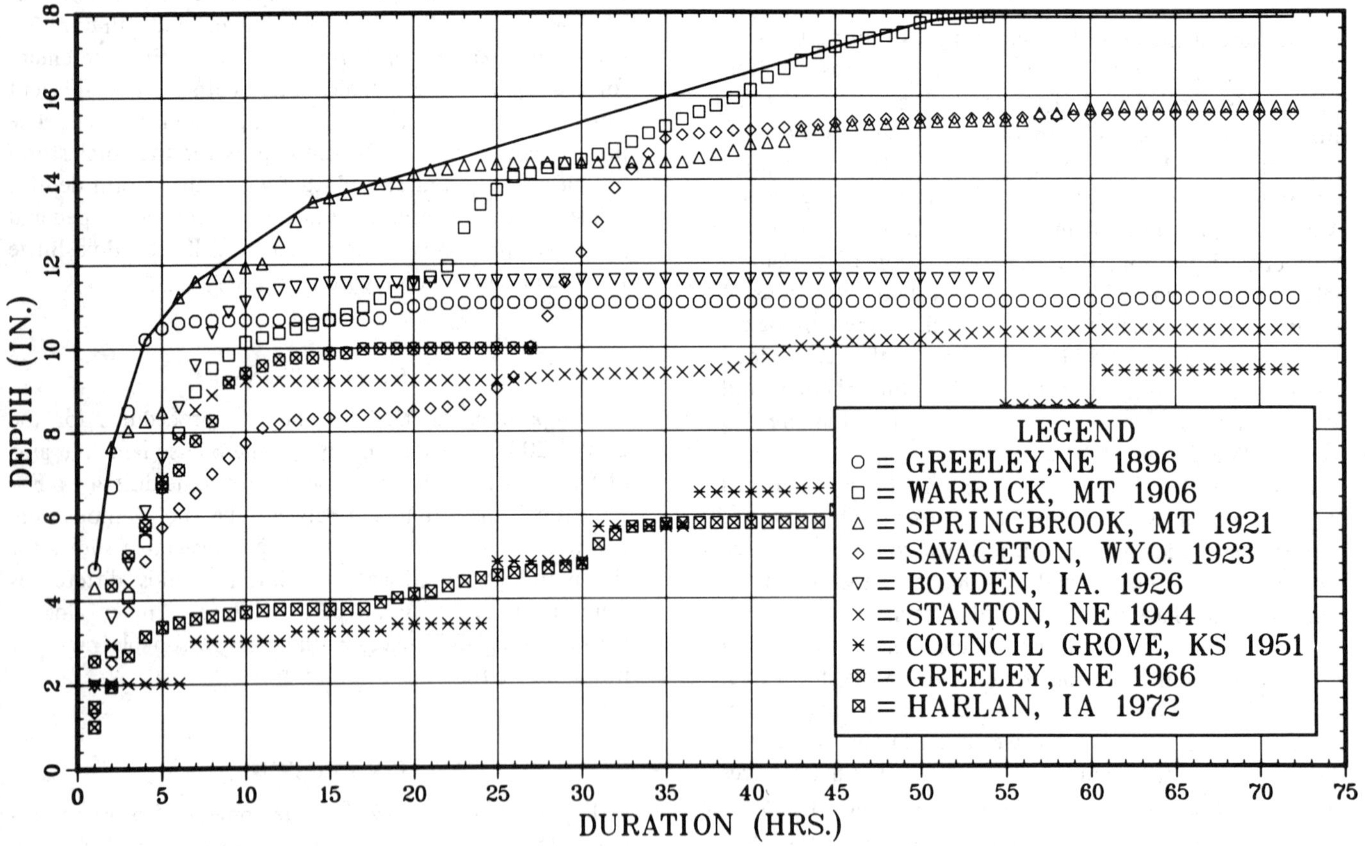

Figure 2. Precipitation Depth-Duration Curve Subbasin I Above Box Butte Dam, Nebraska. Area = 638 square miles.

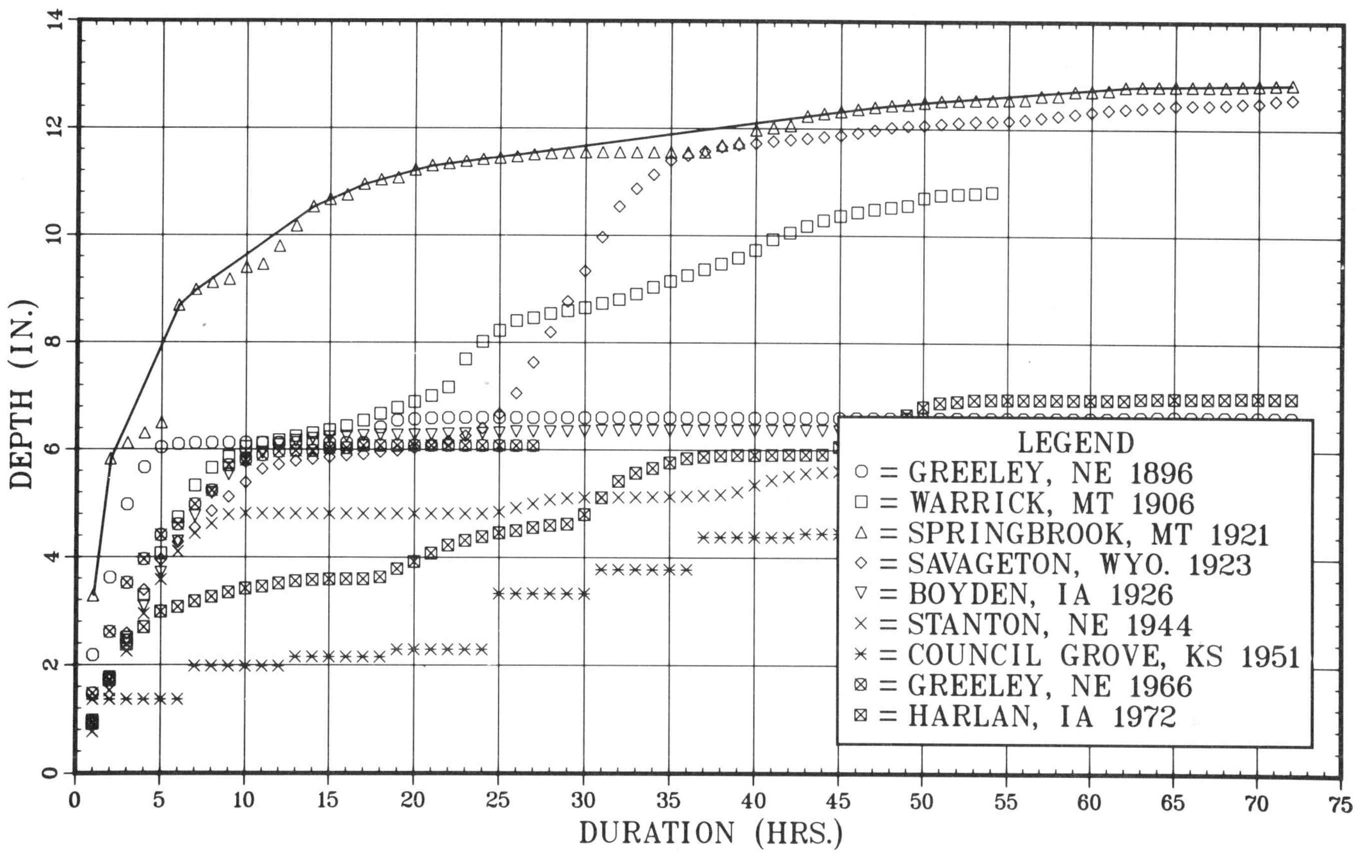

Figure 3. Precipitation Depth-Duration Curve Total Basin Above Box Butte Dam, Nebraska. Area = 1208 square miles.

moisture maximization technique because it provides: (1) consistency with values derived by other state and Federal agencies for the same location, (2) regional consistency, (3) inclusion of all severe storm effects in the same homogeneous area, (4) time savings in calculations, and (5) more objectivity in derivation of the PMP estimates.

LITERATURE CITED

"Flood Hydrology," *In:* Part 6 of Water Studies, Vol. IV of the Reclamation Manual. 1977. U.S. Department of the Interior, Bureau of Reclamation, Washington, D.C.

Hansen, E. M., L. C. Schreiner, and J. F. Miller, 1982. Application of Probable Maximum Precipitation Estimates, United States East of 105th Meridian. U.S. Department of Commerce, National Oceanic and Atmospheric Administration, National Weather Service, Hydrometeorological Report No. 52, Silver Spring, Maryland (in preparation).

Ho, Francis P. and John T. Riedel, 1980. Seasonal Variation of 10-Square-Mile Probable Maximum Precipitation Estimates, United States East of the 105th Meridian. U.S. Department of Commerce, National Oceanic and Atmospheric Administration, National Weather Service and U.S. Nuclear Regulatory Commission, Hydrometeorological Report No. 53, Silver Spring, Maryland.

Riedel, John T. and Louis C. Schreiner, 1980. Comparison of Generalized Estimates of Probable Maximum Precipitation with Greatest Observed Rainfalls. U.S. Department of Commerce, National Oceanic and Atmospheric Administration, National Weather Service, NOAA Technical Report NWS 25, Silver Spring, Maryland.

Schreiner, Louis C. and John T. Riedel, 1978. Probable Maximum Precipitation Estimates, United States East of the 105th Meridian. U.S. Department of Commerce, National Oceanic and Atmospheric Administration, National Weather Service, and U.S. Department of the Army, Corps of Engineers, Hydrometeorological Report No. 51, Washington, D.C.

INTERNATIONAL SYMPOSIUM ON HYDROMETEOROLOGY
JUNE AMERICAN WATER RESOURCES ASSOCIATION 1982

DAM BREACH PARAMETERS, OUTFLOW PEAKS, AND FLOOD STAGES

Krishan P. Singh and Arni Snorrason[1]

ABSTRACT: Important breach parameters were identified and their ranges estimated from a study of earth dam failures. The National Weather Service (NWS) and the U.S. Army Corps of Engineers Hydrologic Engineering Center (HEC) dam breach models were evaluated regarding their theoretical bases. Both models use similar input data and breach descriptions, but the NWS uses the St. Venant equations for routing, whereas the HEC uses hydrologic routing. Information on eight dams in Illinois was taken from the Corps of Engineers inspection reports. Various combinations of breach parameters (failure time, TF; depth of overtopping, h_f; and breach size, B) were used for breach simulations by both methods with the 1.00-, 0.50-, and 0.25-PMF inflow hydrographs. The NWS model performs better than the HEC model in simulating reasonable flood-stage profiles in the downstream reaches. The breach outflow peaks are affected significantly by B but less so by h_f. The ratio of outflow peak to inflow peak and the effect of TF on outflows decrease as the drainage area above the dam and the storage impounded increase. For the range of breach parameters studied, the range of outflow peaks and flood stages downstream from the dam can be determined for regulatory and disaster prevention measures.
(KEY TERMS: dam failure; breach parameters; dam-break models; breach simulations; outflow peaks; flood stages.)

INTRODUCTION

The purpose of this study was threefold: 1) evaluation of breach parameters from a literature survey of historical earthdam failures; 2) evaluation and comparison of the theoretical and practical merits of two dam-break models; and 3) sensitivity analysis of important breach parameters. The information from the literature survey (Singh and Snorrason, 1982) gave the range of values for some important breach parameters. The average breach width, B, depends on the dam height, h_d. Most of the cases fell between two lines, defined by $B = 2h_d$ and $B = 5h_d$. The maximum water depth over the dam during failure by overtopping, h_f, generally lies in the range of 0.5 to 2.0 feet. The failure time, TF, from inception to completion of breach generally lies in the range of 0.25 to 1.0 hour. The historical cases also suggest that the final elevation of the breach bottom, YBMIN, is usually the original ground surface elevation in the channel. No information was available on the side slopes of the breach section.

A complete set of simulations on a single dam consists of the following combinations:

1) Routing of inflow hydrographs for PMF, 0.5 PMF, and 0.25 PMF floods without the existence of the reservoir.
2) Routing of the inflow hydrographs through the reservoir, but with the dam intact, even if overtopped.
3) Routing of inflow hydrographs with the failure of the dam, using eight combinations of breach parameters.

Eight earth dams were selected for breach simulation. The basic data on the dams were taken from Inspection Reports, National Dam Safety Program, published by the Department of the Army, Chicago District, Corps of Engineers (COE). The name of the dam, height of the dam above the streambed, storage at normal pool level, and peak of the probable maximum flood (PMF) inflow hydrograph are given below. The information on size (S = small, IM = intermediate, L = large) as well as hazard (H = high) categories, as given in the Inspection Reports, is also included.

	Height of dam h_d, ft.	Storage S, ac. ft.	Peak of PMF Q, cfs	COE	
				Size	Hazard
Pierce Lake Dam	46.0	2,660	30,500	IM	H
Lake in the Hills Dam #1	40.0	598	8,400	IM	H
Lake in the Hills Dam #2	14.5	78.9	11,318	S	H
Lake Marian Dam	50.0	151	3,164	IM	H
Clinton Lake Dam	65.0	74,200	150,200	L	H
Lake Springfield Dam	48.0	53,504	121,364	L	H
Weslake Dam	48.0	224	1,243	IM	H
Kinkaid Lake Dam	92.0	78,500	71,000	L	H

[1]Respectively, Principal Hydrologist, Illinois State Water Survey, 605 East Springfield, Champaign, Illinois 61820; and Graduate Research Assistant, Illinois State Water Survey, 605 East Springfield, Champaign, Illinois 61820.

THEORETICAL EVALUATION OF SELECTED DAM-BREAK MODELS

The models chosen for evaluation were the U.S. Army Corps of Engineers HEC-1 dam-break version, referred to as HEC, and the National Weather Service dam-break model, referred to as NWS. The HEC program uses the modified-Puls (MP) method for the routing of a flood wave due to dam breach. The MP method is based on the continuity equation transformed into a finite difference equation of the form (Chow, 1964):

$$\frac{1}{2}(I_1 + I_2)\,\Delta t + (S_1 + \frac{1}{2}Q_1\,\Delta t) - Q_1\,\Delta t = (S_2 + \frac{1}{2}Q_2\Delta t) \qquad (1)$$

in which I is the inflow to the reach, S is the storage in the reach, and Q is the outflow from the reach. The subscripts 1 and 2 refer to the beginning and end of a time interval, respectively.

The HEC model computes storage-discharge and discharge elevation relationships for each routing reach using normal depth assumptions, cross-sectional data, channel and overbank roughness, channel slope, and length of the reach. The relationships are then used in the MP method. For reservoir routing, the storage-outflow relationship for the reservoir may be input directly, or generated based on the reservoir elevation-area-capacity data and the hydraulic characteristics of the outlet works and the dam itself (COE, 1978b). The boundary conditions for the MP method are given by the inflow hydrograph for the reservoir routing and by the breach hydrograph for the channel routing. The basic assumption of the MP method is that it uses only the continuity equation and the normal flow equation, neglecting all dynamic effects which are very important for dam breach flood waves.

The routing method of the NWS model is based on the Saint-Venant equations (Fread, 1977), the continuity equation

$$\frac{\partial(A + A_o)}{\partial t} + \frac{\partial Q}{\partial x} = q \qquad (2)$$

and a conservation of momentum equation

$$\frac{\partial Q}{\partial t} + \frac{\partial(Q^2/A)}{\partial x} + gA\,(\frac{\partial h}{\partial x} + S_f + S_e) = 0 \qquad (3)$$

where h is the water surface elevation; A is the active cross-sectional area of flow; A_o is the inactive (off-channel storage) cross-sectional area; x is the longitudinal distance along the channel; t is the time; q is the lateral inflow or outflow per linear distance along the channel (inflow is positive and outflow is negative in sign); g is the acceleration due to gravity; S_f is the friction slope, evaluated by the Manning equation for uniform steady flow; and S_e is the expansion-contraction slope. The partial differential equations 2 and 3 are represented by a corresponding set of non-linear finite difference algebraic equations solved by the Newton-Raphson method (Fread, 1977 and 1979). The initial conditions are given by known steady discharge at the dam, from which the discharge and water surface elevation at each cross section are calculated by solving the steady state nonuniform flow equation. The boundary conditions at the upstream boundary are the reservoir inflow hydrograph, which is routed either by dynamic or storage routing through the reservoir. The breach hydrograph is the upstream boundary condition for the channel routing. At the downstream boundary, the Manning equation is used if the flow is controlled by the channel. Otherwise, a single valued rating curve is used for the downstream boundary.

The basic assumptions of the NWS model are those of the Saint-Venant equations, discussed by Yen (1973) when the equations are derived from the complete one-dimensional open channel flow equations. According to Yen, equations (2) and (3) do not give accurate results if the assumption of hydrostatic pressure distribution is seriously violated.

BREACH HYDROGRAPH COMPUTATION

Both HEC and NWS use the same breach parameters for the description of the breach formation. The five basic breach parameters are the side slope of the breach section, z; the final bottom width of the breach, BBW; the final elevation of the bottom of the breach, YBMIN; the time from inception to completion of breach, TF; and the failure elevation, HF. The elevation of the breach bottom, YBMIN, was usually taken to be the channel bottom or the dominant ground elevation at the dam, except when this was not physically justifiable due to backwater effects. The slope, z, was considered to be 0.5, except for piping failures where it was set at 0.0. The final breach bottom width, BBW, was chosen to be $2h_d$ and $4h_d$, where h_d is the height of the dam, which combined with a side slope of 0.5 gave a maximum width of $5h_d$ at the top. The overtopping depth (or the water depth over the dam when failure starts), h_f or HF-HD, was taken as 0.5 or 2.00 ft. The time of failure was chosen to be 0.25, 0.50, and 1.00 hour. These ranges of breach parameters cover reasonably well the values obtained from historical earthdam failures.

Both programs use breach formation based on a procedure by Fread and Harbaugh (1973). The NWS model assumes that the breach starts at a point and both the breach width and depth increase at a linear rate over the failure time, TF, until the terminal width is attained and the breach bottom has eroded to the elevation YBMIN. If TF is less than ten minutes, the width of the breach starts at a value of BBW rather than at a point. This latter version is the one used by the HEC program for all breach formations. This difference in procedures results in different breach outflow hydrographs with the two models, usually resulting in a higher peak discharge for the NWS model since at maximum breach size more water is retained in the reservoir and thus higher water levels.

The NWS program requires cross-sectional information immediately downstream of the dam in order to calculate tailwater elevations for any needed correction for partial submergence. The program also corrects for the velocity of approach. The HEC ignores both the effects of partial submergence and velocity of approach. In our studies, submergence effects were much more significant than the approach velocity effects. The input data for both the models are essentially the same and can be divided into two groups. The first data group pertains to the inflow hydrograph, the reservoir, the dam, and the breach. The second group of data pertains to the routing of the breach hydrograph through the downstream channel and valley.

EXAMPLE

Out of the eight dams studied, the Lake in the Hills Dam #1 is selected for illustrative purposes. It is located on Woods Creek, a tributary to the Fox River, McHenry County, Illinois. It is an earthfill structure, approximately 40 ft. high and 780 ft. long. The dam is classified in the intermediate size and high hazard potential category. Pertinent data (COE, 1978a) about the dam, spillway, and reservoir follow.

Drainage area	8.52 sq. mi.
Dam	
Elevation, top of dam	827.0 ft. msl
Height above streambed	40.0 ft.
Length	780.0 ft.
Type	Earth embankment
Reservoir	
Elevation, normal pool*	822.0 ft. msl
Area, normal pool	53.0 ac.
Capacity, normal pool	598.0 ac. ft.
Length, normal pool	0.8 mi.
Spillway	
Elevation, weir crest	822.0 ft. msl
Length, crest	29.8 ft.
Type	Concrete broad-crested weir
Freeboard	
With normal pool	5.0 ft.
With 10-year flood event	1.0 ft.

*Top of spillway crest.

The Manning's roughness coefficient, n, was estimated by visual observation to be 0.05 for the channel and 0.07 for the overbank flow (COE, 1978a).

The breach parameters were chosen on the basis of the guidelines established previously. The value of the bottom elevation of the breach, YBMIN, was set at 789.0 ft. which is about 2 ft. above the channel bottom elevation. The bottom breach width, BBW, was set at 80 and 160 ft. for the small and large breach, respectively. The time from the inception of the breach to its completion, TF, has been taken as 0.25, 0.50, or 1.00 hour. The depth of overtopping when the breach starts, or h_f, equal to HF-HD, has been taken as 0.5 or 2.0 ft.; the HF and HD denote the failure elevation and elevation of the top of the dam, respectively.

The peak discharges for both methods and all combinations of breach parameters are shown in Table 1. The peak discharges for the no-reservoir condition are essentially the same as for the no-failure condition due to small storage in the reservoir. The peak outflows with the NWS method due to the failure of the dam are about 7 to 38 percent higher than with the HEC method. This is mainly due to the difference in the mode of breach formation. Increase in peak outflows due to higher failure elevations is about 14 percent with the NWS method and 20 percent with the HEC method, whereas it is 25 to 65 percent with a 50 percent reduction in failure time. Bigger breach size results in an increase of about 7 to 51 percent in peak outflow. The peak values are slightly lower with the 0.50 PMF and 0.25 PMF floods than with the PMF flood. The peak outflow in general varies from about 3.5 to about 11.5 times the PMF inflow peak and up to 25 times the 0.25 PMF inflow peak.

The peak flows and maximum water stages in the 1.7-mi. downstream channel are shown in Figure 1 for TF = 0.5 hr, BBW = 160, and HF-HD = 0.5 ft. The peak flows are higher with the NWS model than with the HEC model, and the difference is approximately constant along the channel. The maximum flood stages with the NWS model are usually higher than with the HEC model, but they agree fairly well for the no-reservoir condition. The profiles due to breach with 0.50 PMF and 0.25 PMF behave in much the same way as with 1.00 PMF, but these have been left out for clarity.

The maximum and minimum flood stages with the breach and with the no-reservoir condition for the 1.00 PMF are shown in Figure 2 for four selected cross sections along the downstream channel, as calculated with the NWS method. The elevation difference between the maximum and minimum flood stages with the breach decreases from about 6.5 ft. at station 0.057-mile to 2 ft. at the end of the 1.7-mile reach. However, even the minimum flood stages with the breach are about 4.0 to 6.0 ft. higher than with the no-reservoir condition.

CONCLUSIONS

The following conclusions are drawn from this study:

1) The review of historical earthdam failures gives valuable information on the range of important breach parameters.

2) The NWS model has sounder theoretical basis than the HEC, but still requires essentially the same input data. The results show that the NWS model performs better than the HEC model in predicting reasonable (and not erratic) flood stages in the downstream reaches.

TABLE 1. Peak Outflows: Lake in the Hills Dam #1

Table II-	Item	Inflow Flood*	YBMIN	BBW	TF	HF	Peak Outflow, cfs NWS	HEC
A	No-reservoir conditions	1.00	–	–	–	–	8,399	8,399
B	No-reservoir conditions	0.50	–	–	–	–	4,199	4,200
C	No-reservoir conditions	0.25	–	–	–	–	2,099	2,110
E	No-failure conditions	1.00	–	–	–	–	8,392	8,392
F	No-failure conditions	0.50	–	–	–	–	4,194	4,195
G	No-failure conditions	0.25	–	–	–	–	2,094	2,093
H	Failure conditions	1.00	789	80	0.50	827.5	45,980	35,607
I	Failure conditions	1.00	789	80	0.50	829.0	51,977	43,187
J	Failure conditions	1.00	789	80	0.25	829.0	64,742	60,423
K	Failure conditions	1.00	789	80	1.00	829.0	36,036	29,014
L	Failure conditions	1.00	789	160	0.50	827.5	55,018	41,230
M	Failure conditions	1.00	789	160	0.50	829.0	63,541	49,638
N	Failure conditions	1.00	789	160	0.25	829.0	97,494	77,333
O	Failure conditions	1.00	789	160	1.00	829.0	38,595	32,270
P	Failure conditions	0.50	789	80	0.50	827.5	44,134	34,055
Q	Failure conditions	0.50	789	160	0.50	827.5	53,692	39,550
R	Failure conditions	0.25	789	80	0.50	827.5	44,135	32,175
S	Failure conditions	0.25	789	160	0.50	827.5	51,868	37,450

*Inflow flood hydrograph corresponds to 0.25, 0.50, or 1.00 times the probable maximum flood, PMF, hydrograph.

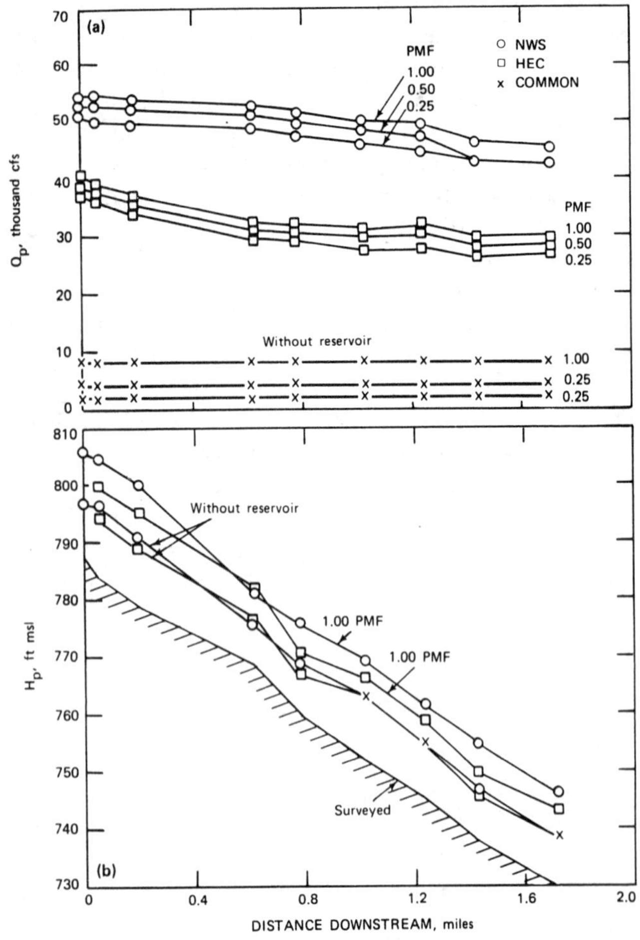

Figure 1. Peak Flows and Flood Stages Downstream of Lake in the Hills Dam #1 (BBW = $4h_d$; TF = 0.5 hour; h_f = 0.5 ft.).

3) For the Lake in the Hills Dam #1, all parameters (h_f, TF, and BBW) affected the breach outflow peaks, whereas the size of the inflow hydrograph did not matter as long as it failed the dam.

4) The ratio of outflow peak with dam breach to the inflow peak for the PMF decreases with increases in the reservoir storage capacity.

5) For the eight dams studied, the outflow peak increased by 13-83 percent for small storage and 1 to 5 percent for large storage reservoirs when the time to failure, TF, was reduced by 50 percent.

LITERATURE CITED

Chow, V. T., 1964. Handbook of Applied Hydrology. McGraw-Hill Book Company, New York, NY, 25:38-40.

COE, 1978a. Lake in the Hills Inspection Report, National Dam Safety Program. U.S. Army Corps of Engineers, Chicago, IL.

COE, 1978b. Flood Hydrograph Package (HEC-1), Users Manual for Dam Safety Investigations. The Hydrologic Engineering Center, U.S. Army Corps of Engineers, Davis, CA.

Fread, D. L. and T. E. Harbaugh, 1973. Transient Hydraulic Simulation of Breached Earth Dams. Journal of Hydraulics Division, ASCE, 99-HY1:139-154.

Fread, D. L., 1977. The Development and Testing of a Dam-Break Flood Forecasting Model. Proceedings, Dam-Break Flood Modeling Workshop, U.S. Water Resources Council, Washington, D.C., 1977, 164-197.

Fread, D. L., 1979. DAMBRK, The NWS Dam-Break Flood Forecasting Model. Office of Hydrology, National Weather Service (NWS), Silver Spring, MD.

Singh, K. P. and A. Snorrason, 1982. Sensitivity of Outflow Peaks and Flood Stages to the Selection of Dam Breach Parameters and Simulation Models. Illinois State Water Survey Contract Report 289, 179 p.

Yen, B. C., 1973. Open-Channel Flow Equations, Revisited. Journal of Engineering Mechanics Division, ASCE, 99-EMS:979-1009.

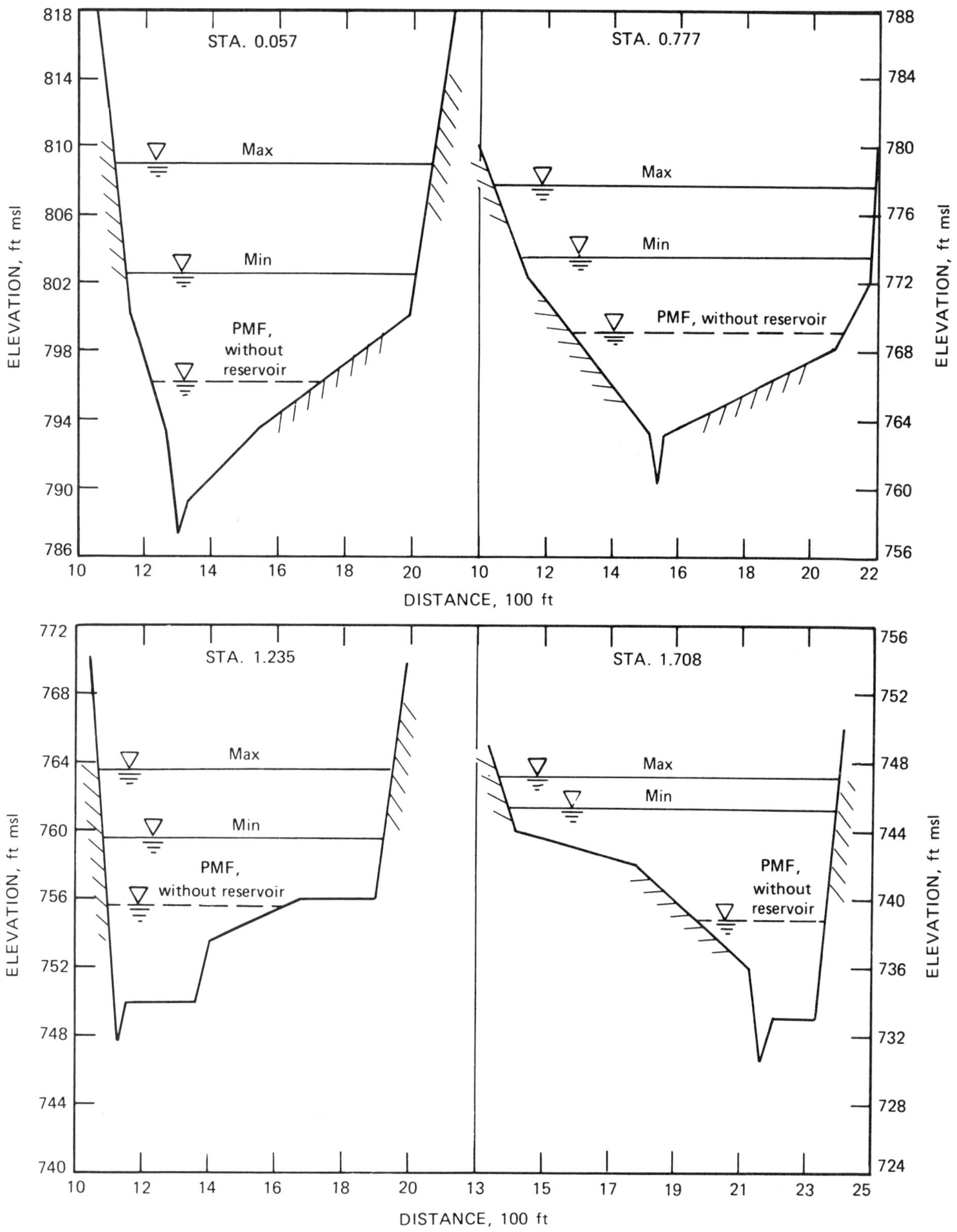

Figure 2. Flood Stages Downstream of Lake in the Hills Dam No. 1.

FLOOD HYDROLOGY OF STREAMS IN NORTHERN UTAH

Purushottam Dass, Ronald L. Lambert, and David D. Clement[1]

ABSTRACT: The hydrologic investigations reported in this publication relate to streams through 20 communities in Northern Utah. In general, these streams develop in the Wasatch mountains and flow through the mountainous canyons to the populated valley areas downstream, eventually terminating into the Great Salt Lake. The study area watersheds are subjected to high-intensity short-duration cloudburst type thunderstorms. The stream gage records available in the area are separated into the snowmelt and rainfall induced flood peaks which are analyzed for Log Pearson Type III distribution for both rainfall and snowmelt events. Due to lack of sufficient gage data related to rainfall caused flood events, 16 of the watersheds are modeled to simulate 10-, 50- and 100-year recurrence interval rainfall caused flood events. The model used for simulation is the Storm Water Management Model developed by the Environmental Protection Agency. The model is calibrated by using information available on several historical floods in the area. Regression equations are developed to predict 10-, 50- and 100-year recurrence interval flood flows for both snowmelt and rainfall caused flood events. The discharge-frequency distributions for these events are statistically combined to give discharge-frequency distributions for the combined events. The regression equations presented here may be used for streams along the Wasatch Front Range in Northern Utah and for other hydrologically similar areas.
(KEY TERMS: models; flood recurrence interval; Utah; discharge frequency; regression analysis.)

INTRODUCTION

Flood damages in the United States of America are estimated to be about four billion dollars each year. An accurate estimate of the flood damage potential is a key element for any flood damage abatement program. To estimate discharge-frequency information for any flooding source requires consideration of climatic, meteorological, physiological and land use factors. Any rational approach to estimate flood flows, therefore, will be more accurate and efficient when applied at local or state level rather than regional or national level.

The hydrologic approach presented in this study is applicable for streams along the Wasatch Mountains in Northern Utah. The study area lies between the Wasatch range mountains on the east and the Great Salt Lake on the west and extends from Salt Lake City in the south up to Brigham City in the north. Most of the streams flow through steep canyons that discharge abruptly onto the flatter areas formed by the recession of ancient Lake Bonneville (Great Salt Lake).

A good understanding of the general terrain of the area can be obtained from Figure 1. Flood flows in Parrish Creek in the picture were caused by a localized cloudburst type rainfall. The adjoining Deuel Creek is dry.

FLOOD HISTORY

According to Woolley (Woolley, 1946), more than 500 reports of outstanding "cloudburst" floods in Utah from 1847 to 1938 indicate that 112 different times farm crops, usually in small patches and varying in amounts, have been destroyed. Stretches of highway have been washed away or covered with mud and boulders, and bridges destroyed about 140 different times; canals and diversion dams have been damaged more than 70 times; railroad tracks have been washed out or covered up more than 60 times, and at least 26 persons have lost their lives by drowning.

During the period 1939-1969, a total of 836 cloudburst floods were reported in Utah (Butler and Marsell, 1972). The largest number was in 1967 when 88 floods were reported in the state; 63 floods were reported in 1968 and 56 were reported in 1961. The extent of damage varied with location and a total of nine persons lost their lives by drowning in the cloudburst floods.

PRECIPITATION

The precipitation characteristics in the study area can be divided into three distinct patterns: (1) Pacific frontal storms; (2) continental cyclone storms, and (3) convective storms associated with moist air from the Gulf of Mexico (Utah State Univ., 1968). The first two types of storms occur throughout the year with varying strength and frequency. Snowfall during the winter months is primarily from Pacific frontal storms. Convective "Gulf" storms are

[1]Respectively, Principal, Senior Project Engineer and Hydraulic Engineer, Gingery Associates, Inc., 2840 S. Vallejo Street, Englewood, Colorado 80110.

Figure 1. Aerial View of Parrish (at left) and Deuel Creeks After the 1930 Flood. (Photo by U.S. Forest Service.)

confined entirely to the summer season and are the cloudburst type storms. It is estimated that 80 to 90 percent of the storms during July through September in the study area are of this type.

ANALYSIS OF STREAM GAGING RECORDS

All the stream gaging stations in the vicinity of the study area, as well as several gages on other streams in Northern Utah exhibiting similar basin characteristics are used in this analysis. A total of 16 gaging stations with seven or more years of record are considered in this study. The stage recording charts from the U.S.G.S. archives were obtained for the selected gaging stations. The charts were inspected along with climatological data for temperature and precipitation to identify the snowmelt caused peaks and the rainfall caused peaks. Stage-Discharge Rating Tables for the gages are used to calculate the discharges for the stages obtained from the strip charts.

The mean, standard deviation and skew coefficient for each of the gages are determined for both rainfall and snowmelt flood peaks. For rainfall events, a weighted average skew of −0.22 is calculated. For the snowmelt events, regionalized skew coefficients for the area are taken from the Water Resources Council's Bulletin #17A (U.S. Water Resources Council, 1977). The 10-, 50-, 100- and 500-year frequency discharges are evaluated at each gage location for snowmelt and rainfall events.

A regression analysis is used to develop equations predicting the 10-, 50-, 100-, and 500-year frequency discharges. The significant basin parameters considered in this analysis are: drainage area, elevation, basin slope, seasonal precipitation, basin shape, aspect, and vegetative cover. The regression analysis identifies the drainage area as the only key parameter and indicates that additional variables do not significantly increase the accuracy of the predictions. The resultant equations are as follows.

For Snowmelt

$$Q_{10} = 14.13\, A^{0.94};\ \bar{R} = 0.84 \quad (1)$$

$$Q_{50} = 20.44\, A^{0.92};\ \bar{R} = 0.84 \quad (2)$$

$$Q_{100} = 22.57\,A^{0.91};\,\bar{R} = 0.84 \quad (3)$$

$$Q_{500} = 28.09\,A^{0.90};\,\bar{R} = 0.83 \quad (4)$$

For Rainfall

$$Q_{10} = 2.8\,A^{1.21};\,\bar{R} = 0.61 \quad (5)$$

$$Q_{50} = 4.6\,A^{1.31};\,\bar{R} = 0.58 \quad (6)$$

$$Q_{100} = 5.2\,A^{1.34};\,\bar{R} = 0.57 \quad (7)$$

$$Q_{500} = 7.3\,A^{1.42};\,\bar{R} = 0.56 \quad (8)$$

Where Q is the discharge in cubic feet per second, A is the drainage area in square miles, and R is the correlation coefficient. Subscripts 10, 50, 100 and 500 indicate the flood recurrence interval in years.

A graphical display of how well the regression equations fit the original data is shown in Figure 2 for the snowmelt caused flood peaks. The regression equations (5) through (8) for rainfall caused flood peaks show poor correlation coefficients. Also since most of the gaging stations are located near the canyon mouths, floods measured were generated from the forested mountainous watershed upstream and do not, therefore, reflect the effect of developed watershed below. It was necessary, therefore, to utilize alternate methods to evaluate rainfall caused flood peaks from the forested mountainous watershed as well as from the developed gently sloping watershed below.

RAINFALL-RUNOFF SIMULATION

The Storm Water Management Model commonly known as SWMM (U.S. Environmental Protection Agency) is used in this study to simulate hydrographs for varying frequency rainfall storms. The rainfall depth-frequency data for the mountainous watersheds is obtained from Precipitation-Characteristics of Summer Storms at High Stations in Utah (Farmer and Fletcher, 1971), which gives the results of analysis of 14 recording precipitation gages in the area. For the valley areas similar data is obtained from Rainfall Intensity Duration Frequency Curves (Sonntag and Glenne, 1978). A storm duration of one hour is chosen for use because the basins studied have small drainage areas and relatively low times of concentration.

The correction factors to change the point rainfall to area rainfall are used from the National Weather Service (1973) publication. The temporal distribution of the rainfall for the region is analyzed by Farmer and Fletcher (1971) up to 10 year frequency events. This information is extended for the higher frequency events by analyzing the major storm events recorded at the precipitation gage in Farmington.

In SWMM, each watershed subarea is described by its location, drainage area, slope, width, roughness value, imperviousness, surface storage and infiltration rate. The location, area, slope and width are taken directly from 7.5 minute U.S.G.S. topographic maps. Roughness values and imperviousness are estimated based on topographic maps and field inspection. Surface storage and infiltration rates are determined based on available references by U.S. Army Corps of Engineers (1973), U.S. Environmental Protection Agency (1971), and Utah State University (1976) and judgment, and are calibrated against the gage records where possible.

A total of sixteen streams within the study area are simulated to yield runoff hydrographs for 10-, 50-, and 100-year frequency storms. These streams are chosen to cover the typical range of drainage areas. The intention was to model a limited number of streams by the SWMM program and then to develop the necessary regression equations of a regional nature to predict discharges for additional streams not modeled. The hydrographs are generally obtained at two selected locations for each stream; the canyon mouth of the stream and the downstream limit of the developed valley area. A stepwise regression approach is used to develop regression equations using the rainfall-runoff modeling results.

The key parameters considered are the total drainage area, total impervious area, drainage area at the canyon mouth, basin shape and channel slope. The drainage area at the canyon mouth is noted to be the key parameter to predict discharge at the canyon mouth location. At locations downstream from the canyon mouth, significant flows are generated from the impervious developed area. At such locations, both the total drainage area and the total impervious area are noted as the key parameters. The derived regression equations are (9) through (14). The subscripts F, I and T refer to the forested mountainous area, the impervious area, and the total drainage area, respectively. The other variables are as explained earlier. A visual display showing the variation of actual values from the perfect fit line is shown in Figure 3 for the 100-year frequency events.

Discharge Due to Rainfall at the Canyon Mouth

$$Q_{10} = 7.97\,A_F^{1.18};\,\bar{R} = 0.91 \quad (9)$$

$$Q_{50} = 21.2\,A_F^{1.22};\,\bar{R} = 0.97 \quad (10)$$

$$Q_{100} = 32.7\,A_F^{1.22};\,\bar{R} = 0.97 \quad (11)$$

Discharge Due to Rainfall in Valley Areas

$$Q_{10} = 282\,A_I^{0.61}\,A_T^{0.20};\,\bar{R} = 0.88 \quad (12)$$

$$Q_{50} = 250\,A_I^{0.31}\,A_T^{0.48};\,\bar{R} = 0.84 \quad (13)$$

$$Q_{100} = 268\,A_I^{0.23}\,A_T^{0.54};\,\bar{R} = 0.82 \quad (14)$$

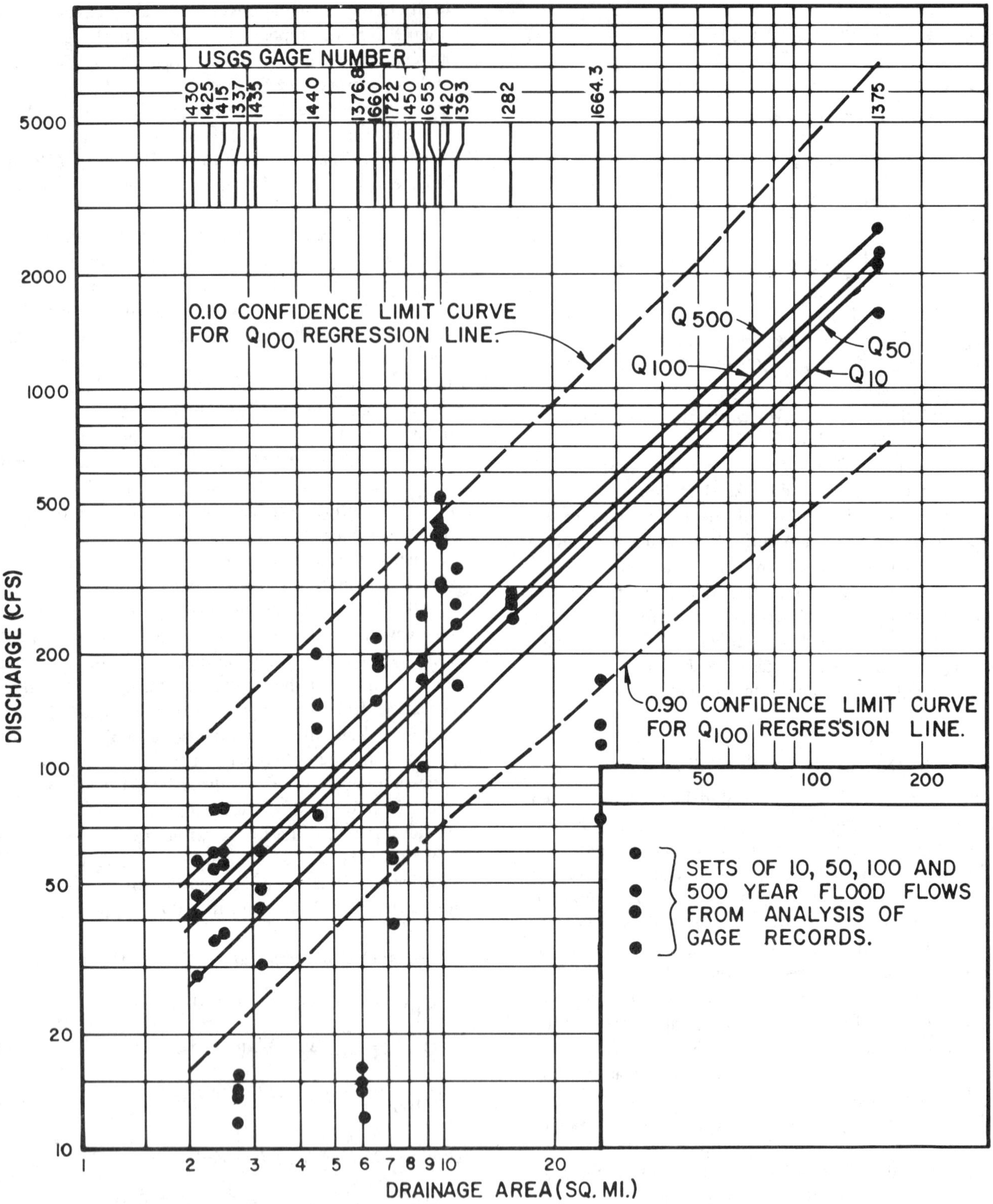

Figure 2. Discharge – Drainage Area Curves for Snowmelt Events.

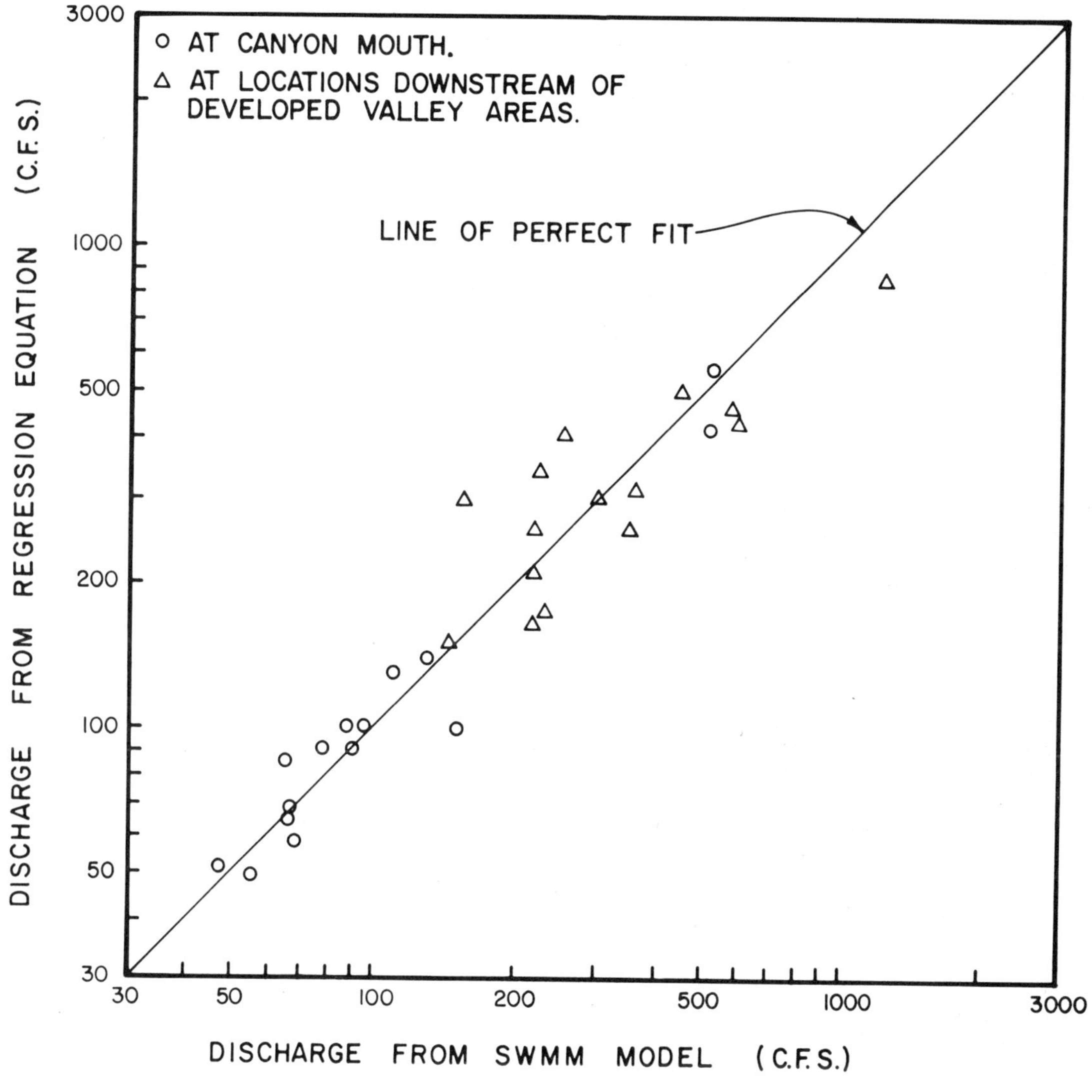

Figure 3. Comparison of 100-Year Frequency Discharge as Calculated by the Regression Equations (11) and (14) and by the SWMM Model.

COMBINING SNOWMELT AND RAINFALL CAUSED FLOOD PEAKS

Snowmelt and rainfall caused floods are assumed to be independent hydrologic events. They can, therefore, be statistically combined to reflect the true probability of flooding. These calculations are carried out based on the following equation:

$$P = P_s + P_r - (P_s P_r)$$

where:

P = exceedence probability of combined flood event
P_s = exceedence probability of snowmelt flood of the same discharge
P_r = exceedence probability of rainfall flood of the same discharge

SUMMARY AND CONCLUSIONS

The stage recording chargs for 16 gaging stations are analyzed to calculate the annual peak flows caused individually by snowmelt and by rainfall. The 10-, 50-, 100- and 500-year frequency flood flows are calculated from the analysis of the annual peak flows. The stepwise regression analysis for the 10-, 50-, 100- and 500-year flood flow values resulted in Equation (1) through (4) for the snowmelt events and (5) through (8) for the rainfall events. Equations (5) through (8) show poor correlation and, in addition, do not reflect the runoff generated in the urbanizing areas below the gage locations. Instead, rainfall regression equations (9) through (14) are developed based on the results of computer modeling with the SWMM program. The 500-year frequency flows are obtained by extrapolation procedures. The snowmelt and rainfall events are statistically combined to obtain the discharge-frequency curve. The regression equations presented here can be used directly for streams along the Wasatch Front Range in Northern Utah and other hydrologically similar areas.

ACKNOWLEDGMENTS

The work upon which this paper is based was supported in part by the Federal Emergency Management Agency (FEMA) to complete flood insurance studies for 20 communities in Utah. Dr. John Liou of FEMA Region VIII provided constructive discussions as the study was progressing.

LITERATURE CITED

Butler, Elmer and Ray E. Marsell, 1972. Cloudburst Floods in Utah, 1939-1969, prepared by U.S. Geological Survey in cooperation with the Utah Department of Natural Resources, Utah.

Croft, A. R., and R. G. Marston, 1950. Summer Rainfall Characteristics in Northern Utah. Transactions, American Geophysical Union, 13(1).

Farmer, Eugene and Joel E. Fletcher, 1971. Precipitation Characteristics of Summer Storms at High Stations in Utah. Intermountain Forest and Range Experiment Station, Ogden, Utah.

National Weather Service, 1973. Precipitation-Frequency Atlas of the Western United States, Volume VI-Utah. National Oceanic and Atmospheric Administration, Silver Spring, Maryland.

Sonntag, Gary and Bard Glenne, 1978. Rainfall Intensity Duration Frequency Curves. University of Utah, Salt Lake City, Utah.

U.S. Army Corps of Engineers, 1973. Hydrology for the Farmington-Centerville FPI Study. A technical addendum, Sacramento District.

U.S. Environmental Protection Agency, 1971. Storm Water Management Model, User's Manual, Version II. National Environmental Research Center, Cincinnati, Ohio.

U.S. Water Resources Council, 1977. Guidelines for Determining Flood Flow Frequency. Bulletin #17A of the Hydrology Committee.

Utah State University, 1968. Hydrologic Atlas of Utah. Utah Water Research Laboratory in cooperation with Utah Department of Natural Resources.

Utah State University, 1976. Application of a Hydrologic Model to the Planning and Design of Storm Drainage Systems for Urban Areas. Utah Water Research Laboratory, Logan, Utah.

Wooley, Ralf R., 1946. Cloudburst Floods in Utah, 1850-1938. U.S. Geological Survey, Water Supply Paper 994, Washington, D.C.

INTERNATIONAL SYMPOSIUM ON HYDROMETEOROLOGY
JUNE AMERICAN WATER RESOURCES ASSOCIATION 1982

MULTIDISCIPLINARY APPROACH TO THE FLOOD HYDROLOGY OF FOOTHILL STREAMS IN COLORADO

Robert D. Jarrett and John E. Costa[1]

ABSTRACT: Conventional hydrologic analyses fail to adequately describe the flood hydrology of foothill streams in Colorado. Annual peak flows are derived from snowmelt, rainfall, or a combination of rain on snow. When snowmelt- and rain-generated peaks are examined separately, flood-frequency analyses show different trends based upon elevation. Above about 7,500 feet (2,300 meters), snowmelt dominates; rain generally does not contribute to the flood potential for recurrence intervals less than the 100-year flood. Where rainfall does significantly contribute to flooding above about 7,500 feet (2,300 meters), unit discharges are small when compared with lower elevation floods resulting from rainfall. Below about 7,500 feet (2,300 meters), rainfall-produced floods predominate. Stratigraphic and geomorphic evidence supports these conclusions and indicates that catastrophic cloudbursts floods in foothill zones have recurrence intervals on the order of thousands of years at any given site. In small basins above 7,500 feet (2,300 meters), many large floods attributed to intense rainfall are, in fact, debris flows and not waterfloods. A paleohydraulic method using maximum boulder sizes was developed to reconstruct historic maximum flood peaks in small ungaged foothill basins.
(KEY TERMS: Colorado; debris flows; flood frequency; floods; geomorphology; mixed population.)

INTRODUCTION

More than three-fourths of Colorado's population is concentrated along or near foothills at the base of high mountains. Many foothill streams flow through scenic canyons before flowing onto much flatter plains or plateaus. Because of the esthetic setting, development is increasing rapidly in these canyon areas. However, foothill streams are subject to frequent and often destructive floods. Foothill streams flood as a result of both excessive snowmelt and rainfall, but historically the most destructive flooding results from "cloudburst-type" rainfall associated with severe thunderstorms during the spring and summer.

The concentration of population, coupled with the highly destructive nature of flash floods, dictates that sound scientific methods be employed in flood studies relating to land use planning, establishment of flood-insurance rates, and the design of transportation facilities. Underestimation of flood potential can lead to tragic loss of life and destruction of property. Overestimation of flood potential virtually precludes additional development in the canyons and reduces property values. Consequently, public agencies engaged in land use planning and regulation seek an optimal balance between protection of life and property and recognition of individual property rights.

Techniques used in deriving flood characteristics for foothill streams are inadequate or unverified; thus, vastly differing results can be obtained from independent analyses of the same flood record. Because some floods result from snowmelt at higher elevations whereas others result from intense rainfall at lower elevations, available flood records for foothill streams are applicable only to the collection sites or to nearby sites on the same stream.

Purpose

As a result of the above problems and the potentially serious consequences, a comprehensive, multidisciplinary project was undertaken to evaluate the flood hydrology of foothill streams in Colorado. On July 31, 1976, the need for a study of this type was dramatically reinforced by the occurrence of the devastating Big Thompson flood along the Front Range foothills (McCain and others, 1979). As much as 12 inches of rain fell on about 60 square miles of the drainage area. The flood lasted only a few hours, but, during that time, it caused the loss of at least 139 lives and property damage estimated at $35 million. The flood crest on the Big Thompson River moved through the 7.6 miles between Drake and the canyon mouth in about 30 minutes, an average speed of about 15 miles per hour. This paper provides an overview of conclusions of the multidisciplinary study.

[1]Respectively, Hydrologist, U.S. Geological Survey, Water Resources Division, Colorado District, Building 53, Mail Stop 415, Box 25046, Denver Federal Center, Lakewood, Colorado 80225; and Associate Professor, Department of Geography, University of Denver, Denver, Colorado 80208, and U.S. Geological Survey, Mail Stop 415, Box 25046, Denver Federal Center, Lakewood, Colorado 80225.

Approach

This comprehensive study of streams in the foothills region of Colorado concentrated on: (1) analysis of available streamflow and precipitation data, (2) the use of geomorphic and stratigraphic techniques in flood-hydrology studies, (3) flood information transfer techniques, and (4) additional data needs. A multidisciplinary approach in which the analysis of available streamflow and precipitation data is complemented with geomorphic and stratigraphic techniques can answer questions about the flood hydrology of an area with little data that no study limited to a single discipline can provide.

ANALYSES OF EXISTING DATA

Typically, records from streamflow-gaging stations have been used where available to derive flood-frequency characteristics for gaged sites. The predominant computation method is to fit the log-Pearson Type-III frequency distribution (U.S. Water Resources Council, 1981) to the observed annual flood series. This conventional hydrologic analysis fails to account for the mixed population of rainfall-runoff and snowmelt-runoff peaks contributing to the total population of flood peaks in the foothills region.

Streamflow records of 69 unregulated streams in the South Platte River, the Arkansas River, and the Colorado River basins were examined to determine peak discharges from snowmelt and rainfall runoff during each water year. Hydrograph shape and weather records were used to determine the meteorologic source of runoff. Snowmelt-runoff peaks result from the seasonal ablation of snowpack and appear on the hydrographs as low peaks that fluctuate diurnally. Rainfall runoff from convective storms or frontal systems affects streamflow much more quickly and has a rapidly peaked hydrograph.

Methods described in the U.S. Water Resources Council Bulletin 17B (1981) generally were followed to determine the station flood-frequency curves including development of generalized skew coefficients for the rainfall-runoff and snowmelt-runoff peak-flow arrays. If the peak flows are classified according to meteorologic cause, separate frequency curves of each population can be constructed and a composite flood-frequency curve can be created. The separate snowmelt-runoff and rainfall-runoff flood-frequency curves for each station can be combined, if the populations are independent, by using the equation:

P (composite) = P (snowmelt) + P (rainfall) – P (snowmelt x rainfall),

where P is the probability of occurrence (Crippen, 1978). The flood-frequency curves for one of the 69 stations analyzed, Clear Creek near Golden, Colorado, are shown in Figure 1.

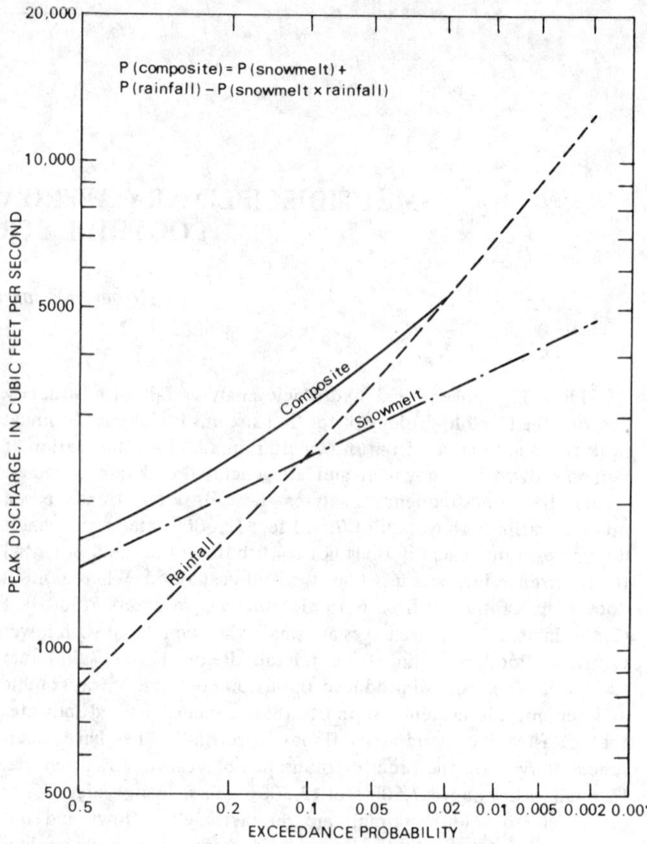

Figure 1. Flood Frequency Curves for Clear Creek Near Golden, Colorado.

Inspection of the plotted rainfall and snowmelt flood-frequency curves indicates which meteorologic cause predominates at a station. The following comparison between a low- and high-elevation station helps demonstrate that the change from snowmelt- to rainfall-dominated flooding is abrupt over a small range in elevation. Clear Creek near Golden (Figure 1) has a gage elevation of 5,735 feet (1,748 meters). The snowmelt-runoff peak flows predominate (based on where the snowmelt and rainfall frequency curves cross) to the 0.12-probability flood; then rainfall-runoff peak flows predominate. Clear Creek near Lawson (Figure 2) has a gage elevation of 8,080 ft. (2,963 m). In this case, the snowmelt-runoff peak flows predominate to the 0.002-probability flood; then rainfall-runoff peak flows predominate.

Similar comparisons were made for all 69 stations and indicate that snowmelt runoff dominates above about 7,500 ft. (2,300 m) (the elevation varies with major river basins). Above this elevation, rainfall generally does not contribute to floods having a recurrence interval of less than 100 years (0.01-probability) of occurrence. Below about 7,500 ft. (2,300 m), rainfall-produced peak flows predominate.

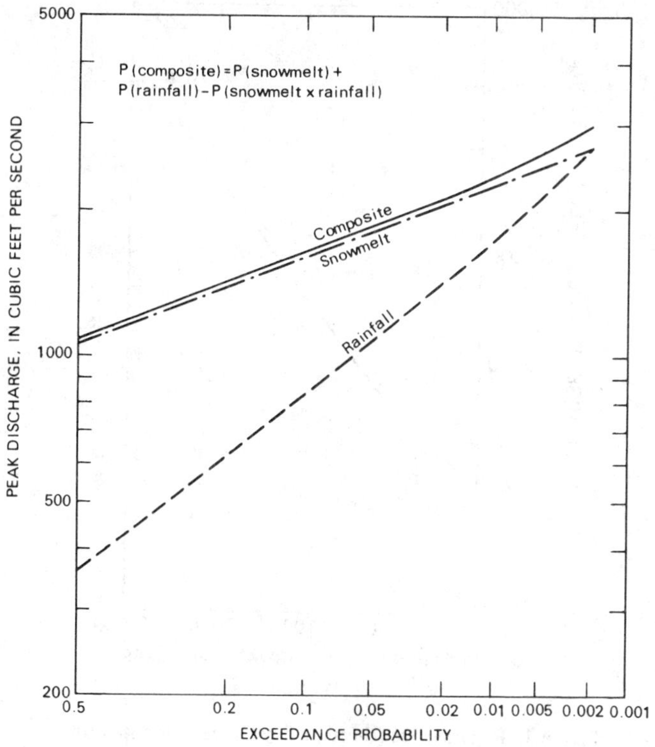

Figure 2. Flood Frequency Curves for Clear Creek Near Lawson, Colorado.

A few of the higher elevation stations exhibit flood-frequency relations in which rainfall runoff predominates, but this probably is due to low snowmelt peak flows. The important point to note is that the rainfall peak flows at these higher elevations have a relatively low unit discharge when compared to lower elevation rainfall peak discharges. For example, Fish Creek near Estes Park, Colorado, at an elevation of 7,476 ft. (2,279 m) with a drainage area of 16 square miles (41.4 square kilometers) has a maximum recorded rainfall-runoff peak flow of 400 cubic feet per second ($ft.^3/s$) in 33 years, or 25 cubic feet per second per square mile ($ft.^3/s/mi.^2$). Even the 100-year rainfall-runoff discharge is only 603 $ft.^3/s$ or 37.7 $ft.^3/s/mi.^2$. Basins of comparable size at lower elevations in the foothills and plains regions have unit peak-discharge values of hundreds to thousands of cubic feet per second per square mile.

One task of the project was to compile a record of recent floods resulting from intense rainfall in the Colorado mountains and to review literature describing historic mountain flooding. Gaging-station records and on-site investigations indicate that debris flows in small mountain streams have been confused with waterfloods. A debris flow is a rapidly flowing, structurally coherent slurry of water and sediment, resembling wet concrete. Geomorphic, sedimentologic, and hydraulic evidence attributed to debris flows were found in five of seven sites studied, each with an area of less than 4 $mi.^2$ (10 km^2) and an altitute greater than 7,500 ft. (2,300 m), where floods resulting from intense rainfall had been reported (Costa and Jarrett, 1981).

The distinction between debris flows and waterfloods is important because: (1) mitigating procedures for waterfloods, such as channelization and damming, may not be effective for debris flows, (2) peak-discharge measurements based on evidence of debris flows lead to excessive estimates of the magnitude of floods in the Rocky Mountains, and (3) because of sparse rainfall data in mountain regions, indirect-discharge estimates have been used by some hydrologists and engineers to determine the amount of rainfall that occurred during a storm. This may lead to inaccurate estimates of rainfall and flood discharges that are used in the design of flood-control structures. It is probable other documented waterfloods in mountainous areas are also debris flows.

An understanding of the flood hydrology of an area requires sufficient information on the processes causing the floods. This would require a network of precipitation gages to document the types of storms and type, volume, intensity, and areal extent of precipitation. The existing precipitation network is too sparse to record the very localized, intense, severe rainstorms that are the most frequent cause of flash floods in the United States (Changnon and Vogel, 1981). This is particularly true of the foothills region. Precipitation studies by Hansen, and others (1978) and Henz (1974) and documented large storms in Colorado indicate that intense cloudburst storms are not as common above 7,000 ft. (2,130 m) as below 7,000 ft. (2,130 m) in Colorado.

GEOMORPHIC AND STRATIGRAPHIC TECHNIQUES

Historic and prehistoric floods in foothill streams in Colorado are recorded in valleys and channels as distinctive deposits and landforms. The interpretation of these deposits and landforms can provide important supplemental information about the spatial occurrence, magnitude, and frequency of large floods.

The maximum size of boulders in coarse boulder deposits from historic catastrophic flash floods in the foothills region was used as the independent variable to reconstruct mean flood velocity and average depth (Costa, 1981). Several valley cross sections in reaches with bedrock controls were used to reconstruct discharges. When tested at eight sites, reconstructed discharges ranged from +31 to −75 percent and averaged −28 percent of the discharges determined from conventional slope-area measurements. Because flood-deposited boulder bars persist in foothill valleys for many centuries, this paleohydraulic reconstruction technique is widely applicable in small foothill streams where stratigraphic and geomorphic evidence of severe floods exist, but no hydrologic data are available.

Old historic and prehistoric flood deposits have been dated by botanical techniques. The age of the largest trees (determined from counting tree rings) growing on flood deposits or landforms gives a minimum time since the flood occurred. On historic flood deposits, lichen cover percentages have been estimated. This may be useful as a supplemental dating method. For example, a 5- to 15-percent cover of lichens on stable flood boulder bars on foothills flood plains would indicate the boulders were deposited about 40-50 years ago.

Radiocarbon dating of landforms and deposits has also been used to estimate the recurrence interval of floods (Costa, 1978). Figure 3 shows the stratigraphy of a large, truncated debris fan deposited in the Big Thompson Canyon that was extensively eroded by the flood of 1976. Deposits as old as 7,000 to 10,000 C^{14} years were eroded for the first time. Figure 4 shows how the geomorphic and stratigraphic data can be used in flood-frequency analysis. Figure 4 is a regionalized flood-frequency curve for the site shown in Figure 3 (Costa, 1978). Extrapolation of the curve to the discharge of the 1976 flood at this site (28,200 $ft.^3/s$) would give a recurrence interval greater than

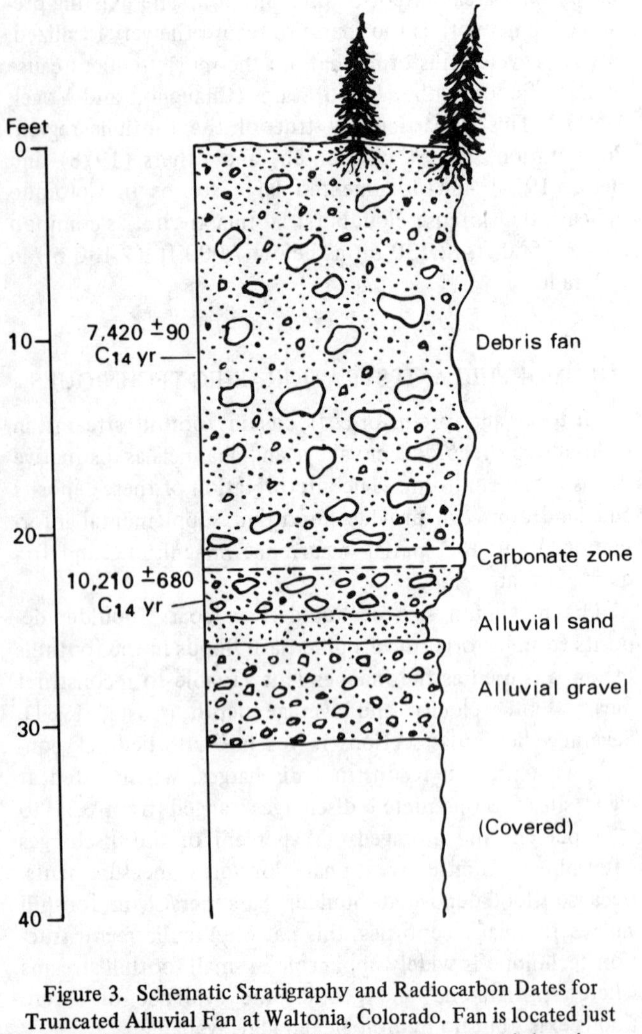

Figure 3. Schematic Stratigraphy and Radiocarbon Dates for Truncated Alluvial Fan at Waltonia, Colorado. Fan is located just below area of maximum runoff. Peak discharge at this location is estimated to be 28,200 cubic feet per second. Radiocarbon dates from the Radiocarbon Laboratory, University of Texas, Austin, Texas (Tx 2836, Tx 2837.)

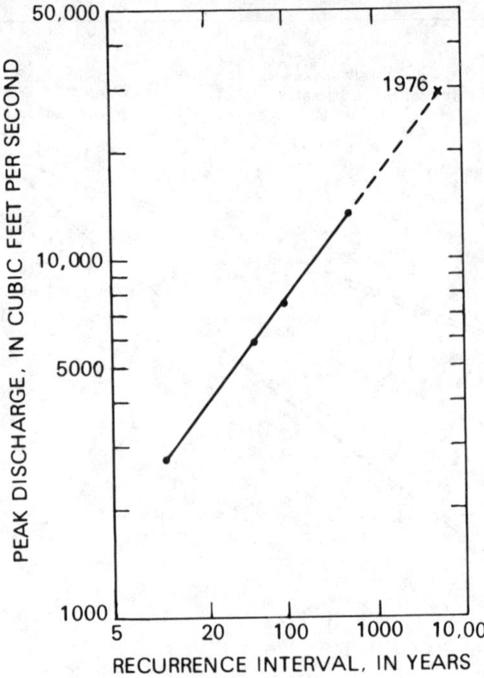

Figure 4. Regional Flood-Frequency Curve Computed for Ungaged Cross Section on Big Thompson River at Alluvial Fan Shown in Figure 3. Curve developed from multiple regression techniques of McCain and Jarrett (1976). Relation was extended to the estimated flood peak at this location (28,200 cubic feet per second).

5,000 years at this cross section, which is in excellent agreement with the radiocarbon dates (7,000-10,000 C-14 years) from the truncated alluvial fan shown in Figure 3.

INVESTIGATIONS OF FLOOD INFORMATION TRANSFER TECHNIQUES

The analysis of available data shows how to evaluate the flood-frequency relations at gaging stations. However, flood characteristics are also required at ungaged sites. This information can be obtained through flood information transfer techniques. Two general methods traditionally have been used for deriving flood characteristics of streams in Colorado. One is the regionalization of gaging-station flood-frequency analyses on basin physical or climatic parameters. The second is to derive flood characteristics through synthetic hydrologic or rainfall-runoff modeling methods. Both of these methods failed to adequately describe the flood hydrology of foothill streams (McCain and Ebling, 1979). Aside from the fact that there are limited rainfall and streamflow data, all transfer methods assume the total basin area contributes runoff during cloudburst rainstorms. However, available data indicate that floods in the foothills region are caused by cloudbursts of very limited areal extent. Geomorphic, sedimentologic, and botanic techniques have been successfully applied to identify contributing drainage areas following severe rainstorms. This is done by mapping the spatial extent of landforms, deposits, and damaged vegetation in a watershed following a severe rainstorm to identify the contributing area of runoff. This study shows that with existing data the most promising and practical method to determine flood characteristics is use of multiple-regression techniques of gaging-station flood-frequency analyses on the basis of basin physical, climatic, and geomorphic parameters. Special emphasis is needed to define the contributing drainage area.

ADDITIONAL DATA NEEDS

The existing streamflow data and data network are a reflection of the needs of current and past water users. Most data are collected to provide streamflow information for the users of agricultural, domestic, and industrial water supplies and not specifically for the determination of flood characteristics. Thus, there is a lack of data on small drainage basins in the foothills area. Consequently 17 crest-stage gages were established to provide peak-flow data on small watersheds of foothill streams and to evaluate this type of gage for a more extensive data-collection program.

An urgent need exists for additional hydrologic information in the foothills region in Colorado. One of the objectives of this project was to provide input for the design of a data-collection network. This study provides insight into the shortcomings of the existing data network and the types of data required. To understand the flood hydrology of foothill streams will require additional streamflow and precipitation data. A streamflow-data network requires additional continuous-recording and crest-stage gages, and methods need to be applied to measure floods that do not occur at gaging stations. Recording precipitation gages and radar and satellite data are needed to provide information on precipitation intensities, volumes, areal distribution, and topographic effects. Geomorphic and stratigraphic information is needed to supplement hydrologic data. An additional data-collection program does not have to be undertaken by one agency; however, the data-collection program should be coordinated. The important aspects of the additional data-collection program are: (1) that it meet the needs of as many users as possible, (2) that the data are collected by generally approved methods, and (3) that the data are readily available to all users.

SUMMARY

A multidisciplinary study was undertaken to analyze the flood hydrology of foothill streams in Colorado because of the inadequacy of existing techniques, lack of precipitation and streamflow data, and complexity of the hydrologic processes involved. Geomorphic and stratigraphic techniques were incorporated into the analyses to extend the spatial and temporal framework of foothills flood hydrology enough to immediately answer questions, without waiting for the conventional collection and analysis of additional data. A more meaningful and useful precipitation and streamflow data-collection network based on the results of this study would greatly enhance meaningful data collection.

LITERATURE CITED

Changnon, S. A. and J. L. Vogel, 1981. Hydroclimatological Characteristics of Isolated Severe Rainstorms. Water Resources Research 17(6):1694-1700.

Costa, J. E., 1978. Holocene Stratigraphy in Flood-Frequency Analysis. Water Resources Research 14:626-632.

Costa, J. E., 1981. Verification of Paleohydraulic Reconstruction. Geological Society of America. Abstract with Programs 13(7):431.

Costa, J. E. and R. D. Jarrett, 1981. Debris Flows in Small Mountain Stream Channels of Colorado and Their Hydrologic Implications. Association of Engineering Geologists Bulletin 18:309-322.

Crippen, J. R., 1978. Composite Log-Type III Frequency-Magnitude Curve of Annual Floods. U.S. Geological Survey Open-File Report 78-352, 5 pp.

Hansen, W. R., John Chronic, and John Matelock, 1978. Climatography of the Front Range Urban Corridor and Vicinity, Colorado. U.S. Geological Survey Professional Paper 1019, 59 pp.

Henz, J. G., 1974. Colorado High Plains Thunderstorm System – A Descriptive Radar-Synoptic Climatology. Fort Collins, Colorado State University, Department of Atmospheric Science, unpublished M.S. thesis, 66 pp.

McCain, J. F. and J. L. Ebling, 1979. A Plan for Study of Flood Hydrology of Foothill Streams in Colorado. U.S. Geological Survey Open-File Report 79-1276, 29 pp.

McCain, J. F., L. R. Hoxit, R. A. Maddox, C. F. Chappell, and Fernando Caracena, 1979. Meteorology and Hydrology in Big Thompson River and Cache la Poudre River Basins, Larimer and Weld Counties, Colorado, Part A of Storm and Flood of July 31-August 1, 1976, in the Big Thompson River and Cache la Poudre River Basins, Larimer and Weld Counties, Colorado: U.S. Geological Survey Professional Paper 1115A, pp. 1-85.

McCain, J. F. and R. D. Jarrett, 1976. Manual for Estimating Flood Characteristics of Natural-Flow Streams in Colorado. Colorado Water Conservation Board, Technical Manual No. 1, 68 pp.

U.S. Water Resources Council, 1981. Guidelines for Determining Flood Flow Frequency, Revised Bulletin 17B of the Hydrology Committee. Washington, D.C., Government Printing Office, 28 pp., 14 apps.

INTERNATIONAL SYMPOSIUM ON HYDROMETEOROLOGY
JUNE AMERICAN WATER RESOURCES ASSOCIATION 1982

SYNTHETIC UNIT-HYDROGRAPH PARAMETERS: DAM SAFETY PROGRAM

Krishan P. Singh[1]

ABSTRACT: The National Dam Inspection Act of 1972 authorized the Corps of Engineers to conduct hydrologic, hydraulic, structural, and geologic evaluations of dams with a high or significant hazard potential. Extensive precipitation and unit hydrograph analyses were conducted in the Illinois State Water Survey to help with the hydrologic evaluations. Four to six observed hydrographs for high floods and associated storm hyetographs for each of 131 basins were analyzed to develop unit hydrographs suitable for converting a 100-year rainfall excess to a 100-year flood hydrograph. Eleven unit-hydrograph parameters were evaluated: unit-hydrograph duration, peak, and time to peak; time to 0.25-, 0.50-, and 0.75-peak for both the rising and falling limbs of the unit hydrograph; time base; and a factor relating a small change in duration to the change in the peak. These parameters delineate a unit hydrograph very satisfactorily. The state was divided into eight homogeneous regions. The parameters in each region were regressed with easily measurable basin factors. The correlation coefficients were highly significant.
(KEY TERMS: unit hydrograph; synthetic hydrology; regression analyses; dam safety; basin factors; regional analysis.)

INTRODUCTION

The National Dam Inspection Act of 1972 authorized the Secretary of the Army, acting through the Chief of Engineers, to initiate an inventory program for all dams satisfying certain size criteria, and a safety inspection of all non-federal dams in the United States that are classified as having a high or significant hazard potential. The dam safety inspection reports contain hydraulic and hydrologic evaluations which require information on storms of various frequency and the soil moisture conditions at the beginning of a design storm, as well as suitable unit hydrographs for converting the design storms into flood hydrographs. The unit hydrographs are derived by a number of methods, mostly untested for Illinois conditions.

1) The watershed area is subdivided into smaller subareas. The subarea synthetic hydrographs are lagged by appropriate travel time and accumulated at the basin outlet to obtain the total watershed synthetic unit hydrograph.

2) A synthetic unit hydrograph is derived with the Snyder's (1938) equations developed from an analysis of a large number of hydrographs from drainage basins in the Appalachian Mountain region.

3) Unit hydrographs can be derived from model hydrographs (Mitchell, 1972) which are developed from a dimensionless translation hydrograph, having a base time of T hours and a reservoir storage $S = k\,Q^x$.

4) Clark (1945) suggested the derivation of a unit hydrograph by routing the time-area diagram of a basin through storage $S = k\,Q$. The difficulty lies in estimating the time base of the time-area-diagram and the routing constant k.

PROBLEM

Unit-hydrograph procedures presently in use yield unit hydrographs (from analyses of storms and flood hydrographs) which may be suitable for deriving 2- to 5-year or 10-year floods because of the averaging processes inherent in these procedures. For evaluation of the adequacy of spillway capacity for dam safety, unit hydrographs suitable for deriving 100-year flood and probable maximum flood (PMF) hydrographs are needed. Notwithstanding the principle of linearity of the unit hydrograph, the unit hydrographs derived from very high floods yield higher peaks and shorter times to the peak than those derived from small- to medium-size floods although the degree of increase in peak and decrease in time to peak varies from basin to basin and region to region, depending on the physiographic, channel, and basin factors.

Because of the errors associated with estimation of linear storage routing coefficients and time base of the area-time-diagrams, an approach similar to that of Snyder is considered in this study. Not only the information on the unit hydrograph widths at 75, 50, and 25 percent of peak discharge but also the time to these discharges from the beginning of the unit hydrograph need to be known for better definition of the unit hydrograph. Snyder's ratio of t_p to t_r is given a constant value of 5.5, whereas it depends on drainage area and other factors. The objectives of this study were to derive a sufficient number of unit hydrograph parameters for satisfactorily defining unit hydrographs, suitable for determining 100-year flood hydrographs, from

[1] Principal Hydrologist, Illinois State Water Survey, 605 East Springfield, Champaign, Illinois 61820.

easily measurable basin factors and to divide the state into homogeneous regions in terms of similar hydrologic responses to imposed storm inputs.

METHODOLOGY FOR DERIVING UNIT HYDROGRAPHS

The methodology, detailed herein for one basin, was applied to all the 131 basins considered in this study.

1) Stage hydrographs for six to seven floods were converted to discharge hydrographs with the appropriate rating tables. The number of discharge values, spaced at equal time intervals, varied from about 40 to 100.

2) For each hydrograph, hourly and daily rainfall recorded at raingage stations in and around the basins, were tabulated for the storm causing the flood. The daily rainfall values, for 10 days prior to the storm and 5 days after it, were analyzed to define the baseflow separation satisfactorily as well as to modify the falling limb of the hydrograph slightly if some small rainfall (after the major storm) interfered with the usual flow recession.

3) On the basis of the information developed above, three to four flood hydrographs were selected which were well-defined, were sharp-peaked, and had low baseflow, and for which the temporal and spatial distribution of rainfall over the basin was relatively uniform.

4) The baseflow was approximated by a straight line with an upward trend from the beginning of surface runoff to its end.

5) The surface runoff hydrograph ordinates (obtained by subtracting the baseflow from the total flow) were tabulated and stored in the computer. The duration of effective rainfall was estimated from the basin hyetograph.

6) A computer program calculated the unit hydrograph and the S-curve (Chow, 1964) with the duration of effective rainfall determined in step 5, as well as with two durations somewhat higher and two durations somewhat lower than the effective rainfall duration.

7) A suitable unit-hydrograph duration was selected from the information obtained in step 6; the selection was based primarily on closeness to the duration determined in step 5 and the smoothness of S-curves derived by assuming shorter or longer durations.

8) Unit hydrographs for the four flood events, as obtained in step 7, were examined to determine a suitable duration for such high floods. All four unit hydrographs were transformed to unit hydrographs with this suitable duration as well as with two somewhat higher and two somewhat lower durations. The use of different durations helped in defining the rate of change in the unit hydrograph peak with a small change in duration. It also helped in defining such changes in time to the peak and in time to various percent flows, both on the rising and falling limb of the unit hydrograph.

9) The unit hydrograph peaks were plotted with respect to the recurrence interval of the floods (from which they were derived) to discern any trend of increase in peak with the recurrence interval. The expected unit-hydrograph peak and time to the peak for a 100-year recurrence interval flood were estimated.

10) The final unit hydrograph, suitable for developing a 100-year flood hydrograph with the expected peak and time to peak, was drawn, conforming to the general shape of the four unit hydrographs and with runoff equal to 1 inch.

11) The following unit-hydrograph parameters (Figure 1) were estimated.

t_r = duration of effective rainfall or of unit hydrograph, hours
U_p = peak discharge, cfs
$t_{.25}$ = time to 0.25 U_p, hours
$t_{.50}$ = time to 0.50 U_p, hours
$t_{.75}$ = time to 0.75 U_p, hours
t_p = time to peak, hours
$d_{.75}$ = unit hydrograph width at 0.75 U_p, hours
$d_{.50}$ = unit hydrograph width at 0.50 U_p, hours
$d_{.25}$ = unit hydrograph width at 0.25 U_p, hours
t_b = time base of unit hydrograph, hours

In addition, the value of a was computed which gives the rate of increase/decrease in U_p with a small decrease/increase in t_r, in cfs per hour.

REGIONALIZATION OF PARAMETERS

The state was divided into 12 regions, demarcated on the basis of physiography, model flow duration (Singh, 1971), and hydrologic and climatologic homogeneity. Regression analyses with each of the unit hydrograph parameters as a dependent variable and one or more of the basin factors (A, L, and S) as independent variables were made for the basins in each of these regions. The basins from adjacent regions were added to a region or transferred from a region to another region in a systematic way so as to obtain the minimum number of regions, providing maximum homogeneity of unit hydrograph relations for the basins within each region. In these analyses, a small time shift in the beginning of the unit hydrographs of some basins significantly improved the correlations between unit hydrograph parameters and basin factors. This small shift is attributed to imprecise information on the exact time of the beginning of rainfall excess or effective rainfall. As a result of stepwise multiple correlations for each of the parameter sets from about 200 combinations of basins, the state was divided into eight regions. The number of study basins in each region are:

Region	1	2	3	4	5	6	7	8
Basins	15	20	11	12	26	11	19	17

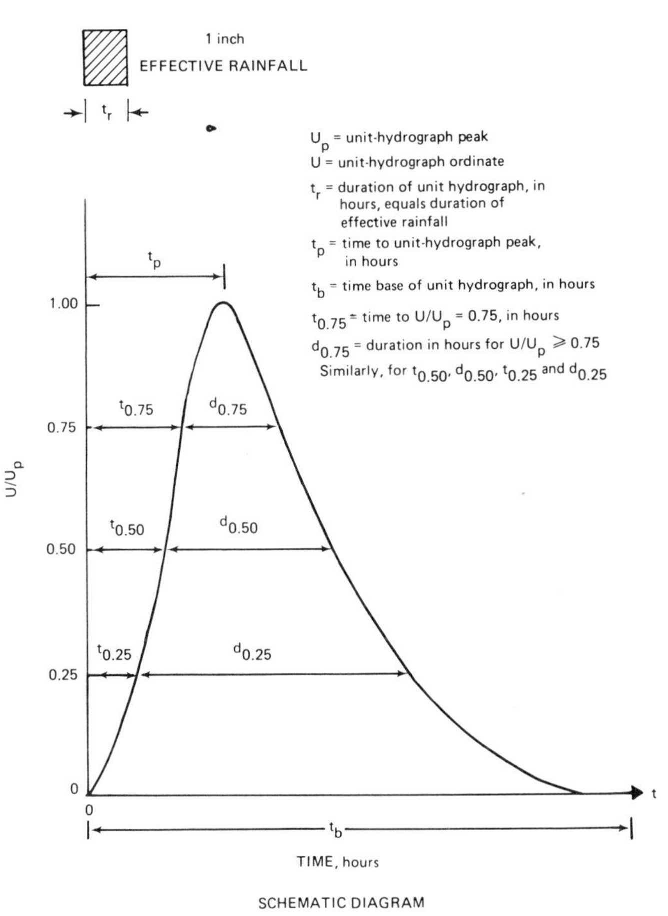

Figure 1. Unit Hydrograph Parameters – Definition Sketch.

The effect of using a unit hydrograph duration t_R, rather than the t_r, on the values of various unit hydrograph parameters is generalized by the following equations on the basis of evaluations made in this study.

$$U_{pR} = U_p - a(t_R - t_r)$$

$$t_{pR} = t_p + 0.5(t_R - t_r)$$

$$t_{.25R} = t_{.25} + 0.5(t_R - t_r)$$

$$t_{.50R} = t_{.50} + 0.5(t_R - t_r)$$

$$t_{.75R} = t_{.75} + 0.5(t_R - t_r)$$

$$t_{bR} = t_b + (t_R - t_r)$$

$$d_{.25R} = d_{.25} + (t_R - t_r)$$

$$d_{.50R} = d_{.50} + 0.75(t_R - t_r)$$

$$d_{.75R} = d_{.75} + 0.5(t_R - t_r)$$

In the above equations, t_R is only slightly higher or lower than t_r. These equations serve the purpose of adjusting the parameters for a change in t_r to a desirable value of t_R from the consideration of a satisfactory time interval between the unit hydrograph ordinates; e.g., for $t_r = 2.76$, a t_R of 3.0 may be more suitable, and for $t_r = 0.43$, a t_r of 0.50 may be more suitable.

The basin factors (A, L, and S) are defined as follows: A = drainage area, in square miles, above the gaging station, as shown in the latest USGS streamflow reports; L = main channel length, in miles, from the gaging station to the basin divide as measured on topographic maps; and S = main channel slope, in feet per mile, determined from elevations at 10 and 85 percent of the distance along the main channel from the gaging station to the basin divide.

The following information was developed for each of the eight regions:

1) Number of basins, stream and gaging stations and their USGS number, drainage area, and length and slope of main channel.

2) Simple regression equations with log L and log S as dependent variables and log A as an independent variable.

3) Derived unit hydrograph parameters: t_r, t_p, U_p, $t_{.75}$, $d_{.75}$, $t_{.50}$, $d_{.50}$, $t_{.25}$, $d_{.25}$, t_b, and a for all the basins; and results of stepwise multiple regressions with log t_r and log a as dependent variables and logs of A, L, and S as independent variables.

4) Modified unit hydrograph parameters for the fitted t_r and a values for all basins in a region.

5) Most significant regression equations, with logarithms of modified unit hydrograph parameters as dependent variables, together with values of standard error, S_e, and correlation coefficient r or R.

6) A summary of basins with fitted U_p within ±10 and ±25 percent of derived U_p and any basins with more than ±25 percent variation.

EXAMPLE: REGION 5

The rivers, streams, and tributaries included in this region are shown in Figure 2, together with the location of the 26 gaging stations used for deriving the unit-hydrograph parameters. These gaging stations; their drainage area, A, above the gaging station; main-channel length, L; and main-channel slope, S, are given in Table 1. The stepwise multiple correlation analyses yielded the following results with the derived t_r and a values at each of the 26 stations:

$$\log t_r = -0.324 + 0.482 \log A - 0.100 \log S\ ;$$
$$(S_e = 0.060;\ R = 0.993)$$

$$\log a = 0.577 + 0.256 \log A + 1.151 \log S\ ;$$
$$(S_e = 0.053;\ R = 0.983)$$

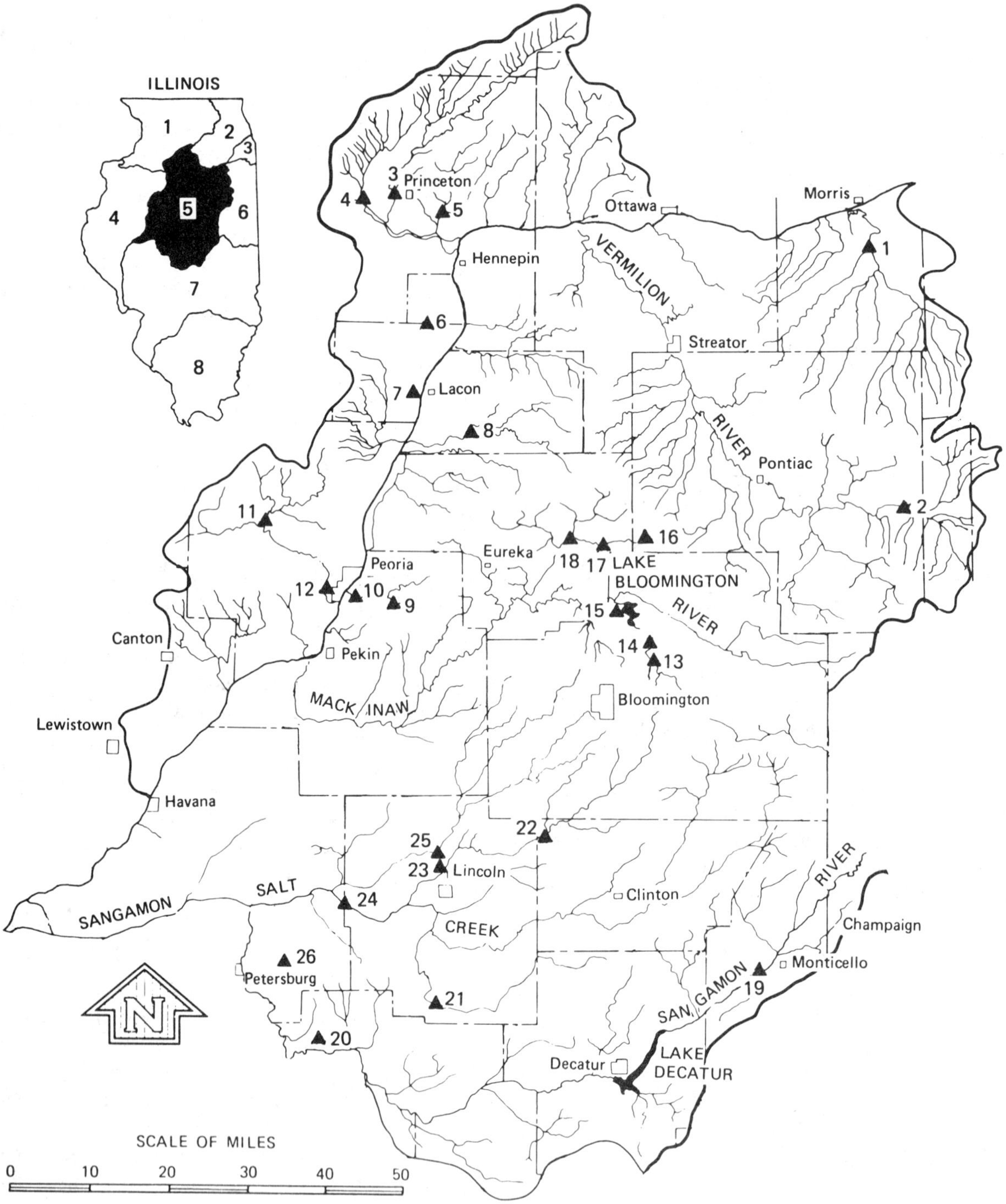

Figure 2. Study Basins in Region 5.

TABLE 1. Basin Factors and Derived, Modified, and Computed Values of Unit Hydrograph Parameters.

	Basin Factors			Derived			Modified			Computed*		
No.	Area	Length	Slope	t_r	t_p	U_p	t_r	t_p	U_p	t_r	t_p	U_p
1	455.00	36.27	4.33	8.00	24.00	10,400	7.84	23.92	10,416	7.84	25.05	9,026
2	186.00	23.00	5.39	5.00	15.50	4,300	4.98	15.49	4,302	4.98	16.04	5,632
3	196.00	54.59	6.07	5.00	14.00	7,000	5.05	14.02	6,994	5.05	15.51	6,356
4	86.70	22.54	9.03	3.00	9.00	5,000	3.27	9.14	4,959	3.27	9.45	4,757
5	99.00	23.50	12.72	3.00	7.50	6,800	3.37	7.69	6,714	3.37	8.52	6,657
6	56.20	27.49	10.24	3.00	9.00	3,350	2.62	8.81	3,408	2.62	7.55	3,838
7	5.66	4.81	53.86	1.00	1.50	2,400	0.73	1.37	2,554	0.73	1.46	2,494
8	115.00	27.68	6.07	4.00	12.50	5,000	3.90	12.45	5,009	3.90	12.62	4,374
9	11.20	6.72	39.86	1.00	2.33	3,000	1.05	2.36	2,974	1.05	2.18	3,250
10	61.20	18.60	18.90	2.00	5.00	7,000	2.57	5.28	6,818	2.57	5.91	6,293
11	119.00	22.18	10.93	4.00	8.00	10,000	3.74	7.87	10,052	3.74	9.80	6,800
12	297.00	39.36	7.50	5.00	14.50	13,000	6.04	15.02	12,828	6.04	16.56	9,883
13	49.00	25.78	5.25	3.00	9.00	2,200	2.62	8.81	2,225	2.62	9.68	2,170
14	53.10	29.20	4.91	3.00	9.00	2,200	2.75	8.87	2,216	2.75	10.29	2,189
15	9.81	6.74	11.88	1.25	3.50	1,200	1.11	3.43	1,215	1.11	3.59	1,254
16	6.30	3.11	11.14	1.00	3.50	700	0.90	3.45	709	0.91	3.11	879
17	30.50	8.47	4.54	2.00	8.50	1,600	2.12	8.56	1,593	2.12	8.60	1,404
18	93.90	13.59	4.22	4.00	13.00	3,200	3.67	12.83	3,220	3.67	13.75	2,932
19	0.10	0.37	34.11	0.08	0.33	130	0.11	0.34	126	0.11	0.38	106
20	1.50	1.36	40.13	0.42	1.12	625	0.40	1.11	631	0.40	1.00	798
21	214.00	37.00	4.65	6.00	24.00	3,600	5.41	23.70	3,652	5.41	18.11	5,595
22	227.00	36.08	6.23	5.00	20.00	6,000	5.40	20.20	5,949	5.40	16.23	7.177
23	306.00	54.48	5.12	6.00	21.00	7,600	6.36	21.18	7,561	6.36	19.91	7,698
24	0.90	1.55	48.94	0.33	0.82	600	0.31	0.81	607	0.31	0.75	642
25	333.00	42.77	5.76	6.00	23.00	8,200	6.55	23.28	8,130	6.55	19.51	8,879
26	0.94	1.57	23.76	0.33	1.09	450	0.34	1.09	449	0.34	1.06	396

*Computed with the regional regression equations.

With the fitted values of t_r and a, the remaining nine unit hydrograph parameters were modified for any difference between the derived and fitted values of the two parameters, using the equations given in the section under "Regionalization of Parameters". The significant regression equations obtained for the modified values are:

$$t_r = 0.474\,A^{0.482}\,S^{-0.100}$$

$$a = 3.777\,A^{0.256}\,S^{1.151}$$

$$t_p = 4.539\,A^{0.388}\,S^{-0.453} \quad (S_e = 0.054;\ R = 0.995)$$

$$U_p = 43.76\,A^{0.701}\,S^{0.709} \quad (S_e = 0.080;\ R = 0.988)$$

$$t_{.75} = 3.600\,A^{0.374}\,S^{-0.469} \quad (S_e = 0.065;\ R = 0.992)$$

$$d_{.75} = 4.561\,A^{0.360}\,S^{-0.593} \quad (S_e = 0.076;\ R = 0.991)$$

$$t_{.50} = 3.005\,A^{0.374}\,S^{-0.481} \quad (S_e = 0.062;\ R = 0.993)$$

$$d_{.50} = 9.184\,A^{0.334}\,S^{-0.645} \quad (S_e = 0.083;\ R = 0.989)$$

$$t_{.25} = 2.223\,A^{0.369}\,S^{-0.493} \quad (S_e = 0.065;\ R = 0.992)$$

$$d_{.25} = 23.40\,A^{0.289}\,S^{-0.763} \quad (S_e = 0.088;\ R = 0.987)$$

$$t_b = 55.73\,A^{0.281}\,S^{-0.709} \quad (S_e = 0.070;\ R = 0.991)$$

The derived, modified, and computed values for t_r, t_p, and U_p are given in Table 1.

APPLICATION OF REGIONALIZED RELATIONS

As an example, a unit hydrograph, suitable for developing a 100-year flood hydrograph, is derived for the Leaf River at Leaf River, Illinois. The following information is available: USGS gaging station number, 05 441000; drainage area, A, 103 square miles; main-channel length, L, 18.27 miles; main-channel, slope, S, 10.45 ft./mi.; and region number, 1. The unit hydrograph parameters computed with regression equations for region 1 are given below, together with parameters with $t_R = 3.0$ hours.

$$t_r = 0.382\,A^{0.415} = 2.61 \text{ hours}$$

$$a = 63.56\,S^{0.246} = 113 \text{ cfs/hour}$$

	for $t_R = 3.0$ hours
$U_p = 118.7\,A^{0.618}\,S^{0.393}$	
$= 5235$ cfs	5191 cfs
$t_p = 1.329\,A^{0.788}\,L^{-0.530}$	
$= 10.99$ hours	11.19 hours
$t_{.75} = 1.240\,A^{0.970}\,L^{-0.904}$	
$= 8.04$ hours	8.24 hours
$t_{.50} = 1.140\,A^{1.044}\,L^{-1.071}$	
$= 6.41$ hours	6.61 hours
$t_{.25} = 0.877\,A^{1.082}\,L^{-1.181}$	
$= 4.27$ hours	4.47 hours
$d_{.75} = 1.658\,A^{0.426}\,S^{-0.266}$	
$= 6.40$ hours	6.60 hours
$d_{.50} = 3.565\,A^{0.402}\,S^{-0.336}$	
$= 10.44$ hours	10.73 hours
$d_{.25} = 7.089\,A^{0.387}\,S^{-0.383}$	
$= 17.35$ hours	17.74 hours
$t_b = 16.87\,A^{0.372}\,S^{-0.326}$	
$= 44.02$ hours	44.41 hours

SUMMARY

Regionalized relations for 11 unit hydrograph parameters, developed in this study, provide an easy and relatively accurate determination of a unit hydrograph suitable for deriving 100-year and probable maximum flood hydrographs with the 100-year and probable maximum storms for an area up to about 340 square miles for drainage basins in Illinois (Singh, 1981).

LITERATURE CITED

Chow, V. T. (editor), 1964. Handbook of Applied Hydrology. McGraw-Hill Book Company, New York.

Clark, C. O., 1945. Storage and the Unit Hydrograph. Transactions, ASCE, Vol. 100:1419-1418.

Mitchell, W. D., 1972. Model Hydrographs. USGS Water Supply Paper 2005.

Singh, K. P., 1971. Model Flow Duration and Streamflow Variability. Water Resources Research, 7(4):1031-1036.

Singh, K. P., 1981. Derivation and Regionalization of Unit Hydrograph Parameters for Illinois (Dam Safety Program). Illinois State Water Survey Contract Report 258, 84 p.

Snyder, F. F., 1938. Synthetic Unit-Graphs. Transactions, American Geophysical Union, 19:447-454.

INTERNATIONAL SYMPOSIUM ON HYDROMETEOROLOGY
JUNE AMERICAN WATER RESOURCES ASSOCIATION 1982

DRAINAGE CLIMATOLOGY FOR CALIFORNIA

J. D. Goodridge[1]

ABSTRACT: Spillway and culvert sizing in California are commonly based on precipitation for various return periods and storm durations. Rainfall records are used because design storms based on runoff records are highly variable. This paper discusses two sets of extreme annual precipitation data, which were compiled for drainage studies. The first is for 700 recording rain gages with 16,000 station years of annual series data for durations ranging from 5 minutes to 24 hours. The second set is for 1700 non-recording rain gages with 41,000 station years of annual series data for durations ranging from 1 to 60 days. A regional frequency analysis is performed on each data set. The method used to select a frequency distribution is by a comparison of the skew and kurtosis of the data sets. The main concept of this paper is the method of regionalizing the standard deviation.
(KEY TERMS: rainfall; depth-duration-frequency; California; design storms; statistics.)

INTRODUCTION

Rainfall is used in hydraulic structure design in California because design storms based on rainfall give more consistent results than design storms based on streamflow records, where the natural variation as measured by the coefficient of variation is larger. There are about 1200 dams in California, where extreme annual rainfall data are used to evaluate the hydraulic safety of spillways. These data are used to develop a unit hydrograph for a spillway design. Design storms are estimates of the magnitude of precipitation events corresponding to the design return period.

The critical storm duration for a design storm at a hydraulic structure is the "time of concentration" This is the time it takes for water in the most remote part of the basin to flow to the outlet or spillway. At two-thirds of the dams in California, the "time of concentration" is an hour or less, and at about half of the dams it is 30 minutes or less. Small culverts and roof top drains have concentration times of only a few minutes.

There are over 700 recording rain gages in California with an average of 23 years of record. In these frequency studies the recording gage records were supplemented with the records of 1700 non-recording gage records, in order to develop regional design values of the coefficients of variation, skew, and kurtosis. The reason for using non-recording gages for regional studies is their length of record. There are over 100 nonrecording rain records in California dating back 100 years or more, whereas there are only a dozen recording gages with as much as 50 years of data.

Managing a large data set for drainage studies would be a very complex procedure except for a few simplifying assumptions. One of these assumptions is that an annual series is adequate for calculating return periods. With the annual series, the highest value for each year is used rather than the highest values of record. This study uses an annual series rather than a partial duration series.

RELATED STUDIES

The data described in this paper are contained in two computer data files, which are contained in a publication of the California Department of Water Resources, called "Rainfall Depth-Duration-Frequency for California," dated February 1981. It includes both short and long duration data on extreme annual storms. The short duration data file contains maximum annual precipitation for 5, 10, 15, 30, 60, 120 and 180 minutes, as well as 6, 12 and 24 hours. The long duration data file contains maximum annual precipitation data for 1, 2, 3, 4, 5, 6, 8, 10, 15, 20, 30 and 60 consecutive days. The short duration data set contains annual precipitation extremes from 700 recording rain gages with 16,000 station years of record. The long duration data set contains data on 1,500 stations with 40,000 station-years of record. This report contains the annual rainfall extremes as well as computed values for a broad range of fixed return periods. The return periods used in each of these reports are for 2, 5, 10, 15, 20, 25, 40, 50, 100, 200, 1000 and 10,000 years as well as the mean value plus 15 standard deviations representing an estimate of the probable maximum precipitation.

DATA

The data for the short-duration rainfall were taken from recording rain gages. About half of the recording gage

[1]State Climatologist, California Department of Water Resources, P.O. Box 388, Sacramento, California 95802.

records are published in "Hourly Precipitation Data" for California and adjacent states, by the National Climatic Center in Ashville, North Carolina 28801. The other half of the data are taken from about forty cooperating agencies located mainly in the urban areas. Local flood control agencies have installed hundreds of gages to develop design data for drainage construction.

The National Weather Service has extracted data on the extreme annual precipitation for 5, 10, 15, 30, 60, 120 and 180 minutes at each of their first order stations. Before 1973 they published only clock hour total rainfall for their recording range network. A part of the effort to develop design data was to examine the original rain gage charts in order to extract short duration annual extreme values for 5 to 180 minutes at all of the National Weather Services recording rain gages in the state. The idea for this project came from a similar one prepared to the Pennsylvania State Highway Department by the Civil Engineering Department at Pennsylvania State University.

The older style of weighing rain gages was ideally suited for measuring short duration rainfall extremes. During the 1970's, the National Weather Service replaced most of these instruments with a newer model, which has only one-tenth of the precision in the depth measurement, and with a 15-minute fixed observation interval. This has eroded the quality of drainage studies, as compared to the data which was collected earlier, because rain depths from the charts from the older gages could be estimated to .01 inch and to five minutes on the time scale.

Only a few of the local agencies have adopted the newer less accurate gages. Some of the local agencies have selected time intervals for reducing rain gage records which are different from those used here. An example is Los Angeles County, which has extracted data on 20-minute annual extremes but not for 15 minutes. The computer program that generates the depth-duration-frequency tables can use a variable format, which accommodates a broad range of storm durations. This report is a collection of the data of many different agencies. Each of these agencies has its own standards of observation as well as its own rain gage chart reduction standards.

ANALYSIS

The analysis of each of the data sets studied here is similar to the precipitation-depth-duration frequency for the 700 recording rain gages in California. This is similar to the example for Sacramento shown on Table 1. The analysis of the extreme annual precipitation data referred to here consists of generalizations made from four basic statistical parameters. These parameters were calculated for each storm duration studied. They are: the mean (M), the standard deviation (S), the coefficient of skew (g) and the coefficient of kurtosis (k), as shown below:

$M = \Sigma X/N$
$S = \Sigma(xx)/(N-1)$
$g = N\Sigma(xxx)/((N-1)(N-2)SSS)$
$k = NN\Sigma(xxxx)/((N-1)(N-2)(N-3)SSSS)$
CV = S/M, the coefficient of variation
X = the magnitude of an event
x = X−M, the departure from the mean
N = the number of years of record

The mean expresses the central tendency of the data set. The standard deviation measures the spread of the data set about the mean. The skew expresses the degree of symmetry, or the lopsideness of the data set due to rainfall being bounded by zero at the lower limit while it lacks an upper limit. The kurtosis is used here to select a frequency distribution. The coefficient of variation is similar to the standard deviation; it measures the variation in the data set.

The actual calculations of precipitation frequency are based on the mean plus a variable number of standard deviations which relate to frequency factor. This was taken from Chow's Handbook of Hydrology as shown below:

$$P(i,j) = M(i) + K(j)S(i) = \text{design storm in inches}$$

where

i refers to return period in years
i refers to storm duration
M(i) = mean precipitation for storm duration (i)
K(j) = frequency factor for return period (j)
S(i) = standard deviation of the maximum annual storm for duration (i)

The design values of the standard deviation used in this paper are the regional values and not calculated ones from the data set. The regional design standard deviation is derived from averaging the coefficients of variation from all of the individual stations in each hydrologic province shown on Figure 1. The individual samples are weighted according to the number of years of record at each station. Regional design values of the coefficients of variation and skew are shown on Figures 2 and 3, as used for storm durations of 5 minutes to 24 hours.

It was assumed that the coefficient of variation is uniform over a range of storm durations from 5 minutes to 60 days as shown on Figure 4. Design values of the coefficients of variation and skew for storm durations of less than 24 hours were assumed to be the same as the values for 24 hours. A major design assumption is that the great number of good-quality daily rain gages are a better basis for adopting a design storm than the relative short records of the recording rain gages. The variation with storm duration of the coefficients of skew and kurtosis are also shown on Figure 4.

FREQUENCY DISTRIBUTIONS

The procedure used here for selecting a frequency distribution is based on the best visual fit of a plot of the

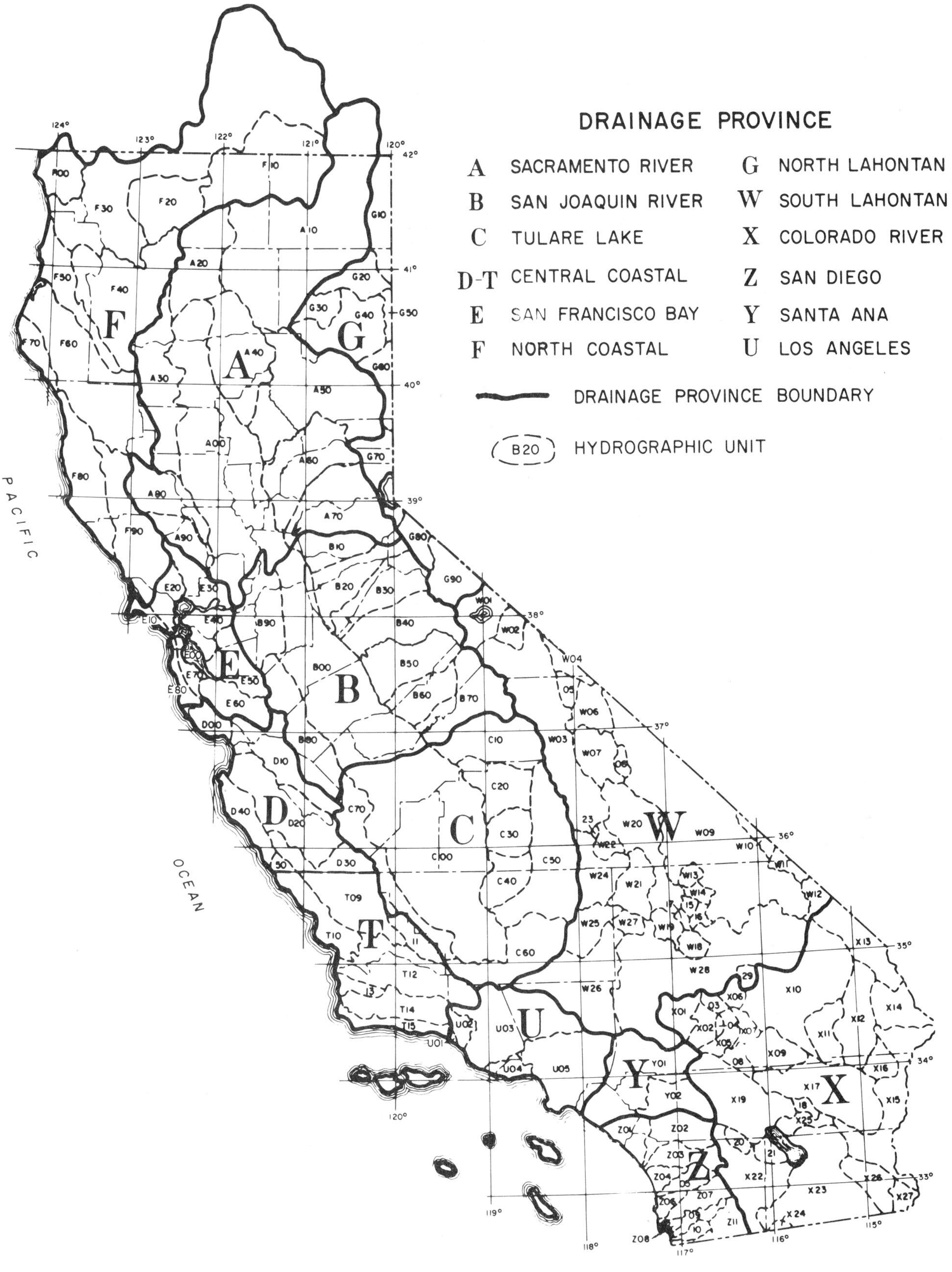

Figure 1. Major Drainage Provinces and Hydrographic Units.

Station No. BSN Order Sub	Station Name	Elev	Sec	TWP	RNG	Lot	BWM	Latit.	Longitude	County Code
A00 7633 0	Sacramento City NWS	25	01	08N	04E	C	M	38.417	121.483	34

Duration
M=Minutes,H=Hours,D=Days,C Yr=Calendar Year, W Yr=Water Year, F Yr=Fiscal Year

Year	5M	10M	15M	30M	1H	2H	3H	6H	12H	24H	C-Yr
1944	.13	.16	.22	.28	.35	.46	.61	1.02	1.67	2.82	22.02
1945	.14	.22	.26	.34	.53	.65	.78	1.28	1.53	2.74	19.45
1946	.17	.30	.39	.46	.46	.61	.67	.90	1.05	1.19	9.05
1947	.17	.26	.28	.36	.55	.66	.67	.96	1.28	1.86	11.09
1948	.12	.17	.21	.32	.50	.67	.83	1.19	1.81	1.93	18.19
1949	.17	.21	.23	.25	.30	.52	.65	.88	.95	1.11	11.03
1950	.36	.44	.47	.52	.62	.78	.91	1.61	1.76	2.15	24.32
1951	.27	.36	.45	.35	.69	.87	1.01	1.40	1.36	1.69	16.22
1952	.18	.22	.23	.27	.40	.70	.99	1.52	1.73	2.72	26.08
1953	.07	.12	.14	.20	.37	.65	.88	1.28	1.59	1.87	12.16
1954	.10	.14	.18	.29	.40	.59	.75	1.17	1.73	2.10	20.99
1955	.20	.29	.33	.37	.42	.62	.85	1.15	1.95	3.27	23.15
1956	.25	.27	.29	.34	.42	.66	.84	1.00	1.49	1.60	15.30
1957	.15	.20	.29	.38	.44	.69	.72	.91	1.13	1.45	18.42
1958	.24	.37	.45	.59	.75	.85	1.12	1.67	1.96	2.66	27.28
1959	.10	.16	.20	.31	.64	.77	.90	1.21	1.44	1.50	11.85
1960	.13	.16	.18	.30	.40	.52	.66	.81	1.24	1.28	14.53
1961	.11	.13	.14	.27	.39	.54	.64	.80	.99	1.32	11.57
1962	.18	.27	.30	.42	.63	.89	1.05	1.55	2.76	5.07	19.88
1963	.19	.24	.28	.39	.49	.70	.83	1.06	1.37	1.92	18.97
1964	.11	.17	.25	.36	.49	.83	1.15	1.97	2.10	2.30	15.07
1965	.11	.20	.27	.35	.41	.63	.74	1.14	1.22	1.34	15.63
1966	.08	.10	.13	.21	.30	.45	.64	.92	1.43	1.54	13.69
1967	.23	.39	.43	.50	.61	.86	1.06	1.23	2.09	3.12	19.63
1968	.13	.20	.23	.25	.26	.37	.45	.79	1.04	1.40	15.28
1969	.17	.26	.32	.39	.63	.99	1.03	1.03	1.47	1.94	26.06
1970	.15	.20	.25	.32	.43	.55	.73	1.00	1.73	2.45	22.83
1971	.13	.17	.21	.31	.47	.74	.84	.94	.96	1.17	9.78
1972	.13	.19	.24	.31	.37	.40	.48	.80	1.06	1.18	13.91
1973	.29	.53	.65	.90	1.01	1.19	1.27	2.01	2.02	2.11	26.48
1974	.17	.22	.24	.26	.37	.62	.65	.79	1.11	1.78	15.16
1975	.16	.20	.21	.28	.38	.59	.79	1.14	1.21	1.85	13.18
1976	.12	.19	.22	.24	.28	.36	.42	.56	.79	.95	6.25
1977	.09	.12	.13	.24	.32	.47	.57	.90	1.15	1.58	11.71
1978	.25	.30	.31	.37	.56	.68	.93	1.27	1.61	1.98	23.74
1979	.28	.35	.42	.53	.59	.61	.75	.95	1.20	1.92	22.33

Figure 1a. Maximum Annual Precipitation (in.)

STATION NO. BSM ORDER SUB	STATION NAME	ELEV	SEC	TWP	RNG	BWM	LATITUDE	LONGITUDE	COUNTY CODE
A00 7633 0	SACRAMENTO CITY NWS	25	01	08N	04E	M	38.417	121.483	34

MAXIMUM PRECIPITATION FOR INDICATED DURATION D-DAYS H-HOURS

RETURN PERIOD IN YEARS	5M	10M	15M	30M	1H	2H	3H	6H	12H	24H	C-YR
2	.14	.21	.25	.33	.44	.62	.75	1.04	1.38	1.85	16.44
5	.19	.28	.34	.45	.59	.83	1.00	1.39	1.85	2.48	21.24
10	.22	.32	.39	.52	.69	.97	1.17	1.62	2.16	2.89	23.96
20	.25	.37	.45	.59	.78	1.10	1.32	1.84	2.44	3.27	26.33
25	.26	.38	.46	.61	.81	1.14	1.37	1.90	2.53	3.39	27.03
40	.27	.39	.47	.62	.82	1.16	1.39	1.93	2.57	3.45	28.46
50	.29	.42	.51	.68	.89	1.26	1.51	2.10	2.80	3.75	29.11
100	.32	.46	.56	.74	.98	1.37	1.65	2.30	3.05	4.09	31.03
200	.34	.50	.60	.80	1.06	1.49	1.79	2.48	3.30	4.43	32.85
1000	.40	.58	.71	.94	1.24	1.74	2.09	2.91	3.87	5.18	36.75
10000	.48	.70	.85	1.13	1.48	2.09	2.51	3.49	4.64	6.22	41.82
PMP	.95	1.38	1.67	2.22	2.93	4.12	4.95	6.88	9.16	12.27	98.44
MEAN	.152	.221	.269	.357	.471	.662	.796	1.107	1.472	1.972	16.798
CLOCK HR. COR.	1.000	1.000	1.000	1.000	1.000	1.000	1.000	1.000	1.000	1.000	1.000
CALCULATED SKEW	1.228	1.549	1.573	1.773	2.791	4.273	3.363	1.839	.997	1.239	.563
REGIONAL SKEW	1.100	1.100	1.100	1.100	1.100	1.100	1.100	1.100	1.100	1.100	.400
SKEW USED	1.100	1.100	1.100	1.100	1.100	1.100	1.100	1.100	1.100	1.100	.400
KURTOSIS	4.770	5.633	5.732	6.620	15.645	30.419	22.023	8.733	4.196	5.464	3.114
N	77	77	77	77	77	77	77	77	77	77	77
RECORD YEAR	1935	1935	1935	1935	1935	1935	1935	1935	1935	1962	1940
RECORD MAXIMUM	.390	.620	.750	.970	1.650	2.620	2.700	2.880	3.080	5.070	34.710
NORMALIZED MAX	3.476	3.677	3.669	3.812	5.644	6.793	6.226	4.666	3.283	4.085	3.008
CALC. COEF. VAR	.450	.490	.487	.450	.443	.435	.384	.343	.333	.384	.355
REGN. COEF. VAR	.348	.348	.348	.348	.348	.348	.348	.348	.348	.348	.324
USED COEF. VAR	.348	.348	.348	.348	.348	.348	.348	.348	.348	.348	.324
MEAN/A	.0091	.0132	.0160	.0213	.0280	.0394	.0474	.0659	.0877	.1174	1.0000
RP10/A	.0133	.0193	.0235	.0312	.0411	.0578	.0695	.0966	.1286	.1722	1.4266
RP25/A	.0156	.0226	.0275	.0365	.0482	.0678	.0815	.1132	.1507	.2018	1.6092
RP50/A	.0172	.0250	.0304	.0404	.0533	.0749	.0900	.1251	.1665	.2230	1.7327
RP100/A	.0188	.0273	.0332	.0441	.0582	.0818	.0983	.1366	.1818	.2435	1.8474
RP1000/A	.0238	.0346	.0421	.0558	.0736	.1036	.1245	.1730	.2302	.3084	2.1878
RP10000/A	.0285	.0415	.0505	.0670	.0884	.1243	.1494	.2077	.2763	.3701	2.4894
PMP/A	.0563	.0819	.0996	.1322	.1744	.2452	.2948	.4098	.5452	.7303	5.8600

PEARSON TYPE III DISTRIBUTION USED
PROBABLE MAXIMUM PRECIPITATION ESTIMATE BASED ON 15 STANDARD DEVIATIONS
WHERE N IS SMALL RESULTS ARE NOT DEPENDABLE

Figure 1b. Precipitation-Duration-Frequency Table.

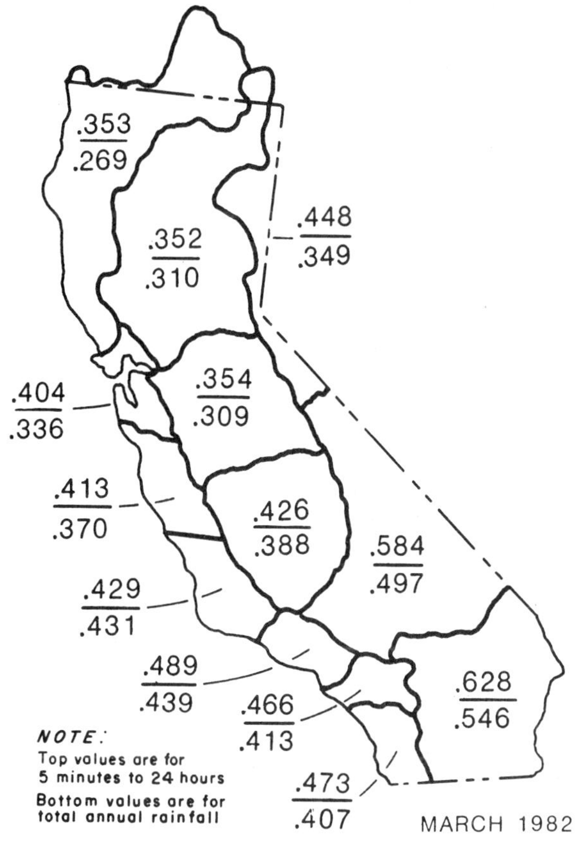

Figure 2. Regional Values of the Coefficient of Variation.

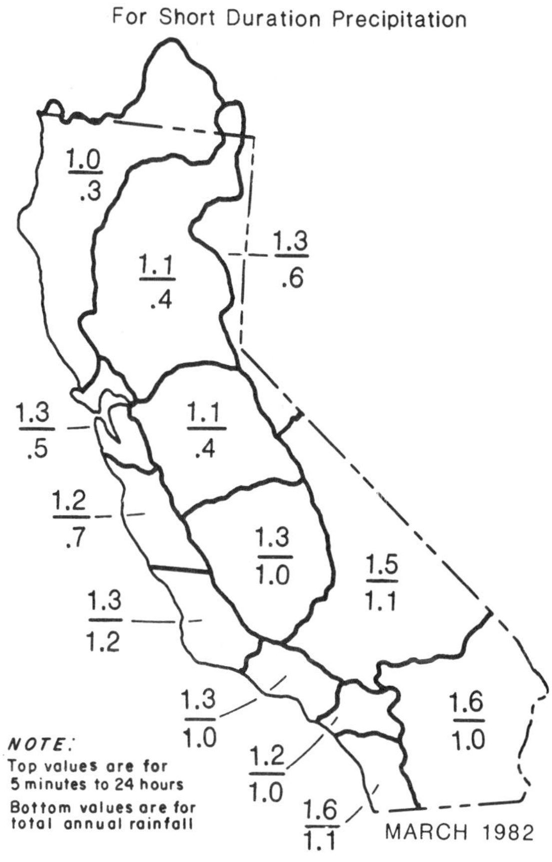

Figure 3. Regional Values of the Coefficient of Skew.

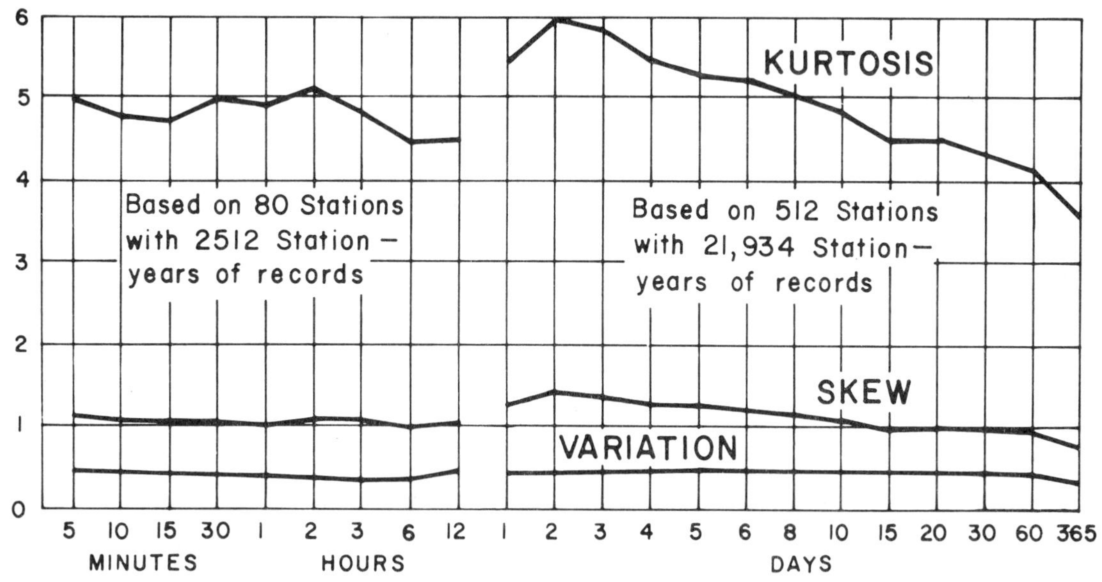

Figure 4. Variation in Extreme Precipitation Statistics with Storm Duration.

coefficients of skew and kurtosis. The averages of skew and kurtosis were weighted by the length of record for each individual station. The statewide averages are plotted on Figure 5.

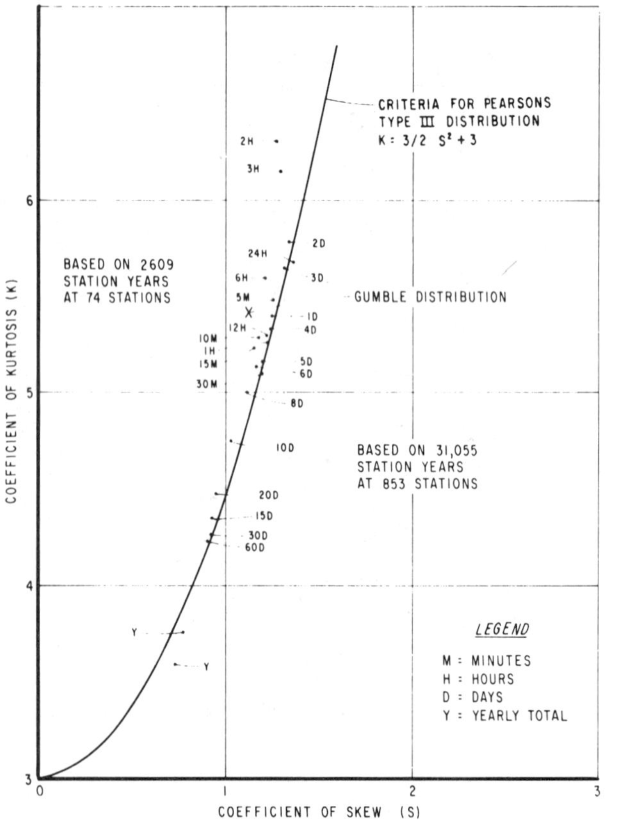

Figure 5. Skew-Kurtosis Relationship of Maximum Annual Precipitation for California.

Several frequency distributions were defined in terms of the mean, standard deviation, and coefficients of skew and kurtosis, by Paul Wu of the Contra Costa County Flood Control District in Martinez, California. A comparison of three of these distributions is shown below for a skew of 1.4:

Return Period (Years)	Frequency Factor (K) Pearson Type III	Weibull	Log-Normal
2	−0.225	−0.244	−0.192
10	1.337	1.359	1.294
100	3.271	2.992	3.260
1000	5.095	4.970	5.352

There is not much difference between these frequency distributions. The one in a thousand year event is five standard deviations more or less above the mean, with whatever distribution might be selected. The more critical statistic is the estimate of the regional coefficient of variation, in which a small difference has a relatively large impact on the size design storm.

The Pearson Type III distribution is the one selected here for assigning return periods to extreme rainfall events in California, based on the skew-kurtosis relationship. The actual frequency factors used in this study were supplied by the Central Technical Unit of the Soil Conservation Service.

The range in skewness for the regions is from 1.0 to 1.6 as shown on Figure 3. A comparison of frequency factors is tabulated here:

Return Period	Frequency Factors Coefficient of Skew 1.0	1.2	1.4	1.6
2	−0.16	−0.20	−0.23	−0.25
10	1.34	1.34	1.34	1.33
100	3.02	3.15	3.27	3.39
1000	4.53	4.81	5.09	5.37

EVALUATING FREQUENCY FACTORS

The frequency factors used here have not been evaluated to determine how many 1000 year storms there have been per 1000 years of record. An alternative test of the appropriateness of the frequency factors was used. This alternative was comparing the maximum storm of record with the length of record. The maximum daily rainfall of record expressed in standard deviations above the mean is plotted against the length of record on Figure 6. Figure 6 also shows a plot of frequency factors for Pearson's Type III distribution, with a skew of 1.1.

Whereas the scatter of points on Figure 5 is quite large, this data when averaged as described here and plotted on Figure 5 shows relatively little scatter.

OUTLIERS

There have been suggestions that very large values of rainfall be eliminated from the data sample as being "outliers," or records from a different population of rare and infrequent storms and not a logical part of a data set where most of the rainfall records have been generated. This would occur where storms are of two types, such as winter time fronts and summer thunderstorms or tropical storms of the summer time. An example of an "outlier" would be the Indio, California, storm of September 1939 where two times the mean annual rainfall (of 3 inches) fell in six hours.

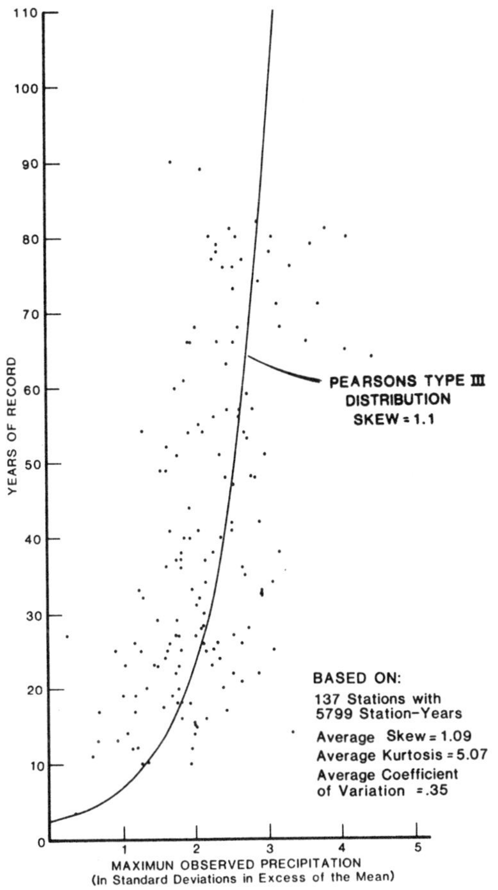

Figure 6. Extreme One Day Precipitation for the Sacramento Valley.

Safety considerations prevail in these studies, and no data have been excluded as "outliers." The apparent effect of the "outliers" is smoothed out in the averaging process. The result is a larger and, hopefully, safer design storm. Public safety aspects are emphasized rather than the course that would result in less costly roadway drainage facilities.

FIXED INTERVAL CORRECTIONS

The average maximum annual rainfall for 1440 consecutive minutes is about 14 percent more than the average annual maximum rainfall based on a once-a-day observation at a fixed time. This is equally true for the maximum 60 minutes and the maximum fixed hourly observation. In 1964, Weiss published a theoretical derivation of the ratio of true to fixed interval maximum precipitation (f) based on a probability model. Values of the fixed interval correction (f), as calculated by Weiss, are tabulated here for various consecutive time intervals (n):

n	1	3	6	12	24
f	1.14	1.04	1.02	1.01	1.00

These intervals are independent of the time interval under study. They are equally valid for data sets representing minutes, hours or days. This ratio applies only to the averages of the extreme events and not to the individual yearly amounts. These "clock hour corrections" are used throughout this study whenever fixed interval data are appropriate.

TRANSPOSING DESIGN STORMS

In California the climate varies abruptly over short distances. Rain gages are almost never available exactly where they are needed. Transferring hydrologic data from one point to another is a major activity. An important problem facing a hydrologist is selecting a suitable precipitation record in the vicinity of a project, for use in modeling storm runoff. The most common factor used for transposing design storms is the mean annual rainfall. Maps of the mean annual rainfall are available in great detail for the state. One of the more useful assumptions in drainage studies is that nearby locations having a similar mean annual rainfall have similar storm frequency characteristics.

The design storms developed in this study are expressed as a fraction of the mean annual precipitation on the bottom of each of the precipitation depth frequency tables, as shown on Figure 1b. As an example, the 10 year 1 hour storm at Sacramento is 0.0411 of the mean annual rainfall. A design storm for a location between rainfall stations can be selected based on interpolating between near bystations.

LIMITATIONS OF PRECIPITATION FREQUENCY STUDIES

Design storms are more sensitive to estimates of the coefficient of variation than to the choice of frequency distribution. The coefficient of variation is definitely a climatic variable. A big improvement in future studies can probably be made by defining regions of climatic homogeneity. The regions used for averaging the coefficients of variation, skew and kurtosis are watershed boundaries rather than climatic zones. Restructuring this study along climatic zones would undoubtedly improve it.

Another of the problems of these studies is the abrupt changes in the mean values of the regional coefficients of variation, which are plotted on Figure 2. It is apparent that there should be a continuous variation in this coefficient rather than abrupt changes as used here. Similarly, the design coefficient of skew would be more logically derived from a contoured map rather than from averages plotted for broad regions as shown on Figure 3. When the computed values of the coefficients of variation and skew are plotted on a map, there is too much interstation variation to draw lines of equal values. A smoothing device is needed. These problems have not been addressed in this paper.

The assumption that the coefficients of variation is adequately uniform for storm durations of 5 minutes to 24 hours needs further study.

Analytical procedures have their limitations, but some of the biggest limitations are probably in the data. It is only the most conscientious weather observer who will measure the rainfall at precisely the same time each day, as needed for studies like this paper. Not all rain gages are located where they will be the most representative of the region's rainfall; gages have historically been located where they will be convenient for the observer. The accuracy of a rain gage is seldom questioned, since rain records are generally used only as indexes of the true rain. Designers seldom question the records, yet in high wind conditions as well as with snow, the rain gages will not catch all of the precipitation that occurs. There are major problems with precipitation gages in the snow zones, where over-caping of gages has been occurring for years even in the published records.

Those who design drainage facilities must consider that the data from these precipitation depth-duration-frequency studies represent a lower limit of the amount of rain or snow that might occur rather than the most likely amount that might occur.

LITERATURE CITED

Goodridge, J. D., February 1981. Rainfall Depth-Duration-Frequency for California. Department of Water Resources, P.O. Box 388, Sacramento, California 95814.

National Climatic Center. Hourly Precipitation Data. Issued monthly since July 1948.

Wu, Baolin and J. D. Goodridge, June 1976. Selection of Frequency Distributions, with tables for Pearson Type III, Log-Normal, Weibell, Normal and Gumbel Distributions. Department of Water Resources, P.O. Box 388, Sacramento, California 95814.

2-MINUTE RAINFALL EXTREMES

David M. Hershfield[1]

ABSTRACT: The analysis of extreme-value rainfall data is rarely extended to durations less than five minutes even though there are many applications for these durations in hydrology and agriculture. The general absence of readily available data for the very short durations is the primary reason for this shortcoming. Breakpoint rainfall data from 32 stations at 11 agricultural watersheds for durations of 2-min to 60-min served as the basis for this study. Relationships were established between the 2-min and longer duration data which were then used as an aid in constructing several generalized maps for the eastern United States, including a 2-yr 2-min map. Unlike the extreme-value analyses of 60-min and 24-hr data, the variability of the annual maximum 2-min values, as determined by the coefficient of variation for the individual stations, exhibits an irregular geographical pattern.
(KEY TERMS: precipitation; extreme-values; frequency maps.)

Figure 1. Station Location Map.

INTRODUCTION

Extreme-value analysis of rainfall amounts for durations less than 5-min have applications in both hydrology and agriculture. Yet, very few such studies appear to have been produced. The lack of available extreme-value data is probably the main reason for the paucity of such studies.

There are two major generalized frequency reports for durations as short as 5-min. Yarnell's paper (1935) presents frequency maps for the conterminous United States whereas Frederick, *et al.* (1979), covers the eastern United States. Several reports containing rainfall-intensity-duration frequency curves include durations from 5-min to 24-hr (Chow, 1953; USWB, 1955). The primary purpose of this paper is to build on the Frederick study by developing relationships between 2-min and longer-duration rainfall data and use this information to construct generalized frequency maps.

DATA

Breakpoint data from recording gages for 32 stations at 11 locations served as the basis for this study. Figure 1 shows the station locations and Table 1 presents the periods of record and gage identification numbers. The annual maximum 2-min rainfalls were extracted from the breakpoint data. Breakpoint refers to the point in time that the rainfall changes intensity. Thus, these data can be used to reproduce the original rainfall trace of the recording chart. Resolution of the annual maximum data was for the beginning and ending on the clock-minute. This provides excellent definition for the 24-hr period containing the maximum rainfall and probably is adequate for durations as short as 5-min. However, for the 2-min duration, the estimated amount is probably closer to the clock 2-min than the 2-min period that contains the maximum rainfall, i.e., the 120-second period. Better definition of the 2-min value would require an instrument with finer resolution than a 6-hr or 24-hr recording gage. Bodtmann and Ruthroff (1976) suggest an analytical method for determining the 1-min rates from weighing raingage recordings. However, the mechanical limitations of the instrument are the main problem in accurately estimating short-duration rainfalls.

DURATION AND FREQUENCY ANALYSES

The relation between the 2-yr 2-min and 2-yr 5-min values for 11 stations, one at each location, is shown in

[1] Research Meteorologist, Hydrology Laboratory, Plant Physiology Institute, USDA-ARS, Beltsville, Maryland 20705.

TABLE 1. Station Location and Period of Record.

Station	State	Raingage No.	Period of Record
Bentonville	AR	000005	1939-47
Safford	AZ	000002	1939-74
		000005	1940-71
		000011	1942-71
		000012	1939-70
		000014	1942-66
		000015	1942-66
Treynor	IA	000101	1963-73
		000105	1963-73
		000112	1964-75
		000113	1964-75
		000115	1964-75
		000116	1964-75
		000117	1964-75
Hastings	NE	RGPB36	1940-67
Albuquerque	NM	000003	1939-72
Coshocton	OH	000100	1938-71
		00Y102	1937-78
		00Y103	1939-78
		00Y108	1938-78
		000109	1940-78
		000113	1940-78
		000115	1939-78
Chickasha	OK	000195	1967-74
Danville	VT	000003	1961-78
		000011	1959-73
		000015	1959-73
Blacksburg	VA	P08R02	1958-72
		P20R02	1958-72
Moorefield	WV	R01P01	1959-67
		R03P03	1959-67
Fennimore	WI	000006	1939-69

Figure 2. The 2-yr values were computed by the Gumbel procedure (1958). It should be noted that the differences in 2-yr estimates from any of the more common extreme-value procedures are insignificant. The geographical distribution of the 2-yr 2-min/2-yr 5-min ratio, which exhibits a slight increase from south to north, is portrayed in Figure 3. In order to make this paper compatible with the results of the Frederick, *et al.*, study, the 2-yr annual series values were adjusted upward by 13 percent for the partial-duration series (Hershfield, 1961). The 2-yr 2-min isopluvial pattern of Figure 4 is quite similar to that for durations more than one order of magnitude longer; i.e., the rainfalls generally decrease from south to north.

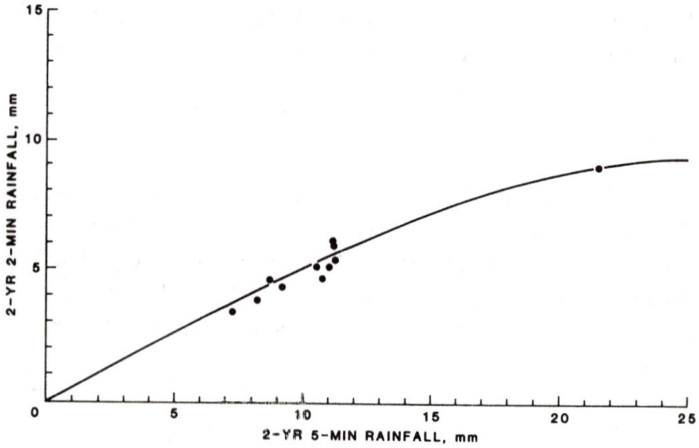

Figure 2. Relationship Between 2-yr 2-min and 2-yr 5-min Rainfall for One Station at Each of 11 Locations.

2-YR 2-MIN/2-YR 5-MIN RATIO

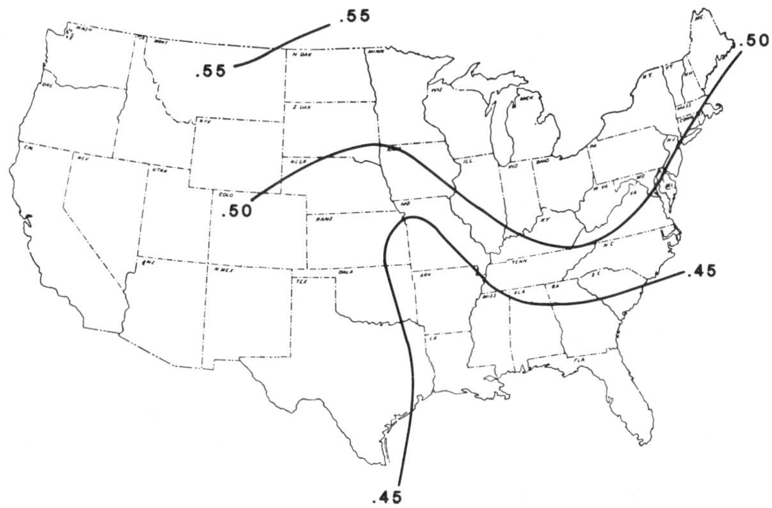

Figure 3. Geographical Distribution of 2-yr 2-min/2-yr 5-min Ratio.

2-YR 2-MIN RAINFALL, mm

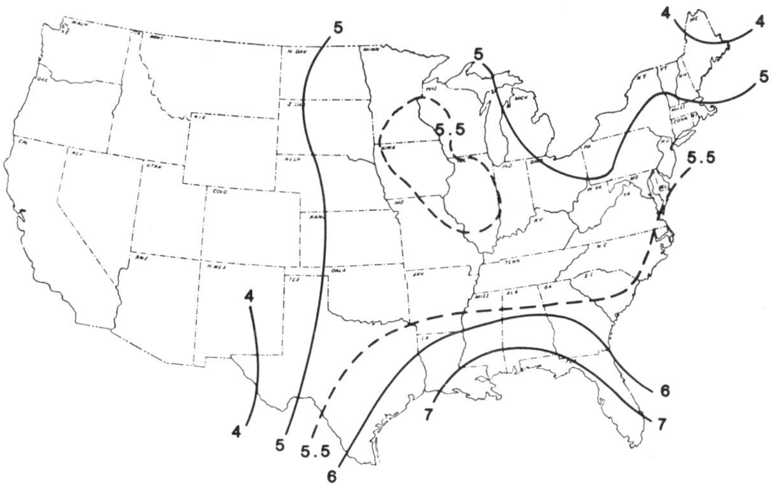

Figure 4. Geographical Distribution of 2-yr 2-min Rainfall (adjusted for partial-duration series).

For most analyses the coefficient of variation (C_V) of an extreme-value series tends to decrease rather uniformly with increasing subsample size, i.e., decreasing duration. However, for the 2-min data, the decrease with duration is rarely present. This may partially be explained by limitations of the 2-min data discussed earlier in this section concerning clock 2-min data versus the maximum amount in a 120-sec period. In lieu of a systematic geographical pattern of the 2-min coefficient of variation, an average C_V of 0.30 can be used with the mean in combination with a common extreme-value procedure to obtain the required frequency. The 2-yr 2-min values on the map of Figure 4 should be reduced by 6 percent to obtain the mean of the annual series.

SPATIAL RELIABILITY

One might ask about the reliability of a 2-min frequency value from one gage to represent a substantial size area. The seven gages, in a relatively flat area of about 10 km^2, at Coshocton, Ohio, provide some variability data. The data of Table 2 displays the 2-yr 2-min values for the seven stations. The mean is 5.40 mm, the standard deviation as computed from the range (Snedecor, 1962) is 0.17 mm, and

TABLE 2. 2-yr 2-min Rainfall Statistics (mm), Coshocton, Ohio.

Station	000100	00Y102	00Y103	000108	000109	000113	000115
2-yr 2-min (adjusted for PD series)	5.49	5.34	5.61	5.98	5.14	5.34	5.40
		Mean = 5.40		SD = .17		C_V = .03	

the coefficient of variation is 0.03. This small variation around the mean is well within the range of normal sampling fluctuations.

Figure 5 shows the rainfall-intensity-frequency-duration curve for six of the seven stations at Coshocton. Division by 30 provides an estimate of the 2-min depth. The Gumbel procedure for fitting to annual maxima was used for the frequency analysis. Dashed lines identify the 25-yr 2-min values and, again, the range of values is relatively small, suggesting considerable reliability for sampling over a substantial size area, i.e., in terms of rainfall depth, the 25-yr values range from 8.7 to 9.7 mm.

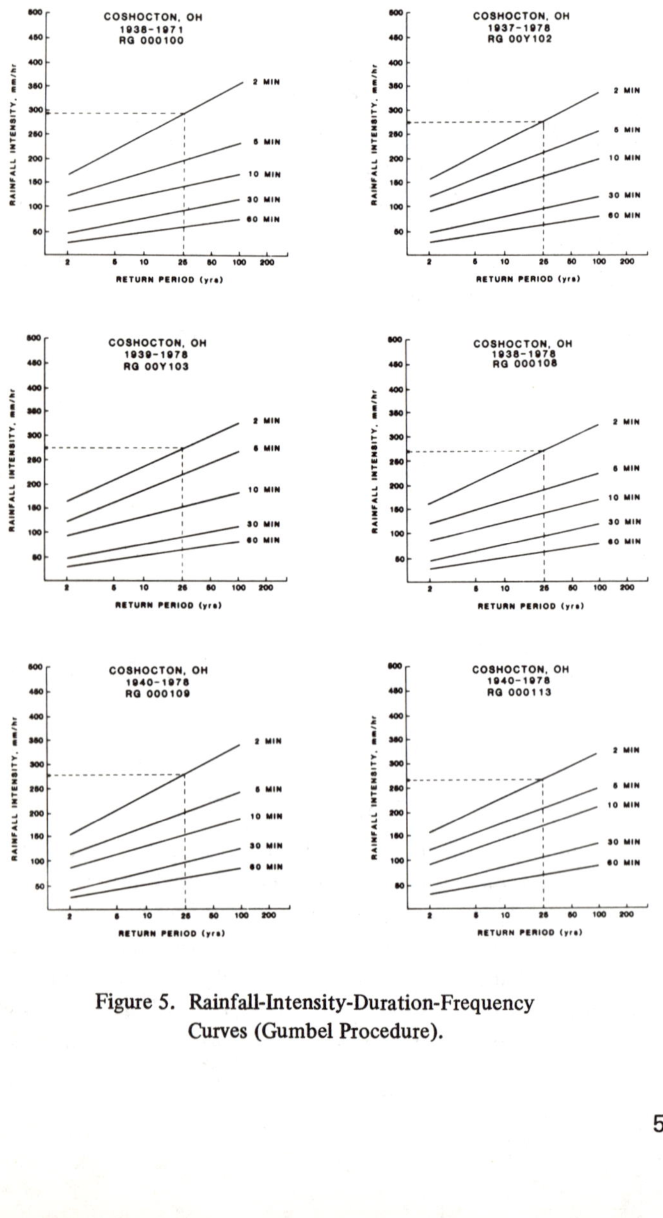

Figure 5. Rainfall-Intensity-Duration-Frequency Curves (Gumbel Procedure).

DISCUSSION

Use has been made of some special data to derive series of 2-min annual maxima. There are limitations to the quality of data because it is unlikely that the instruments, 6-hr and 24-hr recording raingages, respond with the exactness required to produce extremely accurate 2-min amounts. Yet, the quality of the data, taken *in toto* probably represent acceptable estimates. The bias, if there is any, is that the 2-min values are slightly low. The average 2-yr 2-min/2-yr 5-min ratio is about 50 percent with a slight systematic geographical variation. Hershfield (1972) extrapolated rainfall-duration-frequency curves (USWB, 1955) to durations shorter than 5-min and arrived at approximately the same 2-yr 2-min/2-yr 5-min ratio. The curves of the latter study are based on data completely independent of that used here. This lends some support to the quality of the values presented in this investigation.

LITERATURE CITED

Bodtmann, W. F. and C. L. Ruthroff, 1976. The Measurement of 1-min Rates From Weighing Raingage Recordings. J. Appl. Meteorol. 15:1160-1166.

Chow, V. T., 1953. Frequency Analysis of Hydrologic Data With Special Application to Rainfall Intensities. Univ. Illinois, Eng. Exp. Sta. Bull., Series No. 414, 80 pp.

Frederick, R. H., V. A. Myers, and E. P. Auciello, 1977. Five-to-60-Minute Precipitation Frequency for the Eastern and Central United States. NOAA Tech Memo NWS HYDRO-35, 36 pp.

Gumbel, E. J., 1958. Statistics of Extremes. Columbia Univ. Press, 375 pp.

Hershfield, D. M., 1961. Rainfall Frequency Atlas of the United States. U.S. Wea. Bur. Tech. Paper No. 40, 115 pp.

Hershfield, D. M., 1972. Estimating the Extreme-Value 1-min Rainfall. J. Appl. Meteorol. 11:936-940.

Snedecor, G. W. and W. G. Cochran, 1962. Statistical Methods. Iowa State Univ. Press, 534 pp.

U.S. Weather Bureau, 1955. Rainfall-Intensity-Duration-Frequency Curves for Selected Stations in the United States. Hawaiian Islands, and Puerto Rico. Tech. Paper No. 25, 53 pp.

Yarnell, D. L., 1935. Rainfall-Intensity-Frequency Data. U.S. Dept. Agric. Misc. Pub. No. 204, 68 pp.

SOME CHARACTERISTICS OF DAILY SUMMER MONSOON RAINFALL OVER INDIA, SUITABLE FOR HYDROLOGICAL PURPOSES

S. V. Singh and R. H. Kripalani[1]

ABSTRACT: For several hydrological routine operations like flood forecasting, reservoir regulation, etc., climatological knowledge/ forecast of exact quantities of rainfall are required. Apart from the usual climatological information, it is important to know the present state of the rainfall process because the rainfall, up to time scale of at least 1 day, is found to be dependent on its past. In this paper, we study the dependence in daily rainfall of 12 stations and in daily rainfall of 10 meteorological regions during the summer monsoon by (1) analyzing the rainfall as stochastic point process by fitting several types of models like logmodel, Markov chains of order 1 and 2 to station data; and (2) fitting autoregressive and autoregressive moving average models to spatially averaged data. The significant conclusion is that the Markov chain of order 1 fits the runs of wet spells and autoregressive model order 1 fits the spatially averaged data satisfactorily. Further examination of rainfall intensities for spells of various length has revealed that rain intensity is more on spells of 2 or 3 days than on isolated days. These findings will have significant values in simulation of synthetic models of rainfall-runoff process.
(KEY TERMS: persistence; Markov chain; autoregressive models.)

INTRODUCTION

Rainfall climatology is useful for a variety of agricultural and hydrologic operations. For several design and storage problems, certain statistics of rainfall data like extreme distributions, probable maximum precipitation, depth-area-duration characteristics of rain storms can be obtained from past climatic records. Extensive studies have been conducted in India to find these statistics of rainfall from past records. These studies were reviewed at a recent Conference (A.H.I., 1982; also see Dhar, *et al.,* 1981). From this climatic information and variety of operational purposes, the sequences of rainfall can be synthetically generated if parameters of the generating model can be estimated from limited historical records. However, for operational purposes current and forecast information is of maximum value. The forecasts can be made either by a numerical model directly or synoptic/statistical interpretation of numerical output. In India the quantitative forecasts are mainly based on synoptic analogues. Alternatively, the probablistic rainfall forecasts can be made from rainfall itself by using the dependence property of rainfall. We propose to study the dependence in station and spatially averaged daily summer monsoon rainfall in the present study using data of 19 years (1955-73) of 12 stations and 10 meteorological subdivisions (see Figure 1).

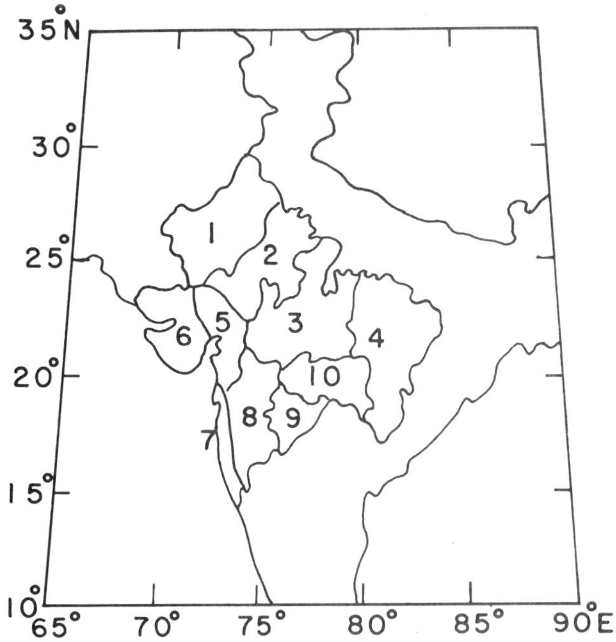

Figure 1. Location Map of 10 Meteorological Subdivisions in Western and Central India.

DEPENDENCE/PERSISTENCE IN STATION RAINFALL

The majority of rain events during monsoon can be attributed to synoptic/subsynoptic systems having life time exceeding 1 day. Thus rain occurs over a place in wet spells

[1] Respectively, Senior Scientific Officer and Senior Scientific Assistant; Indian Institute of Tropical Meteorology, Pashan, Pune-411 008, India.

intermingled with dry spells. In this section we shall describe some properties of this intermittent process.

The Persistence Parameters for Rain Events

The difference between conditional probabilities and the unconditional probabilities is the simplest measure of persistence effect in occurrence of an event (Singh, *et al.*, 1981). Thus the conditional probabilities

PW, PD, PW/W, and PD/D

where W stands for a wet day when rainfall equals or exceeds some threshold (here 2.5 mm) and D stands for a dry day when rainfall is below the threshold. The symbol / is read as "given that" and are required to understand persistence in rain events. Singh, *et al.* (1976), presented charts of these conditional probabilities for each monsoon month (June through September) prepared from data of 221 stations. Singh, *et al.* (1981), presented the charts (reproduced here, see Figure 1) for persistence probability, *viz*. PW/W-PW and PD/D-PD and average length of wet/day days during July-August using data of 22 stations. It is clear from the figure that wet persistence is more in the northwest parts of the country and mid-peninsula, the dry parts and dry persistence is more in wet regions.

Lengths of successive dry and wet spells are found to be independent by correlation analysis. Logarithms of length of each spell is taken before computing correlations, as this variable is in 'counts.' The runs of dry and wet spells thus can be considered as alternative renewal process and distribution functions (models) can be fitted to the frequencies of runs of various length of dry and wet spells separately. Singh, *et al.* (1981), fitted logarithmic, modified logarithmic and geometric models (Markov chains of order 1 and 2) to frequencies of runs of wet and dry days of 22 stations and found that Markov chain order 1 and Markov chain order 2 fit better to the runs of wet and dry days, respectively.

Other Relationships for Station Rainfall

Correlations are also computed between length of dry days and rainfall on first day of the succeeding wet spell and rain intensity in the wet spell (Buishand, 1977; Raudkivi and Lawgun, 1970). These correlations are negligible but have negative signs and thus have no predictive value.

Next, we have examined if the expected values of rain amounts on isolated days, on first, last, and middle days of longer wet spells are different from each other. This kind of analysis is also done by Buishand (1977). He designated the isolated rainly days as Type 0, rainy days either preceeded or followed by a dry days as Type 1 and the rainy days having rain on both preceeding as well as following days as Type 2. We have tested the differences in expected rain amount on these 0, 1, and 2 type rainy days by t-test (Table 1). The results show that expected rain on type 2 days is more than on type 1 which in turn is more than that on Type 0 days.

$$\text{Rain type } 2 > \text{Rain type } 1 > \text{Rain type } 0 \qquad (1)$$

We must however remember that the difference may be solely because part of the rainly days for Type 1 and Type 0 may belong to the adjacent dry spell and rain may occur/begin/end in the middle of the day under consideration. The t-test for difference of rain amount between the first and last day of wet spells shows higher amounts on the first day as compared to the last day. Similar results were found by Smith and Schreiber (1973) and Buishand (1977). We have also presented Trimmed mean (discarding upper 5 percent) values (see Kleiner and Graedel, 1980) and medians in Table 1. A visual representation of these relations are provided in Figure 3 for three stations – Patna, Bangalore, and New Delhi. The figure shows Q-Q plots (Quantile-Quantile plots) (see Gananadesikan, 1977, for further clarifications) between rain on different types of rainy days. We have used decile plot. Under null hypothesis when there is no difference between deciles of two types of rain the Q-Q plot will be a straight line. It is clear from Figure 3 that Equation (1) is satisfied.

Next we have examined the relationship between rainfall intensity and the length of the wet spell. The simple correlations between these parameters are positive and significant for about half the cases. F-test is used to test the significance of the difference in intensity of rainy spells of 1 to 5 day length as the frequency of spells decreases sharply after that. The test values (Table 1) show significant difference between the rain intensities for all stations but Bangalore. Further, tests between intensities of spells of various length, taken in pairs, revealed that it is mainly the intensity of one day spell which is significantly different from other types of spells. This result is also presented in Figure 4 where the median, and quartiles for various lengths of wet spells are plotted. The number of spells of various lengths are also shown in Figure 4. Except for Bangalore, the median intensity of spells increases up to at least 3 days. Trimmed means (see above) for intensity of various lengths of wet spells also suggest that the intensity increases up to 3 day spells in most cases and up to 4 day spells for 7 stations out of 12. However it may solely be due to the increase in Type 2 of rainy days and decrease of Type 0 type of rainy days as the length of the spell increases. Further examination, like significance tests between expected rain on first, second, third day, etc., is needed.

DEPENDENCE IN SPATIALLY AVERAGED RAINFALL

Due to the contribution of localized convection outliers/mavericks are always present in station rainfall. Spatial averaging smooths out these extreme values which can be considered as noise shots as far as response to synoptic forcing is concerned. The spatially averaged rainfall will

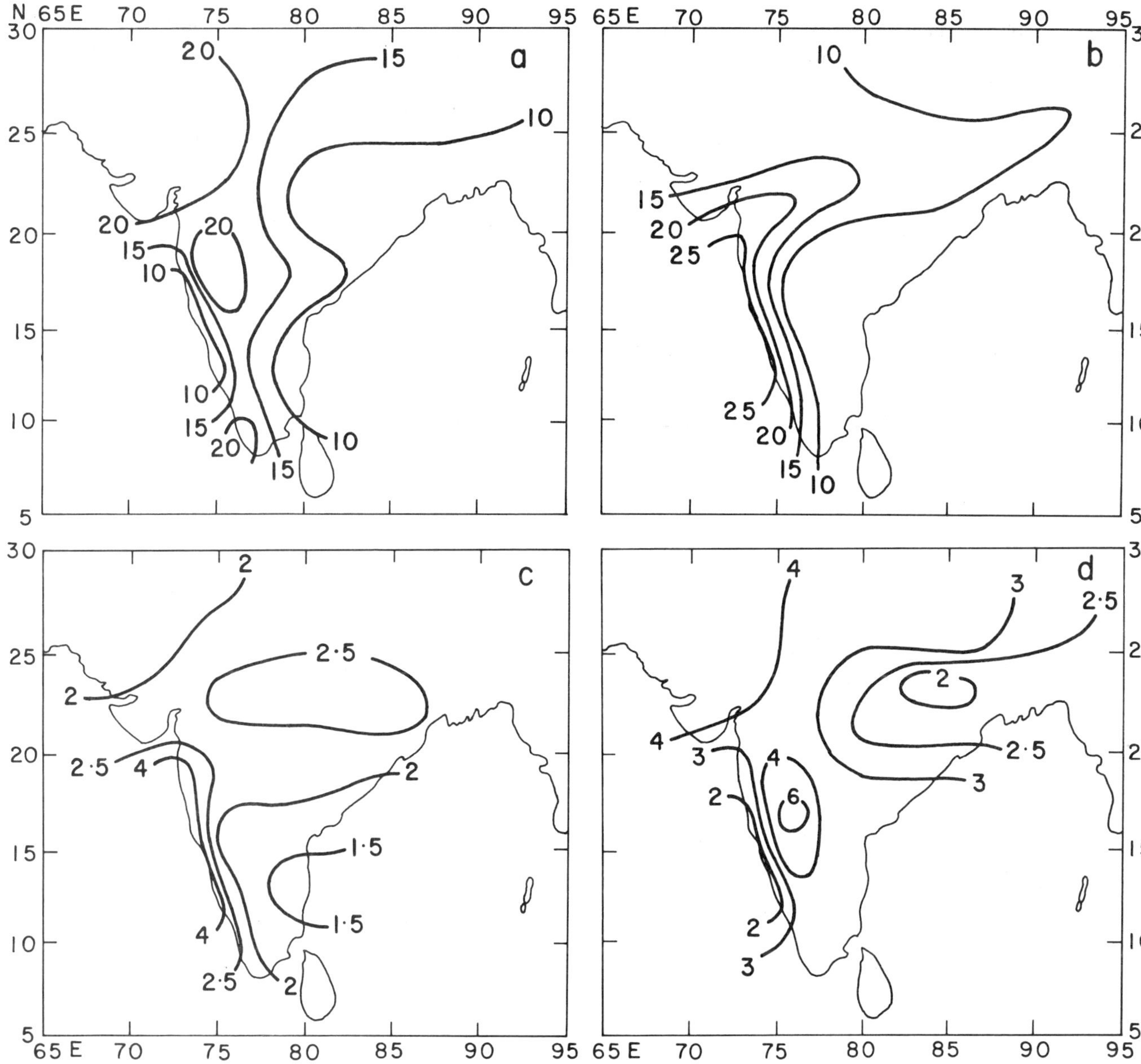

Figure 2. ,(a) Wet Persistence PrW = PW/W-PW; (b) Dry Persistence PrD = PD/D-PD (isolines labeled in terms of percentages are drawn at intervals of 5); (c) Average Length of Wet Spells in Days; (d) Average Length of Dry Spells in Days.

have reduced skewness and will be better amenable to correlation analysis as the outliers will be moderated.

The autocorrelation coefficients are the simplest measure to analyze dependence in measurement (interval or ratio scale) data. Singh, *et al.* (1981), computed autocorrelations in daily subdivisional rainfall up to several lags and found that the regions in western and central India exhibit more dependence than other regions. Here we shall analyze further the dependence in rainfall of 10 meteorological subdivisions by fitting autoregressive (AR) and autoregressive moving average (ARMA) models.

Fitting the Autoregressive Models to Spatially Averaged Data – Method

Autoregressive models are special forms of general class of models known as Autoregressive Integrated Moving Average (ARIMA) models described in some detail by Box and Jenkins (1976). An autoregressive model of order p (AR(p)) is written as

$$\xi_i = \phi_1 \xi_{i-1} + \phi_2 \xi_{i-2} + \ldots + \phi_p \xi_{i-p} + n_i$$

where n_i is a random variate with $E(n_i) = 0$ and

$$\sigma_n^2 = 1 - \phi_1 r_1 - \phi_2 r_2 - \ldots - \phi_p r_p$$

when variance of the stochastic process ξ is unity. $r_1, r_2 \ldots r_p$ are lag 1, 2, . . . p autocorrelation coefficients, and an ARMA model of order (p,q) as

$$\xi_i = \phi_1 \xi_{i-1} + \phi_2 \xi_{i-2} + \ldots + \phi_p \xi_{i-p} + n_i - \theta n_{i-1}$$

$$+ \theta_2 n_{i-2} + \ldots + \theta_2 n_{i-q}$$

ARMA (1,1) reduces to

$$\xi_i = \phi \xi_{i-1} + n_i - \theta n_{i-1} \qquad (2)$$

The choice of type and order of an ARIMA class model is decided by analysis of residuals in several ways. An objective criteria for selection of order of autoregressive models was described by Akaike (1969) (see also Jones, 1975; Katz, 1981). Akaike defined a Final Prediction Error (FPE) criterion as

$$FPE_p = \sigma_p^2 \, (1 + (1+P)/n) \simeq \frac{1}{n} X_p^2 \qquad (3)$$

which can be interpreted as the estimated mean square error multiplied by a factor which gives the average increase in

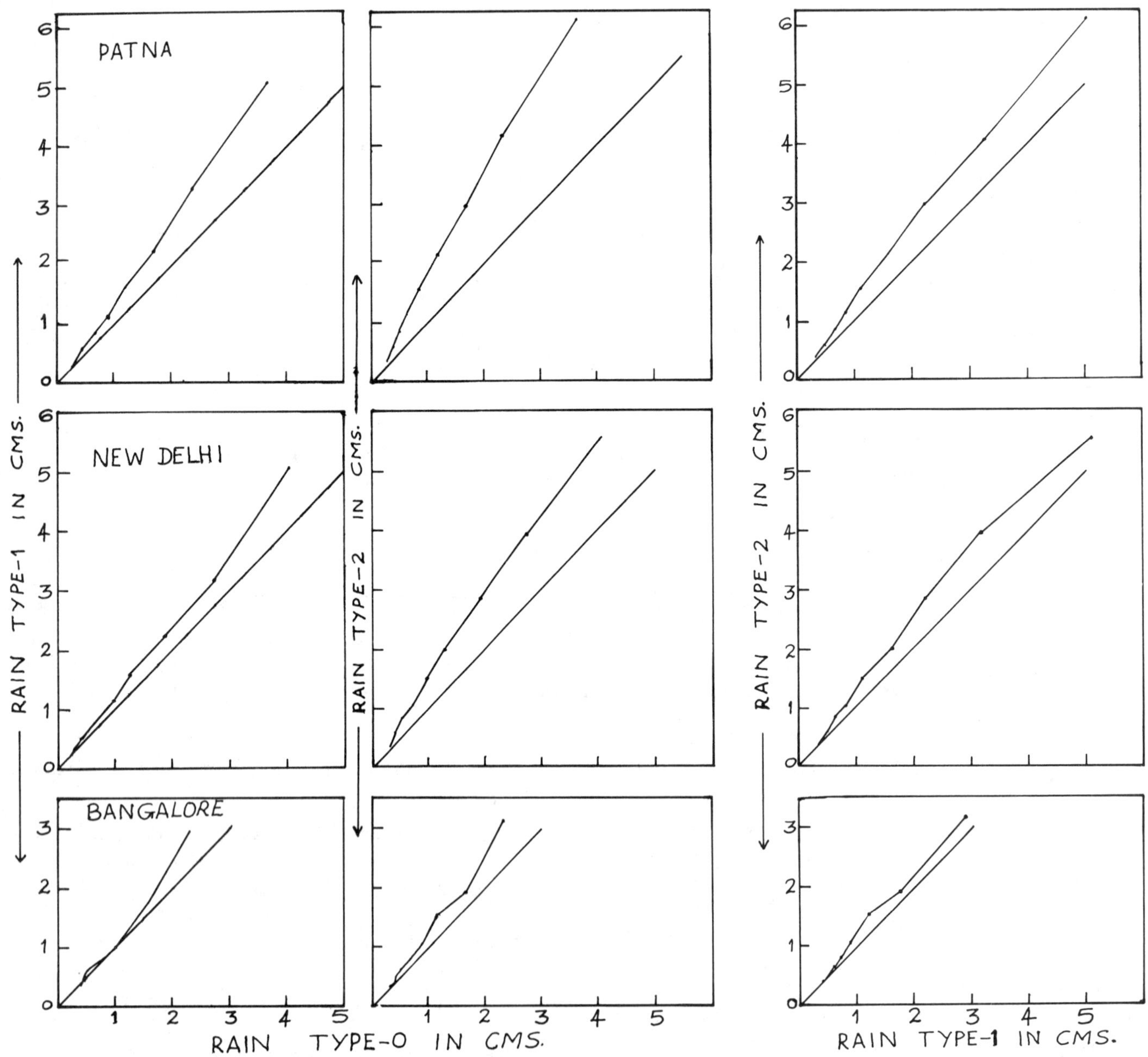

Figure 3. Decile-Decile Plot Between Different Rain Types (*viz.* 0, 1, 2, see text) Taken in Pair for Three Sample Stations – Upper Panel (PATNA), Middle Panel (New Delhi), Lower Panel (Bangalore) (for location of stations, see Table 1).

TABLE 1. Correlation Coefficients Between Rainfall of Successive Days Using 73 Years of Data of July-August.

Station (Lat., Long.) (°N, °E)	Lag-1 Correlation Taking All Days (n=4484)	Lag-1 Correlation Considering Only Rainy Days	Correlation Between Log Length of Dry Spell and Successive Wet Spell	Correlation Between Log of Wet Spell and Its Intensity	t-Tests Between: First and Last Day of Wet Spell	t-Tests Between: Type 0 and Type 1 Rain	t-Tests Between: Type 0 and Type 2 Rain	t-Tests Between: Type 1 and Type 2 Rain
1. Jaipur (27, 76)	0.40	0.07	−0.10	0.08	0.88	−1.24	−2.99	−3.10
2. Patna (26, 85)	0.31	0.11	0.01	0.16	3.11	−2.98	−5.96	−4.32
3. New Delhi (29, 77)	0.33	0.09	−0.01	0.13	2.16	−2.43	−4.06	−1.59
4. Bangalore (13, 78)	0.26	0.11	−0.04	0.05	−0.77	−1.12	−1.72	−0.87
5. Ahmedabad (23, 73)	0.48	0.12	−0.12	0.24	2.23	−3.78	−6.06	−3.63
6. Calcutta (23, 88)	0.22	0.10	0.00	0.15	3.24	−2.74	−5.40	−4.24
7. Ambala (30, 77)	0.29	−0.01	−0.06	0.11	3.64	−1.05	−5.00	−4.87
8. Ambikapur (23, 83)	0.36	0.11	0.02	0.25	2.46	−3.85	−6.73	−5.11
9. Manglore (13, 75)	0.45	0.11	−0.12	0.44	1.07	−1.10	−5.69	−6.95
10. Pune (19, 74)	0.52	0.27	−0.07	0.32	1.48	−4.36	−8.37	−6.90
11. Bombay (19, 73)	0.51	0.39	−0.08	0.48	1.41	−4.44	−8.70	−10.65
12. Hyderabad (17, 78)	0.31	0.12	0.03	0.09	1.93	−2.75	−5.66	−3.73

Station	Trimmed Mean (TM) and Median (Med) in mms for Various Types of Rain: Rain Type 0 TM	Rain Type 0 Med	Rain Type 1 TM	Rain Type 1 Med	Rain Type 2 TM	Rain Type 2 Med	Trimmed Mean (TM) in mms for Intensities of Rainfall During Spell of Various Length of Spell Days: 1	2	3	4	5	F-Test for Testing Significance of Difference Between Intensities of Wet Spell of 1 to 5 Day Length
1. Jaipur	11.90	8.30	14.83	11.60	18.04	13.55	11.90	16.27	17.42	20.42	15.89	3.43
2. Patna	12.69	8.80	16.45	10.90	21.42	15.52	12.69	17.38	17.15	19.22	23.33	7.89
3. New Delhi	14.08	9.70	16.51	10.91	19.90	14.90	14.08	17.40	19.73	20.25	19.98	3.26
4. Bangalore	9.20	6.80	10.12	7.30	11.17	8.00	9.20	10.83	10.60	12.80	13.08	1.18
5. Ahmedabad	9.76	6.20	14.26	9.60	18.35	14.40	9.76	15.18	17.64	18.57	20.12	10.89
6. Calcutta	11.22	8.10	13.46	10.20	16.63	12.70	11.22	14.74	16.74	16.37	17.37	8.68
7. Ambala	15.82	11.42	17.22	12.70	23.83	20.30	15.82	18.15	21.83	19.37	24.14	2.46
8. Ambikapur	11.86	9.30	16.70	12.60	21.28	17.50	11.86	16.17	19.54	22.00	21.66	11.09
9. Manglore	9.28	7.30	14.32	11.60	24.01	19.30	9.28	12.59	15.93	22.93	14.52	8.71
10. Pune	5.14	4.00	7.92	5.80	12.53	9.20	5.14	8.52	10.73	9.44	12.98	20.53
11. Bombay	6.75	5.50	10.37	7.30	19.88	13.40	6.75	9.05	11.24	14.35	16.73	18.21
12. Hyderbad	9.65	7.10	11.42	9.10	14.90	12.10	9.65	11.63	14.48	13.68	15.69	5.96

the prediction variance caused by using estimated coefficients rather than true coefficients. For selected order FPE_p is minimum.

We first fitted for each SDAR models of order 1 to 5 to the daily subdivisional monsoon rainfall series of length 62 day (July-August) for each of the 19 years separately and examined the Akaike's FPE and residual variance after each successive model fitted. The analysis of these results (not presented here) showed that AR(1) fits the rainfall series satisfactorily and higher order models contribute but insignificantly to the variance explained.

We thus fit only three of the simplest types of models, *viz.*, AR(1), AR(2), and ARMA (1,1) in terms of number of parameters to the rainfall data using all the years of data.

We compute autocorrelation up to lag 20 by pooling all the data with due consideration given to the break in series occurring after each season. The autocorrelation for lag ℓ is computed by

$$r_\ell = \frac{\sum_{1}^{NYR} \sum_{i=1}^{N-\ell} n_i \, n_{i+\ell}}{\left(\sum_{1}^{NYR} \sum_{i=1}^{N-\ell} n_i^2\right)^{1/2} \left(\sum_{1}^{NYR} \sum_{i=\ell+1}^{N} n_i^2\right)^{1/2}}$$

where NYR is number of years (19), N is number of days in a season (62). The n_i's are rainfall values with mean subtracted out.

These autocorrelations are then used to estimate the parameters and variance explained by AR(1), AR(2) in the following way:

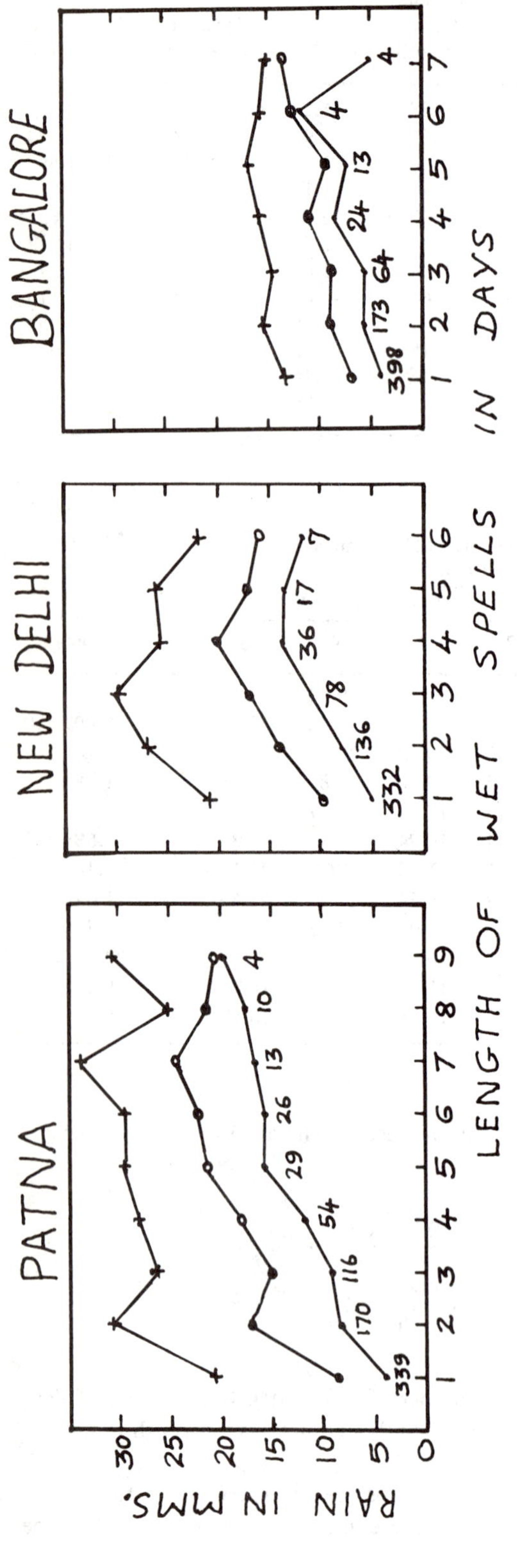

Figure 4. Median (——o——), Upper Quartile (——x——), and Lower Quartile (——●——) Plots for Intensity of Rain for Spells of Various Length, for Patna, New Delhi, and Bangalore (number of spells for various duration are also indicated).

a. AR(1)

$$\phi = r_1, \sigma_n^2 = 1 - r_1^2$$

b. AR(2)

$$\phi_2 = (r_2 - r_1^2) / (1 - r_1^2)$$

$$\phi_1 = r_2(1 - r_2) / (1 - r_1^2)$$

$$\sigma_n^2 = 1 - \phi_1 r_1 - \phi_2 r_2$$

c. ARMA (1, 1)

$$\xi_i = \phi\, \xi_{i-1} + n_i - \theta\, n_{i-1}$$

For estimation of ϕ, the first 10 positive autocorrelation coefficients are used and a function

$$\sum_{i=2}^{10} (r_i - r_1\, \phi^{(i-1)})^2$$

is computed for various values of ϕ within 0 to 1. The value of ϕ is chosen which minimizes the above function. θ is then estimated from:

$$\theta = \frac{(1 - 2\phi r_1 + \phi^2) - \sqrt{(1 - 2\phi r_1 + \phi^2)^2 - 4(\phi - r_1)^2}}{2(\phi - r_1)}$$

and

$$\sigma_n^2 = \frac{(1 - \phi^2)}{(1 + \theta^2 - 2\phi\theta)}$$

Although other nonlinear estimation procedures are more accurate, we feel this procedure estimates, at little cost, the parameters satisfactorily.

Results of Fitting AR and ARMA Models

The fitted parameters are given in Table 2 and we see that the residual variances for two parameter models (AR(2) and ARMA (1,1)) are only negligibly less than that due to AR(1), suggesting that AR(1) is an adequate stochastic model for spatially averaged rainfall. Actual autocorrelations and autocorrelations for AR(1) and ARMA(1,1) process are shown in Figure 5, up to lag 10 for the 10 regions. It is obvious that the ARMA(1,1) process represents the higher lag correlations better than the AR(1) process. This does not contradict the results cited in the preceeding discussion as only residual variance of predictions up to

TABLE 2. Parameters and Residual Variance Due to AR(1), AR(2), and ARMA(1,1) Type Models.

Subdivision No.	Model Type AR(1)		Model Type AR(2)			Model Type ARMA(1,1)		
	$\phi=r_1$	ϕn^2	ϕ_i	ϕ_2	σn^2	ϕ	θ	σn^2
1	0.56	0.69	0.5275	0.0579	0.68	0.6440	0.1237	0.68
2	0.65	0.58	0.6052	0.0689	0.57	0.7520	0.1778	0.57
3	0.63	0.60	0.6405	−0.0167	0.60	0.7440	0.1888	0.59
4	0.54	0.71	0.5704	0.0563	0.68	0.6600	0.1767	0.68
5	0.64	0.59	0.6074	0.0508	0.59	0.7440	0.1744	0.58
6	0.73	0.47	0.7766	−0.0213	0.44	0.6800	−0.0955	0.47
7	0.69	0.53	0.7160	0.0317	0.49	0.7200	0.0659	0.52
8	0.59	0.66	0.5632	0.0454	0.65	0.7200	0.2061	0.65
9	0.45	0.80	0.4050	0.1000	0.79	0.8040	0.4713	0.78
10	0.53	0.72	0.5079	0.0417	0.72	0.7384	0.2994	0.71

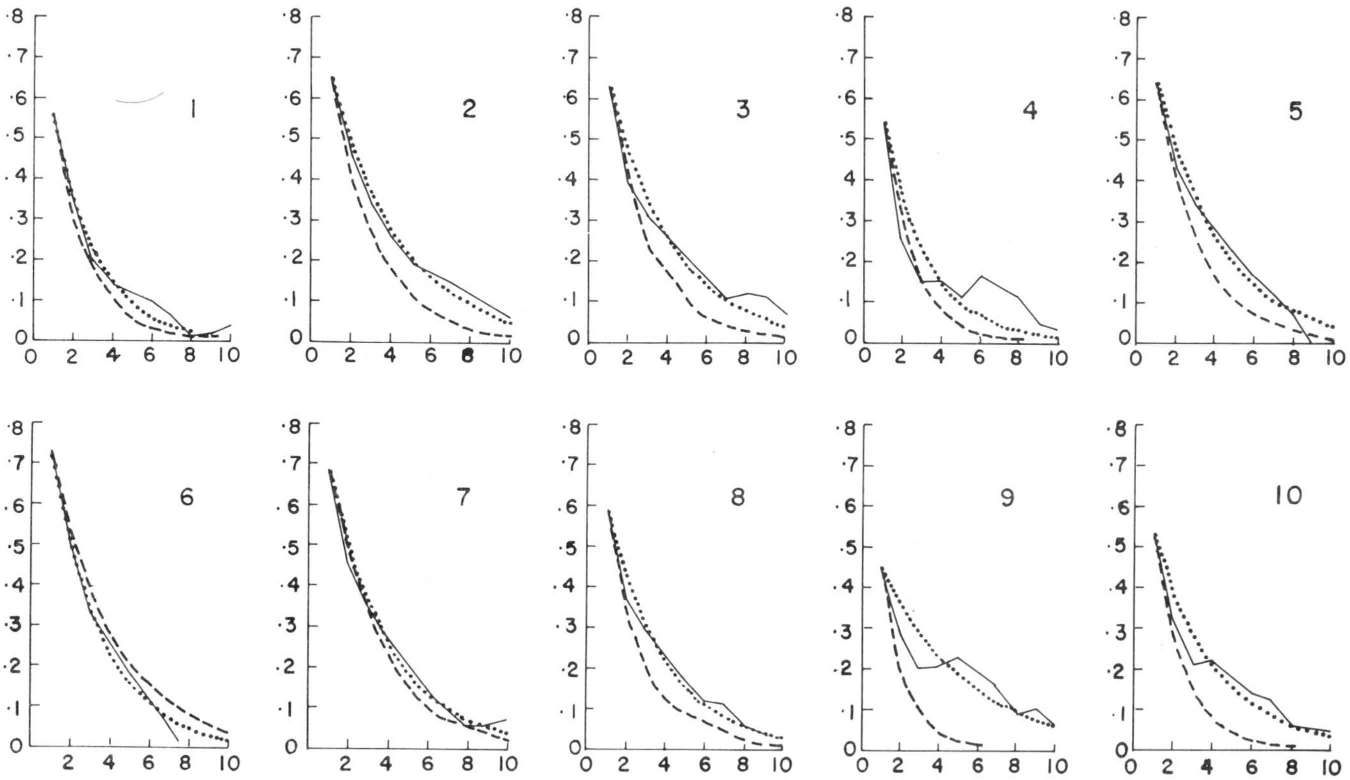

Figure 5. Autocorrelation Coefficient (solid line ——) Up to 10 Days Lag for 10 Meteorological Subdivisions. The correlation coefficients generated by AR(1) and ARMA(1,1) are represented by dashed (- - - -) and dotted line (. . . .), respectively.

time lag 1 only were considered, for which all the three models fair equally well. For higher lead time ARMA(1,1) may be more suitable.

CONCLUSIONS

The following conclusions are drawn:

1. The occurrence of rain events shows dependence which can be well represented by geometric models of Markov type.

2. Rainfall amounts on successive days within a rainy spell are uncorrelated but the rain amount on the second day of a rainy spell is more than on first or last day which is in turn more than rain on isolated days.

3. Rain intensity of wet spells increase as the length increases.

4. The autoregressive model of order 1 fits the time series of spatially averaged daily rainfall satisfactorily and higher order models may not be required to generate sequences of catchment rainfall. On the basis of these results sequences of rainy/dry day in desired period (say a week) and rainfall amount on each day can be generated which will be of substantial value in planning hydrologic operations.

ACKNOWLEDGMENTS

The authors are extremely grateful to Sh. D. R. Sikka, Head of Forecasting Research Division for the encouragement and the facilities provided.

LITERATURE CITED

Akaike, H., 1969. Fitting Autoregressive Models for Prediction. Annals of the Inst. of Statis. Maths. 21:243-247.

Association of Hydrologists of India, 1982. Abstracts of Papers presented at the 'First Annual Convention and Seminar on Hydrological Investigations During the Last 25 Years in India,' May 23-24, 1982, at Waltair, India.

Box, G. E. P. and G. M. Jenkins, 1976. Time Series Analysis: Forecasting and Control. Holden Day, 575 pp.

Buishand, T. A., 1977. Stochastic Modelling of Daily Rainfall Sequences. Mededelingen Landbouwhogescholl Wageningen, KNMI, Nederland, pp. 17-35.

Dhar, O. N., P. R. Rakhecha, and B. N. Mandal, 1981. Some Facts About Indian Rainfall, A Brief Appraisal from Hydrological Considerations. Indian Journal of Power and River Valley Development, Vol. 26:117-125.

Gananadesikan, R., 1977. Methods of Statistical Data Analysis of Multivariant Observations. John Wiley, New York, New York.

Jones, Richard H., 1975. Fitting Autoregressions. J. Amer. Statis. Assn. 70:590-592.

Katz, R. W., 1981. On the Use of Autoregressive Moving Average Processes to Model Meteorological Time Series. Month. Wea. Rev. 109:479-484.

Kleiner, B. and T. E. Gradel, 1980. Exploratory Data Analysis in the Geophysical Sciences. Reviews of Geophysics and Space Physics 18:699-717.

Raudkivi, A. J. and N. Lawgun, 1970. A Markov Chain Model for Rainfall Generation. *In:* Symposium on the Results of Research on Representative and Experimental Basins. Wellington, IAHS Pub. No. 96, pp. 2.69-2.78.

Singh, S. V., P. P. Kamte, S. D. Dahale, and P. V. Upasani, 1976. Rainfall Persistency Over India During Summer Monsoon. Proc. of Symp. on Tropical Monsoons. Indian Institute of Tropical Meteorology, pp. 418-425.

Singh, S. V., R. H. Kripalani, Priya Shaha, P. M. M. Ismail, and S. D. Dahale, 1981. Persistence in Daily and 5-Day Summer Monsoon Rainfall Over India. Arch. Met. Geoph. Biok. Ser. A 30:261-277.

Smith, R. E. and N. A. Schreiber, 1973. Point Process of Seasonal Thunderstorm Rainfall. Water Res. Res. 9:871-884.

AUTHOR INDEX